Nikolaos S. Papageorgiou, Patrick Winkert

Applied Nonlinear Functional Analysis

Also of Interest

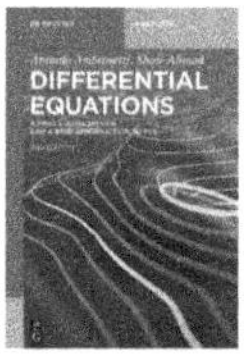

Differential Equations. A First Course on ODE and a Brief Introduction to PDE
Antonio Ambrosetti, Shair Ahmad, 2023
ISBN 978-3-11-118524-8, e-ISBN (PDF) 978-3-11-118567-5,
e-ISBN (EPUB) 978-3-11-118578-1

Scientific Computing. For Scientists and Engineers
Timo Heister, Leo G. Rebholz, 2023
ISBN 978-3-11-099961-7, e-ISBN (PDF) 978-3-11-098845-1,
e-ISBN (EPUB) 978-3-11-098875-8

Nonlinear Functional Analysis and Applications
Jesús Garcia-Falset, Khalid Latrach, 2023
ISBN 978-3-11-103096-8, e-ISBN (PDF) 978-3-11-103181-1,
e-ISBN (EPUB) 978-3-11-103208-5

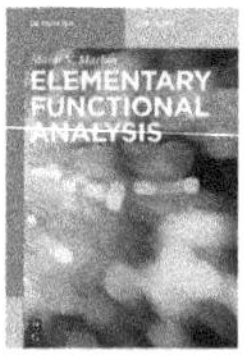

Elementary Functional Analysis
Marat V. Markin, 2018
ISBN 978-3-11-061391-9, e-ISBN (PDF) 978-3-11-061403-9,
e-ISBN (EPUB) 978-3-11-061409-1

Variational Methods in Nonlinear Analysis. With Applications in Optimization and Partial Differential Equations
Dimitrios C. Kravvaritis, Athanasios N. Yannacopoulos, 2020
ISBN 978-3-11-064736-5, e-ISBN (PDF) 978-3-11-064738-9,
e-ISBN (EPUB) 978-3-11-064745-7

Nikolaos S. Papageorgiou, Patrick Winkert

Applied Nonlinear Functional Analysis

An Introduction

2nd edition

DE GRUYTER

Mathematics Subject Classification 2020
26-XX, 28-XX, 46-XX, 47-XX, 49-XX

Authors
Nikolaos S. Papageorgiou
National Technical University of Athens
Department of Mathematics
Zografou Campus
157 80 Athens
Greece
npapg@math.ntua.gr

Patrick Winkert
Technische Universität Berlin
Institut für Mathematik
Straße des 17. Juni 136
10623 Berlin
Germany
winkert@math.tu-berlin.de

ISBN 978-3-11-128421-7
e-ISBN (PDF) 978-3-11-128695-2
e-ISBN (EPUB) 978-3-11-128832-1

Library of Congress Control Number: 2024931997

Bibliographic information published by the Deutsche Nationalbibliothek
The Deutsche Nationalbibliothek lists this publication in the Deutsche Nationalbibliografie; detailed bibliographic data are available on the Internet at http://dnb.dnb.de.

Cover image: By the authors
Typesetting: VTeX UAB, Lithuania
Printing and binding: CPI books GmbH, Leck

www.degruyter.com

This book is dedicated in memory of the first author's mother

M. S. Papageorgiou

and in memory of the second author's father

Wolfgang Winkert

who both passed away during the preparation of the first edition of the book.

Preface to the Second Edition

In this second edition we have added four new sections that we think bring up to date the contents of this volume. More precisely in Chapter 2 we have added Section 2.8 on Hausdorff Measures. These measures are a basic tool in Geometric Analysis and a first encounter with them will be useful to the reader. In Chapter 4 we have added Section 4.8 on Semigroups of Operators, which play a central role in the study of evolution equations. Also, in Chapter 4 we have added Section 4.9 on Generalized Orlicz Spaces. Double phase problems and in general problems with nonstandard growth cannot be studied in the framework of the classical Lebesgue and Sobolev spaces, we need to introduce and use Generalized Orlicz Spaces. In Section 4.9 we have a first look at the structure and properties of these spaces. In Chapter 6 we have added Section 6.8, in which we prove certain useful properties of differential operators that we encounter in boundary value problems (including the double phase operator), we present the Picone identity (useful in proving uniqueness results) and also the Hardy inequality (a valuable tool in the study of singular problems) and we discuss some additional properties of the variational convergence of sets. Furthermore, we use this opportunity to correct some misprints that we found in the first edition.

Finally the authors wish to thank Dr. Ranis N. Ibragimov and Nadja Schedensack of De Gruyter for their kind support and help during the preparation of this second edition of the book.

May 2024

Nikolaos S. Papageorgiou, Athens, Greece
Patrick Winkert, Berlin, Germany

https://doi.org/10.1515/9783111286952-201

Preface to the First Edition

The aim of this book is to present the foundations of modern Nonlinear Functional Analysis and equip the reader with all the necessary tools to continue with theoretical and/or applied research in the field. Nonlinear Functional Analysis is a very broad subject and has applications in many different areas of physics, mechanics, engineering, and economics. In fact, it emerged as a distinct discipline within mathematical analysis specifically as a way to address these needs in a mathematically rigorous way. This way Nonlinear Functional Analysis distinguished itself from the classical Linear Functional Analysis and acquired an interdisciplinary character.

The present book provides a starting point to follow some of the main paths of Nonlinear Functional Analysis, especially those leading to applications. The goal is to present the theories and techniques to the newcomer, which will allow him/her to proceed to more specialized topics. The first three chapters present the main elements of topology, measure theory, and Banach space theory, which are needed to proceed further. In the last three chapters we present more advanced and specialized topics that are motivated by the applications. In Chapter 4 we examine certain spaces of functions and measures that provide the functional framework in the applied problems. We deal with Lebesgue, Lebesgue-Bochner, and Sobolev spaces, which are the basic tools in the study of boundary valued problems. We also study spaces of absolutely continuous functions, of functions of bounded variation, and of measures that eventually lead to Young measures. All these constitute the modern tools in dealing with problems of the calculus of variations, control theory and optimization, as well as mathematical economics. In Chapter 5 we deal with nonsmooth and multivalued analysis, two fields of mathematical analysis that emerged simultaneously in the early 1960's and developed in parallel, feeding each other with new notions and methods. As a result, we deal with convex functions and their duality and subdifferential theory. We also examine the approximation properties of sets and extend the subdifferential theory to the nonconvex one in terms of locally Lipschitz functions in the sense of Clarke. Furthermore, we present the main topological and measure theoretic aspects of set-valued maps with applications to integral functionals. In Chapter 6 we finally study topics that are traditionally associated with what is called "Nonlinear Analysis." These are operators of monotone type, degree theory, fixed point theory, variational principles such as Ekeland's Variational Principle, and variational convergence such as Γ- or epigraphical convergence. With this choice of material, we believe that the reader will be properly equipped at the end to do research in this exciting field of mathematical analysis. Each chapter is followed by at least 50 problems. We encourage the reader to try them in order to test his/her understanding of the material. The solutions to the problems will be posted on the personal site of the second author. Our hope is that the reader, with the help of the material in this book, can proceed with confidence in the many different parts of this field.

https://doi.org/10.1515/9783111286952-202

Finally the authors wish to thank Dr. Apostolos Damialis, Maria Dassing, and Nadja Schedensack of De Gruyter for their kind support and help during the preparation of this book.

January 2018

Nikolaos S. Papageorgiou, Athens, Greece
Patrick Winkert, Berlin, Germany

Contents

1 Basic Topology

Topology, as its name suggests,[1] deals with geometric properties of objects that depend only on their relative positions and not on notions such as size or magnitude. The properties studied by topology are preserved by certain continuous transformations. Discontinuous transformations destroy topological properties. In this chapter we present the basic items of point-set topology that are needed to examine certain topics of applied analysis. We do not claim to have an exhaustive presentation of the subject.

1.1 Basic Notions

We start with the definition of topology.

Definition 1.1.1. Let X be a set and let $\tau \subseteq 2^X$ be such that the following hold:
(a) X and $\emptyset$ both belong to τ;
(b) τ is closed under arbitrary unions, that is, if $\{U_i\}_{i\in I} \subseteq \tau$ is any family of sets in τ, then $\bigcup_{i\in I} U_i \in \tau$;
(c) τ is closed under finite intersections, that is, if $\{U_i\}_{i\in I} \subseteq \tau$ is a finite family of sets in τ, then $\bigcap_{i\in I} U_i \in \tau$.

Then we say that τ is a **topology** on X. The sets in τ are called **open sets**. The complements of the elements of τ are called **closed sets**. In addition we say that the pair (X, τ) is a **topological space**.

Remark 1.1.2. When the topology τ is clearly understood from the context, then we drop it and simply say that X is a topological space. From the definition above it is clear that the family of closed sets contains X and $\emptyset$ and it is closed under finite unions and arbitrary intersections. If X is a set with two topologies τ_1 and τ_2 such that $\tau_1 \subseteq \tau_2$, then we say that τ_1 is **weaker** than τ_2 or that τ_2 is **stronger** than τ_1. The intersection of any family of topologies on X is also a topology that is weaker than every member of the family but stronger than any other topology with this property. Note that for any set X there is a **strongest topology** on X, namely $\tau = 2^X$ known as the **discrete topology**. Moreover, there also exists a **weakest topology** on X, namely $\tau = \{X, \emptyset\}$ known as the **trivial topology**.

In general, a topology is a very large collection of subsets. So it is useful to have a smaller collection of elements of τ, which generates the topology by taking unions.

Definition 1.1.3. Let (X, τ) be a topological space. A **basis** (or **base**) for the topology τ is a subfamily $\mathcal{B}$ of τ such that every member of τ is the union of elements in $\mathcal{B}$. The elements

1 It comes from the Greek word τόπος = location or position.

https://doi.org/10.1515/9783111286952-001

of $\mathcal{B}$ are called **basic open sets** and τ is the topology **generated** by $\mathcal{B}$. A subfamily $\mathcal{L}$ of τ is a **subbasis** of the topology τ if the family of finite intersections of elements in $\mathcal{L}$ is a basis for τ. The elements of $\mathcal{L}$ are called **subbasic open sets.**

In the definition above, we have assumed a topology on X and defined a basis for it. On the other hand, one might start with a basis and using it, generates a topology on X by taking unions. However, not every family in 2^X is a basis for a topology. The next proposition gives necessary and sufficient conditions for a family to generate a topology.

Proposition 1.1.4. *A family $\mathcal{B} \subseteq 2^X$ is a basis for a topology on X if and only if*
(a) *$\bigcup \mathcal{B} = X$, that is, the union of the elements of $\mathcal{B}$ is X;*
(b) *if $B_1, B_2 \in \mathcal{B}$ and $x \in B_1 \cap B_2$, then there exists $B \in \mathcal{B}$ such that $x \in B \subseteq B_1 \cap B_2$.*

Proof. $\Longrightarrow$: The assertion in (a) follows from the fact that X is open; see Definition 1.1.3. Let us prove (b). We know that $B_1 \cap B_2$ is open. So, according to Definition 1.1.3, $B_1 \cap B_2$ is the union of elements in $\mathcal{B}$. Hence we can find $B \in \mathcal{B}$ such that $x \in B \subseteq B_1 \cap B_2$.

$\Longleftarrow$: Let τ be all unions of elements of $\mathcal{B}$. We need to show that τ is a topology on X; see Definition 1.1.1. Evidently $\emptyset \in \tau$ and $X \in \tau$; see (a). In addition, from its definition, τ is closed under arbitrary unions. We have to show that τ is closed under finite intersections. So, let $U_1, U_2 \in \tau$. Then $U_1 \cap U_2 \in \tau$. Given $x \in U_1 \cap U_2$, there exist $B_1, B_2 \in \mathcal{B}$ such that $x \in B_1 \subseteq U_1$ and $x \in B_2 \subseteq U_2$. Therefore, $x \in B_1 \cap B_2 \subseteq U_1 \cap U_2$. By (b) there is $B(x) \in \mathcal{B}$ such that $x \in B(x) \subseteq U_1 \cap U_2$. Obviously, $U_1 \cap U_2 = \bigcup_x B_x \in \tau$. Thus τ is a topology on X. □

Remark 1.1.5. We say that τ is the topology generated by $\mathcal{B}$ and we often write $\tau(\mathcal{B})$ to emphasize the basis generating the topology.

Corollary 1.1.6. *If (X, τ) is a topological space and $\mathcal{B}$ is a subfamily of τ such that for each $U \in \tau$ and $x \in U$, we can find $V \in \mathcal{B}$ such that $x \in V \subseteq U$, then $\mathcal{B}$ is a basis for the topology τ.*

Proposition 1.1.7. *If (X, τ) is a topological space and $\mathcal{B}$ is a basis for τ, then $U \in \tau$, that is, U is open, if and only if for every $x \in U$ there exists $V_x \in \mathcal{B}$ such that $x \in V_x \subseteq U$.*

Proof. $\Longrightarrow$: This follows from (b) of Proposition 1.1.4.

$\Longleftarrow$: We have $U = \bigcup_x V_x \in \tau$. □

Definition 1.1.8. Two bases $\mathcal{B}$ and $\mathcal{B}'$ of X are said to be **equivalent** if $\tau(\mathcal{B}) = \tau(\mathcal{B}')$.

Directly from Propositions 1.1.4 and 1.1.7 we have the following characterization of equivalent topological bases.

Proposition 1.1.9. *Two bases $\mathcal{B}$ and $\mathcal{B}'$ in X are equivalent if and only if*
(a) *for every $B \in \mathcal{B}$ and $x \in B$, there exists $B' \in \mathcal{B}'$ such that $x \in B' \subseteq B$;*
(b) *for every $B' \in \mathcal{B}'$ and $x \in B'$, there exists $B \in \mathcal{B}$ such that $x \in B \subseteq B'$.*

Example 1.1.10. In $\mathbb{R}^N$ with $N \in \mathbb{N}$, let $\mathcal{B} = \{B_r(x): x \in \mathbb{R}^N, r > 0\}$ with $B_r(x) = \{u \in \mathbb{R}^N: |u - x| < r\}$. Then $\mathcal{B}$ is a basis for the so-called **Euclidean topology** (or **standard topology**) on $\mathbb{R}^N$. So, every open set in $\mathbb{R}^N$ is the union of open balls. More generally this is also true for every metric space.

There is a local version of the notion of topological basis.

Definition 1.1.11. Let (X, τ) be a topological space and $x \in X$. We say that $\mathcal{B}(x) \subseteq \tau$ is a **local basis** (or a **local base**) at x if the following hold:
(a) $x \in V$ for every $V \in \mathcal{B}(x)$;
(b) if $x \in U \in \tau$, then there exists $V \in \mathcal{B}(x)$ such that $x \in V \subseteq U$.

Definition 1.1.12. Let (X, τ) be a topological space and $A \subseteq X$.
(a) A **neighborhood** of $x \in X$ is any open set U such that $x \in U$.
(b) We say that $x \in A$ is an **interior point** of A if we can find $U \in \tau$ such that $x \in U \subseteq A$. The **interior** of A, denoted by $\operatorname{int} A$ (or by $\mathring{A}$), is the set of all interior points of A.
(c) We say that $x \in X$ is a **cluster point** (or a **limit point** or an **accumulation point**) of A if every open set containing x contains a point of A distinct from x. The set of all cluster points of A is called the **derived set** of A and is denoted by A'. The **closure** of A, denoted by $\overline{A}$ (or $\operatorname{cl} A$), is the union of A with its set of cluster points, that is, $\overline{A} = A \cup A'$.
(d) We say that $x \in X$ is a **boundary point** of A if $x \in \overline{A} \cap \overline{(X \setminus A)}$. The set of boundary points of A is called the **boundary** of A and is denoted by $\operatorname{bd} A$ (or by ∂A).

Remark 1.1.13. Note that a cluster point or a boundary point of A need not belong to A. In the sequel we denote by $\mathcal{N}(x)$ the family of all neighborhoods of $x \in X$.

Proposition 1.1.14. *If (X, τ) is a topological space and $A, C \subseteq X$, then the following hold:*
(a) $\operatorname{int} A = \bigcup\{U \in \tau: U \subseteq A\}$, *that is,* $\operatorname{int} A$ *is the largest open set contained in A;*
(b) *A is open if and only if* $A = \operatorname{int} A$;
(c) $A \subseteq C$ *implies* $\operatorname{int} A \subseteq \operatorname{int} C$;
(d) $\operatorname{int}(A \cap C) = \operatorname{int} A \cap \operatorname{int} C$.

Proof. (a) Let $\tilde{A} = \bigcup\{U \in \tau: U \subseteq A\}$. Then $\tilde{A}$ is open and by Definition 1.1.12 (b) it is clear that $\operatorname{int} A \subseteq \tilde{A}$. On the other hand, if $x \in \tilde{A}$, then there is $U \in \tau$, $U \subseteq A$ such that $x \in U$. Hence, x is an interior point of A, therefore $\tilde{A} \subseteq \operatorname{int} A$. We conclude that $\tilde{A} = \operatorname{int} A$.

(b) This is an immediate consequence of (a).

(c) We have $\operatorname{int} A \subseteq A \subseteq C$ and since $\operatorname{int} A$ is open, it follows that $\operatorname{int} A \subseteq \operatorname{int} C$, see part (a).

(d) We have $A \cap C \subseteq A$ and $A \cap C \subseteq C$. Then $\operatorname{int}(A \cap C) \subseteq \operatorname{int} A$ and $\operatorname{int}(A \cap C) \subseteq \operatorname{int} C$ because of part (c). This gives

$$\operatorname{int}(A \cap C) \subseteq \operatorname{int} A \cap \operatorname{int} C\,. \tag{1.1.1}$$

On the other hand, $\operatorname{int} A \cap \operatorname{int} C$ is an open subset of $A \cap C$. Hence, because of (a),

$$\operatorname{int} A \cap \operatorname{int} C \subseteq \operatorname{int}(A \cap C)\,. \tag{1.1.2}$$

From (1.1.1) and (1.1.2) we conclude that $\operatorname{int} A \cap \operatorname{int} C = \operatorname{int}(A \cap C)$. □

Remark 1.1.15. In general it is not true that $\operatorname{int}(A \cup C) = \operatorname{int} A \cup \operatorname{int} C$. Indeed let $X = \mathbb{R}$ with the Euclidean topology, see Example 1.1.10, and let $A = [0,1]$ and $C = [1,2]$. Then

$$\operatorname{int} A = (0,1)\,, \quad \operatorname{int} C = (1,2) \quad \text{and} \quad \operatorname{int}(A \cup C) = (0,2)\,.$$

In general we can easily show that if $\{A_i\}_{i \in I}$ is an arbitrary family of subsets of X, then

$$\bigcup_{i \in I} \operatorname{int} A_i \subseteq \operatorname{int} \bigcup_{i \in I} A_i\,.$$

There is an analogous proposition for the closure.

Proposition 1.1.16. *If (X, τ) is a topological space and $A, C \subseteq X$, then the following hold:*
(a) $\overline{A} = \bigcap\{D : D$ *closed*, $D \supseteq A\}$, *that is, $\overline{A}$ is the smallest closed set containing A;*
(b) *A is closed if and only if $A = \overline{A}$;*
(c) *$A \subseteq C$ implies $\overline{A} \subseteq \overline{C}$;*
(d) $\overline{A \cup C} = \overline{A} \cup \overline{C}$.

Proof. (a) Let $A^* = \bigcap\{D : D \text{ closed}, D \supseteq A\}$. Evidently, A^* is closed and so $X \setminus A^*$ is open. Hence, if $x \notin A^*$, then we find $U \in \mathcal{N}(x)$ such that $U \cap A = \emptyset$. Therefore, $x \notin (A \cup A') = \overline{A}$ and so $\overline{A} \subseteq A^*$. Now suppose that $x \in A^* \setminus \overline{A}$. Then there exists $U \in \mathcal{N}(x)$ such that $U \cap A = \emptyset$. Let $C = X \setminus U$. Then C is closed and $C \supseteq A$. Hence $A^* \subseteq C$ and so $x \in C$, a contradiction. Therefore $\overline{A} = A^*$.

(b) This is an immediate consequence of (a).

(c) We have $A \subseteq C \subseteq \overline{C}$ and since $\overline{C}$ is closed, it follows that $\overline{A} \subseteq \overline{C}$, see part (a).

(d) Note that $\overline{A} \cup \overline{C}$ is closed and contains $A \cup C$. Hence

$$\overline{A \cup C} \subseteq \overline{A} \cup \overline{C}\,. \tag{1.1.3}$$

Since $A, C \subseteq A \cup C$, we have $\overline{A}, \overline{C} \subseteq \overline{A \cup C}$, see part (c). Hence

$$\overline{A} \cup \overline{C} \subseteq \overline{A \cup C}\,. \tag{1.1.4}$$

From (1.1.3) and (1.1.4) we conclude that $\overline{A \cup C} = \overline{A} \cup \overline{C}$. □

Remark 1.1.17. In general it is not true that $\overline{A \cap C} = \overline{A} \cap \overline{C}$. To see this, let $X = \mathbb{R}$ with the Euclidean topology and let $A = (0,1)$ as well as $C = (1,2)$. Then $\overline{A \cap C} = \emptyset$ and $\overline{A} \cap \overline{C} = [0,1] \cap [1,2] = \{1\}$. In general we can easily show that if $\{A_i\}_{i \in I}$ is an arbitrary family of subsets of X, then

$$\overline{\bigcap_{i\in I} A_i} \subseteq \bigcap_{i\in I} \overline{A_i}\,.$$

In addition, the following formulas are easy to verify:
- $x \in A'$ if and only if $x \in \overline{(A' \setminus \{x\})}$;
- $(A \cup C)' = A' \cup C'$, $A' \setminus C' \subseteq (A \setminus C)'$, $A'' \subseteq A'$;
- $(\bigcap_{i\in I} A_i)' \subseteq \bigcap_{i\in I} A_i'$ with an arbitrary index set I;
- $\bigcup_{i\in I} A_i' \subseteq (\bigcup_{i\in I} A_i)'$ with an arbitrary index set I;
- $\overline{A}' = A'$;
- $A \subseteq C$ implies $A' \subseteq C'$;
- $(A \setminus \{x\})' = A' = (A \cup \{x\})'$.

The last formula means that the derived set remains unchanged if we add or remove a finite number of elements. If $x \in A \setminus A'$, then we say that x is **isolated**.

Proposition 1.1.18. *If (X, τ) is a topological space and $A \subseteq X$, then the following hold:*
(a) $\operatorname{bd} A = \overline{A} \cap \overline{(X \setminus A)} = \operatorname{bd}(X \setminus A)$;
(b) $\operatorname{bd} A$, $\operatorname{int} A$, $\operatorname{int}(X \setminus A)$ *are pairwise disjoint sets whose union is X;*
(c) $\operatorname{bd} A$ *is a closed set;*
(d) $\overline{A} = \operatorname{int} A \cup \operatorname{bd} A$;
(e) *A is open if and only if* $\operatorname{bd} A \subseteq X \setminus A$;
(f) *A is closed if and only if* $\operatorname{bd} A \subseteq A$;
(g) *A is closed and open (usually called **clopen**) if and only if* $\operatorname{bd} A = \emptyset$.

Proof. (a)–(d) These are immediate consequences of Definition 1.1.12.

(e) $\Longrightarrow$: Since A is open we have $A = \operatorname{int} A$ due to Proposition 1.1.14 (b). From part (b) we know that $\operatorname{int} A$ and $\operatorname{bd} A$ are disjoint sets. Therefore $\operatorname{bd} A \subseteq X \setminus A$.

$\Longleftarrow$: Since $\operatorname{bd} A \subseteq X \setminus A$, no point of A is a boundary point. Hence, every point of A is an interior point, see part (d). Therefore, $A = \operatorname{int} A$, that is, A is open.

(f) This follows from (e) by taking complements.

(g) Combine (e) and (f). □

Definition 1.1.19. A subset A of a topological space X is said to be **dense** if $\overline{A} = X$. We say that the topological space X is **separable** if it has a countable, dense subset.

Remark 1.1.20. It is easy to see that A is dense in the topological space (X, τ) if and only if for every $U \in \tau$, $U \neq \emptyset$ we have $U \cap A \neq \emptyset$. Clearly $\mathbb{R}^N$ is separable since we can take the set of vectors with rational coordinates as a countable, dense set.

Definition 1.1.21. A subset A of a topological space X is said to be **nowhere dense** if $\operatorname{int} \overline{A} = \emptyset$.

Remark 1.1.22. From the definition above we see that $A \subseteq X$ is nowhere dense if and only if $X \setminus \overline{A}$ is dense in X. It follows that $A \subseteq X$ is nowhere dense if and only if $X \setminus \overline{(X \setminus \overline{A})} = \emptyset$ or that A is nowhere dense if and only if $A \subseteq \overline{(X \setminus \overline{A})}$. Any set A that contains a dense set is itself dense. Similarly, any subset of a nowhere dense set is nowhere dense. The closure of a nowhere dense set is nowhere dense.

Proposition 1.1.23. *If X is a topological space and $A \subseteq X$ is open or closed, then* $\operatorname{bd} A$ *is nowhere dense.*

Proof. Suppose that A is open. Then $\operatorname{bd} A = \overline{A} \setminus A$, see Proposition 1.1.18 (d). Hence, $\operatorname{int}\operatorname{bd} A = \operatorname{int}(\overline{A} \setminus A) = \emptyset$, which shows that $\operatorname{bd} A$ is nowhere dense.

Similarly, if A is closed, then $\operatorname{bd} A = A \cap \overline{(X \setminus A)}$, see Definition 1.1.12 (d). Therefore, by Proposition 1.1.14 (d), $\operatorname{int}\operatorname{bd} A = \operatorname{int} A \cap \operatorname{int}\overline{(X \setminus A)}$. Hence, $\operatorname{int}\operatorname{bd} A = \emptyset$ and so $\operatorname{bd} A$ is nowhere dense in X. □

Definition 1.1.24. Let (X, τ) be a topological space and $A \subseteq X$. The **subspace or relative topology** on A is the family

$$\tau(A) = \{U \cap A \colon U \in \tau\} .$$

It is also called the **trace** of τ on A. It is easy to see that $\tau(A)$ is a topology on A.

Proposition 1.1.25. *If (X, τ) is a topological space, $\mathcal{B}$ is a basis for the topology τ and $A \subseteq X$, then $\mathcal{B}(A) = \{U \cap A \colon U \in \mathcal{B}\}$ is a basis for $\tau(A)$.*

Proof. Let $U \in \tau$ and $u \in U \cap A$. We can find $V \in \mathcal{B}$ such that $u \in V \subseteq U$. Then $u \in V \cap A \subseteq U \cap A$. This implies that $\mathcal{B}(A)$ is a basis for $\tau(A)$; see Corollary 1.1.6. □

Proposition 1.1.26. *If (X, τ) is a topological space, $A \in \tau$ and $V \in \tau(A)$, then $V \in \tau$.*

Proof. Since $V \in \tau(A)$ we have $V = U \cap A$ with $U \in \tau$. But $U \cap A \in \tau$ since $A \in \tau$. □

Proposition 1.1.27. *If (X, τ) is a topological space and $A \subseteq X$, then $D \subseteq A$ is $\tau(A)$-closed if and only if $D = C \cap A$ with closed $C \subseteq X$.*

Proof. $\Longrightarrow$: Since $D \subseteq A$ is $\tau(A)$-closed, that is, relatively closed, we have $A \setminus D = U \cap A$ with $U \in \tau$. Then $D = A \setminus (A \setminus D) = A \setminus (U \cap A) = (X \setminus U) \cap A = C \cap A$ with closed $C = X \setminus U$.

$\Longleftarrow$: Let $U = X \setminus C$. Then $U \in \tau$ and we have

$$A \setminus D = A \setminus (C \cap A) = (X \setminus C) \cap A = U \cap A ,$$

which implies that $A \setminus D$ is $\tau(A)$-open and so D is $\tau(A)$-closed. □

As a consequence of Proposition 1.1.26 we have the following observation concerning neighborhoods of a point $x \in A$.

Corollary 1.1.28. *If (X, τ) is a topological space, $A \subseteq X$, $x \in A$ and $V \subseteq A$, then $V \in \mathcal{N}_A(x)$, where $\mathcal{N}_A(x)$ denotes the $\tau(A)$-neighborhoods of x, if and only if $V = U \cap A$ with $U \in \mathcal{N}(x)$.*

This discussion on relativization of topologies leads naturally to the following notion, which will be used in the sequel.

Definition 1.1.29. A property of topological spaces is said to be **hereditary** if every subset with the relative (subspace) topology exhibits this property.

The notion of continuity is central in point-set topology. It is the main tool that allows us to determine which mathematical properties are intrinsic to a particular topological space.

Definition 1.1.30. Let X, Y be topological spaces. We say that a map $f: X \to Y$ is **continuous** at $x \in X$ if for every $U \in \mathcal{N}(f(x))$ we can find $V \in \mathcal{N}(x)$ such that $f(V) \subseteq U$. We say that $f: X \to Y$ is **continuous** if it is continuous at every $x \in X$.

Remark 1.1.31. From the last definition it is clear that continuity is a local property. The next proposition provides a useful global characterization of continuity.

Proposition 1.1.32. *If (X, τ_X) and (Y, τ_Y) are two topological spaces and $f: X \to Y$, then f is continuous if and only if $f^{-1}(\tau_Y) \subseteq \tau_X$, that is, f returns open sets in Y to open sets in X.*

Proof. $\Longrightarrow$: Let $U \in \tau_Y$. Then U is a neighborhood of each of its points. So, $f^{-1}(U)$ contains a neighborhood of everyone of its points. Hence $f^{-1}(U) \in \tau_X$.

$\Longleftarrow$: This is immediate from Definition 1.1.30. □

Remark 1.1.33. Since f^{-1} preserves all set theoretic operations, in the proposition above we may replace τ_Y by a basis $\mathcal{B}_Y$ or even better by a subbasis $\mathcal{L}_Y$.

We have a counterpart of Proposition 1.1.32 with closed sets instead of open sets.

Proposition 1.1.34. *If X and Y are topological spaces and $f: X \to Y$, then f is continuous if and only if for every closed $C \subseteq Y$, $f^{-1}(C)$ is closed in X.*

Proposition 1.1.35. *If X and Y are topological spaces and $f: X \to Y$, then the following statements are equivalent.*
(a) *f is continuous;*
(b) *$f(\overline{A}) \subseteq \overline{f(A)}$ for every $A \subseteq X$;*
(c) *$\overline{f^{-1}(C)} \subseteq f^{-1}(\overline{C})$ for every $C \subseteq Y$.*

Proof. (a) $\Longrightarrow$ (b): Let $A \subseteq X$ and $x \in \overline{A}$. Consider $U \in \mathcal{N}(f(x))$ and choose $V \in \mathcal{N}(x)$ such that $f(V) \subseteq U$, see Definition 1.1.30. We have

$$\begin{aligned} x \in \overline{A} \quad &\Longrightarrow \quad V \cap A \neq \emptyset \quad \Longrightarrow \quad f(V \cap A) \neq \emptyset \\ &\Longrightarrow \quad f(V) \cap f(A) \neq \emptyset \quad \Longrightarrow \quad U \cap f(A) \neq \emptyset\,. \end{aligned}$$

Since $U \in \mathcal{N}(f(x))$ is arbitrary it follows that $x \in \overline{f(A)}$. Hence $f(\overline{A}) \subseteq \overline{f(A)}$.

(b) $\Longrightarrow$ (c): Let $A = f^{-1}(C)$. Then by hypothesis $f(\overline{A}) \subseteq \overline{f(A)} = \overline{f(f^{-1}(C))} \subseteq \overline{C}$ and so $\overline{A} = \overline{f^{-1}(C)} \subseteq f^{-1}(\overline{C})$.

(c) $\Longrightarrow$ (a): Let $C \subseteq Y$ be closed. Then by hypothesis $\overline{f^{-1}(C)} \subseteq f^{-1}(C)$ and so $\overline{f^{-1}(C)} = f^{-1}(C)$, that is, $f^{-1}(C)$ is closed. From Proposition 1.1.34 it follows that f is continuous. □

Proposition 1.1.36. *Let X, Y and Z be topological spaces.*
(a) *If $f: X \to Y$ and $g: Y \to Z$ are continuous maps, then $g \circ f: X \to Z$ is continuous.*
(b) *If $f: X \to Y$ is a continuous map and $A \subseteq X$, then $f|_A: A \to Y$ is continuous for the subspace topology of A.*
(c) *If $X = \bigcup_{i \in I} U_i$ with U_i open and $f: X \to Y$ is a map such that $f|_{U_i}$ is continuous, then $f: X \to Y$ is continuous.*

Proof. (a) If U is open in Z, then $g^{-1}(U)$ is open in Y and $f^{-1}(g^{-1}(U))$ is open in X, see Proposition 1.1.32. But recall that $f^{-1}(g^{-1}(U)) = (g \circ f)^{-1}(U)$. So, by Proposition 1.1.32, $g \circ f$ is continuous.

(b) Let $i: A \to X$ be the inclusion map where A is endowed with the subspace topology. Evidently i is continuous and since $f|_A = f \circ i$ we derive the conclusion using part (a).

(c) Let $V \subseteq Y$ be open. Then $f^{-1}(V) \cap U_i = (f|_{U_i})^{-1}(V)$ is open in X for all $i \in I$. Therefore $f^{-1}(V) = \bigcup_{i \in I} f^{-1}(V) \cap U_i$ is open in X. Taking Proposition 1.1.32 into account yields the continuity of f. □

Continuing in the same way, we prove the so-called "Pasting Lemma".

Proposition 1.1.37 (Pasting Lemma). *If X and Y are topological spaces, $X = A \cup B$ with closed subsets A and B of X, $f: A \to Y$ and $g: B \to Y$ are continuous maps where A and B are endowed with the subspace topology and $f(x) = g(x)$ for all $x \in A \cap B$. Then $h: X \to Y$ defined by*

$$h(x) = \begin{cases} f(x) & \text{if } x \in A \\ g(x) & \text{if } x \in B \end{cases},$$

is continuous.

Proof. Let C be a closed subset of Y. Then

$$h^{-1}(C) = f^{-1}(C) \cup g^{-1}(C) . \tag{1.1.5}$$

By hypothesis $f^{-1}(C)$ is closed in A and since A is closed in Y, from Proposition 1.1.27, we have that $f^{-1}(C)$ is closed in X. Similarly $g^{-1}(C)$ is closed in X. From (1.1.5) it follows that $h^{-1}(C)$ is closed in X. Hence, by Proposition 1.1.34, h is continuous. □

In general the direct image of an open (resp. closed) set by a map need not be open (resp. closed) even if the map is continuous. For this reason we introduce the following definition.

Definition 1.1.38. Let X and Y be two topological spaces. We say that a map $f: X \to Y$ is **open** (respectively, **closed**) if the image of every open (respectively, closed) set in X is open (respectively, closed) in Y.

Remark 1.1.39. It is easy to see that the notions of continuous map, open map, and closed map are independent.

Proposition 1.1.40. *Let (X, τ_X) and (Y, τ_Y) be topological spaces and $f : X \to Y$, then the following statements are equivalent:*
(a) *f is open;*
(b) *$f(\operatorname{int} A) \subseteq \operatorname{int} f(A)$ for every $A \subseteq X$;*
(c) *if $\mathcal{B}_X$ is a basis for τ_X, then $f(\mathcal{B}_X) \subseteq \tau_Y$.*

Proof. (a) $\Longrightarrow$ (b): We have $f(\operatorname{int} A) \subseteq f(A)$ and by hypothesis $f(\operatorname{int} A)$ is open. By Proposition 1.1.14 (a) it follows that $f(\operatorname{int} A) \subseteq \operatorname{int} f(A)$.

(b) $\Longrightarrow$ (c): Let $V \in \mathcal{B}_X$. Then by hypothesis $f(V) = f(\operatorname{int} V) \subseteq \operatorname{int} f(V)$. Hence, $f(V) = \operatorname{int} f(V)$, that is, $f(V) \in \tau_Y$.

(c) $\Longrightarrow$ (a): Let $V \subseteq X$ be open. Then $V = \bigcup_{i \in I} V_i$ with $V_i \in \mathcal{B}_X$. We have

$$f(V) = f\Big(\bigcup_{i \in I} V_i\Big) = \bigcup_{i \in I} f(V_i) \in \tau_Y .$$

Therefore, f is open. □

Next we identify a subfamily of continuous functions that is in the core of point-set topology.

Definition 1.1.41. Let X and Y be two topological spaces and $f : X \to Y$ is a bijection. We say that f is a **homeomorphism** if both f and f^{-1} are continuous. Then we say that the spaces X and Y are **homeomorphic**. Instead of homeomorphism we also say that f is **bicontinuous**.

As an easy consequence of this definition and of Proposition 1.1.40 we have the following proposition.

Proposition 1.1.42. *Let X and Y be topological spaces and let $f : X \to Y$ be a bijection, then the following statements are equivalent:*
(a) *f is a homeomorphism;*
(b) *f is continuous and open;*
(c) *f is continuous and closed;*
(d) *$f(\overline{A}) = \overline{f(A)}$ for every $A \subseteq X$.*

Remark 1.1.43. Given a homeomorphism $f : X \to Y$, $U \subseteq X$ is open if and only if $f(U) \subseteq Y$ is open. Thus a homeomorphism gives a bijection between the topologies of X and Y. Hence, any property of X that is expressed using only the topology of X, yields the same property on Y. Such a property of X is said to be a **topological property** of X.

1.2 Separation and Countability Properties – Convergence

The so-called separation properties determine how rich the supply is of open sets in a given topological space. This is important because the supply of open sets determines the supply of continuous functions. We need to have a rich enough supply of continuous functions in order to produce interesting results.

We start with a notion, which for analysis, is the minimal requirement for a topological space.

Definition 1.2.1. A topological space X is said to be **Hausdorff** (or **T_2-space**) if for every pair $x, u \in X$ we can find $U \in \mathcal{N}(x)$ and $V \in \mathcal{N}(u)$ such that $U \cap V = \emptyset$.

Since our aim is to use topology to investigate problems in analysis, from now on all topological spaces considered are Hausdorff. Let us give an example of a space that is important in algebraic geometry and that is not Hausdorff.

Example 1.2.2. Let $n \in \mathbb{N}$ and let $\mathcal{P}$ denote the set of all polynomials in n variables $\{x_1, \ldots, x_n\}$. Given $p \in \mathcal{P}$, let

$$Z(p) = \{(x_1, \ldots, x_n) \in \mathbb{R}^n : p(x_1, \ldots, x_n) = 0\} .$$

Let $\mathcal{B}$ be the family of all complements of the set $Z(p)$ with $p \in \mathcal{P}$. One can show that $\mathcal{B}$ is a basis for a topology of $\mathbb{R}^n$. This topology is called the **Zariski topology** on $\mathbb{R}^n$ and it turns out that it is not Hausdorff.

Proposition 1.2.3. *The Hausdorff property is hereditary and topological.*

Proof. Let (X, τ) be the topological space and $A \subseteq X$ endowed with the subspace topology $\tau(A)$. Consider two distinct points $x, u \in A$. We can find $U, V \in \tau$ with $x \in U$ and $u \in V$ such that $U \cap V = \emptyset$. Then $U \cap A \in \tau(A)$, $V \cap A \in \tau(A)$ and $(U \cap A) \cap (V \cap A) = \emptyset$. Hence, $(A, \tau(A))$ is Hausdorff.

Let X be a Hausdorff topological space, Y a topological space, and $f: X \to Y$ a homeomorphism. If $y, v \in Y$ are distinct points, then $f^{-1}(y), f^{-1}(v) \in X$ are distinct as well. Since X is Hausdorff we can find $U, V \in \tau$ such that $f^{-1}(y) \in U, f^{-1}(v) \in V$ and $U \cap V = \emptyset$. This implies that $y \in f(U)$, $v \in f(V)$ are both open sets in Y and $f(U) \cap f(V) = \emptyset$. Therefore, Y is Hausdorff as well. ☐

Proposition 1.2.4. *If X is a Hausdorff topological space and $A \subseteq X$ is finite, then A is closed.*

Proof. It suffices to show that every singleton $\{x\}$ is closed. So let $u \in X$ with $u \neq x$. Then we can find $U \in \mathcal{N}(x)$ and $V \in \mathcal{N}(u)$ such that $U \cap V = \emptyset$. This means that $x \notin \overline{\{u\}}$. Therefore $\overline{\{x\}} = \{x\}$ and so every singleton $\{x\}$ is closed. ☐

Proposition 1.2.5. *If X is a Hausdorff topological space and $A \subseteq X$, then $x \in A'$, that is, x is a cluster point of A, if and only if every $U \in \mathcal{N}(x)$ contains infinitely many points of A.*

Proof. $\Longrightarrow$: Arguing by contradiction, suppose that we can find $U \in \mathcal{N}(x)$ such that $U \cap A$ is a finite set. Then $U \cap (A \setminus \{x\})$ is finite. Let $U \cap (A \setminus \{x\}) = \{x_k\}_{k=1}^n$. From Proposition 1.2.4 we know that $\{x_k\}_{k=1}^n$ is a closed subset of X. Hence $X \setminus \{x_k\}_{k=1}^n$ is open. Then

$$V = U \cap (X \setminus \{x_k\}_{k=1}^n) \in \mathcal{N}(x)$$

and $V \cap A = \emptyset$, a contradiction to the fact that $x \in A'$.

$\Longleftarrow$: By hypothesis, every $U \in \mathcal{N}(x)$ intersects A at infinitely many points. Then according to Definition 1.1.12 (c), we have $x \in A'$. □

Proposition 1.2.6. *For a topological space X the following statements are equivalent:*
(a) *X is Hausdorff;*
(b) *Given $x \in X$ and $u \neq x$ we can find $U \in \mathcal{N}(x)$ such that $u \notin \overline{U}$;*
(c) *For every $x \in X$ we have $\{x\} = \bigcap\{\overline{U} : U \in \mathcal{N}(x)\}$.*

Proof. (a) $\Longrightarrow$ (b): Let $x \in X$ and $u \neq x$. Since by hypothesis X is Hausdorff we can find $U \in \mathcal{N}(x)$ and $V \in \mathcal{N}(u)$ such that $U \cap V = \emptyset$. This means that $u \notin \overline{U}$.

(b) $\Longrightarrow$ (c): Let $u \neq x$. By hypothesis we can find $U \in \mathcal{N}(x)$ such that $u \notin \overline{U}$. Therefore we conclude that $\{x\} = \bigcap\{\overline{U} : U \in \mathcal{N}(x)\}$.

(c) $\Longrightarrow$ (a): Let $x \neq u$. We can find $U \in \mathcal{N}(x)$ such that $u \notin \overline{U}$ and $V \in \mathcal{N}(u)$ such that $x \notin \overline{V}$. We set $U' = U \cap (X \setminus \overline{V}) \in \mathcal{N}(x)$ and $V' = V \cap (X \setminus \overline{U}) \in \mathcal{N}(u)$. Evidently $U' \cap V' = \emptyset$ and this shows that X is Hausdorff. □

Now we strengthen the separation property.

Definition 1.2.7. A Hausdorff topological space X is said to be **regular** (or **T_3-space**) if for each closed set $C \subseteq X$ and each $x \notin C$ we can find open sets U and V such that $x \in U$, $C \subseteq V$ and $U \cap V = \emptyset$.

Proposition 1.2.8. *A Hausdorff topological space X is regular if and only if for every point $x \in X$ and every $U \in \mathcal{N}(x)$ we can find $W \in \mathcal{N}(x)$ such that $\overline{W} \subseteq U$.*

Proof. $\Longrightarrow$: Let $x \in X$ and $U \in \mathcal{N}(x)$. Then $X \setminus U$ is a closed set not containing x. Since by hypothesis X is regular, we can find open sets W, V such that

$$x \in W\,, \quad X \setminus U \subseteq V \quad \text{and} \quad W \cap V = \emptyset\,. \tag{1.2.1}$$

We have $W \subseteq X \setminus V$ and so $\overline{W} \subseteq X \setminus V$ since $X \setminus V$ is closed. Then, because of (1.2.1),

$$\overline{W} \subseteq X \setminus V \subseteq X \setminus (X \setminus U) = U\,.$$

This means that $W \in \mathcal{N}(x)$ is the desired neighborhood of x.

$\Longleftarrow$: Let $x \in X$ and let $C \subseteq X$ be closed such that $x \notin C$. Then $X \setminus C \in \mathcal{N}(x)$ and so by hypothesis we can find $W \in \mathcal{N}(x)$ such that $\overline{W} \subseteq X \setminus C$. Then W and $X \setminus \overline{W}$ are open sets such that $x \in W$, $C \subseteq X \setminus \overline{W}$ and $W \cap (X \setminus \overline{W}) = \emptyset$ which by Definition 1.2.7 means that X is regular. □

Proposition 1.2.9. *A Hausdorff topological space X is regular if and only if for every point $x \in X$ and every closed set $C \subseteq X$ such that $x \notin C$ we can find open sets U, V for which we have $\overline{U} \cap \overline{V} = \emptyset$, $U \in \mathcal{N}(x)$ and $C \subseteq V$.*

Proof. $\Longrightarrow$: Let $x \in X$ and let $C \subseteq X$ be a closed set such that $x \notin C$. Since by hypothesis, X is regular, invoking Proposition 1.2.8, we can find $W \in \mathcal{N}(x)$ such that $\overline{W} \subseteq X \setminus C$. A new application of Proposition 1.2.8 produces $U \in \mathcal{N}(x)$ such that $\overline{U} \subseteq W$. Let $V = X \setminus \overline{W}$, which is open. Then we obtain $\overline{U} \subseteq W \subseteq \overline{W} \subseteq X \setminus C$, which gives $C \subseteq X \setminus \overline{W} = V$. Therefore, U and V is the desired pair of open sets.

$\Longleftarrow$: This is obvious from Definition 1.2.7. □

Proposition 1.2.10. *The regularity property is hereditary and topological.*

Proof. Let $A \subseteq X$ and let $D \subseteq A$ be relatively closed and let $x \in A \setminus D$. From Proposition 1.1.27 we have $D = C \cap A$ with closed $C \subseteq X$. Since $x \notin C$ and X is regular, we can find open subsets U, V of X such that $x \in U$, $C \subseteq V$ and $U \cap V = \emptyset$. Then $U \cap A$, $V \cap A$ are relatively open in A, $x \in U \cap A$ and $D \subseteq V \cap A$. This shows that A with the relative (subspace) topology is regular.

Let $f: X \to Y$ be a homeomorphism and $y \in Y$, $C \subseteq Y$ closed with $y \notin C$. Let $x = f^{-1}(y)$ and $\hat{C} = f^{-1}(C)$. Evidently $\hat{C} \subseteq X$ is closed and $x \notin \hat{C}$. Since X is regular we can find open subsets $\hat{U}, \hat{V}$ of X such that $x \in \hat{U}$, $\hat{C} \subseteq \hat{V}$ and $\hat{U} \cap \hat{V} = \emptyset$. This gives $y \in f(\hat{U}) = U$, $f(\hat{C}) = C \subseteq f(\hat{V}) = V$ and $f(\hat{U}) \cap f(\hat{V}) = \emptyset$ since f is a homeomorphism. But from Proposition 1.1.42 we have that U, V are open subsets of Y. Hence we conclude that Y is regular. □

We further strengthen the separation property.

Definition 1.2.11. A Hausdorff topological space X is said to be **normal** (or **T_4-space**) if for each pair A, C of disjoint closed sets in X, we can find open sets U, V such that $A \subseteq U$, $C \subseteq V$ and $U \cap V = \emptyset$.

Remark 1.2.12. The definition above can be equivalently stated as follows: "If U_1, U_2 are open sets in X such that $X = U_1 \cup U_2$, then we can find closed subsets C_1, C_2 of X such that $C_1 \subseteq U_1$, $C_2 \subseteq U_2$ and $X = C_1 \cup C_2$".

The next two propositions characterize normality and are proven with arguments similar to the ones used in Propositions 1.2.8 and 1.2.9.

Proposition 1.2.13. *A Hausdorff topological space X is normal if and only if for each closed set $C \subseteq X$ and each open set $U \subseteq X$ such that $C \subseteq U$ we can find an open set $V \subseteq X$ for which we have $C \subseteq V \subseteq \overline{V} \subseteq U$.*

Proposition 1.2.14. *A Hausdorff topological space X is normal if and only if for each pair A, C of disjoint closed sets in X we can find open sets U, V in X such that $A \subseteq U$, $C \subseteq V$ and $\overline{U} \cap \overline{V} = \emptyset$.*

Proposition 1.2.15. (a) *A closed subset of a normal space is normal.*
(b) *Normality is preserved under continuous, closed surjections.*

Proof. (a) Let X be a normal topological space and $A \subseteq X$ a closed set. Suppose that $C \subseteq A$ is relatively closed. Then $C \subseteq X$ is closed by Proposition 1.1.27. This observation leads immediately to the normality of A.

(b) Let X be a normal topological space, Y a topological space, and $f: X \to Y$ a continuous, closed surjection. Suppose that U_1, U_2 are open subsets of Y such that $Y = U_1 \cup U_2$. Then $\hat{U}_1 = f^{-1}(U_1)$, $\hat{U}_2 = f^{-1}(U_2)$ are open in X and $X = \hat{U}_1 \cup \hat{U}_2$. The normality of X implies that we can find closed subsets $\hat{C}_1, \hat{C}_2$ of X such that $\hat{C}_1 \subseteq \hat{U}_1$, $\hat{C}_2 \subseteq \hat{U}_2$ and $X = \hat{C}_1 \cup \hat{C}_2$; see Remark 1.2.12.

Since f is closed we have that $C_1 = f(\hat{C}_1)$, $C_2 = f(\hat{C}_2)$ are closed subsets of Y and $C_1 \subseteq U_1$, $C_2 \subseteq U_2$ as well as $Y = C_1 \cup C_2$. According to Remark 1.2.12, this means that Y is normal as well. □

Remark 1.2.16. Part (a) of Proposition 1.2.15 fails if the subset is not closed. For a counterexample we refer to Dugundji [102, p. 145].

As we already mentioned in the beginning of this section, richness in open sets implies richness in continuous functions. This is illustrated in the theorem that follows. The result is known as "Urysohn's Lemma".

Theorem 1.2.17 (Urysohn's Lemma). *A Hausdorff topological space X is normal if and only if for each pair A, C of disjoint closed subsets of X we can find a continuous function $f: X \to [0,1]$ such that $f|_A = 0$ and $f|_C = 1$.*

Proof. $\Longrightarrow$: Let D be the set of all rationals r of the form $r = k/2^n$ with $0 \le k/2^n \le 1$, that is, $k = 0, 1, \ldots, 2^n$ dyadic fractions. We show that for every $r \in D$ we can assign an open set $U(r)$ such that

(a) $A \subseteq U(0) \subseteq \overline{U(0)} \subseteq X \setminus C$, $U(1) = X \setminus C$.
(b) $r < r'$ implies $\overline{U(r)} \subseteq U(r')$.

We proceed by induction on the exponent $n \in \mathbb{N}$. So, let

$$E_n = \left\{ U\left(\frac{k}{2^n}\right) : k = 0, 1, \ldots, 2^n \right\}, \quad n \in \mathbb{N}.$$

Then $E_0 = \{U(0), U(1) = X \setminus C\}$ and (a) is satisfied by Proposition 1.2.13. Suppose that E_{n-1} have been constructed. Clearly we need to define $U(k/2^n)$ for $k =$ odd. For $k =$ odd, from the induction hypothesis, we have

$$\overline{U\left(\frac{k-1}{2^n}\right)} \subseteq U\left(\frac{k+1}{2^n}\right),$$

see (b). So we define $U(k/2^n) = U$ with U being an open set such that, due to Proposition 1.2.13,

$$\overline{U\left(\frac{k-1}{2^n}\right)} \subseteq U \subseteq \overline{U} \subseteq U\left(\frac{k+1}{2^n}\right).$$

This completes the induction and we have defined the collection

$$\left\{U\left(\frac{k}{2^n}\right): k = 0, 1, \ldots, 2^n, n \in \mathbb{N}\right\}.$$

We define the desired function f by setting

$$f(x) = \begin{cases} 0 & \text{if } x \in U(r) \text{ for every } r = \text{dyadic fraction as above}, \\ \sup\{r : x \notin U(r)\} & \text{otherwise}. \end{cases}$$

Then f has values in $[0, 1]$ and $f|_A = 0, f|_C = 1$. So it remains to show that f is continuous. Note that the intervals $\{[0, a), (a, 1] : 0 < a < 1\}$ form a subbasis for $[0, 1]$ with the Euclidean topology. So, according to Remark 1.1.33 it suffices to show that $f^{-1}([0, a))$ and $f^{-1}((a, 1])$ are open. Note that $f(x) < a$ if and only if $x \in U(r)$ for some $r < a$. It follows that $f^{-1}([0, a)) = \bigcup_{r<a} U(r)$, which is open. Similarly, $f(x) > a$ if and only if $x \notin \overline{U(r)}$ for some $r > a$. Therefore $f^{-1}((a, 1]) = \bigcup_{r>a}(X \setminus \overline{U(r)})$, which is open. This proves the continuity of f.

$\Longleftarrow$: Let $A, C \subseteq X$ be disjoint closed sets. By hypothesis we can find a continuous function $f : X \to [0, 1]$ such that

$$f|_A = 0 \quad \text{and} \quad f|_C = 1. \tag{1.2.2}$$

Let $U = \{x \in X : f(x) < 1/2\}$ and $V = \{x \in X : f(x) > 1/2\}$. Then $U, V \subseteq X$ are open, $U \cap V = \emptyset, A \subseteq U, C \subseteq V$, see (1.2.2), which implies that X is normal. □

Remark 1.2.18. We can have a form of this result that is a little more flexible. To be more precise, we can replace $[0, 1]$ by $[a, b]$ with $a, b \in \mathbb{R}$, $a \le b$ and $f|_A = a, f|_C = b$. Indeed, let f_0 be the continuous separating function postulated by Theorem 1.2.17. Then set $f = (b - a)f_0 + a$. Evidently this function has the desired properties.

There is another such functional characterization of normality, namely the so-called "Tietze Extension Theorem". We state this result at the end of this section and for its proof, which is rather technical, we refer to Dugundji [102].

Evidently we have

$$\text{Normal} \implies \text{Regular} \implies \text{Hausdorff}.$$

None of these implications is in general reversible. Between regular and normal spaces we can fit another class given in the next definition.

Definition 1.2.19. A Hausdorff topological space X is said to be **completely regular** if for each $x \in X$ and each closed set $C \subseteq X$ with $x \notin C$, we can find a continuous function $f: X \to [0,1]$ such that $f(x) = 0$ and $f|_C = 1$.

Now we pass to the countability properties of a topological space.

Definition 1.2.20. (a) A topological space X is said to be **first countable** if it has a countable local basis at each point of X.
(b) A topological space X is said to be **second countable** if it has a countable basis.

Remark 1.2.21. Evidently a second countable space is also first countable. The converse is not true. Every metric space (X, d) is first countable. Indeed for every $x \in X$, $\mathcal{B}(x) = \{B_r(x): r \in \mathbb{Q}\}$ with $B_r(x) = \{u \in X: d(u,x) < r\}$ is a countable local basis at $+x$ and so X is first countable.

Proposition 1.2.22. *Every second countable space is separable.*

Proof. Let X be a second countable space and let $\mathcal{B}$ be the countable basis of X. Let D be the countable set formed by choosing an element from each nonempty basic open set. Then Corollary 1.1.6 implies that $\overline{D} = X$. □

Remark 1.2.23. The converse of the proposition above is not true. Consider the space $X = \mathbb{R}$ topologized with the topology that has as its basis intervals of the form $(a, b]$ with $a, b \in \mathbb{R}$. This topology is known as the **upper limit topology** and is denoted by τ_u. We can easily check that the Euclidean topology on $X = \mathbb{R}$ is weaker than τ_u. The space $(\mathbb{R}, \tau_u)$ is first countable. To see this, consider $\mathcal{B}(x) = \{(r, x]: r \in \mathbb{Q}\}$ for each $x \in \mathbb{R}$.

In addition, $(\mathbb{R}, \tau_u)$ is separable. Indeed, the rationals are a countable dense subset. However, $(\mathbb{R}, \tau_u)$ is not second countable. To see this, note that if $\{(a_n, b_n]\}_{n\in\mathbb{N}}$ is a countable collection in τ_u, then by choosing $a, b \neq b_n$ for all $n \in \mathbb{N}$, the open set $(a, b]$ cannot be expressed as a union of sets in the countable collection. The proposition above also says that every nonseparable metric space is first countable but not second countable.

Proposition 1.2.24. (a) *Second countability is preserved by continuous open surjections.*
(b) *Second countability is hereditary.*
(c) *Separability is preserved by continuous surjections.*

Proof. (a) Let X be a second countable topological space, Y another topological space, and $f: X \to Y$ a continuous open surjection. Consider a basis $\{U_n\}_{n\in\mathbb{N}}$ for the topology of X, and using Corollary 1.1.6, we see that $\{f(U_n)\}_{n\in\mathbb{N}}$ is a countable basis for Y.

(b) This is obvious.

(c) Let X be a separable topological space, Y another topological space and $f: X \to Y$ a continuous surjection. Consider $D \subseteq X$ as being a countable dense subset. From Proposition 1.1.35 (b) we have $Y = f(X) = f(\overline{D}) \subseteq \overline{f(D)}$. Hence, $Y = \overline{f(D)}$ and $f(D)$ is countable. □

Remark 1.2.25. Clearly, an open subset of a separable topological space is separable for the subspace topology. If X is a second countable topological space, then every subset of X endowed with the subspace topology is separable.

Definition 1.2.26. Let (X, τ) be a topological space.
(a) An **open cover** of X is a collection $\mathcal{D} \subseteq \tau$ such that $X = \bigcup\{U : U \in \mathcal{D}\}$. A **subcover** of an open cover $\mathcal{D}$ is a subfamily $\mathcal{D}'$ of $\mathcal{D}$ such that $X = \bigcup\{U : U \in \mathcal{D}'\}$.
(b) We say that X is a **Lindelöf space** if every open cover contains a countable subcover.

The next result relates the Lindelöf property with second countability. It is known as “Lindelöf's Theorem”.

Theorem 1.2.27 (Lindelöf's Theorem). *Every second countable space is Lindelöf.*

Proof. Let X be a second countable topological space and $\{U_n\}_{n\geq 1}$ a countable basis of X. Consider an open cover $\mathcal{D} = \{V_i\}_{i\in I}$ of X. For each $x \in X$, let $V_{i(x)} \in \{V_i\}_{i\in I}$ be such that $x \in V_{i(x)}$. Let $U_{n(x)} \in \{U_n\}_{n\geq 1}$ be such that $x \in U_{n(x)} \subseteq V_{i(x)}$. Then the family $\{U_{n(x)}\}_{x\in X}$ is a countable open cover of X. For each $U_{n(x)}$ let $V'_{i(x)} \in \mathcal{D}$ be such that $U_{n(x)} \subseteq V'_{i(x)}$. Then the collection $\{V'_{i(x)}\}_{x\in X}$ is a countable subcover of $\mathcal{D}$. Therefore, X is Lindelöf. □

Remark 1.2.28. The converse of the Theorem above is not true. Consider the space $(\mathbb{R}, \tau_u)$; see Remark 1.2.23. Then we can show that it is Lindelöf (see Dugundji [102]), but it is not second countable; see again Remark 1.2.23.

Proposition 1.2.29. (a) *The Lindelöf property is preserved by continuous surjections.*
(b) *A closed subset of a Lindelöf space is Lindelöf for the subspace topology.*

Proof. (a) Let X be a Lindelöf space, Y another topological space, and $f : X \to Y$ a continuous surjection. Consider an open cover $\{U_i\}_{i\in I}$ of Y. Then $\{V_i\}_{i\in I} = \{f^{-1}(U_i)\}_{i\in I}$ is an open cover of X. Since X is Lindelöf, we can find a countable subcover $\{V_n\}_{n\in\mathbb{N}} = \{f^{-1}(U_n)\}_{n\in\mathbb{N}}$. Then $\{U_n\}_{n\in\mathbb{N}}$ is a countable subcover of $\{U_i\}_{i\in I}$ and so we conclude that Y is Lindelöf.

(b) Let X be a Lindelöf space and $C \subseteq X$ a closed subset. Consider an open cover $\{V_i\}_{i\in I}$ of C with the subspace topology. Then $V_i = U_i \cap C$ with $U_i \subseteq X$ open. Then $\{U_i, X \setminus C\}_{i\in I}$ is an open cover of X. Since X is Lindelöf we can find a countable subcover $\{U_n\}_{n\in\mathbb{N}}$. Then $\{U_n \cap C\}_{n\in\mathbb{N}}$ is a countable subcover of $\{V_i\}_{i\in I}$. So, we conclude that C with the subspace topology is Lindelöf. □

We know that a sequence is a map from $\mathbb{N}$ into X but it is more convenient to think of a sequence as a subset of X indexed by $\mathbb{N}$. We generalize this notion by replacing $\mathbb{N}$ with a more general index set.

Definition 1.2.30. Let X be a set.
(a) A **relation** is any subset $R \subseteq X \times X$. Given a relation, it is more suggestive to write xRy instead of $(x, y) \in R$. We say that R is **reflexive** if xRx for all $x \in X$. We say that R is **symmetric** if xRy implies yRx. We say that R is **antisymmetric** if xRy and yRx imply $x = y$. We say that R is **transitive** if xRy and yRz imply xRz.

(b) A relation R is called an **equivalence relation** if it is reflexive, symmetric, and transitive.
(c) A relation R is called a **partial order** if it is antisymmetric and transitive. In this case we write $x \leq y$ if and only if xRy or $x = y$ (a reflexive partial order) and $x < y$ if and only if xRy and $x \neq y$ (a strict partial order). A **linear order** R is a partial order such that for all $x, u \in X$, either xRu or uRx. A **chain** is a linearly ordered subset of a partially ordered set.
(d) A **directed set** is a partially ordered set $(I, \leq)$ such that for any $\alpha, \beta \in I$ we can find $k \in I$ such that $\alpha \leq k$ and $\beta \leq k$.

Remark 1.2.31. Many authors require that a partial order is also reflexive. Definition 1.2.30 (c) is more flexible and allows both "≤" and "<" as partial orders. For any set V let $X = 2^V$ be the collection of all subsets of V. We write $A \leq C$ if and only if $C \subseteq A$ for $C, A \in 2^V$. This is a partial order, the reverse ordering of the sets.

Definition 1.2.32. Let X be a set. A **net** in X is a map $x: D \to X$ with a directed set D. The directed set D is known as the **index set** of the net.

Remark 1.2.33. As for sequences, we denote the map $x: D \to X$ simply by $\{x_\alpha\}_{\alpha\in D}$.

Definition 1.2.34. Let (X, τ) be a topological space. We say that a net $\{x_\alpha\}_{\alpha\in I}$ converges to some $x \in X$ if for every $U \in \mathcal{N}(x)$ there exists $\alpha_0 = \alpha_0(U)$ such that $x_\alpha \in U$ for all $\alpha \geq \alpha_0$, that is, $\{x_\alpha\}$ is eventually in every neighborhood of x. We say that x is the **limit** of the net $\{x_\alpha\}_{\alpha\in I}$ and we write $x_\alpha \to x$ or $x_\alpha \xrightarrow{\tau} x$ if we want to emphasize the topology τ.

Proposition 1.2.35. *A topological space X is Hausdorff if and only if every convergent net has a unique limit.*

Proof. $\Longrightarrow$: Arguing by contradiction, suppose that for a net $\{x_\alpha\}_{\alpha\in I} \subseteq X$ we have

$$x_\alpha \to x \quad \text{and} \quad x_\alpha \to \hat{x} \quad \text{with } x \neq \hat{x}\,.$$

Due to the Hausdorff property we can find $U \in \mathcal{N}(x)$ and $\hat{U} \in \mathcal{N}(\hat{x})$ such that $U \cap \hat{U} = \emptyset$. Furthermore we can find $\alpha_0, \hat{\alpha}_0 \in I$ such that $x_\alpha \in U$ for all $\alpha \geq \alpha_0$ and $x_\alpha \in \hat{U}$ for all $\alpha \geq \hat{\alpha}_0$. Since I is a directed set we can find $\alpha_* \in I$ such that $\alpha_* \geq \alpha_0$, $\alpha_* \geq \hat{\alpha}_0$. Then, $x_\alpha \in U$ and $x_\alpha \in \hat{U}$ for all $\alpha \geq \alpha_*$, a contradiction since $U \cap \hat{U} = \emptyset$. This proves the uniqueness of the limit.

$\Longleftarrow$: We argue again indirectly. To this end, suppose that X is not Hausdorff. Then we can find $x, u \in X$ with $x \neq u$ such that for every $U \in \mathcal{N}(x)$ and every $V \in \mathcal{N}(u)$ there holds $U \cap V \neq \emptyset$. For each $(U, V) \in \mathcal{N}(x) \times \mathcal{N}(u)$, let $x_{UV} \in U \cap V$ and note that the net $\{x_{UV}\}$ converges to both x and u, contradicting our hypothesis. □

Proposition 1.2.36. *If X is a topological space and $A \subseteq X$, then $x \in \overline{A}$ if and only if we can find a net $\{x_\alpha\}_{\alpha\in I} \subseteq A$ such that $x_\alpha \to x$.*

Proof. $\Longrightarrow$: If $U \in \mathcal{N}(x)$, then $U \cap A \neq \emptyset$. Let $x_U \in U \cap A$. Then $\{x_U\}_{U \in \mathcal{N}(x)}$ is a net in A with $\mathcal{N}(x)$ ordered by reverse inclusion, that is, $U_1 \leq U_2$ if and only if $U_2 \subseteq U_1$, and $x_U \to x$.

$\Longleftarrow$: This is obvious. □

Proposition 1.2.37. *If X and Y are topological spaces, $x \in X$ and $f: X \to Y$, then f is continuous at x if and only if for every net $x_\alpha \to x$ we have $f(x_\alpha) \to f(x)$.*

Proof. $\Longrightarrow$: According to Definition 1.1.30, given $U \in \mathcal{N}(f(x))$, we can find $V \in \mathcal{N}(x)$ such that $f(V) \subseteq U$. If $x_\alpha \to x$, then we can find $\alpha_0 \in I$ such that $x_\alpha \in V$ for all $\alpha \geq \alpha_0$. Then $f(x_\alpha) \in f(V) \subseteq U$ for all $\alpha \geq \alpha_0$. Since $U \in \mathcal{N}(f(x))$ is arbitrary, we conclude that $f(x_\alpha) \to f(x)$.

$\Longleftarrow$: Arguing by contradiction, suppose that f is not continuous at x. Then there is $V \in \mathcal{N}(f(x))$ such that $f^{-1}(V) \notin \mathcal{N}(x)$. Then $x \in \overline{X \setminus f^{-1}(V)}$ and so by Proposition 1.2.36 we can find a net $\{x_\alpha\}_{\alpha \in I} \subseteq X \setminus f^{-1}(V)$ such that $x_\alpha \to x$. By hypothesis we have $f(x_\alpha) \to f(x)$. Since $Y \setminus V$ is closed and $f(x_\alpha) \in Y \setminus V$ for all $\alpha \in I$, it follows that $f(x) \in Y \setminus V$, a contradiction. This proves the continuity of f at x. □

The next notion generalizes that of a subsequence.

Definition 1.2.38. Let X be a topological space. A net $\{u_\beta\}_{\beta \in J} \subseteq X$ is a **subnet** of a net $\{x_\alpha\}_{\alpha \in I}$ if there exists a map $\vartheta: J \to I$ such that
(a) $u_\beta = x_{\vartheta(\beta)}$ for every $\beta \in J$;
(b) for each $\alpha_0 \in I$, there exists $\beta_0 \in J$ such that $\beta \geq \beta_0$ implies $\vartheta(\beta) \geq \alpha_0$.

Remark 1.2.39. A subsequence of a sequence is a subnet. But we can have subnets of a sequence that are not subsequences. Indeed, let $\{x_n\}_{n \in \mathbb{N}} = \{n^2 + 1\}_{n \in \mathbb{N}}$ and $\{y_{m,n}\}_{(m \times n) \in \mathbb{N} \times \mathbb{N}} = \{m^2 + 2mn + n^2 + 1\}_{(m \times n) \in \mathbb{N} \times \mathbb{N}}$. Then $\{y_{m,n}\}_{(m \times n) \in \mathbb{N} \times \mathbb{N}}$ is a subnet of $\{x_n\}_{n \in \mathbb{N}}$. To see this, let $\vartheta: \mathbb{N} \times \mathbb{N} \to \mathbb{N}$ be defined by $\vartheta(m, n) = m + n$. However, note that $\{y_{m,n}\}_{(m \times n) \in \mathbb{N} \times \mathbb{N}}$ is not a subsequence of $\{x_n\}_{n \in \mathbb{N}}$.

For the next result, the limits need not be unique.

Proposition 1.2.40. *If X is a topological space, then a net in X converges to a point if and only if every subnet converges in X to the same point.*

Proof. $\Longrightarrow$: This is obvious.

$\Longleftarrow$: Suppose that $\{x_\alpha\}_{\alpha \in I}$ is a net in X and assume that every subnet of $\{x_\alpha\}_{\alpha \in I}$ converges to the same limit x. Arguing by contradiction, suppose that the net $\{x_\alpha\}_{\alpha \in I}$ does not converge to x. Then we can find $U \in \mathcal{N}(x)$ such that for any $\alpha \in I$ we can find $\vartheta(\alpha) \geq \alpha$ such that $x_{\vartheta(\alpha)} \notin V$. If we set $u_\alpha = x_{\vartheta(\alpha)}$, then $\{u_\alpha\}_{\alpha \in I}$ is a subnet of $\{x_\alpha\}_{\alpha \in I}$, which does not converge to x. This contradicts our hypothesis. So, $x_\alpha \to x$. □

Remark 1.2.41. For a bounded real net $\{x_\alpha\}_{\alpha \in I}$ we can define the **limit superior** and the **limit inferior** by setting

$$\liminf_\alpha x_\alpha = \sup_{\alpha \in I} \inf_{\vartheta \geq \alpha} x_\vartheta \quad \text{and} \quad \limsup_\alpha x_\alpha = \inf_{\alpha \in I} \sup_{\vartheta \geq \alpha} x_\vartheta .$$

Evidently, $x_\alpha \to x$ if and only if $x = \liminf_\alpha x_\alpha = \limsup_\alpha x_\alpha$.

Nets were introduced because in general, sequences are not enough to describe a given topology τ.

Definition 1.2.42. Let (X, τ) be a topological space. We denote by τ_{seq} the topology on X whose closed sets are the sequentially τ-closed sets in X.

Remark 1.2.43. From the definition above, it follows directly that τ_{seq} is the strongest, that is, the largest topology on X for which the converging sequences are the τ-converging sequences. Hence, $\tau \subseteq \tau_{\text{seq}}$ and $\tau = \tau_{\text{seq}}$ if and only if (X, τ) is first countable.

In Theorem 1.2.17 we produced a characterization of normal spaces. We conclude this section by providing an alternative characterization of normality. The result is known as the "Tietze Extension Theorem". We do not prove it, since later we will prove a more general result known as the "Dugundji Extension Theorem"; see Theorem 1.7.29.

Theorem 1.2.44 (Tietze Extension Theorem). *A Hausdorff topological space X is normal if and only if for every closed $C \subseteq X$ and for every continuous $f: C \to \mathbb{R}$ there exists a continuous function $\hat{f}: X \to \mathbb{R}$ such that $\hat{f}|_C = f$. Moreover, if $|f(x)| \leq M$ for some $M > 0$ and for all $x \in A$, then $\hat{f}$ can be chosen so that $|\hat{f}(x)| \leq M$ for all $x \in X$.*

1.3 Weak, Product, and Quotient Topologies

Let X and $\{Y_i\}_{i \in I}$ be topological spaces and $f_i: X \to Y_i$ be continuous functions. From Proposition 1.1.32 we see that if we strengthen (enrich) the topology on X, we preserve the continuity of the f_i's. Thus it is natural to inquire what the smallest topology on X is, which preserves the continuity of the f_i's. This leads to the notions of weak and product topologies, which occur in a prominent position in many areas of analysis such as functional analysis.

Definition 1.3.1. Let X be a nonempty set, let $\{(Y_i, \tau_i)\}_{i \in I}$ be a family of Hausdorff topological spaces and let $f_i: X \to Y_i$ with $i \in I$ be a family of functions. The **weak topology** or **initial topology** on X generated by the family of functions $\{f_i\}_{i \in I}$ is the weakest topology on X that makes all f_i's continuous. The weak topology is denoted by $\mathrm{w}(X, \{f_i\})$ or simply by w if X and $\{f_i\}$ are clearly understood.

Remark 1.3.2. Simple set theory reveals that the weak topology is generated, that is, it has as subbasis, the sets of the form

$$\{f_i^{-1}(V): V \in \tau_i, i \in I\}. \tag{1.3.1}$$

Recalling that to check continuity it suffices to consider the inverse image of subbasic sets, another more economical subbasis is given by

$$\{f_i^{-1}(V): V \in \mathcal{L}_i, i \in I\} \tag{1.3.2}$$

with a subbasis $\mathcal{L}_i$ for the topology τ_i. Then a basis for the weak topology is produced by taking finite intersections of the sets above; see (1.3.1) and (1.3.2). An important special case is when $Y_i = \mathbb{R}$ for all $i \in I$. This is the case of the weak topology in functional analysis. Then the subbasic elements are of the form

$$U(x; f, \varepsilon) = \{u \in X: |f(u) - f(x)| < \varepsilon\}$$

with $x \in X, f \in \{f_i\}$ and $\varepsilon > 0$.

Proposition 1.3.3. *A net $\{x_\alpha\}_{\alpha\in J}$ converges to x for the weak topology, which is denoted by $x_\alpha \xrightarrow{\mathrm{w}} x$, if and only if $f_i(x_\alpha) \to f_i(x)$ for all $i \in I$.*

Proof. $\Longrightarrow$: This follows from Proposition 1.2.37, since each f_i is w-continuous.

$\Longleftarrow$: Let $V = \bigcap_{k=1}^n f_{i_k}^{-1}(V_{i_k})$ be a basic neighborhood of X where $V_{i_k} \in \tau_{i_k}$. Since by hypothesis $f_{i_k}(x_\alpha) \to f_{i_k}(x)$, we can find $\alpha_{i_k} \in J$ such that

$$x_\alpha \in f_{i_k}^{-1}(V_{i_k}) \quad \text{for all } \alpha \geq \alpha_{i_k} . \tag{1.3.3}$$

Since J is directed we can find $\alpha_0 \geq \alpha_{i_k}$ for all $k \in \{1, \dots, n\}$. Then $x_\alpha \in V$ for all $\alpha \geq \alpha_0$ because of (1.3.3). This implies $x_\alpha \xrightarrow{\mathrm{w}} x$ in X. □

Proposition 1.3.4. *If Z is another topological space and $g: Z \to X$ is a map, then g is continuous for the weak topology on X if and only if $f_i \circ g$ is continuous for all $i \in I$.*

Proof. $\Longrightarrow$: From Proposition 1.1.36 (a) we know that $f_i \circ g$ is continuous for all $i \in I$.

$\Longleftarrow$: Let $U \subseteq X$ be weakly open. Then

$$U = \bigcup_{\text{arbitrary}} \bigcap_{\text{finite}} f_i^{-1}(V_i) \quad \text{with } V_i \in \tau_i .$$

This gives

$$g^{-1}(U) = \bigcup_{\text{arbitrary}} \bigcap_{\text{finite}} g^{-1}(f_i^{-1}(V_i)) = \bigcup_{\text{arbitrary}} \bigcap_{\text{finite}} (f_i \circ g)^{-1}(V_i) ,$$

which is open in Z, and thus g is continuous. □

Consider X endowed with the weak topology $\mathrm{w}(X, \{f_i\})$. Suppose that $A \subseteq X$. Then we can consider on A the subspace topology induced by $\mathrm{w}(X, \{f_i\})$. However, we can also consider the weak topology $\mathrm{w}(A, \{f_i|_A\})$; see Proposition 1.1.36 (b). It is natural to ask what the relation is between these two topologies on A. It is easy to see that the two topologies have the same convergent nets. This leads to the next result.

Proposition 1.3.5. *If X is endowed with the weak topology $\mathrm{w}(X, \{f_i\})$ and $A \subseteq X$, then $\mathrm{w}(X, \{f_i\})|_A = \mathrm{w}(A, \{f_i|_A\})$.*

As we already mentioned, an analyst requires that a topological space is at least Hausdorff. So we need to know the conditions that guarantee that the weak topology is Hausdorff.

Definition 1.3.6. Let X and $\{Y_i\}_{i\in I}$ be sets and let $f_i: X \to Y_i$ be a family of functions. We say that the family $\{f_i\}_{i\in I}$ is **separating** (or **total**) if for every pair $(x, u) \in X \times X$ with $x \neq u$ we can find $i_0 \in I$ such that $f_{i_0}(x) \neq f_{i_0}(u)$.

Proposition 1.3.7. *If* $\mathrm{w}(X, \{f_i\})$ *is the weak topology on* X*, then* $\mathrm{w}(X, \{f_i\})$ *is Hausdorff if and only if* $\{f_i\}_{i\in I}$ *is separating.*

Proof. $\Longrightarrow$: Arguing by contradiction, suppose that the family $\{f_i\}_{i\in I}$ is not separating. So, we can find a pair $(x, u) \in X \times X$ with $x \neq u$ such that $f_i(x) = f_i(u)$ for all $i \in I$. Let $U \in \mathcal{N}_{\mathrm{w}}(x)$ where $\mathcal{N}_{\mathrm{w}}(x)$ is the family of weak neighborhoods of x. Then we can find $\{f_{i_k}\}_{k=1}^n \subseteq \{f_i\}_{i\in I}$ and $V_{i_k} \in \tau_{i_k}$ with $k \in \{1, \ldots, n\}$ such that

$$x \in \bigcap_{k=1}^{n} f_{i_k}^{-1}(V_{i_k}) \subseteq U . \tag{1.3.4}$$

Since $f_i(x) = f_i(u)$ for all $i \in I$, we have

$$u \in \bigcap_{k=1}^{n} f_{i_k}^{-1}(V_{i_k}) .$$

Due to (1.3.4) it follows $u \in U$. We infer that (X, w) is not Hausdorff, a contradiction.

$\Longleftarrow$: As before, we proceed indirectly. Suppose that (X, w) is not Hausdorff. Then according to Proposition 1.2.35 we can find a net $\{x_\alpha\}_{\alpha\in I} \subseteq X$ such that

$$x_\alpha \xrightarrow{\mathrm{w}} x \quad \text{and} \quad x_\alpha \xrightarrow{\mathrm{w}} \hat{x} , \quad x \neq \hat{x} .$$

For every $i \in I$ we have $f_i(x_\alpha) \to f_i(x)$ and $f_i(x_\alpha) \to f_i(\hat{x})$ in Y_i, which is Hausdorff. Hence, $f_i(x) = f_i(\hat{x})$ for all $i \in I$, see Proposition 1.2.35. This means that the family $\{f_i\}_{i\in I}$ is not separating, a contradiction. □

Next we derive some useful results concerning the weak topology. Let (X, τ) be a Hausdorff topological space. We will use the following notations:

- $C(X, \mathbb{R}) = \{f: X \to \mathbb{R}: f \text{ is continuous}\}$;
- $C_{\mathrm{b}}(X, \mathbb{R}) = \{f: X \to \mathbb{R}: f \text{ is bounded and continuous}\}$.

Proposition 1.3.8. *If* (X, τ) *is a Hausdorff topological space, then* $\mathrm{w}(X, C(X, \mathbb{R})) = \mathrm{w}(X, C_{\mathrm{b}}(X, \mathbb{R}))$.

Proof. Since $C_{\mathrm{b}}(X, \mathbb{R}) \subseteq C(X, \mathbb{R})$ we infer that $\mathrm{w}(X, C_{\mathrm{b}}(X, \mathbb{R})) \subseteq \mathrm{w}(X, C(X, \mathbb{R}))$. So we need to show that the opposite inclusion also holds. Let U be a subbasic open set in $\mathrm{w}(X, C(X, \mathbb{R}))$. Then we have

$$U(x;f,\varepsilon) = \{u \in X\colon |f(u) - f(x)| < \varepsilon\}$$

with $x \in X, f \in C(X,\mathbb{R})$ and $\varepsilon > 0$. Let

$$g(u) = \min\{f(x) + \varepsilon, \max\{f(x) - \varepsilon, f(u)\}\}\,.$$

Evidently we have $g \in C_b(X,\mathbb{R})$ and $U(x;g,\varepsilon) = U(x;f,\varepsilon)$, which implies that $\mathrm{w}(X, C(X,\mathbb{R})) \subseteq \mathrm{w}(X, C_b(X,\mathbb{R}))$. This proves the assertion. □

The next theorem characterizes completely regular spaces (see Definition 1.2.19) via the weak topologies of the previous proposition.

Theorem 1.3.9. *A Hausdorff topological space (X,τ) is completely regular if and only if $\tau = \mathrm{w}(X, C(X,\mathbb{R})) = \mathrm{w}(X, C_b(X,\mathbb{R}))$.*

Proof. $\Longrightarrow$: Let $U \in \tau$ and $x \in U$. Since X is completely regular, we can find $f \in C(X,\mathbb{R})$ such that $f(x) = 0$ and $f|_{X\setminus U} = 1$. Let $V = \{u \in X\colon f(u) < 1\}$. Then V is $\mathrm{w}(X, C(X,\mathbb{R}))$-open, $V \subseteq U$, and $x \in V$. Therefore, U is $\mathrm{w}(X, C(X,\mathbb{R}))$-open and so we infer that

$$\tau \subseteq \mathrm{w}(X, C(X,\mathbb{R}))\,. \tag{1.3.5}$$

From Definition 1.3.1 it is clear that we always have $\mathrm{w}(X, C(X,\mathbb{R})) \subseteq \tau$. This along with (1.3.5) and Proposition 1.3.8 yields $\tau = \mathrm{w}(X, C(X,\mathbb{R})) = \mathrm{w}(X, C_b(X,\mathbb{R}))$.

$\Longleftarrow$: Let $C \subseteq X$ be closed and $x \notin C$. Then $U = X \setminus C \in \mathcal{N}_w(x)$ where $\mathcal{N}_w(x)$ is the family of weak neighborhoods of x. So we can find $V = \bigcap_{i=1}^n \{u \in X\colon |f_i(u) - f_i(x)| < 1\}$, $f_i \in C(X,\mathbb{R})$ for all $i \in \{1,\dots,n\}$, such that $x \in V \subseteq U$. For each $i \in \{1,\dots,n\}$ we define $g_i(u) = \min\{1, |f_i(u) - f_i(x)|\}$ and set $g = \max_{1\le i\le n} g_i$. Obviously $g\colon X \to [0,1]$ is continuous and $g(x) = 0$ as well as $g|_C = 1$. This proves that X is completely regular. □

A weak topology of special interest is the product topology. So, let $\{(X_i,\tau_i)\}_{i\in I}$ be a family of Hausdorff topological spaces. Let $X = \prod_{i\in I} X_i$. The generic element $x \in X$ is denoted by $x = (x_i)$. For every $i \in I$ let $p_i\colon X \to X_i$ be defined by $p_i(x) = x_i$ where p_i is the projection map in the i-th component of the Cartesian product.

Definition 1.3.10. The **product topology** on X is the weak topology $\mathrm{w}(X, \{p_i\})$.

Remark 1.3.11. A basic element for the product topology has the form $V = \prod_{i\in I} V_i$ with $V_i \in \tau_i$ for all $i \in I$ and $V_i = X_i$ for all but a finite number of i's. In addition, note that $x^\alpha = (x_i^\alpha) \to x = (x_i)$ in $X = \prod_{i\in I} X_i$ if and only if $x_i^\alpha \to x_i$ for all $i \in I$. Note that if $A_i \subseteq X_i$ then $\overline{\prod_{i\in I} A_i} = \prod_{i\in I} \overline{A_i}$ and each projection map p_i is open.

Proposition 1.3.12. *$X = \prod_{i\in I} X_i$ with the product topology is Hausdorff.*

Proof. Recall that each X_i is Hausdorff. Let $x = (x_i) \in X$ and $u = (u_i) \in X$ with $x \neq u$. Then we can find at least one $i_0 \in I$ such that $x_{i_0} \neq u_{i_0}$. We can find $U_{i_0}, V_{i_0} \in \tau_{i_0}$ such that $x_{i_0} \in U_{i_0}$, $u_{i_0} \in V_{i_0}$ and $U_{i_0} \cap V_{i_0} = \emptyset$. Let $U = p_{i_0}^{-1}(U_{i_0})$ and $V = p_{i_0}^{-1}(V_{i_0})$. Then both are open in the product topology and $x \in U$, $u \in V$ and $U \cap V = \emptyset$. This implies that X is Hausdorff with the product topology. □

Proposition 1.3.13. *If $\{(X_i, \tau_i)\}_{i \in I}$ is a family of Hausdorff topological spaces, then $X = \prod_{i \in I} X_i$ endowed with the product topology is regular if and only if (X_i, τ_i) is regular for each $i \in I$.*

Proof. $\Longrightarrow$: Each X_i is homeomorphic to a slice of $X = \prod_{i \in I} X_i$. Hence, the implication follows from Proposition 1.2.10.

$\Longleftarrow$: Let $x = (x_i) \in X = \prod_{i \in I} X_i$ and let U be any subbasic neighborhood of x. Then $U = \prod_{i \in I} V_i$ with $V_i = X_i$ for all $i \in I \setminus \{i_0\}$, $V_{i_0} \in \tau_{i_0}$. Exploiting the regularity of X_{i_0} we can find $W_{i_0} \in \tau_{i_0}$ such that

$$x_{i_0} \in W_{i_0} \subseteq \overline{W}_{i_0} \subseteq V_{i_0}\,, \tag{1.3.6}$$

see Proposition 1.2.8. Let $W = \prod_{i \in I} W_i$ with $W_i = X_i$ for all $i \in I \setminus \{i_0\}$ and W_{i_0} as above. Then W is open in the product topology and because of Remark 1.3.11 as well as (1.3.6), it follows that

$$x \in W \subseteq \overline{W} = \prod_{i \in I} \overline{W}_i \subseteq \prod_{i \in I} V_i = V\,.$$

This proves that $X = \prod_{i \in I} X_i$ is regular with the product topology; see Proposition 1.2.8. □

The Cartesian product of normal spaces need not be normal. For a counterexample, see Dugundji [102, p. 145]. However, we have the following result.

Proposition 1.3.14. *If $\{(X_i, \tau_i)\}_{i \in I}$ is a family of Hausdorff topological spaces and $X = \prod_{i \in I} X_i$ endowed with the product topology is normal, then (X_i, τ_i) is normal for each $i \in I$.*

Proof. Note that for each $i \in I$, X_i is homeomorphic to a slice of $X = \prod_{i \in I} X_i$, which is closed, and hence normal due to Proposition 1.2.15 (a). Then the result follows from Proposition 1.2.15 (b). □

Next we will consider the complementary situation to the one that led to the weak topology. So, let X, Y be topological spaces and $f: X \to Y$ be a continuous map. If we weaken the topology on Y we preserve the continuity of f. Hence, we want to identify the largest topology on Y for which f remains continuous.

Definition 1.3.15. Let (X, τ) be a topological space, Y a set, and $f: X \to Y$ a surjection. The **quotient topology** on Y induced by f is $\tau_q = \{U \subseteq Y : f^{-1}(U) \in \tau\}$. When Y is endowed with the quotient topology, then we say that f is a **quotient map**.

Remark 1.3.16. The quotient topology on Y makes f continuous and it is clearly the largest topology on Y that does this.

Proposition 1.3.17. *If (X, τ_X), (Y, τ_Y) are topological spaces and $f: X \to Y$ is supposed to be a continuous, open surjection, then f is a quotient map, that is $\tau_Y = \tau_q$.*

Proof. By definition $\tau_Y \subseteq \tau_q$. On the other hand, if $U \in \tau_q$, then $f^{-1}(U) \in \tau_X$ and since f is open, we have $U = f(f^{-1}(U)) \in \tau_Y$ and so $\tau_q \subseteq \tau_Y$. Therefore $\tau_Y = \tau_q$. □

Corollary 1.3.18. *If $\{(X_i, \tau_i)\}_{i \in I}$ are Hausdorff topological spaces and $X = \prod_{i \in I} X_i$ is endowed with the product topology, then $\tau_i = \tau_q$ for each $i \in I$.*

Proof. Just recall that each projection map $p_i: X = \prod_{i \in I} X_i \to X_i$ is a continuous open surjection. □

Proposition 1.3.19. *If (X, τ_X), (Y, τ_Y) are topological spaces and $f: X \to Y$ is supposed to be a continuous, closed surjection, then f is a quotient map, that is $\tau_Y = \tau_q$.*

Proof. Recall that $\tau_Y \subseteq \tau_q$. Let $U \in \tau_q$. Then $f^{-1}(U) \in \tau_X$ and so $X \setminus f^{-1}(U) =: C \subseteq X$ is closed. Since f is closed, we have that $f(C) \subseteq Y$ is τ_Y-closed. Note that $U = Y \setminus f(C) \in \tau_Y$. Hence $\tau_q \subseteq \tau_Y$ and we conclude that $\tau_Y = \tau_q$. □

The next proposition gives a criterion to recognize when a function defined on a quotient space is continuous.

Proposition 1.3.20. *If (X, τ_X), (Y, τ_Y), and (Z, τ_Z) are topological spaces, $f: X \to Y$ is a quotient map and $g: Y \to Z$, then g is continuous if and only if $g \circ f$ is continuous.*

Proof. $\Longrightarrow$: This follows from Proposition 1.1.36 (a).

$\Longleftarrow$: Let $U \in \tau_Z$. Then $(g \circ f)^{-1}(U) = f^{-1}(g^{-1}(U)) \in \tau_X$. Hence $g^{-1}(U) \in \tau_Y$ since f is a quotient map, see Definition 1.3.15. This proves the continuity of g. □

Now we will show that the whole topic of the quotient topology can be covered by considering Y to be X/R with R being an equivalence relation; see Definition 1.2.30 (b). Suppose $f: X \to Y$ is a surjection and define the relation $R \subseteq X \times X$ by setting xRx' if and only if $f(x) = f(x')$.

Let $e(x)$ be the equivalence class for x. Evidently $f|_{e(x)}$ is constant. Then the map $\hat{f}: X/R \to Y$ defined by $\hat{f}(e(x)) = f(x)$ is actually well-defined and a bijection. Note that if $e(x) = e(x')$, then $f(x) = f(x')$. In order to topologize X/R consider the standard quotient map $e: X \to X/R$ and consider the quotient topology induced by e. Then we have the following result.

Proposition 1.3.21. *If X is a topological space, Y is a set, $f: X \to Y$ is a surjection and R is the equivalence relation defined above, then X/R and Y are homeomorphic when both are endowed with the quotient topology.*

Remark 1.3.22. Instead of using the equivalence relation we may assume that X is partitioned by a collection C of disjoint subsets. Then we define an equivalence relation by setting xRu if and only if x, u are in the same element of C. Then we can consider X/R. The simplest kind of quotient space can be obtained by the equivalence relation R in which only one equivalence class has more than one element $e(x_0) = A$ and for all other equivalence classes we have $e(x) = \{x\}$ with $x \in X \setminus A$. Then X/R is denoted by X/A and we obtain the quotient (identification) space by collapsing A to a single element $\{x_0\}$.

Example 1.3.23. (a) The quotient space of $[0,1]$ obtained by identifying 0 and 1 is homeomorphic to a circle.

(b) The quotient space of $I^2 = [0,1] \times [0,1]$ by identifying the boundary with a single point is homeomorphic to a sphere in $\mathbb{R}^3$.

(c) The quotient space of $I^2 = [0,1] \times [0,1]$ by identifying the points $(0, x_2)$ and $(1, 1-x_2)$ with $0 \le x_2 \le 1$ is homeomorphic to the **Möbius strip**.

(d) Let $X = I^2 = [0,1] \times [0,1]$ and consider an equivalence relation $R \subseteq X \times X$ defined as follows:

$$(x_1, 0)R(x_1, 1) \quad \text{for every } 0 \le x_1 \le 1\,, \tag{1.3.7}$$

$$(0, x_2)R(1, x_2) \quad \text{for every } 0 \le x_2 \le 1\,. \tag{1.3.8}$$

Then the quotient space is realized in two steps and gives a space homeomorphic to the torus. The first step is determined by (1.3.7), which produces a cylinder and then in the second step determined by (1.3.8), where we identify the two bases of the cylinder to generate the torus.

(e) If we replace (1.3.8) in the example above by

$$(0, x_2)R(1, 1-x_2) \quad \text{for every } 0 \le x_2 \le 1\,,$$

then the resulting quotient space X/R is the **Klein bottle**.

(f) Let $D \subseteq \mathbb{R}^2$ be the unit disc, that is, $D = \{(x_1, x_2) \in \mathbb{R}^2 : x_1^2 + x_2^2 \le 1\}$ and consider the equivalence relation

$$xR(-x) \quad \text{for all } \partial D = \{(x_1, x_2) \in \mathbb{R}^2 : x_1^2 + x_2^2 = 1\}\,.$$

That means, diametrically opposite points are identified. Then the quotient space D/R is called the **projective plane** and is denoted by P^2. One can proceed similarly to define P^n for any $n \in \mathbb{N}_0$ as the space obtained from $S^n = \{x \in \mathbb{R}^{n+1} : |x| = 1\}$ by identifying each point x with its antipode $-x$. The space P^n is known as the **projective n-space**.

1.4 Connectedness and Compactness

The property of connectedness says that the space has only one piece. It is a very important topological invariant with important applications in many other branches of mathematics. It is not difficult to come up with a definition of this very intuitive notion.

Definition 1.4.1. Let X be a topological space. A **separation** of X is a pair (U, V) of disjoint, nonempty, open sets of X such that $X = U \cup V$. If such a separation exists, we say that the space is **disconnected.** If there is no such separation for X, then we say that the space is **connected.** A set $A \subseteq X$ is connected, if it is a connected space when endowed with the subspace topology. Note that in a separation the two sets are both open and closed. We say that they are **clopen.**

Example 1.4.2. (a) The space $(\mathbb{R}, \tau_u)$, see Remark 1.2.23, is disconnected and the sets

$$\{x \in \mathbb{R}: x > \lambda\} \quad \text{and} \quad \{x \in \mathbb{R}: x \leq \lambda\}$$

with $\lambda \in \mathbb{R}$ form a separation of $\mathbb{R}$.

(b) The rationals $\mathbb{Q}$ with the relative Euclidean topology form a disconnected space. The sets

$$\{x \in \mathbb{R}: x > \pi\} \cap \mathbb{Q} \quad \text{and} \quad \{x \in \mathbb{R}: x < \pi\} \cap \mathbb{Q}$$

form a separation of $\mathbb{Q}$.

(c) A discrete space that is not a singleton is disconnected and the empty set is disconnected since there are no open sets to form a separation of it.

(d) $\mathbb{R} \setminus \{0\}$ is disconnected since $(-\infty, 0)$ and $(0, +\infty)$ form a separation. Similarly, $\mathbb{R}^2 \setminus \mathbb{R}$ is disconnected and we can have a separation using the sets

$$U = \{(x_1, x_2) \in \mathbb{R}: x_2 > 0\} \quad \text{and} \quad V = \{(x_1, x_2) \in \mathbb{R}: x_2 < 0\}.$$

Here, U is called the **upper half plane** and V is said to be the **lower half plane.**

(e) $\mathbb{R}$, endowed with the Euclidean topology, is connected. To show this we argue by contradiction. So, suppose that $\mathbb{R}$ is disconnected and (U, V) is a separation of $\mathbb{R}$. Let $x \in U$ and $y \in V$ and assume without loss of generality that $x < y$. Then $\hat{U} = U \cap [x, y]$ is closed and bounded in $\mathbb{R}$. Hence, $\hat{u} = \sup \hat{U} \in \hat{U}$. Furthermore, $\hat{u} \notin V$ since U and V are disjoint. Therefore $\hat{u} < y$ and $(\hat{u}, y] \subseteq V$. Thus, $\hat{u} \in \overline{V}$ and so $\hat{u} \in V$. It follows that $\hat{u} \in U \cap V$, a contradiction. This proves the connectedness of $\mathbb{R}$.

Remark 1.4.3. From the examples 1.4.2 (b) and (e) we see that connectedness is not a hereditary property.

Proposition 1.4.4. *The connected subsets of $\mathbb{R}$ are singletons and intervals (open, closed, or half-open).*

Proof. Clearly singletons are connected. In addition, the argument in Example 1.4.2 (e) shows that intervals are connected. It remains to show if $A \subseteq \mathbb{R}$ is connected, then A is an interval. If A is not an interval, then we can find $x, y \in A$ and $u \notin A$ such that $x < u < y$. Then $U = A \cap \{v \in \mathbb{R}: v < c\}$ and $V = A \cap \{v \in \mathbb{R}: v > c\}$ are a separation of A, a contradiction. □

Proposition 1.4.5. *Let X be a topological space. The following statements are equivalent:*
(a) *X is disconnected.*
(b) *There is a nonempty, proper subset of X, which is both open and closed.*
(c) *There is a continuous function from X into the two-point space $\{a, b\}$.*
(d) *X has a nonempty, proper subset A such that $\overline{A} \cap \overline{(X \setminus A)} = \emptyset$.*

Proof. (a) $\Longrightarrow$ (b): Since X is disconnected, it admits a separation (U, V). Then U as well as V are nonempty clopen.

(b) $\Longrightarrow$ (a): Suppose A is a proper, nonempty subset of X that is clopen. Let $C = X \setminus A$. Then (A, C) is a separation of X and so X is disconnected.

(a) $\Longrightarrow$ (c): Let (U, V) be a separation of X. Then the function $f: X \to \{a, b\}$ defined by

$$f(x) = \begin{cases} a & \text{if } x \in U, \\ b & \text{if } x \in V \end{cases}$$

is continuous.

(c) $\Longrightarrow$ (a): Since $f: X \to \{a, b\}$ is continuous, then $U = f^{-1}(a)$ and $V = f^{-1}(b)$ are disjoint, open sets in X such that $X = U \cup V$. So, (U, V) is a separation of X and we conclude that X is disconnected.

(a) $\Longrightarrow$ (d): Let (U, V) be a separation of X. Then $\overline{U} \cap \overline{V} = U \cap V = \emptyset$. So $\overline{U} \cap \overline{(X \setminus U)} = \emptyset$.

(d) $\Longrightarrow$ (a): We have that $\overline{A}$ and $\overline{(X \setminus A)}$ are disjoint, closed sets whose union is X. Hence $\overline{A}$ and $\overline{(X \setminus A)}$ are also open and form a separation of X. □

Corollary 1.4.6. *Let X be a topological space. The following statements are equivalent:*
(a) *X is connected.*
(b) *The only subsets of X that are open and closed are $\emptyset$ and X.*
(c) *There is no continuous function from X onto the two-point space $\{a, b\}$.*
(d) *X has no nonempty, proper subset A such that $\overline{A} \cap \overline{(X \setminus A)} = \emptyset$.*

Proposition 1.4.7. *If X, Y are topological spaces, X is connected and $f: X \to Y$ is continuous, then $f(X)$ is connected.*

Proof. Since $f: X \to f(X)$ is continuous we may assume that f is a continuous surjection. Arguing by contradiction, suppose that $Y = f(X)$ is disconnected and let (U, V) be a separation of Y. Then $f^{-1}(U)$ and $f^{-1}(V)$ are disjoint, open sets in X such that $X = f^{-1}(U) \cup f^{-1}(V)$. Hence X is disconnected, a contradiction. □

Remark 1.4.8. The last proposition gives at once that all open intervals in $\mathbb{R}$ are connected. Indeed recall that every open interval is homeomorphic to $\mathbb{R}$ and that $\mathbb{R}$ is connected; see Example 1.4.2 (e).

If on a connected set A we adjoin some of its limit points we preserve connectedness.

Proposition 1.4.9. *If X is a topological space, $A \subseteq X$ is connected and $A \subseteq C \subseteq \overline{A}$, then C is connected.*

Proof. Arguing by contradiction, suppose that C is disconnected. Hence by Proposition 1.4.5, there exists a continuous surjection $f: C \to \{0, 1\}$. Since A is connected from Corollary 1.4.6 we have that $f(A) = \{0\}$ or $f(A) = \{1\}$. To fix things assume that $f(A) = \{0\}$. From Proposition 1.1.35 we have $f(\overline{A}) \subseteq \overline{f(A)} = \{0\}$. Hence, $f(C) = \{0\}$, a contradiction. □

Corollary 1.4.10. *If X is a topological space and $A \subseteq X$ is connected, then $\overline{A}$ is connected as well.*

Another useful result in determining whether or not a given subset is connected, is the following one.

Proposition 1.4.11. *If (X, τ) is a topological space and $A \subseteq X$, then A is disconnected if and only if there exist open sets $U, V \in \tau$ such that*

$$U \cap A \neq \emptyset\,, \quad V \cap A \neq \emptyset\,, \quad U \cap V \cap A = \emptyset\,, \quad \text{and} \quad A \subseteq U \cup V\,.$$

Proof. $\Longrightarrow$: We have $A = \hat{U} \cup \hat{V}$ with $\hat{U}, \hat{V} \in \tau(A)$ with the subspace topology $\tau(A)$ and

$$\hat{U} = U \cap A \quad \text{as well as} \quad \hat{V} = V \cap A \quad \text{with } U, v \in \tau\,.$$

Then we can easily check that U and V have the desired properties.

$\Longleftarrow$: Let $\hat{U} = U \cap A \neq \emptyset$ and $\hat{V} = V \cap A \neq \emptyset$. We have that $\hat{U}, \hat{V} \in \tau(A)$ and they are disjoint with $A = \hat{U} \cup \hat{V}$. Therefore, A is disconnected. □

It is obvious that connectedness is not preserved by arbitrary unions. Additional restrictions are needed.

Proposition 1.4.12. *If X is a topological space and $\{A_i\}_{i \in I}$ is any family of connected subsets of X such that $\bigcap_{i \in I} A_i \neq \emptyset$, then $\bigcup_{i \in I} A_i$ is connected.*

Proof. Let $C = \bigcup_{i \in I} A_i$. Suppose that C is disconnected. Then by Proposition 1.4.5 we can find a continuous map $f: C \to \{0, 1\}$. Since each A_i is connected, $f|_{A_i}$ is not surjective for all $i \in I$. Let $x_0 \in \bigcap_{i \in I} A_i$. Then $f(x) = f(x_0)$ for all $x \in A_i$ and for all $i \in I$. So, f is not surjective, a contradiction. □

Connectedness is preserved by arbitrary Cartesian products.

Proposition 1.4.13. *If $\{X_i\}_{i \in I}$ is an arbitrary family of nonempty, connected topological spaces, then $X = \prod_{i \in I} X_i$, endowed with the product topology, is connected as well.*

Proof. Arguing by contradiction, suppose that X is disconnected. So there is a continuous map $f: X \to \{0, 1\}$. Fix $u = (u_i)_{i \in I} \in X$ and let $i_1 \in I$. We define $f_{i_1}: X_{i_1} \to X$ by setting $f_{i_1}(x_i) = y = (y_i)_{i \in I}$ with $y_i = u_i$ for $i \neq i_1$ and $y_{i_1} = x_{i_1}$. Evidently f_{i_1} is continuous, which implies the continuity of $f \circ f_{i_1} : X_{i_1} \to \{0, 1\}$. By hypothesis, X_{i_1} is connected. So, $f \circ f_{i_1}$ is constant and $(f \circ f_{i_1})(x_{i_1}) = f(u)$ for every $x_{i_1} \in X_{i_1}$. Hence $f(x) = f(u)$ for all $x \in X$, which are equal to u except for the i_1-component. We repeat this process with another index $i_2 \in I$. Continuing this way we see that $f(x) = f(u)$ for all $x \in X$, which are equal to u except on a finite number of coordinates. This set is dense in X and so by Proposition 1.1.35 (b), f is constant, a contradiction. This proves that X is connected. □

Corollary 1.4.14. *The space $\mathbb{R}^n$ with $n \in \mathbb{N}$ is connected.*

Example 1.4.15. Let $A = \{(0, y) \in \mathbb{R}^2 : 0 \le y \le 1\}$ and $C = \{(x, y) \in \mathbb{R}^2 : 0 < x \le 1, y = \sin(\pi/x)\}$. Evidently C is connected because of Propositions 1.4.7 and 1.4.13. Furthermore, $S = \overline{C} = A \cup C$ is connected; see Corollary 1.4.10. The set S is known as the **topologist's sine curve**.

Remark 1.4.16. It is clear that intersection of even two connected spaces need not be connected. Furthermore, suppose that $\{A_n\}_{n \in \mathbb{N}}$ is a decreasing sequence of connected spaces. Then $\bigcap_{n \ge 1} A_n$ need not be connected. To see this, let $X = I^2 \setminus \{(x, 0) : 1/2 \le x \le 2/3\}$ with $I = [0, 1]$ and $A_n = \{(x, y) \in X : y \le 1/n\}$ with $n \in \mathbb{N}$.

A disconnected space can be decomposed in a unique way into connected **components** and the number of components can be viewed as an indication of how disconnected the space is.

Definition 1.4.17. A **component** of a topological space X is a maximal connected subset C of X. That is, C is connected and it is not properly contained in a connected subset of X.

Remark 1.4.18. A component is necessarily closed. Indeed, from Corollary 1.4.10 we know that $\overline{C}$ is connected. The maximality of C implies that $C = \overline{C}$. Hence, C is closed. The family of distinct components of X form a partition of X. To see this, note if C, C' are two distinct components of X and $C \cap C' \ne \emptyset$, then from Proposition 1.4.12 we have that $C \cup C'$ is connected, contradicting the maximality of the components. Moreover, for the same reason, each $x \in X$ belongs in a unique component. Given $x \in X$ let $C(x)$ denote the component of X containing x. Then, for points $x, u \in X$, $C(x)$ and $C(u)$ are either identical or disjoint. Every connected subset of X is contained in one component and X is connected if and only if it has only one component. Finally if $\{U, V\}$ is a separation of X and C is a component of X, then $C \subseteq U$ or $C \subseteq V$.

Taking into account the remarks above and Proposition 1.4.19, we infer the following result.

Proposition 1.4.19. *If X, Y are topological spaces and $f : X \to Y$ is continuous, then the image of each component of X lies in a component of Y.*

Remark 1.4.20. In particular, a homeomorphism f induces a 1-1 correspondence between the components of X and Y with $C(x)$ being homeomorphic to $C(f(x))$ for all $x \in X$.

Definition 1.4.21. (a) A topological space X is **totally disconnected** provided that each component of X is a singleton.

(b) A point $x \in X$ is a **cut point** of a connected topological space X provided that $X \setminus \{x\}$ is disconnected. We say that $x \in X$ is an n**-cut point** provided that $X \setminus \{x\}$ has n-components.

From Proposition 1.4.19 we infer the following result.

Proposition 1.4.22. *Homeomorphic spaces have the same number of cut points of each type.*

From an analytical point of view, the notion of path-connectedness is more natural. Path-connectedness is a topological property stronger than connectedness and it is useful in many applications. It is a very intuitive notion that in a path-connected space any two distinct points can be joined by a continuous path in the space.

Definition 1.4.23. (a) A **path** in a topological space X is a continuous map $\sigma: [0,1] \to X$. We say that $\sigma(0)$ is the **initial point** of the path and $\sigma(1)$ is the **final point** of the path. The set $\sigma([0,1]) \subseteq X$ is called a **curve** in X. If σ is a path in X, then $\overline{\sigma}(t) = \sigma(1-t)$ for all $t \in [0,1]$ is the **reverse path**.

(b) A topological space X is said to be **path-connected** provided that for each pair of points $x, u \in X$ there is a path in X with initial point x and final point u. A subset C of X is **path-connected** if C has this property for the subspace topology.

The next proposition compares connectedness and path-connectedness.

Proposition 1.4.24. *Every path-connected topological space is connected.*

Proof. Suppose that X is path-connected and let $u \in X$. For each $x \in X$, let σ_x be the path in X with initial point u and final point x. Let $C_x = \sigma_x([0,1])$ be the corresponding curve. From Proposition 1.4.7 we know that $C_x \subseteq X$ is connected. Note that $u \in \bigcap_{x\in X} C_x$. So, from Proposition 1.4.12, it follows that $\bigcup_{x\in X} C_x = X$ is connected. □

Remark 1.4.25. The converse of the above is not true in general. As a counterexample, consider the topologist's sine curve $S = A \cup C$ introduced in Example 1.4.15. Then S is connected but not path-connected. To prove that S is not path-connected, we show that it is not possible to join a point in A to a point in C by a path in S. To this end, let $a \in A$ and let $\sigma: [0,1] \to X$ be a path with initial point a. Note that A is closed in S (see Proposition 1.1.27), and so $\sigma^{-1}(A) \subseteq [0,1]$ is closed and nonempty, since $0 \in \sigma^{-1}(A)$. Let $t \in \sigma^{-1}(A)$ and choose a small $\varepsilon > 0$ such that $\sigma((t-\varepsilon, t+\varepsilon)) \subseteq \overline{B}_{1/2}(\sigma(t)) = \{u \in \mathbb{R}^2 : |u-\sigma(t)| \leq 1/2\}$, which is possible since σ is continuous. Note that $S \cap \overline{B}_{1/2}(\sigma(t))$ consists of a closed interval on the y-axis of $\mathbb{R}^2$ together with parts of the curve $y = \sin(\pi/x)$, each of which is homeomorphic to a closed interval. Moreover, any two of these parts are disjoint in $S \cap \overline{B}_{1/2}(\sigma(t))$. So $A \cap \overline{B}_{1/2}(\sigma(t))$ is a component of $S \cap \overline{B}_{1/2}(\sigma(t))$. Since $\sigma(t) \in A \cap \overline{B}_{1/2}(\sigma(t))$ and $(t-\varepsilon, t+\varepsilon)$ is connected, we must have $\sigma((t-\varepsilon, t+\varepsilon)) \subseteq A \cap \overline{B}_{1/2}(\sigma(t))$. This shows that

$\sigma^{-1}(A) \subseteq [0,1]$ is open. Hence $\sigma^{-1}(A) = [0,1]$ being both closed and open. So, $\sigma([0,1]) \subseteq A$ and this proves that S cannot be path-connected.

Proposition 1.4.26. *If X is a topological space and $u \in X$, then X is path-connected if and only if each $x \in X$ can be joined to u by a path.*

Proof. $\Longrightarrow$: This is obvious.

$\Longleftarrow$: Let $x, x' \in X$ and consider the paths $\sigma, \sigma' : [0,1] \to X$ such that σ has initial point x and final point u as well as σ' having initial point u and final point x'. We define $\hat{\sigma}: [0,1] \to X$ by

$$\hat{\sigma}(t) = \begin{cases} \sigma(2t) & \text{if } t \in [0, \frac{1}{2}], \\ \sigma'(2t-1) & \text{if } t \in [\frac{1}{2}, 1]. \end{cases}$$

This is a continuous path since $\sigma(1) = \sigma'(0) = u$; see Proposition 1.1.37. Moreover, $\hat{\sigma}(0) = x$ and $\hat{\sigma}(1) = x'$. Therefore, X is path-connected. ☐

Definition 1.4.27. Let σ_1, σ_2 be two paths in X such that $\sigma_1(1) = \sigma_2(0)$. The **path composition** of σ_1 and σ_2 denoted by $\sigma_1 * \sigma_2$ is the path in X defined by

$$(\sigma_1 * \sigma_2)(t) = \begin{cases} \sigma_1(2t) & \text{if } t \in [0, \frac{1}{2}], \\ \sigma_2(2t-1) & \text{if } t \in [\frac{1}{2}, 1]. \end{cases}$$

The next result is a straightforward consequence of Definition 1.4.23 (b) and of Proposition 1.1.36 (a).

Proposition 1.4.28. *If X, Y are topological spaces, X is path-connected, and $f: X \to Y$ is continuous, then $f(X)$ is path-connected.*

Remark 1.4.29. It follows that path-connectedness is a topological invariant. In contrast to connectedness, see Corollary 1.4.10, the closure of a path-connected set need not be path-connected. We consider the topologist's sine curve from Example 1.4.15. We have $S = \overline{C}$ and C is path-connected; see Proposition 1.4.28. However, we proved that S is not path-connected; see Remark 1.4.25.

Many results about connectedness have analogues for path connectedness.

Proposition 1.4.30. *If X is a topological space and $\{A_i\}_{i \in I}$ is any family of path-connected subsets of X such that $\bigcap_{i \in I} A_i \neq \emptyset$, then $\bigcup_{i \in I} A_i$ is path-connected.*

Proof. Let $x \in \bigcup_{i \in I} A_i$ and pick $u \in \bigcap_{i \in I} A_i$. Since $x \in A_{i_0}$ for some $i_0 \in I$, we can join x and u by a path in X since A_{i_0} is path-connected. Proposition 1.4.26 implies that $\bigcup_{i \in I} A_i$ is path-connected. ☐

Proposition 1.4.31. *If $\{X_i\}_{i \in I}$ is an arbitrary family of nonempty, path-connected topological spaces, then $X = \prod_{i \in I} X_i$ endowed with the product topology is path-connected as well.*

Proof. Let $x = (x_i)$, $u = (u_i) \in X$. For each $i \in I$, X_i is path-connected so we can find a path σ_i with initial point x_i and final point u_i. Then $\sigma = (\sigma_i)$ is a path in X joining x and u; see Proposition 1.3.4. Hence, X is path-connected as well. □

Definition 1.4.32. A **path component** of a topological space is a maximal path-connected subset C of X. That is, C is path-connected and it is not properly contained in a path-connected subset of X.

Remark 1.4.33. Path components have almost the same properties as components. So every $x \in X$ belongs to exactly one path component denoted by $P(x)$. If $x \neq x'$, then $P(x) \cap P(x') = \emptyset$ or $P(x) = P(x')$. Every path-connected set $C \subseteq X$ is contained in a path component and X is path-connected if and only if X has only one path component. Note that we said almost the same properties. The reason for this, in contrast to components, is that path components need not be closed. Consider the topologist's sine curve $S = A \cup C$, see Example 1.4.15. Then A and C are the path components of S but C is not closed; recall that $\overline{C} = S$. A path component of X is a subset of some component of X.

Connectedness and path-connectedness are global topological properties since they concern the whole topological space. Local topological properties concern the structure of the space near a particular point, if we recall the notion of first countability; see Definition 1.2.20 (a). In the next definition we provide local versions of the notions of connectedness and of path-connectedness.

Definition 1.4.34. A topological space X is said to be **locally connected** (resp. **locally path-connected**) if for every $x \in X$ and every $U \in \mathcal{N}(x)$ we can find a connected (resp. path-connected) $V \in \mathcal{N}(x)$ such that $V \subseteq U$.

Remark 1.4.35. Equivalently X is locally connected (resp. locally path-connected) if and only if every $x \in X$ has a local basis consisting of connected (resp. path-connected) sets. A space can be connected (resp. path-connected) without being locally connected (resp. locally path-connected). Consider the topologist's sine curve (see Example 1.4.15), which is connected but not locally connected. Of course local connectedness (resp. local path-connectedness) does not imply connectedness (resp. path-connectedness). Consider the union of two disjoint, closed balls in $\mathbb{R}^N$.

Proposition 1.4.36. *A topological space X is connected if and only if for each open set $U \subseteq X$ each component of U is open.*

Proof. $\Longrightarrow$: Let C be a component of the open set $U \subseteq X$. Given $x \in C$ we can find a connected open set $V_x \subseteq U$ with $x \in V_x$. We have $V_x \subseteq C$ and since $x \in C$ was arbitrary, we conclude that C is open.

$\Longleftarrow$: Let $x \in X$ and let $U \in \mathcal{N}(x)$. Then by hypothesis the component C of U containing x is open and so X is locally connected. □

Corollary 1.4.37. *If a topological space X is locally connected then every component of X is open (and closed).*

Proposition 1.4.38. *If X is a topological space, then the following statements are equivalent:*
(a) *Every path component of X is open, hence closed as well.*
(b) *Every point of X has a path-connected neighborhood.*

Proof. (a) $\Longrightarrow$ (b): Let $x \in X$ and let $C(x)$ be the path component containing x. By hypothesis $C(x)$ is open and so X is locally path-connected.

(b) $\Longrightarrow$ (a): Let C be a path component and $x \in C$. By hypothesis we can find a path-connected $U \in \mathcal{N}(x)$. Hence, $U \subseteq C$ and since $x \in C$ is arbitrary we conclude that C is open. Note that $X \setminus C$ is the union the remaining open path components, as we just proved, and it is open, so C is closed. □

We saw that path-connectedness is stronger than connectedness; see Proposition 1.4.24. The next proposition provides conditions for the two notions to be equivalent.

Proposition 1.4.39. *A topological space X is path-connected if and only if X is connected and every $x \in X$ has a path-connected neighborhood.*

Proof. $\Longrightarrow$: This follows from Proposition 1.4.24 and the fact that X is a neighborhood of every $x \in X$, and by hypothesis it is path-connected.

$\Longrightarrow$: According to Proposition 1.4.38 every path component of X is open and closed in X. Since X is connected, it follows that it has only one path component, and hence X is path-connected. □

Corollary 1.4.40. *An open subset of $\mathbb{R}^n$ is connected if and only if it is path-connected.*

Remark 1.4.41. The corollary above fails for nonopen sets in $\mathbb{R}^n$. To see this, consider the topologist's sine curve.

Now we pass to another fundamental topological notion, namely the notion of **compactness**. This concept is an abstraction to general topological spaces of a property of closed and bounded intervals, cf. the Heine–Borel Theorem. Compactness does not mean only small in size. It is more than that. For example the intervals $[0,1]$ and $(0,1)$ have the same size but $[0,1]$ is compact while $(0,1)$ is not. Compactness is important in analysis since it combines well with continuity.

Definition 1.4.42. Let X be a Hausdorff topological space. We say that X is **compact** if every open cover admits a finite subcover; see Definition 1.2.26. A subset $A \subseteq X$ is compact provided A, endowed with the relative subspace topology, is compact.

Remark 1.4.43. Since compact subsets of a non-Hausdorff space need not be closed (a rather awkward situation), we have included in the definition of compactness that X is Hausdorff. Since relatively open sets in A are of the form $U \cap A$ with $U \subseteq X$ open, the definition of compactness of $A \subseteq X$ takes the following form: "$A \subseteq X$ is compact if and only if every open cover of A by open sets in X admits a finite subcover".

Definition 1.4.44. Let X be a set and $\mathcal{L} \subseteq 2^X \setminus \{\emptyset\}$. We say that $\mathcal{L}$ has the **finite intersection property** if every finite subcollection of $\mathcal{L}$ has a nonempty intersection.

Proposition 1.4.45. *Let X be a Hausdorff topological space. The following statements are equivalent:*
(a) *X is compact.*
(b) *Every family of nonempty, closed subsets of X with the finite intersection property has a nonempty intersection.*
(c) *Every net in X has a convergent subnet in X.*

Proof. (a) $\Longrightarrow$ (b): Let $\mathcal{L}$ be a family of nonempty, closed subsets of X with the finite intersection property. If $\bigcap_{C\in\mathcal{L}} C = \emptyset$, then $X = \bigcup_{C\in\mathcal{L}}(X \setminus C)$ and so $\{X \setminus C\}_{C\in\mathcal{L}}$ is an open cover of X. The compactness of X implies that we can find a finite subcover such that $X = \bigcup_{k_1}^n (X \setminus C_k)$ with $n \in \mathbb{N}$. Then $\bigcap_{k=1}^n C_k = \emptyset$, contradicting the fact that $\mathcal{L}$ has the finite intersection property.

(b) $\Longrightarrow$ (a): Let $\mathcal{D}$ be an open cover of X. Then $X = \bigcup_{U\in\mathcal{D}} U$ and so $\bigcap_{U\in\mathcal{D}}(X \setminus U) = \emptyset$. This means that the finite intersection property does not hold for the collection $\{X \setminus U\}_{U\in\mathcal{D}}$ and so we can find $\{U_k\}_{k=1}^n \subseteq \mathcal{D}$ such that $\bigcap_{k=1}^n(X \setminus U_k) = \emptyset$. Hence, $X = \bigcup_{k=1}^n U_k$ and so we conclude that X is compact.

(b) $\Longrightarrow$ (c): Let $\{x_i\}_{i\in I}$ be a net in X. Let $A_\alpha = \overline{\{x_i\}_{i\geq\alpha}}$ with $\alpha \in I$. Then $\{A_\alpha\}_{\alpha\in I}$ is a family of nonempty, closed subsets of X with the finite intersection property. So, by hypothesis we can find $x \in \bigcap_{\alpha\in I} A_\alpha$. Evidently, x is a cluster point of $\{x_i\}_{i\in I}$. So, using Proposition 1.2.36 we can find a subnet of $\{x_i\}_{i\in I}$ converging to $x \in X$.

(c) $\Longrightarrow$ (b): Let $\mathcal{L}$ be a family of nonempty, closed subsets of X with the finite intersection property. Let $\mathcal{F}$ be the family of all finite intersections of members of $\mathcal{L}$. Then $\mathcal{F}$ has the finite intersection property and since $\mathcal{L} \subseteq \mathcal{F}$ it suffices to show that $\bigcap_{D\in\mathcal{F}} D \neq \emptyset$. Since the intersection of two elements in $\mathcal{F}$ is again an element of $\mathcal{F}$, we see that $\mathcal{F}$ is directed. Let $x_D \in D$ with $D \in \mathcal{F}$. Then $\{x_D\}_{D\in\mathcal{F}} \subseteq X$ is a net and so by hypothesis it has a cluster point x. Then $x \in D$ for all $D \in \mathcal{F}$ and so $\bigcap_{D\in\mathcal{F}} D \neq \emptyset$. □

Proposition 1.4.46. *If X is a compact topological space and $C \subseteq X$ is closed, then C is compact.*

Proof. Let $\mathcal{L}$ be a cover of C by sets open in X. Then $\mathcal{L}_0 = \mathcal{L} \cup (X \setminus C)$ is an open cover of X. Since X is compact, $\mathcal{L}_0$ has a finite subcover $\{U_k, X \setminus A\}_{k=1}^n$ with $U_k \in \mathcal{L}$. Then $C \subseteq \bigcup_{k=1}^n U_k$ and so C is closed; see Remark 1.4.43. □

Proposition 1.4.47. *If X is a Hausdorff topological space and $C \subseteq X$ is compact, then C is closed.*

Proof. Let $\{x_i\}_{i\in I} \subseteq C$ be a net such that $x_i \to x$. Since X is compact, we can find a subnet $\{u_\alpha\}_{\alpha\in I}$ such that $u_\alpha \to x \in C$; see Propositions 1.4.45 and 1.2.40. Therefore, we conclude that $C \subseteq X$ is compact. □

Corollary 1.4.48. *If X is a compact topological space and $A \subseteq X$, then A is compact if and only if A is closed.*

Proposition 1.4.49. *If X is Hausdorff topological space and K_1, K_2 are compact, disjoint subsets of X, then we can find open $U, V \subseteq X$ such that $K_1 \subseteq U$, $K_2 \subseteq V$ and $U \cap V = \emptyset$.*

Proof. First assume that $K_1 = \{u\}$ is a singleton. Then for each $x \in K_2$ we can find open sets $U_x, V_x \subseteq X$ such that $u \in U_x$, $x \in V_x$ and $U_x \cap V_x = \emptyset$ because X is Hausdorff. Then $\{V_x\}_{x\in K_2}$ is an open cover of K_2. The compactness of K_2 implies that we can find a finite subcover $\{V_{x_k}\}_{k=1}^n$. Let

$$U = \bigcap_{k=1}^{n} U_{x_k} \quad \text{and} \quad V = \bigcup_{k=1}^{n} V_{x_k}\,.$$

Both are open sets in X, $u \in U$ and $K_2 \subseteq V$. So, we have proven the proposition when K_1 is a singleton.

Now consider the case of a general compact set $K_1 \subseteq X$. From the previous part of the proof we know that for every $u \in K_1$ we can find open $U_u, V_u \subseteq X$ such that $u \in U_u$, $K_1 \subseteq V_u$ and $U_u \cap V_u = \emptyset$. Note that $\{U_u\}_{u\in K_1}$ is an open cover of K_1 and so by the compactness we can find a finite subcover $\{U_{u_k}\}_{k=1}^n$. Set

$$U = \bigcup_{k=1}^{n} U_{x_k} \quad \text{and} \quad V = \bigcap_{k=1}^{n} V_{x_k}\,.$$

Then both are open sets in X, $K_1 \subseteq U$, $K_2 \subseteq V$ and $U \cap V = \emptyset$. □

Corollary 1.4.50. *A compact topological space is normal.*

The next result is one of the main theorems on compactness.

Theorem 1.4.51. *If X, Y are Hausdorff topological spaces, $K \subseteq X$ is compact, and $f: X \to Y$ is continuous, then $f(K) \subseteq Y$ is compact.*

Proof. Let $\{V_i\}_{i\in I}$ be an open cover of $f(K)$. Then $\{f^{-1}(V_i)\}_{i\in I}$ is an open cover of K. The compactness of K implies the existence of a finite subcover $\{f^{-1}(V_{i_k})\}_{k=1}^n$, that is $K \subseteq \bigcup_{k=1}^n f^{-1}(V_{i_k})$. Hence

$$f(K) \subseteq f\left(\bigcup_{k=1}^{n} f^{-1}(V_{i_k})\right) = \bigcup_{k=1}^{n} f(f^{-1}(V_{i_k})) \subseteq \bigcup_{k=1}^{n} V_{i_k}\,.$$

Therefore, $f(K)$ is compact. □

In $\mathbb{R}$ the compact sets are closed and bounded; see the Heine–Borel Theorem. So, Theorem 1.4.51 yields the following result known as the "Weierstraß Theorem".

Theorem 1.4.52 (Weierstraß Theorem). *If X is a compact topological space and $f: X \to \mathbb{R}$ is continuous, then there exist $x_0, \hat{x} \in X$ such that*

$$f(x_0) = \inf[f(x): x \in X] \quad \textit{and} \quad f(\hat{x}) = \sup[f(x): x \in X] .$$

Remark 1.4.53. In addition, Theorem 1.4.51 implies that compactness is a topological property.

Theorem 1.4.54. *If X, Y are Hausdorff topological spaces, X is compact and $f: X \to Y$ is a continuous bijection, then f is a homeomorphism.*

Proof. Let $C \subseteq X$ be closed. Then C is compact because of Corollary 1.4.48. Taking into account Theorem 1.4.51, we conclude that $f(C) \subseteq Y$ is compact, hence closed as well; see Proposition 1.4.47. Therefore, f is a closed function and then by Proposition 1.1.42, f is a homeomorphism. □

Compactness is preserved by Cartesian products. This is the celebrated "Tychonoff's Product Theorem". To prove this result, we need some preliminary material. First we present three statements of set theory that are equivalent.

Axiom of Choice: Let K be any set-valued map on a set X such that $K(x) \neq \emptyset$ for all $x \in X$. Then there is a function k on X such that $k(x) \in K(x)$ for all $x \in X$.

Zorn's Lemma: Let $(X, \leq)$ be a partially ordered set such that for every chain $C \subseteq X$ there is an upper bound $\hat{u} \in X$, that is, $x \leq \hat{u}$ for all $x \in C$. Then X has a maximal element, that is, there exists $x_0 \in X$ such that there is no $v \in X$ with $x_0 < v$; see Definition 1.2.30 (c).

Hausdorff Maximal Principle: For every partially ordered set $(X, \leq)$ there is a maximal chain $C \subseteq X$.

Lemma 1.4.55. *If (X, τ) is a Hausdorff topological space and $\mathcal{L}_0$ is a collection of subsets of X with the finite intersection property, then there exists a maximal collection $\mathcal{L}$ of subsets of X with the finite intersection property and containing $\mathcal{L}_0$. Moreover, finite intersections of elements in $\mathcal{L}$ are again in $\mathcal{L}$ and every subset of X intersecting every set in $\mathcal{L}$ is in $\mathcal{L}$.*

Proof. The family of all collections of sets in X with the finite intersection property and containing $\mathcal{L}_0$ is partially ordered by inclusion. Therefore, the Hausdorff Maximal Principle implies the existence of a maximal chain $\mathcal{C}$. Let $\mathcal{L} = \bigcup_{a \in \mathcal{C}} a$.

Let $\{A_k\}_{k=1}^n \subseteq \mathcal{L}$. It belongs to at most n-collections a_k and $\{a_k\}_{k=1}^n$ is linearly ordered. So, there is a collection a_n that contains the others. Hence, $A_k \in a_n$ for all $k = 1, \ldots, n$ and $\bigcap_{k=1}^n A_k \neq \emptyset$ because of the finite intersection property. Thus, $\mathcal{L}$ has the finite intersection property. Note again that $\mathcal{L}$ is maximal.

Let $\mathcal{L}'$ be the collection of all finite intersections of sets in $\mathcal{L}$. Then $\mathcal{L}_0 \subseteq \mathcal{L}'$ and it has the finite intersection property. Hence, by maximality $\mathcal{L}' = \mathcal{L}$.

Finally, let $A \subseteq X$ be such that $A \cap D \neq \emptyset$ for all $D \in \mathcal{L}$. Then the collection $\mathcal{L}' = \mathcal{L} \cup \{A\}$ has the finite intersection property and contains $\mathcal{L}_0$. Therefore, by the maximality, $A \in \mathcal{L}$. □

We will use this lemma to prove "Tychonoff's Product Theorem".

Theorem 1.4.56 (Tychonoff's Product Theorem). *If $\{(X_i, \tau_i)\}_{i\in I}$ are compact topological spaces, then $X = \prod_{i\in I} X_i$ endowed with the product topology is compact.*

Proof. Let $\mathcal{L}_0$ be a collection of closed sets in X with the finite intersection property and let $\mathcal{L}$ be the maximal collection postulated by Lemma 1.4.55. Note that while the elements of $\mathcal{L}_0$ are closed, those of $\mathcal{L}$ need not be closed. We will show that

$$\bigcap_{D\in\mathcal{L}} \overline{D} \neq \emptyset . \tag{1.4.1}$$

For each $i \in I$, let $\mathcal{L}_i$ be the i-projection of $\mathcal{L}$, that is, $\mathcal{L}_i = \{p_i(D) : D \in \mathcal{L}\}$. The elements of this collection need not be open nor closed. However, since $\mathcal{L}$ has the finite intersection property, it follows that so does $\mathcal{L}_i$. Then $\overline{\mathcal{L}_i} = \{\overline{p_i(D)} : D \in \mathcal{L}\}$ has a nonempty intersection; see Proposition 1.4.45. Let $x_i \in \bigcap_{D\in\mathcal{L}} \overline{p_i(D)} \subseteq X_i$ and $x = (x_i) \in X$. We claim that $x \in \overline{D}$ for all $D \in \mathcal{L}$.

Let $U \in \mathcal{N}(x)$. Then from the definition of the product topology we know that we can find $i_1, \dots, i_n \in I$, and $U_{i_k} \in \tau_{i_k}$ with $k = 1, \dots, n$ such that

$$x \in \bigcap_{k=1}^{n} p_{i_k}^{-1}(U_{i_k}) \subseteq U .$$

Note that $x_{i_k} \in U_{i_k} \cap \overline{\mathcal{L}}_{i_k}$, hence $U_{i_k} \cap \mathcal{L}_{i_k} \neq \emptyset$. Therefore, $p_{i_k}^{-1}(U_{i_k}) \cap \mathcal{L} \neq \emptyset$. Thus, Lemma 1.4.55 implies that $p_{i_k}^{-1}(U_{i_k}) \in \mathcal{L}$. Hence $\bigcap_{k=1}^{n} p_{i_k}^{-1}(U_{i_k}) \in \mathcal{L}$. We conclude that (1.4.1) holds and this implies that X is compact; see Proposition 1.4.45. □

Let us now introduce some generalizations of the notion of compactness.

Definition 1.4.57. Let (X, τ) be a Hausdorff topological space.
(a) We say that X is **countably compact** if every countable open cover has a finite subcover.
(b) We say that X is **limit point compact** (or that is has the **Bolzano–Weierstraß property**) if every sequence $\{x_n\}_{n\geq 1} \subseteq X$ has at least one cluster point.
(c) We say that X is **sequentially compact** if every sequence has a τ-convergent subsequence.

Remark 1.4.58. Clearly, "Compactness" implies "Countable Compactness" and "Sequential Compactness" implies "Limit Point Compactness". In general both implications are not reversible.

Combining Definition 1.4.57 and Proposition 1.4.45 gives the following result.

Proposition 1.4.59. *A Hausdorff topological space (X, τ) is countably compact if and only if every countable family of closed sets with the finite intersection property has a nonempty intersection.*

Proposition 1.4.60. *A Hausdorff topological space (X, τ) is countably compact if and only if it is limit point compact.*

Proof. $\Longrightarrow$: Let $\{x_n\}_{n\geq 1} \subseteq X$ and define $A_m = \overline{\{x_n\}_{n\geq m}}$ with $m \in \mathbb{N}$. Then $\{A_m\}_{m\geq 1}$ are closed sets with the finite intersection property. So, $\bigcap_{m\geq 1} A_m \neq \emptyset$ by Proposition 1.4.59. Any $x \in \bigcap_{m\geq 1} A_m \neq \emptyset$ is a cluster point of the sequence. Therefore, X is limit point compact.

$\Longleftarrow$: Let $\{C_n\}_{n\geq 1}$ be closed sets in X with the finite intersection property. Let $x_n \in \bigcap_{k=1}^n C_k$ with $n \in \mathbb{N}$. The limit point compactness of X implies that $\{x_n\}_{n\geq 1}$ has at least one cluster point x. Then $x \in \overline{\{x_n\}_{n\geq 1}} \subseteq \overline{\bigcap_{n\geq 1} C_n} = \bigcap_{n\geq 1} C_n \neq \emptyset$. Using Proposition 1.4.59, this implies that X is countably compact. □

Corollary 1.4.61. *"Sequential Compactness" implies "Countable Compactness".*

The reverse assertion is true under some additional assumptions.

Proposition 1.4.62. *If (X, τ) is a Hausdorff topological space that is first countable and countably compact, then X is sequentially compact.*

Proof. Let $\{x_n\}_{n\geq 1} \subseteq X$ and $x \in \overline{\{x_n\}_{n\geq 1}}$. Let $\{U_k\}_{k\in\mathbb{N}} \subseteq \mathcal{N}(x)$ such that $U_{k+1} \subseteq U_k$ for all $k \in \mathbb{N}$. Recall that X is first countable. Choose $x_m \in U_m \cap \{x_n\}_{n\geq 1}$ with $m \in \mathbb{N}$. Then $\{x_m\}_{m\geq 1}$ is a subsequence of $\{x_n\}_{n\geq 1}$ τ-converging to x. Therefore, X is sequentially compact. □

This proposition together with Lindelöf's Theorem (see Theorem 1.2.27), gives the following result.

Theorem 1.4.63. *If (X, τ) is a Hausdorff topological space that is second countable, then the following statements are equivalent:*
(a) *X is compact.*
(b) *X is countably compact.*
(c) *X is limit point compact.*
(d) *X is sequentially compact.*

Next we introduce a modification of compactness to a local property.

Definition 1.4.64. A Hausdorff topological space (X, τ) is said to be **locally compact** if for every $x \in X$ there exists $U \in \mathcal{N}(x)$ such that $\overline{U}$ is compact.

Remark 1.4.65. A set $A \subseteq X$ such that $\overline{A}$ is compact is said to be **relatively compact** (or **precompact**). The space $\mathbb{R}^N$ with the Euclidean topology is locally compact but not compact. Recall the Heine–Borel Theorem, which says that $A \subseteq \mathbb{R}^N$ is compact if and only if A is closed and bounded. Bounded means that there exists $r > 0$ such that $A \subseteq \overline{B}_r = \{u \in \mathbb{R}^N : |u| \le r\}$.

Proposition 1.4.66. *Let (X, τ) be a Hausdorff topological space. The following statements are equivalent:*
(a) *X is locally compact.*
(b) *For every $x \in X$ and every $U \in \mathcal{N}(x)$ there is a relatively compact $V \in \mathcal{N}(x)$ such that $x \in V \subseteq \overline{V} \subseteq U$.*
(c) *For every compact K and $U \in \tau$ such that $U \supseteq K$, there exists a relatively compact $V \in \tau$ such that $K \subseteq V \subseteq \overline{V} \subseteq U$.*
(d) *X has a basis consisting of relatively compact open sets.*

Proof. (a) $\Longrightarrow$ (b): Let $x \in X$ and $U \in \mathcal{N}(x)$. Taking into account the local compactness of X we find $W \in \mathcal{N}(x)$ such that $\overline{W}$ is compact. Corollary 1.4.50 implies that $\overline{W}$ endowed with the relative topology is regular. Then $\overline{W} \cap U$ is a neighborhood of x in $\overline{W}$. Proposition 1.2.8 implies the existence of an open set $D \subseteq \overline{W}$ such that

$$x \in D \subseteq \overline{D}^{\overline{W}} \subseteq \overline{W} \cap U ,$$

where $\overline{D}^{\overline{W}}$ denotes the closure of D in the relative topology of $\overline{W}$. We have $D = S \cap \overline{W}$ with $S \in \tau$. Let $V = S \cap W \in \mathcal{N}(x)$. This is the desired neighborhood of x.

(b) $\Longrightarrow$ (c): Let $K \subseteq X$ be compact and $U \in \tau$ such that $U \supseteq K$. For every $x \in K$ we can find $V_x \in \mathcal{N}(x)$ relatively compact such that $x \in V_x \subseteq \overline{V}_x \subseteq U$. Evidently $\{V_x\}_{x \in K}$ is an open cover of K and so using compactness we can find a finite subcover $\{V_x\}_{k=1}^n$. Then $V = \bigcup_{k=1}^n V_{x_k} \in \tau$, $\overline{V}$ is compact and $K \subseteq V \subseteq \overline{V} \subseteq U$.

(c) $\Longrightarrow$ (d): Let $\mathcal{B} = \{U \in \tau : \overline{U}$ is compact$\}$. Then since $\{x\}$ is compact, assertion (c) implies that $\mathcal{B}$ is a basis; see Corollary 1.1.6.

(d) $\Longrightarrow$ (a): This is obvious. □

Proposition 1.4.67. *If (X, τ) is a Hausdorff, second countable, locally compact topological space, then X has a countable basis consisting of relatively compact open sets.*

Proof. Let $\{U_n\}_{n \ge 1}$ be a basis of X. Fix $n \in \mathbb{N}$ and let $\{V_x\}_{x \in U_n}$ be an open cover of U_n such that $\overline{V}_x$ is compact and $\overline{V}_x \subseteq U_n$ for all $x \in U_n$; see Proposition 1.4.66. From Proposition 1.2.24 (b) we know that U_n is second countable. So, Lindelöf's Theorem (see Theorem 1.2.27) implies that we can find a countable subcover $\{V_k^n\}_{k \ge 1}$ of U_n. Then the family $\mathcal{B} = \{V_k^n : n, k \in \mathbb{N}\}$ is a countable basis of X consisting of relatively compact open sets. □

The next proposition places more precisely locally compact spaces in the chart of topological spaces.

Proposition 1.4.68. *Every locally compact topological space is completely regular; see Definition* 1.2.19.

Proof. Let $x \in X$ and $C \subseteq X$ be a closed set such that $x \notin C$. Applying Proposition 1.4.66 (c) yields the existence of relatively compact sets $V_1, V_2 \in \tau$ such that

$$x \in V_1 \subseteq \overline{V}_1 \subseteq V_2 \subseteq \overline{V}_2 \subseteq U = X \setminus C .$$

The set $\overline{V}_2$ is compact, and hence normal; see Corollary 1.4.50. Then, Urysohn's Lemma on normality (see Theorem 1.2.17) implies the existence of a continuous function $f: \overline{V}_2 \to [0,1]$ such that $f|_{\overline{V}_2 \setminus V_1} = 0$ and $f(x) = 1$. Let

$$\hat{f}(x) = \begin{cases} f(x) & \text{if } x \in \overline{V}_2 , \\ 0 & \text{if } x \in X \setminus \overline{V}_2 . \end{cases}$$

According to Proposition 1.1.37, $\hat{f}$ is continuous and $\hat{f}|_A = 0$ and $\hat{f}(x) = 1$. Hence, X is completely regular. □

Proposition 1.4.69. *Local compactness is preserved by continuous open surjections.*

Proof. Let X, Y be Hausdorff topological spaces with X locally compact and $f: X \to Y$ being a continuous, open surjection. Let $y \in Y$ and choose $x \in X$ such that $f(x) = y$. Then there exists $U \in \mathcal{N}(x)$ being relatively compact. Since f is open, $f(U) \in \mathcal{N}(y)$ and $f(\overline{U}) \subseteq Y$ is compact; see Theorem 1.4.51. Finally we have $y \in f(U) \subseteq \overline{f(U)} = f(\overline{U})$ with $f(\overline{U})$ being compact. Therefore, Y is locally compact as well. □

Of course every compact space is locally compact. In fact the following proposition is easy to prove.

Proposition 1.4.70. *If (X, τ) is a locally compact topological space, $U \in \tau$ and $C \subseteq X$ is closed, then $U \cap C$ endowed with the relative topology is locally compact.*

Proof. Let $x \in U \cap C$. Choose $V \in \mathcal{N}(x)$ relatively compact such that $x \in V \subseteq \overline{V} \subseteq U$. Then $V \cap (U \cap C)$ is a neighborhood of x in the relative topology of $U \cap C$. It holds

$$\overline{V \cap (U \cap C)}^{\tau(U \cap C)} = \overline{V} \cap (U \cap C) = \overline{V} \cap C$$

and the latter is closed in $\overline{V}$, hence compact. Therefore, $U \cap C$ is locally compact. □

Corollary 1.4.71. *Every open subset and every closed subset of a locally compact space is locally compact for the relative topology.*

We ask the natural question of when we can consider a Hausdorff topological space as a subspace of a compact topological space. Local compactness is the right concept for answering this question.

Definition 1.4.72. Let X be a Hausdorff topological space. A **compactification** of X is a compact topological space Y such that X is homeomorphic to a dense subset of Y. So we may think that X is an actual dense subset of Y.

Proposition 1.4.73. *If (X, τ) is a Hausdorff topological space and $(\hat{X}, \hat{\tau})$ a compactification of X, then X is locally compact if and only if $X \in \hat{\tau}$.*

Proof. $\Longrightarrow$: Let $x \in X$ and choose $U \in \mathcal{N}_X(x)$ relatively compact. We can find $V \in \mathcal{N}_X(x)$ such that $x \in V \subseteq U$. We have $V = W \cap X$ with $W \in \mathcal{N}_{\hat{X}}(x)$ and

$$W = W \cap \hat{X} = W \cap \overline{X} \subseteq \overline{W \cap X} = \overline{V} \subseteq \overline{U} = U \subseteq X .$$

This implies that x is $\hat{\tau}$-interior in X, hence $X \in \hat{\tau}$.

$\Longleftarrow$: We know that $(\hat{X}, \hat{\tau})$ is compact, hence locally compact. Since $X \in \hat{\tau}$ we conclude from Corollary 1.4.71 that X must be locally compact. □

The simplest compactification of noncompact, locally compact topological spaces is the so-called "Alexandrov one-point compactification".

Definition 1.4.74. Let X be a Hausdorff topological space and ∞ an object not in X, called the **point at infinity**. Let $\hat{X} = X \cup \{\infty\}$ and define a topology $\hat{\tau}$ on $\hat{X}$ specifying the following open sets:
(a) $\tau \subseteq \hat{\tau}$;
(b) $\hat{X} \setminus K$ with $K \subseteq X$ compact;
(c) $\hat{X}$.

Then we say that $(\hat{X}, \hat{\tau})$ is the **one-point compactification** of X.

Theorem 1.4.75. *If $\hat{X} = X \cup \{\infty\}$ is as in Definition 1.4.74 and is endowed with the topology $\hat{\tau}$ and (X, τ) is not compact, then $(\hat{X}, \hat{\tau})$ is a compactification of X and $\hat{X}$ is Hausdorff if and only if X is locally compact.*

Proof. First we show that $(\hat{X}, \hat{\tau})$ is compact. So, let $\mathcal{L}$ be an open cover of $\hat{X}$. Then $\mathcal{L}$ must have a member U such that $\infty \in U$. Then by Definition 1.4.74, $\hat{X} \setminus U$ is compact and so it has a finite subcover $\{U_k\}_{k=1}^n \subseteq \mathcal{L}$. Evidently $\{U_k, U\}_{k=1}^n \subseteq \mathcal{L}$ is a finite open cover of $\hat{X}$ and so we conclude that $(\hat{X}, \hat{\tau})$ is compact. It is easy to see from Definition 1.4.74 that $\hat{\tau}|_X = \tau$, that is, the subspace topology of $X \subseteq \hat{X}$ is τ. Since X is not compact, each $\hat{\tau}$-neighborhood of ∞, $\hat{X} \setminus K$ with K compact must intersect X. Hence ∞ is a limit point of X and so $\hat{X} = \overline{X}$. This proves that $(\hat{X}, \hat{\tau})$ is a compactification of X.

Suppose now that $\hat{X}$ is Hausdorff and let $x \in X$. We can find $U, V \in \hat{\tau}$ such that $\infty \in U$, $x \in V$ and $U \cap V = \emptyset$. This implies $V \subseteq \hat{X} \setminus U = K$ with K compact; see Definition 1.4.74. Therefore, X is locally compact.

Conversely, suppose that X is locally compact. Let $x \in X$ and choose $V \in \tau$ such that $x \in V \subseteq \overline{V}$ with $\overline{V}$ compact. Let $U = \hat{X} \setminus \overline{V}$. Then $\infty \in U$, $x \in V$ and $U \cap V = \emptyset$. Hence, $\hat{X}$ is Hausdorff. □

Example 1.4.76. The Alexandrov compactification of $\mathbb{R}^n$ is the n-sphere $S^n = \{u \in \mathbb{R}^{n+1} : |u| = 1\}$. To see this, let $N = (0, 0, \ldots, 0, 1) \in \mathbb{R}^{n+1}$ be the **north pole**. We define the **stereographic projection** $h: S^n \setminus \{N\} \to \mathbb{R}^n$ by

$$h((u_k)_{k=1}^{n+1}) = \frac{(u_k)_{k=1}^{n+1}}{1 - u_{n+1}}.$$

This map sends a point $u \in S^n \setminus \{N\}$ to a point $x \in \mathbb{R}^n$ where the line from N to x intersects $\mathbb{R}^n$. It is a homeomorphism with inverse map

$$h^{-1}((x_k)_{k=1}^{n}) = \frac{((2x_k)_{k=1}^{n}, |x|^2 - 1)}{|x|^2 + 1}.$$

Therefore, $S^n \setminus \{N\}$ is homeomorphic to $\mathbb{R}^n$. Then h extends to a homeomorphism of S^n with the Alexandrov compactification $\hat{\mathbb{R}}^n$ of $\mathbb{R}^n$. We can easily visualize the stereographic projection when $n = 1$ (Fig. 1.1).

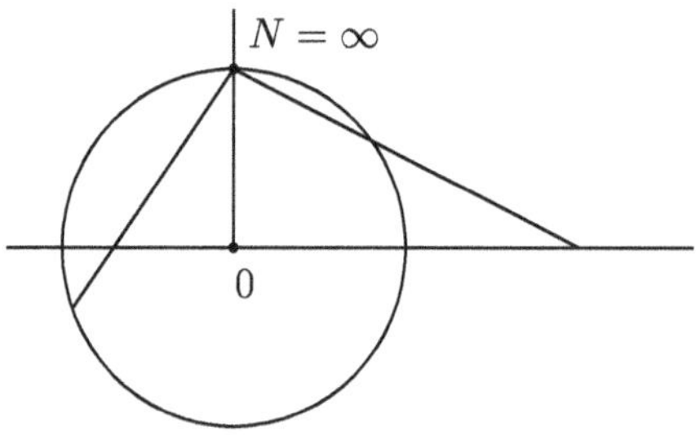

Figure 1.1: Alexandrov one-point compactification of $\mathbb{R}^n$.

This map was known to map makers long ago. From the discussion above we see that by removing a single point from S^n we obtain a space homeomorphic to $\mathbb{R}^n$. Which point we remove is irrelevant because we can rotate any point of S^n into any other. For convenience we remove the north pole N.

Definition 1.4.77. A Hausdorff topological space X is said to be **σ-compact** if it can be expressed as the union of at most countably many compact spaces.

Proposition 1.4.78. *Let (X, τ) be a Hausdorff topological space. The following statements are equivalent:*
(a) *X is locally compact and σ-compact.*
(b) *$X = \bigcup_{k\geq 1} U_k$ with U_k open, relatively compact such that $\overline{U}_k \subseteq U_{k+1}$ with $k \in \mathbb{N}$.*
(c) *X is locally compact and Lindelöf.*

Proof. (a) $\Longrightarrow$ (b): By hypothesis we have $X = \bigcup_{k\geq 1} K_k$ with $K_k \subseteq X$ compact. Proposition 1.4.66 (c) says that we can find $U_1 \supseteq K_1$ open and relatively compact. By induction we can find U_k open, relatively compact such that $U_k \supseteq \overline{U}_{k-1} \cup K_k$. Then $\{U_k\}_{k\geq 1}$ is the desired sequence of open sets.

(b) $\Longrightarrow$ (c): Let $\mathcal{L} = \{U_i\}_{i\in I}$ be an open cover of X. For each $m \in \mathbb{N}$ we can find a finite subfamily $\{U_m^k\}_{k=1}^{n(m)} \subseteq \mathcal{L}$ that covers $\overline{U}_i$ = compact. The family $\{U_m^k : 0 \le k \le n(m), m \in \mathbb{N}\} \subseteq \mathcal{L}$ is a countable subcover; thus X is Lindelöf.

(c) $\Longrightarrow$ (a): Let $\mathcal{L} = \{U_x\}_{x\in X}$ be a cover by relatively compact open sets; see Proposition 1.4.66 (c). The Lindelöf property implies that we can extract a countable subcover. Therefore, X is σ-compact. □

We introduce a generalization of σ-compactness that is determined by some requirement on the behavior of their coverings.

Definition 1.4.79. Let X be a Hausdorff topological space.
(a) Given two covers $\mathcal{L} = \{U_i\}_{i\in I}$ and $\mathcal{L}' = \{V_j\}_{j\in J}$ of X. We say that $\mathcal{L}$ is a **refinement** of $\mathcal{L}'$ if for each $i \in I$ there is a $j \in J$ such that $U_i \subseteq V_j$. We write $\mathcal{L} \prec \mathcal{L}'$.
(b) We say that a cover $\mathcal{L} = \{U_i\}_{i\in I}$ of X is **locally finite** if for every $x \in X$ there exists $V \in \mathcal{N}(x)$ that intersects a finite number of U_i's.
(c) We say that the cover $\mathcal{L} = \{U_i\}_{i\in I}$ of X is **point finite** if for every $x \in X$ there are at most finitely many indices $i \in I$ such that $x \in U_i$.

Remark 1.4.80. Given two covers $\mathcal{L} = \{U_i\}_{i\in I}$ and $\mathcal{L}' = \{V_j\}_{j\in J}$ of X we can define $\mathcal{L}_0 = \{U_i \cap V_j : (i,j) \in I \times J\}$, which is also a cover of X refining both $\mathcal{L}$ and $\mathcal{L}'$. Moreover, if both $\mathcal{L}$ and $\mathcal{L}'$ are locally finite (resp. point finite), then so is $\mathcal{L}_0$. A common refinement of both $\mathcal{L}$ and $\mathcal{L}'$ is also a refinement of $\mathcal{L}_0$.

A refinement of a cover may contain more elements than the given cover.

Definition 1.4.81. A refinement $\mathcal{L} = \{U_i\}_{i\in I}$ of the cover $\mathcal{L}' = \{V_j\}_{j\in J}$ is said to be **precise** if $I = J$ and $U_i \subseteq V_i$ for all $i \in I$.

Proposition 1.4.82. *If X is a Hausdorff topological space and the cover $\mathcal{L}' = \{V_j\}_{j\in J}$ of X has a locally finite (resp. point finite) refinement $\mathcal{L} = \{U_i\}_{i\in I}$, then it has a precise locally finite (resp. point finite) refinement $\hat{\mathcal{L}} = \{\hat{U}_j\}_{j\in J}$. Moreover, if $\mathcal{L}$ is open, then so is $\hat{\mathcal{L}}$.*

Proof. Let $\xi : I \to J$ be the map that assigns to each $i \in I$ a $j \in J$ such that $U_i \subseteq V_j$; see Definition 1.4.79 (a). For every $j \in J$ let $\hat{U}_j = \bigcup\{U_i : \xi(i) = j\}$ (some $\hat{U}_j$ may be empty). Then $\hat{U}_j \subseteq V_j$ for every $j \in J$ and $\hat{\mathcal{L}} = \{\hat{U}_j\}_{j\in J}$ is a cover of X. Clearly, $\hat{\mathcal{L}}$ is locally finite (resp. point finite) if $\mathcal{L}$ is and it is open if $\mathcal{L}$ is open. □

Definition 1.4.83. A Hausdorff topological space X is said to be **paracompact** if each open cover of X admits a locally finite refinement.

An immediate consequence of this definition is the following result.

Proposition 1.4.84. *Every compact topological space is paracompact.*

Closely related to paracompactness is the notion of **partition of unity**, which is essentially a **variable convex combination**.

Definition 1.4.85. Let X be a Hausdorff topological space and $f: X \to \mathbb{R}$ a function.

(a) The **support** of f is the closed set $\operatorname{supp} f := \overline{\{x \in X : f(x) \neq 0\}}$.

(b) A **partition of unity** on X is a family $\{f_i\}_{i \in I}$ of continuous functions $f_i: X \to [0,1]$ such that

 (i) $\{\operatorname{supp} f_i\}_{i \in I}$ form a locally finite closed cover of X;

 (ii) $\sum_{i \in I} f_i(x) = 1$ (the sum is well-defined because of (i)).

 If $\mathcal{L}' = \{V_j\}_{j \in J}$ is an open cover of X, then we say that a partition of unity $\{f_j\}_{j \in J}$ is subordinated to $\mathcal{L}'$ if $\operatorname{supp} f_j \subseteq V_j$ for each $j \in J$.

There is a close relation between paracompactness and partition of unity. The proof of the following theorem is very technical and so it is omitted. We refer to Dugundji [102, Theorem 4.2, p. 170].

Theorem 1.4.86. *A Hausdorff topological space is paracompact if and only if every open cover on X admits a locally finite partition of unity subordinated to the open cover.*

This theorem allows us to fix the place of paracompactness in the chart of topological spaces.

Proposition 1.4.87. *Every paracompact space is normal.*

Proof. Let C_1 and C_2 be two disjoint, closed subspaces of X. We consider the open cover $\mathcal{L} = \{X \setminus C_1, X \setminus C_2\}$. Then Theorem 1.4.86 implies that there is a partition of unity $\{f_1, f_2\}$ subordinated to $\mathcal{L}$. Then $f_1|_{C_2} = 1$ and $f_1|_{C_1} = 0$ and so by Urysohn's Normality Lemma (see Theorem 1.2.17) we conclude that X is normal. □

Closing this section, we mention that there is a "locally compact" version of the Tietze Extension Theorem; see Theorem 1.2.44. This version of the Tietze result reads as follows; see Hewitt–Stromberg [160, Theorem 7.40, p. 99].

Theorem 1.4.88. *If X is locally compact, $K \subseteq X$ is a nonempty, compact set and $U \subseteq K$ is open and $K \subseteq U$, then for every $f \in C(K, \mathbb{R})$ there exists $\hat{f} \in C(X, \mathbb{R})$ with compact support such that $\hat{f}|_K = f$ and f vanishes on $X \setminus U$.*

1.5 Metric Spaces – Baire Category

Metric spaces are a very important class of topological spaces. In fact the development of metric spaces led to the more general notion of topological space. In metric spaces the metric leads to an analysis that is primarily based in the properties of the real line.

Definition 1.5.1. Let X be a set. A **metric** on X is a map $d: X \times X \to \mathbb{R}$ such that the following hold:

(a) $d(x,u) = 0$ if and only if $x = u$;

(b) $d(x,u) = d(u,x)$ for all $x, u \in X$ (symmetry);

(c) $d(x,u) \leq d(x,v) + d(v,u)$ for all $x, u, v \in X$ (triangle inequality).

The pair (X, d) of a set X and of a metric d on X is said to be a **metric space**. If d does not satisfy (a), then d is called a **semimetric** (in French "ecart") and (X, d) is a **semimetric space**.

Remark 1.5.2. If d is a metric, then, based on (a)–(c), it is clear that $d(x,y) \geq 0$ for all $x, y \in X$. If d is a semimetric and $\sim$ is the equivalence relation defined by $x \sim u$ if and only if $d(x,u) = 0$, then $X/\sim$ is a metric space with metric $\hat{d}([x],[u]) = d(x,u)$. Here, for $x \in X$, $[x]$ is the corresponding equivalence class.

Definition 1.5.3. (a) Let (X, d) be a metric space and $A \subseteq X$. The **diameter** of A is defined by

$$\operatorname{diam} A = \sup\left[d(x,u) : x, u \in A\right].$$

If $\operatorname{diam} A < \infty$, then we say that A is **bounded**. Otherwise A is **unbounded**. When $\operatorname{diam} X < \infty$, then we say that d is a **bounded metric**. In addition, for $x \in X$ and $r > 0$, the **open ball** with center x and radius r is defined by

$$B_r(x) = \left\{u \in X : d(u,x) < r\right\}.$$

The corresponding **closed ball** with center x and radius r is defined by

$$\overline{B}_r(x) = \left\{u \in X : d(u,x) \leq r\right\}.$$

(b) Let (X, d) be a metric space. A set $A \subseteq X$ is said to be d-**open** (or simply **open**) if for every $x \in A$ we can find $r = r(x) > 0$ such that $B_r(x) \subseteq A$. The collection

$$\tau_d = \{A \subseteq X : A \text{ is } d\text{-open}\}$$

is a topology on X called the **metric topology** on (X, d).

(c) A topological space (X, τ) is said to be **metrizable** if $\tau = \tau_d$ for some metric d on X. This metric is then said to be **compatible** with the topology. If for two metrics d_1 and d_2 on X, we have $\tau_{d_1} = \tau_{d_2}$, then we say that d_1 and d_2 are **equivalent**.

Remark 1.5.4. The distinction between metric and metrizable spaces is a subtle one. In the case of a metric space we already have a fixed metric. For a metrizable space we have not yet decided from the multitude of equivalent metrics. Note that if d is compatible, then so is kd with $k \in \mathbb{N}$ or $\hat{d}(x,u) = (d(x,u))(1 + d(x,u))$ and $\hat{d}_0(x,u) = \min\{1, d(x,u)\}$. The last two metrics are bounded even if d is not. From the triangle inequality we have

$$\left|d(x,u) - d(y,v)\right| \leq d(x,y) + d(u,v) \quad \text{for all } x, u, y, v \in X. \tag{1.5.1}$$

It follows that d is jointly continuous. Of course τ_d is Hausdorff and first countable and $u_n \xrightarrow{\tau_d} u$ if and only if $d(u_n, u) \to 0$.

In Proposition 1.2.22 we saw that second countability implies separability. For metrizable spaces the two notions are equivalent.

Proposition 1.5.5. *A metrizable space is second countable if and only if it is separable.*

Proof. $\Longrightarrow$: This follows from Proposition 1.2.22.

$\Longleftarrow$: Let (X, τ) be a separable metrizable space and d a compatible metric, that is, $\tau_d = \tau$. Let $D \subseteq X$ be a countable dense set and consider the collection $\mathcal{L} = \{B_{1/n}(x) : x \in D, n \in \mathbb{N}\}$. Clearly, $\mathcal{L}$ is a countable basis for the topology τ; see Corollary 1.1.6. □

Combining this proposition with Proposition 1.2.24 (b) we have the following result.

Corollary 1.5.6. *If X is a separable metrizable space and $A \subseteq X$, then A is separable.*

Definition 1.5.7. Let (X, τ) be a topological space. A set A is said to be an **F_σ-set** if it is the union of at most countably many closed sets. A set C is said to be a **G_δ-set** if it is the intersection of at most countably many open sets.

Proposition 1.5.8. *If X is a metrizable space, then every closed set is G_δ and every open set is F_σ.*

Proof. Let $C \subseteq X$ be closed. Then $U_n = \{x \in X : d(x, C) < 1/n\}$ is open because of the continuity of d. Furthermore $C = \bigcap_{n \geq 1} U_n$. So C is G_δ. Next let $U \subseteq X$ be open. Since $X \setminus U$ is closed, the first part yields that $X \setminus U = \bigcap_{n \geq 1} U_n$ with U_n open. Hence, $U = \bigcup_{n \geq 1} (X \setminus U_n)$ and so U is F_σ. □

Definition 1.5.9. (a) Let (X, d) be a metric space. A sequence $\{x_n\}_{n \geq 1} \subseteq X$ is said to be a **Cauchy sequence** if for any given $\varepsilon > 0$ there exists $n_0 = n_0(\varepsilon) \geq 1$ such that $d(x_n, x_m) \leq \varepsilon$ for all $n, m \geq n_0$, that is, $d(x_n, x_m) \to 0$ as $n, m \to +\infty$. We say that (X, d) is **complete** if every Cauchy sequence in X converges in X.

(b) Let (X, τ) be a topological space. We say that X is **topologically complete** if there is a compatible complete metric d, that is, $\tau_d = \tau$.

Remark 1.5.10. The property of completeness is metric dependent. So it can happen that two metrics are equivalent, that is, they generate the same topology, but one is complete and the other not. On the other hand, topological completeness is a topological property.

Example 1.5.11. The interval $(-1, 1)$ with the usual metric is not a complete metric space but it is topologically complete since it is homeomorphic to $\mathbb{R}$, which is complete. The function $h: (-1, 1) \to \mathbb{R}$ defined by $h(x) = x/(1 - x^2)$ for all $x \in (-1, 1)$ is a homeomorphism between the two spaces.

Definition 1.5.12. Let (X, d) and (Y, ρ) be two metric spaces. A map $f: X \to Y$ is said to be an **isometry** if $d(x, u) = \rho(f(x), f(u))$ for all $x, u \in X$. If f is a surjective isometry, then we say that X and Y are **isometric** spaces. Otherwise we say that f is an isometric embedding.

Remark 1.5.13. Thus an isometric surjection is a distance preserving homeomorphism. In the case of an isometric embedding $f: X \to Y$ we may think of X as a subspace of Y.

Every metric space can be isometrically and densely embedded in a complete metric space.

Theorem 1.5.14. *If (X, d) is any metric space, then there is a complete metric space (Y, ρ) and an isometry $f: X \to Y$ such that $f(X)$ is dense in Y. We say that Y is the* ***completion*** *of X.*

Proof. Let $f_x(u) = d(x, u)$ for all $x, u \in X$. Choose a point $v \in X$ and let

$$S(X, d) = \{f_v + h : h \in C_b(X, \mathbb{R})\} .$$

On $S(X, d)$ we consider the supremum metric d_∞ defined by

$$d_\infty(f_v + h, f_v + \hat{h}) = \sup[|h(x) - \hat{h}(x)| : x \in X] .$$

For any $x, u, y \in X$ we have

$$|d(x, y) - d(u, y)| \leq d(x, u)$$

(see (1.5.1)) and equality holds if $y = x$ or $y = u$. Therefore, for any $u \in X$, taking $x = v$, we have

$$f_u - f_x \in C_b(X, \mathbb{R}) , \quad d_\infty(f_x, f_u) = d(x, u) .$$

In addition we have $f_u \in S(X, d)$ and $S(X, d)$ does not depend on the choice of $v \in X$. Hence, the map $x \to f_x$ from X into $S(X, d)$ is an isometry for d and d_∞. Let Y be the d_∞-closure of the range of this map into $S(X, d)$. But $(C_b(X, \mathbb{R}), d_\infty)$ is complete; recall that the uniform limit of continuous functions is continuous. Hence (Y, d_∞) is complete and this is the completion of (X, d). □

Now we can provide a necessary and sufficient condition for the completeness of a metric space. The necessary part of the result is known as “Cantor’s Intersection Theorem”.

Theorem 1.5.15. *A metric space (X, d) is complete if and only if every decreasing sequence $\{C_n\}_{n\geq 1}$ of nonempty, closed subsets of X such that* $\operatorname{diam} C_n \to 0$ *as $n \to \infty$, has a singleton intersection.*

Proof. $\Longrightarrow$: Let $C = \bigcap_{n\geq 1} C_n$. Then $\operatorname{diam} C \leq \operatorname{diam} C_n$ for all $n \in \mathbb{N}$. Hence, $\operatorname{diam} C = 0$. This means that C is empty or a singleton. We show that $C \neq \emptyset$. For each $n \in \mathbb{N}$ we pick $u_n \in C_n$. Then for $n \geq m$ we have $d(u_n, u_m) \leq \operatorname{diam} C_m \to 0$ as $m \to \infty$. So $\{u_n\}_{n\geq 1} \subseteq X$ is a Cauchy sequence and the completeness of X implies that there exists $u \in X$ such that $u_n \to u$. Evidently $u \in C$ and so $C = \bigcap_{n\geq 1} C_n = \{u\}$.

$\Longleftarrow$: Let $\{u_n\}_{n\geq1} \subseteq X$ be a Cauchy sequence. Set $C_n = \overline{\{u_k : k \geq n\}}$. Since $\{u_n\}_{n\geq1}$ is a Cauchy sequence, we have $\operatorname{diam} C_n \to 0$. By hypothesis $\bigcap_{n\geq1} C_n = \{u\}$ and so we have $u_n \to u$ in X, which means that X is complete. □

Now we consider the Cartesian product of metric spaces. To this end, let $\{X_n\}_{n\geq1}$ be a sequence of nonempty Hausdorff topological spaces and let $X = \prod_{n\geq1} X_n$ be furnished with the product topology.

Proposition 1.5.16. *The product topology on $X = \prod_{n\geq1} X_n$ is metrizable if and only if the space X_n is metrizable for each $n \in \mathbb{N}$.*

Proof. $\Longrightarrow$: Let d be a compatible metric for X. For each $n \in \mathbb{N}$ we fix a $y_n \in X_n$. Then for $u \in X_m$ we define $\hat{u} = (u_k)_{k\geq1} \in X$ by setting $u_k = y_k$ for $k \neq m$ and $u_m = u$. Now we define a metric d_m on X_m by setting $d_m(u, v) = d(\hat{u}, \hat{v})$. It is easy to see that d_m is indeed a metric on X_m. Note that d-convergence in X is equivalent to componentwise convergence. From this it follows easily that τ_{d_m} coincides with the topology of X_m.

$\Longleftarrow$: Assume that each X_n is metrizable and let d_n be a compatible metric. We define a metric d on the product X by setting

$$d((u_n), (v_n)) = \sum_{n\geq1} \frac{1}{2^n} \frac{d_n(u_n, v_n)}{1 + d_n(u_n, v_n)} .$$

It is straightforward that d is a metric. Let $\{\hat{u}_\alpha\}_{\alpha\in J} = \{(u_n^\alpha)\}_{\alpha\in J} \subseteq X$ be a net. We have

$$d(\hat{u}_\alpha, \hat{u}) \to 0 \quad \text{with } \hat{u} = (u_n) \quad \text{if and only if} \quad \lim_{\alpha\in J} d_n(u_n^\alpha, u_n) = 0 , \tag{1.5.2}$$

for all $n \in \mathbb{N}$. From (1.5.2) we infer that the product topology and the τ_d-topology on X coincide. □

In a similar fashion we can also have the following result.

Proposition 1.5.17. *The product topology on X is topologically complete if and only if the space X_n is topologically complete for each $n \in \mathbb{N}$.*

Proposition 1.5.18. *If $\{X_n\}_{n\geq1}$ is a sequence of metrizable spaces and $X = \prod_{n\geq1} X_n$, then X is separable if and only if X_n is separable for each $n \in \mathbb{N}$.*

Proof. $\Longrightarrow$: This is a consequence of the fact that the continuous image of a separable space is separable as well; see Proposition 1.2.24 (c). In our case the continuous map is the projection to the nth factor.

$\Longleftarrow$: From the proof of Proposition 1.5.16 we know that the product topology on X is generated by the metric

$$d(\hat{u}, \hat{v}) = \sum_{n\geq1} \frac{1}{2^n} \frac{d_n(u_n, v_n)}{1 + d_n(u_n, v_n)} \quad \text{for all } \hat{u} = (u_n),\ \hat{v} = (v_n) \in X .$$

For each $n \in \mathbb{N}$ let D_n be a countable, dense subset of X_n. Fix $u_n \in D_n$ for each $n \in \mathbb{N}$ and consider the set $D \subseteq X$ defined by

$$D = \{(y_n) \in X : y_n \in D_n \text{ for each } n \in \mathbb{N} \text{ and } y_n = u_n \text{ eventually}\} .$$

Evidently $D \subseteq X$ is countable and dense. Therefore X is separable. □

Definition 1.5.19. The **Hilbert cube** is the space $\mathbb{H} = [0,1]^{\mathbb{N}}$, that is, the space of all real sequences with values in $[0,1]$.

Remark 1.5.20. Evidently $\mathbb{H}$ is topologically complete, separable, and compact, which follows from the Propositions 1.5.17 and 1.5.18 as well as Theorem 1.4.56.

The next theorem, known as "Urysohn's Theorem", says that in a sense $\mathbb{H}$ is the canonical separable metrizable space.

Theorem 1.5.21 (Urysohn's Theorem). *Every separable metrizable space is homeomorphic to a subset of* $\mathbb{H}$.

Proof. Let (X,d) be a separable metric space and $D = \{y_n\}_{n\geq 1}$ a countable dense subset. We define $\xi_n(u) = \min\{1, d(u,y_n)\}$ for all $n \in \mathbb{N}$ and consider $\xi : X \to \mathbb{H}$ defined by $\xi(u) = (\xi_n(u))_{n\geq 1}$ for all $u \in X$. Each ξ_n is continuous, hence so is ξ. Suppose that $\xi(u) = \xi(v)$ and let $\{y_{n_k}\}_{k\geq 1} \subseteq \{y_n\}_{n\geq 1}$ such that $y_{n_k} \to u$. We have $\lim_{k\to\infty} d(v, y_{n_k}) = 0$, hence $d(v,u) = 0$, which means that $u = v$ and so ξ is $1-1$. Finally we need to show that ξ^{-1} is continuous. To this end, let $\xi(v_n) \to \xi(v)$. Pick $\varepsilon > 0$ and u_m such that $d(v,u_m) < \varepsilon$. Note that

$$d(v_n, u_m) \to d(v, u_m) \quad \text{as } n \to \infty ,$$

which means $d(v_n, u_m) < \varepsilon$ for all $n \geq n_0$. Hence, by the triangle inequality we derive $d(v_n, v) < 2\varepsilon$ for all $n \geq n_0$. Therefore, $v_n \to v$ and so ξ^{-1} is continuous. □

Some features of metrizable spaces are not topological and depend on the particular compatible metric. Such are Cauchy sequences (see Definition 1.5.9 (a)) and uniform continuity, which we are about to introduce.

Definition 1.5.22. Let (X,d) and (Y,ρ) be two metric spaces and $f : X \to Y$ a map.

(a) We say that f is **uniformly continuous** if for every given $\varepsilon > 0$ there exists $\delta = \delta(\varepsilon) > 0$ such that

$$d(x,u) < \delta \quad \text{implies} \quad \rho(f(x), f(u)) < \varepsilon \quad \text{for all } x, u \in X .$$

(b) We say that f is k-**Lipschitz** if

$$\rho(f(x), f(u)) \leq k d(x,u) \quad \text{for all } x, u \in X \text{ with } k > 0 .$$

Remark 1.5.23. A continuous function need not be uniformly continuous. For example, the function $f(x) = x^2$ for $x \in \mathbb{R}$ is continuous but not uniformly continuous. Indeed, note that for $\varepsilon > 0$ the $\delta > 0$ gets smaller as $|x|$ increases. A k-Lipschitz map is uniformly continuous. A 1-Lipschitz map is called **nonexpansive** and if $k \in (0,1)$ we say that f is a **contraction**.

Proposition 1.5.24. *If (X,d) is a metric space and $\varphi\colon \mathbb{R}_+ \to \mathbb{R}_+$ is continuous satisfying*
(a) *φ is nondecreasing, that is, $x \le u$ implies $\varphi(x) \le \varphi(u)$ for all $x, u \ge 0$;*
(b) *φ is subadditive, that is, $\varphi(x+u) \le \varphi(x) + \varphi(u)$ for all $x, u \ge 0$;*
(c) *$\varphi(x) = 0$ if and only if $x = 0$,*

then $\varphi \circ d$ is a metric on X and the identity maps

$$i_1\colon (X,d) \to (X, \varphi \circ d) \quad \text{and} \quad i_2\colon (X, \varphi\circ d) \to (X,d)$$

are both uniformly continuous.

Proof. Applying (a)–(c) it is straightforward to check that $\varphi \circ d$ is a metric on X. Moreover, for given $\varepsilon > 0$ there exists $\delta > 0$ such that $0 \le t < \delta$ implies $0 \le \varphi(t) < \varepsilon$ as well as $0 \le \varphi(t) < \eta = \varphi(\varepsilon)$ implies $0 \le t < \delta$. Here we have used the continuity and monotonicity of φ. Thus we have uniform continuity for both i_1 and i_2. □

Proposition 1.5.25. *If (X,d) and (Y,ρ) are two metric spaces and $f\colon X \to Y$ is uniformly continuous, then f maps Cauchy sequences in X to Cauchy sequences in Y.*

Proof. Let $\{u_n\}_{n\ge1}$ be a Cauchy sequence in X, and for $\varepsilon > 0$ choose $\delta = \delta(\varepsilon) > 0$ such that $d(x,v) < \delta$ implies $\rho(f(x), f(v)) < \varepsilon$ for all $x, v \in X$.

Let $B \subseteq X$ be a ball of radius less than $\delta/2$, which contains $\{u_n\}_{n\ge n_0}$ for some $n_0 \in \mathbb{N}$. Then $f(B)$ contains $\{f(u_n)\}_{n\ge n_0}$. Note that $\operatorname{diam} B < \delta$. Hence $\operatorname{diam} f(B) < \varepsilon$. Thus $f(B)$ is included in a ball $D \subseteq Y$ of radius $\varepsilon > 0$ and so $D \supseteq \{f(u_n)\}_{n\ge\hat{n}}$ for some $\hat{n} \in \mathbb{N}$. Since $\varepsilon > 0$ is arbitrary, we conclude that $\{f(u_n)\}_{n\in\mathbb{N}} \subseteq Y$ is a ρ-Cauchy sequence. □

Remark 1.5.26. The result above fails if f is only continuous. To see this consider the function $f(x) = 1/x$ for all $x \in (0,1)$, which is continuous but not uniformly continuous. Let $u_n = 1/n$ with $n \in \mathbb{N}$. This is a Cauchy sequence in $(0,1)$ but $f(u_n) = n$, which is not a Cauchy sequence.

Theorem 1.5.27. *If (X,d) is a metric space, $D \subseteq X$ a set, (Y,ρ) is a complete metric space and $f\colon D \to Y$ is uniformly continuous, then there exists a unique uniformly continuous map $\hat{f}\colon \overline{D} \to Y$ such that $\hat{f}|_D = f$. In particular, if $Y = \mathbb{R}$ then $\sup_D |f| = \sup_{\overline{D}} |\hat{f}|$.*

Proof. Let $\tilde{u} \in \overline{D}$. Then we find a sequence $\{u_n\}_{n\ge1} \subseteq D$ such that $u_n \to \tilde{u}$ in (X,d). The sequence $\{u_n\}_{n\ge1}$ is a d-Cauchy sequence and then $\{f(u_n)\}_{n\ge1} \subseteq Y$ is a ρ-Cauchy sequence because of Proposition 1.5.25. The completeness of Y implies that $f(u_n) \to y \in Y$. This y is independent of the particular sequence in D approaching $\tilde{u} \in \overline{D}$. Indeed, let $\{x_n\}_{n\ge1} \subseteq D$

be another sequence such that $x_n \to \tilde{u}$ in (X, d). We define

$$h_n = \begin{cases} x_n & \text{if } n = \text{odd} \\ u_n & \text{if } n = \text{even} \end{cases} \quad \text{with } n \in \mathbb{N}.$$

We see that $h_n \to \tilde{u}$ and then $f(h_n) \to y$. Note that $\{f(h_n)\}_{n\geq 1}$ is a Cauchy sequence and for the subsequence $\{f(u_n)\}_{n\geq 1}$ we have that it converges to y in (Y, ρ). Hence, we have shown that y is independent of the sequence $u_n \to \tilde{u} \in \overline{D}$. Therefore, we can set $\hat{f}(\tilde{u}) = y$.

Now we show that $\hat{f}$ is uniformly continuous. From the uniform continuity of f we know that for given $\varepsilon > 0$ there exists $\delta > 0$ such that

$$d(x, u) < \delta \quad \text{implies} \quad \rho(f(x), f(u)) < \varepsilon \quad \text{for all } x, u \in D. \tag{1.5.3}$$

Suppose $x, v \in \overline{D}$ with $d(x, v) < \delta$. Then there exist $\{x_n\}_{n\geq 1}, \{u_n\}_{n\geq 1} \subseteq D$ such that $x_n \to x$ and $v_n \to v$ in (X, d). Hence, $d(x_n, v_n) \to d(x, v)$ and so $d(x_n, v_n) < \delta$ for all $n \geq n_0$. Taking (1.5.3) into account we conclude that $\rho(f(x_n), f(v_n)) < \varepsilon$ for all $n \geq n_0$. Hence, $\rho(f(x), f(v)) \leq \varepsilon$. This proves the uniform continuity of the extension $\hat{f}$. Clearly this extension is unique and we have $\sup_D |f| = \sup_{\overline{D}} |\hat{f}|$. □

Definition 1.5.28. Let (X, d) be a metric space. Recall that

$$C_b(X, \mathbb{R}) = \{f : X \to \mathbb{R} \mid f \text{ is bounded and continuous}\}.$$

We also introduce the subspace

$$U_b(X, \mathbb{R}) = \{f : X \to \mathbb{R} \mid f \text{ is bounded and uniformly continuous}\}$$

of $C_b(X, \mathbb{R})$. On them we consider the **supremum metric** defined by

$$d_\infty(f, g) = \sup_{x\in X} |f(x) - g(x)|.$$

Remark 1.5.29. If X is a metrizable space and d, e are two compatible metrics, then in general we have $U_d(X, \mathbb{R}) \neq U_e(X, \mathbb{R})$. For example, the function $x \to 1/x$ on $(0, 1)$ is not uniformly continuous for the usual metric on $(0, 1)$, but it is uniformly continuous for the metric $\rho(x, u) = |1/x - 1/u|$ for all $x, u \in (0, 1)$.

Proposition 1.5.30. *If (X, d) is a metric space, then X is isometrically embedded into $U_d(X, \mathbb{R})$.*

Proof. We fix $u_0 \in X$ and then for each $x \in X$, let $\eta_x : X \to \mathbb{R}$ be the function defined by $\eta_x(u) = d(x, u) - d(u_0, u)$ for all $u \in X$. We have

$$|\eta_x(u) - \eta_x(v)| \leq |d(x, u) - d(x, v)| + |d(u_0, u) - d(u_0, v)| \leq 2d(u, v),$$

which shows that η_x is 2-Lipschitz. In addition we have $\eta_x(u) \le d(x, u_0)$ for all $u \in X$. Thus, η_x is bounded. Consequently we have $\eta_x \in U_d(X, \mathbb{R})$. Note that

$$|\eta_x(u) - \eta_v(u)| \le d(x, v) \quad \text{for all } u \in X\,,$$

implying $d_\infty(\eta_x, \eta_v) \le d(x, v)$. Moreover, we have $|\eta_x(v) - \eta_v(v)| = d(x, v)$. Therefore, $d_\infty(\eta_x, \eta_v) = d(x, v)$, which means that $x \to \eta_x$ is an isometry. This proves that X is isometrically embedded into $U_d(X, \mathbb{R})$. □

Now we turn our attention to compact metric spaces.

Definition 1.5.31. Let (X, d) be a metric space and $\varepsilon > 0$. An **ε-net** in X is a finite set A in X such that $X = \bigcup_{a \in A} B_\varepsilon(a)$. That is, for every $x \in X$ there exists $a \in A$ such that $d(x, a) < \varepsilon$. We say that (X, d) is **totally bounded** if for every $\varepsilon > 0$ it has an ε-net.

Remark 1.5.32. Clearly a compact metric space is totally bounded.

Proposition 1.5.33. *If the metric space (X, d) is totally bounded, then it is separable.*

Proof. For each $n \in \mathbb{N}$, let $A_n \subseteq X$ be a finite set such that $X = \bigcup_{x \in A_n} B_{1/n}(x)$. Let $D = \bigcup_{n \ge 1} A_n$. Then D is countable and dense in X. □

Proposition 1.5.34. *If (X, d) is a sequentially compact metric space and let $\mathcal{L}$ be an open cover of X, then there is a $\delta > 0$ such that every $A \subseteq X$ with $\operatorname{diam} A < \delta$ is contained in some $U \in \mathcal{L}$.*

Proof. Arguing by contradiction, suppose that we cannot find such a $\delta > 0$. Then for every $n \in \mathbb{N}$ choose $A_n \subseteq X$ with $\operatorname{diam} A_n < 1/n$ and A_n is not contained in any $U \in \mathcal{L}$. Choose $x_n \in A_n$. Since X is sequentially compact, by passing to a subsequence if necessary, we may assume that $x_n \to x$. Let $U \in \mathcal{L} \cap \mathcal{N}(x)$ and choose $\varrho > 0$ such that $B_\varrho(x) \subseteq U$. Then $x_n \in B_{\varrho/2}(x)$ for all $n \ge n_0$ with $1/n_0 < \varrho/2$. Since $\operatorname{diam} A_{n_0} < 1/n_0 < \varrho/2$, we have $A_{n_0} \subseteq B_\varrho(x) \subseteq U$, a contradiction. This proves the proposition. □

Remark 1.5.35. A $\delta > 0$ satisfying the property above is called the **Lebesgue number** of the cover $\mathcal{L}$.

The next theorem provides a complete characterization of compact metric spaces.

Theorem 1.5.36. *If (X, d) is a metric space, then the following statements are equivalent:*
(a) *X is compact;*
(b) *X is complete and totally bounded;*
(c) *X is sequentially compact.*

Proof. (a) $\Longrightarrow$ (b): Since (X, d) is compact, every Cauchy sequence $\{x_n\}_{n \ge 1}$ has a cluster point $x \in X$, see Remark 1.4.58. We claim that $x_n \to x$ in X. Since $\{x_n\}_{n \ge 1}$ is a Cauchy sequence, there exists $n_0 \in \mathbb{N}$ for every given $\varepsilon > 0$ such that

$$d(x_n, x_m) < \varepsilon \quad \text{for all } n, m \ge n_0\,. \tag{1.5.4}$$

Since x is a cluster point of the Cauchy sequence, we can find $k \geq n_0$ such that

$$d(x_k, x) < \varepsilon . \tag{1.5.5}$$

Then, combining (1.5.4) and (1.5.5), we have for $n \geq n_0$

$$d(x_n, x) \leq d(x_n, x_k) + d(x_k, x) < 2\varepsilon ,$$

which means that $x_n \to x$ in X and so X is complete.

For every $\varepsilon > 0$ we have $X = \bigcup_{x\in X} B_\varepsilon(x)$. The compactness of X implies that we can find $x_1, \ldots, x_m$ such that $X = \bigcup_{n=1}^{m} B_\varepsilon(x_n)$. Thus, X is totally bounded.

(b) $\Longrightarrow$ (c): Let $\{x_n\}_{n\geq 1}$ be a sequence in X. Since X is totally bounded, a subsequence S_1 of $\{x_n\}_{n\geq 1}$ must be in a set $B_1 = \{u \in X : d(y_1, u) < 1\}$. Evidently, B_1 is totally bounded. Hence, there exists a subsequence S_2 of S_1, which will be in $B_2 = \{u \in B_1 : d(y_2, u) < 1/2\}$. By induction for each $n \in \mathbb{N}$ we can have a subsequence S_{n+1} of S_n, which is in $B_{n+1} = \{u \in X : d(y_{n+1}, u) < 1/(n+1)\}$. Let $i_1 < i_1 < \cdots < i_n < \cdots$ be such that $x_{i_n} \in S_n$. Then $\{x_n\}_{n\in\mathbb{N}}$ is Cauchy sequence and thus converges. This proves that X is sequentially compact.

(c) $\Longrightarrow$ (a): Let $\mathcal{L}$ be an open cover of X and let $\delta > 0$ be the Lebesgue number of $\mathcal{L}$; see Proposition 1.5.34 and Remark 1.5.35. First we show that X is totally bounded. If this is not the case, then we can find $\varepsilon > 0$ such that no finite family of balls of radius $\varepsilon > 0$ cover X. Inductively we can generate a sequence $\{x_n\}_{n\geq 1} \subseteq X$ such that for all $n \in \mathbb{N}$, $x_n \notin \bigcup_{k<n} B_\varepsilon(x_k)$. For $n \neq m$, we have $d(x_n, x_m) \geq \varepsilon$. This sequence cannot have a convergent subsequence and this contradicts the hypothesis that X is sequentially compact. Therefore X is totally bounded. We choose a $\delta/3$-net $\{x_n\}_{n=1}^{m}$. For each $n \leq m$ let $U_n \in \mathcal{L}$ such that $B_{\delta/3}(x_n) \subseteq U_n$. Then $\{U_n\}_{n=1}^{m}$ is a finite subcover of $\mathcal{L}$ and this proves the compactness of X. □

Corollary 1.5.37. *A metric space is totally bounded if and only if its completion is compact.*

Since bounded sets in $\mathbb{R}^N$ are totally bounded we can state the following characterization of compact sets in $\mathbb{R}^N$. The result is known as the "Heine–Borel Theorem".

Theorem 1.5.38 (Heine–Borel Theorem). *A set $C \subseteq \mathbb{R}^N$ is compact if and only if C is closed and bounded.*

Proposition 1.5.39. *If (X, d) and (Y, ρ) are metric spaces with X being sequentially compact and if $f : X \to Y$ is continuous, then f is uniformly continuous.*

Proof. Given $\varepsilon > 0$ and $x \in X$, let $V_x = f^{-1}(B_{\varepsilon/2}(f(x))) \in \mathcal{N}(x)$. Then, for $u, v \in V_x$ we have

$$\rho(f(u), f(v)) < \varepsilon . \tag{1.5.6}$$

We know that X is sequentially compact because of Theorem 1.5.36. By Proposition 1.5.34 there exists $\delta > 0$ such that for every $v \in X$

$$B_\delta(v) \subseteq V_x \quad \text{for some } x \in X\,. \tag{1.5.7}$$

Recall that this δ is called the Lebesgue number of the cover $\mathcal{L} = \{V_x\}_{x\in X}$; see Proposition 1.5.34 and Remark 1.5.35. Then, because of (1.5.6) and (1.5.7), $u \in B_\delta(v)$ implies $\rho(f(u), f(v)) < \varepsilon$. Hence, f is uniformly continuous. □

The next proposition is an easy consequence of the relevant definitions.

Proposition 1.5.40. (a) *Every metric space X is first countable.*
(b) *For a metric space X the notions of separability, second countability, and Lindelöf are all equivalent.*

Proof. (a) For every $x \in X$, let $\mathcal{B}(x) = \{B_r(x): r \in \mathbb{Q}\}$. Then $\mathcal{B}$ is a countable local basis at $x \in X$. Therefore X is first countable.

(b) First we show that "separability" implies "second countability". Let $\{u_n\}_{n\geq 1}$ be dense in X. Then $\mathcal{B} = \{B_r(u_n): r \in \mathbb{Q}, n \in \mathbb{N}\}$ is a countable basis of X, hence X is second countable. Theorem 1.2.27 says that "second countable" implies "Lindelöf". Finally we show that "Lindelöf" implies "separable". Consider the open cover $\{B_\varepsilon(x)\}_{x\in X}$ with $\varepsilon > 0$ of X. By the Lindelöf property there exists a countable subcover $\{B_\varepsilon(x_k)\}_{k\in\mathbb{N}}$. Let $A(\varepsilon) = \{x_k\}_{k\in\mathbb{N}}$. Then $D = \bigcup_{n\geq 1} A(1/n)$ is a countable dense subset of X. Therefore X is separable. □

Remark 1.5.41. In contrast to general topological spaces (see Proposition 1.2.22), for metric spaces, separability and second countability are equivalent notions.

Combining Proposition 1.5.40 with Theorem 1.4.63 we have the following result.

Theorem 1.5.42. *Let (X, d) be a metric space. Then the following assertions are equivalent:*
(a) *X is compact.*
(b) *X is countably compact.*
(c) *X is limit point compact.*
(c) *X is sequentially compact.*

Definition 1.5.43. A Hausdorff topological space (X, τ) is said to be **Polish** if it is separable and there exists a compatible metric d, that is $\tau = \tau_d$, for which X is complete.

Remark 1.5.44. In a Polish space the compatible metric is not a priori fixed. We know that it exists and generates the topology of X and that the space furnished with this metric is complete. There are many topological spaces that are Polish, but the corresponding complete metric is not particularly simple or natural. However, many constructions and facts depend only on the existence of a complete metric and not on the exact choice.

Proposition 1.5.45. *If X is a Polish space and $A \subseteq X$ is open or closed, then A is Polish.*

Proof. From Corollary 1.5.6 we know that A is separable. First suppose that A is open. We assume that $A \neq X$ and let d be the compatible metric on X for which X is complete. Let

$$\hat{d}(x,u) = d(x,u) + \left|\frac{1}{d(x,A^c)} - \frac{1}{d(u,A^c)}\right| \quad \text{for all } x,u \in A\,. \tag{1.5.8}$$

It is easy to see that $\hat{d}$ is a metric on A. We show that $\hat{d}$ metrizes the subspace topology on A. From the triangle inequality we have

$$|d(x,A^c) - d(u,A^c)| \le d(x,u)\,,$$

which implies that $x \to d(x,A^c)$ is 1-Lipschitz, equivalently nonexpansive. Therefore, $u_n \xrightarrow{\hat{d}} u$ if and only if $u_n \xrightarrow{d} u$. Hence, $\hat{d}$ metrizes the subspace topology on A.

Suppose that $\{u_n\}_{n\ge1} \subseteq A$ is a $\hat{d}$-Cauchy sequence. Then, from (1.5.8) it is clear that $\{u_n\}_{n\ge1}$ is also a d-Cauchy sequence. Therefore, $u_n \xrightarrow{d} u \in X$. If $u \in A^c$, then $d(u_n,A^c) \to 0$ and so from (1.5.8) we have $\hat{d}(u_n,u_m) \to +\infty$ as $n,m \to +\infty$, a contradiction. Thus, $u \in A$ and so $u_n \xrightarrow{d} u$, which proves the completeness of $(A,\hat{d})$.

Now suppose that A is closed. Then $d_A = d|_{A\times A}$ is complete and so A is Polish. □

Proposition 1.5.46. *Countable products and countable intersections of Polish spaces are Polish spaces.*

Proof. For the products the result follows from Propositions 1.5.16, 1.5.17 and 1.5.18. For the intersections let

$$\Delta = \left\{(u_n) \in \prod_{n\ge1} X_n : u_j = u_k \text{ for all } j,k\right\}.$$

Then Δ is closed, hence Polish; see Proposition 1.5.45. But Δ is homeomorphic to $\bigcap_{n\ge1} X_n$. □

The next result is known as "Alexandrov's Theorem" and gives a characterization of Polish spaces.

Theorem 1.5.47 (Alexandrov's Theorem). *If (X,τ) is a Polish space, then $A \subseteq X$ is Polish if and only if A is a G_δ-subset of X.*

Proof. $\Longrightarrow$: Let d be a compatible metric for X and d_0 a compatible complete metric for A. For each $n \in \mathbb{N}$, let V_n be the union of the open subsets U of X for which $U \cap A \neq \emptyset$ and d_0-diam$(U \cap A) < 1/n$, where d_0-diam denotes the diameter for the metric d_0. Since d and d_0 induce the same topology on A we have

$$A \subseteq \overline{A}^{\tau} \cap \left(\bigcap_{n\ge1} V_n\right). \tag{1.5.9}$$

Let $u \in \overline{A}^{\tau} \cap (\bigcap_{n\geq 1} V_n)$. Since $u \in \bigcap_{n\geq 1} V_n$ we can find a sequence $\{U_n\}_{n\geq 1}$ of neighborhoods of x such that

$$U_n \cap A \neq \emptyset \quad \text{and} \quad d_0\text{-}\operatorname{diam}(U_n \cap A) < \frac{1}{n}.$$

Evidently, by replacing U_n with a small neighborhood of u, we may assume that $\{U_n\}_{n\geq 1}$ is decreasing and d-diam $U_n \leq 1/n$. Since (A, d_0) is complete, from Theorem 1.5.15, we have that

$$\{u_0\} = \bigcap_{n\geq 1} \overline{U_n \cap A}^{\tau(A)}. \tag{1.5.10}$$

For every $n \in \mathbb{N}$ we have d-diam $\overline{U_n}^{\tau} \leq 1/n$ and $u, u_0 \in \overline{U_n}^{\tau}$. Hence, because of (1.5.10), $u = u_0$. Therefore, $\overline{A}^{\tau} \cap (\bigcap_{n\geq 1} V_n) \subseteq A$ and due to (1.5.9) it holds that $A = \overline{A}^{\tau} \cap (\bigcap_{n\geq 1} V_n)$. Invoking Proposition 1.5.8 for the closed $\overline{A}^{\tau}$, we conclude that A is a G_δ-subset of X.

$\Longleftarrow$: By hypothesis $A = \bigcap_{n\geq 1} U_n$ with $U_n \subseteq X$ open for all $n \in \mathbb{N}$. From Proposition 1.5.45 we know that each U_n is Polish and so Proposition 1.5.46 implies that $\bigcap_{n\geq 1} U_n = A$ is Polish. □

Remark 1.5.48. From the last theorem we recover the part of Proposition 1.5.45 concerning open sets.

Corollary 1.5.49. *The set of irrational numbers with the topology induced by* $\mathbb{R}$ *is Polish.*

Remark 1.5.50. We mention some more Polish spaces:
- Every locally compact, σ-compact metrizable space is Polish.
- Every locally compact and second countable Hausdorff space is Polish. This is a consequence of the so-called "Urysohn Metrization Theorem", which says that every regular, second countable space is metrizable.
- $\mathbb{N}^\infty$ is Polish (see Proposition 1.5.46) and in fact every Polish space is a continuous image of $\mathbb{N}^\infty$. More precisely every Polish space is a one-to-one continuous image of a closed subset of $\mathbb{N}^\infty$. On $\mathbb{N}^\infty$ we consider the tree metric defined by

$$t(\hat{p}, \hat{q}) = \begin{cases} 0 & \text{if } \hat{p} = \hat{q} \\ \frac{1}{k} & \text{if } \hat{p} \neq \hat{q} \text{ and } k = \min\{n \in \mathbb{N} : p_n \neq q_n\} \end{cases}$$

  for all $\hat{p} = (p_n)$, $\hat{q} = (q_n) \in \mathbb{N}^\infty$. This is a complete metric on $\mathbb{N}^\infty$ compatible with the product topology.
- Every Polish space is a G_δ in some metrizable compactification.

Definition 1.5.51. A Hausdorff space X is said to be a **Souslin space** if there exist a Polish space Y and a continuous surjection $f: Y \to X$.

Remark 1.5.52. Equivalently we can say that the Hausdorff topological space (X, τ) is Souslin if and only if there is a topology $\tau_0 \supseteq \tau$ on X such that (X, τ_0) is homeomorphic to a quotient of a Polish space. A Souslin space is always separable but need not be metrizable. Anticipating some basic material from Chapter 3, we mention that an infinite dimensional separable Banach space with the weak topology is Souslin, but not metrizable. Similarly for the dual X^* of an infinite dimensional separable Banach space endowed with the w*-topology.

Definition 1.5.53. The Souslin subspaces of a Polish space are called **analytic sets**.

Souslin spaces have nice stability properties.

Proposition 1.5.54. (a) *Closed and open subsets of Souslin spaces are Souslin spaces.*
(b) *Countable products of Souslin spaces are Souslin.*
(c) *Countable intersections and countable unions of Souslin subspaces of a Hausdorff topological space V are Souslin.*

Proof. (a): Let X be a Souslin space. Then according to Definition 1.5.51 there exists a Polish space Y and a continuous surjection $f: Y \to X$. Let $E \subseteq X$ be a closed (resp. open) set. Then $f^{-1}(E) \subseteq Y$ is closed (resp. open) and so by Proposition 1.5.45 $f^{-1}(E)$ is Polish. Also $f|_{f^{-1}(E)}$ is continuous and surjective onto $f(f^{-1}(E)) = E$ since f is a surjection. Therefore, by Definition 1.5.51, E is Souslin.

(b): Let $\{X_n\}_{n\geq 1}$ be a family of Souslin spaces. For every $n \in \mathbb{N}$ there exists a Polish space Y_n and a continuous surjection $f_n: Y_n \to X_n$. Set $Y = \prod_{n\geq 1} Y_n$, $X = \prod_{n\geq 1} X_n$ and $\hat{f} = (f_n)_{n\geq 1}: X \to Y$ defined by $\hat{f}(\{y_n\}) = (f_n(y_n))_{n\geq 1}$. Then Y is Polish by Proposition 1.5.46 and $\hat{f}$ is a continuous surjection. So, X is a Souslin space.

(c): Let $\{X_n\}_{n\geq 1}$ be a family of Souslin subspaces of V and let $X = \prod_{n\geq 1} X_n$. We introduce $\hat{V} = V^{\mathbb{N}}$ and $\hat{\Delta}$ the diagonal of $\hat{V}$, that is, $\hat{\Delta} = \{\hat{u} = (u_n)_{n\geq 1}: u_n = u \text{ for all } n \in \mathbb{N}\}$. From Proposition 1.3.12 we know that $\hat{V}$ is Hausdorff and so Problem 1.1 implies that $\hat{\Delta} \subseteq \hat{V}$ is closed. Let $\hat{f}: V \to \hat{\Delta}$ be the canonical map of V onto $\hat{\Delta}$ defined by $\hat{f}(u) = (u, u, \ldots, u, \ldots)$. Then $\hat{f}(X) = \hat{\Delta} \cap (\prod_{n\geq 1} X_n)$ and $\hat{f}$ is a homeomorphism of X onto a closed subspaces of $\prod_{n\geq 1} X_n$. But by part (b) $\prod_{n\geq 1} X_n$ is Souslin, hence by part (a) $\hat{f}(X)$ is Souslin. Therefore X is Souslin.

Now we consider the union $\bigcup_{n\geq 1} X_n$. For every $n \in \mathbb{N}$ we can find a Polish space Y_n and a continuous surjection $f_n: Y_n \to X_n$. Let $\tilde{X}_n = \{n\} \times X_n$ and $\tilde{Y}_n = \{n\} \times Y_n$. Note that both are Polish spaces. Now we consider the map $\tilde{f}_n: \tilde{Y}_n \to \tilde{X}_n$ defined by $\tilde{f}_n(n, y) = (n, f_n(y))$ for all $n \in \mathbb{N}$ and for all $y \in Y_n$. Evidently $\tilde{f}_n$ is a continuous surjection. Let $\tilde{Y} = \bigcup_{n\geq 1} \tilde{Y}_n$ (this set is known as the free or disjoint union of the Y_n's and sometimes it is denoted by $\sum_{n\geq 1} \tilde{Y}_n$) and similarly we set $\tilde{X} = \bigcup_{n\geq 1} \tilde{X}_n$. The function $\tilde{f}: \tilde{Y} \to \tilde{X}$ defined by $\tilde{f}|_{\tilde{Y}_n} = \tilde{f}_n$ for all $n \in \mathbb{N}$ is a continuous surjection. The space $\tilde{Y}$ is Polish; see Proposition 1.5.46. Let $h: \tilde{X} \to \bigcup_{n\geq 1} X_n$ be the canonical projection, that is, $h(n, u) = u$ for all $n \in \mathbb{N}$ and for all $u \in X_n$. This is a homeomorphism onto $\bigcup_{n\geq 1} X_n$. Then $g = h \circ \tilde{f}: \tilde{Y} \to \bigcup_{n\geq 1} X_n$ is a continuous surjection, hence $\bigcup_{n\geq 1} X_n$ is Souslin. □

Directly from Definition 1.5.51, we have the following useful property of Souslin spaces. It shows that although Souslin spaces are not necessarily metrizable, they are sequentially determined.

Proposition 1.5.55. *If X is a Souslin space and $A \subseteq X$, then there exists a countable set $D \subseteq A$ such that D is sequentially dense in A.*

Proof. Let Y be a Polish space and $f: Y \to X$ a continuous surjection. Let $B = f^{-1}(A) \subseteq Y$. Then B is separable and so there exists a countable dense subset $D_0 \subseteq B$, that is, $\overline{D_0}^Y \supseteq B$. Since f is surjective we know that $D = f(D_0) \subseteq A$ is countable and sequentially dense in A. □

Definition 1.5.56. A Hausdorff topological space X is said to be **strongly Lindelöf** if every open subset of X with the subspace topology is Lindelöf; see Definition 1.2.26 (b).

Proposition 1.5.57. *Every Souslin space X is strongly Lindelöf.*

Proof. Let Y be a Polish space and $f: Y \to X$ a continuous surjection. Evidently Y is strongly Lindelöf; see Propositions 1.5.40 (b) and 1.5.45. We can easily check that the continuous image of a strongly Lindelöf space is strongly Lindelöf. Hence X must be strongly Lindelöf. □

Definition 1.5.58. Let X, $\{Y_\alpha\}_{\alpha \in I}$ be sets and $f_\alpha: X \to Y_\alpha$ a family of functions. We say that the family $\{f_\alpha\}_{\alpha \in I}$ is **separating** (or **total**) if for every pair $(x, u) \in X \times X$ with $x \neq u$ we have $f_\alpha(x) \neq f_\alpha(u)$ for some $\alpha \in I$.

Lemma 1.5.59. *If X is a Souslin space, $\{Y_\alpha\}_{\alpha \in I}$ is a family of Hausdorff topological spaces and $f_\alpha: X \to Y_\alpha$ with $\alpha \in I$ is a separating family of continuous maps, then we can find a countable subset $D \subseteq I$ such that $\{f_\alpha\}_{\alpha \in D}$ remains separating.*

Proof. Replacing the Y_α's by their free union (see the proof of Proposition 1.5.54 (c)), we see that without any loss of generality we may assume that $Y_\alpha = Y$ for all $\alpha \in I$. Let $\Delta_X \subseteq X \times X$ and $\Delta_Y \subseteq Y \times Y$ be the diagonals. If $(x, u) \in \Delta_X^c$, then we can find $\alpha \in I$ such that $(f_\alpha(x), f_\alpha(u)) \in \Delta_Y^c$. So, the open sets $(f_\alpha, f_\alpha)^{-1}(\Delta_Y^c)$ with $\alpha \in I$ form an open cover of Δ_X^c. The space $X \times X$ is strongly Lindelöf; see Propositions 1.5.54 (b) and 1.5.57. Therefore we can find a countable $D \subseteq I$ such that $\{(f_\alpha, f_\alpha)^{-1}(\Delta_Y)\}_{\alpha \in D}$ is a countable open cover of Δ_X^c. This means that $\{f_\alpha\}_{\alpha \in D}$ remains separating. □

Combining this lemma with Problem 1.41 we can state the following result concerning compact Souslin spaces.

Theorem 1.5.60. *Every compact Souslin space is metrizable, hence Polish.*

Remark 1.5.61. An improvement of this theorem can be found in Problem 1.42.

The Baire category notion gives a topological meaning to the notion of the size of a set. It is based on density. So, according to Baire, a subset A of a Hausdorff topological space X is considered to be very small (sparse) if there is no nonempty open set $U \subseteq X$ such that $A \cap U$ is dense in U, that is, $\overline{A}$ has an empty interior. Then large sets are those that are not countable unions of sparse sets.

Definition 1.5.62. Let X be a Hausdorff topological space and $A \subseteq X$.
(a) We say that A is **nowhere dense** if $\operatorname{int}\overline{A} = \emptyset$.
(b) We say that A is of **first category** if it is the countable union of nowhere dense sets.
(c) We say that A is of **second category** if it is not of first category.

Remark 1.5.63. Note that $\mathbb{Q}$ is of first category and at the same time dense in $\mathbb{R}$. The set $A \subseteq X$ is nowhere dense if and only if $\operatorname{int}(X \setminus A)$ is dense in X.

Definition 1.5.64. A Hausdorff topological space X is said to be a **Baire space** if the intersection of each countable family of dense, open sets in X is dense.

Proposition 1.5.65. *A Hausdorff topological space X is of second category in itself if and only if every countable family of dense open sets in X has nonempty intersection.*

Proof. $\Longrightarrow$: Let $\{U_n\}_{n\geq 1}$ be dense, open sets. Then $\{U_n^c\}_{n\geq 1} = \{X \setminus U_n\}_{n\geq 1}$ are nowhere dense, closed sets and so $\bigcup_{n\geq 1} U_n^c$ is of first category. Since by hypothesis X is of second category we have

$$X \setminus \left(\bigcup_{n\geq 1} U_n^c\right) = \bigcap_{n\geq 1} U_n \neq \emptyset .$$

$\Longleftarrow$: Arguing by contradiction, suppose that X is of first category. Then $X = \bigcup_{n\geq 1} C_n$ with C_n being nowhere dense and closed for each $n \in \mathbb{N}$. We have

$$X \setminus \left(\bigcup_{n\geq 1} C_n\right) = \bigcap_{n\geq 1} (X \setminus C_n) \neq \emptyset$$

since each $X \setminus C_n = U_n$ with $n \in \mathbb{N}$ is dense and open, a contradiction. This shows that X must be of second category. □

Proposition 1.5.66. *If X is a compact Hausdorff topological space and $A \subseteq X$ is a G_δ-set, then A is a Baire space.*

Proof. First we show that X is a Baire space. Let $\{U_n\}_{n\geq 1}$ be dense, open sets in X and let $V \subseteq X$ be a nonempty, open set. We have $U_1 \cap V \neq \emptyset$ and $U_1 \cap V$ is open. From Corollary 1.4.50 we know that X is normal, hence regular as well. So, we can find an open $W_1 \subseteq X$ such that $\overline{W_1} \subseteq U_1 \cap V$; see Proposition 1.2.8. Similarly, for $n \in \{2, 3, \ldots\}$ there exists open $W_n \subseteq X$ such that $\overline{W}_n \subseteq U_n \cap W_{n-1}$. Evidently $\{\overline{W}_n\}_{n\geq 1}$ is a decreasing sequence of compact sets, hence $\bigcap_{n\geq 1} \overline{W}_n \neq \emptyset$. But $\bigcap_{n\geq 1} \overline{W}_n \subseteq (\bigcap_{n\geq 1} U_n) \cap V$. So, every open set $V \subseteq X$ has a nonempty intersection with $\bigcap_{n\geq 1} U_n$ and this shows that $\bigcap_{n\geq 1} U_n$ is dense in X. Hence, X is a Baire space.

Without loss of generality we may assume that A is dense in X since we can always replace X by $\overline{A}$. Let $\{U_n\}_{n\geq 1}$ be dense, open subsets of A. Then $U_n = V_n \cap A$ with a dense and open $V_n \subseteq X$ for every $n \in \mathbb{N}$. Then

$$\bigcap_{n\geq 1}(V_n \cap A) = \Big(\bigcap_{n\geq 1} V_n\Big) \cap A\,.$$

From the first part of the proof we know that $\bigcap_{n\geq 1} V_n \subseteq X$ is dense. Therefore $\bigcap_{n\geq 1} U_n = \bigcap_{n\geq 1}(V_n \cap A)$ is dense in A. This proves that A is a Baire space. □

Corollary 1.5.67. *If X is a complete metric space and $X = \bigcup_{n\geq 1} C_n$ with closed $C_n \subseteq X$ for all $n \in \mathbb{N}$, then there exists a number $n_0 \in \mathbb{N}$ such that* $\operatorname{int} C_0 \neq \emptyset$.

Now Theorems 1.4.75 and 1.5.47 lead to the so-called "Baire Theorem".

Theorem 1.5.68 (Baire Theorem). (a) *Every locally compact Hausdorff topological space is a Baire space.*
(b) *Every topologically complete Hausdorff space is a Baire space.*

We conclude this section with an important result known as "Stone's Theorem". For the proof we refer to Dugundji [102, p. 186].

Theorem 1.5.69 (Stone's Theorem). *Every metrizable space is paracompact.*

1.6 Function Spaces

Let (X, τ_X) and (Y, τ_Y) be two Hausdorff topological spaces. By $C(X, Y)$ we denote the space of continuous functions $f: X \to Y$. In this section we topologize this space and study its properties.

Definition 1.6.1. Let $K \subseteq X$ be compact and $U \subseteq Y$ be open. We set

$$W(K, U) = \{f \in C(X, Y): f(K) \subseteq U\}\,.$$

The **compact-open topology** (or **c-topology**) on $C(X, Y)$ is the topology τ_ζ on $C(X, Y)$ having as subbasis the family

$$\{W(K, U): K \subseteq X \text{ is compact and } U \subseteq Y \text{ is open}\}\,.$$

Remark 1.6.2. A basic element for the τ_ζ-topology is given by

$$\bigcap_{n=1}^{m} W(K_n, U_n)$$

with compact $K_n \subseteq X$ and open $U_n \subseteq Y$ for all $n \in \{1, \ldots, m\}$. Note that $C(X, Y) \subseteq Y^X$. So, we can consider on $C(X, Y)$ the relative product topology that is the topology of pointwise convergence and is denoted by τ_p. Since $W(\{x\}, U) \in \tau_\zeta$ for all $x \in X$ and all open $U \subseteq Y$, it follows that

$$\tau_p \subseteq \tau_\zeta \,. \tag{1.6.1}$$

Note that we have

$$\bigcap_{n=1}^{m} W(K_n, U) = W\left(\bigcup_{n=1}^{m} K_n, U\right), \quad \bigcap_{n=1}^{m} W(K, U_n) = W\left(K, \bigcap_{n=1}^{m} U_n\right),$$
$$\bigcap_{n=1}^{m} W(K_n, U_n) \subseteq W\left(\bigcup_{n=1}^{m} K_n, \bigcup_{n=1}^{m} U_n\right), \quad \overline{W(K,U)}^{\tau_\zeta} \subseteq W(K, \overline{U}^{\tau_Y})\,.$$

Proposition 1.6.3. *If (X, τ_X) and (Y, τ_Y) are Hausdorff topological spaces and the function space $C(X, Y)$ is endowed with the τ_ζ-topology, then the following hold:*
(a) *$C(X, Y)$ is Hausdorff;*
(b) *$C(X, Y)$ is regular if and only if Y is regular.*

Proof. (a) Let $f, g \in C(X, Y)$ such that $f \neq g$. We can find $x \in X$ such that $f(x) \neq g(x)$. Because Y is Hausdorff, we can find $U \in \mathcal{N}(f(x))$ and $V \in \mathcal{N}(g(x))$ such that $U \cap V = \emptyset$. Then

$$W(\{x\}, U) \in \tau_\zeta \text{ contains } f\,,$$
$$W(\{x\}, V) \in \tau_\zeta \text{ contains } g\,,$$
$$W(\{x\}, U) \cap W(\{x\}, V) = \emptyset\,.$$

This proves that $(C(X, Y), \tau_\zeta)$ is Hausdorff.

(b) $\Longrightarrow$: Evidently, $Y \subseteq C(X, Y)$ (the subspace of constant functions) and $\tau_\zeta(Y) = \tau_Y$. Then the regularity of Y follows from the fact that the property is hereditary; see Proposition 1.2.10.

$\Longleftarrow$: Let $f \in W(K, U)$. The set $f(K) \subseteq Y$ is compact. So, by Problem 1.52 we can find $V \in \tau_Y$ such that $f(K) \subseteq V \subseteq \overline{V} \subseteq U$. Then $f \in W(K, U) \subseteq \overline{W(K,U)}^{\tau_\zeta} \subseteq W(K, \overline{U}^{\tau_Y})$; see Remark 1.6.2. This proves that $(C(X, Y), \tau_\zeta)$ is regular. □

Remark 1.6.4. If Y is normal or first countable or second countable, then $(C(X, Y), \tau_\zeta)$ need not have the same properties.

Let (X, τ_X), (Y, τ_Y) and (Z, τ_Z) be three Hausdorff topological spaces. We can define the map $\eta \colon C(X, Y) \times C(Y, Z) \to C(X, Z)$ given by

$$\eta(f, g) = g \circ f\,. \tag{1.6.2}$$

On $C(X, Y)$, $C(Y, Z)$ and $C(X, Z)$ we consider the corresponding ζ-topologies.

Proposition 1.6.5. *The maps $f \to \eta(f,g)$ and $g \to \eta(f,g)$ are both continuous.*

Proof. We fix $f_1 \in C(X,Y)$ and prove the continuity of $g \to \eta(f_1,g)$ on $C(Y,Z)$. Let $W(K,U)$ be a subbasic neighborhood of $g \circ f_1$. Note that $g \circ f_1 \in W(K,U)$ if and only if $g \in W(f_1(K),U)$. But the set $f_1(K) \subseteq Y$ is compact. Hence, $W(f_1(K),U)$ is a subbasic neighborhood of g. Therefore, $\eta(f_1, W(f_1(K),U)) = W(K,U)$ and this proves the continuity of $g \to \eta(f_1,g)$.

Next we fix $g_1 \in C(Y,Z)$ and consider the map $f \to \eta(f,g_1)$ from $C(X,Y)$ into $C(X,Z)$. The proof of the continuity of this map is similar to the previous part. Note that in this case $g_1 \circ f \in W(K,U)$ if and only if $f \in W(K, g_1^{-1}(U))$ and $g^{-1}(U) \in \tau_Y$. ☐

To have joint continuity of the map η we need to strengthen the conditions on the space Y.

Proposition 1.6.6. *If (Y,τ_Y) is locally compact, then the map η is jointly continuous.*

Proof. Let $(f_1,g_1) \in C(X,Y) \times C(Y,Z)$ and let $W(K,U)$ be a subbasic neighborhood of (f_1,g_1). Note that $f_1(K) \subseteq g_1^{-1}(U)$, $f_1(K) \subseteq Y$ is compact and $g_1^{-1}(U) \subseteq Y$ is open. Since by hypothesis Y is locally compact, we can find relatively compact $V \in \tau_Y$ such that

$$f_1(K) \subseteq V \subseteq \overline{V} \subseteq g_1^{-1}(U)\,;$$

see Proposition 1.4.66 (c). Then we have

$$W(K,V) \subseteq \mathcal{N}(f_1)\,, \quad W(\overline{V},U) \in \mathcal{N}(g_1)\,,$$
$$\eta(W(K,V), W(\overline{V},U)) \subseteq W(K,U)\,.$$

Hence, η is jointly continuous. ☐

Definition 1.6.7. The map $e\colon X \times C(X,Y) \to Y$ defined by $e(x,f) = f(x)$ is called the **evaluation map**. If we fix $x \in X$, the map $e_x\colon C(X,Y) \to Y$ defined by $e_x(f) = f(x)$ is called the **evaluation at** x **map**.

The next proposition establishes the continuity properties of these maps.

Proposition 1.6.8. (a) *If Y is locally compact, then $e\colon X \times C(X,Y) \to Y$ is continuous.*
(b) *For every $x \in X$, the map $e_x\colon C(X,Y) \to Y$ is continuous.*

Proof. Note that when Z is a singleton and $\eta\colon C(Z,X) \times C(X,Y) \to C(Z,Y)$ is the composition map (see (1.6.2)) then $\eta = e$. So, (a) follows from Proposition 1.6.6 while (b) follows from Proposition 1.6.5. ☐

We want to characterize the τ_ζ-compact subsets of $C(X,Y)$. The next definition introduces notions that are crucial in this direction.

Definition 1.6.9. Let (X,τ_X) be a Hausdorff topological space and (Y,d) be a metric space.

(a) A set $\mathcal{F} \subseteq C(X, Y)$ is said to be **equicontinuous** at x if for a given $\varepsilon > 0$ there exists $U \in \mathcal{N}(x)$ such that $d(f(u), f(x)) < \varepsilon$ for all $u \in U$ and for all $f \in \mathcal{F}$. We say that $\mathcal{F}$ is **equicontinuous** if it is equicontinuous at every $x \in X$.
(b) Given $f \in C(X, Y)$ with compact $K \subseteq X$ and $\varepsilon > 0$, we define

$$B_{K,\varepsilon}(f) = \{g \in C(X, Y) : \sup[d(g(x), f(x)) : x \in K] < \varepsilon\} .$$

The sets $B_{K,\varepsilon}(f)$ form a basis for a topology τ_u on $C(X, Y)$ known as the **topology of uniform convergence on compacta.**

Remark 1.6.10. The τ_ζ-topology (see Definition 1.6.1) and the τ_p-topology (see Remark 1.6.2) on $C(X, Y)$ are defined without requiring that Y is a metric space. In contrast, the τ_u-topology (see Definition 1.6.9) explicitly requires that Y must be a metric space. Nevertheless, we can prove the following remarkable result.

Theorem 1.6.11. *If (X, τ_X) is a Hausdorff topological space and (Y, d) is a metric space, then $\tau_\zeta = \tau_u$.*

Proof. First we show that $\tau_\zeta \subseteq \tau_u$. To this end let $f \in W(K, U)$. Then $f(K) \subseteq Y$ is compact and $f(K) \subseteq U$.

Claim: There exists $\varepsilon > 0$ such that

$$f(K)_\varepsilon = \{y \in Y : d(y, f(K)) < \varepsilon\} \subseteq U .$$

Arguing by contradiction, suppose that the claim is not true. Then we can find $\{y_n\}_{n\geq 1} \subseteq Y \setminus U$ such that $d(y_n, f(K)) < 1/n$. Recall that $f(K) \subseteq Y$ is compact. So, for every $n \in \mathbb{N}$ there exists $v_n \in f(K)$ such that $d(y_n, v_n) = d(y_n, f(K)) < 1/n$ for all $n \in \mathbb{N}$. The compactness o $f(K)$ implies that by passing to a subsequence if necessary, we have $v_n \xrightarrow{d} v \in f(K)$ in Y. Since $d(y_n, v_n) < 1/n$ for all $n \in \mathbb{N}$, it follows that $y_n \xrightarrow{d} v \in (X \setminus U) \cap f(K)$, a contradiction, since $f(K) \subseteq U$. This proves that the claim is true.

The claim implies that $B_{K,\varepsilon}(f) \subseteq W(K, U)$, that is

$$\tau_\zeta \subseteq \tau_u . \tag{1.6.3}$$

Next we show that the opposite inclusion holds as well. Let $f \in C(X, Y)$ and let $B_{K,\varepsilon}(f) \subseteq W(K, U)$, see (1.6.3). For every $x \in X$ there exists $V_x \in \mathcal{N}(x)$ such that $f(\overline{V}_x) \subseteq U_x$ with $U_x \subseteq Y$ open and diam $U_x < \varepsilon$. Since K is compact we find $x_1, \ldots, x_n \in K$ such that $K \subseteq \bigcup_{k=1}^n V_{x_k}$. Let $K_{x_k} = \overline{V}_{x_k} \cap K$ for $k \in \{1, \ldots, n\}$. Then $f \in \bigcap_{k=1}^n W(K_{x_k}, U_{x_k}) \subseteq B_{K,\varepsilon}(f)$ and so

$$\tau_u \subseteq \tau_\zeta . \tag{1.6.4}$$

From (1.6.3) and (1.6.4) it follows that $\tau_\zeta = \tau_u$. ☐

We know that $\tau_p \subseteq \tau_\zeta$ ($= \tau_u$ if Y is a metric space); see (1.6.1) and Theorem 1.6.11. However, on equicontinuous sets, the two topologies coincide.

Proposition 1.6.12. *If (X, τ_X) is a Hausdorff topological space, (Y, d) is a metric space and $\mathcal{F} \subseteq C(X, Y)$ is equicontinuous, then $\tau_p(\mathcal{F}) = \tau_\zeta(\mathcal{F})$, that is, the two topologies restricted on $\mathcal{F}$ coincide.*

Proof. Evidently $\tau_p(\mathcal{F}) \subseteq \tau_\zeta(\mathcal{F})$. Moreover, Theorem 1.6.11 yields that $\tau_\zeta = \tau_u$. Therefore, it suffices to find a basic element B for the τ_p-topology such that

$$f \in B \cap \mathcal{F} \subseteq B_{K,\varepsilon}(f) \cap \mathcal{F}\,.$$

Let $\varepsilon_1, \varepsilon_2 > 0$ be such that $2\varepsilon_1 + \varepsilon_2 \leq \varepsilon$. Since $\mathcal{F}$ is equicontinuous and $K \subseteq X$ is compact, we find open sets $\{U_k\}_{k=1}^n$ in X such that $K \subseteq \bigcup_{k=1}^n U_k$ and for each $k \in \{1, \dots, n\}$, each $x, u \in U_k$ and $f \in \mathcal{F}$, $d(f(x), f(u)) < \varepsilon_1$.

We choose $x_k \in U_k$ with $k \in \{1, \dots, n\}$ and let

$$B = \{g \in C(X, Y) \colon d(g(x_k), f(x_k)) < \varepsilon_2 \text{ for all } k \in \{1, \dots n\}\}\,.$$

Let $g \in B \cap \mathcal{F}$. Given $x \in K$, we find $k \in \{1, \dots, n\}$ such that $x \in U_k$. Then we have

$$d(g(x), g(x_k)) \leq \varepsilon_1\,, \quad d(g(x_k), f(x_k)) < \varepsilon_2\,, \quad d(f(x_k), f(x)) \leq \varepsilon_1\,,$$

which implies, by the triangle inequality and the choice of $\varepsilon_1, \varepsilon_2 > 0$, that $d(g(x), f(x)) < \varepsilon$. Hence $g \in B_{K,\varepsilon}(f)$, thus $B \cap \mathcal{F} \subseteq B_{K,\varepsilon} \cap \mathcal{F}$. This proves that $\tau_p(\mathcal{F}) = \tau_\zeta(\mathcal{F})$. □

Proposition 1.6.13. *If (X, τ_X) is a Hausdorff topological space, (Y, d) is a metric space, and $\mathcal{F} \subseteq C(X, Y)$ is equicontinuous, then $\overline{\mathcal{F}}^{\tau_p}$ is equicontinuous as well.*

Proof. Let $x \in X$ and $\varepsilon > 0$. Since $\mathcal{F}$ is equicontinuous, there exists $U \in \mathcal{N}(x)$ such that $d(f(u), f(x)) < \varepsilon$ for all $u \in U$ and for all $f \in \mathcal{F}$.

Let $g \in \overline{\mathcal{F}}^{\tau_p}$. For $v \in U$ we introduce

$$V_v = \left\{h \in C(X, Y) \colon d(h(v), g(v)) < \frac{\varepsilon}{3}, d(h(x), g(x)) < \frac{\varepsilon}{3}\right\} \in \tau_p\,.$$

We have $V_v \cap \mathcal{F} \neq \emptyset$. Let $f \in V_v \cap \mathcal{F}$. We have

$$d(g(v), g(x)) \leq d(g(v), f(v)) + d(f(v), f(x)) + d(f(x), g(x)) \leq 3\frac{\varepsilon}{3} = \varepsilon\,.$$

Hence, $\overline{\mathcal{F}}^{\tau_p}$ is equicontinuous. □

The next theorem is the main result of this section and characterizes the τ_ζ-compact sets in $C(X, Y)$. The result is known as the "Arzela–Ascoli Theorem".

Theorem 1.6.14 (Arzela–Ascoli Theorem). *If (X, τ_X) is a locally compact space, (Y, d) is a metric space, and $\mathcal{F} \subseteq C(X, Y)$, then $\overline{\mathcal{F}}^{\tau_\zeta}$ is τ_ζ-compact if and only if $\mathcal{F}$ is equicontinuous and for every $x \in X$, $\mathcal{F}(x) = \{f(x): f \in \mathcal{F}\} \subseteq Y$ is relatively compact.*

Proof. $\Longrightarrow$: For every $x \in X$, there holds $\mathcal{F}(x) \subseteq \overline{\mathcal{F}}^{\tau_\zeta}(x) = e_x(\overline{\mathcal{F}}^{\tau_\zeta})$ and Proposition 1.6.8 (b) gives that $e_x(\overline{\mathcal{F}}^{\tau_\zeta})$ is compact in Y. We need to show that $\overline{\mathcal{F}}^{\tau_\zeta}$ is equicontinuous. Let $x \in X$ and choose a compact set K such that $K \supseteq V \in \mathcal{N}(x)$. This is possible since X is supposed to be locally compact. Let $\mathcal{L}_\zeta = \{\hat{f} = f|_K : f \in \overline{\mathcal{F}}^{\tau_\zeta}\}$. It suffices to show that $\mathcal{L}_\zeta$ is equicontinuous. Let $r: C(X, Y) \to C(K, Y)$ be defined by $r(f) = f|_K$. Evidently $\mathcal{L}_\zeta = r(\overline{\mathcal{F}}^{\tau_\zeta})$ and r is continuous when both $C(X, Y)$ and $C(K, Y)$ are endowed with their respective τ_ζ-topologies. Note that on $C(X, Y)$ the τ_ζ-topology coincides with the metric topology generated by the uniform metric $\hat{d}_K(f, g) = \max\{d(f(x), g(x)): x \in K\}$. Hence $\mathcal{L}_\zeta$ is $\hat{d}_K$-totally bounded.

Let $\varepsilon > 0$ be given and choose $\varepsilon_1, \varepsilon_2 > 0$ such that $2\varepsilon_1 + \varepsilon_2 \le \varepsilon$. We can find $x_1, \ldots, x_n \in K$ such that $\mathcal{L}_\zeta \subseteq \bigcup_{k=1}^n B_\varepsilon(\hat{f}_k)$. Since each $\hat{f}_K$ is continuous, we can find $U \in \mathcal{N}(x)$ such that

$$d(\hat{f}_k(u), \hat{f}_k(x)) < \varepsilon_2 \quad \text{for all } u \in U \text{ and for all } k \in \{1, \ldots, n\}. \tag{1.6.5}$$

Let $\hat{f} \in \mathcal{L}_\zeta$. Then $\hat{f} \in B_{\varepsilon_1}(\hat{f}_k)$ for some $k \in \{1, \ldots, n\}$. For every $u \in U$ we have

$$d(\hat{f}(u), \hat{f}_k(u)) < \varepsilon_1, \quad d(\hat{f}_k(u), \hat{f}_k(x)) < \varepsilon_2, \quad d(\hat{f}_k(x), \hat{f}(x)) < \varepsilon_1;$$

see (1.6.5). This gives $d(\hat{f}(u), \hat{f}(x)) < \varepsilon$ for all $u \in U$, which implies that $\mathcal{L}_\zeta$ is equicontinuous, and hence, so is $\mathcal{F}$.

$\Longleftarrow$: From Proposition 1.6.13 we know that $\overline{\mathcal{F}}^{\tau_p}$ is equicontinuous. Then Proposition 1.6.12 implies that $\overline{\mathcal{F}}^{\tau_p} = \overline{\mathcal{F}}^{\tau_\zeta}$. Recall that τ_p is the relative product topology on $C(X, Y) \subseteq Y^X$. Using Tychonoff's Product Theorem (see Theorem 1.4.56), we have that $\prod_{x \in X} \mathcal{F}(x)$ is compact in the product topology and so $\overline{\mathcal{F}}^{\tau_p}$ is compact. Therefore, $\overline{\mathcal{F}}^{\tau_\zeta}$ is compact. □

A careful inspection of the second part of the proof above reveals that for that part of the result, the local compactness of X is not needed. So, we can state the following version of the Arzela–Ascoli Theorem.

Theorem 1.6.15. *If (X, τ_X) is a Hausdorff topological space, (Y, d) is a metric space, and $\mathcal{F} \subseteq C(X, Y)$ is a set with the following two properties:*
(a) *$\mathcal{F}$ is equicontinuous;*
(b) *for every $x \in X$, $\mathcal{F}(x) = \{f(x): f \in \mathcal{F}\} \subseteq Y$ is relatively compact,*

then $\overline{\mathcal{F}}^{\tau_\zeta}$ is τ_ζ-compact and equicontinuous on X.

When $Y = \mathbb{R}^N$, exploiting the Heine–Borel Theorem, we can have the following particular version of the Arzela–Ascoli Theorem; see Theorem 1.6.14.

Theorem 1.6.16. *If (X, τ_X) is a compact topological space and $\mathcal{F} \subseteq C(X, \mathbb{R}^N)$, then $\mathcal{F}$ is compact for the supremum metric topology $\tau_{\hat{d}}$ if and only if $\mathcal{F}$ is equicontinuous, $\hat{d}$-closed, and bounded, that is, $|f(u)| \leq M$ for all $u \in X$ and for some $M > 0$.*

Remark 1.6.17. If X is a compact space and (Y, d) is a metric space, then recall that the supremum metric $\hat{d}$ or d_∞ is defined by

$$\hat{d}(f, g) = d_\infty(f, g) = \max\{d(f(x), g(x)) : x \in X\}.$$

Evidently, $f_n \xrightarrow[d_\infty]{\hat{d}}$ if and only if $f_n \to f$ uniformly on X, that is, for given $\varepsilon > 0$, we can find $n_0 = n_0(\varepsilon) \in \mathbb{N}$ such that $d(f_n(u), f(u)) \leq \varepsilon$ for all $u \in X$ and for all $n \geq n_0$.

It is easy to see that uniform limits of continuous maps are again continuous maps. According to Theorem 1.6.11, the $\hat{d}$-metric topology depends only on the topology of Y and on the particular metric d. So, if d_1, d_2 are two compatible metrics on Y, then the corresponding sup-metrics $\hat{d}_1$, $\hat{d}_2$ are compatible as well. Hence we can view $C(X, Y)$ as a topological space without specifying the particular sup-metric and refer to the topology of uniform convergence on $C(X, Y)$.

Proposition 1.6.18. *If X is a compact metrizable space and Y is a separable metrizable space, then the space $C(X, Y)$ with the $\tau_\zeta = \tau_u$-topology is separable and metrizable.*

Proof. On account of Proposition 1.5.40 (b) and Remark 1.6.17, it suffices to show that $C(X, Y)$ is second countable.

Let $D = \{x_n\}_{n \geq 1} \subseteq X$ be a dense set and $\{U_n\}_{n \geq 1}$ a countable basis for X. Let $\{\overline{B}_n\}_{n \geq 1}$ be an enumeration of the countable set of all closed balls with center D and a rational radius. For $n, m \in \mathbb{N}$ let $W_{n,m} = W(\overline{B}_n, U_m)$.

We claim that $\{W_{n,m}\}_{n,m \geq 1}$ is a countable subbasis for $C(X, Y)$. To this end, let $V \subseteq C(X, Y)$ be open and let $f \in V$. We choose $\delta > 0$ such that

$$B_{2\delta}(f) = \{g \in C(X, Y) : \hat{d}(g, f) < 2\delta\} \subseteq V.$$

Let d_Y be a compatible metric on Y and let $Y = \bigcup_{k \geq 1} V_k$ with $V_k \in \{U_n\}_{n \geq 1}$ and $\operatorname{diam} V_k < \delta$. Moreover, let d_X be a compatible metric on X and write the open set $f^{-1}(V_k)$ as a union of d_X-balls with center $u_k \in X$, a rational radius, and closure in $f^{-1}(V_k)$. We have $X = \bigcup_{k \geq 1} f^{-1}(V_k)$ and the compactness of X implies that there exists a finite number of the balls $\overline{B}_n$ with $n \in \mathbb{N}$ such that $\bigcup_{i=1}^k \overline{B}_{n_i} = X$. For each i, choose m_i such that $\overline{B}_{n_i} \subseteq f^{-1}(U_{m_i})$. Let $g \in \bigcap_{i=1}^k W(\overline{B}_{n_i}, U_{m_i})$. If $x \in X$, we choose i such that $x \in \overline{B}_{n_i}$ and note that $f(x), g(x) \in U_{m_i}$. Since $\operatorname{diam} U_{m_i} < \delta$, we have $d_Y(g(x), f(x)) < \delta$, which gives $\hat{d}(g, f) < \delta < 2\delta$. Hence $g \in B_{2\delta}(f) \subseteq V$. Therefore, $f \in \bigcap_{i=1}^k W(\overline{B}_{n_i}, W_{n_i}) \subseteq V$ and this proves the second countability of $C(X, Y)$. □

Remark 1.6.19. Combining Proposition 1.6.18 with Problem 1.21, we conclude that if Y is a Polish space, then so is $C(X, Y)$ equipped with the $\tau_\zeta = \tau_u$-topology.

1.7 Semicontinuous Functions – Miscellaneous Notions

In this section we examine semicontinuous extended real-valued functions and at the end we introduce some topological notions that arise in various parts of nonlinear analysis.

Semicontinuous $\mathbb{R}^*$-valued functions, where $\mathbb{R}^* = \mathbb{R} \cup \{\pm\infty\}$, provide a natural framework to study minimization or maximization problems with constraints. Here we will focus on lower semicontinuous $\overline{\mathbb{R}} = \mathbb{R} \cup \{+\infty\}$-valued functions. Of course with a minus sign all results can be reformulated for upper semicontinuous $\tilde{\mathbb{R}} = \mathbb{R} \cup \{-\infty\}$-valued functions.

So, let X be a set and let $\varphi: X \to \overline{\mathbb{R}} = \mathbb{R} \cup \{+\infty\}$ be a function. We introduce the following sets:

$$\operatorname{epi}\varphi = \{(u,\lambda) \in X \times \mathbb{R}: \varphi(u) \le \lambda\} \text{ is the } \textbf{epigraph of } \varphi\,,$$
$$\varphi^\lambda = \{u \in X: \varphi(u) \le \lambda\} \text{ with } \lambda \in \mathbb{R} \text{ is the } \lambda\textbf{-sublevel set of } \varphi\,,$$
$$\operatorname{dom}\varphi = \{u \in X: \varphi(u) < +\infty\} \text{ is the } \textbf{effective domain of } \varphi\,.$$

To avoid trivial situations, we will always consider functions with $\operatorname{dom}\varphi \ne \emptyset$. In the optimization literature such functions are called **proper**. However, in nonlinear analysis, this name is reserved for maps that have the property where the inverse image of a compact set is compact.

Note that if $\{\varphi_\alpha\}_{\alpha\in I}$ is a family of $\overline{R}$-valued functions then

$$\operatorname{epi}\Big(\sup_{\alpha\in I}\varphi_\alpha\Big) = \bigcap_{\alpha\in I}\operatorname{epi}\varphi_\alpha\,, \tag{1.7.1}$$

$$\operatorname{epi}\Big(\inf_{\alpha\in I}\varphi_\alpha\Big) = \bigcup_{\alpha\in I}\operatorname{epi}\varphi_\alpha\,. \tag{1.7.2}$$

Definition 1.7.1. Let (X,τ) be a Hausdorff topological space and $\varphi: X \to \overline{\mathbb{R}} = \mathbb{R} \cup \{+\infty\}$. We say that φ is **τ-lower semicontinuous at** $x \in X$ if for every $\lambda < \varphi(x)$ there exists $U_\lambda \in \mathcal{N}(x)$ such that $\lambda < \varphi(u)$ for all $u \in U_\lambda$. We say that φ is **τ-lower semicontinuous** if it is τ-lower semicontinuous at every $x \in X$.

Proposition 1.7.2. *If (X,τ) is a Hausdorff topological space and $\varphi: X \to \overline{\mathbb{R}}$ a function, then the following statements are equivalent:*
(a) *φ is τ-lower semicontinuous;*
(b) $\operatorname{epi}\varphi \subseteq X \times \mathbb{R}$ *is closed (we consider the product topology on $X \times \mathbb{R}$);*
(c) *for every $\lambda \in \mathbb{R}$, $\varphi^\lambda \subseteq X$ is closed;*
(d) $\varphi(x) \le \liminf_{u\to x}\varphi(u) = \sup_{U\in\mathcal{N}(x)}\inf_{u\in U}\varphi(u)$ *for all $x \in X$.*

Proof. (a) $\Longrightarrow$ (b): Let $(u,\mu) \notin \operatorname{epi}\varphi$. Then $\mu < \varphi(u)$. Let $\eta \in (\mu, \varphi(u))$. Then by Definition 1.7.1, there exists $U_\eta \in \mathcal{N}(u)$ such that $\mu < \eta < \varphi(v)$ for all $v \in U_\eta$. Then

$$(U_\eta \times (-\infty,\eta)) \cap \operatorname{epi}\varphi = \emptyset\,.$$

Since $U_\eta \times (-\infty,\eta)$ is a neighborhood of (u,λ) in $X \times \mathbb{R}$, we conclude that $(X \times \mathbb{R}) \setminus \operatorname{epi}\varphi$ is open, hence $\operatorname{epi}\varphi$ is closed in $X \times \mathbb{R}$ with the product topology.

(b) $\Longrightarrow$ (c): Note that $\varphi^\lambda \times \{\lambda\} = \operatorname{epi}\varphi \cap (X \times \{\lambda\})$. Therefore $\varphi^\lambda \times \{\lambda\}$ is closed in $X \times \mathbb{R}$. But the map $u \to (u,\lambda)$ is a homeomorphism from X onto $X \times \{\lambda\}$. Therefore φ^λ is closed.

(c) $\Longrightarrow$ (d): Let $\lambda < \varphi(x)$. Since by hypothesis $X \setminus \varphi^\lambda$ is open, we can find $U \in \mathcal{N}(x)$ such that $U \subseteq (X \setminus \varphi^\lambda)$. So, we have $\lambda \le \inf_U \varphi$, which implies $\lambda \le \sup_{U\in\mathcal{N}(x)} \inf_{u\in U} \varphi(u) = \liminf_{u\to x} \varphi(u)$. Since $\lambda < \varphi(x)$ is arbitrary we let $\lambda \nearrow \varphi(x)$ to conclude that $\varphi(x) \le \liminf_{u\to x} \varphi(u)$.

(d) $\Longrightarrow$ (a): Let $\lambda < \varphi(x)$. By hypothesis $\lambda < \sup_{U\in\mathcal{N}(x)} \inf_{u\in U} \varphi(u)$ and thus $\lambda < \inf_{u\in U_0} \varphi(u)$ for some $U_0 \in \mathcal{N}(x)$. Hence, φ is τ-lower semicontinuous at any $x \in X$. □

Remark 1.7.3. If $\varphi: X \to \tilde{\mathbb{R}} = \mathbb{R} \cup \{-\infty\}$, then instead we use the **hypograph** $\operatorname{hyp}\varphi = \{(u,\lambda) \in X \times \mathbb{R}: \lambda \le \varphi(u)\}$ and the **λ-superlevel set** $\varphi_\lambda = \{u \in X: \varphi(u) \ge \lambda\}$. We have that φ is upper semicontinuous if and only if $\operatorname{hyp}\varphi$ is closed if and only if for $\lambda \in \mathbb{R}$, φ_λ is closed if and only if $\varphi(x) \ge \limsup_{u\to x} \varphi(u) = \inf_{U\in\mathcal{N}(x)} \sup_{u\in U} \varphi(u)$ for all $x \in X$.

Proposition 1.7.2 leads to some useful stability properties for lower semicontinuous functions.

Proposition 1.7.4. *If (X,τ) is a Hausdorff topological space and $\varphi_\alpha: X \to \overline{\mathbb{R}}$ with $\alpha \in I$, is a family of τ-lower semicontinuous functions, then the following hold:*
(a) *$\sup_{\alpha\in I} \varphi_\alpha$ is τ-lower semicontinuous;*
(b) *if I is finite, then $\inf_{\alpha\in I} \varphi_\alpha$ is τ-lower semicontinuous.*

Proof. (a) This follows from (1.7.1) and Proposition 1.7.2.

(b) Since I is finite and the finite union of closed sets is closed, the result follows from (1.7.2) and Proposition 1.7.2. □

Similarly, using Proposition 1.7.2, we have the following result.

Proposition 1.7.5. *If (X,τ) is a Hausdorff topological space and $\varphi, \psi: X \to \overline{\mathbb{R}}$ are τ-lower semicontinuous functions, then $\varphi + \psi$ is τ-lower semicontinuous.*

On metric spaces semicontinuous functions can be realized as monotone limits of Lipschitz functions.

Proposition 1.7.6. *If (X,d) is a metric space and $\varphi: X \to \overline{\mathbb{R}}$ is bounded from below, then φ is lower semicontinuous if and only if there exists an increasing sequence of Lipschitz continuous bounded functions $\hat{\varphi}_n: X \to \mathbb{R}$ such that $\hat{\varphi}_n(u) \nearrow \varphi(u)$ for all $u \in X$.*

Proof. $\Longrightarrow$: For every $n \in \mathbb{N}$ let $\varphi_n: X \to \mathbb{R}$ be defined by

$$\varphi_n(u) = \inf[\varphi(x) + nd(x,u): x \in X]\,. \tag{1.7.3}$$

Clearly $\{\varphi_n\}_{n\geq 1}$ is increasing and $\varphi_n \leq \varphi$ for every $n \in \mathbb{N}$. Moreover, for every $v \in X$ we have

$$\varphi_n(u) \leq \varphi(x) + nd(x,u) \leq \varphi(x) + nd(x,v) + nd(v,u) \quad \text{for all } x \in X\,.$$

This gives $\varphi_n(u) \leq \varphi_n(v) + nd(v,u)$, hence $|\varphi_n(u) - \varphi_n(v)| \leq nd(v,u)$. Thus each φ_n is Lipschitz.

We have $\varphi_n(u) \nearrow \tilde{\varphi}(u) \leq \varphi(u)$ for all $u \in X$. Given $\varepsilon > 0$, from (1.7.3), we see that there exists $x_n \in X$ such that

$$\varphi(x_n) + nd(x_n,u) \leq \varphi_n(u) + \varepsilon\,. \tag{1.7.4}$$

Let $\eta \leq \varphi(x)$ for all $x \in X$. So, from (1.7.4), we have

$$d(x_n,u) \leq \frac{1}{n}[\varphi_n(u) + \varepsilon - \eta]\,. \tag{1.7.5}$$

Hence, if $u \in \operatorname{dom}\varphi$, then $d(x_n,u) \leq 1/n[\varphi(u) + \varepsilon - \eta]$, which shows that

$$x_n \xrightarrow{d} u\,. \tag{1.7.6}$$

Hence if we pass to the limit as $n \to \infty$ in (1.7.4) and use (1.7.6), then $\varphi(u) \leq \tilde{\varphi}(u) + \varepsilon$. Since $\varepsilon > 0$ is arbitrary, we let $\varepsilon \searrow 0$ and obtain $\varphi(u) \leq \tilde{\varphi}(u)$, which implies $\varphi(u) = \tilde{\varphi}(u)$ for all $u \in \operatorname{dom}\varphi$.

If $u \notin \operatorname{dom}\varphi$, then we claim that $\tilde{\varphi}(u) = +\infty$. Indeed if $\tilde{\varphi}(u) \in \mathbb{R}$, then from (1.7.5) we have

$$d(x_n,u) \leq \frac{1}{n}[\tilde{\varphi}(u) + \varepsilon - \eta]\,.$$

Hence, $x_n \xrightarrow{d} u$. So, as above we obtain $+\infty = \varphi(u) \leq \tilde{\varphi}(u) < +\infty$, a contradiction. Thus $\varphi_n(u) \nearrow +\infty$ for all $u \notin \operatorname{dom}\varphi$. Finally let $\hat{\varphi}_n = \min\{\varphi_n, n\}$. Then $\hat{\varphi}_n$ is bounded as well. ☐

Remark 1.7.7. If $\varphi: X \to \tilde{\mathbb{R}} = \mathbb{R} \cup \{-\infty\}$ is upper semicontinuous and bounded above, then we can find a decreasing sequence of Lipschitz continuous bounded functions $\hat{\varphi}_n: X \to \mathbb{R}$ such that $\hat{\varphi}_n(u) \to \varphi(u)$ for all $u \in X$ as $n \to \infty$.

From Proposition 1.7.6 and Remark 1.7.7, we infer the following useful result.

Corollary 1.7.8. *If (X,d) is a metric space and $\varphi \in C_b(X,\mathbb{R})$, then there exist two sequences of Lipschitz continuous bounded functions $\xi_n, \eta_n: X \to \mathbb{R}$ such that*
(a) *$\{\xi_n\}_{n\geq 1}$ is increasing and $\xi_n(u) \nearrow \varphi(u)$ for all $u \in X$;*
(b) *$\{\eta_n\}_{n\geq 1}$ is decreasing and $\eta_n(u) \searrow \varphi(u)$ for all $u \in X$.*

In general pointwise convergence of functions does not imply uniform convergence. However, with additional hypotheses we can have this. The result is known as "Dini's Theorem".

Theorem 1.7.9 (Dini's Theorem). *If (X, τ) is a countably compact Hausdorff topological space, $\varphi_n : X \to \mathbb{R}$ with $n \in \mathbb{N}$ is an increasing (resp. decreasing) sequence of lower (resp. upper) semicontinuous functions and $\varphi_n(u) \to \varphi(u)$ for all $u \in X$ with $\varphi : X \to \mathbb{R}$ upper (resp. lower) semicontinuous, then φ is continuous and $\varphi_n \to \varphi$ uniformly, that is $\hat{d}(\varphi_n, \varphi) = \sup_{x \in X} |\varphi_n(x) - \varphi(x)| \to 0$ as $n \to \infty$.*

Proof. We do the case of a lower semicontinuous sequence. The other case is obtained by multiplying with -1. From Proposition 1.7.4 (a), we have that φ is lower semicontinuous as well, hence continuous. Then, for all $n \in \mathbb{N}$, $\varphi_n - \varphi \le 0$ and it is lower semicontinuous. Given $\varepsilon > 0$, let $U_n = \{u \in X : (\varphi_n - \varphi)(u) > -\varepsilon\}$. Then $\{U_n\}_{n \ge 1}$ is an open cover of X and so by countable compactness we can find a finite subcover; see Definition 1.4.57 (a). Since $\{U_n\}_{n \ge 1}$ are increasing, then for some $n \in \mathbb{N}$, $U_n = X$. Hence $-\varepsilon < (\varphi_m - \varphi)(u) \le 0$ for all $m \ge n$. Therefore, $\varphi_n \to \varphi$ uniformly on X. □

Remark 1.7.10. The hypotheses in Theorem 1.7.9 can not be relaxed. Let $\varphi_n(x) = x^n$ for all $x \in [0, 1)$. Then $\varphi_n \searrow 0$ but the convergence is not uniform. The domain $[0, 1)$ is not compact. Moreover, if $X = [0, 1]$, then $\varphi_n(x) = x^n \to \chi_{\{1\}}(x)$ and again the convergence is not uniform since $\chi_{\{1\}}$ is not lower semicontinuous. Note that the characteristic function

$$\chi_C(x) = \begin{cases} 1 & \text{if } x \in C, \\ 0 & \text{if } x \notin C \end{cases}$$

of a closed set C is only upper semicontinuous.

Next we introduce some topological notions that are used often in problems of nonlinear analysis.

Definition 1.7.11. Let (X, τ) be a Hausdorff topological space and $A \subseteq X$. We say that A is a **retract** of X if there is a continuous map $r : X \to A$ such that $r|_A = \mathrm{id}|_A$. The map $r : X \to A$ is called a **retraction**.

Remark 1.7.12. Equivalently we can say that $A \subseteq X$ is a retract of X if $\mathrm{id}|_A$ is continuously extendable to X. The concept of retracts is a topological notion, that is, if $h : X \to Y$ is a homeomorphism and $A \subseteq X$ is a retract of X, then $h(A)$ is a retract of Y.

Example 1.7.13. (a) X and for $u \in X$, the singletons $\{u\}$ are retracts of X.

(b) If $\overline{B}_1^n = \{u \in \mathbb{R}^n : |u| \le 1\}$ and $S^{n-1} = \{u \in \mathbb{R}^n : |u| = 1\}$, then $\overline{B}_1^n$ is a retract of $\mathbb{R}^n$ with a retraction given by

$$r(u) = \begin{cases} \frac{u}{|u|} & \text{if } |u| \ge 1, \\ u & \text{if } |u| < 1, \end{cases}$$

while S^{n-1} is a retract of $\mathbb{R}^n \setminus \{0\}$ with a retraction given by $r(u) = u/|u|$ for all $u \in \mathbb{R}^n \setminus \{0\}$.

(c) Every nonempty closed subset of the Polish space $\mathbb{N}^\infty$ is a retract of $\mathbb{N}^\infty$.

Proposition 1.7.14. *If (X, τ) is a Hausdorff topological space and A is a retract of X, then A is closed.*

Proof. Arguing by contradiction, suppose that A is not closed and let $x \in \overline{A} \setminus A$. Then, for a retraction r, we have $r(x) \neq x$ and so we can find $U \in \mathcal{N}(x)$, $V \in \mathcal{N}(r(x))$ such that $U \cap V = \emptyset$ since X is assumed to be Hausdorff. Because of the continuity of r, there holds $r(U) \subseteq V$. Let $u \in A \cap U$, recall $x \in \overline{A}$, then $r(u) = u \in V$, a contradiction. □

Proposition 1.7.15. *If X is a Hausdorff topological space and $A \subseteq X$, then A is a retract of X if and only if for every Hausdorff topological space Y every continuous map $f: A \to Y$ is continuously extendable on all of X.*

Proof. $\Longrightarrow$: Let $r: X \to A$ be a retraction. Then $f \circ r: X \to Y$ is a continuous extension of f.

$\Longleftarrow$: Let $Y = A$. Then, according to Remark 1.7.12, A is a retract of X. □

Definition 1.7.16. Let X, Y be two Hausdorff topological spaces and $f, g: X \to Y$ two continuous maps. A **homotopy** from f to g is a continuous map $h: [0,1] \times X \to Y$ such that $h(0,\cdot) = f(\cdot)$ and $h(1,\cdot) = g(\cdot)$. Then we say that f and g are **homotopic** and write $f \simeq g$ (or $f \simeq g$ (h) if we need to emphasize the homotopy).

Remark 1.7.17. We can think of the homotopy as a time dependent deformation, with the parameter $t \in [0,1]$ being the time, of f into g as time moves from 0 to 1. This deformation is continuous. So there are no breaks or jumps.

Proposition 1.7.18. *$\simeq$ is an equivalence relation on $C(X, Y)$.*

Proof. First, we see that $f \simeq f$ via the constant homotopy $h(t,\cdot) = f(\cdot)$ for all $t \in [0,1]$. Now let $f, g \in C(X, Y)$ and suppose that $f \simeq g$. Denote by $h: [0,1] \times X \to Y$ the corresponding homotopy. Then $\tilde{h}(t,x) = h(1-t,x)$ for all $t \in [0,1]$ and for all $x \in X$ is a homotopy from g to f. Therefore $g \simeq f$. Finally if $f \simeq g$ (h_1) and $g \simeq k(h_2)$, then

$$h(t,x) = \begin{cases} h_1(2t,x) & \text{if } x \in [0,\frac{1}{2}], \\ h_2(2t-1,x) & \text{if } x \in [\frac{1}{2},1] \end{cases}$$

for all $t \in [0,1]$ and for all $x \in X$ is a homotopy from f to k; see Proposition 1.1.37. Hence, $f \simeq k$. □

Definition 1.7.19. Let X, Y be two Hausdorff topological spaces.

(a) If $f \in C(X, Y)$ is homotopic to a constant map, then we say that f is **nullhomotopic** and we write that $f \simeq 0$.

(b) We say that the space X is **contractible** if id_X is nullhomotopic.

(c) If $\varphi \in C(X,Y)$ and $\psi \in C(Y,X)$, then we say that ψ is a **homotopy inverse** of φ if $\psi \circ \varphi \simeq \mathrm{id}_X$ and $\varphi \circ \psi \simeq \mathrm{id}_Y$. If φ has a homotopy inverse, then φ is said to be a **homotopy equivalence.** In this case we say that X and Y are **homotopy equivalent** (or of the same **homotopy type**).

Remark 1.7.20. It is easy to check by applying Proposition 1.7.18 that homotopy equivalence is an equivalence relation. Note that every convex set in $\mathbb{R}^N$ is contractible and, more generally, every star-shaped set in $\mathbb{R}^N$ is contractible. Recall that a set $A \subseteq \mathbb{R}^N$ is **star-shaped**, if there exists $u_0 \in A$ such that for every $u \in A$, the line segment $[u_0, u] = \{(1-t)u_0 + tu : 0 \leq t \leq 1\}$ is contained in A. In general, a contractible space is one that can be continuously shrunk to a point. Indeed, according to Definition 1.7.19 (b), there exists a continuous map $h: [0,1] \times X \to X$ such that $h(0,x) = x$ for all $x \in X$ and $h(1,x) = x_0$ for all $x \in X$ with $x_0 \in X$.

Definition 1.7.21. Let X be a Hausdorff topological space.

(a) A continuous map $h: [0,1] \times X \to X$ is a **deformation** of X if $h(0,\cdot) = \mathrm{id}_X$. Moreover, if $h(1,X) \subseteq A \subseteq X$, then we say that h is a deformation of X onto A.

(b) A closed set $A \subseteq X$ is a **(resp. strong) deformation retract** of X if there exists a deformation $h: [0,1] \times X \to X$ of X onto A such that $h(1,\cdot)|_A = \mathrm{id}_A$ (resp. such that $h(t,\cdot)|_A = \mathrm{id}_A$ for all $t \in [0,1]$). The deformation h is called a (resp. strong) deformation retraction.

Remark 1.7.22. Note that $A \subseteq X$ is a deformation retract if and only if there exists a retraction $r: X \to A$ (see Definition 1.7.11), such that $i_A \circ r \simeq \mathrm{id}_X$; see Definition 1.7.16. Then, since $r \circ i_A = \mathrm{id}_A$, we infer that the inclusion map $i_A: A \to X$ is a homotopy equivalence.

Example 1.7.23. From Example 1.7.13 (b), we know that S^n is a retract of $\mathbb{R}^{n+1} \setminus \{0\}$. In fact it is a strong deformation retract. Indeed, consider the deformation $h: [0,1] \times (\mathbb{R}^{n+1} \setminus \{0\}) \to \mathbb{R}^{n+1}$ defined by

$$h(t,x) = (1-t)x + t\frac{x}{|x|} \quad \text{for all } t \in [0,1] \text{ and for all } x \in \mathbb{R}^{n+1} \setminus \{0\}\,.$$

Directly from the previous definitions we have the following result.

Proposition 1.7.24. *If X is a Hausdorff topological space, then the following statements are equivalent:*

(a) *X is contractible.*

(b) *X is homotopy equivalent to a singleton.*

(c) *Any point of X is a deformation retract of X.*

Proposition 1.7.25. *If Y is a Hausdorff topological space, then $f \in C(S^n, Y)$ is nullhomotopic if and only if there exists a $\hat{f} \in C(\overline{B}_1^n, Y)$ such that $\hat{f}|_{S^n} = f$, that is, $\hat{f}$ is a continuous extension of f on $\overline{B}_1^n$.*

Proof. $\Longrightarrow$: Since $0 \simeq f$, there exists a homotopy $h: [0,1] \times S^n \to Y$ such that $h(0,\cdot) = u_0$ and $h(1,\cdot) = f$. Let

$$\hat{f}(x) = \begin{cases} u_0 & \text{if } 0 \le |x| \le \frac{1}{2}\,, \\ h(2|x| - 1, \frac{x}{|x|}) & \text{if } \frac{1}{2} \le |x| \le 1\,. \end{cases}$$

Then $\hat{f} \in C(\overline{B}_1^n, Y)$ and $\hat{f}|_{S^n} = f$.

$\Longleftarrow$: Let $h(t,x) = \hat{f}(tx)$ for all $t \in [0,1]$ and for all $x \in \overline{B}_1^n$. Then, using this homotopy, we see that $0 \simeq f$. □

The next notion is related to the Tietze Extension Theorem; see Theorem 1.2.44.

Definition 1.7.26. A Hausdorff topological space X is said to be an **absolute retract** (AR for short) if the following are true:
(a) X is metrizable;
(b) for any metrizable space Y and any closed set $A \subseteq Y$ each $f \in C(A,X)$ can be extended to a $\hat{f} \in C(Y,X)$, that is, $\hat{f}|_A = f$.

Remark 1.7.27. So an AR can replace $\mathbb{R}$ in the Tietze Extension Theorem, see Theorem 1.2.44, for metric spaces.

Proposition 1.7.28. *If X is an AR and C is a retract of X, then C is an AR.*

Proof. Let Y be a metrizable space, $A \subseteq Y$ a closed set, and $f \in C(A, C)$. Let $r: X \to C$ be a retraction. Since X is an AR, there exists $\hat{f} \in C(Y,X)$ such that $\hat{f}|_A = f$. Then $\hat{f}_0 = r \circ \hat{f} \in C(Y, C)$ is the desired extension of f. □

Now we will identify some useful spaces that are AR. The first result is known as "Dugundji's Extension Theorem".

Theorem 1.7.29 (Dugundji's Extension Theorem). *If X is a metrizable space, $A \subseteq X$ is closed, Y is a locally convex space, and $f \in C(A, Y)$, then there exists $\hat{f} \in C(X, Y)$ such that $\hat{f}|_A = f$ and $\hat{f}(X) \subseteq \operatorname{conv} f(A)$.*

Proof. Let d be a compatible metric on X. For $x \in X$ and $r > 0$, let $B(x,r) = \{u \in X : d(u,x) < r\}$. We consider the family $\{B(x, 1/2d(x,A)) : x \in X \setminus A\}$. This is an open cover of $X \setminus A$. Since $X \setminus A$ is paracompact (see Theorem 1.5.69), there exists a locally finite refinement $\{U_\alpha\}_{\alpha\in I}$. For U_α choose $B(x_\alpha, 1/2d(x_\alpha, A))$ such that

$$U_\alpha \subseteq B\left(x_\alpha, \frac{1}{2} d(x_\alpha, A)\right)\,; \tag{1.7.7}$$

see Definition 1.4.79 (a). We choose $u_\alpha \in A$ such that

$$d(x_\alpha, u_\alpha) \le 2d(x_\alpha, A)\,. \tag{1.7.8}$$

We have

$$d(x_\alpha, A) \leq 2d(x, A) \quad \text{for all } x \in U_\alpha \,. \tag{1.7.9}$$

To see (1.7.9) note that for all $x \in U_\alpha$

$$d(x_\alpha, A) \leq d(x_\alpha, x) + d(x, A) \leq \frac{1}{2} d(x_\alpha, A) + d(x, A) \, ;$$

see (1.7.7). Hence, (1.7.9) holds.

Moreover we have

$$d(u, u_\alpha) \leq 6d(u, x) \quad \text{for all } u \in A \text{ and all } x \in U_\alpha \,. \tag{1.7.10}$$

Again, to see (1.7.10), note that, because of (1.7.7) and (1.7.8), for all $u \in A$ and for all $x \in U_\alpha$,

$$\begin{aligned} d(u, u_\alpha) &\leq d(u, x) + d(x, x_\alpha) + d(x_\alpha, u_\alpha) \\ &\leq d(u, x) + \frac{1}{2} d(x_\alpha, A) + 2d(x_\alpha, A) \\ &\leq d(u, x) + d(x, A) + 4d(x, A) \\ &\leq 6d(u, x) \,. \end{aligned}$$

Thus, (1.7.10) holds.

Invoking Theorem 1.4.86, there exists a partition of unity $\{\xi_\alpha\}_{\alpha \in I}$ subordinated to the cover $\{U_\alpha\}_{\alpha \in I}$. We define

$$\hat{f}(u) = \begin{cases} f(u) & \text{if } u \in A \,, \\ \sum_{\alpha \in I} \xi_\alpha(u) f(u_\alpha) & \text{if } u \in X \setminus A \,. \end{cases} \tag{1.7.11}$$

Clearly, $\hat{f}|_A = f$ and $\hat{f}$ is continuous on the open set $X \setminus A$. We need to show the continuity of f at the points of A.

Let $u \in A$ and $V \in \mathcal{N}(f(u))$. Since Y is locally convex and f is continuous at u, we can find a convex set C and a $\delta > 0$ such that

$$f(A \cap B_{\frac{\delta}{6}}(u)) \subseteq C \subseteq V \,. \tag{1.7.12}$$

Let x be any point of $B_{\delta/6}(u) \setminus A$. Since the cover $\{U_\alpha\}_{\alpha \in I}$ is locally finite, it belongs to finitely many sets $U_{\alpha_1}, \ldots, U_{\alpha_n}$. Then $d(x, u) < \delta/6$ and since $x \in U_\alpha$ we have $d(u, u_{\alpha_i}) < \delta$ for all $i \in \{1, \ldots, n\}$; see (1.7.10). This implies that $u_{\alpha_i} \in A \cap B_\delta(u)$ for all $i \in \{1, \ldots, n\}$. Because of (1.7.11) and since C is convex it follows that $\hat{f}(u) \in C$. Therefore, due to (1.7.12), $\hat{f}(B_{\delta/6}(u)) \subseteq V$. Hence $\hat{f}|_A$ is continuous. □

Corollary 1.7.30. *If C is a convex subset of a locally convex space X and C is metrizable, then C is an* AR.

Next we show that in an infinite dimensional normed space X, the unit sphere $\partial B_1 = \{u \in X : \|u\| = 1\}$ is an AR. To do this we will need the following remarkable result due to Klee [192].

Theorem 1.7.31. *If X is an infinite dimensional normed space and $K \subseteq X$ is compact, then $X \setminus C$ and X are homeomorphic.*

Using this theorem, we can prove the following important result.

Theorem 1.7.32. *If X is an infinite dimensional normed space, then $\partial B_1 = \{u \in X : \|u\| = 1\}$ is an* AR *and a retract.*

Proof. By Theorem 1.7.31 X and $X \setminus \{0\}$ are homeomorphic. Due to Corollary 1.7.30, X is an AR. Hence $X \setminus \{0\}$ is an AR as well. Applying the radial retraction $r: X \setminus \{0\} \to \partial B_1$ defined by $r(u) = u/\|u\|$ for all $u \in X \setminus \{0\}$, we see that ∂B_1 is a retract of $X \setminus \{0\}$, hence an AR; see Proposition 1.7.25. Therefore we conclude that ∂B_1 is an AR and a retract of X. □

Remark 1.7.33. The result fails if X is finite dimensional. We will show this in Section 6.4 by using fixed point theory.

1.8 Remarks

(1.1) Point set topology emerged as a coherent field of mathematics with the appearance of Hausdorff's book [155] in 1914. Hausdorff found the right set of axioms to introduce the notion of topology in a general setting. He provided a unified framework for all previous topological research. Abstract spaces were first introduced by Fréchet [128] and Riesz [265]. The notion of a subbasis (see Definition 1.1.3) is due to Bourbaki [46]. The books of Choquet [72], Dugundji [102], Kelley [188], Kuratowski [199, 200], Munkres [247], Nagata [250], and Willard [334] are excellent references for all topics of point-set topology discussed here.

(1.2) The Hausdorff property (see Definition 1.2.1) was among the axioms for a topology used by Hausdorff. Before Hausdorff spaces, there was a more general class, the T_1-spaces introduced by Fréchet and Riesz.

Definition 1.8.1. A topological space X is a T_1**-space** if and only if for every distinct $x, u \in X$, there is a neighborhood of each not containing the other.

Remark 1.8.2. In such spaces singletons are closed sets.

Regular spaces (see Definition 1.2.7) were introduced by Vietoris [319] and the normality property is due to Tietze [310]. Many authors define regularity and normality of T_1-spaces (see Definition 1.8.1): for example, Kelley [188] and Munkres [247]. Here we follow Dugundji [102]. Urysohn's Lemma (see Theorem 1.2.17) was proven by Urysohn [314]. The companion Theorem 1.2.17 (Tietze Extension Theorem) was proven by Tietze [309]. The notion of complete regularity (see Definition 1.2.19) is due to Urysohn [314].

The notions of first and second countability (see Definition 1.2.20) were defined by Hausdorff [155] while the notion of separability is due to Fréchet [128]. The Lindelöf property (see Definition 1.2.26 (b)) goes back to Lindelöf [216] for Euclidean spaces. The general study of Lindelöf spaces started with the paper of Kuratowski–Sierpinski [198].

E. H. Moore [238] and E. H. Moore–Smith [239] developed the general theory of convergence using nets, although the term is due to Kelley [187]. Subnets (see Definition 1.2.38) were introduced by E. H. Moore [240] and studied in detail by Kelley [187]. There is an alternative approach using filters instead of nets. This approach is used by Bourbaki [49].

(1.3) Weak topologies are discussed in Bourbaki [49] under the name "initial topologies". Moreover, quotient topologies were first studied by Alexandrov [5] and R. L. Moore [242]. Weak topologies are important in Banach space theory.

(1.4) The notion of connectedness (see Definition 1.4.23 (b)) is even older and appears in the work of Weierstraß. Locally connected spaces (see Definition 1.4.34) were introduced by Hahn [146] and are discussed in detail in the books of Dugundji [102] and Kuratowski [200].

Here is another notion of "connectedness" for metric spaces that can traced back to the work of Cantor.

Definition 1.8.3. A metric space (X, d) is said to be **well-chained** (or **well-linked**) if for every pair $(x, u) \in X \times X$ and every $\varepsilon > 0$ there exists a finite sequence $v_1, \dots, v_n$ of points in X such that $v_1 = x$, $v_n = u$ and $d(v_k, v_{k+1}) \le \varepsilon$ for all $k \in \{1, \dots, n-1\}$. That means x and u can be joined by a chain of steps at most equal to ε.

Proposition 1.8.4. *Every connected metric space is well-chained. For compact metric spaces we have "connected $\Longleftrightarrow$ well-chained".*

The term "compact space" is due to Fréchet [128] who used it to describe sequential compactness of metric spaces. Hausdorff [155] observed that the sequential definition of compactness is equivalent to the general definition (see Definition 1.4.42) for metric spaces. Alexandrov–Urysohn [6] used Definition 1.4.42 to describe compact spaces and called them "bicompact spaces". The Product Theorem of Tychonoff (see Theorem 1.4.56) was proven by Tychonoff [313] and showed that Definition 1.4.42 is the right one, that is, more general for compactness since it passes to arbitrary products.

Local compactness was introduced by Alexandrov [4] and Tietze [310]. For a topological vector space, local compactness is equivalent to finite dimensionality.

Local compactness is important in integration theory and in the theory of topological groups.

The problem of compactification was initiated by Alexandrov [4] who introduced the one-point compactification; see Definition 1.4.74. Paracompactness was defined by Dieudonne [92] with important contributions of Michael [231, 233], [234].

(1.5) The extension of topological considerations beyond the realm of Euclidean spaces was achieved by Fréchet [128] who introduced metric spaces and allowed the "points" under consideration to be abstract objects and not real numbers or real vectors. The idea of completion of metric spaces can be traced back to Cauchy who tried to define irrational numbers as the limits of Cauchy sequences of rational numbers. The notion of complete metric space can be found in Fréchet [128] and the general completion construction is due to Hausdorff [155]. The supremum metric (see Definition 1.5.28) although attributed to Fréchet, was first used by Weierstraß back in 1885. The systematic study of continuous maps and homeomorphisms started with Fréchet [128] although the idea of homeomorphism (but in a less general context) was used by Poincaré back in 1895.

Next we present an important theorem that gives us conditions under which a Hausdorff topological space is metrizable. The result is due to Urysohn [315] and is known as the "Urysohn's Metrization Theorem".

Theorem 1.8.5 (Urysohn's Metrization Theorem). *Every second countable regular topological space is metrizable.*

Polish spaces are discussed in Bourbaki [49] and Souslin spaces in L. Schwartz [293]. More about them in the Remarks of Chapter 2.

The notions of first and second category spaces (see Definition 1.5.62 (b), (c)) were introduced by Baire [23] who also proved Theorem 1.5.68 (b). Theorem 1.5.68 (a) is due to R. L. Moore [241] and Theorem 1.5.69 is due to A. H. Stone [303].

(1.6) The compact-open topology (see Definition 1.6.1) was defined and studied in detail by Arens [13] and Fox [127]. The Arzela–Ascoli Theorem (see Theorem 1.6.14) was first proven for $C[0,1]$ by Arzela [14] (the necessary part) and by Ascoli [15] (the sufficient part).

Definition 1.8.6. A Hausdorff topological space X is a **k-space** (or a **compactly generated space**) if the following condition hold:

"$C \subseteq X$ is closed if and only if $C \cap K$ is closed for every $K \subseteq X$ compact".

Theorem 1.8.7. (a) *Every locally compact space is a k-space.*
(b) *Every first countable space is a k-space.*

Remark 1.8.8. In particular a metric space is a k-space.

This leads us to the following generalization of Theorem 1.6.14.

Theorem 1.8.9. *Theorem* 1.6.14 *remains true if Y is only a k-space (not necessarily metric space).*

In this general form the result is due to Kelley [188, pp. 233–234].

(1.7) For further results on semicontinuous functions we refer to Dal Maso [80].

The next notion is important in variational problems.

Definition 1.8.10. A function $\varphi: X \to \overline{\mathbb{R}} = \mathbb{R} \cup \{+\infty\}$ is said to be **coercive (sequentially coercive)** if for every $\lambda \in \mathbb{R}$ the sublevel set $\varphi^\lambda = \{x \in X: \varphi(x) \leq \lambda\}$ is relatively compact (relatively sequentially compact).

Remark 1.8.11. Sequentially coercivity implies coercivity. Another name for coercivity is **inf-compactness (sequential inf-compactness)**. Note that lower semicontinuity and coercivity are antagonistic notions. More precisely, let τ_1, τ_2 be two Hausdorff topologies on X and assume that $\tau_2 \subseteq \tau_1$. Then for a function $\varphi: X \to \overline{\mathbb{R}} = \mathbb{R} \cup \{+\infty\}$ we have that "φ is τ_2-lower semicontinuous" implies "φ is τ_1-lower semicontinuous" as well as "φ is τ_1-coercive" implies "φ is τ_2-coercive".

A balance between these two properties leads to the choice of a good topology for variational analysis. For additional information on retracts, absolute retracts, homotopies, etc. we refer to Borsuk [44], Hu [175] and Granas–Dugundji [144].

Problems

Problem 1.1. Suppose that X, Y are Hausdorff topological spaces and $f: X \to Y$ is a continuous map. Show that the set $C = \{(x,u) \in X \times X: f(x) = f(u)\}$ is closed in $X \times X$ with the product topology.

Problem 1.2. Suppose that X, Y are Hausdorff topological spaces and $f, g: X \to Y$ are continuous maps. Show that $\{x \in X: f(x) = g(x)\}$ is closed in X.

Problem 1.3. Show that every subspace of a completely regular space is completely regular. Moreover show that $X = \prod_{\alpha\in I} X_\alpha$ with the product topology is completely regular if and only if each factor space X_α is completely regular.

Problem 1.4. Show that X is completely regular if and only if it is homeomorphic to a subspace of some cube.

Problem 1.5. Show that a topological space X is Hausdorff if and only if the diagonal $D = \{(u,u) \in X \times X: u \in X\}$ is closed in $X \times X$ with the product topology.

Problem 1.6. Suppose that X is a Hausdorff topological space and let $\{u_n\}_{n\geq 1} \subseteq X$ be a sequence such that $u_n \to u \in X$. Show that the set $K = \{u_n\}_{n\geq 1} \cup \{u\}$ is compact. Is the result true for nets? Justify your answer.

Problem 1.7. Show that a regular Lindelöf space is normal.

Problem 1.8. Suppose that X, Y are Hausdorff topological spaces, Y is compact and $f: X \to Y$. Show that f is continuous if and only if $\mathrm{Gr} f = \{(u, y) \in X \times Y : y = f(u)\}$ is closed in $X \times Y$ with the product topology.

Problem 1.9. Suppose that $\{X_\alpha\}_{\alpha \in I}$ are Hausdorff topological spaces and $K_\alpha \subseteq X_\alpha$ with $\alpha \in I$ are compact sets. Let $U \subseteq X = \prod_{\alpha \in I} X_\alpha$ be an open set for the product topology such that $\prod_{\alpha \in I} K_\alpha \subseteq U$. Show that there exists a basic open set V (for the product topology) such that $\prod_{\alpha \in I} K_\alpha \subseteq V \subseteq U$.

Problem 1.10. Let X, Y be Hausdorff topological spaces and let $f: X \to Y$ be a map with $\mathrm{Gr} f = \{(u, y) \in X \times Y : y = f(u)\}$, which is closed in $X \times Y$ with the product topology. Show that for every compact $K \subseteq Y, f^{-1}(K) \subseteq X$ is closed.

Problem 1.11. Let X be a locally compact topological space. Show that X is second countable if and only if it is separable and metrizable.

Problem 1.12. Let X, Y be Hausdorff topological spaces and $A \subseteq X, B \subseteq Y$ are nonempty sets. Show that $A \times B$ is closed (resp. open, dense) in $X \times Y$ with the product topology if and only if A and B are closed (resp. open, dense) in X and Y, respectively.

Problem 1.13. Suppose that X is a normal topological space and $A \subseteq X$ closed. Show that the following statements are equivalent:
(a) A is a G_δ-set.
(b) There exists a continuous map $f: X \to Y$ such that $A = f^{-1}(0)$.
(c) For every closed $C \subseteq X$ with $A \cap C = \emptyset$, there exists a continuous function $f: X \to [0,1]$ such that $f^{-1}(0) = A$ and $f(B) = 1$.

Problem 1.14. Let (X, τ) be a Hausdorff topological space and $\mathcal{L} \subseteq \tau$ be a subbasis of the topology. Assume that every $\mathcal{L}$-cover of X admits a finite subcover. Show that (X, τ) is compact. Remark: This result is known as "Alexandrov's Subbasis Theorem".

Problem 1.15. Let (X, d) be a metric space. Show that there exists a normed space V and an isometry $\xi: X \to V$ such that $\xi(X) \subseteq V$ is closed. Remark: this result is known as the "Arens–Eells Embedding Theorem".

Problem 1.16. Let $A \subseteq \mathbb{R}^N$ be connected and let $A_\varepsilon = \{u \in \mathbb{R}^N : d(u, A) < \varepsilon\}$. Show that A_ε is connected and path-connected.

Problem 1.17. Let X be a Hausdorff topological space that is connected and A is a proper nonempty subset of X. Show that $\mathrm{bd}\, A \neq \emptyset$.

Problem 1.18. Let X be a Hausdorff topological space that is connected and $A \subseteq X$. Assume that $\mathrm{bd}\, A$ is connected. Show that $\overline{A}$ is connected as well.

Problem 1.19. Let X be a Hausdorff topological space and $A \subseteq X$ a connected set. Consider a set $D \subseteq X$ such that $A \cap D \neq \emptyset$ and $A \cap (X \setminus D) \neq \emptyset$. Show that $A \cap \mathrm{bd}\, D \neq \emptyset$.

Problem 1.20. Let X be a Hausdorff topological space, $\{K_\alpha\}_{\alpha\in I}$ is a family of compact subsets of X and $U \subseteq X$ is an open set such that $\cap_{\alpha\in I} K_\alpha \subseteq U$. Show that there exists a finite $F \subseteq I$ such that $\bigcap_{\alpha\in F} K_\alpha \subseteq U$.

Problem 1.21. Let X be a compact topological space and (Y, d) a metric space. On $C(X, Y)$ we consider the supremum metric d_∞; see Definition 1.5.28. Show that $C(X, Y)$ is d_∞-complete if and only if Y is d-complete.

Problem 1.22. Show that a compact metric space cannot be isometric to a proper subset of itself.

Problem 1.23. Let (X, d) be a compact metric space. Show that:
(a) Every nonexpansive map $f: X \to X$ (see Remark 1.5.23) is an isometry.
(b) If $f: X \to X$ satisfies $d(x, u) \le d(f(x), f(u))$ for all $x, u \in X$, then f is an isometry.

Problem 1.24. Let X be a noncompact, locally compact Hausdorff topological space and $\hat{X}$ is its one-point Alexandrov compactification; see Theorem 1.4.75. Show that $\hat{X}$ is metrizable if and only if X is second countable.

Problem 1.25. Let (X, d) and (Y, ρ) be two metric spaces. Show the following two statements:
(a) If $f: X \to Y$ is continuous, then there exists an equivalent metric $\hat{d}$ on X such that $f: (X, \hat{d}) \to (Y, \rho)$ is Lipschitz continuous.
(b) If $\mathcal{L}$ is a countable family of continuous functions from X into Y, then there exists an equivalent metric $\hat{d}$ on X and an equivalent metric $\hat{\rho}$ on Y such that each $f \in \mathcal{L}$ with $f: (X, \hat{d}) \to (Y, \hat{\rho})$ is Lipschitz continuous.

Problem 1.26. Let X be a Hausdorff topological space, (Y, d) a metric space, $f: X \to Y$ a continuous map, and $D_f = \{x \in X : f \text{ is not continuous at } x\}$. Show that D_f is an F_σ-set.

Problem 1.27. Is there a function $f: [0, 1] \to \mathbb{R}$ with D_f being the irrational numbers in $[0, 1]$ (see Problem 1.26)? Justify your answer.

Problem 1.28. Let X be a Hausdorff topological space and $\varphi: X \to \overline{\mathbb{R}} = \mathbb{R} \cup \{+\infty\}$ a coercive and lower semicontinuous (resp. sequentially coercive and sequentially lower semicontinuous) function. Show that there exists $u_0 \in X$ such that $\varphi(u_0) = \inf[\varphi(u): u \in X]$.

Problem 1.29. Let $\varphi: \mathbb{R}^N \to \mathbb{R}$ be a function such that $\lim_{|u|\to\infty} \varphi(u)/|u| > 0$. Show that φ is coercive in the sense of Definition 1.8.10.

Problem 1.30. Let X, Y be metrizable spaces with Y compact and $\varphi: X \times Y \to \overline{\mathbb{R}} = \mathbb{R} \cup \{+\infty\}$ lower semicontinuous. Let $m(u) = \inf[\varphi(u, y): y \in Y]$. Show that $m: X \to \overline{\mathbb{R}}$ is lower semicontinuous and for every $u \in X$ there exists $y_0 \in Y$ such that $m(u) = \varphi(u, y_0)$.

Problem 1.31. Suppose that X is a k-space (see Definition 1.8.6) and Y is a Hausdorff topological space. Show that $f: X \to Y$ is continuous if and only if $f|_K$ is continuous for every compact $K \subseteq X$.

Problem 1.32. Let X be a metric space, $A \subseteq X$ closed, and $V \subseteq [0,1] \times X$ an open set such that $[0,1] \times A \subseteq V$. Show that there exists an open set $U \subseteq X$ such that $A \subseteq U$ and $[0,1] \times U \subseteq V$.

Problem 1.33. Let X be a Hausdorff topological space and $A \subseteq X$ closed. Show that A is a deformation retract of X if and only if A is a retract of X and X is deformable into A.

Problem 1.34. Let X be an AR. Show that any open set $U \subseteq X$ is also an AR.

Problem 1.35. Show that $\mathbb{Q}$ is not topologically complete.

Problem 1.36. Let

$$\chi_{\mathbb{Q}}(x) = \begin{cases} 1 & \text{if } x \in \mathbb{Q}\,, \\ 0 & \text{if } x \notin \mathbb{Q} \end{cases}$$

being the characteristic function of the rationals. Show that $\chi_{\mathbb{Q}}$ is not the pointwise limit of a sequence of continuous functions.

Problem 1.37. Let (X,d) be a compact metric space and $f: X \to X$ an isometry. Show that f is surjective.

Problem 1.38. Is the pointwise limit of lower semicontinuous functions a lower semicontinuous function? How about the uniform limit? Justify your answer.

Problem 1.39. Show that the set of irrational numbers $\mathbb{R} \setminus \mathbb{Q}$ is topologically complete.

Problem 1.40. Let X, Y be Hausdorff topological spaces and $f: X \to Y$. Show that $\operatorname{Gr} f = \{(u,y) \in X \times Y : y = f(u)\}$ is a retract of $X \times Y$.

Problem 1.41. Let (X, τ) be a compact topological space and suppose that there exists a countable, separating family $\mathcal{F}$ of continuous functions $f: X \to Y$ with (Y,d) a metric space. Show that τ is metrizable.

Problem 1.42. Show that every locally compact Souslin space is Polish.

Problem 1.43. Let (X,d) be a metric space and $C_1, C_2 \subseteq X$ nonempty, disjoint, closed sets with C_2 compact. Show that $d(C_1, C_2) = \inf[d(u,v) : u \in C_1, v \in C_2] > 0$.

Problem 1.44. Let X be a locally compact and σ-compact topological space. Show that every open cover $\mathcal{L}$ of X has a locally finite open refinement $\{V_n\}_{n\geq 1}$ such that $\overline{V}_n$ is compact for all $n \in \mathbb{N}$.

Problem 1.45. Let X be a metrizable, locally compact, σ-compact topological space. Show the following:

(a) Every open set $U \subseteq X$ can be written as $U = \bigcup_{n\geq 1} K_n$ with compact K_n and $K_n \subseteq \operatorname{int} K_{n+1}$ for all $n \in \mathbb{N}$.

(b) Every compact set $K \subseteq X$ can be written as $K = \bigcap_{n\geq 1} U_n$ with U_n open, $\overline{U}_n$ compact and $U_n \supseteq U_{n+1}$ for all $n \in \mathbb{N}$.

Problem 1.46. Let X be a locally compact space and $\hat{X}$ its one-point Alexandrov compactification. Set $V = \{\hat{f} \in C(\hat{X}, \mathbb{R}) : f(\infty) = 0\}$. For every $\hat{f} \in V$, let $\tilde{f} = \hat{f}|_X$. Show that $\hat{f} \to \tilde{f}$ is an isometry of V onto $C_0(X, \mathbb{R}) = \{f \in C(X, \mathbb{R})$: for every $\varepsilon > 0$ there exists compact $K \subseteq X$ such that $|f(x)| < \varepsilon$ for all $x \in X \setminus K\}$ being the space of continuous functions on X vanishing at infinity.

Problem 1.47. Let X, Y be Hausdorff topological spaces, $\{V_\alpha\}_{\alpha\in I}$ an open cover of Y, and $f: X \to Y$ a continuous map such that $f_\alpha = f|_{f^{-1}(V_\alpha)}: f^{-1}(V_\alpha) \to V_\alpha$ is a homeomorphism for every $\alpha \in I$. Show that f is a homeomorphism.

Problem 1.48. Let X, Y be Hausdorff topological space, $f: X \to Y$ a map, and $G = \operatorname{Gr} f = \{(u, y) \in X \times Y : y = f(u)\}$. Let $g: X \to G$ be defined by $g(u) = (u, f(u))$. Show that f is continuous if and only if g is a homeomorphism.

Problem 1.49. Let X be a Baire space, Y a separable metric space and $f: X \to Y$ a map such that the inverse image of any open set is a F_σ-set. Show that f is continuous at every point of a dense G_δ-set.

Problem 1.50. Let X be a second countable regular topological space and $U \subseteq X$ an open set. Show that there exists a continuous function $f: X \to [0, 1]$ such that $f(u) > 0$ for all $u \in U$ and $f(u) = 0$ for all $u \in X \setminus U$.

Problem 1.51. Let X, Y be Hausdorff topological spaces, $f: X \to Y$ a continuous map, $C_n \subseteq X$ closed for all $n \in \mathbb{N}$, $C_n \searrow C$ being nonempty compact, and for every $U \supseteq C$ open, there is $n \in \mathbb{N}$ such that $C_n \subseteq U$. Show that $f(C) = \bigcap_{n\geq 1} f(C_n) = \bigcap_{n\geq 1} \overline{f(C_n)}$.

Problem 1.52. Let X be a regular topological space, $K \subseteq X$ compact and $U \subseteq X$ open such that $K \subseteq U$. Show that there exists an open set $V \subseteq X$ such that $K \subseteq V \subseteq \overline{V} \subseteq U$.

Figure 1.2 shows the relations between various spaces introduced in this chapter.

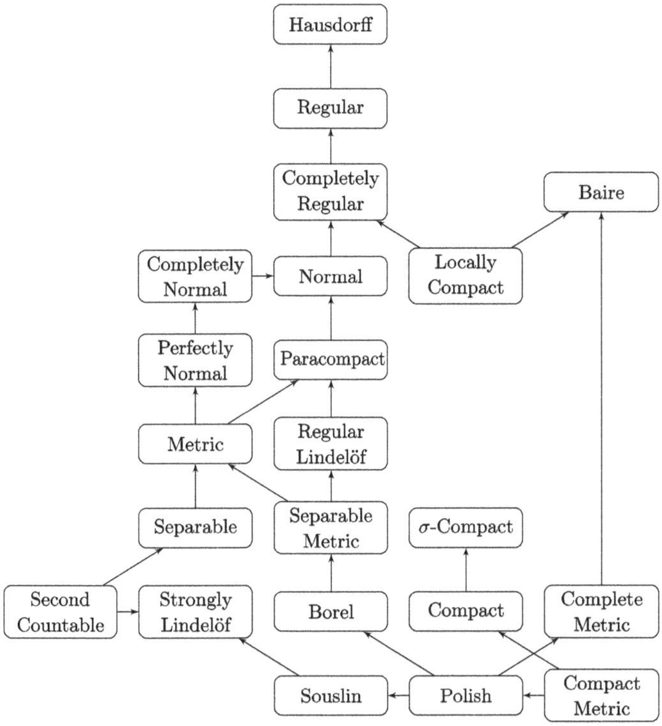

Figure 1.2: Topological spaces: From Compact Metric to Hausdorff.

2 Measure Theory

Measure Theory is the part of mathematical analysis that deals with the development of a precise way to measure large classes of sets and how to integrate functions. It started at the end of the 19th century with the works of Jordan, Borel, Young, and Lebesgue. By that time it was evident that the Riemann integral had serious limitations and had to be replaced by a new integral that was more general (that is, more functions could be integrated) and more flexible (that is, it led to more efficient calculus rules and in particular convergence theorems). The construction of Lebesgue turned out to be extremely fruitful and launched "Measure Theory". The idea of Lebesgue to partition the $f(x)$-axis (instead of the x-axis as it is done in the Riemann integral) was a remarkable conceptual insight, which allowed the full power of measure theory to reveal itself. In this chapter we present some basic aspects of this theory, which are needed to deal with the topics that follow.

2.1 Basic Notions, Measures, and Outer Measures

We start by defining algebras and σ-algebras. These are families of subsets of a given set. On σ-algebras, the theory exhibits its full strength.

Definition 2.1.1. Let X be a set and $\mathcal{L} \subseteq 2^X$ a nonempty family of subsets.
(a) We say that $\mathcal{L}$ is an **algebra** (or a **field**) if $A, B \in \mathcal{L}$ implies $A \cup B \in \mathcal{L}$ and $A^c = X \setminus A \in \mathcal{L}$. That is, $\mathcal{L}$ is closed under finite unions and complementation.
(b) We say that $\mathcal{L}$ is a **σ-algebra** (or a **σ-field**) if $\mathcal{L}$ is an algebra and it is closed under countable unions, that is, if $\{A_n\}_{n\geq 1} \subseteq \mathcal{L}$, then $\bigcup_{n\geq 1} A_n \in \mathcal{L}$.

Remark 2.1.2. Note that if $\mathcal{L}$ is an algebra, then $\emptyset, X \in \mathcal{L}$. Indeed, let $A \in \mathcal{L}$. Then $A^c \in \mathcal{L}$ and so $X = A \cup A^c \in \mathcal{L}$. Hence $\emptyset = X^c \in \mathcal{L}$. Moreover, by de Morgan's law, every algebra (resp. σ-algebra) is closed under finite (resp. countable) intersections. If $E \subseteq X$, then the **restriction** (or **trace**) of $\mathcal{L}$ on E is defined by $\mathcal{L}_E = \{E \cap A : A \in \mathcal{L}\}$.

Example 2.1.3. (a) There are two extreme cases: $\mathcal{L}_1 = \{\emptyset, X\}$ and $\mathcal{L}_2 = 2^X$. Both are σ-algebras with $\mathcal{L}_1$ being the smallest with respect to inclusion and $\mathcal{L}_2$ being the greatest one.
(b) Let $X = [0,1)$ and let $\mathcal{L}$ be the finite union of intervals $[a,b) \subseteq [0,1)$. Then $\mathcal{L}$ is an algebra but not an σ-algebra since $E = \bigcap_{n\geq 1}[0,1/n) = \{0\} \notin \mathcal{L}$.

Evidently the intersection of σ-algebras is again a σ-algebra. This leads to the following definitions.

Definition 2.1.4. (a) Let X be a set and let $\mathcal{F} \subseteq 2^X$ be nonempty. The σ-algebra generated by $\mathcal{F}$, denoted by $\sigma(\mathcal{F})$, is defined by

$$\sigma(\mathcal{F}) = \bigcap \{\mathcal{L} \subseteq 2^X : \mathcal{F} \subseteq \mathcal{L}, \mathcal{L} \text{ is a } \sigma\text{-algebra}\} .$$

https://doi.org/10.1515/9783111286952-002

(b) Let (X, τ) be a Hausdorff topological space. The **Borel σ-algebra** is defined by $\mathcal{B}(X) = \sigma(\tau)$.

As we will see later in our discussion of measures it is often more convenient to start with families that have less structure than σ-algebras and eventually pass to the σ-algebra they generate.

Definition 2.1.5. Let X be a set and let $\mathcal{L} \subseteq 2^X$ be a nonempty family of subsets.
(a) We say that $\mathcal{L}$ is a **ring** if $A, B \in \mathcal{L}$ implies $A \cup B \in \mathcal{L}$ and $A \setminus B \in \mathcal{L}$. That is, $\mathcal{L}$ is closed under finite unions and relative complementation.
(b) We say that $\mathcal{L}$ is a **σ-ring** if $\mathcal{L}$ is a ring and it is closed under countable unions, that is, if $\{A_n\}_{n\geq 1} \subseteq \mathcal{L}$, then $\bigcup_{n\geq 1} A_n \in \mathcal{L}$.
(c) We say that $\mathcal{L}$ is a **semiring** if the following hold:
   (i) $\emptyset \in \mathcal{L}$;
   (ii) $A, B \in \mathcal{L}$ implies $A \cap B \in \mathcal{L}$;
   (iii) $A, B \in \mathcal{L}$ implies $A \setminus B = \bigcup_{k=1}^n C_k$ for some $n \in \mathbb{N}$ and disjoint $\{C_k\}_{k=1}^n \subseteq \mathcal{L}$.

Remark 2.1.6. Note that if $\mathcal{L}$ is a ring and $A \in \mathcal{L}$, then $\emptyset = A \setminus A \in \mathcal{L}$. So, the empty set is always an element of a ring. Hence if $\mathcal{L}$ is a ring and $X \in \mathcal{L}$, then $\mathcal{L}$ is an algebra. Thus we see that the collection of all finite subsets of X is a ring but not an algebra unless X is a finite set. On the other hand the collection of all finite subsets of X and of their complements is an algebra but not a σ-algebra unless X is a finite set. If $\mathcal{L}$ is a ring and $A, B \in \mathcal{L}$, then $A \cap B = A \setminus (A \setminus B) \in \mathcal{L}$. So, a ring is also closed under finite intersections. Similarly $A \Delta B = (A \setminus B) \cup (B \setminus A) \in \mathcal{L}$ and so a ring is also closed under symmetric differences.

We have the following relations among the notions introduced thus far:

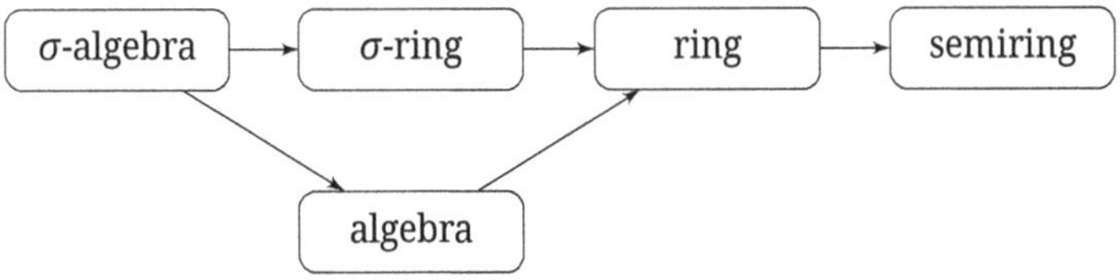

Apart from trivial cases, $\sigma(\mathcal{L})$ (see Definition 2.1.4 (a)) cannot be constructively obtained from $\mathcal{L}$. In order to overcome this difficulty, we introduce the following notions.

Definition 2.1.7. Let X be a set and $\mathcal{D} \subseteq 2^X$. We say that $\mathcal{D}$ is a **Dynkin system** (or a **λ-system**) if the following conditions hold:
(i) $X \in \mathcal{D}$;
(ii) $A, B \in \mathcal{D}$ with $B \subseteq A$ implies $A \setminus B \in \mathcal{D}$;
(iii) $\{A_n\}_{n\geq 1} \subseteq \mathcal{D}$ increasing implies $A = \bigcup_{n\geq 1} A_n \in \mathcal{D}$.

Remark 2.1.8. Evidently (ii) implies that $\emptyset$ is in every Dynkin system and $\{\emptyset, X\}$ as well as 2^X are both Dynkin systems. Consider also the following conditions on the family $\mathcal{D} \subseteq 2^X$:
(iv) $A \in \mathcal{D}$ implies $A^c \in \mathcal{D}$;
(v) for every disjoint sequence $\{A_n\}_{n\geq 1} \subseteq \mathcal{D}$ we have $\bigcup_{n\geq 1} A_n \in \mathcal{D}$.

It is easy to show that $\mathcal{D}$ is a Dynkin system if and only if (i), (iv), and (v) hold if and only if (i), (ii), and (v) hold.

Definition 2.1.9. Let X be a set and $\mathcal{L} \subseteq 2^X$ a nonempty family of subsets of X. We say that $\mathcal{L}$ is a **monotone class** if $\{A_n\}_{n\geq 1} \subseteq \mathcal{L}$ is increasing or decreasing, then

$$A = \bigcup_{n\geq 1} A_n \in \mathcal{L} \quad \text{or} \quad A = \bigcap_{n\geq 1} A_n \in \mathcal{L} .$$

Remark 2.1.10. Any σ-algebra is a monotone class but a topology is not in general. Of course 2^X is always a monotone class and the intersection of a family of monotone classes is a monotone class. So, there is a smallest monotone class containing a nonempty family $\mathcal{L} \subseteq 2^X$. A monotone class that is also an algebra is also a σ-algebra.

The next result is known as the "Dynkin System Theorem". The name "Dynkin's π–λ Theorem" can be also found in the literature.

Theorem 2.1.11 (Dynkin System Theorem). *If X is a set, $\mathcal{L} \subseteq 2^X$ is a nonempty family of subsets that is closed under finite intersections, and $\mathcal{D}$ is a Dynkin system such that $\mathcal{D} \supseteq \mathcal{L}$, then $\mathcal{D} \supseteq \sigma(\mathcal{L})$.*

Proof. Let $\mathcal{D}_0$ be the smallest Dynkin system containing $\mathcal{L}$. Evidently $\mathcal{D}_0 \subseteq \mathcal{D}$. Moreover, $\sigma(\mathcal{L})$ is a Dynkin system. So, we also have $\mathcal{D}_0 \subseteq \sigma(\mathcal{L})$. Let

$$\mathcal{R} = \{A \in \mathcal{D}_0 : A \cap B \in \mathcal{D}_0 \text{ for every } B \in \mathcal{L}\} .$$

Since $\mathcal{L}$ is closed under finite intersections we have $\mathcal{L} \subseteq \mathcal{R}$ and since $\mathcal{D}_0$ is a Dynkin system, we have that $\mathcal{R}$ is a Dynkin system as well. Therefore

$$\mathcal{D}_0 = \mathcal{R} . \tag{2.1.1}$$

Let $\mathcal{R}' = \{E \in \mathcal{D}_0 : E \cap D \in \mathcal{D}_0 \text{ for all } D \in \mathcal{D}_0\}$. Because of (2.1.1), it holds that $\mathcal{D}_0 = \mathcal{R}$ and so we have that $\mathcal{L} \subseteq \mathcal{R}'$, and $\mathcal{R}'$ is a Dynkin system. Hence, $\mathcal{D}_0 = \mathcal{R}'$, which means that $\mathcal{D}_0$ is closed under finite intersections. Thus, $\mathcal{D}_0$ is a σ-algebra; see Remark 2.1.8. Hence,

$$\sigma(\mathcal{L}) = \mathcal{D}_0 \subseteq \mathcal{D} .$$ ☐

Monotone classes are closely related to σ-algebras and by Theorem 2.1.11 are also related to Dynkin systems. The next result illustrates this and is known as the "Monotone Class Theorem".

Theorem 2.1.12 (Monotone Class Theorem). *If X is a set, $\mathcal{L} \subseteq 2^X$ is an algebra and $\mathcal{M} \subseteq 2^X$ is a nonempty, monotone class such that $\mathcal{M} \supseteq \mathcal{L}$, then $\mathcal{M} \supseteq \sigma(\mathcal{L})$.*

Proof. Let $\Sigma = \sigma(\mathcal{L})$ and let $\mathcal{M}_0$ be the smallest monotone class containing $\mathcal{L}$. Evidently $\mathcal{M}_0 \subseteq \mathcal{M}$. If we show that $\Sigma = \mathcal{M}_0$, then we are done.

To this end, we fix $A \in \mathcal{M}_0$ and let

$$\mathcal{M}_0^A = \{B \in \mathcal{M}_0 : A \cap B, B \setminus A \in \mathcal{M}_0\} .$$

Then $\mathcal{M}_0^A$ is a monotone class. If $A \in \mathcal{L}$, then since $\mathcal{L}$ is an algebra, we have $\mathcal{M}_0 \subseteq \mathcal{M}_0^A$, hence $\mathcal{M}_0 = \mathcal{M}_0^A$. So, for any $B \in \mathcal{M}_0$ we have

$$A \cap B, A \setminus B, B \setminus A \in \mathcal{M}_0 \quad \text{for any } A \in \mathcal{L} .$$

Thus, $\mathcal{L} \subseteq \mathcal{M}_0^B$, which implies $\mathcal{M}_0 = \mathcal{M}_0^B$.

Then we see that $\mathcal{M}_0$ is an algebra and so it follows that $\mathcal{M}_0$ is a σ-algebra; see Remark 2.1.10. It follows that $\Sigma \subseteq \mathcal{M}_0$ and because Σ is also a monotone class containing $\mathcal{L}$ we conclude that $\Sigma = \mathcal{M}_0 \subseteq \mathcal{M}$. □

Remark 2.1.13. From the proof above we see that if $\mathcal{L} \subseteq 2^X$ is an algebra, then $\sigma(\mathcal{L})$ coincides with the smallest monotone class generated by $\mathcal{L}$. Therefore, the algebra $\mathcal{L}$ is a monotone class if and only if $\mathcal{L}$ is a σ-algebra.

Since the Borel σ-algebra (see Definition 2.1.4 (b)) is an important σ-algebra, we state some easy but useful facts concerning its generation. The first result is an immediate consequence of Theorem 2.1.11.

Proposition 2.1.14. *If X is a Hausdorff topological space, then the Borel σ-Algebra is the smallest Dynkin system containing the open sets or the closed sets.*

In the context of metric spaces we can state a little different characterization of the Borel sets.

Proposition 2.1.15. *If X is a metrizable space, then the Borel σ-Algebra $\mathcal{B}(X)$ is the smallest family of subsets of X that includes the open sets and it is closed under countable intersections and under countable disjoint unions.*

Proof. From Proposition 1.5.8 we know that every closed set is G_δ. Hence, every family of sets that contains the open sets and is also closed under countable intersections, must contain the closed sets. Then the result follows from Problem 2.1. □

For a similar result for families containing the closed sets, we need to require that we have closure under arbitrary unions, not just disjoint ones.

Proposition 2.1.16. *If X is a metrizable space, then the Borel σ-Algebra $\mathcal{B}(X)$ is the smallest family of subsets of X that includes the closed sets and it is closed under countable intersections and under countable unions.*

Proof. Recall again from Proposition 1.5.8 that every open set is F_σ. Hence every family of sets that contains the closed sets and is closed under countable unions, must contain the open sets as well. Again an appeal to Problem 2.1 concludes the proof. □

Remark 2.1.17. In a Hausdorff topological space the closure of any set belongs to the Borel σ-algebra being closed. Similarly for the interior of any set being open and the boundary of any set being closed. Recalling that singletons are closed sets, we infer that countable sets are Borel. Finally, compact sets are also Borel being closed.

For the real line $\mathbb{R}$ we can choose among many different generators of the Borel σ-algebra. So let

$$\begin{aligned}
&\mathcal{L}_1 = \{(a,b): a < b\}, &&\mathcal{L}_2 = \{[a,b): a < b\}, &&\mathcal{L}_3 = \{(a,b]: a < b\}\,,\\
&\mathcal{L}_4 = \{[a,b]: a < b\}, &&\mathcal{L}_5 = \{(a,\infty): a \in \mathbb{R}\}, &&\mathcal{L}_6 = \{(-\infty,b): b \in \mathbb{R}\}\,,\\
&\mathcal{L}_7 = \{[a,+\infty): a \in \mathbb{R}\}, &&\mathcal{L}_8 = \{(-\infty,b]: b \in \mathbb{R}\}, &&\mathcal{L}_9 = \text{ open sets of } \mathbb{R}\,,\\
&\mathcal{L}_{10} = \text{ closed sets of } \mathbb{R}\,.
\end{aligned}$$

Moreover, by $\mathcal{L}_k^r, k \in \{1,\dots,8\}$ we denote the collection of intervals in $\mathcal{L}_k$ with rational endpoints.

The next result is straightforward.

Proposition 2.1.18. $\mathcal{B}(\mathbb{R}) = \sigma(\mathcal{L}_k)$ *for all* $k \in \{1,\dots,10\}$ *and* $\mathcal{B}(\mathbb{R}) = \sigma(\mathcal{L}_k^r)$ *for all* $k \in \{1,\dots,8\}$.

In many cases we will deal with the extended real line $\mathbb{R}^* = \mathbb{R} \cup \{\pm\infty\}$. In this case we have the following.

Definition 2.1.19. It holds that $\mathcal{B}(\mathbb{R}^*) = \sigma(\mathcal{B}(\mathbb{R}) \cup \{\{+\infty\},\{-\infty\}\})$.

Remark 2.1.20. Evidently $\mathcal{B}(\mathbb{R}^*) = \{$the $\mathcal{B}(\mathbb{R})$-sets or the $\mathcal{B}(\mathbb{R})$-sets with $+\infty$ or $-\infty$ or both attached to them$\}$.

From Proposition 2.1.18 and Definition 2.1.19 we obtain the following.

Proposition 2.1.21. *It holds that* $\operatorname{card}(\mathcal{B}(\mathbb{R})) = \operatorname{card}(\mathcal{B}(\mathbb{R}^*)) = \mathfrak{c}$ *being the cardinality of the continuum.*

Now we pass to set functions.

Definition 2.1.22. Let X be a set, $\emptyset \in \mathcal{L} \subseteq 2^X$ and $\mu\colon \mathcal{L} \to \mathbb{R}^*$ is a set function.

(a) We say that μ is **monotone** if

$$A \subseteq B \quad \text{with } A, B \in \mathcal{L} \quad \text{implies} \quad \mu(A) \le \mu(B)\,.$$

(b) We say that μ is **additive** (or **finitely additive**) if $\{A_k\}_{k=1}^n \subseteq \mathcal{L}$ are pairwise disjoint and $\bigcup_{k=1}^n A_k \in \mathcal{L}$ implies $\mu(\bigcup_{k=1}^n A_k) = \sum_{k=1}^n \mu(A_k)$.

(c) We say that μ is **σ-additive** (or **countably additive**) if $\{A_k\}_{k\geq1} \subseteq \mathcal{L}$ are pairwise disjoint and $\bigcup_{k\geq1} A_k \in \mathcal{L}$ implies $\mu(\bigcup_{k\geq1} A_k) = \sum_{k\geq1} \mu(A_k)$.
(d) We say that μ is **subadditive** if $\{A_k\}_{k=1}^n \subseteq \mathcal{L}$ and $\bigcup_{k=1}^n A_k \in \mathcal{L}$ imply $\mu(\bigcup_{k=1}^n A_k) \leq \sum_{k=1}^n \mu(A_k)$.
(e) We say that μ is **σ-subadditive** if $\{A_k\}_{k\geq1} \subseteq \mathcal{L}$ and $\bigcup_{k\geq1} A_k \in \mathcal{L}$ imply $\mu(\bigcup_{k\geq1} A_k) \leq \sum_{k\geq1} \mu(A_k)$.
(f) When $\mathcal{L} = \Sigma$ is a σ-algebra, then we say that the set function $\mu\colon \Sigma \to \mathbb{R}^* = \mathbb{R} \cup \{\pm\infty\}$ is a **signed-measure** if it takes only one of the values $+\infty$ and $-\infty$, $\mu(\emptyset) = 0$, and it is σ-additive. If μ takes only nonnegative values, then we say that μ is a **measure**.
(g) A pair (X, Σ) with X being a set and $\Sigma \subseteq 2^X$ being a σ-algebra is said to be a **measurable space**. If μ is a measure on (X, Σ), then (X, Σ, μ) is said to be a **measure space**. We say that μ is **finite** (or that the measure space (X, Σ, μ) is finite) if $\mu(X) < \infty$. We say that μ is **σ-finite** if $X = \bigcup_{n\geq1} X_n$ with $X_n \in \Sigma$ and $\mu(X_n) < +\infty$ for all $n \in \mathbb{N}$.

Example 2.1.23. (a) Let X be a nonempty set and $\Sigma = 2^X$. The set function $\mu\colon \Sigma \to [0, +\infty]$ defined by

$$\mu(A) = \begin{cases} \operatorname{card}(A) & \text{if } A \text{ is finite}\,, \\ +\infty & \text{otherwise}\,, \end{cases}$$

is a measure known as the **counting measure**. If X is finite (resp. countable), then $\mu\colon \Sigma \to [0, +\infty]$ is finite (resp. σ-finite). More generally, let $f\colon X \to [0, +\infty)$ be a function and define $\mu\colon 2^X \to [0, +\infty]$ by setting

$$\mu(A) = \sum_{x\in A} f(x) = \sup\left[\sum_{x\in F} f(x)\colon F \subseteq A \text{ is finite}\right].$$

Then $\mu\colon 2^X \to [0, +\infty]$ is a measure that is σ-finite if $\{x \in X\colon f(x) > 0\}$ is countable. Evidently, if $f(x) = 1$ for all $x \in X$, then we have the counting measure. If $f(x_0) = 1$ and $f(x) = 0$ if $x \neq x_0$, then $\mu\colon 2^X \to [0, +\infty]$ is called the **Dirac measure** at x_0 and is denoted by δ_{x_0}.
(b) Let X be an uncountable set and let

$$\Sigma = \{A \subseteq X\colon A \text{ is countable or } A^c \text{ is countable}\}\,.$$

Then Σ is a σ-algebra being the σ-algebra of countable or co-countable sets. The set function $\mu\colon \Sigma \to [0, 1]$ defined by

$$\mu(A) = \begin{cases} 0 & \text{if } A \text{ is countable}\,, \\ 1 & \text{if } A^c \text{ is countable, that is, } A \text{ is co-countable} \end{cases}$$

is a finite measure.

The next proposition summarizes the main properties of measures.

Proposition 2.1.24. *Let (X, Σ, μ) be a measure space. Then the following hold:*
(a) $\mu(A \cup B) + \mu(A \cap B) = \mu(A) + \mu(B)$ *for all* $A, B \in \Sigma$.
(b) $\mu(A) = \mu(B) + \mu(A \setminus B)$ *for all* $A, B \in \Sigma$ *with* $B \subseteq A$.
(c) $\mu(B) \le \mu(A)$ *for all* $A, B \in \Sigma$ *with* $B \subseteq A$ *(monotonicity).*
(d) $\mu(\bigcup_{k\ge1} A_k) \le \sum_{k\ge1} \mu(A_k)$ *for all* $\{A_k\}_{k\ge1} \subseteq \Sigma$ *(σ-subadditivity).*
(e) *If* $\{A_k\}_{k\ge1} \subseteq \Sigma$ *is increasing, then* $\mu(\bigcup_{k\ge1} A_k) = \lim_{k\to\infty} \mu(A_k)$ *(continuity from below).*
(f) *If* $\{A_k\}_{k\ge1} \subseteq \Sigma$ *is decreasing and* $\mu(A_1) < +\infty$*, then* $\mu(\bigcap_{k\ge1} A_k) = \lim_{k\to\infty} \mu(A_k)$ *(continuity from above).*

Proof. (a) By additivity we have

$$\mu(A) = \mu(A \cap B) + \mu(A \setminus B) \quad \text{and} \quad \mu(B) = \mu(A \cap B) + \mu(B \setminus A)\,.$$

Adding these two equations gives

$$\begin{aligned}\mu(A) + \mu(B) &= \mu(A \cap B) + [\mu(A \cap B) + \mu(A \setminus B) + \mu(B \setminus A)] \\ &= \mu(A \cap B) + \mu(A \cup B)\end{aligned}$$

again by the additivity.

(b) Let $A = B \cup (A \setminus B)$ and use the additivity we obtain $\mu(A) = \mu(B) + \mu(A \setminus B)$.

(c) Since μ is nonnegative, the assertion follows from (b).

(d) Let $B_1 = A_1$ and $B_k = A_k \setminus \bigcup_{i=1}^{k-1} A_i$ for $k \ge 2$. Then the sets $\{B_k\}_{k\ge1}$ are disjoint and $\bigcup_{k\ge1} B_k = \bigcup_{k\ge1} A_k$. Then, taking the σ-additivity and part (c) into account it follows

$$\mu\Big(\bigcup_{k\ge1} A_k\Big) = \mu\Big(\bigcup_{k\ge1} B_k\Big) = \sum_{k\ge1} \mu(B_k) \le \sum_{k\ge1} \mu(A_k)\,.$$

(e) Let $A_0 = \emptyset$. Then

$$\mu\Big(\bigcup_{k\ge1} A_k\Big) = \sum_{k\ge1} \mu(A_k \setminus A_{k-1}) = \lim_{n\to\infty} \sum_{k=1}^{n} \mu(A_k \setminus A_{k-1}) = \lim_{n\to\infty} \mu(A_n)\,.$$

(f) Let $B_k = A_1 \setminus A_k$. Then $\{B_k\}_{k\ge1} \subseteq \Sigma$ is increasing, $\mu(A_1) = \mu(A_k) + \mu(B_k)$ for all $k \in \mathbb{N}$, see part (b), and $\bigcup_{k\ge1} B_k = A_1 \setminus \bigcap_{k\ge1} A_k$. By parts (e) and (b) there holds

$$\mu(A_1) = \mu\Big(\bigcap_{k\ge1} A_k\Big) + \lim_{k\to\infty} \mu(B_k) = \mu\Big(\bigcap_{k\ge1} A_k\Big) + \lim_{k\to\infty} [\mu(A_1) - \mu(A_k)]\,.$$

Hence, subtracting $\mu(A_1) < \infty$ from both sides gives $\mu(\bigcap_{k\ge1} A_k) = \lim_{k\to\infty} \mu(A_k)$. □

Remark 2.1.25. Clearly, the condition $\mu(A_1) < +\infty$ in Proposition 2.1.24 (f) can be replaced by the hypothesis that $\mu(A_n) < +\infty$ for some $n \in \mathbb{N}$ since the first $(n-1)$ sets do not affect the intersection.

It turns out that continuity from below (see Proposition 2.1.24 (e)) for an additive set function is equivalent to σ-additivity.

Proposition 2.1.26. *If X is a set, $\mathcal{L} \subseteq 2^X$ is an algebra of sets in X and $\mu: \mathcal{L} \to [0, +\infty]$ is an additive set function, then μ is σ-additive if and only if μ is continuous from below, that is, if $\{A_n\}_{n\geq 1} \subseteq \mathcal{L}$ is increasing, $\bigcup_{n\geq 1} A_n \in \mathcal{L}$, then $\mu(\bigcup_{n\geq 1} A_n) = \lim_{n\to\infty} \mu(A_n)$.*

Proof. $\Longrightarrow$: This follows from the proof of Proposition 2.1.24 (e).

$\Longleftarrow$: Suppose we have continuity from below. Let $\{B_k\}_{k\geq 1} \subseteq \mathcal{L}$ be a sequence of pairwise disjoint sets such that $\bigcup_{k\geq 1} B_k \in \mathcal{L}$. We set $A_n = \bigcup_{k=1}^{n} B_k$. From the continuity from below hypothesis, it follows

$$\mu\Big(\bigcup_{k\geq 1} B_k\Big) = \mu\Big(\bigcup_{k\geq 1} A_k\Big) = \lim_{n\to\infty} \mu(A_n) = \lim_{n\to\infty} \sum_{k=1}^{n} \mu(B_k) = \sum_{k\geq 1} \mu(B_k)\,.$$

This shows that $\mu: \mathcal{L} \to [0, +\infty]$ is σ-additive. □

We get a similar result when we suppose continuity from above at the empty set.

Proposition 2.1.27. *If X is a set, $\mathcal{L} \subseteq 2^X$ is an algebra of sets in X and $\mu: \mathcal{L} \to [0, +\infty]$ is an additive set function with $\mu(X) < +\infty$, then μ is σ-additive if and only if μ is continuous from above at the empty set, that is, if $\{A_k\}_{k\geq 1} \subseteq \mathcal{L}$ is a decreasing sequence such that $\bigcap_{k\geq 1} A_k = \emptyset$, then $\lim_{k\to\infty} \mu(A_k) = 0$.*

Proof. $\Longrightarrow$: This implication follows again from the proof of Proposition 2.1.24 (f).

$\Longleftarrow$: Let $\{A_k\}_{k\geq 1} \subseteq \mathcal{L}$ be an increasing sequence such that $\bigcup_{k\geq 1} A_k \in \mathcal{L}$. Let $B_n = (\bigcup_{k\geq 1} A_k) \setminus A_n$ for all $n \in \mathbb{N}$. Then $\{B_n\}_{n\geq 1} \subseteq \mathcal{L}$ is decreasing and $\bigcap_{n\geq 1} B_n = \emptyset$. Therefore, by hypothesis, we have

$$0 = \lim_{n\to\infty} \mu(B_n) = \mu\Big(\bigcup_{k\geq 1} A_k\Big) - \lim_{n\to\infty} \mu(A_n)\,.$$

Hence, $\mu(\bigcup_{k\geq 1} A_k) = \lim_{k\to\infty} \mu(A_k)$ and so μ is continuous from below. Then Proposition 2.1.26 implies that μ is σ-additive. □

The next result gives a necessary and sufficient condition for two finite measures to be equal. It suffices to know that they coincide on a generating family that is closed under finite intersections.

Proposition 2.1.28. *If (X, Σ) is a measurable space, $\Sigma = \sigma(\mathcal{L})$ with $\mathcal{L}$ closed under finite intersections, μ_1, μ_2 are two finite measures on Σ and $\mu_1(X) = \mu_2(X)$ as well as $\mu_1|_{\mathcal{L}} = \mu_2|_{\mathcal{L}}$, then $\mu_1 = \mu_2$.*

Proof. Let $\mathcal{D} = \{A \in \Sigma: \mu_1(A) = \mu_2(A)\}$. Applying Proposition 2.1.24 (b) and (c), we see that $\mathcal{D}$ is a Dynkin system; see Definition 2.1.7. Moreover, by hypothesis, $\mathcal{L} \subseteq \mathcal{D}$. Then, invoking Theorem 2.1.11, we infer that $\Sigma = \sigma(\mathcal{L}) = \mathcal{D}$, which means that $\mu_1 = \mu_2$. □

Corollary 2.1.29. *If X is a Hausdorff topological space, $\mathcal{B}(X)$ is its Borel σ-field and μ_1, μ_2 are two finite measures on $\mathcal{B}(X)$, which coincide on the open or closed sets, then $\mu_1 = \mu_2$.*

In the next definition we introduce a notion that will lead us to a property reminiscent of the intermediate value property.

Definition 2.1.30. Let (X, Σ, μ) be a measure space.

(a) We say that the measure $\mu: \Sigma \to [0, +\infty]$ is **semifinite** if for every $A \in \Sigma$ with $\mu(A) > 0$, there exists $B \in \Sigma$ with $B \subseteq A$ such that $0 < \mu(B) < +\infty$.

(b) We say that $A \in \Sigma$ is an **atom** of μ if $0 < \mu(A) < +\infty$ and for every $B \subseteq A$ with $B \in \Sigma$ either $\mu(B) = 0$ or $\mu(B) = \mu(A)$. A measure without any atoms is called **nonatomic.**

Remark 2.1.31. The measure μ on Σ is nonatomic if for every set $A \in \Sigma$ with $\mu(A) > 0$, there exists $B \in \Sigma$ with $B \subseteq A$ such that $0 < \mu(B) < \mu(A)$. For the Dirac measure

$$\delta_{x_0}(A) = \begin{cases} 1 & \text{if } x_0 \in A\,, \\ 0 & \text{otherwise}\,, \end{cases} \qquad \text{with } x_0 \in X,\ A \in \Sigma\,,$$

we see that $\{x_0\}$ is an atom. The main examples of atoms are singletons $\{x\}$ with positive measure.

Here is the result that recalls the intermediate value property.

Proposition 2.1.32. *If (X, Σ, μ) is a nonatomic measure space, then the range of μ is the interval $[0, \mu(X)]$.*

Proof. We fix $\lambda \in (0, \mu(X))$ and define $\mathcal{L} = \{A \in \Sigma : 0 < \mu(A) \le \lambda\}$. First we show that $\mathcal{L} \neq \emptyset$. The nonatomicity of μ implies the existence of $B \in \Sigma$ such that $0 < \mu(B) < \mu(X)$. The same argument (nonatomicity of μ) implies that we can find $E_1, E_2 \in \Sigma$ such that $B = E_1 \cup E_2$, $E_1 \cap E_2 = \emptyset$ and $\mu(E_1), \mu(E_2) \in (0, \mu(B))$. It follows that at least one of the sets E_1, E_2 satisfies $\mu(E_1) \in (0, 1/2\mu(B)]$. Proceeding inductively, suppose that we produced $E_1, \ldots, E_n \in \Sigma$ such that

$$\mu(E_n) \in \left(0, \frac{1}{2^n}\mu(B)\right]. \tag{2.1.2}$$

Applying again the nonatomicity of μ there exists $E_{n+1} \in \Sigma$ with $E_{n+1} \subseteq E_n$ such that $\mu(E_{n+1}) \in (0, 1/2\mu(E_n)]$. Evidently, because of (2.1.2) we have $\mu(E_{n+1}) \le 1/2^{n+1}\mu(B)$. Therefore, (2.1.2) holds for all $n \in \mathbb{N}$. Moreover, for a large enough $n \in \mathbb{N}$, we have $\mu(E_n) \le \lambda$. Hence, $E_n \in \mathcal{L}$ for a large enough $n \in \mathbb{N}$, thus yielding $\mathcal{L} \neq \emptyset$.

Next we show that there exists a Σ-set with measure equal to λ. To this end, let $D_0 = \emptyset$ and suppose that $D_n \in \Sigma$ is given. Let

$$\lambda_n = \sup[\mu(C) : C \in \Sigma, D_n \subseteq C, \mu(C) \le \lambda]\,.$$

Choose $D_{n+1} \in \Sigma$ such that

$$D_n \subseteq C_{n+1} \quad \text{and} \quad \lambda_n - \frac{1}{n} \leq \mu(D_{n+1}) \leq \lambda_n . \tag{2.1.3}$$

It holds $0 < \lambda_{n+1} \leq \lambda_n \leq \lambda$ and so $\lim_{n\to\infty} \lambda_n = \hat{\lambda}$ exists and $\hat{\lambda} \leq \lambda$. We define

$$\hat{D} = \bigcup_{n\geq 1} D_n . \tag{2.1.4}$$

This implies, due to (2.1.3) and Proposition 2.1.24 (e), that

$$\mu(\hat{D}) = \lim_{n\to\infty} \mu(D_n) = \hat{\lambda} . \tag{2.1.5}$$

We need to show that $\hat{\lambda} = \lambda$. If $\hat{\lambda} < \lambda$, then $\mu(X \setminus \hat{D}) = \mu(X) - \mu(\hat{D}) > \lambda - \hat{\lambda} > 0$; see Proposition 2.1.24 (b). Reasoning as in the first part of the proof with X replaced by $X \setminus \hat{D}$ and λ replaced by $\lambda - \hat{\lambda} > 0$, we produce

$$C \in \Sigma , \quad C \subseteq X \setminus \hat{D} \quad \text{and} \quad 0 < \mu(C) < \lambda - \hat{\lambda} . \tag{2.1.6}$$

Then, the subadditivity yields $\hat{\lambda} = \mu(\hat{D}) < \mu(C \cup \hat{D}) \leq \lambda$, which gives, because of (2.1.5) and (2.1.6), that $\lambda_n < \mu(C \cup \hat{D})$ for all sufficiently large $n \in \mathbb{N}$. But $D_n \subseteq C \cup \hat{D}$ for all $n \in \mathbb{N}$; see (2.1.4). This contradicts the definition of λ_n for large enough $n \in \mathbb{N}$. We conclude that $\hat{\lambda} = \lambda$ and the proof is finished. □

The notion of outer measure is an abstract generalization of the "outer area" when we apply the exhaustion method of Archimedes to calculate the area of a bounded region in $\mathbb{R}^2$.

Definition 2.1.33. Let X be a nonempty set and $\mu^*: 2^X \to [0, +\infty]$ be a set function. We say that μ^* is an **outer measure** if it satisfies the following conditions:
(a) $\mu^*(\emptyset) = 0$;
(b) μ^* is monotone, that is, $A \subseteq B$ implies $\mu^*(A) \leq \mu^*(B)$;
(c) μ^* is σ-subadditive, that is, $\mu^*(\bigcup_{n\geq 1} A_n) \leq \sum_{n\geq 1} \mu^*(A_n)$.

We say that the outer measure μ^* is finite (resp. σ-finite) if $\mu^*(X) < +\infty$ (resp. $X = \bigcup_{n\geq 1} X_n$ and $\mu^*(X_n) < +\infty$ for all $n \in \mathbb{N}$).

A way to produce an outer measure is to start with a family of elementary sets on which a measure is naturally defined (for example intervals in $\mathbb{R}$ and rectangles in $\mathbb{R}^2$) and approximate any set from above by countable unions of such elementary sets. This process is formalized in the proposition that follows.

Proposition 2.1.34. *If X is a nonempty set, $\mathcal{L} \subseteq 2^X$ is such that $\emptyset, X \in \mathcal{L}$, $\vartheta: \mathcal{L} \to [0, +\infty]$ satisfies $\vartheta(\emptyset) = 0$ and for any $A \in \mathcal{L}$ we set*

$$\mu^*(A) = \inf\left[\sum_{n\geq 1} \vartheta(E_n) : E_n \in \mathcal{L}, A \subseteq \bigcup_{n\geq 1} E_n\right], \tag{2.1.7}$$

then μ^* is an outer measure.

Proof. First note that in (2.1.7) the infimum is taken over by a nonempty set since $A \subseteq X$ and by hypothesis, $X \in \mathcal{L}$. Moreover, $\mu^*(\emptyset) = 0$ and it is clear from (2.1.7) that $A \subseteq B$ implies $\mu^*(A) \leq \mu^*(B)$. Finally we show the σ-additivity of μ^*. So, let $\{A_k\} \subseteq 2^X$ and $\varepsilon > 0$. For each $k \in \mathbb{N}$ we can find $\{E_n^k\}_{n\geq 1} \subseteq \mathcal{L}$ such that

$$A_k \subseteq \bigcup_{n\geq 1} E_n^k \quad \text{and} \quad \sum_{n\geq 1} \vartheta(E_n^k) \leq \mu^*(A_k) + \frac{\varepsilon}{2^k} .$$

Let $A = \bigcup_{k\geq 1} A_k$. Then we have

$$A \subseteq \bigcup_{k,n\geq 1} E_n^k \quad \text{and} \quad \sum_{k,n\geq 1} \vartheta(E_n^k) \leq \sum_{k\geq 1} \mu^*(A_k) + \varepsilon .$$

This gives, due to (2.1.7), $\mu^*(A) \leq \sum_{k\geq 1} \mu^*(A_k) + \varepsilon$. Letting $\varepsilon \searrow 0$, we conclude that μ^* is σ-subadditive. Therefore μ^* is an outer measure. ☐

Example 2.1.35. Let $f: \mathbb{R} \to \mathbb{R}$ be an increasing function. Let $\mathcal{L}$ be the family of all intervals $(a, b]$ with $a, b \in \mathbb{R}$ and set $\vartheta((a, b]) = f(b) - f(a)$. Then the conditions in Proposition 2.1.34 are satisfied and by applying (2.1.7) we can define an outer measure μ^*. This outer measure is called the **Lebesgue–Stieltjes outer measure** and if $f(x) = x$ for all $x \in \mathbb{R}$ it is called the **Lebesgue outer measure.** Note that

$$\mu^*((a, b]) = f(b) - \lim_{x\to a^+} f(x) \leq f(b) - f(a) = \vartheta((a, b]) .$$

Thus, the inequality is strict at those points where f is not continuous from the right.

Now we will pass from outer measures to measures. Outer measures, although defined on the entire power set 2^X have the disadvantage that they are not σ-additive. However, when restricted to a particular subset of 2^X, they become σ-additive. In this direction we need the following remarkable definition due to Carathéodory.

Definition 2.1.36. Let X be a nonempty set and μ^* is an outer measure on 2^X. We say that $A \subseteq X$ is **μ^*-measurable,** if $\mu^*(B) = \mu^*(B \cap A) + \mu^*(B \cap A^c)$ for all $B \subseteq X$, that is, A splits additively all sets in X.

Remark 2.1.37. From Definition 2.1.33 we know that it holds that

$$\mu^*(B) \leq \mu^*(B \cap A) + \mu^*(B \cap A^c) \quad \text{for all } B \subseteq X ,$$

due to the subadditivity property of the outer measure. In order to check the μ^*-measurability of a set $A \subseteq X$ it suffices to show that

$$\mu^*(B) \geq \mu^*(B \cap A) + \mu^*(B \cap A^c) \quad \text{for all } B \subseteq X \text{ with } \mu^*(B) < +\infty\,.$$

This definition of Carathéodory essentially says that the outer measure $\mu^*(A)$ of A is equal to its **inner measure** $\mu^*(X) - \mu^*(A^c)$. For this reason Definition 2.1.36 is the right one and leads to a σ-algebra on which μ^* is σ-additive, hence a measure. This is shown in the next theorem known as the "Carathéodory Theorem".

Theorem 2.1.38 (Carathéodory Theorem). *If X is a nonempty set and $\mu^*: 2^X \to [0, +\infty]$ is an outer measure, then the family Σ^* of all μ^*-measurable sets is a σ-algebra and $\mu = \mu^*|_{\Sigma^*}$ is a measure.*

Proof. The symmetric character of Definition 2.1.36 implies that Σ^* is closed under complementation.

Next let $A, E \in \Sigma^*$ and let $B \subseteq X$. We have

$$\begin{aligned}\mu^*(B) &= \mu^*(B \cap A) + \mu^*(B \cap A^c)\\ &= \mu^*(B \cap A \cap E) + \mu^*(B \cap A \cap E^c) + \mu^*(B \cap A^c \cap E) + \mu^*(B \cap A^c \cap E^c)\,.\end{aligned}$$

Note that $A \cup E = (A \cap E) \cup (A \triangle E) = (A \cap E) \cup (A \cap E^c) \cup (A^c \cap E)$. Hence, by the subadditivity,

$$\mu^*(B \cap (A \cup E)) \leq \mu^*(B \cap A \cap E) + \mu^*(B \cap A \cap E^c) + \mu^*(B \cap A^c \cap E)\,.$$

This implies

$$\mu^*(B \cap (A \cup E)) + \mu^*(B \cap (A \cup E)^c) \leq \mu^*(B)\,.$$

Hence, see Remark 2.1.37, $A \cup E \in \Sigma^*$ and thus, Σ^* is an algebra.

In addition, if $A, E \in \Sigma^*$ and $A \cap E = \emptyset$, then

$$\mu^*(A \cup E) = \mu^*((A \cup E) \cap A) + \mu^*((A \cup E) \cap A^c) = \mu^*(A) + \mu^*(E)$$

where we recall that $\mu^*(A \cap E) = 0$. This means that μ^* is additive on Σ^*.

Now we show that Σ^* is a σ-algebra. Let $\{A_n\}_{n\geq 1} \subseteq \Sigma^*$ and let $D = \bigcup_{n\geq 1} A_n$. Since from the first part of the proof, we have

$$D_k = \bigcup_{n=1}^{k} A_n \in \Sigma^* \quad \text{and} \quad D_k \setminus \bigcup_{n=1}^{k-1} A_n \in \Sigma^* \quad \text{for all } k \in \mathbb{N}\,,$$

without any loss of generality we may assume that the sets $\{A_n\}_{n\geq 1} \subseteq \Sigma^*$ are mutually disjoint. For any $B \subseteq X$, since $D_n, A_n \in \Sigma^*$, we have for all $n \in \mathbb{N}$

$$\begin{aligned}\mu^*(B) &= \mu^*(B \cap D_n) + \mu^*(B \cap D_n^c)\\ &= \mu^*(B \cap A_n) + \mu^*\Big(B \cap \Big(\bigcup_{i \leq n-1} A_i\Big)\Big) + \mu^*(B \cap D_n^c)\,.\end{aligned}$$

Then, by induction on $n \in \mathbb{N}$, we show that

$$\mu^*(B) = \sum_{i=1}^{n} \mu^*(B \cap A_i) + \mu^*(B \cap D_n^c) \geq \sum_{i=1}^{n} \mu^*(B \cap A_i) + \mu^*(B \cap D^c)$$

since μ^* is additive and since $D_n \subseteq D$ for all $n \in \mathbb{N}$. We let $n \to \infty$ and obtain

$$\mu^*(B) \geq \sum_{i \geq 1} \mu^*(B \cap A_i) + \mu^*(B \cap D^c) \geq \mu^*(B \cap D) + \mu^*(B \cap D^c)$$

by the σ-subadditivity; see Definition 2.1.36. This implies that $D \in \Sigma^*$ (see Remark 2.1.37) and $\mu^*(B) = \sum_{i \geq 1} \mu(B \cap A_i) + \mu(B \cap D^c)$.

Let $B = D \subseteq X$. Then $\mu^*(D) = \sum_{i \geq 1} \mu^*(A_i)$ and so we conclude that Σ^* is a σ-algebra and $\mu = \mu^*|_{\Sigma^*}$ is a measure. □

Definition 2.1.39. Let (X, Σ, μ) be a measure space.

(a) A set $A \in \Sigma$ is said to be **μ-null** (or simply **null** if μ is clearly understood) if $\mu(A) = 0$.

(b) We say that μ is **complete** if Σ contains all subsets of null sets.

Remark 2.1.40. If A is μ-null and $B \subseteq A$, then $\mu(B) = 0$, provided $B \in \Sigma$. But in general it need not be the case that $B \in \Sigma$. For example this is the case with the Borel σ-algebra $\mathcal{B}(\mathbb{R})$. However, completeness can always be achieved by simply extending the domain of the measure. This is done in the next proposition whose proof is straightforward and so it is omitted.

Proposition 2.1.41. *If (X, Σ, μ) is a measure space, $\mathcal{N} = \{D \in \Sigma : \mu(D) = 0\}$, $\Sigma_\mu = \{A \cup E : A \in \Sigma, E \subseteq D \in \mathcal{N}\}$ and $\overline{\mu}(A \cup E) = \mu(A)$ for all $A \cup E \in \Sigma_\mu$, then Σ_μ is a σ-algebra and $\overline{\mu}$ is a complete measure on Σ_μ.*

Let (X, Σ^*, μ) be the measure space produced in Theorem 2.1.38.

Proposition 2.1.42. *(X, Σ^*, μ) is a complete measure space.*

Proof. Assume that $\mu^*(A) = 0$. Then, by the subadditivity, the monotonicity, and since $\mu^*(A) = 0$, for any $B \subseteq X$, we have

$$\mu^*(B) \leq \mu^*(B \cap A) + \mu^*(B \cap A^c) \leq \mu^*(B \cap A^c) \leq \mu^*(B).$$

This gives $A \in \Sigma^*$ and so $\mu = \mu^*|_{\Sigma^*}$ is complete. □

Now let X be a set and let $\mathcal{L} \subseteq 2^X$ be a semiring. We consider a σ-additive set function $\mu: \mathcal{L} \to [0, +\infty]$. Applying Proposition 2.1.34, we can define the outer measure $\mu^*: 2^X \to [0, +\infty]$ corresponding to μ. It holds that $\mu^*(A) = \mu(A)$ for all $A \in \mathcal{L}$. We have the following result.

Proposition 2.1.43. *If $\mathcal{D}$ is a semiring satisfying $\mathcal{L} \subseteq \mathcal{D} \subseteq \Sigma^*$, then μ^* is the unique extension of μ to a σ-additive set function on $\mathcal{D}$.*

Proof. Let $\lambda\colon \mathcal{D} \to [0, +\infty]$ be a σ-additive extension of μ on $\mathcal{D}$ and let λ^* be the corresponding outer measure; see Proposition 2.1.34. If $A \subseteq X$ and $\{E_n\}_{n\geq 1} \subseteq \mathcal{L}$ are such that $A \subseteq \bigcup_{n\geq 1} E_n$, then

$$\lambda^*(A) \leq \sum_{n\geq 1} \lambda^*(E_n) = \sum_{n\geq 1} \lambda(E_n) = \sum_{n\geq 1} \mu(E_n) .$$

This implies

$$\lambda^*(A) \leq \mu^*(A) \quad \text{for every } A \subseteq X . \tag{2.1.8}$$

In order to show that $\lambda = \mu^*$ on $\mathcal{D}$, it suffices to show that $\mu^*(A) \leq \lambda(A)$ for all $A \in \mathcal{D}$ with $\mu^*(A) < +\infty$. Recall that μ is σ-additive. Fix $A \in \mathcal{D}$ with $\mu^*(A) < +\infty$ and $\varepsilon > 0$. Consider $\{E_n\}_{n\geq 1} \subseteq \mathcal{L}$ such that

$$A \subseteq \bigcup_{n\geq 1} E_n \quad \text{and} \quad \sum_{n\geq 1} \mu(E_n) \leq \mu^*(A) + \varepsilon ; \tag{2.1.9}$$

see Proposition 2.1.34. Taking Problem 2.2 into account we find pairwise disjoint $\{C_n\}_{n\geq 1} \subseteq \mathcal{L}$ such that

$$\hat{E} = \bigcup_{n\geq 1} E_n = \bigcup_{n\geq 1} C_n \in \sigma(\mathcal{D}) .$$

We know that $\mu^*|_{\sigma(\mathcal{D})}$ and $\lambda^*|_{\sigma(\mathcal{D})}$ are both measures that coincide with μ on $\mathcal{L}$. Therefore

$$\mu^*(\hat{E}) = \sum_{n\geq 1} \mu^*(C_n) = \sum_{n\geq 1} \mu(C_n) = \sum_{n\geq 1} \lambda(C_n) = \lambda^*(\hat{E}) . \tag{2.1.10}$$

Moreover, because of (2.1.8) and (2.1.9) as well as the σ-subadditivity of μ^* and since $\mu^*|_{\mathcal{L}} = \mu$, we have

$$\lambda^*(\hat{E} \setminus A) \leq \mu^*(\hat{E} \setminus A) = \mu^*(\hat{E}) - \mu^*(A) \leq \sum_{n\geq 1} \mu(E_n) - \mu^*(A) \leq \varepsilon . \tag{2.1.11}$$

Hence $\mu^*(A) \leq \mu^*(\hat{E}) = \lambda^*(\hat{E}) = \lambda(A) + \lambda^*(\hat{E} \setminus A) \leq \lambda(A) + \varepsilon$; see (2.1.10) and (2.1.11). Letting $\varepsilon \searrow 0$, we obtain $\mu^*(A) \leq \lambda(A)$. Therefore, $\lambda(A) = \mu^*(A)$ for all $A \in \mathcal{D}$. ☐

The Lebesgue measure on $\mathbb{R}$ was the starting point of "Measure Theory". So, let us look in some detail at how we can produce it using the previous abstract theory. To this end, we introduce

$$\mathcal{L} = \{(a, b] : a \leq b, a, b \in \mathbb{R}\}$$

with $(a, a] = \emptyset$. This is a semiring of subsets of $\mathbb{R}$. Let $\lambda\colon \mathcal{L} \to [0, +\infty]$ be the set function defined by $\lambda((a, b]) = b - a$. This set function is σ-additive and σ-finite. Using Proposition 2.1.43, we know that λ has a unique extension to $\Sigma^* = \Sigma_\lambda$ being the σ-field of λ^*-measurable sets; see Definition 2.1.36. We continue to denote this extension by λ. Then

- λ is the Lebesgue measure on $\mathbb{R}$.
- $\Sigma^* = \Sigma_\lambda$ is the σ-algebra of the Lebesgue measurable subsets of $\mathbb{R}$.

Note that λ is translation invariant, that is $\lambda(A) = \lambda(A + x)$ for all $A \in \Sigma_\lambda$ and for all $x \in \mathbb{R}$. Moreover, we have $\lambda(\theta A) = |\theta|\lambda(A)$ for all $A \in \Sigma_\lambda$ and for all $\theta \in \mathbb{R}$.

From the previous discussion it is not clear if $\Sigma_\lambda = 2^{\mathbb{R}}$. In fact the next theorem shows that this is not the case. Indeed there are subsets of $\mathbb{R}$ that are not Lebesgue measurable.

Theorem 2.1.44. *There is no translation invariant measure defined on all of $2^{\mathbb{R}}$, which assigns to every interval its length.*

Proof. We will define a subset of $\mathbb{R}$, which is not Lebesgue measurable. On $\mathbb{R}$ we consider the following equivalence relation

$$x \sim u \quad \text{if and only if} \quad x - u \in \mathbb{Q} .$$

Choose a single element $x \in [0, 1]$ from every equivalence class formed by $\sim$. Here we assume that the Axiom of Choice holds. Let $A \subseteq [0, 1]$ be the set formed by these representatives. Suppose that $A \in \Sigma_\lambda$. Then by translation invariance we have that $\{A + r\}_{r \in \mathbb{Q}}$ is a countable, Lebesgue measurable partition of $\mathbb{R}$ with $\lambda(A + r) = \eta$ independent of $r \in \mathbb{Q}$. If $\eta = 0$, then we have a contradiction to the fact that $\lambda(\mathbb{R}) = +\infty$. If $\eta > 0$, then, with $D = \mathbb{Q} \cap [0, 1]$, we obtain $2 = \lambda([0, 2]) = \sum_{r \in D} \lambda(A + r) = +\infty$, again a contradiction. Hence, $A \notin \Sigma_\lambda$. □

In general the measure theoretic and topological properties of sets in $\mathbb{R}$ differ.

Example 2.1.45. Singletons have a Lebesgue measure of zero. Hence, $\lambda(\mathbb{Q}) = 0$. Let $\{r_n\}_{n \ge 1} \subseteq [0, 1]$ be an enumeration of the rationals in $[0, 1]$. Let $I_n = (r_n - \varepsilon/2^n, r_n + \varepsilon/2^n)$ and let $U = (0, 1) \cap (\bigcup_{n \ge 1} I_n)$. Evidently, $U \subseteq [0, 1]$ is open and dense, so topologically "large". On the other hand we have $\lambda(U) \le \sum_{n \ge 1} \varepsilon/2^n = \varepsilon$. Hence, U is measure theoretically "small". Similarly, $C = [0, 1] \setminus U$ is nowhere dense and closed, thus topologically small, but $\lambda(C) \ge 1 - \varepsilon$, thus it is measure theoretically "large".

The Cantor set will help us to get an idea on what the relation is between $\mathcal{B}(\mathbb{R})$ and Σ_λ.

Example 2.1.46. The **Cantor set** is constructed as follows. Let $C_0 = [0, 1]$. We trisect $[0, 1]$ and remove the open middle third $(1/3, 2/3)$. We set $C_1 = [0, 1/3] \cup [2/3, 1]$. Then we trisect each of the two intervals of C_1 and remove the open middle thirds. We obtain $C_2 = [0, 1/9] \cup [2/9, 1/3] \cup [2/3, 7/9] \cup [8/9, 1]$. We proceed inductively. So, suppose we have C_n. This consists of 2^n closed intervals. We trisect each one of them and remove the open middle thirds. The remaining part of C_n is the set C_{n+1}, which is the union of 2^{n+1} disjoint closed intervals. Evidently $\{C_n\}_{n\geq 1}$ is decreasing. Then the Cantor set C of $[0,1]$ is defined by $C = \bigcap_{n\geq 1} C_n$. This set consists of those points $x \in [0,1]$, which in base -3 have an expansion $x = \sum_{k\geq 1} a_k 1/3^k$ with $a_k \neq 1$ for all $k \in \mathbb{N}$.

Proposition 2.1.47. *The Cantor set C has the following properties:*
(a) *C is compact and nowhere dense.*
(b) $\lambda(C) = 0$.
(c) $\text{card}(C) = \mathfrak{c} =$ *the cardinality of the continuum*.

Proof. (a) Clearly C is closed since it is the intersection of closed sets. Hence C is compact. Moreover, $\text{int}\, C = \emptyset$ as it contains no interval since at each stage, each interval has length $1/3^n$. Therefore, C is nowhere dense.

(b) At each stage we remove 2^{n-1} open intervals each one of length $1/3^n$. Therefore the total measure of the removed set at the nth step is $2^{n-1}/3^n$. Hence, we have

$$\lambda([0,1] \setminus C) = \sum_{n\geq 1} \frac{2^{n-1}}{3^n} = \frac{1}{2} \sum_{n\geq 1} \left(\frac{2}{3}\right)^n = 1 .$$

Thus, $\lambda(C) = 0$.

(c) Let $x \in C$. Then $x = \sum_{k\geq 1} a_k/3^k$ with $a_k = 0$ or $a_k = 2$ for all $k \in \mathbb{N}$. Let $f(x) = \sum_{k\geq 1} c_k/2^k$ with $c_k = a_k/2$ for all $k \in \mathbb{N}$, the base -2 expansion of $x \in C$. Hence, $f\colon C \to [0,1]$ is onto, thus $\text{card}(C) = \mathfrak{c}$. □

Remark 2.1.48. The Cantor set is interesting because it is "large" from the cardinality point of view but negligible from the measure theoretic point of view. We can generalize the above construction and have "Cantor-like sets" that still satisfy (a) and (c) from Proposition 2.1.47. So, let I be a bounded interval and $\vartheta \in (0,1)$. We call the open interval with the same midpoint as I and length $\vartheta\lambda(I)$ the **open middle** ϑ. Now let $\{\vartheta_k\}_{k\geq 1} \subseteq (0,1)$ and produce a decreasing sequence $\{\hat{C}_k\}_{k\geq 1}$ of closed sets in $[0,1]$ as follows: $\hat{C}_0 = [0,1]$ and $\hat{C}_k$ is produced by removing the open middle ϑ_k from each component interval of $\hat{C}_{k-1}$. We set $\hat{C} = \bigcap_{k\geq 1} \hat{C}_k$. We still have that $\hat{C}$ is compact and nowhere dense and $\text{card}(\hat{C}) = \mathfrak{c}$. Concerning the Lebesgue measure, note that $\lambda(\hat{C}_k) = (1-\vartheta_k)\lambda(\hat{C}_{k-1})$ for all $k \geq 2$. So, $\lambda(\hat{C}) = \prod_{k\geq 1}(1-\vartheta_k) = \lim_{n\to\infty} \prod_{k=1}^{n}(1-\vartheta_k)$. If $\vartheta_k = \vartheta \in (0,1)$ for all $k \in \mathbb{N}$, then $\lambda(\hat{C}) = 0$. Note that the Cantor set corresponds to the particular case of $\vartheta = 1/3$. If $\vartheta_k \to 0$ sufficiently fast as $k \to \infty$, then $\lambda(\hat{C}) > 0$. In particular, $\prod_{k\geq 1}(1-\vartheta_k) > 0$ if and only if $\sum_{k\geq 1} \vartheta_k < +\infty$. We point out that part (c) of the proposition above implies that there are $2^{\mathfrak{c}}$ Lebesgue measurable subsets of $\mathbb{R}$. On the other hand $\text{card}(\mathcal{B}(\mathbb{R})) = \mathfrak{c}$. So, there are many more Lebesgue measurable sets than Borel sets in $\mathbb{R}$ although it is not easy to produce a set that is Lebesgue measurable but not a Borel set. For such a concrete set we refer to Federer [120, p. 68].

2.2 Measurable Functions – Integration

The Lebesgue integral is defined for measurable functions. For this reason we start this section with a discussion of measurable functions.

Definition 2.2.1. Let (X, Σ) and $(Y, \mathcal{L})$ be two measurable spaces and $f: X \to Y$ be a map. We say that f is $(\Sigma, \mathcal{L})$**-measurable** if $f^{-1}(A) \in \Sigma$ for all $A \in \mathcal{L}$. If X, Y are Hausdorff topological spaces, then they become measurable spaces by considering their Borel σ-algebras $\mathcal{B}(X)$, $\mathcal{B}(Y)$ and then f is said to be **Borel measurable** (or simply a **Borel function**). When $Y = \mathbb{R}$ or $Y = \mathbb{R}^*$ we always use the Borel σ-field of Y.

Remark 2.2.2. The reason that we use the Borel σ-algebra on $\mathbb{R}$ as range space is that the Lebesgue σ-algebra Σ_λ, as the completion of $\mathcal{B}(\mathbb{R})$, is in general too large for the Lebesgue measure; see Remark 2.1.48. In particular, there exists a continuous, nondecreasing function $h: [0,1] \to [0,1]$ and a Lebesgue measurable set $C \subseteq [0,1]$ such that $h^{-1}(C)$ is not Lebesgue measurable (assuming the Axiom of Choice). In fact $h(x) = 1/2[\hat{f}(x) + x]$ with $\hat{f}$ being the function from the proof of Proposition 2.1.47 (c) extended to all of $[0,1]$ by declaring it to be constant on each interval missing from C. Then $\hat{f}$ is nondecreasing and continuous and is known as the **Cantor function.**

Proposition 2.2.3. *If* (X, Σ) *and* $(Y, \mathcal{L})$ *are measurable spaces,* $\mathcal{L} = \sigma(\mathfrak{a})$ *and* $f: X \to Y$*, then* f *is* $(\Sigma, \mathcal{L})$*-measurable if and only if* $f^{-1}(A) \in \Sigma$ *for all* $A \in \mathfrak{a}$.

Proof. $\Longrightarrow$: This is immediate from Definition 2.2.1.

$\Longrightarrow$: Let $\mathcal{D} = \{A \subseteq Y : f^{-1}(A) \in \Sigma\}$. Evidently $\mathcal{D} \supseteq \mathfrak{a}$ and $\mathcal{D}$ is a σ-algebra. Therefore, $\mathcal{D} \supseteq \sigma(\mathfrak{a}) = \mathcal{L}$ and this proves the $(\Sigma, \mathcal{L})$-measurability of f. □

Combining Propositions 2.1.18 and 2.2.3 we have the following result.

Proposition 2.2.4. *If* (X, Σ) *is a measurable space and* $f: X \to \mathbb{R}$*, then the following statements are equivalent:*
(a) f *is* Σ*-measurable;*
(b) $f^{-1}((a, +\infty)) \in \Sigma$ *for all* $a \in \mathbb{R}$*;*
(c) $f^{-1}([a, +\infty)) \in \Sigma$ *for all* $a \in \mathbb{R}$*;*
(d) $f^{-1}((-\infty, a]) \in \Sigma$ *for all* $a \in \mathbb{R}$*;*
(e) $f^{-1}((-\infty, a)) \in \Sigma$ *for all* $a \in \mathbb{R}$.

Remark 2.2.5. In case f is $\mathbb{R}^*$-valued, we need to add the requirement that $f^{-1}(\pm\infty) \in \Sigma$ in the statements (b)–(e). Evidently we can take $a \in \mathbb{Q}$ in (b)–(e).

Immediately from Definition 2.2.1, we have that the composition preserves measurability.

Proposition 2.2.6. *If* (X, Σ)*,* $(Y, \mathcal{L})$*,* $(Z, \mathcal{D})$ *are measurable spaces and* $f: X \to Y$*,* $g: Y \to Z$ *are measurable maps, then* $h = g \circ f: X \to Z$ *is measurable as well.*

Moreover, we have the following as a consequence of Proposition 2.2.3.

Proposition 2.2.7. *If X, Y are Hausdorff topological spaces and $f: X \to Y$ is continuous, then f is Borel measurable.*

Proposition 2.2.8. *If (X, Σ) is a measurable space and $f, g: X \to \mathbb{R}$ are Σ-measurable functions, then $f \pm g$ and fg are both Σ-measurable.*

Proof. If $f(x) + g(x) < a$, then $f(x) < a - g(x)$. Let $c \in \mathbb{Q}$ be such that $f(x) < c < a - g(x)$. So, we have that

$$\begin{aligned} &\{x \in X: f(x) + g(x) < a\} \\ &= \bigcup_{c \in \mathbb{Q}} \Big[\{x \in X: f(x) < c\} \bigcap \{x \in X: g(x) < a - c\}\Big] \in \Sigma . \end{aligned}$$

Hence $f + g$ is Σ-measurable.

Since $-g$ is Σ-measurable, if g is, it follows that $f - g$ is Σ-measurable as well.

For any $h: X \to \mathbb{R}$ being Σ-measurable and $a \geq 0$, we have

$$\{x \in X: h(x)^2 > a\} = \{x \in X: h(x) > a^{\frac{1}{2}}\} \bigcup \{x \in X: h(x) < -a^{\frac{1}{2}}\} \in \Sigma .$$

Therefore h^2 is Σ-measurable.

Since $fg = 1/2[(f+g)^2 - f^2 - g^2]$ using the fact above and the Σ-measurability of $f+g$, we conclude that fg is Σ-measurable. □

Remark 2.2.9. The result above is also valid for R^*-valued functions, provided we always take the same value for $f \pm g$ at the points where it is undefined, that is, of the form $\infty - \infty$. In addition, recalling that we always define $0(\pm\infty) = 0$, the function fg is Σ-measurable for R^*-valued f and g.

Proposition 2.2.10. *If (X, Σ) is a measurable space and $f_n: \Sigma \to \mathbb{R}^*$ with $n \in \mathbb{N}$ are Σ-measurable, then*

$$\sup\{f_n\}_{n=1}^m, \quad \inf\{f_n\}_{n=1}^m, \quad \sup_{n\geq 1} f_n, \quad \inf_{n \geq 1} f_n, \quad \liminf_{n\to\infty} f_n, \quad \limsup_{n\to\infty} f_n$$

are all Σ-measurable.

Proof. Let $g(x) = \sup_{1\leq n\leq m} f_n(x)$. Then for all $a \in \mathbb{R}$, we have

$$\{x \in X: g(x) > a\} = \bigcup_{n=1}^{m} \{x \in X: f_n(x) > a\} \in \Sigma .$$

Thus g is Σ-measurable. Similarly, if $\hat{g}(x) = \sup_{n\geq 1} f_n(x)$, then for all $a \in \mathbb{R}$, we have

$$\{x \in X: \hat{g}(x) > a\} = \bigcup_{n\geq 1} \{x \in X: f_n(x) > a\} \in \Sigma .$$

In a similar fashion we also show that $\inf_{1\le n\le m} f_n$ and $\inf_{n\ge 1} f_n$ are both Σ-measurable.

Finally, recall that $\liminf_{n\to\infty} f_n = \sup_{k\ge 1}\inf_{n\ge k} f_n$ and $\limsup_{n\to\infty} f_n = \inf_{k\ge 1}\sup_{n\ge k} f_n$, to conclude that both are Σ-measurable. □

When a sequence of measurable functions does not converge pointwise, we can still have the measurability of the set of points where pointwise convergence occurs.

Proposition 2.2.11. *If (X, Σ) is a measurable space and $f_n: X \to \mathbb{R}$ with $n \ge 1$ is a sequence of Σ-measurable functions, then the set $C = \{x \in X: \lim_{n\to\infty} f_n(x)$ exists$\} \in \Sigma$.*

Proof. Given $x \in C$, we have that $\{f_n(x)\}_{n\ge 1} \subseteq \mathbb{R}$ is a Cauchy sequence. So, for $\varepsilon = 1/n$ with $n \in \mathbb{N}$ we can find $m = m(\varepsilon) \in \mathbb{N}$ such that

$$|f_{m+k}(x) - f_m(x)| < \frac{1}{n} \quad \text{for all } k \in \mathbb{N}\,.$$

Therefore it follows

$$\begin{aligned} C &= \left\{x \in X: \forall n \in \mathbb{N}\ \exists m \in \mathbb{N} \text{ such that } |f_{m+k}(x) - f_m(x)| < \frac{1}{n}\ \forall k \in \mathbb{N}\right\} \\ &= \bigcap_{n\ge 1}\bigcup_{m\ge 1}\bigcap_{k\ge 1}\left\{x \in X: |f_{m+k}(x) - f_m(x)| < \frac{1}{n}\right\} \in \Sigma\,. \end{aligned}$$ □

In Proposition 2.2.10 we saw that the pointwise limit of Σ-measurable, $\mathbb{R}^*$-valued functions is Σ-measurable as well. This result can be extended to maps with values in a metric space.

Proposition 2.2.12. *If (X, Σ) is a measurable space, Y is a metrizable space and $f_n: X \to Y$ with $n \in \mathbb{N}$ is a sequence of Σ-measurable functions such that $f_n(x) \to f(x)$ in Y for all $x \in X$, then f is Σ-measurable as well.*

Proof. Let $C \subseteq Y$ be a closed set. According to Proposition 2.2.3 it suffices to show that $f^{-1}(C) \in \Sigma$. Let d be a compatible metric on Y. Let $U_n = \{y \in Y: d(y, C) < 1/n\}$ with $n \in \mathbb{N}$. These sets are open and $C = \bigcap_{n\ge 1} U_n$; see Proposition 1.5.8. Let $x \in f^{-1}(C)$. Then $f(x) \in C$ and $f_n(x) \to f(x)$ in Y. Since for each $n \in \mathbb{N}$, U_n is a neighborhood of $f(x)$ there exists $m \in \mathbb{N}$ such that $f_k(x) \in U_n$ for all $k \ge m$, which implies

$$x \in \bigcap_{n\ge 1}\bigcup_{m\ge 1}\bigcap_{k\ge m} f_k^{-1}(U_n)\,.$$

This yields

$$f^{-1}(C) \subseteq \bigcap_{n\ge 1}\bigcup_{m\ge 1}\bigcap_{k\ge m} f_k^{-1}(U_n)\,. \tag{2.2.1}$$

Next suppose that $x \in \bigcap_{n\ge 1}\bigcup_{m\ge 1}\bigcap_{k\ge m} f_k^{-1}(U_n)$. So for every $n \in \mathbb{N}$, $f_k(x)$ is eventually in U_n, hence $f(x) = \lim_{k\to\infty} f_k(x) \in \overline{U}_n$. Therefore $f(x) \in \bigcap_{n\ge 1}\overline{U}_n$. But $\overline{U}_{n+1} \subseteq U_n$. Hence $f(x) \in \bigcap_{n\ge 1} U_n = C$, which gives $x \in f^{-1}(C)$. Hence

$$\bigcap_{n\geq 1}\bigcup_{m\geq 1}\bigcap_{k\geq m} f_k^{-1}(U_n) \subseteq f^{-1}(C)\,. \tag{2.2.2}$$

From (2.2.1) and (2.2.2) it follows that

$$f^{-1}(C) = \bigcap_{n\geq 1}\bigcup_{m\geq 1}\bigcap_{k\geq m} f_k^{-1}(U_n) \in \Sigma\,.$$

Thus, f is Σ-measurable. □

Remark 2.2.13. The result above fails if Y is not metrizable. To see this let $Y = I^I$ with $I = [0, 1]$ furnished with the product topology. Then Y is compact by Tychonoff's Theorem (see Theorem 1.4.56), but it is not metrizable. Let $f_n: I \to Y$ with $n \in \mathbb{N}$ be the sequence of maps defined by

$$f_n(x)(t) = [1 - n|x - t|]^+ \quad \text{for all } x, t \in I\,.$$

Note that each $f_n: I \to Y$ is continuous, thus Borel measurable. In addition, $f_n(x)(t) \to \chi_{\{x\}}(t)$ for all $t \in I$. Here

$$\chi_{\{x\}}(t) = \begin{cases} 1 & \text{if } t = x\,,\\ 0 & \text{if } t \neq x \end{cases}$$

is the indicator function of the singleton $\{x\}$.

For each $x \in I$ there exists an open set $U_x \subseteq Y$ such that $f^{-1}(U_x) = \{x\}$ (for example, let $U_x = \{f \in Y = I^I : f(x) > 0\}$). Let $D \subseteq I$ be a non-Borel set and let $V = \bigcup_{x\in D} U_x$. Evidently $V \subseteq I^I$ is open and $f^{-1}(V) = D$. This shows that f is not measurable.

Definition 2.2.14. Let (X, Σ, μ) be a measure space. A statement about $x \in X$ is said to hold **almost everywhere** or **a. e.** (for **almost all** x or **a. a.** $x \in X$) if it holds for all $x \notin D$ with $\mu(D) = 0$. Note that the set of all $x \in X$ for which the statement holds will be in Σ_μ but not necessarily in Σ.

Measurability is not affected by changing the function on a μ-null set.

Proposition 2.2.15. *If (X, Σ, μ) is a complete measure space, $(Y, \mathcal{L})$ is a measurable space, $f: X \to Y$ is $(\Sigma, \mathcal{L})$-measurable and $g: X \to Y$ satisfies $f(x) = g(x)$ for μ-a. a. $x \in X$, then g is $(\Sigma, \mathcal{L})$-measurable as well.*

Next we will introduce the functions, which are the building blocks for the theory of integration.

Definition 2.2.16. Let (X, Σ) be a measurable space.

(a) Given $A \subseteq X$, the **characteristic function** χ_A of A is defined by

$$\chi_A(x) = \begin{cases} 1 & \text{if } x \in A\,,\\ 0 & \text{if } x \notin A\,. \end{cases}$$

(b) A **simple function** is a measurable function $s\colon X \to \mathbb{R}$, which has finite range. So, if $a_1, \dots, a_n$ are the distinct values of s, then we can write $s(x) = \sum_{k=1}^{n} a_k \chi_{A_k}(x)$ with $A_k = \{x \in X\colon s(x) = a_k\} \in \Sigma$. We call this the **standard representation** of s.

Remark 2.2.17. Since in probability theory a characteristic function is a Fourier transform, probabilists use the name indicator function and denote it by i_A. On the other hand, in nonsmooth analysis and optimization, this name and symbol are reserved for another function, namely

$$i_A(x) = \begin{cases} 0 & \text{if } x \in A\,, \\ +\infty & \text{if } x \notin A\,. \end{cases}$$

A simple function is a linear combination with distinct coefficients of characteristic functions of disjoint sets whose union is X. One of the coefficients a_k may well be zero, but still the term $a_k \chi_{A_k}$ is implicitly understood in the standard representation so as to have $X = \bigcup_{k=1}^{n} A_k$. If s and τ are simple functions, then so are $s + \tau$ and $s\tau$.

Simple functions approximate measurable functions.

Proposition 2.2.18. *If (X, Σ) is a measurable space and $f\colon \to [0, +\infty]$ is a Σ-measurable function, then there exists a sequence $\{s_n\}_{n\geq 1}$ of simple functions on X such that*

$$0 \leq s_1(x) \leq s_2(x) \leq \cdots \leq s_n(x) \to f(x) \quad \textit{for all } x \in X \textit{ as } n \to \infty\,.$$

Moreover the convergence is uniform on any set on which f is bounded from above.

Proof. Given $n \in \mathbb{N}$ we partition the interval $[0, n)$ into $n2^n$ half-open intervals of length $1/2^n$. Then for each $1 \leq k \leq n2^n$ with $k \in \mathbb{N}$ we define

$$D_{n,k} = \left\{x \in X\colon \frac{k-1}{2^n} \leq f(x) < \frac{k}{2^n}\right\}, \quad D_n = \{x \in X\colon f(x) \geq n\}\,.$$

The Σ-measurability of f implies that $D_{n,k}, D_n \in \Sigma$. We set

$$s_n = \sum_{k=1}^{n2^n} \frac{k-1}{2^n} \chi_{D_{n,k}} + n\chi_{D_n}\,.$$

Evidently this is a simple function for every $n \in \mathbb{N}$. Let $x \in D_{n,k}$. Then

$$\frac{2k-2}{2^{n+1}} \leq f(x) < \frac{2k}{2^{n+1}}\,,$$

which implies that $s_{n+1}(x) = (2k-2)/2^{n+1}$ or $s_{n+1}(x) = (2k-1)/2^{n+1}$. Hence $s_n(x) \leq s_{n+1}(x)$.

Now let $x \in D_n$. Then $f(x) \geq n$ and we have $f(x) \geq n+1$ or $n \leq f(x) < n+1$. If the first case holds, then $s_{n+1}(x) \geq n+1 > n = s_n(x)$. In the second case, let $k \in \{1, \dots, (n+1)2^{n+1}\}$

such that $(k-1)/2^{n+1} \le f(x) < k/2^{n+1}$. Since $f(x) > n$ it follows that $k/2^{n+1} > n$, hence $k = (n+1)2^{n+1}$. Therefore, $s_{n+1}(x) = n + 1 - 1/2^{n+1} > n = s_n(x)$. This proves that $s_n \le s_{n+1}$.

Now we prove the pointwise convergence. So, fix $x \in X$ such that $f(x) \in [0, +\infty)$ and let $n > f(x)$. Then

$$0 \le f(x) - f_n(x) < \frac{1}{2^n},$$

which gives $f_n(x) \to f(x)$ as $n \to \infty$.

On the other hand, if $f(x) = +\infty$, then $f_n(x) = n \to +\infty$. Finally if $0 \le f(x) \le M$ for some $M > 0$ and for all $x \in X$, then (2.2) holds for every $x \in X$ provided $n > M$. Therefore $f_n \to f$ uniformly. □

If $f^+ = \max\{f, 0\}$ and $f^- = \min\{-f, 0\}$, then $f = f^+ - f^-$ as well as $|f| = f^+ + f^-$ and if $f: X \to \mathbb{R}$ is Σ-measurable, then so are f^+ and f^-; see Proposition 2.2.10. So using Proposition 2.2.18 on each of the functions f^+ and f^- we have the following.

Corollary 2.2.19. *If (X, Σ) is a measurable space and $f: X \to \mathbb{R}$ is Σ-measurable, then there exists a sequence $\{s_n\}_{n\ge1}$ of simple functions on X such that*

$$|s_1| \le |s_2| \le \cdots \le |s_n| \le \ldots |f| \ldots, \quad s_n(x) \to f(x) \quad \textit{for all } x \in X.$$

Moreover if f is bounded, then the convergence is uniform.

We can extend these results to maps with values in a separable metric space. This is useful when studying integration of Banach space-valued maps; see the Lebesgue–Bochner integral in Section 4.2.

Proposition 2.2.20. *If (X, Σ) is a measurable space, (Y, d) is a separable metric space and $f: X \to Y$, then the following hold:*

(a) *If (Y, d) is in addition totally bounded, then f is Σ-measurable if and only if it is the d-uniform limit of a sequence of simple functions with values in Y.*

(b) *f is Σ-measurable if and only if f is the d-pointwise limit of a sequence of simple functions with values in Y.*

Proof. (a) $\Longrightarrow$: Suppose that $f: X \to Y$ is Σ-measurable and let $\varepsilon > 0$. Since Y is by hypothesis totally bounded, there exists $y_1, \ldots, y_m \in Y$ such that $Y = \bigcup_{k=1}^m B_\varepsilon(y_k)$ with $B_\varepsilon(y_k) = \{y \in Y : d(y, y_k) < \varepsilon\}$. We set $A_1 = B_\varepsilon(y_1)$ and $A_{k+1} = B_\varepsilon(y_{k+1}) \setminus \bigcup_{i=1}^k B_\varepsilon(y_i)$ for all $k \in \{1, \ldots, m-1\}$. Then $\{A_k\}_{k=1}^m$ are mutually disjoint Borel sets in Y whose union is Y. We have

$$X = \bigcup_{k=1}^m f^{-1}(A_k) \quad \text{and} \quad f^{-1}(A_k) \cap f^{-1}(A_n) = \emptyset \quad \text{if } k \ne n.$$

We define $s: X \to Y$ by $s(x) = y_k$ if $x \in f^{-1}(A_k)$. Evidently s is a simple function and $d(s(x), f(x)) < \varepsilon$ for all $x \in X$. Therefore f is the d-uniform limit of a sequence of simple functions with values in Y.

$\Longleftarrow$: This is a consequence of Proposition 2.2.12.

(b) By Theorem 1.5.21 there is a homeomorphism (embedding) $\xi: Y \to \mathbb{H}$ onto a subset of the Hilbert cube $\mathbb{H} = [0,1]^{\mathbb{N}}$. Let $e(u,y) = d_{\mathbb{H}}(\xi(u), \xi(y))$ for all $u, y \in Y$. Then e is a metric on Y, compatible with d and (Y, e) is totally bounded. By part (a) we know that f is the e-uniform limit of a sequence of simple functions. Since e and d are topologically equivalent, we have that the sequence of simple functions is d-pointwise convergent to f. □

Definition 2.2.21. Let $\{(Y_\alpha, \mathcal{L}_\alpha)\}_{\alpha\in I}$ be a family of measurable spaces and $f_\alpha: X \to Y_\alpha$ be a map for each $\alpha \in I$. There is a unique σ-algebra on X with respect to which the f_α's are all measurable and this is the σ-algebra generated by the sets $f_\alpha^{-1}(A_\alpha)$ for all $A_\alpha \in \mathcal{L}_\alpha$ and all $\alpha \in I$. It is called the σ-algebra generated by $\{f_\alpha\}_{\alpha\in I}$ and is denoted by $\sigma(\{f_\alpha\})$.

Proposition 2.2.22. *If $(Y, \mathcal{L})$ is a measurable space, $f: X \to Y$ and $g: X \to \mathbb{R}$ are given maps, then g is $\sigma(f)$-measurable if and only if there exists a $\mathcal{L}$-measurable $h: Y \to \mathbb{R}$ such that $g = h \circ f$.*

Proof. $\Longrightarrow$: First we assume that g is a $\sigma(f)$-simple function. Then $g = \sum_{k=1}^{n} a_k \chi_{A_k}$ with $a_k \in \mathbb{R}$ and $A_k \in \sigma(f)$. For $k \in \{1, \dots, n\}$ let $C_k \in \mathcal{L}$ be such that $A_k = f^{-1}(C_k)$. We set $h = \sum_{k=1}^{n} a_k \chi_{C_k}$. Then h is a $\mathcal{L}$-simple function on Y and clearly $g = h \circ f$.

Now suppose that g is a general $\sigma(f)$-measurable function. Then by Corollary 2.2.19 there exists a sequence $\{s_n\}_{n\ge1}$ of $\sigma(f)$-simple functions such that $s_n(x) \to g(x)$ for all $x \in X$. From the first part of the proof we can find $h_n: Y \to \mathbb{R}$ with $n \in \mathbb{N}$ being $\mathcal{L}$-measurable functions such that $s_n = h_n \circ f$ with $n \in \mathbb{N}$. Let $E = \{y \in Y: \lim_{n\to\infty} h_n(y)$ exists in $\mathbb{R}\}$. Since $h_n(f(x)) = s_n(x) \to g(x)$ it follows that $f(X) \subseteq E$. Define

$$h(y) = \lim_{n\to\infty} h_n(y) \quad \text{if } y \in E \quad \text{and} \quad h(y) = 0 \quad \text{if } y \notin E\,.$$

From the inclusion $f(X) \subseteq E$ it follows that $g = h \circ f$. Moreover, from Proposition 2.2.11 we know that $E \in \mathcal{L}$. Hence $h_n\chi_E$ is $\mathcal{L}$-measurable and since $h_n\chi_E \to h\chi_E$ it follows that h is $\mathcal{L}$-measurable.

$\Longleftarrow$: This follows from Proposition 2.2.6. □

Definition 2.2.23. Let $\{(X_\alpha, \Sigma_\alpha)\}_{\alpha\in I}$ be a family of measurable spaces. Set $X = \prod_{\alpha\in I} X_\alpha$ and let $p_\alpha: X \to X_\alpha$ with $\alpha \in I$ be the corresponding projection (coordinate) maps. Then the **product σ-algebra** on X denoted by $\bigotimes_{\alpha\in I} \Sigma_\alpha$ is defined by $\bigotimes_{\alpha\in I} \Sigma_\alpha = \sigma(\{p_\alpha\})$.

Remark 2.2.24. Let (X, Σ), $(Y, \mathcal{L})$ be two measurable spaces. A set of the form $A \times B$ with $A \in \Sigma$, $B \in \mathcal{L}$ is said to be a **measurable rectangle** . By $\mathcal{R}$ we denote the family of measurable rectangles in $X \times Y$. It is easy to see that $\mathcal{R}$ is an algebra. Then $\Sigma \otimes \mathcal{L} = \sigma(\mathcal{R})$. More generally if the index set I is countable, then

$$\bigotimes_{\alpha\in I} \Sigma_\alpha = \sigma\Big(\prod_{\alpha\in I} A_\alpha : A_\alpha \in \Sigma_\alpha\Big)\,.$$

Proposition 2.2.25. *If $\{(X_\alpha, \Sigma_\alpha)\}_{\alpha\in I}$ are measurable spaces and each Σ_α is generated by $\mathfrak{a}_\alpha$, then $\bigotimes_{\alpha\in I}\Sigma_\alpha$ is generated by $\hat{\mathfrak{a}} = \{p_\alpha^{-1}(B_\alpha): B_\alpha \in \mathfrak{a}_\alpha, \alpha \in I\}$. Moreover, if the index set I is countable, then $\bigotimes_{\alpha\in I}\Sigma_\alpha$ is generated by $\tilde{\mathfrak{a}} = \{\prod_{\alpha\in I} B_\alpha: B_\alpha \in \mathfrak{a}_\alpha\}$.*

Proof. From Definition 2.2.23 it is clear that $\sigma(\hat{\mathfrak{a}}) \subseteq \bigotimes_{\alpha\in I}\Sigma_\alpha$. Let

$$\mathcal{D}_\alpha = \{B \subseteq X_\alpha: p_\alpha^{-1}(B) \in \sigma(\hat{\mathfrak{a}})\}, \quad \alpha \in I .$$

It is easy to see that $\mathcal{D}_\alpha$ is a σ-algebra and $\mathfrak{a}_\alpha \subseteq \mathcal{D}_\alpha$. Therefore $\Sigma_\alpha \subseteq \mathcal{D}_\alpha$ for all $\alpha \in I$. Hence $\bigotimes_{\alpha\in I}\Sigma_\alpha \subseteq \sigma(\hat{\mathfrak{a}})$ and so equality holds.

The second assertion follows from Remark 2.2.24. □

Proposition 2.2.26. *If $\{X_k\}_{k=1}^n$ are Hausdorff topological spaces, then the following hold:*
(a) $\bigotimes_{k=1}^n \mathcal{B}(X_k) \subseteq \mathcal{B}(\prod_{k=1}^n X_k)$;
(b) *If $\{X_k\}_{k=1}^n$ are second countable, then $\bigotimes_{k=1}^n \mathcal{B}(X_k) = \mathcal{B}(\prod_{k=1}^n X_k)$.*

Proof. (a) By Proposition 2.2.25, $\bigotimes_{k=1}^n \mathcal{B}(X_k)$ is generated by the sets $p_k^{-1}(U_k)$ with open $U_k \subseteq X_k$ for all $k \in \{1, \dots, n\}$. These sets are open in $X = \prod_{k=1}^n X_k$ and so, we infer that $\bigotimes_{k=1}^n \mathcal{B}(X_k) \subseteq \mathcal{B}(X)$.

(b) Let $\mathcal{D}_k$ be a countable basis of $X_k, k \in \{1, \dots, n\}$. Recall that every open set in X_k is a countable union of elements in $\mathcal{D}_k$. Therefore $\mathcal{B}(X)$ is generated by $\mathcal{D}_k$ and $\mathcal{B}(X)$ is generated by $\hat{\mathcal{D}} = \{\prod_{k=1}^n B_k: B_k \in \mathcal{D}_k\}$. Hence, we conclude that $\bigotimes_{k=1}^n \mathcal{B}(X_k) = \mathcal{B}(X)$. □

Definition 2.2.27. Let X, Y be nonempty sets and $A \subseteq X \times Y$. For each $x \in X$ and each $y \in Y$, the x**-section** of A (resp. the y**-section** of A) are defined by

$$A_x = \{y \in Y: (x, y) \in A\} \quad (\text{resp. } A^y = \{x \in X: (x, y) \in A\}) .$$

Clearly for every $x \in X$ and every $y \in Y$ we have $\emptyset_x = \emptyset^y = \emptyset$ and $(X \times Y)_x = Y$ as well as $(X \times Y)^y = X$.

Remark 2.2.28. If $\{A_\alpha\}_{\alpha\in I} \subseteq X \times Y$, then for all $x \in X$ and for all $y \in Y$ we have

$$\Big(\bigcup_{\alpha\in I} A_\alpha\Big)_x = \bigcup_{\alpha\in I}(A_\alpha)_x , \quad \Big(\bigcap_{\alpha\in I} A_\alpha\Big)_x = \bigcap_{\alpha\in I}(A_\alpha)_x ,$$
$$\Big(\bigcup_{\alpha\in I} A_\alpha\Big)^y = \bigcup_{\alpha\in I}(A_\alpha)^y , \quad \Big(\bigcap_{\alpha\in I} A_\alpha\Big)^y = \bigcap_{\alpha\in I}(A_\alpha)^y .$$

So, it follows that if $\mathcal{L}$ is a σ-algebra on X and $\mathcal{D} = \{A \subseteq X \times Y: A^y \in \mathcal{L} \text{ for all } y \in Y\}$, then $\mathcal{D}$ is a σ-algebra on $X \times Y$. Similarly for $\mathcal{F}$ being a σ-algebra on Y. Finally, if (X, Σ) and $(Y, \mathcal{L})$ are measurable spaces and $A \subseteq X \times Y$, then we say that A has **measurable sections** if for all $x \in X$ and for all $y \in Y$, $A_x \in \mathcal{L}$ and $A^y \in \Sigma$.

Proposition 2.2.29. *If (X, Σ) and $(Y, \mathcal{L})$ are measurable spaces and $A \in \Sigma \bigotimes \mathcal{L}$, then A has measurable sections.*

Proof. Let

$$\hat{\mathcal{D}} = \{A \subseteq X \times Y : A_x \in \mathcal{L} \text{ and } A^y \in \Sigma \text{ for all } x \in X \text{ and for all } y \in Y\}.$$

Then $\hat{\mathcal{D}}$ is a σ-algebra that contains measurable rectangles. Note that

$$(A \times B)_x = \begin{cases} B & \text{if } x \in A \\ \emptyset & \text{if } x \notin A \end{cases} \quad \text{and} \quad (A \times B)^y = \begin{cases} A & \text{if } y \in B \\ \emptyset & \text{if } y \notin B. \end{cases}$$

Therefore, we have that $\sigma(\mathbb{R}) = \Sigma \bigotimes \mathcal{L} \subseteq \hat{\mathcal{D}}$, see Remark 2.2.24. □

Definition 2.2.30. Let (X, Σ) be a measurable space, Y and V are two Hausdorff topological spaces and $f : X \times Y \to V$. We say that f is a **Carathéodory function** if the following properties hold:
(a) $x \mapsto f(x, y)$ is Σ-measurable for every $y \in Y$;
(b) $y \mapsto f(x, y)$ is continuous for every $x \in X$.

Proposition 2.2.31. *If (X, Σ) is a measurable space, Y is a separable metrizable space, V is a metrizable space and $f : X \times Y \to V$ is a Carathéodory function, then f is jointly measurable, that is, f is $(\Sigma \bigotimes \mathcal{B}(Y), \mathcal{B}(V))$-measurable.*

Proof. Let d be a compatible metric for Y and e a compatible metric for V. Recall that Y is separable. So, let $D = \{y_k\}_{k \geq 1}$ be dense in Y. Moreover, let $C \subseteq V$ be a closed set. Then $f(x, u) \in C$ if and only if for every $n \in \mathbb{N}$ there exists $y_k \in D$ such that

$$d(u, y_k) < \frac{1}{n} \quad \text{and} \quad e(f(z, y_k), C) < \frac{1}{n}.$$

Therefore we have

$$f^{-1}(C) = \bigcap_{n \geq 1} \bigcup_{k \geq 1} \{x \in X : f(z, y_k) \in C_{\frac{1}{n}}\} \times B_{\frac{1}{n}}(y_k)$$

with $C_{1/n} = \{v \in V : e(v, C) < 1/n\}$. The measurability of $f(\cdot, y_k)$ and the openness of $C_{1/n}$ imply that $\{x \in X : f(z, y_k) \in C_{1/n}\} \in \Sigma$ for all $n, k \in \mathbb{N}$. Thus $f^{-1}(C) \in \Sigma \bigotimes \mathcal{B}(Y)$. □

The next theorem, known as "Egorov's Theorem", says that in a finite measure space, pointwise convergence of a sequence of measurable functions is in fact "almost" uniform.

Theorem 2.2.32 (Egorov's Theorem). *If (X, Σ, μ) is a finite measure space, (Y, d) is a metric space and $f_n : X \to Y$ with $n \in \mathbb{N}$ is a sequence of Σ-measurable functions such that $f_n(x) \xrightarrow{d} f(x)$ for μ-a. a. $x \in X$, then for any given $\varepsilon > 0$ there exists $A_\varepsilon \in \Sigma$ with $\mu(A_\varepsilon) < \varepsilon$ such that $f_n \xrightarrow{d} f$ uniformly on $X \setminus A_\varepsilon$. That is,* $\limsup_{n \to \infty} [d(f_n(x), f(x)) : x \in A_\varepsilon] = 0$.

Proof. From Proposition 2.2.12 we know that f is Σ-measurable. For $m, k \in \mathbb{N}$ let

$$A_{m,k} = \left\{x \in X : d(f_n(x), f(x)) \leq \frac{1}{m} \text{ for all } n \geq k\right\}.$$

For every $m \in \mathbb{N}$ we have $\mu(X \setminus A_{m,k}) \searrow 0$ as $k \to +\infty$. We choose $k(m) \in \mathbb{N}$ such that $\mu(X \setminus A_{m,k(m)}) < \varepsilon/2^m$ and $D_\varepsilon = \bigcap_{m\geq 1} A_{m,k(m)} \in \Sigma$. Then for $A_\varepsilon = X \setminus D_\varepsilon$ we have $\mu(A_\varepsilon) < \varepsilon$ and $f_n \xrightarrow{d} f$ uniformly on $D_\varepsilon = X \setminus A_\varepsilon$. □

From Chapter 1 we know that a continuous function for the subspace (relative) topology on $A \subseteq X$ cannot always be extended in a continuous fashion to all of X. Think of $f_1(x) = 1/x$ for $x \in (0, 1]$ and $f_2(x) = \sin(1/x)$ for $x \in (0, 1]$ (being bounded as well), which cannot be extended continuously to $[0, 1]$. In contrast, a measurable function from $A \subseteq X$ with the trace σ-algebra can be extended measurably to all of X. The point that we want to emphasize is that A need not be measurable, otherwise the result is obvious. We start with an easy observation that is useful in many circumstances.

Lemma 2.2.33. *If (X, Σ) and $(Y, \mathcal{L})$ are measurable spaces, $\{A_n\}_{n\geq 1} \subseteq \Sigma$ are mutually disjoint sets such that $X = \bigcup_{n\geq 1} A_n$ and $f_n : A_n \to Y$ with $n \in \mathbb{N}$ are $(\Sigma_{A_n}, \mathcal{L})$-measurable functions, then $f : X \to Y$ defined by $f|_{A_n} = f_n$ for all $n \in \mathbb{N}$ is $(\Sigma, \mathcal{L})$-measurable.*

Proof. For every $B \in \mathcal{L}$ we have $f_n^{-1}(B) \in \Sigma_{A_n} = \{A_n \cap D : D \in \Sigma\}$; see Remark 2.1.2. So, $f_n^{-1}(B) = A_n \cap D_n$ with $D_n \in \Sigma$. Note that $f^{-1}(B) = \bigcup_{n\geq 1} f_n^{-1}(B) = \bigcup_{n\geq 1}(A_n \cap D_n) \in \Sigma$. □

Theorem 2.2.34. *If (X, Σ) is a measurable space, $A \subseteq X$ (not necessarily in Σ), and $f : A \to \mathbb{R}$ is Σ_A-measurable (see Remark 2.1.2), then there exists a Σ-measurable function $\hat{f} : X \to \mathbb{R}$ such that $\hat{f}|_A = f$.*

Proof. Let V be the set of all functions $f : A \to \mathbb{R}$ that are Σ_A-measurable and admit a Σ-measurable extension on X. Evidently V is a vector space and it contains the simple functions. Recall that $f = f^+ - f^-$, so we may assume that $f \geq 0$. Proposition 2.2.18 implies that there exist Σ_A-simple functions $\{s_n\}_{n\geq 1}$ such that $0 \leq s_n \nearrow f$. Let $\hat{s}_n$ be the Σ-measurable extension of s_n and recall that $s_n \in V$ for all $n \in \mathbb{N}$. Let $\hat{f}(x) = \lim_{n\to\infty} \hat{s}_n(x)$ when this limit exists and it is finite. Otherwise we set $\hat{f}(x) = 0$. Evidently $\hat{f}|_A = f$. If C is the set of $x \in X$ where the sequence $\{\hat{s}_n(x)\}$ converges, then from Proposition 2.2.11 we have that $C \in \Sigma$. We define

$$\hat{h}_n = \hat{s}_n \quad \text{on } C \quad \text{and} \quad \hat{h}_n = 0 \quad \text{on } X \setminus C \quad \text{for all } n \in \mathbb{N}.$$

From Lemma 2.2.33 we know that for each $n \in \mathbb{N}$, $\hat{h}_n$ is Σ-measurable and $\hat{h}_n(x) \to \hat{f}(x)$ for all $x \in X$. Therefore by Proposition 2.2.11, $\hat{f}$ is Σ-measurable. □

Now we are ready to define the Lebesgue integral of a measurable function.

Definition 2.2.35. Let (X, Σ, μ) be a measure space.

(a) If $s: X \to [0, +\infty]$ is a simple function with standard representation $s = \sum_{k=1}^{n} a_k \chi_{A_k}$, then the integral of s with respect to the measure μ is defined by

$$\int_X s \, d\mu = \sum_{k=1}^{n} a_k \mu(A_k).$$

(b) If $f: X \to [0, +\infty]$ is Σ-measurable, then the integral of f with respect to the measure μ is defined by

$$\int_X f \, d\mu = \sup \left[\int_X s \, d\mu : 0 \le s \le f \text{ and } s \text{ is simple} \right].$$

(c) If $f: X \to \mathbb{R}^*$ is Σ-measurable and at least one of $\int_X f^+ \, d\mu$ and $\int_X f^- \, d\mu$ is finite, then the integral of f with respect to the measure μ is defined by

$$\int_X f \, d\mu = \int_X f^+ \, d\mu - \int_X f^- \, d\mu.$$

If both $\int_X f^+ \, d\mu$ and $\int_X f^- \, d\mu$ are finite, then we say that f is (μ)-**integrable**.

Remark 2.2.36. Since $|f| = f^+ + f^-$ we see that f is integrable if and only if $\int_X |f| \, d\mu < \infty$. Moreover, we have $| \int_X f \, d\mu| \le \int_X |f| \, d\mu$.

Definition 2.2.37. Let (X, Σ, μ) be a measure space and $f: X \to \mathbb{R}^*$ a μ-integrable function. The **integral of** f **over** A with respect to the measure μ is defined by

$$\int_A f \, d\mu = \int_X f\chi_A \, d\mu.$$

Remark 2.2.38. Recalling that any set $A \in \Sigma$ defines in a natural way a measure space with the trace σ-algebra $\Sigma_A = \{A \cap D : D \in \Sigma\}$ (see Remark 2.1.2), we see that it suffices to define the integral over the whole space X and we have it automatically defined over $A \in \Sigma$.

Some straightforward observations concerning the integral are listed below.

Proposition 2.2.39. *If (X, Σ, μ) is a measure space and V is the set of all μ-integrable functions, then V is a vector space, the integral is a linear functional on V and $f \le g$ μ-a. e. implies $\int_X f \, d\mu \le \int_X g \, d\mu$.*

Proposition 2.2.40. *If (X, Σ, μ) is a measure space and $f, g: X \to R^*$ are μ-integrable functions, then the following hold:*
(a) *$f \ge 0$ and $\int_X f \, d\mu = 0$ imply $f = 0$ μ-a. e.;*
(b) *the set $A = \{x \in X : f(x) \ne 0\}$ is σ-finite;*
(c) *$\int_C f \, d\mu = \int_C g \, d\mu$ for all $C \in \Sigma$ if and only if $f = g$ μ-a. e. if and only if $\int_X |f - g| \, d\mu = 0$.*

Proof. (a) Let $A = \{x \in X: f(x) > 0\}$ and $A_n = \{x \in X: f(x) \geq 1/n\}$ with $n \in \mathbb{N}$. Then $A_n \nearrow A$ and so $\mu(A_n) \nearrow \mu(A)$; see Proposition 2.1.26. If $\mu(A) > 0$, then there exists $n \in \mathbb{N}$ such that $\mu(A_n) > 0$. We have

$$0 < \frac{1}{n}\mu(A_n) \leq \int_{A_n} f \, d\mu \leq \int_X f \, d\mu = 0 ,$$

which is a contradiction. Therefore $\mu(A) = 0$ and so $f(x) = 0$ for μ-a. a. $x \in X$.

(b) As above, let $A_n = \{x \in X: |f(x)| \geq 1/n\}$ with $n \in \mathbb{N}$. Then $A_n \in \Sigma$ and $A = \bigcup_{n\geq 1} A_n$. Moreover

$$\frac{1}{n}\mu(A_n) \leq \int_{A_n} |f| \, d\mu \leq \int_X |f| \, d\mu < +\infty ,$$

which gives $\mu(A_n) \leq cn$ for all $n \in \mathbb{N}$ and for some $c > 0$. Hence A is σ-finite.

(c) The second equivalence is obvious. Moreover, if $f = g$ μ-a. e., then $\int_C f \, d\mu = \int_C g \, d\mu$ for all $C \in \Sigma$. So, it remains to show that $\int_C f \, d\mu = \int_C g \, d\mu$ for all $C \in \Sigma$ implies that $f = g$ μ-a. e. To this end let $C = \{x \in X: (f - g)(x) \neq 0\} \in \Sigma$. Suppose that $\mu(C) > 0$. Setting $C_n = \{x \in X: |(f - g)(x)| \geq 1/n\} \in \Sigma$. As above there exists $n \in \mathbb{N}$ such that $\mu(C_n) > 0$. We have $C_n = C_n^+ \cup C_n^-$ with

$$C_n^+ = \left\{x \in X: (f - g)(x) \geq \frac{1}{n}\right\} \in \Sigma$$

and

$$C_n^- = \left\{x \in X: (f - g)(x) \leq -\frac{1}{n}\right\} \in \Sigma .$$

So, at least one of C_n^+, C_n^- has positive μ-measure. To fix things, suppose that $\mu(C_n^+) > 0$. Then

$$0 = \int_{C_n^+} (f - g) \, d\mu \geq \frac{1}{n}\mu(C_n^+) > 0 ,$$

a contradiction. Therefore $\mu(C) = 0$ and so $f = g$ μ-a. e. as in the assertion. □

The next result is known as "Markov inequality".

Proposition 2.2.41 (Markov inequality). *If (X, Σ, μ) is a measure space and $f: X \to \mathbb{R}^*$ is μ-integrable, then for any $\lambda \in (0, +\infty)$ we have*

$$\mu(\{x \in X: |f(x)| \geq \lambda\}) \leq \frac{1}{\lambda}\int_X |f| \, d\mu .$$

Proof. Let $A_\lambda = \{x \in X : |f(x)| \geq \lambda\} \in \Sigma$. Then

$$\infty > \int_X |f| \, \mathrm{d}\mu \geq \int_{A_\lambda} |f| \, \mathrm{d}\mu \geq \lambda\mu(A_\lambda) \quad \text{implies} \quad \mu(A_\lambda) \leq \frac{1}{\lambda} \int_X |f| \, \mathrm{d}\mu . \qquad \square$$

Proposition 2.2.42. *If (X, Σ, μ) is a measure space and $f : X \to \mathbb{R}^*$ is μ-integrable, then the following hold:*
(a) *$\mu(\{x \in X : |f(x)| = +\infty\}) = 0$, that is, f is μ-a. e. $\mathbb{R}$-valued;*
(b) *if $B \in \Sigma$ and $\mu(B) = 0$, then $\int_B f \, \mathrm{d}\mu = 0$.*

Proof. (a) From Proposition 2.2.41 we see that for all $\lambda > 0$, $\mu(\{x \in X : |f(x)| \geq \lambda\}) < +\infty$ and $\lim_{\lambda \to +\infty} \mu(\{x \in X : |f(x)| \geq \lambda\}) = 0$. Note that

$$\{x \in X : |f(x)| \geq n\} \searrow \{x \in X : |f(x)| = +\infty\} \quad \text{as } n \to \infty .$$

This gives, due to Proposition 2.1.24 (f),

$$\mu(\{x \in X : |f(x)| = +\infty\}) = \lim_{n \to \infty} \mu(\{x \in X : |f(x)| \geq n\}) = 0 .$$

(b) We may assume that $f \geq 0$ since $f = f^+ - f^-$. If f is a simple function, then clearly from Definitions 2.2.35 (a) and 2.2.37 we have $\int_B f \, \mathrm{d}\mu = 0$. Then Definition 2.2.35 (b) implies that $\int_B f \, \mathrm{d}\mu = 0$. $\square$

2.3 Convergence Theorems and L^p-Spaces

We start with certain convergence theorems that reveal the continuity properties of the Lebesgue integral.

The first such result is the so-called “Beppo Levi Theorem”.

Theorem 2.3.1 (Beppo Levi Theorem). *If (X, Σ, μ) is a measure space and $f_n : X \to \mathbb{R}^*_+$ with $n \in \mathbb{N}$ is an increasing sequence of Σ-measurable functions such that $f_n \nearrow f$, then $\lim_{n \to \infty} \int_X f_n \, \mathrm{d}\mu = \int_X f \, \mathrm{d}\mu$.*

Proof. From Proposition 2.2.10 we have that f is Σ-measurable. The monotonicity of the integral function implies that

$$\lim_{n \to \infty} \int_X f_n \, \mathrm{d}\mu \leq \int_X f \, \mathrm{d}\mu . \tag{2.3.1}$$

Claim: If s is a simple function and $s \leq f$, then $\int_X s \, \mathrm{d}\mu \leq \lim_{n \to \infty} \int_X f_n \, \mathrm{d}\mu$.

For every $x \in X$ and every $\eta \in (0, 1)$ there exists $n_0 = n_0(x, \eta) \in \mathbb{N}$ such that $\eta s(x) \leq f_n(x)$ for all $n \geq n_0$.

If we set $B_n = \{x \in X : \eta s(x) \le f_n(x)\}$, then $\{B_n\}_{n\ge1} \subseteq \Sigma$ and $B_n \nearrow X$. We have $\eta\chi_{B_n} s \le \chi_{B_n} f_n \le f_n$.

Let $s = \sum_{k=1}^m a_k\chi_{A_k}$ be the standard representation of the simple function s. Then one gets

$$\begin{aligned}\eta\sum_{k=1}^m a_k\mu(A_k\cap B_n) &= \eta\int_X \chi_{B_n} s\,\mathrm{d}\mu \le \int_X f_n\,\mathrm{d}\mu \le \sup_{n\ge1}\int_X f_n\,\mathrm{d}\mu \\ &= \lim_{n\to\infty}\int_X f_n\,\mathrm{d}\mu\,.\end{aligned} \tag{2.3.2}$$

Note that for every $k \in \{1,\dots,m\}$, due to Proposition 2.1.24 (e), it holds that $\mu(A_k \cap B_n) \nearrow \mu(A_k)$ as $n \to \infty$. This implies, because of (2.3.2), that

$$\eta\sum_{k=1}^m a_k\mu(A_k) = \eta\int_X s\,\mathrm{d}\mu \le \lim_{n\to\infty}\int_X f_n\,\mathrm{d}\mu\,.$$

Recall that $\eta \in (0,1)$ is arbitrary. So, let $\eta \to 1^-$. Then $\int_X s\,\mathrm{d}\mu \le \lim_{n\to\infty}\int_X f_n\,\mathrm{d}\mu$. This proves the claim.

From the claim and Definition 2.2.35 (b), we derive

$$\int_X f\,\mathrm{d}\mu \le \lim_{n\to\infty}\int_X f_n\,\mathrm{d}\mu\,. \tag{2.3.3}$$

From (2.3.1) and (2.3.3) we conclude that $\int_X f_n\,\mathrm{d}\mu \nearrow \int_X f\,\mathrm{d}\mu$. ☐

Corollary 2.3.2. *If (X,Σ,μ) is a measure space and $f: X \to \mathbb{R}_+^*$ is Σ-measurable, then $\int_X f\,\mathrm{d}\mu = \lim_{n\to\infty}\int_X s_n\,\mathrm{d}\mu$ for every increasing sequence of simple functions $s_n \nearrow f$.*

Now we can prove the famous “Monotone Convergence Theorem”.

Theorem 2.3.3 (Monotone Convergence Theorem). *If (X,Σ,μ) is a measure space and $f_n: X \to \mathbb{R}^*$ with $n \in \mathbb{N}$ is a sequence of Σ-measurable functions such that $f_n \nearrow f$ and $\int_X f_1\,\mathrm{d}\mu > -\infty$, then $\int_X f_n\,\mathrm{d}\mu \nearrow \int_X f\,\mathrm{d}\mu$ as $n \to \infty$.*

Proof. Just let $g_n = f_n - f_1 \ge 0$ for all $n \in \mathbb{N}$ and apply Theorem 2.3.1 to this sequence. ☐

Remark 2.3.4. The hypothesis that $\int_X f_1\,\mathrm{d}\mu > -\infty$ cannot be removed. To see this, consider the sequence $f_n = -\chi_{[n,\infty)}$ with $n \in \mathbb{N}$. Then $f_n \nearrow 0$ but $\int_X f_n\,\mathrm{d}\mu = -\infty$ for all $n \in \mathbb{N}$. Moreover, there is a “decreasing” version of the theorem, namely $f_n \searrow f$ and $\int_X f_1\,\mathrm{d}\mu < +\infty$ imply that $\int_X f_n\,\mathrm{d}\mu \searrow \int_X f\,\mathrm{d}\mu$.

We can also formulate Theorem 2.3.3 in a series form.

Theorem 2.3.5. *If (X,Σ,μ) is a measure space and $f_n: X \to \mathbb{R}_+^*$ with $n \in \mathbb{N}$ is a sequence of Σ-measurable functions, then*

$$\int_X \Big(\sum_{n\ge 1} f_n\Big)\,\mathrm{d}\mu = \sum_{n\ge 1}\int_X f_n\,\mathrm{d}\mu\,.$$

The next convergence theorem is known as “Fatou’s Lemma”.

Theorem 2.3.6 (Fatou’s Lemma). *If (X,Σ,μ) is a measure space and $f_n, h\colon X \to \mathbb{R}^*$ with $n \in \mathbb{N}$ are Σ-measurable functions, then the following hold:*

(a) *If $h \le f_n$ μ-a. e. for all $n \in \mathbb{N}$ and $-\infty < \int_X h\,\mathrm{d}\mu$, then*

$$\int_X \liminf_{n\to\infty} f_n\,\mathrm{d}\mu \le \liminf_{n\to\infty}\int_X f_n\,\mathrm{d}\mu\,.$$

(b) *If $f_n \le h$ μ-a. e. for all $n \in \mathbb{N}$ and $\int_X h\,\mathrm{d}\mu < +\infty$, then*

$$\limsup_{n\to\infty}\int_X f_n\,\mathrm{d}\mu \le \int_X \limsup_{n\to\infty} f_n\,\mathrm{d}\mu\,.$$

Proof. (a) Let $g_n = \inf_{k\ge n} f_k$ with $n \in \mathbb{N}$. Then $g_n \ge h$ for all $n \in \mathbb{N}$ and $g_n \nearrow \liminf_{n\to\infty} f_n$. Invoking the Monotone Convergence Theorem (see Theorem 2.3.3) we have

$$\int_X g_n\,\mathrm{d}\mu \nearrow \int_X \liminf_{n\to\infty} f_n\,\mathrm{d}\mu\,.$$

It follows $\int_X g_n\,\mathrm{d}\mu \le \int_X f_n\,\mathrm{d}\mu$ for all $n \in \mathbb{N}$ which implies

$$\int_X \liminf_{n\to\infty} f_n\,\mathrm{d}\mu \le \liminf_{n\to\infty}\int_X f_n\,\mathrm{d}\mu\,.$$

(b) Just apply (a) to the sequence $\{-f_n\}_{n\ge 1}$. □

Remark 2.3.7. The bound by h cannot be removed. To see this, consider $X = \mathbb{R}$ and $\mu = \lambda$ being the Lebesgue measure. Let $f_n = -1/n\chi_{[0,n]}$ for all $n \in \mathbb{N}$. Then $\liminf_{n\to\infty}\int_{\mathbb{R}} f_n\,\mathrm{d}\lambda = -1 < 0 = \int_X \liminf_{n\to\infty} f_n\,\mathrm{d}\mu$ and so Fatou’s Lemma fails.

Now we will present the main convergence theorem for the Lebesgue integral known as the “Lebesgue Dominated Convergence Theorem”. It allows us to interchange limits and integrals under general conditions and is the main reason why the Lebesgue integral is more powerful than the Riemann integral.

Theorem 2.3.8 (Lebesgue Dominated Convergence Theorem). *If (X,Σ,μ) is a measure space and $f_n\colon X \to \mathbb{R}^*$ with $n \in \mathbb{N}$ is a sequence of Σ-measurable functions such that*

- *$f_n(x) \to f(x)$ for μ-a. a. $x \in X$;*
- *$|f_n(x)| \le h(x)$ for μ-a. a. $x \in X$ and for all $n \in \mathbb{N}$*

with h being a μ-integrable function, then f is μ-integrable and $\int_X |f_n - f| \, d\mu \to 0$. In particular there holds

$$\int_X f_n \, d\mu \to \int_X f \, d\mu \quad \text{as } n \to \infty .$$

Proof. From Proposition 2.2.12 we know that f is Σ-measurable. Moreover, $|f(x)| \le h(x)$ for μ-a. a. $x \in X$. Therefore, f is μ-integrable.

Note that $0 \le |f_n - f| \le 2h$ μ-a. e. for all $n \in \mathbb{N}$. Applying Fatou's Lemma, Theorem 2.3.6, gives

$$0 \le \liminf_{n\to\infty} \int_X |f_n - f| \, d\mu \le \limsup_{n\to\infty} \int_X |f_n - f| \, d\mu \le 0 ,$$

which implies $\int_X |f_n - f| \, d\mu \to 0$ as $n \to \infty$. Hence,

$$\left| \int_X (f_n - f) \, d\mu \right| \to 0 \quad \text{and so} \quad \int_X f_n \, d\mu \to \int_X f \, d\mu \quad \text{as } n \to \infty . \qquad \square$$

Remark 2.3.9. If the dominating function h is not μ-integrable, then the theorem fails in general. To see this, consider $X = [0,1]$ and $\mu = \lambda$ being the Lebesgue measure. Let $f_n = n\chi_{[0,1/n]}$ with $n \in \mathbb{N}$. Then $\lim_{n\to\infty} \int_0^1 f_n \, d\lambda = 1 \neq 0 = \int_0^1 \lim_{n\to\infty} f_n \, d\lambda$.

We have already seen in Proposition 2.2.42 (b) that integration is insensitive to changes on null sets. Hence, we can integrate functions f that are only defined on a measurable set A with a null complement by simply setting $f|_{A^c} = 0$. This also implies that if f is $\mathbb{R}^*$-valued and it is a. e. $\mathbb{R}$-valued, then for the purposes of integration we can treat f as $\mathbb{R}$-valued. With this in mind we are led to the introduction of the following spaces of integrable functions.

Definition 2.3.10. Let (X, Σ, μ) be a measure space and let $1 \le p < \infty$. For any Σ-measurable function $f\colon X \to \mathbb{R}^*$ we define

$$\|f\|_p = \left(\int_X |f|^p \, d\mu \right)^{\frac{1}{p}} .$$

Let

$$\mathcal{L}^p(X) = \{ f\colon X \to \mathbb{R}^* \colon f \text{ is } \Sigma\text{-measurable}, \|f\|_p < +\infty \} .$$

Evidently $\mathcal{L}^p(X)$ is a vector space. However in order to have a vector space on which $\|\cdot\|_p$ is a norm, we need to take care of functions that differ only on a μ-null set. So, we consider the following equivalence relation on $\mathcal{L}^p(X)$

$$f \sim h \quad \text{if and only if} \quad f(x) = h(x) \quad \text{for } \mu\text{-a. a. } x \in X .$$

Then we define $L^p(X) = \mathscr{L}^p(X)/\sim$.

Next let $f: X \to \mathbb{R}^*$ be Σ-measurable and define the **essential supremum** $\|f\|_\infty$ by

$$\|f\|_\infty = \inf\{\vartheta \geq 0 : \mu(\{x \in X : |f(x)| \geq \vartheta\}) = 0\}$$

with the convention that $\inf \emptyset = +\infty$. We define

$$\mathscr{L}^\infty(X) = \{f: X \to \mathbb{R}^* : f \text{ is } \Sigma\text{-measurable}, \|f\|_\infty < +\infty\}$$

and $L^\infty(X) = \mathscr{L}^\infty(X)/\sim$.

Given $1 \leq p < \infty$ we say that $1 < p' \leq \infty$ is the **conjugate** of p if $1/p + 1/p' = 1$. Note that $p' = p/(p-1)$.

Recall the following elementary inequality known as "Young's inequality". It is a very special case of the so-called "Young–Fenchel inequality", which we discuss in Section 5.3.

Lemma 2.3.11 (Young's inequality). *If $p, p' \in (1, \infty)$ are conjugate exponents and $a, b \geq 0$, then $ab \leq 1/pa^p + 1/p'b^{p'}$ with equality if and only $b = a^{p-1}$.*

Next we will present three inequalities that are very basic in the theory of L^P-spaces. The first inequality is known as "Hölder's inequality".

Theorem 2.3.12 (Hölder's inequality). *If (X, Σ, μ) is a measure space, $1 \leq p < \infty$, $1 < p' \leq \infty$ are conjugate exponents and $f \in L^p(X)$, $h \in L^{p'}(X)$, then $fh \in L^1(X)$ and $\|fh\|_1 \leq \|f\|_p \|h\|_{p'}$.*

Moreover, for $1 < p < \infty$, equality holds if and only if

$$\frac{|f(x)|^p}{\|f\|_p^p} = \frac{|h(x)|^{p'}}{\|h\|_{p'}^{p'}} \quad \text{for } \mu\text{-a. a. } x \in X .$$

Proof. First assume that $p \in (1, \infty)$, hence $p' \in (1, \infty)$. Let $a = |f(x)|/\|f\|_p$ and $b = |h(x)|/\|h\|_{p'}$. Then by applying Young's inequality (see Lemma 2.3.11) it follows

$$\frac{|f(x)h(x)|}{\|f\|_p \|h\|_{p'}} \leq \frac{1}{p} \frac{|f(x)|^p}{\|f\|_p^p} + \frac{1}{p'} \frac{|h(x)|^{p'}}{\|h\|_{p'}^{p'}} \tag{2.3.4}$$

with equality if and only if $|f(x)|^p/\|f\|_p^p = |h(x)|^{p'}/\|h\|_{p'}^{p'}$ for μ-a. a. $x \in X$.

Integrating (2.3.4) it follows

$$\frac{1}{\|f\|_p \|h\|_{p'}} \int_X |fh| \, d\mu \leq \frac{1}{p} + \frac{1}{p'} = 1 ,$$

which implies $\|fh\|_1 \leq \|f\|_p \|h\|_{p'}$.

If $p = 1$, then $p' = +\infty$ and from the definition of the L^∞-norm, we have

$$\|fh\|_1 = \int_X |fh| \, d\mu \le \|h\|_\infty \int_X |f| \, d\mu = \|f\|_1 \|h\|_\infty . \qquad \square$$

When $p = p' = 2$, the inequality is usually called the "Cauchy–Bunyakowsky–Schwarz inequality".

Corollary 2.3.13 (Cauchy–Bunyakowsky–Schwarz inequality). *If (X, Σ, μ) is a measure space and $f, h \in L^2(X)$, then $fh \in L^1(X)$ and $\|fh\|_1 \le \|f\|_2\|h\|_2$. Moreover, equality holds if and only if $f(x)^2/\|f\|_2^2 = h(x)^2/\|h\|_2^2$ for μ-a. a. $x \in X$.*

The second inequality is known as the "Minkowski inequality". In fact it is a consequence of Hölder's inequality.

Theorem 2.3.14 (Minkowski inequality). *If (X, Σ, μ) is a measure space and $f, h \in L^p(X)$ with $1 \le p \le \infty$, then $\|f + h\|_p \le \|f\|_p + \|h\|_p$.*

Proof. Via the triangle inequality the result is clear if $p = 1$ or $p = +\infty$.

So, assume that $1 < p < \infty$ and that $f + h \neq 0$, otherwise the result is clear. We estimate

$$|f(x) + h(x)|^p \le (|f(x)| + |h(x)|)|f(x) + h(x)|^{p-1},$$

which gives

$$\|f + h\|_p^p \le \int_X |f(x)||f(x) + h(x)|^{p-1} \, d\mu + \int_X |h(x)||f(x) + h(x)|^{p-1} \, d\mu .$$

Recall that $p - 1 = p/p'$. So, let $|f + h|^{p-1} \in L^{p'}(X)$ and apply Hölder's inequality (see Theorem 2.3.12) to get

$$\|f + h\|_p^p \le (\|f\|_p + \|h\|_p)\|f + h\|_p^{p-1} .$$

This implies $\|f + h\|_p \le \|f\|_p + \|h\|_p$. $\square$

The third inequality is the so-called "Jensen inequality".

Theorem 2.3.15 (Jensen inequality). *If (X, Σ, μ) is a finite measure space, $f \in L^1(X)$ and $\varphi\colon \mathbb{R} \to \mathbb{R}$ is a convex function, then*

$$\varphi\Big(\frac{1}{\mu(X)} \int_X f \, d\mu\Big) \le \frac{1}{\mu(X)} \int_X (\varphi \circ f) \, d\mu .$$

Moreover, if φ is strictly convex, then equality holds if and only if f is a constant function.

Proof. It is well-known that φ is continuous. See Section 5.1 for more general continuity results for convex functions. In what follows for notational economy we set

$$(f)_X = \frac{1}{\mu(X)} \int_X f \, d\mu \tag{2.3.5}$$

being the average of f over X.

The convexity of φ implies that there exists $\eta \in \mathbb{R}$ such that

$$\eta(t - (f)_X) \le \varphi(t) - \varphi((f)_X) \quad \text{for all } t \in \mathbb{R}\,. \tag{2.3.6}$$

So, if $t = f(x)$, then, due to (2.3.5),

$$\eta\Big(\int_X f \, d\mu - (f)_X \mu(X)\Big) = 0 \le \int_X (\varphi \circ f) \, d\mu - \varphi((f)_X)\mu(X)\,.$$

This yields

$$\varphi\Big(\frac{1}{\mu(X)} \int_X f \, d\mu\Big) \le \frac{1}{\mu(X)} \int_X (\varphi \circ f) \, d\mu\,.$$

Finally, if φ is strictly convex, then (2.3.6) is a strict inequality for all $t \ne (f)_X$. If f is not constant, then $f(x) - (f)_X$ takes on both positive and negative values on sets of positive measure. Therefore, we cannot have equality. □

Now let us state some consequences of theses inequalities. The first is a consequence of Hölder's inequality; see Theorem 2.3.12.

Proposition 2.3.16. *If (X, Σ, μ) is a measure space, $1 \le p_k \le \infty$ for all $k = 1, \dots, n$, $\sum_{k=1}^n 1/p_k = 1/r \le 1$ and $f_k \in L^{p_k}(X)$ for all $k = 1, \dots, n$, then $\prod_{k=1}^n f_k \in L^r(X)$ and $\|\prod_{k=1}^n f_k\|_r \le \prod_{k=1}^n \|f_k\|_{p_k}$.*

Proof. Let $F = \{k \in \{1, \dots, n\} : p_k < \infty\}$ and assume that $F \ne \emptyset$ or otherwise the result is clear. Then

$$\Big\|\prod_{k=1}^n f_k\Big\|_r \le \Big\|\prod_{k \in F} f_k\Big\|_r \prod_{k \notin F} \|f_k\|_\infty \quad \text{and} \quad \sum_{k \in F} \frac{1}{p_k} = \frac{1}{r}\,.$$

So we may assume that $F = \{1, \dots, n\}$. First consider the case $n = 2$. By hypothesis one obtains

$$\frac{r}{p_1} + \frac{r}{p_2} = 1\,.$$

Applying Hölder's inequality for $p = p_1/r$ and $p' = p_2/r$ to the functions $|f_1|^r, |f_2|^r$ leads to

$$\|f_1 f_2\|_r^r \le \|f_1\|_{p_1}^r \|f_2\|_{p_2}^r\,.$$

That shows the proof for $n = 2$. When $n > 2$, we argue by induction. So let $1/\vartheta = \sum_{k=2}^n 1/p_k$. Hence $1/r = 1/p_1 + 1/\vartheta$. Assuming that the result holds for $n-1$, we have, by the induction assumptions and the validity of the case $n = 2$, that

$$\left\|\prod_{k=1}^n f_k\right\|_r \le \|f_1\|_{p_1}\left\|\prod_{k=2}^n f_k\right\|_\vartheta \le \|f_1\|_{p_1}\prod_{k=2}^n \|f_k\|_{p_k} = \prod_{k=1}^n \|f_k\|_{p_k}\,. \qquad \square$$

Another useful consequence of Hölder's inequality (see Theorem 2.3.12) is the so-called "Interpolation inequality".

Proposition 2.3.17 (Interpolation inequality). *If (X, Σ, μ) is a measure space, $1 \le p \le q \le \infty$ and $f \in L^p(X) \cap L^q(X)$, then $f \in L^r(X)$ for all $p \le r \le q$ and $\|f\|_r \le \|f\|_p^t \|f\|_q^{1-t}$ with*

$$\frac{1}{r} = \frac{t}{p} + \frac{1-t}{q} \quad \textit{with } t \in [0,1]\,. \tag{2.3.7}$$

Proof. If $q = \infty$, then $t = p/r$ and $|f|^r \le \|f\|_\infty^{r-p}|f|^p$. Hence

$$\|f\|_r \le \|f\|_\infty^{1-\frac{p}{r}}\|f\|_p^{\frac{p}{r}} = \|f\|_p^t\|f\|_\infty^{1-t}\,.$$

So, suppose now that $q < \infty$. Consider the conjugate exponents $p/(tr)$, $q/((1-t)r)$; see (2.3.7). Then by applying Hölder's inequality (see Theorem 2.3.12), it follows

$$\|f\|_r^r = \int_X |f|^r \, d\mu = \int_X |f|^{tr}|f|^{(1-t)r} \, d\mu \le \|f\|_p^{tr}\|f\|_q^{(1-t)r}\,,$$

which gives $\|f\|_r \le \|f\|_p^t\|f\|_q^{1-t}$. $\square$

In finite measure spaces, by using Hölder's inequality, we can show that the L^p-spaces decrease as p increases.

Proposition 2.3.18. *If (X, Σ, μ) is a finite measure space and $1 \le p \le q \le \infty$, then $L^q(X) \subseteq L^p(X)$ and $\|f\|_p \le \|f\|_q \mu(X)^{1/p-1/q}$.*

Proof. First assume that $q = \infty$. Then for $f \in L^\infty(X)$ we have

$$\|f\|_p^p = \int_X |f|^p \, d\mu \le \|f\|_\infty^p \mu(X)\,.$$

Next assume that $q < \infty$. Consider the conjugate exponents q/p and $q/(q-p)$ and apply Hölder's inequality for them and $f \in L^p(X)$ as well as 1. This gives

$$\|f\|_p^p = \int_X |f|^p \, d\mu \le \||f|^p\|_{\frac{q}{p}}\|1\|_{\frac{q}{p-q}} = \|f\|_q^p \mu(X)^{\frac{1}{p}-\frac{1}{q}} < +\infty\,. \qquad \square$$

Now we turn our attention to the Minkowski inequality; see Theorem 2.3.14. Evidently this inequality implies that $(L^p(X), \|\cdot\|_p)$ with $1 \le p \le \infty$ is a normed space. In fact, it is a complete normed space, that is, a Banach space.

Theorem 2.3.19. *If (X, Σ, μ) is a measure space and $1 \le p \le \infty$, then $(L^p(X), \|\cdot\|_p)$ is a Banach space.*

Proof. First assume that $p = \infty$. Let $\{f_n\}_{n\ge1} \subseteq L^\infty(X)$ be a Cauchy sequence. From Definition 2.3.10 we obtain

$$|f_n(x) - f_m(x)| \le \|f_n - f_m\|_\infty \quad \text{for } \mu\text{-a. a. } x \in X \text{ and for all } n, m \in \mathbb{N}\,.$$

This gives that $\{f_n(x)\}_{n\ge1} \subseteq \mathbb{R}$ is a Cauchy sequence for all $x \in X \setminus A$ with $\mu(A) = 0$. Then, for all $x \in X \setminus A$, $f_n(x) \to f(x)$. Let $f(x) = 0$ for $x \in A$. From Proposition 2.2.12 we know that f is Σ-measurable and

$$|f(x) - f_m(x)| \le \sup_{n\ge m} \|f_n - f_m\|_\infty \le 1$$

for $m \in \mathbb{N}$ large enough and for all $x \in X \setminus A$. This yields $\|f\|_\infty \le \|f_m\|_\infty + 1$ for $m \in \mathbb{N}$ large enough. Hence, $f \in L^\infty(X)$ and so $L^\infty(X)$ is a Banach space.

Next assume that $1 \le p < \infty$. Let $\{f_n\}_{n\ge1} \subseteq L^p(X)$ be a Cauchy sequence. Recall that a Cauchy sequence is convergent if it has a convergent subsequence. So we may assume that

$$\|f_m - f_n\|_p < \frac{1}{2^n} \quad \text{for all } n \in \mathbb{N} \text{ and for all } m > n \text{ with } m \in \mathbb{N}\,. \tag{2.3.8}$$

Let $A(n) = \{x \in X : |f_n(x) - f_{n+1}(x)| \ge 1/n^2\}$. Then $\chi_{A(n)} 1/n^2 \le |f_n - f_{n+1}|$ for all $n \in \mathbb{N}$. Thus, because of (2.3.8),

$$\mu(A(n)) \frac{1}{n^{2p}} \le \int_X |f_n - f_{n+1}|^p \, d\mu < 2^{-np} \quad \text{for all } n \in \mathbb{N}\,.$$

Therefore

$$\sum_{n\ge1} \mu(A(n)) \le \sum_{n\ge1} \frac{n^{2p}}{2^{np}} < +\infty\,.$$

Let $C(n) = \bigcup_{m\ge n} A(m)$. Then $\{C(n)\}_{n\ge1}$ is decreasing and $\mu(C(n)) \to 0$ as $n \to \infty$. Hence, if $C = \bigcap_{n\ge1} C(n)$, then $\mu(C) = 0$ and for $x \in X \setminus C$ we have

$$|f_n(x) - f_m(x)| \le \frac{1}{n^2} \quad \text{for all } n \in \mathbb{N} \text{ large enough}\,.$$

Then for any $m > n$ it holds that $|f_m(x) - f_n(x)| \le \sum_{k\ge n} 1/k^2 \to 0$ as $n \to \infty$. So it follows that, for μ-a. a. $x \in X$, $\{f_n\}_{n\ge1}$ is a Cauchy sequence and so it converges to some $f(x)$. On the exceptional μ-null set, we put $f(x) = 0$. Clearly f is measurable and by Fatou's Lemma (see Theorem 2.3.6), one gets

$$\int_X |f|^p \, d\mu \leq \liminf_{n\to\infty} \int_X |f_n|^p \, d\mu < \infty$$

since a Cauchy sequence is bounded. Hence, $f \in L^p(X)$.

Similarly, we obtain

$$\int_X |f - f_n|^p \, d\mu \leq \liminf_{m\to\infty} \int_X |f_m - f_n|^p \, d\mu ,$$

which implies that $f_n \to f$ in $L^p(X)$. □

A useful consequence of the result above is the following corollary.

Corollary 2.3.20. *If (X, Σ, μ) is a measure space, $\{f_n\}_{n\geq 1} \subseteq L^p(X)$ with $1 \leq p \leq \infty$, and $f_n \to f$ in $L^p(X)$, then there is a subsequence $\{f_{n_k}\}_{k\geq 1}$ of $\{f_n\}_{n\geq 1}$ such that $f_{n_k}(x) \to f(x)$ μ-a. e.*

Example 2.3.21. We have to pass to a subsequence to get pointwise convergence. To see this, consider the sequence $f_k = \chi_{[(i-1)/n, i/n]}$ for $k = i + (n(n-1))/2$ with $n \in \mathbb{N}$ and $i = 1, \ldots, n$. Then $\int_0^1 f_k^p \, d\lambda = 1/n \to 0$, that is, $f_n \to 0$ in $L^p[0,1]$. However, $\liminf_{k\to\infty} f_k(x) = 0 < 1 = \limsup_{k\to\infty} f_k(x)$ for all $x \in [0,1]$ and so we do not have pointwise convergence.

The next result provides a useful dense subset of the Banach space $L^p(X)$. It is a straightforward consequence of Proposition 2.2.18.

Proposition 2.3.22. *If (X, Σ, μ) is a measure space, then the set of simple functions in $L^p(X)$ is dense in $L^p(X)$ for $1 \leq p \leq \infty$.*

We continue with the examination of the Banach spaces $L^p(X)$ for $1 \leq p \leq \infty$. Next we examine under what conditions we can have separability of $L^p(X)$. We start with a definition.

Definition 2.3.23. Let (X, Σ, μ) be a measure space. On Σ we define the semimetric

$$d_\mu(A, B) = \mu(A \triangle B) \quad \text{for all } A, B \in \Sigma .$$

According to Remark 1.5.2 if we introduce on Σ the equivalence relation $\sim$ defined by $A \sim B$ if and only if $\mu(A \triangle B) = 0$, then, on $\Sigma(\mu) = \Sigma/\sim$, d_μ is a metric. Clearly we have

$$d_\mu(A, B) = \|\chi_A - \chi_B\|_1 \quad \text{for all } A, B \in \Sigma(\mu) .$$

Proposition 2.3.24. *If (X, Σ, μ) is a measure space, then $(\Sigma(\mu), d_\mu)$ is a separable metric space if and only if the Banach space $L^1(X)$ is separable.*

Proof. $\Longrightarrow$: Let $\{A_k\}_{k\geq 1} \subseteq \Sigma(\mu)$ be a countable d_μ-dense subset. Then the set of all functions that are finite linear combinations of $\{\chi_{A_k}\}_{k\geq 1}$ with rational coefficients is a countable dense subset of $L^1(X)$. Hence $L^1(X)$ is separable.

$\Longleftarrow$: By identifying an element of Σ with its characteristic function, we see that $\Sigma(\mu)$ can be viewed as a subset of $L^1(X)$. Then the separability of $L^1(X)$ implies the separability of $\Sigma(\mu)$. ☐

The next proposition provides a condition for the separability of $(\Sigma(\mu), d_\mu)$.

Proposition 2.3.25. *If (X, Σ, μ) is a finite measure space and $\Sigma = \sigma(\mathcal{L})$ with $\mathcal{L}$ being countable, then $(\Sigma(\mu), d_\mu)$ is separable.*

Proof. Note that the ring generated by $\mathcal{L}$ is still countable. So we may assume that $\mathcal{L}$ is a ring. Then, using Problem 2.3, for every $A \in \Sigma(\mu)$ we can find $B \in \mathcal{L}$ such that $d_\mu(A, B) = \mu(A \triangle B) \leq \varepsilon$. Hence $\mathcal{L}$ is d_μ-dense in $\Sigma(\mu)$ and so $(\Sigma(\mu), d_\mu)$ is separable. ☐

Corollary 2.3.26. *If X is a separable metric space, $\Sigma = \mathcal{B}(X)$ and μ is a finite measure on Σ, then $(\Sigma(\mu), d_\mu)$ is separable.*

In fact combining Propositions 2.3.18, 2.3.24, and 2.3.25, we can state the following result.

Proposition 2.3.27. *If (X, Σ, μ) is a σ-finite measure space, $\Sigma = \sigma(\mathcal{L})$ with $\mathcal{L}$ countable and $\mathfrak{a}$ is the smallest algebra containing $\mathcal{L}$, then the simple functions of the form $s = \sum_{k=1}^n a_k\chi_{A_k}$ with $n \in \mathbb{N}$, $a_k \in \mathbb{Q}$, $A_k \in \mathfrak{a}$, $\mu(A_k) < \infty$, $k = 1, \ldots, n$ form a countable dense subset of $L^p(X)$ for $1 \leq p < \infty$. In particular, $L^p(X)$ is separable for $1 \leq p < \infty$.*

For the space $L^\infty(X)$ we show that it is not separable. In order to show this first we mention the following decomposition result, which can be found in Dudley [101, p. 82].

Proposition 2.3.28. *If (X, Σ, μ) is a σ-finite measure space, then $\mu = \mu_a + \mu_d$ with μ_a purely atomic and μ_d nonatomic. Moreover the atoms on which μ_a is defined are at most countable.*

We can use this result to establish the nonseparability of $L^\infty(X)$.

Proposition 2.3.29. *If (X, Σ, μ) is a σ-finite measure space, then the Banach space $L^\infty(X)$ is not separable.*

Proof. Applying Proposition 2.3.28, we split X into its atomic part X_a and its nonatomic (diffuse) part X_d. We consider two distinct cases: (a) X_d is not μ-null. (b) X_d is μ-null.

Suppose that (a) holds. Then for each $\eta \in (0, \mu(X_d))$ there exists $A_\eta \in \Sigma$ such that $\mu(A_\eta) = \eta$; see Proposition 2.1.32. Then $\{A_\eta\}_{\eta\in(0,\mu(X_d))}$ is an uncountable set of distinct Σ-sets, that is, $\mu(A_\eta \triangle A_{\eta'}) > 0$ if $\eta \neq \eta'$. Let

$$U_\eta = \left\{f \in L^\infty(X) \colon \|f - \chi_{A_\eta}\|_\infty < \frac{1}{2}\right\}, \quad \eta \in (0, \mu(X_d)) = I .$$

Then $\{U_\eta\}_{\eta\in I}$ is an uncountable family of nonempty, open, and mutually disjoint sets in $L^\infty(X)$. This means that $L^\infty(X)$ is not separable. Indeed, if $L^\infty(X)$ were separable, then there would be a countable dense set $\{f_n\}_{n\geq 1} \subseteq L^\infty(X)$. For each $\eta \in I$ we have $U_\eta \cap \{f_n\}_{n\geq 1} \neq \emptyset$. So we can choose $n(\eta) \in \mathbb{N}$ such that $f_{n(\eta)} \in U_\eta$. The map $\eta \to n(\eta)$ is injective; recall that the sets are mutually disjoint. Therefore I is countable, a contradiction. The case (b) follows from Proposition 2.3.28. □

The main convergence theorem in the theory of Lebesgue integration is the "Lebesgue Dominated Convergence Theorem"; see Theorem 2.3.8. Two of the main ingredients in that result are:

- $f_n(x) \to f(x)$ μ-a. e. as $n \to \infty$ (the pointwise convergence of the sequence);
- $|f_n(x)| \leq h(x)$ for μ-a. a. $x \in X$ and for all $n \in \mathbb{N}$ with $h \in L^1(X)$ (existence of a dominating integrable function).

Both can be weakened. To weaken the pointwise convergence requirement we introduce the following convergence concept.

Definition 2.3.30. Let (X, Σ, μ) be a measure space. A sequence $f_n\colon X \to \mathbb{R}^*$ with $n \in \mathbb{N}$ of Σ-measurable functions **converges in measure** to a Σ-measurable function f if for every $\varepsilon > 0$

$$\mu(\{x \in X\colon |f_n(x) - f(x)| \geq \varepsilon\}) \to 0 \quad \text{as } n \to \infty .$$

We denote the convergence in measure by $f_n \xrightarrow{\mu} f$.

If μ is a probability measure, that is, $\mu(X) = 1$, then we say that the sequence $\{f_n\}_{n\geq 1}$ **converges in probability to** f.

We say that the sequence $\{f_n\}_{n\geq 1}$ is a **Cauchy sequence in measure** if for every $\varepsilon > 0$,

$$\lim_{n,m\to\infty} \mu(\{x \in X\colon |f_n(x) - f_m(x)| \geq \varepsilon\}) = 0 .$$

The following proposition is a straightforward consequence of the definition above.

Proposition 2.3.31. *If (X, Σ, μ) is a measure space, then the following hold:*
(a) *$f_n \xrightarrow{\mu} f$ and $h_n \xrightarrow{\mu} h$ imply $\eta f_n + \vartheta h_n \xrightarrow{\mu} \eta f + \vartheta h$ for all $\eta, \vartheta \in \mathbb{R}$;*
(b) *$f_n \xrightarrow{\mu} f$ implies $f_n^\pm \xrightarrow{\mu} f^\pm$ and $|f_n| \xrightarrow{\mu} |f|$;*
(b) *$f_n \xrightarrow{\mu} f$ and $f_n \xrightarrow{\mu} g$ imply $f = g$ μ-a. e.*

Proposition 2.3.32. *If (X, Σ, μ) is a finite measure space and $f_n \to f$ μ-a. e., then $f_n \xrightarrow{\mu} f$.*

Proof. For every $n \in \mathbb{N}$, let

$$\begin{aligned} A_n &= \{x \in X\colon |f_n(x) - f(x)| \geq \varepsilon\} \\ &= \left\{x \in X\colon \frac{|f_n(x) - f(x)|}{1 + |f_n(x) - f(x)|} \geq \frac{\varepsilon}{1+\varepsilon}\right\} . \end{aligned} \tag{2.3.9}$$

This gives $\mu(A_n) \leq (1+\varepsilon)/\varepsilon \int_X (|f_n - f|)/(1 + |f_n - f|)\, d\mu$ by the Markov inequality; see Proposition 2.2.41. But from the Lebesgue Dominated Convergence Theorem (see Theorem 2.3.8), it follows

$$\frac{1+\varepsilon}{\varepsilon} \int_X \frac{|f_n - f|}{1 + |f_n - f|}\, d\mu \to 0 \quad \text{as } n \to \infty .$$

Hence $\mu(A_n) \to 0$ and so $f_n \xrightarrow{\mu} f$; see (2.3.9). ☐

In fact in finite measure spaces convergence in measure is strictly weaker than pointwise convergence.

Example 2.3.33. Let $X = [0,1]$, $\Sigma = \mathcal{B}([0,1])$, $\mu = \lambda|_{[0,1]}$ with λ being the Lebesgue measure on $\mathbb{R}$. Consider the sequence of Σ-measurable functions

$$f_n(x) = \chi_{[\frac{i}{2^k}, \frac{i+1}{2^k}]}(x) \quad \text{for all } i \in \{0, 1, \dots, 2^k - 1\},\ n = i + 2^k .$$

It follows that

$$\lambda(\{x \in [0,1] : |f_n(x)| \geq \varepsilon\}) = \frac{1}{2^k} \to 0 \quad \text{as } n = n(k) \to +\infty .$$

Hence, $f_n \xrightarrow{\mu} 0$. But the pointwise limit of the f_n's does not exist at any $x \in [0,1]$.

The following is a variant of the Markov inequality (see Proposition 2.2.41) and is known as the "Chebyshev inequality".

Proposition 2.3.34 (Chebyshev inequality). *If (X, Σ, μ) is a measure space, $f \in L^p(X)$, $1 \leq p < \infty$, and $\lambda > 0$, then*

$$\mu(\{x \in X : |f(x)| \geq \lambda\}) \leq \frac{1}{\lambda^p} \|f\|_p^p .$$

Proof. Let $A_\lambda = \{x \in X : |f(x)| \geq \lambda\}$. Then $\|f\|_p^p \geq \int_{A_\lambda} |f|^p\, d\mu \geq \lambda^p \mu(A_\lambda)$. ☐

Using the Chebyshev inequality we can compare convergence in $L^p(X)$ for $1 \leq p < \infty$ with convergence in measure.

Proposition 2.3.35. *If (X, Σ, μ) is a measure space, $\{f_n\}_{n\geq 1} \subseteq L^p(X)$ with $1 \leq p < \infty$, and $\|f_n - f\|_p \to 0$, then $f_n \xrightarrow{\mu} f$.*

Proof. Applying the Chebyshev inequality (see Proposition 2.3.34) yields the assertion of the proposition. ☐

Although convergence in measure is strictly weaker than pointwise convergence, we can always extract from any convergent sequence in measure a pointwise convergent subsequence.

Proposition 2.3.36. *If (X, Σ, μ) is a measure space and $f_n \xrightarrow{\mu} f$, then there exists a subsequence $\{f_{n_k}\}_{k\geq 1} \subseteq \{f_n\}_{n\geq 1}$ such that $f_{n_k} \to f$ μ-a. e.*

Proof. Since $f_n \xrightarrow{\mu} f$ there is a strictly increasing sequence $\{k_n\}_{n\geq 1} \subseteq \mathbb{N}$ such that

$$\mu\left(\left\{x \in X : |f_k(x) - f(x)| \geq \frac{1}{n}\right\}\right) < \frac{1}{2^n} \quad \text{for all } k \geq k_n .$$

For each $n \in \mathbb{N}$, let $A_n = \{x \in X : |f_{k_n}(x) - f(x)| \geq 1/n\} \in \Sigma$. We set $A = \bigcap_{k\geq 1} \bigcup_{n\geq k} A_n \in \Sigma$. Then we have

$$\mu(A) \leq \mu\left(\bigcup_{n\geq k} A_n\right) \leq \sum_{n\geq k} \mu(A_n) \leq \frac{1}{2^{k+1}} \quad \text{for every } k \in \mathbb{N} .$$

Hence, $\mu(A) = 0$.

If $x \notin A$, then there exists $k_0 \in \mathbb{N}$ such that $x \notin \bigcup_{n\geq k_0} A_n$ and so $|f_{k_n}(x) - f(x)| < 1/n$ for all $n \geq k_0$. Thus $f_{k_n}(x) \to f(x)$ for all $x \notin A$ with $\mu(A) = 0$. □

Definition 2.3.37. Let (X, Σ, μ) be a measure space and let $\mathcal{M}(X) = \{f : X \to \mathbb{R}^* : f$ is Σ-measurable$\}$. As before, we define $f \sim h$ if and only if $f = h$ μ-a. e. Then we set $L^0(X) = \mathcal{M}(X)/\sim$. When $\mu(X) < \infty$ on $L^0(X)$ we introduce the translation invariant metric

$$d_\mu(f, h) = \int_X \frac{|f - h|}{1 + |f - h|}\, d\mu \quad \text{for all } f, h \in L^0(X) . \tag{2.3.10}$$

Remark 2.3.38. It is easy to check that d_μ is a metric on $L^0(X)$. For the triangle inequality, use the elementary inequality that says that

$$a, b, c \in \mathbb{R}_+, \quad a \leq b + c \quad \text{implies} \quad \frac{a}{1+a} \leq \frac{b}{1+b} + \frac{c}{1+c} .$$

In the next proposition we show that in finite measure spaces, convergence in measure is in fact a metric convergence.

Proposition 2.3.39. *If (X, Σ, μ) is a finite measure space and $\{f_n\}_{n\geq 1} \subseteq L^0(X)$, $f \in L^0(X)$, then $f_n \xrightarrow{\mu} f$ if and only if $f_n \xrightarrow{d_\mu} f$ in $L^0(X)$; see* (2.3.10).

Proof. In what follows for a given $\varepsilon > 0$ let

$$\begin{aligned} A_n &= \{x \in X : |f_n(x) - f(x)| \geq \varepsilon\} \\ &= \left\{x \in X : \frac{|f_n(x) - f(x)|}{1 + |f_n(x) - f(x)|} \geq \frac{\varepsilon}{1+\varepsilon}\right\}, \quad n \in \mathbb{N} . \end{aligned} \tag{2.3.11}$$

Suppose that $f_n \xrightarrow{\mu} f$. Then we can find $n_0 \in \mathbb{N}$ such that

$$\mu(A_n) \leq \varepsilon \quad \text{for all } n \geq n_0 . \tag{2.3.12}$$

Then, because of (2.3.11) and (2.3.12), it follows

$$\begin{aligned} d_\mu(f_n, f) &= \int_{A_n} \frac{|f_n - f|}{1 + |f_n - f|} \, d\mu + \int_{X \setminus A_n} \frac{|f_n - f|}{1 + |f_n - f|} \, d\mu \\ &\le \mu(A_n) + \frac{\varepsilon}{1+\varepsilon} \mu(X \setminus A_n) \le (1 + \mu(X))\varepsilon \end{aligned}$$

for all $n \ge n_0$. This gives $d_\mu(f_n, f) \to 0$ as $n \to \infty$.

Now assume that $f_n \xrightarrow{d_\mu} f$. Then $\varepsilon/(1+\varepsilon)\chi_{A_n} \le (f_n - f)/(1 + |f_n - f|)$ for all $n \in \mathbb{N}$; see (2.3.11). This implies $\mu(A_n) \le (1+\varepsilon)/(\varepsilon) d_\mu(f_n, f) \to 0$ as $n \to \infty$. Hence $f_n \xrightarrow{\mu} f$. □

The next notion will allow us to relax the dominating function requirement in the Lebesgue Dominated Convergence Theorem; see Theorem 2.3.8.

Definition 2.3.40. Let (X, Σ, μ) be a measure space and $\mathcal{F} \subseteq L^0(X)$. We say that $\mathcal{F}$ is **uniformly integrable** if for every $\varepsilon > 0$ there exists $D_\varepsilon \in \Sigma$ with $\mu(D_\varepsilon) < \infty$ and $\sup_{f \in \mathcal{F}} \int_{X \setminus D_\varepsilon} |f| \, d\mu \le \varepsilon$ as well as $\lim_{c \to \infty} \sup_{f \in \mathcal{F}} \int_{\{|f| \ge c\}} |f| \, d\mu = 0$.

Remark 2.3.41. In the literature one can find other definitions of uniform integrability that are equivalent to the definition above when $\mu(X) < \infty$. Some of these alternative definitions are examined in the exercises. In particular we mention the following equivalent definition for a set $\mathcal{F} \subseteq L^1(X)$ to be uniformly integrable:

(UI)' (a) $\mathcal{F} \subseteq L^1(X)$ is bounded, that is $\sup_{f \in \mathcal{F}} \|f\|_1 < \infty$;

(b) for every $\varepsilon > 0$ there exists $D_\varepsilon \in \Sigma$ with $\mu(D_\varepsilon) < \infty$ such that $\sup_{f \in \mathcal{F}} \int_{X \setminus D_\varepsilon} |f| \, d\mu \le \varepsilon$;

(c) for every $\varepsilon > 0$ there exists $\delta > 0$ such that $\mu(A) \le \delta$ implies $\sup_{f \in \mathcal{F}} \int_A |f| \, d\mu \le \varepsilon$.

The next result is a key property of the Lebesgue integral and will help us identify uniformly integrable subsets of $L^1(X)$. The result is referred to as the **absolute continuity** property of the integral.

Proposition 2.3.42. *If (X, Σ, μ) is a measure space and $f \in L^1(X)$, then for any given $\varepsilon > 0$ there exists $\delta = \delta(\varepsilon) > 0$ such that*

$$A \in \Sigma, \quad \mu(A) \le \delta \quad \textit{implies} \quad \int_A |f| \, d\mu \le \varepsilon .$$

Proof. Since $f = f^+ - f^-$, without any loss of generality, we may assume that $f \ge 0$. Let $f_n = \min\{f, n\}$ with $n \in \mathbb{N}$. Then $f_n \nearrow f$ and so by the Monotone Convergence Theorem (Theorem 2.3.3), we have $\int_X f_n \, d\mu \nearrow \int_X f \, d\mu$. So, given $\varepsilon > 0$ there exists $n_0 = n_0(\varepsilon) \in \mathbb{N}$ such that

$$0 \le \int_X (f - f_n) \, d\mu \le \frac{\varepsilon}{2} \quad \text{for all } n \ge n_0 . \tag{2.3.13}$$

If $\delta = \varepsilon/(2n_0)$ and $A \in \Sigma$ satisfies $\mu(A) \leq \delta$, then, due to (2.3.13),

$$\int_A f \, d\mu \leq \int_A f_{n_0} \, d\mu + \int_X (f - f_{n_0}) \, d\mu \leq \varepsilon . \qquad \square$$

Corollary 2.3.43. *If (X, Σ, μ) is a measure space and $\mathcal{F} \subseteq L^0(X)$ satisfies*

$$|f(x)| \leq h(x) \quad \text{for } \mu\text{-a. a. } x \in X \text{ and for all } f \in \mathcal{F} \text{ with } h \in L^1(X) ,$$

then $\mathcal{F}$ is uniformly integrable. In particular, every finite set $\mathcal{F} \subseteq L^1(X)$ is uniformly integrable.

Now we can state the generalization of the Lebesgue Dominated Convergence Theorem; see Theorem 2.3.8. The result is known as the "Vitali Convergence Theorem" or "Extended Dominated Convergence Theorem".

Theorem 2.3.44 (Vitali Convergence Theorem). *If (X, Σ, μ) is a measure space, $\{f_n\}_{n \geq 1} \subseteq L^1(X)$ is uniformly integrable and $f_n \xrightarrow{\mu} f$ as $n \to \infty$, then $f \in L^1(X)$ and $\|f_n - f\|_1 \to 0$. In particular, we have $\int_X f_n \, d\mu \to \int_X f \, d\mu$.*

Proof. On account of Proposition 2.3.36, we may assume that $f_n \to f$ μ-a. e. Given $\varepsilon > 0$, let $\delta > 0$ and $D_\varepsilon \in \Sigma$ be as postulated by (UI)′; see Remark 2.3.41. Moreover, thanks to Egorov's Theorem, Theorem 2.2.32, we know that there exists $A_\varepsilon \in \Sigma$ with $A_\varepsilon \subseteq D_\varepsilon$ and $\mu(A_\varepsilon) \leq \delta$ such that

$$f_n \to f \quad \text{uniformly on } D_\varepsilon \setminus A_\varepsilon . \tag{2.3.14}$$

We have

$$\begin{aligned} \int_{D_\varepsilon} |f_n - f| \, d\mu &= \int_{A_\varepsilon} |f_n - f| \, d\mu + \int_{D_\varepsilon \setminus A_\varepsilon} |f_n - f| \, d\mu \\ &\leq \int_{A_\varepsilon} |f_n| \, d\mu + \int_{A_\varepsilon} |f| \, d\mu + \|f_n - f\|_{L^\infty(D_\varepsilon \setminus A_\varepsilon)} \mu(D_\varepsilon) . \end{aligned} \tag{2.3.15}$$

Note that according to (UI)′ (see also Definition 2.3.40), it holds that

$$\int_{A_\varepsilon} |f_n| \, d\mu \leq \varepsilon , \quad \int_{X \setminus D_\varepsilon} |f_n| \, d\mu \leq \varepsilon \quad \text{for all } n \in \mathbb{N} . \tag{2.3.16}$$

Moreover, by Fatou's Lemma, one gets

$$\int_{A_\varepsilon} |f| \, d\mu \leq \varepsilon , \quad \int_{X \setminus D_\varepsilon} |f| \, d\mu \leq \varepsilon . \tag{2.3.17}$$

Taking (2.3.15), (2.3.16) and (2.3.17) into account it follows that

$$\int_X |f_n - f| \, d\mu \le \int_{X\setminus D_\varepsilon} |f_n| \, d\mu + \int_{X\setminus D_\varepsilon} |f| \, d\mu + \int_{D_\varepsilon} |f_n - f| \, d\mu$$
$$\le 4\varepsilon + \|f_n - f\|_{L^\infty(D_\varepsilon \setminus A_\varepsilon)} \mu(D_\varepsilon) \quad \text{for all } n \in \mathbb{N} .$$

Hence, because of (2.3.14) and since $\mu(D_\varepsilon)$ is finite and $\varepsilon > 0$ is arbitrary, it follows that $f_n \to f$ in $L^1(X)$. □

Now that once we have the convergence theorems for the Lebesgue integral, we can establish the existence and uniqueness of the product measure.

So, let (X, Σ, μ) and $(Y, \mathcal{L}, \nu)$ be two measure spaces. Suppose that $\Sigma = \sigma(\mathfrak{a})$ and $\mathcal{L} = \sigma(\mathfrak{b})$. We want to define a measure ξ on rectangles of the form $A \times B$ with $A \in \mathfrak{a}$ and $B \in \mathfrak{b}$ such that

$$\xi(A \times B) = \mu(A)\nu(B) \quad \text{for all } A \in \mathfrak{a},\ B \in \mathfrak{b} . \tag{2.3.18}$$

If the generators $\mathfrak{a}$ and $\mathfrak{b}$ are rich enough, we can have the uniqueness of the measure ξ satisfying (2.3.18).

Proposition 2.3.45. *If (X, Σ, μ) and $(Y, \mathcal{L}, \nu)$ are two measure spaces, $\Sigma = \sigma(\mathfrak{a})$, $\mathcal{L} = \sigma(\mathfrak{b})$ and*

(i) *$\mathfrak{a}$ and $\mathfrak{b}$ are closed under finite intersections;*

(ii) *there exists sequences $\{A_n\}_{n\ge1} \subseteq \mathfrak{a}$, $\{B_n\}_{n\ge1} \subseteq \mathfrak{b}$ with $A_n \nearrow X$, $B_n \nearrow Y$ and $\mu(A_n) < \infty$, $\nu(B_n) < \infty$ for all $n \in \mathbb{N}$,*

then there is at most on measure ξ on $\Sigma \bigotimes \mathcal{L}$ satisfying (2.3.18).

Proof. From Proposition 2.2.25 we know that $\Sigma \bigotimes \mathcal{L} = \sigma(\mathfrak{a} \times \mathfrak{b})$. Moreover we have

$$A_n \times B_n \nearrow X \times Y \quad \text{and} \quad \xi(A_n \times B_n) = \mu(A_n)\nu(B_n) < \infty \quad \text{for all } n \in \mathbb{N} .$$

Proposition 2.1.28 implies the uniqueness of ξ. □

Now we examine the issue of the existence of the product measure.

Theorem 2.3.46. *If (X, Σ, μ) and $(X, \mathcal{L}, \nu)$ are two σ-finite measure spaces, then the set function $\xi\colon \Sigma \times \mathcal{L} \to [0, +\infty]$ defined by $\xi(A \times B) = \mu(A)\nu(B)$ for all $A \in \Sigma$, $B \in \mathcal{L}$, extends uniquely to a σ-finite measure on $\Sigma \bigotimes \mathcal{L}$ such that*

$$\xi(C) = \int_Y \int_X \chi_C(x,y) \, d\mu d\nu = \int_X \int_Y \chi_C(x,y) \, d\nu d\mu \quad \textit{for all } C \in \Sigma \bigotimes \mathcal{L}$$

and $x \to \chi_C(x,y)$, $y \to \chi_C(x,y)$, $x \to \int_Y \chi_C(x,y) \, d\nu$ and $y \to \int_X \chi_C(x,y) \, d\mu$ are measurable.

Proof. The uniqueness follows from Proposition 2.3.45. Consider sequences $\{A_n\}_{n\geq 1} \subseteq \Sigma$ and $\{B_n\}_{n\geq 1} \subseteq \mathcal{L}$ such that

$$A_n \nearrow X\,, \quad B_n \nearrow Y \quad \text{and} \quad \mu(A_n) < \infty\,, \quad \nu(B_n) < \infty \quad \text{for all } n \in \mathbb{N}\,.$$

Note that $C_n = A_n \times B_n \nearrow X \times Y$. For every $n \in \mathbb{N}$, let D_n be the family of all subsets $E \subseteq X \times Y$ such that

- $x \to \chi_{E\cap C_n}(x,y)$ and $y \to \chi_{E\cap C_n}(x,y)$ are measurable;
- $x \to \int_Y \chi_{E\cap C_n}(x,y)\, d\nu$ and $y \to \int_X \chi_{E\cap C_n}(x,y)\, d\mu$ are measurable;
- $\int_Y \int_X \chi_{E\cap C_n}(x,y)\, d\mu d\nu = \int_X \int_Y \chi_{E\cap C_n}(x,y)\, d\nu d\mu$.

It is a straightforward procedure to check that D_n is a Dynkin system; see Definition 2.1.7, which contains $\Sigma \times \mathcal{L}$. So, applying the Dynkin System Theorem (see Theorem 2.1.11) yields that $\Sigma \otimes \mathcal{L} \subseteq D_n$ for all $n \in \mathbb{N}$. Since $C_n \nearrow X \times Y$, Proposition 2.2.10 implies the measurability of $x \to \chi_C(x,y)$ and $y \to \chi_C(x,y)$ and then the Monotone Convergence Theorem (see Theorem 2.3.3) gives the measurability of $x \to \int_Y \chi_C(x,y)\, d\nu$ and of $y \to \int_X \chi_C(x,y)\, d\mu$.

Finally, if $E = X \times Y$, then we have that

$$C \to \xi(C) = \int_Y \int_X \chi_C(x,y)\, d\mu d\nu = \int_X \int_Y \chi_C(x,y)\, d\nu d\mu$$

is indeed a measure on $\Sigma \otimes \mathcal{L}$ and $\xi(A\times B) = \mu(A)\nu(B)$ for all $A \in \Sigma$ and for all $B \in \mathcal{L}$. □

Definition 2.3.47. Let (X, Σ, μ) and $(X, \mathcal{L}, \nu)$ be two σ-finite measure spaces. The unique measure ξ on $\Sigma \otimes \mathcal{L}$ produced in Theorem 2.3.46 is called the **product measure** of μ and ν and is denoted by $\mu \times \nu$. The measure space $(X \times Y, \Sigma \otimes \mathcal{L}, \mu \times \nu)$ is called the **product measure space**.

Remark 2.3.48. Now we can define the Lebesgue measure λ^n on $(\mathbb{R}^n, \mathcal{B}(\mathbb{R}^n))$ such that

$$\lambda^n(R) = \prod_{k=1}^{n} (b_k - a_k) \quad \text{for all rectangles } R = \prod_{k=1}^{n} [a_k, b_k)\,.$$

The next two theorems enable us to interchange the order of integration and to calculate integrals with respect to product measures using iteration. Their proofs are straightforward. Indeed, the results are true for characteristic functions, hence for simple functions. Then exploit the density of the simple functions to pass to the general case.

The first result is known as "Tonelli's Theorem".

Theorem 2.3.49 (Tonelli's Theorem). *If (X, Σ, μ) and $(X, \mathcal{L}, \nu)$ are two σ-finite measure spaces and if $f : X \times Y \to [0, \infty]$ is $\Sigma \otimes \mathcal{L}$-measurable, then the following hold:*

(a) *for all $y \in Y$, $x \to f(x,y)$ is Σ-measurable and for all $x \in X$, $y \to f(x,y)$ is $\mathcal{L}$-measurable;*

(b) $x \to \int_Y f(x,y)\,\mathrm{d}\nu$ *is* Σ*-measurable and* $y \to \int_X f(x,y)\,\mathrm{d}\mu$ *is* $\mathcal{L}$*-measurable;*
(c) $\int_{X\times Y} f\,\mathrm{d}(\mu\times\nu) = \int_Y \int_X f(x,y)\,\mathrm{d}\mu\mathrm{d}\nu = \int_X \int_Y f(x,y)\,\mathrm{d}\nu\mathrm{d}\mu$.

The second is known as "Fubini's Theorem".

Theorem 2.3.50 (Fubini's Theorem). *If* (X,Σ,μ) *and* $(X,\mathcal{L},\nu)$ *are two* σ*-finite measure spaces,* $f\colon X\times Y \to \mathbb{R}^*$ *is* $\Sigma\bigotimes\mathcal{L}$*-measurable and at least one of the following three integrals is finite*

$$\int_{X\times Y} |f|\,\mathrm{d}(\mu\times\nu)\,, \quad \int_Y\int_X |f|\,\mathrm{d}\mu\mathrm{d}\nu\,, \quad \int_X\int_Y |f|\,\mathrm{d}\nu\mathrm{d}\mu\,,$$

then all three integrals are finite, $f \in L^1(X\times Y)$ *and*
(a) $x \to f(x,y) \in L^1(X)$ *for* ν*-a. a.* $y \in Y$;
(b) $y \to f(x,y) \in L^1(Y)$ *for* μ*-a. a.* $x \in X$;
(c) $y \to \int_X f(x,y)\,\mathrm{d}\mu \in L^1(Y)$;
(d) $x \to \int_Y f(x,y)\,\mathrm{d}\nu \in L^1(X)$;
(e) $\int_{X\times Y} f\,\mathrm{d}(\mu\times\nu) = \int_Y \int_X f(x,y)\,\mathrm{d}\mu\mathrm{d}\nu = \int_X \int_Y f(x,y)\,\mathrm{d}\nu\mathrm{d}\mu$.

2.4 Signed Measures and Radon–Nikodym Theorem

In this section we examine the notion of differentiating a measure ν with respect to another measure μ defined on the same σ-algebra. This differentiation theory can be developed more precisely if we extend the notion of measure and allow also negative values. This leads us to the concept of signed measure already introduced in Definition 2.1.22 (f). For convenience, let us recall the definition here.

Definition 2.4.1. Let (X,Σ) be a measurable space and $\mu\colon\Sigma\to\mathbb{R}^*$ be a set function. We say that μ is a **signed measure** if the following hold:
(a) $\mu(\emptyset) = 0$;
(b) μ takes at most one of the values $+\infty$ and $-\infty$, that is, either $\mu\colon\Sigma\to(-\infty,+\infty]$ or $\mu\colon\Sigma\to[-\infty,+\infty)$;
(c) for every sequence $\{A_n\}_{n\geq1}\subseteq\Sigma$ of pairwise disjoint sets, we have

$$\mu\Big(\bigcup_{n\geq1} A_n\Big) = \sum_{n\geq1}\mu(A_n)\,. \tag{2.4.1}$$

Remark 2.4.2. If $\mu(\bigcup_{n\geq1} A_n)$ is finite in (2.4.1), then the sum on the right-hand side must converge independently of any rearrangement since the left-hand side is independent of the order of the terms. So the sum in (2.4.1) converges absolutely. Note that if μ_1, μ_2 are two measures on Σ and at least one of them is finite, then $\mu = \mu_1 - \mu_2$ is a signed measure.

Straightforward modifications in the proofs of Propositions 2.1.26 and 2.1.27 lead to the following characterization of signed measures.

Proposition 2.4.3. *If (X, Σ) is a measurable space and $\mu: \Sigma \to \mathbb{R}$ is an additive set function such that $\mu(\emptyset) = 0$, then μ is a signed measure if and only if one of the following equivalent properties holds:*
(a) *$\{A_n\}_{n\geq 1} \subseteq \Sigma$ and $A_n \nearrow A$ imply $\mu(A_n) \to \mu(A)$;*
(b) *$\{A_n\}_{n\geq 1} \subseteq \Sigma$ and $A_n \searrow A$ imply $\mu(A_n) \to \mu(A)$;*
(c) *$\{A_n\}_{n\geq 1} \subseteq \Sigma$ and $A_n \searrow \emptyset$ imply $\mu(A_n) \to 0$.*

As we will see in the sequel, in order to study signed measures it is convenient to write them as differences of measures. For this reason we state the following definition.

Definition 2.4.4. Let (X, Σ) be a measurable space and $\mu: \Sigma \to \mathbb{R}^*$ be a signed measure. A set $A \in \Sigma$ is said to be a **positive** (resp. **negative**) set for μ, if $\mu(B) \geq 0$ (resp. $\mu(B) \leq 0$) for all $B \in \Sigma$, $B \subseteq A$.

Example 2.4.5. Suppose that (X, Σ, μ) is a measure space and let $f: X \to \mathbb{R}^*$ be a Σ-measurable function such that at least one of $\int_X f^+ \, d\mu$ and $\int_X f^- \, d\mu$ is finite. Then the set function $\nu: \Sigma \to \mathbb{R}^*$ defined by $\nu(A) = \int_A f \, d\mu = \int_X f\chi_A \, d\mu$ is a signed measure and a set $A \in \Sigma$ is positive (resp. negative, null) for ν if $f \geq 0$ (resp. $f \leq 0$, $f = 0$) μ-a. e. on A.

It can happen that a set has positive μ-measure with μ being a signed measure but the set is not positive for μ.

Example 2.4.6. Let $X = \mathbb{R}$ and $\Sigma = \mathcal{B}(X)$. Consider $f: \mathbb{R} \to \mathbb{R}$ to be an odd function that is λ-integrable where λ denotes the Lebesgue measure. Assume that $f(x) > 0$ for all $x > 0$. Then $\nu(A) = \int_A f \, d\lambda$ is a signed measure (see Example 2.4.5), and any set of the form $[-a, b]$ with $0 < a < b$ has positive ν-measure without being a positive set for ν.

Next we will describe the structure of signed measures. We will show that X is the union of two disjoint sets, one positive and the other one negative. We start with a proposition for positive sets.

Proposition 2.4.7. *If (X, Σ) is a measurable space, $\mu: \Sigma \to \mathbb{R}^*$ is a signed measure and $A \in \Sigma$ is a positive set for μ, then any $B \in \Sigma$, $B \subseteq A$ is also a positive set for μ. Moreover, the union of any countable family of positive sets for μ is a positive set for μ.*

Proof. The first part of the conclusion is an immediate consequence from Definition 2.4.4.

Suppose that $\{A_n\}_{n\geq 1} \subseteq \Sigma$ are positive sets for μ. Let $C_n = A_n \setminus \bigcup_{k=1}^{n-1} A_k$. Then $C_n \in \Sigma$, $C_n \subseteq A_n$ and so from the first part C_n is positive for μ. Note that $\bigcup_{n\geq 1} A_n = \bigcup_{n\geq 1} C_n$ and the C_n's are mutually disjoint. So, if $B \in \Sigma$, $B \subseteq \bigcup_{n\geq 1} A_n$, then, by the σ-additivity of μ, $\mu(B) = \sum_{n\geq 1} \mu(B \cap C_n)$. Hence, $\mu(B) \geq 0$. So, we conclude that $\bigcup_{n\geq 1} A_n \in \Sigma$ is a positive set for μ. □

Now we can state the following important theorem for signed measures. The result is known as the "Hahn Decomposition Theorem".

Theorem 2.4.8 (Hahn Decomposition Theorem). *If (X, Σ) is a measurable space and $\mu: \Sigma \to \mathbb{R}^*$ is a signed measure, then there exists a positive set $P \in \Sigma$ and a negative set $N \in \Sigma$ such that $X = P \cup N$ and $P \cap N = \emptyset$. Moreover, if P', N' is another such positive-negative decomposition of X, then $P \triangle P' = N \triangle N'$ is μ-null.*

Proof. Without any loss of generality we may assume that μ has values in $[-\infty, +\infty)$; see Definition 2.4.1. We define

$$\eta = \sup[\mu(A) : A \in \Sigma, A \text{ is a positive set for } \mu] \geq 0 . \tag{2.4.2}$$

Let $\{A_n\}_{n\geq1} \subseteq \Sigma$ be a sequence of positive sets such that $\mu(A_n) \to \eta$. Let $P = \bigcup_{n\geq1} A_n$. Then Propositions 2.4.7 and 2.4.3 imply that

$$P \text{ is positive for } \mu \text{ and } \mu(P) = \eta < +\infty . \tag{2.4.3}$$

Let $N = X \setminus P$. We claim that N is a negative set for μ. Arguing by contradiction, suppose that N is not negative for μ.

First we show that N cannot contain a positive set that is not μ-null. Indeed, if $A \subseteq N$ is positive and $\mu(A) > 0$, then $A \cup P$ is positive (see Proposition 2.4.7), and $\mu(A \cup P) = \mu(A) + \mu(P) \geq \eta$ (see (2.4.3)), a contradiction to the definition of $\eta \geq 0$ (see (2.4.2)).

Second, if $A \subseteq N$ and $\mu(A) > 0$, then there exists $B \in \Sigma$, $B \subseteq A$ with $\mu(B) > \mu(A)$. Indeed, since A is not positive, we can find $C \in \Sigma$, $C \subseteq A$ with $\mu(C) < 0$. Then if $B = A \setminus C$, we have $\mu(B) = \mu(A) - \mu(C) > \mu(A)$.

Since we have assumed that N is not a negative set for μ, we can produce a sequence $\{A_n\}_{n\geq1} \subseteq \Sigma$ with $A_n \subseteq N$ for all $n \in \mathbb{N}$ and a sequence $\{k_n\}_{n\geq1} \subseteq \mathbb{N}$ as follows:

k_1 is the smallest natural number for which we can find $B \in \Sigma$, $B \subseteq N$ with $\mu(B) > 1/k_1$. We set $A_1 = B$. Continuing inductively, let k_n be the smallest natural number for which we can find $B \in \Sigma$, $B \subseteq A_{n-1}$ with $\mu(B) \geq \mu(A_{n-1}) + 1/k_n$. We set $A_n = B$. Let $A = \bigcap_{n\geq1} A_n$. Then by Proposition 2.4.3, it follows that $\infty > \mu(A) = \lim_{n\to\infty} \mu(A_n) \geq \sum_{n\geq1} 1/k_n$, which gives $k_n \to \infty$. But as before, there exists $B \in \Sigma$, $B \subseteq A$ with $\mu(B) \geq \mu(A) + 1/k$ for some $k \in \mathbb{N}$. Then for large enough $n \in \mathbb{N}$, we have $k < k_n$ and $B \subseteq A_{n-1}$, a contradiction to the construction of the sequences $\{A_n\}_{n\geq1} \subseteq \Sigma$ and $\{k_n\}_{n\geq1} \subseteq \mathbb{N}$. It follows that N is negative for μ.

Finally suppose that P', N' is another such positive-negative pair. We have $P \setminus P' \subseteq P$ and $P \setminus P' \subseteq N'$, which yields that $P \setminus P'$ is both positive and negative for μ; see Proposition 2.4.7. This gives $\mu(P \setminus P') = 0$. Similarly we can show this for the set $P' \setminus P$. This completes the proof of the theorem. □

Remark 2.4.9. The pair (P, N) is called a **Hahn decomposition** for the signed measure μ.

The Hahn decomposition will lead us to a canonical decomposition of a signed measure. First we state a definition that is central in our considerations in this section.

Definition 2.4.10. Let (X, Σ) be a measurable space and $\mu, \nu\colon \Sigma \to [0, +\infty]$ be two measures.

(a) We say that μ and ν are **mutually singular**, denoted by $\mu \perp \nu$, if there exists two disjoint sets $X_\mu, X_\nu \in \Sigma$ such that $X = X_\mu \cup X_\nu$ and for every $A \in \Sigma$, it holds that

$$\mu(A) = \mu(A \cap X_\mu) \quad \text{and} \quad \nu(A) = \nu(A \cap X_\nu)\,.$$

(b) We say that ν is **absolutely continuous** with respect to μ, denoted by $\nu \ll \mu$, if for every $A \in \Sigma$ with $\mu(A) = 0$ it holds that $\nu(A) = 0$.

Proposition 2.4.11. *If (X, Σ) is a measurable space and $\mu, \nu\colon \Sigma \to [0, +\infty]$ are two measures with ν being finite, then $\nu \ll \mu$ if and only if for every $\varepsilon > 0$ there exists $\delta > 0$ such that*

$$A \in \Sigma \quad \textit{and} \quad \mu(A) \le \delta \quad \textit{imply} \quad \nu(A) \le \varepsilon\,. \tag{2.4.4}$$

Proof. $\Longrightarrow$: Arguing by contradiction suppose that the implication is not true. Then there exist $\varepsilon > 0$ and a sequence $\{A_n\}_{n\ge1} \subseteq \Sigma$ such that

$$\mu(A_n) \le \frac{1}{2^n} \quad \text{and} \quad \nu(A_n) \ge \varepsilon \quad \text{for all } n \in \mathbb{N}\,. \tag{2.4.5}$$

Set $B_k = \bigcup_{n\ge k} A_n \in \Sigma$ and $B = \bigcap_{k\ge1} B_k \in \Sigma$. Then

$$\mu(B) \le \mu(B_k) \le \sum_{n\ge k} \frac{1}{2^n} = \frac{1}{2^{k+1}} \to 0 \quad \text{as } k \to +\infty\,.$$

Hence,

$$\mu(B) = 0\,. \tag{2.4.6}$$

On the other hand, since ν is finite, Proposition 2.1.24 (f) gives

$$\nu(B) = \lim_{n\to\infty} \nu(B_n) \ge \lim_{n\to\infty} \nu(A_n) \ge \varepsilon\,;$$

see (2.4.5). This contradicts the hypothesis that $\nu \ll \mu$; see (2.4.6).

$\Longleftarrow$: If $A \in \Sigma$ with $\mu(A) = 0$, then $\nu(A) \le \varepsilon$ for all $\varepsilon > 0$ and so $\nu(A) = 0$. Therefore $\nu \ll \mu$. ☐

Remark 2.4.12. From the proposition above, we infer that if ν is finite, then $\nu \ll \mu$ if and only if $\lim_{\mu(A)\to0} \nu(A) = 0$.

If ν is not finite, then only the implication "$\Longleftarrow$" is valid in Proposition 2.4.11.

Example 2.4.13. Let $X = (0,1)$, $\Sigma = \mathcal{B}((0,1))$ and $\mu = \lambda$ be the Lebesgue measure on $(0,1)$. Define $\nu(A) = \int_A 1/x \, d\lambda(x)$ for all $A \in \mathcal{B}((0,1))$. Then $\nu \ll \mu$, but (2.4.4) fails.

Now we will use the Hahn decomposition of X to produce a canonical representation of a signed measure as the difference of two measures. The result is known as the "Jordan Decomposition Theorem".

Theorem 2.4.14 (Jordan Decomposition Theorem). *If (X,Σ) is a measurable space and $\mu: \Sigma \to \mathbb{R}^*$ is a signed measure, then there exist unique positive measures $\mu_+, \mu_-: \Sigma \to [0,+\infty]$ with at least one of them finite such that $\mu = \mu_+ - \mu_-$ and $\mu_+ \perp \mu_-$.*

Proof. Let (P,N) be a Hahn decomposition for μ; see Theorem 2.4.8. We define

$$\mu_+(A) = \mu(A \cap P) \quad \text{and} \quad \mu_-(A) = -\mu(A \cap N) \quad \text{for all } A \in \Sigma\,.$$

Then we have $\mu = \mu_+ - \mu_-$ and $\mu_+ \perp \mu_-$.

Suppose that (ξ_+, ξ_-) is another pair of measures such that $\mu = \xi_+ - \xi_-$ and $\xi_+ \perp \xi_-$. Let $A, B \in \Sigma$ be such that $A \cap B = \emptyset$, $A \cup B = X$ and $\xi_+(B) = \xi_-(A) = 0$. Then $X = A \cup B$ is another Hahn decomposition for μ and so $\mu(P \triangle A) = 0$; see Theorem 2.4.8. Therefore for any $D \in \Sigma$ it follows that

$$\xi_+(D) = \xi_+(D \cap A) = \mu(D \cap A) = \mu(D \cap P) = \mu_+(D)\,,$$

which gives $\xi_+ = \mu_+$.

Similarly we show that $\xi_- = \mu_-$ and this proves the uniqueness of the difference decomposition. □

Definition 2.4.15. The measures μ_+ and μ_- from the proposition above are called the **positive** and **negative variations** of μ and $\mu = \mu_+ - \mu_-$ is called the **Jordan decomposition** of μ. The **total variation** of μ is the measure $|\mu|$ defined by $|\mu| = \mu_+ + \mu_-$.

Remark 2.4.16. For every $A \in \Sigma$ we have

$$\begin{aligned}
\mu_+(A) &= \sup[\mu(C): C \in \Sigma, C \subseteq A, C \text{ is positive}] = \sup[\mu(C): C \in \Sigma, C \subseteq A]\,,\\
\mu_-(A) &= -\inf[\mu(C): C \in \Sigma, C \subseteq A, C \text{ is negative}] = -\inf[\mu(C): C \in \Sigma, C \subseteq A]\,,\\
|\mu|(A) &= \sup\left[\sum_{k=1}^{n} |\mu(A_k)| : n \in \mathbb{N}, \{A_k\}_{k=1}^{n} \subseteq \Sigma \text{ are disjoint and } A = \bigcup_{k=1}^{n} A_k\right].
\end{aligned}$$

Moreover, using the Jordan decomposition, we can define the Lebesgue integral with respect to a signed measure. So, let (X,Σ) be a measurable space and let $\mu: \Sigma \to \mathbb{R}^*$ be a signed measure. Consider $f: X \to \mathbb{R}^*$ a Σ-measurable function and $A \in \Sigma$. Suppose that at least one of the integrals $\int_A df\mu_+$ and $\int_A f \, d\mu_-$ is finite. Then the Lebesgue integral of f over A is defined as

$$\int_A f \, d\mu = \int_A f \, d\mu_+ - \int_A f \, d\mu_-\,.$$

If both integrals $\int_A f \, d\mu_+$, $\int_A f \, d\mu_-$ are finite, then we say that f is **Lebesgue integrable** with respect to μ over the set $A \in \Sigma$.

The Jordan decomposition established in Theorem 2.4.14 is minimal in the following sense.

Proposition 2.4.17. *If (X, Σ) is a measurable space, $\mu: \Sigma \to \mathbb{R}^*$ is a signed measure and $\mu = \xi_1 - \xi_2$ with $\xi_1, \xi_2: \Sigma \to [0, +\infty]$ being measures, then $\xi_1 \geq \mu_+$ and $\xi_2 \geq \mu_-$.*

Proof. We have $\mu \leq \xi_1$. Hence, for all $A \in \Sigma$,

$$\mu_+(A) = \mu(A \cap P) \leq \xi_1(A \cap P) \leq \xi_1(A) .$$

Therefore $\mu_+ \leq \xi_1$. Similarly we show that $\mu_- \leq \xi_2$. □

We extend the notions introduced in Definition 2.4.10 to signed measures.

Definition 2.4.18. Let (X, Σ) be a measurable space and $\mu, \nu: \Sigma \to \mathbb{R}^*$ be two signed measures.

(a) We say that μ and ν are **mutually singular**, denoted by $\mu \perp \nu$, if $|\mu| \perp |\nu|$; see Definition 2.4.10 (a).

(b) We say that ν is **absolutely continuous** with respect to μ, denoted by $\nu \ll \mu$, if $|\nu| \ll |\mu|$; see Definition 2.4.10 (b).

Remark 2.4.19. If μ is a signed measure, then $\mu_+ \perp \mu_-$.

The notion of mutual singularity is the antithesis of the notion of absolutely continuity.

Proposition 2.4.20. *If (X, Σ) is a measurable space and $\mu, \nu: \Sigma \to \mathbb{R}^*$ are signed measures, then $\mu \perp \nu$ and $\nu \ll \mu$ imply $\nu = 0$.*

Proof. Since by hypothesis $\mu \perp \nu$, there exist $A, B \in \Sigma$ with $A \cap B = \emptyset$, $X = A \cup B$, and $|\mu|(A) = |\nu|(B) = 0$; see Definition 2.4.18 (a). By hypothesis we also have that $\nu \ll \mu$ and so $|\nu|(A) = 0$; see Definition 2.4.18 (b). For every $C \in \Sigma$, it holds that

$$\begin{aligned} |\nu|(C) &= |\nu|(C \cap A) + |\nu|(C \cap B) \geq |\nu(C \cap A)| + |\nu(C \cap B)| \\ &\geq |\nu(C \cap A) + \nu(C \cap B)| = |\nu(C)| , \end{aligned}$$

by the additivity of ν. Hence, $|\nu(C)| = 0$ for all $C \in \Sigma$ and so $\nu \equiv 0$. □

Proposition 2.4.21. *If (X, Σ) is a measurable space and $\mu, \nu: \Sigma \to \mathbb{R}^*$ are signed measures, then $\nu \ll \mu$ if and only if $\nu_+ \ll \mu$ and $\nu_- \ll \mu$.*

Proof. $\Longrightarrow$: Suppose that $A \in \Sigma$ satisfies $|\mu|(A) = 0$. Then for $B \in \Sigma$, $B \subseteq A$ it follows $|\mu|(B) = 0$ and so $|\nu(B)| \leq |\nu|(B) = 0$. From Remark 2.4.16 we have

$$\nu_+(A) = \sup\left[\nu(B): B \in \Sigma, B \subseteq A\right] = 0 .$$

Hence $\nu_+ \ll \mu$. Similarly we show that $\nu_- \ll \mu$.

$\Longleftarrow$: Suppose that $A \in \Sigma$ satisfies $|\mu|(A) = 0$. By hypothesis one gets $\nu_+(A) = \nu_-(A) = 0$. Recall that $|\nu| = \nu_+ + \nu_-$; see Definition 2.4.15. Therefore $|\nu|(A) = 0$ and we have proved that $\nu \ll \mu$. □

Remark 2.4.22. Evidently $\nu \ll \mu$ if and only if $A \in \Sigma$ with $|\nu|(A) = 0$ imply $\mu(A) = 0$.

In a similar fashion we also show the following facts about singular and absolutely continuous signed measures.

Proposition 2.4.23. *If (X, Σ) is a measurable space and $\mu, \nu, \xi : \Sigma \to \mathbb{R}^*$ are signed measures, then the following hold:*
(a) *$\mu \ll \xi$ and $\nu \ll \xi$ imply $|\mu| + |\nu| \ll \xi$;*
(b) *$\mu \perp \xi$ and $\nu \perp \xi$ imply $|\mu| + |\nu| \perp \xi$;*
(c) *$\mu \ll \xi$ and $\nu \ll \mu$ imply $\nu \ll \xi$;*
(d) *$\mu \perp \xi$ and $\nu \ll \mu$ imply $\nu \perp \xi$.*

Definition 2.4.24. Let (X, Σ) be a measurable space and $\mu : \Sigma \to \mathbb{R}^*$ be a signed measure.
(a) We say that μ is **finite** if $\mu(A) \in \mathbb{R}$ for every $A \in \Sigma$.
(b) We say that μ is **σ-finite** if there exists a sequence $\{A_n\}_{n\geq 1} \subseteq \Sigma$ such that $X = \bigcup_{n\geq 1} A_n$ and $\mu(A_n) \in \mathbb{R}$ for all $n \in \mathbb{N}$.

Remark 2.4.25. A signed measure μ is finite if and only if $|\mu(X)| < +\infty$. Moreover, we can assume in Definition 2.4.24 (b) that the A_n's are mutually disjoint.

Proposition 2.4.26. *If (X, Σ) is a measurable space, $\nu : \Sigma \to \mathbb{R}^*$ is a finite signed measure and $\mu : \Sigma \to [0, +\infty]$ is a measure, then $\nu \ll \mu$ if and only if for every $\varepsilon > 0$ there exists $\delta > 0$ such that $A \in \Sigma$, $\mu(A) \leq \delta$ imply $|\nu(A)| \leq \varepsilon$.*

Proof. According to Definition 2.4.18 (b), $\nu \ll \mu$ if and only if $|\nu| \ll \mu$, and recall that $|\nu(A)| \leq |\nu|(A)$ for all $A \in \Sigma$. Then the conclusion follows from Proposition 2.4.11. □

Corollary 2.4.27. *If (X, Σ, μ) is a measure space and $f \in L^1(X)$, then for a given $\varepsilon > 0$ there exists $\delta = \delta(\varepsilon) > 0$ such that $A \in \Sigma$, $\mu(A) \leq \delta$ imply $|\int_A f \, d\mu| \leq \varepsilon$.*

The technical result, which we prove next, will be used in the proof of the main structural result concerning signed measures, the so-called "Radon–Nikodym Theorem".

Lemma 2.4.28. *If (X, Σ) is a measurable space, μ, ν are measures on Σ with μ being σ-finite, $\nu \not\equiv 0$ and $\nu \ll \mu$, then there exist $\varepsilon > 0$ and $B \in \Sigma$ with $0 < \mu(B) < +\infty$ such that $\varepsilon\mu(C) \leq \nu(C)$ for all $C \in \Sigma$, $C \subseteq B$, that is, B is a positive set for $\mu - \varepsilon\nu$.*

Proof. Let $\{A_n\}_{n\geq 1} \subseteq \Sigma$ be disjoint sets such that $X = \bigcup_{n\geq 1} A_n$ and $\mu(A_n) < +\infty$ for all $n \in \mathbb{N}$. Since $\nu \not\equiv 0$ we can find $m \in \mathbb{N}$ such that $\nu(A_m) > 0$. We choose $\varepsilon > 0$ small such that

$$\nu(A_m) - \varepsilon\mu(A_m) = (\nu - \varepsilon\mu)(A_m) > 0\,.$$

From Problem 2.53 we know that there exists $B \in \Sigma$, $B \subseteq A_m$ such that

$$(\nu - \varepsilon\mu)(B) > 0 \quad \text{and} \quad B \text{ is a positive set for } \nu - \varepsilon\mu\,. \tag{2.4.7}$$

Evidently $(\nu - \varepsilon\mu)(B) < +\infty$. Moreover, if $\mu(B) = 0$, then from (2.4.7) we have $\nu(B) > 0$, which contradicts the hypothesis that $\nu \ll \mu$. Therefore $\mu(B) > 0$. In addition, (2.4.7) implies that $\varepsilon\mu(C) \le \nu(C)$ for all $C \in \Sigma$, $C \subseteq B$. □

We saw in Example 2.4.5 that for a given measure space (X, Σ, μ) and $f \in L^1(X)$, the set function $\Sigma \ni A \xrightarrow{\nu} \int_A f\,d\mu$ is a signed measure. It is natural to ask whether the converse is true as well. Namely, if $\nu \ll \mu$, then can we find $f \in L^1(X, \mu)$ such that $d\nu = f\,d\mu$? The answer to this fundamental question is given by the so-called "Radon–Nikodym Theorem".

Theorem 2.4.29 (Radon–Nikodym Theorem). *If (X, Σ) is a measurable space, $\mu: \Sigma \to [0, +\infty]$ is a σ-finite measure, $\nu: \Sigma \to \mathbb{R}$ is a σ-finite signed measure and $\nu \ll \mu$, then there exists a unique up to equality μ-a. e. Σ-measurable function $f: X \to \mathbb{R}^*$ such that $\nu(A) = \int_A f\,d\mu$ for all $A \in \Sigma$.*

Proof. We know that ν_+, ν_- are finite measures on Σ and from Proposition 2.4.21, we know that $\nu_+ \ll \mu$ and $\nu_- \ll \mu$. Moreover, one has $\nu = \nu_+ - \nu_-$. Therefore without any loss of generality we may assume that ν is a σ-finite measure. It holds that $\Sigma \subseteq \Sigma_\mu \subseteq \Sigma_\nu$.

First assume that ν is finite. We introduce the set

$$\mathfrak{L} = \left\{h \in L^1(X): h \ge 0\ \mu\text{-a. e. and} \int_A h\,d\mu \le \nu(A) \text{ for all } A \in \Sigma_\mu\right\}. \tag{2.4.8}$$

We have $0 \in \mathfrak{L}$ and so $\mathfrak{L} \neq \emptyset$. Let $h_1, h_2 \in \mathfrak{L}$ and $A \in \Sigma_\mu$ and let

$$B = \{x \in A: h_1(x) \ge h_2(x)\}\,, \quad C = A \setminus B = \{x \in A: h_2(x) > h_1(x)\}\,.$$

Evidently $B, C \in \Sigma_\mu$, $A = B \cup C$ and $B \cap C = \emptyset$. Hence

$$\begin{aligned}\int_A \max\{h_1, h_2\}\,d\mu &= \int_B \max\{h_1, h_2\}\,d\mu + \int_C \max\{h_1, h_2\}\,d\mu \\ &= \int_B h_1\,d\mu + \int_C h_2\,d\mu \le \nu(B) + \nu(C) = \nu(A)\,.\end{aligned}$$

Thus, $\max\{h_1, h_2\} \in \mathfrak{L}$. We define

$$\eta = \sup\left[\int_X h\,d\mu: h \in \mathfrak{L}\right] \le \nu(X) < +\infty\,;$$

see (2.4.8). Let $\{h_n\}_{n\geq 1} \subseteq \mathfrak{L}$ be such that $\lim_{n\to\infty} \int_X h_n \, d\mu = \eta$. We set $g_n = \max\{h_k\}_{k=1}^n$. Then from the previous part of the proof we have that $\{g_n\}_{n\geq 1} \subseteq \mathfrak{L}$ is increasing and $\int_X g_n \, d\mu \nearrow \eta$. From the Monotone Convergence Theorem (see Theorem 2.3.3) we know that there exists $g \in L^1(X,\mu)$ such that $g_n \nearrow g$ and $\int_X g \, d\mu = \eta$. We have

$$0 \leq g_n\chi_A \nearrow g\chi_A \quad \text{and} \quad \int_X g_n\chi_A \, d\mu = \int_A g_n \, d\mu \leq \nu(A) \quad \text{for all } n \in \mathbb{N},$$

which implies $\int_A g \, d\mu \leq \nu(A)$ for all $A \in \Sigma_\mu$ and so $g \in \mathfrak{L}$.

Finally we show that $\nu(A) = \int_A g \, d\mu$ for all $A \in \Sigma_\mu$. Let

$$\xi(A) = \nu(A) - \int_A g \, d\mu \quad \text{for all } A \in \Sigma_\mu . \tag{2.4.9}$$

Then ξ is a measure on Σ_μ and $\xi \ll \mu$. Suppose that $\xi \not\equiv 0$. Then Lemma 2.4.28 implies that there exist $\varepsilon > 0$ and $B \in \Sigma_\mu$ such that

$$0 < \mu(B) < \infty \quad \text{and} \quad \varepsilon\mu(C) \leq \xi(C) \quad \text{for all } C \in \Sigma_\mu,\ C \subseteq B . \tag{2.4.10}$$

Let $h = g + \varepsilon\chi_B$. Then $h \geq 0$ μ-a. e. and $h \in L^1(X,\mu)$. We have $\eta = \int_X g \, d\mu < \int_X h \, d\mu$, which gives

$$h \notin \mathfrak{L} . \tag{2.4.11}$$

On the other hand, for every $A \in \Sigma_\mu$, we derive, combining (2.4.8), (2.4.9), (2.4.10),

$$\begin{aligned}
\int_A h \, d\mu &= \int_A [g + \varepsilon\chi_B] \, d\mu = \int_A g \, d\mu + \varepsilon\mu(B \cap A) \leq \int_A g \, d\mu + \xi(B \cap A) \\
&\leq \int_A g \, d\mu + \nu(B \cap A) - \int_{B\cap A} g \, d\mu = \int_{A\setminus B} g \, d\mu + \nu(B \cap A) \\
&\leq \nu(A \setminus B) + \nu(B \cap A) = \nu(A) .
\end{aligned}$$

This yields

$$h \in \mathfrak{L} . \tag{2.4.12}$$

Comparing (2.4.11) and (2.4.12), we reach a contradiction. Therefore

$$\nu(A) = \int_A g \, d\mu \quad \text{for all } A \in \Sigma .$$

Proposition 2.2.40 (c) implies that $g \in L^1(X,\mu)$ is unique.

Now suppose that ν is σ-finite. Then we find a sequence $\{A_n\}_{n\geq 1} \subseteq \Sigma$ of disjoint sets such that $X = \bigcup_{n\geq 1} A_n$ with $\nu(A_n) < +\infty$ for all $n \in \mathbb{N}$. Let $\nu_n = \nu|_{A_n}$ for every $n \in \mathbb{N}$, that is, $\nu_n(B) = \nu(B \cap A_n)$ for all $n \in \mathbb{N}$. Evidently, ν_n is a finite measure on Σ and $\nu_n \ll \mu$. So, from the first part of the proof there exists a unique $g_n \in L^1(X,\mu)$ such that $\nu_n(B) = \int_B g_n \, d\mu$ for all $B \in \Sigma$. Recall that the A_n's are disjoint. We define $g = \sum_{n\geq 1} g_n \chi_{A_n}$ and we have that $g: X \to \mathbb{R}$ is Σ-measurable as well as

$$\nu(B) = \sum_{n\geq 1} \nu(B \cap A_n) = \sum_{n\geq 1} \int_B g_n \chi_{A_n} \, d\mu = \int_B g \, d\mu ,$$

see Theorem 2.3.5. □

Definition 2.4.30. The unique (up to equality μ-a. e.) function $g: X \to \mathbb{R}^*$ postulated by Theorem 2.4.29 is called the **Radon–Nikodym derivative** of ν with respect to μ and is denoted by $d\nu/d\mu = g$ or by $d\nu = g \, d\mu$. If ν is finite, then $g \in L^1(X,\mu)$ and if ν is a measure then $g \geq 0$ μ-a. e.

Theorem 2.4.29 leads to an interesting decomposition of ν. This result is known as the "Lebesgue Decomposition Theorem".

Theorem 2.4.31 (Lebesgue Decomposition Theorem). *If (X,Σ) is a measurable space, $\mu: \Sigma \to [0,+\infty]$ a σ-finite measure and $\nu: \Sigma \to \mathbb{R}^*$ is a σ-finite signed measure, then $\nu = \nu_a + \nu_s$ with $\nu_a \ll \mu$, $\nu_s \perp \mu$ and this decomposition is unique.*

Proof. Let $\xi = \mu + \nu$. Then ξ is a σ-finite measure on Σ and $\mu \ll \xi$, $\nu \ll \xi$. Applying Theorem 2.4.29, we can find Σ-measurable functions $g, h: X \to [0,+\infty]$ such that

$$\mu(A) = \int_A g \, d\xi \quad \text{and} \quad \nu(A) = \int_A h \, d\xi \quad \text{for all } A \in \Sigma . \tag{2.4.13}$$

Let $B = \{x \in X : g(x) > 0\}$ and $C = \{x \in X : g(x) = 0\}$. Then $B, C \in \Sigma$, $B \cap C = \emptyset$, $X = B \cup C$ and $\mu(C) = 0$; see (2.4.13). Let $\hat{\nu} = \nu|_C$, that is, $\hat{\nu}(E) = \nu(E \cap C)$ for all $E \in \Sigma$. Then $\hat{\nu}(B) = 0$ and so it follows that $\hat{\nu} \perp \mu$. Let $\tilde{\nu} = \nu|_B$, that is, $\tilde{\nu}(E) = \nu(E \cap B)$ for all $E \in \Sigma$. We obtain $\tilde{\nu}(E) = \nu(E \cap B) = \int_{E\cap B} h \, d\xi$; see (2.4.13) and $\nu = \tilde{\nu} + \hat{\nu}$.

We need to show that $\tilde{\nu} \ll \mu$. To this end, let $E \in \Sigma$ be such that $\mu(E) = 0$. Then $0 = \mu(E) = \int_E g \, d\xi$ (see (2.4.13)) and so, since $g \geq 0$ ξ-a. e., $g(x) = 0$ for ξ-a. a. $x \in E$. As $g|_{E\cap B} > 0$, we must have $\xi(E \cap B) = 0$, hence $\nu(E \cap B) = 0$ since $\nu \ll \xi$. Therefore $\tilde{\nu}(E) = \nu(E \cap B)$ and this shows that $\tilde{\nu} \ll \mu$.

Finally we show the uniqueness of this decomposition. So, suppose that (ν_a, ν_s) and (ν_a', ν_s') are two such decompositions. Then

$$\nu_a - \nu_a' = \nu_s' - \nu_s . \tag{2.4.14}$$

From Proposition 2.4.23 we have

$$\nu_a - \nu'_a \ll \mu \quad \text{and} \quad (\nu'_s - \nu_s) \perp \mu\,. \tag{2.4.15}$$

From (2.4.14), (2.4.15) and Proposition 2.4.20, we conclude that $\nu_a = \nu'_a$ and $\nu_s = \nu'_s$. Hence, the decomposition is unique. □

Definition 2.4.32. The decomposition $\nu = \nu_a + \nu_s$ provided by the previous theorem with $\nu_a \ll \mu$ as well as $\nu_s \perp \mu$ is called the **Lebesgue decomposition** of ν with respect to μ.

We conclude this section with two useful results concerning setwise limits of sequences of finite measures.

The first result is known as the "Vitali–Hahn–Saks Theorem".

Theorem 2.4.33 (Vitali-Hahn-Saks Theorem). *If (X, Σ) is a measurable space, $\{\nu_n\}_{n \geq 1}$ are finite signed measures, μ is a finite measure, $\nu_n \ll \mu$ for all $n \in \mathbb{N}$ and for all $A \in \Sigma$, the limit $\nu(A) = \lim_{n\to\infty} \nu_n(A)$ exists, then $\nu\colon \Sigma \to \mathbb{R}$ is a signed measure such that $\nu \ll \mu$.*

Proof. On account of the Jordan Decomposition Theorem (see Theorem 2.4.14) we may assume that the ν_n's are measures. First we show that $\{\nu_n\}_{n\geq 1}$ is in fact uniformly absolutely continuous with respect to μ, that is, for given $\varepsilon > 0$ there exists $\delta = \delta(\varepsilon) > 0$ such that $\mu(A) \leq \delta$ implies $\nu_n(A) \leq \varepsilon$ for all $n \in \mathbb{N}$; see Proposition 2.4.11.

Let $\Sigma(\mu)$ and d_μ be as in Definition 2.3.23. We claim that $(\Sigma(\mu), d_\mu)$ is a complete metric space. Indeed, let $S = \{\chi_A : A \in \Sigma_\mu\} \subseteq L^1(X, \mu)$. Let $\{\chi_{A_n}\}_{n\geq 1} \subseteq S$ and assume that $\chi_{A_n} \to f$ in $L^1(X, \mu)$. Then according to Corollary 2.3.20, there exists a subsequence $\{\chi_{A_{n_k}}\}_{k\geq 1}$ of $\{\chi_{A_n}\}_{n\geq 1}$ such that $\chi_{A_{n_k}}(x) \to f(x)$ for μ-a. a. $x \in X$. Therefore, range$(f) = \{0, 1\}$ and since f is measurable, there exists $A \in \Sigma_\mu$ such that $f = \chi_A$. This implies that S is a closed subset of $L^1(X, \mu)$, hence a complete metric space in its own right. But S is isometrically isomorphic to $(\Sigma(\mu), d_\mu)$. Therefore the latter is a complete metric space.

Note that for every $n \in \mathbb{N}$

$$|\nu_n(A) - \nu_n(B)| \leq \nu_n(A \triangle B) \quad \text{for all } A, B \in \Sigma \text{ and } \nu_n \ll \mu\,.$$

So, the map $\nu_n\colon \Sigma \to [0, +\infty)$ with $n \in \mathbb{N}$ is well-defined and continuous. We introduce the sets

$$D_k = \{A \in \Sigma : |\nu_n(A) - \nu_m(A)| \leq \varepsilon \text{ for all } n, m \geq k\}, \quad k \in \mathbb{N}\,.$$

These sets are closed and $\Sigma = \bigcup_{k\in\mathbb{N}} D_k$. So, according to Theorem 1.5.68 (b), we can find $k \in \mathbb{N}$ such that $\operatorname{int} D_k \neq \emptyset$. This means that there exist $\tilde{A} \in D_k$ and $\delta_1 > 0$ such that $A \in \Sigma$ and $\mu(A \triangle \tilde{A}) \leq \delta_1$ imply $A \in D_k$. By hypothesis, $\nu_i \ll \mu$ for all $i \in \{1, \ldots, k\}$. So using Proposition 2.4.11 there is a $\delta \in (0, \delta_1]$ such that $A \in \Sigma$, $\mu(A) \leq \delta$ imply $\nu_i(A) \leq \varepsilon$ for all $i \in \{1, \ldots, k\}$.

If $A \in \Sigma$ and $\mu(A) \leq \delta$, then $\mu((A \cup \tilde{A}) \triangle \tilde{A}) \leq \mu(A) \leq \delta \leq \delta_1$ and so

$$\begin{aligned} |\nu_n(A) - \nu_k(A)| &= |(\nu_n - \nu_k)(A \cup \tilde{A}) - (\nu_n - \nu_k)(\tilde{A} \setminus A)| \\ &\leq |(\nu_n - \nu_k)(A \cup \tilde{A})| + |(\nu_n - \nu_k)(\tilde{A} \setminus A)| \leq 2\varepsilon \end{aligned}$$

for all $n \geq k$. Therefore it follows that $A \in \Sigma$, $\mu(A) \leq \delta$ imply $\nu_n(A) \leq 2\varepsilon + \nu_k(A) \leq 3\varepsilon$ for all $n \in \mathbb{N}$, which is the uniform absolute continuity of $\{\nu_n\}_{n\geq1}$ with respect to μ.

Now let $\{A_n\}_{n\geq1} \subseteq \Sigma$ be mutually disjoint sets and $\varepsilon > 0$. We set $A = \bigcup_{n\geq1} A_n \in \Sigma$. Let $\delta > 0$ be as postulated by the uniform absolute continuity with respect to μ established in the first part of the proof. We choose $k \in \mathbb{N}$ such that $\mu(A \setminus \bigcup_{i=1}^k A_i) \leq \delta$; see Proposition 2.1.24 (e). This implies

$$\left|\nu_n(A) - \sum_{i=1}^m \nu_n(A_i)\right| = \left|\nu_n\left(A \setminus \bigcup_{i=1}^m A_i\right)\right| \leq \varepsilon \quad \text{for all } n, m \geq k\,.$$

Hence

$$\left|\nu(A) - \sum_{i=1}^m \nu(A_i)\right| \leq \varepsilon \quad \text{for all } m \geq k\,.$$

Since $\varepsilon > 0$ is arbitrary, it follows that $\nu(A) = \sum_{i\in\mathbb{N}} \nu(A_i)$ and so ν is a measure. Moreover, from the first part of the proof and Proposition 2.4.11 we have $\nu \ll \mu$. □

The next theorem, known as "Nikodym's Theorem", is an easy consequence of the theorem above.

Theorem 2.4.34 (Nikodym's Theorem). *Let (X, Σ) be a measurable space. If $\{\nu_n\}_{n\geq1}$ is a sequence of nonzero finite measures defined on Σ such that the limit $\lim_{n\to\infty} \nu_n(A)$ exists for all $A \in \Sigma$, then $\nu(A) = \lim_{n\to\infty} \nu_n(A)$ with $A \in \Sigma$ is a finite measure.*

Proof. Consider the set function $\mu\colon \Sigma \to [0, +\infty)$ defined by

$$\mu(A) = \sum_{n\in\mathbb{N}} \frac{1}{2^n} \frac{\nu_n(A)}{\nu_n(X)} \quad \text{for all } A \in \Sigma\,.$$

Evidently μ is a finite measure on Σ and $\nu_n \ll \mu$ for all $n \in \mathbb{N}$. So, invoking Theorem 2.4.33, we conclude that ν is a finite measure on Σ. □

2.5 Regular and Radon Measures

In this section we investigate the connections between measure theory and topology. When we combine the measure theoretic and topological structures, we obtain stronger and more interesting results.

Throughout this section (X, τ) is a Hausdorff topological space. Additional conditions on X will be introduced as needed. By $C_c(X)$ we denote the space of all continuous functions $f\colon X \to \mathbb{R}$ with compact support. Recall that the support of f, denoted by $\operatorname{supp} f$, is defined to be the closure of the set $\{x \in X : f(x) \neq 0\}$.

Definition 2.5.1. The **Baire σ-algebra** of X, denoted by $\mathrm{Ba}(X)$, is defined to be the smallest σ-algebra on X, which makes all functions in $C_c(X)$ measurable. So, $\mathrm{Ba}(X)$ has as generators the sets $\{x \in X : f(x) \geq \eta\}$ with $f \in C_c(X)$ and $\eta \in \mathbb{R}$. These sets are known as **Baire sets.**

This new σ-algebra is most useful within the framework of locally compact spaces.

Lemma 2.5.2. *If X is locally compact, $K \subseteq X$ is compact and $W \subseteq X$ is open such that $K \subseteq W$, then we can find $U \in \tau \cap \mathrm{Ba}(X)$ and a compact G_δ-set C such that $K \subseteq U \subseteq C \subseteq W$.*

Proof. Proposition 1.4.66 (c) says that there exists $D \in \tau$ being relatively compact such that $K \subseteq D \subseteq \overline{D} \subseteq W$. Then Proposition 1.4.68 implies that there is $f \in C_c(X)$ such that $f|_K = 1$ and $f|_{D^c} = 0$. Let $C = \{x \in X : f(x) \geq 1/2\}$. Then $C \subseteq X$ is compact, G_δ, $U = \{x \in X : f(x) > 1/2\} \in \tau$ and we have $K \subseteq U \subseteq C \subseteq W$. □

Corollary 2.5.3. *If X is locally compact, then $\tau \cap \mathrm{Ba}(X)$ is a basis for τ.*

Proof. Let $x \in X$ and $U \in \mathcal{N}(x)$. Then Lemma 2.5.2 implies that there exists $f \in C_c(X)$ such that $f(x) = 1$ and $f|_{U^c} = 0$. Consider the set $V = \{x \in X : f(x) > 1/2\}$. Then $V \in \tau \cap \mathrm{Ba}(X)$ and $V \subseteq U$. □

Now we can give an alternative characterization of $\mathrm{Ba}(X)$ when X is locally compact.

Theorem 2.5.4. *If X is locally compact, then*

$$\mathrm{Ba}(X) = \sigma(\{C \subseteq X : C \text{ is compact and a } G_\delta\text{-set}\}) .$$

Proof. Let $\mathcal{L} = \sigma(\{C \subseteq X : C \text{ is compact and a } G_\delta\text{-set}\})$. For every $f \in C_c(X)$ and $\eta > 0$, the set $\{x \in X : f(x) \geq \eta\}$ is compact and G_δ. Note that $\{f \geq \eta\} = \bigcap_{n\geq1}\{f > \eta - 1/n\}$. Therefore $\{x \in X : f(x) \geq \eta\} \in \mathcal{L}$ for all $f \in C_c(X)$ and for all $\eta > 0$. For $\eta < 0$, we have $0 < -\eta + \eta/(2n) < -\eta$ and

$$\{f \geq \eta\} = \{f < \eta\}^c = \{-f > -\eta\}^c = \Big(\bigcap_{n\geq1}\Big\{-f \geq -\eta + \frac{\eta}{2n}\Big\}\Big)^c \in \mathcal{L} .$$

Moreover, note that $\{f \geq 0\} = \bigcap_{n\geq1}\{f \geq -1/n\} \in \mathcal{L}$. So, every set $\{x \in X : f(x) \geq \eta\}$ for $f \in C_c(X)$ and $\eta \in \mathbb{R}$, belongs to $\mathcal{L}$ and we have

$$\mathrm{Ba}(X) \subseteq \mathcal{L} ; \tag{2.5.1}$$

see Definition 2.5.1. Now suppose that $K = \bigcap_{n\geq1} W_n$ with $W_n \in \tau$ being compact. Lemma 2.5.2 implies that we can find $U_n \in \tau \cap \mathrm{Ba}(X)$ such that $K \subseteq U_n \subseteq W_n$ for all $n \in \mathbb{N}$. Then $K = \bigcap_{n\geq1} U_n \in \mathrm{Ba}(X)$, which gives

$$\mathcal{L} \subseteq \mathrm{Ba}(X) . \tag{2.5.2}$$

From (2.5.1) and (2.5.2) we conclude that $\mathcal{L} = \mathrm{Ba}(X)$. □

Next we compare the Baire and Borel σ-algebras.

Theorem 2.5.5. (a) $\mathrm{Ba}(X) \subseteq \mathcal{B}(X)$
(b) *If X is locally compact, separable and metrizable, then* $\mathrm{Ba}(X) = \mathcal{B}(X)$.

Proof. (a) Just recall that every continuous function $f: X \to \mathbb{R}$ is Borel measurable.

(b) From Proposition 1.4.78 (see also Proposition 1.5.40), we know that X is σ-compact. Therefore, every closed subset of X is likewise σ-compact. It follows that it suffices to show that every compact set belongs to $\mathrm{Ba}(X)$. But Proposition 1.5.8 says that every compact set in X is G_δ. So, according to Theorem 2.5.4, it belongs to $\mathrm{Ba}(X)$ and we conclude that $\mathrm{Ba}(X) = \mathcal{B}(X)$. □

Using Proposition 1.4.66 (d) we have at once the following result.

Proposition 2.5.6. *If X is locally compact and $\hat{\mathcal{B}}$ is a basis for τ, then* $\mathrm{Ba}(X) \subseteq \sigma(\hat{\mathcal{B}}) \subseteq \mathcal{B}(X)$.

The next theorem is the Baire counterpart of Proposition 2.2.26 (b).

Theorem 2.5.7. *If X and Y are second countable, locally compact spaces, then* $\mathrm{Ba}(X\times Y) = \mathrm{Ba}(X) \bigotimes \mathrm{Ba}(Y)$.

Proof. Note that $X \times Y$ is locally compact. We define

$$\mathcal{M}(A) = \{B \subseteq Y : A \times B \in \mathrm{Ba}(X \times Y)\} .$$

It is routine to check that $\mathcal{M}(A)$ is a σ-ring for any A. Suppose that $C \subseteq X$ is compact and a G_δ-set. Then if $E \subseteq Y$ is compact and G_δ, then so is $C\times E \subseteq X\times Y$ and we infer that $\mathcal{M}(C)$ contains every compact G_δ-set in Y. Moreover, we have $Y \in \mathcal{M}(C)$; see Proposition 1.4.78 and Theorem 1.2.27. It follows that $\mathcal{M}(C)$ is a σ-algebra containing $\mathrm{Ba}(Y)$.

Let $\mathcal{L} = \{A \subseteq X : \mathrm{Ba}(Y) \subseteq \mathcal{M}(A)\}$. This family is closed under countable intersections and under complementation and we have seen above it contains every compact G_δ. Therefore

$$\mathrm{Ba}(X) \bigotimes \mathrm{Ba}(Y) \subseteq \mathrm{Ba}(X \times Y) . \tag{2.5.3}$$

On the other hand, from Corollary 2.5.3, we know that the family

$$\mathcal{B} = \{U \times V : U \subseteq X \text{ Baire open}, V \subseteq Y \text{ Baire open}\}$$

is a basis for $X \times Y$. Since $U \times V \in \mathrm{Ba}(X) \bigotimes \mathrm{Ba}(Y)$ it follows that $\sigma(\mathcal{B}) \subseteq \mathrm{Ba}(X) \bigotimes \mathrm{Ba}(Y)$. Then Proposition 2.5.6 gives

$$\mathrm{Ba}(X \times Y) \subseteq \mathrm{Ba}(X) \bigotimes \mathrm{Ba}(Y) . \tag{2.5.4}$$

From (2.5.3) and (2.5.4), we conclude that $\mathrm{Ba}(X \times Y) = \mathrm{Ba}(X) \bigotimes \mathrm{Ba}(Y)$. □

Definition 2.5.8. (a) A **(signed) Borel measure** is a (signed) measure defined on $\mathcal{B}(X)$.
(b) We say that a Borel measure μ is **regular** if for every $A \in \mathcal{B}(X)$

$$\begin{aligned}\mu(A) &= \inf[\mu(U): U \subseteq X \text{ is open}, A \subseteq U] \quad \text{(outer regularity)}\\ &= \sup[\mu(C): C \subseteq X \text{ is closed}, C \subseteq A] \quad \text{(inner regularity)}.\end{aligned}$$

(c) We say that a Borel measure μ is **compact regular** if for every $A \in \mathcal{B}(X)$

$$\mu(A) = \sup[\mu(K): K \subseteq X \text{ is compact}, K \subseteq A].$$

(d) We say that a Borel measure is a **Radon measure** if the following hold:
- $\mu(K) < +\infty$ for every compact $K \subseteq X$;
- $\mu(A) = \inf[\mu(U): U \subseteq X$ is open, $A \subseteq U]$ for all $A \in \mathcal{B}(X)$;
- $\mu(A) = \sup[\mu(K): K \subseteq X$ is compact, $K \subseteq A]$ for all $A \in \mathcal{B}(X)$.

For a signed Borel measure μ we say that μ is regular (resp. compact regular, Radon) if $|\mu|$ is such a measure or equivalently if μ_+ and μ_- have the corresponding properties.

Remark 2.5.9. Evidently two regular Borel measures are equal if and only if they coincide on the open or closed subsets. Similarly two compact regular measures are equal if and only if they coincide on the compact sets.

Proposition 2.5.10. *For finite Borel measures μ, outer and inner regularity are equivalent properties.*

Proof. Suppose that for all $A \in \mathcal{B}(X)$

$$\mu(A) = \inf[\mu(U): U \subseteq X \text{ is open}, A \subseteq U]. \tag{2.5.5}$$

Taking Proposition 2.1.24 (b) and (2.5.5) into account yields

$$\begin{aligned}\mu(X) - \mu(A) = \mu(A^c) &= \inf[\mu(U): U \subseteq X \text{ is open}, A^c \subseteq U]\\ &= \mu(X) - \sup[\mu(C): C \subseteq X \text{ is closed}, C \subseteq A].\end{aligned}$$

Therefore, $\mu(A) = \sup[\mu(C): C \subseteq X$ is closed, $C \subseteq A]$. Hence, outer regularity implies inner regularity.

In a similar way we show that the opposite implication holds as well. So, the two notions are equivalent. □

Theorem 2.5.11. *If $\mu: \mathcal{B}(X) \to [0, +\infty)$ is a finite, compact regular Borel measure, then μ is a Radon measure.*

Proof. Since every compact subset of X is closed, for every $A \in \mathcal{B}(X)$ we derive

$$\begin{aligned}\mu(A) &= \sup[\mu(K): K \subseteq X \text{ is compact}, K \subseteq A]\\ &\le \sup[\mu(C): C \subseteq X \text{ is closed}, C \subseteq A] \le \mu(A).\end{aligned}$$

Hence,

$$\mu(A) = \sup[\mu(C) \colon C \subseteq X \text{ is closed}, C \subseteq A] . \tag{2.5.6}$$

From (2.5.6) and Proposition 2.5.10, we conclude that μ is a Radon measure. □

Theorem 2.5.12. *If X is metrizable and $\mu \colon \mathcal{B}(X) \to [0, +\infty)$ is a finite Borel measure, then μ is regular.*

Proof. Let $\mathcal{M} = \{A \in \mathcal{B}(X) \colon A$ is both outer and inner regular$\}$; see Definition 2.5.8 (a). We are going to show that $\mathcal{M}$ is a σ-algebra containing all the open sets. Therefore $\mathcal{M} = \mathcal{B}(X)$.

Fact 1: $A \in \mathcal{M}$ implies $A^c \in \mathcal{M}$.

This is immediate from the definition of $\mathcal{M}$. Recall that μ is finite and that $\mu(X) - \mu(A) = \mu(A^c)$; see Proposition 2.1.24 (b).

Fact 2: $\{A_n\}_{n\ge1} \subseteq \mathcal{M}$ implies $A = \bigcup_{n\ge1} A_n \in \mathcal{M}$.

For every $n \in \mathbb{N}$ there exist an open $U_n \subseteq X$ and a closed $C_n \subseteq X$ such that

$$C_n \subseteq A_n \subseteq U_n \quad \text{and} \quad \mu(U_n) \le \mu(C_n) + \frac{\varepsilon}{2^n} . \tag{2.5.7}$$

Let $U = \bigcup_{n\ge1} U_n$. Then $U \subseteq X$ is open and $A \subseteq U$. We know that $U \setminus A \subseteq \bigcup_{n\ge1}(U_n \setminus A_n)$. Then, due to (2.5.7), this gives

$$\begin{aligned} 0 \le \mu(U) - \mu(A) = \mu(U \setminus A) &\le \sum_{n\ge1} \mu(U_n \setminus A_n) \\ &= \sum_{n\ge1} (\mu(U_n) - \mu(A_n)) \le \sum_{n\ge1} \frac{\varepsilon}{2^n} = \varepsilon . \end{aligned}$$

Hence,

$$\mu(A) = \inf[\mu(U) \colon U \subseteq X \text{ is open}, A \subseteq U] \quad (\text{outer regularity of } A) .$$

Let $C = \bigcup_{n\ge1} C_n$. Arguing as above, we show that

$$\mu(A) \le \mu(C) + \varepsilon . \tag{2.5.8}$$

For every $m \in \mathbb{N}$, let $\tilde{C}_m = \bigcup_{n=1}^m C_n$. Evidently $\tilde{C}_m$ is closed and $\tilde{C}_m \nearrow C$. Invoking Proposition 2.1.24 (e), there exists $m \in \mathbb{N}$ such that $\mu(C) \le \mu(\tilde{C}_m) + \varepsilon$ which gives, thanks to (2.5.8), that $\mu(A) \le \mu(\tilde{C}_m) + 2\varepsilon$. This finally yields

$$\mu(A) = \sup[\mu(C) \colon C \subseteq X \text{ is closed}, C \subseteq A] \quad (\text{inner regularity of } A) .$$

Hence, $A \in \mathcal{M}$.

Fact 3: $\mathcal{M}$ contains all open sets.

Let $U \subseteq X$ be open. Proposition 1.5.8 says that U is a F_σ-set. So, we can find closed subsets $\{C_n\}_{n\geq1}$ of X such that $C_n \nearrow U$. Then $\mu(C_n) \nearrow \mu(U)$; see Proposition 2.1.24 (e). Hence

$$\mu(U) = \sup[\mu(C) : C \subseteq X \text{ is closed}, C \subseteq U] ,$$

which gives $U \in \mathcal{M}$ since U is open.

Combining Facts 1–3 shows that $\mathcal{M} = \mathcal{B}(X)$. □

Proposition 2.5.13. *If X is metrizable and $\mu: \mathcal{B}(X) \to [0, +\infty)$ is a finite Borel measure, then μ is compact regular if and only if for every $\varepsilon > 0$ there exists a compact $K_\varepsilon \subseteq X$ such that $\mu(X) - \varepsilon \leq \mu(K_\varepsilon)$.*

Proof. $\Longrightarrow$: This is immediate from Definition 2.5.8 (c).

$\Longleftarrow$: From Theorem 2.5.12 we know that μ is regular. So, it suffices to show that for every closed $C \subseteq X$, we have

$$\mu(C) = \sup[\mu(K) : K \subseteq X \text{ is compact}, K \subseteq C] . \tag{2.5.9}$$

Arguing by contradiction suppose that there exists a closed $C \subseteq X$ such that (2.5.9) is not true. So we can find $\varepsilon > 0$ such that

$$\sup[\mu(K) : K \subseteq X \text{ is compact}, K \subseteq C] \leq \mu(C) - \frac{\varepsilon}{2} . \tag{2.5.10}$$

For $K \subseteq X$ compact we have that $K \cap C \subseteq C$ is compact and, because of (2.5.10),

$$\mu(K) = \mu(K \cap C) + \mu(K \cap C^c) \leq \mu(C) - \frac{\varepsilon}{2} + \mu(C^c) = \mu(X) - \frac{\varepsilon}{2} .$$

Since $K \subseteq X$ is arbitrary, we get a contradiction to our hypothesis. □

On Polish spaces all finite Borel measures are Radon measures.

Theorem 2.5.14. *If X is a Polish space and $\mu: \mathcal{B}(X) \to [0, +\infty)$ is a finite Borel measure, then μ is a Radon measure.*

Proof. On account of Theorem 2.5.11 we only need to show that μ is compact regular. Suppose that $D = \{x_k\}_{k\geq1} \subseteq X$ is dense. We consider the closed balls $\overline{B}_n(x_k) = \{x \in X : d(x, x_k) \leq 1/n\}$ with $n, k \in \mathbb{N}$. Obviously $X = \bigcup_{k\geq1} \overline{B}_n(x_k)$ for every $n \in \mathbb{N}$. Given $\varepsilon > 0$, for every $n \in \mathbb{N}$, we can find $m_n \in \mathbb{N}$ such that

$$\mu\left(X \setminus \bigcup_{k=1}^{m_n} \overline{B}_n(x_k) \right) \leq \frac{\varepsilon}{2^n} . \tag{2.5.11}$$

Let $K = \bigcap_{n\geq1} \bigcup_{k=1}^{m_n} \overline{B}_n(x_k)$. The set K is closed and totally bounded, hence K is compact;

see Theorem 1.5.36. Taking (2.5.11) into account it follows

$$\mu(X) - \mu(K) = \mu(X \setminus K) = \mu\left[\bigcup_{n\geq 1}\left(X \setminus \bigcup_{k=1}^{m_n} \overline{B}_n(x_k)\right)\right]$$
$$\leq \sum_{n\geq 1} \mu\left(X \setminus \bigcup_{k=1}^{m_n} \overline{B}_n(x_k)\right) \leq \sum_{n\geq 1} \frac{\varepsilon}{2^n} = \varepsilon\,.$$

Hence, μ is compact regular (see Proposition 2.5.13), and so, μ is a Radon measure. □

In the next proposition we produce another useful dense subset of $L^p(X)$ for $1 \leq p < \infty$.

Proposition 2.5.15. *If X is locally compact and $\mu\colon \mathcal{B}(X) \to [0, +\infty]$ is a Radon measure, then $C_c(X)$ is dense in $L^p(X)$ for $1 \leq p < \infty$ where $C_c(X)$ is the space of all continuous functions $f\colon X \to \mathbb{R}$ that have a compact support.*

Proof. From Proposition 2.3.22, we know that simple functions are dense in $L^p(X)$. So, it suffices to show that for every $A \in \mathcal{B}(X)$ with $\mu(A) < +\infty$ we can approximate χ_A in the L^p-norm by $C_c(X)$-functions. Given $\varepsilon > 0$ there exist an open set $U \subseteq X$ and a compact set $K \subseteq X$ such that

$$K \subseteq A \subseteq U \quad \text{and} \quad \mu(U \setminus K) \leq \varepsilon^p\,. \tag{2.5.12}$$

Since X is locally compact, combining Urysohn's Lemma (see Theorem 1.2.17) and Proposition 1.4.66 (c), we can find $f \in C_c(X)$ such that $\chi_K \leq f \leq \chi_U$. Then, using (2.5.12), $\|\chi_A - f\|_p \leq \mu(U \setminus K)^{1/p} \leq \varepsilon$, which demonstrates that $C_c(X)$ is dense in $L^p(X)$ for $1 \leq p < \infty$. □

Remark 2.5.16. Since $L^\infty(X)$ contains noncontinuous functions, the density result above fails for $p = +\infty$.

The next theorem is another remarkable result in the spirit of Egorov's Theorem; see Theorem 2.2.32. It asserts that a Borel measurable map between certain metric spaces is "almost" continuous. The result is known as "Lusin's Theorem".

Theorem 2.5.17 (Lusin's Theorem). *If X is a Polish space, Y is a separable metric space, $f\colon X \to Y$ is Borel measurable, and $\mu\colon \mathcal{B}(X) \to [0, +\infty)$ is a finite Borel measure, then given any $\varepsilon > 0$, there exists $K_\varepsilon \subseteq X$ being compact such that $\mu(X \setminus K_\varepsilon) \leq \varepsilon$ and $f|_{K_\varepsilon}$ is continuous.*

Proof. We know that Y is second countable; see Proposition 1.5.5. So, let $\{V_n\}_{n\geq 1}$ be a countable basis for the metric topology of Y. We have $f^{-1}(V_n) \in \mathcal{B}(X)$ for all $n \in \mathbb{N}$ and so using Theorem 2.5.12 there exists an open set $U_n \subseteq X$ such that

$$f^{-1}(V_n) \subseteq U_n \quad \text{and} \quad \mu(U_n \setminus f^{-1}(V_n)) \leq \frac{\varepsilon}{2^{n+1}} \quad \text{for all } n \in \mathbb{N}\,. \tag{2.5.13}$$

The set $f^{-1}(V_n)$ is relatively open in $(X \setminus U_n) \cup f^{-1}(V_n)$. Note that $f^{-1}(V_n) = [(X \setminus U_n) \cup f^{-1}(V_n)] \cap U_n$, see (2.5.13). Let

$$A_\varepsilon = X \setminus \bigcup_{n\geq 1}(U_n \setminus f^{-1}(V_n)) = \bigcap_{n\geq 1}((X \setminus U_n) \cup f^{-1}(V_n)) .$$

Thanks to (2.5.13), one gets

$$\mu(X \setminus A_\varepsilon) \leq \frac{\varepsilon}{2} . \tag{2.5.14}$$

Using Theorem 2.5.14 there exists $K_\varepsilon \subseteq A_\varepsilon$ being compact such that $\mu(A_\varepsilon \setminus K_\varepsilon) \leq \varepsilon/2$, which gives $\mu(X \setminus K_\varepsilon) \leq \varepsilon$; see (2.5.14).

For every $n \in \mathbb{N}$, $f^{-1}(V_n)$ is relatively open in K_ε. Since $\{V_n\}_{n\geq 1}$ is a basis for the metric topology of Y, it follows that for all open $V \subseteq Y$, $f^{-1}(V)$ is relatively open in K_ε. Hence $f|_{K_\varepsilon}$ is continuous. □

In addition there is also a second version of Lusin's Theorem.

Theorem 2.5.18 (Lusin's Theorem, Second Version). *If X is locally compact, μ is a Radon measure and $f: X \to \mathbb{R}$ is a Borel measurable function that vanishes outside a set of finite μ-measure, then for given $\varepsilon > 0$, there exist $A \in \mathcal{B}(X)$ and $h \in C_c(X)$ such that $\mu(A) \leq \varepsilon$ and $f|_{X\setminus A} = h|_{X\setminus A}$. Moreover if f is bounded, then it holds that $\|h\|_\infty \leq \|f\|_\infty$.*

Proof. First assume that f is bounded. Let $A = \{x \in X: f(x) \neq 0\} \in \mathcal{B}(X)$. By hypothesis, $\mu(A) < +\infty$. So, we can use Proposition 2.5.15 and find $\{h_n\}_{n\geq 1} \subseteq C_c(X)$ such that $h_n \to f$ in $L^1(X)$. So, by passing to a suitable subsequence, if necessary we may assume that $h_n(x) \to f(x)$ for μ-a. a. $x \in X$; see Corollary 2.3.20. Invoking Egorov's Theorem (see Theorem 2.2.32), there exists $B \subseteq A$ such that

$$\mu(A \setminus B) \leq \frac{\varepsilon}{3} \quad \text{and} \quad h_n \xrightarrow{\mu} f \quad \text{on } B . \tag{2.5.15}$$

Exploiting the fact that μ is a Radon measure, we find a compact set $K \subseteq B$ and an open set $U \supseteq B$ such that

$$\mu(B \setminus K) \leq \frac{\varepsilon}{3} \quad \text{and} \quad \mu(U \setminus A) \leq \frac{\varepsilon}{3} . \tag{2.5.16}$$

Since $h_n \xrightarrow{\mu} f$ on K, it follows that $f|_K$ is continuous. Invoking the locally compact version of the Tietze Extension Theorem (see Theorem 1.4.88), there exists $\hat{h} \in C_c(X)$ such that $\hat{h}|_K = f|_K$ and $\operatorname{supp} \hat{h} \subseteq U$. Hence, $D = \{x \in X: \hat{h}(x) \neq f(x)\} \subseteq U \setminus K$, which demonstrates, due to (2.5.15) and (2.5.16), that $\mu(D) \leq \mu(U \setminus K) \leq \varepsilon$.

Now let $\xi: \mathbb{R} \to \mathbb{R}$ be defined by

$$\xi(t) = \begin{cases} t & \text{if } |t| \leq \|f\|_\infty , \\ \|f\|_\infty \operatorname{sgn} t & \text{if } |t| > \|f\|_\infty . \end{cases}$$

Evidently $\xi(0) = 0$, and so ξ is continuous. So, if we define $h = \xi \circ \hat{f}$, then $h \in C_c(X)$, $h = f$ on the set $\{\hat{h} = f\}$ and $\|h\|_\infty \le \|f\|_\infty$.

Finally we consider the general case in which f is unbounded. In this case we define $A_n = \{x \in X: 0 < |f(x)| \le n\} \in \mathcal{B}(X)$. Then $A_n \nearrow A$ and for large enough $n \ge 1$, we have that $\mu(A \setminus A_n) \le \varepsilon/2$. Then from the first part of the proof there exists $h \in C_c(X)$ such that $h = f\chi_{A_n}$ outside a set $D \in \mathcal{B}(X)$ with $\mu(D) \le \varepsilon/2$. Then finally we have $h = f$ outside a set $D_0 \in \mathcal{B}(X)$ with $\mu(D_0) \le \varepsilon$. ☐

There is a parametric variant of Lusin's Theorem concerning Carathéodory functions; see Definition 2.2.30. The result is known as "Scorza–Dragoni Theorem".

Theorem 2.5.19 (Scorza–Dragoni Theorem). *If T and X are Polish spaces, Y is a separable metric space, $\mu: \mathcal{B}(T) \to [0, +\infty)$ is a finite compact regular Borel measure, and $f: T \times X \to Y$ is a Carathéodory function, then for every $\varepsilon > 0$ there exists a compact set $K_\varepsilon \subseteq T$ with $\mu(T \setminus K_\varepsilon) \le \varepsilon$ such that $f|_{K_\varepsilon \times X}$ is continuous.*

Proof. From Theorem 1.5.21 we know that Y is homeomorphic to a subset of the Hilbert cube $\mathbb{H} = [0,1]^{\mathbb{N}}$. Let $h = (h_n)_{n\in\mathbb{N}}: Y \to \mathbb{H}$ be this homeomorphism. Then f is a Carathéodory function if and only if for every $n \in \mathbb{N}$, $h_n \circ f: T \times X \to [0,1]$ is a Carathéodory function. Therefore without any loss of generality we may assume that $Y = [0,1]$.

Let $\{U_n\}_{n\ge1}$ be a basis for the topology of X and let $\{x_m\}_{m\ge1} \subseteq X$ be dense. For every $q \in [0,1] \cap \mathbb{Q}$ let $\xi_{nq}: X \to [0,1]$ be defined by $\xi_{nq}(x) = q\chi_{U_n}(x)$. Since U_n is open, χ_{U_n} is lower semicontinuous (see Definition 1.7.1), and if $\varphi: X \to Y = [0,1]$ is lower semicontinuous, then $\varphi(x) = \sup[\xi_{nq}(x): \xi_{nq} \le \varphi]$ with $x \in X$. So, we define

$$A_{nqm} = \{t \in T: \xi_{nq}(x_m) \le f(t, x_m)\} \in \mathcal{B}(T) .$$

Let $A_{nq} = \bigcap_{m\in\mathbb{N}} A_{nqm} \in \mathcal{B}(T)$. The density of $\{x_m\}_{m\ge1}$ in X, the continuity of $f(t,\cdot)$, and the lower semicontinuity of ξ_{nq} imply that

$$A_{nq} = \{t \in T: \xi_{nq}(x) \le f(t,x) \text{ for all } x \in X\} .$$

We set $\eta_{nq}(t,x) = \chi_{A_{nq}}(t)\xi_{nq}(x)$. Then $\eta_{nq} \le f$ and for all $(t,x) \in T \times X$ we have $f(t,x) = \sup_{n,q} \eta_{nq}(t,x)$. Note that $\mathbb{N} \times ([0,1] \cap \mathbb{Q})$ is countable. So we can write that

$$f = \sup_{k\in\mathbb{N}} \chi_{B_k} h_k \quad \text{with } B_k \in \mathcal{B}(T) , \quad h_k \text{ is lower semicontinuous on } X .$$

Since by hypothesis μ is a finite, compact regular measure on T, there exist an open set $V_k \subseteq T$ and a compact set $K_k \subseteq T$ such that

$$K_k \subseteq B_k \subseteq V_k \quad \text{and} \quad \mu(V_k \setminus K_k) \le \frac{\varepsilon}{2^{k+2}} \quad \text{for all } k \in \mathbb{N} . \tag{2.5.17}$$

Let $E_k = K_k \cup (X \setminus V_k)$ for all $k \in \mathbb{N}$. Then $\chi_{B_k}|_{E_k}$ is continuous (see (2.5.17)), and this implies that $\chi_{B_k} h_k$ is lower semicontinuous. Let $E = \bigcap_{k\in\mathbb{N}} E_k \subseteq T$ be compact. We see that $\mu(T \setminus E) \leq \varepsilon/2$ and $f|_{E \times X}$ is lower semicontinuous as the upper envelope of lower semicontinuous functions; see Proposition 1.7.4 (a). The same argument applied to $1 - f$ produces another compact set $\tilde{E} \subseteq T$ with $\mu(T \setminus \tilde{E}) \leq \varepsilon/2$ and $(1-f)|_{\tilde{E} \times X}$ is lower semicontinuous. We set $T_\varepsilon = E \cap \tilde{E} \subseteq T$, which is compact. Then we see that $\mu(T \setminus T_\varepsilon) \leq \varepsilon$ and $f|_{T_\varepsilon \times X}$ continuous. □

Next we introduce an extension of the notion of a Carathéodory function (see Definition 2.2.30), which is important in calculus of variation, optimal control, and optimization.

Definition 2.5.20. Let (X, Σ) be a measurable space, Y a Hausdorff topological space, and $f: X \times Y \to \overline{\mathbb{R}} = \mathbb{R} \cup \{+\infty\}$. We say that f is a **normal integrand** if the following hold:
(a) f is $\Sigma \otimes \mathcal{B}(Y)$-measurable;
(b) $y \to f(x, y)$ is lower semicontinuous for all $x \in X$.

Proposition 2.5.21. *If (X, Σ, μ) is a complete measure space, Y is a Polish space, and $f: X \times Y \to \overline{\mathbb{R}} = \mathbb{R} \cup \{+\infty\}$ is a normal integrand such that there is a Carathéodory function $\xi: X \times Y \to \mathbb{R}$ satisfying $\xi(x, y) \leq f(x, y)$ for all $(x, y) \in X \times Y$, then there is a sequence of Carathéodory functions $f_n: X \times Y \to \mathbb{R}$ such that $\xi(x, y) \leq f_n(x, y) \leq f(x, y)$ for all $(x, y) \in X \times Y$ and $f_n \nearrow f$ as $n \to \infty$.*

Proof. We reason as in the proof of Proposition 1.7.6. So, we define

$$f_n(x, y) = \inf[f(x, y) + nd(y, z): z \in Y] \quad \text{for all } n \in \mathbb{N}$$

with d being the metric on Y. If $\{z_m\}_{m \geq 1} \subseteq Y$ is dense in Y, then

$$f_n(x, y) = \inf_{m \in \mathbb{N}} [f(x, y) + nd(y, z_m)] \quad \text{for all } n \in \mathbb{N}.$$

This shows that f_n is $\Sigma \otimes \mathcal{B}(X)$-measurable; see Proposition 2.2.31. Clearly we have $\xi(x, y) \leq f_n(x, y)$ for all $(x, y) \in X \times Y$, for all $n \in \mathbb{N}$ and as in the proof of Proposition 1.7.6, we show that $f_n \nearrow f$. □

Using this proposition we can have the following extension of the Scorza–Dragoni Theorem; see Theorem 2.5.19.

Theorem 2.5.22. *If T and Y are Polish spaces, μ is a finite, compact regular Borel measure on T and $f: T \times X \to \overline{\mathbb{R}} = \mathbb{R} \cup \{+\infty\}$ is a normal integrand bounded below by a Carathéodory function ξ, then for given $\varepsilon > 0$ there is a compact set $T_\varepsilon \subseteq T$ such that $\mu(T \setminus T_\varepsilon) \leq \varepsilon$ and $f|_{T_\varepsilon \times X}$ is lower semicontinuous.*

Proof. Using Proposition 2.5.21, there exist Carathéodory functions f_n such that $\xi \leq f_n \leq f$ for all $n \in \mathbb{N}$ and $f_n \nearrow f$. We apply the Scorza–Dragoni Theorem (see Theorem 2.5.19), and for each $n \in \mathbb{N}$ there is a compact set $T_n \subseteq T$ with $\mu(T \setminus T_n) \leq \varepsilon/(2^n)$ and $f_n|_{T_n \times X}$

is continuous. Let $T_\varepsilon = \bigcap_{n\geq 1} T_n \subseteq T$ being compact. Then, of course, $\mu(T \setminus T_\varepsilon) \leq \varepsilon$ and $f|_{T_\varepsilon \times X}$ is lower semicontinuous. □

Definition 2.5.23. Let (X, Σ, μ) be a measure space, $(Y, \mathcal{L})$ a measurable space, and $f: X \to Y$ a $(\Sigma, \mathcal{L})$-measurable map. Then μ induces an **image measure** $\mu \circ f^{-1}$ on Y by $(\mu \circ f^{-1})(A) = \mu(f^{-1}(A))$ for all $A \in \mathcal{L}$.

Since f^{-1} preserves all the set theoretic operations, we see that indeed $\mu \circ f^{-1}$ is a measure on $(Y, \mathcal{L})$.

Proposition 2.5.24. *If (X, Σ, μ) is a measure space, $(Y, \mathcal{L})$ is a measurable space, $f: X \to Y$ is a $(\Sigma, \mathcal{L})$-measurable map, and $h: Y \to \mathbb{R}$ is a $\mathcal{L}$-measurable function, then*

$$\int_Y h \, \mathrm{d}(\mu \circ f^{-1}) = \int_X (h \circ f) \, \mathrm{d}\mu$$

whenever either side exists.

Proof. If $h = \chi_A$ with $A \in \mathcal{L}$, then the result follows from Definition 2.5.23. So, the result is also true for simple functions that are linear combinations of characteristic functions. Finally we use Proposition 2.2.18 to pass to the general case. □

Image measures via continuous maps preserve the property of being a Radon measure

Proposition 2.5.25. *If X, Y are Hausdorff topological spaces, X is compact, $f: X \to Y$ is continuous, and $\mu: \mathcal{B}(X) \to [0, +\infty]$ is a Radon measure, then $\mu \circ f^{-1}: \mathcal{B}(Y) \to [0, +\infty]$ is a Radon measure as well.*

Proof. According to Theorem 2.5.11, it suffices to show that $\mu \circ f^{-1}$ is compact regular. Since μ is a Radon measure, for every $A \in \mathcal{B}(Y)$ one gets

$$(\mu \circ f^{-1})(A) = \sup[\mu(K): K \subseteq X \text{ is compact}, K \subseteq f^{-1}(A)]\,; \tag{2.5.18}$$

see Definition 2.5.23. For a compact $K \subseteq f^{-1}(A)$ it follows $f(K) \subseteq A$ and so $K \subseteq f^{-1}(f(K)) \subseteq f^{-1}(A)$. Hence

$$\mu(K) \leq \mu(f^{-1}(f(K))) \leq (\mu \circ f^{-1})(A)\,. \tag{2.5.19}$$

The continuity of f implies that $\tilde{K} = f(K) \subseteq Y$ is compact. Then from (2.5.18) and (2.5.19) it follows that

$$(\mu \circ f^{-1})(A) = \sup[(\mu \circ f^{-1})(\tilde{K}): \tilde{K} \subseteq Y \text{ is compact}, \tilde{K} \subseteq A]\,,$$

which shows that $\mu \circ f^{-1}$ is compact regular, hence a Radon measure. □

2.6 Analytic (Souslin) Sets

In Definition 1.5.51 we introduced the notion of a Souslin space. Souslin spaces are of fundamental importance in measure theory since they give to the theory of Borel sets and Borel functions depth and power.

Let us start by recalling the definition of Souslin space.

Definition 2.6.1. A Hausdorff topological space X is said to be a **Souslin space** if it is the continuous image of a Polish space, that is, there exists a Polish space Y and a continuous surjection $f: Y \to X$. A subset of a Hausdorff topological space that is a Souslin space is called a **Souslin set.** A Souslin subset of a Polish space is called **analytic set** as well. The complement of a Souslin set is called **co-Souslin set** (or **coanalytic set**).

Remark 2.6.2. We have that a Souslin space is always separable but need not to be metrizable; see Remark 1.5.52. Moreover, using Remark 1.5.50, we see that a nonempty subset of a Hausdorff space is a Souslin set if it is the image of the Polish space $\mathbb{N}^\infty$ under a continuous map.

Given a set B, by B^f we denote the set of all finite sequences with terms in the set B. That is, $B^f = \bigcup_{n\geq 1} B_n^f$ with B_n^f being the set of n-sequences.

Of special interest to us is the set $\mathbb{N}^f$. Note that $\mathbb{N}^f$ is countable in contrast to $\mathbb{N}^\infty$, which is uncountable. Using $\mathbb{N}^f$ we introduce the following definition.

Definition 2.6.3. Let X be a nonempty set and $\mathcal{L} \subseteq 2^X$. An **$\mathcal{L}$-Souslin scheme** is a map $A: \mathbb{N}^f \to \mathcal{L}$. Let $\mathcal{D}$ be the family of all $\mathcal{L}$-Souslin schemes. The **Souslin operation** (or **A-operation**) over the class $\mathcal{L}$ is a map $\mathfrak{a}: \mathcal{D} \to \mathcal{L}$ such that

$$\mathfrak{a}(A) = \bigcup_{p\in\mathbb{N}^\infty} \bigcap_{k\in\mathbb{N}} A(p_1, \dots, p_k) \quad \text{for all } A \in \mathcal{D}. \tag{2.6.1}$$

The collection of all sets of this form is denoted by $S(\mathcal{L})$. The elements of $S(\mathcal{L})$ are called **$\mathcal{L}$-Souslin** (or **$\mathcal{L}$-analytic**) sets. A Souslin scheme A is said to be **regular** (or **monotone**) if $A(p_1, \dots, p_{k+1}) \subseteq A(p_1, \dots, p_k)$ with $p \in \mathbb{N}^\infty$.

Remark 2.6.4. If $\emptyset \in \mathcal{L}$ (or if $\mathcal{L}$ contains disjoint sets), then $\emptyset \in S(\mathcal{L})$. Note that in (2.6.1) the union is uncountable. So, if $\mathcal{L}$ is a σ-algebra and A is an $\mathcal{L}$-Souslin scheme, then $\mathfrak{a}(A)$ may be outside of $\mathcal{L}$. In what follows we will use the following notation. Given $s = (s_k)_{k=1}^n \in \mathbb{N}^f$ and $p \in \mathbb{N}^\infty$, we write $s < p$ if and only if $s_1 = p_1, \dots, s_n = p_n$.

In the next proposition we collect some basic properties of the operator S.

Proposition 2.6.5. *If X is a nonempty set and $\mathcal{L}, \mathcal{L}' \subseteq 2^X$, then the following hold:*
(a) $S(\mathcal{L}) \subseteq S(\mathcal{L}')$ *if $\mathcal{L} \subseteq \mathcal{L}'$, that is, S is monotone;*
(b) $S(\mathcal{L})_\delta = S(\mathcal{L})$, *that is, S is closed under countable intersections;*
(c) $S(\mathcal{L})_\sigma = S(\mathcal{L})$, *that is, S is closed under countable unions;*
(d) $\mathcal{L} \subseteq S(\mathcal{L})$.

Proof. (a) This is an immediate consequence of Definition 2.6.3.

(b) Clearly we have $S(\mathcal{L}) \subseteq S(\mathcal{L})_\delta$. Suppose that $\bigcap_{k\geq 1} \mathfrak{a}(A_k) \in S(\mathcal{L})_\delta$. We need to produce an $\mathcal{L}$-Souslin scheme $A\colon \mathbb{N}^f \to \mathcal{L}$ such that $\mathfrak{a}(A) = \bigcap_{k\geq 1} \mathfrak{a}(A_k)$. To this end for every $k \in \mathbb{N}$, let $T_k = \{(2m-1)2^{k-1}\colon m \in \mathbb{N}\}$. Then $\{T_k\}_{k\geq 1}$ is a partition of $\mathbb{N}$ into infinitely many infinite sets. For each $k \in \mathbb{N}$, let $\xi_k\colon \mathbb{N}^\infty \to \mathbb{N}^\infty$ be defined by

$$\xi_k((p_n)) = (p_{2^{k-1}}, p_{3\cdot 2^{k-1}}, p_{5\cdot 2^{k-1}}, \dots),$$

that is, ξ picks from the sequence $(p_n)_{n\in\mathbb{N}}$ those elements with index in T_k. We will produce an $\mathcal{L}$-Souslin scheme A such that

$$\bigcap_{s<p} A(s) = \bigcap_{k\geq 1} \bigcap_{s<\xi_k(p)} A_k(s) \quad \text{for all } p \in \mathbb{N}^\infty. \tag{2.6.2}$$

We rewrite (2.6.1) as

$$\bigcap_{n\geq 1} A(p_1, \dots, p_n) = \bigcap_{k\geq 1} \bigcap_{m\geq 1} A_k(p_{2^{k-1}}, p_{3\cdot 2^{k-1}}, \dots, p_{(2m-1)\cdot 2^{k-1}}) \tag{2.6.3}$$

for all $p \in \mathbb{N}^\infty$. If $(p_1, \dots, p_n) \in \mathbb{N}^f$, then $n = (2m-1)2^{k-1}$ for exactly one pair $(m, k) \in \mathbb{N} \times \mathbb{N}$. Let

$$A(p_1, p_2, \dots, p_n) = A_k(p_{2^{k-1}}, p_{3\cdot 2^{k-1}}, \dots, p_{(2m-1)\cdot 2^{k-1}}). \tag{2.6.4}$$

Then (2.6.4) defines an $\mathcal{L}$-Souslin scheme, which satisfies (2.6.3) and consequently (2.6.2) as well.

Let $x \in \mathfrak{a}(A) = \bigcup_{p\in\mathbb{N}^\infty} \bigcap_{s<p} A(s)$; see (2.6.1). So, for some $p_0 \in \mathbb{N}^\infty$ we have

$$x \in \bigcap_{s<p_0} A(s) = \bigcap_{k\geq 1} \bigcap_{s<\xi_k(p_0)} A_k(s);$$

see (2.6.2). Hence

$$x \in \bigcap_{s<\xi_k(p_0)} A_k(s) \subseteq \bigcup_{p\in\mathbb{N}^\infty} \bigcap_{s<p} A_k(s) = \mathfrak{a}(A) \quad \text{for all } k \in \mathbb{N},$$

which implies that $x \in \bigcap_{k\geq 1} \mathfrak{a}(A_k)$. Hence

$$\mathfrak{a}(A) \subseteq \bigcap_{k\geq 1} \mathfrak{a}(A_k). \tag{2.6.5}$$

Next suppose that $x \in \bigcap_{k\geq 1} \mathfrak{a}(A_k)$. Then, from (2.6.1), one gets $x \in \bigcup_{p\in\mathbb{N}^\infty} \bigcap_{s<p} A_k(s)$ for all $k \in \mathbb{N}$, which implies $x \in \bigcap_{s<p_k} A_k(s)$ for some $p_k \in \mathbb{N}^\infty$ and for all $k \in \mathbb{N}$.

Let $\hat{p} \in \mathbb{N}^\infty$ such that $\xi_k(\hat{p}) = p_k$ for all $k \in \mathbb{N}$. Then $x \in \bigcap_{k\geq 1} \bigcap_{s<\xi_k(\hat{p})} A_k(s)$, which implies, due to (2.6.2),

$$x \in \bigcap_{s<\hat{p}} A(s) \subseteq \bigcup_{p\in\mathbb{N}^\infty} \bigcap_{s<p} A(s) = \mathfrak{a}(A)\,.$$

Hence,

$$\bigcap_{k\geq 1} \mathfrak{a}(A_k) \subseteq \mathfrak{a}(A)\,. \tag{2.6.6}$$

From (2.6.5) and (2.6.6) we conclude that $\mathfrak{a}(A) = \bigcap_{k\geq 1} \mathfrak{a}(A_k)$.

(c) Clearly we have $S(\mathcal{L}) \subseteq S(\mathcal{L})_\sigma$. Consider $\bigcup_{k\geq 1} \mathfrak{a}(A_k) \in S(\mathcal{L})_\sigma$. We need to generate an $\mathcal{L}$-Souslin scheme A such that $\mathfrak{a}(A) = \bigcup_{k\geq 1} \mathfrak{a}(A_k)$.

If $s = (s_k)_{k=1}^n \in \mathbb{N}^f$, then $p_1 = (2m-1)2^{k-1}$ for exactly one pair $(m,k) \in \mathbb{N} \times \mathbb{N}$. We define

$$A(s_1,\ldots,s_n) = A\big((2m-1)2^{k-1}, s_2,\ldots,s_n\big) = A_k(m,s_2,\ldots,s_n)\,.$$

This is an $\mathcal{L}$-Souslin scheme for which we have

$$\bigcap_{n\geq 1} A((2m-1)2^{k-1}, s_2,\ldots,s_n) = \bigcap_{n\geq 1} A_k(m,s_2,\ldots,s_n) \tag{2.6.7}$$

for all $k \in \mathbb{N}$ and for all $(m, s_2, s_2, \ldots) \in \mathbb{N}^\infty$. Let $x \in \mathfrak{a}(A) = \bigcup_{p\in\mathbb{N}^\infty} \bigcap_{s<p} A(s)$; see (2.6.1). Then $x \in \bigcap_{n\geq 1} A(p_1,\ldots,p_n)$ for some $p \in \mathbb{N}^\infty$ which gives, choosing $(m,k) \in \mathbb{N}\times\mathbb{N}$ such that $p_1 = (2m-1)2^{k-1}$, $x \in \bigcap_{n\geq 1} A_k(m,p_2,\ldots,p_n) \subseteq \mathfrak{a}(A_k)$. Hence

$$\mathfrak{a}(A) \subseteq \bigcup_{k\geq 1} \mathfrak{a}(A_k)\,. \tag{2.6.8}$$

Next let $x \in \bigcup_{k\geq 1} \mathfrak{a}(A_k) = \bigcup_{k\geq 1} \bigcup_{p\in\mathbb{N}^\infty} \bigcap_{s<p} A_k(s)$. Then for some $k \in \mathbb{N}$ and some $(m, s_2, s_3, \ldots) \in \mathbb{N}^\infty$, one gets $x \in \bigcap_{n\geq 1} A_k(m,s_2,\ldots,s_n)$. Then, because of (2.6.7), it follows that

$$x \in \bigcap_{n\geq 1} A((2m-1)2^{k-1}, s_2,\ldots,s_n) \subseteq \mathfrak{a}(A)\,.$$

This finally gives

$$\bigcup_{k\geq 1} \mathfrak{a}(A_k) \subseteq \mathfrak{a}(A)\,. \tag{2.6.9}$$

From (2.6.8) and (2.6.9) we conclude that $\mathfrak{a}(A) = \bigcup_{k\geq 1} \mathfrak{a}(A_k)$.

(d) For $B \in \mathcal{L}$ we set $A(s) = B$ for all $s \in \mathbb{N}^f$. Then $\mathfrak{a}(A) = B$. □

In fact S is an idempotent operator. For a proof of this result we refer to Klein–Thompson [194, Theorem 12.2.3, p. 143].

Proposition 2.6.6. *If X is a nonempty set and $\mathcal{L} \subseteq 2^X$, then $S(S(\mathcal{L})) = S(\mathcal{L})$.*

Concerning complementation, it is not true in general that $S(\mathcal{L})$ is closed under complementation. Hence, we cannot say in general that $S(\mathcal{L})$ is a σ-algebra. In order for $S(\mathcal{L})$ to contain $\sigma(\mathcal{L})$, we need additional hypotheses.

Proposition 2.6.7. *If X is a nonempty set, $\mathcal{L} \subseteq 2^X$ and for every $B \in \mathcal{L}$ we have that $X \setminus B \in S(\mathcal{L})$, then $\sigma(\mathcal{L}) \subseteq S(\mathcal{L})$.*

Proof. We know that the smallest algebra containing $\mathcal{L}$ is produced by taking finite intersections of finite unions of elements of $\mathcal{L}$ and of complements of elements of $\mathcal{L}$. Then Propositions 2.6.5 and 2.6.6 and the hypothesis imply that $S(S(S(\mathcal{L}))) = S(\mathcal{L})$. But $S(\mathcal{L})$ is a monotone class; see Proposition 2.6.5. So, using Theorem 2.1.12, we conclude that $\sigma(\mathcal{L}) \subseteq S(\mathcal{L})$. □

In Definition 2.6.1 we mentioned that a Souslin space that is a subset of a Polish space is called **analytic**. Next we give an alternative definition of analytic sets in terms of the Souslin operation and subsequently we show that the two notions of analyticity are in fact equivalent.

Definition 2.6.8. Let X be a Polish space and let $\mathcal{F}_X$ denote the family of closed subsets of X. The **analytic sets** of X are the elements of $S(\mathcal{F}_X)$.

Therefore we have two definitions of analytic sets; see Definition 2.6.1 and Definition 2.6.8. Next we show that they are equivalent and we also provide some other useful characterizations of analytic sets.

Proposition 2.6.9. *If X is a Polish space and $E \subseteq X$ is nonempty, then the following statements are equivalent:*

(a) *there exists a continuous function $f: \mathbb{N}^\infty \to X$ such that $E = f(\mathbb{N}^\infty)$;*
(b) *there exists a closed set $C \subseteq \mathbb{N}^\infty \times X$ such that $E = \operatorname{proj}_X C$;*
(c) *E is a Souslin space; see Definition 1.5.51;*
(d) *E is an analytic set and more precisely there is a regular Souslin scheme A consisting of closed subsets of X with a vanishing diameter such that $\mathfrak{a}(A) = E$.*

Proof. (a) $\Longrightarrow$ (b): Since $f: \mathbb{N}^\infty \to X$ is continuous, $\operatorname{Gr} f = C \subseteq \mathbb{N}^\infty \times X$ is closed and $\operatorname{proj}_X C = E$.

(b) $\Longrightarrow$ (c): We know that $\mathbb{N}^\infty \times X$ is Polish; see Remark 1.5.50 and Proposition 1.5.46. The set $C \subseteq \mathbb{N}^\infty \times X$ being closed is itself Polish; see Proposition 1.5.45. The projection map $\operatorname{proj}_X: C \to E$ is a continuous open surjection. Therefore, by Definition 1.5.51, we conclude that E is a Souslin space.

(c) $\Longrightarrow$ (a): According to Definition 1.5.51, there is a Polish space Y and a continuous surjection $h: Y \to E$. Moreover, from Remark 1.5.50 we know that there is a continuous surjection $g: \mathbb{N}^\infty \to Y$. Let $f = h \circ g: \mathbb{N}^\infty \to E$. Then f is a continuous surjection.

(a) $\Longrightarrow$ (d): By hypothesis there is a continuous surjection $f: \mathbb{N}^\infty \to E$. Consider the Souslin scheme defined by

$$A(p_1,\ldots,p_n) = \overline{f(U_{p_1,\ldots,p_n})} = f(\{p_1\} \times \cdots \times \{p_n\} \times \mathbb{N} \times \mathbb{N} \times \ldots)\,.$$

Clearly this Souslin scheme is regular (see Definition 2.6.3), and consists of closed sets. Moreover, the scheme $\{U_s : s \in \mathbb{N}^f\}$ has a vanishing diameter for the tree metric t; see Remark 1.5.50. Note that if $B \subseteq X$ is an F_σ-set and $\varepsilon > 0$, then we can write $B = \bigcup_{n\geq 1} B'_n$ with $\{B'_n\}$ pairwise disjoint F_σ-sets each having diameter less than $\varepsilon > 0$. Using this fact and an induction argument, we show that $E = \mathfrak{a}(A)$.

(d) $\Longrightarrow$ (a): By hypothesis we have $E = \bigcup_{p\in\mathbb{N}^\infty} \bigcap_{k\geq 1} A(p_1,\ldots,p_k)$. Since X is complete, in order for $\bigcap_{k\geq 1} A(p_1,\ldots,p_k)$ to be empty is that for some $k \in \mathbb{N}$, $A(p_1,\ldots,p_k) = \emptyset$. We define

$$\mathfrak{L} = \{p \in \mathbb{N}^\infty : A(p_1,\ldots,p_k) \neq \emptyset \text{ for all } k \in \mathbb{N}\}\,.$$

Using the definition of the tree metric (see Remark 1.5.50), we can easily see that $\mathfrak{L} \subseteq \mathbb{N}^\infty$ is closed. Hence Example 1.7.13 (c) implies that $\mathfrak{L}$ is a retract of $\mathbb{N}^\infty$. We have

$$E = \bigcup_{p\in\mathfrak{L}} \bigcap_{k\geq 1} A(p_1,\ldots,p_k)\,.$$

For each $p \in \mathfrak{L}$ let $g(p)$ be the unique element of $\bigcap_{k\geq 1} A(p_1,\ldots,p_k)$. Recall that a Souslin scheme has a vanishing diameter, and apply Theorem 1.5.15. The map $g : \mathfrak{L} \to E$ is bijective and continuous. Let $r : \mathbb{N}^\infty \to \mathfrak{L}$ be a retraction map. Then $f = g \circ r : \mathbb{N}^\infty \to E$ is a continuous surjection. □

From Proposition 2.6.5, we have the following.

Proposition 2.6.10. *If X is a Polish space, then countable intersections and countable unions of analytic sets are analytic.*

Next we are going to show that the analytic sets contain the Borel sets.

Proposition 2.6.11. *If X is a Polish space and $B \in \mathcal{B}(X)$, then B is analytic.*

Proof. From Proposition 1.5.8, we know that every open set of X is F_σ. Hence, every open set is analytic; see Definition 2.6.8. Then Proposition 2.6.7 implies that $\mathcal{B}(X) \subseteq S(\mathcal{F}_X)$. Using Propositions 2.6.5 and 2.6.6 it follows that

$$S(\mathcal{F}_X) \subseteq S(\mathcal{B}(X)) \subseteq S(S(\mathcal{F}_X)) = S(\mathcal{F}_X)\,.$$ □

Remark 2.6.12. From the proof above we see that $S(\mathcal{F}_X) = S(\mathcal{B}(X))$. If X is countable, then $\mathcal{B}(X) = S(\mathcal{F}_X)$, that is, Borel and analytic sets coincide. If X is uncountable, then the class of analytic sets $S(\mathcal{F}_X)$ is strictly larger than the Borel σ-algebra $\mathcal{B}(X)$. In fact we can have an analytic set whose complement is not analytic.

We want to have a closer look at the relation between Borel and analytic sets. We start with a definition.

Definition 2.6.13. Let X be a Polish space and let $A_1, A_2 \subseteq X$ be nonempty. We say that A_1 and A_2 can be **separated by Borel sets** if there are disjoint Borel sets $B_1, B_2 \subseteq X$ such that $A_1 \subseteq B_1$ and $A_2 \subseteq B_2$.

Lemma 2.6.14. *Let X be a Polish space.*
(a) *If $\{A_n\}_{n\geq 1}$ and C are nonempty subsets of X such that for every $n \in \mathbb{N}$ the sets A_n and C can be separated by Borel sets, then $\bigcup_{n\geq 1} A_n$ and C can be separated by Borel sets.*
(b) *If $\{A_n\}_{n\geq 1}$ and $\{C_n\}_{n\geq 1}$ are nonempty subsets of X such that for each $(n, m) \in \mathbb{N} \times \mathbb{N}$ the sets A_n and C_m can be separated by Borel sets, then the sets $\bigcup_{n\geq 1} A_n$ and $\bigcup_{n\geq 1} C_n$ can be separated by Borel sets.*

Proof. (a) By hypothesis, for each $n \in \mathbb{N}$ there exist disjoint Borel sets B_n and D_n such that $A_n \subseteq B_n$ and $C \subseteq D_n$. Then $\bigcup_{n\geq 1} B_n$ and $\bigcap_{n\geq 1} D_n$ are disjoint Borel sets and $\bigcup_{n\geq 1} A_n \subseteq \bigcup_{n\geq 1} B_n$ and $C \subseteq \bigcap_{n\geq 1} D_n$.

(b) From part (a) above for each $n \in \mathbb{N}$, the sets A_n and $\bigcup_{m\geq 1} C_m$ can be separated by Borel sets. A second application of part (a) implies that $\bigcup_{n\geq 1} A_n$ and $\bigcup_{m\geq 1} C_m$ can be separated by Borel sets. □

Now we show that disjoint analytical sets can be separated by Borel sets. The result is known as the "Separation Theorem" and has important consequences, some of which we explore here.

Theorem 2.6.15 (Separation Theorem). *If X is a Polish space and $A_1, A_2 \subseteq X$ are nonempty disjoint analytical sets, then A_1 and A_2 can separated by Borel sets.*

Proof. Invoking Proposition 2.6.9, there exist continuous surjections

$$f_1 : \mathbb{N}^\infty \to A_1 \quad \text{and} \quad f_2 : \mathbb{N}^\infty \to A_2 .$$

For any $s \in \mathbb{N}^f$, we set $U_s = \{s_1\} \times \cdots \times \{s_k\} \times \mathbb{N} \times \mathbb{N} \times \ldots$ and then define $A_1^s = f_1(U_s)$ as well as $A_2^s = f_2(U_s)$.

Arguing indirectly, suppose that A_1 and A_2 cannot be separated by Borel sets. Since it holds that $A_1 = \bigcup_{n\geq 1} A_1^n$ and $A_2 = \bigcup_{n\geq 1} A_2^n$, using Lemma 2.6.14, there exist $n_1, m_1 \in \mathbb{N}$ such that the sets $A_1^{n_1}$ and $A_2^{m_1}$ cannot be separated by Borel sets. Note that

$$A_1^{n_1} = \bigcup_{n\geq 1} A_1^{n_1,n} \quad \text{and} \quad A_2^{m_1} = \bigcup_{n\geq 1} A_2^{m_1,n} .$$

Hence, a new application of Lemma 2.6.14 gives $n_2, m_2 \in \mathbb{N}$ such that $A_1^{n_1,n_2}$ and $A_2^{m_1,m_2}$ cannot be separated by Borel sets. Continuing this way, we produce $p(1) = (n_k)$ and $p(2) = (m_k) \in \mathbb{N}^\infty$ such that

$$A_1^{n_1,\ldots,n_k} \quad \text{and} \quad A_2^{m_1,\ldots,m_k} , \qquad k \in \mathbb{N}$$

cannot be separated by Borel sets. Let $x = f_1(p(1)) \in A_1$ and $u = f_2(p(2)) \in A_2$. We have $x \neq u$ since the sets A_1 and A_2 are disjoint. Let $U_1 \in \mathcal{N}(x)$ and $U_2 \in \mathcal{N}(u)$ such that

$U_1 \cap U_2 = \emptyset$. The continuity of f_1 and f_2 implies that for $k \in \mathbb{N}$ large enough we have

$$A_1^{n_1,\dots,n_k} = f_1(U_{n_1,\dots,n_k}) \subseteq U_1 \quad \text{and} \quad A_2^{m_1,\dots,m_k} = f_2(U_{m_1,\dots,m_k}) \subseteq U_2 \, .$$

Therefore the open sets U_1 and U_2, which are Borel as well, separate $A_1^{n_1,\dots,n_k}$ and $A_2^{m_1,\dots,m_k}$, a contradiction. □

Corollary 2.6.16. *If X is a Polish space and $\{A_n\}_{n\geq 1}$ are pairwise disjoint analytic sets, then there exists a sequence $\{B_n\}_{n\geq 1}$ of pairwise disjoint Borel sets such that $A_n \subseteq B_n$ for every $n \in \mathbb{N}$.*

Corollary 2.6.17. *If X is a Polish space and $A \subseteq X$ is both analytic and coanalytic, that is, $X \setminus A$ is analytic as well, then $A \in \mathcal{B}(X)$.*

Proof. Using Theorem 2.6.15 there are disjoint Borel sets B_1, B_2 such that $A \subseteq B_1$ and $X \setminus A \subseteq B_2$. Evidently $A = B_1$ and $X \setminus A = B_2$. Therefore $A \in \mathcal{B}(X)$. □

Remark 2.6.18. Clearly the converse of the corollary above is true as well. Namely, every Borel set in X is both analytic and coanalytic.

Applying Corollary 2.6.17 we obtain the following characterizations of Borel measurable maps between Polish spaces.

Proposition 2.6.19. *If X, Y are Polish spaces and $f: X \to Y$, then the following statements are equivalent:*
(a) *f is Borel measurable;*
(b) *$\operatorname{Gr} f \in \mathcal{B}(X \times Y) = \mathcal{B}(X) \otimes \mathcal{B}(Y)$;*
(c) *$\operatorname{Gr} f \subseteq X \times Y$ is analytic.*

Proof. (a) $\Longrightarrow$ (b): Let $\varphi: X \times Y \to Y \times Y$ be defined by $\varphi(x,y) = (f(x), y)$. Since by hypothesis f is Borel measurable, for every $B, C \in \mathcal{B}(X)$ we have $\varphi^{-1}(B \times C) \in \mathcal{B}(X) \otimes \mathcal{B}(Y) = \mathcal{B}(X \times Y)$; see Proposition 2.2.26 (b). Therefore φ is Borel measurable. Let $D = \{(y,z) \in Y \times Y : y = z\}$. Then $D \subseteq Y \times Y$ is closed and $\operatorname{Gr} f = \varphi^{-1}(D) \in \mathcal{B}(X \times Y) = \mathcal{B}(X) \otimes \mathcal{B}(Y)$.

(b) $\Longrightarrow$ (c): This implication is a consequence of Proposition 2.6.11.

(c) $\Longrightarrow$ (a): Let $B \in \mathcal{B}(Y)$. Then $X \times B \in \mathcal{B}(X \times Y)$ and so it is analytic. It follows that $\operatorname{Gr} f \cap (X \times B) \subseteq X \times Y$ is analytic. Note that

$$f^{-1}(B) = \operatorname{proj}_X(\operatorname{Gr} f \cap (X \times B)) \tag{2.6.10}$$

with $\operatorname{proj}_X : X \times Y \to X$ being the projection map defined by $\operatorname{proj}_X(x,y) = x$ for all $(x,y) \in X \times Y$. We know that proj_X is continuous. Since $\operatorname{Gr} f \cap (X \times B)$ is analytic, we find a continuous surjection $h: \mathbb{N}^\infty \to \operatorname{Gr} f \cap (X \times B)$; see Proposition 2.6.9. Then $\operatorname{proj}_X \circ h: \mathbb{N}^\infty \to f^{-1}(B)$ (see (2.6.10)) is a continuous surjection. Hence $f^{-1}(B) \subseteq X$ is analytic; see Proposition 2.6.9. In a similar way we show that $f^{-1}(Y \setminus B) \subseteq X$ is analytic. But

$f^{-1}(Y \setminus B) = X \setminus f^{-1}(B)$. Therefore $f^{-1}(B) \subseteq X$ is coanalytic. Invoking Corollary 2.6.17, we conclude that $f^{-1}(B) \in \mathcal{B}(X)$ and so f is Borel measurable. □

Definition 2.6.20. Let (X, Σ) and $(Y, \mathcal{L})$ be two measurable spaces. A bijection $f: X \to Y$ is said to be an **isomorphism** if f is $(\Sigma, \mathcal{L})$-measurable and f^{-1} is $(\mathcal{L}, \Sigma)$-measurable. Then the measurable spaces (X, Σ) and $(Y, \mathcal{L})$ are said to be **isomorphic**. If X, Y are Hausdorff topological spaces and $\Sigma = \mathcal{B}(X)$, $\mathcal{L} = \mathcal{B}(Y)$, then we use the term **Borel isomorphism**.

Proposition 2.6.21. *If X, Y are Polish spaces and $f: X \to Y$ is a Borel isomorphism, then $E \subseteq X$ is analytic if and only if $f(E) \subseteq Y$ is analytic.*

Proof. $\Longrightarrow$: Since $E \subseteq X$ is analytic, we have $E = \mathfrak{a}(A)$ with A being a $\mathcal{F}_X$-Souslin scheme. Then $f(E) = S(f \circ A)$ with $f \circ A$ being the $\mathcal{B}(Y)$-Souslin scheme defined by $(f \circ A)(x) = f(A(x))$. Hence, $f(E)$ is analytic; see Remark 2.6.12.

$\Longleftarrow$: This is proven in a similar way. □

Corollary 2.6.22. *If X, Y are Polish spaces, $f: X \to Y$ is Borel measurable, $E \in \mathcal{B}(X)$ and $f|_E$ is one-to-one, then $f(E) \in \mathcal{B}(Y)$.*

Now we examine the measurability of analytic sets. Although analytic sets need not be Borel, it turns out that they will always be measurable for the completion of any probability measure defined on the Borel sets.

Definition 2.6.23. Let X be a Polish space and let $M_1^+(X)$ be the set of probability measures on X. Given $\mu \in M_1^+(X)$ let $\mathcal{B}(X)_\mu$ be the completion of the Borel σ-algebra $\mathcal{B}(X)$. Recall that $\mathcal{B}(X)_\mu$ can be described as the family of all sets of the form $B \cup N$ with $B \in \mathcal{B}(X)$ and N is a subset of a μ-null set. The **universal σ-algebra** $\hat{\Sigma}_X$ is defined by

$$\hat{\Sigma}_X = \bigcap_{\mu \in M_1^+(X)} \mathcal{B}(X)_\mu .$$

The elements of $\hat{\Sigma}_X$ are said to be **universally measurable sets**.

Next we will see that analytic sets are universally measurable.

Theorem 2.6.24. *If X is a Polish space and $E \subseteq X$ is analytic, then $E \in \hat{\Sigma}_X$, that is, E is universally measurable.*

Proof. According to Proposition 2.6.9 there exists $f: \mathbb{N}^\infty \to X$ being a continuous map such that $f(\mathbb{N}^\infty) = E$. Let $\mu \in M_1^+(X)$ and for any $k, m \in \mathbb{N}$ let

$$N(k, m) = \{p = (p_k) \in \mathbb{N}^\infty : p_k \le m\} .$$

We see that $f(N(k, m)) \nearrow f(\mathbb{N}^\infty) = E$ as $m \to +\infty$. So, for a given $\varepsilon > 0$ there exists $m_1 \in \mathbb{N}$ such that $\mu^*(f(N(1, m_1))) \ge \mu^*(E) - \varepsilon/2$ with μ^* being the outer measure corresponding to μ; see Proposition 2.1.34.

Similarly, for all $k \in \mathbb{N}$, we can find $m_k \in \mathbb{N}$ such that

$$\mu(\overline{f(C_k)}) \geq \mu^*(f(C_k)) \geq \mu^*(E) - \sum_{i=1}^{k} \frac{\varepsilon}{2^i} \geq \mu^*(E) - \varepsilon$$

with $C_k = \bigcap_{i=1}^k N(i, m_i)$. Letting $k \to \infty$ we see that $C_k \searrow C = \bigcap_{i\geq 1} N(i, m_i)$. Note that each C_k is closed and C is compact. Let $U \supseteq C$ be open. Then U is a union of basic open sets and the compactness of C implies that this union is finite. Each basic open set depends on only finitely many coordinates. Let $j \in \mathbb{N}$ be the largest index of any coordinate in the definition of the sets of this finite subcover. We have $C_j \subseteq U$ and according to Problem 1.51 it holds that $\mu(f(C)) \geq \mu^*(E) - \varepsilon$. The set $f(X) \subseteq X$ is compact. Taking $\varepsilon = 1/n$ with $n \in \mathbb{N}$ we have a countable union of compact sets that is a Borel set $B \subseteq E$ with $\mu(B) = \mu^*(E)$. Therefore $\mu^*(E \setminus B) = 0$ and $E \in \mathcal{B}(X)_\mu$; see Proposition 2.1.41. We conclude that $E \in \hat{\Sigma}_X$. □

The following characterization of the universal σ-algebra $\hat{\Sigma}_X$ is immediate from Definition 2.6.23 and the proof of Theorem 2.6.24.

Proposition 2.6.25. *If X is a Polish space and $E \subseteq X$, then $E \in \hat{\Sigma}_X$ if and only if for any $\mu \in M_1^+(X)$ there exists $B \in \mathcal{B}(X)$ such that $\mu(E \triangle B) = 0$.*

There is a third σ-algebra that we can define for a Polish space X.

Definition 2.6.26. Let X be a Polish space. The **analytic σ-algebra** α_X is the smallest σ-algebra containing the analytic subsets of X, that is, $\alpha_X = \sigma(S(\mathcal{F}_X))$.

If $E \in \alpha_X$, then we say that E is **analytically measurable**. Therefore on any Polish space X we can define three important σ-algebras:
- $\mathcal{B}(X)$ = the Borel σ-algebra.
- α_X = the analytic σ-algebra.
- $\hat{\Sigma}_X$ = the universal σ-algebra.

These σ-algebras are related as follows

$$\mathcal{B}(X) \subseteq S(\mathcal{F}_X) \subseteq \alpha_X \subseteq \hat{\Sigma}_X . \tag{2.6.11}$$

If X is countable, then all classes in (2.6.11) are equal to 2^X. If X is uncountable, then all inclusions in (2.6.11) are strict.

Definition 2.6.27. Let X, Y be Polish spaces, $C \subseteq X$ be nonempty, and $f: C \to Y$. We say that f is **analytically** (resp. **universally**) **measurable** if $C \in \alpha_X$ (resp. $C \in \hat{\Sigma}_X$) and $f^{-1}(E) \in \alpha_X$ (resp. $f^{-1}(E) \in \hat{\Sigma}_X$) for all $E \in \mathcal{B}(Y)$.

The composition of functions preserves universal measurability.

Proposition 2.6.28. *If X, Y, Z are Polish spaces, $C \in \hat{\Sigma}_X$, $E \in \hat{\Sigma}_Y$, $f: C \to Y$, $g: E \to Y$, and $f(C) \subseteq E$, then $g \circ f: C \to Z$ is universally measurable.*

Proof. Let $B \in \mathcal{B}(Z)$. The universal measurability of g implies that $g^{-1}(B) \in \hat{\Sigma}_Y$. Since $(g \circ f)^{-1}(B) = f^{-1}(g^{-1}(B))$ we need to show that for every $D \in \hat{\Sigma}_Y, f^{-1}(D) \in \hat{\Sigma}_X$. Given $\mu \in M_1^+(X)$ we consider the image measure $\mu \circ f^{-1}$ on Y; see Definition 2.5.23. Let $F \in \mathcal{B}(Y)$ be such that $(\mu \circ f^{-1})(F \triangle D) = 0$. The universal measurability of f implies that $f^{-1}(F) \in \hat{\Sigma}_X$. Hence, by applying Proposition 2.6.25, there exists $G \in \mathcal{B}(X)$ such that $\mu(G \triangle f^{-1}(F)) = 0$. Therefore $\mu(G \triangle f^{-1}(D)) = 0$ and this implies, due to Proposition 2.6.25, that $f^{-1}(D) \in \hat{\Sigma}_X$. □

From the proof above, we deduce the following corollary.

Corollary 2.6.29. *If X, Y are Polish spaces, $C \in \hat{\Sigma}_X$, and $f: C \to Y$ is universally measurable, then for every $E \in \hat{\Sigma}_Y$ we have $f^{-1}(E) \in \hat{\Sigma}_X$.*

Remark 2.6.30. Composition of functions does not preserve analytic measurability. The composition of two analytically measurable functions is universally measurable.

2.7 Selection and Projection Theorems

In this section we prove some results, which in addition to being interesting from a purely theoretical viewpoint, are used in many applied fields such as calculus of variations, optimization, optimal control, and mathematical economics.

The mathematical setting is the following: We are given a measurable space (Ω, Σ), a separable metric space (X, d), and a multifunction (so-called set-valued map) $F: \Omega \to 2^X$. The first basic question we want to study is whether we can find a single-valued, Σ-measurable map $f: \Omega \to X$ such that $f(w) \in F(w)$ for all $w \in \Omega$. Such a map is called a **measurable selection** of F. Its existence is not straightforward. First we need to introduce and discuss some notions of measurability for the multifunction F.

In what follows, (Ω, Σ) is a measurable space and (X, d) is a separable metric space. Additional hypotheses will be introduced as needed.

Definition 2.7.1. Let $F: \Omega \to 2^X$ be a multifunction.
(a) We say that F is **measurable** if for every open $U \subseteq X$,

$$F^-(U) = \{w \in \Omega: F(w) \cap U \neq \emptyset\} \in \Sigma .$$

(b) We say that F is **graph measurable** if

$$\operatorname{Gr} F = \{(w, x) \in \Omega \times X: x \in F(w)\} \in \Sigma \bigotimes \mathcal{B}(X) .$$

Remark 2.7.2. Note that in the definitions above we do not require that F be nonempty valued. By **domain of** F we mean the set $\operatorname{dom} F = \{w \in \Omega: F(w) \neq \emptyset\}$. If F is measurable, then clearly $\operatorname{dom} F \in \Sigma$ and so for measurable multifunctions, there is no loss of generality in assuming that $\operatorname{dom} F = X$. If F is single-valued, then measurability coincides with Σ-measurability. Evidently both notions make sense even if X is a general

Hausdorff topological space. However, the most interesting properties and results can be established for X being a Polish space in the case of measurable multifunctions and for X being a Souslin space in the case of graph measurable multifunctions. Therefore, we see that the theory of measurable multifunctions requires separability of the ambient space. Without it we cannot go far. For economy in the presentation we have fixed X to be a separable metric space.

Proposition 2.7.3. *If $F: \Omega \to 2^X$ and for all closed $C \subseteq X$, $F^-(C) = \{w \in \Omega: F(w) \cap C \neq \emptyset\} \in \Sigma$, then F is measurable.*

Proof. From Proposition 1.5.8 we know that every open set $U \subseteq X$ is F_σ. So, $U = \bigcup_{n\geq1} C_n$ with closed $C_n \subseteq X$ for all $n \in \mathbb{N}$. Then, by hypothesis,

$$F^-(U) = F^-\Big(\bigcup_{n\geq1} C_n\Big) = \bigcup_{n\geq1} F^-(C_n) \in \Sigma\,.$$

Hence, F is measurable. □

Remark 2.7.4. The converse of the proposition above is not true in general.

The measurability of F can be characterized functionally.

Proposition 2.7.5. *The multifunction $F: \Omega \to 2^X$ is measurable if and only if for all $x \in X$, the $\overline{\mathbb{R}}_+$-valued function $w \to d(x, F(w))$ is Σ-measurable.*

Proof. $\Longrightarrow$: Given $x \in X$ and $\eta > 0$, let $L_\eta(x) = \{w \in \Omega: d(x, F(w)) < \eta\}$. Then we see that $L_\eta(x) = F^-(B_\eta(x))$ with $B_\eta(x) = \{u \in X: d(u, x) < \eta\}$. Hence, $L_\eta(x) \in \Sigma$ and this implies the Σ-measurability of $w \to d(x, F(w))$.

$\Longleftarrow$: Given $x \in X$ and $\eta > 0$, by hypothesis, it holds that

$$F^-(B_\eta(x)) = L_\eta(x) \in \Sigma\,. \tag{2.7.1}$$

Let $U \subseteq X$ be open. The separability of X implies that $U = \bigcup_{n\geq1} B_{\eta_n}(x_n)$. Then

$$F^-(U) = \bigcup_{n\geq1} F^-(B_{\eta_n}(x_n)) \in \Sigma\,;$$

see (2.7.1). Thus, F is measurable. □

Let us introduce some notation:

$$P_f(X) = \{A \subseteq X: A \text{ is nonempty and closed}\}\,, \quad \hat{P}_f(X) = P_f(X) \cup \{\emptyset\}\,,$$
$$P_k = \{A \subseteq X: A \text{ is nonempty and compact}\}\,.$$

Proposition 2.7.6. *If $F: \Omega \to \hat{P}_f(X)$ is measurable, then F is graph measurable.*

Proof. Since F is closed valued, we have that

$$\operatorname{Gr} F = \big\{(w, x) \in \Omega \times X: d\big(x, F(w)\big) = 0\big\}\,. \tag{2.7.2}$$

But using Proposition 2.7.5 we see that $(w, x) \to d(x, F(w))$ is a Carathéodory function; see Definition 2.2.30. Then Proposition 2.2.3 implies that it is jointly measurable and so from (2.7.2) it follows that $\mathrm{Gr}\, F \in \Sigma \bigotimes \mathcal{B}(X)$, that is, F is graph measurable. □

Recall that if $U \subseteq X$ is open, then $A \cap U \neq \emptyset$ if and only if $\overline{A} \cap U \neq \emptyset$. This straightforward observation leads to the following useful result.

Proposition 2.7.7. *The multifunction $F: \Omega \to 2^X$ is measurable if and only if $w \to \overline{F}(w) = \overline{F(w)}$ is measurable.*

For $P_k(X)$-valued multifunctions we obtain the converse of Proposition 2.7.3.

Proposition 2.7.8. *If $F: \Omega \to P_k(X)$ is measurable, then for all closed $C \subseteq X$, it holds that $F^-(C) = \{w \in \Omega: F(w) \cap C \neq \emptyset\} \in \Sigma$.*

Proof. In what follows for every $E \subseteq X$, we set

$$F^+(E) = \{w \in \Omega: F(w) \subseteq E\}. \tag{2.7.3}$$

Let $C \subseteq X$ be nonempty and closed and let $U_n = \{x \in X: d(x, C) > 1/n\}$ with $n \in \mathbb{N}$. Then U_n is open for each $n \in \mathbb{N}$ and $\{U_n\}_{n\geq 1}$ is increasing. We set $D_n = \overline{U}_n$ with $n \in \mathbb{N}$. Then

$$X \setminus C = \bigcup_{n\geq 1} U_n = \bigcup_{n\geq 1} D_n. \tag{2.7.4}$$

Let $w \in F^+(X \setminus C)$. Then $F(w) \subseteq X \setminus C$; see (2.7.3). Due to (2.7.4) and recalling that $\{U_n\}_{n\geq 1}$ is increasing as well as F is $P_k(X)$-valued, we see that there exists $n \in \mathbb{N}$ such that $F(w) \subseteq U_n \subseteq D_n$. Then, due to (2.7.3), it follows $F^+(X \setminus C) = \bigcup_{n\geq 1} F^+(D_n)$. Since F is measurable we derive

$$F^-(C) = X \setminus (F^+(X \setminus C)) = X \setminus \bigcup_{n\geq 1} F^+(D_n) = \bigcap_{n\geq 1} F^-(X \setminus D_n) \in \Sigma. \qquad \square$$

Proposition 2.7.9. *If $F: \Omega \to P_f(X)$ is measurable, then $F^-(K) \in \Sigma$ for all compact $K \subseteq X$.*

Proof. On account of Theorem 1.5.21 we may assume that X is dense in a compact metric space (Y, d_Y). Consider the multifunction $G: \Omega \to P_k(Y)$ defined by $G(w) = \overline{F(w)}^{d_Y}$. Proposition 2.7.7 guarantees the measurability of G. Now let $K \subseteq X$ compact. We have

$$F^-(K) = \{w \in \Omega: F(w) \cap K \neq \emptyset\} = \{w \in \Omega: G(w) \cap K \neq \emptyset\} = G^-(K) \in \Sigma$$

by Proposition 2.7.8. □

When we introduce extra structure on the space, we can say more. To be more precise, we have the following result.

Proposition 2.7.10. *If X is σ-compact and $F: \Omega \to P_f(X)$, then the following statements are equivalent:*

(a) $F^-(C) \in \Sigma$ *for every closed* $C \subseteq X$.
(b) F *is measurable.*
(c) $F^-(K) \in \Sigma$ *for every compact* $K \subseteq X$.

Proof. (a) $\Longrightarrow$ (b): This implication follows from Proposition 2.7.3.

(b) $\Longrightarrow$ (c): This implication follows from Proposition 2.7.9.

(c) $\Longrightarrow$ (a): By hypothesis, $X = \bigcup_{n\geq 1} K_n$ with compact K_n. Then for closed $C \subseteq X$ it holds that

$$F^-(C) = \bigcup_{n\geq 1}(C \cap K_n) \in \Sigma$$

since $C \cap K_n \subseteq X$ is compact for every $n \in \mathbb{N}$. □

The next theorem summarizes the measurability properties of closed valued multifunctions.

Theorem 2.7.11. *Let* (Ω, Σ) *be a measurable space,* (X, d) *a separable metric space, and* $F: \Omega \to P_f(X)$ *a multifunction. Consider the following statements:*
(a) $F^-(C) \in \Sigma$ *for every closed* $C \subseteq X$.
(b) F *is measurable.*
(c) *For every* $x \in X$, $w \to d(x, F(w))$ *is* Σ*-measurable.*
(d) F *is graph measurable.*

Then (a) $\Longrightarrow$ *(b)* $\Longleftrightarrow$ *(c)* $\Longrightarrow$ *(d) and if* X *is* σ*-compact, then (a)* $\Longleftrightarrow$ *(b)* $\Longleftrightarrow$ *(c)* $\Longrightarrow$ *(d).*

Now we are ready for the first existence theorem for measurable selections. The result is known as the "Kuratowski–Ryll Nardzewski Selection Theorem".

Theorem 2.7.12 (Kuratowski–Ryll Nardzewski Selection Theorem). *If* (Ω, Σ) *is a measurable space,* X *is a Polish space, and* $F: \Omega \to P_f(X)$ *is a measurable multifunction, then* F *admits a measurable selection, that is, there exists a* Σ*-measurable function* $f: \Omega \to X$ *such that* $f(w) \in F(w)$ *for all* $w \in \Omega$.

Proof. Let d be a bounded compatible metric on X. We may assume that the d-diameter of X is strictly less than 1. Let $\{x_n\}_{n\geq 1}$ be dense in X. We produce inductively a sequence of Σ-measurable maps $f_n: \Omega \to X$ with $n \in \mathbb{N}_0$, which satisfy

$$d(f_n(w), F(w)) < \frac{1}{2^n} \quad \text{for all } n \in \mathbb{N}_0 \text{ and for all } w \in \Omega\,, \tag{2.7.5}$$

$$d(f_n(w), f_{n-1}(w)) < \frac{1}{2^{n-1}} \quad \text{for all } n \in \mathbb{N} \text{ and for all } w \in \Omega\,. \tag{2.7.6}$$

Let us start with f_0. We define $f_0: \Omega \to X$ by $f_0(w) = x_1$ for all $w \in \Omega$. Since by hypothesis $\operatorname{diam} X < 1$, inequality (2.7.5) holds for $n = 0$. For the induction hypothesis, we assume that we have already produced $f_0, f_1, \ldots, f_{n-1}$, which satisfy (2.7.5) as well as (2.7.6). For every $k \in \mathbb{N}$, we define

$$A_k^n = \left\{w \in \Omega : d(x_k, F(w)) < \frac{1}{2^n}\right\}, \quad C_k^n = \left\{w \in \Omega : d(x_k, f_{n-1}(w)) < \frac{1}{2^{n-1}}\right\}$$

and $E_k^n = A_k^n \cap C_k^n$. First we show that $\Omega = \bigcup_{k\geq 1} E_k^n$. So, let $w \in \Omega$. The induction hypothesis says that there exists $u \in F(w)$ such that $d(f_{n-1}(w), u) < 1/(2^{n-1})$; see (2.7.5). The density of $\{x_n\}_{n\geq 1}$ in X implies that there is $k \in \mathbb{N}$ such that $d(x_k, u) < 1/2^n$ and $d(x_k, u)+d(u, f_{n-1}(w)) < 1/2^{n-1}$. By the triangle inequality we have $d(x_k, f_{n-1}(w)) < 1/2^{n-1}$. Hence, we see that $w \in E_k^n$, thus $\Omega = \bigcup_{k\geq 1} E_k^n$. The measurability of F and Proposition 2.7.5 imply that $A_k^n \in \Sigma$. Taking the induction hypothesis into account, the Σ-measurability of f_{n-1} implies that $C_k^n \in \Sigma$. Therefore $E_k^n \in \Sigma$. We define a function $f_n : \Omega \to X$ by setting $f_n(w) = x_k$ for all $w \in E_k^n \setminus \bigcup_{i=1}^{k-1} E_i^n$. Hence f_n is Σ-measurable and satisfies (2.7.5) and (2.7.6). This completes the induction.

From (2.7.6) we infer that for every $w \in \Omega$, $\{f_n(w)\}_{n\geq 0} \subseteq X$ is a Cauchy sequence. Therefore

$$f_n(w) \xrightarrow{d} f(w) \quad \text{for all } w \in \Omega \text{ as } n \to \infty .$$

Proposition 2.2.12 implies that f is Σ-measurable and $d(f(w), F(w)) = 0$ for all $w \in \Omega$. Since $F(w) \in P_f(X)$ for all $w \in \Omega$, we conclude that $f(w) \in F(w)$ for all $w \in \Omega$. Therefore $f : \Omega \to X$ is a Σ-measurable selection of F. □

In fact we can produce a whole sequence of dense Σ-measurable selections of F.

Theorem 2.7.13. *If (Ω, Σ) is a measurable space, X is a Polish space and $F : \Omega \to P_f(X)$, then the following statements are equivalent:*
(a) *F is measurable;*
(b) *there exists a sequence of Σ-measurable selections $f_n : \Omega \to X$ of F such that $F(w) = \overline{\{f_n(w)\}_{n\geq 1}}$ for all $w \in \Omega$.*

Proof. (a) $\Longrightarrow$ (b): Let $\{U_n\}_{n\geq 1}$ be a countable basis for the metric topology of X. For every $n \in \mathbb{N}$, we define the multifunction

$$F_n(w) = \begin{cases} F(w) \cap U_n & \text{if } F(w) \cap U_n \neq \emptyset , \\ F(w) & \text{otherwise} , \end{cases}$$

for all $w \in \Omega$. Let $\Omega_n = F^-(U_n) \in \Sigma$ with $n \in \mathbb{N}$. Then for every open set $V \subseteq X$ we obtain

$$F_n^-(V) = \{w \in \Omega_n : F(w) \cap U_n \neq \emptyset\} \cup \{w \in (\Omega \setminus \Omega_n) : F(w) \cap V \neq \emptyset\} \in \Sigma ,$$

which implies that F_n is measurable for all $n \in \mathbb{N}$. Then, thanks to Proposition 2.7.7, it follows that $\overline{F}_n$ is measurable for all $n \in \mathbb{N}$.

Invoking Theorem 2.7.12 there exists a sequence $f_n\colon \Omega \to X$ with each f_n being a Σ-measurable selection of F_n. Note that $\overline{F}_n(w) \subseteq F(w)$ for all $n \in \mathbb{N}$ and for all $w \in \Omega$. Hence, f_n is a Σ-measurable selection of F. Evidently, $F(w) = \overline{\{f_n(w)\}_{n\geq 1}}$ for all $w \in \Omega$.

(b) $\Longrightarrow$ (a): For every $x \in X$, it holds that

$$d(x, F(w)) = \inf_{n\geq 1} d(x, f_n(w)) \quad \text{for all } w \in \Omega\,,$$

which demonstrates, because of Proposition 2.2.10, that $w \to d(x, F(w))$ is Σ-measurable. Hence, due to Proposition 2.7.5, we get that F is measurable. □

We can state another measurable selection theorem for graph measurable multifunctions. First we start with a definition.

Definition 2.7.14. (a) A family $\mathcal{L}$ of subsets of a set X is said to **separate points** in X if for every two distinct points $x, u \in X$ there is $A \in \mathcal{L}$ such that $x \in A$, $u \notin A$ or $x \notin A$, $u \in A$.

(b) A family $\mathcal{D}$ of $\mathbb{R}$-valued functions on X is said to **separate points** in X if for every two distinct points $x, u \in X$ there is $f \in \mathcal{D}$ such that $f(x) \neq f(w)$.

(c) A σ-algebra $\mathcal{L}$ of subsets of a set X is said to be **countably generated** if there is a countable family $\{A_n\}_{n\geq 1} \subseteq \mathcal{L}$ such that $\mathcal{L} = \sigma(\{A_n\}_{n\geq 1})$.

(d) A σ-algebra $\mathcal{L}$ of subsets of a set X is said to be **countably separated** if there is a countable family $\{A_n\}_{n\geq 1} \subseteq \mathcal{L}$ that separates points in X, see (a).

Example 2.7.15. Suppose X is a separable metric space and $\mathcal{L} = \mathcal{B}(X)$ being the Borel σ-algebra. Then $\mathcal{B}(X)$ is countably generated and countably separated. To see this consider $\{U_n\}_{n\geq 1}$ being a countable basis for the metric topology. Then $\sigma(\{U_n\}_{n\geq 1}) = \mathcal{B}(X)$, that is, $\mathcal{B}(X)$ is countably generated and clearly, $\{U_n\}_{n\geq 1}$ separates points in X, that is, $\mathcal{B}(X)$ is countably separated.

Proposition 2.7.16. *If (Ω, Σ) is a measurable space, Y is a Hausdorff topological space and $D \in \Sigma \bigotimes \mathcal{B}(Y)$, then there exists $\Sigma_0 \subseteq \Sigma$ being a countably generated sub-σ-algebra of Σ such that $D \in \Sigma_0 \bigotimes \mathcal{B}(Y)$.*

Proof. Let $\mathcal{L} = \{C \in \Sigma \bigotimes \mathcal{B}(X)\colon$ the conclusion of the proposition holds$\}$. Clearly $\mathcal{L}$ includes all measurable rectangles; see Remark 2.2.24. Moreover $\mathcal{L}$ is closed under complementation. Let $\{C_n\}_{n\geq 1} \subseteq \mathcal{L}$. Then $C_n \in \Sigma_{on} \bigotimes \mathcal{B}(X)$ with $\Sigma_{on} \subseteq \Sigma$ being a countably generated sub-σ-algebra. Then $\bigcup_{n\geq 1} C_n \in \sigma(\bigcup_{n\geq 1} \Sigma_{on}) \bigotimes \mathcal{B}(X)$ and $\sigma(\bigcup_{n\geq 1} \Sigma_{on})$ is countably generated. Therefore $\mathcal{L}$ is a σ-algebra and so we must have $\mathcal{L} = \Sigma \bigotimes \mathcal{B}(X)$. □

Extending the notion of universal σ-algebra (see Definition 2.6.23) to arbitrary measurable spaces, we state the following definition.

Definition 2.7.17. Let (Ω, Σ) be a measurable space. The **universal σ-algebra** corresponding to Σ is defined by $\hat{\Sigma} = \bigcap_{\mu \in M_1^+(\Omega)} \Sigma_\mu$ where $M_1^+(\Omega)$ denotes the set of all probability measures on Ω and Σ_μ is the μ-completion of Σ. We say that the measurable space (Ω, Σ) is complete if $\Sigma = \hat{\Sigma}$.

Using this definition and Corollary 2.6.29 (see also the proof of Proposition 2.6.28) we have the following result.

Proposition 2.7.18. *If (Ω_1, Σ_1) and (Ω_2, Σ_2) are measurable spaces and $f: \Omega_1 \to \Omega_2$ is a (Σ_1, Σ_2)-measurable map, then f is $(\hat{\Sigma}_1, \hat{\Sigma}_2)$-measurable.*

The next result is the original version of the so-called "Yankov–von Neumann Selection Theorem". For its proof we refer to Klein–Thompson [194, Theorem 14.3.2, p. 166].

Theorem 2.7.19 (Yankov–von Neumann Selection Theorem). *If X, Y are Polish spaces, $F: X \to 2^Y \setminus \{\emptyset\}$, and $\operatorname{Gr} F \in a_{X \times Y}$, then there exists an analytically measurable function $f: X \to Y$ such that $f(x) \in F(x)$ for all $x \in X$.*

Recalling that a Souslin space is the continuous image of a Polish space (see Definition 1.5.51), from Theorem 2.7.19 we easily deduce the following result.

Theorem 2.7.20. *If X is a Borel subset of a Polish space, Y is a Souslin space, $F: X \to 2^Y \setminus \{\emptyset\}$, and $\operatorname{Gr} F \subseteq X \times Y$ is a Souslin subset, then there exists an analytically measurable map $f: X \to Y$ such that $f(x) \in F(x)$ for all $x \in X$.*

Remark 2.7.21. Note that Borel sets of Polish spaces are usually called **Borel spaces**.

Proposition 2.7.22. *If (Ω, Σ) is a measurable space such that Σ is countably generated and countably separated, then there is a subset E of $\{0,1\}^{\mathbb{N}}$ such that (Ω, Σ) and $(E, \mathcal{B}(E))$ are isomorphic; see Definition 2.6.20.*

Proof. Let $\{A_n\}_{n \geq 1}$ be the generators of Σ. We are going to show that they separate points in Ω. Arguing by contradiction, suppose that for some $w, w' \in \Omega$, $w \neq w'$ it holds that $\chi_{A_n}(w) = \chi_{A_n}(w')$ for all $n \in \mathbb{N}$. Let $\Sigma_0 = \{A \subseteq \Omega : \chi_{A_n}(w) = \chi_{A_n}(w')\}$. Evidently Σ_0 is a σ-algebra and $A_n \in \Sigma_0$ for all $n \in \mathbb{N}$, thus $\Sigma \subseteq \Sigma_0$, which contradicts the fact that Σ is countably separated. Let $f: \Omega \to \{0,1\}^{\mathbb{N}}$ be defined by $f(w) = \{\chi_{A_n}(w)\}_{n \geq 1}$. Clearly f is one-to-one and Σ-measurable. We need to show that $f^{-1}: E = f(\Omega) \to \Omega$ is measurable. So, we want to show that if $A \in \Sigma$, then $f(A) \in \mathcal{B}(E)$. Let $\Sigma_1 = \{A \subseteq \Omega : f(A) \in \mathcal{B}(E)\}$. This is a σ-algebra and $A_n \in \Sigma_1$ for all $n \in \mathbb{N}$ since $f(A_n) = \{(e_k) \in \{0,1\}^{\mathbb{N}} : e_n = 1\} \cap E$. Therefore, $\Sigma \subseteq \Sigma_1$ and we have proven the measurability of f^{-1}. Hence, we have that (Ω, Σ) and $(E, \mathcal{B}(E))$ are isomorphic. □

Remark 2.7.23. Recall that $\{0,1\}^{\mathbb{N}}$ and $\mathbb{N}^\infty$ are isometrically isomorphic. Hence, $\{0,1\}^{\mathbb{N}}$ is Polish.

Proposition 2.7.24. *If (Ω, Σ) is a measurable space such that Σ is countably generated and countably separated, X is a Souslin space and $F: \Omega \to 2^X \setminus \{\emptyset\}$ is a graph measurable multifunction, then F admits a $\hat{\Sigma}$-measurable selection.*

Proof. Invoking Proposition 2.7.22 we know that there exists $E \subseteq \{0,1\}^{\mathbb{N}}$ such that (Ω, Σ) and $(E, \mathcal{B}(E))$ are isomorphic. Let $h: \Omega \to E$ be this isomorphism. The measurable spaces $(\Omega \times X, \Sigma \bigotimes \mathcal{B}(X))$ and $(E \times X, \mathcal{B}(E) \bigotimes \mathcal{B}(X))$ are isomorphic. Moreover, from Proposition 2.2.26 (b) we know that $\mathcal{B}(E) \bigotimes \mathcal{B}(X) = \mathcal{B}(E \times X)$.

We introduce the multifunction $F_1: E \to 2^X \setminus \{\emptyset\}$ defined by $F_1 = F \circ h^{-1}$. We have $\operatorname{Gr} F_1 = (h, \mathrm{id}_X)(\operatorname{Gr} F)$ with id_X being the identity map on X. Therefore $\operatorname{Gr} F_1 \in \mathcal{B}(E) \bigotimes \mathcal{B}(X) = \mathcal{B}(E \times X)$.

Hence, there exists $D_1 \in \mathcal{B}(P \times X)$ with $P = \{0,1\}^{\mathbb{N}}$ such that $\operatorname{Gr} F_1 = D_1 \cap (E \times X)$. Then $E = \operatorname{proj}_P \operatorname{Gr} F_1 \subseteq E_1 = \operatorname{proj}_P D_1$. Let $h': \Omega \to E_1$ be defined by $h'(w) = h(w)$ for all $w \in \Omega$. Then h' is injective and Σ-measurable. Let $F_2: E_1 \to 2^X \setminus \{\emptyset\}$ be the multifunction defined by $\operatorname{Gr} F_2 = D_1$. We claim that

$$F_2(h'(w)) = F_1(h(w)) \quad \text{for all } w \in \Omega . \tag{2.7.7}$$

To this end, note that for every $u \in E$ we have

$$F_1(u) = \operatorname{proj}_X[\operatorname{Gr} F_1 \cap (\{u\} \times X)] \quad \text{and} \quad F_2(u) = \operatorname{proj}_X[\operatorname{Gr} F_2 \cap (\{u\} \times X)] .$$

Recall that $\operatorname{Gr} F_1 = \operatorname{Gr} F_2 \cap (E \times X)$. So

$$\operatorname{Gr} F_1 \cap (\{u\} \times X) = \operatorname{Gr} F_2 \cap (\{u\} \times X) ,$$

which gives $F_1(u) = F_2(u)$ for all $u \in E$ and this proves (2.7.7).

Since $D_1 \in \mathcal{B}(E \times X)$, D_1 is a Souslin subset of $E \times X$. Hence, we can apply Theorem 2.7.20 and obtain $f_2: E_1 \to X$ being an analytically measurable map such that $f_2(u) \in F_2(u)$ for all $u \in E_1$. Since h' is $(\Sigma, \mathcal{B}(E_1))$-measurable, using Proposition 2.7.18 we have that h' is $(\hat{\Sigma}, \hat{\mathcal{B}}(E_1))$-measurable. Let $f = f_2 \circ h'$. Then $f: \Omega \to X$ is $\hat{\Sigma}$-measurable and $f(w) \in F(w)$ for all $w \in \Omega$. □

Now we are ready for the second measurable selection theorem which is graph conditioned. The result is usually known as the "Yankov–von Neumann–Aumann Selection Theorem".

Theorem 2.7.25 (Yankov–von Neumann–Aumann Selection Theorem). *If (Ω, Σ) is a complete measurable space, X is a Souslin space, and $F: \Omega \to 2^X \setminus \{\emptyset\}$ is graph measurable, then F admits a Σ-measurable selection.*

Proof. Using Proposition 2.7.16 there is a countably generated sub-σ-algebra $\Sigma_0 \subseteq \Sigma$ such that $\operatorname{Gr} F \in \Sigma_0 \bigotimes \mathcal{B}(X)$. On Ω we define an equivalence relation $\sim$ by

$$w \sim w' \quad \text{if and only if} \quad \chi_A(w) = \chi_A(w') \quad \text{for all } A \in \Sigma_0 . \tag{2.7.8}$$

Let $\Omega_* = \Omega/\sim$ and let $p: \Omega \to \Omega_*$ be the canonical projection on the quotient space, that is, $p(w) = \dot{w}$ being the equivalence class of $w \in \Omega$. Let $\Sigma_* = p(\Sigma_0) = \{p(A): A \in \Sigma_0\}$.

It is easy to see that Σ_* is a σ-algebra and if $\{A_n\}_{n\geq 1}$ are the generators of Σ_0, that is, $\Sigma_0 = \sigma(\{A_n\}_{n\geq 1})$, then $\Sigma_* = \sigma(\{p(A_n)\}_{n\geq 1})$. Therefore Σ_* is countably generated.

Next suppose that $\dot{w} \neq \dot{w}'$. Then we can find $A \in \Sigma_0$ such that $\chi_A(w) \neq \chi_A(w')$; see (2.7.8). This is equivalent saying that $\chi_{p(A)}(\dot{w}) = \chi_{p(A)}(\dot{w}')$. It follows that Σ_* is also countably separated. Moreover, note that p is a one-to-one correspondence between Σ_0 and Σ_*. Let id_X be the identity map on X and let $\eta: \Omega \times X \to \Omega_* \times X$ be defined by $\eta = (p, \mathrm{id}_X)$. Then $\operatorname{Gr} F \in \Sigma \otimes \mathcal{B}(X)$ implies that $\eta(\operatorname{Gr} F) \in \Sigma_* \otimes \mathcal{B}(X)$. Let $F_1: \Omega_* \to 2^X \setminus \{\emptyset\}$ defined by $\operatorname{Gr} F_1 = \eta(\operatorname{Gr} F)$. We can now apply Proposition 2.7.24 and produce a $\hat{\Sigma}_*$-measurable selection $f_1: \Omega_* \to X$ of F_1, that is, $f_1(w) \in F_1(w)$ for all $w \in \Omega$. Let $f = f_1 \circ p$ and for $w \in \Omega$ we define $\mathcal{D}(w) = \{A \in \Sigma_0 \otimes \mathcal{B}(X): A_{w'} = A_w \text{ for all } w' \in \dot{w}\}$. Recall that A_w is the w-section of A; see Definition 2.2.27. Note that $\mathcal{D}(w)$ is an algebra and a monotone class. Hence, Theorem 2.1.12 implies that $\mathcal{D}(w)$ is a σ-algebra. It follows that $\mathcal{D}(w) = \Sigma_0 \otimes \mathcal{B}(X)$. Since $\operatorname{Gr} F_w = F(w)$, we see that F is constant on $\dot{w}$ and we have $F(w') = F_1(\dot{w})$ for all $w' \in \dot{w}$. Because $f(w') = f(\dot{w})$ we obtain that $f(w) \in F(w)$ for all $w \in \Omega$. Proposition 2.7.18 implies that f is Σ-measurable. This finishes the proof. □

As for the Kuratowski–Ryll Nardzewski Selection Theorem (see Theorem 2.7.12), we can improve the result above and produce a whole dense sequence of measurable selections. To do this, we will need the following result due to Leese [210, p. 407].

Proposition 2.7.26. *If (Ω, Σ) is a complete measurable space, X is a Souslin space, and $F: \Omega \to 2^X \setminus \{\emptyset\}$ is graph measurable, then there exists a Polish space Y, a measurable multifunction $G: \Omega \to P_f(Y)$, and a continuous map $h: Y \to X$ such that $F(w) = h(G(w))$ for all $w \in \Omega$.*

Remark 2.7.27. Using this proposition and the Kuratowski–Ryll Nardzewski Selection Theorem (see Theorem 2.7.12), we have at once the Yankov–von Neumann–Aumann Selection Theorem; see Theorem 2.7.25. The conclusion of this proposition looks similar to the definition of Souslin spaces; see Definition 1.5.51. For this reason graph measurable multifunctions into a Souslin space are also called multifunctions of **Souslin-type**.

Theorem 2.7.28. *If (Ω, Σ) is a complete measurable space, X is a Souslin space, and $F: \Omega \to 2^X \setminus \{\emptyset\}$ is graph measurable, then there exists a sequence of Σ-measurable selections $f_n: \Omega \to X$ of F such that $F(w) \subseteq \overline{\{f_n(w)\}}_{n\geq 1}$ for all $w \in \Omega$.*

Proof. Applying Proposition 2.7.26 there is a Polish space Y, a measurable multifunction $G: \Omega \to P_f(Y)$, and a continuous map $h: Y \to$ such that

$$F(w) = h(G(w)) \quad \text{for all } w \in \Omega\,. \tag{2.7.9}$$

Invoking Theorem 2.7.13 there is a sequence of Σ-measurable selections $g_n: \Omega \to Y$ of G such that

$$G(w) = \overline{\{g_n(w)\}}_{n\geq 1} \quad \text{for all } w \in \Omega\,. \tag{2.7.10}$$

The continuity of h implies that $f_n = h \circ g_n : \Omega \to X$ with $n \in \mathbb{N}$ is a sequence of Σ-measurable selections of F (see (2.7.9)), and using Proposition 1.1.35 (b) as well as (2.7.10) we derive that

$$F(w) \subseteq \overline{\{f_n(w)\}_{n\geq 1}} \quad \text{for all } w \in \Omega . \qquad \square$$

Given a Borel subset in a Cartesian product it is natural to ask whether its projection on a factor is Borel as well. The next example shows that the answer to this question is negative. This fact was the starting point for Souslin to develop the theory of analytic sets; see Remarks 2.9.

Example 2.7.29. We show that the projection of a Borel set in $\mathbb{R}^2$ need not be Borel. So, let $X = [0,1]$, $Y = [0,1] \cap (\mathbb{R} \setminus \mathbb{Q})$ being the set of the irrationals in $[0,1]$. From Corollary 1.5.49 we know that Y is a Polish space. Let $A \subseteq X$ be analytic but not Borel and let $f: Y \to A$ be a continuous function. Then $\operatorname{Gr} f \in \mathcal{B}(X \times Y) = \mathcal{B}(X) \otimes \mathcal{B}(Y)$ but $\operatorname{proj}_X \operatorname{Gr} f = A \notin \mathcal{B}(X)$.

Next we will show that the projection of a Borel set is universally measurable. We will need two auxiliary lemmata.

Lemma 2.7.30. *If*

$$K_n = \left\{ \sum_{k\geq 1} \frac{s_k}{4^k} : s \in \{0,1\}^{\mathbb{N}}, s_n = 1 \right\},$$

then $K_n \subseteq \mathbb{R}$ is compact and for every $s \in \{0,1\}^{\mathbb{N}}$, it holds that $\sum_{k\geq 1} s_k/4^k \in K_n$ if and only if $s_n = 1$.

Proof. We know that $\{0,1\}^{\mathbb{N}}$ is compact. Let $C_n = \{s \in \{0,1\}^{\mathbb{N}} : s_n = 1\}$. This set is closed, hence compact. Consider the function $f: \{0,1\}^{\mathbb{N}} \to \mathbb{R}$ defined by $f(s) = \sum_{k\geq 1} s_k/4^k$. Then f is the uniform limit of continuous functions, hence it is continuous. It follows that $f(C_n) = K_n$ is compact. Note that f is injective, hence it is a homeomorphism (see Theorem 1.4.54), and $f(s) \in K_n$ if and only if $s \in C_n$. $\square$

Lemma 2.7.31. *If (Ω, Σ) is a measurable space, Y is a Hausdorff topological space, and $D \in \Sigma \otimes \mathcal{B}(Y)$, then there exists $C \in \mathcal{B}(\mathbb{R} \times Y)$ and a Σ-measurable function $f: \Omega \to \mathbb{R}$ such that $D = \{(w,y) \in \Omega \times Y : (f(w), y) \in C\}$.*

Proof. Invoking Proposition 2.7.16 there exists a countably generated sub-σ-algebra $\Sigma_0 \subseteq \Sigma$ such that $D \in \Sigma_0 \otimes \mathcal{B}(Y)$. Suppose $\Sigma_0 = \sigma(\{A_n\}_{n\geq 1})$ and consider the function $f: \Omega \to \mathbb{R}$ defined by $f(w) = \sum_{k\geq 1} 1/4^k \chi_{A_k}(w)$. Lemma 2.7.31 says that for every $n \in \mathbb{N}$ and every $w \in \Omega$ we have $f(w) \in K_n$ if and only if $\chi_{A_n}(w) = 1$ if and only if $w \in A_n$. Hence

$$f^{-1}(K_n) = A_n . \tag{2.7.11}$$

Evidently f is Σ-measurable and we define $\xi(w,y) = (f(w),y)$ and $\mathcal{L} = \{\xi^{-1}(E): E \in \mathcal{B}(\mathbb{R} \times Y)\}$. Clearly $\mathcal{L}$ is a σ-algebra and from (2.7.11) we see that $\xi^{-1}(K_n \times B) = f^{-1}(K_n) \times B = A_n \times B$ with $B \in \mathcal{B}(Y)$. This implies $A_n \times B \in \mathcal{L}$ for all $n \in \mathbb{N}$ and for all $B \in \mathcal{B}(Y)$. Therefore $D \in \Sigma_0 \times \mathcal{B}(Y) \subseteq \mathcal{L}$. So, there is a set $C \in \mathcal{B}(\mathbb{R} \times Y)$ such that $D = \xi^{-1}(C)$ and this proves the lemma. □

Now we are ready for the measurable projection theorem known as the "Yankov–von Neumann–Aumann Projection Theorem".

Theorem 2.7.32 (Yankov–von Neumann–Aumann Projection Theorem). *If (Ω, Σ) is a complete measurable space, X is a Souslin space, and $D \in \Sigma \otimes \mathcal{B}(X)$, then* $\mathrm{proj}_\Omega D \in \Sigma$.

Proof. Lemma 2.7.31 says that there exist $C \in \mathcal{B}(\mathbb{R} \times X)$ and a Σ-measurable function $f: \Omega \to \mathbb{R}$ such that $D = \{(w,x) \in \Omega \times X: (f(w),x) \in C\}$. Then $\mathrm{proj}_\Omega D = f^{-1}(\mathrm{proj}_\mathbb{R} C)$. The space $X \times \mathbb{R}$ is Souslin (see Proposition 1.5.54 (b)), and since $C \in \mathcal{B}(\mathbb{R} \times X)$ it follows that C is Souslin; see Proposition 2.6.11. The set $\mathrm{proj}_\mathbb{R} C$ is the continuous image of a Souslin space, therefore it is a Souslin space as well. As f is Σ-measurable, invoking Proposition 2.7.18, we conclude that $D \in \hat{\Sigma}$. □

We mention two more measurable projection theorems. The first is due to Brown–Purves [63].

Theorem 2.7.33. *If X, Y are Polish spaces, $D \in \mathcal{B}(X \times Y) = \mathcal{B}(X) \otimes \mathcal{B}(Y)$ and for every $x \in D$, $D_x \subseteq Y$ is σ-compact, then* $\mathrm{proj}_X D \in \mathcal{B}(X)$.

For the second projection theorem, we need to introduce a special class of spaces.

Definition 2.7.34. Let Y be a Hausdorff topological space. We say that Y is of **class** σ MK, if $Y = \bigcup_{n \geq 1} K_n$ with each K_n with $n \in \mathbb{N}$ large enough, being metrizable compact.

Remark 2.7.35. Recall that every metrizable compact space is the continuous image of a Cantor set; see Kuratowski [199, p. 444]. Therefore X is σ MK if and only if X is the continuous image of a closed set in $\mathbb{R}$. A separable, metrizable, locally compact space belongs to the class σ MK. But the space need not be metrizable. Again anticipating some material from Chapter 3, let X be a separable Banach space and let X^* be its topological dual. We have $X^* = \bigcup_{n \geq 1} n\overline{B}_1^*$ with $\overline{B}_1^* = \{x^* \in X^*: \|x^*\|_* \leq 1\}$ being the closed unit ball in X^*. We know that $\overline{B}_1^*$ equipped with the relative w*-topology is metrizable compact; see Section 3.3. So, $X^*_{\mathrm{w}^*}$, that is, X^* furnished with the w*-topology, is a σ MK-space.

The next measurable projection theorem is due to Levin [215].

Theorem 2.7.36. *If X is a Borel subset of a Polish space, that is, a Borel space, Y is a σ MK-space and $D \in \mathcal{B}(X \times Y) = \mathcal{B}(X) \otimes \mathcal{B}(Y)$ with $D_x \in P_f(Y)$ for every $x \in X$, then* $\mathrm{proj}_X D \in \mathcal{B}(X)$.

Remark 2.7.37. Note that in this case the projection of a Borel set is Borel.

Comparable Souslin topologies on a set X generate the same Borel σ-algebras.

Proposition 2.7.38. *If τ_1 and τ_2 are two comparable Souslin topologies on X, then $\mathcal{B}(X_{\tau_1}) = \mathcal{B}(X_{\tau_2})$.*

Proof. To fix things we assume that $\tau_2 \subseteq \tau_1$. Then $\mathcal{B}(X_{\tau_2}) \subseteq \mathcal{B}(X_{\tau_1})$. Let $A \in \mathcal{B}(X_{\tau_1})$. Then A is τ_1-Souslin; see Propositions 2.6.11 and 2.6.9. Hence, there exist a Polish space Y and a continuous surjection $f: Y \to (A, \tau_1(A))$; see Definition 1.1.24. Then $f: Y \to (A, \tau_2(A))$ is continuous as well and so A is τ_2-Souslin. The same argument applied to $A^c = X \setminus A$ shows that A^c is τ_2-Souslin as well. Invoking Corollary 2.6.17 we conclude that $A \in \mathcal{B}(X_{\tau_2})$. Hence $\mathcal{B}(X_{\tau_1}) \subseteq \mathcal{B}(X_{\tau_2})$ and so finally we conclude that $\mathcal{B}(X_{\tau_1}) = \mathcal{B}(X_{\tau_2})$. □

Remark 2.7.39. More generally if τ_1 and τ_2 are two Souslin topologies on X and $\tau_1 \cap \tau_2$ is Hausdorff, then $\mathcal{B}(X_{\tau_1}) = \mathcal{B}(X_{\tau_2}) = \mathcal{B}(X_{\tau_1 \cap \tau_2})$.

Proposition 2.7.40. *If (Ω, Σ) is a complete measurable space, X is a Polish space, and $F: \Omega \to 2^X \setminus \{\emptyset\}$ is graph measurable, then $F^-(D) \in \Sigma$ for all $D \in \mathcal{B}(X)$.*

Proof. Note that $F^-(D) = \operatorname{proj}_\Omega[\operatorname{Gr} F \cap (\Omega \times B)] \in \Sigma$; see Theorem 2.7.32. □

Therefore, we can state the following theorem, which summarizes the measurability properties of closed valued multifunctions.

Theorem 2.7.41. *Let (Ω, Σ) be a measurable space, (X, d) is a separable metric space and $F: \Omega \to P_f(X)$. Consider the following statements:*
(a) *$F^-(D) \in \Sigma$ for all $D \in \mathcal{B}(X)$;*
(b) *$F^-(C) \in \Sigma$ for all closed $C \subseteq X$;*
(c) *F is measurable;*
(d) *for every $x \in X$, $w \to d(x, F(w))$ is Σ-measurable;*
(e) *there exists a sequence of Σ-measurable selections $f_n: \Omega \to X$ such that $F(w) = \overline{\{f_n(w)\}_{n\geq 1}}$ for all $w \in \Omega$;*
(f) *F is graph measurable.*

We have the following implications:
(1) *(a) $\Longrightarrow$ (b) $\Longrightarrow$ (c) $\Longleftrightarrow$ (d) $\Longrightarrow$ (f).*
(2) *If X is complete, that is, X is a Polish space, then (c) $\Longleftrightarrow$ (d) $\Longleftrightarrow$ (e).*
(3) *If X is σ-compact, then (b) $\Longleftrightarrow$ (c).*
(4) *If $\Sigma = \hat{\Sigma}$, that is, the measurable space is complete, and X is complete, then (a) to (f) are all equivalent.*

2.8 Hausdorff Measures

In this section we introduce and study certain lower dimensional measures in $\mathbb{R}^N$ known as "Hausdorff measures". These measures allow us to measure very small sets. So, for example, the Hausdorff measures offer the possibility on measuring fractal sets and to define a new notion of dimension for a given set extending the standard topological dimension.

The next example shows that the classical approach described in Section 2.1 is not adequate. Consider the set

$$C = \left\{ \left(t, \sin\left(\frac{1}{t}\right) \right) : 0 \le t \le 1 \right\} .$$

Following Proposition 2.1.34, we define

$$\mu(C) = \inf\left\{ \sum_{k\in\mathbb{N}} \operatorname{diam} D_k : C \subseteq \bigcup_{k\in\mathbb{N}} D_k \right\} .$$

Then $\mu(C) < \infty$, whereas it should be $\mu(C) = \infty$.

The problem with the classical definition is that the approximating sets D_k are not required to follow the geometry of the curve. For this reason we adopt a different approach.

Let

$$\Gamma(s) = \int_0^\infty t^{s-1} e^{-t}\, dt \quad \text{for } 0 < s < \infty$$

be the classical Euler gamma function. For $s \ge 0$ we define

$$\alpha(s) = \frac{\pi^{\frac{s}{2}}}{\Gamma(1 + \frac{s}{2})} .$$

When $s \in \mathbb{N}$, then $\alpha(s)$ is the volume of the unit ball of $\mathbb{R}^s$.

Definition 2.8.1. For each $s \ge 0$, $\varepsilon > 0$ and $C \subseteq \mathbb{R}^N$, we define

$$H^s_\varepsilon(C) = \inf\left\{ \frac{\alpha(s)}{2^s} \sum_{k\in\mathbb{N}} (\operatorname{diam} D_k)^s : \operatorname{diam} D_k \le \varepsilon,\ C \subseteq \bigcup_{k\in\mathbb{N}} D_k \right\} .$$

Clearly, $\varepsilon \to H^s_\varepsilon(C)$ is nondecreasing and so we can define the s-**dimensional Hausdorff measure** of C by

$$H^s(C) = \lim_{\varepsilon\to 0^+} H^s_\varepsilon(C) = \sup_{\varepsilon>0} H^s_\varepsilon(C) .$$

Remark 2.8.2. The reason that we take $\alpha(s)$ to be the volume of the unit ball when $s \in \mathbb{N}$, is for $H^s(C)$ to agree with the intuitive notion of s-**dimensional area** when C is a well behaved set. So, if $C \subseteq \mathbb{R}^{n+k}$ is an n-dimensional C^1-submanifold, then $H^n(C)$ is the n-**dimensional surface area** of C. Also, note that $\lambda^n(B_r(x)) = \alpha(n) r^n$ for all $B_r(x) \subseteq \mathbb{R}^n$ with λ^n being the n-dimensional Lebesgues measure on $\mathbb{R}^n$. The requirement that $\varepsilon \to 0^+$ forces the sets D_k of the ε-covering to follow the local geometry of C.

Proposition 2.8.3. *For $s \geq 0$, $H^s: 2^{\mathbb{R}^N} \to \overline{\mathbb{R}}_+ = [0, +\infty]$ is an outer measure; see Definition* 2.1.33.

Proof. For all $C \in 2^{\mathbb{R}^N}$ with $\operatorname{diam} C \leq \delta$ we have

$$H^s_\varepsilon(\emptyset) \leq (\operatorname{diam} C)^s \leq \delta^s ,$$

which shows that $H^s(\emptyset) = 0$. The monotonicity of H^s is an immediate consequence of Definition 2.8.1. Finally, let $\{C_m\}_{m\in\mathbb{N}} \subseteq 2^{\mathbb{R}^N}$ and for every $m \in \mathbb{N}$, let $\{D^n_m\}_{n\in\mathbb{N}}$ be an ε-covering of C_m such that

$$\frac{\alpha(s)}{2^s} \sum_{n\in\mathbb{N}} \operatorname{diam} D^n_m \leq \frac{\varepsilon}{2^m} + H^s_\varepsilon(C_m) .$$

Evidently, $\{D^n_m\}_{(m,n)\in\mathbb{N}^2}$ is an ε-covering of $\bigcup_{m\in\mathbb{N}} C_m$ and

$$H^s_\varepsilon\Big(\bigcup_{m\in\mathbb{N}} C_m\Big) \leq 2\varepsilon + \sum_{m\in\mathbb{N}} H^s_\varepsilon(C_m) .$$

Letting $\varepsilon \to 0^+$, we obtain the σ-additivity of H^s. Therefore, H^s is an outer measure in the sense of Definition 2.1.33. □

Now we can proceed as in Section 2.1; see Definition 2.1.36.

Definition 2.8.4. We say that $C \subseteq \mathbb{R}^N$ is **H^s-measurable** if

$$H^s(E) = H^s(E \cap C) + H^s(E \cap C^c) \quad \text{for all } E \subset \mathbb{R}^N ,$$

that is, C splits additively all sets in $\mathbb{R}^N$. We denote the family of all H^s-measurable sets by Σ^*_s.

Then using Carathéodory's Theorem stated in Theorem 2.1.38 we obtain the following result.

Theorem 2.8.5. *Σ^*_s is a σ-algebra and H^s is a complete measure on Σ^*_s.*

Remark 2.8.6. More generally, the theory can be developed on a general metric space. For a presentation of this more general framework, we refer to the book of Ambrosio–Tilli [12].

Let $\mathcal{B}(\mathbb{R}^N)$ be the Borel σ-algebra of $\mathbb{R}^N$. We will show that $\mathcal{B}(\mathbb{R}^N) \subseteq \Sigma^*_s$ and so H^s is a Borel measure. We will need the following result known in the literature as the "Carathéodory Criterion".

Proposition 2.8.7 (Carathéodory Criterion). *If (X, d) is a metric space and μ^* is an outer measure on X, then $\mathcal{B}(X) \subseteq \Sigma^*$ if and only if*

$$\mu^*(A \cup C) \geq \mu^*(A) + \mu^*(C) \quad \textit{for all } A, C \subseteq X$$

with

$$d(A,C) = \inf\{d(a,c) : a \in A,\ c \in C\} > 0\,,$$

where Σ^* *is the* σ*-algebra of all* μ^**-measurable sets; see Definition* 2.1.36 *and Theorem* 2.1.38.

Proof. $\Longrightarrow$: This is obvious.

$\Longleftarrow$: It suffices to show that for any set $E \subseteq X$ with $\mu^*(E) < \infty$ and every open set $U \subseteq X$, we have

$$\mu^*(E) \geq \mu^*(E \cap U) + \mu^*(E \cap U^c)\,.$$

Let $D = E \cap U$ and

$$D_k = D \cap \left\{x \in X : d(x, U^c) \geq \frac{1}{k}\right\}.$$

Then $d(D_k, E \cap U^c) > 0$, hence $d(E \cap D_k, E \cap U^c) > 0$. By hypothesis we have

$$\mu^*((E \cap D_k) \cup (E \cap U^c)) \geq \mu^*(E \cap D_k) + \mu^*(E \cap U^c)\,,$$

and so

$$\mu^*(E) \geq \mu^*(E \cap D_k) + \mu^*(E \cap U^c)\,. \tag{2.8.1}$$

Let $D_0 = \emptyset$ and $S_n = D_{n+1} \cap D_n^c$ for $n \in \mathbb{N}$. Then

$$\mu^*(D_{2k+1}) \geq \sum_{n=1}^{k} \mu^*(S_{2n})\,.$$

But

$$D = \bigcup_{k \in \mathbb{N}} D_k = D_{2k} + \bigcup_{n \geq k} S_{2n} + \bigcup_{n \geq k+1} S_{2n-1}$$

and

$$\sum_{n \in \mathbb{N}} \mu^*(S_{2n}) \leq \mu^*(E) < \infty, \quad \sum_{n \in \mathbb{N}} \mu^*(S_{2n+1}) \leq \mu^*(E) < \infty\,.$$

Therefore,

$$\mu^*(D) \leq \mu^*(D_{2k}) + \sum_{n \geq k} \mu^*(S_n) + \sum_{n \geq k+1} \mu^*(S_{2n-1})\,.$$

This implies that $\mu^*(D_k) \to \mu^*(D)$ and so $\mu^*(E \cap D_k) \to \mu^*(E \cap D) = \mu^*(E \cap U)$.

So, if we pass to the limit as $k \to \infty$ in (2.8.1), we obtain

$$\mu^*(E) \geq \mu^*(E \cap U) + \mu^*(E \cap U^c),$$

which shows that $U \in \Sigma^*$ and so $\mathcal{B}(X) \subseteq \Sigma^*$. □

Definition 2.8.8. (a) Let X be a set and μ^* an outer measure on X. We say that μ^* is **regular**, if for each set $E \subseteq X$, we can find $C \in \Sigma^*$ and $E \subseteq C$ such that $\mu^*(E) = \mu^*(C)$.
(b) An outer measure μ^* on $\mathbb{R}^N$ is said to be **Borel**, if $\mathcal{B}(\mathbb{R}^N) \subseteq \Sigma^*$.

Theorem 2.8.9. *If $0 \leq s < \infty$, then H^s is a Borel regular outer measure.*

Proof. Let $A, C \subseteq \mathbb{R}^N$ be such that $d(A, C) > 0$. Let $\mathcal{D} = \{D_k\}_{k\in\mathbb{N}}$ be a cover of $A \cup C$ with $\operatorname{diam} D_k \leq \varepsilon < d(A, C)$. Let $\mathcal{D}_1 = \{D \in \mathcal{D} : D \cap A \neq \emptyset\}$ and $\mathcal{D}_2 = \{D \in \mathcal{D} : D \cap C \neq \emptyset\}$. Then $\mathcal{D}_1, \mathcal{D}_2$ are disjoint and $\mathcal{D} \in \mathcal{D}_1 \cup \mathcal{D}_2$. So we have

$$\sum_{k\in\mathbb{N}} \frac{\alpha(s)}{2^s}(\operatorname{diam} D_k)^s = \sum_{D\in\mathcal{D}_1} \frac{\alpha(s)}{2^s}(\operatorname{diam} D)^s + \sum_{D\in\mathcal{D}_2} \frac{\alpha(s)}{2^s}(\operatorname{diam} D)^s .$$

Therefore, $H^s_\varepsilon(A \cup C) \geq H^s_\varepsilon(A) + H^s_\varepsilon(C)$. Letting $\varepsilon \to 0^+$, we conclude that $H^s(A \cup C) \geq H^s(A) + H^s(C)$. So, by Proposition 2.8.7, $\mathcal{B}(\mathbb{R}^N) \subseteq \Sigma^*_s$ and so H^s is Borel.

Finally, we show the regularity of H^s. We know that $\operatorname{diam} C = \operatorname{diam} \overline{C}$ and so we can say that

$$H^s_\varepsilon(C) = \inf\left\{\frac{\alpha(s)}{2^s}\sum_{k\in\mathbb{N}} \operatorname{diam} D_k : C \subseteq \bigcup_{k\in\mathbb{N}} D_k,\ \operatorname{diam} C_k \leq \delta,\ C_k \text{ closed}\right\}.$$

Let $C \subseteq \mathbb{R}^N$ with $H^s(C) < \infty$. Then $H^s_\varepsilon(C) < \infty$ for all $\varepsilon > 0$. For each $k \in \mathbb{N}$, let $\{D^n_k\}_{n\in\mathbb{N}}$ be closed sets such that

$$\operatorname{diam} D^n_k \leq \frac{1}{n}, \quad C \subseteq \bigcup_{k\in\mathbb{N}} D^n_k$$

and

$$\sum_{k\in\mathbb{N}} \frac{\alpha(s)}{2^s}(\operatorname{diam} C^n_k)^s \leq H^s_{\frac{1}{n}}(C) + \frac{1}{n}. \tag{2.8.2}$$

We set $C_n = \cup_{k\in\mathbb{N}} D^n_k$ and $E = \cap_{k\in\mathbb{N}} C_n$. Then $E \in \mathcal{B}(\mathbb{R}^N)$ and $C \subseteq E$. Moreover, we have

$$H^s_{\frac{1}{n}}(E) \leq \sum_{k\in\mathbb{N}} \frac{\alpha(s)}{2^s}(\operatorname{diam} D^n_k)^s \leq H^s_{\frac{1}{n}}(C) + \frac{1}{n},$$

due to (2.8.2). Letting $n \to \infty$, we obtain $H^s(E) \leq H^s(C)$ and so, since $C \subseteq E$, $H^s(E) = H^s(C)$. Thus, H^s is regular. □

Observe that $\alpha(0) = 1$. Then $H^0(\{u\}) = 1$ for all $u \in \mathbb{R}^N$. Therefore, we can state the following result.

Proposition 2.8.10. *H^0 is the counting measure and $H^1(\cdot) = \lambda^1(\cdot)$ is the Lebesgue measure.*

The Hausdorff measures have the scaling property, well-known for the classical length, area and volume.

Proposition 2.8.11. *If $0 \le s < \infty$, $C \subseteq \mathbb{R}^N$ and $\lambda > 0$, the $H^s(\lambda C) = \lambda H^s(C)$.*

Proof. Let $\{D_k\}_{k\in\mathbb{N}}$ be an ε-cover of C. Then $\{\lambda D_k\}_{k\in\mathbb{N}}$ is a $\lambda\varepsilon$-cover of λC. Hence

$$H^s_{\lambda\varepsilon}(\lambda C) \le \sum_{k\in\mathbb{N}} \frac{\alpha(s)}{2^s}(\lambda \operatorname{diam} D_k)^s = \frac{\lambda^s \alpha(s)}{2^s} \sum_{k\in\mathbb{N}} (\operatorname{diam} D_k)^s$$

and so $H^s_{\lambda\varepsilon}(\lambda C) \le \lambda^s H^s_\varepsilon(C)$. Letting $\varepsilon \to 0$, we obtain $H^s(\lambda C) \le \lambda H^s(C)$.

Replacing λ with λ^{-1} and C with λC, we get the opposite inequality and so we conclude that

$$H^s(\lambda C) = \lambda H^s(C) \quad \text{for all } \lambda > 0\,.$$

□

Proposition 2.8.12. *If $0 \le s < \infty$, $C \subseteq \mathbb{R}^N$ and $\varphi\colon C \to \mathbb{R}^m$ is Lipschitz continuous with Lipschitz constant $\vartheta > 0$, then $H^s(\varphi(C)) \le \vartheta^s H^s(C)$.*

Proof. Let $\{D_k\}_{k\in\mathbb{N}}$ be an open ε-cover. Then $\{\varphi(C \cap D_k)\}_{k\in\mathbb{N}}$ is an $\vartheta\varepsilon$-cover of $\varphi(C)$ and

$$H^s(\vartheta\varepsilon)(\varphi(C)) = \frac{\alpha(s)}{2^s} \sum_{k\in\mathbb{N}} (\operatorname{diam} \varphi(C \cap D_k))^s \le \vartheta^s \frac{\alpha(s)}{2^s} \sum_{k\in\mathbb{N}} (\operatorname{diam} D_k)^s\,.$$

Hence $H^s_{\vartheta\varepsilon}(\varphi(C)) \le \vartheta^s H^s_\varepsilon(C)$. Letting $\varepsilon \to 0^+$ yields $H^s(\varphi(C)) \le \vartheta^s H^s(C)$. □

Corollary 2.8.13. *If $0 \le s < \infty$, $C \subseteq \mathbb{R}^N$ and $\varphi\colon C \to \mathbb{R}^m$ is an isometry, then $H^s(\varphi(C)) = H^s(C)$.*

Proposition 2.8.14. *If $s > N$, then $H^s|_{\mathbb{R}^N} = 0$.*

Proof. It suffices to show that $H^s(Q) = 0$ with Q being the unit cube in $\mathbb{R}^N$. Fix $m \in \mathbb{N}$. Then we can decompose Q into m^N cubes with side $\frac{1}{m}$ and diameter $\frac{\sqrt{N}}{m}$. We have

$$H^s_{\frac{\sqrt{N}}{m}}(Q) \le \sum_{k=1}^{m^N} \frac{\alpha(s)}{m^s} N^{\frac{s}{2}} \frac{1}{m^{s-N}} \to 0 \quad \text{as } m \to \infty\,,$$

since $s > N$. Therefore, $H^s(Q) = 0$ and so $H^s(\mathbb{R}^N) = 0$. □

The next proposition provides a convenient way to identify an H^s-null set.

Proposition 2.8.15. *If $0 \le s < 1$ and $C \subseteq \mathbb{R}^N$ satisfies $H^s_\delta = 0$ for some $\delta > 0$, then $H^s(C) = 0$.*

Proof. The result is obvious if $s = 0$. So we assume that $s > 0$. Fix $\vartheta > 0$. Then we can find a cover $\{D_k\}_{k\in\mathbb{N}}$ of C, that is, $C \subseteq \cup_{k\in\mathbb{N}} D_k$, such that

$$\sum_{k\in\mathbb{N}} \frac{\alpha(s)}{2^s} (\operatorname{diam} D_k)^s \leq \vartheta .$$

For each $k \in \mathbb{N}$ we have

$$\operatorname{diam} D_k \leq s\left(\frac{\vartheta}{\alpha(s)}\right)^{\frac{1}{s}} = \varepsilon . \tag{2.8.3}$$

Then it follows

$$H^s_\varepsilon(C) \leq \vartheta . \tag{2.8.4}$$

Note that $\vartheta \to 0^+$ implies $\varepsilon \to 0^+$; see (2.8.3). So, in the limit as $\varphi \to 0^+$, we obtain, due to (2.8.4), that $H^s(C) = 0$. □

Another result in this direction is the following one.

Proposition 2.8.16. *If $0 \leq s < t < \infty$ and $C \subseteq \mathbb{R}^N$, then the following hold:*
(a) $H^s(C) < \infty$ *implies* $H^t(C) = 0$.
(b) $H^t(C) > 0$ *implies* $H^s(C) = \infty$.

Proof. (a) Suppose $H^s(C) < \infty$ and let $\delta > 0$. Then we can find $\{D_k\}_{k\in\mathbb{N}}$ such that $C \subseteq \cup_{k\in\mathbb{N}} D_k$, $\operatorname{diam} D_k \leq \delta$ and

$$\frac{\alpha(s)}{2^s} \sum_{k\in\mathbb{N}} (\operatorname{diam} D_k)^s \leq H^s_\delta(C) + 1 \leq H^s(C) + 1 . \tag{2.8.5}$$

From Definition 2.8.1 we have

$$\begin{aligned} H^t_\delta(C) &\leq \frac{\alpha(t)}{2^t} \sum_{k\in\mathbb{N}} (\operatorname{diam} D_k)^t \\ &\leq \frac{\alpha(t)}{\alpha(s)} 2^{s-t} \frac{\alpha(s)}{2^s} \sum_{k\in\mathbb{N}} (\operatorname{diam} D_k)^s (\operatorname{diam} D_k)^{t-s} \\ &\leq \frac{\alpha(t)}{\alpha(s)} 2^{s-t} \delta^{t-s} (H^s(C) + 1) , \end{aligned}$$

because of (2.8.5). Letting $\delta \to 0^+$ gives us $H^t(C) = 0$.

(b) This is an immediate consequence of part (a). □

This leads us to the following definition.

Definition 2.8.17. Let $C \subseteq \mathbb{R}^N$. The **Hausdorff dimension of** C is defined to be

$$H_{\dim}(C) = \inf\{0 \leq s < \infty : H^s(C) = 0\} .$$

Remark 2.8.18. If $C \subseteq \mathbb{R}^N$, then $H_{\mathrm{dim}}(C) \le N$. Also, note that

$$\begin{cases} H^t(C) = 0 & \text{for all } t > H_{\mathrm{dim}}(C)\,, \\ H^s(C) = \infty & \text{for all } s < H_{\mathrm{dim}}(C)\,. \end{cases} \tag{2.8.6}$$

We also mention that H^s with $0 \le s < N$ is not a Radon measure since $\mathbb{R}^N$ is not σ-finite with respect to H^s. Note that $H_{\mathrm{dim}}(C)$ can be any number in $[0, \infty]$ and so it need not be an integer.

Example 2.8.19. Let $C \subseteq \mathbb{R}$ be the Cantor ternary set; see Example 2.1.46. Then

$$H_{\mathrm{dim}}(C) = \frac{\ln(2)}{\ln(3)}\,.$$

For the proof we refer to Gasiński–Papageorgiou [135, Proposition 1.3.9, p. 27]. Also, every countable set C has zero Hausdorff dimension.

Proposition 2.8.20. *If $C \subseteq \mathbb{R}^N$ and $\varphi\colon C \to \mathbb{R}^m$ with $m \in \mathbb{N}$ satisfies*

$$|\varphi(u) - \varphi(v)| \le \vartheta|u - v| \quad \textit{for all } u, v \in \mathbb{R}^n \textit{ with } \vartheta > 0\,,$$

then $H_{\mathrm{dim}}(\varphi(C)) \le H_{\mathrm{dim}}(C)$.

Proof. If $t > H_{\mathrm{dim}}(C)$, then $H^t(C) = 0$; see Remark 2.8.18. From Proposition 2.8.12, we have

$$H^t(\varphi(C)) \le \vartheta^t H^t(C)\,,$$

which implies $H^t(\varphi(C)) = 0$. Then, due to (2.8.6), $H_{\mathrm{dim}}(\varphi(C)) \le t$. Letting $t \to H_{\mathrm{dim}}(C)$ gives us $H_{\mathrm{dim}}(\varphi(C)) \le H_{\mathrm{dim}}(C)$. ☐

The Hausdorff dimension provides information about the topological structure of the set.

Proposition 2.8.21. *If $C \subseteq \mathbb{R}^N$ and $H_{\mathrm{dim}}(C) < 1$, then C is totally disconnected, that is, its connected components are the points.*

Proof. Let $u, v \in C$ with $u \neq v$ and let $\varphi_u(y) = |y - u|$. Evidently, φ_u is 1-Lipschitz (nonexpansive) and so from Proposition 2.8.10 it follows

$$H_{\mathrm{dim}}(\varphi_u(C)) \le H_{\mathrm{dim}}(C) < 1\,.$$

Then $H^1(\varphi_u(C)) = 0$, so $\varphi_u(C)$ is Lebesgue null; see Proposition 2.8.10. Therefore, there exists $\eta < \varphi_u(v)$ such that $\eta \notin \varphi_u(C)$. Hence,

$$C = \{c \in C\colon \varphi_u(c) < \eta\} \cup \{c \in C\colon \varphi_u(c) > \eta\}\,.$$

This shows that u, v are in different connected components. ☐

Let λ^N denote the Lebesgue measure on $\mathbb{R}^N$. We want to compare λ^N and H^N. If $N = 1$, then we know that $\lambda^1 = H^1$. In order to show that this equality remains valid also in higher dimensions, we will need the so-called "isodiametric inequality".

Theorem 2.8.22 (Isodiametric inequality). *If $C \subseteq \mathbb{R}^N$, then $\lambda^N(C) \leq \frac{\alpha(N)}{2^N}(\operatorname{diam} C)^N$.*

Proof. Clearly, we may assume that $C \subseteq \mathbb{R}^N$ is closed and $\operatorname{diam} C < \infty$. Let $e \in \mathbb{R}^N$ be a unit vector. We have

$$\mathbb{R}^N = \mathbb{R}e \oplus V \quad \text{with } V = (\mathbb{R}e)^\perp .$$

Let p_V be the orthogonal projection on V and let $C_V = p_V(C)$. Consider the line ℓ_u passing from $u \in C_V$ in the direction e and let

$$\eta(u) = \frac{1}{2}\lambda^1(C \cap \ell_u) \quad \text{for } u \in C_V . \tag{2.8.7}$$

We define

$$K_e(C) = \{(v, u) \in \mathbb{R} \times V : |v| \leq \eta(u),\ u \in C_V\} .$$

Then, using Fubini's Theorem given in Theorem 2.3.50 we have that η is measurable and

$$\int_{C_V} \eta(u)\, du = \frac{1}{2}\lambda^N(C) .$$

Claim: $\operatorname{diam} K_e(C) \leq \operatorname{diam} C$.
Let $u, u' \in C_V$ and set

$$\begin{aligned} \mu &= \sup(C \cap \ell_u) & \beta &= \inf(C \cap \ell_u), \\ \mu' &= \sup(C \cap \ell_{u'}) & \beta' &= \inf(C \cap \ell_{u'}) . \end{aligned}$$

Without any loss of generality we may assume that $\mu - \beta' \geq \mu' - \beta$. Then we have

$$2\eta(u) \leq \mu - \beta \quad \text{and} \quad 2\eta(u') \leq \mu' - \beta' .$$

So, from the definition of η in (2.8.7) it follows that

$$\begin{aligned} (|u - u'|^2 + (\eta(u) - \eta(u'))^2)^{\frac{1}{2}} &\leq (|u - u'|^2 + |\mu - \beta'|^2)^{\frac{1}{2}} \\ &\leq \operatorname{diam}((C \cap \ell_u) \cup (C \cap \ell_{u'})) . \end{aligned}$$

This implies that $\operatorname{diam} K_e(C) \leq \operatorname{diam} C$ and so the Claim is proved.

Let $\{e_k\}_{k=1}^N \subseteq \mathbb{R}^N$ be an orthogonal basis. We define

$$\hat{C} = K_{e_N} \circ \cdots \circ K_{e_1}(C) .$$

Then, by the Claim, we have $\operatorname{diam}\hat{C} \le \operatorname{diam} C$ and so $\lambda^N(\hat{C}) \le \lambda^N(C)$.

The set $\hat{C}$ is symmetric with respect to the basis, thus it is symmetric with respect to the origin. Therefore, $u \in \hat{C}$ implies $-u \in \hat{C}$. Then $2|u| \le \operatorname{diam}\hat{C}$ for all $u \in \hat{C}$ and so $u \in \overline{B}_{\frac{1}{2}\operatorname{diam}\hat{C}}$. This shows that $\hat{C} \subseteq \overline{B}_{\frac{1}{2}\operatorname{diam}\hat{C}}$.

Finally we have

$$\lambda^N(C) = \lambda^N(\hat{C}) \le \frac{\alpha(N)}{2^N}(\operatorname{diam}\hat{C})^N \le \frac{\alpha(N)}{2^N}(\operatorname{diam}\hat{C})^N . \qquad \square$$

Remark 2.8.23. Hidden in the above proof is the so-called "Steiner symmetrization" of a set $C \subseteq \mathbb{R}^N$. Let $e \in \mathbb{R}^N$ with $|e| = 1$ and let $V = (\mathbb{R}e)^{\perp}$. Then the Steiner symmetrization of C with respect to V is the set

$$\hat{C} = \bigcup_{\substack{u \in V \\ C \cap \ell_v^e}} \left\{ v + \vartheta e : |\vartheta| \le \frac{1}{2} H^1(C \cap \ell_v) \right\}$$

with $\ell_v^e = \{v + \vartheta e : \vartheta \in \mathbb{R}\}$, so the line through v in the direction e. We have seen that $\operatorname{diam}\hat{C} \le \operatorname{diam} C$ and if C is measurable, then so is $\hat{C}$ and $\lambda^N(C) = \lambda^N(\hat{C})$.

We can use Theorem 2.8.22 to establish the equality $\lambda^N = H^N$ for any dimension N.

Theorem 2.8.24. *It holds $\lambda^N = H^N$ on $\mathbb{R}^N$.*

Proof. Let $C \subseteq \mathbb{R}^N$. First we show that $\lambda^N(C) \le H^N(C)$. For this purpose, let $\delta > 0$ and consider sets $\{D_k\}_{k\in\mathbb{N}}$ such that $C \subseteq \cup_{k\in\mathbb{N}} D_k$ with $\operatorname{diam} D_k \le \delta$. From the isodiametric inequality given in Theorem 2.8.22 we have

$$\lambda^N(C) \le \sum_{k\in\mathbb{N}} \lambda^N(D_k) \le \frac{\alpha(N)}{2^N} \sum_{k\in\mathbb{N}} (\operatorname{diam} D_k)^N .$$

From this we conclude $\lambda^N(C) \le H_\delta^N(C) \le H^N(C)$.

Next, we show that $H^N \ll \lambda^N$. To this end, let $C \subseteq \mathbb{R}^N$ and $\delta > 0$. We have

$$H_\delta^N(C) \le \inf\left\{ \frac{\alpha(N)}{2^N} \sum_{k\in\mathbb{N}} (\operatorname{diam} Q_k)^N : Q_k \text{ is a cube, } C \subseteq \bigcup_{k\in\mathbb{N}} Q_k, \operatorname{diam} Q_k \le \delta \right\} .$$

For each cube $Q \subseteq \mathbb{R}^N$ we have

$$\frac{\alpha(N)}{2^N}(\operatorname{diam} Q)^N = c_N \lambda^N(Q)$$

with $c_N = \frac{\alpha(N)}{2^N}(N^N)^{\frac{1}{2}}$. It follows that $H_\delta^N(C) \le c_N \lambda^N(C)$ and so $H^N(C) \le c_N \lambda^N(C)$. Therefore,

$$H^N \ll \lambda^N . \tag{2.8.8}$$

Consider cubes $\{Q_k\}_{k\in\mathbb{N}}$ such that

$$C \subseteq \bigcup_{k\in\mathbb{N}} Q_k, \quad \operatorname{diam} Q_k \leq \delta$$

and

$$\sum_{k\in\mathbb{N}} \lambda^N(Q_k) \leq \lambda^N(C) + \varepsilon \tag{2.8.9}$$

with $\varepsilon > 0$. We can find disjoint balls $\{B_k^m\}_{m\in\mathbb{N}} \subseteq Q_k$ such that

$$\operatorname{diam} B_k^m \leq \delta \quad \text{and} \quad \lambda^N\Big(Q_k \setminus \bigcup_{m\in\mathbb{N}} B_k^m\Big) \leq \frac{\varepsilon}{2^k}\,.$$

Using (2.8.8) we obtain

$$H^N\Big(Q_k \setminus \bigcup_{m\in\mathbb{N}} B_k^m\Big) \leq \frac{c\varepsilon}{2^k} \quad \text{with } c > 1\,. \tag{2.8.10}$$

Therefore, applying (2.8.10) and (2.8.9), leads to

$$\begin{aligned} H_\delta^N(C) &\leq \sum_{k\in\mathbb{N}} H_\delta^N(Q_k) \leq \sum_{k\in\mathbb{N}}\sum_{m\in\mathbb{N}} \frac{\alpha(N)}{2^N}(\operatorname{diam} B_k^m)^N + c\varepsilon \leq \sum_{k\in\mathbb{N}}\sum_{m\in\mathbb{N}} \lambda^N(B_k^m) + c\varepsilon \\ &\leq \sum_{k\in\mathbb{N}} \lambda^N(Q_k) + c\varepsilon \leq \lambda^N(C) + (c+1)\varepsilon\,. \end{aligned}$$

Since $\delta, \varepsilon > 0$ were arbitrary, we conclude that $H^N(C) \leq \lambda^N(C)$ and so $\lambda^N = H^N$. □

Remark 2.8.25. It can happen that the set $C \subseteq \mathbb{R}$ is Lebesgue-null, but $H_{\dim}(C) = 1$.

Closely related with the theory of Hausdorff measures and the fine properties of functions are the so-called "covering theorems".

Definition 2.8.26. A collection $\mathfrak{B}$ of closed balls in $\mathbb{R}^N$ is said to be a **fine cover** of $C \subseteq \mathbb{R}^N$ if

$$C \subseteq \bigcup_{B\in\mathfrak{B}} B \quad \text{and} \quad \inf\{\operatorname{diam} B : u \in B,\ B \in \mathfrak{B}\} = 0 \quad \text{for all } u \in C\,.$$

We present two covering theorems. Both can be found in Evans–Gariepy [116]. The first is known as "Vitali's Covering Theorem".

Theorem 2.8.27 (Vitali's Covering Theorem). *If $\mathfrak{B}$ is a collection of nonempty closed balls in $\mathbb{R}^N$ such that*

$$\sup\{\operatorname{diam} B : B \in \mathfrak{B}\} < \infty\,,$$

then there exists a countable subfamily $\mathcal{L} \subseteq \mathfrak{B}$ such that

$$\bigcup_{B\in\mathfrak{B}} B \subseteq \bigcup_{B\in\mathcal{L}} \hat{B}$$

with $\hat{B}$ being the concentric closed ball with 5 times the radius of B.

The second covering theorem is due to Besicovitch [35] and for the proof we refer again to Evans–Gariepy [116, p. 30] and Ziemer [342, p. 9].

Theorem 2.8.28 (Besicovitch's Covering Theorem). *There exists a positive number $M = M(N) > 1$ such that for any family $\mathfrak{B}$ of closed balls in $\mathbb{R}^N$ whose cardinality is at least M and $r_* = \sup\{r > 0 : B_r(u) \in \mathfrak{B}\} < \infty$, contains disjoints subfamilies $\{\mathfrak{B}_k\}_{k=1}^M$ such that if C is the set of centers of balls in $\mathfrak{B}$, then*

$$C \subseteq \bigcup_{k=1}^{M} \bigcup_{B\in\mathfrak{B}_k} B .$$

Corollary 2.8.29. *If $C \subseteq \mathbb{R}^N$, $\mathfrak{B}$ is a family of closed balls which are a fine cover of C and $U \subseteq \mathbb{R}^N$ is open, then $\mathfrak{B}$ contains a countable disjoint subfamily $\mathcal{L}$ such that*

$$\bigcup_{B\in\mathcal{L}} B \subseteq U \quad \text{and} \quad \lambda^N\Big((C \cap U) \setminus \bigcup_{B\in\mathcal{L}} B\Big) = 0 .$$

2.9 Remarks

(2.1) Cantor [67] was one of the first to give a general definition of the measure of a set. However, the definition he gave produced a nonadditive measure. Then came the French mathematician Jordan [184] who defined a set to be measurable if its topological boundary has zero measure. So, the set of rational numbers in an interval is not measurable. Moreover, there are open sets that are not measurable. Finally, the measure that Jordan defined is only finitely additive. Then came Borel [43] who showed that the length of intervals can be extended to a σ-additive set function on the σ-algebra generated by intervals, the Borel σ-algebra. The Borel measure is based on the fact that any open set $U \subseteq \mathbb{R}$ is the union of countably many disjoint intervals. However, we should mention that Borel did not use the terminology of open sets. At that time mathematicians focused on closed – even more specifically on perfect – sets. The notion, together with the name of open set, was introduced by Baire [23] in his thesis. Borel did not use his theory of measure to develop a corresponding theory of integration. Borel sets are produced by infinite applications of certain set-theoretic operations and so we cannot have a good insight concerning their structure. This led to an axiomatic definition of measurable sets. An important contribution to this came from Carathéodory [69] who introduced the notion of outer measure in the sense of Definition 2.1.33. Carathéodory worked on $\mathbb{R}^N$. Moreover,

Definition 2.1.36 about μ^*-measurable sets is also due to Carathéodory [69]. It is a rather strange definition, not that intuitive. It singles out as measurable those sets which split all sets in X in two parts on which μ is additive. It is not clear how Carathéodory came up with this definition. Nevertheless, it turned out to be a very fruitful one. It gives a σ-algebra – in general not the largest possible – which contains the Borel sets and on which μ is a measure. Vitali [320] was the first to establish the existence of a nonmeasurable set in $\mathbb{R}$; see Theorem 2.1.44. A detailed account of the historical development of measurable sets can be found in Chapter 4 of Hawkins [156]. Concerning the atoms of a measure (see Definition 2.1.30 (b)), we mention the following result known as "Saks Lemma", see Dunford–Schwartz [105, Lemma IV.9.7, p. 308].

Lemma 2.9.1 (Saks Lemma). *If (X, Σ, μ) is a finite measure space, then for every $\varepsilon > 0$ there exists a finite partition of X into pairwise disjoint sets $\{A_k\}_{k=1}^n \subseteq \Sigma$ such that either $\mu(A_k) \leq \varepsilon$ for all $k \in \{1, \dots, k\}$ or A_k is an atom with $\mu(A_k) > \varepsilon$ for all $k \in \{1, \dots, k\}$.*

Proposition 2.1.32 is a particular case of a more general result due to Lyapunov [226] known as the "Lyapunov Convexity Theorem". The result has important applications in many applied areas such as optimal control and mathematical economics; see Hermes–LaSalle [159] and Klein–Thompson [194].

Theorem 2.9.2 (Lyapunov Convexity Theorem). *If (X, Σ) is a measurable space and $\mu_1, \dots, \mu_n : X \to \mathbb{R}$ are nonatomic measures, then the set $R = \{(\mu_k(A))_{k=1}^n : A \in \Sigma\} \subseteq \mathbb{R}^n$ is compact and convex.*

The Cantor set (see Example 2.1.46) plays an important role in foundational work and it is also a useful tool in topology.

Further details on measure theory can be found in the books of Bogachev [40, 41], Dudley [101], Folland [125], Halmos [151], Hewitt–Stromberg [160], Royden [283], and Rudin [284].

(2.2) There is no doubt that Lebesgue's theory of integration is one of the major mathematical breakthroughs in the 20 $\underline{\text{th}}$-century. Lebesgue was influenced by the ideas of Borel, but his theory of measure is more general. His theory was first presented in his thesis [205]. Many of the questions left open in his thesis were resolved in his book [206] published two years later. It was based on lectures he gave to the College de France in the period 1902–1903. With his integral, Lebesgue was able to overcome a number of difficulties that were associated with Riemann's theory of integration. In particular the limit theorems for the new integral are substantially more general and helped in the dissemination of Lebesgue's theory. Proposition 2.2.12 goes back to Hausdorff [155] while the example produced in Remark 2.2.13 is due to Dudley [100]; see also Dudley [101, Proposition 4.2.3, p. 96]. Theorem 2.2.32 was proven by Egorov [109]. Egorov was the mathematical mentor of Lusin. We mention that Egorov's Theorem as well as Lusin's Theorem (see Theorem 2.5.17) were stated without proof in Lebesgue [206]. Theorem 2.2.34 is due to von Alexits [324] and Sierpinski [296]. The use of simple functions in the definition of the

Lebesgue integral (see Definition 2.2.35) underlines the main difference with Riemann's method. More precisely, in contrast to Riemann, Lebesgue does not consider partitions of the domain $[a, b]$ of f. Instead he considers partitions of the range of f. A detailed discussion of the development of Lebesgue's method can be found in Hawkins [156].

We conclude our remarks on this subsection with two useful observations. The first concerns Egorov's Theorem (see Theorem 2.2.32) and indicates when we can drop the hypothesis that $\mu(X) < \infty$.

Proposition 2.9.3. *If (X, Σ, μ) is a measure space, $f_n: X \to \mathbb{R}$ with $n \in \mathbb{N}$ is a sequence of Σ-measurable functions such that*

$$f_n \to f \quad \mu\text{-a. e.} \quad \text{and} \quad |f_n(x)| \le h(x) \quad \mu\text{-a. e. with } h \in L^1(X),$$

then given $\varepsilon > 0$ there exists $A_\varepsilon \in \Sigma$ with $\mu(A_\varepsilon) < \varepsilon$ such that $f_n \to f$ uniformly in $X \setminus A_\varepsilon$.

The second observation shows how the Lebesgue measure changes under nonsingular linear transformations.

Proposition 2.9.4. *If $L: \mathbb{R}^N \to \mathbb{R}^N$ is linear and nonsingular, then the following hold:*
(a) *$L(A) \in \mathcal{B}(\mathbb{R}^N)$ for all $A \in \mathcal{B}(\mathbb{R}^N)$;*
(b) *$\lambda^N(L(A)) = |\det(L)|\lambda^N(A)$ for all $A \in \mathcal{B}(\mathbb{R}^N)$.*

(2.3) Theorem 2.3.1 – and consequently Theorems 2.3.3 as well as 2.3.5 – are due to Beppo Levi [213]. Theorem 2.3.6 is due to Fatou [119]. Theorem 2.3.8 is the "crown jewel" of Lebesgue's theory and was proved by Lebesgue [208]. The L^p-spaces were defined by Riesz [268] when $p = 2$, [269] when $1 < p < 2$ and [270] when $2 < p < \infty$. Riesz [269, 270] proved the completeness of $L^p, p \neq 2$ while the completeness of L^2 was proved by Fischer [122]. The Cauchy–Bunyakowsky–Schwarz inequality (see Corollary 2.3.13) was first proven by Cauchy (1821) for finite sums, then by Bunyakowsky (1859) for Riemann integrals and finally by Schwarz (1885) for double integrals. Hölder's inequality (see Theorem 2.3.12) can be found in Rogers [279] and Hölder [170]. Of course the inequalities proven by Rogers and Hölder do not have the form of Theorem 2.3.12, but it can be shown that they imply Theorem 2.3.12. Note that Hölder acknowledges that he was inspired by the work of Rogers. For this reason Dudley [101] calls the result "Rogers–Hölder inequality". Theorem 2.3.14 was proven by Minkowski [235] for finite sums and by Riesz [270] for integrals. Jensen's inequality (see Theorem 2.3.15) was obtained by Jensen [182]. Convergence in measure, initially called also **asymptotic convergence**, can be found in early works of Borel and Lebesgue but a systematic study of it can be found in Riesz [269], who pointed out a gap in the book of Lebesgue concerning this mode of convergence and in Fréchet [130, 131]. In fact Fréchet [130] showed that convergence in measure is metrizable by the metric

$$d_F(f, h) = \inf_{\varepsilon > 0}[\varepsilon + \mu\{x \in X : |f(x) - h(x)| > \varepsilon\}].$$

Another metric was introduced by Fan [118] who defined

$$d_K(f,h) = \inf[\varepsilon \geq 0 \colon \mu\{x \in X \colon |f(x) - h(x)| > \varepsilon\} < \varepsilon] .$$

The metric in (2.3.10) was first introduced by Nikodym [253].

The notion of uniform integrability and the main results concerning it go back to the works of Lebesgue, Vitali, and de la Vallee Poussin. Additional equivalent formulations of this notion can be found in Gasiński–Papageorgiou [136, see Problems 1.7, 1.15, 1.16, 1.17].

Lebesgue [206] was the first to establish for bounded measurable functions of two variables the reduction of multiple integrals to repeated ones. Later Fubini [133] proved Theorem 2.3.50 and the appearance of his result marked a real triumph for Lebesgue's method. As Fubini pointed out, the Lebesgue integral is necessary for this kind of study. Theorem 2.3.49 is due to Tonelli [311].

We conclude the remarks of this subsection with a result on the existence of the essential supremum for a family of functions. The result is useful in probability theory and elliptic partial differential equations.

Proposition 2.9.5. *If (X, Σ, μ) is a σ-finite measure space and $\mathcal{F}$ is a family of Σ-measurable, $\mathbb{R}$-valued functions, then there exists a unique (up to μ-a. e. equality) Σ-measurable function $h \colon X \to \mathbb{R}$ such that $f(x) \leq h(x)$ for μ-a. a. $x \in X$ and for all $f \in \mathcal{F}$.*

If h' is another Σ-measurable function such that $f(x) \leq h'(x)$ for μ-a. a. $x \in X$ and for all $f \in \mathcal{F}$, then $h(x) \leq h'(x)$ for μ-a. a. $x \in X$.

We call $h = \operatorname{ess\,sup} \mathcal{F}$. In addition there is a sequence $\{f_n\}_{n\geq 1} \subseteq \mathcal{F}$ such that $\operatorname{ess\,sup} \mathcal{F} = \sup_{n\geq 1} f_n$. Finally if $\mathcal{F}$ is upward directed, that is, if $f_1, f_2 \in \mathcal{F}$, then there exists $f \in \mathcal{F}$ such that $f_1 \leq f, f_2 \leq f$, then $\{f_n\}_{n\geq 1}$ can be chosen to be increasing.

(2.4) Signed measures were first considered by Lebesgue [208] who studied such measures of the form

$$\mu(A) = \int_A f(x)\,\mathrm{d}\nu(x) \quad \text{with}\, f \in L^1(\nu) .$$

The Hahn Decomposition Theorem (see Theorem 2.4.8) was proven by Hahn [147]. Concerning the Jordan Decomposition Theorem (see Theorem 2.4.14), we mention that Jordan (1881) introduced functions of bounded variation on an interval $[a, b]$ and proved that such a function can be written as the difference of two nondecreasing functions; see also Section 4.3.

The more general Theorem 2.4.14 was named after Jordan as a tribute of his important contributions on the subject. Note that if μ is a finite signed measure on $[a, b]$, then $f(x) = \mu([a, x])$ with $x \in [a, b]$ is a function of bounded variation and $f = g - h$ with $g(x) = \mu_+([a, x])$ and $h(x) = \mu_-([a, x])$ for all $x \in [a, b]$.

The Radon–Nikodym Theorem (see Theorem 2.4.29) started with Lebesgue who obtained the special case of absolute continuity with respect to the Lebesgue measure. The case of Borel measures on $\mathbb{R}^N$ was proven by Radon [263] and a little later by Daniell [82] as well. The general form of the theorem is due to Nikodym [253]. The Lebesgue decomposition in the general abstract setting (see Theorem 2.4.31) can be found in Saks [287]. There is a unifying short proof of Theorems 2.4.29 and 2.4.33 due to von Neumann [329]; see also Dudley [101, p. 134] and Rudin [284, p. 130]. Although Theorem 2.4.33 is called the Vitali–Hahn–Saks Theorem, others also contributed to its formulation, like Lebesgue and Nikodym. It appears the general form was proven by Saks [287]. Theorems 2.4.33 and 2.4.34 are very useful in general measure theory.

(2.5) The definition of the Baire σ-algebra (see Definition 2.5.1) is not the same in all authors. For example, Dudley [101, p. 174] defines the Baire σ-algebra to be the smallest σ-algebra for which all $f \in C_b(X)$ are measurable. Recall that $C_b(X)$ is the space of all $\mathbb{R}$-valued, continuous, and bounded functions. Other definitions of $\mathrm{Ba}(X)$ are provided by Bogachev [41, p. 12] and Halmos [151, p. 220]. Here we follow Royden [283, p. 301]. We should point out that for the Borel σ-algebra, there are some different definitions. More precisely, some of the older texts define the Borel σ-algebra to be the σ-algebra generated by the compact sets. This is in general smaller than the Borel σ-algebra of Definition 2.1.4 (b).

Similarly the terminology introduced in Definition 2.5.8 is not uniform. People use other names for the same notions, see, for example Aliprantis–Border [7, pp. 434–435]. Topological measure theory started with the seminal paper of Radon [263] who worked on $\mathbb{R}^N$. A classical reference on Radon measures is the book of Schwartz [293].

The topological structure of the ambient space leads to the definition of the support of a measure.

Definition 2.9.6. Let X be a Hausdorff topological space and $\mu: \mathcal{B}(X) \to [0, \infty]$ a Borel measure. The **support of** μ is the set

$$\operatorname{supp} \mu = \{x \in X : \mu(U) > 0 \text{ for all } U \in \mathcal{N}(x)\} .$$

Remark 2.9.7. Evidently $\operatorname{supp} \mu$ is closed and if $A \in \mathcal{B}(X), A \subseteq X \setminus \operatorname{supp} \mu$, then $\mu(A) = 0$. Every Radon measure has a unique support.

We have a regularity result for functions that are integrable with respect to a Radon measure. The result is known as the “Vitali–Carathéodory Theorem”; see Rudin [284, p. 57].

Theorem 2.9.8 (Vitali–Carathéodory Theorem). *If X is a locally compact topological space, $\mu: \mathcal{B}(X) \to [0, \infty]$ is a Radon measure, $f \in L^1(X, \mu)$ and $\varepsilon > 0$, then there exist $g: X \to \mathbb{R}$ being upper semicontinuous, bounded above and $h: X \to \mathbb{R}$ being lower semicontinuous, bounded below such that $g(x) \leq f(x) \leq h(x)$ for μ-a. a. $x \in X$ and $\int_X (h - g)\, d\mu \leq \varepsilon$.*

Remark 2.9.9. There is an alternative approach to Lebesgue integration due to Daniell [81] based on the extension of positive linear functionals. Within that theory, the

Vitali–Carathéodory Theorem is essentially the definition of the measurability and integrability of f.

As was the case with Egorov's Theorem (see Theorem 2.2.32), Lusin's Theorem (see Theorem 2.5.17) was first stated without proof by Lebesgue [206]. Lusin [223] proved the result later. There is a category analog to Lusin's Theorem.

Theorem 2.9.10. *If X is a separable metric space and $f: X \to \mathbb{R}$ is Borel measurable, then there is a set D of first category such that $f|_{X\setminus D}$ is continuous.*

Theorem 2.5.19 is due to Scorza Dragoni [294]. Normal integrands (see Definition 2.5.20) is a basic tool in many applied fields such as calculus of variations, optimization and optimal control; see Buttazzo [65], Ekeland–Temam [114], and Papageorgiou–Kyritsi [256].

(2.6) The theory of Souslin or analytic or A-sets started when Souslin, a student of Lusin, discovered an error in Lebesgue [207]. Lebesgue claimed that the projection of a Borel set in $\mathbb{R}^2$ onto the x-axis is again a Borel set. Souslin realized that this is not true and went on to introduce analytic sets and started their study. Souslin [300] also produced an analytic set in the real line whose complement is not analytic and so it is not Borel; see Proposition 2.6.11 and Remark 2.6.12. Lusin [224] proved that analytic sets in $\mathbb{R}$ are Lebesgue measurable. Unfortunately, Souslin died very young at the age of 25 in 1919. The work on analytic sets was continued initially by Lusin and subsequently by many other mathematicians. Theorem 2.6.15 is due to Lusin [225] and is one of the most important results in the theory of analytic sets with far-reaching consequences. In addition to the σ-algebras $\mathcal{B}(X)$, α_X, $\hat{\Sigma}_X$ there is a fourth σ-algebra known as the **limit σ-algebra** denoted by $\mathcal{L}_X$ and it is between α_X and $\hat{\Sigma}_X$. For a discussion of this σ-algebra see Bertsekas–Shreve [34, Appendix B4]. Analytic (Souslin) sets are discussed in the books of Aliprantis–Border [7], Bertsekas–Shreve [34], Bogachev [41], Cohn [76], Dudley [101], Klein–Thompson [194], and Srivastava [301].

(2.7) Measurable multifunctions are an important tool in many applied areas. Detailed studies of measurable multifunctions can be found in the books of Aliprantis–Border [7], Aubin–Frankowska [20], Castaing–Valadier [71], Denkowski–Migórski–Papageorgiou [87], Hu–Papageorgiou [173], and Klein–Thompson [194]. Theorem 2.7.12 was proven by Rohlin [280] and later by Kuratowski–Ryll Nardzewski [201]. There is a gap in the proof of Rohlin and for this reason the result is attributed to Kuratowski–Ryll Nardzewski. Theorem 2.7.25 as stated is due to Sainte-Beuve [286]. Earlier versions of it were proven by Yankov [335], von Neumann [330] and Aumann [21]. The same can be said for Theorem 2.7.32.

(2.8) Hausdorff measures were first considered by Carathéodory [68] in the context of his general theory of outer measures. Hausdorff [154] extended the definition to general $s > 0$ not necessarily an integer and proved that the Hausdorff dimension of the Cantor set is $\frac{\ln(2)}{\ln(3)}$. For the theory of Hausdorff measures and the related covering theorems we refer to the books of Ambrosio–Tilli [12], Evans–Gariepy [116], Gasiński–Papageorgiou [135] and Ziemer [342]

Problems

Problem 2.1. Let X be a set and let $\mathcal{L} \subseteq 2^X$ be nonempty. Show that $\sigma(\mathcal{L})$ is the smallest family $\mathfrak{L} \subseteq 2^X$, which contains $\mathcal{L}$ and satisfies the following assertions:
(a) $A \in \mathcal{L}$ implies $A^c \in \mathfrak{L}$;
(b) $\mathfrak{L}$ is closed under countable intersections;
(c) $\mathfrak{L}$ is closed under countable disjoint unions.

Problem 2.2. Let X be a set and let $\mathcal{L} \subseteq 2^X$ be a semiring. Show that:
(a) If $A, A_1, \ldots, A_n \in \mathcal{L}$, then there exist $\{B_i\}_{i=1}^m \subseteq \mathcal{L}$ pairwise disjoint such that $A \setminus \bigcup_{k=1}^n A_k = \bigcup_{i=1}^m B_i$.
(b) If $\{A_n\}_{n\geq1} \subseteq \mathcal{L}$, then there exist $\{C_k\}_{k\geq1} \subseteq \mathcal{L}$ pairwise disjoint such that $\bigcup_{n\geq1} A_n = \bigcup_{k\geq1} C_k$ and for each $k \geq 1$ there exists $n \geq 1$ such that $C_k \subseteq A_n$.

Problem 2.3. Let (X, Σ, μ) be a finite measure space and $\{A_i\}_{i\in I} \subseteq \Sigma$ are pairwise disjoint with an arbitrary index set I. Show that $\mu(A_i) = 0$ for all $i \in T \setminus I_0$ with I_0 is at most countable.

Problem 2.4. Let (X, Σ, μ) be a finite nonatomic measure space and let $\{\eta_n\}_{n\geq1} \subseteq (0, +\infty)$ be such that $\sum_{n\geq1} \eta_n \leq \mu(X)$. Show that there is $\{A_n\}_{n\geq1} \subseteq \Sigma$ pairwise disjoint such that $\mu(A_n) = \eta_n$ for all $n \in \mathbb{N}$.

Problem 2.5. Let (X, Σ, μ) be a measure space with μ being semifinite (see Definition 2.1.30) and $A \in \Sigma$, $\mu(A) = +\infty$. Show that there exists $C \in \Sigma$, $C \subseteq A$ with $\mu(C) = +\infty$ and that C is σ-finite.

Problem 2.6. Let (X, Σ, μ) be a measure space. Show that μ is semifinite (see Definition 2.1.30) if and only if for all $A \in \Sigma$ with $\mu(A) > 0$ there holds

$$\mu(A) = \sup\left[\mu(C) : C \in \Sigma, C \subseteq A, 0 < \mu(C) < \infty\right].$$

Problem 2.7. Let X be a σ-compact metric space, $\mathcal{B}(X)$ is the Borel σ-algebra of X, and μ_1, μ_2 are two finite measures on $\mathcal{B}(X)$, which are equal on compact sets. Show that $\mu_1 = \mu_2$.

Problem 2.8. Let (X, Σ, μ) be a measure space and μ^* the outer measure defined in (2.1.7) with $\mathcal{L} = \Sigma$ and $\vartheta = \mu$. Show that:
(a) $\mu^*(A) = \inf[\mu(B) : B \in \Sigma, A \subseteq B]$ for every $A \subseteq X$.
(b) For every $A \subseteq X$ there exists $B \in \Sigma_{\mu^*}$ such that $A \subseteq B$ and $\mu^*(A) = \mu(B)$.

Problem 2.9. Let (Ω, Σ, μ) be a measure space, $\{A_n\}_{n\geq 1} \subseteq \Sigma$ with $\sum_{n\geq 1} \mu(A_n) < \infty$, and $\liminf_{n\to\infty} \mu(A_n) \geq \vartheta \geq 0$. Let D_∞ be the set of elements in Ω that belong to an infinity of sets A_n. Show that $D_\infty \in \Sigma$ and $\mu(D_\infty) \geq \vartheta$.

Problem 2.10. Let X be a nonempty set, $\mathcal{L} \subseteq 2^X$ is an algebra, and $\mu: \mathcal{L} \to [0, \infty]$ is an additive set function. Let μ^* be the outer measure defined in (2.1.7) with $\mathcal{L} = \Sigma$ and $\vartheta = \mu$. Show that every element in $\mathcal{L}$ is μ^*-measurable; see Definition 2.1.36. Moreover, show that if μ is σ-additive, then $\mu^*|_{\mathcal{L}} = \mu$.

Problem 2.11. Let $\mathcal{L}$ be a σ-algebra of sets in $\mathbb{R}$. Show that $\mathcal{B}(\mathbb{R}) \subseteq \mathcal{L}$ if and only if any continuous function $f: \mathbb{R} \to \mathbb{R}$ is $\mathcal{L}$-measurable.

Problem 2.12. Let (X, Σ, μ) be a measure space, $f: X \to [0, \infty]$ a Borel function, and let $d_f(t) = \mu(\{x \in X: f(x) > t\})$. Show that:
(a) d_f is right continuous.
(b) If $\mu(X) < \infty$, then for every $t_0 > 0$ it holds that $\lim_{t\to t_0^-} d_f(t) = \mu(\{x \in X: f(x) \geq t_0\})$.

Problem 2.13. Given $\varepsilon > 0$, produce a dense open set $U \subseteq \mathbb{R}$ such that $\lambda(U) \leq \varepsilon$, where λ is the Lebesgue measure on $\mathbb{R}$.

Problem 2.14. Suppose that $1 \leq p < \infty$ and let $f \in L^p(\mathbb{R}^N)$ for the Lebesgue measure on $\mathbb{R}^N$. Show that

$$\lim_{h\to 0} \int_{\mathbb{R}^N} |f(x+h) - f(x)| \, d\lambda = 0 .$$

Problem 2.15. (a) Suppose that $f: \mathbb{R}^N \to \mathbb{R}$ is integrable and $K \subseteq \mathbb{R}^N$ is nonempty and compact. Show that $\lim_{|y|\to\infty} \int_{K+y} |f(x)| \, dx = 0$.
(b) Suppose that $f: \mathbb{R}^N \to \mathbb{R}$ is uniformly continuous and $f \in L^p(\mathbb{R}^N)$ for some $1 \leq p < \infty$. Show that $\lim_{|x|\to\infty} f(x) = 0$.

Problem 2.16. Let X be a nonempty set, Y is a metrizable space and $f: X \to Y$ is a map that is the pointwise limit of simple functions. Show that $f(X) \subseteq Y$ is separable.

Problem 2.17. Let (X, Σ) be a measurable space, Y a second countable Hausdorff topological space, and $f: X \to Y$ a Σ-measurable multifunction. Show that $\operatorname{Gr} f \in \Sigma \otimes \mathcal{B}(Y)$.

Problem 2.18. Let (Ω, Σ, μ) be a measure space and $\mathcal{L} \subseteq \Sigma$ a countable subset such that if $A \in \Sigma$, $\mu(A) < \infty$, then there exists $B \in \mathcal{L}$ with $\mu(A \triangle B) \leq \varepsilon$. Show that $L^p(\Omega)$ is separable for all $1 \leq p < \infty$.

Problem 2.19. Let (Ω, Σ, μ) be a σ-finite measure space and assume that $f \in L^p(\Omega)$ for all $p \geq p_0 \geq 1$. Show that $\lim_{p\to+\infty} \|f\|_p = \|f\|_\infty$.

Problem 2.20. Let (X, Σ), $(Y, \mathcal{L})$, and $(V, \mathcal{D})$ be measurable spaces, $f: X \to Y$, $g: X \to V$, and let $h: X \to Y \times V$ be defined by $h(x) = (f(x), g(x))$ for all $x \in X$. Show that h is $(\Sigma, \mathcal{L} \otimes \mathcal{D})$-measurable if and only if f is $(\Sigma, \mathcal{L})$-measurable and g is $(\Sigma, \mathcal{D})$-measurable.

Problem 2.21. Let (X, Σ) be a measurable space, Y, Y_1, Y_2 separable metrizable spaces, and V a Hausdorff topological space. Suppose that

$$\begin{aligned} &f_k : X \times Y \to Y_k, \quad k = 1,2 \quad \text{are Carathéodory functions}\,, \\ &g: Y_1 \times Y_2 \to V \quad \text{is Borel measurable}\,. \end{aligned}$$

Show that $h: X \times Y \to V$ defined by $h(x,y) = g(f_1(x,y), f_2(x,y))$ is $\Sigma \bigotimes \mathcal{B}(X)$-measurable.

Problem 2.22. Let $E \subseteq \mathbb{R}$ be Lebesgue measurable with $\lambda(E) > 0$. Show that there exists a nonmeasurable subset of E.

Problem 2.23. Let (X, Σ, μ) be a finite measure space and $f_{nm}: X \to \mathbb{R}$ with $n, m \in \mathbb{N}$ a family of Σ-measurable functions such that

$$f_{nm}(x) \to f_n(x) \quad \mu\text{-a.e. as } m \to \infty \quad \text{and} \quad f_n(x) \to f(x) \quad \mu\text{-a.e. as } n \to \infty\,.$$

Show that there exists an increasing sequence $m_n \in \mathbb{N}$ with $n \geq 1$ such that

$$f_{nm_n}(x) \to f(x) \quad \mu\text{-a.e. as } n \to \infty\,.$$

Problem 2.24. Let X be a compact metrizable space and Y be a separable metrizable space, and consider the function space $C(X, Y)$ with the τ_u-topology; see Remark 1.6.17. Let

$$\mathcal{L} = \{e_x^{-1}(C), C \subseteq Y \text{ is closed}\}\,;$$

see Definition 1.6.7. Show that $\mathcal{B}(C(X, Y)) = \sigma(\mathcal{L})$.

Problem 2.25. Let (X, Σ) be a measurable space, V a compact metrizable space, Y a separable metrizable space, and consider the function space $C(V, Y)$ endowed with the τ_u-topology; see Remark 1.6.17.

(a) Given a Carathéodory function $f: X \times V \to Y$, show that $\hat{f}: X \to C(V, Y)$ defined by $\hat{f}(x)(\cdot) = f(x, \cdot)$ is Σ-measurable.

(b) If $h: X \to C(V, Y)$ is Σ-measurable, show that $\tilde{h}: X \times V \to Y$ defined by $\tilde{h}(x, \cdot) = h(x)(\cdot)$ is a Carathéodory function.

Problem 2.26. Let (X, Σ, μ) be a measure space and $f: X \to \mathbb{R}$ is a μ-integrable function. Show that the set $C = \{x \in X : f(x) \neq 0\}$ has σ-finite μ-measure.

Problem 2.27. Suppose that X and Y are Hausdorff topological spaces such that

$$D(Y) = \{(y, v) \in Y \times Y : y = v\} \in \mathcal{B}(Y) \bigotimes \mathcal{B}(Y)\,.$$

Show that the graph of any Borel function $f: X \to Y$ belongs to $\mathcal{B}(X) \otimes \mathcal{B}(Y)$.

Problem 2.28. Let (X, Σ, μ) be a finite measure space. Show that there exists an at most countable family $\{A_n\}_{n\geq 1} \subseteq \Sigma$ of atoms such that $X \setminus \bigcup_{n\geq 1} A_n$ is nonatomic.

Problem 2.29. Let (X, Σ, μ) be a measure space with μ being semifinite (see Definition 2.1.30 (a)), and let $f, g: X \to [0, +\infty]$ be two Σ-measurable functions such that

$$\int_A f \, d\mu \leq \int_A g \, d\mu \quad \text{for all } A \in \Sigma \text{ with } \mu(A) < \infty .$$

Show that $f(x) \leq g(x)$ for μ-a. a. $x \in X$.

Problem 2.30. Let $A \subseteq \mathbb{R}$ be a set of finite Lebesgue measure and let $f: \mathbb{R} \to \mathbb{R}$ be defined by $f(x) = \lambda(A \cap (-\infty, x])$ for all $x \in \mathbb{R}$. Here λ denotes the Lebesgue measure on $\mathbb{R}$. Show that f is continuous.

Problem 2.31. Let $A \subseteq \mathbb{R}$ be a Lebesgue measurable set with $\lambda(A) > 0$ with λ being the Lebesgue measure on $\mathbb{R}$. Show that $A - A$ contains an open set.

Problem 2.32. Let (X, Σ, μ) be a measure space and $f: X \to [0, \infty]$ is a Σ-measurable function. Show that $\int_X f \, d\mu = \int_0^\infty \mu(\{x \in X : f(x) > s\}) \, ds$.

Problem 2.33. Let (X, Σ, μ), $(Y, \mathcal{L}, \nu)$ be two σ-finite measure spaces. Show that $(X \times Y, \Sigma \bigotimes \mathcal{L}, \mu \times \nu)$ is σ-finite as well.

Problem 2.34. Let (X, Σ, μ) be a measure space, $f_n, f: X \to [0, +\infty)$ with $n \geq 1$ are Σ-measurable functions and suppose that $f_n \xrightarrow{\mu} f$. Show that for every $\vartheta > 0$, $f_n^\vartheta \xrightarrow{\mu} f^\vartheta$.

Problem 2.35. Let (X, Σ, μ) be a nonatomic measure space and $f: X \to [0, \infty]$ is a Σ-measurable function. Show that the measure $\Sigma \ni A \to \xi(A) = \int_A f \, d\mu$ is nonatomic if and only if $\mu(\{x \in X : f(x) = +\infty\}) = 0$.

Problem 2.36. Let X be a Hausdorff topological space, $\mu: \mathcal{B}(X) \to [0, +\infty)$ be a finite Borel measure, and $f: X \to \mathbb{R}$ be a continuous function. Show that there exists an at most countable set $D \subseteq \mathbb{R}$ such that $\mu(\{x \in X : f(x) = \eta\}) > 0$ for all $\eta \in D$.

Problem 2.37. Let X, Y be two metric spaces and $f: X \to Y$. Let $C_f = \{x \in X : f \text{ is continuous}\}$. Show that $C_f \in \mathcal{B}(X)$.

Problem 2.38. Does the Lebesgue Dominated Convergence Theorem (see Theorem 2.3.8) hold for nets? Justify your answer.

Problem 2.39. Let X be a Polish space and $A \subseteq X$. Show that A is analytic if and only if $A = \text{proj}_X B$ with $B \in \mathcal{B}(X \times X) = \mathcal{B}(X) \bigotimes \mathcal{B}(X)$.

Problem 2.40. Let (X, Σ) be a measurable space and Y a metric space. Show that $f: X \to Y$ is Σ-measurable if and only if for all continuous $\varphi: Y \to \mathbb{R}$ we have that $\varphi \circ f$ is Σ-measurable.

Problem 2.41. Let (Ω, Σ) be a measurable space, X a separable metrizable space, Y a Hausdorff topological space, $f: \Omega \times X \to Y$ a Carathéodory map, and $U \subseteq Y$ be open. Show that the multifunction $w \to G(w) = \{x \in X : f(w, x) \in U\}$ is measurable.

Problem 2.42. Let (Ω, Σ) be a measurable space, X is a Polish space and $F_n: \Omega \to P_f(X)$ with $n \in \mathbb{N}$ are measurable multifunctions such that for every $w \in \Omega$, there exists $n \in \mathbb{N}$ such that $F_n(w) \in P_k(X)$. Show that $w \to \bigcap_{n\geq 1} F_n(w)$ is measurable.

Problem 2.43. Let $\{X_n\}_{n\geq 1}$ be a sequence of Polish spaces and for each $n \in \mathbb{N}$, $A_n \subseteq X_n$ is analytic. Show that $\prod_{n\geq 1} A_n$ is an analytic subset of $\prod_{n\geq 1} X_n$.

Problem 2.44. Let X, Y be a Polish spaces, $A \in \mathcal{B}(X)$, $f: A \to Y$ is a Borel measurable map, and $E = f(A)$. Assume that f is injective and $B \in \mathcal{B}(Y)$. Show that f^{-1} is Borel measurable.

Problem 2.45. Let X, Y be Polish spaces and $f: X \to Y$ be Borel measurable.
(a) Show that if $A \subseteq X$ is analytic, then $f(A) \subseteq Y$ is analytic.
(b) Show that if $B \subseteq Y$ is analytic, then $f^{-1}(B) \subseteq X$ is analytic.

Problem 2.46. Let X, Y be Hausdorff topological spaces and $f: X \to Y$ be a map that has a graph that is a Souslin subset of $X \times Y$. Show that f is Borel measurable.

Problem 2.47. Let (X, Σ, μ) be a finite measure space, $K \subseteq L^1(X)$ be uniformly integrable, and K^* be the sequential closure for the μ-almost everywhere convergence in K. Show that K^* is uniformly integrable as well.

Problem 2.48. Let (X, Σ, μ) be a measure space and $C \subseteq L^1(X)$ a uniformly integrable set. Show that for given $\varepsilon > 0$ there exist $\xi_\varepsilon \in L^1(X)_+$ and $\delta > 0$ such that $A \in \Sigma$, $\int_A \xi_\varepsilon \, d\mu \leq \delta$ implies $\sup_{f\in C} \int_A |f| \, d\mu \leq \varepsilon$.

Problem 2.49. Let (X, Σ, μ) be a measure space and $C \subseteq L^1(X)$ a uniformly integrable set. Show that for given $\varepsilon > 0$ there is $\xi_\varepsilon \in L^1(X)_+$ such that $\sup_{f\in C} \int_{\{|f|\geq \xi_\varepsilon\}} |f| \, d\mu \leq \varepsilon$.

Problem 2.50. Let (X, Σ, μ) be a measure space and $C \subseteq L^1(X)$. Assume that for every $\varepsilon > 0$ we can find $\xi_\varepsilon \in L^1(\Omega)_+$ such that

$$\sup_{f\in C} \int_{\{|f|\geq \xi_\varepsilon\}} |f| \, d\mu \leq \varepsilon .$$

Show that C is uniformly integrable.

Problem 2.51. Let (X, Σ, μ) be a measure space and $C \subseteq L^1(X)$ be a bounded set, and suppose that for every $\varepsilon > 0$ we can find $\xi_\varepsilon \in L^1(\Omega)_+$ and $\delta > 0$ such that $A \in \Sigma$, $\int_A h_\varepsilon \, d\mu \leq \delta$ implies that $\sup_{f\in C} \int_A |f| \, d\mu \leq \varepsilon$. Show that C is uniformly integrable.

Problem 2.52. Let (Ω, Σ) be a measurable space, X a separable metrizable space, $f\colon \Omega \times X \to \mathbb{R}$ a Carathéodory function, and $F\colon \Omega \to P_k(X)$ a measurable multifunction. Let $m(w) = \min[f(w,x)\colon x \in F(w)]$ and $M(w) = \{x \in F(w)\colon m(w) = f(w,x)\}$. Show that m and M are both measurable.

Problem 2.53. Let (X, Σ) be a measurable space and μ, ν be finite measures on (X, Σ). Show that either $\mu \perp \nu$ or that there exist $\varepsilon > 0$ and $B \in \Sigma$ with $\mu(B) > 0$ and $\nu \geq \varepsilon\mu$ on B, that is, B is a positive set for $\nu - \varepsilon\mu$.

3 Basic Functional Analysis

Functional Analysis emerged as a coherent field of mathematics in the first four decades of the 20th century. It provided a unified framework to treat different objects using abstraction and axiomatization. The main idea is to view functions as points, respectively elements, of an abstract space endowed with certain structures that are axiomatically defined. This way mathematicians were able to "escape" from the usual finite dimensional Euclidean spaces and consider infinite dimensional function spaces. The starting point was the thesis of Fréchet in 1906 who introduced the abstract notion of "metric space" – a concept that was influential in the development of both functional analysis and point set topology. The work of Fréchet was the culmination of the efforts and contributions of many prominent mathematicians from France, Germany, and Italy. Combined with the revolution of measure theory this provided a fertile ground for the development of functional analysis. The prominent figure in the story is that of the Polish mathematician Stefan Banach (1892–1945).

In this chapter, we review the basic notions and results of "Linear Functional Analysis". Moreover, we touch on "Operator Theory" and in particular, we discuss the spectral properties of compact self-adjoint operators on a Hilbert space.

3.1 Topological Vector Spaces, Hahn–Banach Theorem

We start with the basic notion of a topological vector space. Recall that a **vector space** or **linear space** is a set X equipped with two operations $+ : X \times X \to X$ defined by $(x, u) \to x + u$ called the vector addition and $\cdot : \mathbb{K} \times X \to X$ defined by $(\lambda, x) \to \lambda \cdot x$ called the scalar multiplication where $\mathbb{K} = \mathbb{R}$ or $\mathbb{K} = \mathbb{C}$.

Definition 3.1.1. A **topological vector space** is a vector space endowed with a Hausdorff topology τ, which makes the two vector space operations above continuous. Then we say that τ is a vector topology on X.

Remark 3.1.2. Continuity of vector addition means that if $x, u \in X$ and $V \in \tau$ is a neighborhood of $x + u$, that is, $V \in \mathcal{N}(x + u)$, then there exist $U_x \in \mathcal{N}(x)$ and $U_u \in \mathcal{N}(u)$ such that $U_x + U_u \subseteq V$. Similarly the continuity of the scalar multiplication implies that if $(\lambda, x) \in \mathbb{K}\times X$ with $\mathbb{K} = \mathbb{R}$ or $\mathbb{K} = \mathbb{C}$ and $V \in \mathcal{N}(\lambda x)$, then there exist $\varepsilon > 0$ and $U_x \in \mathcal{N}(x)$ such that $\mu U_x \subseteq V$ for all $|\mu - \lambda| < \varepsilon$. Moreover, for a given $x \in X$ and a given $\lambda \in \mathbb{K}$ we introduce

$$\hat{T}_x(u) = x + u \quad \text{for all } u \in X \quad \text{(the translation operator)}\,,$$
$$\hat{M}_\lambda(u) = \lambda u \quad \text{for all } u \in X \quad \text{(the scalar multiplication operator)}\,.$$

Clearly, these operators are homeomorphisms of X onto X. It follows that the vector topology τ is **translation invariant**, that is, $U \in \tau$ if and only if $x + U \in \tau$ for all $x \in X$.

https://doi.org/10.1515/9783111286952-003

Hence, τ is completely determined by any local basis, in particular by the local basis at the origin. If the vector topology is induced by a metric d, then the metric is invariant, that is, $d(x+v, u+v) = d(x, u)$ for all $x, u, v \in X$.

An immediate consequence of these observations is the following simple lemma.

Lemma 3.1.3. *Let (X, τ) be a topological vector space.*
(a) *For all $U, V \in \tau$ and for all $\lambda \in \mathbb{K}$ it follows $U + V \in \tau$ and $\lambda U \in \tau$.*
(b) *If $A \subseteq X$ and $U \in \tau$, then $\overline{A} + U = A + U$ and it is open.*
(c) *If $K \subseteq X$ is compact and $C \subseteq X$ is closed, then $K + C \subseteq X$ is closed.*
(d) *If $K_1, K_2 \subseteq X$ are compact sets, then $K_1 + K_2 \subseteq X$ is compact.*
(e) *If $\varphi: X \to \mathbb{R}$ is linear, then φ is continuous if and only if φ is continuous at $x = 0$.*

Proof. (a), (b), and (e) are clear. (c) Let $\{v_\alpha\}_{\alpha \in I} \subseteq K + C$ be a net such that $v_\alpha \to v$. We have that $v_\alpha = x_\alpha + u_\alpha$ with $x_\alpha \in K$ and $v_\alpha \in C$ for all $\alpha \in I$. The compactness of K implies that there exists a subnet $\{x_\beta\}_{\beta \in J}$ of $\{x_\alpha\}_{\alpha \in I}$ such that $x_\beta \to x \in K$; see Proposition 1.4.45 (c). Then $u_\beta = v_\beta - x_\beta \to v - x = u \in C$ since C is closed; see Proposition 1.2.36. Therefore $v = x + u$ with $x \in K$ and $u \in C$. Hence, we conclude that $K + C$ is closed.

(d) Since "+" is continuous on $X \times X$ and $K_1 \times K_2 \subseteq X \times X$ is compact (see Theorem 1.4.56) we conclude that $+(K_1 \times K_2) = K_1 + K_2 \subseteq X$ is compact; see Theorem 1.4.51. □

Remark 3.1.4. The algebraic sum of two closed sets need not be closed. In $\mathbb{R}^2$ equipped with the usual Euclidean metric, we consider the sets

$$C_1 = \left\{ \left(x, \frac{1}{x}\right) : x \in \mathbb{R} \setminus \{0\} \right\} \quad \text{and} \quad C_2 = \{(u, 0) : u \in \mathbb{R}\} .$$

Then both are closed in $\mathbb{R}^2$ but $C_1 + C_2 = \{(x+u, 1/x) : x \in \mathbb{R} \setminus \{0\}, u \in \mathbb{R}\}$ is not closed in $\mathbb{R} \times \mathbb{R}$.

Remark 3.1.5. Let X, Y be two vector spaces. Recall that a map $A: X \to Y$ is called a linear function if it is additive and homogeneous, that is,

$$\begin{aligned} A(x+y) &= A(x) + A(y) \quad \text{for all } x, y \in X , \\ A(\lambda x) &= \lambda A(x) \quad \text{for all } \lambda \in \mathbb{K} \text{ and for all } x \in X . \end{aligned}$$

By $N(A)$ we denote the kernel of A, that is, $N(A) = \{x \in X : A(x) = 0\}$ and by $R(A)$ the range of A, that is, $R(A) = \{A(x) : x \in X\}$.

Now we introduce certain classes of sets that are important in the study of topological vector spaces.

Definition 3.1.6. Let X be a vector space and $A \subseteq X$.
(a) We say that A is **convex** if for all $x, u \in A$ and $\lambda \in [0, 1]$, it holds $(1 - \lambda)x + \lambda u \in A$.
(b) We say that A is **absorbing** if for any $x \in X$ there is $t = t(x) > 0$ such that $x \in tA$. So every absorbing set contains the origin.

(c) We say that A is **balanced** if $\lambda A \subseteq A$ for all $\lambda \in \mathbb{K}$ with $|\lambda| \le 1$.
(d) We say that A is **symmetric** if $A = -A$.

Lemma 3.1.7. *If (X, τ) is a topological vector space and $V \in \mathcal{N}(0)$, then there exists a symmetric set $U \in \mathcal{N}(0)$ such that $U + U \subseteq V$.*

Proof. The continuity of the vector addition operation implies that there exist $U_1, U_2 \in \mathcal{N}(0)$ such that $U_1 + U_2 \subseteq V$. Let $U = U_1 \cap (-U_1) \cap U_2 \cap (-U_2)$. Then $U \in \mathcal{N}(0)$ is symmetric and $U + U \subseteq V$. □

Proposition 3.1.8. *If (X, τ) is a topological vector space, $K \subseteq X$ is compact, $C \subseteq X$ is closed, and $K \cap C = \emptyset$, then there exists $U \in \mathcal{N}(0)$ such that $(K + U) \cap (C + U) = \emptyset$.*

Proof. We assume that $K \neq \emptyset$ or otherwise the result is obvious. Let $x \in K$. Applying Lemma 3.1.7 there is a symmetric $U_x \in \mathcal{N}(0)$ such that $(x+U_x+U_x+U_x)\cap C = \emptyset$. Exploiting the symmetry of U_x it follows that $(x+U_x+U_x)\cap(C+U_x) = \emptyset$. The compactness of K implies that there exist $\{x_n\}_{n=1}^m \subseteq K$ such that $K \subseteq \bigcup_{n=1}^m (x_n + U_{x_n})$. Let $U = \bigcap_{n=1}^m U_{x_n} \in \mathcal{N}(0)$. Then

$$K + U \subseteq \bigcup_{n=1}^{m} (x_n + U_{x_n} + U) \subseteq \bigcup_{n=1}^{m} (x_n + U_{x_n} + U_{x_n}) .$$

We conclude that $(K + U) \cap (C + U) = \emptyset$. □

Note that $K + U$ is an open set containing K and $C + U$ is an open set containing C; see Lemma 3.1.3 (b). Taking K to be a singleton we obtain the following result.

Corollary 3.1.9. *Every topological vector space is regular; see Definition* 1.2.7.

Proposition 3.1.10. *Let (X, τ) be a topological vector space.*
(a) *If $A \subseteq X$, then $\overline{A} = \bigcap_{U\in\mathcal{N}(0)}(A + U)$.*
(b) *If $A, C \subseteq X$, then $\overline{A} + \overline{C} \subseteq \overline{A + C}$.*
(c) *If $A \subseteq X$ is convex, then $\operatorname{int} A$ and $\overline{A}$ are convex.*
(d) *If $A \subseteq X$ is balanced, then $\overline{A}$ is balanced and when $0 \in \operatorname{int} A$, then $\operatorname{int} A$ is balanced.*

Proof. (a) We know that $x \in \overline{A}$ if and only if $(x + U) \cap A \neq \emptyset$ for all $U \in \mathcal{N}(0)$. Hence, $x \in \overline{A}$ if and only if $x \in A - U$ for every $U \in \mathcal{N}(0)$. But $U \in \mathcal{N}(0)$ if and only if $-U \in \mathcal{N}(0)$.

(b) Let $x \in \overline{A}$, $u \in \overline{C}$ and let $V \in \mathcal{N}(x + u)$. Then there exist $V_x \in \mathcal{N}(x)$, $V_u \in \mathcal{N}(u)$ such that $V_x + V_u \subseteq V$. Then choose $x' \in A \cap V_x$ and $u' \in C \cap V_u$. The existence follows since $x \in \overline{A}$ and $u \in \overline{C}$. Then $x' + u' \in (A + C) \cap V$. Since $V \in \mathcal{N}(x + u)$ we conclude that $x + u \in \overline{A + C}$, thus $\overline{A} + \overline{C} \subseteq \overline{A + C}$.

(c) Since $\operatorname{int} A \subseteq A$ and A is convex, it follows that

$$(1 - \lambda) \operatorname{int} A + \lambda \operatorname{int} A \subseteq A \quad \text{for all } \lambda \in (0, 1) . \tag{3.1.1}$$

Note that the left-hand side in (3.1.1) is an open set and so

$$(1-\lambda)\operatorname{int}A + \lambda\operatorname{int}A \subseteq \operatorname{int}A \quad \text{for all } \lambda \in (0,1)\,.$$

Hence, $\operatorname{int}A$ is convex. For $\lambda \in (0,1)$, due to part (b) and since A is convex, one gets

$$(1-\lambda)\overline{A} + \lambda\overline{A} = \overline{(1-\lambda)A} + \overline{\lambda A} \subseteq \overline{(1-\lambda)A + \lambda A} \subseteq \overline{A}\,.$$

Therefore $\overline{A}$ is convex.

(d) The proof that $\overline{A}$ is balanced is similar to the proof of part (c).

Let $\lambda \in \mathbb{K}$ be such that $0 < |\lambda| \leq 1$. Since A is balanced, we derive $\lambda\operatorname{int}A = \operatorname{int}\lambda A \subseteq \lambda A \subseteq A$, which shows that $\lambda\operatorname{int}A \subseteq A$. Moreover, since $0 \in \operatorname{int}A$, for $\lambda = 0$, it follows that $\lambda\operatorname{int}A \subseteq \operatorname{int}A$ and so $\operatorname{int}A$ is balanced. □

This leads to the following structural result for the topology of X.

Proposition 3.1.11. *Let (X,τ) be a topological vector space.*
(a) *Every $V \in \mathcal{N}(0)$ contains a balanced $U \in \mathcal{N}(0)$.*
(b) *Every convex $V \in \mathcal{N}(0)$ contains a balanced convex $U \in \mathcal{N}(0)$.*

Proof. (a) Let $V \in \mathcal{N}(0)$. Exploiting the continuity of the scalar multiplication operation, there exist $\delta > 0$ and $\tilde{U} \in \mathcal{N}(0)$ such that $\lambda\tilde{U} \subseteq V$ for all $\lambda \in \mathbb{K}$ with $|\lambda| < \delta$. Let U be the union of all these sets $\lambda\tilde{U}$. Evidently, $U \in \mathcal{N}(0)$, U is balanced and $U \subseteq V$.

(b) Let $V \in \mathcal{N}(0)$ be convex. Let $A = \bigcap_{|\lambda|=1}\lambda V$. Applying part (a), let $\hat{U} \in \mathcal{N}(0)$ be balanced such that $\hat{U} \subseteq V$. We have $\lambda^{-1}\hat{U} = \hat{U}$ for all $\lambda \in \mathbb{K}$ with $|\lambda| = 1$. Hence $\hat{U} \subseteq \lambda V$ and thus $\hat{U} \subseteq A$. This means that $\hat{U} \subseteq \operatorname{int}A \in \mathcal{N}(0)$. Moreover, $\operatorname{int}A \subseteq V$. The set A is convex, being the intersection of convex sets. Hence, $\operatorname{int}A$ is convex; see Proposition 3.1.10 (c). We claim that $\operatorname{int}A$ is balanced. According to Proposition 3.1.10 (d) it suffices to show that A is balanced. To this end, let $t \in [0,1]$ and $\mu \in \mathbb{K}$ with $|\mu| = 1$. Then, since $\lambda V \in \mathcal{N}(0)$ is convex,

$$t\mu A = \bigcap_{|\lambda|=1} t\mu\lambda V = \bigcap_{|\lambda|=1} t\lambda V \subseteq \bigcap_{|\lambda|=1} \lambda V\,.$$

Therefore, $t\mu A \subseteq A$ and so A is balanced. We conclude that $U = \operatorname{int}A \in \mathcal{N}(0)$ is the desired balanced and convex neighborhood of the origin. □

Corollary 3.1.12. *Every topological vector space has a local basis consisting of balanced sets.*

We introduce some particular types of topological vector spaces depending on the structure of the local basis.

Definition 3.1.13. Let (X,τ) be a topological vector space.
(a) A set $A \subseteq X$ is said to be **bounded** if for every $U \in \mathcal{N}(0)$ there is a $t_U > 0$ such that $A \subseteq tU$ for all $t > t_U$.
(b) We say that X is **locally convex** if it has a local basis $\mathcal{B}$ consisting of convex sets.
(c) We say that X is **locally bounded** if it has a bounded set in $\mathcal{N}(0)$.

(d) We say that X is **Fréchet** if it is locally convex and the topology τ is induced by a complete translation invariant metric d.
(e) A **norm** on X is a real function $\|\cdot\|$ such that
   $(\mathrm{e})_1$ $\|x\| \geq 0$ for all $x \in X$ and $\|x\| = 0$ if and only if $x = 0$;
   $(\mathrm{e})_2$ $\|\lambda x\| = |\lambda|\|x\|$ for all $(\lambda, x) \in \mathbb{K} \times X$;
   $(\mathrm{e})_3$ $\|x + u\| \leq \|x\| + \|u\|$ for all $x, u \in X$, which is called triangle inequality.
   X equipped with a norm is called a **normed space**. The norm defines a translation invariant metric $d(x, u) = \|x - u\|$. If (X, d) is complete, then X is a **Banach space**.
(f) We say that X is **normable** if τ is generated by the metric induced by a norm.

Remark 3.1.14. If X is locally bounded, then it is first countable. Indeed, if $U \in \mathcal{N}(0)$ is bounded and $r_n \to 0^+$, then $\{r_n U\}_{n \in \mathbb{N}}$ is a local basis for the origin.

Finite dimensional vector spaces exhibit some distinguishing properties. The **Euclidean norm** on X being finite dimensional with $\dim X = n$ is defined by

$$\|x\|_2 = \left(\sum_{k=1}^{n} |x_k|^2\right)^{\frac{1}{2}} \quad \text{for all } x = (x_k)_{k=1}^n \in X\,.$$

The topology on X induced by $\|\cdot\|_2$ is known as the **Euclidean topology**. It turns out that the Euclidean space is the prototype of a n-dimensional vector space.

Definition 3.1.15. Let X be a vector space and let $\|\cdot\|, |\cdot|$ be two norms on X. We say that these norms are **equivalent** if there exist constants $\eta > m > 0$ such that

$$m\|x\| \leq |x| \leq \eta\|x\| \quad \text{for all } x \in X\,.$$

Remark 3.1.16. Equivalence of norms is an equivalence relation and equivalent norms generate the same topology on X.

Proposition 3.1.17. *In a finite dimensional vector space any two norms are equivalent.*

Proof. Let X be the n-dimensional vector space with norm $\|\cdot\|$ and consider $\mathbb{R}^n$ equipped with the norm $\|\cdot\|_2$. Let $\{e_k\}_{k=1}^n \subseteq X$ be a basis for X and consider the linear map $A\colon \mathbb{R}^n \to X$ defined by

$$A(\lambda) = \sum_{k=1}^{n} \lambda_k e_k \quad \text{for all } \lambda = (\lambda_k)_{k=1}^n \in \mathbb{R}^n\,.$$

It is easy to see that A is an isomorphism. Moreover, we obtain the estimate

$$\|A(\lambda)\| \leq \sum_{k=1}^{n} |\lambda_k|\|e_k\| \leq \left(\sum_{k=1}^{n} |\lambda_k|^2\right)^{\frac{1}{2}} \left(\sum_{k=1}^{n} \|e_k\|^2\right)^{\frac{1}{2}} \leq \eta\|\lambda\|_2 \tag{3.1.2}$$

with $\eta = (\sum_{k=1}^n \|e_k\|^2)^{1/2}$. Therefore, A is continuous.

In addition, let $\xi = \|\cdot\| \circ A \colon \mathbb{R}^N \to \mathbb{R}$, that is,

$$\xi(\lambda) = \left\| \sum_{k=1}^{n} \lambda_k e_k \right\| \quad \text{for all } \lambda = (\lambda_k)_{k=1}^{n} \in \mathbb{R}^N . \tag{3.1.3}$$

Of course, ξ is continuous. Moreover, $\partial B_1 = \{\lambda \in \mathbb{R}^N \colon \|\lambda\|_2 = 1\}$ is closed and bounded, and thus compact; see Theorem 1.5.38. Hence, there exists $\lambda^* \in \partial B_1$ such that

$$\xi(\lambda^*) = \inf_{\lambda \in \partial B_1} \xi(\lambda) = m \geq 0 ;$$

see Theorem 1.4.52. If $m = 0$, then $\| \sum_{k=1}^{n} \lambda_k^* e_k \| = 0$ (see (3.1.3)), a contradiction since $\lambda^* \in \partial B_1$. Hence, $m > 0$ and we get

$$m\|\lambda\|_2 \leq \|A(\lambda)\| \quad \text{for all } \lambda \in \mathbb{R}^n . \tag{3.1.4}$$

From (3.1.2) and (3.1.4) we infer that X and $\mathbb{R}^n$ are linearly homeomorphic and so we conclude that any two norms on X are equivalent. □

Corollary 3.1.18. *Every finite dimensional normed space is complete, thus a Banach space.*

Corollary 3.1.19. *Every finite dimensional subspace of a normed space is closed.*

Next we will give a characterization of finite dimensional normed spaces in terms of the topological properties of the closed unit ball $\overline{B}_1 = \{x \in X \colon \|x\| \leq 1\}$. First we need an auxiliary result known as the "Riesz Lemma".

Lemma 3.1.20 (Riesz Lemma). *If X is a normed space, $Y \subseteq X$ is a proper, closed vector subspace, and $0 < \vartheta < 1$, then there exists $x_\vartheta \in (X \setminus Y) \cap \partial B_1$ such that $d(x_\vartheta, Y) \geq \vartheta$.*

Proof. Let $u \in X \setminus Y$. Since Y is closed it holds that $d(u, Y) = m > 0$. We choose $y \in Y$ such that $\|u - y\| \leq m/\vartheta$ and set $x_\vartheta = (u - y)/(\|u - y\|) \in \partial B_1$. Then for every $v \in Y$ it follows that

$$\|x_\vartheta - v\| = \frac{1}{\|u - y\|} \|u - (y + v\|u - y\|)\| . \tag{3.1.5}$$

Note that $y + v\|u - y\| \in Y$. Therefore, from (3.1.5) and the choice of $y \in Y$, it results in $\|x_\vartheta - v\| \geq m/(m/\vartheta) = \vartheta$. □

Applying this lemma, we have the following characterization of finite dimensional normed spaces.

Theorem 3.1.21. *A normed space X is finite dimensional if and only if $\overline{B}_1$ is compact.*

Proof. $\Longrightarrow$: This direction follows from Theorem 1.5.38.

$\Longleftarrow$: The set $\overline{B}_1$ is totally bounded; see Remark 1.5.32. Hence, there is $\{x_k\}_{k=1}^{n} \subseteq \overline{B}_1$ such that

$$\overline{B}_1 \subseteq \bigcup_{k=1}^{n}(x_k + B_{\frac{1}{2}}) \tag{3.1.6}$$

with $B_{1/2} = \{x \in X: \|x\| < 1/2\}$. Let $Y = \operatorname{span}\{x_k\}_{k=1}^n$. The Corollary 3.1.19 implies that $Y \subseteq X$ is closed. Suppose that $Y \neq X$. Then by Lemma 3.1.20 we find $\hat{x} \in (X \setminus Y) \cap \partial B_1$ such that

$$d(\hat{x}, Y) \geq \vartheta > \frac{1}{2}\,. \tag{3.1.7}$$

Comparing (3.1.6) and (3.1.7) we reach a contradiction. Hence $X = Y$ and so X is finite dimensional. □

Proposition 3.1.22. *If X is a finite dimensional normed space, Y is a normed space and $L: X \to Y$ is a linear map, then L is continuous.*

Proof. Suppose $\dim X = n$ and let $\{e_k\}_{k=1}^n$ be a basis for X. Since L is linear, we derive for $x = \sum_{k=1}^n \lambda_k e_k \in X$ with $\lambda_k \in \mathbb{K}$ that $L(x) = \sum_{k=1}^n \lambda_k L(e_k)$. Hence $\|L(x)\|_Y \leq \sum_{k=1}^n |\lambda_k| \|e_k\|_X \leq M(\sum_{k=1}^n |\lambda_k|^2)^{1/2}$ by the Bunyakowsky–Cauchy–Schwarz inequality for finite sums with $M = (\sum_{k=1}^n \|L(e_k)\|_Y^2)^{1/2}$. On the other hand we know from Proposition 3.1.17 the existence of $m > 0$ such that $\|\lambda\|_2 \leq 1/m\|x\|_X$ where $\lambda = (\lambda_k)_{k=1}^n \in \mathbb{R}^n$. Therefore, it follows $\|L(x)\|_Y \leq M/m\|x\|_X$. Hence L is continuous. □

Remark 3.1.23. In particular, if X is a finite dimensional normed space, then every linear functional $f: X \to \mathbb{R}$ is continuous. In fact the converse is true as well.

We conclude our discussion of finite dimensional topological vector spaces with a result closely related to Theorem 3.1.21. It says that there are no infinite dimensional locally compact topological vector spaces.

Proposition 3.1.24. *A topological vector space (X, τ) is locally compact if and only if X is finite dimensional.*

Proof. $\Longrightarrow$: Let $U \in \mathcal{N}(0)$ be relatively compact. So there is $\{x_k\}_{k=1}^n \subseteq U$ such that

$$\overline{U} \subseteq \bigcup_{k=1}^{n}\left(x_k + \frac{1}{2}U\right) = \{x_1, \ldots, x_n\} + \frac{1}{2}U\,. \tag{3.1.8}$$

Let $Y = \operatorname{span}\{x_k\}_{k=1}^n$. Then from (3.1.8) it follows

$$\frac{1}{2}\overline{U} \subseteq \frac{1}{2}\left[Y + \frac{1}{2}U\right] = Y + \frac{1}{2^2}U\,.$$

By induction we have

$$\overline{U} \subseteq Y + \frac{1}{2^n}U \quad \text{for all } n \in \mathbb{N}\,. \tag{3.1.9}$$

We fix $x \in \overline{U}$. Then from (3.1.9) we see that $x = y_n + 1/2^n u_n$ with $y_n \in Y$, $u_n \in U$ and $n \in \mathbb{N}$. Since U is relatively compact we find a subnet $\{u_\beta\}_{\beta \in J}$ of $\{u_n\}_{n \in \mathbb{N}}$ such that $u_\beta \to u$. Moreover, $1/2^\beta \to 0$. Hence, $y_\beta = x - (1/2^\beta)u_\beta \to x \in Y$. Therefore $\overline{U} \subseteq Y$ and since U is absorbing, we conclude that $X = Y$. Hence, X is finite dimensional.

$\Longleftarrow$: Since X is finite dimensional, we see that X is linearly homeomorphic to $(\mathbb{R}^n, \|\cdot\|_2)$. As X is a normed space, invoking Theorem 3.1.21, we get that $\overline{B}_1$ is compact. Thus, X is locally compact. □

Proposition 3.1.25. *If (X, τ) is a topological vector space and $A \subseteq X$, then the following statements are equivalent:*
(a) *A is bounded; see Definition* 3.1.13 *(a).*
(b) *If $\{x_n\}_{n\geq 1} \subseteq A$ and $\{\lambda_n\}_{n\geq 1} \subseteq \mathbb{K}$ with $\lambda_n \to 0$, then $\lambda_n x_n \to 0$ in X.*

Proof. (a) $\Longrightarrow$ (b): Let $U \in \mathcal{N}(0)$ be balanced; see Corollary 3.1.12. Then $A \subseteq tU$ for some $t > 0$. Suppose $\{x_n\}_{n\geq 1} \subseteq A$ and $\{\lambda_n\}_{n\geq 1} \subseteq \mathbb{K}$ such that $\lambda_n \to 0$. Then there exists $n_0 \in \mathbb{N}$ such that $|\lambda_n| t < 1$ for all $n > n_0$. Since U is balanced, it follows $\lambda_n x_n = \lambda_n t 1/t x_n \in U$ for all $n > n_0$. We conclude that $\lambda_n x_n \to 0$ in X as $n \to \infty$.

(b) $\Longrightarrow$ (a): Arguing by contradiction suppose that A is not bounded. Then there exist $t_n \to +\infty$ and $U \in \mathcal{N}(0)$ such that $(X \setminus t_n U) \cap A \neq \emptyset$ for all $n \in \mathbb{N}$. Let $x_n \in A$ with $x_n \notin t_n U$ for all $n \in \mathbb{N}$. We have $1/t_n x_n \notin U$ for all $n \in \mathbb{N}$. Hence $1/t_n u_n$ does not converge to 0, a contradiction to our hypothesis. □

Next we take a closer look at convex sets. In Proposition 3.1.10 (c) we saw that the interior and the closure of a convex set remain convex. In fact we can say more.

Proposition 3.1.26. *If X is a topological vector space, $C \subseteq X$ is a convex set and $0 \leq t < 1$, then $(1 - t) \operatorname{int} C + t\overline{C} \subseteq \operatorname{int} C$.*

Proof. For $t = 0$, the result is trivially true. So, suppose that $0 < t < 1$ and let $x \in \operatorname{int} C$ and $u \in \overline{C}$. Then there exists $U \in \mathcal{N}(0)$ such that $x + U \subseteq C$. Note that $u - (1 - t)/tU \in \mathcal{N}(u)$ and so there exists $y \in C \cap (u - (1 - t)/tU)$. Therefore $t(u - y) \in (1 - t)U$. Let $V = (1 - t)(x + U) + ty = (1 - t)x + (1 - t)U + ty$. This is a nonempty open set and $V \subseteq C$ due to the convexity of C. One gets

$$(1 - t)x + tu = (1 - t)x + t(u - y) + ty \in (1 - t)x + (1 - t)U + ty = V \subseteq C,$$

which gives $(1 - t)x + tu \in \operatorname{int} C$. □

Proposition 3.1.27. *If X is a topological vector space and $C \subseteq X$ is convex, then $\overline{\operatorname{int} C} = \overline{C}$ and $\operatorname{int} \overline{C} = \operatorname{int} C$.*

Proof. From Proposition 3.1.26 it follows $(1 - t) \operatorname{int} C + t\overline{C} \subseteq \operatorname{int} C$ for all $0 \leq t \leq 1$. Letting $t \to 1^-$ gives $\overline{C} = \overline{\operatorname{int} C}$.

Let $u \in \operatorname{int} C$ and $x \in \operatorname{int} \overline{C}$. Then there exists $U \in \mathcal{N}(0)$ such that $x + U \subseteq \overline{C}$. Since U is absorbing there exists $\vartheta \in (0, 1)$ such that $\vartheta(x - u) \in U$. Then $x + \vartheta(x - u) \in \overline{C}$. Applying

Proposition 3.1.26 gives $x - \vartheta(x-u) = (1-\vartheta)x + \vartheta u \in \operatorname{int} C$. Applying Proposition 3.1.26 again yields

$$x = \frac{1}{2}[x - \vartheta(x-u)] + \frac{1}{2}[x + \vartheta(x-u)] \in \operatorname{int} C\,.$$

This shows $\operatorname{int}\overline{C} \subseteq \operatorname{int} C \subseteq \operatorname{int}\overline{C}$ and so $\operatorname{int}\overline{C} = \operatorname{int} C$. □

Remark 3.1.28. Usually, sets C satisfying $\overline{\operatorname{int} C} = \overline{C}$ and $\operatorname{int}\overline{C} = \operatorname{int} C$ are called **regular**.

Clearly the intersection of any family of convex sets is again convex. So we can state the following definition.

Definition 3.1.29. Let X be a vector space and $A \subseteq X$ a nonempty set. The **convex hull** of A, denoted by conv A, is the intersection of all convex sets that contain A. Therefore, conv A is the smallest convex set containing A. An alternative description is given by

$$\operatorname{conv} A = \left\{x \in X : \exists x_k \in A, t_k \geq 0, k = 1, \ldots, n \text{ with } \sum_{k=1}^{n} t_k = 1, x = \sum_{k=1}^{n} t_k x_k\right\}.$$

That is, conv A is the set of all convex combinations of elements in A. If X is a topological vector space, the **closed convex hull** of A, denoted by $\overline{\operatorname{conv}}A$, is the set $\overline{\operatorname{conv} A}$.

For finite dimensional vector spaces, the convex hull of a set is described more precisely by the so-called “Carathéodory Convexity Theorem”.

Theorem 3.1.30 (Carathéodory Convexity Theorem). *If X is an m-dimensional vector space, $A \subseteq X$, and $x \in \operatorname{conv} A$, then x is the convex combination of at most $(m+1)$-elements of A.*

Proof. From Definition 3.1.29 we know that $x = \sum_{k=1}^{n} t_k x_k$ with $t_k \geq 0, x_k \in A, k = 1, \ldots, n$ and $\sum_{k=1}^{n} t_k = 1$. Without any loss of generality we may assume that $t_k > 0$ for all $k = 1, \ldots, n$.

Suppose that $n > m+1$, then $\{x_k - x_1\}_{k=2}^{n}$ must be linearly dependent. Hence, there exist $\beta_2, \ldots, \beta_m \in \mathbb{R}$ not all of them equal to zero such that

$$\sum_{k=2}^{n} \beta_k x_k - \left(\sum_{k=2}^{n} \beta_k\right) x_1 = 0\,.$$

Thus, there are $\eta_1, \ldots, \eta_n \in \mathbb{R}$ not all of them equal to zero such that $\sum_{k=1}^{n} \eta_k x_k = 0$ and $\sum_{k=1}^{n} \eta_k = 0$.

We set

$$I_+ = \{k \in \{1, \ldots, n\} : \eta_k > 0\}\,, \quad I_- = \{k \in \{1, \ldots, n\} : \eta_k < 0\}\,,$$
$$\mu = \min_{k \in I_+} \frac{t_k}{\eta_k}\,, \quad J = \{k \in I_+ : t_k - \mu\eta_k = 0\}\,.$$

The sets I_+, I_- and J are nonempty and $\mu > 0$. One obtains

$$x = \sum_{k=1}^{n} t_k x_k = \sum_{k=1}^{n} (t_k - \mu\eta_k) x_k = \sum_{k \notin J} (t_k - \mu\eta_k) x_k \,. \tag{3.1.10}$$

If $k \in I_+$, then $t_k - \mu\eta_k \geq 0$. If $k \in I_-$, then $t_k - \mu\eta_k > 0$. If $k \in I_+ \setminus J$, then $t_k - \mu\eta_k > 0$. Moreover, we get

$$\sum_{k=1}^{n} (t_k - \mu\eta_k) = \sum_{k=1}^{n} t_k - \mu \sum_{k=1}^{n} \eta_k = 1 \,. \tag{3.1.11}$$

From (3.1.10) and (3.1.11) we see that x is written as a convex combination with positive weights of n' elements with $n' < n$. We repeat this process until $n' \leq m + 1$. □

Corollary 3.1.31. *If X is an m-dimensional topological vector space and $K \subseteq X$ is compact, then* conv $K \subseteq X$ *is compact as well.*

Proof. Let $D = \{(t_1, \ldots, t_{m+1}) : t_k \geq 0, k = 1, \ldots, m + 1, \sum_{k=1}^{m+1} t_k = 1\} \subseteq \mathbb{R}^{m+1}$ and consider the map $\xi : \mathbb{R}^{m+1} \times (\prod_{k=1}^{m+1} X_k = X) \to X$ defined by

$$\xi((t_k)_{k=1}^{m+1}, x_1, \ldots, x_{m+1}) = \sum_{k=1}^{m+1} t_k x_k \,.$$

It is easy to see that ξ is continuous. Since $D \subseteq \mathbb{R}^{m+1}$ and $\prod_{k=1}^{m+1}(C_k = K) \subseteq \prod_{k=1}^{m+1}(X_k = X)$ are both compact, we get that $Dx(\prod_{k=1}^{m+1} C_k = K)$ is compact as well and so $\xi(D, K, \ldots, K) \subseteq X$ is also compact. But according to Theorem 3.1.30, $\xi(D, K, \ldots, K) = \operatorname{conv} K$. Hence, conv $K \subseteq X$ is compact. □

The corollary fails in infinite dimensional topological vector spaces.

Example 3.1.32. Let $c_0 = \{(x_n)_{n\geq 1} : x_n \in \mathbb{R} \text{ for all } n \in \mathbb{N} \text{ with } x_n \to 0\}$ furnished with the norm $\|(x_n)_{n\geq 1}\| = \sup\{|x_n| : n \in \mathbb{N}\}$. Then c_0 is a Banach space. Let $\hat{u}_n = (\delta_{k,n} 1/n)$ with $\delta_{k,n}$ being the Kronecker delta. Evidently $\hat{u}_n \in c_0$ for all $n \in \mathbb{N}$. Let $K = \{\hat{u}_n\} \cup \{0\}$. Then $K \subseteq c_0$ is compact, but

$$\hat{u} = \sum_{n\geq 1} \frac{1}{2^k} \hat{u}_n \in \overline{\operatorname{conv} K} \,, \quad \hat{u} \notin \operatorname{conv} K \,.$$

Thus, conv K is not closed, hence it is not compact.

In the next definition we extend the notion of total boundedness (see Definition 1.5.31), to general topological vector spaces that are not necessarily metrizable.

Definition 3.1.33. Let X be a topological vector space with a local basis $\mathcal{B}$. A set $A \subseteq X$ is said to be **totally bounded** if for every $U \in \mathcal{B}$ there exists a finite subset $F \subseteq X$ such that $A \subseteq F + U$.

Remark 3.1.34. The following assertions are easy to see:

(a) A totally bounded set is bounded; see Definition 3.1.13 (a).

(b) The closure of a totally bounded set is totally bounded.

(c) Compact sets are totally bounded.

Proposition 3.1.35. *If X is a locally convex space and $A \subseteq X$ is totally bounded, then* conv A *is totally bounded.*

Proof. Let $U \in \mathcal{N}(0)$ be convex. Since A is totally bounded, there exists a finite $F \subseteq X$ such that $A \subseteq F + 1/2U$. Corollary 3.1.31 implies that conv F is compact. Let $x \in$ conv A. Then

$$x = \sum_{k=1}^{n} t_k x_k \quad \text{with } t_k \geq 0, \ x_k \in A, \ k = 1, \ldots, n, \ \sum_{k=1}^{n} t_k = 1.$$

For every $k \in \{1, \ldots, n\}$ there is $u_k \in F$ such that $x_k \in u_k + 1/2U$. Then one gets

$$x = \sum_{k=1}^{n} t_k(x_k - u_k) + \sum_{k=1}^{n} t_k u_k \in \frac{1}{2}U + \text{conv } F$$

since U is convex. Hence

$$\text{conv } A \subseteq \text{conv } F + \frac{1}{2}U. \tag{3.1.12}$$

As we already remarked, conv $F \subseteq X$ is compact. Thus, we find a finite set $E \subseteq$ conv F such that conv $F \subseteq E + 1/2U$, which gives, due to (3.1.12) and the fact that U is convex, that conv $A \subseteq E + U$ and so we see that conv A is totally bounded in the sense of Definition 3.1.33. $\square$

From this proposition we deduce the following useful result.

Theorem 3.1.36. *If X is a Fréchet space and $A \subseteq X$ is compact, then $\overline{\text{conv}}\,A \subseteq X$ is compact as well.*

Proof. Since $A \subseteq X$ is compact, it is totally bounded; see Theorem 1.5.36. Then Proposition 3.1.35 implies that conv A is totally bounded, hence $\overline{\text{conv}}\,A$ is totally bounded; see Remark 3.1.34. Then Theorem 1.5.36 implies that $\overline{\text{conv}}\,A$ is compact. $\square$

Next we introduce an important class of convex functionals that describes locally convex spaces.

Definition 3.1.37. Let X be a vector space. A function $\rho: X \to \mathbb{R}$ is a **seminorm** if the following hold:

(a) ρ is subadditive, that is, $\rho(x + u) \leq \rho(x) + \rho(u)$ for all $x, u \in X$.

(b) ρ is absolutely homogeneous, that is, $\rho(\lambda x) = |\lambda|\rho(x)$ for all $\lambda \in \mathbb{K}$ and for all $x \in X$.

If $\rho(x) \neq 0$ for $x \neq 0$, then the seminorm is a norm; see Definition 3.1.13 (e). A family $\mathcal{P}$ of seminorms on X is said to be **separating** if for each $x \neq 0$ there exists $\rho \in \mathcal{P}$ such that $\rho(x) \neq 0$. Given an absorbing set $A \subseteq X$, the real functional $\rho_A: X \to \mathbb{R}$ defined by $\rho_A(x) = \inf[t > 0: x \in tA]$ is the **Minkowski functional of** A (or **gauge of** A).

Proposition 3.1.38. *If X is a vector space and $\rho: X \to \mathbb{R}$ is a seminorm, then the following hold:*
(a) $\rho(0) = 0$, $|\rho(x) - \rho(u)| \leq \rho(x-u)$ *for all* $x, u \in X$, $\rho(x) \geq 0$ *for all* $x \in X$;
(b) $N(\rho) = \{x \in X: \rho(x) = 0\}$ *is a vector subspace of* X;
(c) $B_1 = \{x \in X: \rho(x) < 1\}$ *is convex, absorbing, balanced, and* $\rho = \rho_{B_1}$.

Proof. (a) From Definition 3.1.37 we have $\rho(0) = \rho(\lambda 0) = |\lambda|\rho(0)$ for all $\lambda \in \mathbb{K}$, hence $\rho(0) = 0$. Moreover,

$$\rho(x) = \rho(x - u + u) \leq \rho(x-u) + \rho(u) \quad \text{for all } x, u \in X ,$$

hence, $|\rho(x) - \rho(u)| \leq \rho(x-u)$ by interchanging the roles of x and u. If $u = 0$, then we see that $\rho(x) \geq 0$ for all $x \in X$.

(b) Let $\lambda \in \mathbb{K}$ and $x, u \in N(p)$. Then

$$0 \leq \rho(\lambda x + u) \leq |\lambda|\rho(x) + \rho(u) = 0 .$$

Hence $\lambda x + u \in N(p)$ and so $N(p)$ is a vector subspace of X.

(c) Let $x, u \in B_1$ and $t \in (0,1)$. Then $\rho((1-t)x + tu) \leq (1-t)\rho(x) + t\rho(u) < 1$, which implies that B_1 is convex.

If $x \in X$ and $\vartheta > \rho(x)$, then $\rho(1/\vartheta x) = 1/\vartheta \rho(x) < 1$ and so B_1 is absorbing. Moreover, it is clear that B_1 is balanced. From the previous argument we see that $\rho_{B_1} \leq \rho$. Next let $0 < \eta \leq \rho(x)$. Then $1 \leq \rho(1/\eta x)$ and so $1/\eta x \notin B_1$. Therefore, $\rho \leq \rho_{B_1}$ and we conclude that $\rho = \rho_{B_1}$. ☐

For the Minkowski functional we obtain the following result.

Proposition 3.1.39. *If X is a vector space and $A \subseteq X$ is convex and absorbing, then the following hold:*
(a) ρ_A *is subadditive and positively homogeneous, that is,* ρ_A *is sublinear;*
(b) ρ_A *is a seminorm if* A *is in addition balanced;*
(c) *if* $B = \{x \in X: \rho_A(x) < 1\}$ *and* $C = \{x \in X: \rho_A(x) \leq 1\}$, *then* $B \subseteq A \subseteq C$ *and* $\rho_B = \rho_A = \rho_C$.

Proof. (a) For every $x \in X$, let $A(x) = \{t > 0: x \in tA\}$. Pick $t \in A(x)$ and $\vartheta > t$. Since $0 \in A$ and A is convex, it holds $\vartheta \in A(x)$. Therefore, $A(x)$ is a half-line starting at $\rho_A(x)$. Suppose that $\rho_A(x) < \vartheta$ and $\rho_A(u) < \mu$. Let $\tau = \vartheta + \mu$. Then it follows that $1/\vartheta x \in A$, $1/\mu u \in A$ and since A is convex

$$\frac{1}{\tau}(x+u) = \left(\frac{\vartheta}{\tau}\right)\frac{1}{\vartheta}x + \left(\frac{\mu}{\tau}\right)\frac{1}{\mu}u \in A .$$

This gives $\rho_A(x + u) \le \tau$ and so ρ_A is subadditive. Of course, ρ_A is also positively homogeneous.

(b) This is immediate from Definition 3.1.6 (c) and Definition 3.1.37.

(c) Suppose $\rho_A(x) < 1$. Then $1 \in A(x)$ and so $x \in X$. On the other hand if $x \in A$, then $\rho_A(x) \le 1$ and so we conclude that $B \subseteq A \subseteq C$. It follows that $B(x) \subseteq A(x) \subseteq C(x)$ for every $x \in X$ and so $\rho_B(x) \le \rho_A(x) \le \rho_C(x)$. Suppose $\rho_C(x) < \vartheta < \mu$. Then $1/\vartheta x \in C$ and so $\rho_A(1/\vartheta x) \le 1$, hence

$$\rho_A\left(\frac{1}{\mu}x\right) = \rho_A\left(\frac{\vartheta}{\mu}\frac{1}{\vartheta}x\right) = \frac{\vartheta}{\mu}\rho_A\left(\frac{1}{\vartheta}x\right) \le \frac{\vartheta}{\mu} < 1 .$$

Therefore, $1/\mu x \in B$, $\rho_B(1/\mu x) \le 1$, hence $\rho_B(x) \le \mu$. We conclude that $\rho_B = \rho_A = \rho_C$. □

Seminorms characterize locally convex topologies. The following theorem can be found in Yosida [336, p. 26].

Theorem 3.1.40. *If X is a vector space and $\{\rho_\alpha\}_{\alpha\in I}$ is a separating family of seminorms on X, then there is a weakest locally convex topology on X making all the seminorms continuous. Conversely, any locally convex space is topologized by the seminorms defined by the Minkowski functionals of the convex, absorbing, balanced sets. Such sets are often called* ***barrels****. Moreover, if $\mathcal{F} \subseteq \mathbb{R}^X$ is a set of $\mathbb{R}$-valued linear functionals on X, then the weakest topology on X making all elements of $\mathcal{F}$ continuous is locally convex.*

What about normable spaces, the next more restrictive class of vector spaces after locally convex spaces? We have the following theorem known as “Kolmogorov’s Normability Criterion”.

Theorem 3.1.41 (Kolmogorov’s Normability Criterion). *A topological vector space X is normable if and only if it is locally convex and locally bounded, that is, it possesses a bounded convex neighborhood of the origin.*

Proof. $\Longrightarrow$: The open unit ball $B_1 = \{x \in X : \|x\| < 1\}$ is a bounded convex neighborhood of the origin.

$\Longleftarrow$: Let $U \in \mathcal{N}(0)$ be bounded convex. We may also assume that U is balanced; see Corollary 3.1.12. Let $\|x\| = \rho_U(x)$ for all $x \in X$ with ρ_U being the Minkowski functional of U. Note that $\{tU\}_{t>0}$ is a local basis for the topology of X. If $x \neq 0$, then there exists $t > 0$ such that $x \notin tU$ and so $\|x\| = \rho_U(x) \ge t$. Then, from Proposition 3.1.39 (b) we infer that $\|\cdot\|$ is a norm on X. Moreover, from Proposition 3.1.39 (c) we conclude that $\{x \in X : \|x\| < t\} = tU$ for all $t > 0$. Therefore the norm topology coincides with the initial locally convex topology on X. □

Now we are ready for one of the most important results in analysis with far-reaching consequences. This is the celebrated “Hahn–Banach Extension Theorem”.

Theorem 3.1.42 (Hahn–Banach Extension Theorem). *If X is a vector space, $\rho: X \to \mathbb{R}$ is subadditive and positively homogeneous, that is, sublinear, $V \subseteq X$ is a vector subspace,*

$f: V \to \mathbb{R}$ is linear and $f(x) \le \rho(x)$ for all $x \in V$, then there exists $\hat{f}: X \to \mathbb{R}$ being linear such that $\hat{f}|_V = f$ and $\hat{f}(x) \le \rho(x)$ for all $x \in X$.

Proof. We assume that $V \neq X$ and let $u \in X \setminus V$. Let $Y = \text{span}\{V \cup \{u\}\}$. Then each $y \in Y$ can be written in a unique way as $y = x + \lambda u$ with $x \in V$ and $\lambda \in \mathbb{R}$. Then any extension $\hat{f}$ of f on Y must be of the form $\hat{f}(x + \lambda u) = f(x) + \lambda \hat{f}(u)$. So, the main problem is to define $\hat{f}(u)$. Recall that the extension $\hat{f}$ must satisfy $\hat{f} \le \rho$ on Y. Therefore

$$f(x) + \lambda \hat{f}(u) \le \rho(x + \lambda u) . \tag{3.1.13}$$

Taking $\lambda = 1$ in (3.1.13) yields $\hat{f}(u) \le \rho(x + u) - f(x)$. Similarly, if we take $\lambda = -1$ and replace x by $-x$ in (3.1.13) we infer $-f(x) - \hat{f}(u) \le \rho(-x - u)$ because the subadditivity of ρ implies $-f(x) \le f(-x)$. It follows that

$$-f(v) - \rho(-v - u) \le \hat{f}(u) \le -f(x) + \rho(x + u) \quad \text{for all } v, x \in V . \tag{3.1.14}$$

Therefore the value $\hat{f}(u)$ cannot be chosen arbitrarily but it must satisfy (3.1.14). However, in order to make (3.1.14) possible, we need to have

$$-f(v) - \rho(-v - u) \le -f(x) + \rho(x + u) \quad \text{for all } v, x \in V . \tag{3.1.15}$$

But note that $f(x) - f(v) = f(x - v) \le \rho(x - v) = \rho(x + u + (-v - u)) \le \rho(x + u) + \rho(-v - u)$ and so (3.1.15) holds. Now we can define the extension $\hat{f}$ of f on Y. We can take for example $\hat{f}(u) = \inf[-f(x) + \rho(x + u) : x \in V]$ and obtain $\hat{f}(x + \lambda u) = f(x) + \lambda \hat{f}(u)$. Clearly $\hat{f}$ is linear on Y and $\hat{f}|_V = f$. We need to show that $\hat{f} \le \rho$. Since $\hat{f}|_V = f$ we get $\hat{f} \le \rho$ when $\lambda = 0$. So, let $\lambda \neq 0$. Then we replace $v, x \in V$ by $1/\lambda x \in V$ in (3.1.14). This gives

$$-f\left(\frac{1}{\lambda} x\right) - \rho\left(-\frac{1}{\lambda} x - u\right) \le \hat{f}(u) \le -f\left(\frac{1}{\lambda} x\right) + \rho\left(\frac{1}{\lambda} x + u\right),$$

which implies

$$f\left(\frac{1}{\lambda} x\right) + \hat{f}(u) \le \rho\left(\frac{1}{\lambda} x + u\right) \quad -f\left(\frac{1}{\lambda} x\right) - \hat{f}(u) \le \rho\left(-\frac{1}{\lambda} x - u\right). \tag{3.1.16}$$

If $\lambda > 0$, then if we multiply the first inequality in (3.1.16) with λ, we obtain $\hat{f}(x + \lambda u) \le \rho(x + \lambda u)$. If $\lambda < 0$, then multiplying the second inequality in (3.1.16) with $-\lambda$ gives $\hat{f}(x + \lambda u) \le \rho(x + \lambda u)$. In summary we have showed that $\hat{f} \le \rho$.

Now let

$$\mathcal{L} = \{(Y, \hat{f}) : Y \text{ is a subspace of } X \text{ containing } V \text{ and } \hat{f} \text{ is a linear extension of } f \text{ on } Y \text{ with } \hat{f} \le \rho\} .$$

We order $\mathcal{L}$ as follows: $(Y, \hat{f}) \le (Y', \hat{f}')$ if $Y \subseteq Y'$ and $\hat{f}'|_Y = \hat{f}$. Then every chain $\mathcal{D}$ of $\mathcal{L}$ has an upper bound in $\mathcal{L}$ namely if $\mathcal{D} = \{(Y_\alpha, \hat{f}_\alpha)_{\alpha \in I}\}$, then $Y = \bigcup_{\alpha \in I} Y_\alpha$ is a linear subspace of

X and $\hat{f}(x) = \hat{f}_\alpha(x)$ for $x \in Y_\alpha$ is a well-defined linear functional on Y. Evidently, $(Y, \hat{f}) \in \mathcal{L}$ and $(Y_\alpha, \hat{f}_\alpha) \le (Y, \hat{f})$ for all $\alpha \in I$. By Zorn's Lemma (see Section 1.4), $\mathcal{L}$ admits a maximal element $(Y, \hat{f})$. We must have $Y = X$ or otherwise we repeat the construction in the first part of the proof and contradict the maximality of $(Y, \hat{f})$. □

Remark 3.1.43. It should be noted that the extension $\hat{f}$ is in general not unique.

A careful reading of the proof of Theorem 3.1.42 reveals that the complex variant of the result requires a modification of the condition on ρ since positive homogeneity of ρ makes in that case no sense.

Theorem 3.1.44 (Hahn–Banach Extension Theorem (Complex Variant)). *If X is a complex vector space, $\rho: X \to \mathbb{R}$ is a seminorm, $V \subseteq X$ is a vector subspace, $f: V \to \mathbb{C}$ is linear and $|f(x)| \le \rho(x)$ for all $x \in X$, then there exists $\hat{f}: X \to \mathbb{C}$ being linear such that $\hat{f}|_V = f$ and $|\hat{f}(x)| \le \rho(x)$ for all $x \in X$.*

From now on, unless otherwise stated, all vector spaces will be over the reals.

Definition 3.1.45. Let X, Y be normed spaces. A linear operator $A: X \to Y$ is **bounded** if $\|A(x)\|_Y \le M\|x\|_X$ for some $M > 0$ and for all $x \in X$. The smallest $M \ge 0$ for which the inequality above holds, is called the **operator norm** of A and it is denoted by

$$\|A\|_L = \sup\left[\frac{\|A(x)\|_Y}{\|x\|_X} : x \in X, x \neq 0\right].$$

By $L(X, Y)$ we denote the vector space of all bounded, linear operators from X into Y. Evidently $(L(X, Y), \|\cdot\|_L)$ is a normed space and the resulting norm topology is called the **uniform operator topology**. If $Y = \mathbb{R}$, then $L(X, \mathbb{R}) = X^*$ is the **topological dual** and its elements are called **bounded linear functionals**. If $x \in X$ and $x^* \in X^*$ we usually write $\langle x^*, x\rangle$ instead of $x^*(x)$ and call $\langle \cdot, \cdot \rangle$ the **duality brackets** for the pair (X^*, X).

The proof of the next proposition is straightforward and so its proof is omitted.

Proposition 3.1.46. *If X, Y are normed spaces and $A: X \to Y$ is a linear operator, then the following properties are equivalent:*

(a) *A is bounded.*

(b) *A is continuous.*

(c) *A is continuous at $x = 0$.*

Proposition 3.1.47. *If X is a normed space and Y is a Banach space, then $(L(X, Y), \|\cdot\|_L)$ is a Banach space.*

Proof. Suppose that $\{A_n\}_{n\ge1} \subseteq L(X, Y)$ is a $\|\cdot\|_L$-Cauchy sequence. Then it follows

$$\left\|(A_n - A_m)(x)\right\|_Y \le \|A_n - A_m\|_L \|x\| \quad \text{for all } n, m \in \mathbb{N} \text{ and for all } x \in X.$$

Since Y is complete, one gets that $A(x) = \lim_{n\to\infty} A_n(x)$ exists for all $x \in X$. Of course, $A: X \to Y$ is linear and

$$\|A(x) - A_n(x)\|_Y = \lim_{m\to\infty} \|A_m(x) - A_n(x)\|_Y \le \limsup_{m\to\infty} \|A_m - A_n\|_L \|x\| .$$

So, for given $\varepsilon > 0$, there exists $n_0 = n_0(\varepsilon) \in \mathbb{N}$ such that

$$\|A(x) - A_n(x)\|_Y \le \varepsilon\|x\| \quad \text{for all } x \in X \text{ and for all } n \ge n_0 . \tag{3.1.17}$$

Hence

$$\|A(x)\|_Y = \|A(x) - A_{n_0}(x)\|_Y + \|A_{n_0}(x)\|_Y \le (\varepsilon + \|A_{n_0}\|_L)\|x\|_X .$$

This implies that $A \in L(X, Y)$ and $\|A_n - A\|_L \to 0$ as $n \to \infty$; see (3.1.17). □

Corollary 3.1.48. *If X is a normed space, then X^* is a Banach space and $\|x^*\|_* = \sup\{|\langle x^*, x\rangle| : \|x\| \le 1\} = \sup\{\langle x^*, x\rangle : \|x\| \le 1\}$.*

Proposition 3.1.49. *If X is a normed space, $V \subseteq X$ is a vector subspace, and $u^* \in V^*$, then there exists $x^* \in X^*$ such that $x^*|_V = u^*$ and $\|x^*\|_* = \|u^*\|_{V^*}$.*

Proof. Applying Theorem 3.1.42 with $\rho(x) = \|u^*\|_{V^*}\|x\|$ for all $x \in X$ yields the assertion. □

Proposition 3.1.50. *If X is a normed space and $x_0 \in X$, then there exists $x_0^* \in X^*$ such that $\|x_0^*\|_* = \|x_0\|$ and $\langle x_0^*, x_0\rangle = \|x_0\|^2$.*

Proof. Applying Proposition 3.1.49 with $V = \mathbb{R}x_0$ and $x_0^*(tx_0) = \langle x_0^*, tx_0\rangle = t\|x_0\|^2$ gives the desired result $\|x_0^*\|_* = \|x_0\|$. □

Remark 3.1.51. The element $x_0^* \in X^*$ is not unique in general. In order to have uniqueness we need additional structure on X^*, for example, strict convexity; see Section 3.4. The multivalued map $\mathcal{F}: X \to 2^{X^*} \setminus \{\emptyset\}$ defined by $\mathcal{F}(x) = \{x^* \in X^* : \|x^*\|_* = \|x\| \text{ and } \langle x^*, x\rangle = \|x\|^2\}$ is called the **duality map** from X into X^*. It is important in Nonlinear Analysis and we will encounter it again in Section 6.1.

Proposition 3.1.52. *If X is a normed space and $x \in X$, then*

$$\|x\| = \sup[|\langle x^*, x\rangle| : x^* \in X^*, \|x^*\|_* \le 1] = \sup[\langle x^*, x\rangle : x^* \in X^*, \|x^*\|_* \le 1] .$$

Proof. We assume that $x \ne 0$. Note that

$$\sup[|\langle x^*, x\rangle| : x^* \in X^*, \|x^*\|_* \le 1] \le \|x\| . \tag{3.1.18}$$

On the other hand from Proposition 3.1.50 we know that there is $x_0^* \in X^*$ such that $\|x_0^*\|_* = \|x\|$ and $\langle x_0^*, x\rangle = \|x\|^2$. Let $\hat{x}_0^* = x_0^*/\|x\|$. Then $\|\hat{x}_0^*\|_* = 1$ and $\langle \hat{x}_0^*, x\rangle = \|x\|$. This combined with (3.1.18) implies the assertion of the proposition. □

Now we will produce some important geometric interpretations of the Hahn–Banach Extension Theorem; see Theorem 3.1.42. These are the well-known “Separation Theorems” for convex sets.

Definition 3.1.53. Let X be a vector space. A **hyperplane** is a set of the form $\{f = \vartheta\} = \{x \in X : f(x) = \vartheta\}$ with $f : X \to \mathbb{R}$ being linear and $\vartheta \in \mathbb{R}$. A hyperplane determines two **half-spaces**, namely $\{f \geq \vartheta\} = \{x \in X : f(x) \geq \vartheta\}$ and $\{f \leq \vartheta\} = \{x \in X : f(x) \leq \vartheta\}$. Given two sets $A, C \subseteq X$, we say that the hyperplane $H = \{f = \vartheta\}$ **separates** A **and** C, if $A \subseteq H_- = \{f \leq \vartheta\}$ and $C \subseteq H_+ = \{f \geq \vartheta\}$. We say that H **strongly separates** A and C if there exists $\varepsilon > 0$ such that

$$A \subseteq H_-^\varepsilon = \{f \leq \vartheta - \varepsilon\} \quad \text{and} \quad C \subseteq H_+^\varepsilon = \{f \geq \vartheta + \varepsilon\}\,.$$

Proposition 3.1.54. *If X is a topological vector space, then a hyperplane $H = \{f = \vartheta\}$ is either closed or dense. H is closed if and only if f is continuous while H is dense if and only if f is discontinuous.*

Proof. Due to the linearity of f, we may assume that $\vartheta = 0$. If f is continuous, then H is closed while if H is dense, then clearly f is not continuous. Now assume that H is closed. Suppose that $\{x_\alpha\}_{\alpha \in I} \subseteq X$ and $x_\alpha \to 0$. In addition, let $u \in X$ with $f(u) = 1$. Arguing by contradiction, suppose that $f(x_\alpha) \nrightarrow 0$; see Proposition 3.1.46. Then, for at least a subnet, we have $|f(x_\alpha)| \geq \varepsilon$ for all $\alpha \in I$. Let $v_\alpha = u - f(u)/(f(x_\alpha))x_\alpha$. Then $v_\alpha \in H$ since $\vartheta = 0$ and $v_\alpha \to u$. So, $u \in H$, a contradiction. Therefore $f(x_\alpha) \to 0$ and so f is continuous; see Proposition 3.1.46.

Now suppose that f is discontinuous. Then there exist a net $\{x_\alpha\}_{\alpha \in I} \subseteq X$ and $\varepsilon > 0$ such that

$$x_\alpha \to 0 \quad \text{and} \quad |f(x_\alpha)| \geq \varepsilon \quad \text{for all } \alpha \in I\,.$$

Given any $u \in X$, let $v_\alpha = u - f(u)/(f(x_\alpha))x_\alpha \in H$ for all $\alpha \in I$. We have $v_\alpha \to u$ and so we conclude that H is dense. ☐

Definition 3.1.55. Let X be a vector space and $A \subseteq X$. A point $x \in A$ is said to be an **absorbing point** of A, if $A - x \subseteq X$ is absorbing; see Definition 3.1.6 (b).

Remark 3.1.56. If X is a topological vector space and $\operatorname{int} A \neq \emptyset$, then every $x \in \operatorname{int} A$ is an absorbing point. However, the set A can have absorbing points even if $\operatorname{int} A = \emptyset$. Suppose that X is a normed space and $A = \partial B_1 \cup \{0\}$ where $\partial B_1 = \{x \in X : \|x\| = 1\}$. Then $x = 0$ is absorbing but $\operatorname{int} A = \emptyset$.

Next we present the “First Separation Theorem”.

Theorem 3.1.57 (First Separation Theorem). *If X is a vector space, $A, C \subseteq X$ are two nonempty convex sets, $A \cap C = \emptyset$ and one of them has an absorbing point, then they can be separated by a hyperplane $H = \{f = \vartheta\}$ with $f \neq 0$ and $A \cup C$ is not included in H.*

Proof. Suppose A has an absorbing point. Then $A - C$ has an absorbing point x. Since $A \cap C = \emptyset$, we see that $x \neq 0$. Moreover, the set $E = A - C - x$ is nonempty, convex, and absorbing, and $-x \notin E$ since $A \cap C = \emptyset$. Then Proposition 3.1.39 implies that ρ_E is sublinear.

Suppose that $\rho_E(-x) < 1$. Then there exist $0 \leq t < 1$ and $e \in E$ such that $x = te$. Note that $0 \in E$ being absorbing. So we have $-x = te + (1-t)0 \in E$, a contradiction. Therefore

$$\rho_E(-x) \geq 1 . \tag{3.1.19}$$

Let $V = \mathbb{R}(-x)$ and let $f\colon V \to \mathbb{R}$ be defined by $f(t(-x)) = t$. Clearly, f is linear and $f \leq \rho_E$ on V. Indeed, if $t \geq 0$, then $\rho_E(t(-x)) = t\rho_E(-x) \geq t$; see (3.1.19). If $t < 0$, then $f(t(-x)) < 0 \leq \rho_E(t(-x))$. Invoking Theorem 3.1.42 implies the existence of $\hat{f}\colon X \to \mathbb{R}$ being linear such that $\hat{f}|_V = f$ and $\hat{f} \leq \rho_E$. Note that $\hat{f}(x) = -1$ and so $\hat{f} \neq 0$.

We claim that $\hat{f}$ separates A and C. To see this, let $a \in A$ and $c \in C$. It holds

$$\begin{aligned}\hat{f}(a) &= \hat{f}(a-c-x) + \hat{f}(x) + \hat{f}(c) \leq \rho_E(a-c-x) + \hat{f}(x) + \hat{f}(c)\\ &= \rho_E(a-c-x) - 1 + \hat{f}(c) \leq 1 - 1 + \hat{f}(c) = \hat{f}(c) .\end{aligned}$$

Since $a \in A$ and $c \in C$ are arbitrary, we see that $\hat{f}$ separates A and C. Finally, since $0 \in E$, we have $x = a - c$ with $a \in A$ and $c \in C$. Recall that $\hat{f}(x) = -1$. Then $\hat{f}(a) \neq \hat{f}(c)$ and so we cannot have A and C to be subsets of the same hyperplane. □

Lemma 3.1.58. *If X is a topological vector space, $f\colon X \to \mathbb{R}$ is linear, and f is bounded above or bounded below on a neighborhood of the origin, then f is continuous.*

Proof. Let $U \in \mathcal{N}(0)$ be symmetric and assume that $f \leq M$ on U. Then, for given $\varepsilon > 0$, one gets, since U is symmetric, that $x - u \in \varepsilon/MU$ implies $|f(x) - f(u)| = |f(x-u)| \leq \varepsilon/MM = \varepsilon$. Hence, f is continuous. □

Using this lemma, we can state a topological version of Theorem 3.1.57.

Theorem 3.1.59. *If X is a topological vector space, $A, C \subseteq X$ are nonempty convex sets, $A \cap C = \emptyset$ and one of them has nonempty interior, then they can be separated by a closed hyperplane H and $A \cup C$ is not included in H.*

Proof. Applying Theorem 3.1.57, we obtain a separating hyperplane $H = \{f = \vartheta\}$ with $f \neq 0$. We only need to show that f is continuous. Suppose that $\operatorname{int} A \neq \emptyset$. Then $f(a) \leq \vartheta \leq f(c)$ for all $a \in A$ and for all $c \in C$. Note that if $x \in \operatorname{int} A$, then $U = \operatorname{int} A - x \in \mathcal{N}(0)$ and so $f|_U$ is bounded above, hence f is continuous; see Lemma 3.1.58. □

Next we present the "Second Separation Theorem" called "Strong Separation Theorem".

Theorem 3.1.60 (Strong Separation Theorem). *If X is a locally convex space and $A, C \subseteq X$ are nonempty, disjoint, convex sets, then A and B can be strongly separated by a closed hyperplane if and only if there exists $U \in \mathcal{N}(0)$ being convex such that $(A + U) \cap C = \emptyset$.*

Proof. $\Longrightarrow$: Let f be the linear functional associated with the closed separating hyperplane. Then f is continuous; see Proposition 3.1.54. Moreover, taking $\varepsilon > 0$ from the strong separation (see Definition 3.1.53), $U = \{x \in X : |f(x)| < \varepsilon\}$ is a convex neighborhood of the origin and $(A + U) \cap C = \emptyset$.

$\Longleftarrow$: The set $A + U$ is convex and open. So, we can apply Theorem 3.1.59 and find a linear, continuous functional $f: X \to \mathbb{R}$ and $\vartheta \in \mathbb{R}$ as well as $\varepsilon > 0$ such that $f(a) \le \vartheta - \varepsilon$ for all $a \in A$ and $f(c) \ge \vartheta + \varepsilon$ for all $c \in C$. Hence A and C are strongly separated by $H = \{f = \vartheta\}$. □

Corollary 3.1.61. *If X is locally convex, $A, C \subseteq X$ are nonempty, disjoint, convex sets and A is compact as well as C is closed, then A and C can be strongly separated by a closed hyperplane.*

Proof. The set $X \setminus C$ is open and $A \subseteq X \setminus C$. The compactness of A implies that there exists a convex neighborhood $U \in \mathcal{N}(0)$ such that $A + U \subseteq X \setminus C$. Hence $(A + U) \cap C = \emptyset$. Applying Theorem 3.1.60 gives the assertion. □

Proposition 3.1.62. *If X is a normed space, $V \subseteq X$ is a vector subspace, and $\overline{V} \neq X$, then there exists $x^* \in X^*$ with $x^* \neq 0$ such that $\langle x^*, v\rangle = 0$ for all $v \in V$.*

Proof. Let $u \in X \setminus \overline{V}$. Then apply Corollary 3.1.61 with $A = \{x_0\}$ and $C = \overline{V}$. Thus, we find $x^* \in X^*$ with $x^* \neq 0$ and $\vartheta \in \mathbb{R}$ such that $\langle x^*, x_0\rangle < \vartheta < \langle x^*, v\rangle$ for all $v \in \overline{V}$. But since $\overline{V}$ is a vector space, we see that $\langle x^*, v\rangle = 0$ for all $v \in \overline{V}$ since $\lambda\langle x^*, v\rangle > \vartheta$ for all $\lambda \in \mathbb{R}$, hence $\vartheta < 0$. □

Remark 3.1.63. This proposition is useful for determining whether a linear subspace V is dense in X. We must have that the only element of X^* vanishing on V is $x^* = 0$.

3.2 Three Fundamental Theorems

In this section we present three basic theorems that are the core results of linear functional analysis. These are the “Uniform Boundedness Principle”, the “Open Mapping Theorem”, and the “Closed Graph Theorem”. All three depend on the Baire Category Theorem; see Theorem 1.5.68. We recall that the Baire Category Theorem, roughly speaking, provides conditions for a set to be large in the sense that it has a nonempty interior.

We start with the “Uniform Boundedness Principle”. This theorem asserts that for any family of bounded linear operators, pointwise boundedness implies uniform boundedness, that is, boundedness in the operator norm. As before, we consider real vector spaces.

Theorem 3.2.1 (Uniform Boundedness Principle). *If X is a Banach space, Y is a normed space, and $\mathcal{L} \subseteq L(X, Y)$ satisfies*

$$\sup[\|A(x)\|_Y : A \in \mathcal{L}] = M(x) < \infty\,,$$

then there exists $M_0 > 0$ *such that* $\sup[\|A\|_L : A \in \mathcal{L}] \le M_0$.

Proof. For every $n \in \mathbb{N}$ let $E_n = \{x \in X : \|A(x)\|_Y \le n \text{ for all } A \in \mathcal{L}\}$. The hypothesis implies that

$$X = \bigcup_{n\ge1} E_n\,. \tag{3.2.1}$$

Moreover, we claim that for every $n \in \mathbb{N}$, $E_n \subseteq X$ is closed. To see this, let $\{x_m\}_{m\ge1} \subseteq E_n$ and assume that $x_m \to x$ in X. We obtain $\|A(x_m)\|_Y \le n$ for all $A \in \mathcal{L}$ and for all $m \in \mathbb{N}$. The continuity of A (see Proposition 3.1.46) implies that $\|A(x_m)\|_Y \to \|A(x)\|_Y$ as $m \to \infty$ for every $A \in \mathcal{L}$. Therefore, $\|A(x)\|_Y \le n$ for all $A \in \mathcal{L}$ and so $x \in E_n$, which implies that $E_n \subseteq X$ is closed for every $n \in \mathbb{N}$.

From (3.2.1) and the Baire Category Theorem (see Theorem 1.5.68 and Corollary 1.5.67), we infer that there exists $n_0 \in \mathbb{N}$ such that $\operatorname{int} E_{n_0} \ne \emptyset$. Hence, there exists $\varepsilon > 0$ such that

$$\overline{B}_\varepsilon(x_0) \subseteq E_{n_0} \quad \text{with } \overline{B}_\varepsilon(x_0) = \{x \in X : \|x - x_0\|_X \le \varepsilon\}\,. \tag{3.2.2}$$

Let $x \in X$ with $\|x\|_X \le \varepsilon$ and $A \in \mathcal{L}$. Then, due to (3.2.2),

$$\begin{aligned}\|A(x)\|_Y &= \|A(x + x_0) - A(x_0)\|_Y \le \|A(x + x_0)\|_Y + \|A(x_0)\|_Y \\ &\le n_0 + n_0 = 2n_0\,.\end{aligned} \tag{3.2.3}$$

Thus, for all $u \in X$ with $\|u\|_X = 1$, it follows, because of (3.2.3), that

$$\|A(u)\|_Y = \frac{1}{\varepsilon}\|A(\varepsilon u)\|_Y \le \frac{2n_0}{\varepsilon} \quad \text{for all } A \in \mathcal{L}\,.$$

Hence,

$$\sup[\|A(u)\|_Y : \|u\|_X \le 1] = \|A\|_L \le \frac{2n_0}{\varepsilon} \quad \text{for all } A \in \mathcal{L}\,.$$

□

Theorem 3.2.1 leads to the so-called "Banach–Steinhaus Theorem", which says that the pointwise limit of a sequence of bounded linear operators is a bounded linear operator.

Theorem 3.2.2 (Banach–Steinhaus Theorem). *If* X, Y *are Banach spaces and* $\{A_n\}_{n\ge1} \subseteq L(X, Y)$ *is a sequence such that*

$$A_n(x) \to A(x) \quad \text{in } Y \text{ as } n \to \infty \text{ for all } x \in X\,,$$

then the following hold:

(a) $A \in L(X, Y)$ *and* $\sup_{n\geq 1} \|A_n\|_L < \infty$;
(b) $\|A\|_L \leq \liminf_{n\to\infty} \|A_n\|_L$.

Proof. (a) Clearly, $A: X \to Y$ is linear. Since $\{A_n(x)\}_{n\geq 1} \subseteq Y$ is convergent, it holds that

$$\sup_{n\in\mathbb{N}} \|A_n(x)\|_Y = M(x) < \infty .$$

Applying Theorem 3.2.1, there exists $M_0 > 0$ such that $\sup_{n\in\mathbb{N}} \|A_n\|_L \leq M_0 < \infty$, which implies $\|A_n(x)\|_Y \leq M_0\|x\|_X$ for all $x \in X$ and for all $n \in \mathbb{N}$. Therefore, we derive $\|A(x)\|_Y = \lim_{n\to\infty} \|A_n(x)\|_Y \leq M_0\|x\|_X$ for all $x \in X$, which, due to Proposition 3.1.46, results in $A \in L(X, Y)$.

(b) It holds that $\|A_n(x)\|_Y \leq \|A_n\|_L\|x\|_X$ for all $x \in X$ and for all $n \in \mathbb{N}$. This gives $\|A(x)\|_Y \leq \liminf_{n\to\infty} \|A_n\|_L\|x\|_X$ for all $x \in X$ and so, $\|A\|_L \leq \liminf_{n\to\infty} \|A_n\|_L$. □

Example 3.2.3. (a) Theorems 3.2.1 and 3.2.2 fail if X is only a normed space. To see this, let us define the following subspaces:

$$l^\infty = \left\{\hat{x} = (x_n)_{n\geq 1} \in \mathbb{R}^{\mathbb{N}} : \sup_{n\geq 1} |x_n| < \infty\right\} ,$$
$$c_0 = \{\hat{x} = (x_n)_{n\geq 1} \in \mathbb{R}^{\mathbb{N}} : x_n \to 0 \text{ as } n \to \infty\} ,$$
$$X = \{\hat{x} = (x_n)_{n\geq 1} \in \mathbb{R}^{\mathbb{N}} : \text{ there exists } n_0 \in \mathbb{N} \text{ such that } x_n = 0 \text{ for } n \geq n_0\} .$$

Evidently, $X \subseteq c_0 \subseteq l^\infty$ and we furnish l^∞ with the supremum norm $\|\hat{x}\| = \sup_{n\in\mathbb{N}} |x_n|$. With this norm, l^∞ is a Banach space, c_0 is a closed subspace hence a Banach space itself, but $\overline{X}^{\|\cdot\|} = c_0$. Let $A_n: X \to X$ with $n \geq 1$ and $A: X \to X$ be defined by

$$A_n(\hat{x}) = (x, 2x_2, \ldots, nx_n, 0, 0, \ldots) , \quad A(\hat{x}) = (kx_k)_{k\geq 1} .$$

Then $A_n(\hat{x}) \to A(\hat{x})$ as $n \to \infty$ for all $\hat{x} \in X$ and $\|A_n\|_L = n$ for all $n \in \mathbb{N}$. Precisely, $\{A_n\}_{n\geq 1}$ is pointwise convergent, hence pointwise bounded as well, but $\sup_{n\geq 1} \|A_n\|_L = \infty$ and thus, A is not bounded.

(b) In Theorem 3.2.1 (b) the inequality can be strict. Let

$$l^2 = \left\{\hat{x} = (x_n)_{n\geq 1} \subseteq \mathbb{R}^{\mathbb{N}} : \sum_{n\geq 1} x_n^2 < \infty\right\}$$

furnished with the norm,

$$\|\hat{x}\| = \left(\sum_{n\geq 1} x_n^2\right)^{\frac{1}{2}} .$$

With this norm, l^2 becomes a Banach space. In fact it becomes a Hilbert space; see Section 3.5. Let $X = l^2$, $Y = \mathbb{R}$, and consider the bounded linear operators $A_k: l^2 \to \mathbb{R}$

with $k \geq 1$ defined by $A_k(\hat{x}) = x_k$ for every $\hat{x} = (x_n)_{n\geq 1} \in l^2$ and for every $k \in \mathbb{N}$. Evidently, $A_k(\hat{x}) \to 0$ as $k \to \infty$ for every $\hat{x} \in l^2$ but $\|A_k\|_L = 1$ for all $n \in \mathbb{N}$.

Theorem 3.2.1 leads to interesting characterizations of bounded sets in a Banach space X and in its dual X^*; see Definition 3.1.45. In the next section we will interpret these results in terms of weak and weak* topologies, respectively.

Proposition 3.2.4. *If X is a normed space and $B \subseteq X$ is nonempty, then B is bounded if and only if $x^*(B) = \{\langle x^*, u\rangle : u \in B\} \subseteq \mathbb{R}$ is bounded for every $x^* \in X^*$.*

Proof. $\Longrightarrow$: This follows from the fact that $|\langle x^*, u\rangle| \leq \|x^*\|_* \|u\|$ for every $x^* \in X^*$ and for all $u \in B$. So, if B is bounded, then $\|u\| \leq M$ for some $M > 0$ and for all $u \in B$. Therefore, $x^*(B) \subseteq [-\varrho, \varrho]$ with $\varrho = \|x^*\|_* M$.

$\Longleftarrow$: For every $u \in B$, let $A_u(x^*) = \langle x^*, u\rangle$ for all $x^* \in X^*$ where $\langle \cdot, \cdot \rangle$ denotes the duality brackets for the pair (X^*, X). Then $A_u \in L(X^*, \mathbb{R})$ for all $u \in B$ and by hypothesis,

$$\sup_{u\in B} |A_u(x^*)| = \sup_{u\in B} |\langle x^*, u\rangle| < +\infty .$$

Since X^* is a Banach space (see Corollary 3.1.48), we can apply Theorem 3.2.1 and find $M > 0$ such that

$$|A_u(x^*)| = |\langle x^*, u\rangle| \leq M\|x^*\|_* \quad \text{for all } x^* \in X^* \text{ and for all } u \in B .$$

Because of Proposition 3.1.52 we infer that $\|u\| \leq M$, which shows that B is bounded. □

There is also a "dual" version of this result.

Proposition 3.2.5. *If X is a Banach space and $B^* \subseteq X^*$ is nonempty, then B^* is bounded if and only if $x(B^*) = \{\langle u^*, x\rangle : u^* \in B^*\} \subseteq \mathbb{R}$ is bounded for every $x \in X$.*

Proof. $\Longrightarrow$: This is as in the previous proof.

$\Longleftarrow$: For every $u^* \in B^*$, let $A_{u^*}(x) = \langle u^*, x\rangle$ for all $x \in X$. Then $A_{u^*} \in L(X, \mathbb{R})$ for all $u^* \in B^*$ and by hypothesis,

$$\sup_{u^*\in B^*} |A_{u^*}(x)| = \sup_{u^*\in B^*} |\langle u^*, x\rangle| < \infty .$$

Since X is a Banach space, we can apply Theorem 3.2.1 and find $M > 0$ such that

$$|A_{u^*}(x)| = |\langle u^*, x\rangle| \leq M\|x\| \quad \text{for all } x \in X \text{ and for all } u^* \in B^* .$$

Then, Corollary 3.1.48 implies that $\|u^*\|_* \leq M$ for all $u^* \in B^*$. □

Next we will prove the "Open Mapping Theorem", which asserts that a surjective bounded linear operator between Banach spaces is an open map.

In order to prove this theorem, we will need two auxiliary results.

Lemma 3.2.6. *If X, Y are Banach spaces and $A \in L(X,Y)$ surjective, then there exists $\vartheta > 0$ such that for any $\varepsilon > 0$ and $y \in Y$ we find $x \in X$ such that*

$$\|A(x) - y\|_Y \leq \varepsilon \quad \text{and} \quad \|x\|_X \leq \frac{1}{\vartheta}\|y\|_Y .$$

Proof. Let $B_1^X = \{x \in X : \|x\|_X < 1\}$. The surjectivity of A implies that

$$Y = \bigcup_{n\geq 1} A(nB_1^X) .$$

Then by the Baire Category Theorem there is $n \in \mathbb{N}$ such that $\operatorname{int} \overline{A(nB_1^X)} \neq \emptyset$. This implies $B_\eta(y_0) \subseteq \overline{A(nB_1^X)}$ for some $\eta > 0$ and $y_0 \in Y$. Here $B_\eta(y_0) = \{y \in Y : \|y - y_0\|_Y < \eta\}$. Given $y \in Y$ with $\|y\|_Y < \eta$, let $\{x_k\}_{k\geq 1}, \{u_k\}_{k\geq 1} \subseteq nB_1^X$ such that

$$A(x_k) \to y_0 \quad \text{and} \quad A(u_k) \to y_0 + y \quad \text{in } Y \text{ as } k \to \infty .$$

Let $v_k = u_k - x_k$ for $k \in \mathbb{N}$. Then

$$A(v_k) \to y \quad \text{in } Y \text{ as } k \to \infty \quad \text{and} \quad \|v_k\|_X < 2n \quad \text{for all } k \in \mathbb{N} . \tag{3.2.4}$$

Let $w \in Y \setminus \{0\}$ and let $z = (\eta/2) \cdot (w/\|w\|)$. Then $z \in Y$ and $\|z\|_Y < \eta$. From (3.2.4) we know that there exist $\{\tilde{v}_k\}_{k\geq 1} \subseteq X$ such that

$$A(\tilde{v}_k) \to z = \frac{\eta}{2}\frac{w}{\|w\|_X} \quad \text{in } Y \text{ as } k \to \infty \quad \text{and} \quad \|\tilde{v}_k\|_X < 2n \quad \text{for all } k \in \mathbb{N} .$$

Hence,

$$A\left(\frac{2}{\eta}\|w\|_X \tilde{v}_k\right) \to w \quad \text{in } Y \text{ as } k \to \infty . \tag{3.2.5}$$

Note that

$$\frac{2}{\eta}\|w\|_X\|\tilde{v}_k\|_X < \frac{4n}{\eta}\|w\|_X \quad \text{for all } k \in \mathbb{N} .$$

Finally let $\vartheta = \eta/(4n)$ and apply (3.2.5) to obtain the result of the lemma. □

Using this lemma, we can prove the following proposition.

Proposition 3.2.7. *If X, Y are Banach spaces, $B_1^X = \{x \in X : \|x\|_X < 1\}$, $B_1^Y = \{y \in Y : \|y\|_Y < 1\}$, and $A \in L(X,Y)$ is surjective, then there exists $\delta > 0$ such that $\delta B_1^Y \subseteq A(B_1^X)$.*

Proof. Let $\vartheta > 0$ be as postulated by Lemma 3.2.6. Let $y \in \vartheta B_1^Y$ and $\varepsilon = 1/2\vartheta > 0$. Using Lemma 3.2.6, there exists $x_1 \in X$ such that

$$\|A(x_1) - y\|_Y \leq \frac{\vartheta}{2} \quad \text{and} \quad \|x_1\|_X \leq \frac{1}{\vartheta}\|y\|_Y < 1 . \tag{3.2.6}$$

Now consider $y - A(x_1) \in Y$ and $\varepsilon = \vartheta/4$. A new application of Lemma 3.2.6 gives $x_2 \in X$ such that

$$\|A(x_2) - (y - A(x_1))\|_Y \le \frac{\vartheta}{4} \quad \text{and} \quad \|x_2\|_X \le \frac{1}{\vartheta}\|y - A(x_1)\|_Y < \frac{1}{2},$$

see (3.2.6). Suppose that we have produced $\{x_k\}_{k\ge1}^n \subseteq X$ such that

$$\left\| A\left(\sum_{k=1}^{n} x_k\right) - y \right\|_Y \le \frac{\vartheta}{2^n} \quad \text{and} \quad \|x_k\|_X \le \frac{1}{2^{k-1}} \quad \text{for all } k = 1, \dots, n. \tag{3.2.7}$$

Using Lemma 3.2.6, we obtain $x_{n+1} \in X$ such that

$$\left\| A\left(\sum_{k=1}^{n+1} x_k\right) - y \right\|_Y \le \frac{\vartheta}{2^{n+1}} \quad \text{and} \quad \|x_{n+1}\|_X \le \frac{1}{\vartheta}\left\| A\left(\sum_{k=1}^{n} x_k\right) - y \right\|_Y \le \frac{1}{2^n}; \tag{3.2.8}$$

see (3.2.7). By induction we have a sequence $\{x_n\}_{n\ge1} \subseteq X$ such that (3.2.8) holds.

Let $u_n = \sum_{k=1}^n x_k \in X$ with $n \in \mathbb{N}$. For $m > n$ one gets

$$\|u_m - u_n\|_X = \left\| \sum_{k=n+1}^{m} x_k \right\|_X \le \sum_{k=n+1}^{m} \frac{1}{2^k};$$

see (3.2.8). This implies that $\{u_n\}_{n\ge1} \subseteq X$ is a Cauchy sequence. Since X is a Banach space, we obtain $u_n \to u$ in X. Then

$$\|u\|_X \le \sum_{k\ge1} \|x_k\|_X \le \sum_{k\ge1} \frac{1}{2^{k-1}} = 2,$$

which shows that $u \in 2B_1^X$. From (3.2.8) it follows $\|A(u_n) - y\|_Y \le \vartheta/2^n$, hence $A(u_n) \to y$ in Y. But we also have $A(u_n) \to A(u)$ in Y. Therefore, $y = A(u)$. Recall that $y \in \vartheta B_1^Y$ is arbitrary and $x \in 2B_1^X$. That means $\vartheta/2B_1^Y \subseteq A(B_1^X)$. Choosing $\delta = \vartheta/2 > 0$, we obtain the assertion of the proposition. □

Remark 3.2.8. This proposition provides estimates for the solutions $x \in X$ of $A(x) = y \in Y$ in terms of y. That the equation $A(x) = y$ always has a solution for all $y \in Y$ is a consequence of the surjectivity of A.

Once we have this proposition, we can easily prove the "Open Mapping Theorem".

Theorem 3.2.9 (Open Mapping Theorem). *If X, Y are Banach spaces and $A \in L(X, Y)$ is surjective, then A is an open map, that is, it maps open sets in X to open sets in Y.*

Proof. Let $U \subseteq X$ be nonempty and open, and let $x_0 \in U$. Let $V = U - x_0 \in \mathcal{N}(0)$. Then there exists $\xi > 0$ such that $\xi B_1^X \subseteq V$. Using Proposition 3.2.7 we find $\delta > 0$ such that

$$A(V) \supseteq A(\xi B_1^X) = \xi A(B_1^X) \supseteq \xi\delta B_1^Y,$$

which implies

$$A(U) = A(V + x_0) = A(x_0) + A(V) \supseteq A(x_0) + \xi\delta B_1^Y .$$

The last set is open in Y centered at $A(x_0)$ with a radius of $\xi\delta > 0$. This means that $A(U) \subseteq Y$ is open. □

As an easy consequence of the Open Mapping Theorem we obtain the so-called "Banach Theorem".

Theorem 3.2.10 (Banach Theorem). *If X, Y are Banach spaces and $A \in L(X, Y)$ is a bijection, that is, A is surjective and injective, then $A^{-1} \in L(Y, X)$.*

Proof. First note that $A^{-1}: Y \to X$ is a well-defined linear map. Let $U \subseteq X$ be open. Due to Theorem 3.2.9 it follows that $(A^{-1})^{-1}(U) = A(U) \subseteq Y$ is open. Then Proposition 3.1.46 implies that $A^{-1} \in L(Y, X)$. □

Definition 3.2.11. Let X be a vector space and let $\|\cdot\|$, $|\cdot|$ be two norms on X. We say that the two norms are **equivalent** if there exists a constant $\vartheta \geq 1$ such that

$$\frac{1}{\vartheta}\|x\| \leq |x| \leq \vartheta\|x\| \quad \text{for all } x \in X .$$

Remark 3.2.12. This notion defines an equivalence relation on the set of all possible norms on X. The norms $\|\cdot\|$, $|\cdot|$ on X are equivalent if and only if $\mathrm{id}: (X, \|\cdot\|) \to (X, |\cdot|)$ and $\mathrm{id}: (X, |\cdot|) \to (X, \|\cdot\|)$ are both bounded linear operators. Two norms are equivalent if and only if they generate the same metric topology on X. Finally, if $\|\cdot\|$, $|\cdot|$ are equivalent norms, then $(X, \|\cdot\|)$ is a Banach space if and only if $(X, |\cdot|)$ is a Banach space.

Proposition 3.2.13. *If V is a vector space, $\|\cdot\|$ and $|\cdot|$ are two norms on V with V being a Banach space for both norms and there exists $\eta > 0$ such that*

$$|x| \leq \eta\|x\| \quad \textit{for all } x \in V ,$$

then $\|\cdot\|$ and $|\cdot|$ are equivalent norms on V.

Proof. Let $X = (V, \|\cdot\|)$, $Y = (V, |\cdot|)$, and $A = \mathrm{id}: X \to Y$ with $\mathrm{id}(x) = x$ for all $x \in X$. Then $A \in L(X, Y)$ is bijective and we can apply Theorem 3.2.10 and infer that $A^{-1} = \mathrm{id}: Y = (V, |\cdot|) \to X = (V, \|\cdot\|)$ is continuous. So, it follows that the norms $\|\cdot\|$ and $|\cdot|$ are equivalent; see Remark 3.2.12. □

Recall that a continuous map $f: X \to Y$ has a closed graph $\mathrm{Gr} f = \{(x, y) \in X \times Y : y = f(x)\}$. The converse is not true in general. To see this, let $X = Y = \mathbb{R}_+$ and consider the function $f: \mathbb{R}_+ \to \mathbb{R}_+$ defined by

$$f(x) = \begin{cases} 0 & \text{if } x = 0 , \\ \frac{1}{x} & \text{if } x > 0 . \end{cases}$$

Then $\operatorname{Gr} f$ is closed but f is not continuous at $x = 0$. For linear operators between Banach spaces, the situation changes and we have the third basic theorem of linear functional analysis, which is called the "Closed Graph Theorem".

Theorem 3.2.14 (Closed Graph Theorem). *If X, Y are Banach spaces and $A: X \to Y$ is a linear operator, then $A \in L(X, Y)$ if and only if $\operatorname{Gr} A = \{(x, y) \in X \times Y : y = A(x)\} \subseteq X \times Y$ is closed.*

Proof. $\Longleftarrow$: The graph of any continuous map (linear or not) is closed.

$\Longrightarrow$: On X we consider the following norms

$$\|x\| = \|x\|_X \quad \text{and} \quad |x| = \|x\|_X + \|A(x)\|_Y \quad \text{for all } x \in X\,.$$

Note that $|\cdot|$ is called the **graph norm**. Since $\operatorname{Gr} A \subseteq X \times Y$ is closed, $(X, |\cdot|)$ is a Banach space. Moreover, the inequality $\|x\| \le |x|$ for all $x \in X$ is clearly satisfied. Invoking Proposition 3.2.13, we conclude that $\|\cdot\|$ and $|\cdot|$ are equivalent norms. Thus, there exists $M > 0$ such that $|x| \le M\|x\|$ for all $x \in X$, which implies $\|A(x)\|_Y \le M\|x\|_X$ for all $x \in X$. Then Proposition 3.1.46 finally gives $A \in L(X, Y)$. ☐

We can apply these results to quotient spaces (see Section 1.3), which in turn will lead us to complemented spaces.

So, let X be a normed vector space and $V \subseteq X$ a closed subspace. We define the equivalence relation $\sim$ on X by

$$x \sim u \quad \text{if and only if} \quad x - u \in V\,. \tag{3.2.9}$$

Let $[x]$ denote the equivalence class corresponding to $x \in X$. Then $[x] = x + V = \{x + v : v \in V\}$ and let X/V be the quotient space, that is, the set of all equivalence classes under $\sim$ defined by (3.2.9). So, the whole subspace V is collapsed in the quotient space X/V and identified with the zero vector. The quotient space X/V becomes a vector space under the following operations:

$$\begin{aligned} &\text{vector addition:} && [x_1] + [x_2] = x_1 + V + x_2 + V = x_1 + x_2 + V\,, \\ &\text{scalar multiplication:} && \lambda(x + V) = \lambda x + V\,, \end{aligned}$$

for all $x_1, x_2, x \in X$ and for all $\lambda \in \mathbb{R}$. As we already mentioned, the zero vector in X/V is $0 + V = V$. We can define a norm on X/V by setting

$$\|[x]\| = \inf[\|x + v\| : v \in V]\,.$$

It is easy to check that this is a norm on X/V. Note that

$$\|[x]\| = \inf[\|x + v\| : v \in V] = \inf[\|x - v\| : v \in V] \quad \text{for all } x \in X\,. \tag{3.2.10}$$

Proposition 3.2.15. *If X is a normed space and $V \subseteq X$ is a closed subspace, then the following hold:*

(a) $\|x\| \geq \|[x]\|$ *for all* $x \in X$;
(b) *if* $x \in X$ *and* $\varepsilon > 0$, *then there exists* $u \in X$ *with* $u \sim x$, *that is* $[x] = [u]$, *such that* $\|u\| \leq \|[x]\| + \varepsilon$.

Proof. (a) This is an immediate consequence from (3.2.10).

(b) Let $v \in V$ be such that $\|x-v\| \leq d(x,M)+\varepsilon = \|[x]\|+\varepsilon$; see (3.2.10). Set $u = x-v \in [x]$. Then $\|u\| \leq \|[x]\| + \varepsilon$. □

Remark 3.2.16. Suppose that $x, y \in X$ be such that $\|[x-y]\| < \vartheta$ for some $\vartheta > 0$. Then according to Proposition 3.2.15 (b), there exists $y' \in X$ such that $[x-y] = [x-y']$ and $\|x-y'\| < \vartheta$.

Proposition 3.2.17. *If X is a Banach space and $V \subseteq X$ is a closed subspace, then X/V is a Banach space as well.*

Proof. Suppose that $\{\|[x_n]\|\}_{n\geq 1} \subseteq X/V$ is a Cauchy sequence. By passing to a subsequence if necessary we may assume that

$$\|[x_n - x_{n+1}]\| < \frac{1}{2^n} \quad \text{for all } n \in \mathbb{N}\,.$$

According to Remark 3.2.16 we can find $x_2' \in X$ such that $[x_1 - x_2] = [x_1 - x_2']$ and $\|x_1 - x_2'\| < 1/2$. Then $[x_2] = [x_2']$ and so we may assume that $x_2' = x_2$. Now again by Remark 3.2.16, there exists $x_3' \in X$ such that $[x_2 - x_3] = [x_2 - x_3']$ and $\|x_2 - x_3'\| < 1/2^2$. As for x_2', we may assume that $x_3' = x_3$. Inductively we obtain that $\|x_n - x_{n+1}\| < 1/2^n$ for all $n \in \mathbb{N}$. So, $\{x_n\}_{n\geq 1} \subseteq X$ is a Cauchy sequence and we may say that $x_n \to x \in X$. Then, Proposition 3.2.15 (a) gives

$$\|[x_n] - [x]\| = \|[x_n - x]\| \leq \|x_n - x\|\,.$$

Hence, $[x_n] \to [x]$ and so X/V is a Banach space. □

Remark 3.2.18. In fact there is a kind of converse to the result above. Namely, if X is a normed space, $V \subseteq X$ is a closed subspace, and both V and X/V are complete, then X is a Banach space; see Problem 3.10.

Definition 3.2.19. Let X be a normed space and let $V \subseteq X$ be a closed subspace. The map $p: X \to X/V$ defined by $p(x) = [x]$ is called the **quotient map**.

Proposition 3.2.20. *If X is a normed space and $V \subseteq X$ is a closed subspace, then the quotient map $p \in L(X, X/V)$ is surjective and open, and $N(p) = V$, and if $V \neq X$, then $\|p\|_L = 1$.*

Proof. We only need to show that p is open. Let $U \subseteq X$ be open, $x \in U$, and let $B_1^X = \{u \in X : \|u\| < 1\}$. Then we find $\vartheta > 0$ such that $x + \vartheta B_1^X \subseteq U$, hence $p(x) + \vartheta p(B_1^X) \subseteq p(U)$. We claim that $p(B_1^X) = B_1^{X/V} = \{[x] \in X/V : \|[x]\| < 1\}$. To see this, let $x \in B_1^X$. Then $\|p(x)\| =$

$\|[x]\| \leq \|x\| < 1$; see Proposition 3.2.15 (a). Therefore, $p(B_1^X) \subseteq B_1^{X/V}$. On the other hand if $[u] \in B_1^{X/V}$, then there is $u' \in B_1^X$ such that $p(u') = [u'] = [u]$ (see Proposition 3.2.15 (b)), and so $B_1^{X/V} \subseteq p(B_1^{X/V})$. Thus finally $p(B_1^X) = B_1^{X/V}$ and so $p(x) + \vartheta B_1^{X/V} \subseteq p(U)$. Hence, p is open. □

Proposition 3.2.21. *If X, Z are normed spaces, $V \subseteq X$ is a closed subspace and $A \in L(X,Z)$ satisfies $N(A) = \{x \in X : A(x) = 0\} \supseteq V$, then there exists a unique $\hat{A} \in L(X/V, Z)$ such that $A = \hat{A} \circ p$.*

Proof. The operator $\hat{A}: X/V \to Z$ defined by $\hat{A}([x]) = A(x)$ is well-defined since $V \subseteq N(A)$. Clearly $\hat{A}$ is linear and

$$\|\hat{A}([x])\|_Z = \|A(x+v)\|_Z \leq \|A\|_L \|x+v\|_X \quad \text{for all } v \in V\,,$$

since $V \subseteq N(A)$. Hence,

$$\|\hat{A}([x])\|_Z \leq \|A\|_L \inf[\|x+v\|_X : v \in V] = \|A\|_L \|[x]\|\,.$$

This shows that $\hat{A} \in L(X/V, Z)$ and $A = \hat{A} \circ p$. Clearly $\hat{A}$ is unique. □

Remark 3.2.22. This is a factorization theorem and it can be better remembered if we use the following figure:

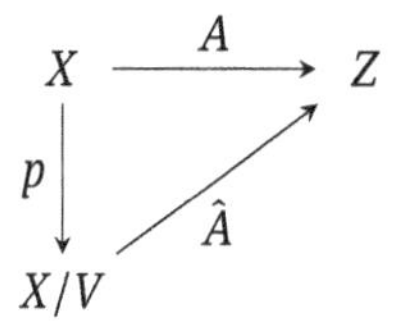

Proposition 3.2.23. *If X, Z are Banach spaces, $A \in L(X,Z)$ is surjective and $V = N(A) = \{x \in X : A(x) = 0\}$, then X/V and Z are isomorphic, that is, there exists $\mathfrak{L}: X/V \to Z$ being a linear, continuous bijection with a continuous inverse.*

Proof. From Proposition 3.2.21 we know that there exists a unique $\hat{A} \in L(X/V, Z)$ such that $A = \hat{A} \circ p$. If $\hat{A}([x]) = \hat{A}([u])$, then $A(x) = A(u)$ and so $x - u \in N(A)$, which means that $\hat{A}$ is one-to-one. Let $z \in Z$ and recall that A is surjective. Then we can find $x \in X$ such that $A(x) = z$. Thus, $\hat{A}([x]) = z$, which implies that $\hat{A}$ is surjective, that is, a bijection. Invoking Theorem 3.2.10, we conclude that $\hat{A}$ is an isomorphism. □

Definition 3.2.24. Let X be a normed space and let $D \subseteq X$. The **annihilator** of D is defined by

$$D^\perp = \{x^* \in X^* : \langle x^*, d \rangle = 0 \text{ for all } d \in D\}\,.$$

Evidently, $D^\perp$ is a closed vector subspace of X^*.

Using this notion we can characterize the dual of a quotient space.

Proposition 3.2.25. *If X is a normed space and $V \subseteq X$ is a closed subspace, then $(X/V)^*$ and $V^\perp$ are isometrically isomorphic.*

Proof. Let $l \in (X/V)^*$ and let $x^* = l \circ p: X \to \mathbb{R}$. Then $x^* \in X^*$ and $x^*|_V = 0$. So, $x^* \in V^\perp$. Conversely, let $x^* \in V^\perp$. Then according to Proposition 3.2.21, there exists a unique $l \in (X/V)^*$ such that $x^* = l \circ p$. So, the linear map $\xi: (X/V)^* \to V^\perp$ defined by $\xi(l) = l \circ p$ is a bijection and

$$l([x]) = \langle \xi(l), x \rangle = \langle \xi(l), x + v \rangle \le \|\xi(l)\|_* \|x + v\| \quad \text{for all } v \in V\,.$$

Thus

$$\|l\|_{(X/V)^*} \le \|\xi(l)\|_*\,. \tag{3.2.11}$$

On the other hand, thanks to Proposition 3.2.15 (a), one gets

$$\langle \xi(l), x \rangle = l([x]) \le \|l\|_{(X/V)^*} \|[x]\| \le \|l\|_{(X/V)^*} \|x\|\,.$$

This gives

$$\|\xi(l)\|_* \le \|l\|_{(X/V)^*}\,. \tag{3.2.12}$$

From (3.2.11) and (3.2.12) we infer that $\|\xi(l)\|_* = \|l\|_{(X/V)^*}$ and so ξ is an isometric isomorphism. □

We present some additional properties of closed subspaces in Banach spaces.

Proposition 3.2.26. *If X is a Banach space and $V, W \subseteq X$ are closed subspaces of X such that $V + W$ is closed, then there exists $\hat{c} > 0$ such that every $u \in V + W$ admits a decomposition $u = v + w$ with $v \in V$ and $w \in W$ as well as*

$$\|v\| \le \hat{c}\|u\| \quad \textit{and} \quad \|w\| \le \hat{c}\|u\|\,.$$

Proof. We consider the Cartesian product $V \times W$ furnished with the norm $\|(v, w)\| = \|v\| + \|w\|$. Moreover, we consider on $V + W$ the norm inherited from X. Let $A: V \times W \to V + W$ be defined by $A((v, w)) = v + w$. Evidently, $A \in L(V \times W, V + W)$ and is surjective. Since $V \times W$ and $V + W$ are Banach spaces, invoking the Open Mapping Theorem (see Theorem 3.2.9), there exists $c > 0$ such that $u \in V + W$ with $\|u\| < c$ implies $u = v + w$ with $v \in V$, $w \in W$ and $\|v\| + \|w\| < 1$. By the homogeneity, there holds for every $u \in V + W$ that $u = v + w$ with $v \in V$, $w \in W$ and $\|v\| + \|w\| \le 1/c\|u\|$. Then for $\hat{c} = c^{-1}$ we have the result. □

Definition 3.2.27. Let X be a normed space. A closed subspace $V \subseteq X$ is called **complemented** (or we say that it admits a **topological complement**), if there exists a closed subspace $W \subseteq X$ such that $V \cap W = \{0\}$ and $X = V + W$ (we write $X = V \oplus W$). Then we say that V and W are **complementary** subspaces of X.

The next results shows that finite dimensional subspaces or subspaces with finite codimension, are complemented.

Proposition 3.2.28. *If X is a normed space and $V \subseteq X$ is a closed subspace such that $\dim V < \infty$ or $\dim(X/V) < \infty$, then V is complemented.*

Proof. Let $n = \dim V < \infty$ and let $\{e_k\}_{k=1}^n$ be a basis of V. According to Proposition 3.1.49, there exists $\{e_m^*\}_{m=1}^n \subseteq X^*$ such that

$$\langle e_m^*, e_k\rangle = \delta_{mk} = \begin{cases} 1 & \text{if } m = k\,, \\ 0 & \text{if } m \neq k\,. \end{cases}$$

Let $W = \{x \in X\colon \langle e_m^*, x\rangle = 0 \text{ for all } m \in \{1, \dots, n\}\}$. Clearly $W \subseteq X$ is a closed subspace and $X = V \oplus W$ since $x - \sum_{m=1}^n \langle e_m^*, x\rangle e_m \in W$ for all $x \in X$.

Next let $n = \dim(X/V) < \infty$. We choose $\{x_k\}_{k=1}^n \subseteq X$ such that $\{[x_k]\}_{k=1}^n$ is a basis of X/V. Then $W = \operatorname{span}\{x_k\}_{k=1}^n \subseteq X$ is closed (see Corollary 3.1.19) and satisfies $X = V \oplus W$. □

Remark 3.2.29. It is not true that every closed subspace of an infinite dimensional Banach space is complemented. For example, $c_0 \subseteq l^\infty$ is a closed subspace, but it is not complemented; see Phillips [262]. In fact a result due to Lindenstrauss–Tzafriri [217] says that every Banach space that is not a Hilbert space admits a closed subspace that is not complemented.

3.3 Weak and Weak* Topologies

In this section we study the weak topology on a normed space X and the weak* topology on X^*, which is always a Banach space; see Corollary 3.1.48. These are locally convex topologies and are special cases of the weak topologies introduced in Definition 1.3.1 when $Y_i = \mathbb{R}$ for all $i \in I$ and $\{f_i\}_{i\in I} = X^*$ (for the weak topology) as well as $\{f_i\}_{i\in I} = X$ (for the weak* topology).

The strong (norm) topology on an infinite dimensional normed space is too strong for many purposes. In particular, note that a strongly compact set in an infinite dimensional normed space has an empty interior. Indeed, if this is not the case, then the space is locally compact, hence by Proposition 3.1.24, it is finite dimensional, a contradiction. The main result of this section is “Alaoglu’s Theorem” (see Theorem 3.3.38), which says that the unit ball in the dual space X^* is compact for the relative weak* topology. This result is reminiscent of the classical Heine–Borel Theorem; see Theorem 1.5.38.

Definition 3.3.1. Let X be a normed space. The **weak topology** on X is the weakest topology on X with respect to which every element $x^* \in X^*$ ($x^*: X \to \mathbb{R}$ being norm continuous and linear) is continuous. We denote the weak topology by $\mathrm{w}(X, X^*)$ or simply by w.

Remark 3.3.2. As we already mentioned, the w-topology is a particular case of the weak (initial) topology introduced in Definition 1.3.1 when the initial space is X (the normed space), $Y_i = \mathbb{R}$ for all $i \in I$, $I = X^*$ and $f_{x^*}: X \to \mathbb{R}$ with $x^* \in X^* = I$ is the linear functional $f_{x^*}(x) = \langle x^*, x\rangle$. Recall that $\langle\cdot,\cdot\rangle$ denotes the duality brackets for the pair (X^*, X). Evidently the weak topology w is weaker than the norm (metric) topology on X.

Proposition 3.3.3. *The weak topology* $\mathrm{w}(X, X^*)$ *is Hausdorff.*

Proof. From Corollary 3.1.61, we know that $\{f_{x^*}\}_{x^* \in X^* = I}$ is separating and so Proposition 1.3.7 implies that $\mathrm{w}(X, X^*)$ is Hausdorff. □

The weak topology on X is clearly linear, that is, both operations, vector addition and scalar multiplication, are continuous. Moreover, it is locally convex; see Theorem 3.1.40. Note that $\mathbb{R}$ is regular and recall that regularity is hereditary and topological (see Proposition 1.2.10), and it is preserved in Cartesian products; see Proposition 1.3.13. Therefore, we can improve Proposition 3.3.3 in the following way.

Proposition 3.3.4. *The weak topology* $\mathrm{w}(X, X^*)$ *is regular; see Definition* 1.2.7.

Remark 3.3.5. In fact for the same reasons, $\mathrm{w}(X, X^*)$ is completely regular; see Definition 1.2.19.

The linearity of the weak topology implies that in order to describe it we only need to specify a local basis at the origin. Then by translation we obtain a local basis at any other point. Remark 1.3.2 allows us to give a precise description of the local basis at the origin.

Proposition 3.3.6. *A typical basic weak neighborhood of the origin is given by*

$$U(0; x_1^*, \dots, x_n^*, \varepsilon) = \{x \in X : |\langle x_k^*, x\rangle| < \varepsilon \textit{ for all } k = 1, \dots, n\}$$

with $\{x_k^*\}_{k=1}^n \subseteq X^*$, $n \in \mathbb{N}$ *and* $\varepsilon > 0$. *As* $\varepsilon > 0$, $n \in \mathbb{N}$ *and* $\{x_k^*\}_{k=1}^n$ *vary, we cover a local basis for the weak topology at the origin. At any other point* $x_0 \in X$ *the local basis consists of sets of the form*

$$x_0 + U(0; x_1^*, \dots, x_n^*, \varepsilon) = \{x \in X : |\langle x_k^*, x - x_0\rangle| < \varepsilon \textit{ for all } k = 1, \dots, n\} .$$

In infinite dimensional normed spaces the weak topology and the strong (norm) topology never coincide. To see this we will need to recall some simple facts from linear algebra. The first is an algebraic variant of the factorization result stated in Proposition 3.2.21.

Lemma 3.3.7. *If X, Y, Z are vector spaces, $f: X \to Z$ and $g: X \to Y$ are linear maps and $N(g) \subseteq N(f)$, where $N(g) = \{x \in X: g(x) = 0\}$, $N(f) = \{x \in X: f(x) = 0\}$, then there exists a linear map $\xi: Y \to Z$ such that $f = \xi \circ g$.*

Proof. Let $\xi: g(X) \to Z$ be defined by $\xi(g(x)) = f(x)$ for all $x \in X$. This linear map is well-defined since if $g(x_1) = g(x_2)$, then $x_1 - x_2 \in N(g) \subseteq N(f)$ and so $f(x_1) = f(x_2)$. Extending ξ to a linear map on all of Y gives $f = \xi \circ g$. □

Using this lemma, we can prove the second auxiliary result from linear algebra.

Lemma 3.3.8. *If X is a vector space, $f, f_1, \ldots, f_n: X \to \mathbb{R}$ are linear maps and $\bigcap_{k=1}^n N(f_k) \subseteq N(f)$, then f is a linear combination of the f_k's.*

Proof. Let $X = X$, $Y = \mathbb{R}^n$, $Z = \mathbb{R}$, $f = f$ and $g = (f_k)_{k=1}^n$ and apply Lemma 3.3.7 to produce a linear functional $\xi: \mathbb{R}^n \to \mathbb{R}$ such that $f = \xi \circ g$. Then $\xi(\hat{y}) = \sum_{k=1}^n \lambda_k y_k$ with $\lambda_1, \ldots, \lambda_n \in \mathbb{R}$, $\hat{y} = (y_k)_{k=1}^n \in \mathbb{R}^n$. It follows that $f(x) = \sum_{k=1}^n \lambda_k f_k(x)$ for all $x \in X$. □

These auxiliary results lead to the following important observations about the weak topology.

Proposition 3.3.9. *If X is an infinite dimensional normed space and $U \subseteq X$ is nonempty and* w*-open, then U is not bounded.*

Proof. Translating U if necessary, we may assume that $0 \in U$. By Proposition 3.3.6 there exist $x_1^*, \ldots, x_n^* \in X^*$ and $\varepsilon > 0$ such that $U(0; x_1^*, \ldots, x_n^*, \varepsilon) \subseteq U$. Note that $V = \bigcap_{k=1}^n N(x_k^*) \subseteq U$. Of course, V is a vector subspace of X and we claim that $V \neq \{0\}$. Indeed, if $V = \{0\}$, then it holds that $V \subseteq N(x^*)$ for all $x^* \in X^*$ and so Lemma 3.3.8 implies that x^* is a linear combination of the x_k^*'s. This means that $X^* = \operatorname{span}\{x_k^*\}_{k=1}^n$ and so X^* is finite dimensional, and hence X is finite dimensional, a contradiction. Therefore U is not bounded since it contains V. □

Remark 3.3.10. This proposition implies that weakly open sets are large. In particular, if $x \in V$ (see the previous proof), $x \neq 0$, then $\mathbb{R}x \subseteq U$. Therefore the open unit ball $B_1 = \{x \in X: \|x\| < 1\}$ is never w-open in an infinite dimensional normed space X.

Corollary 3.3.11. *If X is an infinite dimensional normed space, then the weak and strong (norm) topology do not coincide.*

In finite dimensional normed spaces, which are then of course Banach spaces, the two topologies coincide.

Proposition 3.3.12. *If X is a finite dimensional normed space, then the weak topology and the strong (norm) topology coincide.*

Proof. By definition, the weak topology is smaller than the strong topology. So, in order to prove the proposition, it suffices to show that every strongly open set is weakly open. Let $x_0 \in X$ and let U be a strongly open set containing x_0. Then there exists $\varrho > 0$ such that

$$B_\varrho(x_0) = \{x \in X : \|x - x_0\| < \varrho\} \subseteq U\,. \tag{3.3.1}$$

Let $\{e_k\}_{k=1}^n$ be a basis for X with $\|e_k\| = 1$ for all $k = 1, \dots, n$. Then every $x \in X$ admits an expression $x = \sum_{k=1}^n \lambda_k e_k$ with $\lambda_k \in \mathbb{R}$. For every $k = 1, \dots, n$ the coordinate map $x \to \lambda_k$, denoted by x_k^*, is linear and continuous for every $k = 1, \dots, n$. We consider $U(x_0; x_1^*, \dots, x_n^*, \varrho/n)$ being the basic weak neighborhood of x_0 determined by these coordinate maps. Then it follows

$$\|x - x_0\| \le \sum_{k=1}^n |\langle x_k^*, x - x_0\rangle| \le n\frac{\varrho}{n} = \varrho \quad \text{for all } x \in U\Big(x_0; x_1^*, \dots, x_n^*, \frac{\varrho}{n}\Big),$$

which implies

$$U\Big(x_0; x_1^*, \dots, x_n^*, \frac{\varrho}{n}\Big) \subseteq B_\varrho(x_0) \subseteq U\,,$$

see (3.3.1). That means that U is w-open and so the two topologies coincide. □

In what follows, we denote the convergence in the weak topology by $\xrightarrow{w}$ and the convergence in the strong (norm) topology by $\to$.

Proposition 3.3.13. *If X is a normed space and $\{x_\alpha\}_{\alpha\in I} \subseteq X$ is a net, then the following hold:*

(a) *$x_\alpha \xrightarrow{w} x$ if and only if $\langle x^*, x_\alpha\rangle \to \langle x^*, x\rangle$ for all $x^* \in X^*$;*
(b) *$x_\alpha \to x$ implies $x_\alpha \xrightarrow{w} x$;*
(c) *$x_\alpha \xrightarrow{w} x$ implies $\|x\| \le \liminf_{\alpha\in I} \|x_\alpha\|$ and a weakly convergent sequence is norm bounded;*
(d) *$x_\alpha \xrightarrow{w} x$ in X and $x_\alpha^* \to x^*$ in X^* imply $\langle x_\alpha^*, x_\alpha\rangle \to \langle x^*, x\rangle$.*

Proof. (a) This is a consequence of Proposition 1.3.3.

(b) For every $x^* \in X^*$, we have

$$\big|\langle x^*, x_\alpha\rangle - \langle x^*, x\rangle\big| = \big|\langle x^*, x_\alpha - x\rangle\big| \le \|x^*\|_* \|x_\alpha - x\| \to 0\,.$$

(c) Suppose that there is a sequence $\{x_n\}_{n\in\mathbb{N}} \subseteq X$ such that $x_n \xrightarrow{w} x$. Then, it follows $\langle x^*, x_n - x\rangle \to 0$ for all $x^* \in X^*$, which implies $\sup_{n\in\mathbb{N}} |\langle x^*, x_n - x\rangle| < \infty$. Taking Theorem 3.2.1 into account there exists $M > 0$ such that $\|x_n\| \le M$ for all $n \in \mathbb{N}$.

Evidently we may assume that $x \neq 0$. According to Proposition 3.1.50, there exists $\hat{x}^* \in X^*$ with $\|\hat{x}^*\|_* = 1$ such that $\langle \hat{x}^*, x\rangle = \|x\|$. So, $\|x\| = \lim_{\alpha\in I} |\langle \hat{x}^*, x_\alpha\rangle|$. Then, for given $\varepsilon > 0$ we can find $\alpha_0 = \alpha_0(\varepsilon) \in I$ such that

$$\|x\| - \varepsilon \le \big|\langle x^*, x_\alpha\rangle\big| \le \|x_\alpha\| \quad \text{for all } \alpha \ge \alpha_0\,.$$

Hence, $\|x\| \le \liminf_{\alpha\in I} \|x_\alpha\|$.

(d) Applying part (c), we derive, for some $M > 0$ and for every $\alpha \in I$, that

$$\begin{aligned}|\langle x_\alpha^*, x_\alpha\rangle - \langle x^*, x\rangle| &\le |\langle x_\alpha^* - x^*, x_\alpha\rangle| + |\langle x^*, x_\alpha - x\rangle| \\ &\le \|x_\alpha^* - x^*\|_* M + |\langle x^*, x_\alpha - x\rangle| \to 0 .\end{aligned}$$

Thus, $\langle x_\alpha^*, x_\alpha\rangle \to \langle x^*, x\rangle$. □

Remark 3.3.14. We emphasize that the boundedness in Proposition 3.3.13 (c) holds only for weakly convergent sequences and it fails for nets. Indeed, every infinite dimensional normed space admits a net $\{x_\alpha\}_{\alpha\in I} \subseteq X$ such that $x_\alpha \xrightarrow{w} 0$ in X and $\sup[\|x_\eta\| : \eta \ge \alpha, \eta \in I] = +\infty$. To see this let E denote the collection of all nonempty finite subsets of X^*. This set is directed by the set inclusion, that is, if $\alpha, \eta \in E$, $\alpha \ge \eta$ if and only if $\alpha \supseteq \eta$. For each $\alpha = (x_k^*)_{k=1}^n \in E$ there exist some $x_\alpha \in \bigcap_{k=1}^n N(x_k^*)$ such that $\|x_\alpha\| = \operatorname{card} \alpha$. The net $\{x_\alpha\}_{\alpha\in E}$ has the desired properties.

The weak topology is not metrizable in general and so sequences are not adequate to describe it. In fact we have the following result.

Proposition 3.3.15. *If X is a normed space and the weak topology on X is metrizable, then X is finite dimensional.*

Proof. Since the weak topology is metrizable, it is first countable. Hence, we can find a sequence $\{x_n^*\}_{n\ge 1} \subseteq X^*$ such that for any given $U \in \mathcal{N}_w(0)$ being the filter of weak neighborhoods of the origin, there exist $\varepsilon \in (0,1) \cap \mathbb{Q}$ and $n_U \in \mathbb{N}$ such that

$$U(0; x_1^*, \ldots, x_{n_U}^*, \varepsilon) \subseteq U . \tag{3.3.2}$$

For each $x^* \in X^*$, we have $U(0; x^*, 1) \in \mathcal{N}_w(0)$ and so by (3.3.2) it follows that

$$U(0; x_1^*, \ldots, x_{n(U(0;x^*,1))}^*, \varepsilon) \subseteq U(0; x^*, 1) .$$

Then

$$\bigcap_{k=1}^{n(U(0;x^*,1))} N(x_k^*) \subseteq N(x^*) ,$$

which, due to Lemma 3.3.8, results in

$$x^* \in \operatorname{span}\{x_k^*\}_{k=1}^{n(U(0;x^*,1))} .$$

Since $x^* \in X^*$ is arbitrary, it follows that $X^* = \bigcup_{k\ge 1} V_k$ with each V_k being finite dimensional. Recall that X^* is a Banach space. So, invoking Corollary 1.5.67 we see that $\operatorname{int} V_{k_0} \neq \emptyset$ for some $k_0 \in \mathbb{N}$. This means that $V_{k_0} = X^*$ and so X^* is finite dimensional. Hence X is finite dimensional. □

In what follows, we define for a normed space

$$\overline{B}_1 = \{x \in X : \|x\| \le 1\} \quad \text{and} \quad \partial B_1 = \{x \in X : \|x\| = 1\}.$$

Both sets are strongly closed. However, the situation changes for a weak topology. This is another illustration of the character of weak topology compared with strong (norm) topology, in the case of infinite dimensional normed spaces, of course; see Proposition 3.3.12.

Proposition 3.3.16. *If X is an infinite dimensional normed space, then $\overline{\partial B_1}^w = \overline{B}_1$.*

Proof. First we point out that the set $\overline{B}_1$ is w-closed. Indeed, if $\{x_\alpha\}_{\alpha\in I} \subseteq \overline{B}_1$ is a net such that $x_\alpha \xrightarrow{w} x$, then from Proposition 3.3.13 (c) one gets $\|x\| \le \liminf_{\alpha\in I} \|x_\alpha\| \le 1$. Hence $x \in \overline{B}_1$ and so $\overline{B}_1$ is w-closed. It follows that

$$\overline{\partial B_1}^w \subseteq \overline{B}_1. \tag{3.3.3}$$

Next let $x_0 \in B_1 = \{x \in X : \|x\| < 1\}$ and take $U \in \mathcal{N}_w(x_0)$ being the filter of weak neighborhoods of x_0. We may always assume that U is basic, that is,

$$U = U(x_0; x_1^*, \dots, x_n^*, \varepsilon) \quad \text{with } \{x_k^*\}_{k=1}^n \subseteq X^* \text{ and } \varepsilon > 0.$$

We fix $u \in \bigcap_{k=1}^n N(x_k^*)$, $u \ne 0$ (see the proof of Proposition 3.3.9) and consider the function $\xi : \mathbb{R}_+ \to \mathbb{R}_+$ defined by $\xi(\lambda) = \|x_0 + \lambda u\|$ for all $\lambda \ge 0$. We see that ξ is continuous, $\xi(0) < 1$ and $\lim_{\lambda\to+\infty} \xi(\lambda) = +\infty$. So, by Bolzano's Theorem there exists $\lambda_0 > 0$ such that $\xi(\lambda_0) = \|x_0 + \lambda_0 u\| = 1$, hence $x_0 + \lambda_0 u \in \partial B_1$.

Moreover, for every $k = 1, \dots, n$ we obtain $|\langle x_k^*, x_0 + \lambda_0 u - x_0\rangle| = 0$, which shows that $x_0 + \lambda_0 u \in \partial B_1 \cap U$. Therefore it follows that $B_1 \subseteq \overline{\partial B_1}^w$ and since the weak topology is smaller we infer that $\overline{B_1} \subseteq \overline{B_1}^w \subseteq \overline{\partial B_1}^w$. Finally, because of (3.3.3), we conclude that $\overline{B}_1 = \overline{\partial B_1}^w$. □

Remark 3.3.17. Consider the infinite dimensional Banach space $l^1 = \{\hat{x} = (x_n)_{n\ge1} \in \mathbb{R}^{\mathbb{N}} : \sum_{n\ge1} |x_n| < \infty\}$ which is called the space of all absolutely summable sequences in $\mathbb{R}$. One can show that weak and norm convergent sequences coincide in l^1. This is known as "Schur's Theorem" and its proof can be found in the book of Diestel [90, p. 85].

Our previous discussion of the weak topology has established that in an infinite dimensional normed space there are many more strongly closed sets than there are weakly closed sets. In the next theorem we show that for convex sets both notions agree. This is a remarkable result since a purely algebraic property, namely convexity, leads to a purely topological conclusion, namely that weak and strong closures coincide. The result is known as "Mazur's Theorem".

Theorem 3.3.18 (Mazur's Theorem). *If X is a normed space and $C \subseteq X$ is convex, then $\overline{C} = \overline{C}^w$.*

Proof. Since the strong (norm) topology is larger than the weak topology we directly obtain

$$\overline{C} \subseteq \overline{C}^{w} . \tag{3.3.4}$$

Arguing by contradiction suppose that the inclusion in (3.3.4) is strict. That means there exists $x_0 \in \overline{C}^{w} \setminus \overline{C}$. Invoking the Strong Separation Theorem (see Theorem 3.1.60), we find $x^* \in X^* \setminus \{0\}$ and $\varepsilon > 0$ such that

$$\langle x^*, x_0 \rangle + \varepsilon \le \langle x^*, u \rangle \quad \text{for all } u \in \overline{C} .$$

We set $\vartheta = \inf[\langle x^*, u\rangle : u \in \overline{C}]$ and $U = \{x \in X : \langle x^*, x\rangle < \vartheta\}$. Evidently $U \in \mathcal{N}_w(x_0)$ with $\mathcal{N}_w(x_0)$ being the filter of weak neighborhoods of x_0. Then $U \cap C = \emptyset$ and so $x_0 \notin \overline{C}^{w}$, a contradiction. Therefore from (3.3.4) we conclude that $\overline{C} = \overline{C}^{w}$. □

Corollary 3.3.19. *If X is a normed space and $V \subseteq X$ is a vector subspace, then $\overline{V} = \overline{V}^{w}$.*

Corollary 3.3.20. *If X is a normed space and $x_n \xrightarrow{w} x$, then there exists a sequence $\{u_n\}_{n\ge1} \subseteq X$ consisting of convex combinations of the x_n's such that $u_n \to x$ in X.*

Proof. Let $C = \overline{\operatorname{conv}}\{x_n\}_{n\ge1}$. Theorem 3.3.18 gives $x \in \overline{C}^{w} = \overline{C}$ and so $x \in \overline{\operatorname{conv}}\{x_n\}_{n\ge1}$. The result follows. □

Remark 3.3.21. This corollary known as "Mazur's Lemma" says that if $x_n \xrightarrow{w} x$, then for a given $\varepsilon > 0$ there exist $t_1, \ldots, t_m \ge 0$ such that $\sum_{k=1}^{m} t_k = 1$ and $\|x - \sum_{k=1}^{m} t_k x_k\| < \varepsilon$.

Corollary 3.3.22. *If X is a normed space and $C \subseteq X$ is convex, then C is closed if and only if C is w-closed.*

The next result is a consequence of the projective character of the weak topology.

Proposition 3.3.23. *If X, Y are normed spaces, then $A \in L(X, Y)$ if and only if A is weak-to-weak continuous.*

Proof. Note that $A \in L(X, Y)$ if and only if $A(\overline{B}_1^X) \subseteq Y$ is bounded with $\overline{B}_1^X = \{x \in X : \|x\|_X \le 1\}$; see Proposition 3.1.46. From Proposition 3.2.4 we know that $A(\overline{B}_1^X) \subseteq Y$ is bounded if and only if $y^*(A(\overline{B}_1^X)) \subseteq \mathbb{R}$ is bounded for every $y^* \in Y^*$. But a linear functional on a normed space is continuous if and only if it is weakly continuous. Invoking Proposition 1.3.4 we conclude that A is continuous if and only if it is weak-to-weak continuous. □

From Proposition 3.2.4 we have the following result about bounded sets.

Proposition 3.3.24. *If X is a normed space and $A \subseteq X$, then A is bounded if and only if A is w-bounded.*

Remark 3.3.25. We can formulate this result in a more general form. We say that a locally convex topology τ on X is compatible with the pair (X^*, X) if and only if $(X_\tau)^* = X^*$.

Then $A \subseteq X$ is bounded if and only if A is τ-bounded. In short, we can say that boundedness is duality invariant.

On the dual space X^* we can define two topologies. The first is the usual strong (metric) topology induced by the norm and the second is the weak topology $\mathrm{w} = \mathrm{w}(X^*, X^{**})$. Recall that the weak topology w is the weakest topology on X^* such that $(X^*_{\mathrm{w}})^* = X^{**}$. There is a third topology that we can define known as the **w*-topology**. This topology makes sense only on dual spaces.

Definition 3.3.26. Let X be a normed space and X^* is the topological dual, that is, $X^* = L(X, \mathbb{R})$. The **weak* topology** on X^* is the weakest topology w^* on X^* such that $(X^*_{\mathrm{w}^*})^* = X$. Consider now the linear functional $f_x : X^* \to \mathbb{R}$ defined by $f_x(x^*) = \langle x^*, x \rangle$. Then the weak* topology is the weakest topology on X^* making the collection $\{f_x\}_{x \in X}$ of maps from X^* into $\mathbb{R}$ continuous. The weak* topology on X^* is denoted by w^* or by $\mathrm{w}(X^*, X)$.

Remark 3.3.27. Since $X \subseteq X^{**}$ it is clear that $\mathrm{w}^* \subseteq w$, that is, the weak* topology has fewer open (resp. closed) sets than the weak topology.

Similarly to the weak topology (see Proposition 3.3.4 and Remark 3.3.5), we have the following result.

Proposition 3.3.28. *If X is a normed space, then X^*, equipped with the weak* topology, is a completely regular locally convex space.*

Moreover, we obtain the next two propositions as a consequence from Proposition 3.3.12.

Proposition 3.3.29. *If X is a normed space, then the w^*, the w, and the strong topologies on X^* coincide if and only if X is finite dimensional.*

Proposition 3.3.30. *If X is a normed space, then the basic weak* neighborhood of the origin has the form*

$$U(0; x_1, \dots, x_n, \varepsilon) = \{x^* \in X^* : |\langle x^*, x_k \rangle| < \varepsilon \text{ for all } k = 1, \dots, n\}$$

with $\{x_k\}_{k=1}^n \subseteq X$, $n \in \mathbb{N}$ and $\varepsilon > 0$. Since the weak topology is linear, we obtain the local basis at any other point by translation.*

The proof of Proposition 3.3.13 gives the following result. In what follows we denote the convergence in weak* topology by $\xrightarrow{\mathrm{w}^*}$.

Proposition 3.3.31. *If X is a normed space and $\{x^*_a\}_{a \in I} \subseteq X^*$ is a net, then the following hold:*

(a) $x^*_a \xrightarrow{\mathrm{w}^*} x^*$ *if and only if* $\langle x^*_a, x \rangle \to \langle x^*, x \rangle$ *for all* $x \in X$;

(b) $x^*_a \to x^*$ *or* $x^*_a \xrightarrow{\mathrm{w}} x^*$ *implies* $x^*_a \xrightarrow{\mathrm{w}^*} x^*$;

(c) $x_\alpha^* \xrightarrow{w^*} x^*$ *implies* $\|x^*\|_* \le \liminf_{\alpha\in I} \|x_\alpha^*\|_*$ *and every weakly** *convergent sequence is norm bounded;*

(d) $x_\alpha^* \xrightarrow{w^*} x^*$ *and* $x_\alpha \to x$ *in* X *imply* $\langle x_\alpha^*, x_\alpha\rangle \to \langle x^*, x\rangle$.

Remark 3.3.32. From the definition of the weak* topology, we see that any linear functional $f: X^* \to \mathbb{R}$, which is continuous for the w*-topology, has the form $f(x^*) = \langle x^*, \hat{x}\rangle$ for some $\hat{x} \in X$.

Proposition 3.3.33. *If* X *is a normed space and* $H \subseteq X^*$ *is a* w**-closed hyperplane, then there exist* $\hat{x} \in X$, $\hat{x} \neq 0$, *and* $\vartheta \in \mathbb{R}$ *such that*

$$H = \{x^* \in X^* : \langle x^*, \hat{x}\rangle = \vartheta\} .$$

Proof. We know that $H = \{x^* \in X^* : f(x^*) = \vartheta\}$ with $f: X^* \to \mathbb{R}$ being linear and $\vartheta \in \mathbb{R}$; see Definition 3.1.53. Since by hypothesis H is w*-closed, Proposition 3.1.54 implies that f is w*-continuous. Finally, using Remark 3.3.32, we conclude that there exists $\hat{x} \in X$ such that $H = \{x^* \in X^* : \langle x^*, \hat{x}\rangle = \vartheta\}$. □

Recall that every $x \in X$ defines in a natural way a linear functional $f_x: X^* \to \mathbb{R}$ according to the formula $f_x(x^*) = \langle x^*, x\rangle$. Indeed, we see that $|f_x(x^*)| = |\langle x^*, x\rangle| \le \|x^*\|_*\|x\|$, which shows that f_x is bounded, that is, $f_x \in X^*$, and $\|f_x\|_* \le \|x\|$. Thus we can define the map $j: X \to X^{**}$ by $j(x) = f_x$. Clearly j is linear, injective, and $\|j(x)\|_* \le \|x\|$ for all $x \in X$. Additional information about this map is supplied by the next proposition.

Proposition 3.3.34. *If* X *is a normed space and* $j: X \to X^{**}$ *is the linear map defined above, then* j *is an isometric isomorphism onto* $j(X)$.

Proof. We already proved that j is an isomorphism onto $j(X)$ and $\|j(x)\|_* \le \|x\|$ for all $x \in X$. On the other hand, from Proposition 3.1.50, we know that there exists $x^* \in X^*$ such that $\|x^*\|_* = 1$ and $j(x)(x^*) = \langle x^*, x\rangle = \|x\|$. This shows that $\|j(x)\|_* \ge \|x\|$ for all $x \in X$. Hence, j is an isometry. □

Definition 3.3.35. The isometry $j: X \to X^{**}$ of Proposition 3.3.34 is called the **canonical embedding** of the normed space X into X^{**}.

Remark 3.3.36. Using the canonical embedding we can identify X with a subspace of X^{**}. Moreover, $\overline{j(X)}$ is a closed subspace of the Banach space X^{**}. Hence, $V = \overline{j(X)}$ is a Banach space as well. Therefore j is an isometric isomorphism onto a dense subset of the Banach space V. Hence, the canonical embedding provides a shortcut to the completion of a normed space. Every normed space can be viewed as a dense subspace of a Banach space. When the canonical embedding j is not surjective, then the weak topology $w(X^*, X^{**})$ is strictly larger than the weak* topology. Indeed let $\hat{u} \in X^{**} \setminus j(X)$ and consider the subspace $H = \{x^* \in X^* : \langle \hat{u}, x^*\rangle = 0\}$. Then H is w-closed, but it is not w*-closed; see Proposition 3.3.33. In fact this example shows that Mazur's Theorem (see Theorem 3.3.18) fails for the w*-topology. A strongly closed convex set need not be w*-closed. Moreover, a normed space and its completion have the same dual space; however, their weak* topologies differ. So, one should be careful when dealing with the weak* topology of the dual of a normed space and that of the dual of the Banach space resulting from its completion.

Since X can be viewed as a subspace of X^{**}, it is natural to ask what kind of subspace it is. The answer is given by the so-called "Goldstine's Theorem". In what follows we set

$$B_1^X = \{x \in X : \|x\| < 1\}, \qquad \overline{B}_1^X = \{x \in X : \|x\| \le 1\},$$
$$B_1^{X^{**}} = \{x^{**} \in X^{**} : \|x^{**}\|_{**} < 1\}, \quad \overline{B}_1^{X^{**}} = \{x^{**} \in X^{**} : \|x^{**}\|_{**} \le 1\}.$$

Theorem 3.3.37 (Goldstine's Theorem). *If X is a normed space, then $\overline{j(B_1^X)}^{w^*} = \overline{B}_1^{X^{**}}$ and $\overline{j(X)}^{w^*} = X^{**}$.*

Proof. Clearly, the second equality is a consequence of the first. So, let us prove the first one.

Let $x^{**} \in X^{**} \setminus \overline{j(B_1^X)}^{w^*}$. Since $\overline{j(B_1^X)}^{w^*} \subseteq X^{**}$ is convex and w*-closed, by the Strong Separation Theorem (see Corollary 3.1.61), there exists $x^* \in (X^{**}_{w^*})^* = X^*$ with $x^* \neq 0$ such that

$$\sup[\langle x^*, u^{**}\rangle : u^{**} \in \overline{j(B_1^X)}^{w^*}] < \langle x^*, x^{**}\rangle . \tag{3.3.5}$$

We may always assume that $\|x^*\|_* = 1$. Then, from (3.3.5), we have

$$1 = \|x^*\|_* < \langle x^*, x^{**}\rangle \le \|x^*\|_* \|x^{**}\|_{**} .$$

Hence, $1 < \|x^{**}\|_{**}$ and so $\overline{j(B_1^X)}^{w^*} = \overline{B}_1^{X^{**}}$. □

The weaker a topology is, the more compact sets it has. The next theorem is the most important feature of the weak* topology. It is reminiscent of the Heine–Borel-Theorem and it is the reason why the weak* topology is important in the theory of Banach spaces. The result is known as "Alaoglu's Theorem".

Theorem 3.3.38 (Alaoglu's Theorem). *If X is a normed space, then $\overline{B}_1^{X^*} = \{x^* \in X^* : \|x^*\|_* \le 1\}$ is w*-compact. More generally every w*-closed and bounded subset of X^* is w*-compact.*

Proof. Suppose that $x^* \in \overline{B}_1^{X^*}$. Then for each $x \in \overline{B}_1^X$ it follows $|\langle x^*, x\rangle| \le 1$. Therefore

$$x^*(\overline{B}_1^X) \subseteq I = \{\lambda \in \mathbb{R} : |\lambda| \le 1\} .$$

We can identify each element of $\overline{B}_1^{X^*}$ with a point in $I^{\overline{B}_1^X}$. From Tychonoff's Theorem, see Theorem 1.4.56, $I^{\overline{B}_1^X}$ equipped with the product topology is compact. Since the weak* topology is by definition the topology of pointwise convergence on $\overline{B}_1^X$, the identification of $\overline{B}_1^{X^*}$ with a subset of $I^{\overline{B}_1^X}$ leaves the weak* topology unchanged. So, it remains to show that $\overline{B}_1^{X^*}$ is closed in $I^{\overline{B}_1^X}$. To this end, let $\{x_\alpha^*\}_{\alpha\in I} \subseteq \overline{B}_1^{X^*}$ be a net and assume that it converges pointwise to $g \in I^{\overline{B}_1^X}$. Evidently g is linear and so g is the restriction on $\overline{B}_1^X$ of a linear functional x^* on X. Moreover, since $|g(x)| \leq 1$ for all $x \in \overline{B}_1^X$, it follows that $x^* \in \overline{B}_1^{X^*}$ and this proves that $\overline{B}_1^{X^*}$ is closed in $I^{\overline{B}_1^X}$, and hence w*-compact.

Every bounded set $C \subseteq X^*$ satisfies $C \subseteq r\overline{B}_1^{X^*}$ for some $r > 0$. Since $\overline{B}_1^{X^*}$ is w*-compact and C is w*-closed, we conclude that it is w*-compact. □

Remark 3.3.39. From the theorem above, we derive that if X is a normed space and $C \subseteq X^*$, then C is w*-closed and bounded implies that C is w*-compact.

For the converse to hold, we need to assume that X is a Banach space. To see this, let $X = \{\hat{x} = (a_n)_{n\in\mathbb{N}} : a_n = 0 \text{ for all } n \geq n_0\}$ equipped with the norm $\|\hat{x}\| = \sum_{n\in\mathbb{N}} |a_n|$. Clearly this is a normed space but not a Banach space. Consider a sequence $\{\xi_n\}_{n\in\mathbb{N}} \subseteq \mathbb{R}$ with $\xi_n > 0$ for all $n \in \mathbb{N}$ such that $\xi_n \to +\infty$ as $n \to \infty$. Let $\{\hat{x}_n\}_{n\in\mathbb{N}} \subseteq X^*$ be defined by $\hat{x}_n^*(\hat{x}) = a_n$ for all $n \in \mathbb{N}$.

Let $D = \{0, \xi_1\hat{x}_1, \xi_2\hat{x}_2, \ldots, \xi_n\hat{x}_n, \ldots\} \subseteq X^*$. This set is unbounded since $\|\xi_n\hat{x}_n\|_* = \xi_n \to +\infty$. However, it holds $\xi_n\hat{x}_n(\hat{x}) = \xi_n a_n \to 0$ for all $\hat{x} \in X$. So, $\xi_n\hat{x}_n \xrightarrow{w^*} 0$ and it follows that $D \subseteq X^*$ is w*-bounded.

From the previous remark, we have the following corollary.

Corollary 3.3.40. *If X is a Banach space and $C \subseteq X^*$, then C is bounded if and only if C is w*-bounded, that is, $x(C) \subseteq \mathbb{R}$ is bounded for every $x \in X$.*

We conclude this section with a remarkable result of R. C. James, which provides a necessary and sufficient condition for a set C in a Banach space X to be weakly compact. The result is known as “James' Theorem” and its proof is lengthy and can be found in Holmes [171, p. 157].

Theorem 3.3.41 (James' Theorem). *If X is a Banach space and $C \subseteq X$ is bounded and w-closed, then C is w-compact if and only if every $x^* \in X^*$ attains its supremum over C.*

3.4 Separable and Reflexive Banach Spaces

In this section we examine two special classes of Banach spaces, namely separable and reflexive Banach spaces. They exhibit special properties, which are important in applications.

Definition 3.4.1. (a) A normed space X is **separable** if it contains a countable dense subset.

(b) A normed space X is **reflexive** if the canonical embedding $j\colon X \to X^{**}$ (see Definition 3.3.35) is surjective. A reflexive normed space is necessarily complete, that is, a Banach space.

Remark 3.4.2. Any subset of a separable normed space is a separable metric space. Many important spaces in analysis are separable and/or reflexive. Every finite dimensional Banach space is separable and reflexive. In the definition of reflexivity it is essential to use the canonical embedding j stated in Definition 3.3.35. R. C. James produced in 1951 a remarkable example of a nonreflexive Banach space X that is isometrically isomorphic to X^{**}. In this example, the image of X under the canonical embedding $j\colon X \to X^{**}$ is a closed subspace of codimension one. A detailed construction of this space can be found in Megginson [230]; see Section 4.5. In what follows, for the sake of notational simplicity, we drop the use of the map j. It is understood that X is embedded into X^{**} via the canonical embedding.

Proposition 3.4.3. *If X is a Banach space and X^* is separable, then X is separable.*

Proof. Let $\{x_n^*\}_{n\geq 1} \subseteq X^*$ be dense. Thanks to Corollary 3.1.48 we know that

$$\|x_n^*\|_* = \sup[\langle x_n^*, x\rangle : x \in X, \|x\| \leq 1]$$

for all $n \in \mathbb{N}$. Hence, there exists $x_n \in X$ such that

$$\|x_n\| = 1 \quad \text{and} \quad \frac{1}{2}\|x_n^*\|_* \leq \langle x_n^*, x_n\rangle\,, \qquad n \in \mathbb{N}\,. \tag{3.4.1}$$

Let $V_0 = \operatorname{span}_{\mathbb{Q}}\{x_n\}_{n\in\mathbb{N}}$, that is, V_0 is the set of all finite linear combinations with coefficients in $\mathbb{Q}$ of the vectors $\{x_n\}_{n\in\mathbb{N}}$. This set is countable since $V_0 = \bigcup_{m\geq 1} V_m$ with V_m being the set of linear combinations with coefficients in $\mathbb{Q}$ of $\{x_n\}_{n=1}^m$. Each V_m is countable, and so $V_0 = \bigcup_{m\geq 1} V_m$ is countable as well.

Let $V = \operatorname{span}\{x_n\}_{n\in\mathbb{N}}$. We claim that V is dense in X. To this end, let $x^* \in V^\perp$. Then there exists $\{x_{n_k}^*\}_{k\in\mathbb{N}} \subseteq \{x_n^*\}_{n\in\mathbb{N}}$ such that

$$x_{n_k}^* \to x^* \quad \text{in } X^* \text{ as } k \to \infty\,. \tag{3.4.2}$$

Then, because of (3.4.1) and since $x^* \in V^\perp$, it follows that

$$\|x_{n_k}^*\|_* \leq 2\langle x_{n_k}^*, x_{n_k}\rangle = 2\langle x_{n_k}^* - x^*, x_{n_k}\rangle \leq 2\|x_{n_k}^* - x^*\|_*\|x_{n_k}\| = 2\|x_{n_k}^* - x^*\|_*\,.$$

Hence, thanks to (3.4.2), one gets $x_{n_k}^* \to x^* = 0$ in X^*. This shows that $V^\perp = \{0\}$ and so V is dense in X; see Remark 3.1.63. Since V_0 is countable and dense in V, we conclude that X is separable. □

Remark 3.4.4. The converse of this result is not true. Namely, separability of X does not imply separability of X^*. For example, $X = L^1([0,1])$ is separable (see Proposition 2.3.24), but $X^* = L^\infty([0,1])$ is not separable; see Proposition 2.3.29. In Section 4.1 we will show that $L^\infty([0,1]) = L^1([0,1])^*$.

Theorem 3.4.5. *A Banach space X is reflexive if and only if $\overline{B}_1^X = \{x \in X : \|x\| \le 1\}$ is w-compact.*

Proof. $\Longrightarrow$: The reflexivity of X implies that $X = X^{**}$. Hence $\overline{B}_1^X = \overline{B}_1^{X^{**}}$. By Alaoglu's Theorem (see Theorem 3.3.38), $\overline{B}_1^{X^{**}}$ is w*-compact and from Proposition 1.3.5, we know that

$$\mathrm{w}(X^{**}, X^*)|_X = \mathrm{w}(X, X^*) . \tag{3.4.3}$$

Therefore, $\overline{B}_1^X$ is w-compact.

$\Longleftarrow$: Since by hypothesis, $\overline{B}_1^X$ is w-compact, it is w*-closed in X^{**}; see (3.4.3). The Goldstine's Theorem (see Theorem 3.3.37), gives $\overline{\overline{B}_1^X}^{\mathrm{w}^*} = \overline{B}_1^{X^{**}}$ and since $\overline{B}_1^X$ is w*-closed in X^{**}, we obtain $\overline{B}_1^X = \overline{B}_1^{X^{**}}$. Therefore $X = X^{**}$ and so we conclude that X is reflexive. ☐

Proposition 3.4.6. *A Banach space X is reflexive if and only if X^* is reflexive.*

Proof. $\Longrightarrow$: Since X is reflexive, we know that $X = X^{**}$ and so the weak and weak* topologies on X^* coincide. Alaoglu's Theorem (see Theorem 3.3.38) implies that $\overline{B}_1^{X^*} = \{x^* \in X^* : \|x^*\|_* \le 1\}$ is w-compact and so Theorem 3.4.5 implies that X^* is reflexive.

$\Longleftarrow$: Since X^* is reflexive, by the previous part of the proof we have that X^{**} is reflexive as well. Then, Theorem 3.4.5 implies that $\overline{B}_1^{X^{**}} = \{x^{**} \in X^{**} : \|x^{**}\|_{**} \le 1\}$ is w-compact. The set $\overline{B}_1^X$ is closed, convex, hence a w-closed subset of $\overline{B}_1^{X^{**}}$; see Mazur's Theorem (Theorem 3.3.18). Therefore $\overline{B}_1^X$ is w-compact in X^{**}. Since the w*-topology on X^{**} is weaker than the w-topology, it follows that $\overline{B}_1^X$ is w*-compact in X^{**}. Hence it is w-compact in X; see (3.4.3). We conclude by using Theorem 3.4.5. ☐

Proposition 3.4.7. *If X is a reflexive Banach space and V is a closed subspace of X, then V is a reflexive Banach space.*

Proof. We know that

$$\mathrm{w}(V, V^*) = \mathrm{w}(X, X^*)|_V ; \tag{3.4.4}$$

see Proposition 1.3.5. The set $\overline{B}_1^V = \{x \in V : \|x\| \le 1\}$ is a weakly closed subset of the weakly compact set $\overline{B}_1^X$; see Theorem 3.4.5. Combining this with (3.4.4), we infer that $\overline{B}_1^V$ is w-compact in V. Then invoking Theorem 3.4.5 we conclude that V is reflexive. ☐

Combining Propositions 3.4.3 and 3.4.6, we obtain the following.

Proposition 3.4.8. *If X is a Banach space, then X is separable and reflexive if and only if X^* is separable and reflexive.*

Proposition 3.4.9. *If X is a reflexive Banach space and $V \subseteq X$ is a closed subspace, then X/V is reflexive.*

Proof. From Proposition 3.2.25, we know that $(X/V)^*$ and $V^\perp$ are isometrically isomorphic. Let $\xi\colon (X/V)^* \to V^\perp$ be this isometric isomorphism. If $p\colon X \to X/V$ is the quotient map (see Definition 3.2.19), then from the proof of Proposition 3.2.25 we know that

$$\xi(l) = l \circ p \quad \text{for all } l \in (X/V)^* .$$

Let $l^* \in (X/V)^{**}$. The map $l^* \circ \xi^{-1}\colon V^\perp \to \mathbb{R}$ is a bounded linear functional on a subspace of X^*. Hence, by Proposition 3.1.49 there exists $x^{**} \in X^{**}$ such that $\langle x^{**}, x^* \rangle = \langle l^*, \xi^{-1}(x^*)\rangle$ for all $x^* \in V^\perp$. This implies that

$$\left\langle x^{**}, l \circ p \right\rangle = \left\langle l^*, l \right\rangle \quad \text{for all } l \in (X/V)^* . \tag{3.4.5}$$

The reflexivity of X implies that there exists $x \in X$ such that $j(x) = x^{**}$ with j being the canonical embedding. Let $u = [x] = p(x) \in X/V$. Combining Definition 3.3.35 and (3.4.5), it follows that

$$\left\langle l^*, l \right\rangle = \left\langle x^{**}, l \circ p \right\rangle = \left\langle j(x), l \circ p \right\rangle = \langle l \circ p, x\rangle = \left\langle l, p(x) \right\rangle = \langle l, u\rangle .$$

Hence, $j(u) = l^*$ with j being the canonical embedding for X/V. Since $l^* \in (X/V)^{**}$ is arbitrary, it follows that j is surjective and so X/V is reflexive; see Definition 3.4.1 (b). □

We know that on an infinite dimensional normed space and on its dual, the weak and weak* topologies are never metrizable. Nevertheless, the traces of these topologies on certain subspaces can be metrizable. The results that follow investigate this issue. We start with a general topological result.

Lemma 3.4.10. *If (X, τ) is a compact topological space and $\{f_n\}_{n \ge 1}$ is a separating sequence of continuous functions on X (see Definition 1.3.6), then the topology τ is metrizable.*

Proof. We may assume that $|f_n(x)| \le 1$ for all $x \in X$ and for all $n \in \mathbb{N}$. On X we consider the metric d defined by

$$d(x, u) = \sum_{n \in \mathbb{N}} \frac{1}{2n} \left|f_n(x) - f_n(u)\right| \quad \text{for all } x, u \in X .$$

Let τ_d be the metric topology induced by this metric on X. For every fixed $u \in X$, $x \to d(x, u)$ is τ-continuous as the uniform limit of τ-continuous functions. So, for every $\varepsilon > 0$, it follows that $B_\varepsilon(u) = \{x \in X : d(x, u) < \varepsilon\} \in \tau$, which means that $\tau_d \subseteq \tau$. Using Theorem 1.4.54, we see that the identity map $i_X\colon (X, \tau) \to (X, \tau_d)$ is a homeomorphism. Hence $\tau = \tau_d$. □

Using this lemma, we can state the first metrizability result for the weak* topology.

Theorem 3.4.11. *If X is a separable normed space and $C \subseteq X^*$ is* w**-compact, then C equipped with the* w**-topology is metrizable.*

Proof. Let $\{x_n\}_{n\geq 1} \subseteq X$ be dense in X. If $j: X \to X^{**}$ is the canonical embedding, then

$$\langle j(x_n), x^* \rangle = \langle x^*, x_n \rangle \quad \text{for all } n \in \mathbb{N} \text{ and for all } x^* \in X^* \,;$$

see Definition 3.3.35. So, if $\langle j(x_n), x^* \rangle = 0$ for all $n \in \mathbb{N}$, we derive that $\langle x^*, x_n \rangle = 0$ for all $n \in \mathbb{N}$ and the density of $\{x_n\}_{n\geq 1}$ in X implies that $x^* = 0$. Therefore $\{j(x_n)\}_{n\geq 1} \subseteq X^{**}$ is separating and each $j(x_n)$ is w*-continuous. Applying Lemma 3.4.10, we conclude that (C, w^*) is metrizable. ☐

We can improve this result in the following way.

Theorem 3.4.12. *If X is a normed space, then the following hold:*
(a) *$(\overline{B}_1^{X^*}, w^*)$ is metrizable if and only if X is separable;*
(b) *$(\overline{B}_1^{X}, w)$ is metrizable if and only if X^* is separable.*

Proof. (a) $\Longrightarrow$: Since $(\overline{B}_1^{X^*}, w^*)$ is metrizable we can find a countable basis $\{U_n\}_{n\geq 1}$ at the origin. We obtain

$$U_n = \{x^* \in \overline{B}_1^{X^*} : |\langle x^*, x \rangle| < \varepsilon_n \text{ for all } x \in F_n\}, \quad n \in \mathbb{N}$$

with $F_n \subseteq X$ finite and $\varepsilon_1, \ldots, \varepsilon_n > 0$. Let $E = \bigcup_{n\geq 1} F_n$. Then $E \subseteq X$ is countable and $x^*(E) = 0$ implies $x^* \in U_n$ for all $n \in \mathbb{N}$ and so $x^* = 0$. Moreover, if $x^*(\overline{\operatorname{span}} E) = 0$, then $x^* = 0$. Therefore $\overline{\operatorname{span}} E = X$ and so we conclude that X is separable.

$\Longleftarrow$: This follows from Theorem 3.4.11.

(b) $\Longrightarrow$: As before, let $\{U_n\}_{n\geq 1}$ be a countable local basis at the origin of X. We obtain

$$U_n = \{x \in \overline{B}_1^{X} : |\langle x^*, x \rangle| < \varepsilon_n \text{ for all } x^* \in F_n^*\}, \quad n \in \mathbb{N} \tag{3.4.6}$$

with $F_n^* \subseteq X^*$ finite and $\varepsilon_1, \ldots, \varepsilon_n > 0$. Let $E^* = \bigcup_{n\geq 1} F_n^*$. Then $E^* \subseteq X^*$ is countable and so $\overline{\operatorname{span}} E^*$ is separable. We will show that $X^* = \overline{\operatorname{span}} E^*$. Arguing by contradiction, suppose that there exists $\hat{x}^* \in X^* \setminus \overline{\operatorname{span}} E^*$. Let $d = d(x^*, \overline{\operatorname{span}} E^*)$. Then we can find $\hat{x}^{**} \in X^{**}$ such that

$$\|\hat{x}^{**}\|_{**} = \frac{1}{d}, \quad \hat{x}^{**}(\overline{\operatorname{span}} E^*) = 0 \quad \text{and} \quad \langle \hat{x}^{**}, \hat{x}^* \rangle = 1\,; \tag{3.4.7}$$

see Proposition 3.1.50. We introduce

$$V = \left\{ x \in \overline{B}_1^{X} : |\langle \hat{x}^*, x \rangle| < \frac{d}{2} \right\}. \tag{3.4.8}$$

Then V is a weak neighborhood of the origin in X and so $U_{n_0} \subseteq V$ for some $n_0 \in \mathbb{N}$. Note that $d\hat{x}^{**} \in \overline{B}_1^{X^{**}}$ and so by Goldstine's Theorem (see Theorem 3.3.37), there is $\hat{x} \in \overline{B}_1^X$ such that

$$|\langle d\hat{x}^{**} - \hat{x}, x^*\rangle| < \varepsilon_{n_0} \quad \text{for all } x^* \in F_{n_0}^* \quad \text{and} \quad |\langle d\hat{x}^{**} - \hat{x}, \hat{x}^*\rangle| < \frac{d}{2}.$$

Then, due to (3.4.7),

$$|\langle x^*, \hat{x}\rangle| < \varepsilon_{n_0} \quad \text{for all } x^* \in F_{n_0}^* \quad \text{and} \quad |\langle \hat{x}^*, \hat{x}\rangle| > \frac{d}{2}.$$

This gives, with view to (3.4.6) and (3.4.8), that $\hat{x} \in U_{n_0}$ and $\hat{x} \notin V$, a contradiction to the fact that $U_{n_0} \subseteq V$. Therefore $X^* = \overline{\operatorname{span} E^*}$, and so X^* is separable.

$\Longleftarrow$: According to Alaoglu's Theorem (see Theorem 3.3.38), we know that $\overline{B}_1^{X^{**}}$ is w^*-compact. Since X^* is separable, from part (a) we derive that $(\overline{B}_1^{X^{**}}, w^*)$ is metrizable. Since $\overline{B}_1^X \subseteq \overline{B}_1^{X^{**}}$ via the canonical embedding and $\mathrm{w}(X^{**}, X^*)|_X = w(X, X^*)$, we conclude that $(\overline{B}_1^X, w)$ is metrizable. □

Remark 3.4.13. In particular, this theorem says that if X (resp. X^*) is separable and $C \subseteq X^*$ (resp. $C \subseteq X$) is bounded, then (C, w^*) (resp. (C, w)) is metrizable.

A subset C of a normed space X is said to be **weakly sequentially compact** (resp. **weakly countably compact, weakly limit point compact**) if it is sequentially compact (resp. countably compact, limit point compact) in the weak topology; see Definition 1.4.57.

A remarkable result known as the "Eberlein–Smulian Theorem" says that all these notions are equivalent to weak compactness. The proof of this result is lengthy and can be found in Dunford–Schwartz [105, p. 430] and Megginson [230, p. 248].

Theorem 3.4.14 (Eberlein–Smulian Theorem). *If X is a normed space and $C \subseteq X$, then the following properties are equivalent:*
(a) *C is (relatively) weakly compact.*
(b) *C is (relatively) weakly sequentially compact.*
(c) *C is (relatively) weakly countably compact.*
(d) *C is (relatively) weakly limit point compact.*

Remark 3.4.15. The theorem above is not true for the weak* topology.

Combining Theorems 3.4.5 and 3.4.14, we infer the following sequential characterization of reflexivity.

Theorem 3.4.16. *A Banach space X is reflexive if and only if every bounded sequence in X admits a weakly convergent subsequence.*

Two other consequences of Theorem 3.4.14 are the following two results.

Theorem 3.4.17. *If X is a separable normed space and $C \subseteq X$ is weakly compact, then (C, w) is metrizable.*

Theorem 3.4.18. *If X is a reflexive Banach space, $C \subseteq X$ is bounded, and $x \in \overline{C}^{\mathrm{w}}$, then there exists a sequence $\{x_n\}_{n\geq 1} \subseteq C$ such that $x_n \xrightarrow{\mathrm{w}^*} x$ in X.*

The next proposition provides a way to identify weakly compact sets.

Proposition 3.4.19. *If X is a Banach space, $C \subseteq X$ is w-closed, and for every $\varepsilon > 0$ there is a weakly compact set $K_\varepsilon \subseteq X$ such that $C \subseteq K_\varepsilon + \varepsilon \overline{B}_1^X$, then C is weakly compact.*

Proof. Viewing C as a subset of X^{**} via the canonical embedding, we directly obtain

$$\overline{C}^{\mathrm{w}^*} \subseteq \overline{K_\varepsilon + \varepsilon \overline{B}_1^X}^{\mathrm{w}^*} = \overline{K_\varepsilon}^{\mathrm{w}^*} + \overline{\varepsilon \overline{B}_1^X}^{\mathrm{w}^*} = K_\varepsilon + \varepsilon \overline{B}_1^{X^{**}},$$

since K_ε is w-compact and due to Theorem 3.3.37. Therefore

$$\overline{C}^{\mathrm{w}^*} \subseteq \bigcap_{\varepsilon>0} (K_\varepsilon + \varepsilon \overline{B}_1^{X^{**}}) \subseteq X,$$

which shows that C is w-compact since C is w-closed. □

Continuing with weakly compact sets, we show that this property is preserved if we take the closed convex hull of the set.

Proposition 3.4.20. *If X is a Banach space and $C \subseteq X$ is w-compact, then $\overline{\mathrm{conv}}\, C \subseteq X$ is w-compact as well.*

Proof. Let $x^* \in X^*$. Then

$$\sup[\langle x^*, x\rangle : x \in C] = \sup[\langle x^*, u\rangle : u \in \overline{\mathrm{conv}}\, C]. \tag{3.4.9}$$

Because $C \subseteq X$ is w-compact, there exists $\hat{x} \in C$ such that

$$\langle x^*, \hat{x}\rangle = \sup[\langle x^*, x\rangle : x \in C].$$

This implies, due to (3.4.9), that

$$\langle x^*, \hat{x}\rangle = \sup[\langle x^*, u\rangle : u \in \overline{\mathrm{conv}}\, C].$$

Since $x^* \in X^*$ is arbitrary, invoking James's Theorem (see Theorem 3.3.41), we conclude that $\overline{\mathrm{conv}}\, C$ is w-compact. Note that $\overline{\mathrm{conv}}\, C$ is w-closed by Theorem 3.3.18. □

Next we introduce some new classes of Banach spaces based on some geometric properties of the unit ball.

Definition 3.4.21. Let X be a Banach space.

(a) We say that X is **strictly convex** if for all $x, u \in X$ with $x \neq u$ and $\|x\| = \|u\| = 1$ it holds $\|(1-t)x + tu\| < 1$ for all $t \in (0,1)$.
(b) We say that X is **uniformly convex** if for every $\varepsilon > 0$ there exists $\delta = \delta(\varepsilon) > 0$ such that

$$x, u \in X, \quad \|x\| \le 1, \quad \|u\| \le 1, \quad \|x - u\| \ge \varepsilon \quad \text{imply} \quad \frac{1}{2}\|x + u\| \le 1 - \delta\,.$$

(c) We say that X is **locally uniformly convex** if for every $\varepsilon > 0$ and $x \in X$ with $\|x\| = 1$ there exists $\delta = \delta(\varepsilon, x) > 0$ such that

$$u \in X, \quad \|u\| = 1, \quad \|x - u\| \ge \varepsilon \quad \text{imply} \quad \frac{1}{2}\|x + u\| \le 1 - \delta\,.$$

Remark 3.4.22. Evidently it holds

$$\text{Uniformly convex} \implies \text{Locally uniformly convex} \implies \text{Strictly convex}\,.$$

Note that these implications are not reversible in general. For finite dimensional spaces, the three notions are equivalent.

Proposition 3.4.23. *Let X be a Banach space. The following properties are equivalent:*
(a) *X is strictly convex.*
(b) *The boundary of the unit ball called the unit sphere contains no line segments.*
(c) *$x \neq u$ and $\|x\| = \|u\| = 1$ implies $\|x + u\| < 2$.*
(d) *If $\|x - y\| = \|x - u\| + \|u - y\|$ for $x, u, y \in X$, then there exists $t \in [0,1]$ such that $u = (1-t)x + ty$.*
(e) *Every $x^* \in X^* \setminus \{0\}$ attains its supremum on $\overline{B}_1^X$ on at most one point.*

Proof. (a) $\Longrightarrow$ (b): This is obvious from Definition 3.4.21 (a).

(b) $\Longrightarrow$ (a): Arguing by contradiction suppose that we can find $x, u \in X, x \neq u$, $\|x\| = \|u\| = 1$ and $t_0 \in (0,1)$ such that $\|(1-t_0)x + t_0 u\| = 1$. Let $t \in (0, t_0)$. Then we obtain

$$(1-t_0)x + t_0 u = \frac{1-t_0}{1-t}((1-t)x + tu) + \frac{t_0 - t}{1-t}u\,,$$

which gives

$$1 \le \frac{1-t_0}{1-t}\|(1-t)x + tu\| + \frac{t_0 - t}{1-t}\,.$$

Hence $\|(1-t)x + tu\| \ge 1$ and so $\|(1-t)x + tu\| = 1$.

Similarly we treat the case $t \in (t_0, 1)$. Therefore the line segment $[x, u]$ is on the unit sphere of X, a contradiction to the hypothesis.

(a) $\Longrightarrow$ (c) and (c) $\Longrightarrow$ (b): These implications are obvious.

(a) $\Longrightarrow$ (d): Let $x, u, y \in X$ be such that $\|x - y\| = \|x - u\| + \|u - y\|$. We may assume that $\|x - u\| \neq 0$, $\|u - y\| \neq 0$ and $\|x - u\| \le \|u - y\|$. Then we derive

$$\begin{aligned}
&\left\| \frac{1}{2}\frac{x-u}{\|x-u\|} + \frac{1}{2}\frac{u-y}{\|u-y\|} \right\| \\
&\geq \left\| \frac{1}{2}\frac{x-u}{\|x-u\|} + \frac{1}{2}\frac{u-y}{\|x-u\|} \right\| - \left\| \frac{1}{2}\frac{u-y}{\|x-u\|} - \frac{1}{2}\frac{u-y}{\|u-y\|} \right\| \\
&= \frac{1}{2}\frac{\|x-y\|}{\|x-u\|} - \frac{1}{2}\frac{\|u-y\| - \|x-u\|}{\|x-u\|} \\
&= \frac{1}{2}\frac{1}{\|x-u\|} 2\|x-u\| = 1 .
\end{aligned}$$

Hence we obtain

$$\left\| \frac{x-u}{\|x-u\|} + \frac{u-y}{\|u-y\|} \right\| = 2,$$

which finally gives

$$\frac{x-u}{\|x-u\|} = \frac{u-y}{\|u-y\|} .$$

Therefore $u = (1-t)x + ty$ with $t = (\|x-u\|)/(\|x-y\|) \in (0,1)$.

(d) $\Longrightarrow$ (c): Let $x, y \in X$, $x \neq y$ with $\|x\| = \|y\| = 1/2\|x+y\| = 1$. Then $\|x+y\| = \|x\| + \|y\|$, which gives $u = 0 = (1-t)x - ty$ for some $t \in (0,1)$. Hence $x = t/(1-t)y$ and so $t = 1/2$, that is, $x = y$, a contradiction. Therefore we conclude that $\|x+y\| < 2$.

(a) $\Longrightarrow$ (e): Let $x^* \in X^* \setminus \{0\}$, and suppose that there exist $x, u \in X$ with $\|x\| = \|u\| = 1$ such that $\langle x^*, x\rangle = \langle x^*, u\rangle = \|x^*\|_*$. For $t \in (0,1)$ it follows that

$$\|x^*\|_* = (1-t)\langle x^*, x\rangle + t\langle x^*, u\rangle = \langle x^*, (1-t)x + tu\rangle \leq \|x^*\|_* \|(1-t)x + tu\| ,$$

which implies $1 \leq \|(1-t)x + tu\| < 1$, a contradiction. Thus, $x^* \in X^* \setminus \{0\}$ has at most one maximizer on the closed unit ball of X.

(e) $\Longrightarrow$ (c): Suppose that there are $x, u \in X$, $x \neq u$ with $\|x\| = \|u\| = 1$, $\|x+u\| = 2$. Invoking Proposition 3.1.50, there exists $x^* \in X^*$ such that $\|x^*\|_* = 1$ and $\langle x^*, 1/2(x+u)\rangle = 1/2\|x+u\| = 1$. Hence

$$\langle x^*, x\rangle + \langle x^*, u\rangle = 2 . \tag{3.4.10}$$

It holds $\langle x^*, x\rangle \leq 1$ and $\langle x^*, u\rangle \leq 1$. So, from (3.4.10) it follows that $\langle x^*, x\rangle = \langle x^*, u\rangle = 1$, which contradicts the hypothesis. Therefore $\|x+u\| < 2$. □

From the proposition above and its proof we directly obtain the following corollary.

Corollary 3.4.24. *Let X be a Banach space. The following properties are equivalent:*
(a) *X is strictly convex.*
(b) *If $x, u \in X$, $\|x\| = \|u\| = 1$ and $\|x+u\| = 2$, then $x = u$.*
(c) *If $x, u \in X$ satisfy $2\|x\|^2 + 2\|u\|^2 = \|x+u\|^2$, then $x = u$.*
(d) *If $x, u \in X \setminus \{0\}$ satisfy $\|x+u\| = \|x\| + \|u\|$, then $x = tu$ for some $t > 0$.*

A sequential reformulation of Definition 3.4.21 (b), (c) gives the following characterization of uniform convexity and local uniform convexity.

Proposition 3.4.25. *Let X be a Banach space.*

(a) *X is uniformly convex if and only if for every $\{x_n\}_{n\geq1}, \{u_n\}_{n\geq1} \subseteq \overline{B}_1^X$ such that $\|x_n + u_n\| \to 2$, we have $\|x_n - u_n\| \to 0$ as $n \to \infty$.*

(b) *X is locally uniformly convex if and only if for any $x \in X$, $\|x\| = 1$ and for every sequence $\{x_n\}_{n\geq1} \subseteq X$ with $\|x_n\| = 1$ for all $n \in \mathbb{N}$ such that $\|x_n + x\| \to 2$, we have $\|x_n - x\| \to 0$.*

Remark 3.4.26. In the characterizations above, the sequence can be replaced by nets.

Another characterization of uniform convexity is given by the next proposition.

Proposition 3.4.27. *If X is a Banach space, then X is uniformly convex if and only if for every sequences $\{x_n\}_{n\geq1}, \{u_n\}_{n\geq1} \subseteq X$ with $\{x_n\}_{n\geq1}$ bounded such that*

$$2\|x_n\|^2 + 2\|u_n\|^2 - \|x_n + u_n\|^2 \to 0 \quad \text{as } n \to \infty\,,$$

we have $\|x_n - u_n\| \to 0$ as $n \to \infty$.

Proof. $\Longrightarrow$: Note that

$$\begin{aligned}(\|x_n\| - \|u_n\|)^2 &= 2\|x_n\|^2 + 2\|u_n\|^2 - (\|x_n\| + \|u_n\|)^2 \\ &\leq 2\|x_n\|^2 + 2\|u_n\|^2 - \|x_n + u_n\|^2\end{aligned}$$

for all $n \in \mathbb{N}$. Hence $\|x_n\| - \|u_n\| \to 0$ as $n \to \infty$. Therefore if $\|x_n\| \to 0$ or $\|u_n\| \to 0$, then $\|x_n - u_n\| \to 0$. So we may assume that there exists $\varepsilon > 0$ such that

$$\|x_n\| \geq \varepsilon \quad \text{and} \quad \|u_n\| \geq \varepsilon \quad \text{for all } n \in \mathbb{N}\,.$$

Let $y_n = x_n/\|x_n\|$, $v_n = u_n/\|u_n\|$ with $n \in \mathbb{N}$. Then $\|y_n\| = \|v_n\| = 1$ for all $n \in \mathbb{N}$ and $\|y_n + v_n\| \to 2$. It follows that $\|y_n - v_n\| \to 0$ and so $\|x_n - u_n\| \to 0$.

$\Longleftarrow$: This implication is obvious; see Proposition 3.4.25. □

Uniformly convex Banach spaces are reflexive. The result is known as the "Milman–Pettis Theorem".

Theorem 3.4.28 (Milman–Pettis Theorem). *If X is a uniformly convex Banach space, then X is reflexive.*

Proof. Let $x^{**} \in \overline{B}_1^{X^{**}}$. Invoking the Goldstine's Theorem (see Theorem 3.3.37), we can find a net $\{x_\alpha\}_{\alpha\in I} \subseteq \overline{B}_1^X$ such that $x_\alpha \xrightarrow{w^*} x^{**}$ in X^{**}. Exploiting the w*-lower semicontinuity of the norm $\|\cdot\|_{**}$ on X^{**} (see Proposition 3.3.31 (c)), we see that $\|x_\alpha + x_\beta\| \to 2$. Applying Proposition 3.4.25 (a) gives $\|x_\alpha - x_\beta\| \to 0$, which implies that $\{x_\alpha\}_{\alpha\in I} \subseteq X$ is a

Cauchy net. The completeness of X implies that $x_\alpha \to x^{**} \in X$ and so $X = X^{**}$, that is, X is reflexive. □

In Remark 3.3.17 we mentioned that in the Banach space l^1 for sequences, weak and norm convergences are equivalent. More generally, any Banach space having this property is said to have the **Schur property**.

Example 3.4.29. The Banach space (in fact Hilbert space; see Section 3.5) $l^2 = \{\hat{x} = (x_n)_{n\geq 1} \in \mathbb{R}^{\mathbb{N}} : \sum_{n\geq 1} x_n^2 < \infty\}$ does not have the Schur property. Since l^2 is a Hilbert space, we have $(l^2)^* = l^2$, see Theorem 3.5.21. Let $e_n = (0, \ldots, 1, 0, \ldots)$ with 1 at the n-th spot. Then for every $\hat{x}^* \in (l^2)^* = l^2$ we have $\langle x^*, e_n \rangle \to 0$, that is, $e_n \xrightarrow{w^*} 0$. On the other hand $\|e_n\| = 1$ for all $n \in \mathbb{N}$ and so $e_n \not\to 0$ in the norm topology.

However l^2 as well as every Hilbert space has the following weakened version of the Schur property.

Definition 3.4.30. A normed space X is said to have the **Kadec–Klee property** if it satisfies the following condition:

$$\text{For every sequence } \{x_n\}_{n\geq 1} \subseteq X \text{ such that } x_n \xrightarrow{w} x \text{ in } X$$
$$\text{and } \|x_n\| \to \|x\|, \text{ we have } x_n \to x \text{ in } X.$$

Remark 3.4.31. The names **Radon–Riesz property** or **property (H)** are also used in the literature.

Proposition 3.4.32. *If X is a locally uniformly convex Banach space, then X has the Kadec–Klee property.*

Proof. Consider $x_n \xrightarrow{w} x$ in X. Evidently we may assume that $x \neq 0$. Let $u \in X, u \neq 0$. Let $y_n = x_n/\|x_n\|, y = x/\|x\|$ with $n \in \mathbb{N}$. Then $\|y_n\| = \|y\| = 1$ for all $n \in \mathbb{N}$ and $y_n \xrightarrow{w} y$ in X. Hence

$$2 = 2\|y\| \leq \liminf_{n\to\infty} \|y_n + y\| \leq \limsup_{n\to\infty} \|y_n + y\| \leq \lim_{n\to\infty} \|y_n\| + \|y\| = 2\,;$$

see Proposition 3.3.13 (c). Then $\lim_{n\to\infty} \|y_n + y\| = 2$. Proposition 3.4.25 (b) implies that $\|y_n - y\| \to 0$ since X is locally uniformly convex. □

3.5 Hilbert Spaces

In this section we turn our attention to Hilbert spaces, which are Banach spaces with some additional structure, resulting from the presence of an inner product. The inner product supplies a very rich structure, which leads to important simplifications and makes Hilbert spaces the infinite dimensional analog of Euclidean spaces.

Definition 3.5.1. Let H be a vector space over the field $\mathbb{F}$ with $\mathbb{F} = \mathbb{R}$ or $\mathbb{F} = \mathbb{C}$. An **inner product** on X is a map $(\cdot,\cdot): H \times H \to \mathbb{F}$ such that
(a) $(\lambda x + u, y) = \lambda(x,y) + (u,y)$ for all $x, u, y \in H$ and for all $\lambda \in \mathbb{F}$ (linearity);
(b) $(x,u) = \overline{(u,x)}$ for all $x, u \in H$ (conjugate symmetry);
(c) $(x,x) \geq 0$ and $(x,x) = 0$ if and only if $x = 0$ (positive definiteness).

Remark 3.5.2. Linearity in (a) in fact means linearity in the first argument. In the second argument the map is conjugate linear. Property (b) is sometimes called **Hermitian symmetry**.

The next result is of fundamental importance and is known as the "Cauchy–Bunyakowsky–Schwarz inequality".

Proposition 3.5.3 (Cauchy-Bunyakowsky-Schwarz inequality). *If H is a vector space with inner product $(\cdot,\cdot)$, then $|(x,u)|^2 \leq (x,x)(u,u)$ for all $x, u \in H$.*

Proof. Let $x, u \in H$ and let $\lambda \in \mathbb{F}$. Then it follows that

$$0 \leq (x - \lambda u, x - \lambda u) = (x,x) - \overline{\lambda}(x,u) - \lambda\overline{(x,u)} + |\lambda|^2(u,u)\,.$$

Choosing $\lambda = (x,u)/\vartheta$ with $\vartheta > 0$ results in

$$0 \leq (x,x) - \frac{1}{\vartheta}\left(2 - \frac{(u,u)}{\vartheta}\right)|(x,u)|^2\,.$$

If $u \neq 0$, then choose $\vartheta = (u,u)$ to get the desired inequality. If $u = 0$, then $(x,u) = 0$ and so the inequality holds trivially. □

Proposition 3.5.4. *If H is a vector space with inner product $(\cdot,\cdot)$, then $\|x\| = (x,x)^{1/2}$ for all $x \in H$ defines a norm on H.*

Proof. We only need to verify the triangle inequality. So, let $x, u \in H$. Then, using Proposition 3.5.3, it follows that

$$\begin{aligned}\|x+u\|^2 &= (x+u, x+u) = (x,x) + (x,u) + (u,x) + (u,u)\\ &= \|x\|^2 + 2\,\mathrm{Re}(x,u) + \|u\|^2 \leq \|x\|^2 + 2|(x,u)| + \|u\|^2\\ &\leq \|x\|^2 + 2\|x\|\|u\| + \|u\|^2 = (\|x\| + \|u\|)^2\,.\end{aligned}$$

This shows the assertion. □

Remark 3.5.5. A vector space with an inner product will be referred as an **inner product space**. Usually we will not explicitly mention the inner product unless we want to distinguish between different inner products defined on H. The norm $\|\cdot\|$ defined in Proposition 3.5.4 is the norm defined (induced or generated) by the inner product $(\cdot,\cdot)$.

At this point it is natural to ask when a norm is defined by an inner product. The next proposition will lead to a necessary and sufficient condition for this to happen.

Proposition 3.5.6. *If H is an inner product space, then the following hold:*
(a) *Parallelogram law: For all $x, u \in H$ we have*

$$\|x + u\|^2 + \|x - u\|^2 = 2(\|x\|^2 + \|u\|^2)\,.$$

(b) *Polarization identities: For all $x, u \in H$ we have*

$$(x, u) = \frac{1}{4}[\|x + u\|^2 - \|x - u\|^2 + i\|x + iu\|^2 - i\|x - iu\|^2] \quad \text{if } \mathbb{F} = \mathbb{C}\,,$$
$$(x, u) = \frac{1}{4}[\|x + u\|^2 - \|x - u\|^2] \quad \text{if } \mathbb{F} = \mathbb{R}\,.$$

Proof. (a) For all $x, u \in H$ and for all $\lambda \in \mathbb{F}$ one gets

$$\begin{aligned}\|x + \lambda u\|^2 &= \|x\|^2 + 2\,\mathrm{Re}(\bar{\lambda}(x, u)) + |\lambda|^2\|u\|^2 \\ &= \|x\|^2 + 2[\mathrm{Re}\,\lambda\,\mathrm{Re}(x, u) - \mathrm{im}\,\lambda\,\mathrm{im}(x, u)] + |\lambda|^2\|u\|^2\,.\end{aligned} \tag{3.5.1}$$

Choosing $\lambda = 1$ and $\lambda = -1$ in (3.5.1) and adding these equalities, we obtain the desired parallelogram law.

(b) Choosing $\lambda = 1$ and $\lambda = -1$ in (3.5.1) and subtracting, we get the real polarization identity, that is, the case $\mathbb{F} = \mathbb{R}$. Choosing $\lambda = i$ and $\lambda = -i$ in (3.5.1) and subtracting, we obtain the complex polarization identity, that is, the case $\mathbb{F} = \mathbb{C}$. □

The next theorem provides a necessary and sufficient condition for a norm to be generated by an inner product. For a proof of this result, we refer to Weidmann [332, p. 9].

Theorem 3.5.7. *A norm on a vector space H is defined by an inner product if and only if it satisfies the parallelogram law. Moreover, if the norm on H satisfies the parallelogram law, then the unique inner product defining the norm is given by the polarization identities; see Proposition* 3.5.6 *(b).*

Definition 3.5.8. A **Hilbert space** is a complete inner product space.

Remark 3.5.9. So, according to Theorem 3.5.7, a Hilbert space is a Banach space whose norm satisfies the parallelogram law.

Theorem 3.5.10. *Every Hilbert space H is uniformly convex, hence reflexive; see Theorem* 3.4.28.

Proof. Let $\varepsilon > 0$ and let $x, u \in H$ with $\|x\| \le 1$, $\|u\| \le 1$ and $\|x - u\| \ge \varepsilon$. Using the parallelogram law (see Proposition 3.5.6 (a)), we derive $\|(x + u)/2\|^2 \le 1 - \varepsilon^2/4$, which implies that

$$\frac{1}{2}\|x+u\| \le 1-\delta \quad \text{with } \delta = 1-\left(1-\frac{\varepsilon^2}{4}\right)^{\frac{1}{2}} > 0\,.$$

Therefore H is uniformly convex; see Definition 3.4.21 (b). □

The next notion is particular to inner product spaces and gives them the extra structure with respect to general Banach spaces.

Definition 3.5.11. Let H be an inner product space and $x, u \in H$. We say that x, u are **orthogonal** denoted by $x \perp u$ if $(x, u) = 0$. If $x \in H$ and $C \subseteq H$, then we say that x is orthogonal to C denoted by $x \perp C$ if $x \perp u$ for all $u \in C$. Finally if $C, D \subseteq X$, we say that the two sets are orthogonal, denoted by $C \perp D$ if $x \perp u$ for all $x \in C$ and for all $u \in D$. We say that $C \subseteq X$ is an **orthogonal set** if $x \perp u$ for all $x, u \in C$ with $x \neq u$.

Remark 3.5.12. Clearly, $x \perp u$ if and only if $u \perp x$. Hence $C \perp D$ if and only if $D \perp C$. Moreover, $C \perp D$ implies $C \cap D = \{0\}$.

The next result is an extension of the classical "Pythagorean Theorem".

Theorem 3.5.13 (Generalized Pythagorean Theorem). *If H is an inner product space and $\{x_k\}_{k=0}^n \subseteq H$ is a finite orthogonal set, then*

$$\left\|\sum_{k=0}^{n} x_k\right\|^2 = \sum_{k=0}^{n} \|x_k\|^2\,.$$

Proof. First suppose that $n = 1$, that is, we have a pair $x_0, x_1 \in X$ of orthogonal vectors. Since $x_0 \perp x_1$ we derive

$$\|x_0 + x_1\|^2 = (x_0 + x_1, x_0 + x_1) = \|x_0\|^2 + 2\,\mathrm{Re}(x_0, x_1) + \|x_1\|^2 = \|x_0\|^2 + \|x_1\|^2\,.$$

So, the result holds for $n = 1$. Proceeding by induction, suppose that it holds for some $n \in \mathbb{N}$, that is

$$\left\|\sum_{k=0}^{n} x_k\right\|^2 = \sum_{k=0}^{n} \|x_k\|^2 \quad \text{for every orthogonal set } \{x_k\}_{k=0}^n \subseteq H\,. \tag{3.5.2}$$

Let $\{x_k\}_{k=0}^{n+1} \subseteq H$ be an arbitrary orthogonal set. Since $x_{n+1} \perp \{x_k\}_{k=0}^n$ it follows that $x_{n+1} \perp \sum_{k=0}^n x_k$, and hence

$$\left\|\sum_{k=0}^{n+1} x_k\right\|^2 = \left\|\sum_{k=0}^{n} x_k + x_{n+1}\right\|^2 = \left\|\sum_{k=0}^{n} x_k\right\|^2 + \|x_{n+1}\|^2 = \sum_{k=0}^{n+1} \|x_k\|^2\,;$$

see (3.5.2). So, the induction is complete and the Generalized Pythagorean Theorem holds. □

We can state an infinite version of the Pythagorean Theorem.

Theorem 3.5.14. *If H is an inner product space and $\{x_k\}_{k\geq 1} \subseteq H$ is an orthogonal sequence, then the following hold:*
(a) $\sum_{k\geq 1} x_k$ *exists in X implies that* $\sum_{k\geq 1} \|x_k\|^2 < \infty$ *and* $\| \sum_{k\geq 1} x_k\|^2 = \sum_{k\geq 1} \|x_k\|^2$.
(b) *If H is a Hilbert space and* $\sum_{k\geq 1} \|x_k\|^2 < \infty$, *then* $\sum_{k\geq 1} x_k$ *exists in H.*

Proof. (a) By hypothesis we have

$$\sum_{k=1}^{n} x_k \to \sum_{k\geq 1} x_k \quad \text{in } H \text{ as } n \to \infty\,,$$

which implies that

$$\left\| \sum_{k=1}^{n} x_k \right\|^2 \to \left\| \sum_{k\geq 1} x_k \right\|^2 \quad \text{as } n \to \infty\,. \tag{3.5.3}$$

From Theorem 3.5.13 we obtain

$$\left\| \sum_{k=1}^{n} x_k \right\|^2 = \sum_{k=1}^{n} \|x_k\|^2 \quad \text{for every } n \in \mathbb{N}\,.$$

Hence, due to (3.5.3)

$$\sum_{k=1}^{n} \|x_k\|^2 \to \left\| \sum_{k\geq 1} x_k \right\|^2 \quad \text{as } n \to \infty\,.$$

Therefore

$$\left\| \sum_{k\geq 1} x_k \right\|^2 = \sum_{k\geq 1} \|x_k\|^2 < \infty\,.$$

(b) For $m > n$, it holds

$$\left\| \sum_{k=1}^{m} x_k - \sum_{k=1}^{n} x_k \right\|^2 = \left\| \sum_{k=n+1}^{m} x_k \right\|^2 = \sum_{k=n+1}^{m} \|x_k\|^2\,;$$

see Theorem 3.5.13. Hence

$$\left\| \sum_{k=1}^{m} x_k - \sum_{k=1}^{n} x_k \right\|^2 \to 0 \quad \text{as } n \to \infty\,.$$

Therefore, $\{\sum_{k=1}^{n} x_k\}_{n\in\mathbb{N}} \subseteq H$ is a Cauchy sequence. Since H is a Hilbert space it follows that $\sum_{k=1}^{n} x_k \to \sum_{k\geq 1} x_k$ in H as $n \to \infty$. □

Corollary 3.5.15. *If H is a Hilbert space and $\{x_k\}_{k\geq1} \subseteq H$ is an orthogonal sequence, then $\sum_{k\geq1} \|x_k\|^2 < +\infty$ if and only if $\sum_{k\geq1} x_k$ exists in H and $\|\sum_{k\geq1} x_k\|^2 = \sum_{k\geq1} \|x_k\|^2$.*

Example 3.5.16. Two classical examples of Hilbert spaces are the following ones:

(a) $\mathbb{R}^N$ equipped with the Euclidean inner product

$$(\hat{x}, \hat{u}) = \sum_{k=1}^{N} x_k u_k \quad \text{with } \hat{x} = (x_k)_{k=1}^N,\ \hat{u} = (u_k)_{k=1}^N \in \mathbb{R}^N .$$

(b) The Banach space $l^2 = \{\hat{x} = (x_k)_{k\geq1} \in \mathbb{R}^{\mathbb{N}} : \sum_{k\geq1} x_k^2 < \infty\}$ equipped with the inner product

$$(\hat{x}, \hat{u}) = \sum_{k\geq1} x_k u_k \quad \text{for all } \hat{x}, \hat{u} \in l^2 .$$

Remark 3.5.17. The other sequence Banach spaces $l^p = \{\hat{x} = (x_k)_{k\geq1} \in \mathbb{R}^{\mathbb{N}} : \sum_{k\geq1} |x_k|^p < \infty\}$ with $1 < p < \infty$ and $p \neq 2$ are not Hilbert spaces. We can easily see that the parallelogram law fails; see Theorem 3.5.7.

Now we present a basic property of closed convex sets in a Hilbert space. From now on all Hilbert spaces considered are real, that is, $\mathbb{F} = \mathbb{R}$.

Theorem 3.5.18. *If H is a Hilbert space and $C \subseteq H$ is nonempty, closed, and convex, then for any given $x \in H$ there exists a unique element $p_C(x) \in C$ such that $\|x - p_C(x)\| \leq \|x - u\|$ for all $u \in C$.*

Proof. By translating things if necessary, we assume that $x = 0$. Let $\eta = \inf[\|u\| : u \in C]$ and consider the minimizing sequence $\{u_n\}_{n\geq1} \subseteq C$, that is, $\|u_n\| \searrow \eta$ as $n \to \infty$. From the parallelogram law (see Proposition 3.5.6 (a)), one gets for $m > n$, that

$$\|u_m - u_n\|^2 = 2\|u_m\|^2 + 2\|u_n\|^2 - 4\left\|\frac{u_m + u_n}{2}\right\|^2 \leq 2\|u_m\|^2 + 2\|u_n\|^2 - 4\eta^2 ,$$

since C is convex. Hence $\|u_m - u_n\|^2 \to 0$ as $m, n \to \infty$ and so, $\{u_n\}_{n\geq1} \subseteq C$ is a Cauchy sequence. Thus $u_n \to u$ in H and $\|u\| = \eta$.

Now we show the uniqueness of this best approximation (minimum norm) point u. Suppose that some $v \in C$ satisfies $\|v\| = \eta$. A new application of the parallelogram law gives

$$0 \leq \|u - v\|^2 = 2\|u\|^2 + \|v\|^2 - 4\left\|\frac{u + v}{2}\right\|^2 \leq 4\eta^2 - 4\eta^2 = 0 ,$$

recall again the convexity of C. Then $u = v$. So, $u = p_C(x)$ is the unique best approximation of x in C. □

Definition 3.5.19. The map $p_C: H \to C$ assigning to each $x \in H$ its unique best approximation from C is called the **metric projection of H onto C**.

The next proposition establishes the main properties of the metric projection map.

Proposition 3.5.20. *If H is a Hilbert space, $C \subseteq H$ is nonempty, closed, and convex, and $p_C: X \to C$ is the metric projection map. Then the following hold:*
(a) $p_C|_C = \mathrm{id}_C$;
(b) *if $x \in X \setminus C$, then $p_C(x) \in \mathrm{bd}\, C$;*
(c) $(x - p_C(x), u - p_C(x)) \leq 0$ *for all $u \in C$;*
(d) $\|p_C(x) - p_C(y)\| \leq \|x - y\|$ *for all $x, y \in H$;*
(e) *if C is a closed vector subspace of H, then $x - p_C(x) \perp C$ and $p_C \in L(H)$.*

Proof. (a) This is obvious.

(b) Let $t \in (0, 1)$ and let $x_t = (1 - t)x + tp_C(x)$. We get

$$\|x - x_t\| = t\|x - p_C(x)\| < \|x - p_C(x)\| .$$

So, if $p_C(x) \in \mathrm{int}\, C$, then for $t \in (0, 1)$ close to 1 it follows $x_t \in C$, a contradiction. Hence $p_C(x) \in \mathrm{bd}\, C$.

(c) Let $x \in X$, $u \in C$ and $t \in (0, 1)$. The convexity of C implies

$$\begin{aligned}\|x - p_C(x)\|^2 &\leq \|x - ((1 - t)p_C(x) + tu)\|^2 = \|x - p_C(x) - t(u - p_C(x))\|^2 \\ &= \|x - p_C(x)\|^2 - 2t(x - p_C(x), u - p_C(x)) + t^2\|u - p_C(x)\|^2 ,\end{aligned}$$

which implies $2(x - p_C(x), u - p_C(x)) \leq t\|u - p_C(x)\|^2$. We let $t \searrow 0$ and obtain $(x - p_C(x), u - p_C(x)) \leq 0$ for all $u \in C$.

(d) Let $x, y \in H$. Using part (c) with $u = p_C(y) \in C$ it follows that

$$(x - p_C(x), p_C(y) - p_C(x)) \leq 0 . \tag{3.5.4}$$

Reversing the roles of $x, y \in H$ we also obtain

$$(y - p_C(y), p_C(x) - p_C(y)) \leq 0 . \tag{3.5.5}$$

Adding (3.5.4) and (3.5.5) yields

$$(x - y, p_C(y) - p_C(x)) + (p_C(y) - p_C(x), p_C(y) - p_C(x)) \leq 0 ,$$

which leads to

$$\|p_C(x) - p_C(y)\|^2 \leq \|x - y\|\|p_C(x) - p_C(y)\| ;$$

see Proposition 3.5.3. This finally gives $\|p_C(x) - p_C(y)\| \leq \|x - y\|$ for all $x, y \in H$.

(e) For every $u \in C$ and $\vartheta \in \mathbb{R}$ we get

$$\begin{aligned}\|x - p_C(x)\|^2 &\le \|x - [p_C(x) + \vartheta(\pm u)]\|^2\\ &= \|x - p_C(x)\|^2 \mp 2\vartheta(x - p_C(x), u) + \vartheta^2\|u\|^2 ,\end{aligned}$$

which turns into

$$\pm 2(x - p_C(x), u) \le \vartheta\|u\|^2 .$$

Letting $\vartheta \searrow 0$, it results in $\pm(x - p_C(x), u) \le 0$ for all $u \in C$ and so $(x - p_C(x), u) = 0$ for all $u \in H$ since $C \subseteq H$ is a subspace. This gives

$$x - p_C(x) \perp C . \tag{3.5.6}$$

Finally note that using (3.5.6), for all $u, x \in H$, leads to

$$(p_C(x + y) - (p_C(x) + p_C(y)), u) = 0 \quad \text{for all } u \in C .$$

Hence, $p_C(x + y) = p_C(x) + p_C(y)$, that is, p_C is additive. Clearly $p_C(0) = 0$ and for all $\lambda \in \mathbb{R} \setminus \{0\}$ it follows that $(p_C(\lambda x) - \lambda p_C(x), u) = 0$ for all $u \in C$, which shows that $p_C(\lambda x) = \lambda p_C(x)$, that is, p_C is homogeneous. Therefore $p_C \in L(H)$. □

A remarkable application of this result is a characterization of the topological dual of a Hilbert space. The result is known as the "Riesz–Fréchet Representation Theorem for Hilbert Spaces".

Theorem 3.5.21 (Riesz–Fréchet Representation Theorem for Hilbert Spaces). *If H is a Hilbert space and $x^* \in H^*$, then there exists a unique $x_0 \in H$ such that $\langle x^*, y\rangle = (x_0, y)$ for all $y \in H$ and $\|x^*\|_* = \|x_0\|$.*

Proof. Let $V = (x^*)^{-1}(0)$. This is a closed subspace of H. We may assume that $V \neq H$ otherwise $x^* = 0$ and the result is trivially true with $x_0 = 0$. Let $u_0 \in H \setminus V$, $u_1 = p_V(u_0)$ and $u = (u_0 - u_1)/(\|u_0 - u_1\|)$. Then $\|u\| = 1$ and $(u, x) = 0$ for all $x \in V$; see Proposition 3.5.20 (e). Therefore $u \notin V$. For any $y \in H$, we set

$$z = y - tu \quad \text{with } t = \frac{\langle x^*, y\rangle}{\langle x^*, u\rangle} .$$

Note that $\langle x^*, u\rangle \neq 0$ since $u \notin V$. Then $\langle x^*, z\rangle = 0$ and so $z \in V$. Therefore $(u, z) = 0$, which implies that

$$\langle x^*, y\rangle = \langle x^*, u\rangle(u, y) \quad \text{for all } y \in H . \tag{3.5.7}$$

So if we set $x_0 = \langle x^*, u\rangle u$, then it follows that $\langle x^*, y\rangle = (x_0, y)$ for all $y \in H$. Clearly this x_0 is unique. Evidently, thanks to Proposition 3.5.3 one gets $\|x_0\| \le |\langle x^*, u\rangle|\|u\| =$

$|\langle x^*, u\rangle| \le \|x^*\|_*$. Moreover, from (3.5.7) and Proposition 3.5.3 we conclude that $\|x^*\|_* \le |\langle x^*, u\rangle|\|u\| = \|x_0\|$, which implies $\|x^*\|_* = \|x_0\|$. □

Remark 3.5.22. According to this theorem there is a surjective linear isometry from H^* into H. This means that we can identify H^* with H, that is, a Hilbert space is self-dual. However, it is not always possible to do this identification. This is the case of evolution triples, which we will discuss in Section 4.2.

Definition 3.5.23. Let H be a Hilbert space and $C \subseteq H$. The **orthogonal complement** $C^\perp$ of C is the set

$$C^\perp = \{x \in H : (x, u) = 0 \text{ for all } u \in C\} .$$

On account of Theorem 3.5.21 the orthogonal complement of C is simply the annihilator of C introduced in Definition 3.2.24. Evidently $C^\perp$ is a closed vector subspace of H. Moreover, $C^{\perp\perp} = (C^\perp)^\perp$.

Remark 3.5.24. Clearly $\{0\}^\perp = H, H^\perp = \{0\}$. Moreover $C \perp C^\perp, C \cap C^\perp \subseteq \{0\}$ and if $0 \in C$, then $C \cap C^\perp = \{0\}$. Also, if $C, D \subseteq H$ are nonempty sets, then $C \perp D$ if and only if $C \subseteq D^\perp$. Since $\perp$ is a symmetric relation, that is, $C \perp D$ if and only if $D \perp C$, we also obtain that $D \subseteq C^\perp$. Moreover, $C \perp D$ implies that $C \cap D \subseteq \{0\}$. We can easily see that

$$\begin{aligned} &C \subseteq D \text{ implies that } D^\perp \subseteq C^\perp \text{ and } C^{\perp\perp} \subseteq D^{\perp\perp} , \\ &C^\perp = (\operatorname{span} C)^\perp = (\overline{\operatorname{span}} C)^\perp . \end{aligned} \tag{3.5.8}$$

In addition, since $C \perp C^\perp$ and $C^\perp \perp C^{\perp\perp}$, we derive that $C \subseteq C^{\perp\perp}$ and $C^\perp \subseteq C^{\perp\perp\perp}$, here $C^{\perp\perp\perp} = (C^{\perp\perp})^\perp$. Therefore we have $C \subseteq C^{\perp\perp}$ and $C^\perp = C^{\perp\perp\perp}$. Finally, if $C \subseteq H$ is a vector subspace, then $C^{\perp\perp} = \overline{C}$ and $C^\perp = \{0\}$ if and only if C is dense in H.

Proposition 3.5.25. *If H is a Hilbert space and V is a closed vector subspace of H, then $H = V \oplus V^\perp$; see Definition* 3.2.27.

Proof. It is easy to see that $V \oplus V^\perp$ is a closed vector subspace of H. Suppose that $H \ne V \oplus V^\perp$. Then there exists $u \in H, u \ne 0$ such that $u \perp V \oplus V^\perp$. We have $u \in V^\perp \cap V^{\perp\perp} = \{0\}$, a contradiction. Therefore, $H = V \oplus V^\perp$. □

From Propositions 3.5.20 and 3.5.25 we infer at once the so-called "Projection Theorem".

Theorem 3.5.26 (Projection Theorem). *If H is a Hilbert space and V is a closed vector subspace of H, then there exists a unique pair of continuous linear operators $P: H \to V$ and $Q: H \to V^\perp$ such that*

(a) *$x \in V$ implies that $P(x) = x$, $Q(x) = 0$ and $y \in V^\perp$ implies that $P(y) = 0$, $Q(y) = y$;*
(b) *$P(x) = p_V(x)$ and $Q(x) = p_{V^\perp}(x)$;*
(c) *for all $x \in H$ one has $\|x\|^2 = \|P(x)\|^2 + \|Q(x)\|^2$.*

Now we turn our attention to orthogonal sets that lead to bases for Hilbert spaces. First we recall the following basic notion from linear algebra.

Definition 3.5.27. Let X be a vector space and $C \subseteq X$. We say that C is **linearly independent** if every $x \in C$ is not a linear combination of vectors in $C \setminus \{x\}$, that is, $x \notin \operatorname{span}[C \setminus \{x\}]$. A set $C \subseteq X$ that is not linearly independent is said to be **linearly dependent**.

Remark 3.5.28. The empty set $\emptyset$ is linearly independent. Also, the singleton $C = \{x\}$, $x \neq 0$ is linearly independent. Any set $C \subseteq X$, which contains the origin, is linearly independent. Finally, $C \subseteq X$ is linearly independent if and only if every finite subset of C is linearly independent.

Proposition 3.5.29. *If H is an inner product space and $C \subseteq H$ is an orthogonal set consisting of nonzero vectors, then C is linearly independent.*

Proof. Arguing by contradiction, suppose that there is a sequence $\{x_k\}_{k=0}^n$ with $n \geq 1$ such that $x_0 = \sum_{k=1}^n \lambda_k x_k$ with $\lambda_k \in \mathbb{R}$, $k = 1, \ldots, n$; see Remark 3.5.28. Exploiting the orthogonality of the set C yields $\|x_0\|^2 = \sum_{k=1}^n \lambda_k (x_k, x_0) = 0$, a contradiction. □

Definition 3.5.30. Let H be an inner product space and $C \subseteq X$. We say that C is an **orthonormal set** if it is an orthogonal set consisting of vectors with unit norm, that is, unit vectors.

Remark 3.5.31. Every orthogonal set consisting of nonzero vectors can be normalized. Indeed, if C is an orthogonal set such that $x \neq 0$ for all $x \in C$, then $\{x/\|x\| : x \in C\}$ is an orthonormal set.

From Proposition 3.5.29 we directly obtain the following result.

Proposition 3.5.32. *If H is an inner product space and $C \subseteq H$ is an orthonormal set, then C is linearly independent.*

The next proposition is an immediate consequence of Definition 3.5.30.

Proposition 3.5.33. *If H is an inner product space, $C \subseteq X$ is an orthonormal set, and $x \in H$ with $\|x\| = 1$ and $x \perp C$, then $C \cup \{x\}$ is an orthonormal set as well.*

Definition 3.5.34. Let H be an inner product space and let $\mathcal{L}$ be the family of all orthonormal subsets of H. Evidently $\mathcal{L} \neq \emptyset$ since $C = \{x\}$ with $\|x\| = 1$ is orthonormal. A set $C \in \mathcal{L}$ is **maximal orthonormal** if there is no set $C' \in \mathcal{L}$ such that $C' \neq C$ and $C \subseteq C'$.

Proposition 3.5.35. *If H is an inner product space and $C \subseteq H$ is an orthonormal set, then the following statements are equivalent:*
(a) *C is a maximal orthonormal set.*
(b) *There is no unit vector $x \in X$, that is, $\|x\| = 1$, such that $C \cup \{x\}$ is an orthonormal set.*
(c) $C^{\perp} = \{0\}$.

Proof. (a) $\Longrightarrow$ (b): Otherwise $C \cup \{x\}$ contradicts the maximality of C.

(b) $\Longrightarrow$ (c): Let $y \in H$ be such that $y \perp C$. Let $x = y/\|y\|$. Then $\|x\| = 1$ and $C \cup \{x\}$ is an orthonormal set, a contradiction.

(c) $\Longrightarrow$ (a): Arguing by contradiction, suppose that there exists an orthonormal set $C' \subseteq X$ such that $C' \setminus C \neq \emptyset$. Let $x \in C' \setminus C$. Then $\|x\| = 1$ with $x \perp C$, a contradiction. □

Proposition 3.5.36. *If H is an inner product space and $C \subseteq H$ is an orthonormal set, then there exists a maximal orthonormal set $C_0 \subseteq H$ such that $C \subseteq C_0$.*

Proof. Let $\mathcal{L}_C = \{D \in 2^H : D \text{ is an orthonormal set}, C \subseteq D\}$ and let $\mathcal{D}$ be a chain in $\mathcal{L}_C$. Let $\bigcup \mathcal{D} = \bigcup_{D \in \mathcal{D}} D$ and consider $x, u \in \bigcup \mathcal{D}$ with $x \neq u$. Then $x \in D_x \in \mathcal{D}$ and $u \in D_u \in \mathcal{D}$. Since $\mathcal{D}$ is a chain, we may assume that $D_x \subseteq D_u$. Hence $x, u \in D_u$ and so $\bigcup \mathcal{D} \in \mathcal{L}_C$. Invoking Zorn's Lemma (see Section 1.4), $\mathcal{L}_C$ has a maximal element C_0 such that $C \subseteq C_0$ and C_0 is orthogonal. If we can find a unit vector $x \in H$ such that $C_0 \cup \{x\}$ is orthonormal, then $C_0 \cup \{x\} \in \mathcal{L}_C$ and this contradicts the maximality of C_0. This proves that C_0 is a maximal orthonormal set. □

Now that we have established that maximal orthonormal sets exist, we can show that they span H.

Proposition 3.5.37. *If H is an inner product space and $C \subseteq H$ is an orthonormal set, then the following hold:*
(a) *$\overline{\operatorname{span}} C = H$ implies that C is maximal orthonormal.*
(b) *If H is a Hilbert space and $C \subseteq X$ is maximal orthonormal, then $\overline{\operatorname{span}} C = H$.*

Proof. (a) From (3.5.8) we know that $C^\perp = (\overline{\operatorname{span}} C)^\perp = H^\perp = \{0\}$. Then Proposition 3.5.35 implies that C is maximal orthonormal.

(b) From Proposition 3.5.35 and (3.5.8), we deduce that $0 = C^\perp = (\overline{\operatorname{span}} C)^\perp$. Hence $\overline{\operatorname{span}} C = H$; see Remark 3.5.24. □

Definition 3.5.38. Let H be an inner product space. A set $B \subseteq H$ is an **orthonormal basis** of H if the following hold:
(a) B is an orthonormal set.
(b) $\overline{\operatorname{span}} B = H$.

Remark 3.5.39. According to Proposition 3.5.37, every Hilbert space admits an orthonormal basis. In fact, for Hilbert spaces, the notions of maximal orthonormal set and of orthonormal basis coincide. That is, if H is a Hilbert space, then $B \subseteq H$ is a maximal orthonormal set if and only if $B \subseteq H$ is an orthonormal set. In finite dimensional Hilbert space all orthonormal bases are finite and have cardinality equal to the dimension of the space.

The next proposition establishes a fundamental inequality for inner product spaces known as "Bessel's inequality". First let us see how we interpret summation over an arbitrary index set.

Definition 3.5.40. Let (X, τ) be a Hausdorff topological vector space, I be an arbitrary index set, and $I \ni \alpha \to x_\alpha \in X$ be a map. Then the sum $\sum_{\alpha\in I} x_\alpha$ is defined as follows: Let $\mathcal{F}$ be the family of all finite subsets of I ordered by inclusion. Then $\sum_{\alpha\in I} x_\alpha = x$ if and only if the net $\{\sum_{\alpha\in F} x_\alpha\}_{F\in\mathcal{F}}$ τ-converges to x. This is called **unconditional convergence** since it does not depend on any ordering on the index set I. If $I = \mathbb{N}$, then $\sum_{n\geq 1} x_n = x$ means that $\sum_{n=1}^{m} x_n \xrightarrow{\tau} x$ as $m \to \infty$. Then the series $\sum_{n\geq 1}(-1)^n 1/n$ is convergent but not unconditionally convergent.

Remark 3.5.41. If $X = \mathbb{R}$, then $\sum_{\alpha\in I} x_\alpha = x \in \mathbb{R}$ means that for a given $\varepsilon > 0$ there exists a finite set $F \subseteq I$ such that $|x - \sum_{\alpha\in G} x_\alpha| \leq \varepsilon$ for all finite $F \subseteq G \subseteq I$. On the other hand $\sum_{\alpha\in I} x_\alpha = +\infty$ means that for any given $M > 0$ we can find a finite set $F \subseteq I$ such that $\sum_{\alpha\in G} x_\alpha \geq M$ for all $F \subseteq G \subseteq I$. Also recall that absolutely convergent series can be rearranged (see Amann–Escher [10, p. 201]) and we mention a remarkable result known as the "Orlicz–Pettis Theorem", which says that a series $\sum_{n\geq 1} x_n$ in a Banach space X is weakly unconditionally convergent if and only if it is strongly unconditionally convergent. Finally we mention another important result, the "Dvoretzky–Rogers Theorem", which says that if X is infinite dimensional, then there exists a sequence $\{x_n\}_{n\geq 1} \subseteq X$ such that $\sum_{n\geq 1} x_n$ is unconditional convergent and $\sum_{n\geq 1} \|x_n\| = +\infty$.

Lemma 3.5.42. *If $X = \mathbb{R}$ and $\{x_\alpha\}_{\alpha\in I} \subseteq [0, +\infty)$, then*

$$\sum_{\alpha\in I} x_\alpha = \sup\left[\sum_{\alpha\in F} x_\alpha : F \subseteq I \text{ is finite}\right].$$

Proof. First suppose that $\sum_{\alpha\in I} x_\alpha < +\infty$. Then for a given $\varepsilon > 0$ there exists a finite set $F \subseteq I$ such that

$$\sum_{\alpha\in F} x_\alpha \geq \sum_{\alpha\in I} x_\alpha - \varepsilon .$$

Hence,

$$\sum_{\alpha\in I} x_\alpha \geq \sum_{\alpha\in G} x_\alpha \geq \sum_{\alpha\in F} x_\alpha \geq \sum_{\alpha\in I} x_\alpha - \varepsilon \quad \text{for all finite } F \subseteq G \subseteq I .$$

Therefore

$$\sum_{\alpha\in I} x_\alpha = \sup\left[\sum_{\alpha\in F} x_\alpha : F \subseteq I \text{ finite}\right].$$

Now assume that $\sum_{\alpha\in I} x_\alpha = +\infty$. Then for any given $M > 0$ there exists a finite $F \subseteq I$ such that $\sum_{\alpha\in F} x_\alpha \geq M$, which implies that $\sum_{\alpha\in G} x_\alpha \geq M$ for all finite $F \subseteq G \subseteq I$. Hence,

$$\sup\left[\sum_{\alpha\in F} x_\alpha : F \subseteq I \text{ finite}\right] = +\infty . \qquad \square$$

Remark 3.5.43. If I is uncountable and uncountably many x_α are different from zero, then $\sum_{\alpha\in I} x_\alpha$ cannot converge to a finite limit.

The next result is a fundamental inequality in the theory of Hilbert spaces and is known as "Bessel's inequality".

Proposition 3.5.44 (Bessel's inequality). *If H is an inner product space and $\{x_\alpha\}_{\alpha\in I} \subseteq H$ is an orthonormal set, then $\sum_{\alpha\in I} |(x, x_\alpha)|^2 \le \|x\|^2$ for all $x \in H$.*

Proof. On account of Lemma 3.5.42, we may assume that I is finite. Let $u = \sum_{\alpha\in I}(x, x_\alpha)x_\alpha$. Then $(x, u) = \sum_{\alpha\in I}(x, x_\alpha)^2 = (u, u)$; see Theorem 3.5.13. Therefore $x - u \perp u$ and so $\|x\|^2 = \|x - u\|^2 + \|u\|^2$ due to the Generalized Pythagorean Theorem; see Theorem 3.5.13. Hence, $\|x\|^2 \ge \|u\|^2 = (u, u) = \sum_{\alpha\in I}(x, x_\alpha)^2$. □

Corollary 3.5.45. *If H is an inner product space and $\{x_\alpha\}_{\alpha\in I} \subseteq H$ is an orthonormal set, then for every $x \in H$ the set $\{\alpha \in I: (x, x_\alpha) \ne 0\}$ is countable.*

Remark 3.5.46. We have already mentioned that every Hilbert space has an orthonormal basis. In fact all orthonormal bases of a Hilbert space have the same cardinality, that is, every maximal orthonormal set in an inner product space has the same cardinality.

Proposition 3.5.47. *If H is a separable inner product space, then every orthonormal set in H is countable.*

Proof. Let $B = \{x_\alpha\}_{\alpha\in I} \subseteq H$ be an orthonormal set and let $D = \{u_n\}_{n\ge 1} \subseteq H$ be dense, which is possible since H is separable. Then for any $\alpha \in I$, $\overline{B}_{1/2}(x_\alpha) \cap D \ne \emptyset$. So, there exists $n_\alpha \in \mathbb{N}$ such that $\|x_\alpha - u_{n_\alpha}\| \le 1/2$.

Let $\varphi: I \to \mathbb{N}$ be defined by $\varphi(\alpha) = n_\alpha$. We claim that φ is injective. Using the parallelogram law and the Generalized Pythagorean Theorem, it follows that

$$\begin{aligned}\sqrt{2} = \|x_\alpha - x_\beta\| &= \|x_\alpha - u_{n_\alpha} - x_\beta + u_{n_\beta} + u_{n_\alpha} - u_{n_\beta}\| \\ &\le \|x_\alpha - u_{n_\alpha}\| + \|x_\beta - u_{n_\beta}\| + \|u_{n_\alpha} - u_{n_\beta}\| \le 1 + \|u_{n_\alpha} - u_{n_\beta}\| .\end{aligned}$$

Hence, $\sqrt{2} - 1 \le \|u_{n_\alpha} - u_{n_\beta}\|$ for all $\alpha, \beta \in I$ with $\alpha \ne \beta$. This proves the injectivity of φ, which means that $\operatorname{card} I \le \operatorname{card} \mathbb{N}$ and so I is countable. □

This leads to the following useful characterization of separable Hilbert spaces.

Theorem 3.5.48. *A Hilbert space H is separable if and only if it has a countable orthonormal basis.*

Given a linearly independent sequence one can produce an orthonormal set with the same linear span. The process to achieve this is known as the "Gram–Schmidt Orthonormalization Process".

Proposition 3.5.49 (Gram–Schmidt Orthonormalization Process). *If H is an inner product space and $\{u_n\}_{n\ge1} \subseteq H$ are linearly independent, then there exists an orthonormal sequence $\{x_n\}_{n\ge1} \subseteq H$ such that* $\operatorname{span}\{u_n\}_{n\ge1} = \operatorname{span}\{x_n\}_{n\ge1}$.

Proof. Let $x_1 = u_1/\|u_1\|$. So, the result holds for $n = 1$. Proceeding by induction suppose that we have produced $x_1, \dots, x_{n-1}$. Then we set

$$h_n = u_n - \sum_{k=1}^{n-1} (u_k, x_k) x_k \,.$$

Evidently $h_n \perp x_k$ for all $k = 1, \dots, n-1$ and $h_n \neq 0$ since $u_n \notin \operatorname{span}\{u_k\}_{k=1}^{n-1}$, due to the linear independence of the sequence $\{u_n\}_{n\ge1} \subseteq H$. According to the induction hypothesis, we have $\operatorname{span}\{u_k\}_{k=1}^{n-1} = \operatorname{span}\{x_k\}_{k=1}^{n-1}$. Let $x_n = h_n/\|h_n\|$. Then by induction we have produced the desired orthonormal set $\{x_n\}_{n\ge1} \subseteq H$. □

We conclude this section with a brief look at the notion of the basis for a vector space X. If X is finite dimensional, then it is well-known that a basis is a set $\{e_k\}_{k=1}^n$ such that every $x \in X$ can be written in a unique way as $x = \sum_{k=1}^n \lambda_k e_k$ with $\lambda_k \in \mathbb{R}$ known as the coordinates of x for the given basis. How do we extend this notion to infinite dimensional vector spaces?

Definition 3.5.50. (a) Given a vector space X, a **Hamel basis** is a set $\{e_\alpha\}_{\alpha\in I} \subseteq X$ such that every $x \in X$ can be written in a unique way as $x = \sum_{\alpha\in I} \lambda_\alpha e_\alpha$ with only finite numbers of the real λ_α different from zero. If X is finite dimensional, then a Hamel basis is the usual basis. But in infinite dimensional spaces there are no obvious Hamel bases although they can be shown to exist via Zorn's Lemma.

(b) Let X be a Banach space. A sequence $\{x_n\}_{n\ge1} \subseteq X$ is a **Schauder basis** for X if for each $x \in X$ there exists a unique sequence $\{\lambda_n\}_{n\ge1} \subseteq \mathbb{R}$ such that $x = \sum_{n\ge1} \lambda_n x_n$.

Remark 3.5.51. The Hamel basis is an algebraic notion that does not relate to any topology. A Banach space with a Schauder basis is necessarily separable. Banach [28, p. 111] asked if every infinite dimensional separable Banach space has a Schauder basis. This question was settled in the negative by Enflo [115] who produced a separable reflexive Banach space with no Schauder basis.

3.6 Bounded and Unbounded Linear Operators

Let X, Y be Banach spaces. Recall that by $L(X, Y)$ we denote the Banach space of all bounded linear operators from X into Y. The norm of $L(X, Y)$ is defined by

$$\|A\|_L = \sup\left[\frac{\|A(x)\|_Y}{\|x\|_X} : x \in X \setminus \{0\}\right]; \tag{3.6.1}$$

see Definition 3.1.45. If $X = Y$, we write $L(X,X) = L(X)$.

Definition 3.6.1. (a) The norm (metric) topology induced on $L(X, Y)$ by the norm $\|\cdot\|_L$ (see (3.6.1)) is called the **uniform operator topology** or simply the **norm topology**.

(b) The **strong operator topology** on $L(X, Y)$ is the weakest topology on $L(X, Y)$ for which the maps $e_x : L(X, Y) \to Y$ with $x \in X$ defined by $e_x(A) = A(x)$ for all $A \in L(X, Y)$ are continuous. Then a local basis at the origin consists of the sets

$$\{A \in L(X, Y) : \|A(x_k)\|_Y < \varepsilon \text{ for } k = 1, \dots, n\}$$

with $n \in \mathbb{N}$ and $\varepsilon > 0$. A net $\{A_\alpha\}_{\alpha \in I} \subseteq L(X, Y)$ converges to $A \in L(X, Y)$ in this topology if and only if $\|A_\alpha(x) - A(x)\|_Y \to 0$ for all $x \in X$. We write $A_\alpha \xrightarrow{\text{s}} A$ in $L(X, Y)$.

(c) The **weak operator topology** on $L(X, Y)$ is the weakest topology on $L(X, Y)$ for which the maps $e_{x,y^*} : L(X, Y) \to \mathbb{R}$ with $x \in X$ and $y^* \in Y^*$ defined by $e_{x,y^*}(A) = \langle y^*, A(x)\rangle$ are continuous. Then a local basis at the origin consists of the sets

$$\{A \in L(X, Y) : |\langle y_i^*, A(x_k)\rangle| < \varepsilon \text{ for } k = 1, \dots, n,\ i = 1, \dots, m\}$$

with $n, m \in \mathbb{N}$ and $\varepsilon > 0$. A net $\{A_\alpha\}_{\alpha \in I} \subseteq L(X, Y)$ converges to $A \in L(X, Y)$ in this topology if and only if $|\langle y^*, A_\alpha(x)\rangle - \langle y^*, A(x)\rangle| \to 0$ for all $x \in X, y^* \in Y^*$. We write $A_\alpha \xrightarrow{\text{w}} A$ in $L(X, Y)$.

Remark 3.6.2. Evidently it holds that

$$\text{weak topology} \subseteq \text{strong topology} \subseteq \text{norm topology}.$$

We should not confuse the weak operator topology with the weak topology that we can define on the Banach space $L(X, Y)$. Let V be a third Banach space and consider the map $\vartheta : L(X, Y) \times L(Y, V) \to L(X, V)$ defined by $\vartheta(A, B) = B \circ A$. Then ϑ is jointly continuous for the uniform operator topology but only separately continuous for the strong and weak operator topologies. In general, the strong and weak operator topologies are not first countable and this complicates their study.

Proposition 3.6.3. *If H is a Hilbert space and $\{A_n\}_{n\ge1} \subseteq L(H)$ is a sequence such that $\{(y, A_n(x))\}_{n\ge1}$ is convergent for all $x, y \in H$, then there exists $A \in L(H)$ such that $A_n \xrightarrow{\text{w}} A$.*

Proof. For given $x, y \in H$ we derive $\sup_{n\ge1} |(y, A_n(x))| < \infty$. Invoking Theorem 3.2.1 we obtain that $\sup_{n\ge1} \|A_n(x)\| < \infty$. A second application of Theorem 3.2.1 gives $\sup_{n\ge1} \|A_n\|_L < \infty$.

Let $\xi(x, y) = \lim_{n\to\infty} (y, A_n(x))$. Evidently ξ is bilinear and

$$|\xi(x, y)| \le \limsup_{n\to\infty} |(y, A_n(x))| \le \|y\|\|x\| \Big(\sup_{n\ge1} \|A_n\|_L\Big).$$

Hence ξ is bounded. Then there exists $A \in L(H)$ such that $(y, A(x)) = \xi(x, y)$; see Theorem 3.5.21. Therefore we get $A_n \xrightarrow{\text{w}} A$ in $L(H)$. □

In a similar way we obtain the corresponding result for the strong operator topology.

Proposition 3.6.4. *If X, Y are Banach spaces, $\{A_n\}_{n\geq1} \subseteq L(X,Y)$ and $\{A_n(x)\}_{n\geq1} \subseteq Y$ is a Cauchy sequence for each $x \in X$, then there exists $A \in L(X,Y)$ such that $A_n \xrightarrow{s} A$.*

Remark 3.6.5. Both results fail for nets of operators.

Definition 3.6.6. Let X, Y be normed spaces and $A \in L(X,Y)$. The **adjoint** (or **dual**) operator of A is the unique operator $A^*: Y^* \to X^*$ defined by

$$A^*(y^*) = y^* \circ A \quad \text{for all } y^* \in Y^* .$$

Continuing, the **second adjoint** (or **second dual** or **bidual**) $(A^*)^*$ of A is the unique linear map $A^{**}: X^{**} \to Y^{**}$ such that

$$A^{**}(x^{**}) = x^{**} \circ A^* \quad \text{for all } x^{**} \in X^{**} .$$

The next proposition summarizes the main properties of A^* and A^{**}.

Proposition 3.6.7. *If X, Y are normed spaces and $A, S, T \in L(X,Y)$, then the following hold:*
(a) *$A^* \in L(Y^*, X^*)$ and $\|A^*\|_L = \|A\|_L$.*
(b) *If $\lambda_1, \lambda_2 \in \mathbb{R}$, then $(\lambda_1 S + \lambda_2 T)^* = \lambda_1 S^* + \lambda_2 T^*$.*
(c) *$A^{**}|_X = A$.*
(d) *If V is a third normed space and $B \in L(Y,V)$, then $(B \circ A)^* = A^* \circ B^*$.*
(e) *If A is invertible, that is, A^{-1} exists and $A^{-1} \in L(Y,X)$, then A^* is invertible as well and $(A^*)^{-1} = (A^{-1})^*$.*

Proof. (a) For all $x \in X$ and for all $y^* \in Y^*$ one gets

$$\langle A^*(y^*), x\rangle = \langle y^*, A(x)\rangle \leq \|y^*\|_* \|A(x)\| \leq \|y^*\|_* \|A\|_L \|x\| ,$$

hence $\|A^*(y^*)\| \leq \|y^*\|_* \|A\|_L$, and so $\|A^*\|_L \leq \|A\|_L$. Given $\varepsilon > 0$ there exists $x_0 \in X$ with $\|x_0\| = 1$ such that $\|A\|_L - \varepsilon \leq \|A(x_0)\|$. Let $y^* \in Y^*$ with $\|y^*\|_* = 1$ such that $\langle y^*, A(x_0)\rangle = \|A(x_0)\|$; see Proposition 3.1.50. Then it follows that

$$\langle A^*(y^*), x_0\rangle = \langle y^*, A(x_0)\rangle = \|A(x_0)\| \geq \|A\|_L - \varepsilon ,$$

which gives $\|A^*\|_L \geq \|A\|_L - \varepsilon$. Letting $\varepsilon \searrow 0$, we obtain $\|A^*\|_L \geq \|A\|_L$. Therefore, $A^* \in L(Y^*, X^*)$ and $\|A^*\|_L = \|A\|_L$.

(b) This follows immediately from Definition 3.6.6.

(c) This is also clear from Definition 3.6.6.

(d) For all $x \in X$ and for all $v^* \in V^*$ we derive

$$\langle A^*(B^*(v^*)), x\rangle = \langle B^*(v^*), A(x)\rangle = \langle v^*, B(A(x))\rangle$$

and so we conclude that $A^* \circ B^* = (B \circ A)^*$.

(e) Since A is invertible we have $A^{-1} \circ A = i_X = A \circ A^{-1}$. Then using part (c) we obtain

$$A^* \circ (A^{-1})^* = i_X^* = i_{X^*} = (A^{-1})^* \circ A^* .$$

Hence, A^* is invertible and $(A^*)^{-1} = (A^{-1})^*$. □

Remark 3.6.8. According to this proposition the map $A \to A^*$ from $L(X,Y)$ into $L(Y^*,X^*)$ is an isometric isomorphism. It is also continuous for the weak operator topologies but not for the strong operator topology. When $X = Y = H$ is a complex Hilbert space, that is, over $\mathbb{F} = \mathbb{C}$, then, since H is self-dual, that is, $H = H^*$, we want to define A^* on the space H. From the Riesz–Fréchet Representation Theorem (see Theorem 3.5.21), we know that H is isometric with its dual H^* but the isometry is a conjugate isomorphism $j: H \to H^*$. We set $A' = j^{-1} \circ A^* \circ j$ and get that

$$\begin{aligned}(x, A(y)) &= \langle j(x), A(y)\rangle = \langle A^*(j(x)), y\rangle = (j^{-1}(A^*(j(x))), y) \\ &= (A'(x), y)\end{aligned} \tag{3.6.2}$$

for all $x, y \in H$. Then $A' \in L(H)$ is the Hilbert space adjoint and now the map $A \to A'$ is conjugate linear, that is, $\lambda A \to \bar{\lambda} A'$ for all $\lambda \in \mathbb{C}$ because A' is defined on H rather than on H^* and H is identified with H^* by a conjugate isometric isomorphism. However, in what follows for notational uniformity we denote A' by A^* with the understanding that A^* is defined on H. When H is a real Hilbert space, we define again $A' = A^*$ on H as above.

Proposition 3.6.9. *If H is a Hilbert space over $\mathbb{R}$ or $\mathbb{C}$ and if $A \in L(H)$, then $\|A\|_L^2 = \|A^* \circ A\|_L$.*

Proof. Taking Proposition 3.6.7 (a) and (3.6.2) into account yields

$$\begin{aligned}\|A\|_L^2 &= \sup[\|A(x)\|: \|x\| \le 1] = \sup[(A(x), A(x)): \|x\| \le 1] \\ &= \sup[(A^*(A(x)), x): \|x\| \le 1] \le \|A^* \circ A\|_L \le \|A^*\|_L \|A\|_L = \|A\|_L^2 .\end{aligned}$$ □

Example 3.6.10. Section 4.1 shows that $(l^1)^* = l^\infty$. Consider the right shift operator $A \in L(l^1)$ defined by $A(\hat{x}) = (0, x_1, x_2, \ldots)$ for all $\hat{x} = (x_n)_{n\ge1} \in l^1$. Then $A^*: l^\infty \to l^\infty$ is defined by $A^*(\hat{u}) = (u_2, u_3, \ldots)$ for all $\hat{u} = (u_n)_{n\ge1} \in l^\infty$. In this case we have $\|A\|_L = \|A^*\|_L = 1$.

Proposition 3.6.11. *If X, Y are normed spaces and $A \in L(X,Y)$, then $A^* \in L(Y^*,X^*)$ is weak*-to-weak* continuous.*

Conversely, if $T: Y^* \to X^*$ is a weak*-to-weak* continuous linear operator, then there exists $A \in L(X,Y)$ such that $A^* = T$.

Proof. Let $\{y_\alpha^*\}_{\alpha\in I} \subseteq Y^*$ be a net such that $y_\alpha^* \xrightarrow{w^*} y^*$ in Y^*. Then for every $x \in X$, it follows that

$$\langle A^*(y_\alpha^*), x\rangle = \langle y_\alpha^*, A(x)\rangle \to \langle y^*, A(x)\rangle = \langle A^*(y^*), x\rangle ,$$

hence, $A^*(y_\alpha^*) \xrightarrow{w^*} A^*(y^*)$ and so A^* is weak*-to-weak* continuous.

Let $j_X: X \to X^{**}$ and $j_Y: Y \to Y^{**}$ be the canonical embeddings; see Definition 3.3.35. For every $x \in X$, $j_X(x)T$ is a w*-continuous linear functional on Y^*, hence $j_X(x)T \in j_Y(Y)$. Then $j_Y^{-1}(j_X(x)T) \in Y$. So, we can define an operator $A: X \to Y$ by setting $A(x) = j_Y^{-1}(j_X(x)T)$ for all $x \in X$. Clearly A is linear. Moreover, let $\{x_\alpha\}_{\alpha\in I} \subseteq X$ be a net such that $x_\alpha \xrightarrow{w} x$. Then $j_X(x_\alpha) \xrightarrow{w^*} j_X(x)$; see Proposition 3.3.23. Hence, for all $y^* \in Y^*$, we have

$$(j_X(x_\alpha)T)(y^*) \to (j_X(x)T)(y^*) \quad \text{in } \mathbb{R} ,$$

thus $j_X(x_\alpha)T \xrightarrow{w^*} j_X(x)T$ in Y^{**}. Therefore,

$$A(x_\alpha) = j_Y^{-1}(j_X(x_\alpha)T) \xrightarrow{w} j_Y^{-1}(j_X(x)T) = A(x) \quad \text{in } Y .$$

This means that $A: X \to Y$ is weak-to-weak continuous, hence $A \in L(X, Y)$; see Proposition 3.3.23. Moreover, with view to Definition 3.3.35, we get

$$\langle A^*(y^*), x\rangle = \langle y^*, A(x)\rangle = \langle y^*, j_Y^{-1}(j_X(x)T)\rangle = \langle j_X(x)T, y^*\rangle = \langle T(y^*), x\rangle .$$

Thus, $A^* = T$. □

Corollary 3.6.12. *If X, Y are normed spaces and $S: X^* \to Y^*$ is weak*-to-weak* continuous, then $S \in L(X^*, Y^*)$.*

Next we introduce some important special classes of linear operators.

Definition 3.6.13. (a) Let X be a vector space and let $P: X \to X$ be a linear operator. We say that P is a **projection** if $P^2 = P$, that is, $P(P(x)) = P(x)$ for all $x \in X$.
(b) Let H be a Hilbert space and $A \in L(H)$. We say that A is **self-adjoint** (or **hermitian**) if $A = A^*$, that is, $(A(x), y) = (x, A(y))$ for all $x, y \in H$.
(c) Let H be a Hilbert space and $P \in L(H)$. We say that P is an **orthogonal projection** if P is a projection and P is self-adjoint.

Proposition 3.6.14. *If H is a Hilbert space and $T, S \in L(H)$ are self-adjoint and commuting, that is, $T \circ S = S \circ T$, then $T \circ S \in L(H)$ is self-adjoint as well.*

Proof. For every $x, y \in H$ we see that

$$(T(S(x)), y) = (S(x), T(y)) = (x, S(T(y))) = (x, T(S(y))) .$$

This shows that $T \circ S$ is self-adjoint. □

Proposition 3.6.15. *If H is a Hilbert space and $A \in L(H)$ is self-adjoint, then for every $m \in \mathbb{N}$, A^m is self-adjoint and $\|A^m\|_L = \|A\|^m$.*

Proof. That A^m is self-adjoint for every $m \in \mathbb{N}$ follows from Proposition 3.6.14. From Proposition 3.6.9 we see that

$$\|A\|_L^2 = \|A^* \circ A\|_L = \|A^2\|_L, \quad \|A^4\|_L = \|A^2\|_L^2 = \|A\|_L^4$$

and so on. Therefore we obtain

$$\|A^{2^n}\|_L = \|A\|_L^{2^n}. \tag{3.6.3}$$

If $1 \leq m \leq 2^n$, then

$$\|A^{2^n}\|_L = \|A^m \circ A^{2^n - m}\|_L \leq \|A^m\|_L \|A\|_L^{2^n - m} \leq \|A\|_L^m \|A\|_L^{2^n - m} = \|A\|_L^{2^n},$$

which, due to (3.6.3), results in

$$\|A^m\|_L \|A\|_L^{2^n - m} = \|A\|_L^{2^n}.$$

Thus, $\|A^m\|_L = \|A\|_L^m$. □

Proposition 3.6.16. *If H is a Hilbert space and $A \in L(H)$ is self-adjoint, then $\|A\|_L = \sup[|(A(x), x)|: \|x\| \leq 1]$.*

Proof. For $x \in H$ with $\|x\| \leq 1$ we infer

$$|(A(x), x)| \leq \|A(x)\| \|x\| \leq \|A\|_L \|x\|^2 \leq \|A\|_L,$$

which gives

$$\sup[|(A(x), x)|: \|x\| \leq 1] \leq \|A\|_L. \tag{3.6.4}$$

Let $\eta = \sup[|(A(x), x)|: \|x\| \leq 1]$. Then $|(A(u), u)| \leq \eta \|u\|^2$ for all $u \in H$. For $u \in H$ with $u \neq 0$ let $\lambda = (\|A(u)\|/\|u\|)^{1/2}$ and $y = 1/\lambda A(u)$. Since A is self-adjoint, $(A(\lambda u), y) \in \mathbb{R}$ and, due to the parallelogram law, we obtain

$$\begin{aligned}
\|A(u)\|^2 &= (A(u), A(u)) = \left(A(\lambda u), \frac{1}{\lambda} A(u)\right) = (A(\lambda u), y) \\
&= \frac{1}{4}[(A(\lambda u + y), \lambda u + y) - (A(\lambda u - y), \lambda u - y)] \\
&\leq \frac{1}{4}\eta(\|\lambda u + y\|^2 + \|\lambda u - y\|^2) = \frac{1}{2}\eta(\|\lambda u\|^2 + \|y\|^2) \\
&= \frac{1}{2}\eta\left(\lambda^2 \|u\|^2 + \frac{1}{\lambda^2}\|A(u)\|^2\right) = \eta \|u\| \|A(u)\|,
\end{aligned}$$

where we used the fact that $\lambda\|u\| = 1/\lambda\|A(u)\|$. Hence $\|A(u)\| \le \eta\|u\|$ for every $u \in H$, which gives $\|A\|_L \le \eta$ and so, because of (3.6.4), the result follows. ☐

Next we present a useful factorization result.

Proposition 3.6.17. *If X, Y, V are Banach spaces, $A \in L(X,Y)$, $T \in L(V,Y)$, and A is injective, then the following statements are equivalent:*
(a) $R(T) \subseteq R(A)$.
(b) *There exists $S \in L(V,X)$ such that $A \circ S = T$.*

Proof. (a) $\Longrightarrow$ (b): Let $S = A^{-1} \circ T\colon V \to X$, where we recall that A is injective. Then S is linear and $A \circ S = T$. We claim that $\operatorname{Gr} S \subseteq V \times X$ is closed. To this end, let $\{v_n\}_{n\ge1} \subseteq V$ such that $v_n \to v$ in V and $S(v_n) \to x$ in X. Then $A(x) = \lim_{n\to\infty} A(S(v_n)) = \lim_{n\to\infty} T(v_n) = T(v) = A(S(v))$. Since A is injective it follows that $x = S(v)$ and so $\operatorname{Gr} S \subseteq V \times X$ is closed. Hence, by the Closed Graph Theorem (see Theorem 3.2.14), we conclude that $S \in L(V,X)$.

(b) $\Longrightarrow$ (a): It holds that $R(T) = R(A \circ S) \subseteq R(A)$. ☐

Next we present two theorems relating operators with the same range space and their adjoints. We start with an auxiliary result.

Lemma 3.6.18. *If X, Y are normed spaces, $A \in L(X,Y)$ and $x^* \in X^*$, then the following statements are equivalent:*
(a) $x^* \in R(A^*)$.
(b) $|\langle x^*, x\rangle| \le c\|A(x)\|_Y$ *for all $x \in X$ and for some $c > 0$.*

Proof. (a) $\Longrightarrow$ (b): Of course, $x^* = A^*(y^*)$ for some $y^* \in Y^*$. Then

$$|\langle x^*, x\rangle| = |\langle A^*(y^*), x\rangle| = |\langle y^*, A(x)\rangle| \le \|y^*\|_* \|A(x)\|_Y \quad \text{for all } x \in X\,,$$

which gives $|\langle x^*, x\rangle| \le c\|A(x)\|_Y$ with $c = \|y^*\|_*$.

(b) $\Longrightarrow$ (a): There exists a continuous, linear functional $g\colon R(A) \to \mathbb{R}$ such that $x^* = g \circ A$. According to Proposition 3.1.49, there exists $y^* \in Y^*$ such that $y^*|_{R(A)} = g$. Then $x^* = y^* \circ A = A^*(y^*)$; see Definition 3.6.6. ☐

Theorem 3.6.19. *If X, Y, V are Banach spaces and $A \in L(X,Y)$, $T \in L(V,Y)$ with $R(T) \subseteq R(A)$, then $\|T^*(y^*)\|_* \le c\|A^*(y^*)\|_*$ for all $y^* \in Y^*$ and for some $c > 0$.*

Proof. Let $\hat{X} = X/N(A)$ with $N(A)$ being the kernel of A and $p\colon X \to \hat{X}$ being the quotient map. Then $p^*\colon \hat{X}^* \to X^*$ is an isometric embedding onto $N(A)^\perp \subseteq X^*$; see Proposition 3.2.25. Let $\hat{A}\colon \hat{X} \to Y$ be defined by $\hat{A} \circ p = A$. Then $A^* = p^* \circ \hat{A}^*$, and so

$$\|A^*(y^*)\|_* = \|\hat{A}^*(y^*)\|_* \quad \text{for all } y^* \in Y^*\,.$$

By hypothesis, $R(T) \subseteq R(A) = R(\hat{A})$ and $\hat{A}$ is injective. So, we can use Proposition 3.6.17 and produce $S \in L(V,\hat{X})$ such that $\hat{A} \circ S = T$. Then, since $\hat{A} \circ S = T$,

$$\begin{aligned}
\|T^*(y^*)\|_* &= \sup\left[\frac{\langle T^*(y^*), v\rangle}{\|v\|_V} : v \in V, v \neq 0\right] \\
&= \sup\left[\frac{\langle y^*, T(v)\rangle}{\|v\|_V} : v \in V, v \neq 0\right] \\
&= \sup\left[\frac{\langle \hat{A}^*(y^*), S(v)\rangle}{\|v\|_V} : v \in V, v \neq 0\right] \\
&\leq \sup\left[\frac{\|\hat{A}^*(y^*)\|_* \|S(v)\|_{\hat{X}}}{\|v\|_V} : v \in V, v \neq 0\right] \\
&= \|S\|_L \|\hat{A}^*(y^*)\|_* \quad \text{for all } y^* \in Y^* .
\end{aligned}$$

So, the conclusion of the theorem holds with $c = \|S\|_L$. □

Theorem 3.6.20. *If X, Y, V are normed spaces and $A \in L(X, Y)$, $T \in L(X, V)$, then the following statements are equivalent:*
(a) $R(T^*) \subseteq R(A^*)$.
(b) $\|T(x)\|_V \leq c\|A(x)\|_Y$ *for all* $x \in X$ *and for some* $c > 0$.

Proof. (a) ⟹ (b): Using Theorem 3.6.19, we infer

$$\|T^{**}(x^{**})\|_{**} \leq c\|A^{**}(x^{**})\|_{**} \quad \text{for all } x^{**} \in X^{**} \text{ and for some } c > 0 .$$

Applying Proposition 3.6.7 gives

$$\|T(x)\|_V = \|T^{**}(x)\|_{**} \leq c\|A^{**}(x)\|_{**} = c\|A(x)\|_Y \quad \text{for all } x \in X .$$

(b) ⟹ (a): Let $x^* \in R(T^*) \subseteq X^*$. Using Lemma 3.6.18 yields

$$|\langle x^*, x\rangle| \leq c_0\|T(x)\|_V \quad \text{for all } x \in X \text{ and for some } c_0 > 0 ,$$

which implies

$$|\langle x^*, x\rangle| \leq c_0 c\|A(x)\|_V \quad \text{for all } x \in X .$$

Hence, with view to Lemma 3.6.18 we see that $x^* \in R(A^*)$. Thus, $R(T^*) \subseteq R(A^*)$. □

Theorem 3.6.21. *If X, Y, V are Banach spaces, X is reflexive, $A \in L(X, Y)$, $T \in L(V, Y)$ and*

$$\|T^*(y^*)\|_* \leq c\|A^*(y^*)\|_* \quad \textit{for all } y^* \in Y^* \textit{ and for some } c > 0 ,$$

then $R(T) \subseteq R(A)$.

Proof. Applying Theorem 3.6.20 we obtain $R(T^{**}) \subseteq R(A^{**})$. Let $v \in V$ and let $x \in X^{**} = X$ such that $A(x) = A^{**}(x) = T^{**}(v) = T(v)$; see Proposition 3.6.7 (c). Hence $R(T) \subseteq R(A)$. □

Motivated from Definition 3.2.24, we introduce a similar notion for sets in X^*.

Definition 3.6.22. Let X be a normed space and $E \subseteq X^*$. The **preannihilator** of E is defined by

$$^{\perp}E = \{x \in X : \langle x^*, x\rangle = 0 \text{ for all } x^* \in E\}.$$

Evidently $^{\perp}E$ is a closed linear subspace of X.

Remark 3.6.23. It is easy to see that if $E \subseteq X^*$ is a vector subspace, then $\overline{E}^{w^*} = (^{\perp}E)^{\perp}$, E is w*-closed if and only if $E = (^{\perp}E)^{\perp}$, and $\overline{E}^{w^*} = X^*$ if and only if $^{\perp}E = \{0\}$.

Moreover, if Y, V are closed vector subspaces of X, then

$$V \cap Y = {}^{\perp}(V^{\perp} + Y^{\perp}), \quad (V \cap Y)^{\perp} \supseteq \overline{V^{\perp} + Y^{\perp}},$$
$$V^{\perp} \cap Y^{\perp} = (V + Y)^{\perp}, \quad {}^{\perp}(V^{\perp} \cap Y^{\perp}) = \overline{V + Y}.$$

Proposition 3.6.24. *If X, Y are normed spaces and $A \in L(X, V)$, then the following hold:*
(a) *$R(A)^{\perp} = N(A^*)$ and $^{\perp}R(A^*) = N(A)$.*
(b) *$\overline{R(A)} = Y$ if and only if A^* is injective.*
(c) *A is injective if and only if $\overline{R(A^*)}^{w^*} = X^*$.*

Proof. (a) Note that

$$\begin{aligned} y^* \in R(A)^{\perp} \quad &\text{if and only if} \quad \langle y^*, A(x)\rangle = 0 \quad \text{for all } x \in X \\ &\text{if and only if} \quad \langle A^*(y^*), x\rangle = 0 \quad \text{for all } x \in X \\ &\text{if and only if} \quad A^*(y^*) = 0. \end{aligned}$$

Hence $R(A)^{\perp} = N(A^*)$. Similarly, we have

$$\begin{aligned} x \in {}^{\perp}R(A^*) \quad &\text{if and only if} \quad \langle A^*(y^*), x\rangle = 0 \quad \text{for all } y^* \in Y^* \\ &\text{if and only if} \quad \langle y^*, A(x)\rangle = 0 \quad \text{for all } y^* \in Y^* \\ &\text{if and only if} \quad A(x) = 0. \end{aligned}$$

Thus, $^{\perp}R(A^*) = N(A)$.

(b) $\Longrightarrow$: It holds that $R(A)^{\perp} = \{0\}$ and so with part (a), $N(A^*) = \{0\}$. Hence A^* is injective.

$\Longleftarrow$: It holds that $N(A^*) = \{0\}$ and so with part (a), $R(A)^{\perp} = \{0\}$. Hence $\overline{R(A)} = Y$.

(c) $\Longrightarrow$: It holds that $N(A) = \{0\}$ and so with part (a), $^{\perp}R(A^*) = \{0\}$. Hence, $\overline{R(A^*)}^{w^*} = X^*$; see Remark 3.6.23.

$\Longleftarrow$: It holds that $^{\perp}R(A^*) = \{0\}$ (see Remark 3.6.23), and so with part (a), $N(A) = \{0\}$. Hence, A is injective. ☐

Remark 3.6.25. If X, Y are Banach spaces with X or Y finite dimensional and $A \in L(X, Y)$, we know from linear algebra that

$$\begin{aligned} &A \text{ is surjective if and only if } A^* \text{ is injective}, \\ &A^* \text{ is surjective if and only if } A \text{ is injective}. \end{aligned}$$

Indeed in this case $R(A)$ is closed if $\dim Y < \infty$ and $R(A^*)$ is closed if $\dim X < \infty$ and so the equivalences above follow from Proposition 3.6.24. In the general infinite dimensional case we only have the following implications (see Proposition 3.6.24 (a))

$$\begin{aligned} A \text{ is surjective} &\implies A^* \text{ is injective}, \\ A^* \text{ is surjective} &\implies A \text{ is injective}. \end{aligned}$$

The reverse implications fail. To see this, let $X = Y = H = l^2$, which is a Hilbert space and let $A \in L(H)$ be defined by $A(\hat{x}) = (1/nx_n)_{n\geq 1}$ for all $\hat{x} = (x_n)_{n\geq 1} \in l^2$. Then $A^* = A$ and A is injective but not surjective since $R(A) = R(A^*)$ is only dense in H.

Next we present some results dealing with the basic properties of projections.

Proposition 3.6.26. *If X is a normed space and $P \in L(X)$, then P is a projection if and only if $P^* \in L(X^*)$ is a projection.*

Proof. $\implies$: For all $x \in X$ and for all $x^* \in X^*$ we directly obtain

$$\langle P^*(x^*), x\rangle = \langle x^*, P(x)\rangle = \langle x^*, P(P(x))\rangle = \langle P^*(P^*(x^*)), x\rangle .$$

This shows that $P^*(x^*) = P^*(P^*(x^*))$ for all $x^* \in X^*$. Hence P^* is a projection as well.

$\impliedby$: This is proven in a similar fashion. □

Proposition 3.6.27. *If X is a normed space and $P \in L(X)$, then P is a projection if and only if $I - P$ is a projection.*

Proof. $\implies$: For every $x \in X$ one gets

$$(I - P)(I - P)(x) = x - 2P(x) + P\big(P(x)\big) = x - P(x) = (I - P)(x) .$$

Hence $I - P$ is a projection.

$\impliedby$: Note that $P = I - (I - P)$ and so the implication follows from the previous part. □

Proposition 3.6.28. *If X is a normed space and $P \in L(X)$ is a projection, then $N(P) = R(I - P)$ and $R(P) = N(I - P)$.*

Proof. Let $x \in N(p)$. Then $(I - P)(x) = x$ and so $N(P) \subseteq R(I - P)$. Let $u \in R(I - P)$. Then $u = (I - P)(x)$ with $x \in X$. Then $P(u) = P(x - P(x)) = P(x) - P(P(x)) = P(x) - P(x) = 0$

and so $u \in N(p)$. Therefore we conclude that $N(P) = R(I - P)$. Applying this result to the projection $I - P$ we get $R(P) = N(I - P)$. □

Corollary 3.6.29. *If X is a normed space and $P \in L(X)$ is a projection, then $R(P) = \{x \in X : P(x) = x\}$ and $R(P)$ is closed.*

Corollary 3.6.30. *If X is a Banach space and $P \in L(X)$ is a projection, then $X = N(P) \oplus R(P)$.*

If V and W are complementary subspaces of a Banach space X (see Definition 3.2.27), then we obtain in a unique way, for every $x \in X$, that $x = v + w$ with $v \in V$ and $w \in W$. Let $P_V : X \to V$ be the linear operator such that $P_V(x) = v$. Evidently $P_V^2 = P_V$.

Proposition 3.6.31. *$P_V \in L(X)$, that is, P_V is a projection.*

Proof. Suppose that $x_n \to x$ in X and $P_V(x_n) \to v$ in X. Then $(I - P_V)(x_n) \to x - y$ in X. Note that $v \in V$ and $x - v \in W$. So, $v = P_V(x)$ and by the Closed Graph Theorem (see Theorem 3.2.14), it follows that $P_V \in L(X)$. □

Corollary 3.6.32. *If X is a Banach space and $V \subseteq X$ is a subspace, then V is complemented if and only if $V = R(P)$ with $P \in L(X)$ being a projection.*

Corollary 3.6.33. *If X is a Banach space and $V, W \subseteq X$ are complementary subspaces, then V and X/W are isomorphic.*

Next we use complemented subspaces to obtain a kind of Hahn–Banach Extension Theorem for vector valued maps.

Proposition 3.6.34. *If X is a Banach space and $V \subseteq X$ is a subspace, then the following statements are equivalent:*
(a) *For every Banach space Y and every $A \in L(V, Y)$, there is $\hat{A} \in L(X, Y)$ such that $\hat{A}|_V = A$.*
(b) *$\overline{V}$ is complemented in X.*

Proof. (a) $\Longrightarrow$ (b): Let $i_0 : V \to \overline{V}$ be the bounded linear operator defined by $i_0(v) = v$ for all $v \in V$, that is, the identity map on V. Then by hypothesis there exists $\hat{i}_0 \in L(X, \overline{V})$ such that $\hat{i}_0|_V = i_0$. Due to the continuity of $\hat{i}_0$ we directly obtain that $\hat{i}_0|_{\overline{V}}$ coincides with the identity operator of $\overline{V}$. Therefore, $\hat{i}_0 \in L(X)$ is a projection with $R(\hat{i}_0) = \overline{V}$. Then Corollary 3.6.32 implies that $\overline{V}$ is complemented.

(b) $\Longrightarrow$ (a): Corollary 3.6.32 implies that $\overline{V} = R(P)$ with $P \in L(X)$ being a projection. Let Y be a Banach space and $A \in L(V, Y)$. Then there exists $A_0 \in L(\overline{V}, Y)$ such that $A_0|_V = A$; see Theorem 1.5.27. One gets $A_0 \circ P \in L(X, Y)$ and $A_0 \circ P|_V = A$. So, $\hat{A} = A_0 \circ P \in L(X, Y)$. □

Proposition 3.6.35. *If X is a Banach space, Y is a normed space, and $A \in L(X, Y)$, then $A^{-1} \in L(Y, X)$ if and only if $R(A)$ is dense in Y and there exists $c > 0$ such that $\|A(x)\|_Y \geq c\|x\|_X$ for all $x \in X$.*

Proof. $\Longrightarrow$: This is obvious.

$\Longleftarrow$: Evidently A is injective and $A^{-1} \in L(V, X)$ with $V = R(A)$. Hence $A^{-1} \in L(Y, X)$ since by hypothesis $\overline{V} = Y$. Moreover, note that $\|A^{-1}\|_L \le 1/c$. □

Using this proposition we can improve Proposition 3.6.7 (e).

Proposition 3.6.36. *If X is a Banach space, Y is a normed space, and $A \in L(X, Y)$, then A is invertible if and only if A^* is invertible.*

Proof. $\Longrightarrow$: This follows from Proposition 3.6.7 (e).

$\Longleftarrow$: From Proposition 3.6.24 (a) one has that $R(A)^\perp = N(A^*) = \{0\}$ and so $R(A) \subseteq Y$ is dense. Let $x \in X$ and let $x^* \in X^*$ be such that

$$\langle x^*, x\rangle = \|x\|_X \quad \text{and} \quad \|x^*\| = 1\,;$$

see Proposition 3.1.50. Then

$$\begin{aligned}\|x\|_X = \langle x^*, x\rangle = \langle A^*((A^*)^{-1}(x^*)), x\rangle &= \langle (A^*)^{-1}(x^*), A(x)\rangle \\ \le \|(A^*)^{-1}(x^*)\|_* \|A(x)\|_Y &\le \|(A^*)^{-1}\|_L \|A(x)\|_Y\,.\end{aligned}$$

This implies $\|A(x)\|_Y \ge c\|x\|_X$ with $c = (\|(A^*)^{-1}\|_L)^{-1}$. Now we may apply Proposition 3.6.35 and conclude that $A^{-1} \in L(Y, X)$. □

Corollary 3.6.37. *If X is a Banach space, Y is a normed space, and $A \in L(X, Y)$, then the following statements are equivalent:*

(a) *A is invertible.*

(b) *A^* is invertible.*

(c) *There exist $c, \hat{c} > 0$ such that*

$$\begin{aligned}\|A(x)\|_Y &\ge c\|x\|_X && \text{for all } x \in X\,,\\ \|A^*(x^*)\|_* &\ge \hat{c}\|x^*\|_* && \text{for all } x^* \in X^*\,.\end{aligned}$$

In the last part of this section we deal with unbounded linear operators.

Definition 3.6.38. Let X, Y be Banach spaces. An **unbounded linear operator** is a linear map $A\colon D(A) \subseteq X \to Y$ from a linear subspace $D(A)$ into Y. The subspace $D(A)$ is called the **domain** of A. We say that A is **closed** if $\operatorname{Gr} A \subseteq X \times Y$ is closed. By $N(A)$ we denote the kernel of A, that is, $N(A) = \{x \in D(A)\colon A(x) = 0\}$ and by $R(A)$ the range of A, that is, $R(A) = \{A(x)\colon x \in D(A)\}$.

Remark 3.6.39. In this context, A is closed if and only if for every $\{x_n\}_{n\ge1} \subseteq D(A)$ such that $x_n \to x$ in X and $A(x_n) \to y$ in Y, it follows that $x \in D(A)$ and $A(x) = y$. Note that now it is not enough to check that if $x_n \to 0$ in X and $A(x_n) \to y$ in Y, then $y = 0$. Moreover, if A is closed, then $N(A)$ is closed but $R(A)$ need not be closed. In applications most unbounded linear operators are densely defined, that is, $\overline{D(A)} = X$, and closed.

We can extend the notion of adjoint to unbounded linear operators. So, let $A\colon D(A) \subseteq X \to Y$ be an unbounded linear operator that is densely defined, that is, $\overline{D(A)} = X$. Let

$$D(A^*) = \{y^* \in Y^* : |\langle y^*, A(x)\rangle| \le c\|x\| \text{ for all } x \in D(A) \text{ and for some } c > 0\}. \tag{3.6.5}$$

Evidently $D(A^*) \subseteq Y^*$ is a vector subspace. Let $y^* \in D(A^*)$ and consider the functional $f\colon D(A) \to \mathbb{R}$ defined by $f(x) = \langle y^*, A(x)\rangle$ for all $x \in D(A)$. Because of (3.6.5) it follows that $|f(x)| \le c\|x\|$ for all $x \in D(A)$. Since $D(A)$ is dense in X, extending by continuity, there exists a unique functional $\hat{f}\colon X \to \mathbb{R}$ such that $\hat{f}|_{D(A)} = f$ and $|\hat{f}(x)| \le c\|x\|$ for all $x \in X$. Thus, $\hat{f} \in X^*$. Then we set

$$A^*(y^*) = f^*. \tag{3.6.6}$$

Definition 3.6.40. The unbounded linear operator $A^*\colon D(A^*) \subseteq Y^* \to X^*$ defined by (3.6.6) is called the **adjoint** of A. So, according to the previous construction, we obtain

$$\langle y^*, A(x)\rangle = \langle A^*(y^*), x\rangle \quad \text{for all } x \in D(A) \text{ and for all } y^* \in D(A^*). \tag{3.6.7}$$

Remark 3.6.41. In general, we cannot say that A^* is densely defined. However, if A is also closed, then $D(A^*)$ is w*-dense in Y^*. Therefore, if Y is reflexive and $A\colon D(A) \subseteq X \to Y$ is closed and densely defined, then $A^*\colon D(A^*) \subseteq Y^* \to X^*$ is densely defined as well.

Next we show that A^* is always closed.

Proposition 3.6.42. *If X, Y are Banach spaces and $A\colon D(A) \subseteq X \to Y$ is a densely defined unbounded linear operator, then A^* is closed.*

Proof. Suppose that $y_n^* \to y^*$ in Y^* with $y_n^* \in D(A^*)$ for all $n \in \mathbb{N}$ and $A^*(y_n^*) \to x^*$ in X^*. Thanks to (3.6.7) we have

$$\langle y_n^*, A(x)\rangle = \langle A^*(y_n^*), x\rangle \quad \text{for all } x \in D(A) \text{ and for all } n \in \mathbb{N},$$

which implies

$$\langle y^*, A(x)\rangle = \langle x^*, x\rangle \quad \text{for all } x \in D(A).$$

This gives

$$|\langle y^*, A(x)\rangle| \le \|x^*\|_* \|x\|_X \quad \text{for all } x \in D(A),$$

which yields, because of (3.6.5), that $y^* \in D(A^*)$, which in combination with (3.6.7) results in

$$\langle A^*(y^*), x\rangle = \langle x^*, x\rangle \quad \text{for all } x \in D(A).$$

This implies $x^* = A^*(y^*)$. Hence, A^* is closed; see Remark 3.6.39. ☐

Let $i_0: Y^* \times X^* \to X^* \times Y^*$ be the isomorphism defined by $i_0(y^*, x^*) = (-x^*, y^*)$ for all $y^* \in Y^*$ and for all $x^* \in X^*$.

Proposition 3.6.43. *If X, Y are Banach spaces and $A: D(A) \subseteq X \to Y$ is a densely defined unbounded linear operator, then $i_0(\operatorname{Gr} A^*) = (\operatorname{Gr} A)^\perp$.*

Proof. Let $(y^*, x^*) \in Y^* \times X^*$. Then, thanks to (3.6.7), one has

$$\begin{aligned}
(y^*, x^*) \in \operatorname{Gr} A^* \quad & \text{if and only if} \quad \langle y^*, A(x)\rangle = \langle x^*, x\rangle && \text{for all } x \in D(A) \\
& \text{if and only if} \quad \langle y^*, A(x)\rangle - \langle x^*, x\rangle = 0 && \text{for all } x \in D(A) \\
& \text{if and only if} \quad (-x^*, y^*) \in (\operatorname{Gr} A)^\perp .
\end{aligned}$$

□

The next result is an extension of Proposition 3.6.24 to unbounded linear operators. Its proof can be found in Brézis [52, Theorem 2.19, p. 46].

Proposition 3.6.44. *If X, Y are Banach spaces and $A: D(A) \subseteq X \to Y$ is a closed, densely defined, unbounded linear operator, then the following statements are equivalent:*
(a) *$R(A) \subseteq Y$ is closed.*
(b) *$R(A^*) \subseteq X^*$ is closed.*
(c) *$R(A) = {}^\perp N(A^*)$.*
(d) *$R(A^*) = N(A)^\perp$.*

The next two theorems provide useful characterizations of surjective operators.

Theorem 3.6.45. *If X, Y are Banach spaces and $A: D(A) \subseteq X \to Y$ is a closed, densely defined, unbounded linear operator, then the following statements are equivalent:*
(a) *A is surjective, that is, $R(A) = Y$.*
(b) *$\|y^*\|_* \le c\|A^*(y^*)\|$ for all $y^* \in D(A^*)$ and for some $c > 0$.*
(c) *$R(A^*) \subseteq X^*$ is closed and $N(A^*) = \{0\}$.*

Proof. (a) $\Longrightarrow$ (b): It suffices to show that

$$D^* = \{y^* \in D(A^*) : \|A^*(y^*)\|_* \le 1\}$$

is bounded. Then according to Proposition 3.2.5 we need to show that for all $y \in Y$, $\langle D^*, y\rangle \subseteq \mathbb{R}$ is bounded. Exploiting the surjectivity of A, there exists $x \in D(A)$ such that $y = A(x)$. Then

$$\langle y^*, y\rangle = \langle y^*, A(x)\rangle = \langle A^*(y^*), x\rangle ,$$

which implies $|\langle y^*, y\rangle| \le \|x\|$ for every $y^* \in D^*$. Thus, D^* is bounded.

(b) $\Longrightarrow$ (c): Let $x_n^* \in R(A^*)$ for all $n \in \mathbb{N}$ and assume that $x_n^* \to x^*$ in X^*. We can find $y_n^* \in D(A^*)$ such that $x_n^* = A^*(y_n^*)$ for all $n \in \mathbb{N}$. From (b) we see that

$$\|y_m^* - y_n^*\|_* \le c\|A^*(y_m^* - y_n^*)\|_* = c\|A^*(y_m^*) - A^*(y_n^*)\|_* .$$

This shows that $\{y_n^*\}_{n\geq 1} \subseteq Y^*$ is a Cauchy sequence and so, we conclude that $y_n^* \to y^*$ in Y^*. But from Proposition 3.6.42, we know that A^* is closed. Hence, $x^* = A^*(y^*)$, and so $R(A^*) \subseteq X^*$ is closed. From (b) it is clear that $N(A^*) = \{0\}$.

(c) $\Longrightarrow$ (a): From Proposition 3.6.44 one has $R(A) = {}^{\perp}N(A^*) = Y$. □

In a similar way we can prove a dual version of this theorem.

Theorem 3.6.46. *If X, Y are Banach spaces and $A\colon D(A) \subseteq X \to Y$ is a closed, densely defined, unbounded linear operator, then the following statements are equivalent:*
(a) *A^* is surjective, that is $R(A^*) = X^*$.*
(b) *$\|x\|_X \leq c\|A(x)\|_Y$ for all $x \in D(A)$ and for some $c > 0$.*
(c) *$R(A) \subseteq Y$ is closed and $N(A) = \{0\}$.*

Definition 3.6.47. Let X, Y be Banach spaces and let $A\colon D(A) \subseteq X \to Y$ be an unbounded linear operator. We say that A is **closable** if there is a closed unbounded linear operator $\hat{A}\colon D(\hat{A}) \subseteq X \to Y$ such that

$$D(A) \subseteq D(\hat{A}) \quad \text{and} \quad \hat{A}|_{D(A)} = A\,.$$

Every closable operator A has a smallest closed extension called the **closure** of A denoted by $\overline{A}$.

The next proposition characterizes closable operators.

Proposition 3.6.48. *If X, Y are Banach spaces and $A\colon D(A) \subseteq X \to Y$ is an unbounded linear operator, then the following statements are equivalent:*
(a) *A is closable.*
(b) *If $\{x_n\}_{n\geq 1} \subseteq D(A)$ are such that $x_n \to 0$ in X and $A(x_n) \to y$ in Y, then $y = 0$.*
(c) *The projection map $p_X\colon \overline{\operatorname{Gr} A} \to X$ is injective.*

Proof. (a) $\Longrightarrow$ (b): For every closed extension $\hat{A}$ of A one has $y = \hat{A}(0) = 0$.

(b) $\Longrightarrow$ (c): $\overline{\operatorname{Gr} A} \subseteq X \times Y$ is a vector subspace and so $p_X\colon \overline{\operatorname{Gr} A} \to X$ is linear. By hypothesis, $N(p_X) = \{0\}$ and so p_X is injective.

(c) $\Longrightarrow$ (a): Let $D(\hat{A}) = p_X(\overline{\operatorname{Gr} A}) \subseteq X$. This is a vector subspace. Let $p_Y\colon \overline{\operatorname{Gr} A} \to Y$ be the projection on the second factor. Then $\hat{A} = p_Y \circ p_X^{-1}\colon D(\hat{A}) \to Y$ is an unbounded linear operator with $\operatorname{Gr} \hat{A} = \overline{\operatorname{Gr} A}$ and so $\hat{A}$ is a closed extension of A. □

Proposition 3.6.49. *If X, Y are Banach spaces and $A\colon D(A) \subseteq X \to Y$ is a closable unbounded linear operator, then $\operatorname{Gr} \overline{A} = \overline{\operatorname{Gr} A}$.*

Proof. Let $\hat{A}$ be a closed extension of A. Then $\overline{\operatorname{Gr} A} \subseteq \operatorname{Gr} \hat{A}$ and so if $(0, y) \in \overline{\operatorname{Gr} A}$, then $y = 0$. Let $A_0\colon D(A_0) \to Y$ be defined by $D(A_0) = \{x \in X\colon (x, y) \in \overline{\operatorname{Gr} A} \text{ for some } y \in Y\}$ and $A_0(x) = y$ with $y \in Y$ being the unique element such that $(x, y) \in \overline{\operatorname{Gr} A}$. One has $\operatorname{Gr} A_0 = \overline{\operatorname{Gr} A}$ and so A_0 is a closed extension of A. But $A_0 \subseteq \hat{A}$, which is an arbitrary closed extension of A. Therefore, $A_0 = \overline{A}$. □

Remark 3.6.50. Note that the domain $D(A)$ of an unbounded linear operator $A: D(A) \subseteq X \to Y$ is a normed space with the graph norm defined by $|x| = \|x\|_X + \|A(x)\|_Y$ for all $x \in D(A)$; see the proof of Theorem 3.2.14. Therefore an unbounded linear operator can be viewed also as a bounded linear operator from its domain equipped with the graph norm. It is easy to see that $A: D(A) \subseteq X \to Y$ is closed if and only if $D(A) \subseteq X$ is a Banach space when furnished with the graph norm.

Example 3.6.51. (a) Let $X = C[0,1]$ be equipped with the supremum norm. This is a Banach space. Let $A: D(A) \subseteq X \to X$ be the unbounded linear operator defined by $A(u) = u'$ for all $u \in D(A) = C^1[0,1]$. Evidently A is closed and densely defined. Moreover, the graph norm on $D(A)$ is the usual C^1-norm.

(b) Let H be a separable Hilbert space. From Theorem 3.5.48, we know that H has a countable orthonormal basis $\{e_n\}_{n\geq 1}$. Let $\hat{\lambda} = (\lambda_i)_{i\geq 1} \in \mathbb{R}^{\mathbb{N}}$ and consider the linear operator $A_{\hat{\lambda}}: D(A_{\hat{\lambda}}) \subseteq H \to H$ defined by

$$A_{\hat{\lambda}}(x) = \sum_{n\geq 1} \lambda_n (x, e_n) e_n \quad \text{for all } x \in D(A_{\hat{\lambda}}),$$

where

$$D(A_{\hat{\lambda}}) = \left\{ x \in H: \sum_{n\geq 1} |\lambda_n (x, e_n)|^2 < \infty \right\}.$$

This is a closed, densely defined unbounded linear operator. Note that $A_{\hat{\lambda}} \in L(H)$ if and only if $\hat{\lambda} = (\lambda_n)_{n\geq 1}$ is bounded.

We extend the notion of self-adjoint operator to unbounded linear operators.

Definition 3.6.52. Let H be a Hilbert space and $A: D(A) \subseteq H \to H$ is a densely defined unbounded linear operator. Then the **adjoint** of A is the unbounded linear operator $A^*: D(A^*) \subseteq H \to H$ defined by

$$D(A^*) = \{u \in H: |(u, A(x))| \leq c\|x\| \text{ for all } x \in D(A) \text{ and for some } c > 0\}$$

and

$$(A^*(u), x) = (u, A(x)) \quad \text{for all } x \in D(A) \text{ and for all } u \in D(A^*).$$

We say that A is **symmetric**, if $A \subseteq A^*$, that is, $D(A) \subseteq D(A^*)$ and $A^*|_{D(A)} = A$, so A^* is an extension of A. We say that A is **self-adjoint** if $A = A^*$.

Remark 3.6.53. Evidently A is symmetric if and only if $(A(u), x) = (u, A(x))$ for all $x, u \in D(A)$. A symmetric operator is always closable (see Proposition 3.6.42). Recall that $D(A^*) \supseteq D(A)$ is dense in H. If A is symmetric, then A^* is a closed extension of A. So, we consider the smallest closed extension A^{**} of A. We have $A^{**} \subseteq A^*$. Therefore for symmetric operators we obtain $A \subseteq A^{**} \subseteq A^*$. If A is closed and symmetric, then $A = A^{**} \subseteq A^*$. Finally, if A is self-adjoint, then $A = A^{**} = A^*$. Therefore, a closed symmetric operator A is self-adjoint if and only if A^* is symmetric.

3.7 Compact Operators – Fredholm Operators

In this section we study a class of operators that closely resemble the operators on finite dimensional spaces. These operators are similar to $N \times N$ matrices and so are small in the sense that they map the closed unit ball to a small set.

Definition 3.7.1. Let X, Y be Banach spaces and let $D \subseteq X$ be nonempty subset. A map $f: D \to Y$, not necessarily linear, is said to be **compact** if it is continuous and for every bounded set $B \subseteq D$, the set $\overline{f(B)} \subseteq Y$ is compact. By $K(D, Y)$ we denote the family of all compact maps. If $D = X$, then we define $L_c(X, Y) = K(X, Y) \cap L(X, Y)$.

Remark 3.7.2. If Y is finite dimensional, then every continuous bounded map $f: D \to Y$ is compact. If $A \in L_c(X, Y)$, then $R(A)$ is separable.

Another notion closely related to compactness is the following one.

Definition 3.7.3. Let X, Y be Banach spaces and $D \subseteq X$ is nonempty. A map $f: D \to Y$ is said to be **completely continuous** if for every sequence $\{x_n\}_{n\geq 1} \subseteq D$ such that $x_n \xrightarrow{w} x$ with $x \in D$, it follows $f(x_n) \to f(x)$ in Y.

Remark 3.7.4. Completely continuous operators $A \in L(X, Y)$ are also known as **Dunford–Pettis Operators**. It is easy to see that a linear operator $A: X \to Y$ is completely continuous if and only if $A(C) \subseteq Y$ is compact for every weakly compact $C \subseteq X$.

In general the classes of compact maps and of completely continuous maps are distinct. However, for linear operators we can relate the two classes.

Proposition 3.7.5. *If X, Y are Banach spaces and $A \in L_c(X, Y)$, then A is completely continuous.*

Proof. Let $x_n \xrightarrow{w} x$ in X. Then $\{x_n\}_{n\geq 1} \subseteq X$ is bounded and so $\overline{\{A(x_n)\}_{n\geq 1}} \subseteq Y$ is compact. Thus there exists a subsequence $\{x_{n_k}\}_{k\geq 1}$ of $\{x_n\}_{n\geq 1}$ such that $A(x_{n_k}) \to y$ in Y. From Proposition 3.3.23, one has $A(x_n) \xrightarrow{w} A(x)$ in Y. Therefore $y = A(x)$, and so we conclude that $A(x_n) \to A(x)$ in Y. This proves that A is completely continuous. □

Example 3.7.6. The converse is not true in general. Recall that in l^1, weak and norm convergent sequences coincide; see Remark 3.3.17. Then the identity map $i: l^1 \to l^1$ is a completely continuous linear operator, but clearly it is not compact.

However, if we strengthen the structure of X, then the converse of Proposition 3.7.5 holds. In fact we obtain the following result.

Proposition 3.7.7. *If X is a reflexive Banach space, Y is a Banach space, $D \subseteq X$ is nonempty, w-closed, and $f: D \to Y$ is completely continuous, then $f \in K(D, Y)$.*

Proof. Evidently, f is continuous. Let $B \subseteq D$ be a bounded set. We need to show that $\overline{f(B)} \subseteq Y$ is compact. So, let $\{y_n\}_{n\geq 1} \subseteq f(B) \subseteq Y$. Then $y_n = f(x_n)$ with $\{x_n\}_{n\geq 1} \subseteq B$. The reflexivity of X implies that B is relatively weakly compact. So, the Eberlein–Smulian Theorem, Theorem 3.4.14, says that there exists a subsequence $\{x_{n_k}\}_{k\geq 1}$ of $\{x_n\}_{n\geq 1}$ such that $x_{n_k} \xrightarrow{w} x \in D$. We get $y_{n_k} = f(x_{n_k}) \to f(x) \in \overline{f(B)}$, which means that $\overline{f(B)} \subseteq Y$ is compact. □

Corollary 3.7.8. *If X is a reflexive Banach space, Y is a Banach space, and $A \in L(X, Y)$, then $A \in L_c(X, Y)$ if and only if A is completely continuous.*

The next theorem explains why compact maps resemble maps between finite dimensional spaces. First a simple lemma about relatively compact sets in a Banach space Y.

Lemma 3.7.9. *If Y is a Banach space, $K \subseteq Y$ is nonempty and for every $\varepsilon > 0$, there exists a relatively compact set $K_\varepsilon \subseteq Y$ such that for every $y \in K$ we can find $y_\varepsilon \in K_\varepsilon$ such that $\|y - y_\varepsilon\|_Y < \varepsilon$, then $K \subseteq Y$ is relatively compact.*

Proof. Let $\varepsilon > 0$ be given. There exists a relatively compact set $K_{\varepsilon/2} \subseteq Y$ as postulated by the hypothesis of the lemma. The total boundedness of $K_{\varepsilon/2}$ implies that there exist $\{y_\varepsilon^k\}_{k=1}^m \subseteq K_{\varepsilon/2}$ such that

$$K_{\frac{\varepsilon}{2}} \subseteq \bigcup_{k=1}^{m} B_{\frac{\varepsilon}{2}}(y_\varepsilon^k)\,.$$

By hypothesis, given $y \in K$, there exists $y_{\varepsilon/2} \in K_{\varepsilon/2}$ such that $\|y - y_{\varepsilon/2}\|_Y < \varepsilon/2$. Since $y_{\varepsilon/2} \in B_{\varepsilon/2}(y_\varepsilon^{k_0})$ for some $k_0 \in \{1, \ldots, m\}$ one has $\|y_{\varepsilon/2} - y_\varepsilon^{k_0}\|_Y < \varepsilon/2$. Therefore $\|y - y_\varepsilon^{k_0}\|_Y < \varepsilon$, which implies $K \subseteq \bigcup_{k=1}^m B_\varepsilon(y_\varepsilon^k)$. Hence, K is totally bounded and so relatively compact. □

Theorem 3.7.10. *If X, Y are Banach spaces, $D \subseteq X$ is nonempty, bounded, and $f: D \to Y$, then the following two statements are equivalent:*
(a) *$f \in K(D, Y)$.*
(b) *For every $\varepsilon > 0$ there exists a continuous, bounded map $f_\varepsilon: D \to Y$ such that $\|f(x) - f_\varepsilon(x)\|_Y < \varepsilon$ for all $x \in D$ and $f_\varepsilon(D) \subseteq \overline{\operatorname{conv}} f(D)$ as well as $\dim(\operatorname{span} f_\varepsilon(D)) < \infty$.*

Proof. (a) $\Longrightarrow$ (b): Since f is compact, $f(D) \subseteq Y$ is relatively compact. So, for every $\varepsilon > 0$ there exists a sequence $\{y_k\}_{k=1}^m \subseteq f(D)$ such that

$$\min_{k\in\{1,\ldots,m\}} \|f(x) - y_k\|_Y < \varepsilon \quad \text{for all } x \in D\,. \tag{3.7.1}$$

Recall that $f(D)$ is totally bounded. Let $\lambda_k(x) = \max\{\varepsilon - \|f(x) - y_k\|_Y, 0\}$. Clearly $\lambda_k \colon D \to \mathbb{R}_+$ with $k = 1, \dots, m$ are continuous functions and do not all vanish simultaneously for $x \in D$, see (3.7.1). We introduce the map $f_\varepsilon \colon D \to Y$ defined by

$$f_\varepsilon(x) = \frac{\sum_{k=1}^m \lambda_k(x) y_k}{\sum_{k=1}^m \lambda_k(x)} . \tag{3.7.2}$$

Evidently f_ε is continuous, bounded, and

$$\|f_\varepsilon(x) - f(x)\|_Y = \left\| \frac{\sum_{k=1}^m \lambda_k(x)(y_k - f(x))}{\sum_{k=1}^m \lambda_k(x)} \right\|_Y < \frac{\sum_{k=1}^m \lambda_k(x)\varepsilon}{\sum_{k=1}^m \lambda_k(x)} = \varepsilon$$

for all $x \in D$; see (3.7.1). Then the boundedness of $f(D)$ implies the boundedness of $f_\varepsilon(D)$ while from (3.7.2) we see that $\dim(\operatorname{span} f_\varepsilon(D)) < \infty$. Therefore, f_ε is compact, and it is clear from (3.7.2) that $f_\varepsilon(D) \subseteq \overline{\operatorname{conv}} f(D)$.

(b) $\Longrightarrow$ (a): Let $\varepsilon = 1/n$ with $n \in \mathbb{N}$. Then there exist continuous, bounded maps $f_{1/n} \colon D \to Y$ such that

$$\|f(x) - f_{\frac{1}{n}}(x)\|_Y < \frac{1}{n} \quad \text{for all } x \in D ,$$

which shows that f is continuous since it is the uniform limit of continuous maps.

Let $y = f(x)$ with $x \in D$. Then it follows that

$$\|y - y_n\| < \frac{1}{n} \quad \text{with } y_n = f_{\frac{1}{n}}(x) \in f_{\frac{1}{n}}(D) .$$

The set $f_{1/n}(D)$ is relatively compact. So, invoking Lemma 3.7.9 we conclude that $f(D) \subseteq Y$ is relatively compact, that is, $f \in K(D, Y)$. □

Definition 3.7.11. Let X, Y be Banach spaces and $A \in L(X, Y)$. We say that A is a **finite rank operator** (or **finite dimensional operator** or **degenerate operator**) if $\dim R(A) < \infty$. We denote the space of all finite rank operators by $L_f(X, Y)$. Clearly $L_f(X, Y) \subseteq L_c(X, Y)$. This inclusion is strict in general.

According to Theorem 3.7.10, every $A \in L_c(X, Y)$ can be approximated uniformly on bounded sets by compact maps with finite dimensional range. However, we cannot say that these approximating maps are in $L_f(X, Y)$. So, we are led to the following definition.

Definition 3.7.12. We say that the Banach space Y has the **approximation property** if for every Banach space X, $\overline{L_f(X, Y)}^{\|\cdot\|_L} = L_c(X, Y)$.

Remark 3.7.13. The first example of a Banach space without the approximation property was produced by Enflo [115] who considered a separable reflexive space. A Banach space with a Schauder basis has the approximation property. So, Enflo's example also showed that not every separable reflexive Banach space has a Schauder basis, another long-standing open problem; see Remark 3.5.51.

Proposition 3.7.14. *If X, Y are Banach spaces, then $L_c(X, Y)$ is a Banach space.*

Proof. We only need to show that $L_c(X, Y)$ is a closed subspace of $L(X, Y)$. So, let $A_n \to A$ in $L(X, Y)$. Then $\sup[\|A_n(x) - A(x)\|_Y : \|x\|_X \le 1] \to 0$ as $n \to \infty$. Given $\varepsilon > 0$, there exists $n_0 \in \mathbb{N}$ such that $\|A_n(x) - A(x)\|_Y < \varepsilon/2$ for all $x \in \overline{B}_1^X$ and for all $n \ge n_0$. The set $A_{n_0}(\overline{B}_1^X)$ is totally bounded; recall that $A_{n_0} \in L_c(X, Y)$. Hence, there exists a finite $\varepsilon/2$-net $F \subseteq A_{n_0}(\overline{B}_1^X)$; see Definition 1.5.31. Given $x \in \overline{B}_1^X$ there exists $y \in F$ such that $\|A_{n_0}(x) - y\|_Y < \varepsilon/2$. Then

$$\|A(x) - y\|_Y \le \|A(x) - A_{n_0}(x)\|_Y + \|A_{n_0}(x) - y\|_Y < \frac{\varepsilon}{2} + \frac{\varepsilon}{2} = \varepsilon .$$

Hence, F is an ε-net for $A(\overline{B}_1^X)$, thus, $A(\overline{B}_1^X)$ is relatively compact. Therefore, $A \in L_c(X, Y)$. □

Proposition 3.7.15. *If X, Y, V are Banach spaces, $A \in L(X, Y)$, $T \in L(Y, V)$, and A or T is compact, then $T \circ A \in L_c(X, Y)$.*

Proof. First suppose that A is compact. Then $\overline{A(\overline{B}_1^X)} \subseteq Y$ is compact, hence $T(\overline{A(\overline{B}_1^X)}) \subseteq V$ is compact. This means that $T \circ A \in L_c(X, V)$.

Now suppose that T is compact. The set $A(\overline{B}_1^X) \subseteq Y$ is bounded. Since $T \in L_c(Y, V)$ we have that $T(A(\overline{B}_1^X)) \subseteq V$ is relatively compact. This means that $T \circ A \in L_c(X, V)$. □

Corollary 3.7.16. *If X is a Banach space, then $L_c(X)$ is a closed ideal of $L(X)$.*

The next characterization of operator compactness is very useful in many occasions and is known as “Schauder's Theorem”.

Theorem 3.7.17 (Schauder's Theorem). *If X, Y are Banach spaces and $A \in L(X, Y)$, then $A \in L_c(X, Y)$ if and only if $A^* \in L_c(Y^*, X^*)$.*

Proof. $\Longrightarrow$: Let $K = \overline{A(\overline{B}_1^X)}$. Then $K \subseteq Y$ is compact. Moreover, let $B \subseteq Y^*$ be bounded. Then

$$|\langle y^*, y_1 - y_2\rangle| \le c\|y_1 - y_2\| \quad \text{for all } y^* \in B_1 \text{, for all } y_1, y_2 \in K \text{, for some } c > 0 .$$

This shows that $B \subseteq C(K)$ is bounded and equicontinuous. So, invoking the Arzela–Ascoli Theorem (see Theorem 1.6.16), we infer that B is relatively compact. Then, if $\{y_n^*\}_{n\ge1} \subseteq B$, there exists a subsequence $\{y_{n_k}^*\}_{k\ge1}$ of $\{y_n^*\}_{n\ge1}$, which is a uniformly Cauchy sequence on K. This implies that $\{y_{n_k}^* A\}_{k\ge1}$ is a uniformly Cauchy sequence on $\overline{B}_1^X$. Therefore, $\{y_{n_k}^* A\}_{k\ge1} \subseteq X^*$ is convergent. But by Definition 3.6.6, $y_{n_k}^* A = A^*(y_{n_k}^*)$. Thus, we conclude that $A^* \in L_c(Y^*, X^*)$.

$\Longleftarrow$: From the previous implication we obtain that $A^{**} \in L_c(X^{**}, Y^{**})$. Let $j_X : X \to X^{**}$ and $j_Y : Y \to Y^{**}$ be the corresponding canonical embeddings. Then $A = j_Y^{-1} \circ A^{**} \circ j_X$ and so Proposition 3.7.15 implies that $A \in L_c(X, Y)$. □

Definition 3.7.18. (a) If X is a vector space and V is a vector subspace of X, then the **codimension** of V in X is the dimension of the quotient vector space X/V.
(b) Let X, Y be Banach spaces and $A \in L(X, Y)$. We say that A is a **Fredholm operator** if $N(A)$ is finite dimensional and $R(A)$ has finite codimension. The number $i(A) = \dim N(A) - \operatorname{codim} R(A) = \dim N(A) - \dim(Y/R(A))$ is called the **index of** A.

Remark 3.7.19. If $A \in L(X, Y)$ is a Fredholm operator, then $X = N(A) \oplus V$ and $A|_V$ is an isomorphism of V onto $R(A)$. Moreover, $R(A) \subseteq Y$ is closed.

Lemma 3.7.20. *If X is a Banach space, $A \in L(X)$, $T = i_X - A$, and $V = R(T)$ is a proper closed subspace of X, then for every $\varepsilon > 0$ there exists $x_0 \in \overline{B}_1^X$ such that $d(A(x_0), A(V)) \geq 1 - \varepsilon$.*

Proof. According to the Riesz Lemma (see Lemma 3.1.20), there exists $x_0 \in X$ with $\|x_0\| = 1$ such that $d(x_0, V) \geq 1 - \varepsilon$. One has $T(x_0) \in V$ and $A(V) = (i_X - T)(V) \subseteq Y$. Therefore,

$$d(A(x_0), A(V)) \geq d(A(x_0) + T(x_0), V) = d(x_0, V) \geq 1 - \varepsilon . \qquad \square$$

Using this lemma we can prove the following theorem, which gives an important class of Fredholm operators.

Theorem 3.7.21. *If X is a Banach space, $A \in L_c(X)$, and $\lambda \neq 0$, then $\lambda i_X - A$ is a Fredholm operator.*

Proof. Clearly, we may assume that $\lambda = 1$. Let $N = N(i_X - A)$. For every $x \in N$ one has $A(x) = x$. Therefore $A|_N$ is an isomorphism with a subspace of X and $A|_N$ is compact as well. It follows that N is finite dimensional. Proposition 3.2.28 implies that there is a closed subspace V of X such that $X = N \oplus V$. Let $T = i_X - A$ and $\hat{T} = T|_V$. We obtain that $R(T) = T(V) = R(\hat{T})$ and $N(\hat{T}) = N \cap V = \{0\}$, hence $\hat{T}$ is injective. We claim that

$$\inf[\|\hat{T}(x)\| : x \in V, \|x\| = 1] > 0 . \tag{3.7.3}$$

Arguing by contradiction, suppose that (3.7.3) does not hold. Then there exists $x_n \in V$ with $\|x_n\| = 1$ for all $n \in \mathbb{N}$ such that $\|\hat{T}(x_n)\| \to 0$. Since $A \in L_c(X)$ we may assume that $A(x_n) \to u$ in X. Note that $A(x_n) = x_n$ for all $n \geq 1$, so $\|u\| = 1$. Moreover, $\hat{T}(u) = 0$ and this contradicts the injectivity of $\hat{T}$.

From (3.7.3) we infer that $\|\hat{T}(x)\| \geq c\|x\|$ for all $x \in V$ and for some $c > 0$. Then, Theorem 3.6.45 implies that $R(\hat{T}) = R(T) \subseteq X$ is closed.

We will show that $\operatorname{codim} R(T) < \infty$. Inductively we define

$$T^0 = i_X , \quad T^1 = T \quad \text{and} \quad T^{k+1} = TT^k \quad \text{for all } k \in \mathbb{N}_0 .$$

Moreover we set $N_k = N(T^k)$. Since $T^k = (i_X - T)^k$ and powers of compact operators are again compact operators (see Proposition 3.7.15), we get $T^k = i_X - S_k$ with $S_k \in L_c(X)$. From the first part of the proof we see that $\dim N_k < \infty$ for all $k \in \mathbb{N}_0$.

Let $Z_k = R(T^k) = T^k(V_1)$ with $k \in \mathbb{N}_0$. We have that

$$\{N_k\}_{k\in\mathbb{N}_0} \text{ is increasing and } \{Z_k\}_{k\in\mathbb{N}_0} \text{ is decreasing}. \tag{3.7.4}$$

For some $n \in \mathbb{N}_0$ we obtain $Z_n = Z_{n+1}$. Indeed if all the inclusions $Z_k \supseteq Z_{k+1}$ are strict, then with Lemma 3.7.20 there exists $u_n \in \overline{B}_1^{Z_n}$ such that $d(A(u_n), A(Z_{n+1})) \geq 1/2$. Then $\|A(u_n) - A(u_m)\| \geq 1/2$ for $n \neq m$, a contradiction to the fact that $A \in L_c(X)$.

Similarly, for some $m \in \mathbb{N}_0$, it holds $N_m = N_{m+1}$. Indeed if $x \in N_k$, that is, $T^k(x) = 0$, then $T^{k-1}(T(x)) = 0$ and so $T(x) \subseteq N_{k-1} \subseteq N_k$; see (3.7.4). Therefore, again via Lemma 3.7.20, we conclude that $N_m = N_{m+1}$ for some $m \in \mathbb{N}_0$. Thus, we obtain

$$Z_n = Z_{n'} \quad \text{for all } n' \geq n \quad \text{and} \quad N_m = N_{m'} \quad \text{for all } m' \geq m.$$

Let $i = \max\{n, m\}$. We claim that $X = N_i \oplus Z_i$. Let $x \in X$. Then $T^i(x) \in Z_i$ and $T^i(Z_i) = T^i(T^i(X)) = T^{2i}(X) = T^i(X) = Z_i$. Therefore there exists $u \in Z_i$ such that $T^i(u) = T^i(x)$, hence $T^i(u - x) = 0$. Therefore, $u - x \in N_i$ and $x = x - u + u$. Since $X = N_i \oplus Z_i$, the codimension of Z_i and also of $Z_1 \supseteq Z_i$ is finite. □

Example 3.7.22. (a) If X, Y are finite dimensional Banach spaces, then every linear operator $A: X \to Y$ is a Fredholm operator and $i(A) = \dim X - \dim Y$.

(b) If X, Y are Banach spaces and $A \in L(X, Y)$ is a bijection, then A is a Fredholm operator and $i(A) = 0$.

(c) Let $X = l^p$ with $1 \leq p \leq \infty$ and let $A \in L(l^p)$ be defined by

$$A(\hat{x}) = (x_{n+k})_{n\geq 1} \quad \text{for all } \hat{x} = (x_n)_{n\geq 1} \in l^p \text{ and for some } k \in \mathbb{N}.$$

Recall that for every $n \in \mathbb{N}$, $e_n = (0, \dots, 0, 1, 0 \dots)$ where 1 is located at the n-th entry. We see that $N(A) = \operatorname{span}\{e_n\}_{n=1}^k$, $R(A) = \{e_n\}_{n\geq k+1}$, and $R(A) = l^p$. Therefore A is a Fredholm operator and $i(A) = k$.

Let us consider the case where $X = Y$ are Banach spaces and $A \in L_c(X)$. Then according to Theorem 3.7.21, $i_X - A$ is a Fredholm operator. The next theorem, known as the “Fredholm Alternative Theorem”, asserts that either the nonhomogenous linear equation $x - A(x) = u$ has a solution $x \in X$ for every $u \in X$ or the corresponding homogeneous equation $x - A(x) = 0$ has a nontrivial solution. The result has interesting applications in boundary values problems.

Theorem 3.7.23 (Fredholm Alternative Theorem). *If X is a Banach space, $A \in L_c(X, Y)$ and $\lambda \neq 0$, then the equation $\lambda x - A(x) = u$ has a solution for every $u \in X$ if and only if the equation $x - A(x) = 0$ only has the trivial solution.*

Proof. Again we may assume that $\lambda = 1$. Let $T = i_X - A$. If $A(x) - x = 0$ only has the trivial solution, then $N = N(T) = \{0\}$ and so T is an isomorphism into. We will show that it is surjective.

Let $V_k = R(T^k)$ for all $k \in \mathbb{N}_0$. From the proof of Theorem 3.7.21 we know that there exists $n \in \mathbb{N}_0$ such that $V_k = V_n$ for all $k \geq n$. We claim that $V_1 = V_0 = X$. If this is not the case, let $m \in \mathbb{N}$ be the smallest integer such that $V_{m-1} \neq V_m = V_{m+1}$. We pick $u \in V_{m-1} \setminus V_m$. Then $T(u) \in V_m = V_{m+1}$. Hence, there exists $v \in V_m$ such that $T(u) = T(v)$ and $u \neq v$ since $u \notin V_m$. But this contradicts the injectivity of T.

Next, assume that T is surjective. Let $N_k = N(T^k)$ for $k \in \mathbb{N}$. We need to show that $N_1 = N(T) = \{0\}$. Recall that $\{N_k\}_{k\geq 1}$ is increasing. Arguing by contradiction, suppose that there is $x_1 \neq 0$ such that $x_1 \in N_1$. Inductively we will generate a sequence $\{x_k\}_{k\geq 1} \subseteq X$ such that $T(x_{k+1}) = x_k$ and $x_k \in N_k \setminus N_{k-1}$ for all $k \in \mathbb{N}$. Suppose that $x_1, \ldots, x_k$ have been constructed. Since $R(T) = X$, there exists $x_{k+1} \in X$ such that $T(x_{k+1}) = x_k$. Then $T^k(x_{k+1}) = T^{k-1}(x_k) = \cdots = x_1 \neq 0$ and $T^k(x_{k+1}) = T(x_1) = 0$. This completes the induction. Since $N_m = N_{m+1}$ for some $m \in \mathbb{N}_0$ (see the proof of Theorem 3.7.21), we have proven the assertion of the theorem. □

Next we prove a duality property of Fredholm operators, that is, we show that $A \in L(X, Y)$ is Fredholm if and only if $A^* \in L(Y^*, X^*)$ is Fredholm. We start with a simple lemma.

Lemma 3.7.24. *If X, Y are Banach spaces, $A \in L(X, Y)$ and $\dim(Y/R(A)) < \infty$, then $R(A) \subseteq Y$ is a closed subspace.*

Proof. Let $m = \dim(Y/R(A)) < \infty$. Then there exist vectors $\{y_k\}_{k=1}^m \subseteq Y$ such that

$$[y_k] = y_k + R(A) \in Y/R(A) \quad \text{for all } k \in \{1, \ldots, m\}$$

form a basis of $Y/R(A)$. We introduce the space

$$\hat{X} = X \times \mathbb{R}^m \quad \text{with norm } \|(x, \hat{\lambda})\|_{\hat{X}} = \|x\|_X + |\hat{\lambda}|$$

for all $x \in X$ and for all $\hat{\lambda} = (\lambda_k)_{k=1}^m \in \mathbb{R}^m$. Of course $\hat{X}$ with the norm above is a Banach space. Let $\hat{A} \in L(\hat{X}, Y)$ be defined by

$$\hat{A}(x, \hat{\lambda}) = A(x) + \sum_{k=1}^m \lambda_k y_k .$$

Then $\hat{A}$ is surjective and

$$N(\hat{A}) = \{(x, \lambda) \in X \times \mathbb{R}^m : A(x) = 0, \hat{\lambda} = 0\} = N(A) \times \{0\} .$$

Invoking Theorem 3.8.19, there exists $c > 0$ such that

$$\inf[\|x + u\|_X : u \in N(A)] + |\hat{\lambda}| \leq c \left\| A(x) + \sum_{k=1}^m \lambda_k y_k \right\|_Y \quad \text{for all } x \in X, \hat{\lambda} \in \mathbb{R}^m .$$

Let $\hat{\lambda} = 0$. Then

$$\inf[\|x+u\|_X : u \in N(A)] \le c\|A(x)\|_Y \quad \text{for all } x \in X\,,$$

which shows that $R(A) \subseteq Y$ is closed; see Theorem 3.8.19. □

Using this lemma, we can prove the duality property for Fredholm operators.

Theorem 3.7.25. *If X, Y are Banach spaces and $A \in L(X,Y)$, then the following hold:*
(a) *A is a Fredholm operator if and only if A^* is a Fredholm operator.*
(b) *If A is a Fredholm operator, then $\dim N(A^*) = \dim(Y/R(A))$ and $\dim N(A) = \dim(X^*/R(A^*))$.*

Proof. According to Theorem 3.8.19, $R(A) \subseteq Y$ is closed if and only if $R(A^*) \subseteq X^*$ is closed. So, we may assume that both $R(A) \subseteq Y$ and $R(A^*) \subseteq X^*$ are closed subspaces. Then

$$R(A^*) = N(A)^\perp \quad \text{and} \quad R(A)^\perp = N(A^*)\,; \tag{3.7.5}$$

see Proposition 3.6.44. Applying Proposition 3.2.25, one has

$$N(A)^* = X^*/N(A)^\perp = X^*/R(A^*) \quad \text{and} \quad (Y/R(A))^* = R(A)^\perp = N(A^*)\,;$$

see (3.7.5). This completes the proof of both statements (a) and (b). □

The last part of this section is devoted to the spectral theory of bounded linear operators. First, let us recall some standard results about invertible operators. Recall that $A \in L(X,Y)$ is invertible if and only if it is an isomorphism of X onto Y with X, Y being Banach spaces. Moreover, from Proposition 3.6.7 (e) we know that $A \in L(X,Y)$ with X, Y being Banach spaces is invertible if and only if A^* is invertible and $(A^{-1})^* = (A^*)^{-1}$. In addition, if X, Y, V are Banach spaces and $A \in L(X,Y)$, $T \in L(Y,V)$ are invertible operators, then $T \circ A \in L(X,V)$ is invertible as well and $(T \circ A)^{-1} = A^{-1}T^{-1}$.

Lemma 3.7.26. *If X is a Banach space, $A \in L(X)$ and $\|A\|_L < 1$, then $i_X - A \in L(X)$ is invertible and $(i_X - A)^{-1} = \sum_{n\ge 0} A^n$ with the series being absolutely convergent.*

Proof. Note that

$$\sum_{n\ge 0} \|A^n\|_L \le \sum_{n\ge 0} \|A\|_L^n < \infty$$

since by hypothesis $\|A\|_L < 1$. Hence $\sum_{n\ge 0} A^n$ is absolutely convergent in $L(X)$. Then we obtain

$$(i_X - A)\sum_{n\ge 0} A^n = (i_X - A) + (A - A^2) + \cdots = i_X\,,$$

which is called the **telescoping sum**. Similarly we get $(\sum_{n\ge 0} A^n)(i_X - A) = i_X$. Therefore we conclude that $i_X - A \in L(X)$ is invertible and $(i_X - A)^{-1} = \sum_{n\ge 0} A^n$. □

Lemma 3.7.27. *If X is a Banach space, $A, T \in L(X)$, A is invertible and $\|A-T\|_L < 1/\|A^{-1}\|_L$, then T is invertible as well and $\|T^{-1} - A^{-1}\|_L \le (\|A^{-1}\|_L^2 \|T - A\|_L)/(1 - \|A^{-1}\|_L \|T - A\|_L)$.*

Proof. Note that

$$\|A^{-1}(A - T)\|_L \le \|A^{-1}\|_L \|T - A\|_L < 1 .$$

Using Lemma 3.7.26 it follows that $i_X - A^{-1}(A - T) = A^{-1}T \in L(X)$ is invertible. Hence $T \in L(X)$ is invertible since $T = A(A^{-1}T)$. Moreover, we get

$$\left(i_X - A^{-1}(A - T)\right)^{-1} = \sum_{n\ge 0} (A^{-1}(A - T))^n ;$$

see Lemma 3.7.26. Therefore

$$T^{-1} = \left(A - (A - T)\right)^{-1} = \left(A(i_X - A^{-1}(A - T))\right)^{-1} = \sum_{n\ge 0} (A^{-1}(A - T))^n A^{-1} .$$

Thus,

$$\begin{aligned} \|T^{-1} - A^{-1}\|_L &\le \sum_{n\ge 1} \|(A^{-1}(A - T))^n A^{-1}\|_L \le \|A^{-1}\|_L \sum_{n\ge 1} (\|A^{-1}\|_L \|A - T\|_L)^n \\ &= \frac{\|A^{-1}\|_L^2 \|A - T\|_L}{1 - \|A^{-1}\|_L \|T - A\|_L} . \end{aligned}$$

□

Corollary 3.7.28. *If X is a Banach space and $\mathcal{L} \subseteq L(X)$ is the set of all invertible operators, then $\mathcal{L}$ is an open set in $L(X)$ and the map $A \to A^{-1}$ is a homeomorphism of $\mathcal{L}$ onto $\mathcal{L}$.*

Now we introduce the spectrum of a bounded linear operator. In order to have a complete spectral theory we need to assume that X is a complex Banach space.

Definition 3.7.29. Let X be a complex Banach space and let $A \in L(X)$. The **spectrum** $\sigma(A)$ of A is the set

$$\sigma(A) = \{\lambda \in \mathbb{C} : \lambda i_X - A \text{ is not invertible}\} .$$

The **resolvent set** $\rho(A)$ of A is the complement of $\sigma(A)$, that is, $\rho(A) = \mathbb{C} \setminus \sigma(A)$. The elements of $\rho(A)$ are called **regular values of** A. Moreover, if $\lambda \in \rho(A)$, then $R(\lambda) = (\lambda i_X - A)^{-1} \in L(X)$ is called the **resolvent of** A **at** λ. The spectrum of A is decomposed in the following way:

$$\begin{aligned} P\sigma(A) &= \{\lambda \in \mathbb{C} : \lambda i_X - A \text{ is not injective}\} , \\ R\sigma(A) &= \{\lambda \in \mathbb{C} : \lambda i_X - A \text{ is injective but } R(\lambda i_X - A) \subseteq X \text{ is not dense}\} , \\ C\sigma(A) &= \{\lambda \in \mathbb{C} : \lambda i_X - A \text{ is injective, } R(\lambda i_X - A) \subseteq X \text{ is dense} \\ &\qquad \text{but } \lambda i_X - A \text{ is not surjective}\} . \end{aligned}$$

We call $P\sigma(A)$ the **point spectrum of** A, $R\sigma(A)$ is the **residual spectrum of** A, and $C\sigma(A)$ is the **continuous spectrum of** A. Given $\lambda \in \mathbb{C}$ we see that $\lambda \in P\sigma(A)$ if and only if there exists $x \in X \setminus \{0\}$ such that $A(x) = \lambda x$. These elements are called **eigenvectors** for λ and $N(\lambda i_X - A)$ is the **eigenspace** for λ.

Remark 3.7.30. If X is finite dimensional and $n = \dim X$, then $\sigma(A) = P\sigma(A)$ and $\operatorname{card} \sigma(A) \le n$. If X is infinite dimensional and $A \in L_c(X)$, then $0 \in \sigma(A)$ or otherwise A would be a compact isomorphism, a contradiction.

Proposition 3.7.31. *If X is a Banach space and $A \in L(X)$, then $\sigma(A) = \sigma(A^*)$.*

Proof. From Proposition 3.6.7 (e), we know that $(\lambda i_X - A)$ is invertible if and only if $(\lambda i_X - A)^*$ is invertible. To conclude the proof just note that $(\lambda i_X - A)^* = \lambda i_{X^*} - A^*$. □

On account of Remark 3.6.8, we can state the following corollary concerning operators defined on a Hilbert into itself.

Corollary 3.7.32. *If H is a complex Hilbert space and $A \in L(H)$, then $\sigma(A^*) = \{\bar{\lambda} : \lambda \in \sigma(A)\}$.*

Proposition 3.7.33. *If X is a Banach space and $A \in L(X)$, then $\sigma(A) \subseteq \mathbb{C}$ is compact and if $\lambda \in \sigma(A)$, then $|\lambda| \le \|A\|_L$.*

Proof. Corollary 3.7.28 implies that $\rho(A) \subseteq \mathbb{C}$ is open. Hence, $\sigma(A) = \mathbb{C} \setminus \rho(A)$ is closed. Let $\lambda \in \mathbb{C}$ such that $|\lambda| > \|A\|_L$. Then $\lambda i_X - A = \lambda(i_X - 1/\lambda A)$ and so with Lemma 3.7.26, $\lambda i_X - A$ is invertible. Therefore, if $\lambda \in \sigma(A)$, then $|\lambda| \le \|A\|_L$ and $\sigma(A) \subseteq \mathbb{C}$ is compact. □

The next result is valid only for complex Banach spaces. That is why we said that in order to have a complete theory, we need to consider Banach spaces over $\mathbb{C}$.

Proposition 3.7.34. *If X is a complex Banach space and $A \in L(X)$, then $\sigma(A) \neq \emptyset$.*

Proof. We fix $\lambda_0 \in \sigma(A)$ and consider $\lambda \in \mathbb{C}$ such that $|\lambda - \lambda_0| < \|(\lambda_0 i_X - A)^{-1}\|_L^{-1}$. Using Lemma 3.7.27 for the operators $\lambda_0 i_X - A$ and $\lambda i_X - A$, we get

$$\begin{aligned} R(\lambda) = (\lambda i_X - A)^{-1} &= \sum_{n\ge 0} [(\lambda_0 i_X - A)^{-1}(\lambda_0 - \lambda) i_X]^n (\lambda_0 i_X - A)^{-1} \\ &= \sum_{n\ge 0} (\lambda_0 - \lambda)^n (\lambda_0 i_X - A)^{-(n+1)} = \sum_{n\ge 0} (\lambda_0 - \lambda)^n R(\lambda_0)^{n+1} ; \end{aligned}$$

see the proof of Lemma 3.7.27. Note that the series is absolutely convergent. So $\lambda \to R(\lambda)$ is an analytic function from $\rho(A)$ into $L(X)$. From the proof of Proposition 3.7.33 we know that if $|\lambda| > \|A\|_L$, then $R(\lambda) = \sum_{n\ge 0} 1/\lambda^{n+1} A^n$, hence $\|R(\lambda)\|_L \le 1/(|\lambda| - \|A\|_L)$. Arguing by contradiction, suppose that $\rho(A) = \mathbb{C}$. Then $R(\lambda) \to 0$ as $|\lambda| \to +\infty$. So with Liouville's Theorem, we obtain that $R \equiv 0$, a contradiction since the values of R are invertible operators. Therefore $\rho(A) \neq \mathbb{C}$ and so $\sigma(A) \neq \emptyset$. □

As we already pointed out (see Remark 3.7.30), if $\dim X < \infty$ and $A \in L(X)$, then $\sigma(A) = P\sigma(A)$, just recall that in this case A is injective if and only if A is surjective.

However, it is not true in general that every point of $\sigma(A)$ is an eigenvalue. For compact operators every nonzero element of $\sigma(A)$ is an eigenvalue.

Proposition 3.7.35. *If X is a Banach space, $A \in L_c(X)$ and $\lambda \in \sigma(A) \setminus \{0\}$, then $\lambda \in P\sigma(A)$.*

Proof. Suppose that $\lambda \neq 0$ is not an eigenvalue of A. Then according to Definition 3.7.29 we obtain $N(\lambda i_X - A) = \{0\}$. Then with the Fredholm Alternative Theorem (see Theorem 3.7.23), we have $R(\lambda i_X - A) = X$. Hence, according to Theorem 3.2.10, $\lambda i_X - A$ is invertible, which means that $\lambda \notin \sigma(A)$. □

Lemma 3.7.36. *If X is a Banach space, $A \in L(X)$, $\{\lambda_k\}_{k=1}^n$ are distinct eigenvalues of A and e_k is an eigenvector corresponding to λ_k for each $k = 1, \ldots, n$ with $n \in \mathbb{N}$, then $\{e_k\}_{k=1}^n \subseteq X$ are linearly independent.*

Proof. The proof goes by induction. So, suppose that $\{e_k\}_{k=1}^{n-1}$ are linearly independent. Let $e_n = \sum_{k=1}^{n-1} \vartheta_k e_k$ with $\vartheta_k \in \mathbb{C}$. Then $\sum_{k=1}^{n-1} \lambda_n \vartheta_k e_k = \lambda_n e_n = A(e_n) = \sum_{k=1}^{n-1} \lambda_k \vartheta_k e_k$. Hence, $\sum_{k=1}^{n-1}(\lambda_n - \lambda_k)\vartheta_k e_k = 0$. Since by the induction hypothesis $\{e_k\}_{k=1}^{n-1} \subseteq X$ are linearly independent and $\lambda_n - \lambda_k \neq 0$, we must have $\vartheta_k = 0$ for all $k = 1, \ldots, n$. Therefore $\{e_k\}_{k=1}^n \subseteq X$ are linearly independent. □

Proposition 3.7.37. *If X is a Banach space, $A \in L_c(X)$, and $\varepsilon > 0$, then A has only finitely many eigenvalues $\lambda \in \mathbb{C}$ such that $|\lambda| > \varepsilon$.*

Proof. Arguing by contradiction, suppose that there exist distinct eigenvalues $\{\lambda_k\}_{k \geq 1}$ such that $|\lambda_k| > \varepsilon$ for all $k \in \mathbb{N}$. For every eigenvalue λ_k, we choose an eigenvector e_k. For $n \in \mathbb{N}$ let $X_n = \operatorname{span}\{e_k\}_{k=1}^n$. With Lemma 3.7.36 it follows that $A(X_n) = X_n$ and $X_{n-1} \neq X_n$. Invoking the Riesz Lemma (see Lemma 3.1.20), there is a $u_n \in X_n$ such that

$$d(u_n, u_{n+1}) \geq \frac{1}{2} \quad \text{and} \quad \|u_n\| = 1 \quad \text{for all } n \geq 2\,. \tag{3.7.6}$$

Let $y_n = 1/\lambda_n u_n$ and note that $\|y_n\| \leq 1/\varepsilon$. Then $A(y_n) \in X_n$ and $u_n - A(y_n) \in X_{n-1}$. To see this second inclusion, note that $y_n = \sum_{k=1}^n \vartheta_k e_k$ with $\vartheta_k \in \mathbb{C}$. Then

$$u_n - A(y_n) = \sum_{k=1}^{n} \left(1 - \frac{\lambda_k}{\lambda_n}\right)\vartheta_k e_k = \sum_{k=1}^{n-1} \left(1 - \frac{\lambda_k}{\lambda_n}\right)\vartheta_k e_k \in X_{n-1}\,.$$

Let $n > m$. Then $A(y_m) \in X_m \subseteq X_{n-1}$ and $u_n - A(y_n) \in X_{n-1}$. Therefore one has

$$\begin{aligned} \|A(y_n) - A(y_m)\| &\geq d(A(y_n), X_{n-1}) = d(A(y_n) + u_n - A(y_n), X_{n-1}) \\ &= d(u_n, X_{n-1}) \geq \frac{1}{2}\,; \end{aligned} \tag{3.7.7}$$

see (3.7.6). But $\{y_n\}_{n \geq 1} \subseteq A(\overline{B}_\varepsilon)$ and the latter is relatively compact, a contradiction to (3.7.7). This proves that only finitely many eigenvalues $\lambda \in \mathbb{C}$ satisfy $|\lambda| > \varepsilon$. □

Combining this proposition with Theorem 3.7.21 we obtain the following corollary.

Corollary 3.7.38. *If X is a Banach space and $A \in L_c(X)$, then $\sigma(A) = \{0\} \cup P\sigma(A)$ with $P\sigma(A)$ either a finite set possibly empty or a sequence $\{\lambda_k\}_{k\geq 1} \subseteq \mathbb{C}$ exists such that $\lambda_k \to 0$ as $k \to \infty$ and each λ_k has a corresponding eigenspace that is finite dimensional.*

Now we focus on self-adjoint operators defined on a Hilbert space.

Proposition 3.7.39. *If H is a Hilbert space and $A \in L(H)$ is self-adjoint, then $P\sigma(A) \subseteq \mathbb{R}$ and eigenvectors corresponding to different eigenvalues are orthogonal.*

Proof. Since $A \in L(H)$ is self-adjoint, from Definition 3.6.13 (b) it follows that

$$(A(x), y) = (x, A(y)) \quad \text{for all } x, y \in H\,.$$

Suppose $x = y \in H$. Then

$$(A(x), x) = (x, A(x)) = \overline{(A(x), x)} \quad \text{for all } x \in H\,.$$

Hence

$$(A(x), x) \in \mathbb{R} \quad \text{for all } x \in H\,. \tag{3.7.8}$$

Suppose that $\lambda \in P\sigma(A)$. then $(A(x), x) = (\lambda x, x) = \lambda\|x\|^2$, which implies, because of (3.7.8), that $\lambda = (A(x), x)/\|x\|^2 \in \mathbb{R}$.

Next let $\lambda, \mu \in P\sigma(A)$ with $\lambda \neq \mu$ and suppose that $x, u \in H$ are eigenvectors corresponding to λ, μ, respectively. Then one gets

$$\begin{aligned}(A(x), u) &= (\lambda x, u) = \lambda(x, u)\,,\\ (A(x), u) &= (x, A(u)) = (x, \mu u) = \mu(x, u)\end{aligned}$$

since the eigenvalues are real; see above. It follows that $(\lambda - \mu)(x, u) = 0$. As $\lambda \neq \mu$ we conclude that $(x, u) = 0$. ☐

Proposition 3.7.40. *If H is a Hilbert space and $A \in L(H)$ is self-adjoint, then $\lambda \in \sigma(A)$ if and only if $\inf[\|\lambda x - A(x)\| : \|x\| = 1] = 0$.*

Proof. $\Longrightarrow$: Suppose that $\inf[\|(\lambda i_H - A)(x)\| : \|x\| = 1] > 0$. Then there exists $c > 0$ such that

$$\|(\lambda i_H - A)(x)\| \geq c\|x\| \quad \text{for all } x \in H\,. \tag{3.7.9}$$

We will show that $(\lambda i_H - A)^{-1} \in L(H)$ and so $\lambda \in \rho(A)$. According to Proposition 3.6.35, it suffices to show that $R(\lambda i_H - A)$ is dense in H. If this is not the case, then there exists $\hat{u} \in H \setminus \{0\}$ such that $((\lambda i_H - A)(x), \hat{u}) = 0$ for all $x \in H$. This gives $(x, (\overline{\lambda} i_H - A)\hat{u}) = 0$ for all $x \in H$. Therefore $\overline{\lambda}\hat{u} = A(\hat{u})$, that is, $\overline{\lambda} \in P\sigma(A)$.

But from Proposition 3.7.39 we know that $P\sigma(A) \subseteq \mathbb{R}$. Hence, $\lambda = \overline{\lambda}$ and so $(\lambda i_H - A)(\hat{u}) = 0$, a contradiction to (3.7.9). It follows that $R(\lambda i_H - A)$ is dense in H and so Proposition 3.6.35 implies that $(\lambda i_H - A)^{-1} \in L(H)$, and thus $\lambda \in \rho(A)$.

$\Longleftarrow$: Let $\lambda \in \rho(A)$. Then $(\lambda i_H - A)^{-1} \in L(H)$. So, for $x \in H$ with $\|x\| = 1$ we get

$$1 = \|x\| = \|(\lambda i_H - A)^{-1}(\lambda i_H - A)(x)\| \le \|(\lambda i_H - A)^{-1}\|_L \|(\lambda i_H - A)(x)\|$$
$$\le \|\lambda i_H - A\|_L^{-1} \|(\lambda i_H - A)(x)\| .$$

Hence, $\|\lambda i_H - A\|_L \le \|(\lambda i_H - A)(x)\|$, which gives

$$\|(\lambda i_H - A)^{-1}\|_L^{-1} \le \inf[\|(\lambda i_H - A)(x)\| : \|x\| = 1] .$$

So, if $\inf[\|(\lambda i_H - A)(x)\| : \|x\| = 1] = 0$, then we must have $\lambda \in \sigma(A)$. □

Using this proposition we can conclude that the spectrum of a self-adjoint operator is real; compare with Proposition 3.7.39.

Proposition 3.7.41. *If H is a Hilbert space and $A \in L(H)$ is self-adjoint, then $\sigma(A) \subseteq \mathbb{R}$.*

Proof. Let $\lambda = \eta + i\vartheta$ with $\vartheta \neq 0$. For every $x \in H$ with $\|x\| = 1$ we obtain

$$(\lambda x - A(x), x) - (x, \lambda x - A(x)) = (\lambda - \bar{\lambda})\|x\|^2 = 2i\vartheta .$$

Hence,

$$2|\vartheta| = |(\lambda x - A(x), x) - (x, \lambda x - A(x))| \le |(\lambda x - A(x), x)| + |(x, \lambda x - A(x))|$$
$$\le 2\|(\lambda i_H - A)(x)\| .$$

Therefore,

$$|\vartheta| \le \inf[\|(\lambda i_H - A)(x)\| : \|x\| = 1] . \tag{3.7.10}$$

So, from (3.7.10) and Proposition 3.7.40, we see that $\lambda \in \sigma(A)$ implies that $\vartheta = 0$. Thus, $\sigma(A) \subseteq \mathbb{R}$. □

Using this fact we can locate more precisely the spectrum of a self-adjoint operator.

Proposition 3.7.42. *If H is a Hilbert space, $A \in L(H)$ is self-adjoint, and*

$$m_A = \inf[(A(x), x) : \|x\| = 1] , \quad M_A = \sup[(A(x), x) : \|x\| = 1] ,$$

then $\sigma(A) \subseteq [m_A, M_A]$.

Proof. Note that if $T = A + \mu i_H$, then $T \in L(H)$ is self-adjoint and $m_T = m_A + \mu$ as well as $M_T = M_A + \mu$. So, without any loss of generality we may assume that $0 \le m_A \le M_A$.

From Proposition 3.6.16, we know that $M_A = \|A\|_L$, while from Proposition 3.7.41, we know that $\sigma(A) \subseteq \mathbb{R}$. We will show that, for every $\vartheta > 0$, $\lambda = M_A + \vartheta \notin \sigma(A)$. According to Proposition 3.7.40, it suffices to show that

$$\inf[\|(\lambda i_H - A)(x)\| : \|x\| = 1] > 0 .$$

For every $x \in H$ with $\|x\| = 1$ one has

$$((\lambda i_H - A)(x), x) = (\lambda x, x) - (A(x), x) \geq (\lambda - M_A)\|x\|^2 = \vartheta \|x\|^2 = \vartheta \,.$$

This gives $0 < \vartheta \leq \|(\lambda i_H - A)(x)\|$ for all $x \in H$ with $\|x\| = 1$. Therefore, $0 < \vartheta \leq \inf[\|(\lambda i_H - A)(x)\| : \|x\| = 1]$. Then, due to Proposition 3.7.40, this finally proves that $\lambda \notin \sigma(A)$.

Similarly we show that for every $\vartheta > 0$, $\lambda = m_A - \vartheta \notin \sigma(A)$. Hence, we conclude that $\sigma(A) \subseteq [m_A, M_A]$. □

Proposition 3.7.43. *If H is a Hilbert space and $A \in L(H)$ is self-adjoint, then $m_A, M_A \in \sigma(A)$; see Proposition* 3.7.42.

Proof. As before (see the proof of Proposition 3.7.42), we may assume that $0 \leq m_A \leq M_A$. Recall that $M_A = \|A\|_L$; see Proposition 3.6.16. Let $\{x_n\}_{n\geq 1} \subseteq H$ with $\|x_n\| = 1$ for all $n \in \mathbb{N}$ such that

$$(A(x_n), x_n) \to M_A = \|A\|_L \quad \text{as } n \to \infty \,. \tag{3.7.11}$$

Then, using the fact that $(A(x), x) \geq 0$ for all $x \in H$ and the validity of (3.7.11), it follows that

$$\begin{aligned} 0 &\leq \|(M_A i_H - A)(x_n)\|^2 = (M_A x_n - A(x_n), M_A x_n - A(x_n)) \\ &= M_A^2 + \|A(x_n)\|^2 - 2M_A(A(x_n), x_n) \leq M_A^2 + M_A^2 - 2M_A(A(x_n), x_n) \to 0 \end{aligned}$$

as $n \to \infty$. Hence $\inf[\|(M_A i_H - A)(x)\| : \|x\| = 1] = 0$, which gives $M_A \in \sigma(A)$; see Proposition 3.7.40.

Similarly we show that $m_A \in \sigma(A)$. □

Next we restrict further ourselves to compact self-adjoint operators.

Proposition 3.7.44. *If H is a Hilbert space and $A \in L_c(H)$ is self-adjoint, then $P\sigma(A) \neq \emptyset$.*

Proof. If $A = 0$, then $\lambda = 0 \in P\sigma(A)$. So, suppose that $A \neq 0$. Then Proposition 3.7.43 gives $\|A\|_L \in \sigma(A)$. Since $\|A\|_L \neq 0$, Corollary 3.7.38 implies that $\|A\|_L \in P\sigma(A)$. □

Proposition 3.7.45. *If H is an infinite dimensional Hilbert space and $A \in L_c(H) \setminus \{0\}$ is self-adjoint, then $\sigma(A) = \{0\} \cup \{\lambda_k\}_{k\geq 1}$ with λ_k being distinct nonzero eigenvalues of A, one of these eigenvalues equals $\|A\|_L$ and $\{\lambda_k\}_{k\geq 1}$ is either finite or a countable sequence such that $\lambda_k \to 0$. Moreover, the Hilbert space H admits an orthonormal basis consisting of eigenvectors corresponding to the eigenvalues of A.*

Proof. This is basically Corollary 3.7.38. From Proposition 3.7.43 we also know that one of the eigenvalues equals $\|A\|_L$. It remains to prove the last part of the proposition concerning the basis of H. Let $\lambda \in P\sigma(A)$ and let $N_\lambda = N(\lambda i_H - A)$. From Theorem 3.7.21 we know that $\dim N_\lambda < +\infty$. Let B_λ be an orthonormal basis for N_λ and let $B = \bigcup_{\lambda \in P\sigma(A)} B_\lambda$. From Proposition 3.7.39 we know that $B \subseteq H$ is an orthonormal set and $\overline{\operatorname{span} B}$ contains all the eigenvectors of A. Suppose that $H \neq \overline{\text{span}B}$ and let $V = (\overline{\text{span}B})^{\perp}$. Note that $\overline{\text{span}B}$ is A-invariant. Hence so is V. One has $\sigma(A) = \sigma(A|_{\overline{\text{span}B}}) + \sigma(A|_V)$. But $\sigma(A|_V)$ contains an eigenvalue (see Proposition 3.7.44), and so a corresponding eigenvector u as well. Then u is also an eigenvector of A and so $u \in V \cap \overline{\text{span}B}$, $u \neq 0$, a contradiction. This means that $H = \overline{\text{span}B}$, and so B is an orthonormal basis of H. □

Corollary 3.7.46. *If H is a Hilbert space and $A \in L_c(H)$ is self-adjoint, then $\sigma(A) = \overline{P\sigma(A)}$.*

Proof. If H is finite dimensional, then $\sigma(A) = P\sigma(A)$ and it is compact; see Proposition 3.7.33. If H is infinite dimensional, then $P\sigma(A)$ is a countable sequence or a finite sequence. If it is a countable sequence, then the conclusion follows from Proposition 3.7.44. If it is a finite sequence, then since the eigenspaces for the nonzero eigenvalues are finite dimensional (see Corollary 3.7.38), and H is infinite dimensional, then on account of Proposition 3.7.44 we must have that $\lambda = 0 \in P\sigma(A)$. □

We have reached the main result on the spectral analysis of compact self-adjoint operators defined on a Hilbert space. The result is known as the "Spectral Decomposition Theorem".

Theorem 3.7.47 (Spectral Decomposition Theorem). *If H is an infinite dimensional separable Hilbert space and $A \in L_c(H)$ is self-adjoint, then there exists an orthonormal basis $\{e_k\}_{k\geq 1} \subseteq H$ consisting of eigenvectors corresponding to the distinct eigenvalues $\{\lambda_k\}_{k\geq 1} \subseteq \mathbb{R}$ and*

$$A(x) = \sum_{k\geq 1} \lambda_k (x, e_k) e_k \quad \text{for all } x \in H .$$

Moreover, for every $\lambda \in \rho(A)$ and $x \in H$, it holds that

$$R(\lambda)(x) = \sum_{k\geq 1} \frac{(x, e_k)}{\lambda - \lambda_k} e_k .$$

Proof. Let $\{e_k\}_{k\geq 1} \subseteq H$ be an orthonormal basis of H consisting of eigenvectors; see Propositions 3.7.43 and 3.5.47. Then, for $1 \leq n < m$, one has

$$\left\| \sum_{k=n}^{m} \lambda_k (x, e_k) e_k \right\|^2 = \sum_{k=n}^{m} |\lambda_k (x, e_k)|^2 \leq \|A\|_L \sum_{k=n}^{m} |(x, e_k)|^2 \to 0$$

as $n \to \infty$; see Proposition 3.7.33. Hence, $\sum_{k\geq 1} \lambda_k (x, e_k) e_k$ converges in H.

If $\|x\| \leq 1$, then, for every $n \in \mathbb{N}$, we derive

$$\begin{aligned} \left\| \sum_{k=1}^{n} \lambda_k (x, e_k) e_k \right\|^2 &= \sum_{k=1}^{n} \lambda_k^2 |(x, e_k)|^2 \leq \|A\|_L^2 \sum_{k=1}^{n} |(x, e_k)|^2 \\ &\leq \|A\|_L^2 \sum_{k\geq 1} |(x, e_k)|^2 = \|A\|_L^2 \|x\|^2 . \end{aligned}$$

Consider the operator T defined by $T(x) = \sum_{k\geq 1} \lambda_k (x, e_k) e_k$. Of course, $T \in L(H)$. Hence, $A(e_k) = T(e_k)$ for all $k \in \mathbb{N}$ and so $A = T$.

Now suppose that $\lambda \in \rho(A)$. Recalling that $\sigma(A) = \mathbb{C} \setminus \rho(A)$ is compact, it follows that $d(\lambda, \sigma(A)) > \vartheta > 0$. Hence, $|\lambda - \lambda_k| > \vartheta$ for all $k \in \mathbb{N}$. Therefore,

$$\left\| \sum_{k=n}^{m} \frac{(x, e_k)}{\lambda - \lambda_k} e_k \right\|^2 = \sum_{k=n}^{m} \frac{|(x, e_k)|^2}{|\lambda - \lambda_k|^2} < \frac{1}{\vartheta^2} \sum_{k=n}^{m} |(x, e_k)|^2 .$$

This shows that $\sum_{k\geq 1} (x, e_k)/(\lambda - \lambda_k) e_k$ is convergent in H for all $x \in H$.

Let $T(x) = \sum_{k\geq 1} (x, e_k)/(\lambda - \lambda_k) e_k$. Then, for $\|x\| \leq 1$, we obtain

$$\left\| \sum_{k=1}^{n} \frac{(x, e_k)}{\lambda - \lambda_k} e_k \right\|^2 \leq \frac{1}{\vartheta^2} \sum_{k=1}^{n} |(x, e_k)|^2 = \frac{1}{\vartheta^2} \|x\|^2 \leq \frac{1}{\vartheta^2} .$$

Thus, $T \in L(H)$.

Since $x = \sum_{k\geq 1} (x, e_k) e_k$, we have $A(x) = \sum_{k\geq 1} \lambda_k (x, e_k) e_k$ and

$$(\lambda i_H - A)(x) = \sum_{k\geq 1} (\lambda - \lambda_k)(x, e_k) e_k .$$

As $(e_k, e_i) = \delta_{k,i}$ we then get

$$(\lambda i_H - A)(T(x)) = \sum_{k,i\geq 1} (\lambda - \lambda_k) \frac{(x, e_i)}{\lambda - \lambda_i} (e_k, e_i) e_k = \sum_{k\geq 1} (x, e_k) e_k = x .$$

Similarly we show that $T((\lambda i_H - A)(x)) = x$ for all $x \in H$. Therefore, $T = R(\lambda)$. □

We conclude this section by introducing two more classes of bounded linear operators of Hilbert space into itself.

Definition 3.7.48. Let H be a Hilbert space and $A \in L(H)$.
(a) We say that A is **normal** if $A \circ A^* = A^* \circ A$.
(b) We say that A is **unitary** if A is invertible and $A^{-1} = A^*$.

Remark 3.7.49. Clearly every unitary operator is normal and every self-adjoint operator is normal.

Proposition 3.7.50. *If H is a Hilbert space and $A \in L(H)$, then A is normal if and only if $\|A(x)\| = \|A^*(x)\|$ for all $x \in H$.*

Proof. For every $x \in H$, we derive

$$\begin{aligned} \|A(x)\|^2 - \|A^*(x)\|^2 &= (A(x), A(x)) - (A^*(x), A^*(x)) \\ &= (A^*(A(x)), x) - (A(A^*(x)), x) \\ &= ((A^* \circ A - A \circ A^*)(x), x) . \end{aligned} \tag{3.7.12}$$

From (3.7.12) it follows that A is normal if and only if $\|A(x)\| = \|A^*(x)\|$ for all $x \in H$. □

Proposition 3.7.51. *If H is a Hilbert space and $A \in L(H)$ is surjective, then the following statements are equivalent:*
(a) *A is unitary.*
(b) *$(A(x), A(u)) = (x, u)$ for all $x, u \in H$.*
(c) *A is an isometry.*

Proof. (a) ⟹ (b): For every $x, u \in H$ it holds that

$$(A(x), A(y)) = (A^*(A(x)), u) = (x, u) .$$

(b) ⟺ (c): This follows from the polarization identities; see Proposition 3.5.6 (b).

(c) ⟹ (a): The operator A is an isometry and surjective, and hence, $A^{-1} \in L(H)$; see Theorem 3.2.10. Moreover, for all $x, u \in H$, one has

$$(A^*(A(x)), u) = (A(x), A(u)) = (x, u) ,$$

since (b) is equivalent to (c). Hence, $A^* \circ A = i_H$ and similarly $A \circ A^* = i_H$, and so $A^{-1} = A^*$. □

Remark 3.7.52. So according to the proposition above, $A \in L(H)$ is unitary if and only if it preserves inner products.

3.8 Remarks

(3.1) The major development of mathematics in the twentieth century was the emphasis on the axiomatic method. This abstract tendency with emphasis on the structural properties led to the development of whole new areas such as "Functional Analysis" with the seminal contributions of Banach, von Neumann, and Riesz to mention only a few major figures and to "Modern Algebra" where prominent figures were Noether and van der Waerden. In this approach, the emphasis is not on the objects but on the rules used to handle them, which are the same for many different classes of objects. The power of the axiomatic method can be traced back in the work of Euclid who provided a model for space locally. The first breakthrough in the abstract axiomatic approach was achieved by Fréchet who introduced abstract metric spaces in this thesis [128]. He was the first to go beyond the familiar concrete Euclidean space setting. The normed space axioms (see Definition 3.1.13 (e)) were first introduced by Banach [26] in this thesis. Normed spaces are a subset of metric spaces. The thing that makes normed spaces such a prolific concept is the linkage between the algebraic and the topological structures of the space. This is expressed by the requirement that the two algebraic operations, namely vector addition and scalar multiplication, are continuous. This leads at a higher level of generality to the notion of a topological space. Moreover, the convexity of the balls in a normed space lead to topological vector spaces with a local neighborhood basis consisting of convex sets. These are the locally convex spaces (see Definition 3.1.13 (b)) first introduced by von Neumann [328]. Until the mid-forties, the study of functional analysis focused on normed spaces. The first major paper on the theory of locally convex spaces was that of Dieudonné–Schwartz [95] motivated by Schwartz's construction of the theory of distributions. Lemma 3.1.20, called Riesz Lemma, was proved by Riesz [271] and turned out to be a fruitful result for many occasions. Theorem 3.1.30 is due to Carathéodory [69] and Theorem 3.1.41, due to Kolmogorov [195], seems to be the first theorem about locally convex spaces. The Hahn–Banach Theorem (see Theorem 3.1.42) is crucial in the development of the theory of normed spaces. The first version of it was due to Minkowski, who proved that every boundary point in the closed unit ball of a finite dimensional normed space admits at least one supporting hyperplane through it. Later Helly [158] generalized the ideas of Minkowski to certain separable spaces. Fifteen years later, in 1927, Hahn [149] starting from the work of Helly, proved an extension theorem in a more general form without any separability hypothesis. Soon thereafter, we have the result of Banach [27] (see also Banach [28]), who proved the theorem in general vector spaces apart from any topology. The Hahn–Banach Theorem turned out to be a major tool in the development of the theory of locally convex spaces. Although the original proof uses transfinite induction, this part of the argument was later replaced by use of the Zorn's Lemma. The complex version of the result (see Theorem 3.1.44) is due to Bohnenblust–Sobczyk [42] and Suchomlinov [305]. Theorem 3.1.59, the First Separation Theorem, is due to Edelheit [107]. Theorem 3.1.60, the Strong Separation Theorem, is due to Tukey [312] and Klee [192].

(3.2) Theorem 3.2.1, the Uniform Boundedness Principle, was first proved by Hahn [148] for sequences of linear functionals. A more general form was produced by Hildebrandt [163]. The general version of the result and a proof based on the Baire Category Theorem were provided by Banach–Steinhaus [29]; see also Theorem 3.2.2. Theorem 3.2.9, the Open Mapping Theorem, was proved by Schauder [288] for Banach spaces. Banach [28] extended the result to Fréchet spaces; see Definition 3.1.13 (d). Theorem 3.2.10 and Theorem 3.2.14, the Closed Graph Theorem, are due to Banach [28]. Banach [28] extended both the Open Mapping Theorem and the Closed Graph Theorem to topological groups. The book of Banach [28] turned out to be one of the most influential books in analysis and remains a reference even today.

Definition 3.8.1. Let P be a property of normed spaces. Suppose that if X is a normed space and $V \subseteq X$ is a closed subspace such that if two of the spaces X, V, X/V have property P, then so does the third. Then we say that P is a **three space property**.

Using this notion we can improve Proposition 3.2.17 in the following way.

Proposition 3.8.2. *Completeness is a three space property.*

(3.3) We point out that Banach [28] worked only with weakly convergent sequences and did not use the notion of "weak topology". In certain occasions this led to unnecessary separability assumptions. The first explicity description of weak neighborhoods in a Hilbert space was given by von Neumann [326] who was the first to recognize that the weak topology is indeed a topology. He also realized the nonmetrizability of the weak topology in an infinite dimensional normed space; see Proposition 3.3.15. Further discussion on this issue can be found in Wehausen [331]. Proposition 3.3.16 was first proven for $X = l^2$ by von Neumann [326]. Theorem 3.3.18 is due to Mazur [228]. Earlier particular versions of this result for the Banach space $C[0,1]$ can be found in Gillespie–Hurwitz [139] and Zalcwasser [338]. That bounded linear operators are weakly continuous was first observed by Banach [27]. The converse (see Proposition 3.3.23) is due to Bade [22]. Theorem 3.3.37, Goldstine's Theorem, is naturally due to Goldstine [143] and Theorem 3.3.38, Alaoglu's Theorem, was proved by Alaoglu [3]. For separable Banach spaces the theorem can be found in Banach [28]. For this reason some people call it the "Banach–Alaoglu Theorem"; see, for example, Megginson [230, p. 229]. Theorem 3.3.41 is due to James [180] and is one of the most influential results in Banach space theory.

Another locally convex topology on X^* being the dual of the normed space X is the bounded weak* topology introduced by Dieudonné [94].

Definition 3.8.3. Let X be a normed space. The **bounded weak* topology** (or the **bw*-topology**) is the strongest topology on X^* which coincides with the relative w*-topology on each set $t\overline{B}_1^{X^*} = \{x^* \in X^* : \|x^*\|_* \le t\}$. Therefore a set $U \subseteq X^*$ is bw*-open if and only if $U \cap t\overline{B}_1^{X^*}$ is relatively w*-open in $t\overline{B}_1^{X^*}$ for every $t > 0$ and $C \subseteq X^*$ is bw*-closed if and only if $C \cap t\overline{B}_1^{X^*}$ is relatively w*-closed in $\overline{B}_1^{X^*}$ for all $t > 0$.

Remark 3.8.4. It can be shown (see, for example Dunford–Schwartz [105, Lemma V.5.4, p. 427]) that a local basis at the origin for the bw*-topology is given by the sets

$$B(S) = \{x^* \in X^* : |\langle x^*, x\rangle| < 1 \text{ for all } x \in S\},$$

where $S = \{x_k\}_{k\ge1} \subseteq X$ is a sequence converging to zero. We have $w^* \subseteq \text{bw}^* \subseteq \text{norm}$.

These inclusions are strict if X is an infinite dimensional normed space. Directly from Definition 3.8.3 we see that if $\{x_\alpha^*\}_{\alpha\in I} \subseteq X^*$ is a bounded net and $x^* \in X^*$, then $x_\alpha^* \xrightarrow{\text{w}^*} x^*$ if and only if $x_\alpha^* \xrightarrow{\text{bw}^*} x^*$. Of course $X^*_{\text{bw}^*}$ is a locally convex space.

Proposition 3.8.5. *If X is a Banach space, then $X = (X^*_{\text{w}^*})^* = (X^*_{\text{bw}^*})^*$.*

Using this proposition one can show the following theorem known as the "Krein–Smulian Theorem".

Theorem 3.8.6 (Krein–Smulian Theorem). *If X is a Banach space and $C \subseteq X^*$ is a nonempty convex set, then C is w*-closed if and only if $C \cap t\overline{B}_1^{X^*}$ is w*-closed for every $t > 0$, that is, C is w*-closed if and only if C is bw*-closed.*

Remark 3.8.7. As in Mazur's Theorem (see Theorem 3.3.18), in the theorem above we see that an algebraic property, namely the convexity of C, has topological consequences.

Corollary 3.8.8. *If X is a separable Banach space and $C \subseteq X^*$ is a nonempty convex set, then C is* w*-closed if and only if it is weakly* *sequentially closed.*

We can introduce one more locally convex topology on X^*. Recall that the weak* topology is the weakest topology τ on X^* such that $(X^*_\tau)^* = X$. Suppose we ask for the strongest (finest) topology m on X^* for which $(X^*_{\mathrm{m}})^* = X$ is satisfied.

Theorem 3.8.9. *There exists a strongest topology* m *on X^* such that $(X^*_{\mathrm{m}})^* = X$. This is the topology of uniform convergence on all* w*-compact sets, that is, $x^*_\alpha \xrightarrow{\mathrm{m}} x^*$ in X^* if and only if $\sup[\langle x^*_\alpha - x^*, u\rangle : u \in K] \to 0$ for all* w*-compact $K \subseteq X$. The space X^*_{m} is locally convex and* m *is called the* ***Mackey topology*** *on X^* and is denoted by $\mathrm{m}(X^*, X)$.*

We have already seen how important the notion of convexity is. Next we will see that in some convex sets we can isolate special points of them that in fact generate the set.

Definition 3.8.10. Let X be a topological vector space and $C \subseteq X$ be a nonempty, closed, convex set. A set $E \subseteq C$ is **extremal** in C if E is nonempty, closed, convex and if $x, u \in C$ and $(1-\lambda)x + \lambda u \in E$ for some $\lambda \in (0,1)$, then $x, u \in E$. An **extreme point** of C is an $x \in C$ such that $\{x\}$ is an extremal subset of C, that is, x is an extreme point of C if it does not lie in the interior of any nontrivial closed line segment of C. By $\operatorname{ext} C$ we denote the set of extreme points of C.

The following is the basic theorem about extreme points and it is known as the "Krein–Milman Theorem".

Theorem 3.8.11 (Krein–Milman Theorem). *If X is a locally convex space and $C \subseteq X$ is nonempty, compact, and convex, then $\operatorname{ext} C \neq \emptyset$ and $C = \overline{\operatorname{conv}} \operatorname{ext} C$.*

For more on the structure of convex sets, we refer to Giles [138]. The books of Aliprantis–Border [7], Beauzamy [32], Brézis [52], Denkowski–Migórski–Papageorgiou [87], Diestel [90], Fabian et al. [117], Giles [138], Holmes [171], Megginson [230], Rudin [285], and Yosida [336] discuss in detail the weak and weak* topologies.

(3.4) Reflexive Banach spaces were introduced by Hahn [149]. He called them **regular**. The term "reflexive" is due to Lorch [220] and Theorem 3.4.5 is due to James [179]. There are other useful characterizations of reflexivity. We mention three of them. The first is due to Smulian [298].

Theorem 3.8.12. *If X is a Banach space, then X is reflexive if and only if for every decreasing sequence $\{C_n\}_{n\geq 1}$ of nonempty, bounded, closed, convex subsets of X, it holds that $\bigcap_{n\geq 1} C_n \neq \emptyset$.*

The second is due to James [181].

Theorem 3.8.13. *If X is a Banach space, then the following statements are equivalent:*
(a) *X is not reflexive.*
(c) *For every $\lambda \in (0,1)$ there exists a sequence $\{x_n\}_{n\geq 1} \subseteq X$ with $\|x_n\| = 1$ for all $n \in \mathbb{N}$ such that $d(\operatorname{conv}\{x_k\}_{k=1}^n, \overline{\operatorname{conv}}\{x_k\}_{k\geq n+1}) \geq \lambda$ for every $n \in \mathbb{N}$.*
(c) *For some $\lambda \in (0,1)$ there exists a sequence $\{x_n\}_{n\geq 1} \subseteq X$ with $\|x_n\| = 1$ for all $n \in \mathbb{N}$ such that $d(\operatorname{conv}\{x_k\}_{k=1}^n, \overline{\operatorname{conv}}\{x_k\}_{k\geq n+1}) \geq \lambda$ for every $n \in \mathbb{N}$.*

Remark 3.8.14. The interesting feature of the theorem above for reflexivity is that it is intrinsic. Namely, it does not require any knowledge of X^* or X^{**}.

The third is also due to James [181].

Theorem 3.8.15. *If X is a Banach space, then X is reflexive if and only if every $x^* \in X^*$ is norm attaining, that is, there exists $x_0 \in X$, $\|x_0\| \leq 1$ such that $\|x^*\|_* = \langle x^*, x_0\rangle$.*

It is easy to check the next proposition.

Proposition 3.8.16. *Separability and reflexivity are three space properties; see Definition* 3.8.1.

The direct assertions in Theorem 3.4.12 are due to Banach [28] and concerning Theorem 3.4.14, the Eberlein–Smulian Theorem, Smulian [299] showed that weakly compact sets are weakly sequentially compact. Later Eberlein [106] proved the converse. Whitley [333] provided an elementary proof of the theorem. Theorem 3.4.18 reveals the distinctive character of weakly compact sets. They are sequentially compact and each subset of a weakly compact set has a sequentially determined closure. These properties are a particular instance of a more general class of spaces known as **angelic space**; see Floret [124].

Strict convexity and uniform convexity (see Definition 3.4.21 (a), (b)) were introduced by Clarkson [75]. Local uniform convexity was introduced by Lovaglia [221]. In the paper of Smith [297], we find examples of reflexive Banach spaces that are locally uniformly convex but not uniformly convex, and of reflexive and nonreflexive Banach spaces that are strictly convex but not locally uniformly convex. The Kadec–Klee property is also called the **Radon–Riesz property** or the H-property; see Day [83].

Proposition 3.8.17. *If X is a uniformly convex Banach space and $V \subseteq X$ is a closed subspace, then X/V is uniformly convex as well.*

(3.5) The notion of abstract Hilbert spaces was introduced by von Neumann [325]. His definition is for a separable space and his aim was to develop the spectral theory for classes of operators on this abstract space. Earlier special realizations of Hilbert spaces were examined by many authors. In particular, Hilbert [162] published between 1904 and 1910 a series of six papers collected in book form developing Hilbert space methods to study integral equations. The name Hilbert space was first used by Riesz [266] for what we know today as l^2. Theorem 3.5.21 was stated by Riesz [267] and Fréchet [129] as separate notes in the same issue of the “Comptes Rendus”. In addition to Bessel’s inequality (see Proposition 3.5.44), we should also mention the so-called **Parseval’s identity**.

Proposition 3.8.18 (Parseval’s identity). *If H is a Hilbert space and $\{e_n\}_{n\geq 1} \subseteq H$ is an orthonormal set, then $\{e_n\}_{n\geq 1}$ is an orthonormal basis for H if and only if $\|x\|^2 = \sum_{n\geq 1}(x, e_n)^2$ for all $x \in H$.*

The Gram–Schmidt Orthonormalization Process was first discovered by the Danish statistician Gram. It was elaborated further by Schmidt [290] who demonstrated its usefulness in the study of Hilbert spaces.

(3.6) The operator topologies in Definition 3.6.1 were introduced, in the context of Hilbert spaces, by von Neumann [326]. The notion of adjoint operators (see Definition 3.6.6) was first introduced by Banach [28]. Of course the notion was used earlier in the context of matrix theory. The notion of projection operator (see Definition 3.6.13 (a), (c)) is due to Schmidt [290]. The theory of unbounded linear operators was stimulated by attempts in the late 1920s to give quantum mechanics a rigorous mathematical foundation. The first fundamental works on this subject are those of von Neumann [326, 325], [327], and Stone [304]. A more detailed treatment of unbounded linear operators can be found in the books of Goldberg [142], Hille–Phillips [165], Kato [186], Reed–Simon [264], and Weidmann [332].

We state a theorem related to the material of this section.

Theorem 3.8.19. *If X, Y are Banach spaces and $A \in L(X, Y)$, then the following statements are equivalent:*
(a) *$R(A) \subseteq Y$ is closed;*
(b) *$\inf[\|x + v\|_X : A(v) = 0] \leq c\|Ax\|_Y$ for all $x \in X$ and for some $c > 0$;*
(c) *$R(A^*) \subseteq X^*$ is closed;*
(d) *$\inf[\|y^* + x^*\|_{Y^*} : A^*(v^*) = 0] \leq c\|A^*y^*\|_{X^*}$ for all $y^* \in Y^*$ and for some $c > 0$.*

(3.7) The notion of compact operators (see Definition 3.7.1) is essentially due to Hilbert [162]. However, the general definition was given by Riesz [271]. Theorem 3.7.10 is due to Schauder [288]. It is the starting point of the Leray–Schauder degree theory; see Section 6.2. Theorem 3.7.17 is due to Schauder [288]. The terminology “Fredholm Operator” was introduced in recognition of the pioneering work of E. Fredholm on integral equations. The work of Fredholm influenced Hilbert. Fredholm operators exhibit nice composition and stability properties.

Proposition 3.8.20. *If X, Y, V are Banach spaces and $A \in L(X, Y)$, $T \in L(Y, V)$ are Fredholm operators, then $T \circ A \in L(X, V)$ is a Fredholm operator and $i(T \circ A) = i(A) + i(T)$.*

Proposition 3.8.21. *If X, Y are Banach spaces and $A \in L(X, Y)$ is a Fredholm operator, then the following hold:*
(a) *$A + L$ is a Fredholm operator for every $L \in L_c(X, Y)$ and $i(A + L) = i(A)$;*

(b) *there exists* $\varepsilon > 0$ *such that if* $T \in L(X,Y)$ *with* $\|T\|_L < \varepsilon$, *then* $A + T$ *is a Fredholm operator and* $i(A+T) = i(A)$.

The terminology "spectrum" of $A \in L(X)$ comes from Hilbert who published some papers in book form [162] initiating modern spectral theory. The mathematical setting of self-adjoint operators on a Hilbert space was an important mathematical tool for the development by physicists of the theory of quantum mechanics.

Definition 3.8.22. Let H be a Hilbert space and $A \in L(H)$. We say that A is **positive** (or **monotone**) if $(A(x), x) \geq 0$ for all $x \in H$. Then we write $A \geq 0$. Moreover, if $A, T \in L(H)$, then we write $A \geq T$ if and only if $A - T \geq 0$.

Remark 3.8.23. Every positive $A \in L(H)$ with H being a complex Hilbert space is automatically self-adjoint. This is false for real Hilbert spaces. Moreover, $A^* \circ A \geq 0$ for any $A \in L(H)$.

Proposition 3.8.24. *If* H *is a Hilbert space,* $A \in L(H)$ *and* $A \geq 0$, *then there exists a unique* $T \in L(H)$, $T \geq 0$ *such that* $T^2 = A$. *Moreover,* T *commutes with every bounded linear operator, which commutes with* A. *We denote* T *by* $A^{1/2}$, *the square root of* A.

Definition 3.8.25. Let H be a Hilbert space and $A \in L(H)$. Then $|A| = (A^* \circ A)^{1/2}$; see Proposition 3.8.24.

Finally let us state a result on the usage of unitary operators (see Definition 3.7.48), to identify compact self-adjoint operators.

Proposition 3.8.26. *If* H *is a separable Hilbert space and* $A, T \in L_c(H)$ *is self-adjoint, then there exists a unitary operator* $U \in L(H)$ *such that* $U^* \circ T \circ U = A$ *if and only if* $\dim N(\lambda U - A) = \dim N(\lambda I - T)$ *for all* $\lambda \in \mathbb{C}$. *We say that the operators* A *and* T *are* ***unitarily equivalent***.

Problems

Problem 3.1. Let X be a vector space and let $\rho\colon X \to \mathbb{R}_+$ be a function such that
(a) $\rho(x) = 0$ if and only if $x = 0$;
(b) $\rho(\lambda x) = |\lambda|\rho(x)$ for all $x \in X$ and for all $\lambda \in \mathbb{F}$.

Show that ρ is a norm if and only if $\overline{B}_1 = \{x \in X : \rho(x) \leq 1\}$ is convex.

Problem 3.2. Let X be a vector space and let $\|\cdot\|$, $|\cdot|$ be two equivalent norms on X, that is, they generate the same topology. Show that $(X, \|\cdot\|)$, $(X, |\cdot|)$ are either both Banach spaces or both are noncomplete.

Problem 3.3. Let X be a topological vector space and let $\{C_k\}_{k=1}^n$ be a finite family of compact, convex subsets of X. Show that $\operatorname{conv}(\bigcup_{k=1}^n C_k)$ is compact.

Problem 3.4. Let X be a normed space, $Y \subseteq X$ be a closed subspace, and let $V \subseteq X$ be a finite dimensional subspace. Show that $Y + V = \{y + v : y \in Y, v \in V\} \subseteq X$ is closed.

Problem 3.5. Let X be a normed space and $V \subseteq X$ is a finite dimensional subspace. Show that there exists $x \in X$ with $\|x\| = 1$ such that $1 = d(x, V)$.

Problem 3.6. Let X be a normed space that is a Polish space for the norm topology. Show that X is a Banach space.

Problem 3.7. Show that a normed space X is complete, that is, X is a Banach space, if and only if every absolutely convergent series in X is convergent.

Problem 3.8. Let K be a compact topological space and let $D \subseteq K$ be a closed set. Show that $C(D)$ is isomorphic to a quotient of $C(K)$.

Problem 3.9. Let K, D be compact topological spaces and let $A: C(K) \to C(D)$ be a linear operator such that $f \geq 0$ implies $A(f) \geq 0$, that is, A is positive. Show that A is continuous and $\|A\|_L = \|A(1)\|_{C(D)}$ with $1 \in C(K)$ is the constant function equal to 1.

Problem 3.10. Let $X = C[0,1]$, $u \in X$, and $f: X \to \mathbb{R}$ be a linear function defined by $f(y) = \int_0^1 y(t)u(t)\,dt$ for all $y \in X$. Show that $f \in X^*$ and $\|f\|_* = \int_0^1 |u(t)|\,dt$.

Problem 3.11. Let X be a normed space and $C \subseteq X$ be a nonempty set. Show that $\overline{\operatorname{conv}}\, C = \{x \in X : \langle x^*, x\rangle \leq \sigma_C(x^*)\} = \sup\{\langle x^*, c\rangle : c \in C\}$, whereby $\sigma_C : X^* \to \overline{\mathbb{R}} = \mathbb{R} \cup \{+\infty\}$ is called the **support function of** C.

Problem 3.12. Show that every normed space is isometrically isomorphic to a subspace of $C(K)$ for some compact topological space K.

Problem 3.13. Let X, Y be Banach spaces and let $A \in L(X, Y)$ be surjective. Show that there exists $M > 0$ such that for every $y \in Y$ there is $x \in A^{-1}(y)$ satisfying $\|x\|_X \leq M\|y\|_Y$.

Problem 3.14. Let X, Y be Banach spaces and let $A \in L(X, Y)$ be surjective. Show that Y is isomorphic to $X/N(A)$.

Problem 3.15. Let X be a Banach space and let $C \subseteq X$ be a weakly compact set. Show that C is bounded.

Problem 3.16. Let X be a normed space and $\{x_n^*\}_{n\geq 1} \subseteq X^*$. Suppose that there exists a sequence $\{\varepsilon_n\}_{n\geq 1} \subseteq (0, +\infty)$ with $\varepsilon_n \to 0$ such that for every $x \in X$ there exists $\eta_x > 0$ with $|\langle x_n^*, x\rangle| \leq \eta_x \varepsilon_n$ for all $n \in \mathbb{N}$. Show that $x_n^* \to 0$.

Problem 3.17. Show that separability and reflexivity are three space properties; see Definition 3.8.1.

Problem 3.18. Show that a normed space X is reflexive if and only if each separable, closed subspace $V \subseteq X$ is reflexive.

Problem 3.19. Show that if Y is an infinite dimensional subspace of l^1, then Y is not reflexive.

Problem 3.20. Let X be a separable Banach space. Show that there exists $x_n^* \in X^*$ with $\|x_n^*\|_* = 1$ for all $n \in \mathbb{N}$ such that $\{x_n^*\}_{n\geq 1}$ is separating on X.

Problem 3.21. Let X, Y be Banach spaces with X being reflexive and let $A \in L(X, Y)$ be surjective. Show that Y is reflexive as well.

Problem 3.22. Let X be a Banach space with a separable dual X^*. Show that $\mathcal{B}(X^*) = \mathcal{B}(X^*_{w^*})$. Recall that if Z is a Hausdorff topological space, then $\mathcal{B}(Z)$ denotes the Borel σ-algebra of Z.

Problem 3.23. Let X be a normed space and let $C \subseteq X^*$ be a nonempty, w*-closed set. Show that for any given $x^* \in X^*$ there exists $u_0^* \in C$ such that $\|x^* - u_0^*\|_* = d(x^*, C)$. A set that has this best approximation property for every element in the space is called **proximinal**.

Problem 3.24. Show that a Banach space X is reflexive if and only if every closed convex set is proximinal; see Problem 3.23.

Problem 3.25. Let X, Y be two nontrivial normed spaces and assume that $L(X, Y)$ equipped with the operator norm is a Banach space. Show that Y is a Banach space.

Problem 3.26. Let X be a reflexive Banach space and let Y be another Banach space that is isomorphic to X. Show that Y is reflexive as well.

Problem 3.27. Let X, Y be Banach spaces with X being nonreflexive and Y being reflexive. Suppose that $A \in L(X, Y)$ is injective. Show that $R(A) \subseteq Y$ cannot be closed.

Problem 3.28. Let X be a Banach space and let $C \subseteq X^*$ be a w*-compact set. Show that $\overline{\operatorname{conv}}^{w^*} C$ is w*-compact.

Problem 3.29. Let X be a separable Banach space. Show that X^* is w*-separable.

Problem 3.30. Let H and V be real Hilbert spaces and let $k\colon H \times V \to \mathbb{R}$ be a bilinear form that is bounded, that is, there exists $c > 0$ such that $|k(u, v)| \leq c\|u\|_H \|v\|_V$ for all $u \in H$ and for all $v \in V$. Show that there exists a unique $A \in L(H, V)$ such that $k(u, v) = (A(u), v)_V$ for all $u \in H$ and for all $v \in V$.

Problem 3.31. Let H be a Hilbert space and let $\{e_n\}_{n\geq 1} \subseteq H$ be an orthonormal set. Suppose that $u = \sum_{n\geq 1} a_n e_n$. Show that $a_n = (u, e_n)$ for all $n \in \mathbb{N}$.

Problem 3.32. Let H, V be infinite dimensional separable Hilbert spaces, let $\{e_n\}_{n\geq 1} \subseteq H$ be an orthonormal basis for H, and let $\{\xi_n\}_{n\geq 1} \subseteq V$ be an orthonormal basis for V. Suppose that $A \in L(H, V)$ and $A = (e_n) = \sum_{m\geq 1} \lambda_{nm} \xi_m$ for all $n \in \mathbb{N}$. Show that $\sum_{m\geq 1} |\lambda_{nm}|^2 \leq \|A\|_L^2$ for all $n \in \mathbb{N}$ and $\sum_{n\geq 1} |\lambda_{nm}|^2 \leq \|A\|_L$ for all $m \in \mathbb{N}$.

Problem 3.33. Let H be a Hilbert space and let $A \in L(H)$ be a self-adjoint positive operator. Show that the following statements are equivalent:
(a) $R(A) \subseteq H$ is dense.
(b) $N(A) = \{0\}$.
(c) $(A(x), x) > 0$ for all $x \neq 0$.

Problem 3.34. Let H be a Hilbert space and let $A, T: H \to H$ be two linear operators such that $(A(x), u) = (x, T(u))$ for all $x, u \in H$. Show that $A \in L(H)$ and $T = A^*$.

Problem 3.35. Let H be a Hilbert space and let $\{A_n\}_{n\geq 1} \subseteq L(H)$ be such that $\sup_{n\geq 1} |(A_n(x), u)| < \infty$ for all $x, u \in H$. Show that $\sup_{n\geq 1} \|A_n\|_L < \infty$.

Problem 3.36. Let H be a Hilbert space and let $\{A_n\}_{n\geq 1} \subseteq L(H)$ be such that $\lim_{n\to\infty} |(A_n(x), u)| = 0$ for all $x, u \in H$. Can we say that $\|A_n\|_L \to 0$? Justify your answer.

Problem 3.37. Let K, D be compact spaces, let $g \in C(K, D)$, and let $A: C(K) \to C(D)$ be the operator defined by $A(f)(t) = f(g(s))$ for all $s \in K$ and for all $t \in D$. Show that
(a) $A \in L(C(K), C(D))$ and find $\|A\|_L$.
(b) $R(A) = C(D)$ if and only if g is injective.
(c) A is an isometry if and only if g is surjective.

Problem 3.38. Let X be a Banach space, let V be a normed space, and let $A \in L(X, V)$. Show that: $A^{-1} \in L(V, X)$ if and only if $R(A) \subseteq V$ is dense and $\|A(x)\|_V \geq c\|x\|_X$ for all $x \in X$ and for some $c > 0$.

Problem 3.39. Let H be a Hilbert space and let $A \in L(H)$ be normal. Show that $\lim_{n\to\infty} \|A^n\|_L^{1/n} = \|A\|_L$.

Problem 3.40. Let H be a Hilbert space and let $P \in L(H)$ be a projection, that is, $P^2 = P$. Show that the following properties are equivalent:
(a) P is an orthogonal projection.
(b) P is normal.
(c) $(P(x), x) = \|P(x)\|^2$ for all $x \in H$.

Problem 3.41. Let X, Y be Banach spaces with X being reflexive, $A \in L_c(X, Y)$, $\|\cdot\|_X$ being the norm of X, and $|\cdot|_X$ being another norm on X, which generates a weaker topology on X. Show that for every $\varepsilon > 0$ there exists $c_\varepsilon > 0$ such that

$$\|A(x)\|_Y \leq \varepsilon\|x\|_X + c_\varepsilon|x|_X \quad \text{for all } x \in X .$$

Problem 3.42. Let X be a normed space and let $P \in L_c(X)$ be a projection, that is, $P^2 = P$. Show that $P \in L_f(X)$.

Problem 3.43. Let H be a Hilbert space and let $A \in L(H)$ be self-adjoint. Assume that $A \geq \vartheta i_H$ for some $\vartheta > 0$; see Definition 3.8.22. Show that A is invertible.

Problem 3.44. Let H be a Hilbert space and let $A \in L(H)$ be self-adjoint. Show that the residual spectrum $R\sigma(A)$ of A (see Definition 3.7.29) is empty.

Problem 3.45. Let X be a Banach space and let $A \in L(X)$ and $\lambda \in \mathbb{C}$. Suppose that there exists a sequence $\{x_n\}_{n\geq 1} \subseteq X$ with $\|x_n\| = 1$ for all $n \in \mathbb{N}$ such that $A(x_n) - \lambda x_n \to 0$ in X. Show that $\lambda \in \sigma(A)$.

Problem 3.46. Let H be a Hilbert space and let $P \in L(H)$ be an orthogonal projection. Show that $0 \leq P \leq i_H$; see Definition 3.8.22.

Problem 3.47. Let H be a Hilbert space and let $A \in L(H)$ be such that $(A(x), x) \geq c\|x\|^2$ for all $x \in H$ and for some $c > 0$. Show that A is an isomorphism.

Problem 3.48. Let X be an infinite dimensional Banach space and let $A \in L_c(X)$. Show that there exists $h \in X$ such that there is no $x \in X$ for which we have $A(x) = h$.

Problem 3.49. Let X be a Banach space and let $A: D(A) \subseteq X \to X$ be an unbounded linear operator. Suppose there exists $\lambda \in \mathbb{C}$ such that $(A - \lambda I)^{-1} \in L(X)$. Show that A is closed.

Problem 3.50. Let X, Y be Banach spaces and let $A: D(A) \subseteq X \to Y$ be an unbounded linear operator such that $\|A(x)\|_Y \geq c\|x\|_X$ for all $x \in D(A)$ and for some $c > 0$. Show that A is closed.

Problem 3.51. Let H be a Hilbert space, let $\{u_n\}_{n\geq 1} \subseteq H$ be an orthonormal set and let $A \in L_c(H)$. Show that $A(u_n) \to 0$ in H.

Problem 3.52. Let X, Y be Banach spaces, let $A: X \to Y$ be a linear operator, and suppose that for every $y^* \in Y^*$ one has $y^* \circ A \in X^*$. Show that $A \in L(X, Y)$.

Problem 3.53. Let X be a Banach space and let $P \in L(X)$ be a projection, that is, $P^2 = P$. Show that P^* is a projection in X^*.

Problem 3.54. Let H be a Hilbert space and let $A \in L_c(H)$. Show that there exists $x \in H$ with $\|x\| \leq 1$ such that $\|A(x)\| = \|A\|_L$.

Problem 3.55. Let X, Y be Banach spaces with $Y \neq 0$. Show that X is reflexive if and only if for every $A \in L_c(X, Y)$ there exists $x \in X$ with $\|x\|_X \leq 1$ such that $\|A(x)\|_Y = \|A\|_L$.

Problem 3.56. Let X, Y be Banach spaces and let $A \in L_c(X, Y)$. Show that $R(A) \subseteq Y$ is separable.

Problem 3.57. Let X, Y be Banach spaces and let $A \in L(X, Y)$, which satisfies $\|A(x)\|_Y \geq c\|x\|_X$ for all $x \in X$ and for some $c > 0$. Is it possible for A to be compact? Justify your answer.

Problem 3.58. Let X be an infinite dimensional Banach space and let $A \in L_c(X)$. Show that $0 \in \overline{A(\partial B_1)}$.

Problem 3.59. Let X be a Banach space that is w-separable. Show that X is separable.

Problem 3.60. Let X be an infinite dimensional Banach space and let $K \subseteq X$ be a nonempty, compact set. Show that $\operatorname{int} K = \emptyset$.

Problem 3.61. Let X be a Banach space and assume that there exists an uncountable family $\{U_i\}_{i \in I}$ such that
(a) for each $i \in I$, $U_i \subseteq X$ is nonempty and open;
(b) $U_i \cap U_j = \emptyset$ if $i \neq j$.

Show that X is nonseparable.

4 Banach Spaces of Functions and Measures

Now that we have a reasonable background on measure theory and functional analysis, we can look at concrete spaces of functions and measures that are common in many different fields of analysis. We study them using the abstract tools developed in Chapter 2 (measure theory) and in Chapter 3 (functional analysis). We start with the L^p-spaces, which we already encountered in Section 2.3. Now our emphasis is on the duality theory for such spaces. Then we consider Banach space-valued functions.

Vector valued integration theories were first developed in the 1930s in an attempt to better understand differentiation theorems for Banach space-valued functions. Functions of bounded variation are associated with the early days of real analysis. Recently in the context of geometric measure theory they have found new applications. It is a small natural step to pass from functions of bounded variation to absolutely continuous and Lipschitz functions. Associated with Lipschitz functions are some interesting and useful extension theorems. Subsequently we pass to Sobolev spaces. The theory of Sobolev spaces is one of the most useful tools in modern mathematics with many remarkable applications. The Banach spaces of measures and their modes of convergence are useful in probability theory and stochastic analysis. Finally capacities and Young measures provide useful applications of the previous material.

4.1 L^p-Spaces

Let (X, Σ, μ) be a measure space and suppose that $p, p' \in [1, +\infty]$ are conjugate exponents, that is, $1/p + 1/p' = 1$; see Definition 2.3.10. Given $h \in L^{p'}(X)$, Hölder's inequality (see Theorem 2.3.12) implies that the linear functional $f \to \xi_h(f) = \int_X fh \, d\mu$ is bounded, and hence continuous, and so $\xi_h \in L^p(X)^*$. Moreover, $\|\xi_h\|_* \le \|h\|_{p'}$. Next we more closely examine this functional. It will lead us to a very convenient description of the dual of $L^p(X)$.

Proposition 4.1.1. *If p and p' are conjugate exponents and if $p' = \infty$, μ is semifinite, then*

$$\|h\|_{p'} = \|\xi_h\|_* = \sup\left[\left|\int_X fh \, d\mu\right| : f \in L^p(X), \|f\|_p = 1\right].$$

Proof. As we already mentioned for Hölder's inequality, it holds that $\|\xi_h\|_* \le \|h\|_{p'}$. If $h(z) = 0$ μ-a. e., so clearly we obtain equality. So, suppose now that $h \ne 0$ and $1 < p' < \infty$. Let

$$\hat{f}(x) = \frac{|h(x)|^{p'-1} \operatorname{sgn} h(x)}{\|h\|_{p'}^{p'-1}},$$

where

https://doi.org/10.1515/9783111286952-004

$$\operatorname{sgn} h(x) = \begin{cases} -1 & \text{if } h(x) < 0\,, \\ 0 & \text{if } h(x) = 0\,, \\ 1 & \text{if } h(x) > 0\,. \end{cases}$$

Since $p' - 1 = p'/p$, it follows that

$$\|\hat{f}\|_p^p = \frac{1}{\|h\|_{p'}^{(p'-1)p}} \int_X |h(x)|^{(p'-1)p}\, d\mu = \frac{1}{\|h\|_{p'}^{p'}} \|h\|_{p'}^{p'} = 1\,.$$

Then by Corollary 3.1.48, this implies

$$\|\xi_h\|_* \geq \int_X \hat{f} h\, d\mu = \frac{1}{\|h\|_{p'}^{p'-1}} \int_X |h|^{p'}\, d\mu = \|h\|_{p'}\,.$$

This gives $\|\xi_h\|_* = \|h\|_{p'}$.

If $p' = 1$, then $f = \operatorname{sgn} h$ and $\|f\|_\infty = 1$. Then $\int_X fh\, d\mu = \int_X |h|\, d\mu = \|h\|_1$ and so we have again $\|\xi_h\|_* = \|h\|_1$.

Finally, if $p' = +\infty$, then we need to assume that μ is semifinite; see Definition 2.1.30 (a). Clearly every σ-finite measure is semifinite. Let $\varepsilon > 0$ and consider the set $A_\varepsilon = \{x \in X : |h(x)| \geq \|h\|_\infty - \varepsilon\}$. Then $\mu(A_\varepsilon) > 0$ and since μ is semifinite, there exists $B \in \Sigma$ with $B \subseteq A_\varepsilon$ such that $0 < \mu(B) < \infty$. Let $f = 1/(\mu(B))\chi_B \operatorname{sgn} h$. Then $\|f\|_1 = 1$ and so

$$\|\xi_h\|_* \geq \int_X fh\, d\mu = \frac{1}{\mu(B)} \int_B |h|\, d\mu \geq \|h\|_\infty - \varepsilon\,.$$

Since $\varepsilon > 0$ is arbitrary, we let $\varepsilon \to 0^+$ and obtain $\|\xi_h\|_* \geq \|h\|_\infty$. Therefore, $\|\xi_h\|_* = \|h\|_\infty$. ☐

Next we will show the converse. Namely, if $f \to \int_X fh\, d\mu$ is a bounded linear functional, then $h \in L^{p'}(X)$ in all cases of interest.

Proposition 4.1.2. *If p and p' are conjugate exponents, $h: X \to \mathbb{R}$ is a Σ-measurable function such that $fh \in L^1(X)$ for all f in the space $\mathcal{L}$ of Σ-simple functions, which vanish outside a set of finite measure and*

$$N_{p'}(h) = \sup\left[\left|\int_X fh\, d\mu\right| : f \in \mathcal{L}, \|f\|_p = 1\right] < \infty \tag{4.1.1}$$

and one of the following holds:

(a) *$D_h = \{x \in X : h(x) \neq 0\}$ is σ-finite;*

(b) *μ is semifinite,*

then $h \in L^{p'}(X)$ and $\|h\|_{p'} = N_{p'}(h)$.

Proof. If f is bounded, Σ-measurable and vanishes outside a set $A \in \Sigma$ of finite μ-measure, then according to Corollary 2.2.19, there exists a sequence $\{s_n\}_{n\geq 1}$ of Σ-simple functions such that

$$|s_n| \leq |f| \quad \text{and} \quad s_n \to f \quad \text{uniformly on } X .$$

With the Dominated Convergence Theorem (see Theorem 2.3.8), one has

$$\left|\int_X fh \, d\mu\right| = \lim_{n\to\infty} \left|\int_X s_n h \, d\mu\right| .$$

So, if $\|f\|_p = 1$, then $|\int_X fh \, d\mu| \leq N_{p'}(h)$.

First assume that $p' < \infty$. If μ is semifinite, then, because of (4.1.1), it is easy to see that for every $\varepsilon > 0$ the set $\{x \in X : |h(x)| > \varepsilon\}$ has finite μ-measure and so D_h is σ-finite. Therefore it is enough to consider the case where (a) holds. Let $\{A_n\}_{n\geq 1} \subseteq \Sigma$ be an increasing sequence of sets of finite μ-measure such that $D_h = \bigcup_{n\geq 1} A_n$. As before, let $\{s_n\}_{n\geq 1}$ be a sequence of Σ-simple functions such that $s_n(x) \to h(x)$ μ-a. e. and $|s_n| \leq |h|$. Let $\hat{s}_n = s_n \chi_{E_n}$. Then $\hat{s}_n(x) \to h(x)$ μ-a. e. and $|\hat{s}_n| \leq |h|$ for all $n \in \mathbb{N}$. Let

$$f_n = \frac{|\hat{s}_n|^{p'-1} \operatorname{sgn} h}{\|\hat{s}_n\|_{p'}^{p'-1}} .$$

From the proof of Proposition 4.1.1 one gets $\|f_n\|_p = 1$ and from Fatou's Lemma (see Theorem 2.3.6), it follows that

$$\begin{aligned} \|h\|_{p'} &\leq \liminf_{n\to\infty} \|\hat{s}_n\|_{p'} = \liminf_{n\to\infty} \int_X |f_n \hat{s}_n| \, d\mu \leq \liminf_{n\to\infty} \int_X |f_n h| \, d\mu \\ &= \liminf_{n\to\infty} \int_X f_n h \, d\mu \leq N_{p'}(h) ; \end{aligned} \tag{4.1.2}$$

see the first part of the proof. On the other hand, from Hölder's inequality (see Theorem 2.3.12), we infer that

$$N_{p'}(h) \leq \|h\|_{p'} . \tag{4.1.3}$$

From (4.1.2) and (4.1.3) we conclude that $N_{p'}(h) = \|h\|_{p'}$. Next assume that $p' = +\infty$. For $\varepsilon > 0$, let $A_\varepsilon = \{x \in X : |h(x)| \geq N_\infty(h) + \varepsilon\}$. If $\mu(A_\varepsilon) > 0$, then there exists $E \in \Sigma$ with $E \subseteq A_\varepsilon$ such that $\mu(E) \in (0, +\infty)$. We set

$$f = \frac{1}{\mu(E)} \chi_E \operatorname{sgn} h .$$

We get $\|f\|_1 = 1$ and

$$\int_X fh \, d\mu \geq \frac{1}{\mu(E)} \int_E |h| \, d\mu \geq N_\infty(h) + \varepsilon .$$

But since f is bounded, by the first part of the proof, this cannot happen. Therefore, $\|h\|_\infty \leq N_\infty(h)$. The opposite inequality is clear from (4.1.1). Hence, $\|h\|_\infty = N_\infty(h)$. □

Now we are ready to describe the dual of $L^p(X)$ with $1 < p < \infty$. The result is known as the "Riesz Representation Theorem".

Theorem 4.1.3 (Riesz Representation Theorem). *If (X, Σ, μ) is a measure space and $1 < p < \infty$, then $L^p(X)^*$ is isometrically isomorphic to $L^{p'}(X)$ with $1/p + 1/p' = 1$.*

Proof. First assume that μ is finite. Then all Σ-simple functions belong to $L^p(X)$. Suppose $\xi \in L^p(X)^*$ and $A \in \Sigma$. We set $\vartheta(A) = \xi(\chi_A)$. If $\{A_n\}_{n\geq 1} \subseteq \Sigma$ are pairwise disjoint and $A = \bigcup_{n\geq 1} A_n$, then $\chi_A = \sum_{n\geq 1} \chi_{A_n}$. The series converges in $L^p(X)$ since

$$\left\| \chi_A - \sum_{n=1}^{m} \chi_{A_n} \right\|_p = \left\| \sum_{n\geq m+1} \chi_{A_n} \right\|_p = \mu\Big(\bigcup_{n\geq m+1} A_n \Big)^{\frac{1}{p}} \to 0 \quad \text{as } n \to \infty ,$$

where we recall that $p < \infty$. Therefore, from the linearity and continuity of ξ, we obtain

$$\vartheta(A) = \sum_{n\geq 1} \xi(\chi_{A_n}) = \sum_{n\geq 1} \vartheta(A_n) .$$

This shows that ϑ is a signed measure.

If $\mu(A) = 0$, then $\chi_A = 0$ and so $\vartheta(A) = 0$. Hence $\vartheta \ll \mu$; see Remark 2.4.22. Invoking the Radon–Nikodym Theorem (see Theorem 2.4.29), there exists $h \in L^1(X)$ such that $\xi(\chi_A) = \vartheta(A) = \int_A h \, d\mu$ for all $A \in \Sigma$. This implies

$$\xi(s) = \int_X sh \, d\mu \quad \text{for every simple function } s . \tag{4.1.4}$$

From (4.1.4) it follows

$$\left| \int_X sh \, d\mu \right| = |\xi(s)| \leq \|\xi\|_* \|s\|_p < \infty .$$

Invoking Proposition 4.1.2 we infer that $h \in L^{p'}(X)$. Then from (4.1.4) and the density of simple functions in $L^p(X)$ (see Proposition 2.3.22), we conclude that $\xi(f) = \int_X fh \, d\mu$ for all $f \in L^p(X)$.

Next suppose that μ is σ-finite. Let $\{A_n\}_{n\geq 1} \subseteq \Sigma$ be an increasing sequence such that $0 < \mu(A_n) < \infty$ for all $n \in \mathbb{N}$ and $X = \bigcup_{n\geq 1} A_n$. From the first part of the proof we get that

$$L^p(A_n)^* = L^{p'}(A_n) \quad \text{for all } n \in \mathbb{N} \tag{4.1.5}$$

and that $L^p(A_n)$ (resp. $L^{p'}(A_n)$) is a subspace of $L^p(X)$ (resp. $L^{p'}(X)$), namely those functions which vanish outside of A_n. If $\xi \in L^p(X)^*$, then from (4.1.5) we obtain the existence of $h_n \in L^{p'}(A_n)$ such that

$$\xi(f) = \int_{A_n} fh_n \, d\mu \quad \text{for all } f \in L^p(A_n) \text{ with } n \in \mathbb{N}.$$

The function h_n is unique up to a μ-null set and so $h_n(x) = h_m(x)$ μ-a. e. on A_n for $n < m$. Therefore we can define $h: X \to \mathbb{R}$ by setting $h = h_n$ on A_n for all $n \in \mathbb{N}$. According to the Monotone Convergence Theorem (see Theorem 2.3.3), one has

$$\|h\|_{p'} = \lim_{n\to\infty} \|h_n\|_{p'} \leq \|\xi\|_* < \infty.$$

Hence, $h \in L^{p'}(X)$. For $f \in L^p(X)$, according to the Dominated Convergence Theorem (see Theorem 2.3.8), we obtain $f\chi_{A_n} \to f$ in $L^p(X)$, which implies

$$\xi(f) = \lim_{n\to\infty} \xi(f\chi_{A_n}) = \lim_{n\to\infty} \int_{A_n} fh \, d\mu = \int_X fh \, d\mu.$$

Finally we consider the general case of an arbitrary measure space. If $A \in \Sigma$ is σ-finite, then from the second part of the proof there exists a unique up to μ-null set $h_A \in L^{p'}(A)$ such that $\xi(f) = \int_X fh_A \, d\mu$ for all $f \in L^p(A)$ and $\|h_A\|_{p'} \leq \|\xi\|_*$. If $E \in \Sigma$ with $A \subseteq E$ is σ-finite, then $h_E = h_A$ μ-a. e. and so $\|h_A\|_{p'} \leq \|h_E\|_{p'}$. Let

$$\eta = \sup[\|h_A\|_{p'} : A \in \Sigma \text{ is } \sigma\text{-finite}] \leq \|\xi\|_*.$$

We choose a sequence $\{A_n\}_{n\geq 1} \subseteq \Sigma$ with each A_n σ-finite such that $\|h_{A_n}\|_{p'} \to \eta$. Let $E = \bigcup_{n\geq 1} A_n$. Then $E \in \Sigma$ is σ-finite and $\|h_E\|_{p'} \geq \|h_{A_n}\|_{p'}$ for all $n \in \mathbb{N}$. Therefore $\|h_E\|_{p'} = \eta$. If $D \in \Sigma$ is a σ-finite set with $E \subseteq D$, then

$$\int_X |h_E|^{p'} \, d\mu + \int_X |h_{D\setminus E}|^{p'} \, d\mu = \int_X |h_D|^{p'} \, d\mu \leq \eta^{p'} = \int_X |h_E|^{p'} \, d\mu.$$

This gives $\int_X |h_{D\setminus E}|^{p'} \, d\mu = 0$, that is, $h_{D\setminus E} = 0$ μ-a. e. since $p' < \infty$. Hence $h_D = h_E$ μ-a. e. But if $f \in L^p(X)$, then $D = E \cup \{x \in X : f(x) \neq 0\} \in \Sigma$ is σ-finite; see Problem 2.26. Hence

$$\xi(f) = \int_X fh_D \, d\mu = \int_X fh_E \, d\mu.$$

Therefore we can finally take $h = h_E$. □

Corollary 4.1.4. *If (X, Σ, μ) is a measure space and $1 < p < \infty$, then $L^p(X)$ is a reflexive Banach space.*

Next we consider the dual of $L^1(X)$. To this end we need to restrict ourselves to σ-finite measure spaces. The result is again known as the "Riesz Representation Theorem for L^1".

Theorem 4.1.5 (Riesz Representation Theorem for L^1). *If (X, Σ, μ) is a σ-finite measure space, then $L^1(X)^*$ is isometrically isomorphic to $L^\infty(X)$.*

Proof. First suppose that μ is finite. Let $\xi \in L^1(X)^*$. Reasoning as in the first part of the proof of Theorem 4.1.3 there exists a unique $h \in L^1(X)$ such that

$$\xi(f) = \int_X fh \, d\mu \quad \text{for all } f \in L^1(X) .$$

Invoking Proposition 4.1.2, we infer that $h \in L^\infty(X)$.

Now assume that μ is σ-finite. Then there exists an increasing sequence $\{A_n\}_{n\geq 1} \subseteq \Sigma$ such that $X = \bigcup_{n\geq 1} A_n$ and $0 < \mu(A_n) < \infty$ for all $n \in \mathbb{N}$. From the first part of the proof there exist a unique $h_n \in L^\infty(A_n)$ for each $n \in \mathbb{N}$ such that $\xi(f) = \int_X fh_n \, d\mu$ for all $f \in L^1(A_n)$. Evidently, $h_n = h_m$ μ-a. e. on A_n for $n < m$. So if $h: X \to \mathbb{R}$ is defined by $h = h_n$ on A_n for all $n \in \mathbb{N}$, then $\|h\|_\infty \leq \|\xi\|_*$. Hence, $h \in L^\infty(X)$ and $\xi(f) = \int_X fh \, d\mu$ for all $f \in L^1(X)$ as well as $\|\xi\|_* = \|h\|_\infty$. Propositions 4.1.1 and 4.1.2 imply that $L^1(X)^*$ is isometrically isomorphic to $L^\infty(X)$. □

Proposition 4.1.6. *If (X, Σ, μ) is a measure space and $1 < p < \infty$, then $L^p(X)$ is uniformly convex.*

Proof. Let $f, h \in L^p(X)$ with $\|f\|_p = \|h\|_p \leq 1$, $\varepsilon > 0$, and set $A_\varepsilon = \{x \in X : |(f - h)(x)| \leq \varepsilon|(f + h)(x)|\}$. We obtain

$$\int_{A_\varepsilon} \left|\frac{1}{2}(f - h)\right|^p d\mu \leq \varepsilon^p \int_X \left|\frac{1}{2}(f + h)\right|^p d\mu \leq \varepsilon^p . \tag{4.1.6}$$

The function $t \to |t|^p$ is strictly convex. Hence $s \to 1/2[|s+1|^p + |s-1|^p] - |s|^p$ is continuous and positive on $\mathbb{R}$. So, there exists $r = r(\varepsilon, p) > 0$ such that

$$\frac{1}{2}[|s + 1|^p + |s - 1|^p] - |s|^p \geq r \quad \text{for all } s \in \left[-\frac{1}{\varepsilon}, \frac{1}{\varepsilon}\right] . \tag{4.1.7}$$

Let $s = (f(x) + h(x))/(f(x) - h(x))$ for $x \in X \setminus A_\varepsilon$. Then from (4.1.7) it follows that

$$\frac{1}{2}[|f(x)|^p + |h(x)|^p] \geq r\left|\frac{1}{2}(f - h)(x)\right|^p + \left|\frac{1}{2}(f + h)(x)\right|^p \tag{4.1.8}$$

for all $x \in X \setminus A_\varepsilon$. Moreover, recall that

$$\frac{1}{2}[|f(x)|^p + |h(x)|^p] \geq \left|\frac{1}{2}(f + h)(x)\right|^p \quad \text{for all } x \in X . \tag{4.1.9}$$

We integrate (4.1.8) over $X \setminus A_\varepsilon$, (4.1.9) over A_ε and then add both equations to obtain

$$1 \geq \int_X \left|\frac{1}{2}(f+h)(x)\right|^p d\mu + \int_{X\setminus A_\varepsilon} r\left|\frac{1}{2}(f-h)\right|^p d\mu . \tag{4.1.10}$$

If we choose $\delta = r\varepsilon^p$, then from (4.1.10) we see that

$$\int_{X\setminus A_\varepsilon} \left|\frac{1}{2}(f-h)\right|^p d\mu \leq \varepsilon^p \quad \text{if} \left\|\frac{1}{2}(f+h)\right\|_p^p \geq 1-\delta . \tag{4.1.11}$$

From (4.1.6) and (4.1.11) it follows that

$$\left\|\frac{1}{2}(f+h)\right\|_p^p \geq 1-\delta \quad \text{implies} \quad \left\|\frac{1}{2}(f-h)\right\|_p^p \leq 2\varepsilon^p . \tag{4.1.12}$$

From (4.1.12) and Definition 3.4.21 (b) we conclude that $L^p(X)$ with $1 < p < \infty$ is uniformly convex. □

Remark 4.1.7. From the Milman–Pettis Theorem (see Theorem 3.4.28), it follows that $L^p(X)$ with $1 < p < \infty$ is reflexive. So, we have reached Corollary 4.1.4 following a different route.

Let (X, Σ, μ) be a measure space and $1 \leq p < \infty$ as well as $1 < p' \leq \infty$ be conjugate exponents, that is, $1/p + 1/p' = 1$. If $p = \infty$, then we set $p' = 1$. When $p = 1$ or $p = \infty$ we assume in addition that μ is finite. On account of the Riesz Representation Theorems (see Theorems 4.1.3 and 4.1.5), we have the following modes of convergence.

Definition 4.1.8. Let $\{f_n, f\}_{n\geq 1} \subseteq L^p(X)$.

(a) If $1 \leq p < \infty$, then we say that the f_n's **converge weakly** to f denoted by $f_n \xrightarrow{w} f$ if

$$\int_X f_n h \, d\mu \to \int_X fh \, d\mu$$

is satisfied for all $h \in L^{p'}(X)$.

(b) If $p = +\infty$, then we say that the f_n's **converge weakly*** to f denoted by $f_n \xrightarrow{w^*} f$ if

$$\int_X f_n h \, d\mu \to \int_X fh \, d\mu$$

is satisfied for all $h \in L^1(X)$.

Applying Vitali's Convergence Theorem (see Theorem 2.3.44), we have the following result.

Proposition 4.1.9. *If $\{f_n\}_{n\geq1} \subseteq L^p(X)$ with $1 < p < \infty$ is bounded and $f_n(x) \to f(x)$ μ-a. e. or $f_n \xrightarrow{\mu} f$ as $n \to +\infty$, then $f \in L^p(X)$ and $f_n \xrightarrow{w} f$ in $L^p(X)$.*

From Propositions 3.3.13 and 3.3.31 we conclude the following.

Proposition 4.1.10. *If $\{f_n\}_{n\geq1} \subseteq L^p(X)$ with $1 \leq p \leq \infty$ and $f_n \xrightarrow{w} f$ in $L^p(X)$ for $1 \leq p < \infty$ and $f_n \xrightarrow{w^*} f$ in $L^\infty(X)$, then*

$$\|f\|_p \leq \liminf_{n\to\infty} \|f_n\|_p \leq \sup_{n\geq1} \|f_n\|_p < \infty\,.$$

Recalling that $L^p(X)$ is uniformly convex for $1 < p < \infty$ and since uniformly convex Banach spaces exhibit the Kadec–Klee Property (see Proposition 3.4.32), we can state the following result.

Proposition 4.1.11. *If $\{f_n\}_{n\geq1} \subseteq L^p(X)$, $1 < p < \infty$, $f_n \xrightarrow{w} f$ in $L^p(X)$ and $\|f_n\|_p \to \|f\|_p$, then $\|f_n - f\|_p \to 0$.*

The reflexivity of $L^p(X)$ for $1 < p < \infty$ implies that bounded sets in $L^p(X)$ are relatively weakly compact; see Theorem 3.4.5. Then the Eberlein–Smulian Theorem (see Theorem 3.4.14), gives the following result.

Proposition 4.1.12. *If $\{f_n\}_{n\geq1} \subseteq L^p(X)$ with $1 < p < \infty$ and $\|f_n\|_p \leq M$ for some $M > 0$ and for all $n \in \mathbb{N}$, then there exists a subsequence $\{f_{n_k}\}_{k\geq1}$ of $\{f_n\}_{n\geq1}$ such that $f_{n_k} \xrightarrow{w} f$ in $L^p(X)$ with $f \in L^p(X)$.*

For the case $p = \infty$ instead of the Eberlein–Smulian Theorem, which is not valid for the w*-topology (see Remark 3.4.15), we use Theorem 3.4.12 (a), which imposes additional restrictions on the measure space. So, we assume that the metric space $(\Sigma(\mu), d_\mu)$ (see Definition 2.3.23) is separable. Then, according to Proposition 2.3.24, this is equivalent to saying that $L^1(X)$ is separable. Therefore, applying Theorem 3.4.12 (a), we obtain the following.

Proposition 4.1.13. *If $\{f_n\}_{n\geq1} \subseteq L^\infty(X)$, $\|f_n\|_\infty \leq M$ for some $M > 0$ and for all $n \in \mathbb{N}$ and suppose that $L^1(X)$ is separable, then there exists a subsequence $\{f_{n_k}\}_{k\geq1}$ of $\{f_n\}_{n\geq1}$ such that $f_{n_k} \xrightarrow{w^*} f$ in $L^\infty(X)$ with $f \in L^\infty(X)$.*

Remark 4.1.14. If $X \subseteq \mathbb{R}^N$ is Borel and μ is a Radon measure on $\mathcal{B}(X)$, then $L^1(X)$ is separable.

Recalling that $L^p(X)$-simple functions are dense in $L^p(X)$ for any $p \in [1, \infty]$ (see Proposition 2.3.22), we directly get the following proposition.

Proposition 4.1.15. *If $\{f_n, f\}_{n\geq1} \subseteq L^p(X)$ with $1 \leq p \leq \infty$, then $f_n \xrightarrow{w} f$ in $L^p(X)$ for $1 \leq p < \infty$ and $f_n \xrightarrow{w^*} f$ in $L^\infty(X)$ for $p = +\infty$ if and only if*
(a) $\sup_{n\geq1} \|f_n\|_p < \infty$;

(b) $\int_A f_n \,d\mu \to \int_A f \,d\mu$ *for all* $A \in \Sigma$ *with* $\mu(A) < \infty$.

Remark 4.1.16. The space $L^1(X)$ is not reflexive. To see this, assume that μ is nonatomic; see Definition 2.1.30 (b). Then there exists a decreasing sequence $\{A_n\}_{n\geq 1} \subseteq \Sigma$ such that $0 < \mu(A_n)$ for all $n \in \mathbb{N}$ and $\mu(A_n) \to 0^+$. We set $f_n = \chi_{A_n}\|\chi_{A_n}\|_1^{-1}$ for $n \in \mathbb{N}$. Then $\|f_n\|_1 = 1$ for all $n \in \mathbb{N}$. If $L^1(X)$ is reflexive, then by passing to a subsequence if necessary we may assume that $f_n \xrightarrow{w} f$ in $L^1(X)$ with $f \in L^1(X)$. Then it follows that

$$\int_X f_n h \,d\mu \to \int_X fh \,d\mu \quad \text{for all } h \in L^\infty(X)\,; \tag{4.1.13}$$

see Definition 4.1.8. Fix $k \in \mathbb{N}$ and let $h = \chi_{A_k} \in L^\infty(X)$. One has

$$\int_X f_n \chi_{A_k} \,d\mu = 1 \quad \text{for all } n \geq k\,,$$

which in view of (4.1.13) results in

$$\int_X f\chi_{A_k} \,d\mu = 1 \quad \text{for all } k \in \mathbb{N}\,. \tag{4.1.14}$$

On the other hand, from the Dominated Convergence Theorem, we conclude that

$$\int_X f\chi_{A_k} \,d\mu \to 0 \quad \text{as } k \to \infty\,. \tag{4.1.15}$$

Comparing (4.1.14) and (4.1.15) we have a contradiction, which proves that the space $L^1(X)$ cannot be reflexive.

For $1 < p < \infty$, the space $L^p(X)$ is reflexive while we just saw that $L^1(X)$ is not. It follows that the relatively weakly compact sets are the bounded ones. The situation is different for $L^1(X)$ and the characterization of weakly compact sets is more involved.

Example 4.1.17. Let $X = (-1, 1)$ be equipped with the Lebesgue measure and consider the sequence

$$f_n(x) = \begin{cases} n & \text{if } x \in [-\frac{1}{2n}, \frac{1}{2n}],\ n \in \mathbb{N} \\ 0 & \text{otherwise}\,. \end{cases}$$

Then we easily see the following facts

$$f_n \geq 0\,, \quad \int_X f_n \,dx = 1 \quad \text{for all } n \in \mathbb{N}\,,$$

$$f_n(x) \to 0 \quad \text{a. e.}\,, \quad \int_X f_n h \,dx \to h(0) \quad \text{for all } h \in C(-1, 1)\,.$$

The sequence $\{f_n\}_{n\geq1} \subseteq L^1(-1,1)$ is bounded, but evidently we cannot find a subsequence that converges weakly.

In the case of the space $L^1(X)$, weakly compact sets can be characterized using the notion of uniform integrability; see Definition 2.3.40. The result is known as the "Dunford–Pettis Theorem".

Theorem 4.1.18 (Dunford–Pettis Theorem). *If (X, Σ, μ) is a finite measure space and $\mathcal{F} \subseteq L^1(X)$ is bounded, then $\mathcal{F}$ is relatively weakly compact if and only if it is uniformly integrable.*

Proof. $\Longrightarrow$: We may assume that $\mathcal{F} = \{f_n\}_{n\geq1}$. Since by hypothesis $\mathcal{F} \subseteq L^1(X)$ is relatively weakly compact, according to the Eberlein–Smulian Theorem (see Theorem 3.4.14), we may assume that $f_n \xrightarrow{w} f$ in $L^1(X)$ as $n \to \infty$. Then according to Proposition 4.1.15, we obtain

$$\int_A f_n \, d\mu \to \int_A f \, d\mu \quad \text{for all } A \in \Sigma .$$

If $\nu_n(A) = \int_A f_n \, d\mu$ with $n \in \mathbb{N}$ and $\nu(A) = \int_A f \, d\mu$, then these are signed measures such that $\nu_n \ll \mu$ for all $n \in \mathbb{N}$ and $\nu_n(A) \to \nu(A)$ for all $A \in \Sigma$. From the first part of the proof of the Vitali–Hahn–Saks Theorem (see Theorem 2.4.33), we derive that $\{\nu_n\}_{n\geq1}$ is uniformly absolutely continuous with respect to μ. But $d\nu_n = f_n d\mu$ for all $n \in \mathbb{N}$. Therefore, $\{f_n\}_{n\geq1} \subseteq L^1(X)$ is uniformly integrable.

$\Longleftarrow$: First suppose that Σ is countably generated, that is, $\Sigma = \sigma(\{A_k\}_{k\geq1})$. Then via a diagonal argument on the generators $\{A_k\}_{k\geq1}$ and by passing to a subsequence of $\{f_n\}_{n\geq1}$, we see that $\lim_{n\to\infty} \int_{A_k} f_n \, d\mu$ exists for every $k \in \mathbb{N}$. Let

$$\Sigma_0 = \left\{ E \in \Sigma \colon \lim_{n\to\infty} \int_E f_n \, d\mu \text{ exists} \right\} .$$

Evidently $\{A_k\}_{k\geq1} \subseteq \Sigma_0$ and Σ_0 is a Dynkin system; see Definition 2.1.7. Invoking Theorem 2.1.11, we conclude that $\Sigma_0 = \Sigma$. Therefore,

$$\lim_{n\to\infty} \int_A f_n \, d\mu \quad \text{exists for all } A \in \Sigma .$$

If $\nu_n(A) = \int_A f_n \, d\mu$ for all $A \in \Sigma$ and for all $n \in \mathbb{N}$, then $\nu_n \ll \mu$ for all $n \in \mathbb{N}$ and we have just proven that $\nu_n(A) \to \nu(A)$ for all $A \in \Sigma$. From the Vitali–Hahn–Saks Theorem (see Theorem 2.4.33), we get that ν is a signed measure satisfying $\nu \ll \mu$. Then, by the Radon–Nikodym Theorem (see Theorem 2.4.29), there exists $f \in L^1(X)$ such that $\nu(A) = \int_A f \, d\mu$ for all $A \in \Sigma$. Hence $\int_A f_n \, d\mu \to \int_A f \, d\mu$ for all $A \in \Sigma$ which, due to Proposition 4.1.15, gives that $f_n \xrightarrow{w} f$ in $L^1(X)$. Therefore, $\mathcal{F} \subseteq L^1(X)$ is relatively weakly compact.

Next we remove the hypothesis that Σ is countably generated. In this case we replace Σ with the σ-algebra Σ' generated by the countably many sets

$$\{x \in X: f_n(x) > \eta\} \quad \text{and} \quad \{x \in X: f_n(x) < -\eta\}$$

for all $n \in \mathbb{N}$ and for all $\eta > 0$ with $\eta \in \mathbb{Q}$. Moreover we replace X with

$$V = \bigcup_{n\geq 1}\{x \in X: f_n(x) \neq 0\} .$$

Finally note that by a straightforward application of the Radon–Nikodym Theorem, one has, for any $h \in L^\infty(V, \Sigma)$, the existence of $h' \in L^\infty(V, \Sigma')$ such that

$$\int_V fh \, d\mu = \int_V fh' \, d\mu \quad \text{for all } f \in L^1(V, \Sigma') . \qquad \square$$

Proposition 4.1.19. *If (X, Σ, μ) is a finite measure space and $\{u_n\}_{n\geq 1} \subseteq L^1(X)$ is relatively weakly compact, then $u_n \xrightarrow{w} u$ in $L^1(X)$ as $n \to \infty$ if and only if*

$$\|u + y\|_1 \leq \liminf_{n\to\infty} \|u_n + y\|_1 \quad \textit{for all } y \in L^1(X) .$$

Proof. $\Longrightarrow$: Let $y \in L^1(X)$ be fixed. Evidently $u_n + y \xrightarrow{w} u + y$ in $L^1(X)$. So, invoking Proposition 3.3.13 (c), it follows that

$$\|u + y\|_1 \leq \liminf_{n\to\infty} \|u_n + y\|_1 .$$

$\Longleftarrow$: Since $\{u_n\}_{n\geq 1} \subseteq L^1(X)$ is relatively weakly compact, by passing to a subsequence if necessary, we may assume that $u_n \xrightarrow{w} \hat{u}$ in $L^1(X)$ as $n \to \infty$. Let $A = \{x \in X: \hat{u}(x) > u(x)\}$ and $E = \{x \in X: \hat{u}(x) \leq u(x)\}$. Evidently, $A, E \in \Sigma$. From the Dunford–Pettis Theorem (see Theorem 4.1.18), we conclude that $\{u_n - u\}_{n\geq 1} \subseteq L^1(X)$ is uniformly integrable. So, given $\varepsilon > 0$, there exists $c > 0$ such that

$$\int_{\{|u_n-u|\geq c\}} |u_n - u| \, d\mu \leq \varepsilon \quad \text{for all } n \in \mathbb{N} ; \tag{4.1.16}$$

see Definition 2.3.40. Let $y = -u - c\chi_A + c\chi_E$. Since μ is finite, one gets that $y \in L^1(X)$. Moreover, we obtain

$$|u_n + y| \leq c - (u_n - u)\chi_A + (u_n - u)\chi_E + 2|u_n - u|\chi_{\{|u_n-u|\geq c\}} ,$$

which implies, since $X = A \cup E$ and because of (4.1.16), that

$$
\begin{aligned}
c\mu(X) = \|u + y\|_1 &\le \liminf_{n\to\infty} \|u_n + y\|_1 \\
&\le c\mu(X) - \int_A (\hat{u} - u)\, d\mu + \int_E (\hat{u} - u)\, d\mu + 2\varepsilon \\
&= c\mu(X) - \int_X |\hat{u} - u|\, d\mu + 2\varepsilon .
\end{aligned}
$$

This implies that $\int_X |\hat{u} - u|\, d\mu \le 2\varepsilon$. Since $\varepsilon > 0$ is arbitrary, we let $\varepsilon \searrow 0$ to conclude that $\hat{u} = u$. Therefore, $u_n \xrightarrow{w} u$ in $L^1(X)$. □

Let $\xi\colon \mathbb{R} \to [0, +\infty)$ be a continuous function with $\xi(0) = 0$ which satisfies the following condition:

$$
\begin{aligned}
&\text{For every } \varepsilon > 0, \text{ there exists } c_\varepsilon > 0 \text{ such that} \\
&|\xi(s + t) - \xi(s)| \le \varepsilon\xi(s) + c_\varepsilon\xi(t) \text{ for all } s, t \in \mathbb{R} .
\end{aligned}
\tag{4.1.17}
$$

Remark 4.1.20. Condition (4.1.17) is satisfied by convex functions. Moreover, if $\xi\colon \mathbb{R} \to [0, +\infty)$ is continuous, $\xi(s) > 0$ for all $s \ne 0$ and there exist $p, q \in (1, +\infty)$ such that

$$
\lim_{s\to 0} \frac{\xi(s)}{|s|^p} = \eta_0 > 0 \quad \text{and} \quad \lim_{s\to\pm\infty} \frac{\xi(s)}{|s|^q} = \eta_\infty > 0 ,
$$

then ξ satisfies condition (4.1.17).

Proposition 4.1.21. *If (X, Σ, μ) is a measure space, $\xi\colon \mathbb{R} \to [0, \infty)$ is a continuous function with $\xi(0) = 0$, which satisfies condition (4.1.17), and $f_n\colon X \to \mathbb{R}$ with $n \in \mathbb{N}$ is a sequence of Σ-measurable functions such that*

$$
f_n \to f \quad \mu\text{-a. e.}, \quad \sup_{n\ge 1} \int_X \xi(f_n)\, d\mu < \infty , \quad \textit{and} \quad \int_X \xi(f)\, d\mu < \infty ,
$$

then

$$
\sup_{n\ge 1} \int_X \xi(f_n - f)\, d\mu < \infty \quad \textit{and} \quad \int_X |\xi(f_n) - \xi(f) - \xi(f_n - f)|\, d\mu \to 0
$$

as $n \to \infty$.

Proof. Applying condition (4.1.17) yields

$$
|\xi(f_n) - \xi(f_n - f)| \le \varepsilon\xi(f_n - f) + c_\varepsilon\xi(f) . \tag{4.1.18}
$$

Let $\varepsilon = 1/2$. Then from (4.1.18) we derive $\xi(f_n - f) \le 2[\xi(f_n) + c_{1/2}\xi(f)]$ for all $n \in \mathbb{N}$. Hence, $\sup_{n\ge 1} \int_X \xi(f_n - f)\, d\mu < \infty$. Let

$$
\vartheta_{n,\varepsilon} = [|\xi(f_n) - \xi(f) - \xi(f_n - f)| - \varepsilon\xi(f_n - f)]^+ .
$$

Then

$$\vartheta_{n,\varepsilon} \le (1 + c_\varepsilon)\xi(f) \quad \mu\text{-a. e.} \tag{4.1.19}$$

Moreover we have

$$\vartheta_{n,\varepsilon} \to 0 \quad \mu\text{-a. e. as } n \to \infty\,. \tag{4.1.20}$$

Then (4.1.19), (4.1.20) and the Dominated Convergence Theorem imply that $\int_X \vartheta_{n,\varepsilon}\, d\mu \to 0$ as $n \to \infty$. Note that

$$|\xi(f_n) - \xi(f) - \xi(f_n - f)| \le \vartheta_{n,\varepsilon} + \varepsilon\xi(f_n - f) \quad \mu\text{-a. e.}\,,$$

which gives

$$\limsup_{n\to\infty} \int_X |\xi(f_n) - \xi(f) - \xi(f_n - f)|\, d\mu \le M\varepsilon \quad \text{for some } M > 0\,.$$

Since $\varepsilon > 0$ is arbitrary, we conclude that

$$\int_X |\xi(f_n) - \xi(f) - \xi(f_n - f)|\, d\mu \to 0 \quad \text{as } n \to \infty\,.$$ □

Of special interest is the case when $\xi(s) = |s|^p$ with $1 \le p < \infty$. In this particular form, the result is known as the "Brézis–Lieb Lemma".

Lemma 4.1.22 (Brézis–Lieb Lemma). *If (X, Σ, μ) is a measure space, $\{f_n\}_{n\ge1} \subseteq L^p(X)$ with $1 \le p < \infty$ is bounded, and $f_n \to f$ μ-a. e., then*

$$\lim_{n\to\infty} [\|f_n\|_p^p - \|f_n - f\|_p^p] = \|f\|_p^p\,.$$

Corollary 4.1.23. *If (X, Σ, μ) is a measure space, $\{f_n\}_{n\ge1} \subseteq L^p(X)$ with $1 \le p < \infty$ is bounded and $f_n \to f$ μ-a. e. as well as $\|f_n\|_p \to \|f\|_p$, then $\|f_n - f\|_p \to 0$.*

We have already seen that bounded sequences in L^1 need not have weakly convergent subsequences; see Example 4.1.17. However, if we exclude a decreasing sequence of measurable sets $\{A_k\}_{k\ge1}$ with $\mu(A_k) \to 0$, then we can extract a weakly convergent subsequence. This is the content of the so-called "Bitting Theorem".

Theorem 4.1.24 (Bitting Theorem). *If (X, Σ, μ) is a measure space and $\{f_n\}_{n\ge1} \subseteq L^1(X)$ is a bounded sequence, then there exists a subsequence $\{f_{n_k}\}_{k\ge1}$ of $\{f_n\}_{n\ge1}$, a nonincreasing sequence $\{A_m\}_{m\ge1} \subseteq \Sigma$ with $\mu(A_m) \searrow 0$ and $f \in L^1(X)$ such that $f_{n_k} \xrightarrow{w} f$ in $L^1(X \setminus A_m)$ for all $m \in \mathbb{N}$ as $k \to \infty$.*

Proof. For $n \in \mathbb{N}$ and $c \ge 0$, let $\vartheta_{n,c} = \int_{\{|f_n|\ge c\}} |f_n|\, d\mu \ge 0$. Evidently, $c \to \vartheta_{n,c}$ is nonincreasing. We define $\eta = \lim_{c\to\infty} \sup_{n\ge1} \vartheta_{n,c} \ge 0$. If $\eta = 0$, then $\{f_n\}_{n\ge1} \subseteq L^1(X)$ is uniformly integrable and so on account of the Dunford–Pettis Theorem (see Theorem 4.1.18), the result holds with $A_m = \emptyset$ for all $m \in \mathbb{N}$.

So, assume that $\eta > 0$. For each $k \in \mathbb{N}$, let $n_k \in \mathbb{N}$ be such that $\vartheta_{n_k,2^k} \geq \sup_{n\geq 1} \vartheta_{n,2^k} - 1/k$. Hence,

$$\vartheta_{n_k,2^k} \geq \eta - \frac{1}{k}, \tag{4.1.21}$$

since $\{\sup_{n\geq 1} \vartheta_{n,2^k}\}_{k\geq 1}$ is nonincreasing. Moreover, by monotonicity we know that the following limit exists:

$$\eta' = \lim_{c\to\infty} \sup_{n\geq 1} \int_{\{c\leq |f_{n_k}|<2^k\}} |f_{n_k}|\, d\mu \geq 0\,.$$

So, for sufficiently large $c > 0$, there is a subsequence $\{k(c)\}$ such that

$$\int_{\{c\leq |f_{n_{k(c)}}|<2^{k(c)}\}} |f_{n_{k(c)}}|\, d\mu \geq \frac{\eta'}{2}\,. \tag{4.1.22}$$

Then, because of (4.1.21) and (4.1.22), it follows that

$$\begin{aligned}\eta &= \lim_{c\to\infty} \sup_{n\geq 1} \vartheta_{n,c} \geq \lim_{c\to\infty} \sup_{k\geq 1} \vartheta_{n_k,c} \geq \lim_{c\to\infty} \vartheta_{n_{k(c)},c} \\ &= \lim_{c\to\infty}\left[\int_{\{c\leq |f_{n_{k(c)}}|<2^{k(c)}\}} |f_{n_{k(c)}}|\, d\mu + \int_{\{|f_{n_{k(c)}}|>2^{k(c)}\}} |f_{n_{k(c)}}|\, d\mu\right] \geq \frac{\eta'}{2} + \eta\,.\end{aligned}$$

Thus, $\eta' = 0$ and so

$$\lim_{c\to\infty} \sup_{k\geq 1} \int_{\{c\leq |f_{n_k}|<2^k\}} |f_{n_k}|\, d\mu = 0\,. \tag{4.1.23}$$

Let $A_m = \bigcup_{k\geq m}\{|f_{n_k}| \geq 2^k\}$. Then this sequence has the following properties:

(a) $\{A_m\}_m \subseteq \Sigma$ is nonincreasing;
(b) $\mu(A_m) \searrow 0$ as $m \to \infty$;
(c) for fixed $m \in \mathbb{N}$ we have

$$\lim_{c\to\infty} \sup_{k\geq m} \int_{\{c\leq |f_{n_k}|\}\setminus A_m} |f_{n_k}|\, d\mu = 0\,.$$

Property (a) is clear from the definition of A_m. For property (b), we see that, since $\{f_n\}_{n\geq 1} \subseteq L^1(X)$ is bounded,

$$2^k \mu(\{|f_{n_k}| \geq 2^k\}) \leq \int_{\{|f_{n_k}|\geq 2^k\}} |f_{n_k}|\, d\mu \leq M$$

for some $M > 0$. Hence,

$$\mu(A_m) \leq \sum_{k\geq m} \mu(\{|f_{n_k}| \geq 2^k\}) \leq M \sum_{k\geq m} \frac{1}{2^k} \to 0 \quad \text{as } m \to \infty .$$

Finally for property (c), fix $m \in \mathbb{N}$ and note that

$$0 \leq \lim_{c\to+\infty} \sup_{k\geq m} \int_{\{c\leq|f_{n_k}|\}\setminus A_m} |f_{n_k}| \, d\mu \leq \lim_{c\to+\infty} \sup_{k\geq m} \int_{\{c\leq|f_{n_k}|<2^k\}} |f_{n_k}| \, d\mu \leq 0 ;$$

see (4.1.23). Therefore, the property follows.

From property (c) it follows that, for every $m \in \mathbb{N}$, $\{f_{n_k}\}_{k\geq 1} \subseteq L^1(X \setminus A_m)$ is uniformly integrable.

By a standard diagonal argument, we can find a further subsequence, not relabeled, such that $f_{n_k} \xrightarrow{w} \tilde{f}_m$ in $L^1(X \setminus A_m)$ for all $m \in \mathbb{N}$.

The uniqueness of the weak limit implies that there exists a Σ-measurable function $f: X \to \mathbb{R}$ such that

$$f(x) = \tilde{f}_m(x) \quad \text{for all } x \in X \setminus A_m \text{ with } m \in \mathbb{N} .$$

Moreover, because of

$$\int_{X\setminus A_m} |f| \, d\mu \leq \sup_{n\geq 1} \|f_n\|_1 < \infty ,$$

we conclude that $f \in L^1(X)$. ☐

This leads to the following definition.

Definition 4.1.25. Let (X, Σ, μ) be a measure space, $\{f_n\}_{n\geq 1} \subseteq L^1(X)$ is a bounded sequence and $f \in L^1(X)$. We say that the f_n's converge to f in the **bitting sense**, denoted by $f_n \xrightarrow{b} f$, if there exists a decreasing sequence $\{A_m\}_{m\geq 1} \subseteq \Sigma$ with $\mu(A_m) \searrow 0$ such that $f_n \xrightarrow{w} f$ in $L^1(X \setminus A_m)$ for every $m \in \mathbb{N}$.

Remark 4.1.26. The bitting limit, if it exists, is unique. Indeed, suppose that $f_n \xrightarrow{b} f$ and $f_n \xrightarrow{b} \hat{f}$ as $n \to \infty$. Then it follows that

$$f_n \xrightarrow{w} f \quad \text{in } L^1(X \setminus A_m) \quad \text{and} \quad f_n \xrightarrow{w} \hat{f} \quad \text{in } L^1(X \setminus \hat{A}_m) \quad \text{for all } m \in \mathbb{N} ,$$

where $\{A_m\}_{m\geq 1}, \{\hat{A}\}_{m\geq 1} \subseteq \Sigma$ are decreasing sequences such that

$$\mu(A_m) \to 0 \quad \text{and} \quad \mu(\hat{A}_m) \to 0 \quad \text{as } m \to \infty .$$

The uniqueness of the weak limit implies that $f(x) = \hat{f}(x)$ for μ-a. a. $x \in X \setminus (A_m \cup \hat{A}_{\hat{m}})$ for all $m, \hat{m} \in \mathbb{N}$. Therefore, $f(x) = \hat{f}(x)$ for μ-a. a. $x \in X \setminus D$, where $D =$

$(\bigcap_{m\geq 1} A_m)\bigcup(\bigcap_{m\geq 1} \hat{A}_m)$. Note that $\mu(D) \leq \mu(A_m) + \mu(\hat{A}_m) \to 0$ as $m \to \infty$. Thus, $f = \hat{f}$ μ-a. e.

Very often approximation techniques are based on the regularization method by convolution. This method is realized with the use of mollifiers.

Definition 4.1.27. Let $\Omega \subseteq \mathbb{R}^N$ be an open set.

(a) For $\varepsilon > 0$, let $\Omega_\varepsilon = \{x \in \Omega : d(x, \partial\Omega) > \varepsilon\}$.

(b) By $L^p_{\text{loc}}(\Omega)$ with $1 \leq p < \infty$ we denote the space of all measurable functions $f: \Omega \to \mathbb{R}$ such that $f \in L^p(U)$ for all $U \subset\subset \Omega$, that is, for every open $U \subseteq \Omega$ such that $\overline{U} \subseteq \Omega$ with $\overline{U}$ being compact.

(c) A sequence of **mollifiers** is any sequence of functions on $\mathbb{R}^N$ such that

$$\vartheta_n \in C_c^\infty(\Omega)\,, \quad \operatorname{supp}\vartheta_n \subseteq B_{\frac{1}{n}}(0)\,, \quad \int_{\mathbb{R}^N} \vartheta_n \, dx = 1\,, \quad \vartheta_n \geq 0 \quad \text{for all } n \in \mathbb{N}\,,$$

where $C_c^\infty(\Omega)$ denotes the space of all C^∞-functions on Ω that have compact support. Consider the C^∞-function $\vartheta: \mathbb{R}^N \to \mathbb{R}$ defined by

$$\vartheta(x) = \begin{cases} c\exp(\frac{1}{|x|^2-1}) & \text{if } |x| \leq 1\,, \\ 0 & \text{if } |x| > 1 \end{cases}$$

with $c > 0$ such that $\int_{\mathbb{R}^N} \vartheta(x)\,dx = 1$. For any $\varepsilon > 0$ we set $\vartheta_\varepsilon(x) = 1/\varepsilon^N \vartheta(x/\varepsilon)$ for all $x \in \mathbb{R}^N$. Then $\{\vartheta_\varepsilon\}_{\varepsilon>0}$ is the **standard mollifier**.

(d) Given $f \in L^1_{\text{loc}}(\Omega)$ we define $f^\varepsilon = \vartheta_\varepsilon * f$, where $*$ denotes the **operation of convolution**, that is, $f^\varepsilon(x) = \int_\Omega \vartheta_\varepsilon(x-y)f(y)\,dy$ for all $x \in \Omega_\varepsilon$.

Proposition 4.1.28. *Given $f \in L^1_{\text{loc}}(\Omega)$, it holds that $f^\varepsilon \in C^\infty(\Omega_\varepsilon)$ for every $\varepsilon > 0$.*

Proof. We fix $x \in \Omega_\varepsilon$. For any $k \in \{1, \dots, N\}$, let $e_k = (0, \dots, 1, \dots 0)$ be the $k \overset{\text{th}}{=}$ basic vector in $\mathbb{R}^N$. For $\lambda \in \mathbb{R}$ with a small $|\lambda|$, one has $x + \lambda e_k \in \Omega_\varepsilon$. Then

$$\begin{aligned} \frac{f^\varepsilon(x+\lambda e_k) - f^\varepsilon(x)}{\lambda} &= \frac{1}{\varepsilon^N}\int_\Omega \frac{1}{\lambda}\Big[\vartheta\Big(\frac{x+\lambda e_k - y}{\varepsilon}\Big) - \vartheta\Big(\frac{x-y}{\varepsilon}\Big)\Big]f(y)\,dy \\ &= \frac{1}{\varepsilon^N}\int_U \frac{1}{\lambda}\Big[\vartheta\Big(\frac{x+\lambda e_k - y}{\varepsilon}\Big) - \vartheta\Big(\frac{x-y}{\varepsilon}\Big)\Big]f(y)\,dy \end{aligned}$$

for some $U \subset\subset \Omega$. Note that

$$\begin{aligned} &\lim_{\lambda\to 0}\frac{1}{\lambda}\Big[\vartheta\Big(\frac{x+\lambda e_k - y}{\varepsilon}\Big) - \vartheta\Big(\frac{x-y}{\varepsilon}\Big)\Big] \\ &= \frac{1}{\varepsilon}\frac{\partial\vartheta}{\partial x_k}\Big(\frac{x-y}{\varepsilon}\Big) = \varepsilon^N \frac{\partial\vartheta}{\partial x_k}(x-y) \quad \text{for all } y \in U\,. \end{aligned} \tag{4.1.24}$$

Moreover,

$$\frac{1}{\lambda}\left[\vartheta\left(\frac{x+\lambda e_k-y}{\varepsilon}\right)-\vartheta\left(\frac{x-y}{\varepsilon}\right)\right]f(y)\le\frac{1}{\varepsilon}\|D\vartheta\|_\infty|f(y)| \tag{4.1.25}$$

with $1/\varepsilon\|D\vartheta\|_\infty|f(\cdot)|\in L^1(U)$. From (4.1.24) and (4.1.25) we see that we can apply the Dominated Convergence Theorem to get

$$\frac{\partial f^\varepsilon}{\partial x_k}(x)=\int_\Omega\frac{\partial\vartheta_\varepsilon}{\partial x_k}(x-y)f(y)\,\mathrm{d}y\quad\text{for all }k\in\{1,\dots,N\}.$$

Therefore we conclude that $f^\varepsilon\in C^\infty(\Omega_\varepsilon)$. □

Proposition 4.1.29. *If $f\in C(\mathbb{R})$, then $f^\varepsilon\to f$ as $\varepsilon\to0^+$ uniformly on compact subsets of Ω.*

Proof. Let $U\subset\subset\Omega$ and let $V\subseteq\mathbb{R}^N$ be open such that $U\subseteq V\subseteq\Omega$. For $x\in U$ we obtain

$$f^\varepsilon(x)=\frac{1}{\varepsilon^N}\int_{B_\varepsilon(x)}\vartheta\left(\frac{x-y}{\varepsilon}\right)f(y)\,\mathrm{d}y=\int_{B_1(0)}\vartheta(u)f(x-\varepsilon u)\,\mathrm{d}u\,.$$

Since $\int_{B_1(0)}\vartheta(u)\,\mathrm{d}u=1$ (see Definition 4.1.27 (c)), one has

$$|f^\varepsilon(x)-f(x)|\le\int_{B_1(0)}\vartheta(u)|f(x-\varepsilon u)-f(x)|\,\mathrm{d}u\,. \tag{4.1.26}$$

If $f|_V$ is uniformly continuous, then from (4.1.26) we infer that $f^\varepsilon\to f$ uniformly on V. □

Proposition 4.1.30. *If $f\in L^p_{\mathrm{loc}}(\Omega)$ with $1\le p<\infty$, then $f^\varepsilon\to f$ in $L^p_{\mathrm{loc}}(\Omega)$.*

Proof. Let $U\subset\subset V\subset\subset\Omega$, $x\in U$ and small $\varepsilon>0$. For $1<p<\infty$ and $1/p+1/p'=1$ we derive, using Hölder's inequality, that

$$\begin{aligned}|f^\varepsilon(x)|&\le\int_{B_1(0)}\vartheta(u)^{\frac{1}{p'}}\vartheta(u)^{\frac{1}{p}}|f(x-\varepsilon u)|\,\mathrm{d}u\\&\le\left(\int_{B_1(0)}\vartheta(u)\,\mathrm{d}u\right)^{\frac{1}{p'}}\left(\int_{B_1(0)}\vartheta(u)|f(x-\varepsilon u)|^p\,\mathrm{d}u\right)^{\frac{1}{p}}\\&=\left(\int_{B_1(0)}\vartheta(u)|f(x-\varepsilon u)|^p\,\mathrm{d}u\right)^{\frac{1}{p}}.\end{aligned}$$

Applying Fubini's Theorem we obtain

$$\int_U|f^\varepsilon(x)|^p\,\mathrm{d}x\le\int_{B_1(0)}\vartheta(u)\left(\int_U|f(x-\varepsilon u)|^p\,\mathrm{d}x\right)\mathrm{d}u\le\int_V|f(y)|^p\,\mathrm{d}y \tag{4.1.27}$$

for small $\varepsilon > 0$. Since $f \in L^p(V)$, there exists $h \in C(\overline{V})$ such that

$$\|f - h\|_{L^p(V)} \le \delta \quad \text{with } \delta > 0\,; \tag{4.1.28}$$

see Proposition 2.5.15. Hence, due to (4.1.27),

$$\|f^\varepsilon - h^\varepsilon\|_{L^p(U)} \le \delta\,. \tag{4.1.29}$$

Finally, combining (4.1.28), (4.1.29) and Proposition 4.1.29, we see that, for small $\varepsilon > 0$,

$$\|f^\varepsilon - f\|_{L^p(U)} \le \|f^\varepsilon - h^\varepsilon\|_{L^p(U)} + \|h^\varepsilon - h\|_{L^p(U)} + \|h - f\|_{L^p(U)} \le 3\delta\,.$$

Therefore, $f^\varepsilon \to f$ in $L^p_{\text{loc}}(\Omega)$ as $\varepsilon \to 0^+$. □

Corollary 4.1.31. *If $f \in L^p(\mathbb{R}^N)$ with $1 \le p < \infty$, then $f^\varepsilon \to f$ in $L^p(\mathbb{R}^N)$ as $\varepsilon \to 0^+$.*

Corollary 4.1.32. *If $\Omega \subseteq \mathbb{R}^N$ is open, then $C_c(\Omega)$ is dense in $L^p(\Omega)$ for $1 \le p < \infty$.*

Remark 4.1.33. This corollary is a particular case of Proposition 2.5.15 when $X = \Omega$ and $\mu = \lambda^N$ = the Lebesgue measure on $\mathbb{R}^N$. Finally we mention for future use that if $f \in L^1(\mathbb{R}^N)$ and $h \in L^p(\mathbb{R}^N)$ with $1 \le p \le \infty$, then $f * g \in L^p(\mathbb{R}^N)$ and $\|f * h\|_p \le \|f\|_1 \|h\|_p$, which is a version of Young's inequality. For the proof we refer to Brézis [52, Theorem 4.15, p. 104].

In the last part of this section we will have a quick look at some basic sequence spaces. So, let $\mathbb{R}^{\mathbb{N}}$ be the space of all real sequences. For $1 \le p < \infty$, the l^p-norm of a sequence $\hat{x} = (x_k)_{k\ge1} \in \mathbb{R}^{\mathbb{N}}$ is defined by

$$\|\hat{x}\| = \left(\sum_{k\ge1} |x_k|^p\right)^{\frac{1}{p}}.$$

For $p = \infty$, the l^∞-norm of $\hat{x} = (x_k)_{k\ge1} \in \mathbb{R}^{\mathbb{N}}$ is defined by

$$\|\hat{x}\|_\infty = \sup_{k\ge1} |x_k|\,.$$

These are norms on $L^P(\mathbb{N})$ with $1 \le p \le \infty$ when we consider the counting measure on $\mathbb{N}$.

Definition 4.1.34. We introduce the following sequence spaces

$$\begin{aligned}
c_0 &= \{\hat{x} = (x_k)_{k\ge1} \in \mathbb{R}^{\mathbb{N}} : x_k \to 0 \text{ as } k \to \infty\}\,,\\
c &= \left\{\hat{x} = (x_k)_{k\ge1} \in \mathbb{R}^{\mathbb{N}} : \lim_{n\to\infty} x_n \text{ exists in } \mathbb{R}\right\}\,,\\
l^p &= \{\hat{x} = (x_k)_{k\ge1} \in \mathbb{R}^{\mathbb{N}} : \|\hat{x}\|_p < \infty\}\,, \quad 1 \le p \le \infty\,,\\
s_c &= \{\hat{x} = (x_k)_{k\ge1} \in \mathbb{R}^{\mathbb{N}} : x_k = 0 \text{ for all but a finite number of } k\text{'s}\}\,.
\end{aligned}$$

Remark 4.1.35. We can view s_c as all continuous $\mathbb{R}$-valued functions on $\mathbb{N}$ equipped with the discrete topology, which have compact support. Similarly, l^∞ is the space of all bounded continuous $\mathbb{R}$-valued functions on $\mathbb{N}$ while c_0 is the space of all bounded continuous $\mathbb{R}$-valued functions on $\mathbb{N}$, which vanish at infinity. On s_c, c_0 and c we consider the $\|\cdot\|_\infty$-norm. We easily see that

$$s_c \subseteq l^p \subseteq c_0 \subseteq c \subseteq l^\infty \subseteq \mathbb{R}^{\mathbb{N}} .$$

Proposition 4.1.36. *If $1 \le p < q \le \infty$, then $l^p \subseteq l^q$ and the inclusion is proper.*

Proof. Suppose that $\hat{x} = (x_k)_{k\ge 1} \in l^p$. Then $\{x_k\}_{k\ge 1}$ is a sequence converging to zero, hence it is bounded and so $\hat{x} \in l^\infty$. Therefore, $l^p \subseteq l^\infty$ for all $1 \le p < \infty$.

Now suppose that $1 \le p < q < \infty$ and let $\hat{x} = (x_k)_{k\ge 1} \in l^p$. Since $x_k \to 0$, there exists $m \in \mathbb{N}$ such that $|x_k| \le 1$ for all $k \ge m$. Then $|x_k|^q \le |x_k|^p$ and this proves that $\sum_{k\ge 1} |x|^q < \infty$. Hence $\hat{x} = (x_k)_{k\ge 1} \in l^q$. We conclude that $l^p \subseteq l^q$.

Finally, note that $\hat{x} = (1/k^{1/p})_{k\ge 1} \in l^q$, but $\hat{x} \notin l^p$. So the inclusion $l^p \subseteq l^q$ is proper. □

On account of Remark 4.1.35 and since the l^p-spaces result from $L^p(\mathbb{N})$ with the counting measure, we obtain the following result.

Proposition 4.1.37. *The spaces l^p with $1 \le p \le \infty$ and c_0 as well as c are Banach spaces. The space s_c is not complete.*

From Theorems 4.1.3 and 4.1.5 we get the following proposition.

Proposition 4.1.38. *For $1 \le p < \infty$, $(l^p)^* = l^{p'}$ with $1/p + 1/p'$. Therefore $(l^1)^* = l^\infty$. Moreover, for $1 < p < \infty$, l^p is a reflexive Banach space.*

Taking Hölder's inequality (see Theorem 2.3.12) into account yields the following.

Proposition 4.1.39. *If $1 \le p, p' \le \infty$ are conjugate exponents and $\hat{x} = (x_k)_{k\ge 1} \subseteq l^p$, $\hat{u} = (u_k)_{k\ge 1} \subseteq l^{p'}$, then the series $\langle \hat{x}, \hat{u}\rangle = \sum_{k\ge 1} x_k \hat{u}_k$ converges absolutely and $|\langle \hat{x}, \hat{u}\rangle| \le \|\hat{x}\|_p \|\hat{u}\|_{p'}$.*

From Remark 3.3.17 we know that l^1 has the following property known as the **Schur property**.

Proposition 4.1.40. *The Banach space l^1 has the following property*

$$\hat{x}_n \xrightarrow{\text{w}} \hat{x} \quad \text{in } l^1 \quad \text{implies} \quad \|\hat{x}_n - \hat{x}\|_1 \to 0 ,$$

*which is called the **Schur property**. Hence, every weakly compact subset of l^1 is also norm compact.*

Consider the particular sequences

$$\hat{e} = (1, 1, 1, \ldots) , \quad \hat{e}_k = (0, \ldots, 0, e_k = 1, 0, \ldots) \quad \text{for } k \in \mathbb{N} .$$

Proposition 4.1.41. *The sequence $\{\hat{e}, \hat{e}_k\}_{k\geq 1}$ is a Schauder basis (see Definition 3.5.50(b)) for the Banach space c. Hence, c is separable.*

Proof. Let $\hat{x} = (x_k)_{k\geq 1} \in c$ and let $x_\infty = \lim_{k\to\infty} x_k$. Then

$$\hat{x} = x_\infty \hat{e} + \sum_{k\geq 1} (x_k - x_\infty)\hat{e}_k ,$$

which shows that $\{\hat{e}, \hat{e}_k\}_{k\geq 1}$ is a Schauder basis for c. □

Since c_0 is a closed subspace of c, we infer the following.

Corollary 4.1.42. *The Banach space c_0 is separable.*

Proposition 4.1.43. *If $1 \leq p < \infty$, then $\{\hat{e}_k\}_{k\geq 1}$ is a Schauder basis for l^p. Hence, l^p is separable for $1 \leq p < \infty$.*

Proof. Let $\hat{x} = (x_k)_{k\geq 1} \in l^p$ with $1 \leq p < \infty$. Then it follows that

$$\left\| \hat{x} - \sum_{k=1}^{n} x_k \hat{e}_k \right\|_p = \left(\sum_{k\geq n+1} |x_k|^p \right)^{\frac{1}{p}} \to 0 \quad \text{as } n \to \infty .$$

Hence $\hat{x} = \sum_{k\geq 1} x_k \hat{e}_k$ and so we have that $\{\hat{e}_k\}_{k\geq 1}$ is a Schauder basis for l^p with $1 \leq p < \infty$. □

Proposition 4.1.44. $c_0^* = l^1$.

Proof. Every $\hat{u} = (u_k)_{k\geq 1} \in l^1$ defines a linear functional $\xi_{\hat{u}} \colon c_0 \to \mathbb{R}$ by

$$\xi_{\hat{u}}(\hat{x}) = \sum_{k\geq 1} x_k u_k \quad \text{for all } \hat{x} = (x_k)_{k\geq 1} \in c_0 .$$

Moreover, one has

$$|\xi_{\hat{u}}(\hat{x})| \leq \sum_{k\geq 1} |x_k u_k| \leq \|\hat{x}\|_\infty \sum_{k\geq 1} |u_k| = \|\hat{x}\|_\infty \|\hat{u}\|_1 , \tag{4.1.30}$$

which shows that $\xi_{\hat{u}} \in c_0^*$. Thus, $\hat{u} \to \xi_{\hat{u}}$ is a bounded linear map from l^1 into c_0^*. We claim that this map is an isometric isomorphism. To see this, let $\hat{u} = (u_k)_{k\geq 1} \in l^1$ and define $\lambda_k = \operatorname{sgn} u_k$ as well as $\lambda_k = 0$ if $u_k = 0$ for $k \in \mathbb{N}$. We infer that

$$\lambda_k = 1 \quad \text{if } u_k > 0 , \quad \lambda_k = 0 \quad \text{if } u_k = 0 , \quad \lambda_k = -1 \quad \text{if } u_k < 0 \quad \text{for } k \in \mathbb{N} .$$

For $n \in \mathbb{N}$ we define

$$\hat{y}_n = \sum_{k=1}^{n} \lambda_k \hat{e}_k \in c_0 .$$

This gives $\xi_{\hat{u}}(\hat{y}_n) = \sum_{k=1}^{n} |y_k|$ and $\|\hat{y}_n\|_\infty = 1$, which implies $\|\xi_{\hat{u}}\|_* \geq \sum_{k=1}^{n} |u_k|$ for all $n \in \mathbb{N}$. Thus

$$\|\xi_{\hat{u}}\|_* \geq \|\hat{u}\|_1 \, . \tag{4.1.31}$$

From (4.1.30) and (4.1.31), it follows that $\|\xi_{\hat{u}}\|_* = \|\hat{u}\|_1$. Therefore, $\xi_{\hat{u}}$ is an isometry. We need to show that $\hat{u} \to \xi_{\hat{u}}$ is surjective. So, let $\xi^* \in c_0^*$ and let $u_k = \xi^*(\hat{e}_k)$ for all $k \in \mathbb{N}$. As before, we define $\hat{y}_n = \sum_{k=1}^{n} (\operatorname{sgn} u_k)\hat{e}_k \in c_0$. Then $\|\hat{y}_n\|_\infty = 1$ for all $n \in \mathbb{N}$ large enough and so we derive

$$\sum_{k=1}^{n} |u_k| = \xi^*(\hat{y}_n) \leq \|\xi^*\|_* \quad \text{for all } n \in \mathbb{N} \, ,$$

which directly yields

$$\|\hat{u}\|_1 = \sum_{k \geq 1} |u_k| \leq \|\xi^*\|_* \quad \text{with } \hat{u} = (u_k)_{k \geq 1} \, .$$

Therefore, $\hat{u} \in l^1$. Since $\xi_{\hat{u}}(\hat{e}_k) = u_k = \xi^*(\hat{e}_k)$ for all $k \in \mathbb{N}$ and since $\operatorname{span}\{\hat{e}_k\}_{k \geq 1}$ is dense in c_0 we infer that $\xi_{\hat{u}} = \xi^*$. Thus, the map $\hat{u} \to \xi_{\hat{u}}$ is surjective, hence an isomorphism. □

4.2 Lebesgue–Bochner Spaces

In this section we deal with Banach space-valued functions. We define integrals for such functions and Lebesgue spaces for them which we study in detail. These spaces play an important role in the theory of evolution equations and in the study of Young measures, which in turn are basic tools in the theory of calculus of variations and in optimal control.

We start by introducing some notions of measurability for Banach space-valued functions.

Definition 4.2.1. Let (Ω, Σ, μ) be a measure space and let X be a Banach space.

(a) A **simple function** $s\colon \Omega \to X$ is a function of the form

$$s(w) = \sum_{k=1}^{n} \eta_k \chi_{A_k}(w) \quad \text{for all } w \in \Omega$$

with $n \in \mathbb{N}$, $\{\eta_k\}_{k=1}^{n} \subseteq X$ and $\{A_k\}_{k=1}^{n} \subseteq \Sigma$ mutually disjoint.

(b) A function $f\colon \Omega \to X$ is **strongly measurable** if there exist a sequence $\{s_n\}_{n \geq 1}$ of simple functions such that $\|s_n(w) - f(w)\| \to 0$ μ-a. e.

(c) A function $f\colon \Omega \to X$ (resp. $f\colon \Omega \to X^*$) is said to be **weakly measurable** (resp. **weakly*-measurable**) if $w \to \langle x^*, f(w) \rangle$ is Σ-measurable for all $x^* \in X^*$ (resp.

$w \to \langle f(w), x\rangle$ is Σ-measurable for all $x \in X$). Here by $\langle\cdot,\cdot\rangle$ we denote the duality brackets for the pair (X^*, X).

Proposition 4.2.2. *If $f\colon \Omega \to X$ is strongly measurable, then $w \to \|f(w)\|$ is Σ-measurable from Ω into $\mathbb{R}_+$.*

Proof. By Definition 4.2.1 (b), there exists a sequence of simple functions $\{s_n\}_{n\geq 1}$ such that $\|s_n(w) - f(w)\| \to 0$ μ-a. e., which implies that

$$\big|\|s_n(w)\| - \|f(w)\|\big| \leq \|s_n(w) - f(w)\| \to 0 \quad \mu\text{-a. e.}$$

Hence, $w \to \|f(w)\|$ is Σ-measurable; see Proposition 2.2.12. □

Definition 4.2.3. Let (Ω, Σ, μ) be a measure space and let X be a Banach space. A function $f\colon \Omega \to X$ is said to be **essentially separably valued** if there exists $N \in \Sigma$ with $\mu(N) = 0$ such that $f(\Omega \setminus N) \subseteq X$ is separable.

Using this definition we can state a convenient characterization of strongly measurable functions. The result is known as the "Pettis Measurability Theorem".

Theorem 4.2.4 (Pettis Measurability Theorem). *If (Ω, Σ, μ) is a measure space, X is a Banach space and $f\colon \Omega \to X$, then the following statements are equivalent:*
(a) *f is strongly measurable.*
(b) *f is essentially separably valued and $f^{-1}(U) \in \Sigma$ for all open sets $U \subseteq X$.*
(c) *f is essentially separably valued and weakly measurable.*

Proof. (a) $\Longrightarrow$ (b): The strong measurability of f implies that there exists a sequence of simple functions $s_n\colon \Omega \to X$ with $n \in \mathbb{N}$ such that $s_n(w) \to f(w)$ μ-a. e. Then, for every $x^* \in X^*$, $\langle x^*, s_n(w)\rangle \to \langle x^*, f(w)\rangle$ in R for all $w \in \Omega \setminus N$ with $\mu(N) = 0$. Since $w \to \langle x^*, s_n(w)\rangle$ is Σ-measurable for each $n \in \mathbb{N}$, from Proposition 2.2.12 it follows that $w \to \langle x^*, f(w)\rangle$ is Σ-measurable. So, f is weakly measurable. The union E of the ranges of the s_n's is a countable set. Hence $\overline{E} \subseteq X$ is separable and $f(w) \in \overline{E}$ for all $w \in \Omega \setminus N$ with $\mu(N) = 0$. Hence, f is essentially separably valued.

(b) $\Longrightarrow$ (c): Note that for every open $V \subseteq \mathbb{R}$ and for every $x^* \in X^*$, $(x^*)^{-1}(V) \subseteq X$ is open and so $(x^* \circ f)^{-1}(V) = f^{-1}((x^*)^{-1}(V)) \in \Sigma$. Therefore, f is weakly measurable.

(c) $\Longrightarrow$ (a): Without any loss of generality we may assume that X is separable, otherwise we replace X by $\overline{\operatorname{span}}\, f(\Omega \setminus N)$. Then the separability of X implies that there exists a sequence $\{x_n^*\}_{n\geq 1} \subseteq \overline{B}_1^{X^*}$ such that $\|f(w)\| = \sup_{n\geq 1} |\langle x_n^*, f(w)\rangle|$; see Theorem 3.4.12 (a). Hence $w \to \|f(w)\|$ is Σ-measurable.

Let $A_+ = \{w \in \Omega\colon \|f(w)\| > 0\}$. Then $A_+ \in \Sigma$ and $w \to f(w) - x$ is weakly measurable on A_+ for every $x \in X$. Hence, $w \to \|f(w) - x\|$ is $\Sigma \cap A_+$-measurable. Let $\{x_n\}_{n\geq 1} \subseteq X$ be a dense sequence. Given $\varepsilon > 0$, let

$$D_n = \{w \in A_+\colon \|f(w) - x_n\| < \varepsilon\} \in \Sigma \cap A_+ \quad \text{with } n \in \mathbb{N}\,.$$

We define $C_n = D_n \setminus \bigcup_{k=1}^{n-1} D_k \in \Sigma \cap A_+$ and these sets are disjoint. Note that $A_+ = \bigcup_{n\geq 1} C_n$. We define

$$f_\varepsilon(w) = \begin{cases} x_n & \text{if } w \in C_n,\ n \in \mathbb{N}\,, \\ 0 & \text{if } w \in \Omega \setminus A_+\,. \end{cases}$$

Evidently, f_ε is countably valued and $\|f(w) - f_\varepsilon(w)\| < \varepsilon$ for all $w \in \Omega$. Therefore, f is the uniform limit of countably valued functions. Truncating the f_n's, we produce a sequence $\{s_n\}_{n\geq 1}$ of simple functions such that $s_n(w) \to f(w)$ μ-a. e. Hence f is strongly measurable. ☐

Corollary 4.2.5. *If (Ω, Σ, μ) is a measure space, X is a separable Banach space, and $f\colon \Omega \to X$, then the following statements are equivalent:*
(a) *f is strongly measurable.*
(b) *f is measurable.*
(c) *f is weakly measurable.*

Another useful consequence of Theorem 4.2.4 is the following result.

Corollary 4.2.6. *If $f\colon \Omega \to X$ is the μ-a. e. limit of a sequence of strongly measurable functions, then f is strongly measurable.*

Example 4.2.7. (a) Let $X = l^2[0,1]$ and let $\{e_t\}_{t\in[0,1]}$ be the canonical basis of this nonseparable Hilbert space. Let $f\colon [0,1] \to X$ be defined by $f(t) = e_t$. For $x^* \in X^*$ we see that $\langle x^*, f(t)\rangle = 0$ for all $t \in [0,1]\setminus C$ with C being countable since $\sum_{t\in[0,1]}(\langle x^*, e_t\rangle)^2 < \infty$. Therefore, f is weakly measurable. However, $\|f(s) - f(t)\| = \sqrt{2}$ for $s \neq t$, which implies that f is not essentially separably valued and so it is not strongly measurable.

(b) Let $X = L^\infty[0,1]$ and let $f\colon [0,1] \to X$ be defined by $f(t) = \chi_{[0,t]}$. Note that X^* is the space of finitely additive measures, which are absolutely continuous with respect to the Lebesgue measure; see Dunford–Schwartz [105, IV 8.16]. So, every $x^* \in X^*$ is the difference of two positive elements. For $x^* \geq 0$, the function $t \to \langle x^*, f(t)\rangle$ is increasing and so f is weakly measurable. On the other hand, $\|f(s) - f(t)\|_\infty = 1$ for $s \neq t$ and so f is not essentially separably valued. Thus it is not strongly measurable.

The notion of a Bochner integral is an abstraction of Proposition 2.3.22.

Definition 4.2.8. Let (Ω, Σ, μ) be a σ-finite measure space and X is a Banach space.
(a) A simple function $s\colon \Omega \to X$ is **Bochner integrable** if it has the form

$$s(w) = \sum_{k=1}^{n} \eta_k \chi_{A_k}(w) \quad \text{for all } w \in \Sigma$$

with $n \in \mathbb{N}$, distinct elements $\{\eta_k\}_{k=1}^n \subseteq X$ and $\{A_k\}_{k=1}^n \subseteq \Sigma$ are mutually disjoint and $\eta_k = 0$ if $\mu(A_k) = +\infty$. For any $A \in \Sigma$ the **Bochner integral** of s over A is defined by

$$\int_A s\,\mathrm{d}\mu = \sum_{k=1}^{n} \eta_k \mu(A_k \cap A)$$

with $\eta_k \mu(A_k \cap A) = 0$ if $\eta_k = 0$ and $\mu(A_k \cap A) = +\infty$.

(b) A strongly measurable function $f\colon \Omega \to X$ is **Bochner integrable** if there exists a sequence $\{s_n\}_{n\geq 1}$ of Bochner integrable simple functions such that

$$\|s_n(w) - f(w)\| \to 0 \quad \mu\text{-a.e.} \quad \text{and} \quad \int_\Omega \|s_n - f\|\,\mathrm{d}\mu \to 0 \quad \text{as } n \to \infty\,.$$

Then for any $A \in \Sigma$ the **Bochner integral** of f over A is defined by

$$\int_A f\,\mathrm{d}\mu = \lim_{n\to\infty} \int_A s_n\,\mathrm{d}\mu\,. \tag{4.2.1}$$

Remark 4.2.9. It is easy to see that the limit in (4.2.1) exists and is independent of the sequence $\{s_n\}_{n\geq 1}$. Evidently $\int_A f\,\mathrm{d}\mu = \int_\Omega f\chi_A\,\mathrm{d}\mu$.

An immediate consequence of the definition above is the following result.

Proposition 4.2.10. *If $f\colon \Omega \to X$ is Bochner integrable, then*

$$\Big\|\int_\Omega f\,\mathrm{d}\mu\Big\| \leq \int_\Omega \|f\|\,\mathrm{d}\mu \quad \textit{and} \quad \lim_{\mu(A)\to 0} \int_A f\,\mathrm{d}\mu = 0\,.$$

As in the case with the Lebesgue measure, the Bochner integral defines a vector valued measure as well.

Proposition 4.2.11. *If $\{A_k\}_{k\geq 1} \subseteq \Sigma$ is a disjoint partition of Ω and $f\colon \Omega \to X$ is Bochner integrable over A_k with $k \in \mathbb{N}$ and $\sum_{k\geq 1} \int_{A_k} \|f\|\,\mathrm{d}\mu < \infty$, then f is Bochner integrable over Ω and $\int_\Omega f\,\mathrm{d}\mu = \sum_{k\geq 1} \int_{A_k} f\,\mathrm{d}\mu$.*

Proof. Let $\varepsilon > 0$ and choose $n \in \mathbb{N}$ such that

$$\sum_{k\geq n+1} \int_{A_k} \|f\|\,\mathrm{d}\mu \leq \varepsilon\,. \tag{4.2.2}$$

For each $k \in \{1,\ldots,n\}$, we choose a Bochner integrable simple function $s_k\colon \Omega \to X$ such that

$$\int_{A_k} \|f - s_k\|\,\mathrm{d}\mu \leq \frac{\varepsilon}{2^k}\,. \tag{4.2.3}$$

We set $s = \sum_{k=1}^{n} s_k \chi_{A_k}$. Clearly this is a simple function. By applying (4.2.2) and (4.2.3) we obtain

$$\int_X \|f - s\| \, d\mu = \sum_{k\ge1} \int_{A_k} \|f - s\| \, d\mu = \sum_{k=1}^{n} \int_{A_k} \|f - s_k\| \, d\mu + \sum_{k\ge n+1} \int_{A_k} \|f\| \, d\mu$$
$$\le \sum_{k=1}^{n} \frac{\varepsilon}{2^k} + \varepsilon \le 2\varepsilon \, .$$

Hence, f is Bochner integrable over Ω. Finally note that

$$\left\| \int_\Omega f \, d\mu - \sum_{k=1}^{n} \int_{A_k} f \, d\mu \right\| \le \lim_{n\to\infty} \sum_{k\ge n+1} \int_{A_k} \|f| \, d\mu = 0 \, . \qquad \square$$

The definition of the Bochner integral is not that easy to use. The next proposition provides a very convenient criterion for Bochner integrability.

Proposition 4.2.12. *A function $f:\Omega \to X$ is Bochner integrable if and only if f is strongly measurable and $\|f(\cdot)\| \in L^1(\Omega)$.*

Proof. $\Longrightarrow$: Suppose that f is Bochner integrable. Then there exists a sequence of Bochner integrable simple functions $\{s_n\}_{n\ge1}$ such that $\int_\Omega \|s_n - f\| \, d\mu \to 0$. For $w \in \Omega$ and $n, m \in \mathbb{N}$ we get

$$0 \le \big| \|s_n(w)\| - \|s_m(w)\| \big| \le \|s_n(w) - s_m(w)\|$$
$$\le \|s_n(w) - f(w)\| + \|f(w) - s_m(w)\| \, ,$$

which implies that

$$\int_\Omega \big| \|s_n\| - \|s_m\| \big| \, d\mu \le \int_\Omega \|s_n - f\| \, d\mu + \int_\Omega \|f - s_m\| \, d\mu \to 0 \quad \text{as } n, m \to \infty \, .$$

Hence, $\{\|s_n\|\}_{n\ge1} \subseteq L^1(\Omega)$ is a Cauchy sequence, thus bounded. We finally see that

$$\int_\Omega \|f\| \, d\mu \le \int_\Omega \|f - s_n\| \, d\mu + \int_\Omega \|s_n\| \, d\mu \le M \quad \text{for some } M > 0 \text{ and for all } n \in \mathbb{N} \, .$$

$\Longleftarrow$: From Proposition 4.2.11, we may assume without any loss of generality that μ is finite. From the proof of the Pettis Measurability Theorem (see Theorem 4.2.4), we know that for a given $\varepsilon > 0$ we find a countably valued measurable function $h_\varepsilon: \Omega \to X$ such that

$$\|h_\varepsilon(w) - f(w)\| \le \varepsilon \quad \text{for all } w \in \Omega \setminus N \text{ with } \mu(N) = 0 \, . \tag{4.2.4}$$

This gives

$$\int_\Omega \|h_\varepsilon\| \, d\mu \le \int_\Omega \|h_\varepsilon - f\| \, d\mu + \int_\Omega \|f\| \, d\mu \le \varepsilon\mu(\Omega) + \int_\Omega \|f\| \, d\mu < \infty \, .$$

Hence $\|h_\varepsilon(\cdot)\| \in L^1(\Omega)$ and we find $\delta > 0$ such that

$$\int_A \|h_\varepsilon\| \, d\mu \leq \varepsilon \quad \text{for all } A \in \Sigma \text{ with } \mu(A) \leq \delta \,. \tag{4.2.5}$$

We consider a Σ-Partition $\Omega = E \cup A$ with $\mu(A) \leq \delta$ and such that $\hat{h}_\varepsilon = h_\varepsilon \chi_A$ has finite range, that is, $\hat{h}_\varepsilon$ is a simple function. Then, because of (4.2.4) and (4.2.5), we conclude that

$$\begin{aligned} \int_\Omega \|f - \hat{h}_\varepsilon\| \, d\mu &\leq \int_\Omega \|f - h_\varepsilon\| \, d\mu + \int_\Omega \|h_\varepsilon - \hat{h}_\varepsilon\| \, d\mu \\ &\leq \int_\Omega \|f - h_\varepsilon\| \, d\mu + \int_A \|h_\varepsilon\| \, d\mu \leq \varepsilon\mu(\Omega) + \varepsilon \,. \end{aligned}$$

Therefore, f is Bochner integrable in the sense of Definition 4.2.8 (b). □

The next result follows directly from the definition for Bochner integrable simple functions and by approximation for general Bochner integrable functions.

Proposition 4.2.13. *If (Ω, Σ, μ) is a σ-finite measure space, X, Y are Banach spaces, $T \in L(X, Y)$, $f\colon \Omega \to X$ and $T(f)\colon \Omega \to Y$ are both Bochner integrable, then*

$$T\Big(\int_\Omega f \, d\mu\Big) = \int_\Omega (T \circ f) \, d\mu \,.$$

Corollary 4.2.14. *If (Ω, Σ, μ) is a σ-finite measure space, X is a Banach space, and $f\colon \Omega \to X$ is Bochner integrable, then*

$$\Big\langle x^*, \int_\Omega f \, d\mu \Big\rangle = \int_\Omega \langle x^*, f\rangle \, d\mu \quad \textit{for all } x^* \in X^* \,.$$

The next result is a straightforward consequence of Definition 4.2.8.

Proposition 4.2.15. *If $f, h\colon \Omega \to X$ are Bochner integrable functions and $\vartheta, \lambda \in \mathbb{R}$, then the following hold:*
(a) *$\int_A [\vartheta f + \lambda h] \, d\mu = \vartheta \int_A f \, d\mu + \lambda \int_A h \, d\mu$ for all $A \in \Sigma$.*
(b) *If $f(w) \leq h(w)$ μ-a. e., then $\int_A f \, d\mu \leq \int_A h \, d\mu$ for all $A \in \Sigma$.*

Proposition 4.2.16. *If $f, h\colon \Omega \to X$ are Bochner integrable functions and $\int_A f \, d\mu = \int_A h \, d\mu$ for all $A \in \Sigma$, then $f(w) = h(w)$ μ-a. e.*

Proof. On account of the Pettis Measurability Theorem (see Theorem 4.2.4) without any loss of generality, we may assume that X is separable. Then according to Theorem 3.4.12 (a), there exists a sequence $\{x_n^*\}_{n\geq 1} \subseteq \overline{B}_1^{X^*}$ such that $\overline{\{x_n^*\}_{n\geq 1}}^{w^*} = \overline{B}_1^{X^*}$. Applying

Corollary 4.2.14 yields $\int_A \langle x_n^*, f - h\rangle \, \mathrm{d}\mu = 0$ for all $n \in \mathbb{N}$. Hence $\langle x_n^*, f(w) - h(w)\rangle = 0$ for all $w \in \Omega \setminus N$ with $\mu(N) = 0$ and $n \in \mathbb{N}$. Therefore, we obtain $\|f(w) - h(w)\| = 0$ for all $w \in \Omega \setminus N$ with $\mu(N) = 0$ and so $f = h$ μ-a. e. □

An interesting byproduct of this proof is the following corollary.

Corollary 4.2.17. *If $f, h\colon \Omega \to X$ are strongly measurable and $\langle x^*, f(w)\rangle = \langle x^*, h(w)\rangle$ μ-a. e. for all $x^* \in X^*$, the exceptional μ-null set depending on x^*, then $f(w) = h(w)$ μ-a. e.*

Next we present a version of the mean value theorem for Bochner integrals.

Proposition 4.2.18. *If $f\colon \Omega \to X$ is Bochner integrable and $A \in \Sigma$ with $\mu(A) > 0$, then $1/(\mu(A)) \int_A f \, \mathrm{d}\mu \in \overline{\operatorname{conv}} f(A)$.*

Proof. We argue by contradiction. So, suppose that $1/(\mu(A)) \int_A f \, \mathrm{d}\mu \notin \overline{\operatorname{conv}} f(A)$. Then according to the Strong Separation Theorem (see Corollary 3.1.61), there exists $x^* \in X^* \setminus \{0\}$ such that

$$\left\langle x^*, \frac{1}{\mu(A)} \int_A f \, \mathrm{d}\mu \right\rangle < \vartheta \le \langle x^*, f(w)\rangle \quad \text{for all } w \in A .$$

This implies, thanks to Corollary 4.2.14, that

$$\frac{1}{\mu(A)} \int_A \langle x^*, f\rangle \, \mathrm{d}\mu < \vartheta \le \langle x^*, f(w)\rangle \quad \text{for all } w \in A .$$

Therefore,

$$\int_A \langle x^*, f\rangle \, \mathrm{d}\mu < \vartheta \mu(A) \le \int_A \langle x^*, f\rangle \, \mathrm{d}\mu ,$$

a contradiction. This proves the proposition. □

The Lebesgue Dominated Convergence Theorem has its counterpart for the Bochner integral as stated in the next theorem.

Theorem 4.2.19. *If $\{f_n\}_{n\ge 1}$ is a sequence of Bochner integrable functions, $f_n \to f$ μ-a. e. and there exists $\eta \in L^1(\Omega)$ such that $\|f_n(w)\| \le \eta(w)$ μ-a. e. for all $n \in \mathbb{N}$, then f is Bochner integrable,*

$$\int_\Omega \|f_n - f\| \, \mathrm{d}\mu \to 0 \quad \textit{and} \quad \int_A f_n \, \mathrm{d}\mu \to \int_A f \, \mathrm{d}\mu \quad \textit{for all } A \in \Sigma .$$

Proof. Since $\|f_n(w)\| \le \eta(w)$ μ-a. e. for all $n \in \mathbb{N}$ with $\eta \in L^1(\Omega)$, Proposition 4.2.12 implies that f is Bochner integrable.

By hypothesis, $\|f_n(w) - f(w)\| \to 0$ μ-a. e. as $n \to \infty$ and $\|f_n(w) - f(w)\| \le 2\eta(w)$ μ-a. e. So, by the scalar Dominated Convergence theorem we get that $\int_\Omega \|f_n - f\| \, \mathrm{d}\mu \to 0$

as $n \to \infty$. Moreover, for every $A \in \Sigma$, we obtain that

$$\left\| \int_A f_n \, \mathrm{d}\mu - \int_A f \, \mathrm{d}\mu \right\| = \left\| \int_\Omega (f_n - f)\chi_A \, \mathrm{d}\mu \right\| \le \int_\Omega \|f_n - f\| \chi_A \, \mathrm{d}\mu$$
$$\le \int_\Omega \|f_n - f\| \, \mathrm{d}\mu \to 0 \quad \text{as } n \to \infty . \qquad \square$$

Now we are ready to introduce the analogs of the L^p-spaces for $1 \le p \le \infty$ for Banach space-valued functions. These spaces are known as **Lebesgue–Bochner spaces.**

Definition 4.2.20. Let (Ω, Σ, μ) be a measure space and X a Banach space.

(a) For $1 \le p < \infty$ we define $L^p(\Omega, X)$ to be the space of all equivalence classes for the relation of equality μ-a. e. of Bochner integrable functions $f\colon \Omega \to X$ such that $\int_\Omega \|f\|^p \, \mathrm{d}\mu < \infty$. This is a normed space with the norm defined by

$$\|f\|_p = \left(\int_\Omega \|f\|^p \, \mathrm{d}\mu \right)^{\frac{1}{p}} .$$

(b) For $p = \infty$ we define $L^\infty(\Omega, X)$ to be the space of all equivalence classes of Bochner integrable functions $f\colon \Omega \to X$, which are essentially bounded, that is,

$$\operatorname*{ess\,sup}_\Omega \|f(w)\| = \inf[M > 0\colon \|f(w)\| \le M \ \mu\text{-a. e.}] < \infty .$$

This is a normed space with the norm defined by $\|f\|_\infty = \operatorname{ess\,sup}_\Omega \|f(w)\|$.

Remark 4.2.21. By Proposition 4.2.12, $L^1(\Omega, X)$ coincides with the class of all Bochner integrable functions.

The next proposition is an easy consequence of the definition above and of the properties of the Bochner integral.

Proposition 4.2.22. *If (Ω, Σ, μ) is a measure space and X is a Banach space, then the following hold:*

(a) *$L^p(\Omega, X)$ is a Banach space for every $1 \le p \le \infty$.*

(b) *The set of integrable simple functions is dense in $L^p(\Omega, X)$ for $1 \le p < \infty$ and the countably valued functions in $L^\infty(\Omega, X)$ are dense in $L^\infty(\Omega, X)$.*

(c) *If $(\Sigma(\mu), d_\mu)$ is a separable metric space (see Definition 2.3.23), and X is a separable Banach space, then $L^p(\Omega, X)$ is separable as well for $1 \le p < \infty$.*

(d) *If X is reflexive (resp. uniformly convex), then the same is true for $L^p(\Omega, X)$ for $1 < p < \infty$.*

(e) *If X is continuously embedded into Y, then so does $L^p(\Omega, X)$ into $L^q(\Omega, Y)$ for $1 \le q \le p \le \infty$.*

A basic problem in the theory of Lebesgue–Bochner spaces is the identification of the dual of $L^p(\Omega, X)$ for $1 \le p < \infty$. To do this, we need to introduce some basic definitions from the theory of vector measures.

Definition 4.2.23. Let (Ω, Σ) be a measurable space and X a Banach space.

(a) We say that $\xi: \Sigma \to X$ is a **vector measure** if $\xi(\emptyset) = 0$ and for every $\{A_n\}_{n\ge1} \subseteq \Sigma$ mutually disjoint, one has $\xi(\bigcup_{n\ge1} A_n) = \sum_{n\ge1} \xi(A_n)$ in the norm topology of X.

(b) If μ is a measure on Σ, we say that the vector measure $\xi: \Sigma \to X$ is **μ-continuous** if $\lim_{\mu(A)\to0} \xi(A) = 0$. This is equivalent to saying that $\mu(A) = 0$ implies $\xi(A) = 0$. As in the scalar case, we denote this by $\xi \ll \mu$.

(c) We say that a vector measure $\xi: \Sigma \to X$ is of **bounded variation** if

$$|\xi|(\Omega) = \sup_{\mathcal{P}} \sum_{A\in\mathcal{P}} \|\mu(A)\| < \infty ,$$

where $\mathcal{P}$ runs through the finite Σ-partitions of Ω. Similarly we can define this for $E \in \Sigma$ by

$$|\xi|(E) = \sup_{\mathcal{P}'} \sum_{A\in\mathcal{P}'} \|\mu(A)\| < \infty ,$$

where $\mathcal{P}'$ runs through the finite Σ-partitions of E. We call $|\xi|(\cdot)$ the **variation** of ξ and it is a measure on Σ.

(d) The Banach space X is said to have the **Radon–Nikodym Property** (the RNP for short) if for every probability measure μ on Σ and every vector measure $\xi: \Sigma \to X$ of bounded variation with $\xi \ll \mu$, there exists $f \in L^1(\Omega, X)$ such that $\xi(A) = \int_A f \, d\mu$ for all $A \in \Sigma$.

Remark 4.2.24. If $f \in L^1(\Omega, X)$, then from Proposition 4.2.11 we know that $\Sigma \ni A \to \xi(A) = \int_A f \, d\mu$ is a vector measure. The RNP is not a property that every Banach space X has. For example let $X = c_0$ and let $(\Omega, \Sigma, \mu) = ([0,1], \mathcal{B}([0,1]), \lambda)$ with $\mathcal{B}([0,1])$ being the Borel σ-algebra of $[0,1]$ and λ being the Lebesgue measure. We consider the vector measure $\xi: \Sigma \to c_0$ defined by $\xi(A) = (\int_A \cos(nt) \, dt)_{n\ge1}$, which is well-defined by the Riemann–Lebesgue Lemma; see Hewitt–Stromberg [160, p. 249]. Clearly, ξ is of bounded variation and $\xi \ll \lambda$. But $(\cos(nt))_{n\ge1} \notin c_0$. So ξ does not have a density in $L^1([0,1], c_0)$. Hence c_0 does not have the RNP. The RNP is a hereditary property, that is, every closed subspace of a Banach space with the RNP has the RNP.

The next theorem identifies two major classes of Banach spaces that exhibit the RNP. For a proof of this result we refer to Diestel–Uhl [91, pp. 79, 82].

Theorem 4.2.25. *If X is a reflexive Banach space or X is a separable dual Banach space, then X has the* RNP.

Using this notion, we can state our first result concerning the dual of a Lebesgue–Bochner space.

Theorem 4.2.26. *If (Ω, Σ, μ) is a σ-finite measure space and X is a Banach space such that X^* has the* RNP, *then $L^p(\Omega, X)^* = L^{p'}(\Omega, X^*)$ for all $1 \le p < \infty$ with $1/p + 1/p' = 1$.*

Proof. On account of Proposition 4.2.11, we may assume that μ is finite. Let $h \in L^{p'}(\Omega, X^*)$ and define

$$\eta_h(f) = \int_\Omega \langle f, h\rangle \, d\mu \quad \text{for all } f \in L^p(\Omega, X).$$

Evidently, $\eta_h : L^p(\Omega, X) \to \mathbb{R}$ is linear and

$$|\eta_h(f)| \le \int_\Omega |\langle f, h\rangle| \, d\mu \le \int_\Omega \|f\|_X \|h\|_{X^*} \, d\mu \le \|f\|_p \|h\|_{p'},$$

which implies that

$$\eta_h \in L^p(\Omega, X)^* \quad \text{and} \quad \|\eta_h\|_* \le \|h\|_{p'}. \tag{4.2.6}$$

First suppose that $h = \sum_{k\ge1} x_k^* \chi_{A_k}$ with $x_k^* \in X^*$ and a partition $\{A_k\}_{k\ge1} \subseteq \Sigma$ of Ω with $\mu(A_k) > 0$ for all $k \in \mathbb{N}$. Given $\varepsilon > 0$ we choose $\vartheta \in L^p(\Omega)$ with $\vartheta \ge 0$, $\vartheta \ne 0$ and $\|\vartheta\|_p \le 1$ such that

$$\|h\|_{p'} - \frac{\varepsilon}{2} \le \int_\Omega \|h\|_{X^*} \vartheta \, d\mu. \tag{4.2.7}$$

For each $k \in \mathbb{N}$, let $x_k \in X$ with $\|x_k\|_X = 1$ such that

$$\|x_k^*\|_{X^*} - \frac{\varepsilon}{2\|\vartheta\|_1} \le \langle x_k^*, x_k\rangle. \tag{4.2.8}$$

Let $f = \sum_{k\ge1} x_k \vartheta \chi_{A_k} \in L^p(\Omega, X)$. Then $\|f\|_p = \|\vartheta\|_p \le 1$ and, because of (4.2.7) and (4.2.8),

$$\begin{aligned}\int_\Omega \langle f, h\rangle \, d\mu &= \int_\Omega \vartheta \sum_{k\ge1} \langle x_k^*, x_k\rangle \chi_{A_k} \, d\mu \ge \int_\Omega \vartheta \sum_{k\ge1} \left(\|x_k^*\|_{X^*} - \frac{\varepsilon}{2\|\vartheta\|_1} \right) \chi_{A_k} \, d\mu \\ &= \int_\Omega \vartheta \|h\|_{X^*} \, d\mu - \frac{\varepsilon}{2} \ge \|h\|_{p'} - \varepsilon.\end{aligned}$$

Letting $\varepsilon \searrow 0$ we conclude that in this case $\|\eta_h\|_* = \|h\|_{p'}$; see (4.2.6).

Now suppose that $h \in L^{p'}(\Omega, X^*)$ is general, not necessarily countably valued. Then from the proof of Theorem 4.2.4 we know that there exists a sequence of countably valued functions such that $\|h_n - h\|_{p'} \to 0$. Moreover, we know that $\|\eta_h\|_* = \|h_n\|_{p'}$ for all $n \in \mathbb{N}$ and $\|\eta_{h_n} - \eta_h\|_* \le \|h_n - h\|_{p'} \to 0$ as $n \to \infty$. Therefore,

$$\|\eta_h\|_* = \lim_{n\to\infty} \|\eta_{h_n}\|_* = \lim_{n\to\infty} \|h_n\|_{p'} = \|h\|_{p'}.$$

So, we have proved that $L^p(\Omega, X^*)$ is contained isometrically as a subspace of $L^p(\Omega, X)^*$.

Now suppose that X^* has the RNP. Let $\beta \in L^p(\Omega, X)^*$ and consider $\xi: \Sigma \to X^*$ defined by $\langle \xi(A), x \rangle = \beta(x\chi_A)$ for all $A \in \Sigma$ and for all $x \in X$. Clearly, ξ is a vector valued measure. Let $\{A_k\}_{k=1}^n \subseteq \Sigma$ be a finite partition of Ω and let $\{x_k\}_{k=1}^n \subseteq \overline{B}_1^X$, that is, $\|x_k\|_X \leq 1$ for all $k = 1, \ldots, n$. Then it follows that

$$\left|\sum_{k=1}^n \langle \xi(A_k), x_k \rangle\right| = \left|\beta\left(\sum_{k=1}^n x_k \chi_{A_k}\right)\right| \leq \|\beta\|_* \left\|\sum_{k=1}^n x_k \chi_{A_k}\right\|_p$$
$$\leq \|\beta\|_* \left\|\sum_{k=1}^n \chi_{A_k}\right\|_p \leq \|\beta\|_* \mu(\Omega)^{\frac{1}{p}} .$$

Then $|\xi|(\Omega) < \infty$; see Definition 4.2.23 (c).

Since X^* has the RNP there exists $h \in L^1(\Omega, X^*)$ such that $\xi(A) = \int_A h \, d\mu$ for all $A \in \Sigma$. Note that if $f \in L^p(\Omega, X)$ is simple, then $\xi(f) = \int_\Omega \langle h, f \rangle \, d\mu$. Let $\{A_k\}_{k\geq 1} \subseteq \Sigma$ an increasing sequence such that $\Omega = \bigcup_{k\geq 1} A_k$ and $h|_{A_k}$ is bounded. We fix $k_0 \in \mathbb{N}$ and observe that $f \to \int_{A_{k_0}} \langle h, f \rangle \, d\mu$ is a bounded linear functional on $L^p(\Omega, X)$, which agrees with ξ on the simple functions, which are supported on A_{k_0}. It follows that

$$\beta(f\chi_{A_{k_0}}) = \int_\Omega \langle h\chi_{A_{k_0}}, f \rangle \, d\mu \quad \text{for all } f \in L^p(\Omega, X) .$$

Note that $h\chi_{A_{k_0}}$ is bounded, hence $h\chi_{A_{k_0}} \in L^{p'}(\Omega, X^*)$ and $\|h\chi_{A_{k_0}}\|_{p'} \leq \|\beta\|_*$. Since this is true for all $k_0 \in \mathbb{N}$, by the Monotone Convergence Theorem, we get that $h \in L^{p'}(\Omega, X^*)$. Then

$$\beta(f) = \lim_{n\to\infty} \int_\Omega \langle h, f\chi_{A_k} \rangle \, d\mu = \int_\Omega \langle h, f \rangle \, d\mu \quad \text{for all } f \in L^p(\Omega, X) . \qquad \square$$

Remark 4.2.27. In fact the converse is also true, namely if $L^p(\Omega, X)^* = L^{p'}(\Omega, X^*)$ with $1 \leq p < \infty$, then X^* has the RNP; see Diestel–Uhl [91, p. 99].

We can also state a vector valued version of the Dunford–Pettis Theorem; see Theorem 4.1.18.

Theorem 4.2.28. *Let (Ω, Σ, μ) be a finite measure space, X is a Banach space such that both X and X^* have the* RNP, *and let $\mathcal{F} \subseteq L^1(\Omega, X)$ satisfies the following conditions:*
(a) *$\mathcal{F}$ is bounded.*
(b) *$\mathcal{F}$ is uniformly integrable, that is, $\lim_{\mu(A)\to 0} \sup_{f\in\mathcal{F}} \int_A \|f\| \, d\mu = 0$.*
(c) *The set $\{\int_A f \, d\mu : f \in \mathcal{F}\}$ is relatively weakly compact for every $A \in \Sigma$.*

Then $\mathcal{F} \subseteq L^1(\Omega, X)$ is relatively weakly compact.

Proof. Let $\{f_n\}_{n\geq 1} \subseteq \mathcal{F}$. Invoking the Pettis Measurability Theorem (see Theorem 4.2.4), there exists a countable algebra $\mathcal{L} = \{A_k\}_{k\geq 1} \subseteq \Sigma$ such that if $\Sigma_1 = \sigma(\mathcal{L})$, then each f_n is Σ_1-measurable. By a diagonalization process based on condition (c) and using the Eberlein–Smulian Theorem, we produce a subsequence $\{f_{n_m}\}_{m\geq 1}$ of $\{f_n\}_{n\geq 1}$ such that

$$\mathrm{w}-\lim_{m\to\infty} \int_{A_k} f_{n_m}\, \mathrm{d}\mu \quad \text{exists for all } k \in \mathbb{N}\,.$$

Therefore,

$$\mathrm{w}-\lim_{m\to\infty} \int_{A} f_{n_m}\, \mathrm{d}\mu \quad \text{exists for all } A \in \mathcal{L}\,.$$

The condition (b) implies that $\{\int_A f_{n_m}\, \mathrm{d}\mu\}_{m\geq 1} \subseteq X$ is weakly a Cauchy sequence for all $A \in \Sigma_1$. Hence, condition (c) implies that we can define a set function $\xi\colon \Sigma_1 \to X$ by

$$\xi(A) = \mathrm{w}-\lim_{m\to\infty} \int_{A} f_{n_m}\, \mathrm{d}\mu \quad \text{for all } A \in \Sigma_1\,.$$

We see that $\lim_{\mu(A)\to 0}\langle x^*, \xi(A)\rangle = 0$ for each $x^* \in X^*$. Therefore ξ is weakly countably additive and so by the Orlicz–Pettis Theorem (see Remark 3.5.41), we know that ξ is a vector measure such that $\xi \ll \mu$.

Next we show that ξ is of bounded variation. One has

$$\|\xi(A)\|_X \leq \liminf_{m\to\infty} \left\| \int_A f_{n_m}\, \mathrm{d}\mu \right\| \quad \text{for all } A \in \Sigma_1\,.$$

So, if $\mathcal{P} \subseteq \Sigma_1$ is a finite partition of Ω, then, due to condition (a),

$$\begin{aligned} \sum_{A\in\mathcal{P}} \|\xi(A)\|_X &\leq \sum_{A\in\mathcal{P}} \liminf_{m\to\infty} \left\| \int_A f_{n_m}\, \mathrm{d}\mu \right\| \leq \liminf_{m\to\infty} \sum_{A\in\mathcal{P}} \left\| \int_A f_{n_m}\, \mathrm{d}\mu \right\| \\ &\leq \sup_{m\in\mathbb{N}} \sum_{A\in\mathcal{P}} \int_A \|f_{n_m}\|\, \mathrm{d}\mu = \sup_{m\geq 1} \|f_{n_m}\|_1 \leq \sup_{f\in\mathcal{F}} \|f\|_1 < \infty\,. \end{aligned}$$

Hence, ξ is of bounded variation.

Since X has the RNP there exists $f \in L^1(\Omega, \Sigma_1, X)$ such that

$$\xi(A) = \int_A f\, \mathrm{d}\mu \quad \text{for all } A \in \Sigma_1\,.$$

We need to show that $f_{n_m} \xrightarrow{\mathrm{w}} f$ in $L^1(\Omega, \Sigma, X)$ as $m \to \infty$. Hence we will also have weak convergence in $L^1(\Omega, X)$. Then according to the Eberlein–Smulian Theorem, $\mathcal{F} \subseteq L^1(\Omega, X)$ is relatively weakly compact.

Note that $\int_A f_{n_m}\, d\mu \xrightarrow{w} \int_A f\, d\mu$ for all $A \in \Sigma_1$. Hence, for every countably valued $h \in L^\infty(\Omega, \Sigma_1, X^*)$ we have $\int_\Omega \langle h, f_{n_m}\rangle\, d\mu \to \int_\Omega \langle h, f\rangle\, d\mu$. But countably valued functions are dense in $L^\infty(\Omega, \Sigma_1, X^*)$; see Proposition 4.2.22 (b). So, finally

$$\int_\Omega \langle h, f_{n_m}\rangle\, d\mu \to \int_\Omega \langle h, f\rangle\, d\mu \quad \text{for all } h \in L^\infty(\Omega, \Sigma_1, X^*)\,.$$

Thus, $f_{n_m} \xrightarrow{w} f$ in $L^1(\Omega, X)$. □

Next we examine what is the dual of $L^1(\Omega, X)$ when X is an arbitrary Banach space, that is, no condition is imposed on X^*; see Theorem 4.2.26.

Definition 4.2.29. Let (Ω, Σ, μ) be a σ-finite measure space and let X be a Banach space.

(a) Two functions $f, h\colon \Omega \to X^*$, which are w*-measurable are said to be **equivalent**, denoted by $f \sim h$, if $\langle f(w), x\rangle = \langle h(w), x\rangle$ μ-a. e. for all $x \in X$. The exceptional μ-null set depends on $x \in X$ in general. Evidently $\sim$ is an equivalence relation.

(b) By $L^\infty(\Omega, X^*_{w^*})$ we denote the linear space of the equivalence classes for the relation $\sim$ of w*-measurable functions $f\colon \Omega \to X^*$ such that

$$|\langle f(w), x\rangle| \le c\|x\| \quad \mu\text{-a. e., for all } x \in X \text{ and for some } c > 0\,.$$

The exceptional μ-null set may depend on $x \in X$. The infimum of all $c > 0$ is denoted by $\|f\|_{L^\infty(\Omega, X^*_{w^*})}$ and is a norm on $L^\infty(\Omega, X^*_{w^*})$.

Remark 4.2.30. If X is separable and $f \in L^\infty(\Omega, X^*_{w^*})$, then the function $w \to \|f(w)\|_{X^*}$ belongs to $L^\infty(\Omega)$ and it holds $\|f\|_{L^\infty(\Omega, X^*_{w^*})} = \operatorname{ess\,sup}_\Omega \|f(\cdot)\|_{X^*}$. Some authors denote the space $L^\infty(\Omega, X^*_{w^*})$ by $L^\infty_w(\Omega, X^*)$.

Example 4.2.31. Let (Ω, Σ, μ) be a nonatomic σ-finite measure space and let $X = l^2[0,1] = \{x = (x_\alpha)_{\alpha\in[0,1]} \in \mathbb{R}^{[0,1]}\colon \|x\|^2 = \sum_{0\le\alpha\le1} |x_\alpha|^2 < \infty\}$. This means that $x_\alpha = 0$ except for at most a countable number of indices; see Definition 3.5.40. This is a nonseparable Hilbert space and $L^\infty(\Omega, X^*_{w^*}) = L^\infty(\Omega, X_w)$ consists of all functions $w \to f(w) = (f_\alpha(w))_{\alpha\in[0,1]}$ with each f_α being Σ-measurable and essentially bounded with $\operatorname{ess\,sup}_\Omega |f_\alpha(\cdot)| \le M$ for all $\alpha \in [0,1]$. Consider the function $w \to e(w) = (e_\alpha(w))_{\alpha\in[0,1]}$, where

$$e_\alpha(w) = \begin{cases} 1 & \text{if } w = \alpha\,, \\ 0 & \text{otherwise}\,. \end{cases}$$

Then $e \sim 0$ but $\|e(w)\| = 1$ for all $w \in \Omega$. Therefore a function in the equivalence class of zero in $L^\infty(\Omega, X^*_{w^*})$ may be nonzero everywhere. If we multiply $e(w)$ with a scalar function $\vartheta(w)$, we obtain another element in the same class with norm $|\vartheta(w)|$. Hence $w \to \|f(w)\|_{X^*}$ need not be essentially bounded or even measurable for an element $f \in L^\infty(\Omega, X^*_{w^*})$.

The next remarkable result known as the "Lifting Theorem" eliminates the exceptional μ-null set from all elements in $L^\infty(\Omega)$ at once. For a proof of this result we refer to A. and C. Ionescu–Tulcea [178, Theorem IV. 3, p. 46]. In what follows, by $B(\Omega)$ we denote the space of all bounded functions $f: \Omega \to \mathbb{R}$ with the supremum norm.

Theorem 4.2.32 (Lifting Theorem). *If (Ω, Σ, μ) is a σ-finite space, then there exists a linear map $\rho: L^\infty(\Omega) \to B(\Omega)$ such that:*
(a) $\rho(f) \sim f$;
(b) $\rho(1) = 1$, *where* 1 *is the function identically* 1;
(c) $\rho(f)(w) \geq 0$ *for all* $w \in \Omega$ *if* $f(w) \geq 0$ μ*-a. e.*

*The map ρ is called a **linear lifting**.*

Proposition 4.2.33. *If (Ω, Σ, μ) is a σ-finite measure space, X is a Banach space, and $K \in L(L^1(\Omega), X^*)$, then there exists a unique $f \in L^\infty(\Omega, X^*_{w^*})$ such that*

$$\langle K(h), x\rangle = \int_\Omega h(w)\langle f(w), x\rangle \, d\mu \quad \textit{for all } h \in L^1(\Omega) \textit{ and for all } x \in X . \tag{4.2.9}$$

Moreover, $\|K\|_L = \|f\|_{L^\infty(\Omega, X^*_{w^*})}$. The map $S: L(L^1(\Omega), X^*) \to L^\infty(\Omega, X^*_{w^*})$ defined by $S(K) = f$ is linear and surjective, that is, every $f \in L^\infty(\Omega, X^*_{w^*})$ corresponds to a $K \in L(L^1(\Omega), X^*)$ via (4.2.9).

Proof. Let $x \in X$. Then $\eta_x(h) = \langle K(h), x\rangle$ is a linear functional on $L^1(\Omega)$ since

$$|\eta_x(h)| \leq \|K(h)\|_{X^*}\|x\|_X \leq \|K\|_L\|h\|_1\|x\|_X , \tag{4.2.10}$$

which shows that $\eta_x \in L^1(\Omega)^*$. Theorem 4.1.5 implies that there exists a unique $f_x \in L^\infty(\Omega)$ such that

$$\langle K(h), x\rangle = \int_\Omega hf_x \, d\mu \quad \text{for all } h \in L^1(\Omega) \quad \text{and} \quad \|f_x\|_\infty \leq \|K\|_L\|x\|_X ; \tag{4.2.11}$$

see (4.2.10). Note that $x \to f_x \in L^\infty(\Omega)$ is linear and bounded; see (4.2.11). Let ρ be the linear lifting from Theorem 4.2.32. Then $x \to \rho(f_x)(w)$ belongs to X^* for every $w \in \Omega$; see (4.2.11). Therefore there exists $f(w) \in X^*$ with $\|f(w)\|_{X^*} \leq \|K\|_L$ (see again (4.2.11)) such that

$$\langle f(w), x\rangle = \rho(f_x)(w) \quad \text{for all } x \in X . \tag{4.2.12}$$

Then $f \in L^\infty(\Omega, X^*_{w^*})$ and $\|f(w)\|_{X^*} \leq \|K\|_L$ for all $w \in \Omega$. Hence, because of (4.2.11) and (4.2.12),

$$\|f\|_{L^\infty(\Omega, X^*_{w^*})} \leq \|K\|_L \tag{4.2.13}$$

and

$$\langle K(h), x\rangle = \int_\Omega h(w)\langle f(w), x\rangle \, d\mu . \tag{4.2.14}$$

From (4.2.14) it follows that $\|K\|_L \le \|f\|_{L^\infty(\Omega, X^*_{w^*})}$, which becomes equality because of (4.2.13), that is, $\|K\|_L = \|f\|_{L^\infty(\Omega, X^*_{w^*})}$. Evidently S is linear.

Conversely, consider $f \in L^\infty(\Omega, X^*_{w^*})$. Then (4.2.9) defines a unique $K(h) \in X^*$ for every $h \in L^1(\Omega)$. In addition, $h \to K(h)$ is linear and $\|K\|_L \le \|f\|_{L^\infty(\Omega, X^*_{w^*})}$. Reasoning as in the first part of the proof, we obtain $\hat{f} \in L^\infty(\Omega, X^*_{w^*})$ with $\|\hat{f}\|_{L^\infty(\Omega, X^*_{w^*})} \le \|K\|_L$. Then $f - \hat{f}$ produce the zero operator (see (4.2.9)), hence $f \sim \hat{f}$. □

Remark 4.2.34. In fact from the proof above we have $\sup_{w\in\Omega} \|f(w)\|_{X^*} = \|K\|_L$.

So, we can state the following lifting theorem for $L^\infty(\Omega, X^*_{w^*})$.

Corollary 4.2.35. *If (Ω, Σ, μ) is a σ-finite measure space and X is a Banach space, then there exists a continuous linear map $\hat{\rho}: L^\infty(\Omega, X^*_{w^*}) \to L^\infty(\Omega, X^*_{w^*})$ such that*
(a) $\hat{\rho}(f) \sim f$;
(b) $\sup_{w\in\Omega} \|\hat{\rho}(f)(w)\|_{X^*} = \|f\|_{L^\infty(\Omega, X^*_{w^*})}$.

*This map $\hat{\rho}$ is called a **lifting** on $L^\infty(\Omega, X^*_{w^*})$.*

Remark 4.2.36. Evidently, $\hat{\rho}$ depends on the lifting ρ on $L^\infty(\Omega)$ stated in Theorem 4.2.32. Note that if X is separable, then $\|f\|_{L^\infty(\Omega, X^*)} = \|K\|_L$. In general, $L^\infty(\Omega, X^*_{w^*}) \ne L^\infty(\Omega, X^*)$ even if X is separable.

Now we are ready to characterize $L^1(\Omega, X)^*$ for an arbitrary Banach space X.

Theorem 4.2.37. *If (Ω, Σ, μ) is a σ-finite measure space and X is a Banach space, then $L^1(\Omega, X)^*$ is isometrically isomorphic to $L^\infty(\Omega, X^*_{w^*})$ and the duality pairing is given by*

$$\langle f, g\rangle = \int_\Omega \langle f(w), g(w)\rangle_X \, d\mu \quad \textit{for all } g \in L^1(\Omega, X),\ f \in L^\infty(\Omega, X^*_{w^*}) .$$

Proof. Let $f \in L^\infty(\Omega, X^*_{w^*})$. Then $\eta_f: L^1(\Omega, X) \to \mathbb{R}$ defined by $\eta_f(g) = \int_\Omega \langle f, g\rangle \, d\mu$ is bounded linear, hence $\eta_f \in L^1(\Omega, X)^*$ with $\|\eta_f\|_* \le \|f\|_{L^\infty(\Omega, X^*_{w^*})}$. We show that the opposite inequality also holds. So, let $\varepsilon > 0$. Then there exist $x \in X$ with $\|x\|_X = 1$ and $A \in \Sigma$ with $\mu(A) > 0$ such that

$$\|f\|_{L^\infty(\Omega, X^*_{w^*})} - \varepsilon \le \langle f(w), x\rangle_X \quad \text{for all } w \in A .$$

This implies

$$(\|f\|_{L^\infty(\Omega, X^*_{w^*})} - \varepsilon)\mu(A) \le \eta_f(\chi_A x) .$$

Hence, $\|f\|_{L^\infty(\Omega,X^*_{w^*})} \le \|\eta_f\|_*$ since $\varepsilon > 0$ is arbitrary and thus, $\|\eta_f\|_* = \|f\|_{L^\infty(\Omega,X^*_{w^*})}$.

Next we show that every element in $L^1(\Omega,X)^*$ is of the form η_f for some $f \in L^\infty(\Omega,X^*_{w^*})$. So let $\beta \in L^1(\Omega,X)^*$ and let $h \in L^1(\Omega)$. Then $\vartheta\colon X \to \mathbb{R}$ defined by $\vartheta(x) = \beta(hx)$ is linear and $|\vartheta(x)| = |\beta(hx)| \le \|\beta\|_*\|h\|_1\|x\|_X$, which gives

$$\vartheta \in X^* \quad \text{and} \quad \|\vartheta\|_{X^*} \le \|\beta\|_*\|h\|_1 . \tag{4.2.15}$$

Hence, there exists $x_h^* \in X^*$ such that $\beta(hx) = \langle x_h^*, x\rangle$ and $\|x_h^*\|_{X^*} \le \|\beta\|_*\|h\|_1$; see (4.2.15). Consider the map $K\colon L^1(\Omega) \to X^*$ defined by $K(h) = x_h^*$. Then $K \in L(L^1(\Omega),X^*)$ and $\|K\|_L \le \|\beta\|_*$. Invoking Proposition 4.2.33 there exists a unique $f \in L^\infty(\Omega,X^*_{w^*})$ such that

$$\beta(hx) = \langle x_h^*, x\rangle = \int_\Omega h(w)\langle f(w),x\rangle_X \, d\mu \quad \text{for all } h \in L^1(\Omega) \text{ and for all } x \in X .$$

Thus, $\|\beta\|_* = \|f\|_{L^\infty(\Omega,X^*_{w^*})}$. Finally if we take $\{h_k\}_{k=1}^n \subseteq L^1(\Omega)$ and $\{x_k\}_{k=1}^n \subseteq X$, then

$$\beta\left(\sum_{k=1}^n h_k x_k\right) = \int_\Omega \left\langle f(w), \sum_{k=1}^n h_k(w)x\right\rangle_X d\mu .$$

But such functions are dense in $L^1(\Omega,X)$. Therefore,

$$\beta(g) = \int_\Omega \langle f(w), g(w)\rangle_X \, d\mu \quad \text{for all } g \in L^1(\Omega,X) . \qquad \square$$

Remark 4.2.38. If X^* has the RNP, for example when X^* is separable or when X is reflexive, then $L^\infty(\Omega,X^*_{w^*}) = L^\infty(\Omega,X^*)$.

Next we introduce a notion that is a useful tool in the theory of evolution equations.

Definition 4.2.39. A triple (X,H,X^*) of spaces is called an **evolution triple** (or **Gelfand triple**) if the following properties hold:
(a) $X \subseteq H$ and X^* is the dual of X.
(b) X is a separable reflexive Banach space.
(c) H is a separable Hilbert space that is identified with its dual, pivot space; see Theorem 3.5.21.
(d) X is embedded continuously and densely in H.

Remark 4.2.40. We can easily check that property (d) implies that $H^* = H$ is embedded into X^* continuously. Moreover, the reflexivity of X implies that the embedding $H \hookrightarrow X^*$ is also dense. So, we have $X \hookrightarrow H \hookrightarrow X^*$ with all embeddings being continuous and dense. If $2 \le p < \infty$, then $X = L^p[0,1]$, $H = L^2[0,1]$, $X^* = L^{p'}[0,1]$ with $1/p + 1/p' = 1$ is an evolution triple. Other evolution triples, useful in partial differential equations, can be produced using Sobolev functions; see Section 4.5.

In what follows, we denote by $\langle\cdot,\cdot\rangle$ the duality brackets for the pair (X^*, X) and by $(\cdot,\cdot)$ we denote the inner product of H. Moreover by $\|\cdot\|, |\cdot|, \|\cdot\|_*$ we denote the norms of X, H, X^*, respectively. We easily see that

$$\langle\cdot,\cdot\rangle|_{H\times X} = (\cdot,\cdot), \quad \|\cdot\|_* \le c_1|\cdot| \quad \text{and} \quad |\cdot| \le c_2\|\cdot\| \quad \text{for some } c_1, c_2 > 0\,. \tag{4.2.16}$$

Definition 4.2.41. Let $T = [0, b]$ and let X, Y be Banach spaces with $X \subseteq Y$ and $u \in L^1(T, X)$. Then the **distributional derivative** $(\mathrm{d}u)/(\mathrm{d}t)$ of u is understood as the linear operator $(\mathrm{d}u)/(\mathrm{d}t) \in L(C_c^\infty(0, b), Y)$ defined by

$$\frac{\mathrm{d}u}{\mathrm{d}t}(\varphi) = -\int_0^b u\frac{\mathrm{d}\varphi}{\mathrm{d}t}\,\mathrm{d}t \quad \text{for all } \varphi \in C_c^\infty(0, b)\,.$$

Here $C_c^\infty(0, b)$ denotes the space of all C^∞ functions with compact support. We write $(\mathrm{d}u)/(\mathrm{d}t) = u'$ and if $u' \in L^1(T, Y)$, then $\int_0^b \varphi' u\,\mathrm{d}t = -\int_0^b \varphi u'\,\mathrm{d}t$ for all $\varphi \in C_c^\infty(0, b)$.

The next proposition is an immediate consequence of this definition.

Proposition 4.2.42. *If (X, H, X^*) is an evolution triple, $1 \le p < \infty$ and $u \in L^p(T, X)$, then the distributional derivative $u' \in L^{p'}(T, X^*) = L^p(T, X)^*$ with $1/p + 1/p' = 1$ exists if and only if there exists $v \in L^{p'}(T, X^*)$ such that*

$$\int_0^b (u(t), x)\varphi'(t)\,\mathrm{d}t = -\int_0^b \langle v(t), x\rangle\varphi(t)\,\mathrm{d}t \quad \textit{for all } x \in X \textit{ and for all } \varphi \in C_c^\infty(0, b)\,.$$

The distributional derivative is uniquely defined and $u' = v$.

Definition 4.2.43. Let $T = [0, b]$ and let X, Y be Banach spaces such that $X \subseteq Y$ and let $1 \le p, q$. The space $W_{pq}(T, X, Y)$ is defined by

$$W_{pq}(T, X, Y) = \{u \in L^p(T, X) : u' \in L^q(T, Y)\}\,,$$

where u' denotes the distributional derivative of u; see Definition 4.2.41. The space $W_{pq}(T, X, Y)$ is equipped with the norm

$$\|u\|_W = \|u\|_{L^p(T,X)} + \|u'\|_{L^q(T,Y)}\,.$$

This is clearly a Banach space. If $X = Y$ and $p = q$, then $W_{pp}(T, X, X) = W^{1,p}((0, b), X)$ and this is a vector valued Sobolev space; see Section 4.5. Finally when $Y = X^*$ and $q = p'$ with $1/p + 1/p' = 1$, then we write $W_p(0, b) = W_{pp'}(T, X, X^*)$.

Remark 4.2.44. If X, Y are separable and $1 < p, q < +\infty$, then $W_{pq}(0, b)$ is a separable and reflexive Banach space.

Proposition 4.2.45. *If $T = [0, b]$ and if X, Y are Banach spaces with $X \hookrightarrow Y$ continuously and $1 \leq p, q$, then $W_{pq}(0, b) \hookrightarrow C(T, Y)$ continuously.*

Proof. Let $u \in W_{pq}(0, b)$. Then from Definition 4.2.43 we know that u' is Bochner integrable. We set $v(t) = \int_0^1 u'(s)\,ds$. Then

$$\|v(t) - v(\tau)\|_Y = \left\| \int_\tau^t u'(s)\,ds \right\|_Y \leq \int_\tau^t \|u'(s)\|_Y\,ds\,.$$

Hence $v\colon T \to Y$ is continuous. But $v = u + y$ with $y \in Y$. Therefore $u\colon T \to Y$ is continuous as well. From Hölder's inequality we get

$$\|v(t)\|_Y \leq \int_0^b \|u'(t)\|_Y\,dt = \|u'\|_{L^1(T,Y)} \leq c_1 \|u'\|_{L^q(T,Y)} \tag{4.2.17}$$

for some $c_1 > 0$. Taking into account (4.2.16), (4.2.17) as well as the fact that $X \hookrightarrow Y$ continuously, we obtain

$$\begin{aligned} \|y\|_Y &= \frac{1}{b^{\frac{1}{p}}} \left[\int_0^b \|y\|_Y^p\,dt \right]^{\frac{1}{p}} = \frac{1}{b^{\frac{1}{p}}} \|v - u\|_{L^p(T,Y)} \\ &\leq \frac{1}{b^{\frac{1}{p}}} \left[\|v\|_{L^p(T,Y)} + \|u\|_{L^p(T,Y)} \right] \\ &\leq \frac{1}{b^{\frac{1}{p}}} \left[b^{\frac{1}{b}} c_1 \|u'\|_{L^q(T,Y)} + c_2 \|u\|_{L^p(T,Y)} \right] \quad \text{for some } c_2 > 0\,. \end{aligned} \tag{4.2.18}$$

Then, applying Hölder's inequality in (4.2.17) and using (4.2.18), we derive

$$\begin{aligned} \|u\|_{C(T,Y)} &= \sup_{t\in T} \|v(t) - y\|_Y = \sup_{t\in T} \left\| \int_0^t u'(s)\,ds - y \right\|_Y \leq \int_0^b \|u'(s)\|_Y\,ds + \|y\|_Y \\ &\leq 2c_1 \|u'\|_{L^q(T,Y)} + \frac{c_2}{b^{\frac{1}{b}}} \|u\|_{L^p(T,Y)} \leq c_3 \|u\|_W \quad \text{for some } c_3 > 0\,. \end{aligned}$$

Thus, $W_{pq}(0, b) \hookrightarrow C(T, Y)$ continuously. □

Proposition 4.2.46. *If (X, H, X^*) is an evolution triple, then $C^1(T, X)$ is dense in $W_p(0, b)$ and $W_p(0, b) \hookrightarrow C(T, H)$ continuously and densely.*

Proof. Let $a < 0 < b < d$ and let $u \in W_p(0, b)$. We extend u to $(a, 0)$ and to (b, d) by symmetry. Let $\varphi \in C_c^\infty(a, d)$ with $\varphi|_T = 1$ and set $\hat{u} = \varphi u$. Evidently $\hat{u} \in W_p(a, d)$ and $\hat{u}|_T = u$. Moreover, we get

$$\|u\|_{W_p(0,b)} \leq \|\hat{u}\|_{W_p(a,d)} \leq c(\varphi) \|u\|_{W_p(0,b)}$$

with $c(\varphi) > 0$ depending on the test function φ. Note that $\hat{u}$ vanishes on some neighborhoods of a and d. Let ϑ be the standard mollifier (see Definition 4.1.27 (c)), and let $\{u^m = f_{1/m} * u\}_{m\geq 1}$ be the regularizations of u; see Definition 4.1.27 (d). Then, as in Proposition 4.1.28, one has $u^m \in C_c^\infty((a,d),X)$ for every $m \geq 1$ large enough and $u^m \to \hat{u}$ in $W_p(a,d)$ with $\|u^m\|_{W_p(a,d)} \leq \|\hat{u}\|_{W_p(a,d)}$. It follows that $C^1(T,X)$ is dense in $W_p(a,b)$. Moreover, for every $m, n \in \mathbb{N}$, we infer that

$$\frac{1}{2}\frac{\mathrm{d}}{\mathrm{d}t}|u^m(t) - u^n(t)|^2 = ((u^m)'(t) - (u^n)'(t), u^m(t) - u^n(t))\,.$$

Then, by applying (4.2.16), it results in

$$\begin{aligned}\frac{1}{2}|u^m(t) - u^n(t)|^2 &= \int_a^t \langle (u^m)'(s) - (u^n)'(s), u^m(s) - u^n(s)\rangle \,\mathrm{d}s \\ &\leq \int_a^t \|(u^m)'(s) - (u^n)'(s)\|_{X^*} \|u^m(s) - u^n(s)\|_X \,\mathrm{d}s \\ &\leq \frac{1}{2}\|u^m - u^n\|_{W_p(a,d)} \quad \text{for all } t \in (a,d)\,.\end{aligned}$$

Hence, $\{u^m\}_{m\geq 1}$ is a Cauchy sequence in $C([a,d],H)$. Therefore $u^m \to \hat{u}$ in $C([a,d],H)$ as $m \to \infty$ and $\|\hat{u}\|_{C([a,d],H)} \leq \|\hat{u}\|_{W_p(a,d)}$. So, we conclude that $u \in C(T,H)$. More precisely, there is a class representative with this property, and $W_p(0,b) \hookrightarrow C(T,H)$ continuously and of course densely. □

An interesting byproduct of the proof above is the following **integration by parts** formula.

Corollary 4.2.47. *If (X,H,X^*) is an evolution triple and $u, v \in W_p(0,b)$, then $\mathrm{d}/(\mathrm{d}t)(u(t),v(t)) = \langle u'(t), v(t)\rangle + \langle u(t), v'(t)\rangle$ for a. a. $t \in T$.*

The embedding of $W_p(0,b)$ into $C(T,H)$ is not compact in general. However we can prove compact embedding of $W_p(0,b)$ into $L^p(T,H)$. This is a particular case of Theorem 4.2.49 below. First we need a interpolation-type lemma known as "Ehrling's inequality".

Lemma 4.2.48 (Ehrling's inequality). *If X, Y, V are Banach spaces such that $X \subseteq Y \subseteq V$ with the embedding of X into Y being compact and the embedding of Y into V being continuous, then, for a given $\varepsilon > 0$ there exists $c(\varepsilon) > 0$ such that*

$$\|x\|_Y \leq \varepsilon\|x\|_X + c(\varepsilon)\|x\|_V \quad \text{for all } x \in X\,. \tag{4.2.19}$$

Proof. Suppose that (4.2.19) is false. Then there exists $\varepsilon > 0$ and a sequence $\{x_n\}_{n\geq 1} \subseteq X$ such that

$$\|x_n\|_Y > \varepsilon\|x_n\|_X + n\|x_n\|_V \quad \text{for all } n \in \mathbb{N}\,.$$

We set $u_n = x_n/\|x_n\|_X$ for all $n \in \mathbb{N}$. Then $\|u_n\|_X = 1$ for all $n \in \mathbb{N}$ and

$$\|u_n\|_Y \geq \varepsilon + n\|u_n\|_V \quad \text{for all } n \in \mathbb{N}\,. \tag{4.2.20}$$

The set $\{u_n\}_{n\geq 1} \subseteq X$ is bounded. Since by hypothesis $X \hookrightarrow Y$ compactly and $Y \hookrightarrow V$ continuously, by passing to a suitable subsequence if necessary, we may assume that

$$u_n \to u \quad \text{in } Y \text{ and in } V \text{ as } n \to \infty\,. \tag{4.2.21}$$

From (4.2.20) and (4.2.21) we infer that $\|u\|_V = 0$ and $\|u\|_Y \geq \varepsilon > 0$, a contradiction. ☐

Using this lemma, we can prove the following theorem concerning the embedding of $W_p(0,b)$ into $L^p(T,H)$.

Theorem 4.2.49. *If X, Y, V are Banach spaces, X, V are reflexive, $X \subseteq Y \subseteq V$, the embedding of X into Y is compact, the embedding of Y into V is continuous, and $1 < p < \infty$ as well as $1 \leq q < \infty$, then the embedding $W_{pq}(T,X,V) \hookrightarrow L^p(T,Y)$ is compact.*

Proof. We need to show that bounded sets in $W_{pq}(T,X,V)$ are relatively compact in $L^p(T,Y)$. So let $\{u_n\}_{n\geq 1} \subseteq W_{pq}(T,X,V)$ be bounded. Then $\{u_n\}_{n\geq 1} \subseteq L^p(T,X)$ is bounded and $L^p(T,X)$ is reflexive; see Proposition 4.2.22 (d). Therefore, we may assume that

$$u_n \xrightarrow{\text{w}} u \quad \text{in } L^p(T,X) \text{ as } n \to \infty\,. \tag{4.2.22}$$

Recall that $L^q(T,V) \subseteq L^1(T,V)$. So we obtain

$$\{u_n\}_{n\geq 1} \subseteq L^1(T,V) \quad \text{is bounded}\,. \tag{4.2.23}$$

Without any loss of generality we may assume that $u = 0$; otherwise replace u_n by $u_n - u$. Applying Lemma 4.2.48 we see that for given $\varepsilon > 0$

$$\|u_n\|^p_{L^p(T,Y)} \leq \varepsilon\|u_n\|^p_{L^p(T,X)} + c(\varepsilon)\|u_n\|^p_{L^p(T,V)} \quad \text{for all } n \in \mathbb{N}\,. \tag{4.2.24}$$

Pick $\delta \in (0, b/2]$. For $t \in [0, b/2]$ we can write

$$u_n(t) = \tilde{u}_n(t) + y_n(t) \quad \text{with } \tilde{u}_n(t) = \frac{1}{\delta}\int_0^\delta u_n(t+s)\,\mathrm{d}s\,. \tag{4.2.25}$$

Note that, by using integration by parts,

$$\begin{aligned}\int_0^\delta \left(\frac{s}{\delta} - 1\right)\frac{\mathrm{d}}{\mathrm{d}s}u_n(t+s)\,\mathrm{d}s &= \left[\frac{s}{\delta} - 1\right]u_n(t+s)\Big|_0^\delta - \int_0^\delta \frac{1}{\delta}u_n(t+s)\,\mathrm{d}s \\ &= u_n(t) - \tilde{u}_n(t) = y_n(t) \quad \text{for } t \in \left[0, \frac{b}{2}\right].\end{aligned}$$

Hence,

$$\int_0^{\frac{b}{2}} \|u_n(t)\|_V^p \, dt \le 2^{p-1}\left[\int_0^{\frac{b}{2}} \|\tilde{u}_n(t)\|_V^p \, dt + \int_0^{\frac{b}{2}} \|y_n(t)\|_V^p \, dt\right]. \tag{4.2.26}$$

Moreover, we have the following estimate:

$$\int_0^{\frac{b}{2}} \|y_n(t)\|_V^p \, dt \le \int_0^{\frac{b}{2}} \left[\int_0^{\delta} \left(1 - \frac{s}{\delta}\right) \|u_n'(t+s)\|_V \, ds\right]^p dt = \| \|u_n'\|_V * \eta_\delta \|^p_{L^p[0,\frac{b}{2}]},$$

where $*$ denotes convolution and $\eta_\delta(t) = [t/\delta + 1]\chi_{(-\delta,0]}(t)$. Then, taking Remark 4.1.33 into account, one has

$$\| \|u_n'\|_V * \eta_\delta \|_{L^p[0,\frac{b}{2}]} \le \|u_n'\|_{L^1([0,\frac{b}{2}],V)} \|\eta_\delta\|_p .$$

So, finally we obtain, due to (4.2.23), that

$$\int_0^{\frac{b}{2}} \|y_n(t)\|_V^p \, dt \le \|u_n'\|^p_{L^1([0,b],V)} \sqrt{\delta} \le \hat{c}\sqrt{\delta} \tag{4.2.27}$$

for some $\hat{c} > 0$ and for all $n \in \mathbb{N}$. From (4.2.22), (4.2.25) and recalling that $u = 0$, we have

$$\tilde{u}_n(t) \xrightarrow{w} 0 \quad \text{in } X \text{ for all } t \in T .$$

Since $X \hookrightarrow Y$ is compact and $Y \hookrightarrow V$ is continuous, we get

$$\tilde{u}_n(t) \to 0 \quad \text{in } Y \text{ for all } t \in T \quad \text{and} \quad \tilde{u}_n(t) \to 0 \quad \text{in } V \text{ for all } t \in T .$$

This gives

$$\|\tilde{u}_n(t)\|_V \le c_0 \|\tilde{u}_n(t)\|_X \le \frac{c_0}{\delta} \int_0^{\delta} \|u_n(t+s)\|_X \, ds \le \frac{c_0}{\delta} b^{\frac{1}{p'}} \|u_n\|_{L^p(T,X)}$$

for some $c_0 > 0$ and for all $n \in \mathbb{N}$. Hence $\{\tilde{u}_n(t)\}_{n\ge 1} \subseteq V$ is bounded uniformly for $t \in T$. Then, from the Lebesgue Dominated Convergence Theorem, we obtain

$$\int_0^{\frac{b}{2}} \|\tilde{u}_n(t)\|_V^p \, dt \to 0 \quad \text{as } n \to \infty ,$$

which, because of (4.2.26) and (4.2.27), results in

$$\int_0^{\frac{b}{2}} \|u_n(t)\|_V^p \, dt \to 0 \quad \text{as } n \to \infty .$$

In a similar way we show that

$$\int_{\frac{b}{2}}^{b} \|u_n(t)\|_V^p \, dt \to 0 \quad \text{as } n \to \infty ,$$

which implies

$$\|u_n\|_{L^p(T,V)}^p \to 0 \quad \text{and} \quad \|u_n\|_{L^p(T,Y)}^p \to 0 \quad \text{as } n \to \infty ;$$

see (4.2.24) and (4.2.22), and recall that $\varepsilon > 0$ is arbitrary. □

Proposition 4.2.50. *If X, Y are Banach spaces, X is reflexive, $X \hookrightarrow Y$ is continuous, and $u \in L^\infty(T,X) \cap C(T,Y_w)$, then $u \in C(T,X_w)$, where X_w and Y_w denote the spaces X and Y endowed with their weak topologies, respectively.*

Proof. By replacing Y with $\overline{X}^{\|\cdot\|_Y}$ if necessary we may assume that the embedding of X into Y is also dense. Then $Y^* \hookrightarrow X^*$ is continuous and dense since X is reflexive; see Remark 4.2.40.

Let $t_n \to$ in T. Since $u \in C(T,Y_w)$, we have

$$u(t_n) \xrightarrow{w} u(t) \quad \text{in } Y \text{ as } n \to \infty .$$

First we show that $u(t) \in X$ for all $t \in T$ and that $\|u(t)\|_X \le \|u\|_{L^\infty(T,X)}$ for all $t \in T$. To this end, we extend the function u by zero outside T and denote this extension by $\hat{u}$. Using mollification we regularize $\hat{u}$ and obtain a sequence $\{u_n\}_{n\ge1} \subseteq C^1(T,X)$ such that

$$\begin{aligned} &\|u_n(t)\| \le \|u\|_{L^\infty(T,X)} \quad \text{for all } t \in T \text{ and for all } n \in \mathbb{N} , \\ &u_n(t) \xrightarrow{w} u(t) \quad \text{in } Y \text{ and for all } t \in T . \end{aligned} \tag{4.2.28}$$

Using (4.2.28) one has for all $y^* \in X^*$ and for all $t \in T$

$$\begin{aligned} |\langle y^*, u_n(t)\rangle_Y| = |\langle y^*, u_n(t)\rangle_X| &\le \|y^*\|_{X^*} \|u_n(t)\|_X \\ &\le \|y^*\|_{X^*} \|u_n\|_{L^\infty(T,X)} \le \|y^*\|_{X^*} \|u\|_{L^\infty(T,X)} . \end{aligned} \tag{4.2.29}$$

The density of Y^* in X^* and (4.2.29) imply that

$$u(t) \in X \quad \text{and} \quad \|u(t)\|_X \le \|u\|_{L^\infty(T,X)} \quad \text{for all } t \in T . \tag{4.2.30}$$

Let $x^* \in X^*$. The density of Y^* in X^* implies that there exists $\{y_m^*\}_{m\ge1} \subseteq Y^*$ such that $y_m^* \to x^*$ in X^* as $m \to \infty$. Thanks to (4.2.30) one gets

$$\langle y_m^*, u(t_n)\rangle_X \to \langle y_m^*, u(t)\rangle_X \quad \text{as } n \to \infty \text{ and for all } m \in \mathbb{N}$$

and

$$\langle y_m^*, u(t)\rangle_X \to \langle x^*, u(t)\rangle_X \quad \text{as } m \to \infty\,.$$

So, we can find a sequence $\{m(n)\}_{n\geq 1}$ not necessarily strictly increasing such that $m(n) \to \infty$ as $n \to \infty$ and

$$\langle y_{m(n)}^*, u(t_n)\rangle_X \to \langle x^*, u(t)\rangle_X \quad \text{as } n \to \infty\,. \tag{4.2.31}$$

Finally, because of (4.2.30) and (4.2.31), we infer that

$$\begin{aligned}
&|\langle x^*, u(t_n)\rangle_X - \langle x^*, u(t)\rangle_X| \\
&\leq |\langle x^*, u(t_n)\rangle_X - \langle y_{m(n)}^*, u(t_n)\rangle_X| + |\langle y_{m(n)}^*, u(t_n)\rangle_X - \langle x^*, u(t)\rangle_X| \\
&\leq \|x^* - y_{m(n)}^*\|_{X^*}\|u\|_{L^\infty(T,X)} + |\langle y_{m(n)}^*, u(t_n)\rangle_X - \langle x^*, u(t)\rangle_X| \to 0
\end{aligned}$$

as $n \to \infty$. Hence, $u \in C(T, X_w)$. □

We conclude this section with a brief mention of another weaker integral for Banach space valued functions. This is called the **Pettis integral**.

Definition 4.2.51. Let (Ω, Σ, μ) be a measure space and X a Banach space with dual X^*. Suppose that $f\colon \Omega \to X$ is weakly measurable and $\langle x^*, f(\cdot)\rangle \in L^1(\Omega)$ for all $x^* \in X^*$ We call such functions **weakly integrable**. We say that f is **Pettis integrable** if for all $A \in \Sigma$ there exists $x_A \in X$ such that

$$\langle x^*, x_A\rangle = \int_A \langle x^*, f(w)\rangle \, d\mu \quad \text{for all } x^* \in X^*\,.$$

We write $x_A = \mathrm{P} - \int_A f \, d\mu$ and call it the **Pettis integral** of f.

Remark 4.2.52. Clearly a Bochner integrable function is Pettis integrable and the two integrals coincide. The converse is not true in general.

Proposition 4.2.53. *If (Ω, Σ, μ) is a finite nonatomic measure space and X is a Banach space, then the following statements are equivalent:*
(a) *X is finite dimensional.*
(b) *Every Pettis integrable function is also Bochner integrable.*
(c) *Every strongly measurable and Pettis integrable function is Bochner integrable.*

Proof. (a) $\Longrightarrow$ (b): Let $\xi(A) = \mathrm{P} - \int_A f \, d\mu$ for all $A \in \Sigma$. This is a vector measure of bounded variation, $\xi \ll \mu$ and so by the Radon–Nikodym Theorem it holds that $f \in L^1(\Omega, X)$.

(b) $\Longrightarrow$ (c): This is clear.

(c) $\Longrightarrow$ (a): Suppose that $\dim X = +\infty$. Then according to the Dvoretzky–Rogers Theorem (see Remark 3.5.41), there exists a sequence $\{x_n\}_{n\geq 1} \subseteq X$ such that

$$\sum_{n\geq 1} x_n \text{ is unconditionally convergent and } \sum_{n\geq 1} \|x_n\| = +\infty .$$

Let $\{A_n\}_{n\geq 1} \subseteq \Sigma$ be a partition of Ω with $\mu(A_n) > 0$ for all $n \in \mathbb{N}$. Recall that μ is nonatomic. Then $f = \sum_{n\geq 1} x_n/(\mu(A_n))\chi_{A_n}$ is strongly measurable and Pettis integrable, but $f \notin L^1(\Omega, X)$; see Proposition 4.2.12. ☐

The next proposition describes an essential difference between Bochner and Pettis integrable functions.

Proposition 4.2.54. *If (Ω, Σ, μ) is a finite measure space, X is a Banach space, and $f: \Omega \to X$ is a function such that $f = \sum_{n\geq 1} x_n\chi_{A_n}$ with $\{x_n\}_{n\geq 1} \subseteq X$ and $\{A_n\}_{n\geq 1} \subseteq \Sigma$ pairwise disjoint. Then the following hold:*
(a) *f is Bochner integrable if and only if $\sum_{n\geq 1} x_n\mu(A_n)$ is absolutely convergent in X.*
(b) *f is Pettis integrable if and only if $\sum_{n\geq 1} x_n\mu(A_n)$ is unconditionally convergent in X.*

In both cases the integral over $A \in \Sigma$ equals $\sum_{n\geq 1} x_n\mu(A \cap A_n)$.

Definition 4.2.55. Let (Ω, Σ, μ) be a finite measure space and let X be a Banach space.
(a) Two weakly measurable functions $f, h: \Omega \to X$ are said to be **weakly equivalent** if

$$\langle x^*, f(w)\rangle = \langle x^*, h(w)\rangle \quad \mu\text{-a. e. and for all } x^* \in X^* .$$

The exceptional μ-null set depends on $x^* \in X^*$. By $\sim$ we denote the equivalence relation of weak equivalence of functions as above.
(b) Let $\mathcal{P}(\mu, X)$ be the space of all Pettis integrable functions and set $P(\mu, X) = \mathcal{P}(\mu, X)/\sim$. This is a normed space when equipped with the norm

$$\|f\|_{\mathrm{Pe}} = \sup\left[\int_\Omega |\langle x^*, f\rangle| \, d\mu : x^* \in X^*, \|x\|_* \leq 1\right] .$$

Remark 4.2.56. This norm satisfies

$$\sup_{A\in\Sigma}\left\|\int_A f \, d\mu\right\| \leq \|f\|_{\mathrm{Pe}} \leq 2\sup_{A\in\Sigma}\left\|\int_A f \, d\mu\right\| \quad \text{for all } f \in P(\mu, X) .$$

Moreover we have $\|f\|_{\mathrm{Pe}} \leq \|f\|_1$, and they are not equivalent unless X is finite dimensional.

4.3 Functions of Bounded Variations

We start by examining monotone functions.

Definition 4.3.1. Let $A \subseteq \mathbb{R}$ and consider a function $f: A \to \mathbb{R}$.

(a) We say that f is **increasing** (resp. **decreasing**) if for all $x, u \in A$ with $x < u$ we have $f(x) \le f(u)$ (resp. $f(u) \le f(x)$).

(b) We say that f is **strictly increasing** (resp. **strictly decreasing**) if the inequalities from (a) are strict.

Remark 4.3.2. An increasing or decreasing function is called **monotone**. Similarly a strictly increasing or strictly decreasing function is called **strictly monotone**. Of course strictly monotone functions are monotone.

A monotone function need not be continuous. The next proposition tell us how discontinuous they can be.

Proposition 4.3.3. *If $T \subseteq \mathbb{R}$ is an interval and $f: T \to \mathbb{R}$ is a monotone function, then the set of discontinuity points of f is countable.*

Proof. First suppose that $T = [a, b]$ and that f is increasing. The reasoning is similar if f is decreasing. Let $x \in (a, b)$. Then the following limits exist:

$$f_+(x) = \lim_{u \to x^+} f(u) \quad \text{and} \quad f_-(x) = \lim_{u \to x^-} f(u) .$$

Obviously it holds that $s(x) = f_+(x) - f_-(x) \ge 0$ for all $x \in (a, b)$. This is the "jump" of f at x. Evidently f is continuous at x if and only if $s(x) = 0$. For each $n \in \mathbb{N}$, let $A_n = \{x \in (a, b): s(x) \ge 1/n\}$. If $x_1, \ldots, x_m \in A_n$, then it follows that

$$f(b) - f(a) \ge \sum_{k=1}^{m} [f_+(x_k) - f_-(x_k)] \ge \frac{m}{n} ,$$

which implies that A_n is finite for all $n \in \mathbb{N}$ and so $\bigcup_{n \ge 1} A_n$ is countable. But this union is clearly the set of discontinuity points of f.

Now suppose that T is an arbitrary interval. Let $T_n = [a_n, b_n]$ with $n \in \mathbb{N}$ such that

$$a_n \searrow \inf T \quad \text{and} \quad b_n \nearrow \sup T \quad \text{as } n \to \infty .$$

On each T_n the set D_n of discontinuity points of $f|_{T_n}$ is countable. Hence $\bigcup_{n \ge 1} D_n$ is countable and is the set of discontinuity points of f on T. □

There is a kind of converse to this proposition.

Proposition 4.3.4. *If $D \subseteq \mathbb{R}$ is a countable subset, then there exists a monotone function $f: \mathbb{R} \to \mathbb{R}$, which has D as the set of its discontinuity points.*

Proof. If $D \subseteq \mathbb{R}$ is finite, then the construction of f is evident. So assume that D is countably infinite, that is, $D = \{x_n\}_{n\geq 1}$. For each $n \in \mathbb{N}$ let $f_n\colon \mathbb{R} \to \mathbb{R}$ be defined by

$$f_n(x) = \begin{cases} -\frac{1}{n^2} & \text{if } x < x_n , \\ \frac{1}{n^2} & \text{if } x \geq x_n . \end{cases}$$

Then f has only one discontinuity point at x_n. We define

$$f(x) = \sum_{n\geq 1} f_n(x) \quad \text{for all } x \in \mathbb{R} .$$

Since $|f_n(x)| = 1/n^2$ for all $x \in \mathbb{R}$, by the Weierstrass M-test, $\sum_{n\geq 1} f_n(x)$ converges uniformly. Hence, f is continuous at all $x \in \mathbb{R}$ where each f_n is continuous. Therefore f is continuous on $\mathbb{R} \setminus D$. □

Corollary 4.3.5. *There exists an increasing function $f\colon \mathbb{R} \to \mathbb{R}$ that is continuous at all $x \in \mathbb{R} \setminus \mathbb{Q}$ and discontinuous at all $x \in \mathbb{Q}$.*

The previous considerations lead to the following definition.

Definition 4.3.6. Let $T \subseteq \mathbb{R}$ be an interval and let $f\colon T \to \mathbb{R}$ be an increasing function. The function

$$s_f(x) = \sum_{\substack{u\in T \\ u<x}} [f_+(u) - f_-(u)] + f(x) - f_-(x) \quad \text{for all } x \in T$$

with $f_-(\inf T) = f(\inf T)$ if $\inf T \in T$ and $f_+(\sup T) = f(\sup T)$ if $\sup T \in T$, is the **jump or saltus function** of f.

Remark 4.3.7. Note that s_f is increasing and is constant on any open interval on which f is continuous. Moreover it is easy to check that $h(x) = f(x) - s_f(x)$ is an increasing and continuous function on T.

Next we examine the inverse of an increasing function.

Proposition 4.3.8. *If $T \subseteq \mathbb{R}$ is an interval bounded from below, $f\colon T \to \mathbb{R}$ is an increasing function, $I \subseteq \mathbb{R}$ is the smallest interval containing $f(T)$ and $h\colon I \to \mathbb{R}$ is defined by*

$$h(u) = \inf[x \in T : f(x) \geq u] \quad \textit{for all } u \in I ,$$

then the following hold:
(a) *h is increasing and left continuous;*
(b) *h jumps at some $u_0 \in I \setminus \{\sup_T f\}$ if and only if $f(x) = u_0$ for all $x \in (x_1, x_2) \subseteq T$ with $x_1 < x_2$;*
(c) *$h(f(x)) \leq x$ for all $x \in T$ and the inequality is strict if and only if $f|_{[v,x]}$ is constant for some $v < x$;*

(d) *$h(u) = x_0$ for all $u \in (u_1, u_2) \subseteq I$ with $u_1 < u_2$ and for some $x_0 \in \operatorname{int} T$ if and only if f jumps at x_0 and $(u_1, u_2) \subseteq (f_-(x_0), f_+(x_0))$.*

Proof. (a) Let $u_1, u_2 \in T$ with $u_1 \leq u_2$. Then $\{x \in T: f(x) \geq u_2\} \subseteq \{x \in T: f(x) \geq u_1\}$ and so $h(u_1) \leq h(u_2)$. Therefore h is increasing.
Let $u_0 \in I$ with $u_0 > \inf I = \inf_T f$. Then $f(x) < u_0$ for all $x \in T$ with $x < h(u_0)$. Given $\varepsilon > 0$, let $x_0 \in T \cap [h(u_0) - \varepsilon, h(u_0) + \varepsilon)$. Then for every $u \in (f(x_0), u_0)$ we deduce that

$$h(u) = \inf[x \in T: f(x) \geq u] \geq x_0 \geq h(u_0) - \varepsilon .$$

Hence h is left continuous.

(b) $\Longrightarrow$: Suppose that $h(u_0) < h_+(u_0)$ for some $u_0 \in I$ with $u_0 < \sup I = \sup_T f$. Then we get $f(x) \geq u_0$ for all $x > h(u_0)$. On the other hand if $h(u_0) < x < h_+(u_0)$, then the monotonicity of h (see part (a)) implies that $h(u) > x$ for all $u > u_0$. So, $f(x) < u$ for all $u > u_0$, hence $f(x) \leq u_0$. Therefore, $f(x) = u_0$ for all $x \in (h(u_0), h_+(u_0))$.
$\Longleftarrow$: Suppose that $f(x) = u_0$ for all $x \in (x_1, x_2)$ and $u_0 < \sup I = \sup_T f$. Then $h(u_0) \leq x_1$. Moreover, if $u \in (u_0, \sup I)$, then $f(x) = u_0 < u$ and so $h(u) \geq x$ for every $x \in (x_1, x_2)$. Let $x \to x_2^+$. Then $h(u) \geq x_2$ for all $u \in (u_0, \sup I)$. Hence $h_+(u_0) \geq x_2$ and so we conclude that $h(u_0) \leq x_1 < x_2 \leq h_+(u_0)$.

(c) $\Longrightarrow$: Let $u = f(x)$. Then $h(u) = h(f(x)) \leq x$. Suppose this last inequality is strict, so then there exists $v \in T$ with $v < x$ such that $f(v) \geq f(x)$. Hence, $f|_{[v,x]} = f(x)$.
$\Longleftarrow$: If $f|_{[v,x]} =$ constant, then $h(f(x)) \leq v < x$.

(d) $\Longrightarrow$: If $x \in T$ with $x > x_0$, then $f(x) \geq u$ for all $u \in (u_1, u_2)$ with $u_1 < u_2$. Let $u \to u_2^+$. It holds that $f(x) \geq u_2$, hence $f_+(x_0) \geq u_2$. On the other hand, if $x \in T$ with $x < x_0$, then $f(x) < u$ for all $u \in (u_1, u_2)$. Let $u \to u_1^-$. Then $f(x) \leq u_1$ and so $f_-(x_0) \leq u_1$.
$\Longleftarrow$: Let $u \in (f_-(x_0), f_+(x_0))$. If $x < x_0$, then $f(x) < u$ and so $h(u) \geq x_0$. On the other hand one has

$$x_0 \leq h(u) \leq h(f_+(x_0)) \leq h(f(x)) \leq x .$$

Let $x \to x_0^+$. We obtain that $h(u) = x_0$ for all $u \in (f_-(x_0), f_+(x_0))$. ☐

Proposition 4.3.9. *If $f: T \to \mathbb{R}$ is strictly increasing, then h is the left inverse of f and it is continuous.*

Proof. Since f is strictly increasing, using parts (a) and (b) of Proposition 4.3.8, we infer that h is continuous. Moreover, part (c) of Proposition 4.3.8 implies that $h(f(x)) = x$ for all $x \in T$. Hence, h is the left inverse of f. ☐

Remark 4.3.10. The previous two propositions remain valid if f is decreasing (resp. strictly decreasing). In this case $h: I \to \mathbb{R}$ is defined by

$$h(u) = \inf[x \in T: f(x) \leq u] .$$

Now we turn our attention to the differentiability properties of monotone functions. To do this we will need another version of "Vitali's Covering Theorem", see Theorem 2.8.27, which we state and prove below. First we recall the definition of fine cover stated here in terms of cubes; see Definition 2.8.26.

Definition 4.3.11. Consider the $\mathbb{R}^N$ equipped with the Lebesgue measure λ and $\mathcal{L}$, which denotes a family of nontrivial, closed cubes in $\mathbb{R}^N$. We say that $\mathcal{L}$ is a **fine cover** for a set $A \subseteq \mathbb{R}^N$ if for every $x \in A$ and every $\delta > 0$, there exists a cube $Q \in \mathcal{L}$ such that $x \in Q$ and $\operatorname{diam} Q \leq \delta$.

Example 4.3.12. Let $m \in \mathbb{N}$ and $\hat{\eta} = (\eta_k)_{k=1}^N \in \mathbb{Z}^N$ = the N-tuples of integers. We define

$$Q_{m,\hat{\eta}} = \left\{ \hat{x} = (x_k)_{k=1}^N \in \mathbb{R}^N : \frac{\eta_k - 1}{2^m} \leq x_k \leq \frac{\eta_k}{2^m} \text{ for all } k = 1, \ldots, N \right\}.$$

This is a closed diadic cube. We can partition $\mathbb{R}^N$ into closed diadic cubes with pairwise disjoint interiors using the hyperplane $\{x_i = \eta_{k_i} 1/2^m\}$ where for every $i = 1, \ldots, N$, the numbers $\eta_{k_i} \in \mathbb{Z}$. The collection $\mathcal{L}$ of all N-dimensional closed diadic cubes of diameter $\leq \vartheta$ for some $\vartheta > 0$, form a fine cover for any $A \subseteq \mathbb{R}^N$.

The next result is known as "Vitali's Covering Theorem", see also Theorem 2.8.27.

Theorem 4.3.13 (Vitali's Covering Theorem). *If $A \subseteq \mathbb{R}^N$ is bounded and Lebesgue measurable and $\mathcal{L}$ is a fine cover of A, then there exists a countable subcollection $\{Q_n\}_{n\geq 1} \subseteq \mathcal{L}$ with pairwise disjoint interiors such that $\lambda(A \setminus \bigcup_{n\geq 1} Q_n) = 0$.*

Proof. Evidently we may assume that there exists a closed cube $\hat{Q}$ such that $A \subseteq \hat{Q}$ and $Q \subseteq \hat{Q}$ for all $Q \in \mathcal{L}$.

Let $\mathcal{L}_0 = \mathcal{L}$ and let $Q_0 \in \mathcal{L}_0$. If Q_0 covers A we are done. Otherwise we define $\mathcal{L}_1 = \{Q \in \mathcal{L} : \operatorname{int} Q \cap \operatorname{int} Q_0 = \emptyset\}$. Since Q_0 does not cover A, we see that $\mathcal{L}_1 \neq \emptyset$. Let $\vartheta_1 = \sup[\operatorname{diam} Q : Q \in \mathcal{L}_1]$. Choose $Q_1 \in \mathcal{L}_1$ such that $1/2\vartheta_1 < \operatorname{diam} Q_1$. If $Q_0 \cup Q_1$ covers A, we are done; otherwise, we define $\mathcal{L}_2 = \{Q \in \mathcal{L}_1 : \operatorname{int} Q \cap \operatorname{int} Q_1 = \emptyset\}$. Moreover we set $\vartheta_2 = \sup[\operatorname{diam} Q : Q \in \mathcal{L}_2]$. Choose $Q_2 \in \mathcal{L}_2$ such that $1/2\vartheta_2 < \operatorname{diam} Q_2$. We continue this way and inductively we generate collections $\{\mathcal{L}_n\}_{n\geq 1}$, positive numbers $\{\vartheta_n\}_{n\geq 1}$, and cubes $\{Q_n\}_{n\geq 1}$ such that

$$\begin{aligned} &\mathcal{L}_n = \{Q \in \mathcal{L}_{n-1} : \operatorname{int} Q \cap \operatorname{int} Q_{n-1} = \emptyset\} \quad \vartheta_n = \sup[\operatorname{diam} Q : Q \in \mathcal{L}_n] \\ &Q_n \in \mathcal{L}_n \quad \text{with } \frac{1}{2}\vartheta_n < \operatorname{diam} Q_n . \end{aligned}$$

We have

$$\sum_{n\geq 1} \left(\frac{\operatorname{diam} Q_n}{\sqrt{N}} \right)^N = \sum_{n\geq 1} \lambda(Q_n) \leq \lambda(\hat{Q}) < \infty . \tag{4.3.1}$$

This gives

$$\operatorname{diam} Q_n \to 0 \quad \text{as } n \to \infty\,. \tag{4.3.2}$$

Arguing by contradiction, suppose that the conclusion of the theorem is not true. So, there exists $\varepsilon > 0$ such that

$$\lambda\Big(A \setminus \bigcup_{n\geq 1} Q_n\Big) \geq 2\varepsilon\,. \tag{4.3.3}$$

Let Q'_n be another cube with the same center and parallel faces to Q_n such that

$$\operatorname{diam} Q'_n = (4\sqrt{N}+1)\operatorname{diam} Q_n\,. \tag{4.3.4}$$

From (4.3.1) and (4.3.4) we see that there exists $n_0 = n_0(\varepsilon) \in \mathbb{N}$ such that

$$\lambda\Big(\bigcup_{n\geq n_0+1} Q'_n\Big) \leq \sum_{n\geq n_0+1} \lambda(Q'_n) \leq \varepsilon\,. \tag{4.3.5}$$

Applying (4.3.3) and (4.3.5) yields

$$\lambda\left(\left[A \setminus \bigcup_{n=1}^{n_0} Q_n\right] \setminus \bigcup_{n\geq n_0+1} Q'_n\right) \geq 2\varepsilon - \varepsilon = \varepsilon\,.$$

So, we can find $x \in [A \setminus \bigcup_{n=1}^{n_0} Q_n] \setminus \bigcup_{n\geq n_0+1} Q'_n$. Then, since $\bigcup_{n=1}^{n_0} Q_n \subseteq \mathbb{R}^N$ is closed,

$$2\beta = d\Big(x, \bigcup_{n=1}^{n_0} Q_n\Big) > 0\,.$$

Then according to Definition 4.3.11, there exists $Q_\mu \in \mathcal{L}$ with $\operatorname{diam} Q_\mu = \mu \leq \beta$ and $x \in Q_\mu$. It holds that $Q_\mu \cap \operatorname{int} Q_n = \emptyset$ for all $n = 1, \ldots, n_0$. Therefore $Q_\mu \in \mathcal{L}_{n_0+1}$. We claim that

$$Q_\mu \cap \operatorname{int} Q_n \neq \emptyset \quad \text{for some } n \geq n_0\,. \tag{4.3.6}$$

Indeed, if (4.3.6) is not true, then $Q_\mu \in \mathcal{L}_n$ for all $n \in \mathbb{N}$. Hence, because of (4.3.2) and since $1/2\vartheta_n < \operatorname{diam} Q_n$,

$$0 < \mu \leq \vartheta_n \to 0 \quad \text{as } n \to \infty\,,$$

a contradiction. So, (4.3.6) is true. Let $m \geq n_0 + 1$ be the smallest integer $\geq n_0 + 1$ such that (4.3.6) holds. Then $Q_\mu \notin \mathcal{L}_{m+1}$, $Q_\mu \in \mathcal{L}_m$ and $\mu \leq \vartheta_m$. Recall that $x \notin Q'_m$. Therefore $Q_\mu \cap \operatorname{int} Q_m \neq \emptyset$. If

$$\mu = \operatorname{diam} Q_\mu > \frac{1}{2\sqrt{N}}\left[\operatorname{diam} Q'_m - \operatorname{diam} Q_m\right], \tag{4.3.7}$$

then from (4.3.4) and (4.3.7) it follows that $\vartheta_m \geq \mu > \vartheta_m$, a contradiction. This proves the theorem. □

Corollary 4.3.14. *If $A \subseteq \mathbb{R}^N$ is bounded and Lebesgue measurable and $\mathcal{L}$ is a fine cover of A, then for any given $\varepsilon > 0$ there exists a finite collection $\mathcal{F}_\varepsilon = \{Q_n\}_{n=1}^{n_0} \subseteq \mathcal{L}$ with elements that have pairwise disjoint interiors and*

$$\sum_{n\geq 1} \lambda(Q_n) - \varepsilon \leq \lambda(A) \leq \lambda\left(\bigcup_{n=1}^{n_0} (A \cap Q_n)\right) + \varepsilon .$$

Proof. Exploiting the regularity of the Lebesgue measure, there exists $U_\varepsilon \subseteq \mathbb{R}^N$ such that

$$A \subseteq U_\varepsilon \quad \text{and} \quad \lambda(U_\varepsilon) \leq \lambda(A) + \varepsilon . \tag{4.3.8}$$

Moreover, let $\mathcal{L}_\varepsilon = \{Q \in \mathcal{L} : Q \subseteq U_\varepsilon\}$. Using Theorem 4.3.13, we find $\{Q_n\}_{n\geq 1} \subseteq \mathcal{L}_\varepsilon$ with pairwise disjoint interiors such that

$$\lambda\left(A \setminus \bigcup_{n\geq 1} Q_n\right) = 0 , \tag{4.3.9}$$

which in combination with (4.3.8) implies that

$$\sum_{n\geq 1} \lambda(Q_n) \leq \lambda(U_\varepsilon) \leq \lambda(A) + \varepsilon < \infty . \tag{4.3.10}$$

Hence,

$$\sum_{n\geq n_0+1} \lambda(Q_n) \leq \varepsilon \quad \text{for some } n_0 = n_0(\varepsilon) \in \mathbb{N} .$$

From this and (4.3.9) we obtain

$$\lambda(A) = \lambda\left(\bigcup_{n\geq 1} (A \cap Q_n)\right) \leq \lambda\left(\bigcup_{n=1}^{n_0} (A \cap Q_n)\right) + \varepsilon .$$

Then, due to (4.3.10) it follows that

$$\sum_{n\geq 1} \lambda(Q_n) - \varepsilon \leq \lambda(A) . \qquad \square$$

We will use this corollary to establish the differentiability properties of monotone functions.

Definition 4.3.15. Let $f : [a, b] \to \mathbb{R}$ and for fixed $x \in [a, b]$ we define

$$D_+ f(x) = \liminf_{h\to 0^+} \frac{f(x+h) - f(x)}{h} , \quad D_- f(x) = \liminf_{h\to 0^+} \frac{f(x) - f(x-h)}{h} ,$$
$$D^+ f(x) = \limsup_{h\to 0^+} \frac{f(x+h) - f(x)}{h} , \quad D^- f(x) = \limsup_{h\to 0^+} \frac{f(x) - f(x-h)}{h} .$$

We call $D_{\pm}f(x), D^{\pm}f(x)$ the **Dini derivatives** or **Dini derivates** of f at x. Clearly, $D_{+}f(x) \le D^{+}f(x)$ and $D_{-}f(x) \le D^{-}f(x)$. Moreover f is differentiable at x with derivative $f'(x) \in \mathbb{R}$ if and only if $f'(x) = D^{+}f(x) = D_{+}f(x) = D^{-}f(x) = D_{-}f(x)$.

Proposition 4.3.16. *If $f: [a, b] \to \mathbb{R}$ is increasing, then the functions $x \to D_{\pm}f(x)$ and $x \to D^{\pm}f(x)$ are all measurable.*

Proof. We do the proof for $D^{+}f(x)$; the proofs for the other functions are similar.

For every $n \in \mathbb{N}$ let $\xi_n(x) = \sup[(f(x+h) - f(x))/h : 0 < h \le 1/n]$. We see that $D^{+}f(x) = \lim_{n\to\infty} \xi_n(x)$. So, according to Proposition 2.2.10, it suffices to show that each ξ_n is measurable. Let $Q^n = (0, 1/n] \cap \mathbb{Q}$ and set $\vartheta_n(x) = \sup[(f(x+h) - f(x))/h : h \in Q^n]$. Then $\vartheta_n(x) \le \xi_n(x)$ for all $n \in \mathbb{N}$. We will show that the reverse inequality is also true. So, let $\varepsilon > 0$. We can find $s \in (0, 1/n]$ such that $\xi_n(x) - \varepsilon \le (f(x+s) - f(x))/s$. Having fixed $\varepsilon > 0$ and $s \in (0, 1/n]$ as above, we choose $h \in Q^n$ such that

$$\frac{1}{s} - \frac{1}{h} < \frac{\varepsilon}{|f(x+s)| + |f(x)| + 1},$$

which is equivalent to

$$\left[\frac{1}{s} - \frac{1}{h}\right](|f(x+s)| + |f(x)| + 1) < \varepsilon .$$

Since $s < h$, we obtain

$$\left[\frac{1}{s} - \frac{1}{h}\right](f(x+s) + f(x)) < \varepsilon$$

and because of $f(x+s) \le f(x+h)$, it follows that

$$\frac{f(x+s) - f(x)}{s} < \frac{f(x+h) - f(x)}{h} + \varepsilon .$$

Hence, $\xi_n(x) - 2\varepsilon \le \vartheta_n(x)$ for all $\varepsilon > 0$ and thus $\xi_n = \vartheta_n$. But ϑ_n is measurable. Hence, so is ξ_n, and thus $D^{+}f$ is measurable. Similarly we show this for $D^{-}f$, $D_{+}f$, and $D_{-}f$. □

Theorem 4.3.17. *If $f: [a, b] \to \mathbb{R}$ is increasing, then f is differentiable a. e. on $[a, b]$.*

Proof. We will show that the set where the Dini derivatives are not equal is Lebesgue-null. We show this for the set $\{D^{+}f > D_{-}f\}$. The proofs for the others are similar.

So, let $r, q \in \mathbb{Q}$ and define

$$C_{r,q} = \{x \in [a, b] : D^{+}f(x) > r > q > D_{-}f(x)\} .$$

By Proposition 4.3.16, this set is Lebesgue measurable. Let $\varepsilon > 0$ and choose an open set $U_\varepsilon \subseteq \mathbb{R}$ such that $\lambda(U_\varepsilon) \le \lambda(C_{r,q}) + \varepsilon$, which is possible because of the regularity of the Lebesgue measure λ. For every $x \in C_{r,q}$ there exists an arbitrarily small interval $[x - h, x] \subseteq U_\varepsilon$ such that

$$f(x) - f(x - h) < qh\,. \tag{4.3.11}$$

Invoking Corollary 4.3.14 there exists a finite collection $\{I_n\}_{n=1}^{n_0}$ of such intervals that cover $A \subseteq C_{r,q}$ with $\lambda^*(A) > \lambda(C_{r,q}) - \varepsilon$ where λ^* denotes the Lebesgue outer measure; see Definition 2.1.33 and Example 2.1.35. Summing over these intervals, we obtain, due to (4.3.11), that

$$\sum_{n=1}^{n_0} [f(x_n) - f(x_n - h_n)] < q \sum_{n=1}^{n_0} h_n < q\lambda(U_\varepsilon) \le q(\lambda(C_{r,q}) + \varepsilon)\,. \tag{4.3.12}$$

Each $y \in A$ is the left endpoint of an arbitrarily small interval $(y, y + \mu) \subseteq I_n$ and $r\mu < f(y+\mu)-f(y)$. A new application of Corollary 4.3.14 gives a finite collection $\{I_n\}_{n=1}^{\tilde{n}_0}$ of such intervals such that their union contains a subset of A of outer measure $> \lambda(C_{r,q}) - 2\varepsilon$. Summing over these intervals, one gets

$$\sum_{n=1}^{\tilde{n}_0} [f(y_n + \mu_n) - f(y_n)] \ge r \sum_{n=1}^{\tilde{n}_0} \mu_n > r(\lambda(C_{r,q}) - 2\varepsilon)\,. \tag{4.3.13}$$

Each interval I_n is contained in some interval I_m and if we sum over those n for which $I_n \subseteq I_m$, then, since f is increasing, this gives

$$\sum_{n=1}^{\overline{n}_0} [f(y_n + \mu_n) - f(y_n)] \le f(x_n) - f(x_n - h_n)\,,$$

which implies

$$\sum_{n=1}^{\tilde{n}_0} [f(y_n + \mu_n) - f(y_n)] \le \sum_{n=1}^{n_0} (f(x_n) - f(x_n - h_n))\,.$$

Then, from (4.3.12) and (4.3.13), we see that $r(\lambda(C_{r,q}) - 2\varepsilon) < q(\lambda(C_{r,q}) + \varepsilon)$. Since $\varepsilon > 0$ is arbitrary, we let $\varepsilon \searrow 0$ to obtain $r\lambda(C_{r,q}) < q\lambda(C_{r,q})$. Hence, $\lambda(C_{r,q}) = 0$, since $q < r$. So, we have that $k(x) = \lim_{h\to 0}(f(x+h) - f(x))/h$ exists for a. a. $x \in [a,b]$. We need to show that $k(x) \in \mathbb{R}$ for a. a. $\in [a,b]$. We define $k_n(x) = n[f(x + 1/n) - f(x)]$ with $f(x) = f(b)$ if $x \ge b$. Then $k_n(x) \to k(x)$ for a. a. $x \in [a,b]$. Thus, k is measurable. Moreover $k_n \ge 0$ for all $n \in \mathbb{N}$ since f is increasing. From Fatou's Lemma we infer that

$$\begin{aligned}
\int_a^b k\,dx &\le \liminf_{n\to\infty} \int_a^b k_n\,dx = \liminf_{n\to\infty} n \int_a^b \left[f\left(x + \frac{1}{n}\right) - f(x)\right] dx \\
&= \liminf_{n\to\infty} \left[n \int_b^{b+\frac{1}{n}} f\,dx - n \int_a^{a+\frac{1}{n}} f\,dx\right] = f(b) - \limsup_{n\to\infty} n \int_a^{a+\frac{1}{n}} f\,dx \\
&\le f(b) - f(a)\,.
\end{aligned}$$

Hence, $k \in L^1[a,b]$ and so $k(x) \in \mathbb{R}$ for a. a. $x \in [a,b]$. Therefore f is differentiable almost everywhere and $f' = k$. □

Remark 4.3.18. The result above is sharp in the sense that for any given Lebesgue-null set $D \subseteq \mathbb{R}$ there exists an increasing continuous function that is differentiable at all $\mathbb{R}\setminus D$.

Corollary 4.3.19. *If $T \subseteq \mathbb{R}$ is an interval and $f\colon T \to \mathbb{R}$ is a monotone function, then $f' \in L^1[a,b]$ for every $[a,b] \subseteq T$ and*

$$\int_a^b |f'(x)|\,dx \le |f(b) - f(a)|\,.$$

Hence $f' \in L^1_{\mathrm{loc}}(T)$.

Proposition 4.3.20. *If $T \subseteq \mathbb{R}$ is an interval, $f\colon T \to \mathbb{R}$ is a monotone function, and $h > 0$, $[a,b] \subseteq T$ with $b - a > h$, then*

$$\frac{1}{h}\int_a^{b-h} |f(x+h) - f(x)|\,dx \le |f(b) - f(a)|\,.$$

Moreover if f is bounded, then

$$\frac{1}{h}\int_{T_h} |f(x+h) - f(x)|\,dx \le \sup_T f - \inf_T f$$

with $T_h = \{x \in T : x + h \in T\}$.

Proof. To fix things we may assume that f is increasing. Let

$$g(x) = \begin{cases} f(x) & \text{if } x \le b\,, \\ f(b) & \text{if } b \le x\,. \end{cases}$$

Since f is increasing, we obtain

$$\begin{aligned} \frac{1}{h}\int_a^b [g(x+h) - g(x)]\,dx &= \frac{1}{h}\left[\int_b^{b+h} g(x)\,dx - \int_a^{a+h} g(x)\,dx\right] \le \frac{1}{h}[f(b)h - f(a)h] \\ &= f(b) - f(a)\,. \end{aligned}$$

This gives

$$\frac{1}{h}\int_a^{b-h} [f(x+h) - f(x)]\,dx \le \frac{1}{h}\int_a^b [g(x+h) - g(x)]\,dx \le f(b) - f(a)\,.$$

Now suppose that f is bounded. Let $T_n = [a_n, b_n]$ such that $a_n \searrow \inf T$ and $b_n \nearrow \sup T$. Recalling that f is increasing, we get

$$0 \le \frac{1}{h} \int_{a_n}^{b_n - h} [f(x+h) - f(x)] \, dx \le f(b_n) - f(a_n) \le \sup_T f - \inf_T f \, .$$

If we let $n \to \infty$, using the Lebesgue Monotone Convergence Theorem, it follows that

$$\frac{1}{h} \int_{T_h} |f(x+h) - f(x)| \, dx \le \sup_T f - \inf_T f \, .$$

□

An important consequence of Theorem 4.3.17 is the following result.

Theorem 4.3.21. *If $T \subseteq \mathbb{R}$ is an interval, $f_n \colon T \to \mathbb{R}$ with $n \in \mathbb{N}$ is a sequence of increasing functions, and $\sum_{n \ge 1} f_n(x)$ converges pointwise on T, then $\sum_{n \ge 1} f_n(x)$ converges uniformly on compact sets in T; the function $f(x) = \sum_{n \ge 1} f_n(x)$ is differentiable for a. a. $x \in T$ and $f'(x) = \sum_{n \ge 1} f_n'(x)$ for a. a. $x \in T$.*

Proof. The result is local. So, without any loss of generality we may assume that $T = [a, b]$. Let $s_n(x) = \sum_{k=1}^{n} f_k(x)$ for all $n \in \mathbb{N}$. Then $\{s_n\}_{n \ge 1}$ is a sequence of increasing functions such that $s_n(x) \to f(x)$ for all $x \in [a, b]$. We will show that f is increasing as well. Arguing by contradiction, suppose that f is not increasing. Then there exists $x_1, x_2 \in [a, b]$ such that $x_1 < x_2$ and $f(x_2) < f(x_1)$.

Let $h = f(x_1) - f(x_2)$ and $\varepsilon \in (0, h/2)$. We can find $n_0 = n_0(\varepsilon, x_1, x_2) \in \mathbb{N}$ such that

$$|s_n(x_1) - f(x_1)| \le \varepsilon \quad \text{and} \quad |s_n(x_2) - f(x_2)| \le \varepsilon \quad \text{for all } n \ge n_0 \, .$$

This yields

$$s_n(x_2) - s_n(x_1) < f(x_2) + \frac{h}{2} - f(x_1) + \frac{h}{2} = h - [f(x_1) - f(x_2)] = h - h = 0$$

for all $n \ge n_0$. Therefore $s_n(x_1) - s_n(x_2) > 0$ for all $n \ge n_0$, a contradiction to the fact the s_n is increasing. This proves that f is increasing as well.

Now we will show that $s_n \to f$ uniformly on $[a, b]$. Without any loss of generality we may assume that $f_n(a) \ge 0$ for all $n \in \mathbb{N}$. Hence, $f_n \ge 0$ for all $n \in \mathbb{N}$. We have

$$0 \le f(x) - s_n(x) = \sum_{k \ge n+1} f_k(x) \le \sum_{k \ge n+1} f_k(b) \, ,$$

which results in

$$\sup_{x \in [a,b]} |f(x) - s_n(x)| \le \sum_{k \ge n+1} f_k(b) \to 0 \quad \text{as } n \to \infty \, .$$

This proves the desired uniform convergence on $[a, b]$.

Since $\{f_n\}_{n\geq 1}$ and f are all increasing, by Theorem 4.3.17, they are all differentiable on $D \subseteq [a,b]$ with $\lambda([a,b]\setminus D) = 0$ where λ is the Lebesgue measure on $\mathbb{R}$. Let $x \in D$ and $h > 0$ be small such that $x + h \in [a,b]$. Then

$$\frac{f(x+h)-f(x)}{h} = \sum_{n\geq 1} \frac{f_n(x+h)-f_n(x)}{h} \geq \sum_{k=1}^{n} \frac{f_k(x+h)-f_k(x)}{h} = s_n(x)\,,$$

which shows that

$$f'(x) \geq \sum_{k=1}^{n} f_k'(x) = s_n'(x) \geq 0\,.$$

Hence, $s_n'(x) \to g(x)$ for all $x \in D$ as $n \to \infty$. We will show that $g = f'$. It holds that $s_n(b) \to f(b)$. So, there exists a subsequence $\{s_{n_k}(b)\}_{k\geq 1}$ such that $0 \leq f(b) - s_{n_k}(b) \leq 1/2^k$. Since $f - s_{n_k}$ is increasing, we then obtain

$$0 \leq f(x) - s_{n_k}(x) \leq \frac{1}{2^k} \quad \text{for all } x \in [a,b]\,.$$

Thus, $\{f - s_{n_k}\}_{k\geq 1}$ is a sequence of increasing functions and it is convergent to 0. So, reasoning as above, it follows that $f'(x) - s_{n_k}'(x) \to 0$ for a. a. $x \in [a,b]$. Since $\{s_n'\}_{n\geq 1}$ is increasing we conclude that $f'(x) - s_n'(x) \to 0$ for a. a. $x \in [a,b]$. □

The set of monotone functions is not a vector space because the difference of two monotone functions need not be monotone. So, we pass to the smallest vector space containing the set of monotone functions. This is the space of functions of bounded variation.

Definition 4.3.22. (a) Let $T \subseteq \mathbb{R}$ be an interval. A **partition** of T is a finite set $P = \{x_k\}_{k=0}^n \subseteq T$ such that $x_0 < x_1 < \cdots < x_n$. Let $\mathcal{P}$ be the set of all finite partitions of T.

(b) Let $T \subseteq \mathbb{R}$ be an interval and $f\colon T \to \mathbb{R}$. We say that f is of **bounded variation** if

$$\operatorname*{var}_T f = \sup\left[\sum_{k=0}^{n-1} |f(x_{k+1}) - f(x_k)| : P = \{x_k\}_{k=0}^n \in \mathcal{P}\right] < \infty\,.$$

We denote the space of functions of bounded variation by $\mathrm{BV}(T)$.

(c) We say that $f \in \mathrm{BV}_{\mathrm{loc}}(T)$ if $\operatorname{var}_{[a,b]} f < \infty$ for all $[a,b] \subseteq T$.

Remark 4.3.23. Suppose that $b = \sup T \in T$. Then in the definition above it suffices to consider partitions $P = \{x_k\}_{k=0}^n$ of the form $x_0 < x_1 < \cdots < x_n = b$. Indeed if $P = \{x_k\}_{k=0}^n \in \mathcal{P}$ with $x_n < b$ and $P' = P \cup \{b\} \in \mathcal{P}$, then

$$\sum_{k=0}^{n-1} |f(x_{k+1}) - f(x_k)| \leq \sum_{k=0}^{n-1} |f(x_{k+1}) - f(x_k)| + |f(b) - f(x_n)| \leq \operatorname*{var}_T f\,.$$

Similarly if $a = \inf T \in T$. So, in what follows we will use this fact without further comment. Note that $\mathrm{BV}_{\mathrm{loc}}([a,b]) = \mathrm{BV}([a,b])$. Finally if $U \subseteq \mathbb{R}$ is open, then $U = \bigcup_{n\geq 1} T_n$ with $\{T_n\}_{n\geq 1}$ pairwise disjoint intervals. Then for $f\colon U \to \mathbb{R}$ we define $\operatorname{var}_U f = \sum_{n\geq 1} \operatorname{var}_{T_n} f$.

The following proposition is a straightforward consequence of Definition 4.3.22.

Proposition 4.3.24. *If $f, h \in \mathrm{BV}([a,b])$, then the following hold:*
(a) $f \pm h \in \mathrm{BV}([a,b])$;
(b) $fh \in \mathrm{BV}([a,b])$;
(c) *if $h(x) \geq c > 0$ for all $x \in [a,b]$, then $f/h \in \mathrm{BV}([a,b])$;*
(d) *if f is differentiable on $[a,b]$ and f' is bounded, then $f \in \mathrm{BV}([a,b])$;*
(e) *if f is Lipschitz continuous on $[a,b]$, then $f \in \mathrm{BV}([a,b])$.*

Proposition 4.3.25. *If $T \subseteq \mathbb{R}$ is an interval and $f\colon T \to \mathbb{R}$, then the following hold:*
(a) *for every $u \in T$, $\sup_T |f| \leq |f(u)| + \operatorname{var}_T f$; hence if $f \in \mathrm{BV}(T)$, then f is bounded;*
(b) *for every $u \in T$ it holds that $\operatorname{var}_T f = \operatorname{var}_{T\cap(-\infty,u]} f + \operatorname{var}_{T\cap[u,+\infty)} f$;*
(c) *if T does not contain $\sup T$ (resp. $\inf T$), then $\operatorname{var} f = \lim_{u\to(\sup T)^-} \operatorname{var}_{T\cap(-\infty,u]} f$ (resp. $\operatorname{var} f = \lim_{u\to(\inf T)^+} \operatorname{var}_{T\cap[u,+\infty)} f$).*

Proof. (a): Let $x \neq u$ and let $P = \{u, x\} \in \mathcal{P}$. It holds that

$$|f(x)| \leq |f(u)| + |f(x) - f(u)| \leq |f(u)| + \operatorname*{var}_T f\,,$$

which implies $\sup_T |f| \leq |f(u)| + \operatorname{var}_T f$.

(b): Let $T_1 = T \cap (-\infty, u]$ and $T_2 = T \cap [u, +\infty)$. Consider a partition $P_1 = \{x_k\}_{k=0}^n$ of T_1 and a partition $P_2 = \{y_i\}_{i=0}^m$ of T_2. As we already pointed out (see Remark 4.3.23), we can have $x_n = c = y_0$. Then $P = P_1 \cup P_2$ is a partition of T and so

$$\sum_{k=0}^{n-1} |f(x_{k+1}) - f(x_k)| + \sum_{i=0}^{m-1} |f(y_{i+1}) - f(y_i)| \leq \operatorname*{var}_T f\,.$$

This gives

$$\operatorname*{var}_{T_1} f + \operatorname*{var}_{T_2} f \leq \operatorname*{var}_T f\,. \tag{4.3.14}$$

Next let $P = \{x_k\}_{k=0}^n$ be a partition of T with $u < x_n$. Let $m \in \{1, \dots, n\}$ be such that $x_{m+1} \leq u \leq x_m$ and let $P_1 = \{x_k\}_{k=0}^{m-1} \cup \{u\}$ as well as $P_2 = \{u\} \cup \{x_k\}_{k=m}^n$. These are partitions of T_1 and T_2, respectively. We obtain

$$\begin{aligned}
&\sum_{k=0}^{n-1} |f(x_{k+1}) - f(x_k)| \\
&= \sum_{k=0}^{m-2} |f(x_{k+1}) - f(x_k)| + |f(x_m) + f(u) - f(u) - f(x_{m-1})|
\end{aligned}$$

$$+\sum_{k=m}^{n-1}|f(x_{k+1})-f(x_k)| \tag{4.3.15}$$

$$\le \sum_{k=0}^{m-2}|f(x_{k+1})-f(x_k)| + |f(x_{m-1})-f(u)| + |f(u)-f(x_m)|$$

$$+\sum_{k=m}^{n-1}|f(x_{k+1})-f(x_k)| \le \operatorname*{var}_{T_1} f + \operatorname*{var}_{T_2} f\,.$$

From (4.3.14) and (4.3.15), we conclude that $\mathrm{var}_T f = \mathrm{var}_{T_1} f + \mathrm{var}_{T_2} f$.

(c) Evidently we may assume that $\mathrm{var}_T f > 0$. We consider the case $\sup T \notin T$; the other case is treated similarly. Let $\eta \in (0, \mathrm{var}_T f)$ and consider a partition $P = \{x_k\}_{k=0}^n$ such that $\eta < \sum_{k=0}^{n-1} |f(x_{k+1}) - f(x_k)|$. Consider $u \in (x_n, \sup T)$. Then P is a partition of $T \cap (-\infty, u]$ and so from part (b) we conclude that

$$\eta < \sum_{k=0}^{n-1}|f(x_{k+1})-f(x_k)| \le \operatorname*{var}_{T\cap(-\infty,u]} f \le \operatorname*{var}_{T} f\,.$$

Hence,

$$\eta < \lim_{u\to(\sup T)^-} \operatorname*{var}_{T\cap(-\infty,u]} f \le \operatorname*{var}_{T} f\,.$$

Note that the limit exists from part (b) since $u \to \mathrm{var}_{T\cap(-\infty,u]} f$ is increasing. Finally let $\eta \nearrow \mathrm{var}_T f$ to conclude that $\lim_{u\to(\sup T)^-} \mathrm{var}_{T\cap(-\infty,u]} f = \mathrm{var}_T f$. □

Corollary 4.3.26. *If $T = [a,b]$ with $a < b$ and $c \in [a,b]$, then $\mathrm{var}_{[a,b]} f = \mathrm{var}_{[a,c]} f + \mathrm{var}_{[c,b]} f$.*

What can we say about monotone functions?

Proposition 4.3.27. *If $T \subseteq \mathbb{R}$ is an interval and $f\colon T \to \mathbb{R}$ is monotone, then for every subinterval $I \subseteq T$, we have $\mathrm{var}_I f = \sup_I f - \inf_I f$. Therefore $f \in \mathrm{BV}_{\mathrm{loc}}(T)$ and if f is bounded, then $f \in \mathrm{BV}(T)$.*

Proof. To fix things we assume that f is increasing. Let $P = \{x_k\}_{k=0}^n \subseteq I$ be a partition of I. Then

$$\sum_{k=0}^{n-1}|f(x_{k+1})-f(x_k)| = \sum_{k=0}^{n-1}(f(x_{k+1})-f(x_k)) = f(x_n)-f(x_0) \le \sup_I f - \inf_I f\,.$$

Hence,

$$\operatorname*{var}_I f \le \sup_I f - \inf_I f\,. \tag{4.3.16}$$

Evidently we may assume that I is not a singleton. Let $\inf_I f \le u < v \le \sup_I f$ and consider the partition $P = \{u, v\} \subseteq I$. It holds that $f(v) - f(u) = |f(u) - f(v)| \le \mathrm{var}_I f$. If $\sup I \in I$, then we choose $v = \sup_I$ and obtain $f(v) = \sup_I f$. If $\sup I \notin I$, then considering

$v \to (\sup I)^-$, one has $f(v) \to \sup_I f$. We argue similarly for the left end point. So finally this leads to

$$\sup_I f - \inf_I f \le \operatorname{var}_I f\,. \tag{4.3.17}$$

From (4.3.16) and (4.3.17) we conclude that $\operatorname{var}_I f = \sup_I f - \inf_I f$. Hence, $f \in \mathrm{BV}_{\mathrm{loc}}(T)$. Moreover, if f is bounded, then $f \in \mathrm{BV}(T)$. □

Example 4.3.28. The continuity and/or boundedness of f is not enough to guarantee that a function is of bounded variation. To see this consider the function

$$f(x) = \begin{cases} x\sin(\frac{1}{x}) & \text{if } 0 < x \le 1\,, \\ 0 & \text{if } x = 0\,. \end{cases}$$

Evidently f is continuous and bounded. However, $f \notin \mathrm{BV}([0,1])$. Consider $x_k = 1/((k + 1/2)\pi)$. Then $f(x_k) = (-1)^k/((k+1/2)\pi)$ and so $\sum_{k=1}^n 2/(k\pi) \le \operatorname{var}_{[0,1]} f$. Hence, $\operatorname{var}_{[0,1]} f = +\infty$.

Proposition 4.3.29. *If $T \subseteq \mathbb{R}$ is an interval, $x_0 \in T$, $f \in \mathrm{BV}_{\mathrm{loc}}(T)$, and*

$$V_{x_0}(x) = \begin{cases} -\operatorname{var}_{[x,x_0]} f & \text{if } x < x_0 \\ \operatorname{var}_{[x_0,x]} f & \text{if } x_0 \le x \end{cases}\,, \quad \text{for all } x \in T\,,$$

then for all $x, y \in T$ with $x < y$ we have

$$|f(y) - f(x)| \le V_{x_0}(y) - V_{x_0}(x) = \operatorname{var}_{[x,y]} f\,. \tag{4.3.18}$$

Moreover, V_{x_0} and $V_{x_0} \pm f$ are all increasing.

Proof. Let $x, y \in T$ with $x < y$. Thanks to Corollary 4.3.26 we obtain

$$\operatorname{var}_{[x,y]} f = \begin{cases} \operatorname{var}_{[x_0,y]} f - \operatorname{var}_{[x_0,x]} f = V_{x_0}(y) - V_{x_0}(x) & \text{if } x_0 \le x < y\,, \\ \operatorname{var}_{[x,x_0]} f - \operatorname{var}_{[y,x_0]} f = -V_{x_0}(x) + V_{x_0}(y) & \text{if } x < y \le x_0\,, \\ \operatorname{var}_{[x,x_0]} f + \operatorname{var}_{[x_0,y]} f = -V_{x_0}(x) + V_{x_0}(y) & \text{if } x \le x_0 \le y\,. \end{cases}$$

Since $|f(y) - f(x)| \le \operatorname{var}_{[x,y]} f$, then inequality (4.3.18) follows. Moreover, from (4.3.18), we see that $V_{x_0}(x) \le V_{x_0}(y)$ and $\pm[f(y) - f(x)] \le V_{x_0}(y) - V_{x_0}(x)$. Hence V_{x_0} and $V_{x_0} \pm f$ are all increasing. □

Corollary 4.3.30. *If $T \subseteq \mathbb{R}$ is an interval and $f\colon T \to \mathbb{R}$ is a measurable function, then for $h > 0$ we have*

$$\frac{1}{h}\int_{T_h} |f(x+h) - f(x)|\,dx \le \operatorname{var}_T f\,,$$

where $T_h = \{x \in X : x + h \in T\}$.

Proof. Clearly we may assume that $\operatorname{var}_T f < \infty$. Consider $[a, b] \subseteq T$ with $0 < h \le b - a$. From (4.3.18) and Proposition 4.3.20, we get

$$\begin{aligned}\frac{1}{h}\int_a^{b-h} |f(x+h) - f(x)|\,dx &\le \frac{1}{h}\int_a^{b-h} (V(x+h) - V(x))\,dx \\ &\le V(b) - V(a) = \operatorname*{var}_{[a,b]} f\,,\end{aligned} \tag{4.3.19}$$

since V is increasing; see Proposition 4.3.29. Let $T_n = [a_n, b_n]$ with $n \in \mathbb{N}$ be an increasing sequence of subintervals of T such that $a_n \searrow \inf T$ and $b_n \nearrow \sup T$. Assume that $\lambda(T_h) > 0$, otherwise the result is obvious. Then for large enough $n \in \mathbb{N}$, one has $0 < h < b_n - a_n$ and so from (4.3.19) it follows that

$$\frac{1}{h}\int_{a_n}^{b_n-h} |f(x+h) - f(x)|\,dx \le \operatorname*{var}_{[a_n,b_n]} f \le \operatorname*{var}_T f \quad \text{for all } n \in \mathbb{N} \text{ large enough}\,.$$

Passing to the limit as $n \to \infty$ and using the Lebesgue Monotone Convergence Theorem, we obtain

$$\frac{1}{h}\int_{T_h} |f(x+h) - f(x)|\,dx \le \operatorname*{var}_T f\,. \qquad \square$$

Now we come to the theorem that characterizes functions of bounded variation.

Theorem 4.3.31. *If $T \subseteq \mathbb{R}$ is an interval, then the smallest vector space containing all monotone functions (resp. all bounded monotone functions) is $\mathrm{BV}_{\mathrm{loc}}(T)$ (resp. $\mathrm{BV}(T)$). Moreover every $f \in \mathrm{BV}_{\mathrm{loc}}(T)$ (resp. $f \in \mathrm{BV}(T)$) can be written as the difference of two increasing functions (resp. of two bounded increasing functions).*

Proof. Let $f, h\colon T \to \mathbb{R}$. From Definition 4.3.22 (b) we see that for every subinterval $I \subseteq T$ we have $\operatorname{var}_I(\vartheta f) = |\vartheta| \operatorname{var}_I f$ for all $\vartheta \in \mathbb{R}$ and $\operatorname{var}_I(f + h) \le \operatorname{var}_I f + \operatorname{var}_I h$. Hence, $\mathrm{BV}_{\mathrm{loc}}(T)$ (resp. $\mathrm{BV}(T)$) is a vector space. Proposition 4.3.27 implies that the monotone functions (resp. the bounded monotone functions) belong to $\mathrm{BV}_{\mathrm{loc}}(T)$ (resp. to $\mathrm{BV}(T)$). Moreover, if $f \in \mathrm{BV}_{\mathrm{loc}}(T)$ (resp. $f \in \mathrm{BV}(T)$), then $f = V - (V - f)$ and both V, $V - f$ are increasing; see Proposition 4.3.29. So, $\mathrm{BV}_{\mathrm{loc}}(T)$ (resp. $\mathrm{BV}(T)$) is the smallest vector space containing the monotone (resp. bounded monotone) functions. $\square$

Corollary 4.3.32. *$f \in \mathrm{BV}([a, b])$ if and only if f is the difference of two increasing functions.*

Corollary 4.3.33. *If $T \subseteq \mathbb{R}$ is an interval and $f \in \mathrm{BV}_{\mathrm{loc}}(T)$, then f has countably many discontinuity points, f' exists λ-a. e. and*

$$\int_a^b |f'|\,dx \le \operatorname*{var}_{[a,b]} f \quad \textit{for all } [a, b] \subseteq T\,.$$

Moreover, if $f \in \mathrm{BV}(T)$, *then* $f' \in L^1(T)$ *and for* $x_0 \in T$, *one has*

$$\int_T |f'|\,dx \le \int_T |V'_{x_0}|\,dx \le \sup_T V_{x_0} - \inf_T V_{x_0} - \operatorname*{var}_T f\,.$$

Proposition 4.3.34. *If* $f \in \mathrm{BV}([a,b]) \cap C([a,b])$, *then* $V_a(x) = \operatorname{var}_{[a,x]} f$ *is continuous on* $[a,b]$.

Proof. Let $x_0 \in [a,b)$. We show that V_a is right continuous at x_0. Given $\varepsilon > 0$, let $\{x_k\}_{k=1}^n \subseteq [x_0,b]$ be a partition such that

$$\operatorname*{var}_{[x_0,b]} f - \varepsilon \le \sum_{k=0}^{n-1} |f(x_{k+1}) - f(x_k)| = s_n\,. \tag{4.3.20}$$

Clearly the sum s_n only increases if we add new points to the partition. So we may assume that $|f(x_1) - f(x_0)| \le \varepsilon$. Then from (4.3.20), we infer that

$$\operatorname*{var}_{[x_0,b]} f \le \varepsilon + \sum_{k=0}^{n-1} |f(x_{k+1}) - f(x_k)| \le 2\varepsilon + \sum_{k=1}^{n-1} |f(x_{k+1}) - f(x_k)| \le 2\varepsilon + \operatorname*{var}_{[x,b]} f\,.$$

Hence, $\operatorname{var}_{[x_0,x_1]} f \le 2\varepsilon$ and so $V_a(x_1) - V_a(x_0) \le 2\varepsilon$.

Since $\varepsilon > 0$ is arbitrary we conclude that $\lim_{x \to x_0^+} V_a(x) = V_a(x_0)$ and this proves the right continuity of V_a at $x_0 \in [a,b)$. Similarly we show the left continuity of V_a at $\hat{x} \in (a,b]$. Therefore V_a is continuous. □

Corollary 4.3.35. *If* $f \in \mathrm{BV}([a,b]) \cap C([a,b])$, *then* f *can be written as the difference of two continuous increasing functions.*

Proof. Note that $f = V_a - (V_a - f)$, and use Propositions 4.3.34 and 4.3.29. □

We have already seen that $f \to \operatorname{var}_T f$ is absolutely homogeneous, that is, $\operatorname{var}_T(\vartheta f) = |\vartheta| \operatorname{var}_T f$ for all $\vartheta \in \mathbb{R}$ and subadditive, that is, $\operatorname{var}_T(f + h) \le \operatorname{var}_T f + \operatorname{var}_T h$. So, it is almost a norm. What we miss is that $\operatorname{var}_T f = 0$ does not imply that $f \equiv 0$. Instead we have that f is constant. This can be remedied if we add the term $|f(u)|$ for some $u \in T$. This leads to the next result.

Proposition 4.3.36. *If* $T \subseteq \mathbb{R}$ *is an interval and* $u \in T$, *then* $f \to |f(u)| + \operatorname{var}_T f = \|f\|_{\mathrm{BV}}$ *is a norm on* $\mathrm{BV}(T)$.

So, $\mathrm{BV}(T)$ is a normed space. It is natural to ask whether it is complete. This will be proved using the so-called "Helly's Selection Theorem". To prove this result we will need some auxiliary results that are actually of independent interest.

Proposition 4.3.37. *If* $T \subseteq \mathbb{R}$ *is an interval and* $H = \{f\}$ *is an infinite family of functions* $f: T \to \mathbb{R}$ *such that* $|f(x)| \le M$ *for all* $x \in T$, *for all* $f \in H$ *and for some* $M > 0$, *then for every countable set* $D \subseteq T$ *there exists a sequence* $\{f_n\}_{n \ge 1} \subseteq H$ *such that* $\lim_{n\to\infty} f_n(x)$ *exists in* $\mathbb{R}$ *for all* $x \in D$.

Proof. Let $D = \{x_k\}_{k\geq 1}$. Then $\{f(x_1): f \in H\} \subseteq \mathbb{R}$ is bounded. So, there exists a sequence $\{f_n^1\}_{n\geq 1} \subseteq H$ such that $\eta_1 = \lim_{n\to\infty} f_n^1(x_1)$ exists. For this sequence of functions we consider the real sequence $\{f_n^1(x_2)\}_{n\geq 1} \subseteq \mathbb{R}$. Then we find a subsequence $\{f_n^2\}_{n\geq 1}$ of $\{f_n^1\}_{n\geq 1}$ such that $\eta_2 = \lim_{n\to\infty} f_n^2(x_2)$ exists. Inductively, for every $k \in \mathbb{N}$ with $k \geq 2$, there exists a subsequence $\{f_n^k\}_{n\geq 1}$ of $\{f_n^{k-1}\}_{n\geq 1}$ for which $\eta_k = \lim_{n\to\infty} f_n^k(x_k)$ exists. We form the sequence $\{f_n^n(x_k)\}$ of diagonal elements based on the Cantor diagonalization process. Then, for every $x_k \in D$, we obtain that $\{f_n^n(x_k)\}_{n\geq k}$ is a subsequence of $\{f_n^k(x_k)\}_{n\geq 1}$ and so it converges to η_k. □

Proposition 4.3.38. *If $T \subseteq \mathbb{R}$ is an interval and $H = \{f\}$ is an infinite family of increasing functions $f: T \to \mathbb{R}$ such that $|f(x)| \leq M$ for all $x \in T$, for all $f \in H$ and for some $M > 0$, then there exists a sequence $\{f_n\}_{n\geq 1} \subseteq H$ and an increasing function $f_*: T \to \mathbb{R}$ such that $f_n(x) \to f_*(x)$ for all $x \in T$ as $n \to \infty$.*

Proof. Let D be the rational points of T union with the endpoints of T which belong to T. We apply Proposition 4.3.37 to find a sequence $\{f_n\}_{n\geq 1} \subseteq H$ such that $f(x) = \lim_{n\to\infty} f_n(x)$ for all $x \in D$. If $x, y \in D$ with $x \leq y$, then $f_n(x) \leq f_n(y)$ for all $n \in \mathbb{N}$. Hence, $f(x) \leq f(y)$. We extend f on T by setting

$$f(x) = \sup\left[f(y): y \in D, y < x\right].$$

Then f is clearly increasing and so the discontinuity points of f are at most countable. We will show that $f_n(x) \to f(x)$ at every continuity point x of f. So, let $\varepsilon > 0$ and $x_i, x_k \in D$ be given such that $x_i < x < x_k$ and $f(x_k) - f(x_i) \leq \varepsilon/2$. There exists $n_0 \in \mathbb{N}$ such that $|f_n(x_k) - f(x_k)| \leq \varepsilon/2$ and $|f_n(x_i) - f(x_i)| \leq \varepsilon/2$ for all $n \geq n_0$. Recall that f is increasing. So, we obtain

$$\begin{aligned} f(x_i) &\leq f(x) \leq f(x_k) \leq f(x_i) + \frac{\varepsilon}{2}, \\ f(x_i) - f_n(x_i) &\leq |f(x_i) - f_n(x_i)| \leq \frac{\varepsilon}{2}. \end{aligned}$$

Hence,

$$f(x) \leq f_n(x_i) + \varepsilon \quad \text{for all } n \geq n_0. \tag{4.3.21}$$

Similarly we show that

$$f_n(x_k) \leq f(x) + \varepsilon \quad \text{for all } n \geq n_0. \tag{4.3.22}$$

From (4.3.21) and (4.3.22) it follows that

$$f(x) - \varepsilon \leq f_n(x_i) \leq f_n(x) \leq f_n(x_k) \leq f(x) + \varepsilon \quad \text{for all } n \geq n_0.$$

Thus, $f_n(x) \to f(x)$ as $n \to \infty$ for every continuity point x of f.

Let $E \subseteq [a, b]$ be the countable set where f is not continuous. Then Proposition 4.3.37 says that there exists a subsequence of $\{f_n\}_{n\geq1}$, still denoted by the same index, such that $f_n(y) \to \hat{f}(y)$ for all $y \in E$ as $n \to \infty$. Finally let

$$f_*(x) = \begin{cases} f(x) & \text{if } x \in T \setminus E, \\ \hat{f}(x) & \text{if } x \in E. \end{cases}$$

□

Now we are ready to state and prove "Helly's Selection Theorem".

Theorem 4.3.39 (Helly's Selection Theorem). *If $T \subseteq \mathbb{R}$ is an interval and $H \subseteq \mathrm{BV}(T)$ is an infinite subset such that $\|f\|_{\mathrm{BV}} = |f(u)| + \mathrm{var}_T f \leq M$ for all $f \in H$, for some $u \in T$ and for some $M > 0$, then there exists a sequence $\{f_n\}_{n\geq1} \subseteq H$ and a function $f \in \mathrm{BV}(T)$ such that $f_n(x) \to f(x)$ for all $x \in T$.*

Proof. Corollary 4.3.32 says that $f = V^f - (V^f - f)$ for all $f \in H$ with V^f as in Proposition 4.3.29. It holds that V^f and $V^f - f$ are increasing and we obtain

$$\left|V^f(x)\right| \leq M \quad \text{and} \quad \left|V^f(x) - f(x)\right| \leq \left|V^f(x)\right| + |f(x) - f(v)| + |f(v)| \leq 3M\,,$$

for all $x \in T$. Using Proposition 4.3.38 there is a sequence $\{f_n\}_{n\geq1} \subseteq H$ and an increasing function $h_1 \colon T \to \mathbb{R}$ such that $V^{f_n}(x) \to h_1(x)$ for all $x \in T$.

A new application of Proposition 4.3.38 on $\{V^{f_n} - f_n\}$ implies that there exists a subsequence $\{V^{f_{n_k}} - f_{n_k}\}_{k\geq1}$ of $\{V^{f_n} - f_n\}_{n\geq1}$ and an increasing function $h_2 \colon T \to \mathbb{R}$ such that $(V^{f_{n_k}} - f_{n_k})(x) \to h_2(x)$ for all $x \in T$. Theorem 4.3.31 implies that $f = h_1 - h_2 \in \mathrm{BV}(T)$ and $f_{n_k}(x) \to f(x)$ for all $x \in T$. □

Using this theorem we can show that $(\mathrm{BV}(T), \|\cdot\|_{\mathrm{BV}})$ (see Proposition 4.3.36) is in fact a Banach space.

Theorem 4.3.40. *If $T \subseteq \mathbb{R}$ is an interval, then $(\mathrm{BV}(T), \|\cdot\|_{\mathrm{BV}})$ is a Banach space.*

Proof. Let $\{f_n\}_{n\geq1} \subseteq \mathrm{BV}(T)$ be a Cauchy sequence. Hence it is $\|\cdot\|_{\mathrm{BV}}$-bounded. So, by Theorem 4.3.39 there exists a subsequence $\{f_{n_k}\}_{k\geq1}$ of $\{f_n\}_{n\geq1}$ and a function $f \in \mathrm{BV}(T)$ such that $f_{n_k}(x) \to f(x)$ for all $x \in T$.

Given $\varepsilon > 0$ we find a number $n_0 = n_0(\varepsilon) \in \mathbb{N}$ such that

$$\|f_n - f_m\|_{\mathrm{BV}} = |f_n(u) - f_m(u)| + \operatorname*{var}_T(f_n - f_m) \leq \varepsilon \quad \text{for all } n, m \geq n_0\,.$$

This implies that

$$\|f_n - f_{n_k}\|_{\mathrm{BV}} = |f_n(u) - f_{n_k}(u)| + \operatorname*{var}_T(f_n - f_{n_k}) \leq \varepsilon \quad \text{for all } n, n_k \geq n_0\,.$$

Letting $k \to +\infty$ yields

$$|f_n(u) - f(u)| + \limsup_{k\to\infty} \operatorname*{var}_T(f_n - f_{n_k}) \leq \varepsilon \quad \text{for all } n \geq n_0\,. \tag{4.3.23}$$

Claim: The map $f \to \operatorname{var}_T f$ is lower semicontinuous for the pointwise convergence.

Let $f_n \to f$ pointwise and assume that $\operatorname{var}_T f > 0$, otherwise there is nothing to prove. Let $\eta \in (0, \operatorname{var}_T f)$ and let $\{x_k\}_{k=0}^m$ be a partition of T such that

$$\eta < \sum_{k=0}^{m-1} |f(x_{k+1}) - f(x_k)| .$$

Exploiting the pointwise convergence, there exists a number $n_0 = n_0(\varepsilon) \geq 1$ such that

$$|f_n(x_k) - f(x_k)| \leq \frac{\varepsilon}{2m} \quad \text{for all } n \geq n_0 \text{ and for all } k = 1, \ldots, m .$$

Then it follows that

$$\eta < \sum_{k=0}^{m-1} |f(x_{k+1}) - f(x_k)| \leq \sum_{k=0}^{m-1} \left(|f(x_{k+1}) - f(x_k)| + \frac{\varepsilon}{m} \right) \leq \operatorname*{var}_T f_n + \varepsilon$$

for all $n \geq n_0$. Since $\varepsilon > 0$ is arbitrary, we conclude that $\eta \leq \liminf_{n\to\infty} \operatorname{var}_T f_n$ and so $\operatorname{var}_T f \leq \liminf_{n\to\infty} \operatorname{var}_T f_n$. This proves the claim.

Using the claim in (4.3.23) we obtain

$$|f_n(u) - f(u)| + \operatorname*{var}_T(f_n - f) \leq \varepsilon \quad \text{for all } n \geq n_0 .$$

Hence, $f_n \to f$ in $\mathrm{BV}(T)$, and so the latter is a Banach space. □

Remark 4.3.41. The space $\mathrm{BV}(T)$ is not separable. Indeed, let $f_t = \chi_{\{t\}}$ with $t \in T$ and consider the open balls $B_t = \{h \in \mathrm{BV}(T) : \|h - f_t\|_{\mathrm{BV}} < 1\}$ with $t \in T$. Evidently if $t \neq s$, then $B_t \cap B_s = \emptyset$ and $\{B_t\}_{t\in T}$ is uncountable. So, $\mathrm{BV}(T)$ must be nonseparable; see Problem 3.61.

We conclude with a result characterizing continuous functions of bounded variation.

Definition 4.3.42. Let A, B be nonempty sets, $D \subseteq A$, and $f : A \to B$. For every $b \in B$ we define

$$N_f(b, D) = \begin{cases} \operatorname{card}\{a \in D : f(a) = b\} & \text{if the set is finite ,} \\ +\infty & \text{otherwise .} \end{cases}$$

Then $N_f(\cdot, D)$ is called the **Banach indicatrix of f on D.**

The proof of the next theorem can be found in Leoni [211, p. 68] or Natanson [251, p. 225].

Theorem 4.3.43. *If $T \subseteq \mathbb{R}$ is an interval and $f \in C(T)$, then $N_f(\cdot, T)$ is Borel and $\int_{\mathbb{R}} N_f(u, T)\, du = \operatorname{var}_T f$. Moreover, $f \in \mathrm{BV}(T)$ if and only if $N_f(\cdot, T) \in L^1(\mathbb{R})$.*

4.4 Absolutely Continuous Functions

Functions of bounded variation, although differentiable almost everywhere, fail to satisfy the fundamental theorem of calculus for the Lebesgue integral. The Cantor function h (see Remark 2.2.2) is continuous, increasing, and $h'(x) = 0$ for almost all $x \in [0,1]$. So, we have

$$0 = \int_0^1 h'(x)\,\mathrm{d}x < h(1) - h(0) = 1\,.$$

Hence, we need to go to a smaller space of functions. This smaller class of functions is given in the next definition.

Definition 4.4.1. Let $T \subseteq \mathbb{R}$ be an interval and $f\colon T \to \mathbb{R}$. We say that f is **absolutely continuous** if for every $\varepsilon > 0$ there exists a $\delta > 0$ such that

$$\left|\sum_{k=1}^{n} (f(b_k) - f(a_k))\right| \le \varepsilon \tag{4.4.1}$$

for any family of nonoverlapping open intervals (a_k, b_k), $k = 1, \dots, n$ with $[a_k, b_k] \subseteq T$ and $\sum_{k=1}^{n}(b_k - a_k) \le \delta$. We denote the space of absolutely continuous functions by $\mathrm{AC}(T)$. We say that $f\colon T \to \mathbb{R}$ is **locally absolutely continuous** if it is absolutely continuous in $[a,b]$ for every $[a,b] \subseteq T$. We denote the space of locally absolutely continuous functions by $\mathrm{AC}_{\mathrm{loc}}(T)$. Of course it holds that $\mathrm{AC}([a,b]) = \mathrm{AC}_{\mathrm{loc}}([a,b])$.

Remark 4.4.2. If $U \subseteq \mathbb{R}$ is open, then the definition above is still valid if instead of T we use U. Now we require that $[a_k, b_k] \subseteq U$ for all $k = 1, \dots, n$. In the definition above, $n \in \mathbb{N}$ is arbitrary and in fact we can allow it to be $+\infty$ by replacing the finite series by infinite ones. In the definition of absolute continuity of a function, without altering the definition, we can replace (4.4.1) by the following stronger requirement

$$\sum_{k=1}^{n} |f(b_k) - f(a_k)| \le \varepsilon\,. \tag{4.4.2}$$

Indeed, let $\{(a_k, b_k)\}_{k=1}^{n}$ be a family of pairwise disjoint open intervals such that $[a_k, b_k] \subseteq T$ and $\sum_{k=1}^{n}(b_k - a_k) \le \delta$. Let $\mathcal{L}_1$ = the subintervals $[a_k, b_k]$ for which $f(b_k) - f(a_k) \ge 0$ and $\mathcal{L}_2$ = the subintervals for which $f(b_k) < f(a_k)$. Moreover, let $\delta > 0$ be corresponding to $\varepsilon/2$ in (4.4.1). Then

$$\sum_{\mathcal{L}_1} |f(b_k) - f(a_k)| = \left|\sum_{\mathcal{L}_1} (f(b_k) - f(a_k))\right| \le \frac{\varepsilon}{2}$$

and

$$\sum_{\mathcal{L}_2} |f(b_k) - f(a_k)| = \left| \sum_{\mathcal{L}_2} (f(b_k) - f(a_k)) \right| \leq \frac{\varepsilon}{2} .$$

Therefore (4.4.2) holds.

If we take $n = 1$ in Definition 4.4.1, this leads to the following result.

Proposition 4.4.3. *If $f \in \mathrm{AC}(T)$, then f is uniformly continuous.*

Remark 4.4.4. The converse is not true. The function

$$f(x) = \begin{cases} x \sin(\frac{1}{x}) & \text{if } x \in (0,1] , \\ 0 & \text{if } x = 0 , \end{cases}$$

is uniformly continuous on $[0, 1]$ but not absolutely continuous. On the other hand the function

$$f(x) = \begin{cases} x^{1+\varepsilon} \sin(\frac{1}{x}) & \text{if } x \in (0,1] , \\ 0 & \text{if } x = 0 , \end{cases}$$

is absolutely continuous on $[0, 1]$ for $\varepsilon > 0$. Now if $f \in \mathrm{AC}(T)$, then, on account of the previous proposition, f is uniformly continuous and so it can be extended uniquely to $\overline{T}$ to a uniformly continuous function $\hat{f}$; see Theorem 1.5.27. Then $\hat{f} \in \mathrm{AC}(\overline{T})$.

Another straightforward consequence of Definition 4.4.1 is the following proposition.

Proposition 4.4.5. *If $f, h \in \mathrm{AC}([a, b])$ and $c \in \mathbb{R}$, then the following hold:*
(a) *$f \pm ch \in \mathrm{AC}([a, b])$;*
(b) *$fh \in \mathrm{AC}([a, b])$;*
(c) *if $h > 0$, then $f/h \in \mathrm{AC}([a, b])$.*

Proposition 4.4.6. *If $f \in \mathrm{AC}([a, b])$, $f([a, b]) \subseteq [c, d]$ and $\xi \colon [c, d] \to \mathbb{R}$ is Lipschitz continuous, then $h = \xi \circ f \in \mathrm{AC}([a, b])$.*

Proof. Suppose that $|\xi(x) - \xi(u)| \leq \eta |x - u|$ for all $x, u \in [a, b]$ and with $\eta > 0$. Then for any family $\{(a_k, b_k)\}_{k=1}^n$ of pairwise disjoint open intervals, we obtain

$$\sum_{k=1}^{n} |\xi(f(b_k)) - \xi(f(a_k))| \leq \eta \sum_{k=1}^{n} |f(b_k) - f(a_k)| .$$

Hence, $\xi \circ f \in \mathrm{AC}([a, b])$. ☐

Proposition 4.4.7. *If $T \subseteq \mathbb{R}$ is an interval and $f \in \mathrm{AC}_{\mathrm{loc}}(T)$ (resp. $f \in \mathrm{AC}(T)$), then $f \in \mathrm{BV}_{\mathrm{loc}}(T)$ (resp. $f \in \mathrm{BV}(I)$) for every bounded subinterval $I \subseteq T$.*

Proof. First assume that $f \in \mathrm{AC}_{\mathrm{loc}}(T)$. Let $[a, b] \subseteq T$. Choose $\varepsilon = 1$ and let $\delta > 0$ be as in Definition 4.4.1. Moreover, let $n = [(2(b-a))/\delta]$ be the integer part of $(2(b-a))/\delta$ and partition $[a, b]$ to n intervals $[x_k, x_{k+1}]$ of length $(b-a)/n$ with $k = 0, \ldots, n-1$, that is, $a = x_0 < x_1 < \cdots < x_n = b$. Recall $\varepsilon = 1$. It follows that

$$\operatorname*{var}_{[x_k, x_{k+1}]} f \le 1 \quad \text{for all } k = 0, \ldots, n-1,$$

which implies

$$\operatorname*{var}_{[a,b]} f = \sum_{k=0}^{n-1} \operatorname*{var}_{[x_k, x_{k+1}]} f \le n \le \frac{2(b-a)}{\delta} < +\infty .$$

Hence $f \in \mathrm{BV}_{\mathrm{loc}}(T)$.

Next assume that $f \in \mathrm{BV}(T)$ and let $I \subseteq T$ be a bounded subinterval. We know that f can extended to $\bar{I}$ and for the extension $\hat{f}$ we have $\hat{f} \in \mathrm{AC}(\bar{I})$; see Remark 4.4.4. Then from the first part of the proof it follows that $\operatorname{var}_{\bar{I}} f < +\infty$ and so $\operatorname{var}_I f < +\infty$. □

Corollary 4.4.8. (a) *If $T \subseteq \mathbb{R}$ is an interval and $f \in \mathrm{AC}_{\mathrm{loc}}(T)$, then f is differentiable a. e. on T and $f' \in L^1_{\mathrm{loc}}(T)$.*

(b) *If $T \subseteq \mathbb{R}$ is a bounded interval, then $\mathrm{AC}(T) \subseteq \mathrm{BV}(T)$ and for $f \in \mathrm{AC}(T)$ it holds that $f' \in L^1(T)$.*

Remark 4.4.9. The inclusion above is clearly strict. Consider a monotone discontinuous function. Moreover, there exist continuous monotone functions that are not absolutely continuous. Think of the Cantor function; see Remark 2.2.2. So, it is natural to ask what is missing from a continuous function $f \in \mathrm{BV}_{\mathrm{loc}}(T)$ in order to be absolutely continuous. The property that we seek is given in the next definition.

Definition 4.4.10. Let $T \subseteq \mathbb{R}$ be an interval and $f: T \to \mathbb{R}$ a function. We say that f satisfies **Lusin's Condition (N)** if f maps sets of Lebesgue measure zero to sets of Lebesgue measure zero.

Proposition 4.4.11. *If $T \subseteq \mathbb{R}$ is an interval and $f: T \to \mathbb{R}$ is a continuous function, then f maps Lebesgue measurable sets to Lebesgue measurable sets if and only if f satisfies Lusin's Condition (N).*

Proof. $\Longrightarrow$: Arguing by contradiction, suppose that f does not satisfy Lusin's condition (N). Then we can find a Lebesgue-null set $D_0 \subseteq T$ such that $0 < \lambda^*(f(D_0))$ with λ^* being the Lebesgue outer measure on $\mathbb{R}$. Then $f(D_0)$ contains a nonmeasurable set E. Let $C \subseteq D_0$ such that $f(C) = E$. Since C is a subset of a Lebesgue-null set, it is Lebesgue measurable. But E is not, which is a contradiction to the hypothesis.

$\Longleftarrow$: Let $A \subseteq T$ be a Lebesgue measurable set. Then $A = C \cup D$ with C being σ-compact and D being Lebesgue-null. Here we use the fact that the Lebesgue measure is Radon; see Theorem 2.5.14. Then it follows that

$$f(A) = f(C \cup D) = f(C) \cup f(D) . \qquad (4.4.3)$$

Since $C = \bigcup_{n\geq 1} K_n$ with compact K_n, then $f(C) = f(\bigcup_{n\geq 1} K_n) = \bigcup_{n\geq 1} f(K_n)$ and for each $n \in \mathbb{N}$, $f(K_n) \subseteq \mathbb{R}$ is compact. In addition, by hypothesis, $f(D)$ is Lebesgue-null. So, from (4.4.3) it follows that $f(A)$ is measurable. □

Theorem 4.4.12. *If $T \subseteq \mathbb{R}$ is an interval and $f: T \to \mathbb{R}$, then $f \in \mathrm{AC}_{\mathrm{loc}}(T)$ if and only if*
(a) *f is continuous on T;*
(b) *$f \in \mathrm{BV}_{\mathrm{loc}}(T)$;*
(c) *f satisfies Lusin's Condition (N).*

Proof. Since the result is local we may assume that $T = [a, b]$.

First we suppose that f is absolutely continuous on $[a, b]$. Evidently f is continuous and by Proposition 4.4.7, $f \in \mathrm{BV}_{\mathrm{loc}}(T)$. So it remains to prove statement (c). Let $\varepsilon > 0$. There exists $\delta > 0$ such that for any finite (or countable) collection of mutually disjoint intervals $\{(a_k, b_k)\}_{k\geq 1}$ with $[a_k, b_k] \subseteq [a, b]$ such that

$$\sum_{k\geq 1}(b_k - a_k) \leq \delta \quad \text{implies} \quad \sum_{k\geq 1}|f(b_k) - f(a_k)| \leq \varepsilon . \qquad (4.4.4)$$

Let $D \subseteq T$ be Lebesgue-null and let $U = \bigcup_{k\geq 1}(c_k, d_k)$ be an open set such that $D \subseteq U$ and $\lambda(U) = \sum_{k\geq 1}(d_k - c_k) \leq \delta$. It holds that

$$f(D) \subseteq f(U) \subseteq f\Big(\bigcup_{k\geq 1}[c_k, d_k]\Big) \subseteq \bigcup_{k\geq 1}[f(m_k), f(M_k)]$$

with $m_k, M_k \in [c_k, d_k]$ such that

$$f(m_k) = \min[f(u) : u \in [c_k, d_k]] ,$$
$$f(M_k) = \max[f(u) : u \in [c_k, d_k]] .$$

Taking into account (4.4.4) and recalling that $\sum_{k\geq 1}(M_k - m_k) \leq \delta$, we derive

$$\lambda^*(f(D)) \leq \sum_{k\geq 1}[f(M_k) - f(m_k)] \leq \varepsilon .$$

Because $\varepsilon > 0$ is arbitrary, we let $\varepsilon \to 0^+$ to conclude that $\lambda(f(D)) = 0$, that is, $f(D)$ is Lebesgue-null. Therefore f satisfies statement (c).

Now suppose that properties (a), (b), and (c) hold. Arguing by contradiction, suppose that f is not absolutely continuous. So, there exists $\varepsilon_0 > 0$ such that for every $\delta > 0$ there is a pairwise disjoint family of open intervals $\{(a_k, b_k)\}_{k=1}^n$ in $[a, b]$ such that

$$\sum_{k=1}^{n}(b_k - a_k) \leq \delta \quad \text{and} \quad \sum_{k=1}^{n}(M_k - m_k) \geq \varepsilon_0$$

with $m_k = \min[f(u): u \in [a_k, b_k]\}$ and $M_k = \max[f(u): u \in [a_k, b_k]\}$. Let $\sum_{m\geq 1} \delta_m$ be a convergent series of positive terms. For each δ_m, let (a_k^m, b_k^m) with $k = 1, \ldots, n_m$ be pairwise disjoint open intervals for which

$$\sum_{k=1}^{n_m} (b_k^m - a_k^m) \leq \delta_m \quad \text{and} \quad \sum_{k=1}^{n_m} (M_k^m - m_k^m) \geq \varepsilon_0 \, . \tag{4.4.5}$$

Let

$$C_m = \bigcup_{k=1}^{n_m} (a_k^m, b_k^m) \quad \text{and} \quad B = \bigcap_{n\geq 1} \bigcup_{m\geq n} C_m = \limsup_{m\to\infty} C_m \, .$$

Then $\lambda(B) = 0$, and so by (c) it follows that $\lambda(f(B)) = 0$. For $k = 1, \ldots, n_m$ with $m \in \mathbb{N}$, we define the following functions:

$$\xi_k^m(u) = \begin{cases} 1 & \text{if } f(x) = u \text{ for some } x \in (a_k^m, b_k^m)\,, \\ 0 & \text{otherwise}\,. \end{cases}$$

Then $\xi_k^m(u) = 1$ for all $u \in (m_k^m, M_k^m)$ and $\xi_k^m(u) = 0$ for all $u \notin [m_k^m, M_k^m]$. Therefore with $\hat{m} = \min_T f$, $\hat{M} = \max_T f$, that is $[\hat{m}, \hat{M}] = f([a, b])$, we infer that

$$\int_{\hat{m}}^{\hat{M}} \xi_k^m(u)\, du = M_k^m - m_k^m \, . \tag{4.4.6}$$

We set $N_m(u) = \sum_{k=1}^{n_m} \xi_k^m(u)$. Then $N_m(u)$ is the number of intervals (a_k^m, b_k^m) that contain at least one x with $f(x) = u$. So, $N_m(u) \leq N_f(y, T)$ the latter being indicatrix function of f on $T = [a, b]$; see Definition 4.3.42. From (4.4.5) and (4.4.6) we obtain

$$\int_{\hat{m}}^{\hat{M}} N_m(u)\, du \geq \varepsilon_0 \, . \tag{4.4.7}$$

Let $E = \{u \in T = [a, b]: \lim_{m\to\infty} N_m(u) \neq 0\}$ and $F = \{u \in T = [a, b]: N_f(u, T) = +\infty\}$. From Theorem 4.3.43 one has that $N_f(\cdot, T) \in L^1(T)$. Therefore $\lambda(F) = 0$. Let $u_0 \in E \setminus F = E \cap F^c$. There exists a sequence $\{m_i\}_{i\geq 1} \subseteq \mathbb{N}$ such that $N_{m_i}(u_0) \geq 1$. For every $i \in \mathbb{N}$ there exists x_{m_i} such that

$$f(x_{m_i}) = u_0 \, , \quad x_{m_i} \in C_{m_i} \, . \tag{4.4.8}$$

Since $N_f(u_0, T) < \infty$, there are only finitely many distinct $\{x_{m_i}\}_{i\in\mathbb{N}}$ such that (4.4.8) holds. Therefore one of them, call it x_0, occurs an infinite number of times in $\{x_{m_i}\}_{i\in\mathbb{N}}$ and $f(x_0) = u_0$. Then $x_0 \in B$ and $f(x_0) = u_0 \in f(B)$. Hence, we get $E \setminus F = E \cap F^c \subseteq f(B)$, which shows that $\lambda(E) = 0$ and so $\lim_{m\to\infty} N_m(u) = 0$ for almost all $u \in [m, M]$. By the Lebesgue's Dominated Convergence Theorem, we infer that $\lim_{m\to\infty} \int_{\hat{m}}^{\hat{M}} N_m(u)\,du = 0$. This contradicts (4.4.7). Thus, f is absolutely continuous. □

As a consequence of Proposition 4.4.11 and Theorem 4.4.12, we obtain the following corollary.

Corollary 4.4.13. *If $T \subseteq \mathbb{R}$ is an interval and $f \in \mathrm{AC}_{\mathrm{loc}}(T)$, then f maps Lebesgue measurable sets to Lebesgue measurable sets.*

Next we present some results about differentiable functions, which will help us better understand absolutely continuous functions.

Proposition 4.4.14. *If $T \subseteq \mathbb{R}$ is an interval, $f: T \to \mathbb{R}$ is differentiable at every $x \in A$ with $A \subseteq T$ not necessarily measurable and $|f'(x)| \le M$ for all $x \in A$ and for some $M > 0$, then $\lambda^*(f(A)) \le M\lambda^*(A)$ with λ^* being the Lebesgue outer measure.*

Proof. Without any loss of generality, we assume that $A \subseteq \operatorname{int} T$. For every $n \in \mathbb{N}$, let A_n be the set of points $x \in A$ such that $\lambda^*(f(I)) \le (M+\varepsilon)\lambda(I)$ for all intervals $I \subseteq T$ with $x \in I$ and $\lambda(I) \in (0, 1/n)$. We see that $\{A_n\}_{n\ge1}$ is increasing. In addition, if $x \in A$, then

$$|f'(x)| \le M \quad \text{and} \quad f'(x) = \lim_{u\to x} \frac{f(u) - f(x)}{u - x}\,.$$

So, there exists $\delta > 0$ such that

$$|f(u) - f(x)| \le (M+\varepsilon)(u - x) \quad \text{for all } u \in I \text{ with } |u - x| \le \delta\,. \tag{4.4.9}$$

Let $u, v \in I$ with $|u - v| \le \delta$ and $u < x < u'$. Thanks to (4.4.9), we obtain

$$|f(u) - f(v)| \le |f(u) - f(x)| + |f(x) - f(v)| \le (M+\varepsilon)(v - u)\,.$$

This gives $x \in A_n$ for every $n \in \mathbb{N}$ with $n > 1/\delta$. It follows that

$$A = \bigcup_{n\ge1} A_n\,. \tag{4.4.10}$$

We fix $n \in \mathbb{N}$ and let U_n be open such that $\lambda(U_n) \le \lambda^*(A_n) + \varepsilon$. Replacing U_n with $U_n \cap \operatorname{int} T$ if necessary, we may assume that $U_n \subseteq \operatorname{int} T$. We can write $U_n = \bigcup_{k\ge1} I_k^n$ with $\{I_k^n\}_{k\ge1}$ being pairwise disjoint intervals with $0 < \lambda(I_k^n) < 1/n$ for all $k \in \mathbb{N}$. Let $\mathcal{L} = \{k \in \mathbb{N} : I_k^n \cap A_n \ne \emptyset\}$. If $k \in \mathcal{L}$, then $\lambda^*(f(I_k^n)) \le (M+\varepsilon)\lambda^*(I_k^n)$ and so

$$\begin{aligned}
\lambda^*(f(A_n)) &\le \lambda^*\Big(\bigcup_{k\in\mathcal{L}} f(I_k^n)\Big) \le \sum_{k\in\mathcal{L}} \lambda^*(f(I_k^n)) \le (M+\varepsilon)\lambda\Big(\bigcup_{k\ge1} I_k^n\Big) \\
&\le (M+\varepsilon)\lambda(U_n) \le (M+\varepsilon)(\lambda^*(A_n) + \varepsilon)\,.
\end{aligned}$$

Letting $n \to \infty$ gives

$$\lambda^*(f(A)) \le (M + \varepsilon)(\lambda^*(A) + \varepsilon)\,;$$

see (4.4.10). Finally we let $\varepsilon \searrow 0$ to conclude that

$$\lambda^*(f(A)) \le M\lambda^*(A)\,. \qquad \square$$

Corollary 4.4.15. *If $T \subseteq \mathbb{R}$ is an interval, $f: T \to \mathbb{R}$ is differentiable on $A \subseteq T$ and either A is Lebesgue-null or $f'|_A = 0$, then $\lambda(f(A)) = 0$.*

Proof. First suppose that A is Lebesgue-null. For every $n \in \mathbb{N}$, let $A_n = \{x \in A: |f'(x)| \le n\}$. Then $A = \bigcup_{n\ge1} A_n$ and by Proposition 4.4.14, we have that $\lambda^*(f(A_n)) \le n\lambda^*(A_n) = 0$. Since $f(A) = \bigcup_{n\ge1} f(A_n)$, the countable subadditivity of λ^* gives $\lambda(f(A)) = 0$.

The case of $f'|_A = 0$ follows directly from Proposition 4.4.14 with $M = 0$. $\square$

Remark 4.4.16. If we assume that f is almost everywhere differentiable on A in the corollary above, then the result fails. Think of the Cantor function; see Remark 2.2.2.

Proposition 4.4.17. *If $T \subseteq \mathbb{R}$ is an interval, $f: T \to \mathbb{R}$ is Lebesgue measurable, $A \subseteq T$ is Lebesgue measurable, and f is differentiable on A, then $f(A) \subseteq \mathbb{R}$ is Lebesgue measurable and $\lambda(f(A)) \le \int_A |f'(x)|\,dx$.*

Proof. From Corollary 4.4.15 we see that $f|_A$ satisfies Lusin's Condition (N) and so Proposition 4.4.11 implies that $f(A) \subseteq \mathbb{R}$ is Lebesgue measurable.

First suppose that $\lambda(A) < +\infty$. We fix $n \in \mathbb{N}$ and for every $k \in \mathbb{N}$, we define

$$A_n^k = \left\{x \in A: \frac{k-1}{2^n} \le |f'(x)| < \frac{k}{2^n}\right\}.$$

We see that $\{A_n^k\}_{k\ge1}$ are pairwise disjoint and $A = \bigcup_{k\ge1} A_n^k$. Due to Proposition 4.4.14 and the σ-additivity of λ, we get

$$\begin{aligned}
\lambda(f(A)) &= \lambda\Big(f\Big(\bigcup_{k\ge1} A_n^k\Big)\Big) \le \sum_{k\ge1} \lambda(f(A_n^k)) \le \sum_{k\ge1} \frac{k}{2^n}\lambda(A_n^k) \\
&= \sum_{k\ge1} \frac{k-1}{2^n}\lambda(A_n^k) + \frac{1}{2^n}\sum_{k\ge1}\lambda(A_n^k) \le \sum_{k\ge1} \int_{A_n^k} |f'(x)|\,dx + \frac{1}{2^n}\lambda(A) \\
&= \int_A |f'(x)|\,dx + \frac{1}{2^n}\lambda(A) \quad \text{for all } n \in \mathbb{N}\,.
\end{aligned}$$

Letting $n \to \infty$ we conclude that $\lambda(f(A)) \le \int_A |f'(x)|\,dx$.

Next suppose that $\lambda(A) = +\infty$. For every $k \in \mathbb{Z}$ we set $A_k = A \cap [k, k+1)$. Then we obtain from the previous part that

$$\lambda(f(A)) = \lambda\Big(f\Big(\bigcup_{k\in\mathbb{Z}} A_k\Big)\Big) = \lambda\Big(\bigcup_{k\in\mathbb{Z}} f(A_k)\Big) = \sum_{k\in\mathbb{Z}} \lambda(f(A_k))$$

$$\leq \sum_{k\in\mathbb{Z}} \int_{A_k} |f'(x)|\,\mathrm{d}x = \int_A |f'(x)|\,\mathrm{d}x\,. \qquad \square$$

Corollary 4.4.18. *If $T \subseteq \mathbb{R}$ is an interval and $f\colon T \to \mathbb{R}$ is differentiable on $[a,b] \subseteq T$, then $|f(b) - f(a)| \leq \int_a^b |f'(x)|\,\mathrm{d}x$.*

From this corollary we infer the following result.

Theorem 4.4.19. *If $T \subseteq \mathbb{R}$ is an interval, $f\colon T \to \mathbb{R}$ is differentiable everywhere on T and $f' \in L^1_{\mathrm{loc}}(T)$, then $f \in \mathrm{AC}_{\mathrm{loc}}(T)$.*

Next we will show that absolutely continuous functions are exactly the class of functions for which the fundamental theorem of calculus for the Lebesgue integral holds.

Lemma 4.4.20. *If $h \in L^1[a,b]$ and $f(x) = \int_a^x h(t)\,\mathrm{d}t$ for all $x \in [a,b]$, which is the indefinite Lebesgue integral of h, then $f \in \mathrm{AC}([a,b])$ and $f' = h$.*

Proof. From Proposition 2.3.42, we know that for a given $\varepsilon > 0$ there exists a $\delta > 0$ such that if $\{(a_k, b_k)\}_{k=1}^n$ are pairwise disjoint intervals such that $[a_k, b_k] \subseteq [a,b]$ for all $k = 1,\dots,n$ and $\sum_{k=1}^n (b_k - a_k) \leq \delta$, then $|\sum_{k=1}^n \int_{a_k}^{b_k} h(t)\,\mathrm{d}t| \leq \varepsilon$. Then $|\sum_{k=1}^n (f(b_k) - f(a_k))| \leq \varepsilon$ and so $f \in \mathrm{AC}([a,b])$; see Definition 4.4.1. Clearly if h is simple, then $f' = h$ a. e. For the general case, use Corollary 2.2.19. $\square$

Theorem 4.4.21. *If $T \subseteq \mathbb{R}$ is an interval and $f\colon T \to \mathbb{R}$, then $f \in \mathrm{AC}_{\mathrm{loc}}(T)$ if and only if*
(a) *f is continuous on T;*
(b) *f is differentiable a. e. on T and $f' \in L^1_{\mathrm{loc}}(T)$;*
(c) *$f(x) = f(c) + \int_c^x f'(t)\,\mathrm{d}t$ for all $c, x \in T$.*

Proof. First assume that $f \in \mathrm{AC}_{\mathrm{loc}}(T)$. On account of Corollary 4.4.8 we need to prove (c). Let $[a,b] \subseteq T$ such that $c \in [a,b]$. From Lemma 4.4.20 we know that $f \in \mathrm{AC}([a,b])$ and there exists a Lebesgue-null set $D \subseteq [a,b]$ such that f is differentiable on $[a,b]\backslash D$. If we set $g(x) = f(x) - [f(c) + \int_c^x f'(t)\,\mathrm{d}t]$, then $g'(x) = 0$ for all $x \in [a,b] \setminus D$. Then Corollary 4.4.15 implies that $\lambda(g([a,b] \setminus D)) = 0$. Note that $g \in \mathrm{AC}([a,b])$. So, from Theorem 4.4.12, it maps Lebesgue-null sets to Lebesgue-null sets. Hence, $\lambda(g(D)) = 0$. Therefore it follows that $\lambda(g([a,b])) = 0$. The continuity of g implies that $g([a,b])$ is either a singleton or a nondegenerate interval. Then the second possibility is excluded and so we conclude that g is constant. We have $g(c) = 0$, hence $g|_{[a,b]} = 0$ and so $f(x) = f(c) + \int_c^x f'(t)\,\mathrm{d}t$ for all $x \in [a,b]$.

Conversely if (a), (b), and (c) hold, then from Lemma 4.4.20 we get $f \in \mathrm{AC}_{\mathrm{loc}}(T)$. $\square$

Corollary 4.4.22. *If $T \subseteq \mathbb{R}$ is an interval, $f\colon T \to \mathbb{R}$ is everywhere differentiable and $f' \in L^1_{\mathrm{loc}}(T)$, then $f(x) = f(c) + \int_c^x f'(t)\,\mathrm{d}t$ for all $c, x \in T$.*

An important consequence of Theorem 4.4.21 is the following formula known as "Integration by Parts".

Proposition 4.4.23 (Integration by Parts). *If $T \subseteq \mathbb{R}$ is an interval and $f, h \in \mathrm{AC}_{\mathrm{loc}}(T)$, then*

$$\int_c^x fh' \, dt + \int_c^x f'h \, dt = (fh)(x) - (fh)(c) \quad \text{for all } c, x \in T .$$

Proof. Recall that $fh \in \mathrm{AC}_{\mathrm{loc}}(T)$; see Proposition 4.4.5. So, from Theorem 4.4.21, we obtain

$$(fh)(x) - (fh)(c) = \int_c^x (fh)'(x) \, dx .$$

Since f, h, fh are differentiable almost everywhere, one has $(fh)'(x) = f'(x)h(x) + f(x)h'(x)$ for a. a. $x \in T$ and so we obtain the desired formula. □

Now we will use Theorem 4.4.21 to make a connection between the notions of absolute continuity for functions and for Lebesgue–Stieltjes measures; see Example 2.1.35.

Proposition 4.4.24. *A continuous increasing function $f \colon [a, b] \to \mathbb{R}$ is absolutely continuous if and only if the corresponding Lebesgue–Stieltjes measure $\vartheta_f([a, b]) = f(b) - f(a)$ is absolutely continuous with respect to the Lebesgue measure λ, that is, $\vartheta_f \ll \lambda$.*

Proof. $\Longrightarrow$: From Theorem 4.4.21 we get $\vartheta_f(A) = \lambda(f(A))$ for all Lebesgue measurable $A \subseteq [a, b]$. Then invoking Theorem 4.4.12, we see that $\vartheta_f \ll \lambda$.

$\Longleftarrow$: Since ϑ_f is finite on $[a, b]$ and $\vartheta_f \ll \lambda$, for a given $\varepsilon > 0$ there exists $\delta > 0$ such that

$$\lambda(A) \le \delta \quad \text{implies} \quad \vartheta_f(A) \le \varepsilon \, ; \tag{4.4.11}$$

see Proposition 2.4.11. Suppose that $\{(a_k, b_k)\}_{k=1}^n$ are pairwise disjoint open intervals with $[a_k, b_k] \subseteq [a, b]$ for all $k = 1, \ldots, n$. Then for $A = \bigcup_{k=1}^n (a_k, b_k)$, from (4.4.11), one obtains that

$$\lambda(A) = \sum_{k=1}^n (b_k - a_k) \le \delta \quad \text{implies} \quad \vartheta_f(A) = \sum_{k=1}^n [f(b_k) - f(a_k)] \le \varepsilon .$$

Hence, $f \in \mathrm{AC}([a, b])$. □

Corollary 4.4.25. *If $T \subseteq \mathbb{R}$ is an interval, $f \colon T \to \mathbb{R}$ is a monotone function and $[a, b] \subseteq T$, then $f \in \mathrm{AC}([a, b])$ if and only if $|f(b) - f(a)| = \int_a^b |f'(x)| \, dx$. Moreover, if f is bounded, then $f \in \mathrm{AC}(T)$ if and only if $\int_T |f'(x)| \, dx = \sup_T f - \inf_T f$.*

This corollary leads to the following result.

Theorem 4.4.26. *If $T \subseteq \mathbb{R}$ is an interval, $f \in \mathrm{BV}_{\mathrm{loc}}(T)$, and $[a,b] \subseteq T$, then $f \in \mathrm{AC}([a,b])$ if and only if $\int_a^b |f'(x)|\,\mathrm{d}x = \mathrm{var}_{[a,b]} f$. Moreover, if $f \in \mathrm{BV}(T)$, then $f \in \mathrm{AC}(T)$ if and only $\int_T |f'(x)|\,\mathrm{d}x = \mathrm{var}_T f$.*

Proof. $\Longrightarrow$: Let $\{x_k\}_{k=0}^n$ be a partition of $[a,b]$. Then

$$\sum_{k=0}^{n-1} |f(x_{k+1}) - f(x_k)| = \sum_{k=0}^{n-1} \left| \int_{x_k}^{x_{k+1}} f'(x)\,\mathrm{d}x \right| \le \sum_{k=0}^{n-1} \int_{x_{k-1}}^{x_{k+1}} |f'(x)|\,\mathrm{d}x = \int_a^b |f'(x)|\,\mathrm{d}x\,.$$

Hence,

$$\operatorname*{var}_{[a,b]} f \le \int_a^b |f'(x)|\,\mathrm{d}x\,. \tag{4.4.12}$$

Combining (4.4.12) with Corollary 4.3.33 we conclude that $\mathrm{var}_{[a,b]} f = \int_a^b |f'(x)|\,\mathrm{d}x$.

If $f \in \mathrm{AC}(T)$, then from the previous part, for every $[a,b] \subseteq T$, we obtain

$$\int_a^b |f'(x)|\,\mathrm{d}x = \operatorname*{var}_{[a,b]} f\,. \tag{4.4.13}$$

Let $m = \inf T$ as well as $M = \sup T$ and consider $a_n \searrow m$ and $b_n \nearrow M$. From (4.4.13) it follows that

$$\int_{a_n}^{b_n} |f'(x)|\,\mathrm{d}x = \operatorname*{var}_{[a_n,b_n]} f \quad \text{for all } n \in \mathbb{N}\,.$$

Passing to the limit as $n \to \infty$ and using Proposition 4.3.25 as well as the Monotone Convergence Theorem, we derive

$$\int_T |f'(x)|\,\mathrm{d}x = \operatorname*{var}_T f\,.$$

$\Longleftarrow$: From Corollary 4.3.33 it holds that $\int_a^b |f'(x)|\,\mathrm{d}x = \mathrm{var}_{[a,b]} f = \int_a^b V_a'(x)\,\mathrm{d}x = V_a(b)$. Since V_a is increasing (see Proposition 4.3.29), from Corollary 4.4.25 we get $V_a \in \mathrm{AC}([a,b])$. But from Proposition 4.3.29 it follows that $|f(x) - f(y)| \le V_a(x) - V_a(y)$ for all $a \le x \le y \le b$. Hence, $f \in \mathrm{AC}([a,b])$.

If $f \in \mathrm{BV}(T)$, then Corollary 4.3.33 gives, for $x_0 \in T$, that

$$\int_T |f'(x)|\,\mathrm{d}x \le \int_T |V_{x_0}'(x)|\,\mathrm{d}x \le \sup_T V_{x_0} - \inf_T V_{x_0} = \operatorname*{var}_T f\,.$$

Using the hypothesis we obtain

$$\int_T |V'_{x_0}(x)|\,dx = \sup_T V_{x_0} - \inf_T V_{x_0}\,.$$

Hence, $V_{x_0} \in AC(T)$, see Corollary 4.4.25 and as before, we conclude that $f \in AC(T)$. □

By replacing the absolute value by the norm, the notion of absolute continuity can be extended to vector-valued functions.

Definition 4.4.27. Let $T = [a, b]$ and let X be a Banach space. A function $f: T \to X$ is said to be **absolutely continuous** if for every $\varepsilon > 0$ there exists $\delta = \delta(\varepsilon) > 0$ such that for any family $\{(a_k, b_k)\}_{k=1}^n$ of pairwise disjoint open subintervals of T such that $\sum_{k=1}^n (b_k - a_k) \leq \delta$, we have $\sum_{k=1}^n \|f(b_k) - f(a_k)\| \leq \varepsilon$.

In Theorem 4.4.21 we have seen that if $X = \mathbb{R}$ or more generally $\mathbb{R}^N$, then the fundamental theorem of Lebesgue calculus characterizes absolutely continuous functions. This is no longer true for X-valued functions when X is an infinite dimensional Banach space.

Example 4.4.28. Let $X = L^1[0,1]$ and let $f: [0,1] \to X$ be defined by $f(t) = \chi_{[0,t]}$. Clearly f is absolutely continuous. However f is nowhere differentiable. Indeed, if f was differentiable at $t = t_0$, then for every $h \in L^\infty[0,1] = L^1[0,1]^*$, $t \to \xi(t) = \langle h, f(t)\rangle$ is differentiable at $t = t_0$. This means that $t \to \xi(t) = \int_0^t h(s)\,ds$ is differentiable at $t = t_0$ for every $h \in L^\infty[0,1]$. Choose

$$h(s) = \begin{cases} 1 & \text{if } s < t_0 \\ -1 & \text{if } t_0 < s \end{cases}, \quad \text{then } \xi(t) = \begin{cases} t & \text{if } t < t_0 \\ 2t_0 - t & \text{if } t_0 \leq t, \end{cases}$$

and this function is not differentiable at $t = t_0$.

The problem with the example above is the fact that $X = L^1[0,1]$ does not have the RNP; see Definition 4.2.23 (d). In fact we can state the following result.

Theorem 4.4.29. *If $T = [a, b]$ and X is a Banach space, then the following statements are equivalent:*

(a) *X has the* RNP;
(b) *every absolutely continuous function $f: T \to X$ is almost everywhere differentiable and $f(t) = f(s) + \int_s^t f'(\tau)\,d\tau$ for all $0 \leq s \leq t \leq b$.*

Proof. (a) ⟹ (b): Let $f: T \to X$ be an absolutely continuous function. The variation of f,

$$V_0(t) = \sup\left[\sum_{k=0}^{n-1} \|f(t_{k+1}) - f(t_k)\| : 0 = t_0 < t_1 < \cdots < t_n = b\right],$$

is an $\mathbb{R}_+$-valued absolutely continuous function. Therefore there exists a finite positive measure μ on T such that $V_0(t) = \mu([0,t])$. Evidently $\mu \ll \lambda$.

We introduce a vector measure $m\colon \mathcal{B}(X) \to X$, with $\mathcal{B}(X)$ being the Borel σ-field of T, defined as follows. If $U \subseteq T$ is open, then $U = \bigcup_{k\geq 1}(a_k, b_k)$ with a disjoint union and we set $m(U) = \sum_{k\geq 1}[f(b_k) - f(a_k)]$. For a general $A \in \mathcal{B}(T)$, let $\{U_n\}_{n\geq 1}$ be a decreasing sequence of open sets, $A \subseteq U_n$ for all $n \in \mathbb{N}$ such that $\mu(U_n) \searrow \mu(A)$. Note that $\mu(U_n \setminus U_m) \to 0$ for $n < m$ and $n \to \infty$. Therefore $m(A) = \lim_{n\to\infty} m(U_n)$ exists. Evidently m is well-defined, σ-additive, thus a measure, and $\|m(A)\| \leq \mu(A)$ for all $A \in \mathcal{B}(T)$. Therefore $m \ll \mu$, hence $m \ll \lambda$. Since X has the RNP (see Definition 4.2.23 (d)), there exists $h \in L^1(T, X)$ such that $m([0,t]) = \int_0^t h(s)\,\mathrm{d}s$. Therefore $f(t) = f(0) + m([0,t]) = f(0) + \int_0^t h(s)\,\mathrm{d}s$. This finally gives $f'(t) = h(t)$ for a. a. $t \in T$.

(b) $\Longrightarrow$ (a): Let $m\colon \mathcal{B}(T) \to X$ be a vector measure such that $m \ll \lambda$. Let $f(t) = m([0,t])$ for all $t \in T$. Evidently $f\colon T \to X$ is absolutely continuous. So, by hypothesis, $f'(t)$ exists for almost all $t \in T$ and $f(t) = f(0) + \int_0^t f'(s)\,\mathrm{d}s$ for all $t \in T$. Then, as in the first part, we see that $m(A) = \int_A f'(s)\,\mathrm{d}s$ for all $A \in \mathcal{B}(T)$. Hence, X has the RNP. ☐

4.5 Sobolev Spaces

In this section we present an outline of the theory of Sobolev spaces. These spaces play a key role in the theory of partial differential equations.

Definition 4.5.1. Let $\Omega \subseteq \mathbb{R}^N$ be an open set and let $1 \leq p \leq \infty$.

(a) Given $u \in L^p(\Omega)$, the **distributional derivative** $\partial u/(\partial z_k)$ with $k = 1, \dots, N$ is defined by

$$\frac{\partial u}{\partial z_k}(\varphi) = -\int_\Omega u \frac{\partial \varphi}{\partial z_k}\,\mathrm{d}z \quad \text{for all } \varphi \in C_c^1(\Omega)\,.$$

If $\partial u/(\partial z_k) \in L^p(\Omega)$, then

$$\int_\Omega \frac{\partial u}{\partial z_k}\varphi\,\mathrm{d}z = -\int_\Omega u \frac{\partial \varphi}{\partial z_k}\,\mathrm{d}z \quad \text{for all } k = 1, \dots, N\,.$$

If u is differentiable in the classical sense, then it is also differentiable in the distributional sense and the two are equal.

(b) The **Sobolev space** $W^{1,p}(\Omega)$ is defined by

$$W^{1,p}(\Omega) = \left\{u \in L^p(\Omega)\colon \frac{\partial u}{\partial z_k} \in L^p(\Omega) \text{ for all } k = 1, \dots, N\right\}.$$

The space $W^{1,p}(\Omega)$ is equipped with the norm

$$\|u\|_{W^{1,p}(\Omega)} = \left[\|u\|_p^p + \sum_{k=1}^{N} \left\| \frac{\partial u}{\partial x_k} \right\|_p^p \right]^{\frac{1}{p}} \quad \text{for } 1 \le p < \infty\,,$$

$$\|u\|_{W^{1,\infty}(\Omega)} = \max\left\{ \|u\|_\infty, \left\| \frac{\partial u}{\partial z_1} \right\|_\infty, \dots, \left\| \frac{\partial u}{\partial z_N} \right\|_\infty \right\} \quad \text{for } p = +\infty\,.$$

We say that $u \in W^{1,p}_{\text{loc}}(\Omega)$ if $u \in W^{1,p}(U)$ for every open set U such that $\overline{U}$ is compact and $\overline{U} \subseteq \Omega$, that is, U is compactly contained in Ω denoted by $U \subset\subset \Omega$.

Remark 4.5.2. If $p = 2$, then we usually write $W^{1,2}(\Omega) = H^1(\Omega)$ to denote that the space is a Hilbert space with inner product given by

$$(u, v)_{H^1} = \int_\Omega uv\,\mathrm{d}z + \sum_{k=1}^{N} \int_\Omega \frac{\partial u}{\partial z_k} \frac{\partial v}{\partial z_k}\,\mathrm{d}z\,.$$

Another equivalent norm on $W^{1,p}(\Omega)$ is given by

$$|u|_{W^{1,p}(\Omega)} = \|u\|_p + \sum_{k=1}^{N} \left\| \frac{\partial u}{\partial z_k} \right\|_p \quad \text{for all } u \in W^{1,p}(\Omega)\,.$$

Note that $C_c^\infty(\Omega)$, the space of test functions, also denoted by $D(\Omega)$, is a subspace of $W^{1,p}(\Omega)$ for $1 \le p \le \infty$. So, we may consider its closure in $W^{1,p}(\Omega)$.

Definition 4.5.3. $W_0^{1,p}(\Omega) = \overline{C_c^\infty(\Omega)}^{\|\cdot\|_{W^{1,p}(\Omega)}}$.

Later when we introduce the notion of trace, we will characterize the elements of $W_0^{1,p}(\Omega)$ as those Sobolev functions that vanish on $\partial\Omega$.

Example 4.5.4. Let $N = 1$, $\Omega = (-1, 1)$, and $u(z) = |z|$ for all $z \in (-1, 1)$. The function is not differentiable in the classical sense at $z = 0$. Let us compute its distributional derivative. To this end, let $\varphi \in C_c^\infty(-1, 1)$. Then we derive

$$\int_{-1}^{1} u\varphi'\,\mathrm{d}z = \int_0^1 z\varphi'\,\mathrm{d}z + \int_{-1}^{0} (-z)\varphi'\,\mathrm{d}z = -\int_0^1 \varphi\,\mathrm{d}z + \int_{-1}^{0} \varphi\,\mathrm{d}z = -\int_{-1}^{1} \operatorname{sgn}(z)\varphi\,\mathrm{d}z\,.$$

Recall that

$$\operatorname{sgn} z = \begin{cases} -1 & \text{if } -1 \le z < 0\,, \\ 1 & \text{if } 0 < z \le 1\,. \end{cases}$$

Hence, $\mathrm{d}u/(\mathrm{d}z) = u' = \operatorname{sgn}$ and so $u' \in L^\infty(-1, 1)$. Therefore $u \in W^{1,\infty}(-1, 1)$. In particular then it holds $u \in W^{1,p}(-1, 1)$ for all $1 \le p < \infty$.

Remark 4.5.5. A function with a jump at a point is not a Sobolev function. Consider the Heaviside function $u\colon (-1, 1) \to \mathbb{R}$ defined by $u(z) = 0$ if $z < 0$ and $u(z) = 1$ if $z \ge 0$. Then

$u' = \delta_0$ where δ_0 is the Dirac measure concentrated at $z = 0$. But the distribution δ_0 does not correspond to a function $f \in L^1(-1, 1)$.

Continuing with functions of one variable, that is, $N = 1$, we will show that every such Sobolev function has an absolutely continuous representative.

Lemma 4.5.6. *If $u \in L^1_{\text{loc}}(a, b)$ and $u' = 0$ in the distributional sense, then $u(z) = c \in \mathbb{R}$ for a. a. $z \in (a, b)$.*

Proof. By hypothesis we have $\int_a^b u\varphi' \,\mathrm{d}z = 0$ for all $\varphi \in C_c^\infty(a, b)$. Consider the space $V = \{\varphi' : \varphi \in C_c^\infty(a, b)\}$. Note that if $\varphi \in C_c^\infty(a, b)$, then $\varphi' \in C_c^\infty(a, b)$ and $\int_a^b \varphi'(z)\,\mathrm{d}z = \varphi(b) - \varphi(a) = 0$. On the other hand, if $\vartheta \in C_c^\infty(a, b)$ and $\int_a^b \vartheta(z)\,\mathrm{d}z = 0$, then we set $\varphi(z) = \int_0^z \vartheta(t)\,\mathrm{d}t$ and obtain $\varphi \in C_c^\infty(a, b)$ with $\varphi' = \vartheta$. Therefore $V = \{\vartheta \in C_c^\infty(a, b) : \int_a^b \vartheta(z)\,\mathrm{d}z = 0\}$. In order to produce a function $\vartheta \in V$, let $\eta \in C_c^\infty(a, b)$ such that $\int_a^b \eta(z)\,\mathrm{d}z = 1$. Then for any $\varphi \in C_c^\infty(a, b)$, we set

$$\vartheta(z) = \varphi(z) - \left(\int_a^b \varphi(t)\,\mathrm{d}t\right)\eta(z)\,.$$

Evidently $\vartheta \in V$ and $\int_a^b u\vartheta\,\mathrm{d}z = 0$. We get

$$\int_a^b u\varphi\,\mathrm{d}z = \left(\int_a^b \varphi\,\mathrm{d}z\right)\int_a^b u\eta\,\mathrm{d}z = \int_a^b c\varphi\,\mathrm{d}z \quad \text{with } c = \int_a^b u\eta\,\mathrm{d}z\,.$$

Hence, $\int_a^b (u - c)\varphi\,\mathrm{d}z = 0$ for all $\varphi \in C_c^\infty(a, b)$ and so $u(z) = c$ for a. a. $z \in (a, b)$, (see Corollary 4.1.31) and recall that $C_c^\infty(a, b)$ is dense in $C_c(a, b)$. ☐

Theorem 4.5.7. *If $\Omega = (a, b)$ and $1 \le p \le \infty$, then the following hold:*

(a) *for any $u \in W^{1,p}(a, b)$ there exists a unique $\overline{u} \in \mathrm{AC}([a, b])$ such that $u = \overline{u}$ a. e. on Ω and $\overline{u}(x) - \overline{u}(v) = \int_v^x u'(z)\,\mathrm{d}z$ for all $x, v \in (a, b)$;*

(b) *if $u \in L^p(a, b)$ and $u(x) - u(v) = \int_v^x h(z)\,\mathrm{d}z$ for some $h \in L^p(a, b)$, then $u \in W^{1,p}(a, b)$ and $u' = h$ in the distributional sense.*

Proof. (a) Let $h(x) = \int_a^x u'(z)\,\mathrm{d}z$. Then $h \in \mathrm{AC}([a, b])$; see Theorem 4.4.21. Moreover, for every $\varphi \in C_c^\infty(a, b)$, by using Fubini's Theorem, we obtain

$$\begin{aligned}
h'(\varphi) &= -\int_a^b h\varphi'\,\mathrm{d}z = -\int_a^b\left(\int_a^z u'\,\mathrm{d}t\right)\varphi'\,\mathrm{d}z = -\int_a^b\left(\int_a^b \chi_{[a,z]}u'\,\mathrm{d}t\right)\varphi'\,\mathrm{d}z \\
&= -\int_a^b u'\left(\int_a^b \chi_{[a,z]}(t)\varphi'\,\mathrm{d}z\right)\mathrm{d}t = -\int_a^b u'\left(\int_t^b \varphi'\,\mathrm{d}z\right)\mathrm{d}t = \int_a^b u'\varphi\,\mathrm{d}t\,.
\end{aligned}$$

Hence, $h' = u'$ and so $(h - u)' = 0$ in the distributional sense. Thus, Lemma 4.5.6 gives $h = u + c$ with $c \in \mathbb{R}$. Then $\overline{u} = h \in AC([a, b])$ is the desired absolutely continuous representative of u.

(b) Evidently $u \in AC([a, b])$ and so it is differentiable almost everywhere in the classical sense and $u' = h \in L^p(a, b)$. Hence $u \in W^{1,p}(a, b)$ and $u' = b$ in the distributional sense. □

Remark 4.5.8. If $N \geq 2$, then the result above fails.

Proposition 4.5.9. *If $\Omega \subseteq \mathbb{R}^N$ is an open set, then the following hold:*
(a) *$W^{1,p}(\Omega)$ is a Banach space for all $1 \leq p \leq \infty$;*
(b) *$W^{1,p}(\Omega)$ is reflexive for $1 < p < \infty$;*
(c) *$W^{1,p}(\Omega)$ is separable for $1 \leq p < \infty$ and $W^{1,2}(\Omega) = H^1(\Omega)$ is a separable Hilbert space.*

Proof. (a) Let $\{u_n\}_{n\geq 1} \subseteq W^{1,p}(\Omega)$ be a Cauchy sequence. Let $Du = (\partial u/(\partial z_k))_{k=1}^N$ be the gradient of u. We see that $\{u_n\}_{n\geq 1} \subseteq L^p(\Omega)$ and $\{Du_n\}_{n\geq 1} \subseteq L^p(\Omega, \mathbb{R}^N)$ are Cauchy sequences. Therefore $u_n \to u$ in $L^p(\Omega)$ and $Du_n \to g$ in $L^p(\Omega, \mathbb{R}^N)$. By definition it holds that

$$\int_\Omega (Du_n)\varphi \, dz = (-1)^N \int_\Omega u_n(D\varphi) \, dz \quad \text{for all } \varphi \in C_c^\infty(\Omega) \text{ and for all } n \in \mathbb{N} .$$

This implies

$$\int_\Omega g\varphi \, dz = (-1)^N \int_\Omega u(D\varphi) \, dz \quad \text{for all } \varphi \in C_c^\infty(\Omega) .$$

Hence, $g = Du$ and so $u \in W^{1,p}(\Omega)$. Therefore $u_n \to u$ in $W^{1,p}(\Omega)$ and we conclude that $W^{1,p}(\Omega)$ is a Banach space.

(b) The space $V = L^p(\Omega) \times L^p(\Omega, \mathbb{R}^N)$ is reflexive for $1 < p < \infty$. Let $K \in L(W^{1,p}(\Omega), V)$ be defined by $K(u) = (u, Du)$. Clearly this is an isometry into V and so $K(W^{1,p}(\Omega))$ is a closed subspace of V. Therefore $K(W^{1,p}(\Omega))$ is reflexive and then so is $W^{1,p}(\Omega)$.

(c) The space $V = L^p(\Omega) \times L^p(\Omega, \mathbb{R}^N)$ is separable for $1 \leq p < \infty$. Then $K(W^{1,p}(\Omega)) \subseteq V$ is separable, hence so is $W^{1,p}(\Omega)$. Then $W^{1,2}(\Omega) = H^1(\Omega)$ is a separable Hilbert space. □

Next we will see how we can approximate Sobolev functions with smooth functions. We will use approximation by mollification; see Definition 4.1.27 (c) and (d).

Proposition 4.5.10. *If $\Omega \subseteq \mathbb{R}^N$ is open and $u \in W^{1,p}_{\text{loc}}(\Omega)$ for some $1 \leq p < \infty$, then $u^\varepsilon \to u$ in $W^{1,p}_{\text{loc}}$.*

Proof. Let $\{\vartheta_\varepsilon\}_{\varepsilon>0}$ be the standard mollifier and $u^\varepsilon = \vartheta_\varepsilon * u$ with $\varepsilon > 0$; see Definition 4.1.27 (c) and (d). From the proof of Proposition 4.1.28 we know that

$$\frac{\partial u^\varepsilon}{\partial z_k} = \int\limits_\Omega \frac{\partial \vartheta_\varepsilon}{\partial z_k}(z-y)u(y)\,\mathrm{d}y = -\int\limits_\Omega \frac{\partial \vartheta_\varepsilon}{\partial y_k}(z-y)u(y)\,\mathrm{d}y$$
$$= \int\limits_\Omega \vartheta_\varepsilon(z-y)\frac{\partial u}{\partial z_k}(y)\,\mathrm{d}y = \left(\vartheta_\varepsilon * \frac{\partial u}{\partial y_k}\right)(z) \quad \text{for all } z \in \Omega\,.$$

Then the result follows from Proposition 4.1.30. □

Theorem 4.5.11. *If $\Omega \subseteq \mathbb{R}^N$ is open and $u \in W^{1,p}(\Omega)$ for some $1 \le p < \infty$, then there exists a sequence $\{u_n\}_{n\ge1} \subseteq W^{1,p}(\Omega) \cap C^\infty(\Omega)$ such that $u_n \to u$ in $W^{1,p}(\Omega)$.*

Proof. Let $\varepsilon > 0$ be given and define

$$\Omega_0 = \emptyset\,, \quad \Omega_n = \left\{z \in \Omega \colon d(z, \partial\Omega) > \frac{1}{n}\right\} \cap B_r$$

for $n \in \mathbb{N}$ with $B_r = \{z \in \mathbb{R}^N \colon |z| < r\}$. We set $U_n = \Omega_{n+1} \setminus \overline{\Omega}_n$ for all $n \in \mathbb{N}$ and consider a smooth partition $\{\psi_n\}_{n\ge1}$ of unity subordinate to $\{U_n\}_{n\ge1}$. Then we have

$$\psi_n \in C_c^\infty(U_n)\,, \quad 0 \le \psi_n \le 1 \quad \text{for all } n \in \mathbb{N} \quad \text{and} \quad \sum_{n\ge1} \psi_n(z) = 1 \quad \text{for all } z \in \Omega\,.$$

Then $u\psi_n \in W^{1,p}(\Omega)$ and $\operatorname{supp}(u\psi_n) \subseteq U_n$. Hence, there exists $\varepsilon_n > 0$ such that

$$\begin{aligned}
&\operatorname{supp}(\vartheta_{\varepsilon_n} * (u\psi_n)) \subset U_n\,,\\
&\left(\int\limits_\Omega |\vartheta_{\varepsilon_n} * (u\psi_n) - u\psi_n|^p\,\mathrm{d}z\right)^{\frac{1}{p}} < \frac{\varepsilon}{2^n}\,,\\
&\left(\int\limits_\Omega |\vartheta_{\varepsilon_n} * (D(u\psi_n)) - D(u\psi_n)|^p\,\mathrm{d}z\right)^{\frac{1}{p}} < \frac{\varepsilon}{2^n}\,.
\end{aligned} \tag{4.5.1}$$

We set $u_\varepsilon = \sum_{n\ge1} \vartheta_{\varepsilon_n} * (u\psi_n)$. For every $z \in \Omega$ in some neighborhood of it, only a finite number of terms in this sum are nonzero. Therefore $u_\varepsilon \in C^\infty(\Omega)$. We obtain $u = \sum_{n\ge1}(u\psi_n)$. Then from (4.5.1) it follows that

$$\begin{aligned}
\|u_\varepsilon - u\|_p &\le \sum_{n\ge1}\left(\int\limits_\Omega |\vartheta_{\varepsilon_n} * (u\psi_n) - u\psi_n|^p\,\mathrm{d}z\right)^{\frac{1}{p}} < \varepsilon\,,\\
\|Du_\varepsilon - Du\|_p &\le \sum_{n\ge1}\left(\int\limits_\Omega |\vartheta_{\varepsilon_n} * (D(u\psi_n)) - D(u\psi_n)|^p\,\mathrm{d}z\right)^{\frac{1}{p}} < \varepsilon\,.
\end{aligned} \tag{4.5.2}$$

Then $u_\varepsilon \in W^{1,p}(\Omega) \cap C^\infty(\Omega)$ and $u_\varepsilon \to u$ in $W^{1,p}(\Omega)$ as $\varepsilon \to 0^+$. □

Remark 4.5.12. In this theorem we do not claim that the approximating smooth functions belong to $C^\infty(\overline{\Omega})$. For this reason the result is known as the "Local Approximation Theorem by Smooth Functions".

In order to have a "Global Approximation Theorem by Smooth Functions", that is, for the approximating smooth functions to belong to $C^\infty(\overline{\Omega})$, we need to strengthen the condition on Ω.

Definition 4.5.13. Let $\Omega \subseteq \mathbb{R}^N$ be an open set. We say that $\partial\Omega$ is **Lipschitz** if it can be represented locally as the graph of a Lipschitz function defined on some open ball of $\mathbb{R}^{N-1}$.

The next theorem is the global approximation result and its rather technical proof can be found in Evans–Gariepy [116, Theorem 3, p. 127].

Theorem 4.5.14. *If $\Omega \subseteq \mathbb{R}^N$ is bounded, $\partial\Omega$ is Lipschitz, and $u \in W^{1,p}(\Omega)$ for some $1 \le p < \infty$, then there exists a sequence $\{u_n\}_{n\ge1} \subseteq W^{1,p}(\Omega) \cap C^\infty(\overline{\Omega})$ such that $u_n \to u$ in $W^{1,p}(\Omega)$.*

The approximating theorems permit the extension of the usual calculus rules to Sobolev functions.

Proposition 4.5.15. *If $\Omega \subseteq \mathbb{R}^N$ is an open set and $u, v \in W^{1,p}(\Omega) \cap L^\infty(\Omega)$ for some $1 \le p < \infty$, then $uv \in W^{1,p}(\Omega) \cap L^\infty(\Omega)$ and*

$$\frac{\partial(uv)}{\partial z_k} = \frac{\partial u}{\partial z_k} v + u \frac{\partial v}{\partial z_k} \quad \text{a. e. for all } k \in \mathbb{N}\,.$$

Proof. Let $\psi \in C_c^1(\Omega)$ with $\operatorname{supp}\psi \subseteq U \subset\subset \Omega$. From Proposition 4.5.10 it follows that

$$\begin{aligned}
\int_\Omega uv \frac{\partial \psi}{\partial z_k}\, dz &= \int_U uv \frac{\partial \psi}{\partial z_k}\, dz = \lim_{\varepsilon\to 0} \int_U u^\varepsilon v^\varepsilon \frac{\partial \psi}{\partial z_k}\, dz \\
&= -\lim_{\varepsilon\to 0} \int_U \left[\frac{\partial u^\varepsilon}{\partial z_k} v^\varepsilon + u^\varepsilon \frac{\partial v^\varepsilon}{\partial z_k} \right] \psi\, dz = -\int_U \left[\frac{\partial u}{\partial z_k} v + u \frac{\partial v}{\partial z_k} \right] \psi\, dz \\
&= -\int_\Omega \left[\frac{\partial u}{\partial z_k} v + u \frac{\partial v}{\partial z_k} \right] \psi\, dz\,.
\end{aligned}$$

□

Remark 4.5.16. The result is also true if $W^{1,p}(\Omega)$ is replaced by $W_0^{1,p}(\Omega)$; see Definition 4.5.3.

Before proving additional calculus rules for Sobolev functions, let us state a result that shows when a Sobolev function $u \in W^{1,p}(\Omega)$ actually belongs to $W_0^{1,p}(\Omega)$; see Definition 4.5.3.

Proposition 4.5.17. *If $\Omega \subseteq \mathbb{R}^N$ is an open set, $u \in W^{1,p}(\Omega)$ for some $1 \le p < \infty$ and u vanishes outside of a compact set $K \subseteq \Omega$, then $u \in W_0^{1,p}(\Omega)$.*

Proof. Let $\Omega^* \subseteq \mathbb{R}^N$ be a bounded open set such that $K \subseteq \Omega^* \subset\subset \Omega$ with Lipschitz boundary $\partial\Omega^*$. Choose a cut-off function $\psi \in C_c^\infty(\mathbb{R}^N)$ such that $\psi|_K \equiv 1$. Evidently $u = \psi u$. Applying Corollary 4.1.32 and Theorem 4.5.14, there exists a sequence $\{u_n\}_{n\geq 1} \subseteq C_c^\infty(\Omega)$ such that

$$u_n \to u \quad \text{in } L^p(\Omega) \quad \text{and} \quad Du_n \to Du \quad \text{in } L^p(\Omega^*, \mathbb{R}^N) .$$

Then $\psi u_n \to \psi u = u$ in $W^{1,p}(\Omega)$ and $\psi u_n \in C_c^\infty(\Omega)$. Hence $u \in W_0^{1,p}(\Omega)$; see Definition 4.5.3. □

Now we can prove a chain rule for Sobolev functions.

Theorem 4.5.18 (Chain Rule for Sobolev Functions). *If $\Omega \subseteq \mathbb{R}^N$ is a bounded open set with Lipschitz boundary, $\varphi\colon \mathbb{R} \to \mathbb{R}$ is Lipschitz continuous and $u \in W^{1,p}(\Omega)$ (resp. $u \in W_0^{1,p}(\Omega)$) for some $1 \leq p < \infty$, then $\varphi \circ u \in W^{1,p}(\Omega)$ (resp. $\varphi \circ u \in W_0^{1,p}(\Omega)$) and*

$$\frac{\partial}{\partial z_k}(\varphi \circ u)(z) = \varphi^*(u(z))\frac{\partial u}{\partial z_k}(z) \quad \textit{for a. a. } z \in \Omega \textit{ and for all } k \in \mathbb{N} ,$$

where $\varphi^*\colon \mathbb{R} \to \mathbb{R}$ is any measurable function such that $\varphi^* = \varphi'$ a. e.

Proof. First assume that $u \in W^{1,p}(\Omega)$. Using Theorem 4.5.14 there exists a sequence $\{u_n\}_{n\geq 1} \subseteq C^\infty(\overline{\Omega})$ such that $u_n \to u$ in $W^{1,p}(\Omega)$. We set $\xi_n = \varphi \circ u_n$ with $n \in \mathbb{N}$. It is clear that ξ_n is Lipschitz continuous and $|\partial\xi_n/(\partial z_k)| \leq \operatorname{Lip}(\xi_n)$ for all $k = 1, \ldots, N$ and for all $n \in \mathbb{N}$ where $\operatorname{Lip}(\xi_n)$ denotes the Lipschitz constant of ξ_n. Hence $\partial\xi_n/(\partial z_k) \in L^p(\Omega)$ and so $\xi_n \in W^{1,p}(\Omega)$ for all $n \in \mathbb{N}$. We obtain

$$|\xi_n(z) - \varphi(u(z))| = |\varphi(u_n(z)) - \varphi(u(z))| \leq \operatorname{Lip}(\varphi)|u_n(z) - u(z)| \quad \text{for a. a. } z \in \Omega ,$$

which gives $\xi_n \to \varphi \circ u$ in $L^p(\Omega)$.

Let $\{e_k\}_{k=1}^N$ be the standard orthonormal basis of $\mathbb{R}^N$. Using this yields

$$\frac{|\xi_n(z + te_k) - \xi_n(z)|}{|t|} \leq \frac{\operatorname{Lip}(\varphi)|u_n(z + te_k) - u_n(z)|}{|t|} ,$$

which implies

$$\limsup_{n\to\infty} \left\|\frac{\partial\xi_n}{\partial z_k}\right\|_p \leq \operatorname{Lip}(\varphi)\left\|\frac{\partial u}{\partial z_k}\right\|_p \quad \text{for all } k = 1, \ldots, N .$$

Hence, $\{\partial u/(\partial z_k)\}_{n\geq 1} \subseteq L^p(\Omega)$ is bounded for all $k = 1, \ldots, N$. By passing to a suitable subsequence if necessary we may assume that

$$\frac{\partial\xi_n}{\partial z_k} \xrightarrow{\text{w}} h_k \quad \text{in } L^p(\Omega) \text{ as } n \to \infty \text{ with } h_k \in L^p(\Omega) \text{ and } k = 1, \ldots, N . \tag{4.5.3}$$

From (4.5.2) and (4.5.3) it follows that $h_k = (\partial\varphi(u))/(\partial z_k)$ for all $k = 1, \ldots, N$. Therefore $\xi_n \to \varphi(u)$ in $W^{1,p}(\Omega)$ and so $\varphi \circ u \in W^{1,p}(\Omega)$.

Note that $D\xi_n = \varphi'(u_n)Du_n$ for all $n \in \mathbb{N}$. So, taking the limit as $n \to \infty$ we derive $D(\varphi \circ u) = \varphi^*(u)Du$ with a measurable function $\varphi^*: \mathbb{R} \to \mathbb{R}$ such that $\varphi^* = \varphi'$ a. e.

If $u \in W_0^{1,p}(\Omega)$, then in the proof above we have $\{u_n\}_{n\geq 1} \subseteq C_c^\infty(\Omega)$; see Definition 4.5.3. So, $\xi_n \in W^{1,p}(\Omega)$ has compact support. Thus by Proposition 4.5.17, we infer that $\xi_n \in W_0^{1,p}(\Omega)$, and so the limit as $n \to \infty$ gives $\varphi \circ u \in W_0^{1,p}(\Omega)$. Moreover, $D(\varphi \circ u) = \varphi^*(u)Du$. □

This chain rule has interesting consequences.

Corollary 4.5.19. *If $\Omega \subseteq \mathbb{R}^N$ is a bounded open set with Lipschitz boundary, $X = W^{1,p}(\Omega)$ or $W_0^{1,p}(\Omega)$ with $1 \leq p < \infty$ and $u, v \in X$, then the following hold:*
(a) *$u^+, u^-, |u| \in X$ and*

$$Du^+ = \begin{cases} 0 & \text{a. e. on } \{u \leq 0\} \\ Du & \text{a. e. on } \{0 < u\} \end{cases}, \quad Du^- = \begin{cases} -Du & \text{a. e. on } \{u < 0\} \\ 0 & \text{a. e. on } \{0 \leq u\} \end{cases},$$

$$D|u| = \begin{cases} -Du & \text{a. e. on } \{u < 0\} \\ 0 & \text{a. e. on } \{u = 0\} \\ Du & \text{a. e. on } \{0 < u\} \end{cases}.$$

(b) *$\max\{u, v\} = h \in X$, $\min\{u, v\} = g \in X$ and*

$$Dh = \begin{cases} Du & \text{a. e. on } \{u \geq v\} \\ Dv & \text{a. e. on } \{v \geq u\} \end{cases}, \quad Dg = \begin{cases} Du & \text{a. e. on } \{u \leq v\} \\ Dv & \text{a. e. on } \{v \leq u\} \end{cases}.$$

The proper definition of boundary values for Sobolev functions requires the notion of trace. This notion is produced in the next theorem whose proof can be found in Evans–Gariepy [116, Theorem 1, p. 133] and Kufner–John–Fučík [196, Theorem 6.6.4, p. 325].

Theorem 4.5.20. *If $\Omega \subseteq \mathbb{R}^N$ is a bounded open set with Lipschitz boundary $\partial\Omega$ and $1 \leq p < \infty$, then the following hold:*
(a) *there exists a bounded linear operator $\gamma_0: W^{1,p}(\Omega) \to L^p(\partial\Omega)$, where we consider the $(N-1)$-dimensional* ***Hausdorff surface measure*** *σ on $\partial\Omega$, such that $\gamma_0(u) = u|_{\partial\Omega}$ for all $u \in W^{1,p}(\Omega) \cap C(\overline{\Omega})$;*
(b) *for every $\vartheta \in C(\mathbb{R}^N, \mathbb{R}^N)$ and $u \in W^{1,p}(\Omega)$ it holds that*

$$\int_\Omega u(\operatorname{div} \vartheta)\, dz + \int_\Omega (Du, \vartheta)_{\mathbb{R}^N}\, dz = \int_{\partial\Omega} (\vartheta, n)_{\mathbb{R}^N}\, d\sigma,$$

with n being the outward unit normal on $\partial\Omega$;
(c) *$W_0^{1,p}(\Omega) = \ker \gamma_0$.*

Remark 4.5.21. Since by hypothesis $\partial\Omega$ is Lipschitz, $n(z)$ exists for σ-almost all $z \in \partial\Omega$.

Definition 4.5.22. The function $\gamma_0(u) \in L^p(\partial\Omega)$ is uniquely defined up to σ-null sets. It is called the **trace** of u on $\partial\Omega$ and we interpret $\gamma_0(u)$ as describing the **boundary values** of the Sobolev function $u \in W^{1,p}(\Omega)$ with $1 \le p < \infty$.

Remark 4.5.23. In the definition above we have excluded the case $p = +\infty$. The reason for this is that $u \in W^{1,p}_{\text{loc}}(\Omega)$ if and only if u is locally Lipschitz. So, it admits a continuous extension on $\overline{\Omega}$ and then the values of u on $\partial\Omega$ are defined in the classical sense. The operator $\gamma_0 \colon W^{1,p}(\Omega) \to L^p(\partial\Omega)$ is not surjective. In order to describe the range of γ_0 we need to introduce Sobolev spaces of fractional order that are beyond our scope here.

Finally we mention a compactness result for the trace map, which can be found in Kufner–John–Fučík [196, Theorem 6.10.5, p. 344].

Theorem 4.5.24. *If $\Omega \subseteq \mathbb{R}^N$ is a bounded open set with Lipschitz boundary $\partial\Omega$, then the following hold:*
(a) *if $1 < p < N$, then $\gamma_0 \colon W^{1,p}(\Omega) \to L^r(\partial\Omega)$ is compact for any $1 \le r < (Np - p)/(N - p)$;*
(b) *if $N \ge p$, then $\gamma_0 \colon W^{1,p}(\Omega) \to L^r(\partial\Omega)$ is compact for any $r \ge 1$.*

Another result in this vein is the so-called "Rellich–Kondrachov Theorem", which can be found in Adams [1, Theorem 6.2, p. 144].

Theorem 4.5.25 (Rellich–Kondrachov Theorem). *If $\Omega \subseteq \mathbb{R}^N$ is a bounded open set with Lipschitz boundary and let $1 \le p < \infty$, then the following hold:*
(a) *when $1 \le p < N$, it holds $W^{1,p}(\Omega) \hookrightarrow L^r(\Omega)$ for all $r \in [1, p^* = Np/(N - p)]$ and the embedding is compact if $1 \le r < p^*$;*
(b) *when $p = N$, it holds $W^{1,p}(\Omega) \hookrightarrow L^r(\Omega)$ for all $r \in [1, +\infty)$ and the embedding is compact;*
(c) *when $p > N$, it holds that $W^{1,p}(\Omega) \hookrightarrow C(\overline{\Omega})$ compactly.*

Remark 4.5.26. The embedding $W^{1,p}(\Omega) \hookrightarrow L^{p^*}(\Omega)$ is never compact. Also, if Ω is not bounded, then the embedding $W^{1,p}(\Omega) \hookrightarrow L^p(\Omega)$ is not compact.

As a consequence of Theorem 4.5.25 we have the following proposition providing useful equivalent norms on $W^{1,p}(\Omega)$.

Proposition 4.5.27. *If $\Omega \subseteq \mathbb{R}^N$ is a bounded open set with Lipschitz boundary, then $|u| = \|u\|_r + \|Du\|_p$ is an equivalent norm on $W^{1,p}(\Omega)$ in the following cases:*
(a) $1 \le r \le p^*$ *if* $1 \le p < N$;
(b) $1 \le r < \infty$ *if* $p = N$;
(c) $1 \le r \le \infty$ *if* $p > N$.

For the space $W^{1,p}_0(\Omega)$ we can do better. The result is known as "Poincaré's inequality".

Theorem 4.5.28 (Poincaré's inequality). *If $\Omega \subseteq \mathbb{R}^N$ is a bounded open set and $1 \leq p < \infty$, then there exists a constant $c = c(p, N, \Omega) > 0$ such that $\|u\|_p \leq c\|Du\|_p$ for all $u \in W_0^{1,p}(\Omega)$.*

Proof. Since $\Omega \subseteq \mathbb{R}^N$ is bounded, there exists $\eta \in \mathbb{R}$ such that $\Omega \subseteq (-\eta, \eta)^N$. For $\vartheta \in C_c^\infty(\Omega)$, we have

$$\vartheta(z) = \int_{-\eta}^{z_N} \frac{\partial\vartheta}{\partial z_N}(z_1, \ldots, z_{N-1}, s)\, ds\,.$$

By using Hölder's inequality this results in

$$|\vartheta(z)|^p \leq (2\eta)^{p-1} \int_{-\eta}^{\eta} \left| \frac{\partial\vartheta}{\partial z_N}(z_1, \ldots, z_{N-1}, s) \right|^p ds\,.$$

Therefore

$$\begin{aligned} &\int_{[-\eta,\eta]^{N-1}} |\vartheta(z_1, \ldots, z_{N-1}, z_N)|^p\, dz_1 \ldots dz_{N-1} \\ &\leq (2\eta)^{p-1} \int_{[-\eta,\eta]^{N-1}} \left(\int_{-\eta}^{\eta} \left| \frac{\partial\vartheta}{\partial z_N}(z_1, \ldots, z_{n-1}, s) \right|^p ds \right) dz_1 \ldots dz_{N-1}\,, \end{aligned}$$

which implies

$$\|\vartheta\|_p \leq (2\eta)^p \int_\Omega \left| \frac{\partial\vartheta}{\partial z_N} \right|^p dz \leq (2\eta)^p \|D\vartheta\|_p^p\,.$$

Thus, $\|u\|_p \leq 2\eta\|Du\|_p$ for all $u \in W_0^{1,p}(\Omega)$ where we have used Definition 4.5.3. □

Corollary 4.5.29. *If $\Omega \subseteq \mathbb{R}^N$ is a bounded open set and $1 \leq p < \infty$, then $|\cdot| = \|D(\cdot)\|_p$ is an equivalent norm on $W_0^{1,p}(\Omega)$.*

There are some other related inequalities that are also useful.

Definition 4.5.30. We say that $\Omega \subseteq \mathbb{R}^N$ is a **domain** if it is open and connected.

Theorem 4.5.31. *If $\Omega \subseteq \mathbb{R}^N$ is a bounded domain with Lipschitz boundary and $V \subseteq W^{1,p}(\Omega)$ with $1 < p < \infty$ is a closed subspace such that the only constant function belonging to V is the zero function, then there exists a constant $c > 0$ such that $\|v\|_p \leq c\|Dv\|_p$ for all $v \in V$.*

Proof. Arguing by contradiction, suppose that the conclusion of the theorem is false. Then there exists a sequence $\{v_n\}_{n\geq1} \subseteq V$ such that $\|v_n\|_p > n\|Dv_n\|_p$. Let $y_n = v_n/\|v_n\|_p$ with $n \in \mathbb{N}$. Then $y_n \in V$, $\|y_n\|_p = 1$ for all $n \in \mathbb{N}$ and $\|Dy_n\|_p < 1/n$ for all $n \in \mathbb{N}$. Therefore $\{y_n\}_{n\geq1} \subseteq W^{1,p}(\Omega)$ is bounded and so by passing to a suitable subsequence if necessary we may assume that $y_n \xrightarrow{w} y$ in $W^{1,p}(\Omega)$. Hence, $y_n \to y$ in $L^p(\Omega)$ (see Theorem 4.5.25), and so $y \in V$ and $\|y\|_p = 1$. From the weak lower semicontinuity of the norm (see Proposition 3.3.13 (c)), it follows that

$$\|y\|_p + \|Dy\|_p \le \liminf_{n\to\infty}[\|y_n\|_p + \|Dy_n\|_p] .$$

Hence, $\|Dy\|_p \le \liminf_{n\to\infty} \|Dy_n\|_p = 0$, which implies $y \equiv c \in \mathbb{R}$ as Ω is connected. Since $y \in V$ we obtain $y = 0$, a contradiction to the fact that $\|y\| = 1$. □

Corollary 4.5.32. *If $\Omega \subseteq \mathbb{R}^N$ is a bounded domain with Lipschitz boundary, $\Gamma_* \subseteq \partial\Omega$ with $\sigma(\Gamma_*) > 0$ and $V = \{v \in W^{1,p}(\Omega): \gamma_0(v) = 0 \text{ on } \Gamma_*\}$ with $1 < p < \infty$, then there exists a constant $c > 0$ such that $\|v\|_p \le c\|Dv\|_p$ for all $v \in V$.*

The next consequence of Theorem 4.5.31 is the following result known as the "Poincaré–Wirtinger inequality".

Corollary 4.5.33 (Poincaré–Wirtinger inequality). *If $\Omega \subseteq \mathbb{R}^N$ is a bounded domain with a Lipschitz boundary, then there exists a constant $c > 0$ such that*

$$\left\| u - \frac{1}{\lambda^N(\Omega)} \int_\Omega u \, dz \right\|_p \le c\|Du\|_p \quad \text{for all } u \in W^{1,p}(\Omega) \text{ with } 1 < p < \infty ,$$

where λ^N denotes the Lebesgue measure on $\mathbb{R}^N$.

Proof. Let $V = \{v \in W^{1,p}(\Omega): \int_\Omega v \, dz = 0\}$ and apply Theorem 4.5.31. □

We conclude with another equivalent norm for Sobolev spaces that is useful in the study of boundary value problems.

Proposition 4.5.34. *If $\Omega \subseteq \mathbb{R}^N$ is a bounded domain with a Lipschitz boundary and $\beta \in L^\infty(\partial\Omega)$ with $\beta(z) \ge 0$ σ-a. e. and $\beta \not\equiv 0$, then*

$$u \to |u| = \left[\|Du\|_p^p + \int_{\partial\Omega} \beta(z)|u|^p \, d\sigma \right]^{\frac{1}{p}}$$

is an equivalent norm on the Sobolev space $W^{1,p}(\Omega)$ with $1 < p < \infty$.

Proof. For every $u \in W^{1,p}(\Omega)$ we obtain

$$|u|^p \le \|Du\|_p^p + \|\beta\|_{L^\infty(\partial\Omega)} \|u\|_{L^p(\partial\Omega)}^p \le \|Du\|_p^p + \|\beta\|_{L^\infty(\partial\Omega)} \|\gamma_0\|_L^p \|u\|^p ;$$

see Theorem 4.5.20. Then

$$|u| \le c_1\|u\| \quad \text{for some } c_1 > 0 . \tag{4.5.4}$$

Next we show that there exists a constant $c_2 > 0$ such that

$$\|u\|_p \le c_2|u| \quad \text{for all } u \in W^{1,p}(\Omega)\,. \tag{4.5.5}$$

Arguing by contradiction, suppose that (4.5.5) is not true. Then there exists a sequence $\{u_n\}_{n\ge1} \subseteq W^{1,p}(\Omega)$ such that $\|u_n\|_p > n|u_n|$ for all $n \in \mathbb{N}$. Normalizing in $L^p(\Omega)$, we can say that

$$\|u_n\|_p = 1 \quad \text{and} \quad |u_n| < \frac{1}{n} \quad \text{for all } n \in \mathbb{N}\,. \tag{4.5.6}$$

This gives $|u_n| \to 0$ and so $\|Du_n\|_p \to 0$ as $n \to \infty$. It follows that $\{u_n\}_{n\ge1} \subseteq W^{1,p}(\Omega)$ is bounded and so we may assume that

$$u_n \xrightarrow{w} u \quad \text{in } W^{1,p}(\Omega) \quad \text{and} \quad u_n \to u \quad \text{in } L^p(\Omega) \text{ and in } L^p(\partial\Omega)\,; \tag{4.5.7}$$

see Theorems 4.5.25 and 4.5.24. From (4.5.6) and (4.5.7) it follows that

$$\|Du\|_p^p + \int_{\partial\Omega} \beta(z)|u|^p \, d\sigma \le 0\,. \tag{4.5.8}$$

This shows that $u \equiv \xi \in \mathbb{R}$ since $\Omega \subseteq \mathbb{R}^N$ is connected. If $\xi \ne 0$, then from (4.5.8) and the hypothesis on β we obtain $0 < |\xi|^p \int_{\partial\Omega} \beta(z)\, d\sigma \le 0$, a contradiction. Therefore, (4.5.5) holds. From (4.5.5) it follows that

$$\|u\| \le c_3|u| \quad \text{for all } u \in W^{1,p}(\Omega) \text{ and for some } c_3 > 0\,. \tag{4.5.9}$$

Then (4.5.4) and (4.5.9) imply that $\|\cdot\|$ and $|\cdot|$ are equivalent norms on $W^{1,p}(\Omega)$. □

4.6 Spaces of Measures

Let (X, Σ) be a measurable space. We define

$$\begin{aligned} \mathrm{ca}(\Sigma) &= \text{the set of all } \mathbb{R}\text{-valued signed measures on } \Sigma\,, \\ \mathrm{ca}_+(\Sigma) &= \text{the set of all } \mathbb{R}_+\text{-valued measures on } \Sigma\,. \end{aligned}$$

Given $\mu \in \mathrm{ca}(\Sigma)$, one has the **total variation** $|\mu|: \Sigma \to \mathbb{R}_+$ defined by

$$|\mu|(A) = \sup\left[\sum_{k=1}^{n} |\mu(A_k)| : n \in \mathbb{N}, \{A_k\}_{k=1}^{n} \text{ is a } \Sigma\text{-partition of A}\right];$$

see Definition 2.4.15 and Remark 2.4.16. We know the following.

Proposition 4.6.1. *If $\mu \in \mathrm{ca}(\Sigma)$, then the following hold:*
(a) $\sup[|\mu(C)| : C \in \Sigma, C \subseteq A] \le |\mu|(A) \le 2\sup[|\mu(C)| : C \in \Sigma, C \subseteq A]$;
(b) $|\mu| \in \mathrm{ca}_+(\Sigma)$.

Moreover, for $\mu \in \mathrm{ca}(\Sigma)$ we have the positive and negative variations of μ defined by

$$\mu_+ = \frac{1}{2}[|\mu| + \mu] \quad \text{and} \quad \mu_- = \frac{1}{2}[|\mu| - \mu] .$$

Evidently $\mu = \mu_+ - \mu_-$, $|\mu| = \mu_+ + \mu_-$ and $\mu_+, \mu_- \in \mathrm{ca}_+(\Sigma)$. Moreover $\mu_+ \perp \mu_-$ and there exist $P, N \in \Sigma$, the Hahn decomposition of X, such that for all $A \in \Sigma$

$$\begin{aligned}
&\mu_+(A) = \mu(A \cap P) = \sup[\mu(C): C \in \Sigma, C \subseteq A] ,\\
&\mu_-(A) = -\mu(A \cap N) = -\inf[\mu(C): C \in \Sigma, C \subseteq A] ,\\
&\mu_+(A) \geq 0 \quad \text{for all } A \subseteq P , \quad \mu_-(A) \geq 0 \quad \text{for all } A \subseteq N .
\end{aligned}$$

On $\mathrm{ca}(\Sigma)$ we define the following two quantities

$$\|\mu\| = |\mu|(X) \quad \text{and} \quad \|\mu\|_\infty = \sup[|\mu(A)|: A \in \Sigma] .$$

It is easy to see that both are norms on $\mathrm{ca}(\Sigma)$. Moreover, we have the following result.

Proposition 4.6.2. $\|\cdot\|, \|\cdot\|_\infty$ *are equivalent norms on* $\mathrm{ca}(\Sigma)$ *and* $(\mathrm{ca}(\Sigma), \|\cdot\|)$ *and* $(\mathrm{ca}(\Sigma), \|\cdot\|_\infty)$ *are Banach spaces.*

Proof. From Proposition 4.6.1 (a) it follows that

$$\|\mu\|_\infty \leq \|\mu\| \leq 2\|\mu\|_\infty \quad \text{for all } \mu \in \mathrm{ca}(\Sigma) , \tag{4.6.1}$$

which shows that $\|\cdot\|, \|\cdot\|_\infty$ are equivalent norms on $\mathrm{ca}(\Sigma)$. In addition, from (4.6.1) we see that

$$\begin{aligned}
\mu_n \xrightarrow{\|\cdot\|} \mu \quad &\text{if and only if} \quad \mu_n \xrightarrow{\|\cdot\|_\infty} \mu \\
&\text{if and only if} \quad \sup[|\mu_n(A) - \mu(A)|: A \in \Sigma] \to 0 .
\end{aligned} \tag{4.6.2}$$

Now suppose that $\{\mu_n\}_{n\geq 1} \subseteq \mathrm{ca}(\Sigma)$ is a $\|\cdot\|$-Cauchy sequence, hence also a $\|\cdot\|_\infty$-Cauchy sequence because of their equivalence; see (4.6.1). From (4.6.2) we obtain that $\{\mu_n(A)\}_{n\geq 1} \subseteq \mathbb{R}$ is a Cauchy sequence uniformly in $A \in \Sigma$. Thus, there exists an additive $\mu: \Sigma \to \mathbb{R}$ such that

$$\sup[|\mu_n(A) - \mu(A)|: A \in \Sigma] \to 0 \quad \text{as } n \to \infty . \tag{4.6.3}$$

We need to show that $\mu \in \mathrm{ca}(\Sigma)$. Let $\{A_k\}_{k\geq 1} \subseteq \Sigma$ be pairwise disjoint and let $A = \bigcup_{k\geq 1} A_k \in \Sigma$. From (4.6.3) we see that there exists a number $n_0 \in \mathbb{N}$ such that

$$|\mu_n(C) - \mu(C)| \leq \varepsilon \quad \text{for all } C \in \Sigma \text{ and for all } n \geq n_0 . \tag{4.6.4}$$

Moreover, since $\mu_n \in \mathrm{ca}(\Sigma)$ for $n \in \mathbb{N}$, there exists a number $n_1 \in \mathbb{N}$ with $n_1 \geq n_0$ such that

$$\left|\mu_{n_0}(A) - \sum_{k=1}^{n} \mu_{n_0}(A_k)\right| \leq \varepsilon \quad \text{for all } n \geq n_1 . \tag{4.6.5}$$

Taking (4.6.4) and (4.6.5) into account, we get for $n \geq n_1$ that

$$\begin{aligned}\left|\mu(A) - \sum_{k=1}^{n} \mu(A_k)\right| &= \left|\mu\left(A \setminus \bigcup_{k=1}^{n} A_k\right)\right| \leq \left|\mu\left(A \setminus \bigcup_{k=1}^{n} A_k\right) - \mu_{n_0}\left(A \setminus \bigcup_{k=1}^{n} A_k\right)\right| \\ &+ \left|\mu_{n_0}\left(A \setminus \bigcup_{k=1}^{n} A_k\right)\right| = \left|\mu\left(A \setminus \bigcup_{k=1}^{n} A_k\right) - \mu_{n_0}\left(A \setminus \bigcup_{k=1}^{n} A_k\right)\right| \\ &+ \left|\mu_{n_0}(A) - \sum_{k=1}^{n} \mu_{n_0}(A_k)\right| \leq 2\varepsilon .\end{aligned}$$

Hence, $\mu \in \mathrm{ca}(\Sigma)$. Therefore $(\mathrm{ca}(\Sigma), \|\cdot\|)$ and $(\mathrm{ca}(\Sigma), \|\cdot\|_\infty)$ are Banach spaces. □

Definition 4.6.3. Let X be a locally compact topological space. We introduce the following vector spaces of continuous functions:

(a) $C_c(X) = \{f: X \to \mathbb{R}: f$ is continuous with compact support$\}$.

(b) $C_0(X) = \{f: X \to \mathbb{R}: f$ is continuous and for every $\varepsilon > 0$ there exists a compact set $K_\varepsilon \subseteq X$ such that $|f(x)| \leq \varepsilon$ for all $x \in X \setminus K_\varepsilon\}$. These are the continuous functions on $\mathbb{R}$ that vanish at infinity.

(c) $C_b(X) = \{f: X \to \mathbb{R}: f$ is continuous and bounded$\}$.

Remark 4.6.4. From the definitions above it is clear that we always have $C_c(X) \subseteq C_0(X) \subseteq C_b(X)$. If X is compact, then the three spaces coincide.

We endow the space $C_b(X)$ with the supremum norm

$$\|f\|_\infty = \sup[|f(x)|: x \in X] .$$

Proposition 4.6.5. *If X is a locally compact topological space, then the following hold:*

(a) $(C_b(X), \|\cdot\|_\infty)$ *is a Banach space;*

(b) $C_0(X)$ *is a closed subspace of* $C_b(X)$ *and so a Banach space as well;*

(c) $\overline{C_c(X)}^{\|\cdot\|_\infty} = C_0(X)$.

Proof. (a) Let $\{f_n\}_{n\geq 1} \subseteq C_b(X)$ be a Cauchy sequence. The completeness of $\mathbb{R}$ implies that for every fixed $x \in X$, the sequence $\{f_n(x)\}_{n\geq 1} \subseteq \mathbb{R}$ has a limit $f(x)$. We claim that

$$\|f_n - f\|_\infty \to 0 \quad \text{and} \quad f \in C_b(X) . \tag{4.6.6}$$

Given $\varepsilon > 0$, there exists $n_0 \in \mathbb{N}$ such that

$$\|f_n - f_m\|_\infty < \frac{\varepsilon}{3} \quad \text{for all } n, m \geq n_0 . \tag{4.6.7}$$

For $x \in X$ there exists $m = m(\varepsilon, x) \in \mathbb{N}$ with $m \geq n_0$ such that

$$|f_m(x) - f(x)| < \frac{\varepsilon}{3} . \tag{4.6.8}$$

From (4.6.7) and (4.6.8) we obtain

$$|f_n(x) - f(x)| \le |f_n(x) - f_m(x)| + |f_m(x) - f(x)| < \frac{2\varepsilon}{3} \quad \text{for all } n \ge n_0 ,$$

which gives $\|f_n - f\|_\infty < (2\varepsilon)/3$ for all $n \ge n_0$. Therefore the convergence in (4.6.6) is true. Next we show that $f \in C_b(X)$. Let $n \in \mathbb{N}$ such that $\|f_n - f\|_\infty < 1$. Then $|f(x)| < 1 + |f_n(x)|$ for all $x \in X$, and so f is bounded. In fact $\|f\|_\infty \le 1 + \|f_n\|_\infty$. In order to show the continuity of f, let $x \in X$ and $\varepsilon > 0$. We choose a large enough $n \in \mathbb{N}$ such that $\|f_n - f\|_\infty < \varepsilon/3$. Let $U \in \mathcal{N}(x)$ such that

$$|f_n(u) - f_n(x)| < \frac{\varepsilon}{3} \quad \text{for all } u \in U . \tag{4.6.9}$$

Then for $u \in U$, one gets

$$|f(u) - f(x)| \le |f(u) - f_n(u)| + |f_n(u) - f_n(x)| + |f_n(x) - f(x)| < 3\frac{\varepsilon}{3} = \varepsilon ;$$

see (4.6.9). Hence, $f \in C_b(X)$.

(b) Let $\{f_n\}_{n \ge 1} \subseteq C_0(X)$ and assume that $f_n \xrightarrow{\|\cdot\|_\infty} f$. Then given $\varepsilon > 0$ there exists a large enough $n \in \mathbb{N}$ and a compact set $K_\varepsilon \subseteq X$ such that

$$\|f_n - f\|_\infty < \frac{\varepsilon}{2} \quad \text{and} \quad |f_n(x)| < \frac{\varepsilon}{2} \quad \text{for all } x \in X \setminus K_\varepsilon .$$

This shows that

$$|f(x)| < \frac{\varepsilon}{2} + |f_n(x)| < \varepsilon \quad \text{for all } x \in X \setminus K_\varepsilon .$$

Hence, $f \in C_0(X)$.

(c) Given $f \in C_0(X)$ and $\varepsilon > 0$, there exists a compact set $K_\varepsilon \subseteq X$ such that $|f(x)| < \varepsilon$ for all $x \in X \setminus K_\varepsilon$. Invoking Theorem 1.4.88 we find $\hat{f} \in C_c(X)$ such that $\hat{f}|_{K_\varepsilon} = f$. Then $\|\hat{f} - f\|_\infty < \varepsilon$ and this proves the density of $C_c(X)$ in $C_0(X)$ for the $\|\cdot\|_\infty$-norm. □

We know that X being locally compact admits a one-point compactification $\hat{X}$; see Definition 1.4.74 and Theorem 1.4.75. Moreover, from Problem 1.46, we have the following result.

Proposition 4.6.6. *If X is a locally compact space, $\hat{X}$ its one-point compactification, $V = \{\hat{f} \in C(\hat{X}) : \hat{f}(\infty) = 0\}$, and for every $\hat{f} \in V$ let $\tilde{f} = \hat{f}|_X$, then the map $\hat{f} \to \tilde{f}$ is a linear isometry from V into $C_0(X)$.*

For a locally compact topological space X, we set $\Sigma = \mathcal{B}(X) =$ the Borel σ-field of X. We consider the following subspaces of $\mathrm{ca}(\mathcal{B}(X))$

$$\mathrm{ca}_R(\mathcal{B}(X)) = \{\mu \in \mathrm{ca}(\mathcal{B}(X)) : \mu \text{ is Radon}\} ,$$
$$\mathrm{ca}_r(\mathcal{B}(X)) = \{\mu \in \mathrm{ca}(\mathcal{B}(X)) : \mu \text{ is regular}\} ;$$

see Definitions 2.5.8 (d) and (b). It holds

$$\mathrm{ca_R}(\mathcal{B}(X)) \subseteq \mathrm{ca_r}(\mathcal{B}(X)) \subseteq \mathrm{ca}(\mathcal{B}(X))\,.$$

If X is metrizable, then $\mathrm{ca_r}(\mathcal{B}(X)) = \mathrm{ca}(\mathcal{B}(X))$; see Theorem 2.5.12. If X is Polish, then $\mathrm{ca_R}(\mathcal{B}(X)) = \mathrm{ca}(\mathcal{B}(X))$; see Theorem 2.5.14. In fact this is also true in the more general case when X is Souslin; see Schwartz [293, Proposition 3.3]. If X is compact, then clearly $\mathrm{ca_R}(\mathcal{B}(X)) = \mathrm{ca_r}(\mathcal{B}(X))$.

Proposition 4.6.7. *If X is locally compact space, $\hat{X}$ its one-point compactification, $Y = \{\hat{\mu} \in \mathrm{ca_r}(\mathcal{B}(\hat{X})) : \hat{\mu}(\{\infty\}) = 0\}$, and $\tilde{\mu} = \hat{\mu}|_{\mathcal{B}(X)}$ for all $\hat{\mu} \in Y$, then the map $\hat{\mu} \to \tilde{\mu}$ is a linear isometry from Y onto $\mathrm{ca_R}(X)$.*

Proof. Note that every $B \in \mathcal{B}(\hat{X})$ can be written as $B = A \cup \{+\infty\}$ or as $B = A$ with $A \in \mathcal{B}(X)$. In the first case, using the additivity of $\hat{\mu}$, one has $\hat{\mu}(B) = \hat{\mu}(A) + \hat{\mu}(\{\infty\}) = \tilde{\mu}(A)$. □

Now we can have a representation theorem for $C_0(X)$. First we recall the following classical result when X is a compact topological space; see, for example, Ash [16, Theorem 4.3.13, p. 184].

Theorem 4.6.8. *If X is a compact topological space, then $C(X)^* = \mathrm{ca_r}(\mathcal{B}(X))$ and if $\xi \in C_0(X)^*$, then $\xi(f) = \int_X f(x)\,\mathrm{d}\mu$ for some $\mu \in \mathrm{ca_r}(\mathcal{B}(X))$.*

Using this theorem we can prove the general case for X being locally compact.

Theorem 4.6.9. *If X is a locally compact topological space, then $C_c(X)^* = \mathrm{ca_R}(\mathcal{B}(X))$ and if $\xi \in C_0(X)^*$, then $\xi(f) = \int_X f(x)\,\mathrm{d}\mu$ for some $\mu \in \mathrm{ca_R}(\mathcal{B}(X))$.*

Proof. From Proposition 4.6.6 we know that $C_0(X) \simeq V \subseteq C(\hat{X})$ where $\hat{X}$ is the one-point compactification of X. Let $\hat{h}^* \in C(\hat{X})^*$. Then $h^* = \hat{h}^*|_V \in V^*$ and by the Hahn–Banach Theorem every element of V^* is obtained this way. So, if $N = \{\hat{h}^* \in C(\hat{X})^* : \hat{h}^*|_V = 0\}$, then $V^* \simeq C(\hat{X})^*/N$. We know that $\operatorname{codim} V = 1$. Hence $\dim N = 1$ and if δ_∞ is the Dirac measure at ∞, that is, $\delta_\infty(A) = 1$ if $\infty \in A$ and $\delta_\infty(A) = 0$ otherwise, then $N = \mathbb{R}\delta_\infty$. From Proposition 4.6.7 we get that $Y \simeq \mathrm{ca_r}(\mathcal{B}(\hat{X}))/N$ and from Theorem 4.6.8 it follows that $C(\hat{X})^* = \mathrm{ca_r}(\mathcal{B}(\hat{X}))$ where we recall that $\hat{X}$ is compact. Therefore we derive, thanks to Proposition 4.6.7, that

$$C_0(X)^* \simeq \mathrm{ca_r}(\mathcal{B}(\hat{X}))/N \simeq Y \simeq \mathrm{ca_R}(\hat{X})\,. \tag{4.6.10}$$

□

Remark 4.6.10. In (4.6.10) all isomorphisms are isometric and so, if $\mu \in ca_R(\mathcal{B}(X))$, then

$$\|\mu\|_{ca_R(\mathcal{B}(X))} = \sup\left[\int_X f(x)\,\mathrm{d}\mu : \|f\|_{C_0(X)} \le 1\right].$$

Let X be a locally compact topological space. We will introduce some topologies on the space $\mathrm{ca}(\mathcal{B}(X))$. Let $Z = C_b(X)$ or $C_0(X)$ or $C_c(X)$. Given $f \in Z$, we define

$$\xi_f(\mu) = \int_X f(x)\,\mathrm{d}\mu \quad \text{for all } \mu \in \mathrm{ca}(\mathcal{B}(X)).$$

Definition 4.6.11. Let X be a locally compact topological space.

(a) The weakest topology on $\mathrm{ca}(\mathcal{B}(X))$ making all the maps $\{\xi_f : f \in C_\mathrm{b}(X)\}$ continuous is called the **narrow topology** on $\mathrm{ca}(\mathcal{B}(X))$.

(b) The weakest topology on $\mathrm{ca}(\mathcal{B}(X))$ making all the maps $\{\xi_f : f \in C_0(X)\}$ continuous is called the **weak topology** on $\mathrm{ca}(\mathcal{B}(X))$.

(c) The weakest topology on $\mathrm{ca}(\mathcal{B}(X))$ making all the maps $\{\xi_f : f \in C_\mathrm{c}(X)\}$ continuous is called the **vague topology** on $\mathrm{ca}(\mathcal{B}(X))$.

By $\tau_\mathrm{n}(X)$ (resp. $\tau_{\mathrm{w}^*}(X)$, $\tau_\mathrm{v}(X)$) we denote the narrow (resp. the weak, the vague) topology on $\mathrm{ca}(\mathcal{B}(X))$.

Remark 4.6.12. On account of Theorem 4.6.9, $\tau_{\mathrm{w}^*}(X)$ restricted on $\mathrm{ca}_\mathrm{R}(\mathcal{B}(X)) = C_\mathrm{c}(X)^*$ is the usual w*-topology; see Section 3.3. The term **weak topology** comes from probability theory. From Definition 4.6.11 it is clear that

$$\tau_\mathrm{v}(X) \subseteq \tau_{\mathrm{w}^*}(X) \subseteq \tau_\mathrm{n}(X).$$

If X is compact, then these topologies coincide and they all correspond to the w*-topology on $\mathrm{ca}_\mathrm{r}(\mathcal{B}(X)) = \mathrm{ca}_\mathrm{R}(\mathcal{B}(X)) = C(X)^*$; see Theorem 4.6.8.

Proposition 4.6.13. *$\tau_{\mathrm{w}^*}(X) = \tau_\mathrm{n}(X)$ if and only if X is compact.*

Proof. $\Longrightarrow$: Let $f_0 \equiv 1 \in C_\mathrm{b}(X)$. Then $\xi_{f_0} : \mathrm{ca}(\mathcal{B}(X)) \to \mathbb{R}$ defined by $\xi_{f_0}(\mu) = \int_X f_0(x)\,\mathrm{d}\mu$ for all $\mu \in \mathrm{ca}(\mathcal{B}(X))$, is $\tau_\mathrm{n}(X)$-continuous, and hence by hypothesis, also $\tau_{\mathrm{w}^*}(X)$-continuous. From the definition of the $\tau_{\mathrm{w}^*}(X)$-topology there exists $\delta > 0$ and $\{f_k\}_{k=1}^n \subseteq C_0(X)$ such that

$$\left|\int_X f_k(x)\,\mathrm{d}\mu\right| < \delta \quad \text{implies} \quad |\xi_{f_0}(\mu)| = |\mu(X)| < \frac{1}{2}. \tag{4.6.11}$$

Since $f_k \in C_0(X)$ there exists a compact set $K_\delta^k \subseteq X$ such that $|f_k(x)| < \delta$ for all $x \in X \setminus K_\delta^k$. We set $K_\delta = \bigcup_{k=1}^n K_\delta^k$. Then $K_\delta \subseteq X$ is compact. Suppose that there is $x_0 \in X \setminus K_\delta$. Then $\delta_{x_0} \in \mathrm{ca}_\mathrm{R}(\mathcal{B}(X))$ and $|\int_X f_k(x)\,\mathrm{d}\delta_{x_0}| = |f_k(x_0)| < \delta$ for all $k = 1, \dots, n$. Hence, $\delta_{x_0}(X) < 1/2 < 1$ (see (4.6.11)), a contradiction. This shows that $X = K_\delta$ and so X is compact.

$\Longleftarrow$: If X is compact, then $C_0(X) = C_\mathrm{b}(X)$ and so $\tau_{\mathrm{w}^*}(X) = \tau_\mathrm{n}(X)$. □

Corollary 4.6.14. *$\tau_\mathrm{v}(X) = \tau_\mathrm{n}(X)$ if and only if X is compact.*

Let $P_\mathrm{r}(X)$ be the set of all regular probability measures on X, that is,

$$P_\mathrm{r}(X) = \{\mu \in (\mathrm{ca}_\mathrm{r})_+(X) : \mu(X) = 1\}.$$

Proposition 4.6.15. *If X is a compact topological space, then $P_\mathrm{r}(X)$ is $\tau_\mathrm{n}(X)$-compact.*

Proof. First we show that for every $\eta \in \mathbb{R}$, the set

$$\mathrm{ca}_{\mathrm{r}}^{\eta}(\mathcal{B}(X)) = \{\mu \in \mathrm{ca}_{\mathrm{r}}(\mathcal{B}(X)) \colon |\mu|(X) \le \eta\}$$

is $\tau_{\mathrm{n}}(X)$-compact. From Proposition 4.6.13 we obtain $\tau_{\mathrm{n}}(X) = \tau_{\mathrm{w}^*}(X)$ and by the Alaoglu Theorem, it holds that $\mathrm{ca}_{\mathrm{r}}^{\eta}(\mathcal{B}(X))$ is $\tau_{\mathrm{w}^*}(X)$-compact. Hence, $\mathrm{ca}_{\mathrm{r}}^{\eta}(\mathcal{B}(X))$ is $\tau_{\mathrm{n}}(X)$-compact.

Next let $\{\mu_\alpha\}_{\alpha\in I} \subseteq P_{\mathrm{r}}(X)$ be a net such that $\mu_\alpha \xrightarrow{\tau_{\mathrm{n}}(X)} \mu$. Then $\mu_\alpha(X) \to \mu(X)$, just use ξ_{f_0} with $f_0 \equiv 1$, and so $\mu(X) = 1$. Hence $P_{\mathrm{r}}(X)$ is $\tau_{\mathrm{n}}(X)$-closed and since $P_{\mathrm{r}}(X) \subseteq \mathrm{ca}_{\mathrm{r}}^{1}(\mathcal{B}(X))$, from the first part of the proof we conclude that $P_{\mathrm{r}}(X)$ is $\tau_{\mathrm{n}}(X)$-compact. □

Proposition 4.6.16. *If X is a locally compact and second countable topological space, then $\tau_{\mathrm{n}}(X)|_{P_{\mathrm{r}}(X)} = \tau_{\mathrm{v}}(X)|_{P_{\mathrm{r}}(X)}$.*

Proof. It is easy to see that

$$\tau_{\mathrm{v}}(X)|_{P_{\mathrm{r}}(X)} \subseteq \tau_{\mathrm{n}}(X)|_{P_{\mathrm{r}}(X)} \,. \tag{4.6.12}$$

We will show that the opposite inclusion also holds. First note that the hypotheses on X imply that it is a Polish space; see Remark 1.5.50. Let $\{\mu_\alpha, \mu\}_{\alpha\in I} \subseteq P_{\mathrm{r}}(X)$ and assume that $\mu_\alpha \xrightarrow{\tau_{\mathrm{v}}(X)} \mu$. Since X is Polish, all the probability measures are Radon; see Theorem 2.5.14. So, given $\varepsilon > 0$ there exists a compact set $K_\varepsilon \subseteq X$ such that $\mu(X \setminus K_\varepsilon) < \varepsilon$. Then Theorem 1.4.88 implies that we can find $\hat f \in C_{\mathrm{c}}(X)$ such that $\hat f|_{K_\varepsilon} = 1$ with $\hat f(X) \subseteq [0,1]$. We obtain $\xi_{\hat f}(\mu_\alpha) \to \xi_{\hat f}(\mu)$ and so there is an index $\alpha_0 \in I$ such that

$$\left|\xi_{\hat f}(\mu_\alpha) - \xi_{\hat f}(\mu)\right| < \varepsilon \quad \text{for all } \alpha \in I \text{ with } \alpha \ge \alpha_0 \,. \tag{4.6.13}$$

Let $C_\varepsilon = \operatorname{supp} \hat f \supseteq K_\varepsilon$. We have $\chi_{K\varepsilon} \le \hat f \le \chi_{C_\varepsilon}$ and so for $\alpha \in I$ with $\alpha \ge \alpha_0$, it follows, due to (4.6.13), that

$$\begin{aligned} \mu_\alpha(X \setminus C_\varepsilon) &= 1 - \mu_\alpha(C_\varepsilon) \le 1 - \xi_{\hat f}(\mu_\alpha) = 1 - \xi_{\hat f}(\mu) - \xi_{\hat f}(\mu_\alpha) + \xi_{\hat f}(\mu) \\ &\le 1 - \xi_{\hat f}(\mu) + \varepsilon \le 1 - \mu(K_\varepsilon) + \varepsilon = \mu(X \setminus K_\varepsilon) + \varepsilon < 2\varepsilon \,. \end{aligned} \tag{4.6.14}$$

Applying Theorem 1.4.88 again provides $g \in C_{\mathrm{c}}(X)$ such that $g|_{C_\varepsilon} = 1$ with $g(X) \subseteq [0,1]$. Let $f \in C_{\mathrm{b}}(X)$ and set $\tilde f = gf$. Evidently $\operatorname{supp} \tilde f \subseteq \operatorname{supp} g$ and so $\tilde f \in C_{\mathrm{c}}(X)$. Since $\mu_\alpha \xrightarrow{\tau_{\mathrm{v}}(X)} \mu$, there exists an index $\alpha_1 \in I$ such that

$$\left|\xi_{\tilde f}(\mu_\alpha) - \xi_{\tilde f}(\mu)\right| < \varepsilon \quad \text{for all } \alpha \in I \text{ with } \alpha \ge \alpha_1 \,.$$

Since I is a directed set, there is $\alpha_2 \in I$ with $\alpha_2 \ge \alpha_1$ as well as $\alpha_2 \ge \alpha_0$. For $\alpha \in I$ with $\alpha \ge \alpha_2$ we then derive that

$$\begin{aligned} \left|\xi_f(\mu_\alpha) - \xi_f(\mu)\right| &\le \left|\xi_f(\mu_\alpha) - \xi_{\tilde f}(\mu_\alpha)\right| + \left|\xi_{\tilde f}(\mu_\alpha) - \xi_{\tilde f}(\mu)\right| \\ &\quad + \left|\xi_{\tilde f}(\mu) - \xi_f(\mu)\right| \,. \end{aligned} \tag{4.6.15}$$

Note that

$$|(f-\tilde{f})(x)| = \begin{cases} 0 & \text{if } x \in C_\varepsilon\,, \\ f(x) & \text{if } x \in X \setminus C_\varepsilon\,. \end{cases} \tag{4.6.16}$$

Then, taking (4.6.14) into account, we infer that

$$\begin{aligned} |\xi_f(\mu_\alpha) - \xi_{\tilde{f}}(\mu_\alpha)| &\le \int_X |(f-\tilde{f})(x)|\,\mathrm{d}\mu_\alpha \le \|f\|_\infty \mu_\alpha(X \setminus C_\varepsilon) \\ &\le 2\varepsilon\|f\|_\infty\,. \end{aligned} \tag{4.6.17}$$

Moreover we have

$$|\xi_{\tilde{f}}(\mu_\alpha) - \xi_{\tilde{f}}(\mu)| < \varepsilon \quad \text{for all } \alpha \ge \alpha_2 \tag{4.6.18}$$

and

$$|\xi_{\tilde{f}}(\mu) - \xi_f(\mu)| \le \int_X |(\tilde{f}-f)(x)|\,\mathrm{d}\mu \le \|f\|_\infty \mu(X \setminus C_\varepsilon) \le \varepsilon\|f\|_\infty\,; \tag{4.6.19}$$

see (4.6.16). Returning to (4.6.15) and using (4.6.17), (4.6.18), as well as (4.6.19), we get

$$|\xi_f(\mu_\alpha) - \xi_f(\mu)| < \varepsilon(3\|f\|_\infty + 1) \quad \text{for all } \alpha \in I \text{ with } \alpha \ge \alpha_2\,.$$

Hence, $\xi_f(\mu_\alpha) \to \xi_f(\mu)$ and so $\xi_f|_{P_r(X)}$ is $\tau_v(X)$-continuous. Since $f \in C_b(X)$ is arbitrary, we infer that

$$\tau_n(X)|_{P_r(X)} \subseteq \tau_v(X)|_{P_r(X)}\,. \tag{4.6.20}$$

From (4.6.12) and (4.6.20), we conclude that $\tau_n(X)|_{P_r(X)} = \tau_v(X)|_{P_r(X)}$. □

Using Dirac measures we can embed a completely regular topological space X (see Definition 1.2.19) into $\mathrm{ca}_R(\mathcal{B}(X))$.

Proposition 4.6.17. *If X is a completely regular topological space and $\gamma: X \to \mathrm{ca}_r(\mathcal{B}(X))$ is defined by $\gamma(x) = \delta_x$ for all $x \in X$, then γ is a homeomorphism of X onto $(\gamma(X), \tau_n(X)|_{\gamma(X)})$.*

Proof. First note that γ is injective. Indeed, if $x, u \in X$ with $x \neq u$, then by the complete regularity of X there exists $f \in C_b(X)$ such that $f(x) = 0$ and $f(u) = 1$. Then $\xi_f(\delta_x) = 0$ and $\xi_f(\delta_u) = 1$. Hence, $\delta_x \neq \delta_u$.

Next we show that $\gamma: X \to \gamma(X)$ is bicontinuous. First suppose that $x_\alpha \to x$ in X. Given $f \in C_b(X)$ and $\varepsilon > 0$, we define

$$U(\delta_x; f, \varepsilon) = \{\mu \in \mathrm{ca}_R(\mathcal{B}(X)) : |\xi_f(\mu) - \xi_f(\delta_x)| < \varepsilon\}\,.$$

Evidently this is a $\tau_n(X)$-neighborhood of δ_x. We know that $f(x_\alpha) \to f(x)$ and so we can find an index $\alpha_0 \in I$ such that

$$|\xi_f(\delta_{x_\alpha}) - \xi_f(\delta_x)| = |f(x_\alpha) - f(x)| < \varepsilon \quad \text{for all } \alpha \in I \text{ with } \alpha \geq \alpha_0 .$$

Hence, $\xi_f(\delta_{x_\alpha}) = \gamma(x_\alpha) \in U(\delta_x; f, \varepsilon)$ for all $\alpha \in I$ with $\alpha \geq \alpha_0$. This shows that $\gamma(x_\alpha) \overset{\tau_n(X)}{\longrightarrow} \gamma(x)$ and so γ is $\tau_n(X)$-continuous.

Now assume that $\gamma(x_\alpha) \overset{\tau_n(X)}{\longrightarrow} \gamma(x)$. Then, for every $f \in C_b(X)$ we obtain $\xi_f(\delta_{x_\alpha}) \to \xi_f(\delta_x)$ and so

$$f(x_\alpha) \to f(x) . \tag{4.6.21}$$

Let $V \in \mathcal{N}(x)$ and let $g: X \to [0,1]$ be a continuous map such that $g(x) = 0$ and $g|_{X \setminus V} = 1$. From (4.6.21) we see that there exists an index $\alpha_1 \in I$ such that $|g(x_\alpha) - g(x)| = |g(x_\alpha)| < 1/2$ for all $\alpha \in I$ with $\alpha \geq \alpha_1$. So, $x_\alpha \in V$ for all $\alpha \geq \alpha_1$ and this means that $x_\alpha \to x$ in X. Hence γ is bicontinuous and this completes the proof. □

By strengthening the conditions on X we can have additional results.

Proposition 4.6.18. *If X is a metrizable space, then the following hold:*
(a) *for every lower semicontinuous and bounded below function $h: X \to \overline{\mathbb{R}} = \mathbb{R} \cup \{+\infty\}$, the map $\mu \to \xi_h(\mu) = \int_X h(x)\, d\mu$ is $\tau_n(X)$-lower semicontinuous on $\mathrm{ca}_+(\mathcal{B}(X))$;*
(b) *for every upper semicontinuous and bounded above function $h: X \to \overline{\mathbb{R}} = \mathbb{R} \cup \{-\infty\}$, the map $\mu \to \xi_g(\mu) = \int_X g(x)\, d\mu$ is $\tau_n(X)$-upper semicontinuous on $\mathrm{ca}_+(\mathcal{B}(X))$.*

Proof. (a) Applying Proposition 1.7.6 we see that for all $\mu \in \mathrm{ca}_+(\mathcal{B}(X))$

$$\xi_h(\mu) = \sup\left[\xi_f(\mu): \inf_X h \leq f \leq h, f \in C_b(X)\right] . \tag{4.6.22}$$

Each ξ_f is $\tau_n(X)|_{\mathrm{ca}_+(\mathcal{B}(X))}$-continuous. Then, invoking Proposition 1.7.4 (a) we infer that $\mathrm{ca}_+(\mathcal{B}(X)) \ni \mu \to \xi_h(\mu)$ is $\tau_n(X)|_{\mathrm{ca}_+(\mathcal{B}(X))}$-lower semicontinuous.
(b) This follows from (a) by multiplying with -1. □

Proposition 4.6.19. *If X is metrizable, then $(\mathrm{ca}(\mathcal{B}(X)), \tau_n(X))$ is completely regular.*

Proof. Note that $(\mathrm{ca}(\mathcal{B}(X)), \tau_n(x))$ is a locally convex space with a generating family of seminorms given by the maps $\mu \to \xi_f(\mu) = |\int_X f(x)\, d\mu|$. Given $\mu, \mu' \in \mathrm{ca}(\mathcal{B}(X))$, assume that $\xi_f(\mu) = \xi_f(\mu')$ for all $f \in C_b(X)$. Hence,

$$\xi_f(\mu - \mu') = 0 . \tag{4.6.23}$$

For an open set $U \subseteq X$, χ_U is lower semicontinuous and nonnegative. Then from (4.6.22) and (4.6.23) we obtain $|\mu - \mu'|(U) = 0$, which implies $\mu(U) = \mu'(U)$, that is, $\mu = \mu'$. So, the family $\{\xi_f\}_{f \in C_b(X)}$ generating the $\tau_n(X)$-topology is separating, and hence the topology is completely regular; see Section 1.3. □

Proposition 4.6.20. *If X is metrizable, then $(\mathrm{ca_R})_+(\mathcal{B}(X))$ is $\tau_\mathrm{n}(X)$-closed in $\mathrm{ca_R}(\mathcal{B}(X))$.*

Proof. Let $\{\mu_\alpha\}_{\alpha\in I} \subseteq (\mathrm{ca_R})_+(\mathcal{B}(X))$ be a net such that $\mu_\alpha \xrightarrow{\tau_\mathrm{n}(X)} \mu \in \mathrm{ca_R}(\mathcal{B}(X))$. For every $f \in C_\mathrm{b}(X)$ with $f \geq 0$ we have $\xi_f(\mu_\alpha) \geq 0$ for all $\alpha \in I$, and so $\xi_f(\mu) \geq 0$.

Let $K \subseteq X$ be compact. Since $|\mu|$ is regular, given $\varepsilon > 0$, there exists an open set $U \subseteq X$ such that $|\mu|(U \setminus K) < \varepsilon$. Invoking Urysohn's Lemma (see Theorem 1.2.17), there exists a continuous function $f: X \to [0,1]$ such that $f|_K = 1$ and $f|_{X\setminus U} = 0$. Then

$$\xi_f(\mu) = \int_X f(x)\,\mathrm{d}\mu = \int_U f(x)\,\mathrm{d}\mu = \mu(K) + \int_{U\setminus K} f(x)\,\mathrm{d}\mu\,,$$

which gives

$$\mu(K) = \xi_f(\mu) - \int_{U\setminus K} f(x)\,\mathrm{d}\mu \geq -\int_{U\setminus K} f(x)\,\mathrm{d}\mu \geq -\mu(U \setminus K) \geq -\varepsilon\,.$$

Since $\varepsilon > 0$ is arbitrary we let $\varepsilon \searrow 0$ to conclude that $\mu(K) \geq 0$ for all compact sets $K \subseteq X$. But $\mu \in \mathrm{ca_R}(\mathcal{B}(X))$. Hence it is compact regular; see Definition 2.5.8 (d). Therefore $\mu \in (\mathrm{ca_R})_+(\mathcal{B}(X))$ and we have proven that the latter is $\tau_\mathrm{n}(X)$-closed. □

The next theorem yields several equivalent definitions of the narrow topology on $\mathrm{ca}_+(\mathcal{B}(X))$. In what follows, we denote by $U_\mathrm{b}(X)$ the space of all $\mathbb{R}$-valued, bounded, uniformly continuous functions on X.

Theorem 4.6.21. *If X is metrizable and $\{\mu_\alpha, \mu\}_{\alpha\in I} \subseteq \mathrm{ca}_+(\mathcal{B}(X))$, then the following statements are equivalent:*

(a) $\mu_\alpha \xrightarrow{\tau_\mathrm{n}(X)} \mu$;
(b) $\xi_f(\mu_\alpha) \to \xi_f(\mu)$ *for all* $f \in U_\mathrm{b}(X)$;
(c) $\limsup_{\alpha\in I} \mu_\alpha(C) \leq \mu(C)$ *for all closed* $C \subseteq X$;
(d) $\mu(U) \leq \liminf_{\alpha\in I} \mu_\alpha(U)$ *for all open* $U \subseteq X$;
(e) $\lim_\alpha \mu_\alpha(A) = \mu(A)$ *for all* $A \in \mathcal{B}(X)$ *with* $\mu(\overline{A} \setminus \operatorname{int} A) = \mu(\operatorname{bd} A) = 0$.

Proof. (a) $\Longrightarrow$ (b): This is clear from Definition 4.6.11 (a) since $U_\mathrm{b}(X) \subseteq C_\mathrm{b}(X)$.

(b) $\Longrightarrow$ (c): Let $C \subseteq X$ be closed and let $U_n = \{x \in X: d(x,C) < 1/n\}$. Then U_n is open and $C \cap (X \setminus U_n) = \emptyset$. We define $f_n(x) = (d(x, X \setminus U_n))/(d(x,C) + d(x, X\setminus U_n))$ with $x \in X$. Evidently $f_n \in U_\mathrm{b}(X)$ and $f_n|_C = 1$ as well as $f_n|_{X\setminus U_n} = 0$ and $0 \leq f_n \leq 1$. Moreover, $\bigcap_{n\geq 1} U_n = C$. Then it follows that

$$\limsup_\alpha \mu_\alpha(C) \leq \limsup_\alpha \int_X f_n(x)\,\mathrm{d}\mu_\alpha = \int_X f_n(x)\,\mathrm{d}\mu \leq \mu(U_n) \quad \text{for all } n \geq 1\,.$$

Hence, $\limsup_\alpha \mu_\alpha(C) \leq \mu(C)$.

(c) $\Longleftrightarrow$ (d): Since closed sets are the complements of open sets the result follows.

(d) $\Longrightarrow$ (e): Let $A \in \mathcal{B}(X)$ with $\mu(\operatorname{bd} A) = \mu(\overline{A} \setminus \operatorname{int} A) = 0$. Then

$$\limsup_{\alpha} \mu_{\alpha}(A) \le \limsup_{\alpha} \mu_{\alpha}(\overline{A}) \le \mu(\overline{A}) = \mu(A) \tag{4.6.24}$$

and

$$\liminf_{\alpha} \mu_{\alpha}(A) \ge \liminf_{\alpha} \mu_{\alpha}(\operatorname{int} A) \ge \mu(\operatorname{int} A) = \mu(A)\,. \tag{4.6.25}$$

From (4.6.24) and (4.6.25) we conclude that $\mu_{\alpha}(A) \to \mu(A)$.

(e) $\Longrightarrow$ (a): Let $f \in C_{\mathrm{b}}(X)$ and define $\mu_f(A) = \mu(\{x \in X : f(x) \in A\})$ for all $A \in \mathcal{B}(X)$. Then μ_f can have at most a countable number of mass points; see Saks Lemma (Lemma 2.9.1). Hence, given any $\varepsilon > 0$, there exists $\{\eta_k\}_{k=0}^{m}$ and $a, b \in \mathbb{R}$ such that

$$a < f(x) < b \quad \text{for all } x \in X\,, \quad a = \eta_0 < \eta_1 < \cdots < \eta_m = b\,, \tag{4.6.26}$$

$$\eta_k - \eta_{k-1} < \varepsilon\,, \quad \mu(\{x \in X : f(x) = \eta_k\}) = 0 \quad \text{for all } k = 1, \dots, m\,. \tag{4.6.27}$$

Let $A_k = \{x \in X : \eta_{k-1} \le f(x) < \eta_k\}$ with $k = 1, \dots, m$. Then $\{A_k\}_{k=1}^{m} \subseteq \mathcal{B}(X)$ are disjoint and $X = \bigcup_{k=1}^{m} A_k$; see (4.6.26). Moreover, we obtain

$$\overline{A}_k \setminus \operatorname{int} A_k \subseteq \{x \in X : f(x) = \eta_{k-1}\} \cup \{x \in X : f(x) = \eta_k\}\,.$$

This gives, due to (4.6.27), that $\mu(\overline{A}_k \setminus \operatorname{int} A_k) = \mu(\operatorname{bd} A_k) = 0$ for all $k = 1, \dots, m$. Then

$$\mu_{\alpha}(A_k) \to \mu(A_k) \quad \text{for all } k = 1, \dots, m\,. \tag{4.6.28}$$

Let $s(x) = \sum_{k=1}^{m} \eta_{k-1} \chi_{A_k}(x)$ for all $x \in X$. Then $|s(x) - f(x)| < \varepsilon$ for all $x \in X$, see (4.6.27). It follows that

$$\begin{aligned}
&\left| \int_X f(x)\, d\mu_{\alpha} - \int_X f(x)\, d\mu \right| \\
&\le \int_X |f(x) - s(x)|\, d\mu_{\alpha} + \left| \int_X s(x)\, d\mu_{\alpha} - \int_X s(x)\, d\mu \right| + \int_X |s(x) - f(x)|\, d\mu \\
&\le \varepsilon[\mu_{\alpha}(X) + \mu(X)] + \sum_{k=1}^{m} |\mu_{\alpha}(A_k) - \mu(A_k)| |\eta_{k-1}|\,.
\end{aligned}$$

Hence, thanks to (4.6.28),

$$\limsup_{\alpha} \left| \int_X f(x)\, d\mu_{\alpha} - \int_X f(x)\, d\mu \right| \le 2\varepsilon \mu(X)\,.$$

Letting $\varepsilon \searrow 0$ we conclude that $\mu_{\alpha} \xrightarrow{\tau_{\mathrm{n}}(X)} \mu$. □

Remark 4.6.22. The theorem above fails on all of $\mathrm{ca}(\mathcal{B}(X))$. To see this, let $X = \mathbb{R}$ with the usual metric topology. Let $\mu_n = \delta_{1/n} - \delta_{-1/n}$ for all $n \in N$. Then $\mu_n \xrightarrow{\tau_n(X)} 0$. On the other hand if $C = [0, +\infty)$, then $\mu_n(C) = 1$ for all $n \in \mathbb{N}$. Hence (b) is not satisfied.

Next we consider a metrizable space X and we focus on the set of all probability measures on X denoted by $\mathcal{P}(X)$. Note that all probability measures on X are regular; see Theorem 2.5.12. So, $\mathcal{P}(X) = P_r(X)$. The reason we focus on $\mathcal{P}(X)$ is that $\overline{\mathrm{span}}\, \mathcal{P}(X) = \mathrm{ca}(\mathcal{B}(X))$. It turns out that $\mathcal{P}(X)$ equipped with the relative $\tau_n(X)$-topology inherits many of the properties of the space X. So, in what follows without any further saying, we consider on $\mathcal{P}(X)$ the trace of the $\tau_n(X)$-topology.

Theorem 4.6.23. *X is compact if and only if $\mathcal{P}(X)$ is compact metrizable.*

Proof. $\Longrightarrow$: From Proposition 4.6.13 we know that $\tau_n(X) = \tau_{w^*}(X)$. Since $C_b(X) = C(X)$ is separable, the w*-topology on the closed unit ball of $\mathrm{ca}(\mathcal{B}(X))$ is compact and metrizable; see Theorem 3.4.11. But $\mathcal{P}(X)$ is a w*-closed subset of the closed unit ball in $\mathrm{ca}(\mathcal{B}(X))$. We conclude that $\mathcal{P}(X)$ is compact and metrizable.

$\Longleftarrow$: Applying Proposition 4.6.17, we know that X is a topological subspace of $\mathcal{P}(X)$. So, it follows that X is compact. ☐

Theorem 4.6.24. *X is separable if and only if $\mathcal{P}(X)$ is separable metrizable.*

Proof. $\Longrightarrow$: By Urysohn's Theorem (see Theorem 1.5.21), X is homeomorphic to a subset of the Hilbert cube $\mathbb{H} = [0,1]^{\mathbb{N}}$. Therefore there exists an equivalent totally bounded metrization of X. We use this metric d on X. Then the completion $(\overline{X}, \overline{d})$ of (X, d) is compact and we have an isometry $\vartheta\colon U_d(X) \to C(\overline{X})$ defined by $\vartheta(f) = \overline{f}$ with $\overline{f}$ being the unique continuous extension of f on $\overline{X}$; see Theorem 1.5.27. Here, $U_d(X)$ denotes the space of all $\mathbb{R}$-valued d-uniformly continuous functions on X. Since $C(\overline{X})$ is separable, so is $U_d(X)$. Let $D \subseteq U_d(X)$ be a countable dense subset. Let $S\colon \mathcal{P}(X) \to \mathbb{R}^{\mathbb{N}}$ be defined by

$$S(\mu) = \left\{ \int_X f(x)\,\mathrm{d}\mu : f \in D \right\}.$$

We claim that S is a homeomorphism. First we show that S is injective. So suppose that $S(\mu) = S(\lambda)$. Then $\int_X f(x)\,\mathrm{d}\mu = \int_X f(x)\,\mathrm{d}\lambda$ for all $f \in D$. Exploiting the density of D in $U_d(X)$, we obtain that $\int_X f(x)\,\mathrm{d}\mu = \int_X f(x)\,\mathrm{d}\lambda$ for all $f \in U_d(X)$. Let $C \subseteq X$ be closed and let $U_n = \{x \in X : d(x, C) < 1/n\}$. Then U_n is open, $C \cap (X \setminus U_n) = \emptyset$, and $C = \bigcap_{n\ge1} U_n$. Then as in the proof of Theorem 4.6.21, we produce a function $\hat{f}_n\colon X \to [0,1]$ such that

$$\hat{f}_n|_C = 1\,, \quad \hat{f}_n|_{X\setminus U_n} = 0\,, \quad 0 \le \hat{f}_n \le 1\,, \quad \hat{f}_n \in U_d(X) \quad \text{for all } n \in \mathbb{N}\,.$$

It follows that

$$\mu(C) \le \int_X \hat{f}_n(x)\,\mathrm{d}\mu = \int_X \hat{f}_n(x)\,\mathrm{d}\lambda = \int_{U_n} \hat{f}_n(x)\,\mathrm{d}\lambda \le \lambda(U_n) \quad \text{for all } n \in \mathbb{N}\,,$$

which shows that $\mu(C) \leq \lambda(C)$. Interchanging the roles for μ and λ in the argument above, we also get that $\lambda(C) \leq \mu(C)$; hence $\mu(C) = \lambda(C)$, for all closed $C \subseteq X$. Therefore $\mu = \lambda$; see Remark 2.5.9. We have proven that S is injective. Next we show that S is bicontinuous. So, suppose that $\mu_\alpha \to \mu$ in $\mathcal{P}(X)$. Then $\int_X f(x)\,d\mu_\alpha \to \int_X f(x)\,d\mu$ for all $f \in D$; see Theorem 4.6.21. Hence, $S(\mu_\alpha) \to S(\mu)$ and this proves the continuity of S. In order to show the continuity of S^{-1}, let $\{\mu_\alpha\}_{\alpha\in I} \subseteq \mathcal{P}(X)$ be a net and assume that $S(\mu_\alpha) \to S(\mu)$ in $\mathbb{R}^{\mathbb{N}}$, that is, $\int_X f(x)\,d\mu_\alpha \to \int_X f(x)\,d\mu$ for all $f \in D$. For any $h \in U_d(X)$, one has

$$\left|\int_X h(x)\,d\mu_\alpha - \int_X h(x)\,d\mu\right| \leq 2\|h - f\|_\infty + \left|\int_X f(x)\,d\mu_\alpha - \int_X f(x)\,d\mu\right|$$

with $f \in D$. Then

$$\limsup_\alpha \left|\int_X h(x)\,d\mu_\alpha - \int_X h(x)\,d\mu\right| \leq 2\|h - f\|_\infty \quad \text{for all } f \in D\,.$$

Hence, $\mu_\alpha \to \mu$ in $\mathcal{P}(X)$ since $D \subseteq U_d(X)$ is dense. Therefore S^{-1} is continuous and so S is bicontinuous. Thus, S is a homeomorphism. Since $\mathbb{R}^{\mathbb{N}}$ is separable metrizable, then so is $\mathcal{P}(X)$.

$\Leftarrow$: By Proposition 4.6.17, X can be viewed as a topological subspace of $\mathcal{P}(X)$. Therefore X is separable. □

Theorem 4.6.25. *If X is separable metrizable, then X is Polish if and only if $\mathcal{P}(X)$ is Polish.*

Proof. $\Rightarrow$: As in the proof of Theorem 4.6.24, we may assume that X is totally bounded for some metric d. Its d-completion $\overline{X}$ is compact and X being Polish is a G_δ-subset of $\overline{X}$; see Theorem 1.5.47. From Theorem 4.6.23 we know that $P(\overline{X})$ is compact metrizable. Let $P_0 = \{\overline{\mu} \in P(\overline{X}) : \overline{\mu}(\overline{X} \setminus X) = 0\}$ Then $\mathcal{P}(X)$ and P_0 are homeomorphic. If we show that $P_0 \subseteq P(\overline{X})$ is G_δ, then Theorem 1.5.47 implies that P_0 is Polish, and hence, so is $\mathcal{P}(X)$. Since X is G_δ in $\overline{X}$, there exist open sets $\{U_n\}_{n\geq1}$ in $\overline{X}$ such that $X = \bigcap_{n\geq1} U_n$. Let

$$P_n = \{\overline{\mu} \in P(\overline{X}) : \overline{\mu}(\overline{X} \setminus U_n) = 0\} \quad \text{with } n \in \mathbb{N}\,.$$

Then $P_0 = \bigcap_{n\geq1} P_n$ and $P_n = \bigcap_{k\geq1} P_n^k$ with $P_n^k = \{\overline{\mu} \in P(\overline{X}) : \overline{\mu}(X \setminus U_n) < 1/k\}$. Suppose that $\{\overline{\mu}_\alpha\}_{\alpha\in I} \subseteq P(\overline{X}) \setminus P_n^k$ such that $\overline{\mu}_\alpha \to \overline{\mu}$ in $P(\overline{X})$. Then by Theorem 4.6.21, we obtain

$$\overline{\mu}(X \setminus U_n) \geq \limsup_\alpha \overline{\mu}_\alpha(X \setminus U_n) \geq \frac{1}{k}\,,$$

which implies that $P_n^k \subseteq P(\overline{X})$ is open for all $n, k \in \mathbb{N}$. Hence P_0 is G_δ in $P(\overline{X})$ and so $\mathcal{P}(X)$ is Polish.

$\Leftarrow$: Once again we use Proposition 4.6.17 to view X as a closed subspace of $\mathcal{P}(X)$. Then X is Polish by Proposition 1.5.45. □

Dudley [99] furnished a compatible metric

$$d_D(\mu,\mu^*) = \sup[|\xi_f(\mu) - \xi_f(\mu^*)|: f \in \mathrm{Lip_b}(X), \|f\|_{\mathrm{Lip_b}} \le 1] ,$$

where

$$\mathrm{Lip_b}(X) = \{f: X \to \mathbb{R}: f \text{ is bounded and Lipschitz continuous}\}$$

and

$$\|f\|_{\mathrm{Lip_b}} = \|f\|_\infty + \sup_{\substack{x\neq u\\ x,u\in X}} \frac{|f(x)-f(u)|}{d(x,u)} .$$

Definition 4.6.26. Let X be a completely regular topological space and $C \subseteq \mathrm{ca}(\mathcal{B}(X))$. We say that C is **tight** if

(a) $\sup[|\mu|(X): \mu \in C] < \infty$;

(b) for every $\varepsilon > 0$ there exists a compact set $K_\varepsilon \subseteq X$ such that $\sup[|\mu|(X \setminus K_\varepsilon): \mu \in C] \le \varepsilon$.

Using Proposition 4.6.17, we easily see that the following result is true.

Proposition 4.6.27. *If X is completely regular, $A \subseteq X$ and $C = \{\delta_x : x \in A\}$, then C is tight if and only if A is relatively compact.*

Proposition 4.6.28. *If X is completely regular and $C \subseteq \mathrm{ca}(\mathcal{B}(X))$ is bounded, then C is tight if and only if there exists a function $\varphi: X \to \overline{\mathbb{R}} = \mathbb{R} \cup \{+\infty\}$ such that*

(a) *for all $\eta \ge 0$, it holds that $\varphi^\eta = \{x \in X: \varphi(x) \le \eta\}$ is compact;*

(b) $\sup[\int_X \varphi \, \mathrm{d}|\mu|: \mu \in C] < +\infty$.

Proof. $\Longrightarrow$: On account of Definition 4.6.26, there exists an increasing sequence $\{K_n\}_{n\ge1}$ of compact subsets of X such that $\sup[|\mu|(X \setminus K_n): \mu \in C] \le 1/2^n$. Let $\varphi = \sum_{n\ge1} \chi_{X\setminus K_n}$ and let $[\eta]$ be the integer part of $\eta > 0$. We have $\varphi^\eta = K_{[\eta]+1}$ and so we have satisfied (a). Moreover,

$$\sup\Big[\int_X \varphi \, \mathrm{d}|\mu|: \mu \in C\Big] = \sup\Big[\sum_{n\ge1} |\mu|(X \setminus K_n): \mu \in C\Big]$$
$$\le \sum_{n\ge1} \sup[|\mu|(X \setminus K_n): \mu \in C] \le \sum_{n\ge1} \frac{1}{2^n} < \infty .$$

$\Longleftarrow$: Let $M = \sup[\int_X \varphi \, \mathrm{d}|\mu|: \mu \in C] < \infty$; see (b). Given $\varepsilon > 0$, let $K_\varepsilon = \varphi^{M/\varepsilon} \subseteq X$ be compact. Then for every $\mu \in C$ we get

$$|\mu|(X \setminus K_\varepsilon) = |\mu|\Big(\Big\{x \in X: \varphi(x) > \frac{M}{\varepsilon}\Big\}\Big) \le \frac{\varepsilon}{M} \int_X \varphi \, \mathrm{d}|\mu| \le \varepsilon .$$

Hence, C is tight. ∎

Remark 4.6.29. A function $\varphi: X \to \overline{\mathbb{R}} = \mathbb{R} \cup \{+\infty\}$ such that $\varphi^\eta = \{x \in X: \varphi(x) \le \eta\}$ is compact for every $\eta \in \mathbb{R}$ is said to be **inf-compact** or **coercive**.

Proposition 4.6.30. *If X is metrizable and $C \subseteq \mathrm{ca}_+(\mathcal{B}(X))$ is tight, then $\overline{C}^{\tau_n(X)}$ is tight as well.*

Proof. Let $M = \sup[\mu(X): \mu \in C] < +\infty$; see Definition 4.6.26. Given $\mu \in \overline{C}^{\tau_n(X)}$, there exists $\{\mu_\alpha\}_{\alpha \in I} \subseteq C$ such that $\mu_\alpha \xrightarrow{\tau_n(X)} \mu$. Let $f \equiv 1 \in C_b(X)$. Then $\mu_\alpha(X) = \xi_f(\mu_\alpha) \to \xi_f(\mu) = \mu(X)$, and hence, $\mu(X) \le M$ for all $\mu \in \overline{C}^{\tau_n(X)}$.

For every $\varepsilon > 0$ we can find a compact set $K_\varepsilon \subseteq X$ such that

$$\mu(X \setminus K_\varepsilon) \le \varepsilon \quad \text{for all } \mu \in C\,. \tag{4.6.29}$$

The function $\chi_{X \setminus K_\varepsilon}$ is lower semicontinuous and nonnegative. So, if $\mu \in \overline{C}^{\tau_n(X)}$ and $\{\mu_\alpha\}_{\alpha \in I} \subseteq C$ satisfy $\mu_\alpha \xrightarrow{\tau_n(X)} \mu$, then by Proposition 4.6.18, we infer that

$$\mu(X \setminus K_\varepsilon) \le \liminf_\alpha \mu_\alpha(X \setminus K_\varepsilon) \le \varepsilon \quad \text{for all } \mu \in \overline{C}^{\tau_n(X)}\,;$$

see (4.6.29). Hence, $\overline{C}^{\tau_n(X)}$ is tight. ☐

We conclude with a result characterizing the $\tau_n(X)$-compact subsets of $\mathrm{ca}_R(\mathcal{B}(X))$. The result is known as "Prokhorov's Theorem". Its proof can be found in Bourbaki [47, Theorem 5.2, p. 64] in the case of X being completely regular and in Dudley [101, Theorem 11.5.4, p. 316] in the case of X being Polish.

Theorem 4.6.31 (Prokhorov's Theorem). *If X is completely regular and $C \subseteq \mathrm{ca}_R(\mathcal{B}(X))$ is tight, then C is relatively $\tau_n(X)$-compact.*

Remark 4.6.32. For Polish spaces we know that $C \subseteq \mathrm{ca}(\mathcal{B}(X))$ is tight if and only if it is relatively $\tau_n(X)$-compact; see Dudley [101].

4.7 Young Measures

In this section we present a brief introduction of the theory of Young measures. Such measures generalize measurable functions and in fact can be viewed as the completion in some sense of the set of measurable functions. In this completion, the measurable functions correspond to the Dirac Young measures. Nowadays Young measures are an important tool in many parts of mathematical analysis.

Let (Ω, Σ, μ) be a complete probability space and X a Polish space. By $\mathcal{B}(X)$, we denote the Borel σ-algebra of X and by $\mathcal{P}(X)$, the set of all probability measures on X endowed with the narrow topology $\tau_n(X)$. The next result is a "disintegration theorem" and its proof can be found in Valadier [316].

Theorem 4.7.1. *If $p_\Omega: \Omega \times X \to \Omega$ is the projection map, $\lambda \in \mathrm{ca}(\Sigma \otimes \mathcal{B}(X))$, and $\mu = \lambda \circ p_\Omega^{-1}$, then there exists a Σ-measurable $\hat{\lambda}: \Omega \to \mathcal{P}(X)$ such that $\lambda(A \times C) = \int_A \hat{\lambda}(w)(C)\, d\mu$.*

Remark 4.7.2. The map $\hat{\lambda}: \Omega \to \mathcal{P}(X)$ is unique in the sense that if $\hat{\lambda}_0: \Omega \to \mathcal{P}(X)$ is another such measurable map, then $\hat{\lambda}(w)(C) = \hat{\lambda}_0(w)(C)$ μ-a. e. for every $C \in \mathcal{B}(X)$. Since $\mathcal{B}(X)$ is countably generated, $\hat{\lambda}(w) = \hat{\lambda}_0(w)$ μ-a. e. The map $\hat{\lambda}$ is said to be the **disintegration** of $\lambda \in \mathrm{ca}(\Sigma \otimes \mathcal{B}(X))$ with respect to μ. Using the Monotone Class Theorem (see Theorem 2.1.12), one easily sees that

$$\lambda(E) = \int_\Omega \hat{\lambda}(w)(E_w)\, d\mu \tag{4.7.1}$$

for all $E \in \Sigma \otimes \mathcal{B}(X)$ where $E_w = \{x \in X: (w, x) \in E\}$. Using (4.7.1) and approximating with simple functions, we infer that for every $f \in L^1(\Omega \times X, \lambda)$ (or alternatively for every $\Sigma \otimes \mathcal{B}(X)$-measurable $f: \Omega \times X \to \mathbb{R}_+$), we obtain

$$\int_{\Omega \times X} f(w, x)\, d\lambda = \int_\Omega \left[\int_X f(w, x) \hat{\lambda}_w(dx) \right] d\mu\,.$$

Now we can introduce Young measures.

Definition 4.7.3. A **Young measure** on $\Omega \times X$ is a $\lambda \in \mathrm{ca}_+(\Sigma \otimes \mathcal{B}(X))$ such that $\mu = \lambda \circ p_\Omega^{-1}$, that is, $\mu(A) = \lambda(A \times X)$ for all $A \in \Sigma$. By $\mathcal{Y}(\Omega \times X)$ (or simply by $\mathcal{Y}$ if no ambiguity can occur), we denote the space of Young measures on $\Omega \times X$.

Remark 4.7.4. On account of Theorem 4.7.1 we can identify each $\lambda \in \mathcal{Y}(\Omega \times X)$ with its unique disintegration $\hat{\lambda}(w)$. So, we can say that a Young measure is a Σ-measurable map $\hat{\lambda}: \Omega \to \mathcal{P}(X)$. So, $\mathcal{Y} \subseteq \mathrm{ca}_+(\Sigma \otimes \mathcal{B}(X))$ and $\mathcal{Y} \subseteq L^0(\Omega, \mathcal{P}(X)) = \{\hat{\lambda}: \Omega \to \mathcal{P}(X): \hat{\lambda}$ is Σ-measurable$\}$. Note that if $\lambda \in \mathcal{Y}(\Omega \times X)$, then for every $A \in \Sigma$, $\lambda(A \times \cdot) \in \mathrm{ca}_+(\mathcal{B}(X))$ and so it is a Radon measure; see Theorem 2.5.14.

Proposition 4.7.5. *If $\hat{\lambda}: \Omega \to \mathcal{P}(X)$, then the following two statements are equivalent:*
(a) *$\hat{\lambda}$ is Σ-measurable;*
(b) *The map $w \to \xi_C(w) = \hat{\lambda}(w)(C)$ is Σ-measurable for every $C \in \mathcal{B}(X)$.*

Proof. (a) $\Longrightarrow$ (b): Let $U \subseteq X$ be open and let $\eta_U: \mathcal{P}(X) \to [0, 1]$ be defined by $\eta_U(\vartheta) = \vartheta(U)$ for all $\vartheta \in \mathcal{P}(X)$. According to Theorem 4.6.21 η_U is lower semicontinuous. Recall that we consider the $\tau_n(X)$-topology on $\mathcal{P}(X)$. Then $\xi_U = \eta_U \circ \hat{\lambda}$ and so ξ_U is Σ-measurable. Since the elements of $\mathcal{P}(X)$ are regular (see Theorem 2.5.12), we conclude that ξ_C is Σ-measurable for all $C \in \mathcal{B}(X)$.

(b) $\Longrightarrow$ (a): For every $E \in \Sigma \otimes \mathcal{B}(X)$, let $h_E: \Omega \to [0, 1]$ be defined by $h_E(w) = \hat{\lambda}(w)(E_w)$. We set $\mathcal{M} = \{E \in \Sigma \otimes \mathcal{B}(X): h_E$ is Σ-measurable$\}$. A standard argument shows that $\mathcal{M}$ is a monotone class that contains the algebra of measurable rectangles. Then from the Monotone Class Theorem (see Theorem 2.1.12), we obtain $\Sigma \otimes \mathcal{B}(X) \subseteq \mathcal{M}$; see Remark 2.2.24.

Hence, h_E is Σ-measurable for every $E \in \Sigma \otimes \mathcal{B}(X)$. Now let $C \in \mathcal{B}(X)$ and let $E = \Omega \times C$. We have $h_E(w) = \hat{\lambda}(w)(C) = \int_C \chi_C(x)\hat{\lambda}(w)(dx)$. We have seen that h_E is Σ-measurable. Then $w \to \int_X s(x)\hat{\lambda}(w)(dx)$ is Σ-measurable for every simple function $s(x)$ on X. Finally if $h \in C_b(X)$, then there exists a sequence $\{s_n\}_{n\geq 1}$ of simple functions such that $|s_n(x)| \leq h(x)$ for all $x \in X$, for all $n \in \mathbb{N}$ and by Corollary 2.2.19, we also get that $s_n(x) \to h(x)$ for all $x \in X$ as $n \to \infty$. Then by the Lebesgue Dominated Convergence Theorem (see Theorem 2.3.8), it follows

$$\int_X s_n(x)\hat{\lambda}(w)(dx) \to \int_X h(x)\hat{\lambda}(w)(dx) = \langle \hat{\lambda}(w), h \rangle ,$$

which shows that $w \to \langle \hat{\lambda}(w), h \rangle$ is Σ-measurable for all $h \in C_b(X)$. Hence, $w \to \hat{\lambda}(w)$ is Σ-measurable. ∎

Let $L^0(\Omega, X) = \{u : \Omega \to X : u \text{ is } \Sigma\text{-measurable}\}$. Given $u \in L^0(\Omega, X)$, let $\hat{\lambda}^u : \Omega \to \mathcal{P}(X)$ be defined by $\hat{\lambda}^u(w) = \delta_{u(w)}$. Note that $w \to \hat{\lambda}^u(w)(C) = \chi_{u^{-1}(C)}(w)$ is Σ-measurable for every $C \in \mathcal{B}(X)$. Invoking Proposition 4.7.5, we infer that $\hat{\lambda}^u : \Omega \to \mathcal{P}(X)$ is Σ-measurable and so it is a Young measure $\hat{\lambda}^u$; see Remark 4.7.4.

Definition 4.7.6. Given $u \in L^0(\Omega, X)$, $\hat{\lambda}^u$ defined above is the **Young measure associated to the measurable function** u.

Remark 4.7.7. Evidently, $\hat{\lambda}^u$ is the disintegration of the Young measure λ^u defined by

$$\lambda^u(A \times C) = \mu(A \cap u^{-1}(C)) \quad \text{for all } A \in \Sigma \text{ and for all } C \in \mathcal{B}(X) .$$

Then for every $\Sigma \otimes \mathcal{B}(X)$-measurable function $\varphi : \Omega \times X \to \overline{\mathbb{R}}_+ = \mathbb{R}_+ \cup \{+\infty\}$ or every $\varphi \in L^1(\Omega \times X, \lambda^u)$ we have

$$\int_{\Omega\times X} \varphi(w, x)\, d\lambda^u = \int_\Omega \varphi(w, u(w))\, d\mu .$$

Moreover, the map $u \to \lambda^u$ is an embedding of $L^0(\Omega, X)$ into $\mathcal{Y}(\Omega \times X)$. Therefore, we can view $L^0(\Omega, X)$ as a subspace of $\mathcal{Y}(\Omega \times X)$. Of course this also leads to an identification of X with a subspace of $\mathcal{Y}(\Omega \times X)$, that is, identify $x \in X$ with the constant function $u \equiv x$.

We would like to have conditions that allow us to infer that a Young measure is associated with a measurable function. The next proposition provides a criterion to identify such Young measures.

Proposition 4.7.8. *If $\lambda \in \mathcal{Y}(\Omega \times X)$ and $u \in L^0(\Omega, X)$, then $\lambda = \lambda^u$ if and only if $\lambda((\mathrm{Gr}\, u)^c) = 0$, where $\mathrm{Gr}\, u = \{(w, x) \in \Omega \times X : u(w) = x\}$ denotes the graph of u and $(\mathrm{Gr}\, u)^c = (\Omega \times X) \setminus \mathrm{Gr}\, u$.*

Proof. $\Longrightarrow$: Let $\eta_u : \Omega \to \Omega \times X$ be defined by $\eta_u(w) = (w, u(w))$. For every $E \in \Sigma \otimes \mathcal{B}(X)$ we obtain

$$\lambda(E) = \lambda^u(E) = \mu(\eta_u^{-1}(E)) \,. \tag{4.7.2}$$

Note that $\operatorname{Gr} u \in \Sigma \otimes \mathcal{B}(X)$. Hence, $E = (\Omega \times X) \setminus \operatorname{Gr} u \in \Sigma \otimes \mathcal{B}(X)$ and $\eta_u^{-1}(E) = \emptyset$. Therefore, due to (4.7.2), $\lambda((\Omega \times X) \setminus \operatorname{Gr} u) = 0$.

$\Longleftarrow$: Let $E \in \Sigma \otimes \mathcal{B}(X)$ and $A = \eta_u^{-1}(E) \in \Sigma$. Then

$$\lambda(E) = \lambda(E \setminus (\operatorname{Gr} u)^c) \le \lambda(A \times X) = \mu(A) \,. \tag{4.7.3}$$

Moreover we have

$$\mu(A) = \lambda(A \times X) = \lambda((A \times X) \setminus (\operatorname{Gr} u)^c) \le \lambda(E) \,, \tag{4.7.4}$$

since $A \times X \setminus (\operatorname{Gr} u)^c \subseteq E$. From (4.7.3) and (4.7.4) we infer that

$$\lambda(E) = \mu(A) = \mu(\eta_u^{-1}(E)) = \lambda^u(E) \,.$$

Hence, $\lambda = \lambda^u$. □

Definition 4.7.9. (a) An **integrand** is a $\Sigma \otimes \mathcal{B}(X)$-measurable function $\varphi\colon \Omega \times \mathbb{R} \to \mathbb{R}^* = \mathbb{R} \cup \{\pm\infty\}$.

(b) We say that the integrand φ is L^1**-bounded** if there exists $k \in L^1(\Omega)$ such that $|\varphi(w,x)| \le k(w)$ μ-a. e. on Ω for all $x \in X$.

(c) We say that a function $\varphi\colon \Omega \times X \to \mathbb{R}$ is **Carathéodory** if $w \to \varphi(w,x)$ is Σ-measurable for all $x \in X$ and $w \to \varphi(w,x)$ is continuous for μ-a. e. $w \in \Omega$; see also Definition 2.2.30. A Carathéodory function is an integrand; see Proposition 2.2.31. By $\operatorname{Car}(\Omega \times X)$ (resp. $\operatorname{Car}_b(\Omega \times X)$), we denote the set of all Carathéodory (resp. L^1-bounded Carathéodory) integrands.

(d) We say that an integrand φ is **normal** if it takes values in $\overline{\mathbb{R}} = \mathbb{R} \cup \{+\infty\}$ and $\varphi(w,\cdot)$ is lower semicontinuous for μ-a.e $w \in \Omega$. By $\operatorname{Nor}(\Omega \times X)$ we denote the set of all normal integrands on $\Omega \times X$.

Let d be the metric generating the topology of X. Using d and reasoning as in the proof of Proposition 1.7.6, we obtain the following approximation result.

Proposition 4.7.10. *If $\varphi \in \operatorname{Nor}(\Omega \times X)$, then there exists a sequence $\{\varphi_n\}_{n\ge1} \subseteq \operatorname{Car}(\Omega \times X)$ such that $\varphi_n(w,\cdot)$ is d-Lipschitz for every $n \in \mathbb{N}$ and for μ-a. a. $w \in \Omega$ and $\varphi_n \nearrow \varphi$ on $\Omega \times X$. Moreover, if φ is L^1-bounded, then so are the φ_n's.*

Using L^1-bounded Carathéodory integrands we can define a counterpart of the narrow topology (see Definition 4.6.11 (a)), for the space of Young measures $\mathcal{Y}(\Omega \times X)$.

Definition 4.7.11. The **Young narrow topology** on $\mathcal{Y}(\Omega \times X)$ is the weakest topology on $\mathcal{Y}(\Omega \times X)$ for which the maps

$$\lambda \to I_\varphi(\lambda) = \int_{\Omega\times X} \varphi(w,x)\, d\lambda = \int_\Omega \Big[\int_X \varphi(w,x) \hat{\lambda}(w)(dx) \Big] d\mu$$

with $\hat{\lambda}$ the disintegration of λ (see Theorem 4.7.1), and $\varphi \in \mathrm{Car_b}(\Omega \times X)$, are all continuous. We denote by $\tau_{\mathrm{n}}^{\mathcal{Y}}(\Omega \times X)$ or simply $\tau_{\mathrm{n}}^{\mathcal{Y}}$ this topology on $\mathcal{Y}(\Omega \times X)$.

Remark 4.7.12. From the definition above it is clear that

$$\lambda_n \xrightarrow{\tau_{\mathrm{n}}^{\mathcal{Y}}} \lambda \quad \text{if and only if} \quad \lambda_n(A \times \cdot) \xrightarrow{\tau_{\mathrm{n}}(X)} \lambda(A \times \cdot) \quad \text{for all } A \in \Sigma\,. \tag{4.7.5}$$

In fact, using the Monotone Class Theorem, we can show that we can replace Σ in (4.7.5) with an algebra $\mathfrak{a}$ such that $\Sigma = \sigma(\mathfrak{a})$.

Using Definition 4.7.11, (4.7.5), and Proposition 4.6.18 together with Theorem 4.6.21, we easily reach the following result.

Proposition 4.7.13. *If $\{\lambda_\alpha\}_{\alpha \in I} \subseteq \mathcal{Y}(\Omega \times X)$ is a net and $\lambda \in \mathcal{Y}(\Omega \times X)$, then the following statements are equivalent:*

(a) $\lambda_\alpha \xrightarrow{\tau_{\mathrm{n}}^{\mathcal{Y}}} \lambda$;

(b) *for every $\varphi = \chi_A \otimes f$ with $A \in \Sigma$ and a lower semicontinuous and bounded from below function $f: X \to \mathbb{R}$, it holds that*

$$I_\varphi(\lambda) \le \liminf_\alpha I_\varphi(\lambda_\alpha)\,;$$

(c) *for every $\varphi = \chi_A \otimes f$ with $A \in \Sigma$ and for every $f \in U_d^b(X)$ = bounded, d-uniformly continuous functions on X, it holds $I_\varphi(\lambda_\alpha) \to I_\varphi(\lambda)$;*

(d) $\lambda(A \times U) \le \liminf_\alpha \lambda_\alpha(A \times U)$ *for all $A \in \Sigma$ and for all open sets $U \subseteq X$;*

(e) $\limsup_\alpha \lambda_\alpha(A \times C) \le \lambda(A \times C)$ *for all $A \in \Sigma$ and for all closed sets $C \subseteq X$.*

In the next proposition we establish the metrizability of $\mathcal{Y}(\Omega \times X)$.

Proposition 4.7.14. *If Σ is countably generated, then $(\mathcal{Y}(\Omega \times X), \tau_{\mathrm{n}}^{\mathcal{Y}})$ is metrizable.*

Proof. Let $\mathfrak{a} = \{A_n\}_{n \ge 1}$ be a countable algebra such that $\Sigma = \sigma(\mathfrak{a})$. Also, as before, d is a metric generating the topology on X. For $\lambda, \lambda^* \in \mathcal{Y}(\Omega \times X)$ we have $\lambda(A_n \times \cdot), \lambda^*(A_n \times \cdot) \in \mathcal{P}(X)$ for all $n \in \mathbb{N}$ and by Theorem 4.6.25, $\mathcal{P}(X)$ equipped with the $\tau_{\mathrm{n}}(X)$-topology is Polish. We know that the Dudley metric d_D is compatible and

$$\begin{aligned} & d_D(\lambda(A_n \times \cdot), \lambda^*(A_n \times \cdot)) \\ & = \sup[|I_\varphi(\lambda) - I_\varphi(\lambda^*)| : \varphi = \chi_{A_n} \otimes f, f \in \mathrm{Lip_b}(X), \|f\|_{\mathrm{Lip_b}} \le 1] \le 2\mu(A_n) \le 2 \end{aligned}$$

for all $n \in \mathbb{N}$. So, we can define $e: \mathcal{Y}(\Omega \times X) \times \mathcal{Y}(\Omega \times X) \to \mathbb{R}_+$ by

$$e(\vartheta, \vartheta') = \sum_{n \ge 1} \frac{1}{2^n} d_D(\vartheta(A_n \times \cdot), \vartheta'(A_n \times \cdot)) \quad \text{for all } \vartheta, \vartheta' \in \mathcal{Y}(\Omega \times X)\,.$$

It is easy to check that this is a metric on $\mathcal{Y}(\Omega \times X)$. If $\{\lambda_\alpha\}_{\alpha \in I} \subseteq \mathcal{Y}(\Omega \times X)$ is a net and $\lambda \in \mathcal{Y}(\Omega \times X)$, then according to Remark 4.7.12, we obtain

$$\lambda_\alpha \xrightarrow{\tau_n^{\mathcal{Y}}} \lambda \quad \text{if and only if} \quad \lambda_\alpha(A_n \times \cdot) \xrightarrow{\tau_n(X)} \lambda(A_n \times \cdot) \quad \text{for all } n \in \mathbb{N}\,.$$

Hence,

$$\lambda_\alpha \xrightarrow{\tau_n^{\mathcal{Y}}} \lambda \quad \text{if and only if} \quad d_D(\lambda_\alpha(A_n \times \cdot), \lambda(A_n \times \cdot)) \xrightarrow{\alpha} 0 \quad \text{for all } n \in \mathbb{N}\,,$$

which shows that $e(\lambda_\alpha, \lambda) \xrightarrow{\alpha} 0$. So, e generates $\tau_n^{\mathcal{Y}}$ and we conclude that $(\mathcal{Y}(\Omega \times X), \tau_n^{\mathcal{Y}})$ is metrizable. □

Every $\vartheta \in \mathcal{P}(X)$ generates a Young measure

$$\lambda^\vartheta = \mu \otimes \vartheta\,, \tag{4.7.6}$$

whose disintegration is $\hat{\lambda}^\vartheta(w) = \vartheta$ for all $w \in \Omega$.

Definition 4.7.15. $\lambda^\vartheta \in \mathcal{Y}(\Omega \times X)$ defined by (4.7.6) is called the **Young measure corresponding to** $\vartheta \in \mathcal{P}(X)$. The map $\vartheta \to \lambda^\vartheta$ is injective from $\mathcal{P}(X)$ into $\mathcal{Y}(\Omega \times X)$ and so we can say that $\mathcal{P}(X) \subseteq \mathcal{Y}(\Omega \times X)$.

Proposition 4.7.16. (a) $\tau_n^{\mathcal{Y}}(\Omega \times X)|_{\mathcal{P}(X)} = \tau_n(X)$;
(b) $\mathcal{P}(X)$ *is* $\tau_n^{\mathcal{Y}}$*-closed in* $\mathcal{Y}(\Omega \times X)$.

Proof. (a) We know that if $\{\vartheta_\alpha\}_{\alpha \in I} \subseteq \mathcal{P}(X)$ is a net and $\vartheta \in \mathcal{P}(X)$, then

$$\vartheta_\alpha \xrightarrow{\tau_n(X)} \vartheta \quad \text{if and only if} \quad \xi_f(\vartheta_\alpha) \to \xi_f(\vartheta) \quad \text{for all } f \in C_b(X)\,; \tag{4.7.7}$$

see Definition 4.6.11 (a). Then for $\varphi = \chi_A \otimes f$ with $A \in \Sigma$ and $f \in C_b(X)$ we get from (4.7.7) that

$$I_\varphi(\lambda^{\vartheta_\alpha}) = \mu(A)\xi_f(\vartheta_\alpha) \to \mu(A)\xi_f(\vartheta) = I_\varphi(\lambda)\,. \tag{4.7.8}$$

Invoking Proposition 4.7.13, from (4.7.8) we conclude that

$$\vartheta_\alpha \xrightarrow{\tau_n(X)} \vartheta \quad \text{if and only if} \quad \lambda^{\vartheta_\alpha} \xrightarrow{\tau_n^{\mathcal{Y}}} \lambda^\vartheta\,. \tag{4.7.9}$$

(b) Let $\{\vartheta_\alpha\}_{\alpha \in I}$ be a net in $\mathcal{P}(X)$ such that $\lambda^{\vartheta_\alpha} \xrightarrow{\tau_n^{\mathcal{Y}}} \lambda \in \mathcal{Y}(\Omega \times X)$. Let $\vartheta \in \mathcal{P}(X)$ be defined by

$$\vartheta(C) = \int_\Omega \hat{\lambda}(w)(C)\, d\mu \quad \text{for all } C \in \mathcal{B}(X)\,. \tag{4.7.10}$$

Let $\lambda^\vartheta = \mu \otimes \vartheta$. Then, by using (4.7.10), we obtain for $\varphi = \chi_A \otimes f$ with $A \in \Sigma$ and $f \in C_b(X)$ that

$$I_\varphi(\lambda^\vartheta) = \mu(A)\xi_f(\vartheta) = \mu(A)\int_X f(x)\hat{\lambda}(w)(\mathrm{d}x) = \mu(A)\lim_\alpha \xi_f(\vartheta_\alpha) = \lim_\alpha I_\varphi(\lambda^{\vartheta_\alpha}),$$

which shows that $\lambda^{\vartheta_\alpha} \xrightarrow{\tau_\mathrm{n}^\mathcal{Y}} \lambda$; see Proposition 4.7.13. From (4.7.9) we see that $\lambda = \lambda^\vartheta \in \mathcal{P}(X)$. Therefore $\mathcal{P}(X)$ is $\tau_\mathrm{n}^\mathcal{Y}$-closed in $\mathcal{Y}(\Omega \times X)$. ☐

Proposition 4.7.17. *If X, Y are Polish spaces with X homeomorphic to a subset of Y, then $(\mathcal{Y}(\Omega \times X), \tau_\mathrm{n}^\mathcal{Y})$ is homeomorphic to a subset of $(\mathcal{Y}(\Omega \times Y), \tau_\mathrm{n}^\mathcal{Y})$.*

Proof. Let $j: X \to Y$ be the homeomorphism between X and $j(X)$. We have $\mathcal{B}(X) = j^{-1}(\mathcal{B}(Y))$. Let $e: \mathrm{ca}_+(\mathcal{B}(X)) \to \mathrm{ca}_+(\mathcal{B}(Y))$ be defined by $e(\vartheta)(C) = \vartheta(j^{-1}(C))$. This is a homeomorphism between $\mathrm{ca}_+(\mathcal{B}(X))$ and $e(\mathrm{ca}_+(\mathcal{B}(X))) \subseteq \mathrm{ca}_+(\mathcal{B}(Y))$ when the spaces are equipped with their narrow topologies. Similarly for $e|_{\mathcal{P}(X)}$ with $e(\mathcal{P}(X)) \subseteq \mathcal{P}(Y)$. We know that $\mathcal{B}(\mathcal{P}(X)) = e^{-1}(\mathcal{B}(\mathcal{P}(Y)))$. Note that if $\lambda \in \mathcal{Y}(\Omega \times X)$ and $\hat{\lambda}(w)$ is its disintegration, then $\hat{\eta}(w) = e \circ \hat{\lambda}(w)$ is the disintegration of $\eta \in \mathcal{Y}(\Omega \times Y)$ and for all $E \in \Sigma \otimes \mathcal{B}(Y)$ we derive

$$\begin{aligned} \eta(E) &= \int_\Omega e(\hat{\lambda}(w))(E(w))\,\mathrm{d}\mu = \int_\Omega \hat{\lambda}(w)(j^{-1}(E_w))\,\mathrm{d}\mu \\ &= \int_\Omega \hat{\lambda}(w)(k^{-1}(E)_w)\,\mathrm{d}\mu = \lambda(k^{-1}(E)) \end{aligned} \tag{4.7.11}$$

with $k: \Omega \times X \to \Omega \times Y$ defined by $k(w, x) = (w, j(x))$. Hence, the map $H: \mathcal{Y}(\Omega \times X) \to \mathcal{Y}(\Omega \times Y)$ defined by $H(\lambda) = \eta$ is injective.

Let $\{\lambda_\alpha\}_{\alpha\in I} \subseteq \mathcal{Y}(\Omega \times X)$ be a net and $\lambda \in \mathcal{Y}(\Omega \times X)$. We assume that $\lambda_\alpha \xrightarrow{\tau_\mathrm{n}^\mathcal{Y}} \lambda$. Then by Proposition 4.7.13, this is equivalent to saying that

$$\lambda(A \times U) \le \liminf_\alpha \lambda_\alpha(A \times U) \quad \text{for all } A \in \Sigma \text{ and for all open } U \subseteq X.$$

Due to (4.7.11) this is equivalent to

$$\eta(A \times V) = \lambda(A \times j^{-1}(V)) \le \liminf_\alpha \lambda_\alpha(A \times j^{-1}(V)) = \liminf_\alpha \eta_\alpha(A \times V)$$

with $\eta_\alpha = H(\lambda_\alpha)$ and with $A \in \Sigma$ and $V \subseteq Y$ open. This proves that H is bicontinuous and so a homeomorphism. ☐

The next theorem provides a characterization of the $\tau_\mathrm{n}^\mathcal{Y}$-topology in terms of Carathéodory integrands.

Theorem 4.7.18. *If $\{\lambda_\alpha\}_{\alpha\in I} \subseteq \mathcal{Y}(\Omega \times X)$ is a net and $\lambda \in \mathcal{Y}(\Omega \times X)$, then $\lambda_\alpha \xrightarrow{\tau_\mathrm{n}^\mathcal{Y}} \lambda$ if and only if $I_\varphi(\lambda_\alpha) \to I_\varphi(\lambda)$ for all $\varphi \in \mathrm{Car}_\mathrm{b}(\Omega \times X)$.*

Proof. $\Longrightarrow$: First assume that X is compact. Then from Proposition 4.2.33 we know that

$$L^1(\Omega, C(X)) = L^\infty(\Omega, \mathrm{ca}(\mathcal{B}(X))_{\mathrm{w}^*})$$

and $w^* = \tau_n(X)$; see Proposition 4.6.13. We have $\mathrm{Car}_b(\Omega \times X) = L^1(\Omega, C(X))$ and the set $D = \{\varphi = \chi_A \otimes f : A \in \Sigma, f \in C(X)\}$ is dense in $\mathrm{Car}_b(\Omega \times X) = L^1(\Omega, C(X))$. So, the implication follows from Proposition 4.7.13.

Now consider the general noncompact case. Since X is Polish, it is homeomorphic to a subset of the Hilbert cube $\mathbb{H} = [0,1]^{\mathbb{N}}$. Then by Proposition 4.7.17, $\mathcal{Y}(\Omega \times X)$ is homeomorphic to a subset of $\mathcal{Y}(\Omega \times \mathbb{H})$. Since $\mathbb{H}$ is compact, the result follows from the first part of the proof.

$\Longleftarrow$: This is immediate from Definition 4.7.11. □

Using Proposition 4.7.10 we can state the following alternative characterization of convergence in the Young narrow topology.

Theorem 4.7.19. *If $\{\lambda_\alpha\}_{\alpha \in I} \subseteq \mathcal{Y}(\Omega \times X)$ is a net and $\lambda \in \mathcal{Y}(\Omega \times X)$, then $\lambda_\alpha \xrightarrow{\tau_n^{\mathcal{Y}}} \lambda$ if and only if $I_\varphi(\lambda) \leq \liminf_\alpha I_\varphi(\lambda_\alpha)$ for all bounded from below functions $\varphi \in \mathrm{Nor}(\Omega \times X)$.*

We conclude with a compactness result similar to Prokhorov's Theorem; see Theorem 4.6.31.

Definition 4.7.20. A subset $E \subseteq \mathcal{Y}(\Omega \times X)$ is said to be **$\mathcal{Y}$-tight** if for every $\varepsilon > 0$ there exists a compact set $K_\varepsilon \subseteq X$ such that $\lambda(\Omega \times (X \setminus K_\varepsilon)) < \varepsilon$ for all $\lambda \in E$.

Remark 4.7.21. In fact the definition above is equivalent to each of the following conditions:

(a) There exists $f : X \to \overline{\mathbb{R}}_+ = \mathbb{R}_+ \cup \{+\infty\}$ such that for every $\eta \geq 0$, $f^\eta = \{x \in X : f(x) \leq \eta\}$ is compact for which for $\varphi = \chi_\Omega \otimes f$ it holds that $\sup[I_\varphi(\lambda) : \lambda \in E] < \infty$.

(b) For every $\varepsilon > 0$ there exists a tight set $C_\varepsilon \subseteq \mathcal{P}(X)$ (see Definition 4.6.26), such that for $\lambda \in E$ and with $\hat{\lambda}(w)$ being its disintegration, it holds that

$$\{w \in \Omega : \hat{\lambda}(w) \in \mathcal{P}(X) \setminus C_\varepsilon\} \in \Sigma \quad \text{and} \quad \mu(\{w \in \Omega : \hat{\lambda}(w) \in \mathcal{P}(X) \setminus C_\varepsilon\}) \leq \varepsilon .$$

Using the notion of $\mathcal{Y}$-tightness we can state the following Young counterpart of Theorem 4.6.31. The result can be found in Valadier [318].

Theorem 4.7.22. *A set $E \subseteq \mathcal{Y}(\Omega \times X)$ is relatively sequentially $\tau_n^{\mathcal{Y}}$-compact if and only if E is tight.*

Remark 4.7.23. If $C \subseteq L^1(\Omega, \mathbb{R}^N)$ is bounded and $E = \{\lambda^u : u \in C\}$, then $E \subseteq \mathcal{Y}(\Omega \times \mathbb{R}^N)$ is $\mathcal{Y}$-tight. So, if $\{u_n\}_{n \geq 1} \subseteq C$, then there exist a subsequence $\{u_{n_k}\}_{k \geq 1}$ of $\{u_n\}_{n \geq 1}$ and $\lambda \in \mathcal{Y}(\Omega \times \mathbb{R}^N)$ such that $\lambda^{u_{n_k}} \xrightarrow{\tau_n^{\mathcal{Y}}} \lambda$. Moreover, if $\{u_n\}_{n \geq 1} \subseteq C$ is uniformly integrable, then there exist a subsequence $\{u_{n_k}\}_{k \geq 1}$ of $\{u_n\}_{n \geq 1}$ and $u \in L^1(\Omega, \mathbb{R}^N)$, $\lambda \in \mathcal{Y}(\Omega \times X)$ such that

$$\lambda^{u_{n_k}} \xrightarrow{\tau_n^{\mathcal{Y}}} \lambda \quad \text{and} \quad u_{n_k} \xrightarrow{w} u \quad \text{in } L^1(\Omega, \mathbb{R}^N) \text{ with } u(w) = \int_{\mathbb{R}^N} x d\hat{\lambda}(w)(dx)$$

with $\hat{\lambda}(w)$ being the disintegration of the Young measure λ. The limit function u is known as the **barycenter** of $\hat{\lambda}(w)$. Moreover we mention that if μ is nonatomic, then $D = \{\lambda^u : u \in L^0(\Omega, X)\}$ is dense in $(\mathcal{Y}(\Omega \times X), \tau_n^{\mathcal{Y}})$. Finally it is clear from Definitions 4.6.11 (a), 4.7.6, and 4.7.11 that if $C \in \mathcal{P}(X)$ and $E = \{\lambda^{\vartheta} : \vartheta \in C\} \subseteq \mathcal{Y}(\Omega \times X)$, then E is $\mathcal{Y}$-tight if and only if C is tight.

We conclude by providing a list of alternative equivalent definitions of $\mathcal{Y}$-tightness; see Definition 4.7.20. Recall that $f : X \to [0, +\infty]$ is **inf-compact** if for every $\eta \geq 0$, $f^{-1}([0, \eta]) \subseteq X$ is compact. Evidently f is lower semicontinuous. An integrand $f : \Omega \times X \to [0, +\infty]$ is said to be **inf-compact** if for every $w \in \Omega$, $f(w, \cdot)$ is inf-compact.

Proposition 4.7.24. *If X is Polish and $E \subseteq \mathcal{Y}(\Omega \times X)$, then the following statements are equivalent:*
(a) *E is $\mathcal{Y}$-tight.*
(b) *There exists an inf-compact function $f : X \to [0, +\infty]$ such that*

$$\sup_{\lambda \in E} \int_{\Omega \times X} f(x) \, d\lambda(w, x) < +\infty .$$

(c) *There exists an inf-compact integrand $\psi : \Omega \times X \to [0, +\infty]$ such that*

$$\sup_{\lambda \in E} \int_{\Omega \times X} \psi(w, x) \, d\lambda(w, x) < +\infty .$$

(d) *For every $\varepsilon > 0$ there is a measurable multifunction $F : \Omega \to 2^X \setminus \{\emptyset\}$ with compact values such that*

$$\sup_{\lambda \in E} \int_{\Omega} \hat{\lambda}(w)(X \setminus F(w)) \, d\mu < \varepsilon$$

with $\hat{\lambda}(w)$ being the disintegration of λ.

4.8 Semigroups of Operators

Semigroups of operators are important in the theory of evolution equations (dynamic partial differential equations). They generalize the well-known notion of the exponential matrix which gives the solution of a first order linear ordinary differential system. This notion extends easily to an infinite dimensional Banach space X and gives the solution of the Cauchy problem

$$u'(t) = Au(t), \quad t \in \mathbb{R}, \quad u(0) = x_0$$

with $A \in L(X)$ and $x_0 \in X$. The unique solution of this problem is given by

$$u(t) = e^{At}x_0 = \sum_{k\in\mathbb{N}_0} \frac{t^k}{k!}A^k x_0 \quad \text{for all } t \in \mathbb{R}\,.$$

In most applications though, we cannot expect A to be bounded. Instead we deal with operators which are only densely defined, see Remark 3.6.39. In this case it is preferable to use the solution rather than the Cauchy problem as our starting point. This leads to strongly continuous semigroups of operators and their generators. The generators are the typical examples of unbounded operators; see Definition 3.6.38.

Definition 4.8.1. Let X be a real or complex Banach space. A family $\{L_t\}_{t\geq 0} \subseteq L(X)$ is said to be a **strongly continuous semigroup** (or a C_0**-semigroup**) if the following hold.
(a) **Semigroup property**: $L_0 = \mathrm{id}$ and $L_{t+s} = L_t \circ L_s$ for all $t, s \geq 0$;
(b) for every $u \in X$, we have

$$\lim_{t\to 0} \|L_t(u) - u\|_X = 0\,.$$

Remark 4.8.2. From this definition the continuity of the maps $t \to L_t(u)$ follows for all $u \in X$ on $\mathbb{R}_+ = [0, \infty)$, not just at $t = 0$; see (b) in the above definition. In order to see this, note that

$$\|L_{t+s}(u) - L_t(u)\|_X = \|L_t(L_s(u)) - L_t(u)\|_X \leq \|L_t\|_L \|L_s(u) - u\|_X\,.$$

Example 4.8.3. (a) Let $A \in L(X)$ and define

$$L_t = e^{tA} = \sum_{n\in\mathbb{N}_0} \frac{t^n}{n!}A^n \quad \text{for } t \geq 0\,.$$

This series converges in the operator norm and $\{L_t\}_{t\geq 0}$ is a strongly continuous semigroup which can be seen by using the estimate $\|A^n\|_L \leq \|A\|_L^n$.
(b) Let $X = \{u\colon \mathbb{R} \to \mathbb{R}\colon u \text{ is bounded and uniformly continuous}\}$ be equipped with the supremum norm. We set $L_t(u)(s) = u(t+s)$ for all $u \in X$ and for all $t, s \geq 0$. Then $\{L_t\}_{t\geq 0} \subseteq L(X)$ is a strongly continuous semigroup.
(c) Fix $n \in \mathbb{N}$, $p \in [1, \infty)$ and define the heat kernel

$$K_t(x) = \frac{1}{(4\pi t)^{\frac{n}{2}}} e^{-\frac{\|x\|_2^2}{4t}} \quad \text{for all } x \in \mathbb{R}^n \text{ and for all } t \geq 0$$

with $\|\cdot\|_2$ being the Euclidean norm on $\mathbb{R}^n$. We have

$$\int_{\mathbb{R}^n} K_t(s)\,\mathrm{d}s = 1 \quad \text{and} \quad K_{t+s} = K_t * K_s \quad \text{for all } t, s \geq 0\,,$$

where $*$ denotes the operation of convolution. Let $L_t\colon L^p(\mathbb{R}^n) \to L^p(\mathbb{R}^n)$ with $t \geq 0$ be defined by

$$L_t(u) = \begin{cases} u & \text{if } t = 0 , \\ K_t * u & \text{if } 0 < t . \end{cases}$$

Then $\{L_t\}_{t\geq 0} \subseteq L(L^p(\mathbb{R}^n))$ is a strongly continuous semigroup. For each $h \in L^p(\mathbb{R}^n)$ the function

$$u(t,x) = (K_t * h)(x), \quad t > 0,\ x \in \mathbb{R}^n$$

is the solution of the heat equation

$$u_t - \Delta u = 0 \quad \text{and} \quad \lim_{t\to 0} \|u(t,\cdot) - h\|_p = 0 .$$

Proposition 4.8.4. *If $\{L_t\}_{t\geq 0} \subseteq L(X)$ is a strongly continuous semigroup, then the following hold:*

(a) $\sup_{a\leq t\leq b} \|L_t\|_L < \infty$ *for all $b > 0$;*

(b) *for every $u \in X$, $t \to L_t(u)$ is continuous from $\mathbb{R}_+$ into X.*

Proof. (a) We claim that there exist $\varepsilon > 0$ and $M \geq 1$ such that $\|L_t\|_L \leq M$ for all $0 \leq t \leq \varepsilon$.

If this is not the case, then we could find a sequence $\{t_n\}_{n\in\mathbb{N}} \subseteq (0,\infty)$ such that $t_n \to 0$ and $\|L_{t_n}\|_L \geq n$ for all $n \in \mathbb{N}$. Then, by the Uniform Boundedness Principle (see Theorem 3.2.1), we can find $u \in X$ such that $\{L_{t_n}(u)\}_{n\in\mathbb{N}} \subseteq X$ is unbounded. But this contradicts Definition 3.8.1 (b).

Next, let $t = m\varepsilon + s$ with $m \in \mathbb{N}$ and $s \geq 0$. We have

$$L_t = L_{m\varepsilon} \circ L_s = (L_\varepsilon)^m \circ L_s ,$$

which implies

$$\|L_t\|_L \leq M^{1+m} \leq MM^{\frac{t}{\varepsilon}} = Me^{\omega t}$$

with $\omega = \frac{1}{\varepsilon}\ln(M)$.

(b) This was explained in Remark 4.8.2. □

In order to relate $L_t(u)$ to the solution of an evolution equation, we need the notion of the infinitesimal generator of a strongly continuous semigroup.

Definition 4.8.5. Let $\{L_t\}_{t\geq 0}$ be a strongly continuous semigroup. We define the operator A by

$$Au = \lim_{t\to 0} \frac{L_t(u) - u}{t} \tag{4.8.1}$$

with $D(A) = \{u \in X\colon \text{the limit in (4.8.1) exists}\}$. The operator A with the indicated domain $D(A)$ is called the **generator** or **infinitesimal generator** of $\{L_t\}_{t\geq 0}$.

We have the following basic proposition relating the strongly continuous semigroup and its infinitesimal generator.

Proposition 4.8.6. *If $\{L_t\}_{t\geq 0} \subseteq L(X)$ is a strongly continuous semigroup with infinitesimal generator A and $x_0 \in D(A)$, then the following hold:*
(a) $L_t(x_0) \in D(A)$ *for all* $t \geq 0$;
(b) $\frac{\mathrm{d}}{\mathrm{d}t}(L_t(x_0)) = AL_t(x_0) = L_t(Ax_0)$ *for all* $t > 0$;
(c) $\frac{\mathrm{d}^n}{\mathrm{d}t^n}(L_t(x_0)) = A^n L_t(x_0) = L_t(A^n x_0)$ *for* $x_0 \in D(A^n)$ *with* $n \in \mathbb{N}$ *and for all* $t > 0$;
(d) *for all* $u \in X$, $\int_0^t L_\tau(u)\,\mathrm{d}\tau \in D(A)$ *and* $A(\int_0^t L_\tau(u)\,\mathrm{d}\tau) = L_t(u)u$;
(e) $D(A^n)$ *is dense in* X *for all* $n \in \mathbb{N}$ *and* A *is closed in the sense of Definition* 3.6.38.

Proof. (a) Let $x_0 \in D(A)$. Using the semigroup property, we have for $t \geq 0$ and $s > 0$ that

$$\frac{L_{t+s}(x_0) - L_t(x_0)}{s} = L_t \frac{L_s(x_0) - x_0}{s} = \frac{L_s(x_0) - x_0}{s} L_t(x_0) . \tag{4.8.2}$$

Passing to the limit as $s \to 0^+$, we see that $L_t(x_0) \in D(A)$.

(b) From (4.8.2) above we get that

$$\frac{\mathrm{d}^+}{\mathrm{d}t^+}(L_t(x_0)) = AL_t(x_0) = L_t(Ax_0) . \tag{4.8.3}$$

Moreover, we have

$$\frac{L_t(x_0) - L_{t-s}(x_0)}{x_0} = L_{t-s} \frac{L_s(x_0) - x_0}{s} ,$$

which implies

$$\begin{aligned} &\frac{L_t(x_0) - L_{t-s}(x_0)}{x_0} - L_t(x_0) \\ &= L_{t-s}\left(\frac{L_s(x_0) - x_0}{s} - Ax_0\right) + L_{t-s}(x_0) - L_t(x_0) . \end{aligned} \tag{4.8.4}$$

For $t > 0$ and $0 < s < t$, we have from Proposition 4.8.4

$$\left\| L_{t-s}\left(\frac{L_s(x_0) - x_0}{s} - Ax_0\right)\right\|_X \leq M \left\| \frac{L_s(x_0) - x_0}{s} - Ax_0 \right\|_X .$$

Hence, we obtain

$$\left\| L_{t-s}\left(\frac{L_s(x_0) - x_0}{s} - Ax_0\right)\right\|_X \to 0 \quad \text{as } s \to 0^+ . \tag{4.8.5}$$

Moreover, we have

$$\|L_{t-s}(x_0) - L_t(x_0)\|_X \to 0 \quad \text{as } s \to 0^+ . \tag{4.8.6}$$

Returning to (4.8.4), by applying (4.8.5) as well as (4.8.6) we see that

$$\frac{\mathrm{d}^-}{\mathrm{d}t^-}(L_t(x_0)) = L_t(Ax_0) = \frac{\mathrm{d}^+}{\mathrm{d}t^+}L_t(x_0)\,,$$

see also (4.8.3). This gives

$$\frac{\mathrm{d}}{\mathrm{d}t}(L_t(x_0)) = L_t(Ax_0) = A(L_t(x_0))\,.$$

(c) This follows by induction on $n \in \mathbb{N}$.

(d) For $u \in X$ and $s > 0$ we have

$$\begin{aligned}\left(\frac{L_s - \mathrm{id}}{s}\right)\int_0^t L_\tau(u)\,\mathrm{d}\tau &= \frac{1}{s}\int_0^t (L_{\tau+s}(u) - L_\tau(u))\,\mathrm{d}\tau \\ &= \frac{1}{s}\left(\int_t^{t+s} L_\tau(u)\,\mathrm{d}\tau - \int_0^s L_\tau(u)\,\mathrm{d}\tau\right),\end{aligned}$$

which implies $\int_0^t L_\tau(u)\,\mathrm{d}\tau \in D(A)$ and

$$A\left(\int_0^t L_\tau(u)\,\mathrm{d}\tau\right) = L_t(u) - u\,.$$

(e) For $u \in X$, we obtain from (d) that

$$\int_0^t L_\tau(u)\,\mathrm{d}\tau \in D(A) \quad \text{for } t > 0\,.$$

Further, we have

$$\frac{1}{s}\int_0^s L_\tau(u)\,\mathrm{d}\tau \to u \quad \text{as } s \to 0^+\,.$$

Therefore $D(A)$ is dense in X. By induction on $n \in \mathbb{N}$ we infer that $D(A^n)$ is dense in X.

Next, let $\{u_n\}_{n\in\mathbb{N}} \subseteq D(A)$ and assume that

$$u_n \to u \quad \text{in } X \quad \text{and} \quad Au_n \to y \quad \text{in } X\,. \tag{4.8.7}$$

We need to show that $u \in D(A)$ and $y = Au$; see Remark 3.6.39. From (b) we have

$$\frac{L_s(u) - u}{s} = \lim_{n\to\infty}\frac{L_s(u_n) - u_n}{s} = \lim_{n\to\infty}\frac{1}{s}\int_0^s L_\tau(Au_n)\,\mathrm{d}\tau\,,$$

and by (4.8.7) it follows that

$$\lim_{s\to 0^+} \frac{L_s(u) - u}{s} = y .$$

Hence, $u \in D(A)$ and $Au = y$, that is, A is closed. □

We can state the following corollaries.

Corollary 4.8.7. *If two strongly continuous semigroups have the same infinitesimal generator, then they are identical.*

Corollary 4.8.8. *If $\{L_t\}_{t\geq 0}$ is a strongly continuous semigroup, then there exist $M \geq 1$ and $\omega > 0$ such that*

$$\|L_t\|_L \leq Me^{\omega t} \quad \textit{for all } t \geq 0 .$$

Proof. From Proposition 4.8.4 we know that there exist $M \geq 1$ and $\delta > 0$ such that $\|L_t\|_L \leq M$ for $t \in [0, \delta]$. Let $\omega = \frac{1}{\delta}\ln(M)$. Then, for any $t \geq 0$ we can find $n \in \mathbb{N}_0$ and $0 \leq \vartheta < \delta$ such that $t = n\delta + \vartheta$. This implies $L_t = (L_\delta)^n L_\vartheta$ and so

$$\|L_t\|_L \leq \|L_\delta\|_L^n \|L_\vartheta\|_L \leq M^n M \leq Me^{\omega t} ,$$

where we have used that $\ln(M^n) = n\ln(M) \leq n\omega\delta \leq \omega t$. □

Corollary 4.8.9. *If $\{L_t\}_{t\geq 0}$ is a strongly continuous semigroup with infinitesimal generator $A \in L(X)$, then $L_t = e^{At}$ for all $t \geq 0$.*

So, semigroups play a central role in determining the solutions of evolution equations. We have seen that $u(t) = L_t(x_0)$, $x_0 \in D(A)$, $t \geq 0$ is a solution. Now we reverse the problem and ask what are the conditions that we must assume on A in order for A to be the infinitesimal generator of a strongly continuous semigroup. If $A \in L(X)$, then the situation is simple since $L_t = e^{At}$ for $t \geq 0$ is the desired strongly continuous semigroup. For the unbounded case, we have seen that if A is the infinitesimal generator of a strongly continuous semigroup, then the following hold:
- $\overline{D(A)} = X$ (densely defined);
- A is closed.

The additional conditions required for the converse implication are included in the so-called "Hille–Yosida Theorem" or "Hille–Yosida–Phillips Theorem".

First we prove a result which expresses the resolvent operator $(\lambda \,\mathrm{id} - A)^{-1}$ in terms of the semigroup. At this point we assume that X is a complex Banach space. If X is real, then we complexify it in order to make sense of $\lambda \,\mathrm{id} - A$ for $\lambda \in \mathbb{C}$. The spectrum and the resolvent set of an unbounded operator are defined in the same way as for bounded ones; see Definition 3.7.29.

Definition 4.8.10. Let X be a complex Banach space and $A: D(A) \subseteq X \to X$ be an unbounded linear operator. The **spectrum** $\sigma(A)$ of A is the set

$$\sigma(A) = \{\lambda \in \mathbb{C}: \lambda \operatorname{id} - A: D(A) \subseteq X \to X \text{ is not invertible}\}.$$

As in the bounded case $\sigma(A)$ has the following decomposition

$$\sigma(A) = P\sigma(A) \cup R\sigma(A) \cup C\sigma(A),$$

where

$$\begin{aligned} P\sigma(A) &= \{\lambda \in \mathbb{C}: \lambda \operatorname{id} - A \text{ is not injective}\}, \\ R\sigma(A) &= \{\lambda \in \mathbb{C}: \lambda \operatorname{id} - A \text{ is injective but } \overline{R(\lambda \operatorname{id} - A)} \neq X\}, \\ C\sigma(A) &= \{\lambda \in \mathbb{C}: \lambda \operatorname{id} - A \text{ is injective, } \overline{R(\lambda \operatorname{id} - A)} = X, \\ &\qquad \text{but } \lambda \operatorname{id} - A \text{ is not invertible}\}. \end{aligned}$$

We say that $P\sigma(A)$ is the **point spectrum of** A, $R\sigma(A)$ is the **residual spectrum of** A, and $C\sigma(A)$ is the **continuous spectrum of** A.

The **resolvent set** $\rho(A)$ of A is the set $\rho(A) = \mathbb{C} \setminus \sigma(A)$. For $\lambda \in \rho(A)$, $R(\lambda) = (\lambda \operatorname{id} - A)^{-1} \in L(X)$, and it is called the **resolvent operator**. The map $\lambda \to R(\lambda)$ is analytic, and the resolvent identity holds:

$$R(\lambda) - R(\vartheta) = (\vartheta - \lambda)R(\lambda) \circ R(\vartheta) \quad \text{for all } \lambda, \vartheta \in \rho(A).$$

Remark 4.8.11. Note that $\lambda \in P\sigma(A)$ if and only if we can find $u \in D(A)$, $u \neq 0$, such that $Au = \lambda u$. We call $\lambda \in P\sigma(A)$ an **eigenvalue** of A, and the nonzero vector $u \in (\lambda \operatorname{id} - A)$ is known as an **eigenvector** corresponding to λ. We can show that $\rho(A) \neq \emptyset$ implies A being closed, but the converse is not true. Also, $\rho(A) \subseteq \mathbb{C}$ is open.

First we prove a lemma which complements Corollary 3.8.8.

Lemma 4.8.12. *If $\{L_t\}_{t\geq 0}$ is a strongly continuous semigroup, then*

$$\lim_{t\to\infty} \frac{1}{t} \ln(\|L_t\|_L) = \inf_{t>0} \frac{1}{t} \ln(\|L_t\|_L) = \omega_0 \in \mathbb{R} \cup \{\infty\}.$$

Proof. If $L_t = 0$ for all $t > 0$, then clearly $\omega_0 = \infty$. So, we assume that $L_t \neq 0$ for all $t > 0$. Let $\vartheta(t) = \ln(\|L_t\|_L)$ for all $t \geq 0$. The semigroup property and Proposition 4.8.4 (a) imply that

$$\vartheta(0) = 0, \quad \vartheta(t+s) \leq \vartheta(t) + \vartheta(s), \quad M_b = \sup_{0\leq t\leq b} \vartheta(t) < \infty \tag{4.8.8}$$

for all $t, s \geq 0$. Fix $t_0 > 0$ and let $t > 0$. We have $t = kt_0 + s$ with $k \in \mathbb{N}_0$ and $0 \leq s < t_0$. Using (4.8.8) we have

$$\frac{\vartheta(t)}{t} \le \frac{k\vartheta(t_0) + \vartheta(s)}{t} \le \frac{\vartheta(t_0)}{t_0} + \frac{M_{t_0}}{t},$$

which implies

$$\limsup_{t\to\infty} \frac{\vartheta(t)}{t} \le \frac{\vartheta(t_0)}{t_0}.$$

Since $t_0 > 0$ is arbitrary, we obtain

$$\limsup_{t\to\infty} \frac{\vartheta(t)}{t} \le \inf_{t>0} \frac{\vartheta(t)}{t}.$$

Hence, we have

$$\lim_{t\to\infty} \frac{\vartheta(t)}{t} = \inf_{t>0} \frac{\vartheta(t)}{t} = \omega_0. \qquad \square$$

The next auxiliary result is a generalization of the classical "Variation of Constants" formula. It illustrates the close connection of strongly continuous semigroups and evolution equations.

Lemma 4.8.13. *If A is the infinitesimal generator of a strongly continuous semigroup $\{L_t\}_{t\ge 0}$ and*
(i) $f \in C^1([0,b], X)$,
(ii) $x_0 \in D(A)$,

then the function

$$u(t) = L_t(x_0) + \int_0^t L_{t-s} f(s)\,\mathrm{d}s$$

is continuously differentiable and is a classical solution of the Cauchy problem

$$u'(t) = Au(t) + f(t), \quad t \in [0,b], \quad u(0) = x_0.$$

Proof. Let $y(t) = \int_0^t L_{t-s} f(s)\,\mathrm{d}s$ for $t \in [0,b]$. We have

$$\begin{aligned} y(t) &= \int_0^t L_{t-s}\left(f(0) + \int_0^s f'(\tau)\,\mathrm{d}\tau \right) \mathrm{d}s \\ &= \left(\int_0^t L_{t-s}\,\mathrm{d}s \right) f(0) + \int_0^t \left(\int_\tau^t L_{t-s}\,\mathrm{d}s \right) f'(\tau)\,\mathrm{d}\tau, \end{aligned}$$

by a change of order of integration. From Proposition 4.8.6 (d), we know that

$$L_t(x) - x = \int_0^t L_s(Ax)\,\mathrm{d}s \quad \text{for all } x \in D(A)\,.$$

This implies

$$L_{t-\tau}(x) - x = \int_\tau^t AL_{t-s}(x)\,\mathrm{d}s$$

and since A is closed we obtain

$$L_{t-\tau}(x) - x = A\int_\tau^t L_{t-s}(x)\,\mathrm{d}s\,.$$

Because $\overline{D(A)} = X$, it follows

$$L_{t-\tau}(x) - x = A\int_\tau^t L_{t-s}(x)\,\mathrm{d}s\,.$$

Therefore, we have

$$Ay(t) = (L_t - \mathrm{id})f(0) + \int_0^t (L_{t-\tau} - \mathrm{id})f'(\tau)\,\mathrm{d}\tau$$

and in particular $y(t) \in D(A)$ for all $t \geq 0$. Then we can write

$$y(t) = \int_0^t L_{t-s}f(s)\,\mathrm{d}s = \int_0^t L_s f(t-s)\,\mathrm{d}s\,.$$

From this we see that

$$\frac{\mathrm{d}y}{\mathrm{d}t}(t) = L_t f(0) + \int_s^t L_s f'(t-s)\,\mathrm{d}s\,.$$

Hence, $u'(t) = Au(t) + f(t)$ for all $t \in [0, b]$ with $u(0) = x_0$. □

Now we can state and prove the so-called "Resolvent Identity for Semigroups".

Proposition 4.8.14 (Resolvent Identity for Semigroups). *If $A\colon D(A) \subseteq X \to X$ is the infinitesimal generator of a strongly continuous semigroup $\{L_t\}_{t\geq 0}$ and $\lambda \in \mathbb{C}$ satisfies*

$$\omega_0 = \lim_{t\to\infty} \frac{\ln(\|L_t\|_L)}{t} < \operatorname{Re}\lambda\,,$$

with $\operatorname{Re}\lambda$ *being the real part of* λ*, then* $\lambda \in \rho(A)$ *and*

$$(\lambda \operatorname{id} - A)^{-k} = \frac{1}{(k-1)!} \int_0^\infty t^{k-1} e^{-\lambda t} L_t(x)\, \mathrm{d}t$$

for all $x \in X$ *and for all* $k \in \mathbb{N}$.

Proof. First, let $k = 1$ and consider $\lambda \in \mathbb{C}$ such that $\omega_0 < \operatorname{Re}\lambda$. We choose $\omega \in (\omega_0, \operatorname{Re}\lambda)$. From Lemma 4.8.12, we have

$$\|L_t\|_L \le M e^{\omega t} \quad \text{for some } M \ge 1 \text{ and for all } t \ge 0 .$$

This gives

$$\|e^{-\lambda t} L_t(x)\|_X \le M e^{(\omega - \operatorname{Re}\lambda)t} \|x\|_X \quad \text{for all } x \in X \text{ and for all } t \ge 0 .$$

It follows that if

$$R(\lambda)x = \int_0^\infty e^{-\lambda t} L_t(x)\, \mathrm{d}t = \lim_{b \to \infty} \int_0^b e^{-\lambda t} L_t(x)\, \mathrm{d}t \quad \text{for all } x \in X ,$$

then $R(\lambda) \in L(X)$.

Claim 1: If $x \in X$ and $b > 0$, then

$$\eta_b = \int_0^b e^{-\lambda t} L_t(x)\, \mathrm{d}t \in D(A) \quad \text{and} \quad A(\eta_b) = e^{-\lambda b} L_b(x) - x + \lambda \int_0^b e^{-\lambda t} L_t(x)\, \mathrm{d}t = \vartheta_b .$$

This claim follows from Lemma 4.8.13 with $f(t) = e^{-\lambda(b-t)}x$.

Claim 2: If $x \in D(A)$ and $b > 0$, then

$$\int_0^b e^{-\lambda t} L_t(x)\, \mathrm{d}t = \vartheta_b ;$$

see Claim 1. This claim follows from integration by parts and since $\frac{\mathrm{d}}{\mathrm{d}t} L_t(x) = L_t(Ax)$.

Recall that A is closed. Hence

$$R(\lambda)x \in D(A) \quad \text{and} \quad AR(\lambda)x = \lambda R(\lambda)x - x \quad \text{for all } x \in X .$$

From Claim 2 we know that, for $x \in D(A)$,

$$R(\lambda)Ax = \lim_{b \to \infty} \int_0^b e^{-\lambda t} L_t(x)\, \mathrm{d}t = \lambda R(\lambda)x - x ,$$

which implies

$$(\lambda \,\mathrm{id} - A)R(\lambda)x = x \quad \text{for all } x \in X\,, \tag{4.8.9}$$

$$R(\lambda)(\lambda \,\mathrm{id} - A)x = x \quad \text{for all } x \in D(A)\,. \tag{4.8.10}$$

Therefore, $\lambda \,\mathrm{id} - A$ is a bijection and $(\lambda \,\mathrm{id} - A)^{-1} = R(\lambda)$. So, we have proved the result for $k = 1$.

If $k \geq 2$, then $\lambda \to (\lambda \,\mathrm{id} - A)^{-1}x$ is analytic, and

$$(\lambda \,\mathrm{id} - A)^{-1}x = \frac{(-1)^{k-1}}{(k-1)!}\frac{\mathrm{d}^{k-1}}{d\lambda^{k-1}}(\lambda \,\mathrm{id} - A)^{-1}x = \frac{(-1)^{k-1}}{(k-1)!}\frac{\mathrm{d}^{k-1}}{d\lambda^{k-1}}\int_0^\infty e^{-\lambda t}L_t(x)\,\mathrm{d}t$$
$$= \frac{1}{(k-1)!}\int_0^\infty t^{k-1}e^{-\lambda t}L_t(x)\,\mathrm{d}t$$

for all $x \in X$ and for all $\lambda \in \mathbb{C}$ with $\mathrm{Re}\,\lambda > \omega_0$. □

Remark 4.8.15. It follows that

$$\sup\{\mathrm{Re}\,\lambda : \lambda \in \sigma(A)\} \leq \omega_0 = \lim_{t\to\infty} \frac{\ln(\|L_t\|_L)}{t}\,.$$

In fact this inequality can be strict; see Bühler–Salamon [64, Example 7.2.7, p. 371].

Now we are ready for the main result concerning strongly continuous semigroups which is known as the “Hille–Yosida Theorem” or “Hille–Yosida–Phillips Theorem”. It gives necessary and sufficient conditions for a linear operator to be the infinitesimal generator of a strongly continuous semigroup.

Theorem 4.8.16 (Hille–Yosida Theorem). *If $A : D(A) \subseteq X \to X$ is a densely defined linear operator, $\omega \in \mathbb{R}$ and $M \geq 1$, then the following statements are equivalent:*
(a) *A is the infinitesimal generator of a strongly continuous semigroup $\{L_t\}_{t\geq 0} \subseteq L(X)$ such that $\|L_t\|_L \leq Me^{\omega t}$ for all $t \geq 0$.*
(b) *For every $\lambda \in \mathbb{R}$, $\lambda > \omega$, $\lambda \,\mathrm{id} - A$ is invertible and*

$$\|(\lambda \,\mathrm{id} - A)^{-k}\|_L \leq \frac{M}{(\lambda - \omega)^k} \quad \textit{for all } \lambda > \omega \textit{ and for all } k \in \mathbb{N}\,.$$

Proof. (a) ⟹ (b): Let $\lambda > \omega$ and $k \in \mathbb{N}$. Since $\omega_0 \leq \omega$, by Proposition 4.8.14, we have

$$(\lambda \,\mathrm{id} - A)^{-k}x = \frac{1}{(k-1)!}\int_0^\infty t^{k-1}e^{-\lambda t}L_t(x)\,\mathrm{d}t \quad \text{for all } x \in X\,,$$

which implies

$$\begin{aligned}\|(\lambda \operatorname{id} - A)^{-k} x\|_X &= \frac{1}{(k-1)!} \int_0^\infty t^{k-1} e^{-\lambda t} \|L_t(x)\|_X \, dt \\ &\leq \frac{M\|x\|_X}{(k-1)!} \int_0^\infty k^{t-1} e^{-(\lambda-\omega)t} \, dt = \frac{M\|x\|_X}{(\lambda-\omega)^k} \, .\end{aligned}$$

So, we have proved the statement in (b).

(b) $\Longrightarrow$ (a): The proof of this implication will be done in a sequence of steps.

Step 1: $\lim_{\lambda \to +\infty} \lambda R(\lambda) x = x$ for all $x \in X$.

Using (4.8.9) and (4.8.10), we have for $x \in D(A)$

$$\|\lambda R(\lambda)x - x\|_X = \|AR(\lambda)x\|_X = \|R(\lambda)Ax\|_X \leq \frac{M}{\lambda - \omega} \|Ax\|_X \, ,$$

which gives

$$\|\lambda R(\lambda)x - x\|_X \to 0 \quad \text{as } \lambda \to \infty \text{ for } x \in D(A) \, . \tag{4.8.11}$$

By hypothesis, $D(A)$ is dense in X and $\|\lambda R(\lambda)\|_L \leq M$. For any $x \in X$ and $\varepsilon > 0$, let $u \in B_\varepsilon(x)$ with $u \in D(A)$. Then we have

$$\begin{aligned}\|\lambda R(\lambda)x - x\|_X &\leq \|\lambda R(\lambda)(x-u)\|_X + \|\lambda R(\lambda)u - u\|_X + \|u - x\|_X \\ &\leq (M+1)\|u - x\|_X + \|\lambda R(\lambda)u - u\|_X \, .\end{aligned}$$

Let $\varepsilon = \frac{\delta}{M+1}$ with $\delta > 0$. Then we get from the estimate above

$$\|\lambda R(\lambda)x - x\|_X \leq \delta + \|\lambda R(\lambda)u - u\|_X \, ,$$

hence, by (4.8.11)

$$\limsup_{\lambda \to \infty} \|\lambda R(\lambda)x - x\|_X \leq \delta \, .$$

Since $\delta > 0$ is arbitrary, we conclude that

$$\lim_{\lambda \to \infty} \|\lambda R(\lambda)x - x\|_X = 0 \quad \text{for all } x \in X \, .$$

This completes the proof of Step 1.

Step 2: If for $\lambda > \omega$ and $t \geq 0$ we set

$$A_\lambda = \lambda A R(\lambda) \quad \text{and} \quad (T_\lambda)_t = e^{tA_\lambda} = \sum_{k \in \mathbb{N}_0} \frac{t^k}{k!} A_\lambda^k \, ,$$

then we have $\|(T_\lambda)_T\|_L \leq M e^{\frac{\lambda\omega}{\lambda-\omega} t}$.

By (4.8.9) we have $A_\lambda = \lambda^2 R(\lambda) - \lambda\,\mathrm{id}$, from which we obtain that

$$\|(T_\lambda)_t\|_L = e^{-\lambda t}\|e^{t\lambda^2 R(\lambda)}\|_L \le e^{-\lambda t}\sum_{k\in\mathbb{N}_0}\frac{t^k\lambda^{2k}}{k!}\|R(\lambda)^k\|_L$$
$$\le e^{-\lambda t}\sum_{k\in\mathbb{N}_0}\frac{t^k\lambda^{2k}}{k!}\frac{M}{(\lambda-\omega)^k} = Me^{-\lambda t}e^{\frac{\lambda^2 t}{\lambda-\omega}} = Me^{\frac{\lambda\omega}{\lambda-\omega}t}$$

for all $\lambda > \omega$ and for all $t \ge 0$. This completes the proof of Step 2.

Step 3: If $\lambda > \vartheta > \omega$, then

$$\|(T_\lambda)_t(x) - (T_\vartheta)_t(x)\|_X = Me^{\frac{\vartheta\omega}{\vartheta-\omega}t}t\|A_\lambda x - A_\vartheta x\|_X$$

for all $x \in X$.

Note that $A_\lambda A_\vartheta = A_\vartheta A_\lambda$ and so $A_\lambda(T_\vartheta)_t = (T_\vartheta)_t A_\lambda$. This implies

$$(T_\lambda)_t(x) - (T_\vartheta)_t(x) = \int_0^t \frac{\mathrm{d}}{\mathrm{d}s}(T_\vartheta)_{t-s}(T_\lambda)_s\,\mathrm{d}s$$
$$= \int_0^t (T_\vartheta)_{t-s}(T_\lambda)_s(A_\lambda x - A_\vartheta x)\,\mathrm{d}s$$

for all $x \in X$ and for all $t \ge 0$. Then we have

$$\|(T_\lambda)_t(x) - (T_\vartheta)_t(x)\|_X \le \int_0^t \|(T_\vartheta)_{t-s}\|_L\|(T_\lambda)_s\|_L\,\mathrm{d}s\|A_\lambda x - A_\vartheta x\|_X$$
$$\le M^2 e^{-\frac{\vartheta\omega}{\vartheta-\omega}t}\int_0^t e^{-\frac{\vartheta\omega}{\vartheta-\omega}s}e^{\frac{\lambda\omega}{\lambda-\omega}s}\,\mathrm{d}s\|A_\lambda x - A_\vartheta x\|_X$$
$$\le M^2 e^{\frac{\vartheta\omega}{\vartheta-\omega}t}t\|A_\lambda x - A_\vartheta x\|_X\,,$$

where we have used that $\frac{\lambda\omega}{\lambda-\omega} \le \frac{\vartheta\omega}{\vartheta-\omega}$. This completes the proof of Step 3.

Step 4: $T_t(x) = \lim_{\lambda\to\infty}(T_\lambda)_t(x)$ exists for all $x \in X$, and for all $t \ge 0$ and $\{T_t\}_{t\ge0} \subseteq L(X)$ is a strongly continuous semigroup with $\|T_t\|_L \le Me^{\omega t}$ for all $t \ge 0$.

From Step 1 we know that, for all $x \in D(A)$, $Ax = \lim_{\lambda\to\infty} A_\lambda x$. Then Step 3 implies that $T_t(x) = \lim_{\lambda\to\infty}(T_\lambda)_t(x)$ exists and the convergence is uniform on every bounded interval $(0, b]$. Then, using Step 2 and the Banach–Steinhaus Theorem (Theorem 3.2.2), we infer that $T_t \in L(X)$ for all $t \ge 0$. Also note that $t \to T_t(x)$ is continuous for all $x \in X$. Then, by the semigroup property, we have

$$T_sT_t(x) = \lim_{\lambda\to\infty}(T_\lambda)_s(T_\lambda)_t(x) = \lim_{\lambda\to\infty}(T_\lambda)_{t+s}(x) = T_{t+s}(x)$$

for all $t, s \geq 0$. Hence, $\{T_t\}_{t\geq 0}$ is a strongly continuous semigroup. Finally, using Step 2, we obtain

$$\|T_t(x)\|_X = \lim_{\lambda\to\infty} \|(T_\lambda)_t(x)\|_X \leq \lim_{\lambda\to\infty} Me^{\frac{\lambda\omega}{\lambda-\omega}t}\|x\|_X = Me^{\omega t}\|x\|_X .$$

Therefore, $\|T_t\|_L \leq Me^{\omega t}$ for all $t \geq 0$. This completes the proof of Step 4.

Step 5: A is the infinitesimal generator of $\{T_t\}_{t\geq 0}$.

Let E be the infinitesimal generator of $\{T_t\}_{t\geq 0}$ and $x \in D(A)$. It holds that

$$\|(T_\lambda)_t(A_\lambda x) - T_t(Ax)\|_X \leq \|(T_\lambda)_t\|_L\|A_\lambda x - Ax\|_X + \|(T_\lambda)_t(Ax) - T_t(Ax)\|_X ,$$

which implies

$$\lim_{\lambda\to\infty} \|(T_\lambda)_t(A_\lambda x) - T_t(Ax)\|_X = 0 \quad \text{uniformly} ,$$

see Steps 1 and 2. Therefore, we have for $b > 0$

$$\int_0^b T_t(Ax)\,\mathrm{d}t = \lim_{\lambda\to\infty} \int_0^b (T_\lambda)_t(Ax)\,\mathrm{d}t = \lim_{\lambda\to\infty} (T_\lambda)_b(x) - x = T_b(x) - x .$$

This yields

$$\lim_{b\to 0} \frac{T_b(x) - x}{b} = \lim_{b\to 0} \frac{1}{b} \int_0^b T_t(Ax)\,\mathrm{d}t = Ax .$$

Hence, $D(A) \subseteq D(E)$ and $E|_{D(A)} = A$.

Let $v \in D(E)$ and $\lambda > \omega$. We set $x = (\lambda\,\mathrm{id} - A)^{-1}(\lambda v - Ev)$ and know that

$$x \in D(A) \subseteq D(E) \quad \text{and} \quad \lambda x - Ex = \lambda x - Ax = \lambda v - Ev .$$

But $\lambda\,\mathrm{id} - E$ is injective due to Proposition 4.8.14. So, it follows that $v = x \in D(A)$ and then $A = E$. This completes the proof of Step 5 and the proof of the theorem. □

We have seen in Corollary 4.8.8 that if $\{L_t\}_{t\geq 0}$ is a strongly continuous semigroup, then there exists $\omega \in \mathbb{R}$ and $M \geq 1$ such that

$$\|L_t\|_L \leq Me^{\omega t} \quad \text{for all } t \geq 0 . \tag{4.8.12}$$

Definition 4.8.17. If in (4.8.12), $M = 1$ and $\omega = 0$, then $\|L_t\|_L \leq 1$ and $\{L_t\}_{t\geq 0}$ is said to be a **contraction semigroup**.

Remark 4.8.18. The infinitesimal generator of a contraction semigroup is **dissipative** in the sense that there exists, for every $x \in D(A)$, $x^* \in \mathcal{F}(x)$ (see Remark 3.1.51) such that $\mathrm{Re}\langle x^*, Ax\rangle \le 0$ which is known as the "Lumer–Phillips Theorem". If $\{L_t\}_{t\ge 0}$ is a contraction semigroup with infinitesimal generator A, then

$$L_t(x) = \lim_{n\to\infty}\left(\mathrm{id} - \frac{t}{n}A\right)^{-n} x = \lim_{n\to\infty}\left(\frac{n}{t}R\left(\frac{n}{t}\right)\right)^n \quad \text{for all } x \in X\,.$$

This expression is known as the **exponential formula** for the contraction semigroup. It is useful in numerical analysis, since it leads to discretization schemes.

If A is the infinitesimal generator of a contraction semigroup, as before, we define

$$D(A^2) = \{x \in D(A) : Ax \in D(A)\}\,,$$

and inductively

$$D(A^n) = \{x \in D(A) : Ax \in D(A^{n-1})\} \quad \text{for all } n \in \mathbb{N}\,.$$

The **graph norm** on the vector space $D(A)$ is the norm function $x \to |x|_A$ defined by

$$|x|_A = \|x\|_X + \|Ax\|_X \quad \text{for all } x \in D(A)\,.$$

Proposition 4.8.19. *$D(A^2)$ is $|\cdot|_A$-dense in $D(A)$.*

Proof. For $x \in D(A)$, we set $x_\lambda = \lambda R(\lambda)x$ for all $\lambda > 0$. Clearly, $x_\lambda \in D(A)$ and from Step 1 in the proof of Theorem 4.8.16, we have

$$x_\lambda \to x \quad \text{in } X \text{ as } \lambda \to \infty\,.$$

From the definition of the resolvent operator $R(\lambda)$, we have $x = \frac{1}{\lambda}(\lambda\,\mathrm{id} - A)x$ and so $Au_\lambda = \lambda u_\lambda - \lambda u \in D(A)$. Hence $u_\lambda \in D(A^2)$. As above, we have

$$Au_\lambda = \lambda^2 R(\lambda)u - \lambda u = A_\lambda u\,,$$

and so

$$Au_\lambda \to Au \quad \text{in } X \text{ as } \lambda \to \infty\,.$$

Hence, $u_\lambda \to u$ in $(D(A), |\cdot|_A)$ as $\lambda \to \infty$. □

Using $D(A^2)$ and more generally $D(A^n)$ for $n \in \mathbb{N}$, we can have a regularity result for the Cauchy problem

$$u'(t) = Au(t) \quad \text{for } t \ge 0, \quad u(0) = x_0\,. \tag{4.8.13}$$

Theorem 4.8.20. *If A is the infinitesimal generator of a strongly continuous semigroup $\{L_t\}_{t\ge 0}$ and $x_0 \in D(A^2)$, then the solution u of* (4.8.13) *satisfies*

$$u \in C^2(\mathbb{R}_+, X) \cap C^1(\mathbb{R}_+, D(A)) \cap C(\mathbb{R}_+, D(A^2)) .$$

More generally, if $x_0 \in D(A^n)$ *for* $n \in \mathbb{N}$ *with* $n \geq 3$*, then*

$$u \in \bigcap_{k=0}^{n} C^{n-k}(\mathbb{R}_+, D(A^k)) .$$

Proof. Suppose $x_0 \in D(A^2)$. Then $Ax_0 \in D(A)$. Therefore, the function $v(t) = L_t(Ax_0)$ for $t \geq 0$ is differentiable and

$$v'(t) = Av(t),\ t \geq 0, \quad v(0) = Ax_0, \quad v(t) \in D(A) \quad \text{for all } t \geq 0 ,$$

see Proposition 4.8.6. Note that

$$v(t) = L_t(Ax_0) = AL_t(x_0) = u'(t) = Au(t) \quad \text{for } t \geq 0 .$$

Hence,

$$u(t) \in D(A^2) \quad \text{for all } t \geq 0 \quad \text{and} \quad v \in C^1(\mathbb{R}_+, X) \cap C(\mathbb{R}_+, D(A)) .$$

If $n \in \mathbb{N}$ with $n \geq 3$, then we proceed by induction. Assume that the result is true for $n-1$. Let $x_0 \in D(A^n)$. Then $Ax_0 \in D(A^{n-1})$. By the induction hypothesis, we have that $v(t) = u'(t)$ satisfies

$$v \in \bigcap_{k=0}^{n-1} C^{n-1-k}(\mathbb{R}_+, D(A^k)) ,$$

which implies

$$u \in \bigcap_{k=0}^{n-1} C^{n-k}(\mathbb{R}_+, D(A^k)) .$$

We know that $v = Au \in C(\mathbb{R}_+, D(A^{n-1}))$. Therefore, $u \in C(\mathbb{R}_+, D(A^n))$. □

Corollary 4.8.21. *If A is the infinitesimal generator of a contraction semigroup* $\{L_t\}_{t \geq 0}$, $x_0 \in D(A)$ *and* $u(t) = L_t(x_0)$ *for all* $t \geq 0$*, then*

$$\|u'(t)\|_X \leq \|Ax_0\|_X \quad \textit{for all } t \geq 0 .$$

Proof. If $x_0 \in D(A^2)$, then from Theorem 4.8.20, we know that if $v(t) \in L_t(Ax_0)$, then

$$\|u'(t)\|_X = \|v(t)\|_X \leq \|Ax_0\|_X ,$$

since $\|L_t\|_L \leq 1$. If $x_0 \in D(A)$, then from Proposition 4.8.19, there exists a sequence $\{x_0^n\}_{n \in \mathbb{N}} \subseteq D(A^2)$ such that

$$x_0^n \to x_0 \quad \text{and} \quad Ax_0^n \to Ax_0 \quad \text{in } X\,.$$

Let $u_n(t) = L_t(x_0^n)$ for $n \in \mathbb{N}$. We have

$$\|u_n(t) - u_n(t)\|_X \le \|x_0^n - x_n^m\|_X\,,$$
$$\|u_n'(t) - u_m'(t)\|_X \le \|Ax_0^n - Ax_0^m\|_X$$

for $n, m \in \mathbb{N}$. So we see that

$$u_n \to u \quad \text{and} \quad u_n' \to u' \quad \text{uniformly on } \mathbb{R}_+\,.$$

Hence, $\|u'(t)\|_X \le \|Ax_0\|_X$ for all $t \ge 0$. □

Now we consider the inhomogeneous Cauchy problem

$$u'(t) = Au(t) + h(t) \quad \text{for } t \ge 0, \quad u(0) = x_0\,, \tag{4.8.14}$$

with $h\colon [0, b] \to X$. As before, we assume that A is the infinitesimal generator of a strongly continuous semigroup $\{L_t\}_{t\ge 0}$. If $h \equiv 0$, then the solution of (4.8.14) is unique for every initial value $x_0 \in D(A)$.

Definition 4.8.22. We say that $u\colon [0, b] \to X$ is a **classical solution** of (4.8.14) on $[0, b)$ if the following hold:
- $u \in C([0, b), X) \cap C^1((0, b), X)$;
- $u(t) \in D(A)$ for all $0 < t < b$;
- u satisfies (4.8.14) on $[0, b)$.

Remark 4.8.23. Let $v(s) = L_{t-s}(u(s))$ for $t \in [0, b]$. Then v is differentiable on $(0, t)$ and

$$\begin{aligned}\frac{dv}{ds} &= -AL_{t-s}(u(s)) + L_{t-s}(u'(s))\\ &= -AL_{t-s}(u(s)) + L_{t-s}(Au(s)) + L_{t-s}(h(s)) = L_{t-s}(h(s))\,.\end{aligned} \tag{4.8.15}$$

Suppose $h \in L^1([0, b], X)$. Then $L_{t-s}(h(\cdot))$ is integrable, and integrating (4.8.15), we obtain

$$u(t) = L_t(x_0) + \int_0^t L_{t-s}(h(s))\,ds \quad \text{for all } t \in [0, b]\,. \tag{4.8.16}$$

These observations lead to the following result.

Proposition 4.8.24. *If $h \in L^1([0, b], X)$ and $x_0 \in X$, then problem* (4.8.14) *has at most one solution. If* (4.8.14) *has a solution u, then this solution is given by* (4.8.16).

Note that the right-hand side in (4.8.16) is a continuous function on $[0, b]$. Therefore, we can say that (4.8.16) defines a generalized solution of (4.8.14) even if u is not differentiable and therefore (4.8.14) is not satisfied in the sense of Definition 4.8.22.

Definition 4.8.25. If $x_0 \in X$ and $h \in L^1([0, b], X)$, then the function $u \in C([0, b], X)$ defined by

$$u(t) = L_t(x) + \int_0^t L_{t-s}(h(s))\,\mathrm{d}s \quad \text{for } t \in [0, b]\,,$$

is said to be a **mild solution** of problem (4.8.14).

Remark 4.8.26. Evidently, a mild solution need not be a classical one even if $h \equiv 0$. If $h \in L^1([0, b], X)$, then problem (4.8.14) has a unique mild solution.

It is natural to ask what extra conditions are needed on h and x_0 in order for the mild solution to be a classical one. Continuity of h is not enough for the mild solution to be a classical one for $x_0 \in D(A)$. To see this let $x_0 \in X$ be such that $L_t(x_0) \notin D(A)$ for $t \geq 0$. We set $h(t) = L_t(x_0)$. Then h is continuous. We consider the Cauchy problem (4.8.14) with this h and with $x_0 = 0$. We claim that this problem has no classical solution although $0 \in D(A)$. The mild solution of this problem is given by

$$u(t) = \int_0^t L_{t-s}(L_s(x_0))\,\mathrm{d}s = tL_t(x_0)$$

and $t \to tL_t(x_0)$ is not differentiable for $t > 0$. So, the mild solution is not a classical solution.

Therefore, if we want to have a classical solution, we need to assume more than continuity of h.

Proposition 4.8.27. *If A is the infinitesimal generator of a strongly continuous semigroup $\{L_t\}_{t\geq 0}$, $h \in L^1([0, b], X) \cap C((0, b), X)$ and*

$$v(t) = \int_0^t L_{t-s}(h(s))\,\mathrm{d}s \quad \textit{for all } t \in [0, b]\,,$$

then the Cauchy problem (4.8.14) *has a classical solution for every $x_0 \in D(A)$ if one of the following statements is true:*

(a) $v \in C^1((0, b), X)$;

(b) $v(t) \in D(A)$ *for all* $t \in (0, b)$ *and* $Av \in C((0, b), X)$.

If (4.8.14) *has a classical solution u on $[0, b)$ for some $x_0 \in D(A)$, then v satisfies both (a) and (b).*

Proof. If v is a classical solution of (4.8.14) on $[0, b)$ for some $x_0 \in D(A)$, then we have

$$u(t) = L_t(x_0) + \int_0^t L_{t-s}(h(s))\,\mathrm{d}s \quad \text{for all } 0 \le t \le b\,.$$

Hence, $v(t) = u(t) - L_t(x_0)$ for all $t \in [0, b]$. Therefore, v is differentiable and $v'(t) = u'(t) - L_t(Ax_0)$ for all $t \in (0, b)$. Thus, $v \in C^1((0, b), X)$. So, we have (a) satisfied.

Next, note that if $x_0 \in D(A)$, then $L_t(x_0) \in D(A)$ for $t \ge 0$, see Proposition 4.8.6, and so $v(t) = u(t) - L_t(x_0) \in D(A)$ for all $t > 0$. Also, we have

$$Av(t) = Au(t) - AL_t(x_0) = u'(t) - h(t) - L_t(Ax_0)\,.$$

This shows that $t \to Av(t)$ is continuous on $(0, b)$. So we have (b) satisfied.

Note that for $\vartheta > 0$, we have

$$\frac{L_\vartheta - \mathrm{id}}{\vartheta}(v(t)) = \frac{v(t+\vartheta) - v(t)}{\vartheta} = \frac{1}{\vartheta}\int_t^{t+\vartheta} L_{t+\vartheta-s}(h(s))\,\mathrm{d}s\,, \tag{4.8.17}$$

which implies, since h is continuous on $(0, b]$, that

$$\lim_{\vartheta\to 0}\frac{L_\vartheta - \mathrm{id}}{\vartheta}(v(t)) = h(t)\,.$$

If $v \in C^1((0, b), X)$, then from (4.8.17) we see that $v(t) \in D(A)$ and $Av(t) = u'(t) - h(t)$ for all $t \in (0, b)$. Since $v(0) = 0$, we conclude that $u(t) = L_t(x_0) + v(t)$ for $t \in [0, b)$ is a classical solution (4.8.14) with $x_0 \in D(A)$.

If $v(t) \in D(A)$ for $t \in (0, b)$, then from (4.8.17) we see that $\frac{\mathrm{d}^+}{\mathrm{d}t^+}(v(t))$ exists, and we have

$$\frac{\mathrm{d}^+}{\mathrm{d}t^+}(v(t)) = Av(t) + h(t)\,.$$

Hence, $t \to \frac{\mathrm{d}^+}{\mathrm{d}t^+}(v(t))$ is continuous. This implies that $v \in C^1((0, b), X)$ and $v'(t) = Av(t) + h(t)$ and $v(0) = 0$. Therefore $u(t) = L_t(x_0) + v(t)$ is a classical solution of (4.8.14) with $x_0 \in D(A)$. ☐

Corollary 4.8.28. *If A is the infinitesimal generator of a strongly continuous semigroup $\{L_t\}_{t\ge 0}$ and $h \in C^1([0, b], X)$, then the Cauchy problem* (4.8.14) *has a classical solution u on $[0, b)$ for every $x_0 \in D(A)$.*

Proof. We have

$$v(t) = \int_0^t L_{t-s}(h(s))\,\mathrm{d}s = \int_0^t L_s(h(t-s))\,\mathrm{d}s\,.$$

Therefore v is differentiable for $t > 0$ and

$$v'(t) = L_t(h(0)) + \int_0^t L_t(h'(t-s))\,ds = L_t(h(0)) + \int_0^t L_{t-s}(h(s))\,ds$$

is continuous on $(0, b)$. From Proposition 4.8.27, see statement (a), we conclude that u is a classical solution for the Cauchy problem (4.8.14) on $[0, b)$. □

Now we can relate classical and mild solutions for the Cauchy problem (4.8.14).

Theorem 4.8.29. *If A is the infinitesimal generator of a strongly continuous semigroup $\{L_t\}_{t\geq 0}$, $h \in L^1([0,b],X)$ and u is a mild solution of (4.8.14) with $x_0 \in X$, then, for every $b' \in (0,b)$, u is the uniform limit on $[0,b']$ of classical solutions of (4.8.14).*

Proof. From Corollary 4.8.8, we have $\|L_t\|_L \leq Me^{\omega t}$. Also, from Proposition 4.8.6, we know that there exists $x_0^n \in D(A)$ with $n \in \mathbb{N}$ such that $x_0^n \to x_0$ in X.

Moreover, exploiting the density of $C^1([0,b],X)$ in $L^1([0,b],X)$ we can find $h_n \in C^1([0,b],X)$ with $n \in \mathbb{N}$ such that $h_n \to h$ in $L^1([0,b],X)$.

For every $n \in \mathbb{N}$ we consider the Cauchy problem

$$u_n'(t) = Au_n(t) + h_n(t), \quad u_n(0) = x_0^n\,. \tag{4.8.18}$$

From Corollary 4.8.28 we know that (4.8.18) has a unique classical solution u_n on $[0, b)$ and

$$u_n(t) = L_t(x_0^n) + \int_0^t L_{t-s}(h_n(s))\,ds\,.$$

Then, for all $t \in [0, b']$ with $b' < b$ we have

$$\|u_n(t) - u(t)\|_X \leq Me^{\omega b'}\left(\|x_0^n - x_0\|_X + \int_0^b \|h_n(s) - h(s)\|_X\,ds \right),$$

which shows that $u_n \to u$ in $C([0,b'],X)$. □

There is a third notion of solution for the Cauchy problem (4.8.14).

Definition 4.8.30. We say that $u \in C([0,b],X)$ is a **strong solution** of (4.8.14), if u is differentiable almost everywhere on $[0,b]$, $u' \in L^1([0,b],X)$ and

$$u'(t) = Au(t) + h(t) \quad \text{for a. a. } t \in [a,b], \quad u(0) = x_0\,.$$

Remark 4.8.31. If X has the RNP (Radon–Nikodym Property, see Definition 4.2.23), then a strong solution belongs to $\mathrm{AC}([0,b],X) = W^{1,1}((0,b),X)$; see Definition 4.4.1. If $A = 0$, then (4.8.14) has no classical solution unless $h \in C([0,b],X)$, but it has a strong solution given by $u(t) = x_0 + \int_0^t h(s)\,ds$ for all $t \in [0,b]$.

It is natural to ask when a mild solution is in fact a strong solution. In this direction we have a result analogous to Proposition 4.8.27.

Proposition 4.8.32. *If A is the infinitesimal generator of a strongly continuous semigroup $\{L_t\}_{t\geq 0}$, $h \in L^1([0,b],X)$, and*

$$v(t) = \int_0^t L_{t-s}(h(s))\,ds \quad \text{for all } t \in [0,b]\,,$$

then the Cauchy problem (4.8.14) *has a strong solution $u \in C([0,b],X)$ for every $x_0 \in D(A)$ if one of the following statements holds:*
(a) *v is differentiable at a. a. $t \in [0,b]$ and $v' \in L^1([0,b],X)$;*
(b) *$v(t) \in D(A)$ for a. a. $t \in [0,b]$ and $Av \in L^1([0,b],X)$.*

If $u \in C([0,b],X)$ is a strong solution of (4.8.14)*, then u satisfies both (a) and (b) above.*

Corollary 4.8.33. *If A is the infinitesimal generator of a strongly continuous semigroup $\{L_t\}_{t\geq 0}$ and $h: [0,b] \to X$ is differentiable almost everywhere with $h' \in L^1([0,b],X)$, then the Cauchy problem* (4.8.14) *has a unique strong solution $u \in C([0,b],X)$.*

When the ambient space is a Hilbert space, then we have dissipative or accretive operators.

Definition 4.8.34. Let H be a Hilbert space over $\mathbb{R}$ or $\mathbb{C}$ and $A: D(A) \subseteq H \to H$ be a linear operator.
(a) We say that A is **dissipative** if and only if

$$\operatorname{Re}(Au,u) \leq 0 \quad \text{for all } u \in D(A)\,.$$

(b) We say that A is **maximal dissipative** if and only if

$$R(\operatorname{id} - \lambda A) = H \quad \text{for some } \lambda > 0\,.$$

(d) We say that A is **accretive** if and only if for all $\lambda > 0$

$$(\operatorname{id} + \lambda A) \text{ is injective and } (\operatorname{id} + \lambda A)^{-1} \text{ is nonexpansive}\,.$$

(d) We say that A is **maximal accretive** if and only if

$$A \text{ is accretive and } R(\operatorname{id} + \lambda A) = H \quad \text{for all } \lambda > 0\,.$$

Remark 4.8.35. We can show that a maximal dissipative operator A is closed, densely defined, for every $\lambda > 0$, $(\lambda \operatorname{id} - A)$ is invertible and

$$\|(\lambda \operatorname{id} - A)^{-1}\|_H \le \frac{1}{\lambda} .$$

Then Theorem 4.8.16 takes the following form.

Theorem 4.8.36. *If $A: D(A) \subseteq H \to H$ is a linear operator, then the following statements are equivalent:*
(a) *A is the infinitesimal generator of a contraction semigroup.*
(b) *A is maximal dissipative.*
(c) *$-A$ is maximal accretive.*

We can have a corresponding theory for semigroups of nonlinear maps. Let X be a real Banach space and $C \subseteq X$ nonempty.

Definition 4.8.37. A family of maps $S_t: C \to C$ with $t \ge 0$ is said to be a **semigroup of nonexpansive maps on** C if the following hold:
(a) $S_{t+s} = S_t \circ S_s$ for all $t, s \ge 0$;
(b) $S_0 = \operatorname{id}$;
(c) $\lim_{t\to 0} S_t(u) = u$ for all $u \in C$;
(d) $\|S_t(u) - S_t(v)\|_X \le \|u - v\|_X$ for all $t \ge 0$ and for all $u, v \in C$.

Remark 4.8.38. It is easy to check that the function $(t, u) \to S_t(u)$ is continuous on $\mathbb{R}_+ \times C$ into C.

Recall the duality map $\mathcal{F}: X \to 2^X \setminus \{\emptyset\}$ defined by

$$\mathcal{F}(u) = \{u^* \in X^*: \langle u^*, u\rangle = \|u\|_X^2 = \|u^*\|_*^2\} ,$$

see Definition 6.1.20. Using $\mathcal{F}$ we extend the notion of accretive operator to nonlinear maps defined on a general Banach space X.

Definition 4.8.39. A map $A: D(A) \subseteq X \to X$ is said to be **accretive** if for all $u, v \in D(A)$, there exists $x^* \in \mathcal{F}(u-v)$ such that $\langle x^*, A(u) - A(v)\rangle \ge 0$. If, in addition, $\operatorname{id} + A$ is injective, then we say that A is m**-accretive**.

Remark 4.8.40. Introducing the so-called **semi-inner product** $\langle \cdot, \cdot \rangle_+$ by

$$\langle u, v\rangle_+ = \sup\{\langle x^*, u\rangle : x^* \in \mathcal{F}(v)\} , \tag{4.8.19}$$

the above definition can be equivalently written as $\langle A(u) - A(v), u - v\rangle_+ \ge 0$. Since $\mathcal{F}(v) \subseteq X^*$ is nonempty and w*-compact, the supremum in (4.8.19) is attained. If $X = H$ is a Hilbert space, then an accretive operator is also called **monotone** and a maximal accretive operator is called **maximal monotone**. Monotone operators will be introduced in Chapter 6.

The next theorem concerning the generator of semigroups of nonlinear, nonexpansive maps is known as the “Crandall–Liggett Exponential Formula”.

Theorem 4.8.41 (Crandall–Liggett Exponential Formula). *If $A: D(A) \subseteq X \to X$ is a m-accretive map, then*

$$S_t(u) = \lim_{n\to\infty}\left(\mathrm{id} + \frac{1}{n}A\right)^{-n}(u)$$

exists for all $u \in \overline{D(A)}$ and the limit is uniform in $t \in K$ with $K \subseteq \mathbb{R}_+$ being compact. Moreover, $S_t: \overline{D(A)} \to \overline{D(A)}$ is a semigroup of nonlinear, nonexpansive maps and for all $t > 0$ and for all $u \in D(A)$, we have

$$\|S_t(u) - u\|_X \le t \inf\{\|y\|_Y : y \in A(u)\}\,.$$

4.9 Generalized Orlicz Spaces

The Sobolev spaces that we discussed in Section 4.5 are appropriate for equations in which the nonlinear terms exhibit growth and coercivity conditions of polynomial type. However, there are physical phenomena, where the nonlinearities satisfy general growth conditions, governed by an inhomogeneous (that is, space dependent), anisotropic (that is, of different growth in different directions) convex function with no polynomial growth. Such situations lead to a different function space setting based on the so-called “Generalized Orlicz Spaces”, also known as “Musielak–Orlicz Spaces”.

We start with the following basic definition.

Definition 4.9.1. (a) We say that $\varphi: \mathbb{R}_+ \to \mathbb{R}_+$ with $\mathbb{R}_+ = [0,\infty)$ is an N-**function**, if it is convex, continuous, nondecreasing, $\varphi(0) = 0$, $\varphi(x) > 0$ for all $x > 0$ and

$$\text{superlinear at } +\infty: \quad \lim_{x\to+\infty}\frac{\varphi(x)}{x} = +\infty\,,$$

$$\text{sublinear at } 0^+: \quad \lim_{x\to 0^+}\frac{\varphi(x)}{x} = 0\,.$$

(b) Let (Ω, Σ, μ) be a finite complete measure space. We say that $\varphi: \Omega \times \mathbb{R}_+ \to \mathbb{R}_+$ is a **generalized** N-**function**, if for all $x \in \mathbb{R}_+$, $z \to \varphi(z, x)$ is measurable and for μ-a. a. $z \in \Omega$, $\varphi(z,\cdot)$ is a N-function. We denote the family of generalized N-functions by $\mathrm{N}(\Omega)$.

Remark 4.9.2. If $\varphi \in \mathrm{N}(\Omega)$, then φ is a Carathéodory function, hence jointly measurable; see Proposition 2.2.31. Note that if $\varphi \in \mathrm{N}(\Omega)$, then $\varphi(z,x) = \int_0^x f(z,x)\,ds$ with $f(z,x)$ being the right-hand derivative of $\varphi(z,\cdot)$ for all $z \in \Omega$. We have

$$\varphi(z,x) \le f(z,x)x \le \varphi(z,2x) \quad \text{for all } z \in \Omega \text{ and for all } x \ge 0 .$$

We set

$$\psi(z,u) = \inf\{x \in \mathbb{R}_+ : \varphi(z,x) \ge u\} .$$

Then ψ is a Carathéodory function which is nondecreasing in u. If $\varphi(z,\cdot)$ is strictly increasing, then $\psi(z,\cdot)$ is the inverse of $\varphi(z,\cdot)$ denoted by $\varphi^{-1}(z,u)$.

Example 4.9.3. (a) If $p\colon \Omega \to [1,+\infty)$ is a measurable function and $\varphi(z,x) = x^{p(z)}$, then φ is a generalized N-function which is strictly increasing in x and $\varphi^{-1}(z,u) = u^{\frac{1}{p(z)}}$. This function leads to the variable exponent spaces.

(b) If $a \in L^\infty(\Omega)$, $a(z) \ge 0$ μ-a. a. $z \in \Omega$ and $1 \le q < p < \infty$, then

$$\varphi(z,x) = a(z)x^p + x^q \quad \text{for all } z \in \Omega \text{ and for all } x \ge 0$$

is a generalized N-function which is used in **double phase** problems or problems with **unbalanced growth**. Indeed note that

$$x^q \le \varphi(z,x) \le c(x^p + x^q) \quad \text{for } \mu\text{-a. a. } z \in \Omega, \text{ for all } x \ge 0 \text{ and for some } c > 0 ,$$

which justifies the characterization of unbalanced growth.

The next definition anticipates a notion that we will introduce and study in a more general context in Chapter 5; see Definition 5.3.1.

Definition 4.9.4. Let $\varphi \in N(\Omega)$. The **conjugate** or **complementary** function $\varphi^*\colon \Omega \times \mathbb{R}_+ \to \mathbb{R}_+$ is defined by

$$\varphi^*(z,u) = \sup_{x \ge 0}\{ux - \varphi(z,x)\} \quad \text{for all } z \in \Omega \text{ and for all } x \ge 0 .$$

Remark 4.9.5. Note that $\varphi^* \in N(\Omega)$ and φ is the conjugate of φ^*. Also, we have

$$\varphi^*(z,u) = \int_0^u \hat{f}(z,\tau)\,\mathrm{d}\tau$$

with $\hat{f}(z,\tau) = \sup\{s \ge 0 : f(z,s) \le \tau\}$ for all $z \in \Omega$. Moreover, we can easily check that

$$\hat{f}(z, f(z,s)) \ge s, \quad f(z,\hat{f}(z,\tau)) \ge \tau$$

for all $z \in \Omega$, for all $s, \tau \ge 0$,

$$xu \le \varphi(z,x) + \varphi^*(z,u)$$

for all $z \in \Omega$, for all $x, u \ge 0$ and

$$\hat{f}(z,\tau)\tau = \varphi(z,\hat{f}(z,\tau)) + \varphi^*(z,\tau)$$

for all $z \in \Omega$ and for all $\tau \geq 0$.

The generalized Orlicz functions lead to spaces with nice properties when the following property is satisfied.

Definition 4.9.6. Let $\varphi \in \mathrm{N}(\Omega)$. We say that φ satisfies the **Δ_2-condition**, denoted by $\varphi \in \Delta_2$, if there exist a constant $c > 0$ and a function $\vartheta \in L^1(\Omega)_+$ such that

$$\varphi(z,2x) \leq c\varphi(z,x) + \vartheta(z) \quad \text{for a. a. } z \in \Omega \text{ and for all } x \geq 0\,.$$

Remark 4.9.7. Both functions from Example 4.9.3 satisfy the Δ_2-condition.

Definition 4.9.8. Let $\varphi_1, \varphi_2 \in \mathrm{N}(\Omega)$. We say that φ_1 is **weaker** that φ_2, denoted by $\varphi_1 \prec \varphi_2$, if there exist constants $c_1, c_2 > 0$ and $\vartheta \in L^1(\Omega)_+$ such that

$$\varphi_1(z,x) \leq c_1\varphi_2(z,c_2x) + \vartheta(z) \quad \text{for a. a. } z \in \Omega \text{ and for all } x \geq 0\,.$$

Given $\varphi \in \mathrm{N}(\Omega)$, let ρ_φ be the corresponding modular function defined by

$$\rho_\varphi(u) = \int_\Omega \varphi(z,|u|)\,d\mu$$

with $u \in L^0(\Omega) = \{u\colon \Omega \to \mathbb{R}\colon \text{measurable}\}$ and we identify two measurable functions which differ only on a μ-null set.

Using the modular function ρ_φ we can now define the corresponding generalized Orlicz spaces.

Definition 4.9.9. Given $\varphi \in \mathrm{N}(\Omega)$, the **generalized Orlicz space** $L^\varphi(\Omega)$ is defined to be the smallest linear space generated by the set

$$S(\Omega) = \{u \in L^0(\Omega)\colon \rho_\varphi(u) < \infty\}\,.$$

Proposition 4.9.10. *If* $\varphi \in \mathrm{N}(\Omega)$, *then*

$$L^\varphi(\Omega) = \left\{u \in L^0(\Omega)\colon \rho_\varphi\left(\frac{u}{\lambda}\right) \to 0 \text{ as } \lambda \to \infty\right\}. \tag{4.9.1}$$

Proof. Let $u \in L^\varphi(\Omega)$. Then we have

$$u = \sum_{k=1}^n \lambda_k u_k \quad \text{with } \lambda_k \in \mathbb{R},\ u_k \in S(\Omega) \text{ and } n \in \mathbb{N}\,.$$

Clearly, $S(\Omega)$ is convex and so

$$\frac{1}{\sum_{k=1}^{n} |\lambda_k|} u \in S(\Omega), \tag{4.9.2}$$

and so

$$\int_{\Omega} \varphi\left(z, \frac{u}{\sum_{k=1}^{n} |\lambda_k|}\right) dz < \infty .$$

Then, using the convexity of ρ_φ and (4.9.2), we obtain

$$\lim_{\lambda \to \infty} \rho_\varphi\left(\frac{u}{\lambda}\right) = \lim_{\substack{t \to \infty \\ t \geq 1}} \rho_\varphi\left(\frac{u}{t \sum_{k=1}^{n} |\lambda_k|}\right) \leq \lim_{\substack{t \to \infty \\ t \geq 1}} \frac{1}{t} \rho_\varphi\left(\frac{u}{\sum_{k=1}^{n} |\lambda_k|}\right) = 0 .$$

Therefore, we have

$$L^\varphi(\Omega) \subseteq \left\{u \in L^0(\Omega) : \rho_\varphi\left(\frac{u}{\lambda}\right) \to 0 \text{ as } \lambda \to \infty\right\} . \tag{4.9.3}$$

We need to show that the opposite inclusion is also true. So, suppose $u \in L^0(\Omega)$ is such that

$$\rho_\varphi\left(\frac{u}{\lambda}\right) \to 0 \quad \text{as } \lambda \to \infty .$$

Then we can find $\lambda_0 > 0$ such that $\rho_\varphi(\frac{u}{\lambda_0}) < \infty$ which implies $\frac{u}{\lambda_0} \in S(\Omega)$ and so $u \in L^\varphi(\Omega)$. We have proved that

$$\left\{u \in L^0(\Omega) : \rho_\varphi\left(\frac{u}{\lambda}\right) \to 0 \text{ as } \lambda \to \infty\right\} \subseteq L^\varphi(\Omega) . \tag{4.9.4}$$

From (4.9.3) and (4.9.4) we conclude that

$$L^\varphi(\Omega) = \left\{u \in L^0(\Omega) : \rho_\varphi\left(\frac{u}{\lambda}\right) \to 0 \text{ as } \lambda \to \infty\right\} . \qquad \square$$

Remark 4.9.11. Using this proposition we see that the spaces generated by the variable exponent N(Ω)-function

$$\varphi(z, x) = x^{p(z)} \quad \text{for all } z \in \Omega \text{ and for all } x \geq 0$$

and by the double phase N(Ω)-function

$$\varphi(z, x) = a(z)x^p + x^q \quad \text{for all } z \in \Omega \text{ and for all } x \geq 0$$

are given by (4.9.1).

Proposition 4.9.12. *If $\varphi \in \mathrm{N}(\Omega)$, then the function $\|\cdot\|: L^{\varphi}(\Omega) \to \mathbb{R}_+$ defined by*

$$\|u\| = \inf\left\{\lambda > 0 : \rho_{\varphi}\left(\frac{u}{\lambda}\right) \leq 1\right\}$$

is a norm on $L^{\varphi}(\Omega)$.

Proof. From Proposition 4.9.10 we know that $\rho_{\varphi}(u) < \infty$ for all $u \in L^{\varphi}(\Omega)$. First, we show that

$$\|u\| = 0 \quad \text{implies} \quad u(z) = 0 \quad \text{for } \mu\text{-a. a. } z \in \Omega\,. \tag{4.9.5}$$

We have $\varphi(z, 0) = 0$ if and only if $z = 0$; see Definition 4.9.1. If $u = 0$, then $\rho_{\varphi}(\frac{u}{\lambda}) = 0$ for all $\lambda > 0$ and so $\|u\| = 0$. If $\|u\| = 0$, then $\rho_{\varphi}(\frac{u}{\lambda}) \leq 1$ for all $\lambda > 0$. But for $0 < \lambda \leq 1$, on account of the convexity of the modular function and since $\rho_{\varphi}(0) = 0$, we have

$$\frac{1}{\lambda}\rho_{\varphi}(u) \leq \rho_{\varphi}\left(\frac{u}{\lambda}\right),$$

which implies $\frac{1}{\lambda}\rho_{\varphi}(u) = 0$ and so $\varphi(z, u(z)) = 0$ for μ-a. a. $z \in \Omega$. Hence, $u(z) = 0$ for μ-a. a. $z \in \Omega$.

Next we show that

$$\|\vartheta u\| = |\vartheta|\|u\| \quad \text{for all } \vartheta \in \mathbb{R}\,. \tag{4.9.6}$$

We extend $\varphi(z, \cdot)$ on $\mathbb{R}$ by $\varphi(z, x) = \varphi(z, -x)$ for $x \geq 0$. Then we get

$$\begin{aligned}\|\vartheta u\| &= \inf\left\{\lambda > 0 : \rho_{\varphi}\left(\frac{\vartheta u}{\lambda}\right) \leq 1\right\} = \inf\left\{|\vartheta|\lambda' > 0 : \rho_{\varphi}\left(\frac{\vartheta u}{|\vartheta|\lambda'}\right) \leq 1\right\} \\ &= |\vartheta| \inf\left\{\lambda' > 0 : \rho_{\varphi}\left(\frac{u}{\lambda'}\right) \leq 1\right\} = |\vartheta|\|u\|\,.\end{aligned}$$

This proves (4.9.6).

Finally, we prove the triangle inequality, namely that

$$\|u + v\| \leq \|u\| + \|v\| \quad \text{for all } u, v \in L^{\varphi}(\Omega)\,. \tag{4.9.7}$$

Let $u \in L^{\varphi}(\Omega)$ and let $\{\lambda_n\}_{n\in\mathbb{N}} \subseteq \mathbb{R}$ be a minimizing sequence, that is,

$$\lambda_n \searrow \|u\| \quad \text{and} \quad \rho_{\varphi}\left(\frac{u}{\lambda_n}\right) = \int_{\Omega} \varphi\left(z, \frac{u}{\lambda_n}\right) \mathrm{d}z \leq 1\,.$$

Then, since $\lambda_{n+1} < \lambda_n$ for all $n \in \mathbb{N}$, due to the convexity of $\varphi(z, \cdot)$, we obtain

$$\varphi\left(z, \frac{u}{\lambda_n}\right) \leq \frac{\lambda_n}{\lambda_{n+1}}\varphi\left(z, \frac{u}{\lambda_n}\right) \leq \varphi\left(z, \frac{\lambda_n}{\lambda_{n+1}}\frac{u}{\lambda_n}\right) = \varphi\left(z, \frac{u}{\lambda_{n+1}}\right)$$

for a. a. $z \in \Omega$. Hence $\{\xi_n\}_{n\in\mathbb{N}}$ with $\xi_n(z) := \varphi(z, u(z)/\lambda_n)$ for a. a. $z \in \Omega$ is increasing. Moreover,

$$\xi_n(z) \to \varphi\left(z, \frac{u(z)}{\|u\|}\right) \quad \text{for a. a. } z \in \Omega\,.$$

Then, by the Monotone Convergence Theorem (see Theorem 2.3.3), it follows that

$$\rho_\varphi\left(\frac{u}{\|u\|}\right) = \int_\Omega \varphi\left(z, \frac{u}{\|u\|}\right) \mathrm{d}z \le 1\,. \tag{4.9.8}$$

Now, let $u, v \in L^\varphi(\Omega)$. From the convexity of the modular function ρ_φ and (4.9.8) we have

$$\begin{aligned}\rho_\varphi\left(\frac{u+v}{\|u\| + \|v\|}\right) &= \rho_\varphi\left(\frac{\|u\|}{\|u\| + \|v\|}\frac{u}{\|u\|} + \frac{\|v\|}{\|u\| + \|v\|}\frac{v}{\|v\|}\right) \\ &\le \frac{\|u\|}{\|u\| + \|v\|}\rho_\varphi\left(\frac{u}{\|u\|}\right) + \frac{\|v\|}{\|u\| + \|v\|}\rho_\varphi\left(\frac{v}{\|v\|}\right) \le 1\,.\end{aligned}$$

Therefore, $\|u + v\| \le \|u\| + \|v\|$ and so the triangle inequality (4.9.7) holds.

Finally, from (4.9.5), (4.9.6) and (4.9.7) we conclude that $\|\cdot\|$ is a norm on $L^\varphi(\Omega)$. □

Remark 4.9.13. This norm is known as the **Luxemburg norm**, and it is the Minkowski functional corresponding to the convex, absorbing and balanced set

$$C = \{u \in L^0(\Omega) : \rho_\varphi(u) \le 1\}\,;$$

see Definition 3.1.37.

Corollary 4.9.14. *If $\varphi \in \mathrm{N}(\Omega)$ and $u \in L^0(\Omega)$ satisfies $\|u\| \le \eta$ for some $\eta > 0$, then $u \in L^\varphi(\Omega)$.*

Proof. We have $\frac{1}{\eta}\|u\| \le 1$ and so by (4.9.8)

$$\rho_\varphi\left(\frac{u}{\eta}\right) = \rho_\varphi\left(\frac{\|u\| u}{\|u\| \eta}\right) \le \frac{\|u\|}{\eta}\rho_\varphi\left(\frac{u}{\|u\|}\right) \le \frac{\|u\|}{\eta} \le 1\,.$$

This gives $\frac{u}{\eta} \in S(\Omega)$ and so $u \in L^\varphi(\Omega)$ □

Remark 4.9.15. On $L^\varphi(\Omega)$ one can define another norm by

$$|u| = \sup\left\{\int_\Omega uv\,\mathrm{d}z : \rho_{\varphi^*}(v) \le 1\right\}\,.$$

We can verify that this is a norm on $L^\varphi(\Omega)$, and it is known as the **Orlicz norm**. The two norms are equivalent and we have

$$\|u\| \leq |u| \leq 2\|u\| \quad \text{for all } u \in L^{\varphi}(\Omega)\,. \tag{4.9.9}$$

The next result is known as the **unit ball property** and relates the Luxemburg norm with the modular function.

Proposition 4.9.16. *If $\varphi \in \mathrm{N}(\Omega)$ and $u \in L^{\varphi}(\Omega)$, then $\|u\| \leq 1$ if and only if $\rho_\varphi(u) \leq 1$.*

Proof. From the definition of the Luxemburg norm in Proposition 4.9.12, we see that $\rho_\varphi(u) \leq 1$ implies that $\|u\| \leq 1$. Next, suppose that $\|u\| \leq 1$. Then $\rho_\varphi(\frac{u}{\lambda}) \leq 1$ for all $\lambda > 1$. The continuity of $\varphi(z,\cdot)$ and Fatou's Lemma imply that $\rho_\varphi(u) \leq 1$. ☐

Corollary 4.9.17. *If $\varphi \in \mathrm{N}(\Omega)$ and $u \in L^{\varphi}(\Omega)$, then the following hold:*
(a) $\|u\| \leq 1$ *implies* $\rho_\varphi(u) \leq \|u\|$;
(b) $\|u\| \geq 1$ *implies* $\rho_\varphi(u) \geq \|u\|$;
(c) $\|u\| \leq \rho_\varphi(u) + 1$.

Proof. (a) From (4.9.8) we know that

$$\rho_\varphi\left(\frac{u}{\|u\|}\right) \leq 1\,. \tag{4.9.10}$$

Since $\|u\| \leq 1$, exploiting the convexity of $\varphi(z,\cdot)$, we have

$$\frac{1}{\|u\|}\rho_\varphi(u) \leq \rho_\varphi\left(\frac{u}{\|u\|}\right).$$

From this and (4.9.10) we obtain $\rho_\varphi(u) \leq \|u\|$.

(b) Suppose $\|u\| > 1$ and choose $\varepsilon > 0$ small enough so that

$$\|u\| - \varepsilon > 1\,. \tag{4.9.11}$$

From the definition of the Luxemburg norm, we have

$$1 < \rho_\varphi\left(\frac{u}{\|u\| - \varepsilon}\right). \tag{4.9.12}$$

Then (4.9.11) and the convexity of $\varphi(z,\cdot)$ imply

$$\rho_\varphi\left(\frac{u}{\|u\| - \varepsilon}\right) \leq \frac{1}{\|u\| - \varepsilon}\rho_\varphi(u)\,,$$

which by (4.9.12) gives $\|u\| - \varepsilon < \rho_\varphi(u)$. Letting $\varepsilon \to 0$ yields $\|u\| \leq \rho_\varphi(u)$.

(c) This follows from (a) and (b). ☐

We can have a kind of Hölder's inequality.

Proposition 4.9.18. *If $\varphi \in \mathrm{N}(\Omega)$, $u \in L^{\varphi}(\Omega)$ and $h \in L^{\varphi^*}(\Omega)$, then $u, h \in L^1(\Omega)$ and*

$$\int_\Omega |uh| \,\mathrm{d}\mu \le 2\|u\|\|h\|_*$$

with $\|\cdot\|_*$ denoting the Luxemburg norm of $L^{\varphi^*}(\Omega)$.

Proof. If $u = 0$ or $h = 0$, then the inequality is obvious. So, we assume that $u \neq 0 \neq h$. Using the Young–Fenchel inequality, see Remark 4.9.5, we have

$$\int_\Omega \frac{|u|}{\|u\|}\frac{|h|}{\|h\|_*}\,\mathrm{d}\mu \le \rho_\varphi\left(\frac{|u|}{\|u\|}\right) + \rho_{\varphi^*}\left(\frac{|h|}{\|h\|_*}\right) \le 2\,,$$

see Corollary 4.9.17 (a). Therefore,

$$\int_\Omega |uh| \,\mathrm{d}\mu \le 2\|u\|\|h\|_*\,. \qquad \square$$

Remark 4.9.19. Due to (4.9.9) and Proposition 4.9.18 there exists a constant $c > 0$ such that

$$\int_\Omega |uh| \,\mathrm{d}\mu \le c\|u\||h| \quad \text{and} \quad \int_\Omega |uh| \,\mathrm{d}\mu \le c|u|\|h\|$$

for all $u \in L^{\varphi}(\Omega)$ and for all $h \in L^{\varphi^*}(\Omega)$.

Thus far we know that $L^{\varphi}(\Omega)$ is a normed space. In the next theorem we are going to prove that $L^{\varphi}(\Omega)$ is even a Banach space.

Theorem 4.9.20. *If $\varphi \in \mathrm{N}(\Omega)$, then $L^{\varphi}(\Omega)$ is a Banach space.*

Proof. For $A \in \Sigma$, we define

$$\vartheta_x(A) = \int_A \varphi(z, x) \,\mathrm{d}\mu\,.$$

Note that $\vartheta_x(A) = 0$ implies $\varphi(z, x) = 0$ for μ-a. a. $z \in \Omega$ and so $\mu(A) = 0$. Therefore we have $\mu \ll \vartheta_x$; see Definition 2.4.10.

Invoking Proposition 2.4.11, we see that for given $\varepsilon > 0$, we can find $\delta \in (0, 1)$ such that

$$A \in \Sigma \quad \text{and} \quad \vartheta_x(A) \le \delta \quad \text{imply} \quad \mu(A) \le \varepsilon\,. \tag{4.9.13}$$

Consider the Cauchy sequence $\{u_n\}_{n\in\mathbb{N}} \subseteq L^{\varphi}(\Omega)$. Then

$$\|u_n - u_m\| \le \delta \quad \text{for all } m, n \ge n_0\,. \tag{4.9.14}$$

For $\lambda > 0$, let

$$A_\lambda = \{z \in \Omega: |u_n(z) - u_m(z)| > \lambda\}.$$

Then, from (4.9.14) and Corollary 4.9.17 (a), we have

$$\vartheta_\lambda(A_\lambda) = \int_{A_\lambda} \varphi(z,\lambda)\,d\mu \le \int_{A_\lambda} \varphi(z, |u_n - u_m|)\,d\mu \le \rho_\varphi(u_n - u_m) \le \|u_n - u_m\| \le \delta.$$

Therefore, because of (4.9.13), $\mu(A_\lambda) \le \varepsilon$ and so $\{u_n\}_{n\in\mathbb{N}}$ is a Cauchy sequence in measure. Invoking Proposition 2.3.36, there exist $u \in L^0(\Omega)$ and a subsequence $\{u_{n_k}\}_{k\in\mathbb{N}}$ of $\{u_n\}_{n\in\mathbb{N}}$ such that

$$u_{n_k}(z) \to u(z) \quad \text{for } \mu\text{-a.a. } z \in \Omega \text{ as } k \to \infty.$$

We have

$$\varphi(z, \eta|u_{n_k}(z) - u(z)|) \to 0 \quad \text{for } \mu\text{-a.a. } z \in \Omega \text{ as } k \to \infty$$

and for all $\eta > 0$. Let $\lambda > 0$ and $\varepsilon \in (0,1)$. Then we can find $k_0 \in \mathbb{N}$ such that

$$\|\lambda(u_{n_k} - u_{n_m})\| \le \varepsilon \quad \text{for all } k, m \ge k_0. \tag{4.9.15}$$

Using Fatou's Lemma and (4.9.15) we obtain

$$\begin{aligned} \rho_\varphi(\lambda|u_{n_k} - u|) &= \int_\Omega \varphi(z, \lambda|u_{n_k} - u|)\,d\mu = \int_\Omega \lim_{m\to\infty} \varphi(z, \lambda|u_{n_k} - u_{n_m}|)\,d\mu \\ &\le \liminf_{m\to\infty} \int_\Omega \varphi(z, \lambda|u_{n_k} - u_{n_m}|)\,d\mu = \liminf_{m\to\infty} \rho_\varphi(\lambda|u_{n_k} - u_{n_m}|) \\ &\le \liminf_{m\to\infty} \|\lambda(u_{n_k} - u_{n_m})\| \le \varepsilon. \end{aligned}$$

Therefore, $\rho_\varphi(\lambda|u_{n_k} - u|) \to 0$ as $k \to \infty$ for every $\lambda > 0$ and so

$$\rho_\varphi\left(\frac{|u_{n_k} - u|}{\lambda^{-1}}\right) \le 1 \quad \text{for all } k \ge k_0.$$

This implies $\|u_{n_k} - u\| \le \lambda^{-1}$ for all $k \ge k_0$. Since $\lambda > 0$ is arbitrary, we conclude that $u_{n_k} \to u$ in $L^\varphi(\Omega)$. Hence, $L^\varphi(\Omega)$ is a Banach space. □

Remark 4.9.21. $L^\varphi(\Omega)$ equipped with the Orlicz norm is a Banach space as well.

Corollary 4.9.22. *If $\varphi \in \mathrm{N}(\Omega)$, $\{u_n\}_{n\in\mathbb{N}} \subseteq L^\varphi(\Omega)$ and $u_n \to u$ in $L^\varphi(\Omega)$, then there exists a subsequence $\{u_{n_k}\}_{k\in\mathbb{N}}$ of $\{u_n\}_{n\in\mathbb{N}}$ such that*

$$u_{n_k}(z) \to u(z) \quad \text{for } \mu\text{-a.a. } z \in \Omega \text{ as } k \to \infty.$$

Corollary 4.9.23. *If $\varphi \in \mathrm{N}(\Omega)$ and $\{u_n, u\}_{n\in\mathbb{N}} \subseteq L^\varphi(\Omega)$, then $\|u_n - u\| \to 0$ if and only if $\rho_\varphi(\lambda(u_n - u)) \to 0$ for all $\lambda > 0$.*

We present some more properties of the Luxemburg norm.

Proposition 4.9.24. *If $\varphi \in \mathrm{N}(\Omega)$, $\{u_n\}_{n\in\mathbb{N}} \subseteq L^\varphi(\Omega)$ and $u_n(z) \to u(z)$ for μ-a. a. $z \in \Omega$, then $\|u\| \le \liminf_{n\to\infty} \|u_n\|$.*

Proof. The result is obvious if $\lim_{n\to\infty} \|u_n\| = \infty$. So we assume that $\liminf_{n\to\infty} \|u_n\| < \infty$. Then, let $\lambda > \liminf_{n\to\infty} \|u_n\|$, we have $\|u_n\| \le \lambda$ for all $n \ge n_0$. Using Proposition 4.9.16 we get

$$\rho_\varphi\left(\frac{u_n}{\lambda}\right) \le 1 \quad \text{for all } n \ge n_0 .$$

From Fatou's Lemma we have

$$\rho_\varphi\left(\frac{u}{\lambda}\right) \le 1$$

and again by Proposition 4.9.16 it follows that $\|u\| \le \lambda$. Hence, $\|u\| \le \liminf_{n\to\infty} \|u_n\|$. □

Proposition 4.9.25. *If $\varphi \in \mathrm{N}(\Omega)$, $\{u_n\}_{n\in\mathbb{N}} \subseteq L^\varphi(\Omega)$, $|u_n(z)| \nearrow |u(z)|$ for μ-a. a. $z \in \Omega$ and $\sup_{n\in\mathbb{N}} \|u_n\| < \infty$, then $u \in L^\varphi(\Omega)$ and $\|u_n\| \nearrow \|u\|$.*

Proof. From Proposition 4.9.24 we have

$$\|u\| \le \liminf_{n\to\infty} \|u_n\| \le \sup_{n\in\mathbb{N}} \|u_n\| < \infty \tag{4.9.16}$$

and so $u \in L^\varphi(\Omega)$.

Note that $\{\|u_n\|\}_{n\in\mathbb{N}}$ is increasing and

$$\lim_{n\to\infty} \|u_n\| \le \|u\| . \tag{4.9.17}$$

From (4.9.16) and (4.9.17) we conclude that $\|u_n\| \nearrow \|u\|$. □

Recall that $L^p(\Omega) \hookrightarrow L^q(\Omega)$ continuously when $1 < q \le p \le \infty$, as μ is finite by hypothesis. The same embedding is true for the generalized Orlicz spaces $L^\varphi(\Omega)$ and $L^\psi(\Omega)$ provided ψ dominates φ in the sense of Definition 4.9.8.

Theorem 4.9.26. *If $\varphi, \psi \in \mathrm{N}(\Omega)$ and $\varphi \prec \psi$, then $L^\psi(\Omega) \hookrightarrow L^\varphi(\Omega)$.*

Proof. Let $\varphi_\lambda(z, x) = \varphi(z, \lambda x)$ for $\lambda > 0$.

Claim: $L^\varphi(\Omega) = L^{\varphi_\lambda}(\Omega)$ and $\|\lambda u\| = \|u\|_{\varphi_\lambda}$.

Consider $u \in L^\varphi(\Omega)$. From Definition 4.9.9 we have

$$u = \sum_{k=1}^{n} a_k u_k \quad \text{with } a_k \in \mathbb{R} \text{ and } u_k \in S(\Omega) ,$$

equivalently,

$$u = \sum_{k=1}^{n} a_k \lambda \frac{u}{\lambda} \quad \text{and} \quad \frac{u}{\lambda} \in S_{\varphi_\lambda}(\Omega)\,,$$

which is equivalent to $u \in L^{\varphi_\lambda}(\Omega)$. So we have proved that $L^{\varphi}(\Omega) = L^{\varphi_\lambda}(\Omega)$.

Also, as in the proof of Proposition 4.9.12, we have

$$\begin{aligned}\|u\|_{\varphi_\lambda} &= \inf\left\{\vartheta > 0 : \rho_\varphi\left(\frac{\lambda u}{\vartheta}\right) \le 1\right\} = \inf\left\{\lambda\vartheta' > 0 : \rho_\varphi\left(\frac{\lambda u}{\lambda\vartheta'}\right) \le 1\right\} \\ &= \lambda \inf\left\{\vartheta' > 0 : \rho_\varphi\left(\frac{u}{\vartheta'}\right) \le 1\right\} = \lambda\|u\|\,.\end{aligned}$$

This proves the Claim.

According to this Claim, it suffices to assume that there exist $c > 0$ and $h \in L^1(\Omega)_+$ such that

$$\varphi(z,x) \le c\psi(z,x) + h(z) \quad \text{for } \mu\text{-a. a. } z \in \Omega \text{ and for all } x \ge 0\,. \tag{4.9.18}$$

Suppose $\|u\|_\psi = 1$. Then using (4.9.18), we have

$$\rho_\varphi(u) \le c\rho_\psi(u) + \|h\|_1 \le c\|u\|_\psi + \|h\|_1 \le \hat{c}\|u\|_\psi \tag{4.9.19}$$

with $\hat{c} = 1 + c + \|h\|_1$. Note that $\hat{c}\|u\|_\psi \ge 1$. Applying the convexity of $\varphi(z,\cdot)$, this leads to

$$\rho_\varphi\left(\frac{u}{\hat{c}\|u\|_\psi}\right) \le \frac{1}{\hat{c}\|u\|_\psi}\rho_\varphi(u) \le \frac{1}{\hat{c}\|u\|_\psi}\hat{c}\|u\|_\psi = 1\,,$$

where we have used (4.9.19) in the last inequality. Hence, $\|u\|_\varphi \le \hat{c}\|u\|_\psi$.

If $\|u\|_\psi \ne 1$, then let $\hat{u} = \frac{u}{\|u\|_\psi}$. Then $\|\hat{u}\|_\psi = 1$ and from the previous part of the proof, it follows that $\|\hat{u}\|_\varphi \le \hat{c}\|\hat{u}\|_\psi$ and so $\|u\|_\varphi \le \hat{c}\|u\|_\psi$. Therefore, $L^\psi(\Omega) \hookrightarrow L^\varphi(\Omega)$ continuously. □

Remark 4.9.27. The converse of Theorem 4.9.26 is also true. Namely, if $\varphi, \psi \in N(\Omega)$, the measure μ is nonatomic and $L^\psi(\Omega) \hookrightarrow L^\varphi(\Omega)$ continuously, then $\varphi \prec \psi$. For details concerning this implication we refer to Musielak [248, Theorem 8.4, p. 45].

Definition 4.9.28. We say that $\varphi, \psi \in N(\Omega)$ are **equivalent**, denoted by $\varphi \sim \psi$, if there exists $\eta > 1$ such that

$$\psi\left(z, \frac{1}{\eta}x\right) \le \varphi(z,x) \le \psi(z,\eta x)$$

for μ-a. a. $z \in \Omega$ and for all $x \ge 0$.

Using Theorem 4.9.26 we obtain the following result.

Corollary 4.9.29. *If $\varphi, \psi \in \mathrm{N}(\Omega)$ and $\varphi \sim \psi$, then $L^{\varphi}(\Omega) = L^{\psi}(\Omega)$ and the norms are comparable.*

Given $\varphi \in \mathrm{N}(\Omega)$, a simple function need not belong to $L^{\varphi}(\Omega)$.

Example 4.9.30. Let $\Omega = (0,1)$ be equipped with the Lebesgue measure and with Σ being the σ-algebra of the Lebesgue measurable subsets of Ω. Let $\varphi(z,x) = \frac{x^{1+\varepsilon}}{|z|}$ with $\varepsilon > 0$, then $\varphi \in \mathrm{N}(\Omega)$. Let $u = \chi_{\Omega}$, then we have for all $\lambda > 0$

$$\int_0^1 \varphi(z, \lambda|u|)\,\mathrm{d}z = \int_0^1 \frac{\lambda^{1+\varepsilon}}{z}\,\mathrm{d}z = +\infty\,.$$

So, by Proposition 4.9.10, $u \notin L^{\varphi}(\Omega)$.

This example leads to the following definition which identifies an important subclass of $\mathrm{N}(\Omega)$.

Definition 4.9.31. Let $\varphi \in \mathrm{N}(\Omega)$. We say that φ is **locally integrable**, if

$$\int_{\Omega} \varphi(z,x)\,\mathrm{d}z < \infty$$

for all $x > 0$. We denote the subclass of locally integrable N-functions by $\mathrm{N}_0(\Omega)$.

Remark 4.9.32. Let $\varphi \in \mathrm{N}_0$ and let $\hat{S}_0(\Omega)$ denote the set of all simple functions. Then $\hat{S}_0(\Omega) \subseteq L^{\varphi}(\Omega)$.

Definition 4.9.33. Let $\varphi \in \mathrm{N}(\Omega)$. We define the space $E^{\varphi}(\Omega)$ by

$$E^{\varphi}(\Omega) = \left\{u \in L^0(\Omega) \colon \rho_{\varphi}(\lambda u) < \infty \text{ for all } \lambda > 0\right\}.$$

Remark 4.9.34. Recall that

$$S(\Omega) = \left\{u \in L^0(\Omega) \colon \rho_{\varphi}(u) < \infty\right\}.$$

Then $E^{\varphi}(\Omega)$ is the largest linear space contained in $S(\Omega)$. Therefore, we have $E^{\varphi}(\Omega) \subseteq L^{\varphi}(\Omega)$.

Proposition 4.9.35. *If $\varphi \in \mathrm{N}_0(\Omega)$, then $\hat{S}_0(\Omega)$ is $\|\cdot\|$-dense in $E^{\varphi}(\Omega)$.*

Proof. Let $u \in E^{\varphi}(\Omega)$. Since $u = u^+ - u^-$, we may assume that $u \geq 0$. From Proposition 2.2.18 we know that we can find an increasing sequence $\{u_n\}_{n\in\mathbb{N}} \subseteq \hat{S}_0(\Omega)$ such that

$$0 \leq u_n \quad \text{for all } n \in \mathbb{N} \quad \text{and} \quad u_n(z) \nearrow u(z) \quad \text{for } \mu\text{-a.a. } z \in \Omega\,.$$

Then, by Corollary 4.9.23, we have $u_n \to u$ in $L^{\varphi}(\Omega)$. Therefore, $\hat{S}_0(\Omega)$ is $\|\cdot\|$-dense in $E^{\varphi}(\Omega)$. □

Proposition 4.9.36. *If $\varphi \in N_0(\Omega) \cap \Delta_2$ (see Definition 4.9.6), then $E^\varphi(\Omega) = L^\varphi(\Omega)$.*

Proof. We already know that $E^\varphi(\Omega) \subseteq L^\varphi(\Omega)$. So, we need to show that the opposite inclusion also holds. To this end, let $u \in L^\varphi(\Omega)$. Then we can find $\lambda_0 > 0$ such that $\rho_\varphi(\frac{u}{\lambda_0}) < \infty$.

Given $\eta > 0$, there exists $c_0 > 0$ such that

$$\rho_\varphi(u) \le c_0 + \int_{\{z\in\Omega:|u|\ge\eta\}} \varphi(z,|u|)\,dz\,.$$

If $\lambda > 0$, then we can find $m \in \mathbb{N}$ such that $2^m \ge \frac{\lambda_0}{\lambda}$. From the convexity of $\varphi(z,\cdot)$ and the Δ_2-condition, we have

$$\begin{aligned}\rho_\varphi(\lambda u) &\le c_0 + \frac{\lambda\lambda_0}{2^m} \int_{\{z\in\Omega:|u|\ge\eta\}} \varphi\Big(z, \frac{2^m|u|}{\lambda_0}\Big)\,dz\\ &\le c_0 + \frac{\lambda\lambda_0}{2^m}\Big(c^m \int_{\{z\in\Omega:|u|\ge\eta\}} \varphi\Big(z,\frac{|u|}{\lambda_0}\Big)\,dz + \|h\|_1 \sum_{k=0}^{m-1} c^k\Big) < \infty\,.\end{aligned}$$

Since $\lambda > 0$ is arbitrary, we conclude that $u \in E^\varphi(\Omega)$. Therefore, $E^\varphi(\Omega) = L^\varphi(\Omega)$. □

Corollary 4.9.37. *If $\varphi \in N_0(\Omega) \cap \Delta_2$, then $\hat{S}_0(\Omega)$ is $\|\cdot\|$-dense in $L^\varphi(\Omega)$.*

Corollary 4.9.38. *If $\varphi \in N_0(\Omega) \cap \Delta_2$ and $\Sigma = \sigma(\mathcal{L})$ with $\mathcal{L} \subseteq 2^\Omega$ countable, then $L^\varphi(\Omega)$ is separable.*

Remark 4.9.39. We call $\Sigma = \sigma(\mathcal{L})$ **countably generated.** Note that, by Proposition 2.3.25, (Σ, d_μ) is a separable metric space. This is the case if $\Omega \subseteq \mathbb{R}^N$ is a bounded set equipped with the Lebesgue measure on $\mathbb{R}^N$.

We can identify another dense subset of $L^\varphi(\Omega)$; see Harjulehto–Hästö [152, Theorem 3.7.14, p. 71].

Proposition 4.9.40. *If $\varphi \in N_0(\Omega) \cap \Delta_2$ and $0 < \eta \le \operatorname{ess\,inf}_\Omega \varphi(\cdot,1)$, then $C_c^\infty(\Omega)$ is dense in $L^\varphi(\Omega)$.*

Next we examine the reflexivity of $L^\varphi(\Omega)$.

Definition 4.9.41. (a) Let $\varphi \in N(\Omega)$. We say that φ is **uniformly convex**, if for every $\varepsilon > 0$, there exists $\delta > 0$ such that

$$\varphi\Big(z, \frac{|x+y|}{2}\Big) \le \frac{1-\delta}{2}(\varphi(z,|x|) + \varphi(z,|y|))$$

for a. a. $z \in \Omega$ and for all $x, y \in \mathbb{R}$ with $|x-y| \ge \varepsilon \max\{|x|,|y|\}$.

(b) Let $\varphi \in N(\Omega)$. We say that the modular function $\rho_\varphi(\cdot)$ is **uniformly convex**, if for every $\varepsilon > 0$, there exists $\delta > 0$ such that

$$\rho_\varphi\left(\frac{u+v}{2}\right) \le \frac{1-\delta}{2}(\rho_\varphi(u)+\rho_\varphi(v))$$

for all $u, v \in L^\varphi(\Omega)$ with

$$\rho_\varphi\left(\frac{u-v}{2}\right) > \frac{\varepsilon}{2}(\rho_\varphi(u)+\rho_\varphi(v)) .$$

Remark 4.9.42. If $x, y \in \mathbb{R}$ satisfy

$$|x-y| \le \varepsilon \max\{|x|, |y|\} \quad \text{for } \varepsilon \in (0,1) ,$$

then $\frac{1}{2}|x-y| \le \frac{\varepsilon}{2}|x+y|$. So, using the properties of $\varphi(z,\cdot)$, we have

$$\varphi\left(z, \frac{|x-y|}{2}\right) \le \frac{\varepsilon}{2}(\varphi(z,|x|)+\varphi(z,|y|)) \quad \text{for a. a. } z \in \Omega . \tag{4.9.20}$$

Theorem 4.9.43. *If $\varphi \in \mathrm{N}_0 \cap \Delta_2$ and is uniformly convex, then the Luxemburg norm $\|\cdot\|$ on $L^\varphi(\Omega)$ is uniformly convex.*

Proof. Let $\varepsilon \in (0,1)$ and $u, v \in L^\varphi(\Omega)$ be such that

$$\rho_\varphi\left(\frac{u-v}{2}\right) > \frac{\varepsilon}{2}(\rho_\varphi(u)+\rho_\varphi(v)) . \tag{4.9.21}$$

We introduce the set

$$D = \left\{z \in \Omega\colon |u(z)-v(z)| > \frac{\varepsilon}{2}\max\{|u(z)|, |v(z)|\}\right\}.$$

Using (4.9.20) we have

$$\begin{aligned} \rho_\varphi\left(\frac{u-v}{2}\chi_{\Omega\setminus D}\right) &\le \frac{\varepsilon}{4}(\rho_\varphi(u\chi_{\Omega\setminus D})+\rho_\varphi(v\chi_{\Omega\setminus D})) \\ &\le \frac{\varepsilon}{4}(\rho_\varphi(u)+\rho_\varphi(v)) . \end{aligned} \tag{4.9.22}$$

From (4.9.21) and (4.9.22) it follows that

$$\begin{aligned} \rho_\varphi\left(\frac{u-v}{2}\chi_D\right) &= \rho_\varphi\left(\frac{u-v}{2}\right) - \rho_\varphi\left(\frac{u-v}{2}\chi_{\Omega\setminus D}\right) \\ &> \frac{\varepsilon}{4}(\rho_\varphi(u)+\rho_\varphi(v)) . \end{aligned} \tag{4.9.23}$$

Also, using the uniform convexity of φ, we have

$$\rho_\varphi\left(\frac{u+v}{2}\chi_D\right) \le \frac{1-\delta}{2}(\rho_\varphi(u\chi_D)+\rho_\varphi(v\chi_D)) . \tag{4.9.24}$$

Note that, by (4.9.20)

$$0 \le \frac{1}{2}(\rho_\varphi(u\chi_{\Omega\setminus D}) + \rho_\varphi(v\chi_{\Omega\setminus D})) - \rho_\varphi\Big(\frac{u-v}{2}\chi_{\Omega\setminus D}\Big) .$$

Therefore, using this, the convexity of φ and (4.9.24) as well as (4.9.23), it follows that

$$\begin{aligned}\frac{1}{2}(\rho_\varphi(u) + \rho_\varphi(v)) - \rho_\varphi\Big(\frac{u+v}{2}\Big) &\ge \frac{1}{2}(\rho_\varphi(u\chi_D) + \rho_\varphi(v\chi_D)) - \rho_\varphi\Big(\frac{u+v}{2}\chi_D\Big)\\ \ge \frac{\delta}{2}(\rho_\varphi(u\chi_D) + \rho_\varphi(v\chi_D)) \ge \delta\rho_\varphi\Big(\frac{u-v}{2}\chi_D\Big) &\ge \frac{\delta\varepsilon}{4}(\rho_\varphi(u) + \rho_\varphi(v)) ,\end{aligned}$$

which proves that ρ_φ is uniformly convex.

Now suppose that

$$\|u\|, \|v\| \le 1 \quad \text{and} \quad \|u - v\| > \varepsilon . \tag{4.9.25}$$

This gives $\frac{1}{2}\|u - v\| > \frac{\varepsilon}{2}$ and so $\rho_\varphi(\frac{u-v}{2}) > \eta$ for some $\eta = \eta(\varepsilon) > 0$. Then, by (4.9.25) and Proposition 4.9.16, we infer

$$\rho_\varphi\Big(\frac{u-v}{2}\Big) > \frac{\eta}{2}(\rho_\varphi(u) + \rho_\varphi(v)) .$$

From the first part of the proof we know that ρ_φ is uniformly convex. So there exists $\vartheta = \vartheta(\eta) > 0$ such that

$$\rho_\varphi\Big(\frac{u+v}{2}\Big) \le \frac{1-\vartheta}{2}(\rho_\varphi(u) + \rho_\varphi(v)) \le 1 - \vartheta , \tag{4.9.26}$$

where we have used (4.9.25) in the last inequality.

Using the Δ_2-condition, we show that (4.9.26) implies

$$\Big\|\frac{u+v}{2}\Big\| \le 1 - \delta_0 \quad \text{for some } \delta_0 = \delta_0(c, \vartheta) > 0$$

with $c > 0$ being the constant from the Δ_2-condition. This proves that $\|\cdot\|$ is uniformly convex. □

Applying the Milman–Pettis Theorem given in Theorem 3.4.28, we infer the following corollary.

Corollary 4.9.44. *If $\varphi \in \mathrm{N}_0 \cap \Delta_2$ and is uniformly convex, then $L^\varphi(\Omega)$ is reflexive.*

Remark 4.9.45. We can have reflexivity under weaker conditions. Indeed, note that if $\varphi \in \mathrm{N}(\Omega)$, then $E^\varphi(\Omega)^* = L^{\varphi^*}(\Omega)$. So, if $\varphi \in \Delta_2$, $\varphi^* \in \Delta_2$, since $\varphi = \varphi^{**}$, it follows that $L^\varphi(\Omega)$ is reflexive; see Proposition 4.9.36.

Let

$$p \in L^{\infty}(\Omega) \quad \text{with } 1 < p_- = \operatorname*{ess\,inf}_{\Omega} p \le \operatorname*{ess\,sup}_{\Omega} = p_+ < \infty\,, \tag{4.9.27}$$

and

$$\varphi(z,x) = x^{p(z)} \quad \text{for all } z \in \Omega \text{ and for all } x \ge 0. \tag{4.9.28}$$

Then $\varphi \in \mathrm{N}_0(\Omega) \cap \Delta_2$ and is uniformly convex. We denote the corresponding generalized Orlicz space by $L^{p(\cdot)}(\Omega)$.

Proposition 4.9.46. *If φ is given in* (4.9.28) *such that* (4.9.27) *is satisfied, then $L^{p(\cdot)}(\Omega)$ is uniformly convex (thus reflexive) and separable.*

Remark 4.9.47. Conversely, we can show that if $p \in L^{\infty}(\Omega)$ with $1 \le p_- \le p_+ < \infty$ and $L^{p(\cdot)}(\Omega)$ is reflexive, then $1 < p_- \le p_+ < \infty$.

Next, let

$$a \in L^{\infty}(\Omega) \quad \text{with } a(z) \ge 0 \text{ for } \mu\text{-a. a. } z \in \Omega \quad \text{and} \quad 1 < q < p < \infty\,, \tag{4.9.29}$$

and

$$\varphi(z,x) = a(z)x^p + x^q \quad \text{for all } z \in \Omega \text{ and for all } x \ge 0\,. \tag{4.9.30}$$

Then $\varphi \in \mathrm{N}_0 \cap \Delta_2$. This is the double phase integrand, which, as we pointed out in Example 4.9.3 (b), exhibits unbalanced growth in the $x \in \mathbb{R}_+$-variable. Note that φ is uniformly convex. Therefore, we have the following proposition.

Proposition 4.9.48. *If φ is given in* (4.9.30) *such that* (4.9.29) *is satisfied, then $L^{\varphi}(\Omega)$ is uniformly convex (thus reflexive) and separable.*

Next we present some useful facts about these two special cases of generalized Orlicz spaces generated by the $\mathrm{N}(\Omega)$-functions given in (4.9.28) and (4.9.30), respectively.

In what follows ρ_p denotes the modular function on $L^{p(\cdot)}(\Omega)$ defined by

$$\rho_p(u) = \int_{\Omega} |u|^{p(z)}\,\mathrm{d}z \quad \text{for all } u \in L^{p(\cdot)}(\Omega)\,.$$

Moreover, ρ_{φ} is the modular function corresponding to the double phase integrand given by

$$\rho_{\varphi}(u) = \int_{\Omega} (a(z)|u|^p + |u|^q)\,\mathrm{d}z \quad \text{for all } u \in L^{\varphi}(\Omega)\,.$$

Proposition 4.9.49. *If* $u \in L^{p(\cdot)}(\Omega)$, $\lambda > 0$ *and* (4.9.27) *is satisfied, then the following hold:*
(a) $\|u\| \le 1$ *implies* $\|u\|^{p_+} \le \rho_p(u) \le \|u\|^{p_-}$;
(b) $\|u\| \ge 1$ *implies* $\|u\|^{p_-} \le \rho_p(u) \le \|u\|^{p_+}$;
(c) $\|u\| = \lambda$ *if and only if* $\rho_p(\frac{u}{\lambda}) = 1$.

Proof. (a) Let $\|u\| \le 1$. Then we have

$$\int_\Omega \left(\frac{|u|}{\|u\|}\right)^{p(z)} dz \le 1\,,$$

which implies, due to $\|u\| \le 1$, that

$$\frac{1}{\|u\|^{p_-}}\rho_p(u) \le 1\,.$$

So, we have $\rho_p(u) \le \|u\|^{p_-}$.

We know that $\|u\| \le 1$ if and only if $\rho_p(u) \le 1$; see Proposition 4.9.16. Note that

$$\|u\| \le \rho_p(u)^{\frac{1}{p_+}} \iff \left\|\frac{u}{\rho_p(u)^{\frac{1}{p_+}}}\right\| \le 1 \iff \rho_p\left(\frac{u}{\rho_p(u)^{\frac{1}{p_+}}}\right) \le 1\,; \tag{4.9.31}$$

see Proposition 4.9.16. Indeed, observe that

$$\rho_p(u)^{-\frac{p(z)}{p_+}} \le \rho_p(u)^{-1} \quad \text{for all } z \in \Omega\,.$$

So, the last inequality in (4.9.31) is clear. Therefore, we obtain $\|u\|^{p_+} \le \rho_p(u)$.

(b) The proof is similar to that of (a).

(c) If $\|u\| = \lambda$, then $\rho_p(\frac{u}{\lambda}) \le 1$. Suppose that the inequality is strict, that is, $\rho_p(\frac{u}{\lambda}) < 1$. Exploiting the continuity of $\vartheta \to \rho_p(\frac{u}{\vartheta})$ on $(0,\infty)$, we can find $\lambda' \in (0,\lambda)$ such that $\rho_p(\frac{u}{\lambda'}) \le 1$. Then, by Proposition 4.9.16, we obtain $\|u\| \le \lambda' < \lambda$, a contradiction. Therefore, $\rho_p(\frac{u}{\lambda}) = 1$.

Conversely, suppose that $\rho_p(\frac{u}{\lambda}) = 1$. Then once again Proposition 4.9.16 implies $\|u\| \le \lambda$. If $\|u\| < \lambda$, then let $\lambda' \in (\|u\|,\lambda)$. We have $\rho_p(\frac{u}{\lambda'}) \le 1$ which implies $\rho_p(\frac{u}{\lambda}) < 1$, a contradiction. □

A similar result is also valid for the double phase space $L^\varphi(\Omega)$ with φ given in (4.9.30), the proof is similar to that of Proposition 4.9.49.

Proposition 4.9.50. *If* $u \in L^\varphi(\Omega)$ *with* φ *given in* (4.9.30) *such that* (4.9.29) *is satisfied and* $\lambda > 0$, *then the following hold:*
(a) $\|u\| \le 1$ *implies* $\|u\|^p \le \rho_\varphi(u) \le \|u\|^q$;
(b) $\|u\| \ge 1$ *implies* $\|u\|^q \le \rho_\varphi(u) \le \|u\|^p$;
(c) $\|u\| = \lambda$ *if and only if* $\rho_\varphi(\frac{u}{\lambda}) = 1$.

As a direct consequence of Proposition 4.9.16, we obtain the following result.

Proposition 4.9.51. (a) *Let (4.9.27) be satisfied. If $u \in L^{p(\cdot)}(\Omega)$, then $\|u\| < 1$ (resp. $= 1$, > 1) if and only if $\rho_p(u) < 1$ (resp. $= 1$, > 1).*
(b) *Let (4.9.29) be satisfied. If $u \in L^{\varphi}(\Omega)$ with φ given in (4.9.30), then $\|u\| < 1$ (resp. $= 1$, > 1) if and only if $\rho_\varphi(u) < 1$ (resp. $= 1$, > 1).*

We return to the general case. We recall Corollary 4.9.23 as Proposition and give an alternative proof of it.

Proposition 4.9.52. *If $\varphi \in N(\Omega)$ and $\{u_n, u\}_{n\in\mathbb{N}} \subseteq L^{\varphi}(\Omega)$, then $\|u_n - u\| \to 0$ as $n \to \infty$ if and only if $\rho_\varphi(\frac{u_n-u}{\lambda}) \to 0$ as $n \to \infty$ for all $\lambda > 0$.*

Proof. Considering $v_n = u_n - u$ with $n \in \mathbb{N}$, we see that, without any loss of generality, we may assume that $u = 0$. So we have

$$u_n \to 0 \quad \text{in } L^{\varphi}(\Omega) \text{ as } n \to \infty\,.$$

Then, for any $\lambda > 0$, we have $\|\frac{u_n}{\lambda}\| \to 0$ as $n \to \infty$ and so we can find $n_0 \in \mathbb{N}$ such that

$$\left\|\frac{u_n}{\lambda}\right\| \le 1 \quad \text{for all } n \ge n_0\,. \tag{4.9.32}$$

From (4.9.32) and Corollary 4.9.17 (a), we have

$$\rho_\varphi\left(\frac{u_n}{\lambda}\right) \le \left\|\frac{u_n}{\lambda}\right\| \quad \text{for all } n \ge n_0\,,$$

which implies

$$\lim_{n\to\infty} \rho_\varphi\left(\frac{u_n}{\lambda}\right) = 0\,.$$

Next, we prove the converse. So, suppose that

$$\lim_{n\to\infty} \rho_\varphi\left(\frac{u_n}{\lambda}\right) = 0 \quad \text{for all } \lambda > 0\,.$$

Then there exists a number $n_0 \in \mathbb{N}$ such that

$$\rho_\varphi\left(\frac{u_n}{\lambda}\right) \le 1 \quad \text{for all } n \ge n_0\,,$$

which yields $\|u_n\| \le \lambda$; see Proposition 4.9.16. Since $\lambda > 0$ is arbitrary, we conclude that $\|u\| \to 0$ as $n \to \infty$. □

Definition 4.9.53. Let $\varphi \in N(\Omega)$ and $\{u_n, u\}_{n\in\mathbb{N}} \subseteq L^{\varphi}(\Omega)$. We say that the u_n's **converge modularly** to u, denoted by $u_n \xrightarrow{\text{m}} u$, if there exists $\lambda > 0$ such that

$$\lim_{n\to\infty} \rho_\varphi\left(\frac{u_n - u}{\lambda}\right) = 0\,.$$

Remark 4.9.54. On account of Proposition 4.9.52, we see that norm convergence implies modular convergence. The converse is true if $\varphi \in \Delta_2$; see Proposition 4.9.36.

Recall that both the variable exponent function

$$\varphi(z,x) = x^{p(z)} \quad \text{for all } z \in \Omega \text{ and for all } x \geq 0$$

with $p \in L^\infty(\Omega)$, $1 \leq p_- \leq p_+ < \infty$ and the double phase function

$$\varphi(z,x) = a(z)x^p + x^q \quad \text{for all } z \in \Omega \text{ and for all } x \geq 0$$

with $a \in L^\infty(\Omega)$, $a(z) \geq 0$ for a. a. $z \in \Omega$ and $1 \leq q < p < \infty$, satisfy the Δ_2-condition. Therefore, Remark 4.9.54 leads to the following result.

Proposition 4.9.55. (a) *Let* (4.9.27) *be satisfied. If* $\{u_n, u\}_{n\in\mathbb{N}} \subseteq L^{p(\cdot)}(\Omega)$, *then* $\|u_n - u\| \to 0$ *as* $n \to \infty$ *if and only if* $u_n \xrightarrow{\mathrm{m}} u$.
(b) *Let* (4.9.29) *be satisfied. If* $\{u_n, u\}_{n\in\mathbb{N}} \subseteq L^\varphi(\Omega)$ *with* φ *given in* (4.9.30), *then* $\|u_n - u\| \to 0$ *as* $n \to \infty$ *if and only if* $u_n \xrightarrow{\mathrm{m}} u$.

Proposition 4.9.56. (a) *Let* (4.9.27) *be satisfied. If* $\{u_n\}_{n\in\mathbb{N}} \subseteq L^{p(\cdot)}(\Omega)$, *then* $\|u_n\| \to \infty$ *as* $n \to \infty$ *if and only if* $\rho_p(u_n) \to \infty$ *as* $n \to \infty$.
(b) *Let* (4.9.29) *be satisfied. If* $\{u_n\}_{n\in\mathbb{N}} \subseteq L^\varphi(\Omega)$ *with* φ *given in* (4.9.30), *then* $\|u_n\| \to \infty$ *as* $n \to \infty$ *if and only if* $\rho_\varphi(u_n) \to \infty$ *as* $n \to \infty$.

Proof. (a) Suppose $\|u_n\| \to \infty$. Then we can find $n_0 \in \mathbb{N}$ such that $\|u\| \geq M > 1$ for all $n \geq n_0$. This implies

$$\rho_p(u_n) \geq \|u_n\| \geq M > 1 \quad \text{for all } n \geq n_0 ;$$

see Corollary 4.9.17 (b). Hence, $\rho_p(u_n) \to \infty$.

For the converse implication, assume that $\rho_p(u_n) \to \infty$. Then, for a given $M \geq 1$, we can find $n_0 \in \mathbb{N}$ such that $\rho_p(u_n) \geq M^{p_+}$ for all $n \geq n_0$. Since $M \geq 1$, we obtain

$$\rho_p\left(\frac{u_n}{M}\right) \geq 1 \quad \text{for all } n \geq n_0 .$$

This shows that $\|u_n\| \geq M$ for all $n \geq n_0$ and so $\|u_n\| \to \infty$.

(b) This can be shown as in (a) using similar arguments. □

In the treatment of partial differential equations we often encounter the situation where the gradient of the unknown function belongs to a generalized Orlicz space. This leads to **generalized Orlicz–Sobolev spaces**, also known as **Musielak–Orlicz–Sobolev spaces.**

Now, let $\Omega \subseteq \mathbb{R}^N$, $N \geq 1$, be a bounded domain with a Lipschitz boundary $\partial\Omega$. We will focus on these spaces which correspond to the variable exponent function and the double phase function. Precisely, we consider

$$\begin{aligned}&\varphi(z,x) = x^{p(z)} \quad \text{for all } z \in \Omega \text{ and for all } x \geq 0\,,\\ &\text{with} \quad p \in C(\overline{\Omega}) \quad \text{and} \quad 1 < p_- \leq p_+ < N\,,\end{aligned} \tag{4.9.33}$$

and

$$\begin{aligned}&\varphi(z,x) = a(z)x^p + x^q \quad \text{for all } z \in \Omega \text{ and for all } x \geq 0\,,\\ &\text{with} \quad a \in L^\infty(\Omega), \quad a(z) \geq 0 \text{ for } \mu\text{-a. a. } z \in \Omega \quad \text{and} \quad 1 < q < p < N\,.\end{aligned} \tag{4.9.34}$$

Definition 4.9.57. (a) For φ given in (4.9.33), the variable exponent Sobolev space $W^{1,p(\cdot)}(\Omega)$ is defined by

$$W^{1,p(\cdot)}(\Omega) = \{u \in L^{p(\cdot)}(\Omega) \colon |Du| \in L^{p(\cdot)}(\Omega)\}$$

with Du being the weak gradient of u. We endow this space with the norm

$$\|u\|_{1,p(\cdot)} = \|u\|_{p(\cdot)} + \|Du\|_{p(\cdot)} \quad \text{for all } u \in W^{1,p(\cdot)}(\Omega)\,,$$

where $\|Du\|_{p(\cdot)} = \|\,|Du|\,\|_{p(\cdot)}$. Also, we introduce the space

$$W_0^{1,p(\cdot)}(\Omega) = \overline{C_c^\infty(\Omega)}^{\|\cdot\|_{1,p(\cdot)}}\,.$$

(b) For φ given in (4.9.34), the double phase space $W^{1,\varphi}(\Omega)$ is defined by

$$W^{1,\varphi}(\Omega) = \{u \in L^{\varphi}(\Omega) \colon |Du| \in L^{\varphi}(\Omega)\}\,,$$

where as before Du is the weak gradient of u. We equip this space with the norm

$$\|u\|_{1,\varphi} = \|u\|_{\varphi} + \|Du\|_{\varphi} \quad \text{for all } u \in W^{1,\varphi}(\Omega)\,,$$

with $\|Du\|_{\varphi} = \|\,|Du|\,\|_{\varphi}$. Moreover, we introduce the space

$$W_0^{1,\varphi}(\Omega) = \overline{C_c^\infty(\Omega)}^{\|\cdot\|_{1,\varphi}}\,.$$

Let $V = L^{p(\cdot)}(\Omega) \times L^{p(\cdot)}(\Omega)$ or $V = L^{\varphi}(\Omega) \times L^{\varphi}(\Omega)$ under the assumptions (4.9.33) or (4.9.34), respectively. Then V is separable and reflexive, in fact uniformly convex. We consider the map $K \in L(X, V)$ with $X = W^{1,p(\cdot)}(\Omega)$ or $X = W^{1,\varphi}(\Omega)$, defined by $K(u) = (u, |Du|)$. This is an isometry into V and so $K(X) \subseteq V$ is closed. Therefore $K(X)$ is separable and reflexive, hence so are $W^{1,p(\cdot)}(\Omega)$ and $W^{1,\varphi}(\Omega)$. Also they are uniformly convex. The similar conclusion holds for $W_0^{1,p(\cdot)}(\Omega)$ and $W_0^{1,\varphi}(\Omega)$. Therefore, we can state the following proposition.

Proposition 4.9.58. *Let* (4.9.33) *or* (4.9.34) *be satisfied. Then the spaces* $W^{1,p(\cdot)}(\Omega)$, $W_0^{1,p(\cdot)}(\Omega)$ *and* $W^{1,\varphi}(\Omega)$, $W_0^{1,\varphi}(\Omega)$ *are Banach spaces which are separable and reflexive, in fact uniformly convex.*

Let

$$p^*(z) = \frac{Np(z)}{N - p(z)} \quad \text{for all } z \in \overline{\Omega}\,.$$

Using Theorems 4.9.26 and 4.5.25, we have the following result.

Proposition 4.9.59. *Let* (4.9.33) *or* (4.9.34) *be satisfied.*

(a) *If* $s \in C(\overline{\Omega})$ *with* $1 \le s(z) \le p(z)$ *for all* $z \in \overline{\Omega}$, *then we have the continuous embeddings*

$$L^{p(\cdot)}(\Omega) \hookrightarrow L^{s(\cdot)}(\Omega), \quad W^{1,p(\cdot)}(\Omega) \hookrightarrow W^{1,s(\cdot)}(\Omega), \quad W_0^{1,p(\cdot)}(\Omega) \hookrightarrow W_0^{1,s(\cdot)}(\Omega)\,.$$

(b) *If* $1 \le s \le q$, *then we have the continuous embeddings*

$$L^{\varphi}(\Omega) \hookrightarrow L^{s}(\Omega), \quad W^{1,\varphi}(\Omega) \hookrightarrow W^{1,s}(\Omega), \quad W_0^{1,\varphi}(\Omega) \hookrightarrow W_0^{1,s}(\Omega)\,.$$

(c) *If* $s \in C(\overline{\Omega})$ *with* $1 \le s(z) < p^*(z)$ *for all* $z \in \overline{\Omega}$, *then the embeddings*

$$W^{1,p(\cdot)}(\Omega) \hookrightarrow L^{s(\cdot)}(\Omega) \quad \text{and} \quad W_0^{1,p(\cdot)}(\Omega) \hookrightarrow L^{s(\cdot)}(\Omega)$$

are compact.

(d) *If* $1 \le s < q^* = \frac{Nq}{N-q}$, *then the embeddings*

$$W^{1,\varphi}(\Omega) \hookrightarrow L^{s}(\Omega) \quad \text{and} \quad W_0^{1,\varphi}(\Omega) \hookrightarrow L^{s}(\Omega)$$

are compact.

We also have the compact embedding for the double phase case, see [78, Proposition 2.18].

Proposition 4.9.60. *Let* (4.9.34) *be satisfied and suppose in addition that* $p < q^*$ *holds. Then* $W_0^{1,\varphi}(\Omega) \hookrightarrow L^{\varphi}(\Omega)$ *compactly.*

For the spaces $W_0^{1,p(\cdot)}(\Omega)$ and $W_0^{1,\varphi}(\Omega)$ the Poincaré inequality holds.

Proposition 4.9.61. *Let* (4.9.33) *or* (4.9.34) *be satisfied, and suppose that in* (4.9.34) $p < q^*$ *holds. Then there exists* $c > 0$ *such that*

$$\begin{aligned} \|u\|_{p(\cdot)} &\le c\|Du\|_{p(\cdot)} \quad \text{for all } u \in W_0^{1,p(\cdot)}(\Omega),\\ \|u\|_{\varphi} &\le c\|Du\|_{\varphi} \quad \text{for all } u \in W_0^{1,\varphi}(\Omega)\,. \end{aligned}$$

Proof. The proof is identical in both cases. We do it using the notation of the double phase case. So, suppose that the assertion of the proposition is not true. Then we can find a sequence $\{u_n\}_{n\in\mathbb{N}} \subseteq W_0^{1,\varphi}(\Omega)$ such that

$$\|u_n\|_{\varphi} > n\|Du_n\|_{\varphi} \quad \text{for all } n \in \mathbb{N}\,. \tag{4.9.35}$$

Let $y_n = \frac{u_n}{\|u_n\|_\varphi}$ for all $n \in \mathbb{N}$. Then from (4.9.35) we have

$$\|Dy_n\|_\varphi < \frac{1}{n} \leq 1 \quad \text{and} \quad \|y_n\|_\varphi = 1 \quad \text{for all } n \in \mathbb{N}\,. \tag{4.9.36}$$

From (4.9.36) and Proposition 4.9.58, we see that we may assume that

$$y_n \xrightarrow{\mathrm{w}} y \quad \text{in } W_0^{1,\varphi}(\Omega)\,, \tag{4.9.37}$$

which by Proposition 4.9.60 gives $y_n \to y$ in $L^\varphi(\Omega)$. Using this implies

$$\|y\|_\varphi = 1\,. \tag{4.9.38}$$

From (4.9.36) and (4.9.37) we have

$$\|Dy\|_\varphi \leq \liminf_{n\to\infty} \|Dy_n\|_\varphi = 0\,.$$

But then, by Proposition 4.9.59, we have $y = 0$, a contradiction to (4.9.38). So the Poincaré inequality holds. ☐

The next result is a modular version of the Kadec–Klee property; see Definition 3.4.30. In what follows, let $X = L^{p(\cdot)}(\Omega)$ or $W^{1,p(\cdot)}(\Omega)$ or $W_0^{1,p(\cdot)}(\Omega)$ under (4.9.33) and $X = L^\varphi(\Omega)$ or $W^{1,\varphi}(\Omega)$ or $W_0^{1,\varphi}(\Omega)$ under (4.9.34).

From the proof of Theorem 4.9.43 we know that if $\varphi \in \mathrm{N}(\Omega)$ is uniformly convex, then for a given $\varepsilon > 0$, there exists $\delta > 0$ such that

$$\begin{aligned} &\rho_\varphi\left(\frac{u-v}{2}\right) \leq \frac{\varepsilon}{2}(\rho_\varphi(u) + \rho_\varphi(v)) \quad \text{or} \\ &\rho_\varphi\left(\frac{u+v}{2}\right) \leq \frac{1-\delta}{2}(\rho_\varphi(u) + \rho_\varphi(v)) \end{aligned} \tag{4.9.39}$$

for all $u, v \in L^\varphi(\Omega)$.

In the next proposition, ρ stands for the variable exponent modular function ρ_p or for the double phase modular function ρ_φ.

Proposition 4.9.62. *If* $\{u_n\}_{n\in\mathbb{N}} \subseteq X$, $u_n \xrightarrow{\mathrm{w}} u$ *in* X *and* $\rho(u_n) \to \rho(u)$, *then* $u_n \to u$ *in* X.

Proof. We claim that

$$\rho\left(\frac{u_n - u}{2}\right) \to 0 \quad \text{as } n \to \infty\,. \tag{4.9.40}$$

Arguing by contradiction, suppose that (4.9.40) is not true. Then we can find a subsequence $\{u_{n_k}\}_{k\in\mathbb{N}}$ of $\{u_n\}_{n\in\mathbb{N}}$ and $\varepsilon > 0$ such that

$$\rho\left(\frac{u_{n_k} - u}{2}\right) > \varepsilon \quad \text{for all } k \in \mathbb{N}\,. \tag{4.9.41}$$

Then (4.9.41) implies that there is at least a subsequence satisfying the second inequality in (4.9.39). We have

$$\rho\left(\frac{u_{n_k}+u}{2}\right) \leq \frac{1-\delta}{2}(\rho(u_{n_k})+\rho(u)) . \tag{4.9.42}$$

Exploiting the sequential weak lower semicontinuity of the modular function ρ and the hypothesis that $\rho(u_n) \to \rho(u)$, passing to the limit in (4.9.42) as $n \to \infty$, we obtain

$$\rho(u) \leq \liminf_{n\to\infty} \rho\left(\frac{u_{n_k}+u}{2}\right) \leq (1-\delta)\rho(u)$$

and so

$$\rho(u) = 0 . \tag{4.9.43}$$

The convexity of the modular function ρ implies

$$\rho\left(\frac{u_{n_k}-u}{2}\right) \leq \frac{1}{2}(\rho(u_{n_k})+\rho(u)) \quad \text{for all } n \in \mathbb{N} .$$

Passing to the limit as $n \to \infty$ and using the fact that $\rho(u_{n_k}) \to 0$ because of (4.9.43), we obtain (4.9.40). From (4.9.40) and Proposition 4.9.55, we conclude that $u_n \to u$ in X. □

On account of the Poincaré inequality in Proposition 4.9.62 we have the following corollary.

Corollary 4.9.63. *Let* (4.9.33) *or* (4.9.34) *be satisfied. If* $\{u_n\}_{n\in\mathbb{N}} \subseteq W_0^{1,p(\cdot)}(\Omega)$ *or* $W_0^{1,\varphi}(\Omega)$ *and*

$$Du_n \xrightarrow{w} Du \quad \text{in } L^{p(\cdot)}(\Omega, \mathbb{R}^N) \text{ or } L^{\varphi}(\Omega, \mathbb{R}^N) ,$$
$$\rho_p(Du_n) \to \rho_p(Du) \quad \text{or} \quad \rho_\varphi(Du_n) \to \rho_\varphi(Du) ,$$

then $u_n \to u$ *in* $W_0^{1,p(\cdot)}(\Omega)$ *or in* $W_0^{1,\varphi}(\Omega)$.

Finally, using Corollary 4.5.19, we have the following properties of the spaces $X = W^{1,p(\cdot)}, W_0^{1,p(\cdot)}(\Omega)$ or $X = W^{1,\varphi}(\Omega), W_0^{1,\varphi}(\Omega)$ under (4.9.33) or (4.9.34).

Proposition 4.9.64. (a) *If* $u \in X$, *then* $u^\pm \in X$ *and* $D(u^\pm) = \pm(Du)\chi_{\{\pm u>0\}}$.
(b) *The map* $u \to u^\pm$ *is continuous on* X.

4.10 Remarks

(4.1) As we already mentioned in Section 2.3, $L^p[0,1]$ with $p \neq 2$ is a Banach space, which was proven by Riesz [269, 270] while the completeness of $L^2[0,1]$ was established by Fischer [122] and Riesz [268]. Theorem 4.1.3 for the Hilbert space case, that is, $p = 2$, was established simultaneously by Fréchet [129] and Riesz [267] for $\Omega = [0,1]$, that is, for $L^2[0,1]$. For $L^p[0,1]$ with $1 < p < \infty$, the result was proven by Riesz [270]. For a finite measure space (Ω, Σ, μ) the result was proven by Dunford [103] and for an arbitrary measure space by McShane [229]. Theorem 4.1.5 is due to Schwartz [292]. The uniform convexity of the L^p-spaces with $1 < p < \infty$ stated in Proposition 4.1.6 is usually proven using the Clarkson inequalities; see Clarkson [75] and Hewitt–Stromberg [160, p. 225, p. 227]. Theorem 4.1.18 is due to Dunford–Pettis [104]. The notion of uniform integrability is very important with applications to different parts of mathematical analysis. The following result provides a sufficient condition for uniform integrability of a sequence in $L^1(\Omega)$.

Proposition 4.10.1. *If (Ω, Σ, μ) is a finite measure space and $\{f_n\}_{n\geq 1} \subseteq L^1(\Omega)$ is bounded such that*

$$\lim_{n\to\infty} \int_A f_n \, d\mu = 0 \quad \text{for all } A \in \Sigma \,,$$

then $\{f_n\}_{n\geq 1} \subseteq L^1(\Omega)$ is uniformly integrable.

In fact the proof of the result above leads to the following interesting proposition.

Proposition 4.10.2. *If (Ω, Σ, μ) is a finite measure space and $C \subseteq L^1(\Omega)$ is bounded but not uniformly integrable, then there exist sequences $\{f_n\}_{n\geq 1} \subseteq C$ and $\{A_n\}_{n\geq 1} \subseteq \Sigma$ mutually disjoint and $\varepsilon_0 > 0$ such that*

$$\left| \int_{A_n} f_n \, d\mu \right| \geq \varepsilon_0 \quad \text{for all } n \in \mathbb{N} \,.$$

In the problems of Chapter 2 the reader can find alternative equivalent definitions of the notion of uniform integrability. Proposition 4.1.21 and Lemma 4.1.22 are due to Brézis–Lieb [54]. The notion of bitting convergence (see Definition 4.1.25) is due to Chacon, and Theorem 4.1.24 can be found in Brooks–Chacon [57]. Mollification techniques (see Definition 4.1.27) are standard especially in the theory of Sobolev space; see Adams [1] and Brézis [52]. More detailed accounts on the Banach spaces of sequences can be found in Dunford–Schwartz [105] and Lindenstrauss–Tzafriri [217].

(4.2) Integration theories for vector valued functions appeared in the 1930s as a tool in the study of the differentiation properties of vector-valued functions. A second revival of the subject occurred in the 1970s with the study of the geometry of Banach spaces. Nowadays vector-valued integration and the associated Lebesgue–Bochner spaces constitute a mature subject in mathematical analysis with many applications such as in the theory of infinite dimensional dynamical systems and in the theory of infinite dimensional stochastic processes. The Pettis Measurability Theorem (see Theorem 4.2.4) can be found in Pettis [259]. The Bochner integral (see Definition 4.2.8) was first introduced by Bochner [38]. Various parts of the theory of Bochner integration can be found in the books of Denkowski–Migórski–Papageorgiou [87], Diestel–Uhl [91], Gasiński–Papageorgiou [135], Hille–Phillips [165], and Schwabik–Ye [291]. Two basic references for the theory of vector-valued measures (see Definition 4.2.23) are the books of Diestel–Uhl [91] and Dinculeanu [96]. For the notion of RNP (see Definition 4.2.23 (d)) we refer to Diestel–Uhl [91]. Theorem 4.2.26 is essentially due to Bochner–Taylor [39] although not formulated in terms of the RNP. The book of A. and C. Ionescu Tulcea [178] is a very good reference for the notion of linear lifting; see Theorem 4.2.32. Theorem 4.2.37 is due to Dieudonné [93] and Dinculeanu–Foiaş [97] and can be found in the book of A. and C. Ionescu Tulcea [178, p. 95]. Another structural result for the Lebesgue–Bochner space $L^1(\Omega, X)$ is the following one due to Talagrand [308]. Recall that a Banach space is weakly complete if every weakly Cauchy sequence in X converges.

Proposition 4.10.3. *If (Ω, Σ, μ) is a finite measure space and X is a Banach space, which is weakly sequentially complete, then $L^1(\Omega, X)$ is weakly sequentially complete.*

Evolution triples (see Definition 4.2.39) are a basic tool in the study of evolution equations; see Hu–Papageorgiou [174], Lions [218], Roubíček [282], and Zeidler [340]. Lemma 4.2.48 is due to Ehrling [110] and in various forms it is used extensively in the theory of evolution equations. Theorem 4.2.49 is due to Aubin [19] and Lions [218]. For extensions of this result we refer to Roubíček [282, pp. 208, 211].

The Pettis integral was first introduced by Pettis [259]. Interest for it was revived in the 1970s in order to develop an integration theory for functions that are only weakly measurable or for which $\|f(\cdot)\|$ is not integrable. Detailed accounts on the theory of the Pettis integral can be found in Musial [249] and Talagrand [307].

(4.3) The notion of a function of bounded variation (see Definition 4.3.22) goes back to Jordan [183] who also proved that such a function is the difference of two increasing functions. For the theory of monotone functions we refer to Leoni [211] and Natanson [251]. Theorem 4.3.13 is due to Vitali [321]. Covering Theorems like Theorem 4.3.13 are important in geometric measure theory.

We can extend the notion of bounded variation to functions of many variables.

Definition 4.10.4. Let $\Omega \subseteq \mathbb{R}^N$ be an open set. We say that $f \in L^1(\Omega)$ is of **bounded variation** in Ω if

$$\sup\left[\int_\Omega f \operatorname{div}\vartheta \, dz : \vartheta \in C_c^1(\Omega, \mathbb{R}^N) \text{ with } |\vartheta(x)| \le 1 \text{ for all } z \in \Omega\right] < \infty .$$

By $\mathrm{BV}(\Omega)$ we denote the space of all functions on Ω that are of bounded variation. A function $f \in L^1_{\mathrm{loc}}(\Omega)$ has **locally bounded variation** in Ω if for each open U with $U \subset\subset \Omega$ we have $f \in \mathrm{BV}(U)$. We denote the space of such functions by $\mathrm{BV}_{\mathrm{loc}}(\Omega)$. A Lebesgue measurable set $A \subseteq \Omega$ has a **finite perimeter** (resp. **locally finite perimeter**) if $\chi_A \in \mathrm{BV}(\Omega)$ (resp. $\chi_A \in \mathrm{BV}_{\mathrm{loc}}(\Omega)$).

The next theorem is the basic structural result for the space $\mathrm{BV}_{\mathrm{loc}}(\Omega)$.

Theorem 4.10.5. *If $f \in \mathrm{BV}_{\mathrm{loc}}(\Omega)$, then there exists a Radon measure μ on Ω and a μ-measurable function $\xi: \Omega \to \mathbb{R}^N$ such that*
(a) $|\xi(z)| = 1$ *μ-a. e.;*
(b) $\int_\Omega f \operatorname{div} \vartheta \, dz = -\int_\Omega (\vartheta, \xi)_{\mathbb{R}^N} \, d\mu$ *for all* $\vartheta \in C_c^1(\Omega, \mathbb{R}^N)$.

Remark 4.10.6. We usually write $\mu = \|Df\|$. Then $f \in \mathrm{BV}(\Omega)$ if and only if $\|Df\|(\Omega) < \infty$ and $\|f\|_{\mathrm{BV}} = \|f\|_1 + \|Df\|(\Omega)$ is a norm on $\mathrm{BV}(\Omega)$ making it a Banach space.

For further details on the space $\mathrm{BV}(\Omega)$ we refer to Evans–Gariepy [116], Leoni [211], and Ziemer [342].

(4.4) The notion of absolutely continuous functions is due to Lebesgue [208] and Vitali [321]. Lebesgue proved the fundamental theorem of the calculus for the Lebesgue integral while Vitali showed that a function is absolutely continuous if and only if it is an indefinite integral of an L^1-function; see Theorem 4.4.21. Extensions to vector valued functions can be found in Diestel–Uhl [91].

(4.5) The material on Sobolev spaces is standard and can be found in many books such as Adams [1], Brézis [52], Evans–Gariepy [116], and Leoni [211]. Let us also mention a result describing the dual of $W_0^{1,p}(\Omega)$; see Adams [1, Theorem 3.10, p. 50].

Theorem 4.10.7. *If $\Omega \subseteq \mathbb{R}^N$ is open and $1 \le p < \infty$, then*

$$W_0^{1,p}(\Omega)^* = \left\{ g \in C_c^\infty(\Omega)^* : g = -\sum_{k=1}^N D_k h_k \text{ for some } h = (h_k)_{k=1}^N \subseteq L^{p'}(\Omega, \mathbb{R}^N) \right\}.$$

We write $W^{-1,p'}(\Omega) = W_0^{1,p}(\Omega)^*$ with $1/p + 1/p' = 1$.

(4.6) A good and complete reference for the space $\mathrm{ca}(\mathcal{B}(X))$ and its topologies is the book of Bourbaki [48]. Additional information can be found in the books of Aliprantis–Border [7], Bogachev [41], Dellacherie–Meyer [86], Dunford–Schwartz [105], Florescu–Godet-Thobie [123], and Schwartz [293].

The space of finitely additive set functions describes the bidual of $L^1(\Omega)$.

Definition 4.10.8. Let (Ω, Σ) be a measurable space. We define

$$\mathrm{ba}(\Sigma) = \{\mu : \Sigma \to \mathbb{R} : \mu \text{ is finitely additive}\}.$$

Equipped with the supremum norm $\|\mu\|_\infty = \sup[|\mu(A)| : A \in \Sigma]$, $\mathrm{ba}(\Sigma)$ is a Banach space. Another equivalent norm, making $\mathrm{ba}(\Sigma)$ a Banach space, is the total variation norm $\|\mu\| = |\mu|(\Omega)$. Let $\lambda: \Sigma \to \overline{\mathbb{R}}_+ = \mathbb{R}_+ \cup \{+\infty\}$ be σ-finite. We define

$$\mathrm{ba}_\lambda(\Sigma) = \{\mu \in \mathrm{ba}(\Sigma) : \mu \ll \lambda\}.$$

The next result characterizes the bidual of $L^1(\Omega)$, that is, the dual of $L^\infty(\Omega)$, and can be found in Dunford–Schwartz [105, Theorem IV.8.16, p. 296].

Theorem 4.10.9. *If $(\Omega, \Sigma, \lambda)$ is a σ-finite measure space, then $L^1(\Omega)^{**} = L^\infty(\Omega)^* = \mathrm{ba}_\lambda(\Sigma)$.*

Continuing with the dual of $L^\infty(\Omega)$ we will provide a more detailed description of its elements.

Definition 4.10.10. Let $(\Omega, \Sigma, \lambda)$ be a σ-finite measure space and X a Banach space.

(a) We say that $\eta \in L^\infty(\Omega, X)^*$ is **absolutely continuous with respect to** λ if there exists $u \in L^1(\Omega, X^*_{w^*})$ such that

$$\eta(v) = \int_\Omega \langle u(w), v(w)\rangle \, d\lambda \quad \text{for all } v \in L^\infty(\Omega, X) .$$

We say that u is the density of η and

$$\|\eta\|_* = \|u\|_{L^1(\Omega, X^*_{w^*})} = \int_\Omega \|u(w)\|_{X^*} \, d\lambda .$$

Hence, an absolutely continuous element of $L^\infty(\Omega, X)^*$ can be identified with its λ-density.

(b) We say that $\eta \in L^\infty(\Omega, X)^*$ is **singular with respect to** λ if there exists a decreasing sequence $\{A_n\}_{n\ge1} \subseteq \Sigma$ such that $\lambda(A_n) \searrow 0$ and η is supported by A_n with $n \in \mathbb{N}$, that is

$$v \in L^\infty(\Omega, X) \text{ with } v|_{A_n} = 0 \text{ for some } n \in \mathbb{N} \text{ implies } \eta(v) = 0 .$$

The next result is due to Levin [214].

Theorem 4.10.11. *If $(\Omega, \Sigma, \lambda)$ is a σ-finite measure space and X is a Banach space, then $L^\infty(\Omega, X)^* = L^1(\Omega, X^*_{w^*}) \oplus L_s$ with L_s being the space of λ-singular functions, that is, if $\eta \in L^\infty(\Omega, X)^*$, then*

$$\eta(v) = \int_\Omega \langle u(w), v(w)\rangle \, d\lambda + \eta_s(v) \quad \textit{for all } v \in L^\infty(\Omega)$$

with $u \in L^1(\Omega, X^*_{w^*})$ and $\eta_s \in L_s$. Moreover,

$$\|\eta\|_* = \|u\|_{L^1(\Omega, X^*_{w^*})} + \|\eta_s\|_* .$$

(4.7) For a complete account of the theory of Young measures we refer to Balder [24, 25], Florescu–Godet-Thobie [123], Pedregal [258], Roubíček [281], and Valadier [317, 318]. Applications to mathematical economics, optimal control, and calculus of variations can be found in Balder [25], Pedregal [258], and Roubíček [281].

(4.8) The theory of strongly continuous semigroups started in 1948 with the generation theorem of Hille and Yosida; see Theorem 4.8.16. It was established independently by Hille [164] and Yosida [337] for contractions. Hille's approach was based on the convergence of the exponential formula

$$L_t(u) = \lim_{n\to\infty}\left(\mathrm{id} - \frac{1}{n}A\right)^{-n} u$$

with $u \in D(A^2)$ while Yosida used the bounded linear operator

$$A_\lambda = \lambda A R(\lambda) = \lambda^2 R(\lambda) - \lambda\,\mathrm{id} \quad \text{for } \lambda > 0,$$

see Step 2 of the proof of Theorem 4.8.16. For this reason A_λ is called the **Yosida approximation** of A. We mention that if the semigroup is uniformly continuous, that is,

$$\lim_{t'\to t} \|L_{t'} - L_t\|_L = 0,$$

then the generator is a bounded linear operator A and $L_t = e^{At}$; see Corollary 4.8.9. This result goes back to Nathan [252] and it is used to relate the strongly continuous semigroup with the Yosida approximation.

Proposition 4.10.12. *If A is the infinitesimal generator of a strongly continuous semigroup $\{L_t\}_{t\ge 0}$, then*

$$L_t(u) = \lim_{\lambda\to\infty} e^{tA_\lambda}u \quad \textit{for all } u \in X.$$

The generation theorem based on dissipative operators, see Theorem 4.8.36, is due to Phillips [261]. It was extended to general Banach spaces by Lumer–Phillips [222].

More on strongly continuous semigroups can be found in the books of Bühler–Salamon [64], Butzer–Berens [66], Hille–Phillips [165], Martin [227] and Yosida [336].

For nonlinear semigroups we refer to the books of Barbu [30] and Miyadera [237]. Theorem 4.8.41 is due to Crandall–Liggett [77].

(4.9) Recently, generalized Orlicz spaces have attracted the interest of researchers in order to treat boundary values problems with nonstandard growth and nonuniform ellipticity of the differential operator. An account of the theory of theses spaces can be found in the books of Cruz-Uribe–Fiorenza [79], Diening–Harjulehto–Hästö–Růžička [89] (primarily dealing with variable exponent spaces), Harjulehto–Hästö [152] and Musielak [248]. We also mention the recent work of Crespo-Blanco–Gasiński–Harjulehto–Winkert [78] about variable exponent double phase functions and corresponding generalized Orlicz spaces.

Problems

Problem 4.1. Let (Ω, Σ, μ) be a measure space and for $p \ge 1$ let $L^p(\Omega)_+ = \{f \in L^p(\Omega) : f(w) \ge 0\ \mu\text{-a.e.}\}$. Show that the map $f \to f^\vartheta$ is a homeomorphism from $L^p(\Omega)_+$ onto $L^{p/\vartheta}(\Omega)_+$ for every $\vartheta > 0$.

Problem 4.2. Let (Ω, Σ, μ) be a measure space, $1 \le p, q \le \infty$, and $\overline{B}_1^p = \{f \in L^p(\Omega) : \|f\|_p \le 1\}$. Show that the set $\overline{B}_1^p \cap L^q(\Omega)$ is w-closed in $L^q(\Omega)$.

Problem 4.3. Let (Ω, Σ, μ) be a measure space, $1 \le p \le \infty$, $1/p + 1/p' = 1$, and $f : \Omega \to \mathbb{R}$ be Σ-measurable such that $fg \in L^1(\Omega)$ for all $g \in L^{p'}(\Omega)$. Show that $f \in L^p(\Omega)$.

Problem 4.4. Let (Ω, Σ, μ) be a finite measure space, $1 < p < \infty$, $\{f_n, f\}_{n\ge1} \subseteq L^p(\Omega)$, and assume that $f_n(w) \to f(w)$ μ-a. e. and $\|f_n\|_p \to \|f\|_p$ as $n \to \infty$. Show that $f_n \to f$ in $L^p(\Omega)$.

Problem 4.5. Let (Ω, Σ, μ) be a measure space and $f : \Omega \to X$ is a nonzero Σ-measurable function. We set $T_f = \{p \in [1, \infty] : f \in L^p(\Omega)\}$. Show that T_f is an interval.

Problem 4.6. Let (Ω, Σ, μ) be a measure space, $1 < p < \infty$, $\{f_n\}_{n\ge1} \subseteq L^p(\Omega)$, and assume that $f_n \xrightarrow{w} f$ in $L^p(\Omega)$ and $\limsup_{n\to\infty} \|f_n\|_p \le \|f\|_p$. Show that $f_n \to f$ in $L^p(\Omega)$.

Problem 4.7. Suppose that $\{f_n, f\}_{n\ge1} \subseteq L^1[0,1]$ and assume that $f_n(t) \to f(t)$ a. e. on $[0,1]$. Is it true that $f_n \to f$ in $L^1[0,1]$? Justify your answer.

Problem 4.8. Let (Ω, Σ, μ) be a finite measure space and $f, h : \Omega \to \mathbb{R}_+$ are two Σ-measurable functions such that

$$\mu(\{h > \lambda\}) \le \frac{1}{\lambda} \int_{\{h>\lambda\}} f \, d\mu \quad \text{for all } \lambda > 0 .$$

Show that $\|h\|_p \le p/(p-1)\|f\|_p$ for all $p \in (1, \infty)$.

Problem 4.9. Let (Ω, Σ, μ) be a measure space, $1 \le p \le \infty$ and $\{f_n\}_{n\ge1} \subseteq L^p(\Omega)$ such that $f_n \to f$ in $L^p(\Omega)$. Show that there exists a subsequence $\{f_{n_k}\}_{k\ge1}$ of $\{f_n\}_{n\ge1}$ such that for every $\varepsilon > 0$ there exists a set $A_\varepsilon \in \Sigma$ with $\mu(A_\varepsilon) < \varepsilon$ and $f_{n_k} \to f$ uniformly on $X \setminus A_\varepsilon$ and $f_{n_k}(w) \to f(w)$ μ-a. e.

Problem 4.10. Let $f \in L^p(0, +\infty)$, $1 < p < \infty$ and for all $x > 0$ let $F(x) = 1/x \int_0^x f(s)\, ds$. Show that $F \in L^p(0, \infty)$ and $\|F\|_p \le p/(p-1)\|f\|_p$. This inequality is known as "Hardy's inequality".

Problem 4.11. Let (Ω, Σ, μ) be a measure space, $1 \le p \le \infty$, $\{f_n, f\}_{n\ge1} \subseteq L^p(\Omega)$, $f_n \xrightarrow{w} f$ in $L^p(\Omega)$, and $f_n(w) \to \hat{f}(w)$ μ-a. e. Show that $f(w) = \hat{f}(w)$ μ-a. e.

Problem 4.12. Let (Ω, Σ, μ) be a semifinite measure space and $f, h : \Omega \to \overline{\mathbb{R}}_+ = \mathbb{R}_+ \cup \{+\infty\}$ be two Σ-measurable functions such that

$$\int_A f \, d\mu \le \int_A h \, d\mu \quad \text{for all } A \in \Sigma \text{ with } \mu(A) < +\infty .$$

Show that $f(w) \le h(w)$ μ-a. e.

Problem 4.13. Let (Ω, Σ, μ) be a measure space and $f \in L^1(\Omega)$. Show that $\mu(\{f = \pm\infty\}) = 0$.

Problem 4.14. Let (X, d) be a separable metric space, μ a locally finite Borel measure, that is, for all $x \in X$, there exists $r > 0$ such that $\mu(B_r(x)) > 0$ where $B_r(x) = \{u \in X : d(u, x) < r\}$, V a Banach space, and $1 \le p < \infty$.

(a) Show that $C(X, V) \cap L^P(X, V)$ is dense in $L^p(T, V)$.

(b) If μ is a Radon measure, then show that $C_c(X, V)$ is dense in $L^p(X, V)$.

Problem 4.15. Let (X, d) be a locally compact, separable metric space, μ a locally finite Borel measure on X (see Problem 4.14), V a Banach space, and $1 \le p < \infty$. Show that $C_c(X, V)$ is dense in $L^p(X, V)$.

Problem 4.16. Let $T = [0, b]$, H be a separable Hilbert space, and $\{f_n\}_{n\ge1} \subseteq L^2(T, H)$ such that

(a) $f_n \xrightarrow{w} f$ in $L^2(T, H)$;

(b) $\sup_{n\ge1} |f_n(t)| \le M$ for a. a. $t \in T$;

(c) there exists a countable dense set D of H such that $\{(h, f_n(\cdot))\}_{n\ge1} \subseteq L^1(T)$ is relatively compact. Show that there exists a subsequence $\{f_{n_k}\}_{k\ge1}$ of $\{f_n\}_{n\ge1}$ such that $f_{n_k}(t) \xrightarrow{w} f(t)$ in H for a. a. $t \in T$.

Problem 4.17. Let $\Omega \subseteq \mathbb{R}^N$ be an open set, X a Banach space, $D \subseteq X$ a dense subset, and $1 \le p < \infty$. Show that $C_c(\Omega) \otimes C$ is dense in $L^p(\Omega, X)$.

Problem 4.18. Let (Ω, Σ, μ) be a finite measure space with countably generated Σ, X is a separable Banach space and $1 \le p < \infty$. Show that $L^p(\Omega, X)$ is separable.

Problem 4.19. Let (Ω, Σ, μ) be a σ-finite measure space and X is a Banach space. Show that simple functions are dense in $L^\infty(\Omega, X)$ if and only if X is finite dimensional.

Problem 4.20. Let (Ω, Σ, μ) be a measure space, X a Banach space, and $f : \Omega \to X$. Prove the following:

(a) If f is the μ-a. e. limit of a sequence of countably valued functions $\{h_n\}_{n\ge1}$, then f is essentially separably valued.

(b) If f is as in (a), then there exists a sequence $\{\hat{h}\}_{n\ge1}$ of countably valued functions such that $\hat{h}_n \to f$ uniformly on $\Omega \setminus N$ with $\mu(N) = 0$.

(c) If f is essentially separably valued and $w \to \|f(w) - y\|$ is Σ-measurable for all $x \in X$, then f is strongly measurable.

Problem 4.21. Let (Ω, Σ, μ) be a finite measure space, $\Sigma_0 \subseteq \Sigma$ a sub-σ-algebra, and X a Banach space. Show that there exists a unique operator $E^{\Sigma_0} \in L(L^1(\Omega, \Sigma, X), L^1(\Omega, \Sigma_0, X))$ such that

$$\int_A f \, d\mu = \int_A E^{\Sigma_0} f \, d\mu \quad \text{for all } f \in L^1(\Omega, \Sigma, X) .$$

E^{Σ_0} is the **conditional expectation** of f with respect to Σ_0.

Problem 4.22. Show that $L^1(0,1)$ is not the dual space of any normed space V.

Problem 4.23. Let (Ω,Σ,μ) and $(X,\mathcal{L},\lambda)$ be two measure spaces, $A \in L(L^1(\Omega),L^1(X))$, and $D \subseteq L^1(\Omega)$ be a uniformly integrable set. Show that $A(D) \subseteq L^1(X)$ is uniformly integrable.

Problem 4.24. Let $(X,\|\cdot\|)$ and $(H,|\cdot|)$ be two Hilbert spaces with $X \hookrightarrow H$ compactly and densely and let $\{f_n\}_{n\geq 1} \subseteq L^2(T,X)$ with $T = [0,b]$. Show that the following two statements are equivalent:
(a) $f_n \to f$ in $L^2(T,H)$;
(b) $f_n(t) \xrightarrow{w} f(t)$ a. e. in H and $\lim_{\lambda(A)\to 0} \sup_{n\geq 1} \int_A |f_n(t)|^2 \, dt = 0$.

Problem 4.25. Show that the composition of two functions of bounded variation functions need not be of bounded variation. Similarly show this for absolutely continuous functions.

Problem 4.26. Is the uniform limit of absolutely continuous functions an absolutely continuous function? Justify your answer.

Problem 4.27. Find a condition that guarantees that the pointwise limit of a sequence of absolutely continuous functions is an absolutely continuous function. Hint: Recall the Arzela–Ascoli Theorem.

Problem 4.28. Suppose that $f\colon [a,b] \to \mathbb{R}$ is continuous and of bounded variation. Assume that for every $c \in (a,b)$, $f \in \mathrm{AC}([a,c])$. Show that $f \in \mathrm{AC}([a,b])$.

Problem 4.29. Suppose $f \in \mathrm{AC}([a,b])$ and let $p \geq 1$. Show that $|f|^p \in \mathrm{AC}([a,b])$.

Problem 4.30. Let $A \subseteq [0,1]$ be a measurable set such that $\lambda(A \cap [a,b]) \geq \vartheta(b-a)$ for some $\vartheta > 0$ and for all $0 \leq a \leq b \leq 1$ where λ stands for the Lebesgue measure on $\mathbb{R}$. Show that $\lambda(A) = 1$.

Problem 4.31. Let $A \subseteq \mathbb{R}$ be not necessarily Lebesgue measurable. Show that

$$\lim_{\delta\to 0} \frac{\lambda^*(A \cap [t-\delta, t+\delta])}{2\delta} = 1 \quad \text{for a. a. } t \in \mathbb{R},$$

where λ^* denotes the Lebesgue outer measure.

Problem 4.32. Let $f\colon [a,b] \to \mathbb{R}$ be continuous and $\eta < \mathrm{var}_{[a,b]} f \in [0,+\infty]$. Show that there exists $\delta > 0$ such that for every partition $a = x_0 < x_1 < \cdots < x_n = b$ with $\max\{x_{k+1} - x_k : k = 0,1,\ldots,n-1\} < \delta$ it holds that

$$\sum_{k=0}^{n-1} |f(x_{k+1}) - f(x_k)| > \eta .$$

Problem 4.33. Show that $\mathrm{BV}([a,b])$ is compactly embedded into $L^p([a,b])$.

Problem 4.34. Let $f \in \mathrm{BV}([a,b])$ and let $\bar{f} = 1/(b-a)\int_a^b f\,dx$. Show that

$$\int_a^b |f(x) - \bar{f}|\,dx \le \frac{b-a}{2}\int_a^b |f'|\,dx$$

and that this inequality can be strict.

Problem 4.35. Let $f \in C^1(a,b)$. Show that $f \in \mathrm{BV}([a,b])$ if and only if $f' \in L^1(a,b)$ and we have $\mathrm{var}_{[a,b]} f = \|f'\|_1$.

Problem 4.36. Find a bounded function f such that $f \in \mathrm{BV}_{\mathrm{loc}}(\mathbb{R})$ but $f \notin \mathrm{BV}(\mathbb{R})$.

Problem 4.37. Let $f(x) = x^2\sin(\pi/x)$ if $x \in (0,1]$, $f(0) = 0$, and $\hat{f}(x) = x^2\sin(\pi/x^2)$ if $x \in (0,1]$, $\hat{f}(0) = 0$. Show that f is absolutely continuous, but $\hat{f}$ is not.

Problem 4.38. Let X, Y, Z be Banach spaces with X reflexive, $X \hookrightarrow V$ continuously, and $K \in L_c(X,Y)$. Show that for every $\varepsilon > 0$ there exists $c_\varepsilon > 0$ such that

$$\|K(x)\|_Y \le \varepsilon\|x\|_X + c_\varepsilon\|x\|_V \quad \text{for all } x \in X\,.$$

Problem 4.39. Let $\Omega \subseteq \mathbb{R}^N$ be an open set, $1 \le p \le \infty$, and $u \in W_0^{1,p}(\Omega)$ with $u \ge 0$. Show that there exists a sequence $\{u_n\}_{n\ge1} \subseteq C_c^\infty(\Omega)$ with $u_n \ge 0$ for all $n \in \mathbb{N}$ such that $u_n \to u$ in $W_0^{1,p}(\Omega)$.

Problem 4.40. Let $\Omega \subseteq \mathbb{R}^N$ be an open set, $1 \le p < \infty$, $u \in W^{1,p}(\Omega)$, and u vanishes outside a compact set $K \subseteq \Omega$. Show that $u \in W_0^{1,p}(\Omega)$.

Problem 4.41. Let $\Omega \subseteq \mathbb{R}^N$ be a bounded open set, $1 \le p < \infty$, and $u \in W^{1,p}(\Omega)$ be such that $\lim_{z\to x} u(z) = 0$ for all $x \in \partial\Omega$. Show that $u \in W_0^{1,p}(\Omega)$.

Problem 4.42. Let $\{u_n\}_{n\ge1} \subseteq W^{1,1}(0,b)$ be a sequence such that $u_n(t) \to u(t)$ for a. a. $t \in (0,1)$ and there exists $h \in L^1(0,1)$ such that $|u_n'(t)| \le h(t)$ for a. a. $t \in T$ and for all $n \in \mathbb{N}$. Show that $u_n \to u$ uniformly on $[0,1]$.

Problem 4.43. Let $u \in W^{1,p}(0,1)$ with $1 \le p < \infty$ and let $Z = \{t \in (0,1)\colon u(t) = 0\}$. Show that $u'(t) = 0$ for a. a. $t \in Z$.

Problem 4.44. Let $\Omega \subseteq \mathbb{R}^N$ be a bounded open set with Lipschitz boundary, $p \in (1,N)$, and $r \in (1, ((N-1)p)/(N-p))$. Show that for every $\varepsilon > 0$ there exists $c_\varepsilon > 0$ such that

$$\|\gamma_0(u)\|_{L^r(\partial\Omega)} \le \varepsilon\|Du\|_{L^p(\Omega)} + c_\varepsilon\|u\|_{L^p(\Omega)} \quad \text{for all } u \in W^{1,p}(\Omega)\,.$$

Problem 4.45. Let $\Omega \subseteq \mathbb{R}^N$ be a bounded open set with Lipschitz boundary $\partial\Omega$. Consider $u \in C^1(\overline{\Omega}) \cap W^{1,p}(\Omega)$ with $1 < p < \infty$, and assume that it has finitely many nodal domains where a nodal domain of u is a connected component of $\Omega \setminus Z(u)$ with $Z(u) = \{z \in \Omega\colon u(z) = 0\}$. Show that for any nodal domain Ω_0, $u_0 = \chi_{\overline{\Omega}_0} u \in W^{1,p}(\Omega)$ and

$$u_0|_{\partial\Omega}(z) = \begin{cases} u(z) & \text{if } z \in \partial\Omega \cap \partial\Omega_0\,, \\ 0 & \text{if } z \in \partial\Omega \setminus \partial\Omega_0\,. \end{cases}$$

Problem 4.46. Let $\Omega \subseteq \mathbb{R}^N$ be a bounded open set, $1 \le p < \infty$, and $h^* \in L^{p'}(\Omega) \subseteq W^{-1,p'}(\Omega)$. Show that

$$\langle h^*, u\rangle = \int_\Omega h^* u \,\mathrm{d}z \quad \text{for all } u \in W_0^{1,p}(\Omega)\,.$$

Problem 4.47. Let X be a Hausdorff topological space. Show that $\mathrm{ca}_\mathrm{r}(\mathcal{B}(X))$ is a closed subspace of $\mathrm{ca}(\mathcal{B}(X))$. Hence it is itself a Banach space.

Problem 4.48. Let X be a locally compact topological space, $\{\mu_\alpha\}_{\alpha\in I} \subseteq \mathrm{ca}_\mathrm{R}^+(\mathcal{B}(X))$ be a net, and $\mu \in \mathrm{ca}_\mathrm{R}^+(\mathcal{B}(X))$. Show that

$$\mu_\alpha \xrightarrow{\tau_\mathrm{n}(X)} \mu \quad \text{if and only if} \quad \mu_\alpha \xrightarrow{\tau_\mathrm{v}(X)} \mu \quad \text{and} \quad \mu_\alpha(X) \to \mu(X)\,.$$

Problem 4.49. Let (X, Σ) be a measurable space. Show that the Banach space $\mathrm{ca}(\Sigma)$ is weakly complete.

Problem 4.50. Let X be a separable metric space and $D \subseteq X$ a countable dense subset. Show that the set of all probability measures supported by finite subsets of D is dense in $(\mathcal{P}(X), \tau_\mathrm{n}(X))$.

Problem 4.51. Let X, Y be separable metric spaces, $h_n, h\colon X \to Y$ with $n \in \mathbb{N}$ be Borel maps with h continuous, $h_n \to h$ uniformly on compact subsets of X, and $\{\mu_n\}_{n\ge1} \subseteq \mathcal{P}(X)$ be tight with $\mu_n \xrightarrow{\tau_\mathrm{n}(X)} \mu \in \mathcal{P}(X)$. Show that $\mu_n \circ h_n^{-1} \xrightarrow{\tau_\mathrm{n}(Y)} \mu \circ h^{-1}$.

Problem 4.52. Let (X, Σ) be a measurable space and $\{\mu_n, \mu\}_{n\ge1} \subseteq \mathrm{ca}(\Sigma)$. Show that $\mu_n \xrightarrow{\mathrm{w}} \mu$ in $\mathrm{ca}(\Sigma)$ if and only if $\mu_n(A) \to \mu(A)$ for all $A \in \Sigma$.

Problem 4.53. Let (X, Σ) be a measurable space, $\{\mu_n\}_{n\ge1} \subseteq \mathrm{ca}_+(\Sigma)$, $\mu_n \xrightarrow{\mathrm{w}} \mu$, and $f_n\colon X \to [0, +\infty)$ be a sequence of Σ-measurable functions such that $f_n(x) \to f(x)$ for all $x \in X$. Show that $\int_X f \,\mathrm{d}\mu \le \liminf_{n\to\infty} \int_X f_n \,\mathrm{d}\mu_n$.

Problem 4.54. Let (Ω, Σ, μ) be a complete probability space, X a Polish space, and $u \in L^0(\Omega, X)$. Let λ^u be the Young measure associated with u (see Definition 4.7.6), and define $\lambda^u(\Omega) = \{\delta_{u(w)}\colon w \in \Omega\} \subseteq \mathcal{P}(X)$. Show that $\lambda^u(\Omega)$ is tight if and only if $u(\Omega) \subseteq X$ is relatively compact.

Problem 4.55. Let X be a metric space and $\mu, \lambda \in \mathcal{P}(X)$ such that $\int_X f \,\mathrm{d}\mu = \int_X f \,\mathrm{d}\lambda$ for all $f \in U_\mathrm{b}(X)$. Show that $\mu = \lambda$.

Problem 4.56. Let V, X be separable metric spaces, $\hat\lambda\colon V \to \mathcal{P}(X)$ be a Borel map, and $f \in C(V \times X)$. Show that $v \to g(v) = \int_X f(v, x)\hat\lambda(v)(\mathrm{d}x)$ is continuous on V.

Problem 4.57. Let $\Omega \subseteq \mathbb{R}^N$ be bounded and open, $C \subseteq W^{1,p}(\Omega)$ with $1 < p < \infty$ be closed and convex, $\{u_n\}_{n\geq 1} \subseteq C$, $u \in L^p(\Omega)$, $y \in L^p(\Omega, \mathbb{R}^N)$, and $u_n \xrightarrow{w} u$ in $L^p(\Omega)$, $Du_n \xrightarrow{w} y$ in $L^p(\Omega, \mathbb{R}^N)$. Show that $u \in C$ and $y = Du$.

Problem 4.58. Let $\Omega \subseteq \mathbb{R}^N$ be bounded and open, $\{u_n\}_{n\geq 1} \subseteq W^{1,p}(\Omega)$ with $1 < p < \infty$ be bounded, and $u_n(z) \to u(z)$ a. e. in Ω. Show that $u \in W^{1,p}(\Omega)$ and $u_n \xrightarrow{w} u$ in $W^{1,p}(\Omega)$.

Problem 4.59. Let $\{f_n\}_{n\geq 1} \subseteq L^p(0,b)$ with $1 < p < \infty$ and assume that $f_n \xrightarrow{w} f$ in $L^p(0,b)$, $f_n' \to f'$ in $W^{-1,p}(0,b)$. Show that $f_n \to f$ in $L^p(0,b)$.

Problem 4.60. Let X be a locally compact separable metric space, $f\colon X \to \mathbb{R}_+$ a lower semicontinuous function, and $\{\mu_n, \mu\}_{n\geq 1} \subseteq \mathrm{ca}_+(\mathcal{B}(X))$ such that $\mu_n \xrightarrow{\tau_n(X)} \mu$ and $\mu_n \leq \lambda$ for all $n \in \mathbb{N}$ and some $\lambda \in \mathrm{ca}_+(\mathcal{B}(X))$ such that $\int_X f \, d\lambda < \infty$. Show that $\int_X f \, d\mu_n \to \int_X f \, d\mu$.

5 Convex Functions – Nonsmooth Analysis

Convex sets and convex functions are a basic tool in many parts of mathematical analysis as well as in many applied fields such as optimization, optimal control, game theory, mathematical economics, and others. They exhibit many interesting properties that lead to remarkable results. For convex sets, topological, algebraic, and geometric notions often coincide. Convex functions have properties that lead to many fruitful continuity properties and a coherent differentiability theory. Moreover, local minima turn out to be global ones. Their systematic study started in the early 1960s. This effort led to a rich theory of convex sets and functions known as "Convex Analysis".

One of the main features of this theory is duality, which provides significant insight into convex optimization in the context of applications. In addition, convex analysis permits the treatment of nonsmooth functions and provides a calculus for convex functions that goes well beyond the classical one. In the absence of smoothness and convexity, the situation becomes more complicated. A major step in this direction was made by considering locally Lipschitz functions. An effective calculus as well as powerful optimality conditions were produced for this class of functions, extending the corresponding theory for continuous and convex functions. The body of these results constitute what is known nowadays as "Nonsmooth Analysis".

5.1 Convex Functions – Continuity Properties

In this chapter we often deal with extended real valued functions. So, we set $\mathbb{R}^* = \mathbb{R} \cup \{\pm\infty\}$ and $\overline{\mathbb{R}} = \mathbb{R} \cup \{+\infty\}$. The operations on $\mathbb{R}^*$ and $\overline{\mathbb{R}}$ are defined as usual. In addition, we set $0 \cdot (\pm\infty) = (\pm\infty) \cdot 0 = 0$. However, we do not define $(+\infty) - \infty$.

Definition 5.1.1. Let X be a real vector space and $f: X \to \mathbb{R}^*$.

(a) We say that f is **convex** if

$$f(\lambda x + (1-\lambda)u) \leq \lambda f(x) + (1-\lambda) f(u) \tag{5.1.1}$$

for all $x, u \in X$ and for all $\lambda \in [0,1]$ provided the right-hand side is defined.

(b) We say that f is **strictly convex** if

$$f(\lambda x + (1-\lambda)u) < \lambda f(x) + (1-\lambda) f(u)$$

for all $x, u \in X$ with $x \neq u$ and for all $\lambda \in [0,1]$ provided the right-hand side is defined.

(c) The set $\operatorname{dom} f = \{x \in X : f(x) < +\infty\}$ is called the **effective domain** of f. Moreover, the **epigraph** of f is the set

$$\operatorname{epi} f = \{(x, \eta) \in X \times \mathbb{R} : f(x) \leq \eta\}.$$

https://doi.org/10.1515/9783111286952-005

(d) We say that f is **proper** if $\operatorname{dom} f \neq \emptyset$ and $f(x) > -\infty$ for all $x \in X$.
(e) We say that f is **concave** (resp. **strictly concave**) if $-f$ is convex (resp. strictly convex).

Remark 5.1.2. In Definitions 5.1.1 (a) and (b), the right-hand side is not defined only if $f(x) = \pm\infty$ and $f(u) = \mp\infty$.

Proposition 5.1.3. *If X is a vector space and $f: X \to \mathbb{R}^*$, then the following hold:*
(a) *f is convex if and only if $\operatorname{epi} f \subseteq X \times \mathbb{R}$ is convex;*
(b) *if f is convex, then $\operatorname{dom} f$ is convex.*

Proof. (a) $\Longrightarrow$: Let $(x, \eta), (u, \vartheta) \in \operatorname{epi} f$ and let $\lambda \in (0, 1)$. Since f is convex, we obtain

$$f(\lambda x + (1-\lambda)u) \leq \lambda f(x) + (1-\lambda) f(u) \leq \lambda\eta + (1-\lambda)\vartheta \,.$$

This shows that $(\lambda x + (1-\lambda)u, \lambda\eta + (1-\lambda)\vartheta) \in \operatorname{epi} f$. But $(\lambda x + (1-\lambda)u, \lambda\eta + (1-\lambda)\vartheta) = \lambda(x, \eta) + (1-\lambda)(u, \vartheta)$. Therefore, $\operatorname{epi} f$ is convex.

$\Longleftarrow$: Note that $\operatorname{dom} f = p_X(\operatorname{epi} f)$ where p_X denotes the projection operator. So, $\operatorname{dom} f \subseteq X$ is convex. It suffices to check (5.1.1) on $\operatorname{dom} f$. So, let $x, u \in \operatorname{dom} f$ and consider $\eta, \vartheta \in \mathbb{R}$ such that $f(x) \leq \eta$ and $f(u) \leq \vartheta$, that is, $(x, \eta), (u, \vartheta) \in \operatorname{epi} f$. By hypothesis, one has $\lambda(x, \eta) + (1-\lambda)(u, \vartheta) \in \operatorname{epi} f$ for every $\lambda \in [0, 1]$. Hence, $f(\lambda x + (1-\lambda)u) \leq \lambda\eta + (1-\lambda)\vartheta$. If both $f(x), f(u) \in \mathbb{R}$, then we can take $\eta = f(x)$, $\vartheta = f(u)$. This gives $f(\lambda x + (1-\lambda)u) \leq \lambda f(x) + (1-\lambda) f(u)$. If $f(x) = -\infty$ and $f(u) = -\infty$, then we let $\eta \searrow -\infty$ and $\vartheta \searrow -\infty$ and again inequality (5.1.1) is satisfied.

(b) Since f is convex, the epigraph of f is convex; see part (a). Recall that $\operatorname{dom} f = p_X(\operatorname{epi} f)$. □

Remark 5.1.4. Note that if $f: D \subseteq X \to \mathbb{R}$, then we introduce $\hat{f}: X \to \overline{\mathbb{R}}$ by setting

$$\hat{f}(x) = \begin{cases} f(x) & \text{if } x \in D\,, \\ +\infty & \text{if } x \in X \setminus D\,, \end{cases}$$

that is, we impose an infinite penalty if we violate the constraint D. So by considering $\overline{\mathbb{R}}$-valued functions, we can deal only with functions defined on all of X. In this respect, the following definition is useful.

Definition 5.1.5. Let X be a vector space and let $D \subseteq X$. The indicator function $i_D: X \to \overline{\mathbb{R}}$ is defined by

$$i_D(x) = \begin{cases} 0 & \text{if } x \in D\,, \\ +\infty & \text{if } x \in X \setminus D\,. \end{cases}$$

Evidently, $D \subseteq X$ is convex if and only if i_D is a convex function. Moreover, $\operatorname{dom} i_D = D$. Thus the study of convex sets is reduced to the study of convex functions.

The following proposition is an easy consequence of Definition 5.1.1.

Proposition 5.1.6. *If X is a vector space, $f, h: X \to \overline{\mathbb{R}}$ are convex functions and $\eta, \vartheta \geq 0$, then $\eta f + \vartheta h: X \to \overline{\mathbb{R}}$ is convex as well.*

Proposition 5.1.7. *If X is a vector space and $f_\alpha: X \to \overline{\mathbb{R}}$ with $\alpha \in I$ is a family of convex functions, then $f = \sup_{\alpha \in I} f_\alpha$ is convex as well.*

Proof. Note that $\operatorname{epi} f = \bigcap_{\alpha \in I} \operatorname{epi} f_\alpha$. Therefore, $\operatorname{epi} f$ is convex and so the result follows from Proposition 5.1.3. □

The next proposition shows that it is quite pathological for a convex function to attain the value $-\infty$.

Proposition 5.1.8. *If X is a topological vector space, $f: X \to \mathbb{R}^*$ is convex, and there exists $x_0 \in \operatorname{int} \operatorname{dom} f$, then f is proper.*

Proof. Arguing by contradiction, suppose that we can find $u \in X$ such that $f(u) = -\infty$. Since $x_0 \in \operatorname{int} \operatorname{dom} f$, for small enough $\lambda \in (0, 1)$, we get $\hat{x} = x_0 + \lambda(x_0 - x) \in \operatorname{dom} f$. Note that $x_0 = 1/(1 + \lambda)\hat{x} + \lambda/(1 + \lambda)u$. Then the convexity of f implies that

$$f(x_0) \leq \frac{1}{1+\lambda} f(\hat{x}) + \frac{\lambda}{1+\lambda} f(u) = -\infty ,$$

a contradiction. Therefore, $f(x) > -\infty$ for all $x \in X$, that is, f is proper. □

Proposition 5.1.9. *If X is a vector space, $f: X \to \mathbb{R}$ is convex and $h: \mathbb{R} \to \mathbb{R}$ is convex and nondecreasing, then $h \circ f: X \to \mathbb{R}$ is convex as well.*

Proof. Let $x, u \in X$ and let $\lambda \in [0, 1]$. Since f is convex, that is, $f(\lambda x + (1 - \lambda u)) \leq \lambda f(x) + (1 - \lambda) f(u)$, we obtain

$$h(f(\lambda x + (1 - \lambda)u)) \leq h(\lambda f(x) + (1 - \lambda) f(u)) \leq \lambda h(f(x)) + (1 - \lambda) h(f(u))$$

because h is nondecreasing and convex. Therefore $h \circ f$ is convex. □

Remark 5.1.10. From the proof above it is clear that if f is strictly convex and h is convex and strictly increasing, then $h \circ f$ is strictly convex.

Corollary 5.1.11. *If X is a normed space, then $x \to \|x\|^p$ is convex if and only if $p \geq 1$. Moreover, if X is strictly convex, then $x \to \|x\|^p$ is strictly convex if and only if $p > 1$.*

Definition 5.1.12. Let X be a vector space and let $f_k: X \to \mathbb{R}^*$ for $k = 1, \ldots, n$ be proper functions. We define

$$f(x) = \inf\left[\sum_{k=1}^{n} f_k(x_k) : x = \sum_{k=1}^{n} x_k\right] .$$

Then f is called the **infimal convolution** of the f_k's and is denoted by

$$f = \bigoplus_{k=1}^{n} f_k .$$

We say that the infimal convolution is **exact at** x if there exists a sequence $\{x_k\}_{k=1}^{n} \subseteq X$ such that

$$x = \sum_{k=1}^{n} x_k \quad \text{and} \quad f(x) = \sum_{k=1}^{n} f(x_k) .$$

Remark 5.1.13. If $n = 2$, then $(f_1 \oplus f_2)(x) = \inf[f_1(x-u) + f_2(u) : u \in X]$, which reminds us of the formula for the usual convolution $(f_1 * f_2)(x) = \int_{\mathbb{R}^N} f_1(x-u) f_2(u)\, du$ if $X = \mathbb{R}^N$. Clearly, if the f_k's are convex, then so is f but may fail to be proper. In order to see this, let $f_1 = i_{C_1}$ and $f_2 = i_{C_2}$ with $C_1, C_2 \subseteq X$ be nonempty, disjoint, and convex. Then $f_1 \oplus f_2 \equiv +\infty$. Similarly, if f_1 and f_2 are linear with $f_1 \neq f_2$, then $f_1 \oplus f_2 = -\infty$. Moreover, from Definition 5.1.12 we get

$$(f_1 \oplus f_2)(x) = \inf[\eta \in \mathbb{R} : \text{there exists } x \in X \text{ such that } (x, \eta) \in (\operatorname{epi} f_1 + \operatorname{epi} f_2)] .$$

In addition, it is easily seen that

$$f_1 \oplus f_2 = f_2 \oplus f_1 \quad \text{and} \quad f_1 \oplus (f_2 \oplus f_3) = (f_1 \oplus f_2) \oplus f_3 .$$

Example 5.1.14. (a) Let $f_1 = f$ and $f_2 = \delta_{\{x_0\}}$. Then $(f_1 \oplus f_2)(x) = f(x - x_0)$. So, if $x_0 = 0$, then $(f_1 \oplus f_2)(x) = f(x)$.

(b) Let X be a normed space, $C \subseteq X$ be nonempty and convex, and $f_1(x) = \|x\|$, $f_2(x) = i_C(x)$ for all $x \in X$. Then $(f_1 \oplus f_2)(x) = \inf[\|x-u\| + \delta_C(u) : x \in X] = \inf[\|x-u\| : u \in C] = d(x, C)$, where $d(x, C)$ stands for the distance of x from C. Since f_1 is convex (see Corollary 5.1.11), and f_2 is also convex, because $C \subseteq X$ is convex, it follows that $x \to d(x, C)$ is convex; see Remark 5.1.13.

Definition 5.1.15. Let X be a vector space and $q : X \to \mathbb{R}^*$. We say that q is **sublinear** if the following hold:
(a) q is proper;
(b) q is positively homogeneous, that is, $q(\lambda x) = \lambda q(x)$ for all $\lambda > 0$ and for all $x \in X$;
(c) q is subadditive, that is, $q(x + u) \leq q(x) + q(u)$ for all $x, u \in X$.

Remark 5.1.16. From this definition it follows that a sublinear function q is convex. Moreover, $q(0) = 0$. If X is a normed space, then $q(x) = \|x\|$ is sublinear. In fact, a sublinear function can be seen as a generalization of the norm in a vector space. Given a convex absorbing set $A \subseteq X$, let ρ_A be the Minkowski function of A; see Definition 3.1.37. Then ρ_A is sublinear. Another important sublinear function is given in the next definition.

Definition 5.1.17. Let X be a normed space and $A \subseteq X$ a nonempty set. The **support function** of A is the function $\sigma(\cdot;A): X^* \to \overline{\mathbb{R}}$ defined by

$$\sigma(x^*;A) = \sup[\langle x^*, u\rangle : u \in A] .$$

Evidently, $\sigma(\cdot;A)$ is sublinear and $\sigma(\cdot;A) = \sigma(\cdot;\overline{\operatorname{conv}} A)$.

Now we turn our attention to the continuity properties of convex functions.

Proposition 5.1.18. *If X is a locally convex space, $f: X \to \overline{\mathbb{R}}$ is proper and convex, and for $x_0 \in \operatorname{dom} f$ there exists $U \in \mathcal{N}(x_0)$ such that $f|_U$ is bounded above, then f is continuous at x_0.*

Proof. Replacing f with $\hat{f}(y) = f(x_0 + y) - f(x_0)$ and $y \in X$ if necessary, we assume, without any loss of generality, that $x_0 = 0$ and $f(0) = 0$.

By hypothesis we have $f(x) \le M$ for all $x \in U \in \mathcal{N}(0)$. Let $V = U \cap (-U)$. Then $V \in \mathcal{N}(0)$, and it is symmetric. Let $\lambda \in (0,1)$ and $x \in \lambda V$. From the convexity of f it follows that

$$f(x) \le \lambda f\left(\frac{1}{\lambda}x\right) \le \varepsilon M , \tag{5.1.2}$$

since $x = (1-\lambda)0 + \lambda 1/\lambda x$ and $f(0) = 0$. Note, since V is symmetric, that

$$-\frac{1}{\lambda}x \in V \subseteq U \quad \text{and} \quad 0 = \frac{1}{1+\lambda}x + \frac{\lambda}{1+\lambda}\left(-\frac{1}{\lambda}x\right).$$

This gives

$$0 = f(0) \le \frac{1}{1+\lambda}f(x) + \frac{\lambda}{1+\lambda}f\left(-\frac{1}{\lambda}x\right).$$

Hence,

$$f(x) \ge -\lambda f\left(-\frac{1}{\lambda}x\right) \ge -\lambda M . \tag{5.1.3}$$

From (5.1.2) and (5.1.3) it follows that $|f(x)| \le \lambda x$ for all $x \in \lambda V$ and this implies the continuity of f at $x_0 = 0$. □

Proposition 5.1.19. *If X is a normed space, and $f: X \to \overline{\mathbb{R}}$ is convex and continuous at $x_0 \in \operatorname{int} \operatorname{dom} f$, then f is Lipschitz on some neighborhood of x_0.*

Proof. Since f is continuous at x_0 there exist $M, \delta > 0$ such that

$$|f(x)| \le M \quad \text{for all } x \in B_{2\delta}(x_0) \subseteq \operatorname{int} \operatorname{dom} f . \tag{5.1.4}$$

Let $x, u \in B_\delta(x_0)$ with $x \neq u$ and $\vartheta = \|u - x\| > 0$. We set

$$v = u + \frac{\delta}{\vartheta}(u - x) = u + \delta\frac{u - x}{\|u - x\|} = \left(1 + \frac{\delta}{\vartheta}\right)u - \frac{\delta}{\vartheta}x . \tag{5.1.5}$$

Then, (5.1.5) implies

$$\|v - x_0\| = \left\|u - x_0 + \delta\frac{u - x}{\|u - x\|}\right\| \le \|u - x_0\| + \delta < 2\delta .$$

Hence $v \in B_{2\delta}(x_0)$. Moreover, from (5.1.5) it follows that $u = \vartheta/(\vartheta + \delta)v + \delta/(\vartheta + \delta)x$, which gives, due to the convexity of f, that $f(u) \le \vartheta/(\vartheta + \delta)f(v) + \delta/(\vartheta + \delta)f(x)$. Since $v \in B_{2\delta}(x_0)$ and because of (5.1.4) we then derive

$$f(u) - f(x) \le \frac{\vartheta}{\vartheta + \delta}[f(v) - f(x)] \le \frac{\vartheta}{\vartheta + \delta}2M \le \frac{2M}{\delta}\|u - x\| . \tag{5.1.6}$$

Interchanging the roles of x and u in the argument above, we also get

$$f(x) - f(u) \le \frac{2M}{\delta}\|x - u\| . \tag{5.1.7}$$

From (5.1.6) and (5.1.7) we conclude that

$$|f(x) - f(u)| \le \frac{2M}{\delta}\|x - u\| \quad \text{for all } x, u \in B_\delta(x_0) .$$ □

Proposition 5.1.20. *If X is a normed space and $f: X \to \overline{\mathbb{R}}$ is proper and convex, then the following statements are equivalent:*
(a) *f is Lipschitz on some neighborhood of x_0;*
(b) *f is continuous at x_0;*
(c) *f is bounded above on some neighborhood of x_0.*

Proof. Clearly, the implications (a) $\Longrightarrow$ (b) $\Longrightarrow$ (c) are easy to verify. Suppose that (c) holds. Then $f(x) \le M$ for some $M > 0$ and for all $x \in B_{2\delta}(x_0)$. From the proof of Proposition 5.1.19, we obtain that $f|_{B_\delta(x_0)}$ is $2M/\delta$-Lipschitz. □

Proposition 5.1.21. *If X is a Banach space and $f: X \to \overline{\mathbb{R}}$ is lower semicontinuous, proper, and convex, then the following statements are equivalent:*
(a) *f is continuous at x_0;*
(b) *$x_0 \in \operatorname{int} \operatorname{dom} f$.*

Proof. (a) $\Longrightarrow$ (b): This implication is clear.

(b) $\Longrightarrow$ (a): Replacing f with $\hat{f}(x) = f(x_0 + x)$ if necessary, we may assume that $x_0 = 0$. Let

$$C_n = \{x \in X : \max\{f(x), f(-x)\} \le n\} .$$

The lower semicontinuity of f implies that each C_n is closed and clearly $X = \bigcup_{n \ge 1} nC_n$. Then the Baire Category Theorem (see Theorem 1.5.68) implies that $\operatorname{int} n_0 C_{n_0} \neq \emptyset$ for some $n_0 \in \mathbb{N}$. Therefore, $0 \in \operatorname{int} C_{n_0}$. Then, by Proposition 5.1.20, we see that f is continuous at $x_0 = 0$. ☐

Definition 5.1.22. Let X be a normed space and let $f\colon \to \mathbb{R}$. We say that f is **locally Lipschitz** if for every $x \in X$ there exist $U \in \mathcal{N}(x)$ and $k_U > 0$ such that

$$|f(u) - f(v)| \le k_U \|u - v\| \quad \text{for all } u, v \in U .$$

As a consequence of Propositions 5.1.21 and 5.1.20 and of the definition above, we can state the following corollary.

Corollary 5.1.23. *If X is a Banach space and $f\colon X \to \overline{\mathbb{R}}$ is lower semicontinuous and convex, then $f|_{\operatorname{int}\operatorname{dom} f}$ is locally Lipschitz. In particular, a continuous and convex function $f\colon X \to \mathbb{R}$ is locally Lipschitz.*

Proposition 5.1.24. *If X is a normed space and $f\colon X \to \overline{\mathbb{R}}$ is proper and convex, then the following statements are equivalent:*
(a) *f is bounded above in a neighborhood of x_0;*
(b) *f is continuous at $x_0 \in X$;*
(c) $\operatorname{int}\operatorname{epi} f \neq \emptyset$*;*
(d) $\operatorname{int}\operatorname{dom} f \neq \emptyset$ *and $f|_{\operatorname{int}\operatorname{dom} f}$ is continuous.*

Moreover, if one of the statements above holds, then

$$\operatorname{int}\operatorname{epi} f = \{(x, \eta) \in X \times \mathbb{R} \colon x \in \operatorname{int}\operatorname{dom} f, f(x) < \eta\} .$$

Proof. (a) $\Longleftrightarrow$ (b): This equivalence is Proposition 5.1.20.

(a) $\Longrightarrow$ (c): Let $U \in \mathcal{N}(x_0)$ such that $f|_U \le M$. Then $U \subseteq \operatorname{int}\operatorname{dom} \varphi$ and $\{(x, \eta) \in U \times \mathbb{R} \colon M < \eta\} \subseteq \operatorname{epi} f$. Hence, $\operatorname{int}\operatorname{epi} f \neq \emptyset$.

(c) $\Longrightarrow$ (a): Let $(x_0, \eta) \in \operatorname{int}\operatorname{epi} f$. Then there exist $U \in \mathcal{N}(x_0)$ and $\varepsilon > 0$ such that $U \times [\eta - \varepsilon, \eta + \varepsilon] \subseteq \operatorname{epi} f$. Hence, $U \times \{\eta\} \subseteq \operatorname{epi} f$ and so $f|_U \le \eta$.

(a) $\Longrightarrow$ (d): As before, without any loss of generality, we may assume that $x_0 = 0$. Let $U \in \mathcal{N}(x_0)$ be such that $f|_U \le M$ for some $M > 0$. Then $U \subseteq \operatorname{dom} f$ and so $\operatorname{int}\operatorname{dom} f \neq \emptyset$. Note that the set $\operatorname{dom} f$ is convex. So, if $x \in \operatorname{int}\operatorname{dom} f$, there is $\lambda > 1$ such that $v = \lambda x \in \operatorname{dom} f$; see Proposition 3.1.26. We set $V = x + (1 - 1/\lambda)U \in \mathcal{N}(x)$. If $y \in V$, then we obtain $y = x + (1 - 1/\lambda)u$ with $u \in U$. From the convexity of f we derive

$$\begin{aligned} f(y) &= f\left(\frac{1}{\lambda} v + \left(1 - \frac{1}{\lambda}\right) u\right) \le \frac{1}{\lambda} f(v) + \left(1 - \frac{1}{\lambda}\right) f(u) \\ &\le \frac{1}{\lambda} f(v) + \left(1 - \frac{1}{\lambda}\right) M = \hat{M} . \end{aligned}$$

This shows that $f|_V$ is bounded above and so $f|_{\operatorname{int}\operatorname{dom} f}$ is continuous.

(d) $\Longrightarrow$ (a): This is clear.

Finally let $D = \{(x,\eta) \in X \times \mathbb{R}: x \in \operatorname{int}\operatorname{dom} f, f(x) < \eta\}$. Clearly, $\operatorname{int}\operatorname{epi} f \subseteq D$. Let $x \in \operatorname{int}\operatorname{dom} f$ with $f(x) < \eta$. We choose $\mu \in \mathbb{R}$ such that $f(x) < \mu < \eta$. By hypothesis $f|_{\operatorname{int}\operatorname{dom} f}$ is continuous, so there exists $U \in \mathcal{N}(x)$ with $U \subseteq \operatorname{int}\operatorname{dom} f$ and $f|_U < \mu$. Then $U \times (\mu, +\infty) \subseteq \operatorname{int}\operatorname{epi} f$. Therefore, $D \subseteq \operatorname{int}\operatorname{epi} f$, which means that $D = \operatorname{int}\operatorname{epi} f$. □

In finite dimensional spaces the situation is simpler.

Proposition 5.1.25. *If X is a finite dimensional vector space and $f: X \to \overline{\mathbb{R}}$ is convex, then $f|_{\operatorname{int}\operatorname{dom} f}$ is locally Lipschitz.*

Proof. Let $x \in \operatorname{int}\operatorname{dom} f$. We can find $\delta > 0$ and $\{e_n\}_{n=1}^{N+1} \subseteq X$, where $N = \dim X$, such that

$$B_\delta(x) \subseteq \operatorname{conv}\{e_n\}_{n=1}^{N+1} \subseteq \operatorname{dom} f\,;$$

see Theorem 3.1.30. So, if $u \in B_\delta(x)$, there exists $\{\lambda_n\}_{n=1}^{N+1} \subseteq [0,1]$ such that

$$\sum_{n=1}^{N+1} \lambda_n = 1 \quad \text{and} \quad u = \sum_{n=1}^{N+1} \lambda_n e_n\,.$$

The convexity of f implies that

$$f(u) \le \sum_{n=1}^{N+1} \lambda_n f(e_n) \le \max[f(e_n): 1 \le n \le N+1] = M\,,$$

which shows that $f|_{\operatorname{int}\operatorname{dom} f}$ is locally Lipschitz; see Proposition 5.1.20. □

Remark 5.1.26. Comparing the proposition above with Corollary 5.1.23, we see that in the infinite dimensional case, we need the extra condition that f is lower semicontinuous.

Definition 5.1.27. Let X be a normed space.

(a) Let $x^* \in X^*$ and $\eta \in \mathbb{R}$. A function $a: X \to \mathbb{R}$ of the form

$$a(x) = \langle x^*, x\rangle + \eta \quad \text{for all } x \in X$$

is said to be a **continuous affine function**. We denote the set of such functions by $\operatorname{Aff}(X)$.

(b) We define the following sets

$$\Gamma(X) = \{f: X \to \mathbb{R}^*: f(x) = \sup[a(x): a \in \operatorname{Aff}(X), a \le f]\}$$
$$\Gamma_0(X) = \{f \in \Gamma(X): f \text{ is proper}\}\,.$$

Evidently $\Gamma_0(X) \subseteq \Gamma(X)$ and both are cones, that is, they are closed under positive scalar multiplication.

The next proposition characterizes these cones.

Proposition 5.1.28. *If X is a normed space and $f: X \to \mathbb{R}^*$, then $f \in \Gamma(X)$ if and only if f is lower semicontinuous and convex. Moreover if f attains the value $-\infty$, then $f \equiv -\infty$.*

Proof. $\Longrightarrow$: The pointwise supremum of an empty set of functions is $-\infty$. So, if the set of continuous affine minorants of f is nonempty, the function f does not take the value $-\infty$ and being the supremum of continuous convex, in fact affine, functions, it is lower semicontinuous and convex; see Propositions 1.7.4 (a) and 5.1.7.

$\Longleftarrow$: Suppose that f is lower semicontinuous, convex, and $f \not\equiv -\infty$. If $f \equiv +\infty$, then clearly f is the pointwise supremum of all continuous, affine functions. So, we assume that f is proper.

Let $(\hat{x}, \hat{\eta}) \notin \operatorname{epi} f$. Since $\operatorname{epi} f$ is closed and convex, by the Strong Separation Theorem, there exist $x^* \in X^*$ and $\alpha, \beta \in \mathbb{R}$ such that

$$\langle x^*, \hat{x}\rangle + \alpha\hat{\eta} < \beta < \langle x^*, x\rangle + \alpha\eta \quad \text{for all } (x, \eta) \in \operatorname{epi} f\,. \tag{5.1.8}$$

First assume that $\hat{x} \in \operatorname{dom} f$, that is, $f(\hat{x}) < +\infty$. Then we choose $x = \hat{x}$ and $\eta = f(\hat{x})$ in (5.1.8) to obtain $0 < \alpha[f(\hat{x}) - \hat{\eta}]$, thus $\alpha > 0$. Again from (5.1.8) one has

$$\hat{\eta} < \frac{\beta}{\alpha} - \frac{1}{\alpha}\langle x^*, x\rangle < f(\hat{x})\,.$$

Set $a(x) = \beta/\alpha - 1/\alpha\langle x^*, x\rangle$. Then $a \in \operatorname{Aff}(X)$ with $a \le f$. Since $\hat{\eta} < f(\hat{x})$ is arbitrary, we conclude that $f = \sup[a : a \in \operatorname{Aff}(X), a \le f]$, that is, $f \in \Gamma(X)$.

Now suppose that $f(\hat{x}) = +\infty$. If $\alpha \ne 0$, then we can argue as above and reach the desired conclusion. So, suppose that $\alpha = 0$. We set $a(x) = \beta - \langle x^*, x\rangle$ for all $x \in X$. Then (5.1.8) gives

$$a(\hat{x}) > 0 \quad \text{and} \quad a(x) < 0 \quad \text{for all } x \in \operatorname{dom} f\,.$$

Therefore there exist $u^* \in X^*$ and $\vartheta \in \mathbb{R}$ such that if $\hat{a}(x) = \vartheta - \langle u^*, x\rangle$ for all $x \in X$, then $\hat{a}(x) < f(x)$ for all $x \in X$. For every $m > 0$ we set $\hat{a}_m(x) = \hat{a}(x) + ma(x)$ for all $x \in X$. Hence, $\hat{\eta} \le \hat{a}_m(x) \le f(x)$ for large enough $m > 0$ and for all $x \in X$. Thus, $f \in \Gamma(X)$. □

Remark 5.1.29. So, $\Gamma_0(X)$ is the cone of lower semicontinuous, convex, proper functions and each $f \in \Gamma_0(X)$ is the supremum of all its continuous affine minorants. Moreover, every $f \in \Gamma_0(X)$ admits continuous affine minorants.

Definition 5.1.30. Let X be a normed space and let $f: X \to \overline{\mathbb{R}}$ be a proper function. The largest minorant $\overline{f} \in \Gamma_0(X)$ of f is called the **Γ-regularization** of f and is denoted by $\overline{f}_c$.

From this definition and Proposition 5.1.28 we have the following.

Proposition 5.1.31. *If X is a normed space and $f: X \to \overline{\mathbb{R}}$ is proper, then $\operatorname{epi} \overline{f}_c = \overline{\operatorname{conv}} \operatorname{epi} f$.*

In addition to the Γ-regularization of a proper function $f: X \to \overline{\mathbb{R}}$, we also introduce its lower semicontinuous regularization.

Definition 5.1.32. Let X be a topological space and let $f: X \to \overline{\mathbb{R}}$ be a proper function. The largest lower semicontinuous minorant of f is called the **lower semicontinuous regularization** of f and is denoted by $\overline{f}$.

The next proposition characterizes $\overline{f}$.

Proposition 5.1.33. *If X is a topological space and $f: X \to \overline{\mathbb{R}}$ is a proper function, then $\overline{f}(x) = \sup_{U \in \mathcal{N}(x)} \inf_{u \in U} f(u) = \liminf_{u \to x} f(u)$ for all $x \in X$ and $\operatorname{epi} \overline{f} = \overline{\operatorname{epi} f}$.*

Proof. Evidently, the function $f_0(x) = \sup_{U \in \mathcal{N}(x)} \inf_{u \in U} f(u)$ is lower semicontinuous and $f_0(x) \le f(x)$ for all $x \in X$. Hence $f_0 \le \overline{f}$. On the other hand, if h is a lower semicontinuous minorant of f, then

$$h(x) = \sup_{U \in \mathcal{N}(x)} \inf_{u \in U} h(u) \le \sup_{U \in \mathcal{N}(x)} \inf_{u \in U} f(u) = f_0(x) \quad \text{for all } x \in X\,;$$

see Remark 1.7.3. Hence, $\overline{f}(x) \le f_0(x)$ for all $x \in X$ and so $\overline{f} = f_0$.

Since $\overline{f} \le f$, we have $\operatorname{epi} f \subseteq \operatorname{epi} \overline{f}$, and hence, $\overline{\operatorname{epi} f} \subseteq \operatorname{epi} \overline{f}$. On the other hand $\overline{\operatorname{epi} f} = \operatorname{epi} h$ for some proper function $h: X \to \overline{\mathbb{R}}$. This function is lower semicontinuous (see Proposition 1.7.2) and satisfies $h \le f$. Therefore, $h = \overline{f}$. □

Remark 5.1.34. In addition, it holds that

$$\{x \in X : \overline{f}(x) \le \eta\} = \bigcap_{\vartheta > \eta} \overline{\{x \in X : f(x) \le \vartheta\}} \quad \text{for all } \eta \in \mathbb{R}\,.$$

Example 5.1.35. If $A \subseteq X$ and $f = i_A$, then $\overline{f} = i_{\overline{A}}$.

In first countable spaces, we can state another characterization of $\overline{f}$ in terms of sequences.

Proposition 5.1.36. *If X is a first countable topological space and $f: X \to \overline{\mathbb{R}}$ is proper, then $\overline{f}$ is characterized by the following two properties:*

(a) *for every sequence $x_n \to x$, we have $\overline{f}(x) \le \liminf_{n \to \infty} f(x_n)$;*
(b) *there exists a sequence $x_n \to x$ such that $\limsup_{n \to \infty} f(x_n) \le \overline{f}(x)$.*

Proof. Since $\overline{f}$ is lower semicontinuous and $\overline{f} \le f$, we obtain

$$\overline{f}(x) \le \liminf_{n \to \infty} \overline{f}(x_n) \le \liminf_{n \to \infty} f(x_n)$$

for every sequence $x_n \to x$. This establishes (a). For property (b) we may assume that $x \in \operatorname{dom} \overline{f}$, that is, $\overline{f}(x) < +\infty$. Let $\{U_n\}_{n \ge 1}$ be a decreasing local basis at x, which exists since X is first countable. Let $\eta_n \searrow \overline{f}(x)$ and $\eta_n \ne \overline{f}(x)$. Then by Proposition 5.1.33 it follows that $\inf_{u \in U_n} f(u) < \eta_n$ and so there exists $x_n \in U_n$ such that $f(x_n) < \eta_n$. Evidently $x_n \to x$ and $\limsup_{n \to \infty} f(x_n) \le \overline{f}(x)$. This establishes property (b). □

A direct consequence of Definition 5.1.32 is the following result.

Proposition 5.1.37. *If X is a topological space and $f, h: X \to \overline{\mathbb{R}}$ are proper functions, then $\bar{f} + \bar{h} \leq \overline{f + h}$. Moreover, if h is continuous, then $\overline{f + h} = \bar{f} + h$.*

Another consequence of Definitions 5.1.32 and 5.1.30 is the following one.

Proposition 5.1.38. *If X is a normed space and $f: X \to \overline{\mathbb{R}}$ is a proper function, then $\bar{f}_c \leq \bar{f} \leq f$.*

5.2 Differentiability of Convex Functions

Convex functions exhibit interesting differentiability properties. In this section we explore some of these properties.

Let X be a normed space, $U \subseteq X$ is an open set, $x \in U$, and $f: U \to \mathbb{R}$. We introduce the following **directional derivatives** of f at x in the direction $h \in X$

$$f'_+(x;h) = \lim_{\lambda \to 0^+} \frac{f(x + \lambda h) - f(x)}{\lambda}, \quad f'_-(x;h) = \lim_{\lambda \to 0^-} \frac{f(x + \lambda h) - f(x)}{\lambda}$$

$$f'(x;h) = \lim_{\lambda \to 0} \frac{f(x + \lambda h) - f(x)}{\lambda}$$

From the definitions above it is clear that $f'_+(x;h) = -f'_-(x;-h)$. Moreover, $f'(x;h)$ exists if and only if $f'_+(x;\pm h)$ both exist and $f'_+(x;h) = -f'_+(x;-h)$. In addition, it holds that $f'_+(x;0) = 0$ and $f'_+(x;\lambda h) = \lambda f'_+(x;h)$ for all $\lambda > 0$.

Definition 5.2.1. (a) We say that f is **Gateaux differentiable at** x if $f'(x;\cdot) \in X^*$. We say that f is **Gateaux differentiable** if it is Gateaux differentiable at every $x \in U$.

(b) We say that f is **Fréchet differentiable at** x if there exists $x^* \in X^*$ such that

$$\lim_{h \to 0} \frac{f(x + h) - f(x) - \langle x^*, h \rangle}{\|h\|} = 0 .$$

We say that f is **Fréchet differentiable** if it is Fréchet differentiable at every $x \in U$.

In what follows we denote by $f'(x) \in X^*$ both the Gateaux and Fréchet derivatives of f at x. It will be clear from the context which one is used.

Remark 5.2.2. Note that both notions remain unaffected by equivalent renorming of X. A function that is Fréchet differentiable at x, is continuous at x, but this is not true for functions that are Gateaux differentiable at x. For example, the function $f: \mathbb{R}^2 \to \mathbb{R}$ defined by

$$f(x_1, x_2) = \begin{cases} \frac{x_1^6}{x_1^8 + (x_2 - x_1^2)^2} & \text{if } (x_1, x_2) \neq 0 \\ 0 & \text{if } (x_1, x_2) = 0 \end{cases}$$

is Gateaux differentiable at $(0, 0)$ with $f'(0, 0) = 0$ but f is not continuous at zero. Note that $f(x, x^2) = 1/x^2$.

Directly from Definition 5.2.1 we obtain the following simple facts.

Proposition 5.2.3. (a) *f is Gateaux differentiable at $x \in U$ if and only if there exists $x^* \in X^*$ such that $f(x+\lambda h) = f(x) + \lambda\langle x^*, h\rangle + o(\lambda)$ as $\lambda \to 0$. Then $x^* = f'(x)$.*

(b) *f is Fréchet differentiable at $x \in U$ if and only if there exists $x^* \in X^*$ such that $f(x+h) = f(x) + \langle x^*, h\rangle + o(\|h\|)$ as $h \to 0$. Then $x^* = f'(x)$.*

(c) *f is Fréchet differentiable at x if and only if f is Gateaux differentiable at x and*

$$\limsup_{\lambda\to 0}\left[\frac{f(x+\lambda h)-f(x)}{\lambda} : \|h\| \le 1\right] = \langle f'(x), h\rangle .$$

For a vector space V and a subset $A \subseteq V$, the **core** (or **algebraic interior** of A) denoted by $\operatorname{cor} A$ is the set of points x of A such that for all $h \in V \setminus \{x\}$ there exists $\lambda \in (0,1)$ for which $[x, (1-\lambda)x+\lambda h) = \{tx+(1-t)[(1-\lambda)x+\lambda h] : 0 < t \le 1\} \subseteq A$. If $\operatorname{Aff} A$, the affine hull of A, is not all of X, then $\operatorname{cor} A = \emptyset$. Therefore, for convex sets A in a topological vector space with $\operatorname{int} A \ne \emptyset$, we have $\operatorname{int} A = \operatorname{cor} A$. This fact and Proposition 5.1.21 imply the following result.

Proposition 5.2.4. *If X is a Banach space and $f : X \to \overline{\mathbb{R}}$ is lower semicontinuous, convex, and Gateaux differentiable at $x \in X$, then f is continuous at x.*

Proposition 5.2.5. *If $X = \mathbb{R}^N$ and $f : X \to \mathbb{R}$ is Lipschitz on some neighborhood of $x \in X$, then f is Fréchet differentiable at x if and only if it is Gateaux differentiable at x.*

Proof. $\Longrightarrow$: This implication is always true.

$\Longleftarrow$: Let $f'(x) \in X$ be the Gateaux derivative of f at x. Arguing by contradiction, suppose that f is not Fréchet differentiable at $x \in \mathbb{R}^N$. Then there exists a sequence $\{h_n\}_{n\ge 1} \subseteq \mathbb{R}^N \setminus \{0\}$ such that $\|h_n\| \to 0$ and

$$d_n = \left|\frac{f(x+h_n) - f(x) - (f'(x), h_n)_{\mathbb{R}^N}}{\|h_n\|}\right| \not\to 0 . \tag{5.2.1}$$

We set $v_n = h_n/\|h_n\|$ and have $h_n = \lambda_n v_n$ with $\lambda_n = \|h_n\|$ for all $n \in \mathbb{N}$. Note that $\|v_n\| = 1$ for all $n \in \mathbb{N}$ and so we may assume that $v_n \to v$ in $\mathbb{R}^N$ with $\|v\| = 1$. We obtain

$$\begin{aligned}
d_n &= \left|\frac{f(x+t_n v_n) - f(x)}{t_n} - (f'(x), v_n)_{\mathbb{R}^N}\right| \\
&\le \left|\frac{f(x+t_n v) - f(x)}{t_n} - (f'(x), v)_{\mathbb{R}^N}\right| + \frac{|f(x+t_n v_n) - f(x+t_n v)|}{t_n} \\
&\quad + |(f'(x), v_n - v)_{\mathbb{R}^N}| \\
&\le \left|\frac{f(x+t_n v) - f(x)}{t_n} - (f'(x), v)_{\mathbb{R}^N}\right| + [k + \|f'(x)\|]\|v_n - v\| \to 0 ,
\end{aligned}$$

where $k > 0$ is the Lipschitz constant on a neighborhood of x. This contradicts (5.2.1). □

Combining this with Proposition 5.1.25 gives the following result.

Corollary 5.2.6. *If $f: \mathbb{R}^N \to \mathbb{R}$ is convex and all partial derivatives $\partial f/\partial x_k(x)$ exist, then f is Fréchet differentiable.*

Remark 5.2.7. From multivariable calculus we know that the existence of partial derivatives does not imply the existence of the Fréchet derivative in general. In fact also it does not imply Gateaux differentiability. Consider the function

$$f(x_1, x_2) = \begin{cases} \frac{x_1(x_1^2-3x_2^2)}{x_1^2+x_2^2} & \text{if } (x_1, x_2) \neq 0\,, \\ 0 & \text{if } (x_1, x_2) = 0\,. \end{cases}$$

Then $f'((0,0);(h,v)) = f(h,v)$, which is not linear.

Proposition 5.2.8. *If X is a normed space and $f: X \to \overline{\mathbb{R}}$ is proper and convex, then the following hold:*

(a) *for $x \in \operatorname{dom} f$ and $h \in X$, the function $\lambda \to (f(x+\lambda h) - f(x))/\lambda$ is increasing on $\mathbb{R} \setminus \{0\}$ and so $f'_+(x;h)$ exists and it holds that*

$$f(x) - f(x-h) \le f'_+(x;h) = \inf_{\lambda>0} \frac{f(x+\lambda h) - f(x)}{\lambda} \le f(x+h) - f(x)\,;$$

(b) *for $x \in \operatorname{dom} f$, the function $h \to f'_+(x;h)$ is sublinear;*

(c) *for $x \in \operatorname{int}\operatorname{dom} f$, it holds $f'_+(x;h) \in \mathbb{R}$ for all $h \in X$;*

(d) *if $f: X \to \mathbb{R}$ is continuous and convex, then $f'_+(x,\cdot)$ exists and is continuous on X.*

Proof. (a) Since f is proper and convex, the same is true for $\varphi(\lambda) = f(x+\lambda h)$ with $\lambda \in \mathbb{R}$. If $\lambda_1 < \lambda_2 < \lambda_3$, we set $\mu_{mk} = \lambda_m - \lambda_k$ for $m, k = 1, 2, 3$. Note that $\lambda_2 = \mu_{32}/\mu_{31}\lambda_1 + \mu_{21}/\mu_{31}\lambda_3$. So, from the convexity of φ, we obtain that

$$\varphi(\lambda_2) \le \frac{\mu_{32}}{\mu_{31}}\varphi(\lambda_1) + \frac{\mu_{21}}{\mu_{31}}\varphi(\lambda_3)\,,$$

which results in

$$\frac{\varphi(\lambda_2) - \varphi(\lambda_1)}{\lambda_2 - \lambda_1} \le \frac{\varphi(\lambda_3) - \varphi(\lambda_1)}{\lambda_3 - \lambda_1} \le \frac{\varphi(\lambda_3) - \varphi(\lambda_2)}{\lambda_3 - \lambda_2}\,. \tag{5.2.2}$$

Consider $0 < \lambda < \vartheta$. Then from (5.2.2) it follows that $\varphi(\lambda)/\lambda \le \varphi(\vartheta)/\vartheta$, which implies that

$$\frac{f(x+\lambda h) - f(x)}{\lambda} \le \frac{f(x+\vartheta h) - f(x)}{\vartheta}\,.$$

So, we have proven that $\lambda \to (f(x+\lambda h) - f(x))/\lambda$ is increasing on $(0, +\infty)$. For $\lambda < 0$, note that

$$\frac{f(x+\lambda h) - f(x)}{\lambda} = -\frac{f(x+(-\lambda)(-h)) - f(x)}{-\lambda}\,,$$

which shows that $\lambda \to (f(x+\lambda h)-f(x))/\lambda$ is increasing on $(-\infty, 0)$. It follows that $f'_+(x;h)$ exists in $\mathbb{R}$ and we get

$$f(x) - f(x-h) \le f'(x;h) = \inf_{\lambda>0} \frac{f(x+\lambda h) - f(x)}{\lambda} \qquad (5.2.3)$$
$$\le f(x+h) - f(x) \quad \text{for all } h \in X\,.$$

(b) Since f is convex, for every $h, v \in X$, we obtain

$$f(x+\lambda(h+v)) = f\left(\frac{1}{2}(x+2\lambda h) + \frac{1}{2}(x+2\lambda v)\right)$$
$$\le \frac{1}{2} f(x+2\lambda h) + \frac{1}{2} f(x+2\lambda v)\,.$$

Hence,

$$f'_+(x;h+v) \le f'_+(x;h) + f'_+(x;v)\,.$$

Clearly, $f'_+(x,\cdot)$ is positively homogeneous. Therefore, $f'_+(x,\cdot)$ is sublinear.

(c) If $x \in \operatorname{int}\operatorname{dom} f$, there exists $\delta > 0$ such that $x \pm \delta h \in \operatorname{dom} f$. From (5.2.3) with h replaced by δh, we have $\delta f'_+(x;h) = f'_+(x;\delta h) \in \mathbb{R}$, which gives $f'_+(x;h) \in \mathbb{R}$ for all $h \in X$.

(d) There exists $\delta > 0$ such that $f(x+h) - f(x) \le 1$ for all $h \in B_\delta(0)$. Then, due to (5.2.3), this yields $f'_+(x,\cdot)$, which is bounded above on $B_\delta(0)$. Since $f'_+(x,\cdot)$ is sublinear (see part (b)), it follows that $f'_+(x,\cdot)$ is continuous; see Proposition 5.1.20. □

Remark 5.2.9. Evidently, if $x \in \operatorname{dom} f$, then $h \to f'_-(x;h)$ is superlinear, that is, $f'_-(x;\cdot)$ is positively homogeneous and superadditive, that is, if $h, v \in X$, then $f'_-(x;h+v) \ge f'_-(x;h) + f'_-(x;v)$.

We can state characterizations of Fréchet and Gateaux differentiability without explicit mention of the derivative.

Proposition 5.2.10. *If X is a Banach space and $f: X \to \overline{\mathbb{R}}$ is convex and continuous at $x_0 \in X$, then f is Fréchet differentiable at $x_0 \in X$ if and only if*

$$\lim_{\lambda\to 0} \frac{f(x_0+\lambda h) + f(x_0-\lambda h) - 2f(x_0)}{\lambda} = 0$$

uniformly in $h \in X$ with $\|h\| = 1$.

Proof. $\Longrightarrow$: The Fréchet differentiability of f at x_0 implies that for a given $\varepsilon > 0$ there exists $\delta > 0$ such that

$$\left|f(x_0+h) + f(x_0-h) - \langle f'(x_0), h\rangle\right| \le \frac{\varepsilon}{2}\|h\| \quad \text{for all } h \in X \text{ with } \|h\| \le \delta\,. \qquad (5.2.4)$$

We use (5.2.4) first with h and then with $-h$. Adding these inequalities leads to

$$0 \le f(x_0+h) + f(x_0-h) - 2f(x_0) \le \varepsilon\|h\| \quad \text{for all } h \in X \text{ with } \|h\| \le \delta\,. \qquad (5.2.5)$$

The first inequality in (5.2.5) is a consequence of the convexity of f. For $\|h\| = 1$ and $\lambda \in (0, \delta]$ one has $\|\lambda h\| \le \delta$ and from (5.2.5) it follows that

$$0 \le \frac{f(x_0 + \lambda h) + f(x_0 - \lambda h) - 2f(x_0)}{\lambda} \le \varepsilon \quad \text{for all } \lambda \in (0, \delta] \text{ with } \|h\| = 1\,.$$

This gives

$$\lim_{\lambda \to 0} \frac{f(x_0 + \lambda h) + f(x_0 - \lambda h) - 2f(x_0)}{\lambda} = 0 \quad \text{uniformly in } h \in X \text{ with } \|h\| = 1\,.$$

Moreover, we easily see that

$$\frac{f(x_0 + \lambda h) + f(x_0 - \lambda h) - 2f(x_0)}{\lambda} = \frac{f(x_0 + \lambda h) - f(x_0)}{\lambda} + \frac{f(x_0 - \lambda h) - f(x_0)}{\lambda}\,.$$

Hence, $f'_+(x_0; h) = f'_-(x_0; h) = f'(x_0; h)$ and so $f'(x_0; \cdot) \in X^*$, which shows that f is Gateaux differentiable.

In addition, $\lim_{\lambda \to 0}(f(x_0 + \lambda h) - f(x_0))/\lambda = f'(x_0; h)$ is uniform in $\|h\| = 1$. Therefore, f is Fréchet differentiable by Proposition 5.2.3. □

In a similar way we can state an analogous characterization of Gateaux differentiability for convex functions.

Proposition 5.2.11. *If X is a Banach space and $f: X \to \overline{\mathbb{R}}$ is convex and continuous at $x_0 \in X$, then f is Gateaux differentiable at x_0 if and only if for every $\varepsilon > 0$ and every $h \in X$ with $\|h\| = 1$ there exists $\delta = \delta(\varepsilon, h) > 0$ such that*

$$f(x_0 + th) + f(x_0 - th) - 2f(x_0) \le \varepsilon t \quad \textit{for all } t \in [0, \delta]\,.$$

Convex functions on the real line have many points of differentiability, as the next proposition shows.

Proposition 5.2.12. *If $T \subseteq \mathbb{R}$ is an open interval and $f: T \to \mathbb{R}$ is convex, then f is differentiable at all but at most countably many points of T.*

Proof. From (5.2.2) we see that $x \to f'_+(x) = \lim_{\lambda \to 0^+}(f(x + \lambda) - f(x))/\lambda$ is nondecreasing on T. Note that the points where f fails to be differentiable are the discontinuity jump points of f'_+. But by Proposition 4.3.3 this set is at most countable. □

Next we introduce a notion that will be the main object of interest in the next section.

Definition 5.2.13. Let X be a normed space, $f: X \to \overline{\mathbb{R}}$ a function, and $x_0 \in \operatorname{dom} f$. The **subdifferential** of f at x_0 is the set

$$\partial f(x_0) = \{x^* \in X^* : \langle x^*, x - x_0 \rangle \le f(x) - f(x_0) \text{ for all } x \in X\}\,.$$

When $x_0 \notin \operatorname{dom} f$, we set $\partial f(x_0) = \emptyset$. The elements of the set $\partial f(x_0)$ are called **subgradients** of f at x_0.

Remark 5.2.14. According to the definition above, $x^* \in \partial f(x_0)$ if and only if the affine function $x \to f(x_0) + \langle x^*, x - x_0\rangle$ supports the epigraph of f at $(x_0, f(x_0))$. Thus the subdifferential generalizes the classical notion of derivative. The subdifferential of a lower semicontinuous convex function may be empty at some points in its effective domain. Consider the function $f(x) = -\sqrt{1 - x^2}$ for all $x \in [-1, 1]$. Then $\partial f(\pm 1) = \emptyset$. The domain of ∂f is the set $D(\partial f) = \{x \in X : \partial f(x) \neq \emptyset\}$. For a convex function f, $\operatorname{dom} f$ is always convex. However, $D(\partial f)$ need not be convex.

Next, we give a detailed study of the subdifferential of a convex function. At this point we want to use the subdifferential to characterize the Gateaux differentiability of convex functions. First we relate $f'_+(x, \cdot)$ and $\partial f(x)$.

Proposition 5.2.15. *If X is a Banach space and $f : X \to \overline{\mathbb{R}}$ is a proper, convex function, then the following hold:*
(a) *for $x_0 \in \operatorname{dom} f$ we have*

$$\partial f(x_0) = \{x^* \in X^* : \langle x^*, h\rangle \leq f'_+(x_0; h) \text{ for all } h \in X\} ;$$

(b) *if $x_0 \in \operatorname{dom} f$ and f is continuous at x_0, then $\partial f(x_0) \subseteq X^*$ is nonempty, w^*-compact and convex;*
(c) *if $x_0 \in \operatorname{int} \operatorname{dom} f$ and f is continuous at x_0, then $\sigma(h; \partial f(x_0)) = f'_+(x_0; h)$ for all $h \in X$; see Definition 5.1.17.*

Proof. (a) Let $x^* \in \partial f(x_0)$. Then for $\lambda > 0$ we obtain $\lambda\langle x^*, h\rangle \leq f(x_0 + \lambda h) - f(x_0)$, which gives $\langle x^*, h\rangle \leq f'_+(x_0; h)$ for all $h \in X$. Hence

$$\partial f(x_0) \subseteq \{x^* \in X^* : \langle x^*, h\rangle \leq f'_+(x_0; h) \text{ for all } h \in X\} =: D^* .$$

On the other hand, if $x^* \in D^*$, then from Proposition 5.2.8 (a) it follows that $\langle x^*, h\rangle \leq f(x_0 + h) - f(x_0)$ for all $x \in X$. Therefore, $x^* \in \partial f(x_0)$ and so $\partial f(x_0) = D^*$.

(b) From Proposition 5.1.24 we know that $\operatorname{int} \operatorname{epi} f \neq \emptyset$ and in addition, that $x_0 \in \operatorname{int} \operatorname{dom} f$. Since $(x_0, f(x_0))$ is a boundary point of $\operatorname{epi} f$, by the First Separation Theorem (see Theorem 3.1.59), there exists $(x^*, \eta) \in X^* \times \mathbb{R}$ with $(x^*, \eta) \neq (0, 0)$ such that

$$\eta(f(x_0) - \lambda) \leq \langle x^*, u - x_0\rangle \quad \text{for all } (u, \lambda) \in \operatorname{epi} f . \tag{5.2.6}$$

Note that λ can increase up to $+\infty$. So, from (5.2.6) we see that $\eta \geq 0$. If $\eta = 0$, then

$$\langle x^*, x_0\rangle \leq \langle x^*, u\rangle \quad \text{for all } u \in \operatorname{dom} f . \tag{5.2.7}$$

Since $x_0 \in \operatorname{int} \operatorname{dom} f$, from (5.2.7) we conclude that $x^* = 0$, a contradiction. So $\eta > 0$ and we can take $\eta = 1$. From (5.2.6) with $\lambda = f(u)$ one obtains $f(x_0) - f(u) \leq \langle x^*, u - x_0\rangle$ for all $u \in \operatorname{dom} f$. Thus, $-x^* \in \partial f(x_0)$ and so $\partial f(x_0) \neq \emptyset$.

From Proposition 5.1.20, we know that there exists $\delta > 0$ such that $f|_{B_\delta(x_0)}$ is Lipschitz. Then, for $x^* \in \partial f(x_0)$, it follows that

$$\langle x^*, h \rangle \leq f(x_0 + h) - f(x_0) \leq k\|h\| \quad \text{for all } h \in B_\delta(0) .$$

Hence, $\|x^*\| \leq k$. Therefore, $\partial f(x_0)$ is bounded and clearly w^*-closed and convex. Thus, $\partial f(x_0)$ is nonempty, w^*-compact, and convex.

(c) Let $\rho(h) = f'_+(x_0; h)$. By Proposition 5.2.8, ρ is sublinear, continuous, and in fact, Lipschitz. Let $h \in X$ with $\|h\| = 1$ and $H = \mathbb{R}h$. We introduce the linear functional $l: H \rightarrow \mathbb{R}$ defined by $l(th) = t\rho(h)$ for all $t \in \mathbb{R}$. Then by the Hahn–Banach Theorem, there exists $\hat{l} \in X^*$ such that $\hat{l}|_H = l$. Moreover, we have $\langle \hat{l}, v \rangle \leq p(v) \leq f(x_0 + v) - f(x_0)$ for all $v \in X$. Thus, $\hat{l} \in \partial f(x_0)$ and $\hat{l}(th) = f'_+(x; th)$ for all $t \geq 0$. Hence,

$$f'_+(x_0; h) = \sigma(h; \partial f(x_0)) \quad \text{for all } h \in X .$$ □

Using this proposition we can state the following characterization of the Gateaux differentiability of a convex function in terms of its subdifferential.

Theorem 5.2.16. *If X is a Banach space and $f: X \rightarrow \overline{\mathbb{R}}$ is proper and convex with $x_0 \in \text{dom} f$, then the following hold:*
(a) *if f is Gateaux differentiable at x_0, then $\partial f(x_0) = \{f'(x_0)\}$;*
(b) *if f is continuous at x_0 and $\partial f(x_0)$ is a singleton, then f is Gateaux differentiable at x_0 and $\partial f(x_0) = \{f'(x_0)\}$.*

Proof. (a) The convexity of f implies that

$$\langle f'(x_0), h \rangle \leq \frac{1}{\lambda}[f(x_0 + \lambda h) - f(x_0)] \leq f(x_0 + h) - f(x_0)$$

for all $\lambda \in (0,1)$ and for all $h \in X$. Then, $f'(x_0) \in \partial f(x_0)$; see Definition 5.2.13. Suppose that $x^* \in \partial f(x_0)$. Then

$$\langle x^*, h \rangle \leq \frac{1}{\lambda}[f(x_0 + h) - f(x_0)] \quad \text{for all } \lambda > 0 \text{ and for all } h \in X .$$

This implies $\langle x^*, h \rangle \leq \langle f'(x_0), h \rangle$ for all $h \in X$. Hence, $x^* = f'(x_0)$ and so $\partial f(x_0) = \{f'(x_0)\}$.

(b) Suppose $\partial f(x_0) = \{x^*\}$. From Proposition 5.2.15 we know that $f'_+(x_0; h) = \langle x^*, h \rangle$ for all $h \in X$. Hence, $f'_+(x_0; \cdot) = x^*$ and so, f is Gateaux differentiable at x_0. □

For separable Banach spaces we have generic Gateaux differentiability of continuous, convex functions.

Theorem 5.2.17. *If X is a separable Banach space and $f: X \rightarrow \mathbb{R}$ is continuous and convex, then f is Gateaux differentiable on a dense G_δ-subset of X.*

Proof. Let $\{x_n\}_{n\geq 1}$ be dense in the unit sphere $\partial B_1 = \{x \in X: \|x\| = 1\}$. For $n, m \in \mathbb{N}$ let

$$A_{n,m} = \{x \in X : \text{there exists } x^*, u^* \in \partial f(u) \text{ such that } \langle x^* - u^*, x_n\rangle \geq 1/m\} .$$

According to Theorem 5.2.16, $f'(x)$ does not exist if and only if $\partial f(x)$ is not a singleton if and only if $x \in \bigcup_{n,m\geq 1} A_{n,m}$. We show that each set $A_{n,m}$ is closed. So, let $\{y_k\}_{k\geq 1} \subseteq A_{n,m}$ and assume that $y_k \to y$ in X. For each $k \in \mathbb{N}$ there exist $x_k^*, u_k^* \in \partial f(y_k)$ such that $\langle x_k^* - u_k^*, x_n\rangle \geq 1/m$.

The separability of X implies that bounded sets in X^* endowed with the relative weak* topology are metrizable; see Theorem 3.4.12 and Remark 3.4.13. From the proof of Proposition 5.2.15 (b), it is clear that $\bigcup_{k\geq 1} \partial f(y_k) \subseteq X^*$ is bounded. Therefore, we may assume that

$$x_k^* \xrightarrow{w^*} x^* \quad \text{and} \quad u_k^* \xrightarrow{w^*} u^* \quad \text{in } X^* .$$

For any $h \in X$ we see that

$$\begin{aligned} \langle x^*, h - y\rangle &= \lim_{k\to\infty} \langle x_k^*, h - y_k\rangle \leq \lim_{k\to\infty} [f(h) - f(y_k)] = f(h) - f(y) , \\ \langle u^*, h - y\rangle &= \lim_{k\to\infty} \langle u_k^*, h - y_k\rangle \leq \lim_{k\to\infty} [f(h) - f(y_k)] = f(h) - f(y) . \end{aligned}$$

Hence, $x^*, u^* \in \partial f(y)$ and $\langle x^* - u^*, x_n\rangle = \lim_{k\to\infty}\langle x_k^* - u_k^*, x_n\rangle \geq 1/m$. Thus, $y \in A_{n,m}$, and so $A_{n,m} \subseteq X$ is closed.

The set $X \setminus A_{n,m} = U_{n,m}$ is open for all $n, m \in \mathbb{N}$. We claim that $U_{n,m}$ is also dense. Let $x_0 \in X$ and consider the function $\xi(\lambda) = f(x_0 + \lambda x_n)$ for all $\lambda \in \mathbb{R}$. According to Proposition 5.2.12 we can approximate x_0 by points of the form $x_0 + \lambda x_n$ with ξ being differentiable at λ. If $x^*, u^* \in \partial f(x_0 + \lambda x_n)$, then their restrictions on $x_0 + \mathbb{R}x_n$ give subgradients of ξ at λ. But ξ is differentiable at λ. So, x^* and u^* coincide on the line $x_0 + \mathbb{R}x_n$. In particular, $\langle x^*, x_n\rangle = \langle u^*, x_n\rangle$. Hence, $x_0 + \lambda x_n \in U_{n,m}$ for all $m \in \mathbb{N}$. From the Baire Category Theorem, we see that $\bigcap_{n,m\geq 1} U_{n,m}$ is dense and G_δ in X. □

Definition 5.2.18. A Banach space X is said to be **weak Asplund** if every continuous and convex function $f: X \to \mathbb{R}$ is Gateaux differentiable at a dense G_δ-subset of X.

This definition in combination with Theorem 5.2.17 yields the following corollary.

Corollary 5.2.19. *Every separable Banach space is weak Asplund.*

Proposition 5.2.20. *If X is a Banach space and $f: X \to \mathbb{R}$ is continuous and convex, then f is Fréchet differentiable at a possibly empty G_δ-set.*

Proof. From Proposition 5.2.10 we know that f is Fréchet differentiable at $x \in X$ if and only if for every $\varepsilon > 0$ there exists $\delta > 0$ such that

$$f(x + \lambda h) + f(x - \lambda h) - 2f(x) < \lambda\varepsilon \tag{5.2.8}$$

for all $h \in X$ with $\|h\| = 1$ and for all $\lambda \in (0, \delta)$. For each $n \in \mathbb{N}$ we define

$$U_n = \Big\{x \in X\colon \text{there exists } \delta > 0 \text{ such that} \\ \sup_{\|h\|=1}\left[\frac{f(x+\delta h) + f(x-\delta h) - 2f(x)}{\delta}\right] < \frac{1}{n}\Big\}.$$

If D is the set of points of Fréchet differentiability of f, from (5.2.8), we obtain that

$$D = \bigcap_{n\geq 1} U_n. \tag{5.2.9}$$

So, we need to show that for every $n \in \mathbb{N}$, U_n is open. Let $x \in U_n$. From Corollary 5.1.23 we know that f is locally Lipschitz. So, there exist $\delta_1 > 0$ and $k > 0$ such that

$$|f(u) - f(v)| \leq k\|u - v\| \quad \text{for all } u, v \in B_{\delta_1}(x). \tag{5.2.10}$$

Moreover, because $x \in U_n$, there exist $\delta > 0$ and $\eta > 0$ such that

$$\frac{f(x+\delta h) + f(x-\delta h) - 2f(x_0)}{\delta} \leq \eta < \frac{1}{n} \tag{5.2.11}$$

for all $h \in X$ with $\|h\| = 1$. We choose a small enough $\delta_2 > 0$ such that $\eta + (4k\delta_2)/\delta < 1/n$ and let $y \in B_{\delta_2(x)}$. Taking (5.2.10) and (5.2.11) into account, one has, for any $h \in X$ with $\|h\| = 1$, that

$$\begin{aligned}
&\frac{f(y+\delta h) + f(y-\delta h) - 2f(y)}{\delta} \\
&\leq \frac{f(x+\delta h) + f(x-\delta h) - 2f(x)}{\delta} + \frac{2|f(y) - f(x)|}{\delta} \\
&\quad + \frac{|f(y+\delta h) - f(x+\delta h)|}{\delta} + \frac{|f(y-\delta h) - f(x-\delta h)|}{\delta} \\
&\leq \eta + \frac{4k\|y - x\|}{\delta} \leq \eta + \frac{4k\delta_2}{\delta} < \frac{1}{n}.
\end{aligned}$$

Hence, $y \in U_n$ and so $B_{\delta_2}(x) \subseteq U_n$. This proves that U_n is open for every $n \in \mathbb{N}$ and so from (5.2.9) we conclude that it is G_δ, possibly empty. □

Definition 5.2.21. A Banach space X is said to be **Asplund** if every continuous and convex function $f\colon X \to \mathbb{R}$ is Fréchet differentiable on a dense set.

Remark 5.2.22. On account of Proposition 5.2.20, f is in fact differentiable on a dense G_δ-set.

The following theorem characterizes Asplund spaces and its proof can be found in Phelps [260, p. 23].

Theorem 5.2.23. *If X is a Banach space, then X is an Asplund space if and only if every separable closed subspace of X has a separable dual space. In particular, every Banach space with a separable dual space is an Asplund space.*

Corollary 5.2.24. *Every reflexive Banach space is an Asplund space.*

Proposition 5.2.25. *If H is a Hilbert space and $f(x) = 1/2\|x\|^2$ for all $x \in H$, then f is Fréchet differentiable and $\langle f'(x), h\rangle = (x, h)$ for all $x, h \in H$, where $(\cdot,\cdot)$ is the inner product of H.*

Proof. For every $x, h \in H$ we easily see that

$$\frac{1}{2}\|x+h\|^2 - \frac{1}{2}\|x\|^2 - (x,h) = \|h\|^2 .$$

Hence, $\langle f'(x), h\rangle = (x, h)$. □

Applying the chain rule and the proposition above, we can state the following result.

Corollary 5.2.26. *The norm of a Hilbert space is Fréchet differentiable at every $x \in H$ with $x \neq 0$.*

Let H be a Hilbert space and let $C \subseteq H$ be a nonempty, closed, and convex set. We define the metric projection map $p_C\colon H \to C$ that assigns to each $x \in H$ its unique best approximation from C; see Definition 3.5.19.

Proposition 5.2.27. *If H is a Hilbert space, $C \subseteq H$ is nonempty, closed, convex, and*

$$f(x) = \frac{1}{2}\left[\|x\|^2 - \|x - p_C(x)\|^2\right] \quad \text{for all } x \in H ,$$

then f is convex, Fréchet differentiable and $\langle f'(x), h\rangle = (p_C(x), h)$ for all $x, h \in H$.

Proof. We easily see that

$$2f(x) = \|x\|^2 - \inf[\|x-u\|^2 : u \in C] = \sup[2(x,u) - \|u\|^2 : u \in C] .$$

So, f is the supremum of affine continuous functions. Therefore, f is convex and from Proposition 3.5.20, we know that f is continuous. Moreover, Proposition 5.2.25 implies that f is Fréchet differentiable and

$$\langle f'(x), h\rangle = (x,h) - (x - p_C(x), h) = (p_C(x), h) \quad \text{for all } h \in H . \qquad \square$$

Remark 5.2.28. If H is a Hilbert space and $f\colon H \to \mathbb{R}$ is Gateaux differentiable at $x_0 \in H$, then $f'(x_0) \in H^*$. From Theorem 3.5.21 we know that there exists a unique $u_0 \in H$ such that $\langle f'(x_0), h\rangle = (u_0, h)$ for all $h \in H$. This element is called the **gradient** of f at x_0 and is denoted by $\nabla f(x_0)$. So, we have $\langle f'(x_0), h\rangle = (\nabla f(x_0), h)$ for all $h \in H$.

Proposition 5.2.29. *If $T \subseteq \mathbb{R}$ is an interval, then the following hold:*
(a) *a differentiable function $f\colon T \to \mathbb{R}$ is convex (resp. strictly convex) if and only if f' is increasing (resp. strictly increasing) on T;*
(b) *a twice differentiable function $f\colon T \to \mathbb{R}$ is convex if and only if $f''(t) \geq 0$ for all $t \in T$.*

Proof. (a) $\Longrightarrow$: This is a consequence of (5.2.2).

$\Longleftarrow$: Arguing by contradiction, suppose that f is not convex. Then there exist $t < s < r$ in the interval T such that

$$\frac{r-s}{r-t} f(t) + \frac{s-t}{r-t} f(r) < f(s)\,,$$

which implies

$$\frac{f(r)-f(s)}{r-s} < \frac{f(s)-f(t)}{s-t}\,. \tag{5.2.12}$$

From (5.2.12) and the Mean Value Theorem, we contradict the hypothesis that f' is increasing. Similarly we show the assertion for strictly convex functions.

(b) This follows immediately from (a). □

Recall that if $f\colon \mathbb{R}^N \to \mathbb{R}$ is twice Gateaux differentiable at x_0, then we can identify $f''(x_0)$ with the Hessian matrix

$$H(x_0) = \big(f_{x_k x_i}(x_0)\big)_{k,i=1}^{N} \quad \text{where } f_{x_k x_i}(x_0) = \frac{\partial f}{\partial x_i \partial x_k}(x_0)$$

by setting

$$f''(x_0)(u,v) = (H(x_0)u, v)_{\mathbb{R}^N} \quad \text{for all } u, v \in \mathbb{R}^N\,.$$

Then the second derivative $\nabla^2 f(x_0) = H(x_0)$ is a symmetric matrix.

Proposition 5.2.30. *If $f\colon \mathbb{R}^N \to \mathbb{R}$ is twice Gateaux differentiable, then f is convex if and only if $\nabla^2 f(x) \geq 0$ for all $x \in \mathbb{R}^N$, that is, $(\nabla f(x)h, h)_{\mathbb{R}^N} \geq 0$ for all $x, h \in \mathbb{R}^N$.*

Proof. Let $x, h \in \mathbb{R}^N$ with $h \neq 0$ and define $\xi(t) = f(x + th)$. Then

$$\xi'(t) = (\nabla f(x+th), h)_{\mathbb{R}^N} \quad \text{and} \quad \xi''(t) = (\nabla^2 f(x+th), h)_{\mathbb{R}^N}\,.$$

The result follows from Proposition 5.2.29. □

Remark 5.2.31. If $\nabla^2 f(x) > 0$ for all $x \in \mathbb{R}^N$, that is, $(\nabla^2 f(x)h, h)_{\mathbb{R}^N} > 0$ for all $x, h \in \mathbb{R}^N$ with $h \neq 0$, then f is strictly convex. However, a function can be strictly convex without $\nabla^2 f(x)$ being positive definite at all points. For example, let $f(x) = x^4$. Then f is strictly convex, but $f'(0) = 0$.

5.3 Conjugate Functions – Convex Subdifferential

Conjugate functions play a major role in the duality theory, which is one of the main themes of "Convex Analysis". Conjugate functions have many interesting properties and are also closely related to the notion of convex subdifferentials, which we investigate in the second half of this section.

Definition 5.3.1. Let X be a locally convex space, X^* is its dual space and $f: X \to \mathbb{R}^*$ is a function. The **conjugate** of f is the function $f^*: X^* \to \mathbb{R}^*$ defined by

$$f^*(x^*) = \sup[\langle x^*, x\rangle - f(x): x \in X]. \tag{5.3.1}$$

We can also define the conjugate of f^* which is called the **second conjugate** of f. This is a function defined on X^{**}.

Remark 5.3.2. Clearly we can restrict ourselves to $x \in \operatorname{dom} f$ in (5.3.1). So, if $\operatorname{dom} f \neq \emptyset$, then $f^*(x^*) > -\infty$ for all $x^* \in X^*$. From (5.3.1) we see that f^* is the pointwise supremum of the family of continuous affine functions $x \to \langle x^*, x\rangle - f(x)$. Therefore, $f^* \in \Gamma(X^*)$; see Definition 5.1.27 (b). The conjugate f^* may not be proper, even if f is proper on X.

The next proposition contains some properties of the conjugate function that are an immediate consequence of Definition 5.3.1.

Proposition 5.3.3. *The following hold:*
(a) $f^*(0) = -\inf[f(x): x \in X]$;
(b) $f \leq h$ *implies* $h^* \leq f^*$;
(c) $(\inf_{i\in I} f_i)^* = \sup_{i\in I} f_i^*$ *and* $(\sup_{i\in I} f_i)^* \leq \inf_{i\in I} f_i^*$ *for every family of functions* $f_i: X \to \mathbb{R}^*$ *with* $i \in I$;
(d) $(\lambda f)^*(x^*) = \lambda f^*(x^*/\lambda)$ *for all* $\lambda > 0$ *and for all* $x^* \in X^*$;
(e) $(f+\eta)^* = f^* - \eta$ *for all* $\eta \in \mathbb{R}$;
(f) *if for* $u \in X$, $f_u(x) = f(x-u)$, *then* $(f_u)^*(x^*) = f^*(x^*) + \langle x^*, u\rangle$ *for all* $x^* \in X^*$;
(g) *if* f *is proper, then* $\langle x^*, x\rangle \leq f(x) + f^*(x^*)$ *for all* $x \in X$ *and* $x^* \in X^*$. *This inequality is known as the* ***Young–Fenchel inequality***.
(h) *if* $C \subseteq X$ *is convex and* $f = i_C$, *then* $f^*(\cdot) = \sigma(\cdot; C)$.

Proposition 5.3.4. *If* X *is a normed space and* $f(x) = \|x\|$ *for all* $x \in X$, *then* $f^*(x^*) = i_{\overline{B}_1^*}(x^*)$ *with* $\overline{B}_1^* = \{x^* \in X^*: \|x^*\|_* \leq 1\}$.

Proof. From Definition 5.3.1, we have

$$f^*(x^*) = \sup[\langle x^*, x\rangle - \|x\|: x \in X]. \tag{5.3.2}$$

First suppose that $\|x^*\|_* \leq 1$. Then $\langle x^*, x\rangle \leq \|x\|$ for all $x \in X$ and so $\langle x^*, x\rangle - \|x\| \leq 0$. Hence, $f^*(x^*) = 0$ for all $\|x^*\|_* \leq 1$; see (5.3.2).

Next suppose that $\|x^*\|_* > 1$. Then

$$\|x^*\|_* = \sup\left[\left\langle x^*, \frac{x}{\|x\|}\right\rangle : x \in X, x \neq 0\right] > 1\,.$$

Thus, there exists $\hat{x} \in X$ with $\hat{x} \neq 0$ such that $\langle x^*, \hat{x}/\|\hat{x}\|\rangle > 1$, which implies that $\langle x^*, \hat{x}\rangle > \|\hat{x}\|$. Then we obtain

$$\begin{aligned} f^*(x^*) &= \sup[\langle x^*, x\rangle - \|x\| : x \in X] \geq \sup[\langle x^*, \lambda\hat{x}\rangle - \lambda\|\hat{x}\| : \lambda > 0] \\ &= \Big(\sup_{\lambda>0} \lambda\Big)[\langle x^*, \hat{x}\rangle - \|\hat{x}\|] = +\infty\,. \end{aligned}$$

So, we have proven that

$$f^*(x^*) = \begin{cases} 0 & \text{if } \|x^*\|_* \leq 1\,, \\ +\infty & \text{if } 1 < \|x^*\|_*\,. \end{cases}$$

Hence, $f^* = i_{\overline{B}_1^*}$. □

Proposition 5.3.5. *If X is a normed space, $\xi\colon \mathbb{R} \to \overline{\mathbb{R}}$ is proper, even, and $f(x) = \xi(\|x\|)$ for all $x \in X$, then $f^*(x^*) = \xi^*(\|x^*\|_*)$ for all $x^* \in X^*$.*

Proof. From Definition 5.3.1, since ξ is even, we obtain

$$\begin{aligned} f^*(x^*) &= \sup[\langle x^*, x\rangle - \xi(\|x\|)] = \sup_{t\geq 0}\sup_{\|x\|=t}[\langle x^*, x\rangle - \xi(t)] \\ &= \sup_{t\geq 0}\Big[\sup_{\|x\|=t}\langle x^*, x\rangle - \xi(t)\Big] = \sup_{t>0}[t\|x^*\|_* - \xi(t)] \\ &= \sup_{t\in\mathbb{R}}[t\|x^*\|_* - \xi(t)] = \xi^*(\|x^*\|_*)\,. \end{aligned}$$ □

Proposition 5.3.6. *If X is a normed space and $f \in \Gamma_0(X)$, then $f^* \in \Gamma_0(X^*)$.*

Proof. Since $f \in \Gamma_0(X)$, it admits a continuous affine minorant; see Remark 5.1.29. So, there exist $x^* \in X^*$ and $\eta \in \mathbb{R}$ such that $\langle x^*, x\rangle + \eta \leq f(x)$ for all $x \in X$. Then

$$f^*(x^*) = \sup[\langle x^*, x\rangle - f(x) : x \in X] \leq -\eta\,,$$

that is, $x^* \in \operatorname{dom} f^*$. Hence, f^* is proper and so $f^* \in \Gamma_0(X^*)$; see Remark 5.3.2. □

Proposition 5.3.7. *If X is a normed space and $f\colon X \to \overline{\mathbb{R}}$, then the following hold:*
(a) *$f^{**}|_X = \overline{f} \leq f$;*
(b) *if f is proper, then $f^{**}|_X = f$ if and only if $f \in \Gamma_0(X)$.*

Proof. (a) We know that $\overline{f} \leq f$. Let $x^* \in X^*$ and $\eta \in \mathbb{R}$. Then

$$\langle x^*, x\rangle + \eta \leq f(x) \quad \text{for all } x \in X \quad \text{if and only} \quad f^*(x^*) \leq -\eta\,. \tag{5.3.3}$$

From Definition 5.1.30 and (5.3.3), it follows that

$$\overline{f}(x) = \sup[\langle x^*, x\rangle + \eta : x^* \in X^*, \eta \in \mathbb{R}, \eta \le -f^*(x^*)]\,. \tag{5.3.4}$$

If $f^*(x^*) > -\infty$ for all $x^* \in X^*$, then from (5.3.4) we have

$$\overline{f}(x) = \sup[\langle x^*, x\rangle - f^*(x^*) : x^* \in X^*] = f^{**}(x) \quad \text{for all } x \in X\,.$$

Hence, $f^{**}|_X = \overline{f}$. If $f^*(x^*) = -\infty$ for some $x^* \in X^*$, then $\overline{f} \equiv +\infty = f^{**}|_X$.
(b) $\Longrightarrow$: From Proposition 5.3.6, it follows that $f = f^{**}|_X \in \Gamma_0(X)$.
$\Longleftarrow$: This follows from (a) and Proposition 5.1.38. □

Remark 5.3.8. If X is a normed space and $f(x) = \|x\|$ for all $x \in X$, then f is continuous, convex, and from Proposition 5.3.4 we see that $f^* = i_{\overline{B}_1^*}$. Invoking Proposition 5.3.7 we recover the familiar formula for $\|\cdot\|$, namely

$$\|x\| = \sup[\langle x^*, x\rangle : \|x^*\|_* \le 1] = \max[\langle x^*, x\rangle : \|x^*\|_* \le 1]$$

by Alaoglu's Theorem; see also Proposition 3.1.52.

Recall the operation of infimal convolution introduced in Definition 5.1.12. We have the following conjugation rule for this operation.

Proposition 5.3.9. *If X is a normed space and $f, h : X \to \overline{\mathbb{R}}$ are proper functions, then $(f \oplus h)^* = f^* + h^*$.*

Proof. For $x^* \in X^*$ we obtain

$$\begin{aligned}(f \oplus h)^*(x^*) &= \sup[\langle x^*, x\rangle - \inf(f(x-u) + h(u) : u \in X) : x \in X]\\ &= \sup[\langle x^*, y\rangle - f(y) + \langle x^*, u\rangle - h(u) : y, u \in X] = f^*(x^*) + h^*(x^*)\,.\end{aligned}$$ □

Proposition 5.3.10. *If X is a normed space and $f, h : X \to \overline{\mathbb{R}}$ are proper, then the following hold:*
(a) $(f + h)^* \le f^* \oplus h^*$;
(b) *if f, h are convex and there exists a point in* $\operatorname{dom} f \cap \operatorname{dom} h$, *where one of the two functions is continuous, then $(f + h)^* = f^* \oplus h^*$.*

Proof. (a) From the Young–Fenchel inequality (see Proposition 5.3.3 (g)), we derive

$$\langle x^* + u^*, x\rangle - f(x) - h(x) \le f^*(x^*) + h^*(u^*)$$

for all $x^*, u^* \in X^*$ and for all $x \in X$. This implies

$$(f + h)^*(x^* + u^*) \le f^*(x^*) + h^*(u^*)\,. \tag{5.3.5}$$

In particular, (5.3.5) holds for all $x^*, u^* \in X^*$ such that $y^* = x^* + u^*$. Therefore,

$$(f+h)^* \le f^* \oplus h^* . \tag{5.3.6}$$

(b) Let $x^* \in X^*$ and $\lambda = (f+h)^*(x^*)$. If $\lambda = +\infty$, then from (5.3.6) we see that $(f^* \oplus h^*)(x^*) = +\infty$ and so we have equality. Let us assume that $\lambda < +\infty$. Note that $x \in \operatorname{dom}(f+h)$ and so $\lambda \in \mathbb{R}$. We consider the set

$$D = \{(x,\eta) \in X \times \mathbb{R} : \eta \le \langle x^*, x\rangle - h(x) - \lambda\} . \tag{5.3.7}$$

This set is convex. Assuming that f is continuous at a point of $\operatorname{dom} f \cap \operatorname{dom} h$, we will show that $D \cap (\operatorname{int} \operatorname{epi} f) = \emptyset$; see Proposition 5.1.24. Indeed, suppose $(\eta, x) \in D \cap (\operatorname{int} \operatorname{epi} f)$. Then $f(x) < \eta \le \langle x^*, x\rangle - h(x) - \lambda$. Hence,

$$\lambda < \langle x^*, x\rangle - (f(x) + h(x)) \le (f+h)^*(x^*) = \lambda ,$$

a contradiction. So, $D \cap (\operatorname{int} \operatorname{epi} f) = \emptyset$ and we can apply the First Separation Theorem (see Theorem 3.1.59) and find $(u^*, \vartheta) \in X^* \times \mathbb{R}$ with $(u^*, \vartheta) \neq (0,0)$ such that

$$\sup[\vartheta\eta + \langle u^*, x\rangle : (x,\eta) \in \operatorname{epi} f] \le \inf[\vartheta\eta + \langle u^*, x\rangle : (x,\eta) \in D] . \tag{5.3.8}$$

Since ϑ can increase up to $+\infty$, from (5.3.7) we see that $\vartheta \le 0$. If $\vartheta = 0$, then $u^* \neq 0$ separates $\operatorname{dom} f$ and $\operatorname{dom} h$, but this is a contradiction since $\operatorname{int} \operatorname{dom} f \cap \operatorname{dom} h \neq \emptyset$. Therefore, $\vartheta < 0$. We divide both parts of (5.3.8) by $|\vartheta|$ and set $\hat{u}^* = 1/|\vartheta| u^*$. Taking (5.3.7) and (5.3.8) into account, we obtain

$$\begin{aligned} f^*(\hat{u}^*) &= \sup[\langle \hat{u}^*, x\rangle - f(x) : x \in X] = \sup[\langle \hat{u}^*, x\rangle - \eta : (x,\eta) \in \operatorname{epi} f] \\ &\le \inf[\langle \hat{u}^*, x\rangle - \eta : (x,\eta) \in D] = \inf[\langle \hat{u}^* - x^*, x\rangle + h(x) : x \in \operatorname{dom} h] + \lambda \\ &= -h^*(x^* - \hat{u}^*) + \lambda \end{aligned}$$

Therefore,

$$(f^* \oplus h^*)(x^*) = f^*(\hat{u}^*) + h^*(x^* - \hat{u}^*) \le \lambda = (f+h)^*(x^*) .$$

Then, due to (5.3.6), we get $f^* \oplus h^* = (f+h)^*$. □

Let $C \subseteq X$ be a nonempty set. Recall that

$$\begin{aligned} d(x, C) &= \inf[\|x - c\| : c \in C] \qquad \text{with } x \in X , \\ \sigma(x^*; C) &= \sup[\langle x^*, c\rangle : c \in C] \quad \text{with } x^* \in X^* . \end{aligned}$$

In the next proposition, to simplify the notation, we write $d_A(\cdot) = d(\cdot, A)$ and $\sigma_A(\cdot) = \sigma(\cdot; A)$.

Proposition 5.3.11. *If X is a normed space and $C \subseteq X$ is nonempty and convex, then $d_A = \sigma_A^*|_X$.*

Proof. Consider the proper convex functions $f(x) = \|x\|$ and $h(x) = i_C(x)$. We have

$$(f \oplus h)(x) = \inf[\|x - u\| + i_C(u) : u \in X] = d_C(x) \quad \text{for all } x \in X .$$

The function $x \to d_C(x)$ is continuous and convex, hence, thanks to Propositions 5.3.9, 5.3.3 (h) and 5.3.4,

$$\begin{aligned} d_C(x) &= (f \oplus h)^{**}(x) = (f^* + h^*)^* = \sup[\langle x^*, x\rangle - i_{\overline{B}_1^*}(x^*) - \sigma_C(x^*)] \\ &= \sup[\langle x^*, x\rangle - \sigma_C(x^*) : x^* \in \overline{B}_1^*] . \end{aligned} \tag{5.3.9}$$

In fact, since $\overline{B}_1^*$ is w*-compact by Alaoglu's Theorem and $x^* \to \langle x^*, x\rangle - \sigma_C(x^*)$ is weakly upper semicontinuous, the second supremum in (5.3.9) becomes maximum. □

Using the support function $\sigma(\cdot; C)$ we can characterize some important classes of subsets of X. For a proof we refer to Moreau [244] and Laurent [202].

Proposition 5.3.12. *If X is a normed space and $C \subseteq X$ is nonempty, closed, and convex, then the following hold:*
(a) *C is bounded if and only if $\sigma(\cdot; C)$ is strongly continuous on X^*;*
(b) *C is compact if and only if $\sigma(\cdot; C)|_{\overline{B}_1^*}$ is w*-continuous;*
(c) *C is weakly compact if and only if $\sigma(\cdot; C)$ is m-continuous where m denotes the Mackey topology; see Theorem 3.8.9;*
(d) *C is weakly locally compact and contains no line if and only if there exists $x^* \in X^*$ such that $\sigma(\cdot; C)$ is m-continuous at x^**

Moreover, $\sigma(\cdot; C) \in \Gamma_0(X)$ and

$$C = \{x \in X : \langle x^*, x\rangle \le \sigma(x^*; C) \text{ for all } x^* \in X^*\} .$$

Now we turn our attention to the subdifferential of a convex function. We start by recalling the definition of the subdifferential.

Definition 5.3.13. Let X be a normed space and $f: X \to \overline{\mathbb{R}}$ a proper function. The **subdifferential** of f at x is the set $\partial f(x)$ defined by

$$\partial f(x) = \{x^* \in X^* : \langle x^*, h\rangle \le f(x + h) - f(x) \text{ for all } h \in X\} . \tag{5.3.10}$$

The elements of $\partial f(x)$ are called **subgradients** of f at x. Moreover, we say that f is **subdifferentiable** at x if $\partial f(x) \neq \emptyset$. This set is known as the **domain** of ∂f and is denoted by $D(\partial f)$, that is, $D(\partial f) = \{x \in X : \partial f(x) \neq \emptyset\}$.

Remark 5.3.14. From (5.3.10) it is clear that $\partial f(x) \subseteq X^*$ is always closed and convex. In fact $\partial f(x)$ is weakly* closed. Note that $D(\partial f) \subseteq \operatorname{dom} f$. In the definition above, f need not be convex. However, a coherent theory and a remarkable calculus can be developed for convex functions, the **convex subdifferential** for short; see Proposition 5.3.17.

We present three characteristic and important examples of subdifferentials.

Example 5.3.15. (a) Let X be a Banach space and let $f(x) = \|x\|$ for all $x \in X$. We show that

$$\partial f(x) = \begin{cases} \overline{B}_1^* = \{x^* \in X^* : \|x^*\|_* \leq 1\} & \text{if } x = 0\,, \\ \{x^* \in X^* : \|x^*\|_* = 1, \langle x^*, x\rangle = \|x\|\} & \text{if } x \neq 0\,. \end{cases} \tag{5.3.11}$$

In order to prove (5.3.11), we first assume that $x = 0$. Then

$$x^* \in \partial f(0) \quad \text{if and only if} \quad \langle x^*, x\rangle \leq \|h\| \quad \text{for all } h \in X\,. \tag{5.3.12}$$

From (5.3.12) and the definition of the dual norm, one has $\partial f(0) = \overline{B}_1^*$.
Next suppose that $x \neq 0$. Let $C = \{x^* \in X^* : \|x^*\|_* = 1, \langle x^*, x\rangle = \|x\|\}$ and take $x^* \in C$. Then $\langle x^*, h\rangle \leq \|h\|$ for all $h \in X$. Since $x^* \in C$, we get

$$\langle x^*, x + h\rangle - \langle x^*, x\rangle \leq \|x + h\| - \|x\|\,.$$

Hence, $x^* \in \partial f(x)$ and so $C \subseteq \partial f(x)$.
On the other hand, if $x^* \in \partial f(x)$, then $-\langle x^*, x\rangle \leq -\|x\|$; see (5.3.9). This gives

$$\langle x^*, x\rangle \geq \|x\|\,. \tag{5.3.13}$$

But $\|x\| = \|2x\| - \|x\| \geq \langle x^*, x\rangle$. Hence, $\langle x^*, x\rangle = \|x\|$; see (5.3.13). Moreover, for $h \in X$ and $\lambda > 0$, we have $\langle x^*, \lambda h\rangle \leq \|x + \lambda h\| - \|x\|$. Hence,

$$\langle x^*, h\rangle \leq \left\|\frac{x}{\lambda} + h\right\| - \left\|\frac{x}{\lambda}\right\| \quad \text{for all } \lambda > 0\,.$$

Sending $\lambda \to +\infty$, we obtain $\langle x^*, h\rangle \leq \|h\|$ for all $h \in X$. Therefore, $\|x^*\|_* \leq 1$. Thus, we conclude that $\partial f(x) = C$.

(b) Let X be a Banach space and $f(x) = 1/2\|x\|^2$ for all $x \in X$. We claim that

$$\partial f(x) = \mathcal{F}(x) = \{x^* \in X^* : \|x^*\|_* = \|x\|, \langle x^*, x\rangle = \|x\|^2\} \quad \text{for all } x \in X\,,$$

is the duality map; see Remark 3.1.51. So, let $x^* \in \mathcal{F}(x)$. Then we obtain

$$\langle x^*, u - x\rangle = \langle x^*, u\rangle - \|x\|^2 \leq \|x\|\|u\| - \|x\|^2 \leq \frac{1}{2}[\|u\|^2 - \|x\|^2]$$

for all $u \in X$. Hence,

$$x^* \in \partial f(x) \quad \text{and so} \quad \mathcal{F}(x) \subseteq \partial f(x) \quad \text{for all } x \in X\,. \tag{5.3.14}$$

On the other hand, if $x^* \in \partial f(x)$, then

$$\langle x^*, u - x\rangle \le \frac{1}{2}[\|u\|^2 - \|x\|^2] \quad \text{for all } u \in X\,. \tag{5.3.15}$$

Let $u = x + \lambda h$ with $\lambda > 0$ and $h \in X$. Then

$$\begin{aligned}\langle x^*, h\rangle &\le \frac{1}{2\lambda}[\|x + \lambda h\|^2 - \|x\|^2] \le \frac{1}{2\lambda}[(\|x\| + \lambda\|h\|)^2 - \|x\|^2] \\ &\le \|x\|\|h\| + \frac{1}{2}\lambda^2\|h\| \quad \text{for all } \lambda > 0\,.\end{aligned}$$

This implies

$$\langle x^*, h\rangle \le \|x\|\|h\|\,. \tag{5.3.16}$$

Moreover, if we choose $u = (1 - \lambda)x$ in (5.3.15), divide by $\lambda > 0$, and let $\lambda \to 0^+$, then

$$\langle x^*, x\rangle \ge \|x\|^2\,. \tag{5.3.17}$$

From (5.3.16) and (5.3.17) it follows that $\langle x^*, x\rangle = \|x\|^2 = \|x^*\|_*^2$ and so $\partial f(x) = \mathcal{F}(x)$; see (5.3.14).

(c) Let X be a Banach space and let $C \subseteq X$ be a closed and convex set. The **normal cone** $N_C(x)$ to C at $x \in C$ is defined by

$$N_C(x) = \{x^* \in X^* : \langle x^*, u - x\rangle \le 0 \text{ for all } u \in C\}\,.$$

This is a closed, convex cone with $0 \in N_C(x)$ and

$$N_C(x) = \partial i_C(x) \quad \text{for all } x \in C\,.$$

Note that $D(\partial i_C) = C$ and $\partial i_C(x) = \{0\}$ if $x \in \operatorname{int} C$. If C is a vector subspace of X, then $\partial i_C(x) = C^\perp$ for all $x \in C$; see Definition 3.2.24. In general, the normal cone to C at $x \in C$ is the set of normal vectors to half-spaces, which support C at x.

Remark 5.3.16. If $X = \mathbb{R}$, then from Example 5.3.15 (a), we get for $f(x) = |x|$ with $x \in \mathbb{R}$ that

$$\partial f(x) = \begin{cases} -1 & \text{if } x < 0\,, \\ \{v \in \mathbb{R} : |v| \le 1\} & \text{if } x = 0\,, \\ +1 & \text{if } x > 0\,. \end{cases}$$

Proposition 5.3.17. *If X is a normed space and $f : X \to \mathbb{R}$ is subdifferentiable at every $x \in X$, then $f \in \Gamma_0(X)$.*

Proof. First we show that f is convex. So, let $x^* \in \partial f(x)$ and $u, v \in X$. Then

$$\langle x^*, u - x\rangle \le f(u) - f(x) \quad \text{and} \quad \langle x^*, v - x\rangle \le f(v) - f(x)$$

and

$$\begin{aligned}\langle x^*, \lambda(u-x)\rangle &\le \lambda[f(u)-f(x)]\\ \langle x^*, (1-\lambda)(v-x)\rangle &\le (1-\lambda)[f(v)-f(x)]\end{aligned} \tag{5.3.18}$$

with $\lambda \in (0,1)$. Adding these two inequalities in (5.3.18), we obtain

$$\langle x^*, \lambda u + (1-\lambda)v - x\rangle \le \lambda f(u) + (1-\lambda)f(v) - f(x)\,. \tag{5.3.19}$$

So, if $x = \lambda u + (1-\lambda)v$, then inequality (5.3.19) gives

$$f(\lambda v + (1-\lambda)v) \le \lambda f(u) + (1-\lambda)f(v)\,,$$

that is, f is convex.

Next we show that f is lower semicontinuous. So, let $x_n \to x$ in X and let $x^* \in \partial f(x)$. Then

$$\langle x^*, x_n - x\rangle \le f(x_n) - f(x) \quad \text{for all } n \in \mathbb{N}\,.$$

Therefore, $f(x) \le \liminf_{n\to\infty} f(x_n)$, that is, f is lower semicontinuous. We conclude that $f \in \Gamma_0(X)$. □

One of the main uses of the subdifferential is to detect minimizers. In fact, directly from Definition 5.3.13, one has the following extension of the classical Fermat rule.

Proposition 5.3.18. *If X is a normed space and $f\colon X \to \overline{\mathbb{R}}$ is proper and convex, then $x_0 \in X$ is a (global) minimizer of f if and only if $0 \in \partial f(x_0)$.*

Proposition 5.3.19. *If X is a Banach space and $f\colon X \to \overline{\mathbb{R}}$ is proper and convex, then $x^* \in \partial f(x)$ if and only if $f(x) + f(x^*) = \langle x^*, x\rangle$. Moreover, if $f \in \Gamma_0(X)$, then $x^* \in \partial f(x)$ if and only if $x \in \partial f^*(x^*)$.*

Proof. Let $x^* \in \partial f(x)$. Then $\langle x^*, u - x\rangle \le f(u) - f(x)$ for all $u \in X$. Hence,

$$\langle x^*, u\rangle - f(u) + f(x) \le \langle x^*, x\rangle \quad \text{for all } u \in X\,.$$

This implies $f^*(x^*) + f(x) \le \langle x^*, x\rangle$ and so, $f(x) + f^*(x^*) = \langle x^*, x\rangle$ according to the Young–Fenchel inequality; see Proposition 5.3.3 (g).

Now suppose that $f(x) + f^*(x^*) = \langle x^*, x\rangle$. The continuous affine function

$$u \to a(u) = \langle x^*, u\rangle - f^*(x^*) = \langle x^*, u\rangle + f(x) - \langle x^*, x\rangle$$

is a minorant of f and $a(x) = f(x)$. Therefore, $x^* \in \partial f(x)$.

Now we assume that $f \in \Gamma_0(X)$. Then from the first part we have $x \in \partial f^*(x^*)$ if and only if $f^*(x^*) + f^{**}(x) = f^*(x^*) + f(x) = \langle x^*, x\rangle$ (see Propositions 5.3.6 and 5.3.7) if and only if $x^* \in \partial f(x)$, from the first part of the proof. □

From Propositions 5.3.18 and 5.3.19 we deduce the following corollaries.

Corollary 5.3.20. *If X is a Banach space and $f \in \Gamma_0(X)$, then f attains its infimum on X if and only if $\partial f^*(0) \cap X \neq \emptyset$.*

Corollary 5.3.21. *If X is a reflexive Banach space and $f \in \Gamma_0(X)$, then $\partial f^* = (\partial f)^{-1}$.*

Recall that $D(\partial f) \subseteq \operatorname{dom} f$. Combining Proposition 5.1.24 and Proposition 5.2.15 (b) we obtain the following.

Proposition 5.3.22. *If X is a Banach space and $f \in \Gamma_0(X)$, then* $\operatorname{int} \operatorname{dom} f \subseteq D(\partial f)$.

Directly from Definition 5.3.13 we can state the following proposition.

Proposition 5.3.23. *If X is a normed space and $f, h: X \to \overline{\mathbb{R}}$, then the following hold:*
(a) $\partial(\lambda f) = \lambda \partial f$ *for all* $\lambda > 0$;
(b) $\partial f + \partial h \subseteq \partial(f + h)$.

Next we are going to improve part (b) of the proposition above.

Theorem 5.3.24. *If X is a normed space, $f, h \in \Gamma_0(X)$, and $\operatorname{dom} f \cap \operatorname{dom} h \neq \emptyset$ with one of the two functions being continuous at $x_0 \in \operatorname{dom} f \cap \operatorname{dom} h$, then $\partial(f + h)(x) = \partial f(x) + \partial h(x)$ for all $x \in X$.*

Proof. From Proposition 5.3.23 (b) we already know that

$$\partial f(x) + \partial h(x) \subseteq \partial(f + h)(x) \quad \text{for all } x \in X . \tag{5.3.20}$$

So, let $x^* \in \partial(f + h)(x)$. Then

$$\langle x^*, u - x\rangle + f(x) + h(x) \leq f(u) + h(u) \quad \text{for all } u \in X . \tag{5.3.21}$$

We consider the following two convex subsets of $X \times \mathbb{R}$

$$C_1 = \{(u, \lambda) \in X \times \mathbb{R} : f(u) - \langle x^*, u - x\rangle - h(x) \leq \lambda\}$$
$$C_2 = \{(u, \lambda) \in X \times \mathbb{R} : \lambda \leq h(x) - h(u)\} .$$

We assume that f is continuous at $x_0 \in \operatorname{dom} f \cap \operatorname{dom} h$. Note that if

$$g(u) = f(u) - \langle x^*, u - x\rangle - h(x) \quad \text{for } u \in X ,$$

then $C_1 = \operatorname{epi} g$ and since g is continuous at x_0, we have that $\operatorname{int} C_1 \neq \emptyset$; see Proposition 5.1.24. Moreover, (5.3.21) implies that $\operatorname{int} C_1 \cap C_2 = \emptyset$. We apply now the First Separation Theorem (see Theorem 3.1.59), and find $(u^*, \eta) \in X^* \times \mathbb{R}$ with $(u^*, \eta) \neq 0$ such that

$$h(x) - h(u) \leq \langle u^*, u\rangle + \eta \leq f(u) - \langle x^*, u - x\rangle - f(x) \quad \text{for all } u \in X . \tag{5.3.22}$$

If $u = x$, then $\eta = -\langle u^*, x\rangle$ (see (5.3.22)), and so

$$\langle u - x, -u^* \rangle \le h(u) - h(x) \quad \text{for all } u \in X\,, \tag{5.3.23}$$

$$\langle u - x, u^* + x^* \rangle \le f(u) - f(x) \quad \text{for all } u \in X\,. \tag{5.3.24}$$

From (5.3.23) we see that $-u^* \in \partial h(x)$ and from (5.3.24) we have $u^* + x^* \in \partial f(x)$. Therefore, $x^* \in \partial(f+h)(x)$ has been decomposed as $x^* = (u^*+x^*)+(-u^*)$ with $u^*+x^* \in \partial f(x)$ and $-u^* \in \partial h(x)$. This means that $\partial(f+h)(x) \subseteq \partial f(x) + \partial h(x)$. This and (5.3.20) imply that

$$\partial(f + h)(x) = \partial f(x) + \partial h(x) \quad \text{for all } x \in X\,.$$ □

Remark 5.3.25. By induction we can extend this to any finite set $\{f_k\}_{k=1}^N \subseteq \Gamma_0(X)$ and obtain

$$\partial\left(\sum_{k=1}^N f_k\right)(x) = \sum_{k=1}^N \partial f_k(x) \quad \text{for all } x \in X$$

provided that $\bigcap_{k=1}^N \operatorname{dom} f_k \neq \emptyset$ and all but one of the functions are continuous at a point $x_0 \in \bigcap_{k=1}^N \operatorname{dom} f_k$. If f is lower semicontinuous and f^* is Fréchet differentiable, then $f \in \Gamma_0(X)$.

Another subdifferential rule concerns composite functions.

Theorem 5.3.26. *If X, Y are normed spaces, $A \in L(X,Y)$, $f \in \Gamma_0(Y)$, and there is a point $A(x_0) \in Y$ where f is continuous and finite, then $\partial(f \circ A)(x) = A^*\partial f(A(x))$ for all $x \in X$.*

Proof. Evidently, $f \circ A \in \Gamma_0(X)$. Let $y^* \in \partial f(A(x))$. We obtain

$$\langle y^*, y - A(x)\rangle_Y \le f(y) - f(A(x)) \quad \text{for all } y \in Y\,. \tag{5.3.25}$$

Choosing $y = A(u)$ with $u \in X$ in (5.3.25) yields

$$\langle y^*, A(u) - A(x)\rangle_Y \le (f \circ A)(u) - (f \circ A)(x) \quad \text{for all } x \in X\,,$$

which implies

$$\langle A^*(y^*), u - x\rangle_X \le (f \circ A)(u) - (f \circ A)(x) \quad \text{for all } x \in X\,.$$

Hence $A^*(y^*) \in \partial(f \circ A)(x)$. So, we have proved that

$$A^*(\partial f(A(x))) \subseteq \partial(f \circ A)(x) \quad \text{for all } x \in X\,. \tag{5.3.26}$$

Now, let $x^* \in \partial(f \circ A)(x)$. Then

$$\langle x^*, u - x\rangle_X \le (f \circ A)(u) - (f \circ A)(x) \quad \text{for all } u \in X\,. \tag{5.3.27}$$

In $Y \times \mathbb{R}$ we consider the affine space

$$L = \{(A(u), \langle x^*, u - x\rangle + (f \circ A)(x)) : u \in X\} .$$

From (5.3.27) and the hypotheses of the theorem, we see that $L \cap (\operatorname{int}\operatorname{epi} f) = \emptyset$. So, as before by the First Separation Theorem, there exists $(y^*, \eta) \in Y^* \times \mathbb{R}$ such that the closed hyperplane H, which is the graph of the affine function $y \to \langle y^*, y\rangle_Y + \eta$ separates L and $\operatorname{epi} f$. Since $L \subseteq H$, we get

$$\langle y^*, A(u)\rangle_Y + \eta = \langle x^*, u - x\rangle_X + (f \circ A)(x) \quad \text{for all } u \in X .$$

It follows that

$$\langle y^*, A(u)\rangle_Y = \langle x^*, u\rangle_X \qquad \text{for all } u \in X , \tag{5.3.28}$$

$$\eta = (f \circ A)(x) - \langle x^*, x\rangle_X \quad \text{for all } u \in X . \tag{5.3.29}$$

Equality (5.3.28) implies $A^*(y^*) = x^*$. Moreover, since $H \cap \operatorname{int}\operatorname{epi} f = \emptyset$, we obtain

$$\langle y^*, y - A(x)\rangle \le f(y) - (f \circ A)(x) \quad \text{for all } y \in Y ,$$

see (5.3.29). Hence, $y^* \in \partial f(A(x))$ which implies $x^* = A^*(y^*) \in A^*(\partial f(A(x)))$. Therefore,

$$\partial(f \circ A)(x) \subseteq A^*(\partial f(A(x))) \quad \text{for all } x \in X . \tag{5.3.30}$$

From (5.3.26) and (5.3.30) we conclude that $\partial(f \circ A)(x) = A^*(\partial f(A(x)))$ for all $x \in X$. □

In the next proposition we give some more subdifferential calculus rules. They are easy consequences of the definition of the subdifferential.

Proposition 5.3.27. *If X is a normed space and $f: X \to \overline{\mathbb{R}}$ is proper and convex, then the following hold:*
(a) *for $h(x) = f(x + x_0)$ with $x \in X$, we have $\partial h(x) = \partial f(x + x_0)$;*
(b) *for $h(x) = \lambda f(x)$ with $\lambda > 0$, we have $\partial h(x) = \lambda \partial f(x)$;*
(c) *for $h(x) = f(\lambda x)$ with $\lambda > 0$, we have $\partial h(x) = \lambda \partial f(\lambda x)$.*

We will return to the properties of the convex subdifferential in Section 6.1, where we discuss maximal monotone maps. The convex subdifferential is a prime example of such a map.

In addition to the subdifferential of a proper, convex function, we can also define the **approximate subdifferential** also called **ε-subdifferential**, which is also a useful tool in some occasions.

Definition 5.3.28. Let X be a normed space and $f: X \to \overline{\mathbb{R}}$ is a proper, convex function. For each $\varepsilon \ge 0$, the **ε-subdifferential** of f at $x \in \operatorname{dom} f$ is defined to be the w^*-closed set

$$\partial_\varepsilon f(x) = \{x^* \in X^* : \langle x^*, u - x\rangle - \varepsilon \leq f(u) - f(x) \text{ for all } u \in X\}\,.$$

Remark 5.3.29. When $\varepsilon = 0$, we recover the notion of the convex subdifferential; see Definition 5.3.13. However, there is a basic difference between the subdifferential, that is, $\varepsilon = 0$, and the ε-subdifferential, that is, $\varepsilon > 0$. The usual subdifferential ∂f is a local notion while $\partial_\varepsilon f$ is a global one, that is, the behavior of f on all of X may be relevant to the construction of $\partial_\varepsilon f$. This explains why ∂f and $\partial_\varepsilon f$ have in general different properties. The next proposition presents an important such difference.

Proposition 5.3.30. *If X is a normed space and $f \in \Gamma_0(X)$, then for every $\varepsilon > 0$ and every $x \in \operatorname{dom} f$, we have $\partial_\varepsilon f(x) \neq \emptyset$.*

Proof. Note that $(x, f(x) - \varepsilon) \notin \operatorname{epi} f$. Then by the Strong Separation Theorem (see Theorem 3.1.60), there exists $(u^*, \eta) \in X^* \times \mathbb{R}$ with $(u^*, \eta) \neq 0$ such that

$$\langle u^*, x\rangle + \eta(f(x) - \varepsilon) < \langle u^*, u\rangle + \eta\lambda \quad \text{for all } (u, \lambda) \in \operatorname{epi} f\,. \tag{5.3.31}$$

We choose $u = x$ and $\lambda = f(x)$ in (5.3.31) to get $\eta(-\varepsilon) < 0$, which implies $\eta > 0$. Without any loss of generality, we can assume that $\eta = 1$ by replacing u^* with $1/\eta u^*$. Setting $x^* = -u^*$ and $\lambda = f(u)$ gives

$$\langle x^*, u - x\rangle - \varepsilon \leq f(u) - f(x) \quad \text{for all } u \in X\,.$$

This shows that $x^* \in \partial f_\varepsilon f(x)$. □

The next proposition generalizes the formula in Proposition 5.2.15 (c).

Proposition 5.3.31. *If X is a normed space and $f \in \Gamma_0(X)$, then for any $x \in \operatorname{dom} f$ we have*

$$f'_+(x; h) = \lim_{\varepsilon \searrow 0} \sigma(h; \partial_\varepsilon f(x)) \quad \textit{for all } h \in X\,. \tag{5.3.32}$$

Proof. Clearly, $\{\partial_\varepsilon f(x)\}_{\varepsilon > 0}$ is decreasing with $\varepsilon > 0$. So, the limit on the right-hand side of (5.3.32) exists in $\mathbb{R}^*$. For $\varepsilon > 0$, $x^* \in \partial f_\varepsilon(x)$, $\lambda > 0$, and $h \in X$, we obtain

$$\langle x^*, h\rangle \leq \frac{1}{\lambda}[f(x + \lambda h) - f(x) + \varepsilon]\,.$$

Let $\lambda = \sqrt{\varepsilon}$. Then

$$\langle x^*, h\rangle \leq \frac{1}{\sqrt{\varepsilon}}[f(x + \sqrt{\varepsilon} h) - f(x)] + \sqrt{\varepsilon}\,.$$

This implies

$$\sigma(h; \partial_\varepsilon f(x)) \leq f'_+(x; h) \quad \text{for all } h \in X\,. \tag{5.3.33}$$

Evidently, we may assume that $f'_+(x;h) > -\infty$. Let $\vartheta \in \mathbb{R}$ such that $\vartheta < f'_+(x;h)$ and let $\varepsilon > 0$. For $\lambda \in [0,1]$ we obtain $f(x) + \lambda\vartheta \leq f(x+\lambda h)$. In $X \times \mathbb{R}$ we consider the sets

$$K = \{(x, f(x) - \varepsilon) + \lambda(h, \vartheta) : 0 \leq \lambda \leq 1\} \quad \text{and} \quad \operatorname{epi} f .$$

Both are convex, K is compact, $\operatorname{epi} f$ is closed, and $K \cap \operatorname{epi} f = \emptyset$. So, by the Strong Separation Theorem there exists $(x^*, \eta) \in X^* \times \mathbb{R}$ with $(x^*, \eta) \neq 0$ such that

$$\langle x^*, x + \lambda h\rangle + \eta(f(x) - \varepsilon + \lambda\vartheta) < \langle x^*, u\rangle + \eta f(u) \tag{5.3.34}$$

for all $u \in \operatorname{dom} f$ and $0 \leq \lambda \leq 1$. In (5.3.34) we choose $u = x$ and $\lambda = 0$. Then $\eta(-\varepsilon) < 0$ and so $\eta > 0$. Now let $u = x$ and $\lambda = 1$ and divide with $\eta > 0$. We obtain

$$\vartheta - \varepsilon \leq \left\langle -\frac{1}{\lambda} x^*, h \right\rangle . \tag{5.3.35}$$

For given $v \in X$ with $x + v \in \operatorname{dom} f$, let $u = x + v$ and $\lambda = 0$. Dividing with $\lambda > 0$, we get

$$\left\langle -\frac{1}{\lambda} x^*, v \right\rangle \leq f(x+v) - f(x) + \varepsilon .$$

Hence, $-1/\lambda x^* \in \partial_\varepsilon f(x)$ and so, due to (5.3.35), $\vartheta - \varepsilon \leq \sigma(h; \partial_\varepsilon f(x))$. This gives

$$f'_+(x;h) - \varepsilon \leq \sigma(h; \partial_\varepsilon f(x)) \quad \text{for all } h \in X \text{ and } \varepsilon > 0 . \tag{5.3.36}$$

From (5.3.33) and (5.3.36) we conclude that (5.3.32) holds. □

Next we introduce a notion that is useful in optimization theory.

Definition 5.3.32. Let X be a normed space and $f \in \Gamma_0(X)$. The **recession function** f^∞ of f is defined by

$$f^\infty(h) = \lim_{\lambda \to +\infty} \frac{f(x+\lambda h)}{\lambda} \quad \text{for all } x \in \operatorname{dom} f \text{ and for all } h \in X . \tag{5.3.37}$$

Remark 5.3.33. The function $\xi(\lambda) = f(x+\lambda h)$ is convex on $\mathbb{R}$. Hence, $\lambda \to (\xi(\lambda) - \xi(0))/\lambda$ is nondecreasing and therefore the limit in (5.3.37) exists.

Proposition 5.3.34. *If X is a normed space and $f \in \Gamma_0(X)$, then*

$$f^\infty(h) = \sup\left[\frac{f(x+\lambda h) - f(x)}{\lambda} : \lambda > 0\right] = \sup[f(x+h) - f(x) : x \in \operatorname{dom} f] .$$

Proof. First we prove that

$$f^\infty(h) = \lim_{\lambda \to +\infty} \frac{f(x+\lambda h)}{\lambda} = \sup\left[\frac{f(x+\lambda h) - f(x)}{\lambda} : \lambda > 0\right] \tag{5.3.38}$$

for $x \in \operatorname{dom} f$ and $h \in X$. Note that

$$\begin{aligned}\lim_{\lambda\to+\infty}\frac{f(x+\lambda h)}{\lambda} &= \lim_{\lambda\to+\infty}\frac{f(x+\lambda h)-f(x)}{\lambda}\\ &\le \sup\left[\frac{f(x+\lambda h)-f(x)}{\lambda} : \lambda > 0\right].\end{aligned} \tag{5.3.39}$$

We need to show that the opposite inequality also holds. We fix $\lambda > 0$ and $t > \lambda$. The convexity of f gives

$$f(x+\lambda h) = f\left(\left(1-\frac{\lambda}{t}\right)x + \frac{\lambda}{t}(x+th)\right) \le \left(1-\frac{\lambda}{t}\right)f(x) + \frac{\lambda}{t}f(x+th).$$

Hence,

$$\frac{f(x+\lambda h)-f(x)}{\lambda} \le \frac{f(x+th)-f(x)}{t}.$$

This yields

$$\begin{aligned}\sup\left[\frac{f(x+\lambda h)-f(x)}{\lambda} : \lambda > 0\right] &\le \lim_{t\to+\infty}\frac{f(x+th)-f(x)}{t}\\ &= \lim_{t\to+\infty}\frac{f(x+th)}{t} = f^{\infty}(h).\end{aligned} \tag{5.3.40}$$

From (5.3.39) and (5.3.40) we infer that (5.3.38) holds.

In order to finish the proof of the proposition we need to show that, for every $u \in \operatorname{dom} f$ and $h \in X$, there holds

$$\sup[f(x+h)-f(x) : x \in \operatorname{dom} f] = \sup\left[\frac{f(u+\lambda h)-f(u)}{\lambda} : \lambda > 0\right]. \tag{5.3.41}$$

Let $x, u \in \operatorname{dom} f$ and $h \in X$. Since $f \in \Gamma_0(X)$ we obtain

$$\begin{aligned}f(x+h) &\le \liminf_{\lambda\to+\infty} f\left(\left(1-\frac{1}{\lambda}\right)x + \frac{1}{\lambda}(u+\lambda h)\right)\\ &\le \liminf_{\lambda\to+\infty}\left[\left(1-\frac{1}{\lambda}\right)f(x) + \frac{1}{\lambda}f(u+\lambda h)\right]\\ &= f(x) + \lim_{\lambda\to+\infty}\frac{f(u+\lambda h)-f(u)}{\lambda}.\end{aligned}$$

This implies

$$\sup[f(x+h)-f(x) : x \in \operatorname{dom} f] \le \sup\left[\frac{f(u+\lambda h)-f(u)}{\lambda} : \lambda > 0\right]. \tag{5.3.42}$$

Let $\vartheta = \sup[f(x+h)-f(x) : x \in \operatorname{dom} f]$ and assume that $\vartheta < +\infty$. Then $x+h \in \operatorname{dom} f$ for every $x \in \operatorname{dom} f$ and so from $f(x+h) \le \vartheta + f(x)$ we deduce that

$$f(x+mh) = f(x) + \sum_{k=1}^{m} [f(x+kh) - f(x+(k-1)h)] \le f(x) + m\vartheta \tag{5.3.43}$$

for all $m \in \mathbb{N}_0$. Let $n, m \in \mathbb{N}_0$ with $n > m$. Then, the convexity of f and (5.3.43) imply

$$\begin{aligned} f\left(x + \frac{m}{n}h\right) &= f\left(\left(1 - \frac{1}{n}\right)x + \frac{x+mh}{n}\right) \le \left(1 - \frac{1}{n}\right)f(x) + \frac{1}{n}f(x+mh) \\ &\le \left(1 - \frac{1}{n}\right)f(x) + \frac{1}{n}(f(x) + m\vartheta) = f(x) + \frac{m}{n}\vartheta . \end{aligned}$$

Exploiting the lower semicontinuity of f, we obtain

$$f(x + \lambda h) \le f(x) + \lambda\vartheta \quad \text{for all } \lambda \ge 0 .$$

Hence,

$$\frac{f(x+\lambda h) - f(x)}{\lambda} \le \vartheta \quad \text{for all } \lambda > 0 ,$$

which gives

$$\sup\left[\frac{f(x+\lambda h) - f(x)}{\lambda} : \lambda > 0\right] \le \vartheta . \tag{5.3.44}$$

From (5.3.42) and (5.3.44) it follows that (5.3.41) holds. So, we have proven the proposition. □

Corollary 5.3.35. *If X is a normed space and $f \in \Gamma_0(X)$, then f^∞ is independent of $x \in \operatorname{dom} f$; see Definition* 5.3.32.

Proposition 5.3.36. *If X is a normed space and $f \in \Gamma_0(X)$, then $f^\infty \in \Gamma_0(X)$ and it is positively homogeneous.*

Proof. For every $x \in \operatorname{dom} f$, the function $g(h) = f(x+h) - f(x)$ belongs to $\Gamma_0(X)$. Then invoking Proposition 5.3.34, we conclude that $f^\infty \in \Gamma_0(X)$. For the positive homogeneity, note that for given $h \in X$ and $t > 0$, we have

$$f^\infty(th) = \lim_{\lambda \to +\infty} \frac{f(x + \lambda t h) - f(x)}{\lambda} .$$

We set $s = \lambda t$. Then

$$f^\infty(th) = t \lim_{s \to +\infty} \frac{f(x+sh) - f(x)}{s} = t f^\infty(h) .$$ □

Proposition 5.3.37. *If X is a normed space and $f \in \Gamma_0(X)$, then $f^\infty(h) + f^\infty(-h) \ge 0$ for all $h \in X$.*

Proof. We assume that $f^\infty(h), f^\infty(-h) \in \mathbb{R}$, otherwise the inequality is clearly true. Taking Proposition 5.3.34 into account gives $x+h \in \operatorname{dom} f$ and $x-h \in \operatorname{dom} f$ for all $x \in \operatorname{dom} f$. Therefore, using once more Proposition 5.3.34 leads to

$$\begin{aligned} f^\infty(h) + f^\infty(-h) &\geq \sup[f(x) - f(x-h) : x \in \operatorname{dom} f] \\ &\quad + \sup[f(x-h) - f(x) : x \in \operatorname{dom} f] \geq 0 . \end{aligned}$$ □

Proposition 5.3.38. *If X is a normed space, $f \in \Gamma_0(X)$, and $\inf_X f > -\infty$, then $f^\infty(h) \geq 0$ for all $h \in X$.*

Proof. Let $m = \inf_X f > -\infty$. Let $x \in \operatorname{dom} f$ and $h \in X$. Then

$$f^\infty(h) = \lim_{\lambda \to +\infty} \frac{f(x + \lambda h)}{\lambda} \geq \lim_{\lambda \to +\infty} \frac{m}{\lambda} = 0 .$$ □

Another easy consequence of Definition 5.3.32 is the following formula.

Proposition 5.3.39. *If X is a normed space and $\{f_k\}_{k=1}^n \subseteq \Gamma_0(X)$ with $\bigcap_{k=1}^n \operatorname{dom} f_k \neq \emptyset$, then*

$$\left(\sum_{k=1}^n f_k\right)^\infty = \sum_{k=1}^n f_k^\infty .$$

Remark 5.3.40. If $f \in \Gamma_0(X)$ is positively homogeneous of degree $p > 1$, that is, $f(\lambda x) = \lambda^p f(x)$ for all $\lambda > 0$ and for all $x \in X$, then we have

$$f^\infty(h) = \begin{cases} 0 & \text{if } f(h) = 0 , \\ +\infty & \text{otherwise} . \end{cases}$$

On the other hand, if f is positively homogeneous of degree 1, then $f^\infty = f$.

5.4 Proximinal and Chebyshev Sets

In Theorem 3.5.18 we proven that every nonempty closed convex set in a Hilbert space has the unique best approximation property. In this section we examine such approximation properties for sets in general normed spaces, not necessarily Hilbert spaces.

Definition 5.4.1. Let X be a normed space and let $C \subseteq X$ be a nonempty set. The **best approximation map** $p_C : X \to 2^C$ is defined by

$$p_C(x) = \{c \in C : \|x - c\| = d(x, C)\} \quad \text{for all } x \in X .$$

If $p_C(x) \neq \emptyset$ for every $x \in X$, then C is said to be **proximinal**. If $p_C(x)$ is a singleton for every $x \in X$, then C is said to be a **Chebyshev set**. In that case $p_C : X \to C$ is called **metric projection**.

Proposition 5.4.2. *If X is a reflexive Banach space and $C \subseteq X$ is nonempty, closed, and convex, then $p_C(x) \neq \emptyset$ for every $x \in X$.*

Proof. Let $\{c_n\}_{n\geq1} \subseteq C$ be such that $\|x - c_n\| \searrow d(x, C)$. Evidently, $\{c_n\}_{n\geq1} \subseteq C$ is bounded. So, by passing to a subsequence if necessary, we may assume that $c_n \xrightarrow{w} c \in C$. We obtain $\|x - c\| \leq \liminf_{n\to\infty} \|x - c_n\| = d(x, C)$, which shows that $\|x - c\| = d(x, C)$ and so $c \in p_C(x)$. □

Proposition 5.4.3. *If X is a strictly convex Banach space and $C \subseteq X$ is nonempty, closed, and convex, then $p_C(x) = \emptyset$ for all $x \in X$ or it is a singleton.*

Proof. Let $c, \hat{c} \in p_C(x)$. The strict convexity of X implies that, if $c \neq \hat{c}$, then $\|2x - (c + \hat{c})\| < 2d(x, C)$; see Definition 3.4.21 (a). Hence,

$$\left\| x - \frac{1}{2}(c + \hat{c}) \right\| < d(x, C),$$

a contradiction since $1/2(c + \hat{c}) \in C$. So, $p_C(x)$ is either empty or a singleton. □

Combining Proposition 5.4.2 and 5.4.3 we can state the following result.

Corollary 5.4.4. *If X is a reflexive and strictly convex Banach space and $C \subseteq X$ is nonempty, closed, and convex, then C is Chebyshev.*

Using the notion of proximinality we can characterize reflexive Banach spaces.

Theorem 5.4.5. *If X is a Banach space, then X is reflexive if and only if every nonempty, closed, convex set is proximinal.*

Proof. $\Longrightarrow$: This implication is stated in Proposition 5.4.2.

$\Longleftarrow$: If X is not reflexive, then on account of James' Theorem (see Theorem 3.3.41), there exists $x^* \in X^*$ with $\|x^*\|_* = 1$ such that $\langle x^*, x\rangle < 1$ for all $x \in X$ with $\|x\| \leq 1$. Let $C = (x^*)^{-1}(1)$. Then $d(0, C) = 1 < \|x\|$ for all $x \in C$. But $C = (x^*)^{-1}(1)$ is nonempty, closed, and convex, and thus by hypothesis proximinal, a contradiction. □

Proposition 5.4.6. *If X is a Banach space and $C \subseteq X$ is a Chebyshev set, then the metric projection map $p_C: X \to C$ has a closed graph and is locally bounded.*

Proof. Let $x_n \to x$ in X and $p_C(x_n) \to u \in C$ in C. We have

$$\big|\|x_n - p_C(x_n)\| - \|x - p_C(x)\|\big| = \big|d(x_n, C) - d(x, C)\big| \leq \|x_n - x\|.$$

Therefore,

$$\|x_n - p_C(x_n)\| \leq \|x_n - x\| + \|x - p_C(x)\|. \tag{5.4.1}$$

This yields $\|x - u\| \leq \|x - p_C(x)\|$, hence $u = p_C(x)$ and so $\operatorname{Gr} p_C$ is closed. From (5.4.1) it is clear that $x \to p_C(x)$ is locally bounded. □

Corollary 5.4.7. *If X is a Banach space and $C \subseteq X$ is a locally compact Chebyshev set, then $p_C: X \to C$ is continuous.*

Proof. Arguing by contradiction, suppose that p_C is not continuous at x_0. Then there exist a sequence $\{x_n\}_{n\geq 1} \subseteq X$ and $\varepsilon > 0$ such that

$$x_n \to x_0 \quad \text{in } X \quad \text{and} \quad \|p_C(x_n) - p_C(x_0)\| \geq \varepsilon \quad \text{for all } n \in \mathbb{N}. \tag{5.4.2}$$

From Proposition 5.4.6, p_C is locally bounded. So, it follows that $\{p_C(x_n)\}_{n\geq 1} \subseteq C$ is bounded. Since by hypothesis C is locally compact, we may assume that $p_C(x_n) \to u \in C$ and $u \neq p_C(x_0)$; see (5.4.2). Then we obtain $\|x_0 - u\| \leq \|x_0 - p_C(x_0)\|$. Hence, $u \in p_C(x_0)$ and so $u = p_C(x_0)$ since C is Chebyshev, a contradiction. □

Corollary 5.4.8. *If X is a finite dimensional Banach space and $C \subseteq X$ is a Chebyshev set, then the metric projection map $p_C: X \to C$ is continuous.*

Proof. Note that C is a closed subset of a locally compact space. Hence, C is locally compact and we can apply Corollary 5.4.7. □

Proposition 5.4.9. *If X is a reflexive Banach space and $C \subseteq X$ is a weakly closed Chebyshev set, then the metric projection map $p_C: X \to C$ is norm-to-weak continuous.*

Proof. Suppose that $x_n \to x$ in X. Then $\|x_n - p_C(x_n)\| = d(x_n, C) \to d(x, C) = \|x - p_C(x)\|$. Since $\{p_C(x_n)\}_{n\geq 1} \subseteq C$ is bounded, we may assume that $p_C(x_n) \xrightarrow{w} u \in C$ since C is weakly closed. Then we obtain

$$\|x - u\| \leq \liminf_{n\to\infty} \|x_n - p_C(x_n)\| = \|x - p_C(x)\|,$$

which shows that $u = p_C(x)$. Therefore, we have $p_C(x_n) \xrightarrow{w} p_C(x)$ for the original sequence and so we conclude that $p_C: X \to C$ is norm-to-weak continuous. □

Corollary 5.4.10. *If X is a reflexive Banach space with the Kadec–Klee property (see Definition 3.4.30), and $C \subseteq X$ is a weakly closed Chebyshev set, then the metric projection map $p_C: X \to C$ is continuous.*

Proof. From the proof of Proposition 5.4.9 we know that $x_n \to x$ in X implies $x_n - p_C(x_n) \xrightarrow{w} x - p_C(x)$ and $\|x_n - p_C(x_n)\| \to \|x - p_C(x)\|$. So, by the Kadec–Klee property it follows that $x_n - p_C(x_n) \to x - p_C(x)$ in X, which implies $p_C(x_n) \to p_C(x)$ in X. Hence, p_C is continuous. □

Next we focus on Hilbert spaces and try to characterize the Chebyshev sets. We start with a technical duality formula that will be used in the sequel.

Lemma 5.4.11. *If H is a Hilbert space, $C \subseteq H$ is a nonempty, closed subset, and $f = 1/2\|\cdot\|^2 + i_C$, then $d_C^2 = \|\cdot\|^2 - 2f^*$, where $d_C(x) = \inf[\|x - c\| : c \in C]$.*

Proof. We identify H and H^*; see Theorem 3.5.21. According to Definition 5.3.1 we see that

$$\begin{aligned} f^*(h) &= \sup\left[(h,x) - \frac{1}{2}\|x\|^2 : x \in C\right] = \sup\left[(h,x) - \frac{1}{2}(x,x) : x \in C\right] \\ &= \sup\left[\frac{1}{2}(h,h) + (h,x) - \frac{1}{2}(h,h) - \frac{1}{2}(x,x) : x \in C\right] \\ &= \frac{1}{2}(h,h) + \sup\left[-\frac{1}{2}(x,x) + (h,x) - \frac{1}{2}(h,h) : x \in C\right] \\ &= \frac{1}{2}(h,h) - \frac{1}{2}\inf[(x,x) - 2(h,x) + (h,h) : x \in C] = \frac{1}{2}(h,h) - \frac{1}{2}d_C^2(h) \end{aligned}$$

for all $h \in H$. Hence, $d_C^2 = \|\cdot\|^2 - 2f^*$. □

Using this lemma we can characterize closed and convex sets in a Hilbert space by employing the distance function.

Theorem 5.4.12. *If H is a Hilbert space and $C \subseteq H$ is nonempty and closed, then the following statements are equivalent:*
(a) *C is convex;*
(b) *d_C^2 is Fréchet differentiable;*
(c) *d_C^2 is Gateaux differentiable.*

Proof. (a) $\Longrightarrow$ (b): Let $\varphi(u) = 1/2d_C^2(u)$. We have

$$\begin{aligned} \varphi(u) &= \inf\left[\frac{1}{2}\|u - c\|^2 : c \in C\right] = \inf\left[\frac{1}{2}\|u - c\|^2 + i_C(c) : c \in H\right] \\ &= \left(\frac{1}{2}\|\cdot\|^2 \oplus i_C\right)(u) \quad \text{for all } u \in H . \end{aligned}$$

Then, by Proposition 5.3.9, we obtain

$$\varphi^*(u^*) = \frac{1}{2}\|u^*\|^2 + \sigma(u^*; C) \quad \text{for all } u^* \in H .$$

From the Young–Fenchel inequality (see Proposition 5.3.3 (g)), it follows that

$$(u^*, u) - \frac{1}{2}\|u^*\|^2 - \sigma(u^*; C) \le \varphi(u) \quad \text{for all } u, u^* \in H . \tag{5.4.3}$$

Fix $v \in H$ and let $u^* = v - p_C(v)$. From the properties of the metric projection (see Proposition 3.5.20), one gets $0 \le (v - p_C(v), p_C(v) - c)$ for all $c \in C$. This implies

$$\begin{aligned} \sigma(u^*; C) &= \sup[(u^*, u) : c \in C] = \sup[(v - p_C(v), c) : c \in C] \\ &= (v - p_C(v), p_C(v)) . \end{aligned} \tag{5.4.4}$$

We return to (5.4.3) and use (5.4.4). We obtain

$$(v - p_C(v), u - p_C(v)) - \frac{1}{2}\|v - p_C(v)\|^2 \le \varphi(u),$$

which implies

$$(v - p_C(v), u - p_C(v)) - \varphi(v) \le \varphi(u) .$$

This finally gives

$$\begin{aligned}\varphi(u) - \varphi(v) &\ge (v - p_C(v), u - p_C(v)) - 2\varphi(v)\\ &= (v - p_C(v), u - p_C(v)) - \|v - p_C(v)\|^2 = (v - p_C(v), u - v) .\end{aligned}$$

Therefore,

$$0 \le \varphi(u) - \varphi(v) - (v - p_C(v), u - v) . \tag{5.4.5}$$

Reversing the roles of u and v, we also have

$$0 \le \varphi(v) - \varphi(u) - (u - p_C(u), v - u) . \tag{5.4.6}$$

From (5.4.5) and (5.4.6) it follows that

$$0 \le \varphi(u) - \varphi(v) - (v - p_C(v), u - v) \le 2\|u - v\|^2$$

and so

$$0 \le \frac{\varphi(u) - \varphi(v) - (v - p_C(v), u - v)}{\|u - v\|} \le 2\|u - v\| .$$

This shows that $\varphi'(v) = v - p_C(v)$ and so $\varphi \in C^1(H)$.

(b) $\Longrightarrow$ (c): This is immediate.

(c) $\Longrightarrow$ (a): Let $f = 1/2\|\cdot\|^2 + i_C$. Taking Lemma 5.4.11 into account yields $f^* = 1/2[\|\cdot\|^2 - d_C^2]$. Hence, f^* is Gateaux differentiable. So, by the Kadec–Klee property of Hilbert spaces it follows that the Gateaux derivative of d_C^2 is continuous, hence d_C^2 is Fréchet differentiable. Then so is f^* and this implies the convexity of $C = \operatorname{dom} f$; see Remark 5.3.25. □

Finally we mention a result of Vlasov [322] on Chebyshev sets. For a proof of this theorem we refer to Giles [137, p. 245].

Theorem 5.4.13. *If X is a Banach space with strictly convex dual space, then every Chebyshev set with continuous metric projection is convex.*

Remark 5.4.14. In a Hilbert space a Chebyshev set is convex if and only if the metric projection map is nonexpansive. Moreover, in a Hilbert space a weakly closed Chebyshev set is convex.

5.5 Smoothness of the Norm

In this section we present some basic results on the differentiability properties of the norms of Banach spaces. These properties provide important information about the geometry of the Banach space.

We start with a basic duality theorem from the theory of convex sets. First we state a definition.

Definition 5.5.1. Let X be a topological vector space and $A \subseteq X$. The **polar** of A is the set $A^\circ \subseteq X^*$ defined by

$$A^\circ = \{x^* \in X^* : \langle x^*, a\rangle \leq 1 \text{ for all } a \in A\}\,.$$

Given a subset $C \subseteq X^*$, the **prepolar** of C is the set ${}^\circ C \subseteq X$ defined by

$${}^\circ C = \{x \in X : \langle c, x\rangle \leq 1 \text{ for all } c \in C\}\,.$$

The next theorem is known as the "Bipolar Theorem". It is the basic result concerning polars of sets in Banach spaces.

Theorem 5.5.2 (Bipolar Theorem). *If X is a Banach space and $A \subseteq X$ as well as $C \subseteq X^*$, then the following hold:*

(a) *${}^\circ(A^\circ)$ is the closed balanced convex hull of A, that is*

$${}^\circ(A^\circ) = \overline{\operatorname{conv}}\,[\{0\} \cup A]\,;$$

(b) *$({}^\circ C)^\circ$ is the w^*-closed balanced convex hull of C, that is,*

$$({}^\circ C)^\circ = \overline{\operatorname{conv}}^{\mathrm{w}^*}\,[\{0\} \cup C]\,.$$

Proof. (a) Clearly, it holds that $\{0\} \cup A \subseteq {}^\circ(A^\circ)$. But the set ${}^\circ(A^\circ)$ is closed and convex; see Definition 5.5.1. Therefore,

$$\overline{\operatorname{conv}}\,[\{0\} \cup A] \subseteq {}^\circ(A^\circ)\,. \tag{5.5.1}$$

On the other hand, every closed half-space that contains $\{0\} \cup A$ also contains ${}^\circ(A^\circ)$. Hence,

$$\overline{\operatorname{conv}}\,[\{0\} \cup A] \supseteq {}^\circ(A^\circ)\,. \tag{5.5.2}$$

Recall that $\overline{\operatorname{conv}} D$ is the intersection of all closed half-spaces in X that contain D. From (5.5.1) and (5.5.2), it follows that

$$^{\circ}(A^{\circ}) = \overline{\operatorname{conv}}\,[\{0\} \cup A]\,.$$

(b) This follows from part (a) above. □

Corollary 5.5.3. *If X is a Banach space and $A \subseteq X$, then* span A *is dense in X if and only if $x^* = 0$ is the only functional that vanishes on A.*

Corollary 5.5.4. *If X is a Banach space and $C \subseteq X^*$, then C separates points in X, that is, C is total, if and only if* span C *is* w**-dense in X^*.*

The Bipolar Theorem (see Theorem 5.5.2) is used to recognize dual norms.

Proposition 5.5.5. *If $(X, \|\cdot\|)$ is a Banach space and $|\cdot|_*$ is a norm on X^* equivalent to the dual norm $\|\cdot\|_*$ of $\|\cdot\|$, then $|\cdot|_*$ is a dual norm to some norm $|\cdot|$ on X equivalent to $\|\cdot\|$ if and only if $|\cdot|_*$ is* w**-lower semicontinuous on X^*.*

Proof. $\Longrightarrow$: By hypothesis, $|x^*|_* = \sup[\langle x^*, x\rangle : |x| \le 1]$. So, $|\cdot|_*$ is the supremum of w*-continuous linear functionals. Hence, $|\cdot|_*$ is w*-lower semicontinuous.

$\Longleftarrow$: Let $\overline{B}_1 = \{x^* \in X^* : |x^*|_* \le 1\}$. Then $\overline{B}_1$ is w*-closed and by the Bipolar Theorem (see Theorem 5.5.2), we obtain $\overline{B}_1 = ({}^{\circ}\overline{B}_1)^{\circ}$. Then $|\cdot|_*$ is the dual norm to the equivalent norm given by the Minkowski functional of ${}^{\circ}\overline{B}_1$. □

Proposition 5.5.6. *If $(X, \|\cdot\|)$ is a Banach space and if $x \in X$ with $\|x\| = 1$, then the following hold:*

(a) *$\|\cdot\|$ is Fréchet differentiable at x if and only if $\|x_n^* - u_n^*\|_* \to 0$ whenever $x_n^*, u_n^* \in X^*$ with $\|x_n^*\|_* = \|u_n^*\| = 1$ satisfy*

$$\lim_{n\to\infty} \langle x_n^*, x\rangle = \lim_{n\to\infty} \langle u_n^*, x\rangle = 1\,; \tag{5.5.3}$$

(b) *$\|\cdot\|$ is Gateaux differentiable at x if and only if $x_n^* - u_n^* \xrightarrow{\text{w}^*} 0$ whenever $x_n^*, u_n^* \in X^*$ with $\|x_n^*\|_* = \|u_n^*\| = 1$ fulfill* (5.5.3).

Proof. (a) $\Longrightarrow$: Since by hypothesis $\|\cdot\|$ is Fréchet differentiable at x, for a given $\varepsilon > 0$ there exists $\delta > 0$ such that

$$\|x + h\| + \|x - h\| \le 2 + \varepsilon\|h\| \quad \text{for all } \|h\| \le \delta\,; \tag{5.5.4}$$

see Proposition 5.2.10. Let $x_n^*, u_n^* \in X^*$ with $\|x_n^*\|_* = \|u_n^*\|_* = 1$ such that (5.5.3) is satisfied. We can find a number $n_0 \in \mathbb{N}$ such that

$$\max\{|\langle x_n^*, x\rangle - 1|, |\langle u_n^*, x\rangle - 1|\} \le \varepsilon\delta \quad \text{for all } n \ge n_0\,. \tag{5.5.5}$$

Then, due to (5.5.4) and (5.5.5), we obtain, for $n \ge n_0$ and $\|h\| \le \delta$, that

$$\begin{aligned}\langle x_n^* - u_n^*, h\rangle &= \langle x_n^*, x+h\rangle + \langle u_n^*, x-h\rangle - \langle x_n^* + u_n^*, x\rangle \\ &\le \|x+h\| + \|x-h\| - \langle x_n^* + u_n^*, x\rangle \\ &\le 2 + \varepsilon\|h\| - \langle x_n^* + u_n^*, x\rangle \le 3\varepsilon\delta\,.\end{aligned} \tag{5.5.6}$$

Taking (5.5.6) into account, we get for $n \ge n_0$ that

$$\|x_n^* - u_n^*\|_* = \sup[\langle x_n^* - u_n^*, y\rangle : \|y\| = 1] = \sup\left[\frac{\langle x_n^* - u_n^*, \delta y\rangle}{\delta} : \|y\| = 1\right] \le 3\varepsilon\,.$$

Hence, $\lim_{n\to\infty} \|x_n^* - u_n^*\|_* = 0$.

$\Longleftarrow$: Arguing by contradiction, suppose that $\|\cdot\|$ is not Fréchet differentiable at x. Applying Proposition 5.2.10, there exist $\varepsilon > 0$ and a sequence $\{h_n\}_{n\ge1} \subseteq X$ with $h_n \to 0$ in X such that

$$\|x+h_n\| + \|x-h_n\| \ge 2 + \varepsilon\|h_n\| \quad \text{for all } n \in \mathbb{N}\,.$$

Let us choose $x_n^*, u_n^* \in X^*$ with $\|x_n^*\|_* = \|u_n^*\|_* = 1$ such that

$$\langle x_n^*, x+h_n\rangle = \|x+h_n\| \quad \text{and} \quad \langle u_n^*, x-h_n\rangle = \|x-h_n\| \quad \text{for all } n \in \mathbb{N}\,. \tag{5.5.7}$$

We obtain

$$|\langle x_n^*, h_n\rangle| \le \|h_n\| \quad \text{and} \quad |\|x+h_n\| - \|x\|| \le \|h_n\| \quad \text{for all } n \in \mathbb{N}\,.$$

Hence, $\langle x_n^*, h_n\rangle \to 0$ and $\|x+h_n\| \to 1$ as $n \to \infty$. Because of (5.5.7) we then derive

$$\lim_{n\to\infty} \langle x_n^*, x\rangle = \lim_{n\to\infty} [\langle x_n^*, x+h_n\rangle - \langle x_n^*, h_n\rangle] = \lim_{n\to\infty} [\|x+h_n\| - \langle x_n^*, h_n\rangle] = 1\,.$$

Similarly, we show that $\lim_{n\to\infty} \langle u_n^*, x\rangle = 1$.

(b) This equivalence is proven similarly as (a) using Proposition 5.2.11 this time. □

Remark 5.5.7. Note that the second statement in (a) is equivalent to the following one:

$$\{x_n^*\}_{n\ge1} \subseteq X^* \text{ with } \|x_n^*\|_* = 1 \text{ is convergent whenever } \langle x_n^*, x\rangle \to 1\,.$$

Similarly, the second statement in (b) can be written equivalently as follows:

$$\text{there exists a unique } x^* \in X^* \text{ with } \|x^*\|_* = 1 \text{ such that } \langle x^*, x\rangle = 1\,.$$

Definition 5.5.8. Let $(X, \|\cdot\|)$ be a Banach space. We say that $\|\cdot\|$ is **Fréchet (resp. Gateaux) differentiable** if $\|\cdot\|$ is Fréchet (resp. Gateaux) differentiable at every $x \in X \setminus \{0\}$.

Remark 5.5.9. The norm $\|\cdot\|$ is never differentiable at $x = 0$. Differentiability conditions for a norm are homogeneous, that is, $\|\cdot\|$ is differentiable at x if it is differentiable at λx with $\lambda \in \mathbb{R} \setminus \{0\}$. So, it is enough to check differentiability at points $x \in X$ with $\|x\| = 1$.

Proposition 5.5.10. *If X is a Banach space with a Fréchet differentiable norm $\|\cdot\|$, then $\|\cdot\| \in C^1(X \setminus \{0\})$.*

Proof. Let $\{x_n, x\}_{n\geq 1} \subseteq X \setminus \{0\}$ and assume that $x_n \to x$. Let $\varphi(u) = \|u\|$ for all $u \in X \setminus \{0\}$ and let $x_n^* = \varphi'(x_n)$ as well as $x^* = \varphi'(u)$. We have

$$\|x_n^*\|_* = \|x^*\|_* = 1, \quad \langle x_n^*, x_n\rangle = \|x_n\| \quad \text{and} \quad \langle x^*, x\rangle = \|x\|;$$

see Example 5.3.15 (a). Let $v_n = x_n/\|x_n\|$ and $v = x/\|x\|$. Then

$$\langle x_n^*, v_n\rangle = 1 \quad \text{and} \quad \langle x^*, v\rangle = 1. \tag{5.5.8}$$

Note that

$$\langle x_n^*, v\rangle \to 1 \quad \text{as} \quad n \to \infty. \tag{5.5.9}$$

Then, (5.5.8), (5.5.9) and Proposition 5.5.6 (see also Remark 5.5.7) imply that $x_n^* \to x^*$ in X^*. This proves that $\varphi(\cdot) = \|\cdot\| \in C^1(X \setminus \{0\})$. □

Theorem 5.5.11. *If X is a Banach space and X^* has Fréchet differentiable norm, then X is reflexive.*

Proof. According to James' Theorem (see Theorem 3.3.41), it suffices to show that every $x^* \in X^*$ attains its norm on $\overline{B}_1^X = \{x \in X : \|x\| \leq 1\}$. So, suppose that $x^* \in X^*$ with $\|x^*\|_* = 1$ and choose $\{x_n\}_{n\geq 1} \subseteq \partial B_1^X$ such that $\langle x^*, x\rangle \to 1$. Then from Proposition 5.5.6 and Remark 5.5.7, we see that $\{x_n\}_{n\geq 1} \subseteq X^{**}$ converges to $x \in \partial B_1^X$. Evidently, $\langle x^*, x\rangle = 1$, and so X is reflexive; see Theorems 3.3.41 and 3.4.5. □

For our next result of this kind we will need the "Bishop–Phelps Theorem". We will prove this theorem in Section 6.6 using the Ekeland variational principle. At this point we limit ourselves to the statement of the theorem for easy reference.

Theorem 5.5.12 (Bishop–Phelps Theorem). *If X is a Banach space, then the set of all elements of X^* that attain their norm is dense in X^*.*

Using this result we can prove the following theorem.

Theorem 5.5.13. *If X is a separable Banach space that admits an equivalent Fréchet differentiable norm, then X^* is separable.*

Proof. Let $\{x_n\}_{n\geq 1} \subseteq \partial B_1^X$ be dense. For each $n \in \mathbb{N}$ there exists $x_n^* \in X^*$ with $\|x_n^*\|_* = 1$ such that $\langle x_n^*, x_n\rangle = 1$. Now let $u^* \in X^*$ with $\|u^*\|_* = 1$ be norm attaining. Then $\langle u^*, x\rangle = 1$ with $x \in X$ and $\|x\| = 1$. Consider a subsequence $\{x_{n_k}\}_{k\geq 1} \subseteq \{x_n\}_{n\geq 1}$ such that $x_{n_k} \to x$. Then $\langle x_{n_k}^*, x_{n_k}\rangle = 1$ for all $k \in \mathbb{N}$. Then according to Proposition 5.5.6 and Remark 5.5.7, we obtain $x_{n_k}^* \to u^*$. Therefore, $\overline{\{x_{n_k}^*\}_{n\geq 1}}$ contains all norm attaining elements of X^*. Invoking the Bishop–Phelps Theorem (see Theorem 5.5.12), we conclude that $\{x_n^*\}_{n\geq 1} \subseteq \partial B_1^{X^*}$ is dense. This means that X^* is separable. □

Proposition 5.5.14. *If X is a Banach space and X^* is strictly convex, then the norm $\|\cdot\|$ of X is Gateaux differentiable.*

Proof. According to Remark 5.5.7 we need to show that for $x \in X$ with $\|x\| = 1$ there exists a unique $x^* \in X^*$ with $\|x^*\|_* = 1$ such that $\langle x^*, x\rangle = 1$. Existence is evident, recall that $\overline{B}_1^{X^*} = \{x^* \in X^* : \|x^*\|_* \leq 1\}$ is w*-compact; see Theorem 3.3.38. So, we need to show the uniqueness. Suppose that $x^*, \hat{x}^* \in X^*$ with $\|x^*\|_* = \|\hat{x}^*\|_* = 1$ and $\langle x^*, x\rangle = 1 = \langle \hat{x}^*, x\rangle$. Then

$$2 = \langle x^* + \hat{x}^*, x\rangle \leq \|x^* + \hat{x}^*\|_* \leq 2\,.$$

Hence, $\|x^* + \hat{x}^*\|_* = 2$ and so $x^* = \hat{x}^*$ by the strict convexity of X^*; see Proposition 3.4.23. □

Theorem 5.5.15. *Every separable Banach space admits an equivalent Gateaux differentiable norm.*

Proof. Let $\{x_n\}_{n\geq 1} \subseteq \partial \overline{B}_1^X = \{x \in X : \|x\| = 1\}$ be dense. We define a new norm $|\cdot|_*$ on X^* given by

$$|x^*|_*^2 = \|x^*\|_*^2 + \sum_{n\geq 1} \frac{1}{2^n} \langle x^*, x_n\rangle^2 \tag{5.5.10}$$

with $\|\cdot\|_*$ being the original dual norm on X^*. From Proposition 5.5.5, $|\cdot|_*$ is the dual norm of a norm $|\cdot|$ on X, which is equivalent to the initial norm $\|\cdot\|$. If we can show that $|\cdot|_*$ is strictly convex, then by Proposition 5.5.14, $|\cdot|$ is Gateaux differentiable.

To this end, let $x^*, u^* \in X^*$ such that

$$2|x^*|_*^2 + 2|u^*|_*^2 = |x^* + u^*|_*^2\,. \tag{5.5.11}$$

Since $2|x^*|_*^2 + 2|u^*|_*^2 - |x^* + u^*|_*^2$ and $2\langle x^*, x_n\rangle^2 + 2\langle u^*, x_n\rangle^2 - \langle x^* + u^*, x_n\rangle^2 \geq 0$, we infer from (5.5.10) and (5.5.11) for all $n \in \mathbb{N}$ that

$$0 = 2\langle x^*, x_n\rangle^2 + 2\langle u^*, x_n\rangle^2 - \langle x^* - u^*, x_n\rangle^2 = \langle x^* - u^*, x_n\rangle^2 \quad \text{for all } n \in \mathbb{N}\,.$$

Hence, the density of $\{x_n\}_{n\geq 1}$ implies $x^* = u^*$. Therefore, $|\cdot|_*$ is strictly convex, and thus, $|\cdot|$ is Gateaux differentiable. □

Another such result is given in the next theorem. The proof can be found in Deville–Godefroy–Zizler [88, Theorem 2.6(i), p. 49 and Theorem 3.1.(ii), p. 51].

Theorem 5.5.16. *Every separable Banach space admits an equivalent norm that is both locally uniformly convex and Gateaux differentiable.*

Proposition 5.5.17. *If $(X, \|\cdot\|)$ is a Banach space and X^* equipped with the dual norm $\|\cdot\|_*$ is locally uniformly convex, then $\|\cdot\|$ is Fréchet differentiable.*

Proof. Let $x \in X$ with $\|x\| = 1$. We choose $x^* \in X^*$ with $\|x^*\|_* = 1$ such that $\langle x^*, x\rangle = 1$. Let $\{x_n^*\}_{n\geq 1} \subseteq X^*$ with $\|x_n^*\|_* = 1$ and $n \in \mathbb{N}$ such that $\langle x_n^*, x\rangle \to 1$. We have

$$\langle x^* + x_n^*, x\rangle \leq \|x^* + x_n^*\|_* \leq 2 \quad \text{for all } n \in \mathbb{N}\,. \tag{5.5.12}$$

As $\langle x^* + x_n^*, x\rangle \to 2$ as $n \to \infty$, from (5.5.12) it follows that $\|x^* + x_n^*\|_* \to 2$. Hence, $\|x_n^* - x^*\|_* \to 0$ since X^* is locally uniformly convex; see Proposition 3.4.25 (b). Therefore, taking Proposition 5.5.6 and Remark 5.5.7 into account, we conclude that $\|\cdot\|$ is Fréchet differentiable. □

We conclude this section with the "Fréchet" version of Theorem 5.5.16. For the proof we refer again to Deville–Godefroy–Zizler [88, Theorem 2.6(ii), p. 49 and Theorem 3.1(i), p. 51].

Theorem 5.5.18. *If X is a Banach space and X^* is separable, then X admits an equivalent norm that is both locally uniformly convex and Fréchet differentiable.*

5.6 Multifunctions – Integral Functionals

Let X be a Hausdorff topological space. We introduce the following notation:

$$P_{\mathrm{f}}(X) = \{A \subseteq X : A \text{ is nonempty and closed}\}\,,$$
$$P_{\mathrm{k}}(X) = \{A \subseteq X : A \text{ is nonempty and compact}\}\,.$$

If X is a locally convex space, then we can introduce some additional notation:

$$P_{\mathrm{f_c}}(X) = \{A \subseteq X : A \text{ is nonempty, closed and convex}\}\,,$$
$$P_{\mathrm{(w)kc}}(X) = \{A \subseteq X : A \text{ is nonempty, (weakly-) compact and convex}\}\,,$$
$$P_{\mathrm{bf(c)}}(X) = \{A \subseteq X : A \text{ is nonempty, bounded, closed (and convex)}\}\,.$$

For sets Y, V and a multifunction $F: Y \to 2^V$, we define

$$F^+(A) = \{y \in Y : F(y) \subseteq A\} \quad \text{for all } A \subseteq V\,,$$
$$F^-(A) = \{y \in Y : F(y) \cap A \neq \emptyset\} \quad \text{for all } A \subseteq V\,.$$

The set $F^+(A)$ is called the **strong inverse image** of A and $F^-(A)$ is said to be the **weak inverse image** of A. We directly see that $F^+(A) \subseteq F^-(A) \subseteq Y$.

These notions satisfy the following calculus rules.

Proposition 5.6.1. *Let X, Y, V be nonempty sets.*

(a) *if $F, G: X \to 2^Y$ are multifunctions and $(F\cup G)(x) = F(x)\cup G(x)$, $(F\cap G)(x) = F(x)\cap G(x)$ for all $x \in X$, then*

$$(F \cup G)^+(A) = F^+(A) \cup G^+(A)\,, \quad (F \cup G)^-(A) = F^-(A) \cup G^-(A)\,,$$
$$(F \cap G)^+(A) \supseteq F^+(A) \cap G^+(A)\,, \quad (F \cap G)^-(A) \subseteq F^-(A) \cap G^-(A)$$

for all $A \subseteq Y$;

(b) *if* $F\colon X \to 2^Y$, $G\colon Y \to 2^V$, *and* $(G \circ F)(x) = G(F(x)) = \bigcup_{y \in F(x)} G(y)$ *for all* $x \in X$, *then* $(G \circ F)^+(A) = F^+(G^+(A))$ *and* $(G \circ F)^-(A) = F^-(G^-(A))$ *for all* $A \subseteq V$;

(c) *if* $F\colon X \to 2^Y$ *and* $\{A_i\}_{i \in I} \subseteq 2^Y$ *with* I *being an arbitrary index set, then*

$$\bigcup_{i \in I} F^+(A_i) \subseteq F^+\Big(\bigcup_{i \in I} A_i\Big)\,, \quad \bigcup_{i \in I} F^-(A_i) = F^-\Big(\bigcup_{i \in I} A_i\Big)\,,$$
$$\bigcap_{i \in I} F^+(A_i) \subseteq F^+\Big(\bigcap_{i \in I} A_i\Big)\,, \quad F^-\Big(\bigcap_{i \in I} A_i\Big) \subseteq \bigcap_{i \in I} F^-(A_i)\,.$$

(d) *if* $F\colon X \to 2^Y$, $G\colon X \to 2^V$, *and* $F \times G\colon X \to 2^{Y \times V}$ *is defined by* $(F \times G)(x) = F(x) \times G(x)$ *for all* $x \in X$, *then* $(F \times G)^+(A \times C) = F^+(A) \cap G^+(C)$ *and* $(F \times G)^-(A \times C) = F^-(A) \cap G^-(C)$ *for all* $A \subseteq Y$ *and* $C \subseteq V$.

Now we are ready to introduce the main continuity notions for multifunctions. Recall that all topological spaces are assumed to be Hausdorff.

Definition 5.6.2. Let X, Y be topological spaces and let $F\colon X \to 2^Y$ be a multifunction.

(a) We say that F is **upper semicontinuous at** $x_0 \in X$ (usc at x_0 for short) if for all open $V \subseteq Y$ such that $F(x_0) \subseteq V$, there exists $U \in \mathcal{N}(x_0)$ such that $F(U) \subseteq V$. If this is true at every $x_0 \in X$, then we say that F is **upper semicontinuous** (usc for short).

(b) We say that F is **lower semicontinuous at** $x_0 \in X$ (lsc at x_0 for short) if for all open $V \subseteq Y$ such that $F(x_0) \cap V \neq \emptyset$, there exists $U \in \mathcal{N}(x_0)$ such that $F(x) \cap V \neq \emptyset$ for all $x \in U$. If this is true at every $x_0 \in X$, then we say that F is **lower semicontinuous** (lsc for short).

(c) We say that F is **continuous at** x_0 (or **Vietoris continuous at** x_0) if it is both usc and lsc at x_0. If this is true at every $x_0 \in X$, then we say that F is **continuous** (or **Vietoris continuous**).

Remark 5.6.3. It is clear from this definition that if F is single-valued, then the notions above coincide with the usual continuity of F.

Definition 5.6.2 easily leads to the following results.

Proposition 5.6.4. *If* X, Y *are topological spaces and* $F\colon X \to 2^Y$, *then the following statements are equivalent:*

(a) F *is* usc, *that is,* $F^+(V) \subseteq X$ *is open for all open* $V \subseteq Y$;

(b) $F^-(C) \subseteq X$ *is closed for every closed* $C \subseteq Y$;

(c) *if* $\{x_\alpha\}_{\alpha \in I} \subseteq X$ *is a net,* $x_\alpha \to x$ *and* $V \subseteq Y$ *is open with* $F(x) \subseteq V$, *then there exists an index* $\alpha_0 \in I$ *such that* $F(x_\alpha) \subseteq V$ *for all* $\alpha \geq \alpha_0$.

Proposition 5.6.5. *If X, Y are topological spaces and $F: X \to 2^Y$, then the following statements are equivalent:*
(a) *F is lsc, that is, $F^-(V) \subseteq X$ is open for all open $V \subseteq Y$;*
(b) *$F^+(C) \subseteq X$ is closed for every closed $C \subseteq Y$;*
(c) *if $\{x_\alpha\}_{\alpha \in I} \subseteq X$ is a net, $x_\alpha \to x$, and $V \subseteq Y$ is open with $F(x) \cap V \neq \emptyset$, then there exists an index $\alpha_0 \in I$ such that $F(x_\alpha) \cap V \neq \emptyset$ for all $\alpha \geq \alpha_0$.*
(d) *if $\{x_\alpha\}_{\alpha \in I} \subseteq X$ is a net, $x_\alpha \to x$, and $y \in F(x)$, then there exists a net $\{y_\alpha\}_{\alpha \in I} \subseteq Y$ with $y_\alpha \to y$ and $y_\alpha \in F(x_\alpha)$ for all $\alpha \in I$.*

Remark 5.6.6. Because of Proposition 5.6.1 (c), we see that F is lsc if and only if $F^-(V)$ is open for every basic open $V \subseteq Y$. For multifunctions the notions of upper and lower semicontinuity are in general distinct. Upper semicontinuity allows upward jumps in the sense of inclusion while lower semicontinuity allows downward jumps. In order to see this, consider the following two multifunctions $F_1, F_2: \mathbb{R} \to 2^{\mathbb{R}}$ defined by

$$F_1(x) = \begin{cases} [0,1] & \text{if } x = 0 \\ \{1\} & \text{if } x \neq 0 \end{cases}, \quad F_2(x) = \begin{cases} \{0\} & \text{if } x = 0 \\ [0,1] & \text{if } x \neq 0 \end{cases}.$$

Then F_1 is usc but not lsc while F_2 is lsc but not usc. Suppose that $F: \mathbb{R} \to 2^{\mathbb{R}}$ is defined by $F(x) = [\xi(x), \eta(x)]$ with $\xi, \eta: \mathbb{R} \to \mathbb{R}$. When ξ is lower semicontinuous and η is upper semicontinuous, then F is upper semicontinuous. On the other hand, when ξ is upper semicontinuous and η is lower semicontinuous, then F is lower semicontinuous.

From Propositions 5.6.4 and 5.6.5 we infer the following result.

Proposition 5.6.7. *If X, Y are topological spaces and $F: X \to 2^Y$, then the following statements are equivalent:*
(a) *F is continuous, that is, both $F^+(V)$ and $F^-(V)$ are open in X for all open $V \subseteq Y$;*
(b) *both $F^+(C)$ and $F^-(C)$ are closed in X for every closed $C \subseteq Y$;*
(c) *if $\{x_\alpha\}_{\alpha \in I} \subseteq X$ is a net, $x_\alpha \to x$, and $V \subseteq Y$ is open with $F(x) \subseteq V$ or $F(x) \cap V \neq \emptyset$, then there exists an index $\alpha_0 \in I$ such that $F(x_\alpha) \subseteq V$ or $F(x_\alpha) \cap V \neq \emptyset$ for all $\alpha \geq \alpha_0$.*

Definition 5.6.8. Let X, Y be topological spaces and $F: X \to 2^Y$. The **graph** of F is the set

$$\operatorname{Gr} F = \{(x, y) \in X \times Y : y \in F(x)\}.$$

Proposition 5.6.9. *If X, Y are topological spaces with Y being regular and $F: X \to P_f(Y)$ is usc, then* $\operatorname{Gr} F \subseteq X \times Y$ *is closed.*

Proof. Let $\{(x_\alpha, y_\alpha)\}_{\alpha \in I} \subseteq \operatorname{Gr} F$ be a net such that $(x_\alpha, y_\alpha) \to (x, y)$ in $X \times Y$. We argue by contradiction. So, suppose that $y \notin F(x)$. The regularity of Y implies the existence of open sets $V_1, V_2 \subseteq Y$ such that $V_1 \in \mathcal{N}(y)$, $F(x) \subseteq V_2$ and $V_1 \cap V_2 = \emptyset$. According to Proposition 5.6.4 there is an index $\alpha_0 \in I$ such that

$$y_\alpha \in F(x_\alpha) \subseteq V_2 \quad \text{with } y_\alpha \in V_1 \text{ for all } \alpha \geq \alpha_0 ,$$

a contradiction since $V_1 \cap V_2 = \emptyset$. Therefore, $\operatorname{Gr} F \subseteq X \times Y$ is closed. □

Remark 5.6.10. If F is $P_k(Y)$-valued, then we can drop the regularity condition on Y. Of course, the converse of Proposition 5.6.9 is not true in general. Simple examples of single-valued functions illustrate this.

Proposition 5.6.11. *If X, Y are topological spaces, $F: X \to P_k(Y)$ has a closed graph and is locally compact, that is, for every $x \in X$ there exists $U \in \mathcal{N}(x)$ such that $\overline{F(U)} \subseteq Y$ is compact, then F is* usc.

Proof. According to Proposition 5.6.4, it suffices to show that if $C \subseteq Y$ is closed, then $F^-(C) \subseteq X$ is closed. So, let $\{x_\alpha\}_{\alpha \in I} \subseteq F^-(C)$ be a net and assume that $x_\alpha \to x$ in X. We can find $U \in \mathcal{N}(x)$ such that $\overline{F(U)} \subseteq Y$ is compact. We choose $\alpha_0 \in I$ such that $x_\alpha \in U$ for all $\alpha \geq \alpha_0$. Let $y_\alpha \in F(x_\alpha) \cap C$ for $\alpha \geq \alpha_0$. Then $\overline{\{y_\alpha\}}_{\alpha \geq \alpha_0}$ is compact and so there exists a subnet $\{y_\beta\}_{\beta \in J}$ of $\{y_\alpha\}_{\alpha \in I}$ such that $y_\beta \to y$. Since $\operatorname{Gr} F \subseteq X \times Y$ and $C \subseteq Y$ are closed, we obtain that $(x, y) \in \operatorname{Gr} F \cap (X \times C)$, that is, $x \in F^-(C)$. Hence, $F^-(C)$ is closed and so F is usc. □

Definition 5.6.12. Let X, Y be topological spaces and $F: X \to 2^Y$. We say that F is **closed** (resp. **sequentially closed**) if $\operatorname{Gr} F \subseteq X \times Y$ is closed (resp. sequentially closed).

Remark 5.6.13. Every closed (resp. sequentially closed) multifunction has closed (resp. sequentially closed) values.

Proposition 5.6.14. *If X, Y are topological spaces, $F: X \to P_k(Y)$ is* usc, *and $K \subseteq X$ is compact, then $F(K) = \bigcup_{x \in K} F(x) \subseteq Y$ is compact.*

Proof. Let $\{y_\alpha\}_{\alpha \in I} \subseteq F(K)$ be a net. We have $y_\alpha \in F(x_\alpha)$ with $x_\alpha \in K$ for all $\alpha \in I$. The compactness of K implies that there exists a subnet $\{x_\beta\}_{\beta \in J}$ such that $x_\beta \to x \in K$ in X. We claim that $\{y_\beta\}_{\beta \in J}$ has a cluster point in $F(x)$. Arguing by contradiction, suppose that for every $y \in F(x)$ we can find $\beta_0(y) \in J$ and $V(y) \in \mathcal{N}(y)$ such that $y_\beta \notin V(y)$ for all $\beta \in J$ with $\beta \geq \beta_0(y)$. Evidently, $\{V(y)\}_{y \in F(x)}$ is an open cover of $F(y) \in P_k(Y)$. So, there exists a finite subcover $\{V(y_k)\}_{k=1}^N$. Then $V = \bigcup_{k=1}^N V(y_k) \in \mathcal{N}(y)$. We can find $\beta_1 \in J$ such that $y_\beta \notin V = \bigcup_{k=1}^N V(y_k) \supseteq F(x)$ for all $\beta \in J$ with $\beta \geq \beta_1$. This contradicts Proposition 5.6.4 (c). So, the claim is true and we can find a subnet $\{y_\gamma\}_{\gamma \in S}$ of $\{y_\beta\}_{\beta \in J}$ such that $y_\gamma \to y \in F(x) \subseteq F(K)$. This proves the compactness of $F(K)$. □

We can characterize upper and lower semicontinuity using the distance and support functions.

Proposition 5.6.15. *If X is a topological space, (Y, d) is a metric space and $F: X \to 2^Y \setminus \{\emptyset\}$, then the following hold:*

(a) F is lsc *if and only if $x \to d(y, F(x))$ is upper semicontinuous for every $y \in Y$;*

(b) *if F is usc, then $x \to d(y, F(x))$ is lower semicontinuous for every $y \in Y$; the converse is true if $F: X \to P_f(Y)$ is locally compact; see Proposition* 5.6.11.

Proof. (a) $\Longrightarrow$: We assume that F is lsc. Let $\lambda \in \mathbb{R}$ and consider the upper level set

$$U_\lambda = \{x \in X : k_y(x) = d(y, F(x)) \geq \lambda\} .$$

We need to show that $U_\lambda \subseteq X$ is closed. So, let $\{x_\alpha\}_{\alpha\in I} \subseteq U_\lambda$ be a net and assume that $x_\alpha \to x$ in X. Given $\varepsilon > 0$, there exists $v \in F(x)$ such that $d(y, v) \leq k_y(x) + \varepsilon$. Let $B_\varepsilon(v) = \{u \in Y : d(u, v) < \varepsilon\}$. The lower semicontinuity of F implies that there exists $\alpha_0 \in I$ such that $F(x_\alpha) \cap B_\varepsilon(v) \neq \emptyset$ for all $\alpha \geq \alpha_0$; see Proposition 5.6.5 (c). So, we can find $y_\alpha \in F(x_\alpha)$ such that $d(y_\alpha, y) < k_y(x) + 2\varepsilon$ for all $\alpha \geq \alpha_0$. Hence, $\lambda \leq k_y(x_\alpha) < k_y(x) + 2\varepsilon$. But $\varepsilon > 0$ is arbitrary. Letting $\varepsilon \searrow 0$, we conclude that $\lambda \leq k_y(x)$ and so $x \in U_\lambda$. This proves the upper semicontinuity of the distance function $x \to k_y(x) = d(y, F(x))$.

$\Longleftarrow$: Let $V \subseteq Y$ be open. We need to show that $F^-(V)$ is open. So, let $x \in F^-(V)$. Then we can find $y \in F(x) \cap V$. Let $\varepsilon > 0$ such that $B_\varepsilon(y) \subseteq V$. Since by hypothesis k_y is upper semicontinuous, there exists $U \in \mathcal{N}(x)$ such that $k_y(x') < k_y(x) + \varepsilon = \varepsilon$ for all $x' \in U$. This implies that $F(x') \cap B_\varepsilon(y) \neq \emptyset$ for all $x' \in U$ and so $F(x') \cap V \neq \emptyset$ for all $x' \in U$. Hence, F is lsc; see Proposition 5.6.5 (a).

(b) Consider the lower level set $L_\lambda = \{x \in X : k_y(x) = d(y, F(x)) \leq \lambda\}$. We need to show that L_λ is closed. So, let $\{x_\alpha\}_{\alpha\in I} \subseteq L_\lambda$ be a net such that $x_\alpha \to x$ in X. The upper semicontinuity of F implies that for given $\varepsilon > 0$ there exists $\alpha_0 \in I$ such that

$$F(x_\alpha) \subseteq F(x)_\varepsilon = \{v \in Y : d(v, F(x)) < \varepsilon\} \quad \text{for all } \alpha \geq \alpha_0 .$$

This shows that $k_y(x) < k_y(x_\alpha) + \varepsilon$ and so $k_y(x) < \lambda + \varepsilon$. Since $\varepsilon > 0$ is arbitrary, we let $\varepsilon \searrow 0$ to conclude that $k_y(x) \leq \lambda$. So, $x \in L_\lambda$ and this proves that L_λ is closed, that is, k_y is lower semicontinuous.

Next, for the converse, assume that F is locally compact and k_y is lower semicontinuous. According to Proposition 5.6.11, in order to show that F is usc, it suffices to show that $\operatorname{Gr} F \subseteq X \times Y$ is closed. So, let $\{(x_\alpha, y_\alpha)\}_{\alpha\in I} \subseteq \operatorname{Gr} F$ be a net such that $(x_\alpha, y_\alpha) \to (x, y)$ in $X \times Y$. We have

$$k_y(x_\alpha) = d(y, F(x_\alpha)) \leq d(y, y_\alpha) \xrightarrow{\alpha\in I} 0 . \tag{5.6.1}$$

Since by hypothesis k_y is lower semicontinuous, we obtain $k_y(x) \leq \liminf_{\alpha\in I} k_y(x_\alpha)$. Because of (5.6.1) we see that $k_y(x) = 0$ and so $y \in F(x)$. Therefore, $\operatorname{Gr} F \subseteq X \times Y$ is closed, hence F is usc. □

Proposition 5.6.16. *If X is a topological space, Y is a normed space equipped with the weak topology, and $F: X \to 2^Y \setminus \{\emptyset\}$ is usc, then $x \to \sigma(y^*; F(x))$ is upper semicontinuous for all $y^* \in Y^*$.*

Proof. We fix $y^* \in Y^*$ and $\varepsilon > 0$ and introduce the set

$$W(y^*, \varepsilon) = \{y \in Y: \langle y^*, y\rangle < \varepsilon\}.$$

Evidently, $W(y^*, \varepsilon) \in \mathcal{N}_w(0)$, where $\mathcal{N}_w(0)$ is the filter of weak neighborhoods of the origin. The upper semicontinuity of F implies that there exists $U \in \mathcal{N}(x)$ such that $F(x') \subseteq F(x) + W(y^*, \varepsilon)$ for all $x' \in U$. Hence,

$$\sigma(y^*; F(x')) \leq \sigma(y^*; F(x)) + \varepsilon \quad \text{for all } x' \in U.$$

This shows that $x \to \sigma(y^*; F(x))$ is upper semicontinuous. □

Example 5.6.17. (a) Consider the multifunction $F: \mathbb{R} \to P_f(\mathbb{R}^2)$ defined by

$$F(x) = \{(u, xu): u \in \mathbb{R}\}.$$

Note that $x \to d(y, F(x))$ is continuous for all $y \in \mathbb{R}^2$, but clearly F is not usc. The multifunction F is not locally compact; see Proposition 5.6.15 (b).

(b) Consider the multifunction $F: \mathbb{R}_+ \to P_{kc}(\mathbb{R})$ defined by

$$F(x) = \begin{cases} \{-1, 1\} & \text{if } x = 0, \\ [0, x] & \text{if } x \neq 0. \end{cases}$$

For every $y^* \in \mathbb{R}$, the function $x \to \sigma(y^*; F(x))$ is upper semicontinuous but F is not usc at $x = 0$. So, the converse of Proposition 5.6.16 is not true in general. It is true if F has values in $P_{wkc}(Y)$; see Hu–Papageorgiou [173, pp. 47–48].

Let X, Y be topological spaces, $F: X \to 2^Y \setminus \{\emptyset\}$ a multifunction, and let $\varphi: X \times Y \to \mathbb{R}$ be a function. We consider the following optimization problem

$$v(x) = \sup[\varphi(x, y): y \in F(x)]. \tag{5.6.2}$$

The function v is known as the **value function** of the optimization problem.

Proposition 5.6.18. *If F is* lsc *and φ is lower semicontinuous, then the value function v defined in* (5.6.2) *is lower semicontinuous.*

Proof. We need to show that the lower level set $L_\lambda = \{x \in X: v(x) \leq \lambda\}$ is closed for every $\lambda \in \mathbb{R}$. So, let $\{x_\alpha\}_{\alpha\in I} \subseteq L_\lambda$ be a net and assume that $x_\alpha \to x$ in X. Let $y \in F(x)$. Proposition 5.6.5 (d) implies that there exists a net $\{y_\alpha\}_{\alpha\in I} \subseteq Y$ such that $y_\alpha \to y$ and $y_\alpha \in F(x_\alpha)$ for all $\alpha \in I$. From (5.6.2), we see that $\varphi(x_\alpha, y_\alpha) \leq v(x_\alpha) \leq \lambda$ for all $\alpha \in I$ which implies that

$$\varphi(x, y) \leq \liminf_{\alpha\in I} \varphi(x_\alpha, y_\alpha) \leq \lambda. \tag{5.6.3}$$

Since $y \in F(x)$ is arbitrary, from (5.6.3) it follows that $v(x) \le \lambda$. Hence, $x \in L_\lambda$ and this proves that L_λ is closed, and thus v is lower semicontinuous. □

Proposition 5.6.19. *If $F: X \to P_k(Y)$ is* usc *and φ is upper semicontinuous, then the value function v is upper semicontinuous.*

Proof. We need to show that the upper level set $U_\lambda = \{x \in X : v(x) \ge \lambda\}$ is closed for every $\lambda \in \mathbb{R}$. So, let $\{x_\alpha\}_{\alpha\in I} \subseteq U_\lambda$ be a net and assume that $x_\alpha \to x$ in X. Since F is $P_k(Y)$-valued for every $\alpha \in I$, there exists $y_\alpha \in F(x_\alpha)$ such that $v(x_\alpha) = \varphi(x_\alpha, y_\alpha)$. From the proof of Proposition 5.6.14, we know that $\{y_\alpha\}_{\alpha\in I}$ has a cluster point $y \in F(x)$. So, there exists a subnet $\{y_\beta\}_{\beta\in J}$ of $\{y_\alpha\}_{\alpha\in I}$ such that $y_\beta \to y$ in Y. We have $\lambda \le v(x_\beta) = \varphi(x_\beta, y_\beta)$ for all $\beta \in J$. Since φ is upper semicontinuous we obtain that

$$\lambda \le \limsup_{\beta\in J} \varphi(x_\beta, y_\beta) \le \varphi(x, y)\,.$$

Since $y \in F(x)$ (see (5.6.2)), we get $\lambda \le v(x)$. Therefore, $x \in U_\lambda$ and so $U_\lambda \subseteq X$ is closed, that is, v is upper semicontinuous. □

Definition 5.6.20. Let X, Y be two sets and let $F: X \to 2^Y \setminus \{\emptyset\}$ be a multifunction. A **selection** of F is a single-valued map $f: X \to Y$ such that $f(x) \in F(x)$ for all $x \in X$. When X and Y have topological structures, we look for **continuous selections**. When X has measure theoretic structures, we look for **measurable selections**.

A usc multifunction need not have a continuous selection.

Example 5.6.21. Consider the multifunction $F: \mathbb{R} \to P_{kc}(\mathbb{R})$ defined by

$$F(x) = \begin{cases} \{-1\} & \text{if } x < 0\,, \\ [-1, 1] & \text{if } x = 0\,, \\ \{1\} & \text{if } x > 0\,. \end{cases}$$

Then F is usc and cannot have a continuous selection. Note that if $\varphi(x) = |x|$, then $F = \partial\varphi$.

However, the situation is different for lsc multifunctions and we can state the so-called "Michael Selection Theorem".

Theorem 5.6.22 (Michael Selection Theorem). *If X is paracompact, Y is a Banach space, and $F: X \to P_{fc}(Y)$ is* lsc, *then F admits a continuous selection.*

Proof. We set $B_1 = \{y \in Y : \|y\| < 1\}$. First we produce a continuous approximation selection, that is, a continuous function $f_\varepsilon: X \to Y$ such that

$$f_\varepsilon(x) \in F(x) + \varepsilon B_1 \quad \text{for all } x \in X \text{ with } \varepsilon > 0\,. \tag{5.6.4}$$

To this end, we choose $y_x \in F(x)$ for every $x \in X$. The lower semicontinuity of F implies that $F^-(y_x + \varepsilon B_1) \subseteq X$ is open. So, $\{F^-(y_x + \varepsilon B_1)\}_{x\in X}$ is an open cover of X. Then the paracompactness of X implies that there exists a locally finite refinement $\{F^-(y_i+\varepsilon B_1)\}_{i\in I}$. Let $\{p_i\}_{i\in I}$ be a continuous partition of unity subordinate to this cover; see Theorem 1.4.86. We set

$$f_\varepsilon(x) = \sum_{i\in I} p_i(x)u_i \quad \text{for all } x \in X \tag{5.6.5}$$

with $u_i \in y_i + \varepsilon B_1$. The local finiteness of the cover implies that the sum in (5.6.5) is finite and so $f_\varepsilon(x)$ is well-defined. The convexity of the values of $x \to F(x) + \varepsilon B_1$ implies that (5.6.4) holds. Inductively we generate a sequence of continuous functions $f_n: X \to Y$ with $n \in \mathbb{N}$ such that

$$f_n(x) \in F(x) + \frac{1}{2^n}B_1 \quad \text{for all } x \in X \text{ and for all } n \in \mathbb{N} \tag{5.6.6}$$

and

$$\|f_{n+1}(x) - f_n(x)\| < \frac{1}{2^{n-1}} \quad \text{for all } x \in X \text{ and for all } n \in \mathbb{N}\,. \tag{5.6.7}$$

For $n = 1$ the function f_1 was produced in the first part of the proof. For the induction hypothesis, assume that we already have the continuous functions $f_k: X \to Y$ with $k \in \{1, \ldots, n\}$, which satisfy (5.6.6) and (5.6.7). Let

$$G_n(x) = F(x) \bigcap \left(f_n(x) + \frac{1}{2^n}B_1 \right) \quad \text{for all } x \in X\,.$$

From (5.6.6) we see that G_n has nonempty values that are also convex. Moreover, it is easy to see that G_n is lsc. So, from the first part of the proof there exists a continuous function $f_{n+1}: X \to Y$ such that

$$f_{n+1}(x) \in G_n(x) + \frac{1}{2^{n+1}}B_1 \quad \text{for all } x \in X\,.$$

This yields

$$f_{n+1}(x) \in f_n(x) + \frac{1}{2^n}B_1 + \frac{1}{2^{n+1}}B_1 \subseteq f_n(x) + \frac{1}{2^{n-1}}B_1 \quad \text{for all } x \in X,$$

and so

$$\|f_{n+1}(x) - f_n(x)\| < \frac{1}{2^{n-1}} \quad \text{for all } x \in X\,.$$

This completes the induction process.

From (5.6.7), we see that $\{f_n\}_{n\geq 1} \subseteq C(X, Y)$ is a Cauchy sequence. So, $f_n \to f \in C(X, Y)$ and $f(x) \in F(x)$ for all $x \in X$; see (5.6.6). Then f is the desired continuous selection. □

Remark 5.6.23. In fact, the existence of a continuous selection for F is equivalent to the paracompactness of X, that is, if X is a topological space, Y is a Banach space, and every lsc multifunction $F: X \to P_{fc}(Y)$ admits a continuous selection, then the space X is paracompact.

We can obtain the continuous selection passing from a prescribed point of $\mathrm{Gr}\, F$.

Corollary 5.6.24. *If X is paracompact, Y is a Banach space, $F: X \to P_{fc}(Y)$ is lsc, and $(\hat{x}, \hat{y}) \in \mathrm{Gr}\, F$, then there exists a continuous map $f: X \to Y$ such that $f(x) \in F(x)$ for all $x \in X$ and $f(\hat{x}) = \hat{y}$.*

Proof. Let $\hat{F}: X \to P_{fc}(Y)$ be the multifunction defined by

$$\hat{F}(x) = \begin{cases} F(x) & \text{if } x \neq \hat{x}\,, \\ \{\hat{y}\} & \text{if } x = \hat{x}\,. \end{cases} \tag{5.6.8}$$

Evidently, $\hat{F}$ is lsc; see Remark 5.6.6. So, by Theorem 5.6.22 there exists a continuous map $f: X \to Y$ such that $f(x) \in \hat{F}(x)$ for all $x \in X$. Then from (5.6.8) it follows that $f(x) \in F(x)$ for all $x \in X$ and $f(\hat{x}) = \hat{y}$. □

If we strengthen the conditions on the spaces X and Y we can improve the conclusion of Theorem 5.6.22 and produce a whole sequence $\{f_n\}_{n\geq 1}$ of continuous selections of F such that $\{f_n(x)\}_{n\geq 1}$ is dense in $F(x)$.

Theorem 5.6.25. *If X is a metric space, Y is a separable Banach space, and $F: X \to P_{fc}(Y)$ is* lsc, *then there exists a sequence $\{f_n\}_{n\geq 1}$ of continuous selections of F such that $F(x) = \overline{\{f_n(x)\}_{n\geq 1}}$ for all $x \in X$.*

Proof. Let $\{y_n\}_{n\geq 1} \subseteq Y$ be dense and let $U_{nm} = F^{-}(B_{1/2^m}(y_n))$ for all $n, m \in \mathbb{N}$. The lower semicontinuity of F implies that each U_{nm} is open in X. In a metric space every open set is F_σ; see Proposition 1.5.8. So, we obtain

$$U_{nm} = \bigcup_{k\in\mathbb{N}} C_{nmk} \quad \text{with closed } C_{nmk} \subseteq X \text{ for all } k \in \mathbb{N}\,.$$

We define

$$F_{nmk}(x) = \begin{cases} F(x) & \text{if } x \notin C_{nmk}\,, \\ \overline{F(x) \cap B_{\frac{1}{2^m}}(y_n)} & \text{if } x \in C_{nmk}\,. \end{cases}$$

Evidently, F_{nmk} is lsc and has values in $P_{fc}(Y)$. So, Theorem 5.6.22 provides a continuous selection $f_{nmk}: X \to Y$ of F_{nmk}. Let $y \in F(x)$ and $m \in \mathbb{N}$. We choose $y_n \in y + 1/2^m B_1$. Then $x \in U_{n(m+2)}$ and so $x \in C_{n(m+2)k}$ for some $k \in \mathbb{N}$. Then we have

$$f_{n(m+2)k}(x) \in y_n + \frac{1}{2^{m+2}}\overline{B}_1 \subseteq y_n + \frac{1}{2^m}B_1\,,$$

which implies that $\overline{\{f_{nmk}(x)\}_{n,m,k\in\mathbb{N}}} = F(x)$ for all $x \in X$. □

The measurability properties of multifunctions and the existence of measurable selections were discussed in Section 2.7. Here we present a few more results that complete the theory of measurable multifunctions.

In what follows, we denote by (Ω, Σ) a measurable space and X is assumed to be a separable Banach space.

Definition 5.6.26. Let $F\colon \Omega \to 2^X \setminus \{\emptyset\}$ be a multifunction. We say that F is **scalarly measurable** if the function $w \to \sigma(x^*; F(w))$ is Σ-measurable for every $x^* \in X^*$.

Proposition 5.6.27. *If (Ω, Σ) is a complete measurable space (see Definition 2.7.17), and $F\colon \Omega \to 2^X \setminus \{\emptyset\}$ is graph measurable, then F is scalarly measurable.*

Proof. According to Theorem 2.7.28 there exists a sequence $\{f_n\}_{n\geq 1}$ of Σ-measurable selections of F such that $F(w) \subseteq \overline{\{f_n(w)\}}_{n\geq 1}$ for all $w \in \Omega$. Then we have

$$\sigma(x^*; F(w)) = \sup_{n\geq 1} \langle x^*, f_n(w) \rangle \quad \text{for all } x^* \in X^* .$$

Hence, $w \to \sigma(x^*; F(w))$ is Σ-measurable and so, F is scalarly measurable. □

If F is $P_{\mathrm{wkc}}(X)$-valued, then measurability (see Definition 2.7.1 (a)), and scalar measurability (see Definition 5.6.26) are in fact equivalent.

Proposition 5.6.28. *If $F\colon \Omega \to P_{\mathrm{wkc}}(X)$, then F is measurable if and only if it is scalarly measurable.*

Proof. $\Longrightarrow$: From Theorem 2.7.13, we know that there exists a sequence $\{f_n\}_{n\geq 1}$ of Σ-measurable selections of F such that $F(w) = \overline{\{f_n(w)\}}_{n\geq 1}$ for all $w \in \Omega$. Then

$$\sigma(x^*; F(w)) = \sup_{n\geq 1} \langle x^*, f_n(w) \rangle \quad \text{for all } x^* \in X^* .$$

Hence, F is scalarly measurable.

$\Longleftarrow$: Since X is separable, then $X^*_{w^*}$ is separable. Recall that $X^*_{w^*}$ is the space X^* endowed with the w*-topology. So, if τ is a locally convex topology on X^* such that $(X^*_\tau)^* = X$, then X^*_τ is separable. In particular this is true if $\tau = \mathrm{m}(X^*, X)$ where $\mathrm{m}(X^*, X)$ is the Mackey topology on X^*; see Theorem 3.8.9. From Proposition 5.3.12, we know that $\sigma(\cdot; F(w))$ is m-continuous for every $w \in \Omega$. Moreover, from Proposition 5.3.11 we obtain

$$d(x, F(w)) = \sup[\langle x^*, x \rangle - \sigma(x^*; F(w)) : x^* \in X] \tag{5.6.9}$$

for all $w \in \Omega$ and for all $x \in X$. Let $\{x_n^*\}_{n\geq 1} \subseteq X^*$ be m-dense. Exploiting the m-continuity of $x^* \to \langle x^*, x \rangle - \sigma(x^*; F(w))$, (5.6.9) yields

$$d(x, F(w)) = \sup_{n\geq 1}[\langle x_n^*, x \rangle - \sigma(x_n^*; F(w))] ,$$

which implies that $w \to d(x, F(w))$ is Σ-measurable. Hence F is measurable; see Proposition 2.7.5. □

Given a multifunction $F:\Omega \to 2^X \setminus \{\emptyset\}$, by $\operatorname{ext} F$ we denote the multifunction that assigns at each $w \in \Omega$ the set $\operatorname{ext} F(w)$ of extreme points of $F(w)$. The next proposition examines the measurability of the multifunction $w \to \operatorname{ext} F(w)$.

Proposition 5.6.29. *If $F:\Omega \to P_{\mathrm{wkc}}(X)$ is a measurable multifunction, then $w \to \operatorname{ext} F(w)$ is graph measurable.*

Proof. From the Krein–Milman Theorem (see Theorem 3.8.11), we know that $\operatorname{ext} F(w) \neq \emptyset$ for all $w \in \Omega$. On X^*, we consider the Mackey topology $\mathrm{m}(X^*, X)$; see Theorem 3.8.9. Endowed with this topology, X^* is separable. Let $\{x_n^*\}_{n\geq 1} \subseteq \overline{B}_1^{X^*}$ be m-dense and consider the function $\eta_F:\Omega \times X \to \mathbb{R}$ defined by

$$\eta_F(w,x) = \begin{cases} \sum_{n\geq 1} \frac{\langle x_n^*, x\rangle^2}{2^n} & \text{if } x \in F(w)\,, \\ +\infty & \text{otherwise}\,. \end{cases}$$

Evidently, η_F is $\Sigma \otimes \mathcal{B}(X)$-measurable and $\eta_F(w,\cdot)|_{F(w)}$ is continuous for every $w \in \Omega$. Denote by $\mathcal{L}$ the space of all affine continuous functions $a:X \to \mathbb{R}$. We define

$$\hat{\eta}_F(w,x) = \inf\big[a(x): a \in \mathcal{L}, a(u) > \eta_F(w,u) \text{ for all } u \in F(w)\big]\,.$$

From Theorem 5.6.25, we know that there exists a sequence of Σ-measurable selections $u_n:\Omega \to X$ with $n \in \mathbb{N}$ of F such that $F(w) = \overline{\{u_n(w)\}_{n\geq 1}}$ for all $w \in \Omega$. For every $(w, x^*) \in \Omega \times X^*$ let

$$e_{x^*}(w) = \sup\big[\eta_F(w,x) - \langle x^*, u\rangle : u \in F(w)\big]\,.$$

Note that $e_{x^*}(w) < +\infty$, $x^* \to e_{x^*}(w)$ is continuous on X^* for every $w \in \Omega$ and

$$e_{x^*}(w) = \sup_{n\geq 1}\big[\eta_F(w,u_n(w)) - \langle x^*, u_n(w)\rangle\big]\,.$$

Hence, $(w, x^*) \to e_{x^*}(w)$ is $\Sigma \otimes \mathcal{B}(X^*)$-measurable. We have

$$\hat{\eta}_F(w,x) = \inf\big[\langle x^*, x\rangle + e_{x^*}(w): x^* \in X^*\big]\,.$$

Then, for $\{x_n^*\}_{n\geq 1} \subseteq X^*$ m-dense, we obtain

$$\hat{\eta}_F(w,x) = \inf_{n\geq 1}\big[\langle x_n^*, x\rangle + e_{x_n^*}(w).\big]$$

Hence, $(w,x) \to \hat{\eta}_F(w,x)$ is $\Sigma \otimes \mathcal{B}(X)$-measurable. But we know that

$$\operatorname{ext} F(w) = \big\{x \in X: \hat{\eta}_F(w,x) = \eta_F(w,x)\big\}\,;$$

see Choquet [72]. Therefore, $\operatorname{Gr} F \in \Sigma \otimes \mathcal{B}(X)$. ☐

From this point on, the standing hypotheses are: (Ω, Σ, μ) is a σ-finite measure space and X is a separable Banach space. Additional hypotheses will be introduced as needed. By $L^0(\Omega, X)$ we denote the space of the equivalence classes of Σ-measurable maps from Ω into X for the relation of equality μ-a. e.

Definition 5.6.30. A set $D \subseteq L^0(\Omega, X)$ is said to be **decomposable** if $\chi_A f + \chi_{A^c} g \in D$ for every triple $(A, f, g) \in \Sigma \times D \times D$.

Remark 5.6.31. Since $\chi_{A^c} = 1 - \chi_A$, formally the notion of decomposability is similar to that of convexity. Of course in the definition of decomposability the coefficients are functions. In what follows we will see that decomposability leads to some implications similar to those of convexity.

Definition 5.6.32. Given a multifunction $F: \Omega \to 2^X \setminus \{\emptyset\}$, let $S_F = \{u \in L^0(\Omega, X): u(w) \in F(w)\ \mu\text{-a. e.}\}$. Moreover, let $S_F^p = S_F \cap L^p(\Omega, X)$ for $1 \le p \le \infty$.

Remark 5.6.33. Clearly, the sets S_F and S_F^p with $1 \le p \le \infty$ are all decomposable.

Proposition 5.6.34. *If $F: \Omega \to 2^X \setminus \{\emptyset\}$ is graph measurable and $S_F^p \ne \emptyset$ for $1 \le p \le \infty$, then there exists a sequence $\{u_n\}_{n \ge 1} \subseteq S_F^p$ such that $F(w) \subseteq \overline{\{u_n(w)\}}_{n\ge1}$ μ-a. e.*

Proof. According to Theorem 2.7.28, there exists a sequence $\hat{u}_n: \Omega \to X$ with $n \in \mathbb{N}$ of Σ-measurable selections of F such that $F(w) \subseteq \overline{\{\hat{u}_n(w)\}}_{n\ge1}$ μ-a. e. Since μ is σ-finite, there exists a sequence $\{A_k\}_{k\in\mathbb{N}} \subseteq \Sigma$, which is a partition of Ω such that $\mu(A_k) < +\infty$ for all $k \in \mathbb{N}$. Let $u_0 \in S_F^p$ and define

$$C_{nki} = \{w \in \Omega: i - 1 \le \|u_n(w)\| < i\} \cap A_k \quad \text{for all } n, k, i \in \mathbb{N},$$
$$u_{nki} = \chi_{C_{nki}} u_n + \chi_{C_{nki}^c} u_0 \quad \text{for all } n, k, i \in \mathbb{N}.$$

Evidently $u_{nki} \in S_F^p$ and $F(w) \subseteq \overline{\{u_{nki}(w)\}}_{n,k,i\ge1}$ for μ-a. a. $w \in \Omega$. □

Corollary 5.6.35. *If $F, G: \Omega \to 2^X \setminus \{\emptyset\}$ are both graph measurable and $S_F^p = S_G^p$ for some $1 \le p \le \infty$, then $F(w) = G(w)$ μ-a. e. in Ω.*

Proposition 5.6.36. *If $F: \Omega \to 2^X \setminus \{\emptyset\}$ is graph measurable and $1 \le p \le \infty$, then $S_F^p \ne \emptyset$ if and only if $m(w) = \inf[\|u\|: u \in F(w)] \in L^p(\Omega)$.*

Proof. $\Longrightarrow$: According to Proposition 5.6.34 there exists a sequence $\{u_n\}_{n\ge1} \subseteq L^p(\Omega, X)$ such that $F(w) \subseteq \overline{\{u_n(w)\}}_{n\ge1}$ μ-a. e. Then $m(w) = \inf_{n\ge1} \|u_n(w)\|$ and so $m \in L^p(\Omega)$.

$\Longleftarrow$: Since μ is σ-finite, for given $\varepsilon > 0$ there exists $e_\varepsilon \in L^p(\Omega)_+$ such that

$$e_\varepsilon(w) > 0 \quad \mu\text{-a. e.} \quad \text{and} \quad \int_\Omega e_\varepsilon(w)\, d\mu = \varepsilon.$$

Consider the multifunction $G_\varepsilon(w) = \{u \in F(w): \|u\| \le m(w) + e_\varepsilon(w)\}$. Evidently, G_ε is graph measurable. So, by the Yankov–von Neumann–Aumann Selection Theorem (see Theorem 2.7.25), there exists a Σ-measurable selection $u_\varepsilon\colon \Omega \to X$ of G_ε. Then $u_\varepsilon \in L^p(\Omega, X)$ and so $S_F^p \neq \emptyset$. □

Lemma 5.6.37. *If $F\colon \Omega \to 2^X \setminus \{\emptyset\}$ is graph measurable, $1 \le p \le \infty$, $\{u_n\}_{n\ge1} \subseteq S_F^p$ satisfies*

$$F(w) \subseteq \overline{\{u_n(w)\}_{n\ge1}} \quad \text{for } \mu\text{-a. a. } w \in \Omega\,,$$

$u \in S_F^p$, and $\varepsilon > 0$, then there exists a finite Σ-partition $\{C_k\}_{k=1}^m$ of Ω such that

$$\left\| u - \sum_{k=1}^m \chi_{C_k} u_k \right\|_p < \varepsilon\,.$$

Proof. We may assume that $u(w) \in F(w)$ for all $w \in \Omega$. Consider a function $\xi \in L^1(\Omega)$ such that $\xi(w) > 0$ for all $w \in \Omega$ and $\int_\Omega \xi \, d\mu < \varepsilon^p/3$. We can find a Σ-partition $\{E_n\}_{n\ge1}$ of Ω such that

$$\|u(w) - u_n(w)\|^p < \xi(w) \quad \text{for all } w \in E_n \text{ and for all } n\,.$$

We choose a large enough $m \in \mathbb{N}$ such that

$$\sum_{n\ge m+1} \int_{E_n} \|u(w)\|^p \, d\mu < \frac{\varepsilon^p}{3\cdot 2^p} \quad \text{and} \quad \sum_{n\ge m+1} \int_{E_n} \|u_1(w)\|^p \, d\mu < \frac{\varepsilon^p}{3\cdot 2^p}\,.$$

Let $\{C_k\}_{k=1}^m$ be a Σ-partition of Ω defined by

$$C_1 = E_1 \bigcup \Big(\bigcup_{n\ge m+1} E_n \Big) \quad \text{and} \quad C_k = E_k \quad \text{for } k = 2, \dots, m\,.$$

Then it follows that

$$\begin{aligned} &\left\| u - \sum_{k=1}^m \chi_{C_k} u_k \right\|_p \\ &= \sum_{k=1}^m \int_{E_k} \|u(w) - u_k(w)\|^p \, d\mu + \sum_{k\ge m+1} \int_{E_k} \|u(w) - u_1(w)\|^p \, d\mu \\ &\le \int_\Omega \xi(w) \, d\mu + \sum_{k\ge m+1} 2^{p-1} \int_{E_k} (\|u(w)\|^p + \|u_1(w)\|^p) \, d\mu < \varepsilon^p\,. \end{aligned}$$ □

Using this lemma we can characterize closed and decomposable sets in $L^p(\Omega, X)$ for $1 \le p < \infty$.

Theorem 5.6.38. *If $D \subseteq L^p(\Omega, X)$ for $1 \le p < \infty$ is nonempty and closed, then D is decomposable if and only if $D = S_F^p$ for a measurable $F\colon \Omega \to P_f(X)$.*

Proof. $\Longrightarrow$: From Proposition 5.6.34, we know that there exists $\{u_n\}_{n\geq1} \subseteq L^p(\Omega, X)$ such that $X = \overline{\{u_n(w)\}_{n\geq1}}$ μ-a. e. Let $\xi_n = \inf[\|u_n - h\|_p : h \in D]$ with $n \in \mathbb{N}$ and let $\{h_{nm}\}_{m\geq1} \subseteq D$ such that

$$\|u_n - h_{nm}\|_p \searrow \xi_n \quad \text{as } m \to \infty \text{ for all } n \in \mathbb{N}\,.$$

We set $F(w) = \overline{\{h_{nm}(w)\}_{n,m\in\mathbb{N}}}$ for all $w \in \Omega$. Clearly, $F\colon \Omega \to P_f(X)$ is measurable. We claim that $D = S_F^p$. So, let $u \in S_F^p$ and $\varepsilon > 0$. Invoking Lemma 5.6.37, there exist finite Σ-partition $\{C_k\}_{k=1}^m$ of Ω and $\{v_k\}_{k=1}^m \subseteq \{h_{nm}\}_{n,m\in\mathbb{N}}$ such that

$$\left\| u - \sum_{k=1}^{m} \chi_{C_k} v_k \right\|_p < \varepsilon\,.$$

The decomposability of D implies that $\sum_{k=1}^m \chi_{C_k} v_k \in D$. Hence, $u \in D$ since D is closed. Therefore,

$$S_F^p \subseteq D\,. \tag{5.6.10}$$

Suppose that $D \neq S_F^p$. Then there exist $u \in K$, $A \in \Sigma$ with $\mu(A) > 0$ and $\delta > 0$ such that

$$\|u(w) - h_{nm}(w)\| \geq \delta \quad \text{for all } w \in A \text{ and for all } n, m \geq 1\,. \tag{5.6.11}$$

Fix $n \in \mathbb{N}$ such that $G = A \cap \{w \in \Omega : \|u(w) - u_n(w)\| < \delta/3\}$ has positive μ-measure. We define

$$f_n = \chi_G u + \chi_{G^c} h_{nm} \quad \text{with } m \in \mathbb{N}\,. \tag{5.6.12}$$

Evidently, $f_n \in D$ and, due to (5.6.11), we obtain

$$\begin{aligned} \|h_{nm}(w) - u_n(w)\| &\geq \|h_{nm}(w) - u(w)\| - \|u(w) - u_n(w)\| \\ &\geq \delta - \frac{\delta}{3} = \frac{2\delta}{3} \quad \text{for all } w \in G\,. \end{aligned} \tag{5.6.13}$$

Taking (5.6.12) and (5.6.13) into account yields

$$\begin{aligned} \|u_n - h_{nm}\|_p^p - \xi_n^p &\geq \|u_n - h_{nm}\|_p^p - \|u_n - f_n\|_p^p \\ &\geq \int_G [\|u_n - h_{nm}\|^p - \|u_n - f_n\|^p]\, d\mu \\ &\geq \left[\left(\frac{2\delta}{3}\right)^p - \frac{\delta^p}{3^p}\right]\mu(G) > 0\,. \end{aligned}$$

Letting $m \to +\infty$ gives a contradiction. Thus, $D = S_F^p$; see (5.6.10).

$\Longleftarrow$: This implication follows from Remark 5.6.33. ☐

The next result is very useful in various parts of applied analysis, since it permits the commutation between sup or inf and the integral.

Theorem 5.6.39. *If $\varphi: \Omega \times X \to \overline{\mathbb{R}} = \mathbb{R} \cup \{+\infty\}$ is $\Sigma \otimes \mathcal{B}(X)$-measurable, $F: \Omega \to 2^X \setminus \{\emptyset\}$ is graph measurable, $I_\varphi: L^p(\Omega, X) \to \mathbb{R}^* = \mathbb{R} \cup \{\pm\infty\}$ is the integral functional given by*

$$I_\varphi(u) = \int_\Omega \varphi(w, u(w))\, d\mu \quad \text{for all } u \in L^p(\Omega, X) \text{ with } 1 \le p \le \infty,$$

and there exists $u_0 \in L^p(\Omega, X)$ such that $I_\varphi(u_0) > -\infty$, then

$$\sup[I_\varphi(u): u \in S_F^p] = \int_\Omega \sup[\varphi(w, x): x \in F(w)]\, d\mu\,.$$

Proof. Let $m(w) = \sup[\varphi(w, x): x \in F(w)]$ and let $\lambda \in \mathbb{R}$. Note that

$$m(w) > \lambda \quad \text{if and only if} \quad \varphi(w, x) > \lambda \quad \text{for some } x \in F(w)\,.$$

It follows that

$$\{w \in \Omega: m(w) > \lambda\} = \mathrm{proj}_\Omega[(w, x) \in \mathrm{Gr}\, F: \varphi(w, x) > \lambda] \in \Sigma_\mu$$

with Σ_μ being the μ-completion of Σ; see Theorem 2.2.32. So, $w \to m(w)$ is Σ_μ-measurable. For every $u \in S_F^p$ we have

$$\varphi(w, u(w)) \le m(w) \quad \mu\text{-a. e.} \tag{5.6.14}$$

This shows $\varphi(w, u_0(w)) \le m(w)$ μ-a. e., and hence, $\int_\Omega m\, d\mu$ exists and is possibly $+\infty$. Then, applying (5.6.14), we obtain

$$\sup[I_\varphi(u): u \in S_F^p] \le \int_\Omega m\, d\mu\,. \tag{5.6.15}$$

If $I_\varphi(u_0) = +\infty$, then we are done. So, suppose $I_\varphi(u_0) \in \mathbb{R}$. This means that $\varphi(\cdot, u_0(\cdot)) \in L^1(\Omega)$. Let $\eta < \int_\Omega m\, d\mu$. We will show that $\eta < I_\varphi(u)$ for some $u \in S_F^p$. Since μ is σ-finite, there exists $\{C_n\}_{n\ge1} \subseteq \Sigma$ such that $C_n \nearrow \Omega$ and $\mu(C_n) < +\infty$ for all $n \in \mathbb{N}$. Moreover, let $e \in L^1(\Omega)$ with $e(w) > 0$ μ-a. e. We set

$$E_n = C_n \cap \{w \in \Omega: \varphi(w, u_0(w)) \le n\}$$

and

$$m_n(w) = \begin{cases} m(w) - \frac{e(w)}{n} & \text{if } w \in E_n,\ e(w) \le n\,, \\ n - \frac{e(w)}{n} & \text{if } w \in E_n,\ e(w) > n\,, \\ \varphi(w, u_0(w)) - \frac{e(w)}{n} & \text{if } w \in E_n^c\,. \end{cases}$$

We see that $\{m_n\}_{n\geq 1} \subseteq L^1(\Omega)$ and $m_n \nearrow m$. So, the Monotone Convergence Theorem (see Theorem 2.3.3) implies that there exists $n_0 \in \mathbb{N}$ such that $\eta < \int_\Omega m_{n_0}\,d\mu$. Set $\vartheta = m_{n_0}$. Then $\eta < \int_\Omega \vartheta\,d\mu$ and $\vartheta(w) < m(w)$ μ-a. e. Define

$$H(w) = F(w) \cap \{x \in X\colon \vartheta(w) \leq \varphi(w,x)\} \neq \emptyset \quad \text{for all } w \in \Omega\,. \tag{5.6.16}$$

Clearly, $\operatorname{Gr} H \in \Sigma \otimes \mathcal{B}(X)$ and so according to Theorem 2.7.25, there exists a Σ-measurable function $u\colon \Omega \to X$ such that $u(w) \in H(w)$ μ-a. e. We define

$$L_n = C_n \cap \{w \in \Omega\colon \|u(w)\| \leq n\}$$

and $h_n = \chi_{L_n} u + \chi_{L_n^c} u_0$ for all $n \in \mathbb{N}$. We have $\{h_n\}_{n\geq 1} \subseteq S_F^p$ and, thanks to (5.6.16),

$$\begin{aligned} I_\varphi(h_n) &= \int_{L_n} \varphi(w,u)\,d\mu + \int_{L_n^c} \varphi(w,u_0)\,d\mu \geq \int_{L_n} \vartheta\,d\mu + \int_{L_n^c} \varphi(w,u_0)\,d\mu \\ &= \int_\Omega \vartheta\,d\mu + \int_{L_n^c} [\varphi(w,u_0) - \vartheta]\,d\mu > \eta + \int_{L_n^c} [\varphi(w,u_0) - \vartheta]\,d\mu\,. \end{aligned} \tag{5.6.17}$$

Recall that $L_n \nearrow \Omega$. Therefore, from (5.6.17), it follows that $I_\varphi(h_n) > \eta$ for all sufficiently large $n \in \mathbb{N}$. Hence, because of (5.6.15), we conclude that

$$\sup[I_\varphi(u)\colon u \in S_F^p] = \int_\Omega m\,d\mu\,. \qquad \square$$

Corollary 5.6.40. *If $F\colon \Omega \to 2^X \setminus \{\emptyset\}$ is graph measurable and $S_F^p \neq \emptyset$ for $1 \leq p \leq \infty$, then*

$$\sigma(v^*; S_F^p) = \int_\Omega \sigma(v^*(w); F(w))\,d\mu$$

for every $v^* \in L^{p'}(\Omega, X^*_{w^*})$ with $1/p + 1/p' = 1$.

Proposition 5.6.41. *If $F\colon \Omega \to 2^X \setminus \{\emptyset\}$ is graph measurable and $S_F^p \neq \emptyset$ for $1 \leq p < \infty$, then $\overline{\operatorname{conv}}\, S_F^p = S^p_{\overline{\operatorname{conv}} F}$.*

Proof. Clearly, $S^p_{\overline{\operatorname{conv}} F} \subseteq L^p(\Omega, X)$ is closed and convex. Hence,

$$\overline{\operatorname{conv}}\, S_F^p \subseteq S^p_{\overline{\operatorname{conv}} F}\,. \tag{5.6.18}$$

Suppose that the inclusion (5.6.18) is strict. Then there exists $u \in S^p_{\overline{\operatorname{conv}} F}$ such that $u \notin \overline{\operatorname{conv}}\, S_F^p$. The Strong Separation Theorem implies that there exist $v^* \in L^{p'}(\Omega, X^*_{w^*})$ and $\varepsilon > 0$ such that $\sigma(v^*; S_F^p) \leq \langle v^*, u\rangle - \varepsilon$. Applying Corollary 5.6.40 yields

$$\int_\Omega \sigma(v^*(w); F(w))\,d\mu \leq \int_\Omega \langle v^*(w), u(w)\rangle\,d\mu - \varepsilon\,,$$

which gives

$$\varepsilon \le \int_\Omega [\langle v^*(w), u(w)\rangle - \sigma(v^*(w); F(w))]\, d\mu \le 0\,,$$

a contradiction. Therefore, equality holds in (5.6.18) and so we are done. □

The next weak compactness result is useful in many applications. We start with a definition.

Definition 5.6.42. A multifunction $F: \Omega \to 2^X \setminus \{\emptyset\}$ is said to be L^p**-integrably bounded** with $1 < p \le \infty$ and simply **integrably bounded** for $p = 1$ if there exists $\vartheta \in L^p(\Omega)$ such that

$$|F(w)| = \sup[\|u\| : u \in F(w)] \le \vartheta(w) \quad \mu\text{-a. e.}$$

Theorem 5.6.43. *If $F: \to P_{\mathrm{wkc}}(X)$ is a graph measurable and integrably bounded multifunction, then S_F^1 is nonempty, convex, and* w*-compact in $L^1(\Omega, X)$.*

Proof. Clearly, $S_F^1 \subseteq L^1(\Omega, X)$ is nonempty, closed, and convex; see Theorem 2.7.25. Let $v^* \in L^\infty(\Omega, X^*_{\mathrm{w}^*}) = L^1(\Omega, X)^*$ (see Theorem 4.2.37), and denote by $((\cdot,\cdot))$ the duality brackets for this pair. Applying Corollary 5.6.40 gives

$$\sigma(v^*; S_F^1) = \int_\Omega \sigma(v^*(w); F(w))\, d\mu\,.$$

Let $S(w) = \{u \in F(w) : \langle v^*(w), u\rangle = \sigma(v^*(w); F(w))\}$. Since F is $P_{\mathrm{wkc}}(X)$-valued, we see that $S(w) \neq \emptyset$ for all $w \in \Omega$. Since $(w, x^*) \to \sigma(x^*; F(w))$ is a Carathéodory function, we see that $w \to \sigma(v^*(w); F(w))$ is Σ-measurable and so S is graph measurable. Invoking the Yankov–von Neumann–Aumann Selection Theorem (see Theorem 2.7.25), there exists $u_0 \in S_F^1$ such that $\langle v^*(w), u_0(w)\rangle = \sigma(v^*(w); F(w))$ μ-a. e. Therefore,

$$\sigma(v^*; S_F^1) = ((v^*, u_0)) = \int_\Omega \langle v^*, u_0\rangle\, d\mu\,.$$

Since $v^* \in L^\infty(\Omega, X^*_{\mathrm{w}^*}) = L^1(\Omega, X)^*$ is arbitrary, we infer from James' Theorem (see Theorem 3.3.41) that $S_F^1 \subseteq L^1(\Omega, X)$ is w-compact. □

Now we turn our attention to integral functionals. First we introduce the different kind of integrands that are involved in the study of integral functionals. Our standing hypotheses are the following:

- (Ω, Σ, μ) is a σ-finite complete measure space.
- X is a separable Banach space.

Definition 5.6.44. A function $\varphi: \Omega \times X \to \overline{\mathbb{R}} = \mathbb{R} \cup \{+\infty\}$ is said to be
(a) an **integrand** if φ is $\Sigma \otimes \mathcal{B}(X)$-measurable;
(b) a **normal integrand** if φ is an integrand and $\varphi(w, \cdot) \in \Gamma_0(X)$ for μ-a. a. $w \in \Omega$;
(c) a **convex integrand** if φ is normal and $\varphi(w, \cdot)$ is convex for μ-a. a. $w \in \Omega$;
(d) a **Carathéodory integrand** if $w \to \varphi(w, x)$ is Σ-measurable for all $x \in X$ and $x \to \varphi(w, x)$ is continuous for μ-a. a. $w \in \Omega$.

Remark 5.6.45. According to Proposition 2.2.31, a Carathéodory integrand is indeed an integrand, that is, it is $\Sigma \otimes \mathcal{B}(X)$-measurable.

Definition 5.6.46. Let $\varphi, \psi: \Omega \times X \to \overline{\mathbb{R}} = \mathbb{R} \cup \{+\infty\}$ be functions. We say that
(a) φ is **μ-dominated** by ψ if

$$\varphi(w, x) \le \psi(w, x) \quad \text{for all } w \in \Omega \setminus N \text{ with } \mu(N) = 0 \text{ and for all } x \in X$$

and we write $\varphi \prec \psi$;
(b) φ and ψ are **μ-equivalent** if $\varphi \prec \psi$, $\psi \prec \varphi$ and we write $\varphi = \psi$.

Proposition 5.6.47. *If $\varphi, \psi: \Omega \times X \to \overline{\mathbb{R}} = \mathbb{R} \cup \{+\infty\}$ are two integrands, $\eta: \Omega \times \mathbb{R}_+ \to \mathbb{R}_+$ with $\mathbb{R}_+ = [0, +\infty)$ is also an integrand such that $\eta(w, \cdot)$ is increasing, and $\eta(\cdot, t) \in L^1(\Omega)$, $-\eta(w, \|x\|)$ is μ-dominated by φ or ψ and*

$$\int_A \varphi(w, u(w))\, d\mu \le \int_A \psi(w, u(w))\, d\mu \quad \textit{for all } u \in L^\infty(\Omega, X) \textit{ and for all } A \in \Sigma\,,$$

then $\varphi \prec \psi$.

Proof. For every $k \in \mathbb{N}$, let $\varphi_k = \min\{\varphi, k\}$ and $\psi_k = \min\{\psi, k\}$. It suffices to show that $\varphi_k \prec \psi_k$ for all $k \in \mathbb{N}$. Let $\varepsilon > 0$ and set

$$\begin{aligned} \Gamma(\varepsilon, k) &= \{(w, x) \in \Omega \times X: \|x\| \le k, \varphi_k(w, x) > \psi_k(w, x) + \varepsilon\}\,, \\ \Gamma(\varepsilon, k)(w) &= \{x \in X: (w, x) \in \Gamma(\varepsilon, k)\}\,, \\ \Omega(\varepsilon, k) &= \{w \in \Omega: \Gamma(\varepsilon, k)(w) \ne \emptyset\}\,. \end{aligned}$$

Evidently, $\Gamma(\varepsilon, k) \in \Sigma \otimes \mathcal{B}(X)$ and so $\Omega(\varepsilon, k) \in \Sigma$. From the Yankov–von Neumann–Aumann Selection Theorem (see Theorem 2.7.25), there exists a $\Sigma \cap \Omega(\varepsilon, k)$-measurable selection $\gamma_{\varepsilon,k}: \Omega(\varepsilon, k) \to X$ of $\Gamma(\varepsilon, k)$. Let $\gamma_{\varepsilon,k}(w) = 0$ for all $w \in \Omega \setminus \Omega(\varepsilon, k)$. We obtain

$$\|\gamma_{\varepsilon,k}(w)\| \le k\,,$$

thus $\gamma_{\varepsilon,k} \in L^\infty(\Omega)$. This implies by the hypothesis that

$$\int_{\Omega(\varepsilon,k)} \varphi(w, \gamma_{\varepsilon,k}(w))\, d\mu \le \int_{\Omega(\varepsilon,k)} \psi(w, \gamma_{\varepsilon,k}(w))\, d\mu\,.$$

We have

$$\varphi_k(w, y_{\varepsilon,k}(w)) > \psi_k(w, y_{\varepsilon,k}(w)) + \varepsilon \quad \text{for all } w \in \Omega(\varepsilon, k)\,.$$

Therefore, $\psi_k(w, y_{\varepsilon,k}(w)) < k$ for all $w \in \Omega(\varepsilon, k)$ and so $\psi_k(w, y_{\varepsilon,k}(w)) = \psi(w, y_{\varepsilon,k}(w))$ for all $w \in \Omega(\varepsilon, k)$. This gives

$$\psi(w, y_{\varepsilon,k}(w)) + \varepsilon < \varphi(w, y_{\varepsilon,k}(w)) \quad \text{for all } w \in \Omega(\varepsilon, k)\,.$$

Without any loss of generality, we assume that $-\eta(w, \|x\|)$ is dominated by φ. Then

$$\int\limits_{\Omega(\varepsilon,k)} \varphi(w, y_{\varepsilon,k}(w))\, d\mu + \varepsilon\mu(\Omega(\varepsilon, k)) \le \int\limits_{\Omega(\varepsilon,k)} \varphi(w, y_{\varepsilon,k}(w))\, d\mu\,,$$

which yields $\mu(\Omega(\varepsilon, k)) = 0$.

Let $N_k = \bigcup_{\varepsilon>0} \Omega(\varepsilon, k)$. Then $N_k \in \Sigma$ and $\mu(N_k) = 0$. Moreover, we get

$$\varphi_k(w, x) \le \psi_k(w, x) \quad \text{for all } w \in \Omega \setminus N_k \text{ and for all } x \in X \text{ with } \|x\| \le k\,.$$

Since this is true for all $k \in \mathbb{N}$, we conclude that $\varphi \prec \psi$. □

Corollary 5.6.48. *If $\varphi, \psi: \Omega \times \overline{\mathbb{R}} = \mathbb{R} \cup \{+\infty\}$ are two integrands as in Proposition* 5.6.47, *then $\varphi = \psi$ if and only if*

$$\int\limits_A \varphi(w, u(w))\, d\mu = \int\limits_A \psi(w, u(w))\, d\mu \quad \textit{for all } u \in L^\infty(\Omega, X) \textit{ and for all } A \in \Sigma\,.$$

Proposition 5.6.49. *If $h: \Omega \times X \to \overline{\mathbb{R}}_+ = \mathbb{R}_+ \cup \{+\infty\}$ is a function such that $h(w, \cdot)$ is lower semicontinuous for μ-a. a. $w \in \Omega$, then there exists a normal integrand φ such that $h \prec \varphi$ and for every integrand ψ satisfying $h \prec \psi$, we have $\varphi \prec \psi$.*

Proof. We may always assume that $h(w, \cdot)$ is lower semicontinuous for every $w \in \Omega$. Let $\mathcal{B}$ be a countable base for X, which exists since X is separable. For every $i = (q, A) = \mathbb{Q} \times \mathcal{B} = I$ we consider the function $q_i(x) = q\chi_A(x)$ for all $x \in X$. Since $h(w, \cdot)$ is lower semicontinuous, one has

$$h(w, x) = \sup[\varphi_i(x) : i \in I(w)] = \sup[\chi_{A_i}(w)\varphi_i(x) : i \in I] \tag{5.6.19}$$

with $I(w) \subseteq I$ and $A_i = \{w \in \Omega : i \in I(w)\} \subseteq \Omega$. We choose $C_i \in \Sigma$ such that

$$A_i \subseteq C_i \quad \text{and} \quad \mu(C_i) = \mu^*(A_i) \tag{5.6.20}$$

with μ^* being the outer measure associated with μ. Recall that $\mu^*(A) = \sup[\mu(C) : C \in \Sigma, C \supseteq A]$. We define

$$\varphi(w,x) = \sup\left[\chi_{C_i}(w)\varphi_i(x) : i \in I\right].$$

Then φ is a normal integrand and $h \prec \varphi$.

Now suppose that ψ is an integrand such that $h \prec \psi$. We may assume that $h(w,x) \leq \psi(w,x)$ for all $w \in \Omega$ and for all $x \in X$. Let $u \in L^1(\Omega, X)$ and let

$$D_i = \{w \in \Omega : \varphi_i(u(w)) \leq \psi(w, u(w))\} \quad \text{for every } i \in I.$$

We see that $D_i \in \Sigma$ and from (5.6.19) it follows that $A_i \subseteq D_i$ for all $i \in I$. Moreover, (5.6.20) implies that $\mu(C_i \setminus D_i) = 0$. Hence,

$$\chi_{C_i}(w)\varphi_i(u(w)) \leq \psi(w, u(w)) \quad \mu\text{-a. e.},$$

which shows that $\varphi(w, u(w)) \leq \psi(w, u(w))$ μ-a. e. ☐

Proposition 5.6.50. *If $\varphi : \Omega \times X \to \overline{\mathbb{R}} = \mathbb{R} \cup \{+\infty\}$ is a normal integrand, then $\varphi^* : \Omega \times X^* \to \overline{\mathbb{R}}$ defined by*

$$\varphi^*(w, x^*) = \sup\left[\langle x^*, x\rangle - \varphi(w,x) : x \in X\right]$$

is a convex integrand on $\Omega \times X^*_{w^*}$.

Proof. Let $E : \Omega \to 2^X \setminus \{\emptyset\}$ be the multifunction defined by

$$E(w) = \operatorname{epi} \varphi(w, \cdot) = \{(x, \lambda) \in X \times \mathbb{R} : \varphi(w,x) \leq \lambda\}.$$

The normality of φ implies that $E(w) \in P_f(X \times \mathbb{R})$ for all $w \in \Omega$. Moreover,

$$\begin{aligned} \operatorname{Gr} E &= \{(w, x, \lambda) \in \Omega \times X \times \mathbb{R} : \varphi(w,x) \leq \lambda\} \\ &\in \Sigma \otimes \mathcal{B}(X) \otimes \mathcal{B}(\mathbb{R}) = \Sigma \otimes \mathcal{B}(X \times \mathbb{R}). \end{aligned}$$

Invoking Theorem 2.7.28, there exist two sequences $u_n : \Omega \to X$ and $\lambda_n : \Omega \to \mathbb{R}$ with $n \in \mathbb{N}$ of Σ-measurable functions such that

$$(u_n, \lambda_n) \text{ is a selection of } E \text{ and } E(w) = \overline{\{(u_n(w), \lambda_n(w))\}_{n\geq 1}} \text{ for all } w \in \Omega.$$

Then we have

$$\varphi^*(w, x^*) = \sup_{n\geq 1}\left[\langle x^*, u_n(w)\rangle - \lambda_n(w)\right].$$

Hence, φ^* is a convex integrand on $\Omega \times X^*_{w^*}$. ☐

For an integrand φ we define the integral functional

$$I_\varphi(u) = \int_\Omega \varphi(w, u(w))\, d\mu$$

with $u: \Omega \to X$ belonging to some vector space of functions. We adopt the convention $+\infty + (-\infty) = +\infty$, that is, in our considerations $+\infty$ dominates over $-\infty$. Then, for a normal integrand $\varphi: \Omega \times X \to \overline{\mathbb{R}} = \mathbb{R} \cup \{+\infty\}$, we define the integral functional $I_\varphi: L^1(\Omega, X) \to \mathbb{R}^* = \mathbb{R} \cup \{\pm\infty\}$ by

$$I_\varphi(u) = \begin{cases} \int_\Omega \varphi(w, u(w))\, d\mu & \text{if } \int_\Omega \varphi(w, u(w))^+\, d\mu < +\infty\,, \\ +\infty & \text{otherwise}\,. \end{cases}$$

Using Proposition 5.6.50, we can define the integral functional $I_{\varphi^*}: L^\infty(\Omega, X^*_{w^*}) \to \mathbb{R}^* = \mathbb{R} \cup \{\pm\infty\}$ by

$$I_{\varphi^*}(u^*) = \begin{cases} \int_\Omega \varphi^*(w, u^*(w))\, d\mu & \text{if } \int_\Omega \varphi^*(w, u^*(w))^+\, d\mu < +\infty\,, \\ +\infty & \text{otherwise}\,. \end{cases}$$

From Theorem 4.2.37, we know that $L^1(\Omega, X)^* = L^\infty(\Omega, X^*_{w^*})$ and the duality brackets for this pair are given by

$$\langle u^*, u\rangle_{L^1} = \int_\Omega \langle u^*(w), u(w)\rangle\, d\mu \quad \text{for all } u \in L^1(\Omega, X) \text{ and for all } u^* \in L^\infty(\Omega, X^*_{w^*})\,.$$

We mention that the two theorems that follow can be stated for I_φ defined on $L^p(\Omega, X)$ with $1 \le p < \infty$. Then $L^p(\Omega, X)^* = L^{p'}(\Omega, X^*_{w^*})$ for $1/p + 1/p' = 1$ and if X^* has the RNP, in particular, if X^* is separable, then $L^p(\Omega, X)^* = L^{p'}(\Omega, X^*)$. In this case the duality brackets are denoted by $\langle\cdot,\cdot\rangle_{L^p}$.

Theorem 5.6.51. *If $I_\varphi: L^1(\Omega, X) \to \mathbb{R}^*$ is finite at $u_0 \in L^1(\Omega, X)$, then $(I_\varphi)^* = I_{\varphi^*}$.*

Proof. According to the Young–Fenchel inequality (see Proposition 5.3.3 (g)), it suffices to show that

$$\int_\Omega \varphi^*(w, u^*(w))\, d\mu \le \sup[\langle u^*, u\rangle_{L^1} - I_\varphi(u): u \in L^1(\Omega, X)] \tag{5.6.21}$$

for all $u^* \in L^\infty(\Omega, X^*_{w^*})$. Let $\eta \in \mathbb{R}$ be such that $\eta < I_{\varphi^*}(u^*)$. Then (5.6.21) will follow if we can produce $u \in L^1(\Omega, X)$ such that $\eta \le \langle u^*, u\rangle_{L^1} - I_\varphi(u)$.

By hypothesis, $I_\varphi(u_0) \in \mathbb{R}$ and so there exists $\vartheta_0 \in L^1(\Omega)$ such that

$$\vartheta_0(w) \le \langle u^*(w), u_0(w)\rangle - \varphi(w, u_0(w)) \quad \mu\text{-a. e.}\,,$$

which implies

$$\vartheta_0(w) \le \varphi^*(w, u^*(w)) \quad \mu\text{-a. e.} \tag{5.6.22}$$

We claim that there exists $\xi \in L^1(\Omega)$ such that

$$\eta < \int_\Omega \xi(w)\,\mathrm{d}\mu \quad \text{and} \quad \xi(w) < \varphi^*(w, u(w)) \quad \mu\text{-a. e.}$$

Let $h \in L^1(\Omega)$ with $h(w) > 0$ μ-a. e. If $I_{\varphi^*}(u^*) \in \mathbb{R}$, then let

$$\xi(w) = \varphi^*(w, u^*(w)) - \varepsilon h(w)$$

with small enough $\varepsilon > 0$ so that $\eta < \int_\Omega \xi(w)\,\mathrm{d}\mu$. If $I_{\varphi^*}(u^*) = +\infty$, then we define

$$g_n(w) = \begin{cases} \min\{nh(w), \frac{1}{2}\varphi^*(w, u^*(w))\} & \text{if } \varphi^*(w, u^*(w)) > 0\,, \\ \varphi^*(w, u^*(w)) - h(w) & \text{if } \varphi^*(w, u^*(w)) \le 0\,. \end{cases}$$

We see that

$$g_n(w) \to \frac{1}{2}\varphi^*(w, u^*(w)) \quad \text{for all } w \in \{w \in \Omega\colon \varphi^*(w, u^*(w)) > 0\}\,.$$

From the Monotone Convergence Theorem, we obtain

$$\int_\Omega g_n(w)\,\mathrm{d}\mu \to +\infty\,.$$

Hence, there exists a number $n_0 \in \mathbb{N}$ large enough such that $\eta < \int_\Omega g_{n_0}(w)\,\mathrm{d}\mu$. Thus if $\xi = g_{n_0}$, then $\xi(w) < \varphi^*(w, u^*(w))$ μ-a. e. Let $K\colon \Omega \to 2^X$ be the multifunction defined by

$$K(w) = \{x \in X\colon \xi(w) \le \langle u^*(w), x\rangle - \varphi(w, x)\} \quad \text{for all } w \in \Omega\,.$$

It holds $\operatorname{Gr} K \in \Sigma \otimes \mathcal{B}(X)$ and so we can apply the Yankov–von Neumann–Aumann Selection Theorem (see Theorem 2.7.25), and produce a Σ-measurable function $k\colon \Omega \to X$ such that $k(w) \in K(w)$ for all $w \in \Omega$.

Since μ is σ-finite, there exists $\Omega_0 \in \Sigma$ with $\mu(\Omega_0) < +\infty$ such that $k|_{\Omega_0}$ is bounded and $\eta < \int_{\Omega_0} \xi(w)\,\mathrm{d}\mu + \int_{\Omega\setminus\Omega_0} \vartheta_0(w)\,\mathrm{d}\mu$; see (5.6.22). We define

$$u(w) = \begin{cases} k(w) & \text{if } w \in \Omega_0\,, \\ u_0(w) & \text{if } w \in \Omega \setminus \Omega_0\,. \end{cases}$$

We obtain $u \in L^1(\Omega, X)$ and

$$\begin{aligned} \xi(w) &\le \langle u^*(w), u(w)\rangle - \varphi(w, u(w)) \quad \text{for all } w \in \Omega_0\,, \\ \vartheta_0(w) &\le \langle u^*(w), u(w)\rangle - \varphi(w, u(w)) \quad \text{for all } w \in \Omega \setminus \Omega_0\,. \end{aligned}$$

Finally we can write

$$\eta < \int_{\Omega_0} \xi(w)\,\mathrm{d}\mu + \int_{\Omega\setminus\Omega_0} \vartheta_0(w)\,\mathrm{d}\mu \le \int_\Omega \langle u^*(w), u(w)\rangle\,\mathrm{d}\mu - \int_\Omega \varphi(w, u(w))\,\mathrm{d}\mu\,,$$

which shows that $I_{\varphi^*}(u^*) = (I_\varphi)^*(u^*)$. □

Theorem 5.6.52. *If $\varphi:\Omega\times X \to \overline{\mathbb{R}} = \mathbb{R}\cup\{+\infty\}$ is a convex integrand, I_φ is finite at $u_0 \in L^1(\Omega, X)$ and I_{φ^*} is finite at $u_0^* \in L^\infty(\Omega, X^*_{\mathrm{w}^*})$, then $I_\varphi \in \Gamma_0(L^1(\Omega, X))$, $I_{\varphi^*} \in \Gamma_0(L^\infty(\Omega, X^*_{\mathrm{w}^*})_{\mathrm{w}^*})$, and they are conjugates to each other.*

Proof. Proposition 5.3.7 (b) says that

$$\varphi(w,\cdot) = \varphi^{**}(w,\cdot)|_X \quad \text{for all } w\in\Omega\,.$$

Therefore the result is a consequence of Theorem 5.6.51. □

Proposition 5.6.53. *If μ is nonatomic, $\varphi:\Omega\times X \to \overline{\mathbb{R}} = \mathbb{R}\cup\{+\infty\}$ is a convex integrand, $1 \le p < \infty$, and $I_\varphi: L^p(\Omega, X) \to \overline{\mathbb{R}} = \mathbb{R}\cup\{+\infty\}$ is continuous at a point, then I_φ is continuous everywhere.*

Proof. Let $u_0 \in L^p(\Omega, X)$ be the point of continuity of I_φ. Replacing I_φ by $\hat{I}_\varphi(u) = I_\varphi(u_0 + u) - I_\varphi(u_0)$ if necessary, we see that, without any loss of generality, we may assume that $u_0 = 0$ and $I_\varphi(0) = 0$. Then the continuity hypothesis implies that there exists $\delta > 0$ such that

$$I_\varphi(u) \le 1 \quad \text{for all } u \in L^p(\Omega, X) \text{ with } \|u\|_{L^p(\Omega,X)} \le \delta\,.$$

Let $u \in L^p(\Omega, X)$. The nonatomicity of μ and the absolute continuity of the Lebesgue integral imply the existence of $\delta_1 > 0$ and pairwise disjoint sets $\{A_k\}_{k=1}^N \subseteq \Sigma$ such that

$$\mu(A_k) \le \delta_1 \quad \text{and} \quad \|\chi_{A_k} u\|_{L^p(\Omega,X)} \le \delta \quad \text{for all } k\in\{1,\dots,N\}\,.$$

It holds that

$$\int_{A_k} \varphi(w, u(w))\,\mathrm{d}\mu = \int_\Omega \varphi(w, \chi_{A_k}(w)u(w))\,\mathrm{d}\mu - \int_{A_k^c} \varphi(w, 0)\,\mathrm{d}\mu \le 1 + \eta$$

for some $\eta > 0$ independent of $k \in \{1,\dots,N\}$. Therefore,

$$\sum_{k=1}^N \int_{A_k} \varphi(w, u(w))\,\mathrm{d}\mu = \int_\Omega \varphi(w, u(w))\,\mathrm{d}\mu = I_\varphi(u) < \infty\,.$$

This shows that I_φ is continuous everywhere on $L^p(\Omega, X)$; see Proposition 5.1.24. □

We can describe the subdifferential of I_φ in the following way.

Theorem 5.6.54. *If $\varphi\colon \Omega\times X \to \overline{\mathbb{R}} = \mathbb{R}\cup\{+\infty\}$ is a convex integrand and $I_\varphi\colon L^p(\Omega, X) \to \overline{\mathbb{R}}$ with $1 \le p < \infty$ is finite at $u_0 \in L^p(\Omega, X)$, then, for all $u \in L^p(\Omega, X)$, we have that $u^* \in \partial I_\varphi(u)$ if and only if $u^*(w) \in \partial\varphi(w, u(w))$ μ-a. e.*

Proof. According to Proposition 5.3.19, we see that

$$u^* \in \partial I_\varphi(u) \quad \text{if and only if} \quad I_\varphi(u) + (I_\varphi)^*(u^*) = \langle u^*, u\rangle_{L^p} .$$

Moreover, Theorem 5.6.51 states that $(I_\varphi)^* = I_{\varphi^*}$. That means we can say $u^* \in \partial I_\varphi(u)$ if and only if

$$\int_\Omega [\varphi(w, u(w)) + \varphi^*(w, u^*(w))]\, d\mu = \int_\Omega \langle u^*(w), u(w)\rangle\, d\mu .$$

Invoking the Young–Fenchel inequality (see Proposition 5.3.3 (g)), we infer that

$$\varphi(w, u(w)) + \varphi^*(w, u^*(w)) = \langle u^*(w), u(w)\rangle \quad \text{for } \mu\text{-a. a. } w \in \Omega ,$$

which is equivalent to saying that $u^*(w) \in \partial\varphi(w, u(w))$ μ-a. e.; see Proposition 5.3.19. ☐

Now we state a lower semicontinuity result, which is useful in many applications.

Theorem 5.6.55. *If $L\colon \Omega \times \mathbb{R}^N \times \mathbb{R}^m \to \overline{\mathbb{R}} = \mathbb{R} \cup \{+\infty\}$ is a measurable function such that*
(i) *$(x, y) \to L(w, x, y)$ is lower semicontinuous for μ-a. a. $w \in \Omega$;*
(ii) *$y \to L(w, x, y)$ is convex for μ-a. a. $w \in \Omega$ and for all $x \in \mathbb{R}^N$;*
(iii) *$\xi(w) - c(|x| + |y|) \le L(w, x, y)$ for μ-a. a. $w \in \Omega$, for all $(x, y) \in \mathbb{R}^N \times \mathbb{R}^m$ with $\xi \in L^1(\Omega)$ and some $c > 0$;*

and if $p, r \in [1, +\infty]$, then the integral functional

$$(u, v) \to I_L(u, v) = \int_\Omega L(w, u(w), v(w))\, d\mu$$

is sequentially lower semicontinuous on $L^p(\Omega, \mathbb{R}^N) \times L^r(\Omega, \mathbb{R}^m)_{\mathrm{w}}$. If $r = +\infty$, then we consider the w^*-topology on $L^r(\Omega, \mathbb{R}^m)$.

Proof. We need to show that the sublevel set

$$S_\eta = \{(u, v) \in L^p(\Omega, \mathbb{R}^N) \times L^r(\Omega, \mathbb{R}^m) \colon I_L(u, v) \le \eta\}$$

is sequentially closed in $L^p(\Omega, \mathbb{R}^N) \times L^r(\Omega, \mathbb{R}^m)_{\mathrm{w}}$, resp. in $L^p(\Omega, \mathbb{R}^N) \times L^\infty(\Omega, \mathbb{R}^m)_{\mathrm{w}^*}$ if $r = +\infty$, for every $\eta \in \mathbb{R}$. To this end, we consider a sequence $\{(u_n, v_n)\}_{n\ge1} \subseteq L^p(\Omega, \mathbb{R}^N) \times L^r(\Omega, \mathbb{R}^m)_{\mathrm{w}}$ such that

$$u_n \to u \quad \text{in } L^p(\Omega, \mathbb{R}^N) \quad \text{and} \quad v_n \xrightarrow{\mathrm{w}} v \quad \text{in } L^r(\Omega, \mathbb{R}^m) ,$$

resp. $v_n \xrightarrow{w^*} v$ in $L^\infty(\Omega, \mathbb{R}^m)$ if $r = +\infty$. By passing to a subsequence if necessary we may assume that $u_n(w) \to u(w)$ μ-a. e. According to Remark 4.7.23, we can also assume that $\lambda^{v_n} \xrightarrow{\tau_n^{\mathcal{Y}}} \lambda$ in $\mathcal{Y}(\Omega \times \mathbb{R}^m)$. By Definition 4.7.6, we have for every $A \in \Sigma$

$$\int_A v_n(w)\,d\mu = \int_A \int_{\mathbb{R}^m} y\delta_{v_n(w)}(dy)\,d\mu \to \int_A \int_{\mathbb{R}^m} y\hat{\lambda}(w)(dy)\,d\mu\,. \tag{5.6.23}$$

Moreover, one has

$$\int_A v_n(w)\,d\mu \to \int_A v(w)\,d\mu\,. \tag{5.6.24}$$

From (5.6.23) and (5.6.24), it follows that

$$\int_A v(w)\,d\mu = \int_A \int_{\mathbb{R}^m} y\hat{\lambda}(w)(dy)\,d\mu \quad \text{for all } A \in \Sigma\,,$$

which implies that

$$v(w) = \int_{\mathbb{R}^m} y\hat{\lambda}(w)(dy) \quad \text{for } \mu\text{-a. a. } w \in \Omega\,. \tag{5.6.25}$$

Since by hypothesis L is normal using Proposition 2.5.21 (see also its proof), there exist Carathéodory integrands $L_k : \Omega \times \mathbb{R}^N \times \mathbb{R}^m \to \mathbb{R}$ with $k \in \mathbb{N}$ such that $L_k(w, \cdot, \cdot)$ is k-Lipschitz for all $w \in \Omega$ and $\xi(w) - c(|x| + |y|) \le L_k(w, x, y)$ for μ-a. a. $w \in \Omega$, for all $x \in \mathbb{R}^N$, and for all $y \in \mathbb{R}^m$. In addition, it holds that $L_k \nearrow L$. Let $\{\vartheta_k\}_{k\ge1} \subseteq C_0(\mathbb{R}^N \times \mathbb{R}^m)$ be an increasing sequence converging pointwise to 1. We set

$$\hat{L}_k(w, x, y) = \min\{k\vartheta_k(x, y), L_k(w, x, y)\}\,.$$

Evidently $\hat{L}_k(w, \cdot, \cdot) \nearrow L(w, \cdot, \cdot)$ for μ-a. a. $w \in \Omega$. We know that $C_0(\mathbb{R}^m)^* = \text{ca}(\mathbb{R}^m)$; see Theorem 4.6.9. So, by Theorem 4.2.37, we obtain

$$L^1(\Omega, C_0(\mathbb{R}^m))^* = L^\infty(\Omega, \text{ca}(\mathcal{B}(\mathbb{R}^m))_{w^*})\,.$$

Then, $\tau_n^{\mathcal{Y}}$ (see Definition 4.7.11) coincides with the relative w*-topology if we view $\mathcal{Y}(\Omega, \mathbb{R}^m)$ as a subset of $L^\infty(\Omega, \text{ca}(\mathcal{B}(\mathbb{R}^m))_{w^*})$. By $((\cdot, \cdot))$ we denote duality brackets for the pair $(L^\infty(\Omega, \text{ca}(\mathcal{B}(\mathbb{R}^m))_{w^*}), L^1(\Omega, C_0(\mathbb{R}^m)))$. Note that

$$\sup_{y\in\mathbb{R}^m} |\hat{L}_k(w, u_n(w), y) - \hat{L}_k(w, u(w), y)| \le k|u_n(w) - u(w)| \to 0 \quad \mu\text{-a. e.}\,,$$

which implies that

$$\hat{L}_k(w, u_n(w), \cdot) \to \hat{L}_k(w, u(w), \cdot) \quad \text{in } C_0(\mathbb{R}^m)\ \mu\text{-a. e.}$$

We set $\tilde{L}_k(u_n)(w) = \hat{L}_k(w, u_n(w), \cdot)$ and $\tilde{L}_k(u)(w) = \hat{L}_k(w, u(w), \cdot)$. Then, by the Lebesgue Dominated Convergence Theorem, we have

$$\tilde{L}_k(u_n) \to \tilde{L}_k(u) \quad \text{in } L^1(\Omega, C_0(\mathbb{R}^m)) .$$

This implies

$$((\tilde{L}_k(u_n), \lambda^{v_n})) \to ((\tilde{L}_k, \lambda)) .$$

Hence,

$$\int_\Omega \int_{\mathbb{R}^m} \hat{L}_k(w, u_n(w), y) \lambda^{v_n(w)}(\mathrm{d}y)\, \mathrm{d}\mu \to \int_\Omega \int_{\mathbb{R}^m} \hat{L}_k(w, u(w), y) \hat{\lambda}(w)(\mathrm{d}y)\, \mathrm{d}\mu .$$

Then, from the Monotone Convergence Theorem, it follows that

$$\int_\Omega \int_{\mathbb{R}^m} \hat{L}_k(w, u_n(w), y) \hat{\lambda}(w)(\mathrm{d}y)\, \mathrm{d}\mu \nearrow \int_\Omega \int_{\mathbb{R}^m} L(w, u(w), y) \hat{\lambda}(w)(\mathrm{d}y)\, \mathrm{d}\mu$$

as $k \to \infty$. So, we can find a sequence $k(n) \to +\infty$ as $n \to \infty$ such that

$$\int_\Omega \int_{\mathbb{R}^m} \hat{L}_{k(n)}(w, u_n(w), y) \lambda^{u_n(w)}(\mathrm{d}y)\, \mathrm{d}\mu \to \int_\Omega \int_{\mathbb{R}^m} L(w, u(w), y) \hat{\lambda}(w)(\mathrm{d}y)\, \mathrm{d}\mu$$

as $n \to \infty$. Since $\hat{L}_{k(n)} \le L$, it follows that

$$\int_\Omega \int_{\mathbb{R}^m} L(w, u(w), y) \hat{\lambda}(w)(\mathrm{d}y)\, \mathrm{d}\mu \le \liminf_{n \to \infty} I_L(u_n, v_n) \le \eta .$$

Applying Jensen's inequality yields

$$\int_\Omega L\Big(w, u(w), \int_{\mathbb{R}^m} y \hat{\lambda}(w)(\mathrm{d}y) \Big)\, \mathrm{d}\mu \le \eta .$$

Then, from (5.6.25) we obtain that $I_L(u, v) \le \eta$. Therefore, $(u, v) \in S_\eta$ and this proves the desired sequential lower semicontinuity of I_L. □

A useful consequence of this theorem is the following result.

Corollary 5.6.56. *If $\Omega \subseteq \mathbb{R}^N$ is a bounded open set with Lipschitz boundary and $L: \Omega \times \mathbb{R}^m \times \mathbb{R}^{mN} \to \overline{\mathbb{R}} = \mathbb{R} \cup \{+\infty\}$ is a measurable function such that*

(i) *$(x, y) \to L(z, x, y)$ is lower semicontinuous for μ-a. a. $z \in \Omega$;*
(ii) *$y \to L(z, x, y)$ is convex for μ-a. a. $z \in \Omega$ and for all $x \in \mathbb{R}^m$;*
(iii) *$\xi(z) - c(|x| + |y|) \le L(z, x, y)$ for μ-a. a. $z \in \Omega$, for all $(x, y) \in \mathbb{R}^m \times \mathbb{R}^{mN}$ with $\xi \in L^1(\Omega)$ and some $c > 0$;*

and if $1 < p < \infty$, *then*

$$u \to J(u) = \int_{\Omega} L(z, u(z), Du(z))\,\mathrm{d}z$$

is sequentially lower semicontinuous on $W^{1,p}(\Omega, \mathbb{R}^m)_w$.

5.7 Lipschitz and Locally Lipschitz Functions

Lipschitz and locally Lipschitz functions play a central role in many parts of analysis. Lipschitz functions are important in the theory of Sobolev spaces on general metric measure spaces and provide an effective substitute for smooth functions. Locally Lipschitz functions admit a powerful subdifferential theory that parallels and to a certain point extends the corresponding subdifferential theory for convex functions; see Section 5.3. In this section we address some of these issues.

Definition 5.7.1. (a) Let (X, d_X) and (Y, d_Y) be two metric spaces and $f: X \to Y$. We say that f is **k-Lipschitz** if there exists $k > 0$ such that

$$d_Y(f(u), f(v)) \leq k d_X(u, v) \quad \text{for all } u, v \in X\,. \tag{5.7.1}$$

We also use the term **Lipschitz** for a function f that is k-Lipschitz for some $k > 0$. The smallest $k > 0$ for which (5.7.1) holds is called the **Lipschitz constant of f**.

(b) We say that $f: X \to Y$ is **locally Lipschitz** if every $x \in X$ has a neighborhood U such that $f|_U$ is Lipschitz. We say that f is locally k-Lipschitz if all these restrictions are k-Lipschitz.

Remark 5.7.2. If $f: X \to Y$ is a Lipschitz bijection and the inverse of f is also a Lipschitz map from Y into X, then we say that f is **bilipschitz**. The notion of **k-bilipschitz** is defined in the same way. Note that f is **1-bilipschitz** if and only if it is an isometry.

The next result is known as the "McShance–Whitney Extension Theorem" for Lipschitz maps.

Theorem 5.7.3 (McShance–Whitney Extension Theorem). *If (X, d) is a metric space, $A \subseteq X$, and $f: A \to \mathbb{R}$ is k-Lipschitz, then there exists a k-Lipschitz function $\hat{f}: X \to \mathbb{R}$ such that $\hat{f}|_A = f$.*

Proof. Let $\hat{f}: X \to \mathbb{R}$ be defined by

$$\hat{f}(x) = \inf[f(u) + kd(u, x) : u \in A] \quad \text{for all } x \in X\,. \tag{5.7.2}$$

We fix $u_0 \in A$ and obtain

$$\begin{aligned} f(u) + kd(u,x) &\geq f(u) + kd(u,u_0) - kd(u_0,x) \geq \hat{f}(u_0) - kd(u_0,x) \\ &= f(u_0) - kd(u_0,x)\,. \end{aligned} \tag{5.7.3}$$

This shows that $\hat{f}(x) > \infty$ for all $x \in X$.

Let $x, y \in X$. Given $\varepsilon > 0$, there exists $u_\varepsilon \in A$ such that

$$f(u_\varepsilon) + kd(u_\varepsilon, x) \leq \hat{f}(x) + \varepsilon\,. \tag{5.7.4}$$

Moreover, from (5.7.2) we get

$$\hat{f}(y) \leq f(u_\varepsilon) + kd(u_\varepsilon, y)\,. \tag{5.7.5}$$

From (5.7.4) and (5.7.5) it follows that

$$\hat{f}(y) - \hat{f}(x) \leq k\big[d(u_\varepsilon, y) - d(u_\varepsilon, x)\big] + \varepsilon \leq kd(y,x) + \varepsilon\,.$$

Since $\varepsilon > 0$ is arbitrary, we let $\varepsilon \searrow 0$ to obtain

$$\hat{f}(y) - \hat{f}(x) \leq kd(y,x)\,. \tag{5.7.6}$$

Interchanging the roles of x and y in the argument above gives

$$\hat{f}(x) - \hat{f}(y) \leq kd(x,y)\,. \tag{5.7.7}$$

Then (5.7.6) and (5.7.7) imply that $\hat{f}$ is k-Lipschitz on all of X. Finally, from (5.7.3) with $x = u_0$, we see that $\hat{f}|_A = f$. □

Remark 5.7.4. If X is a normed space, then from (5.7.2) and Definition 5.1.12 we see that $\hat{f} = f \oplus k\|\cdot\|$. Note that this extension $\hat{f}$ is the largest k-Lipschitz extension of f in the sense that if $\hat{g}: X \to \mathbb{R}$ is k-Lipschitz and $\hat{g}|_A = f$, then $\hat{g} \leq \hat{f}$. On the other hand if we consider $\tilde{f}: X \to \mathbb{R}$ defined by

$$\tilde{f}(x) = \sup[f(u) - kd(u,x) : u \in A]\,, \tag{5.7.8}$$

then $\tilde{f}$ is k-Lipschitz as well and $\tilde{f}|_A = f$, that is, $\tilde{f}$ is a Lipschitz extension of f. The extension given by (5.7.8) is the smallest k-Lipschitz extension of f.

Applying Theorem 5.7.3 to every component of an $\mathbb{R}^m$-valued function, we obtain the following result.

Corollary 5.7.5. *If (X, d) is a metric space, $A \subseteq X$ and $f: A \to \mathbb{R}^m$ is k-Lipschitz, then there exists a $k\sqrt{m}$-Lipschitz function $\hat{f}: X \to \mathbb{R}^m$ such that $\hat{f}|_A = f$.*

Remark 5.7.6. If we use the l^∞-norm on $\mathbb{R}^m$, then we can drop the factor $\sqrt{m}$ in the corollary above. In fact, we can state the following more general result in this case.

Proposition 5.7.7. *If (X, d) is a metric space, (Ω, Σ, μ) is a measure space, $A \subseteq X$, and $f: A \to L^\infty(\Omega)$ is a k-Lipschitz map, then there exists a k-Lipschitz function $\hat{f}: X \to L^\infty(\Omega)$ such that $\hat{f}| = f$.*

Proof. For every $x \in X$, let $\hat{f}(x) \in L^\infty(\Omega)$ be defined by

$$\hat{f}(x)(w) = \inf[f(u)(w) + kd(u, x) : u \in A] \quad \text{for all } w \in \Omega . \tag{5.7.9}$$

As in the proof of Theorem 5.7.3, fixing $u_0 \in A$, we obtain

$$\hat{f}(x) \geq -\|f(u_0)\|_\infty - kd(u_0, x) ;$$

see (5.7.3). Moreover, from (5.7.9) we have

$$\hat{f}(x)(w) \leq f(u_0)(w) + kd(u_0, x) ,$$

which implies that

$$\|\hat{f}(x)\|_\infty \leq \|f(u_0)\|_\infty + kd(u_0, x) < \infty .$$

This means that $\hat{f}$ has values in $L^\infty(\Omega)$. From (5.7.9) we can easily check that $\hat{f}$ is k-Lipschitz and $\hat{f}|_A = f$; see the proof of Theorem 5.7.3. □

The next result explains why the proposition above is useful.

Proposition 5.7.8. *Every metric space (X, d) is embedded isometrically into the Banach space* BC(X), *where* BC(X) *denotes the space of all bounded, continuous functions on X equipped with the supremum norm.*

Proof. We fix $x_0 \in X$ and consider, for each $x \in X$, the function $f_x: X \to \mathbb{R}$ defined by

$$f_x(u) = d(u, x) - d(u, x_0) . \tag{5.7.10}$$

Applying the triangle inequality, we see that f_x is bounded. Moreover, one has

$$|f_x(u) - f_{x'}(u)| = |d(u, x) - d(u, x')| \leq d(x, x') \quad \text{for all } u \in X .$$

Choosing $u = x$ yields $\|f_x - f_{x'}\|_\infty = d(x, x')$ and so we have the desired isometric embedding. □

In the proposition above, the target space for the embedding depends on the metric space itself. For separable metric spaces, we can have a universal target space.

Proposition 5.7.9. *Every separable metric space is embedded isometrically into the Banach space l^∞.*

Proof. Let $\{x_k\}_{k\in\mathbb{N}_0}$ be dense in the separable metric space (X,d). Consider the function $h\colon X\to l^\infty$ defined by

$$h(u)=\{d(u,x_k)-d(x_k,x_0)\}_{k\geq 1}\,. \tag{5.7.11}$$

Then we can easily see that this is an isometric embedding of X into l^∞. □

Remark 5.7.10. There is no canonical isometric embedding of a metric space X into $\mathrm{BC}(X)$, or into l^∞ in case X is separable. The embedding $x\to f_x$ in (5.7.10) depends on the choice of x_0 while in the separable case the embedding h defined in (5.7.11) depends on the chosen dense subset $\{x_k\}_{n\in\mathbb{N}_0}$. Proposition 5.7.9 is not entirely satisfactory, since the target space l^∞ is not separable. There is a theorem due to Banach [28], which says that every separable metric space admits an isometric embedding into the separable Banach space $C[0,1]$.

Next we turn our attention to locally Lipschitz functions and extend the subdifferential theory of continuous convex functions to them. We start by recalling the definition of locally Lipschitz functions.

Definition 5.7.11. Let X be a Banach space and $f\colon X\to\mathbb{R}$. We say that f is **locally Lipschitz** if for every $x\in X$ there exist $U\in\mathcal{N}(x)$ and a constant k_U such that

$$|f(u)-f(v)|\leq k_U\|u-v\|\quad\text{for all } u,v\in U\,.$$

Remark 5.7.12. If $f\colon X\to\mathbb{R}$ is Lipschitz continuous on bounded sets, then f is locally Lipschitz. If $\dim X<+\infty$, then the two properties are equivalent.

Of course the directional derivative $f'_+(x,\cdot)$ of convex functions (see Proposition 5.2.8 (a)) need not exist in this case. It is replaced by the following quantity. In what follows, X will always be a Banach space. Additional hypotheses will be introduced as needed.

Definition 5.7.13. Let $f\colon X\to\mathbb{R}$ be a locally Lipschitz function. The **generalized directional derivative** of f at $x\in X$ in the direction $h\in X$ is defined by

$$f^\circ(x;h)=\limsup_{\substack{u\to x\\ \lambda\searrow 0}}\frac{f(u+\lambda h)-f(u)}{\lambda}=\inf_{\varepsilon,\delta>0}\sup_{\substack{\|u-x\|\leq\varepsilon\\ 0<\lambda\leq\delta}}\frac{f(u+\lambda h)-f(u)}{\lambda}\,. \tag{5.7.12}$$

The importance of this notion is a result of its properties listed in the next proposition.

Proposition 5.7.14. *If $\varphi\colon X\to\mathbb{R}$ is locally Lipschitz, then the following hold:*
(a) *$h\to f^\circ(x;h)$ is sublinear and Lipschitz continuous for every $x\in X$;*
(b) *$(x,h)\to f^\circ(x;h)$ is upper semicontinuous;*
(c) *$f^\circ(x;-h)=(-f)^\circ(x;h)$ for all $x,h\in X$.*

Proof. (a) Because of (5.7.12) it is clear that $f^\circ(x;\cdot)$ is positively homogeneous. Let $h_1, h_2 \in X$. Applying the definition in (5.7.12) gives

$$
\begin{aligned}
& f^\circ(x;h_1+h_2) \\
&= \limsup_{\substack{u\to x\\ \lambda\searrow 0}} \frac{f(u+\lambda(h_1+h_2))-f(u)}{\lambda} \\
&= \limsup_{\substack{u\to x\\ \lambda\searrow 0}} \frac{f(u+\lambda(h_1+h_2))-f(u+\lambda h_2)+f(u+\lambda h_2)-f(u)}{\lambda} \\
&\le \limsup_{\substack{u\to x\\ \lambda\searrow 0}} \frac{f(u+\lambda h_2+\lambda h_1)-f(u+\lambda h_2)}{\lambda} + \limsup_{\substack{u\to x\\ \lambda\searrow 0}} \frac{f(u+\lambda h_2)-f(u)}{\lambda} \\
&= f^\circ(x;h_1)+f^\circ(x;h_2)\,.
\end{aligned}
$$

So, $f^\circ(x;\cdot)$ is subadditive, therefore sublinear.

From the fact that f is locally Lipschitz, if u is near x and $\lambda > 0$ is near 0, then

$$\frac{f(u+\lambda h)-f(u)}{\lambda} \le k\|h\| \quad \text{for all } h\in X\,.$$

Hence,

$$f^\circ(x;h) \le k\|h\|\,. \tag{5.7.13}$$

Since $f^\circ(x;\cdot)$ is sublinear, we obtain

$$0 = f^\circ(x;0) = f^\circ(x;h-h) \le f^\circ(x;h)+f^\circ(x;-h)\,.$$

Due to (5.7.13), we then see that $-f^\circ(x;h) \le f^\circ(x;-h) \le k\|h\|$ and so $|f^\circ(x;h)| \le k\|h\|$. This proves the Lipschitz continuity of $f^\circ(x;\cdot)$.

(b) Suppose $(x_n,h_n) \to (x,h)$ in $X\times X$. Given $n\in\mathbb{N}$, let $u_n \in X$ and $\lambda_n > 0$ be such that $\|u_n\| + \lambda_n \le 1/n$ and

$$
\begin{aligned}
& f^\circ(x_n;h_n) \\
&\le \frac{f(x_n+u_n+\lambda_n h_n)-f(x_n+u_n)}{\lambda_n} + \frac{1}{n} \\
&= \frac{f(x_n+u_n+\lambda_n h_n)-f(x_n+u_n+\lambda_n h)}{\lambda_n} + \frac{f(x_n+u_n+\lambda_n h)-f(x_n+u_n)}{\lambda_n} + \frac{1}{n}\,.
\end{aligned}
$$

This implies

$$\limsup_{n\to\infty} f^\circ(x_n;h_n) \le \limsup_{n\to\infty} \frac{f(x_n+u_n+\lambda_n h)-f(x_n+u_n)}{\lambda_n} \le f^\circ(x;h)\,.$$

Hence, $(x,h) \to f^\circ(x;h)$ is upper semicontinuous.

(c) Let $v = u - \lambda h$. Then we obtain

$$f^\circ(x; -h) = \limsup_{\substack{u \to x \\ \lambda \searrow 0}} \frac{f(u + \lambda(-h)) - f(u)}{\lambda} = \limsup_{\substack{v \to x \\ \lambda \searrow 0}} \frac{(-f)(v + \lambda h) - (-f)(v)}{\lambda}$$
$$= (-f)^\circ(x; h) .$$

□

The differentiability notion, which is closer to the subdifferential theory for locally Lipschitz functions, is the following one.

Definition 5.7.15. Let Y be another Banach space and $f: X \to Y$. We say that f is **strictly differentiable at** x if there exists $f_s'(x) \in L(X, Y)$ such that

$$\lim_{\substack{u \to x \\ \lambda \searrow 0}} \frac{f(u + \lambda h) - f(u)}{\lambda} = f_s'(x)(h) \tag{5.7.14}$$

uniformly for h in compact sets. We say that f is **strictly differentiable** if it is strictly differentiable at every $x \in X$.

Remark 5.7.16. If f is Lipschitz continuous near x, then the limit in (5.7.14) is automatically uniform for h in compact sets. Usually this notion of strict differentiability is called **Hadamard strict differentiability**. We can define Fréchet (resp. Gateaux) strict differentiability by requiring that the limit in (5.7.14) is uniform for h in bounded (resp. finite) sets. Alternatively, we say that f is **strictly Fréchet differentiable at** x if there exists $f_s'(x) \in L(X, Y)$ such that

$$\lim_{\substack{u \to x \\ h \to 0}} \frac{f(u + h) - f(u) - f_s'(x)(h)}{\|h\|} = 0 .$$

Proposition 5.7.17. *If Y is another Banach space, $f: X \to Y$ and $A \in L(X, Y)$, then the following two statements are equivalent:*
(a) *f is strictly differentiable at x and $f_s'(x) = A$;*
(b) *f is Lipschitz near x and*

$$\lim_{\substack{u \to x \\ \lambda \searrow 0}} \frac{f(u + \lambda h) - f(u)}{\lambda} = A(h) \quad \textit{for all } h \in X . \tag{5.7.15}$$

Proof. (a) $\Longrightarrow$ (b): Evidently, (5.7.15) holds on account of Definition 5.7.15. So, it remains to show that f is Lipschitz near x. Arguing by contradiction, suppose we could find sequences $\{u_n\}_{n \ge 1}, \{v_n\}_{n \ge 1} \subseteq X$ such that $u_n, v_n \to x$ with $u_n, v_n \in B_{1/n}(x)$ and

$$\|f(u_n) - f(v_n)\|_Y > n\|u_n - v_n\|_X . \tag{5.7.16}$$

We define $\lambda_n > 0$ and h_n by the equations $v_n = u_n + \lambda_n h_n$ and $\|h_n\| = 1/\sqrt{n}$. Evidently, $\lambda_n \to 0^+$. Let $K = \{h_n\}_{n \ge 1} \cup \{0\}$. Then $K \subseteq X$ is compact and so by hypothesis, for a given $\varepsilon > 0$, there exists $n_0 = n_0(\varepsilon) \in \mathbb{N}$ such that

$$\left\|\frac{f(u_n+\lambda_n h)-f(u_n)}{\lambda_n}-f_s'(x)(h)\right\|_Y<\varepsilon \quad \text{for all } n\geq n_0 \text{ and for all } h\in K\,.$$

But this is impossible, since (5.7.16) gives, for $h = h_n$, that

$$\frac{1}{\lambda_n}\|f(u_n+\lambda_n h)-f(u_n)\|_Y>\frac{1}{\sqrt{n}}\,.$$

Therefore, f is Lipschitz near x and we have proven that (a) implies (b).

(b) $\Longrightarrow$ (a): Let $K \subseteq X$ be compact and $\varepsilon > 0$. By hypothesis, for every $h \in K$, there exists $\delta = \delta(h) > 0$ such that

$$\left\|\frac{f(u+\lambda h)-f(u)}{\lambda}-A(h)\right\|_Y<\varepsilon \tag{5.7.17}$$

for all $u \in B_\delta(x)$ and for all $\lambda \in (0,\delta)$. We have

$$\left\|\frac{f(u+\lambda h)-f(u)}{\lambda}-\frac{f(u+\lambda h')-f(u)}{\lambda}\right\|_Y<k\|h-h'\|_X\,.$$

Taking a sufficiently small $\delta > 0$, one obtains

$$\left\|\frac{f(u+\lambda h')-f(u)}{\lambda}-A(h')\right\|_Y<2\varepsilon\,,$$

for all $u \in B_\delta(x)$, $h' \in B_\delta(h)$ and $\lambda \in (0,\delta)$. The compactness of K implies that the open cover $\{h+\delta B_1\}_{h\in K}$ has a finite subcover $\{h_k+\delta B_1\}_{k=1}^n$. We set $\hat{\delta} = \min_{1\leq k\leq n}\delta(h_k)$ and see that (5.7.17) holds with $\varepsilon > 0$ replaced by $2\varepsilon > 0$ and for all $h \in K$, for all $u \in B_{\hat{\delta}}(x)$, and for all $\lambda \in (0,\hat{\delta})$. Therefore, $A = f_s'(x)$. ☐

Definition 5.7.18. Let Y be another Banach space and $f: X \to Y$. We say that f is **continuously differentiable at** x if the Gateaux derivative of f in a neighborhood U of x exists and the function $U \ni x' \to f'(x') \in L(X,Y)$ is continuous.

Remark 5.7.19. Such a function is also Fréchet differentiable and of course the two derivatives coincide. We denote the space of continuously differentiable functions $f: X \to Y$ by $C^1(X,Y)$.

Proposition 5.7.20. *If Y is another Banach space and $f: X \to Y$ is continuously differentiable at $x \in X$, then f is strictly differentiable at x.*

Proof. From the Mean Value Theorem we know that f is Lipschitz near x. So, according to Proposition 5.7.17 we need to show (5.7.17). Let $\{u_n\}_{n\geq1}, \{h_n\}_{n\geq1} \subseteq X$ and $\{\lambda_n\}_{n\geq1} \subseteq (0,+\infty)$. We need to show that

$$\lim_{n\to\infty}\sup_{\|y^*\|_{Y^*}\leq1}\left\langle y^*,\frac{f(u_n+\lambda_n h_n)-f(u_n)}{\lambda_n}-f'(x)(h_n)\right\rangle=0\,.$$

By the Mean Value Theorem, there exists $\hat{x}_n \in [x_n, x_n + \lambda_n h_n]$ such that

$$\left\langle y^*, \frac{f(u_n + \lambda_n h_n) - f(u_n)}{\lambda_n} - f'(x)(h_n) \right\rangle = \langle y^*, (f'(\hat{x}_n) - f'(x))(h_n) \rangle .$$

Note that

$$\sup[\langle y^*, (f'(\hat{x}_n) - f'(x))(h_n) \rangle : \|y^*\|_{Y^*} \leq 1] \to 0 \quad \text{as } n \to \infty .$$

Hence, $x \to f'(x)$ is continuous into $L(X, Y)$. □

Based on Proposition 5.7.14 (a) we state the following definition.

Definition 5.7.21. Let $f: X \to \mathbb{R}$ be locally Lipschitz. The **generalized subdifferential** (or **Clarke subdifferential**) of f at x is defined by

$$\partial f(x) = \{x^* \in X^* : \langle x^*, h \rangle \leq f^\circ(x; h) \text{ for all } h \in X\} .$$

Remark 5.7.22. From Remark 5.1.29, we know that $\partial f(x) \neq \emptyset$ and it is convex, w*-closed, and bounded; thus it is w*-compact. We have $\sigma(\cdot; \partial f(x)) = f^\circ(x; \cdot)$ and so $x^* \in \partial f(x)$ if and only if $\langle x^*, h \rangle \leq f^\circ(x; h)$ for all $h \in X$.

Proposition 5.7.23. *If $f: X \to \mathbb{R}$ is locally Lipschitz, then* $\mathrm{Gr}\, \partial f \subseteq X \times X^*_{\mathrm{w}^*}$ *is closed.*

Proof. Assume that $x_\alpha \to x$ in X, $x^*_\alpha \xrightarrow{\mathrm{w}^*} x^*$ in X^*, and $x^*_\alpha \in \partial f(x_\alpha)$ for all $\alpha \in I$. We know that $\langle x^*_\alpha, h \rangle \leq f^\circ(x_\alpha; h)$ for all $h \in X$ and for all $\alpha \in I$. Applying Proposition 5.7.14 (b) gives

$$\langle x^*, h \rangle = \lim_{\alpha \in I} \langle x^*_\alpha, h \rangle \leq \limsup_{\alpha \in I} f^\circ(x_\alpha; h) \leq f^\circ(x; h) \quad \text{for all } h \in X .$$

Hence, $x^* \in \partial f(x)$. □

From this proposition and Proposition 5.6.11 we infer the following corollary.

Corollary 5.7.24. *If X is finite dimensional and $f: X \to \mathbb{R}$ is locally Lipschitz, then $\partial f: X \to 2^{X^*} \setminus \{\emptyset\}$ is* usc.

Proposition 5.7.25. *If $f: X \to \mathbb{R}$ is locally Lipschitz, then the following hold:*
(a) *if f is Gateaux or Fréchet differentiable at $x \in X$, then $f'(x) \in \partial f(x)$;*
(b) *if f is strictly differentiable at $x \in X$, then $\partial f(x) = \{f'(x)\}$.*

Proof. (a) Evidently, in each case we have $\langle f'(x), h \rangle \leq f^\circ(x; h)$ for all $h \in X$. So, $f'(x) \in \partial f(x)$; see Remark 5.7.22.

(b) From (5.7.14) we have $\langle f'_s(x), h \rangle = f^\circ(x; h)$ for all $h \in X$. Therefore, $\partial f(x) = \{f'_s(x)\}$. □

Remark 5.7.26. A locally Lipschitz function that is simply differentiable at x may contain more elements in $\partial f(x)$. In order to see this, consider the function $f: \mathbb{R} \to \mathbb{R}$ defined by

$$f(x) = \begin{cases} x^2 \sin(\frac{1}{x}) & \text{if } x \neq 0\,, \\ 0 & \text{if } x = 0\,. \end{cases}$$

Then f is differentiable on $[0,1]$, Lipschitz continuous on $[0,1]$, and $f'(0) = 0$. But $f^\circ(0;h) = |h|$ and so $\partial f(0) = [-1,1]$.

It is natural to ask what the relation is between the convex subdifferential (see Definition 5.2.13) and the generalized (Clarke) subdifferential (see Definition 5.7.21). In this direction, we have the following result. In what follows, by $\partial_c f$ we denote the subdifferential in the sense of convex analysis and by ∂f the generalized subdifferential, to distinguish between them.

Proposition 5.7.27. *If $f: X \to \mathbb{R}$ is continuous and convex, and hence locally Lipschitz by Proposition 5.1.20, then $\partial_c f(x) = \partial f(x)$ for all $x \in X$.*

Proof. Clearly, $f'_+(x';\cdot) \le f^\circ(x;\cdot)$ and so by Proposition 5.2.15, we obtain

$$\partial_c f(x) \subseteq \partial f(x) \quad \text{for all } x \in X\,. \tag{5.7.18}$$

For fixed $h \in X$, the function $(u,\lambda) \to (f(u+\lambda h) - f(u))/\lambda$ is continuous from $X \times (0,+\infty)$ into $\mathbb{R}$. Therefore, for a given $\varepsilon > 0$, there exists $\delta > 0$ such that

$$\frac{f(u+th) - f(u)}{t} \le \frac{f(x+\lambda h) - f(x)}{\lambda} + \varepsilon$$

for all $\|u - x\| \le \delta$ and $|t - \lambda| \le \delta$. Hence,

$$\sup_{\substack{\|u-x\|\le\delta \\ |t-\lambda|\le\delta}} \frac{f(u+th) - f(u)}{t} \le \frac{f(x+\lambda h) - f(x)}{\lambda} + \varepsilon\,.$$

Passing to the limit as $\lambda \to 0^+$ and $\delta \to 0^+$ gives $f^\circ(x;h) \le f'_+(x;h) + \varepsilon$. Letting $\varepsilon \searrow 0$ gives $f^\circ(x;h) \le f'_+(x;h)$ for all $h \in X$. Hence, $\partial f(x) \subseteq \partial_c f(x)$ for all $x \in X$ and so, due to (5.7.18), $\partial_c f(x) = \partial f(x)$ for all $x \in X$. □

Proposition 5.7.28. *If $f: X \to \mathbb{R}$ is locally Lipschitz and $\lambda \in \mathbb{R}$, then $\partial(\lambda f)(x) = \lambda \partial f(x)$ for all $x \in X$.*

Proof. If $\lambda \ge 0$, then $(\lambda f)^\circ = \lambda f^\circ$ and so $\partial(\lambda f)(x) = \lambda \partial f(x)$ for all $x \in X$. Therefore, we assume that $\lambda < 0$ and take, without any loss of generality, $\lambda = -1$. We obtain $x^* \in \partial(-f)(x)$ if and only if $\langle x^*, h\rangle \le (-f)^\circ(x;h)$ for all $h \in X$ if and only if $\langle x^*, h\rangle \le f^\circ(x;-h)$ for all $h \in X$ (see Proposition 5.7.14), if and only if $\langle -x^*, -h\rangle \le f^\circ(x;-h)$ for all $h \in X$ if and only if $x^* \in -\partial f(x)$. Therefore, we conclude that $\partial(\lambda f)(x) = \lambda \partial f(x)$ for all $x \in X$. □

Now we will state an extension of the classical Fermat rule for local extrema.

Proposition 5.7.29. *If $f: X \to \mathbb{R}$ is a locally Lipschitz function that has a local extremum (local maximum or local minimum) at $x \in X$, then $0 \in \partial f(x)$.*

Proof. Since $\partial(-f) = -\partial f$ (see Proposition 5.7.28), it suffices to prove the result for the case when x is a local minimizer. From (5.7.12), it follows that $0 \le f^\circ(x; h)$ for all $h \in X$. Hence, $0 \in \partial f(x)$; see Definition 5.7.21. □

The next notion leads to sharper calculus rules for the generalized subdifferential.

Definition 5.7.30. A locally Lipschitz function $f: X \to \mathbb{R}$ is said to be **regular at** x if the following hold:
(a) $f'_+(x; h)$ exists for all $h \in X$;
(b) $f^\circ(x; h) = f'_+(x; h)$ for all $h \in X$.

Remark 5.7.31. Continuous convex functions and strictly differentiable functions, in particular C^1-functions, are regular.

The next calculus rule for the generalized subdifferential is an easy consequence of Definition 5.7.21.

Proposition 5.7.32. *If $f_k: X \to \mathbb{R}$ for $k \in \{1, \ldots, m\}$ are locally Lipschitz functions and $\{\lambda_k\}_{k=1}^m \subseteq \mathbb{R}$, then*

$$\partial\left(\sum_{k=1}^m \lambda_k f_k\right)(x) \subseteq \sum_{k=1}^m \lambda_k \partial f_k(x) \quad \text{for all } x \in X .$$

Equality holds if each f_k is regular at $x \in X$ or if all but one of the f_k's is strictly differentiable at $x \in X$ and $\lambda_k \ge 0$.

Useful in applications is the following chain rule.

Proposition 5.7.33. *If Y is another Banach space, $g \in C^1(X, Y)$, and $f: Y \to \mathbb{R}$ is locally Lipschitz, then the following hold:*
(a) *$f \circ g: X \to \mathbb{R}$ is locally Lipschitz and $\partial(f \circ g)(x) \subseteq \partial f(g(x)) \circ g'(x)$ in the sense that, for every $x^* \in \partial(f \circ g)(x)$, we have*

$$x^* = g'(x)^* u^* \quad \text{for some } u^* \in \partial f(x) ; \tag{5.7.19}$$

(b) *if f or $(-f)$ is regular at $g(x)$, then $f \circ g$ or $(-f) \circ g$ is regular at x and equality holds in* (5.7.19)*;*
(c) *if g maps every $U \in \mathcal{N}(x)$ onto a set that is dense in a neighborhood of $g(x)$, for example, if $g'(x)$ is surjective, then equality holds in* (5.7.19).

Proof. (a) Since f is locally Lipschitz, there exists $V \in \mathcal{N}(g(x))$ such that $f|_V$ is Lipschitz continuous. Let $U = g^{-1}(V) \subseteq X$. Then $U \in \mathcal{N}(x)$ and $g(U) \subseteq V$. So, $f \circ g|_U$ is Lipschitz continuous, proving that $f \circ g$ is locally Lipschitz as well.

From Proposition 5.7.14 (a), we know that for a given $\varepsilon > 0$ there exists $\delta \in (0, \varepsilon]$ such that

$$f^\circ(g(x); h') \le f^\circ(g(x); h) + \varepsilon \quad \text{for all } h', h \in Y \text{ with } \|h' - h\|_Y \le \delta\,. \tag{5.7.20}$$

In addition, from the definition of the generalized directional derivative, we see that there exist $\eta, \vartheta > 0$ such that

$$\frac{f(y + \lambda h') - f(y)}{\lambda} \le f^\circ(g(x), h') + \varepsilon \le f^\circ(g(x); h) + 2\varepsilon \tag{5.7.21}$$

for all $\|y - g(x)\|_Y \le \eta$, $0 < \lambda \le \vartheta$ and $\|h' - h\|_Y \le \delta$; see (5.7.20). We set $h = g'(x)v$ with $v \in X$. Since $g \in C^1(X, Y)$, there exists $0 < \xi \le \vartheta$ such that

$$\left\| \frac{g(u + \lambda v) - g(u)}{\lambda} - g'(x)v \right\|_Y \le \delta \quad \text{and} \quad \|g(u) - g(v)\|_Y \le \eta$$

for all $\|u - x\|_X \le \xi$ and $0 < \lambda \le \xi$. In (5.7.21) we set $y = g(u)$ and $h' = 1/\lambda[g(u + \lambda v) - g(u)] \in Y$. Then

$$\frac{(f \circ g)(u + \lambda v) - (f \circ g)(u)}{\lambda} \le f^\circ(g(x); g'(x)v) + 2\varepsilon$$

for all $\|u - x\|_X \le \xi$ and $0 < \lambda \le \xi$. Since $\varepsilon > 0$ is arbitrary, it follows that

$$(f \circ g)^\circ(x; v) \le f^\circ(g(x); g'(x)v) = \max[\langle y^*, g'(x)v\rangle : y^* \in \partial f(g(x))]\,. \tag{5.7.22}$$

Hence, $\partial(f \circ g)(x) \subseteq \partial f(g(x)) \circ g'(x)$.

(b) We assume that f is regular at $g(x)$. The case where $(-f)$ is regular at $g(x)$ can be derived from the previous one since $\partial(-f)(g(x)) = -\partial f(g(x))$; see Proposition 5.7.28.

Since f is locally Lipschitz, $g \in C^1(X, Y)$, and f is regular at $g(x)$ we infer (see Definition 5.7.30) that

$$\begin{aligned} & f^\circ(g(x); g'(x)v) \\ & = f'_+(g(x); g'(x)v) = \lim_{\lambda \searrow 0} \frac{f(g(x) + \lambda g'(x)v) - f(g(x))}{\lambda} \\ & = \lim_{\lambda \searrow 0} \left[\frac{f(g(x) + \lambda g'(x)v) - f(g(x + \lambda v))}{\lambda} \right. \\ & \qquad \left. + \frac{f(g(x + \lambda v)) - f(g(x))}{\lambda} \right] \\ & = (f \circ g)'_+(x; v) \le (f \circ g)^\circ(x; v)\,. \end{aligned} \tag{5.7.23}$$

From (5.7.22) and (5.7.23) we infer that

$$f^\circ(g(x); g'(x)v) = (f \circ g)^\circ(x; v) \quad \text{for all } v \in X\,,$$

which implies that

$$\partial(f \circ g)(x) = \partial f(g(x)) \circ g'(x)\,. \tag{5.7.24}$$

(c) If $g \in C^1(X,Y)$ maps any neighborhood of x onto a dense subset of a neighborhood of $g(x)$, then we can write, since $g \in C^1(X,Y)$ and the hypothesis on g, that

$$\begin{aligned} &f^\circ(g(x); g'(x)v) \\ &= \limsup_{\substack{y\to g(x)\\ \lambda\searrow 0}} \frac{f(y+\lambda g'(x)v) - f(y)}{\lambda} = \limsup_{\substack{u\to x\\ \lambda\searrow 0}} \frac{f(g(u)+\lambda g'(x)v) - f(g(u))}{\lambda} \\ &= \limsup_{\substack{u\to x\\ \lambda\searrow 0}} \frac{f(g(u+\lambda v)) - f(g(u))}{\lambda} = (f\circ g)^\circ(x; v) \quad \text{for all } v \in X\,. \end{aligned}$$

From this equality it follows that (5.7.24) holds. □

Remark 5.7.34. Note that the inclusion in part (a) can be written as

$$\partial(f \circ g)(x) \subseteq g'(x)^* \partial f(g(x))\,.$$

Another result in the same direction is the following one.

Proposition 5.7.35. *If $T = [0,b]$, $g \in C^1(T,X)$ and $f\colon X \to T$ is locally Lipschitz, then $h = f \circ g\colon T \to \mathbb{R}$ is differentiable a. e. on T and*

$$h'(t) \le \max[\langle x^*, g'(x)\rangle : x^* \in \partial f(g(x))]\,.$$

Proof. Evidently, h is locally Lipschitz and so it is differentiable a. e. on T. Let $t_0 \in T$ be a point of differentiability of h. Recall that $o(\lambda)/\lambda \to 0$ as $\lambda \to 0$. Then, since f is locally Lipschitz, we obtain

$$\begin{aligned} &h'(t_0) \\ &= \lim_{\lambda\to 0^+} \frac{f(g(t_0+\lambda)) - f(g(t_0))}{\lambda} = \lim_{\lambda\to 0^+} \frac{f(g(t_0)+\lambda g'(t_0) + o(\lambda)) - f(g(t_0))}{\lambda} \\ &= \lim_{\lambda\to 0^+} \Bigg[\frac{f(g(t_0)+\lambda g'(t_0)) - f(g(t_0))}{\lambda} \\ &\qquad + \frac{f(g(t_0)+\lambda g'(t_0)+o(\lambda)) - f(g(t_0)+\lambda g'(t_0))}{\lambda}\Bigg] \\ &= \lim_{\lambda\to 0^+} \frac{f(g(t_0)+\lambda g'(t_0)) - f(g(t_0))}{\lambda} \le f^\circ(g(t_0); g'(t_0)) \\ &= \max[\langle x^*, g'(t_0)\rangle : x^* \in \partial f(g(t_0))]\,. \end{aligned}$$

□

We also have a "Mean Value Theorem for locally Lipschitz functions".

Theorem 5.7.36 (Mean Value Theorem for locally Lipschitz functions). *If $f: X \to \mathbb{R}$ is locally Lipschitz and $x, u \in X$, then there exist $\lambda_0 \in (0,1)$ and $v^* \in \partial f((1-\lambda_0)x + \lambda_0 u)$ such that $f(u) - f(x) = \langle v^*, u - x\rangle$.*

Proof. Let $\xi: \mathbb{R} \to \mathbb{R}$ be defined by

$$\xi(\lambda) = f((1-\lambda)x + \lambda u) + \lambda(f(x) - f(u)) \quad \text{for all } \lambda \in \mathbb{R}.$$

Evidently, ξ is locally Lipschitz, and hence $\xi|_{[0,1]}$ is locally Lipschitz as well. We have $\xi(0) = \xi(1) = f(x)$. It follows that ξ attains a local extremum, that is, a local maximum or a local minimum, at some $\lambda_0 \in (0,1)$. Proposition 5.7.29 implies that $0 \in \partial\xi(\lambda_0)$. A straightforward calculation leads to

$$0 \in \partial\xi(\lambda_0) \subseteq \langle \partial f(x + \lambda_0(u-x)), u - x\rangle + [f(x) - f(u)].$$

Hence, $f(u) - f(x) \in \langle \partial f(x + \lambda_0(u - x)), u - x\rangle$. □

Finally, we introduce some geometric objects that are cones that locally approximate a set. These cones are the nonsmooth counterparts of tangent spaces for smooth manifolds, which are a basic tool in differential geometry.

Definition 5.7.37. Let X be a Banach space, $C \subseteq X$ a nonempty set, and $x \in \overline{C}$.
(a) The **contingent cone** to C at x is defined by

$$T_C(x) = \{h \in X: \exists \lambda_k \searrow 0\ \exists h_k \to h \text{ such that } x + \lambda_k h_k \in \overline{C} \text{ for all } k \in \mathbb{N}\}.$$

(b) The **Clarke tangent cone** to C at x is defined by

$$\begin{aligned} T'_C(x) = \{h \in X: \forall x_k \to x\ \forall \lambda_k \searrow 0\ \exists h_k \to h \text{ such that} \\ x_k + \lambda_k h_k \in \overline{C} \text{ for all } k \in \mathbb{N}\}. \end{aligned}$$

Remark 5.7.38. These cones only depend on the local properties of C near x. So, if $U \in \mathcal{N}(x)$, then $T_C(x) = T_{C\cap U}(x)$ and $T'_C(x) = T'_{C\cap U}(x)$. Moreover, it is clear that $T_C(x) = T_{\overline{C}}(x)$ and $T'_C(x) = T'_{\overline{C}}(x)$ and so without any loss of generality we may always take $C \subseteq X$ to be closed. We have

$$\begin{aligned} T_C(x) &= \left\{h \in X: \liminf_{\lambda\searrow 0} \frac{1}{\lambda} d(x + \lambda h, C) = 0\right\}, \\ T'_C(x) &= \{h \in X: d_C^\circ(x; h) \le 0\}, \end{aligned}$$

where $d_C(x) = d(x, C) = \inf[\|x - u\|: u \in C]$.

Proposition 5.7.39. *If $C \subseteq X$ is nonempty, closed, and convex, then $T_C(x) = \overline{\mathbb{R}_+[C - x]} = T'_C(x)$ is a closed, convex cone.*

Proof. First, we show the first equality. Let $S_C(x) = \bigcup_{\lambda>0} 1/\lambda[C - x]$. We prove the existence of $\lambda_0 > 0$ such that $x + \lambda h \in C$ for all $\lambda \in [0, \lambda_0]$ and for all $h \in S_C(x)$. To this end, note that from the definition of $S_C(x)$ we see that there exists $\lambda_0 > 0$ such that $x+\lambda_0 h \in C$. For $\lambda \in [0, \lambda_0]$ we obtain

$$x + \lambda h = \left(1 - \frac{\lambda}{\lambda_0}\right)x + \frac{\lambda}{\lambda_0}(x + \lambda_0 h) \in C\,, \tag{5.7.25}$$

since $C \subseteq X$ is convex. Now let $h_1, h_2 \in S_C(x)$. Then there exist $\lambda_1, \lambda_2 > 0$ such that $x + \lambda_k h_k \in C$ for $k = 1, 2$. We set $\lambda_0 = \min\{\lambda_1, \lambda_2\}$. Taking (5.7.25) into account gives $x + \lambda_0 h_k \in C$ for $k \in \{1, 2\}$. The convexity of the set C implies that

$$x + \lambda_0((1 - t)h_1 + th_2) \in C \quad \text{for all } t \in [0, 1]\,.$$

Hence, $S_C(x)$ is a convex cone and so $\overline{S_C(x)}$ is a closed, convex cone. Clearly, $T_C(x) = \overline{S_C(x)}$.

Now we prove the second equality. Since $C \subseteq X$ is convex, d_C is convex and Lipschitz continuous. Therefore, $d_C^\circ(x; \cdot) = (d_C)'_+(x; \cdot)$. Then Remark 5.7.38 yields $T_C(x) = T'_C(x)$. □

Remark 5.7.40. So, for convex sets, both tangent cones are convex and are equal. However, the Clarke tangent cone is always convex, even when the set C is not convex. Moreover, both tangent cones are closed and $C \subseteq x + T_C(x)$. The contingent cone exhibits good properties for convex sets.

Definition 5.7.41. (a) Let $C \subseteq X$ be a nonempty, closed, convex set. The **normal cone** to C at x is defined by

$$\begin{aligned} N_C(x) &= \{x^* \in X^* : \langle x^*, x\rangle = \sigma(x^*; C)\} \\ &= \{x^* \in X^* : \langle x^*, u - x\rangle \le 0 \text{ for all } u \in C\}\,. \end{aligned}$$

(b) Let $C \subseteq X$ be a nonempty, closed set. The **Clarke normal cone** to C at x is defined by

$$N'_C(x) = T'_C(x)^- = \{x^* \in X^* : \langle x^*, h\rangle \le 0 \text{ for all } h \in T'_C(x)\}\,.$$

Remark 5.7.42. Evidently, $N_C(x) = T_C(x)^-$, which is the negative polar cone of $T_C(x)$. So, $N_C(x)$ is closed and convex. Note that $N_{\{x\}}(x) = X^*$ and if $x \in \operatorname{int} C$, then $N_C(x) = \{0\}$. If i_C is the indicator function of the set C, then $i_C \in \Gamma_0(X)$ and $\partial i_C(x) = N_C(x)$, where ∂i_C is the subdifferential in the sense of convex analysis. Similarly, if $x \in \operatorname{int} C$, then $N'_C(x) = \{0\}$. For convex sets, we obtain $N_C(x) = N'_C(x)$; see Proposition 5.7.39.

Proposition 5.7.43. *If $C \subseteq X$ is nonempty and $x \in \overline{C}$, then*

$$N'_C(x) = \overline{\bigcup_{\lambda\ge 0} \lambda \partial d_C(x)}^{w^*}\,.$$

Proof. Note that $h \in T'_C(x)$ if and only if $\langle x^*, h\rangle \le 0$ for all $x^* \in \partial d_C(x)$, which is the subdifferential in the sense of Clarke. So, the negative polar cone of $T_C(x)$ is the weak*-closed, convex cone generated by the set $\partial d_C(x) \subseteq X^*$. Hence, $N'_C(x) = \overline{\bigcup_{\lambda\ge 0} \lambda d_C(x)}^{w^*}$. □

5.8 Remarks

(5.1) Convex functions play a central role in many parts of mathematical analysis and its applications such as optimization, optimal control, game theory, mathematical economics, and others. They exhibit interesting properties concerning their continuity and differentiability. Moreover, minimizers are automatically global. Together with the corresponding theory of convex sets, they form a body of results that are known as "Convex Analysis". There are many books on the subject and we mention those by Hiriart-Urruty–Lemaréchal [167, 168], Rockafellar [273], Rockafellar–Wets [278] (finite dimensional theory) and Barbu–Precupanu [31], Borwein–Vanderwerff [45], Ekeland–Temam [114], Gasiński–Papageorgiou [135], Giles [137], Ioffe–Tichomirov [177], Laurent [202], and Phelps [260] (infinite dimensional theory). The theories of convex functions and of convex sets are linked via the following theorem from Hörmander [172].

Theorem 5.8.1. *If X is a locally convex space, then there is a bijective correspondence between nonempty, closed, convex sets and sublinear, w^*-lower semicontinuous functions from X^* into $\overline{\mathbb{R}} = \mathbb{R} \cup \{+\infty\}$. This correspondence maps the set C to the support function $\sigma(\cdot; C)$ of C.*

(5.2) As is evident from the results of this section, convex functions exhibit remarkable differentiability properties. Moreover, when they are not differentiable, they admit a useful multivalued analog of the derivative, which is the subdifferential; see Definition 5.2.13. Theorem 5.2.17 is a classic result of Mazur [228]. Additional results on weak Asplund (see Definition 5.2.18) and Asplund (see Definition 5.2.21) can be found in the books of Fabian et al. [117] and Deville–Godefroy–Zizler [88].

(5.3) Duality is in the core of convex analysis. The notion of a conjugate function (see Definition 5.3.1) was first introduced by Fenchel [121] in 1951. He was motivated by the classical Legendre transform and worked on $\mathbb{R}^N$. The extension to dual pairs of locally convex spaces is due to Brøndsted [56], Moreau [244], and Rockafellar [272, 276]. The finite dimensional duality theory can be found in Fenchel [121], Rockafellar [273], and Rockafellar–Wets [278]. The infinite dimensional theory is included in the books of Barbu–Precupanu [31], Borwein–Vanderwerff [45], Ekeland–Temam [114], Gasiński–Papageorgiou [135], Ioffe–Tichomirov [177], and Laurent [202]. Proposition 5.3.7 (b) is probably the most important and useful result in duality theory. In the literature it appears often under the name "Fenchel–Moreau Theorem". According to Proposition 5.3.7 (a), $f^{**}|_X$ is the largest convex and lower semicontinuous function majorized by f, and sometimes this is expressed by writing that $f^{**}|_X \le \overline{\operatorname{conv}} f$. This fact is important in control theory and in the calculus of variations in the so-called "relaxation method". If $X = \mathbb{R}^N$, then, using the Carathéodory Convexity Theorem (see Theorem 3.1.30), we can state the following convenient expression for f^{**}; see Ioffe–Tichomirov [177, p. 189].

Proposition 5.8.2. *If $f\colon \mathbb{R}^N \to \overline{\mathbb{R}}$ is proper and lower semicontinuous and* $\operatorname{dom} f^{**} \subseteq \mathbb{R}^N$ *is closed, then*

$$f^{**}(x) = \inf\left[\sum_{k=1}^{N+1} \lambda_k f(x_k)\colon x_k \in \mathbb{R}^N, \lambda_k \geq 0, \sum_{k=1}^{N+1} \lambda_k = 1, \sum_{k=1}^{N+1} \lambda_k x_k = x\right].$$

The operation of infimal convolution (see Definition 5.1.12) is due to Moreau [243]. The duality properties of this operation were produced by Ioffe–Tichomirov [177].

The systematic study of the subdifferential theory of convex functions started with Moreau [243] and Rockafellar [273, 274]. Moreau works on Hilbert spaces while Rockafellar [274] considers general Banach spaces. The subdifferential theory of convex functions is closely related to the theory of nonlinear operators of monotone type; see Section 6.1. The study of the convex subdifferential can also be found in the books of Barbu–Precupanu [31], Borwein–Vanderwerff [45], Ekeland–Temam [114], Gasiński–Papageorgiou [135], Ioffe–Tichomirov [177], and Laurent [202]. Theorems 5.3.24 and 5.3.26 are due to Rockafellar [274]. The ε-subdifferential (see Definition 5.3.28) was studied by Hiriart-Urruty [166] and Hiriart-Urruty–Phelps [169].

In connection with the notion of recession function (see Definition 5.3.32), we can introduce the **recession cone**.

Definition 5.8.3. If $C \subseteq X$ is a nonempty and convex set, then its **recession cone** is defined by

$$C^\infty = \{h \in X\colon u + \lambda h \in C \text{ for all } u \in C \text{ and for all } \lambda \geq 0\}.$$

Proposition 5.8.4. *If $C \subseteq X$ is nonempty, convex (and closed), then the following hold:*
(a) *C^∞ is convex (and closed) and $C + \lambda C^\infty \subseteq C$ for all $\lambda \geq 0$;*
(b) *$h \in C^\infty$ if and only if $h = \lim_{n\to\infty} \lambda_n h_n$ with $u_n \in C$ and $\lambda_n \to 0^+$.*

Proposition 5.8.5. *If $f\colon X \to \overline{R} = \mathbb{R} \cup \{+\infty\}$ is proper and convex, then* $(\operatorname{epi} f)^\infty = \operatorname{epi} f^\infty$.

The recession function and its use in convex minimization problems can be found in the book of Attouch–Buttazzo–Michaille [18].

(5.4) We can complete Theorem 5.4.5 as follows. The result is known as the "Day–James Theorem"; see Megginson [230, p. 436].

Theorem 5.8.6 (Day–James Theorem). *A normed space X is strictly convex and reflexive if and only if every nonempty, closed, convex set $C \subseteq X$ is Chebyshev.*

More on proximinal and Chebyshev sets can be found in Vlasov [322, 323].

(5.5) The duality between the geometric properties of ∂B_1 and the differentiability properties of the norm functional are discussed in the books of Deville–Godefroy–Zizler [88] and Fabian et al. [117].

A complement to Theorem 5.5.13 is the following result due to Leach–Whitfield [204].

Proposition 5.8.7. *If X is a separable Banach space such that X^* is not separable, then X admits an equivalent norm, which is nowhere Fréchet differentiable.*

The next renorming theorem is useful in the theory of monotone operators and is known as "Troyanski's Renorming Theorem".

Theorem 5.8.8 (Troyanski's Renorming Theorem). *Every reflexive Banach space can be given an equivalent norm so that X and X^* are both locally uniformly convex and have Fréchet differentiable norms.*

(5.6) Multifunctions are a useful tool in many applications such as optimization, optimal control, calculus of variations, operations research, game theory, mathematical economics, and others. Their continuity and measurability properties are discussed in the books of Aliprantis–Border [7], Aubin–Frankowska [20], Castaing–Valadier [71], Denkowski–Migórski–Papageorgiou [87], Hu–Papageorgiou [173], Klein–Thompson [194], and Papageorgiou–Kyritsi-Yiallourou [256]. Theorem 5.6.22 is due to Michael [232]. Theorem 5.6.39 is very useful in many different situations. It illustrates the power of the notion of decomposability and was first proven by Rockafellar [277]. Our formulation and the proof of the result are based on Hiai–Umegaki [161]. Fryszkowski [132], Hiai–Umegaki [161], Hu–Papageorgiou [173], and Olech [254] have detailed studies of decomposable sets and of their applications. Theorem 5.6.43 has a converse due to Klei [193].

Theorem 5.8.9. *If $F\colon \Omega \to P_f(X)$ is graph measurable, integrably bounded, and $S_F^1 \subseteq L^1(\Omega, X)$ is w-compact and convex, then $F(w) \in P_{wkc}(X)$ for μ-a. a. $w \in \Omega$.*

Integrands and integral functionals can be found in Buttazzo [65]. Theorems 5.6.51 and 5.6.54 are due to Rockafellar [277] and Levin [215] while Theorem 5.6.55 is due to Ioffe [176].

Another related result is given in the next proposition; see Ekeland–Temam [114] where $X = \mathbb{R}$.

Theorem 5.8.10. *If μ is nonatomic, $f\colon \Omega \times X \to \overline{\mathbb{R}} = \mathbb{R} \cup \{+\infty\}$ is a normal integrand, there exists $u_0 \in L^1(\Omega, X)$ such that $I_f(u_0) < +\infty$ and I_f is weakly lower semicontinuous on $L^1(\Omega, X)$, then $f(z, \cdot)$ is convex for all $w \in \Omega$.*

(5.7) Theorem 5.7.3 has the following remarkable extension; see Benyamini–Lindenstrauss [33, Section 1.2]

Theorem 5.8.11. *If H_1 and H_2 are Hilbert spaces, $A \subseteq H_1$, and $f\colon A \to H_2$ is k-Lipschitz, then there exists a k-Lipschitz function $\hat{f}\colon H_1 \to H_2$ such that $\hat{f}|_A = f$.*

For locally Lipschitz functions we have the so-called "Rademacher Theorem"; see Evans–Gariepy [116, p. 81].

Theorem 5.8.12 (Rademacher Theorem). *If $f: \mathbb{R}^N \to \mathbb{R}^m$ is locally Lipschitz, then f is differentiable λ^N-a. e., where λ^N is the Lebesgue measure on $\mathbb{R}^N$.*

Using this theorem, when X is finite dimensional, we can state a definition of the generalized subdifferential, which is more intuitive and geometric than the one in terms of the generalized directional derivative; see Definition 5.7.21 and Clarke [74, p. 63].

Proposition 5.8.13. *If $f: \mathbb{R}^N \to \mathbb{R}$ is locally Lipschitz and $D \subseteq \mathbb{R}^N$ is any Lebesgue-null set, then*

$$\partial f(x) = \overline{\operatorname{conv}}\left\{\lim_{n\to\infty} \nabla f(x_n): x_n \to x, x_n \in D \cup E_f^c\right\} \quad \textit{for every } x \in \mathbb{R}^N ,$$

with E_f being the set of points of differentiability of f. Note that Theorem 5.8.12 *implies that $\lambda^N(E_f^c) = 0$.*

The subdifferential theory for locally Lipschitz functions is due to Clarke [74] and the Mean Value Theorem (see Theorem 5.7.36) is due to Lebourg [209]. The Rademacher Theorem can be extended to functions between Banach spaces. The problem that we face when dealing with such a generalization is that we do not have a natural choice of measure as with the Lebesgue measure λ^N on $\mathbb{R}^N$. So, we need an alternative way to come up with negligible sets.

Definition 5.8.14. (a) A **topological group** is a group G together with a Hausdorff topology, which is compatible with the group structure, that is, $(x,y) \to x \cdot y$ and $x \to x^{-1}$ are both continuous. We say that G is an **Abelian topological group** if G is an Abelian group. Then the group operation is denoted by "+".
(b) A Borel set $A \subseteq G$ is said to be **Haar-null** if there is a probability measure μ on G such that $\chi_A * \mu = 0$, that is,

$$\int_G \chi_A(x+y)\mu(\mathrm{d}x) = 0 \quad \text{for all } y \in G .$$

Remark 5.8.15. So, $A \subseteq G$ is Haar-null if and only if there is a probability measure μ such that $\mu(A+y) = 0$ for all $y \in G$. We call μ the **test measure for** A. If G is locally compact then there is a unique translation invariant probability measure μ and the Haar-null sets are the μ-null sets.

The extension of the Rademacher Theorem is the following one.

Theorem 5.8.16. *If X is a separable Banach space, $U \subseteq X$ is open, and Y is a Banach space with the* RNP *and $f: U \to X$ is Lipschitz continuous, then f is Gateaux differentiable on a set D_f with $X \setminus D_f$ being Haar-null in X.*

For details on these and related issues we refer to Christensen [73] and Benyamini–Lindenstrauss [33].

Problems

Problem 5.1. Let X be a Banach space and let $f: X \to \overline{\mathbb{R}} = \mathbb{R} \cup \{+\infty\}$ be a proper, convex function. Show that $L_\lambda = \{x \in X: f(x) \le \lambda\}$ is bounded for every $\lambda \in \mathbb{R}$ if and only if $\liminf_{\|x\| \to \infty} f(x)/\|x\| > 0$.

Problem 5.2. Let X be a Banach space and let $f: X \to \overline{\mathbb{R}} = \mathbb{R} \cup \{+\infty\}$ be a proper, convex function. Show that f is Lipschitz on bounded subsets of X if and only if ∂f maps bounded sets in X into nonempty, bounded sets in X^*.

Problem 5.3. Let X be a Banach space and let $f: X \to \mathbb{R}$ be a convex function satisfying $f(u) \le k\|u\|$ for all $u \in X$. Show that f is k-Lipschitz.

Problem 5.4. Show that on any infinite dimensional Banach space X there is a convex function $f: X \to \mathbb{R}$ such that f is Gateaux differentiable at $u = 0$, f is lower semicontinuous at $u = 0$, but f is discontinuous at $u = 0$.

Problem 5.5. Let X be a Banach space and let $f: X \to \overline{\mathbb{R}} = \mathbb{R} \cup \{+\infty\}$ be a convex function that is Gateaux differentiable at x_0. Show that $f'(x_0) \in \partial f(x_0)$.

Problem 5.6. Let X be a separable, infinite dimensional Banach space. Show that there is a lower semicontinuous, convex function $f: X \to \overline{\mathbb{R}} = \mathbb{R} \cup \{+\infty\}$ such that $\partial f(0)$ is a singleton but f is not Gateaux differentiable at $u = 0$.

Problem 5.7. Let K be a compact topological space, $v_0 \in C(K)$ and for every $u \in C(K)$ let

$$\begin{aligned} L^+(u) &= \{x \in K: u(x) - v_0(x) = \|u - v_0\|_\infty\}, \\ L^-(u) &= \{x \in K: u(x) - v_0(x) = -\|u - v_0\|_\infty\}, \\ L(u) &= L^+(u) \cup L^-(u). \end{aligned}$$

Consider the convex function $\varphi_0: C(K) \to \mathbb{R}$ defined by $\varphi_0(u) = \|u - v_0\|_\infty$. Show that

$$\begin{aligned} \partial\varphi_0(u) = \{\mu \in \mathrm{ca}_r(\mathcal{B}(X)): \|\mu\| = 1,\ \mu^+ \text{ (resp. } \mu^-) \text{ is concentrated on} \\ L^+(u) \text{ (resp. on } L^-(u))\} \end{aligned}$$

for all $u \in C(K)$.

Problem 5.8. Let (X, d) be a compact metric space and let $\xi: C(X) \to \mathbb{R}$ be defined by $\xi(u) = \max\{u(x): x \in X\}$. Show that ξ is continuous, convex, and $\mu \in \partial\xi(u)$ if and only if $\langle \mu, \mathrm{id}_X \rangle = 1$, $\operatorname{supp} \mu \subseteq \{x \in X: u(x) = \xi(u)\}$.

Problem 5.9. Find the subdifferential of the norm of $L^1([0,1], \mathbb{R}^N)$.

Problem 5.10. Let $f: \mathbb{R}^N \to \mathbb{R}$ be a convex function. Show that ∂f maps bounded sets to bounded sets.

Problem 5.11. Let X be a Banach space and let $f: X \to \overline{\mathbb{R}} = \mathbb{R} \cup \{+\infty\}$ be a function which is proper, convex, and continuous at $x_0 \in \operatorname{int} \operatorname{dom} f$. Show that ∂f is locally bounded at x_0, that is, there exist $\eta, r > 0$ such that $\|x^*\|_* \leq \eta$ for all $x^* \in \partial f(x)$ and for all $x \in \overline{B}_r(x_0)$.

Problem 5.12. Let X be a Banach space and let $f: X \to \overline{\mathbb{R}} = \mathbb{R} \cup \{+\infty\}$ be a proper, convex function that is continuous on $\operatorname{int} \operatorname{dom} f \neq \emptyset$. Show that the subdifferential multifunction $\partial f: \operatorname{int} \operatorname{dom} f \to 2^{X^*} \setminus \{\emptyset\}$ is usc from X with the norm topology into X^* with the w*-topology denoted by $X^*_{w^*}$.

Problem 5.13. Let X be a Banach space and let $f: X \to \overline{\mathbb{R}} = \mathbb{R} \cup \{+\infty\}$ be a proper, convex function that is continuous on $\operatorname{int} \operatorname{dom} f \neq \emptyset$ and Fréchet differentiable at $x_0 \in \operatorname{int} \operatorname{dom} f$. Show that the subdifferential multifunction $\partial f: \operatorname{int} \operatorname{dom} f \to 2^{X^*} \setminus \{\emptyset\}$ is norm-to-norm upper semicontinuous at x_0.

Problem 5.14. Let X, Y be locally convex spaces, $\xi: X \times Y \to \overline{\mathbb{R}} = \mathbb{R} \cup \{+\infty\}$ is convex and $m: Y \to \overline{\mathbb{R}} = \mathbb{R} \cup \{+\infty\}$ is defined by $m(y) = \inf[\xi(x, y): x \in X]$. Show that m is convex as well.

Problem 5.15. Let $f: \mathbb{R} \to \mathbb{R}$ be defined by $f(x) = |x|$ for all $x \in \mathbb{R}$. For a given $\varepsilon > 0$ find $\partial_\varepsilon f(x)$.

Problem 5.16. Let X be a reflexive and strictly convex Banach space and let $f \in \Gamma_0(X)$. For $\lambda > 0$ we define

$$f_\lambda(x) = \left(f \oplus \frac{1}{2\lambda} \|\cdot\|^2\right)(x) \quad \text{for all } x \in X .$$

Show that f_λ is continuous, convex, and the infimal convolution is exact at a unique point $\hat{x}_\lambda \in X$.

Problem 5.17. Suppose that everything is as in Problem 5.16 with a Hilbert space $X = H$ and let $J_\lambda: H \to H$ be the map defined by $J_\lambda(x) = \hat{x}_\lambda$. Show that I_λ is nonexpansive, that is, 1-Lipschitz.

Problem 5.18. Suppose that everything is as in Problem 5.17. Show that for every $x \in H$ we have $J_\lambda(x) \to x$ in H as $\lambda \to 0^+$ and $f = \sup_{\lambda > 0} f_\lambda$.

Problem 5.19. Let X be a Banach space and $f \in \Gamma_0(X)$. Show that

$$\liminf_{\|x\| \to \infty} \frac{f(x)}{\|x\|} = \inf[f^\infty(x): \|x\| = 1] .$$

Problem 5.20. Let X be a Banach space and let $C \subseteq X$ be a nonempty, closed, and convex set. We consider $d_C(x) = d(x, C) = \inf[\|x - c\|: c \in C]$. Show that for every $x \notin C$ we have $\|x^*\|_* = 1$ for all $x^* \in \partial d_C(x)$.

Problem 5.21. Let X be a Banach space, $U \subseteq X$ be open, and let $f: U \to \mathbb{R}$ be a continuous and convex function. Show that f is Gateaux differentiable at $x \in U$ if and only if there is a selection s of ∂f that is norm-to-weak* continuous.

Problem 5.22. Let $\hat{x} = (x_n) \in l^\infty$ and let $p(x) = \limsup_{n\to\infty} |x_n|$. Show that p is continuous and convex, but nowhere Gateaux differentiable.

Problem 5.23. Let X be a Banach space, $C \subseteq X$ be nonempty, U be open containing C, and $f: U \to \mathbb{R}$ be a locally k_0-Lipschitz function. Suppose that x_0 is a minimizer of f on C. Show that x_0 is also a minimizer of $f + \lambda d_C$ on U for all $\lambda \geq k_0$, whereby $d_C(x) = d(x, C)$.

Problem 5.24. Let X be a Banach space, $C \subseteq X$ nonempty, U an open set containing C, and $f: U \to \mathbb{R}$ a locally k-Lipschitz function. Suppose that x_0 is a local minimizer of f on C. Show that $0 \leq f^\circ(x_0; h)$ for all $h \in T_C'(x_0)$.

Problem 5.25. Let X be a reflexive Banach space that is continuously and densely embedded into a Hilbert space. Suppose that $f \in \Gamma_0(X)$ and $c\|u\|_X^2 \leq f(u)$ for all $u \in X$. Then define the map $\hat{\varphi}: H \to \overline{\mathbb{R}} = \mathbb{R} \cup \{+\infty\}$ by

$$\hat{\varphi}(u) = \begin{cases} \varphi(u) & \text{if } u \in X\,, \\ +\infty & \text{if } u \in H \setminus X\,. \end{cases}$$

Show that $\hat{\varphi} \in \Gamma_0(X)$.

Problem 5.26. Let X be a Banach space and let $f \in \Gamma_0(X)$. Show that f is k-Lipschitz if and only if $\operatorname{dom} f^* \subseteq k\overline{B}_1^*$, where $\overline{B}_1^* = \{x^* \in X^*: \|x^*\|_* \leq 1\}$.

Problem 5.27. Let X be a Banach space, $U \subseteq X$ a nonempty, convex set, and $f: U \to \mathbb{R}$ a Gateaux differentiable function such that

$$\langle f'(u) - f'(v), u - v\rangle \geq 0 \quad \text{for all } u, v \in U\,.$$

Show that φ is convex.

Problem 5.28. Let X be a Banach space, $C \subseteq X$ a nonempty set, and $f: C \to \mathbb{R}$ a k-Lipschitz function. We define $\hat{f}_{C,k}(x) = \inf[f(u) + kd(u,x): u \in C]$ for all $x \in X$. Show that x_0 is a global minimizer of f on C if and only if x_0 is a global minimizer of $\hat{f}_{C,k}$ on X.

Problem 5.29. Let X be a Banach space and let $f: X \to \mathbb{R}$ be a continuous, convex function. Show that $0 \in \operatorname{int} \partial f(u)$ if and only if there exists $\varepsilon > 0$ such that $f(u) + \varepsilon\|h\| \leq f(u+h)$ for all $h \in X$.

Problem 5.30. Let $f_n: \mathbb{R}^N \to \mathbb{R}$ with $n \in \mathbb{N}$ be a sequence of convex and differentiable functions such that $f_n(u) \to f(u)$ for all $u \in \mathbb{R}^n$ with $f: \mathbb{R}^N \to \mathbb{R}$ being convex and differentiable as well. Show that $f_n' \to f'$ uniformly on compact subsets of $\mathbb{R}^N$.

Problem 5.31. Let X be a Banach space and let $f: X \to \overline{\mathbb{R}} = \mathbb{R} \cup \{+\infty\}$ be a proper, convex function. Show that $f(u) \geq c\|u\| + \beta$ with $c > 0$, $\beta \in \mathbb{R}$ if and only if $\varphi^*|_{c\overline{B}_1^*} \leq -\beta$, where $\overline{B}_1^* = \{x^* \in X^* : \|x^*\|_* \leq 1\}$.

Problem 5.32. Let X be a Banach space and let $f: X \to \overline{\mathbb{R}} = \mathbb{R} \cup \{+\infty\}$ be a proper, convex function that is lower semicontinuous at some point of its domain. Show that f is coercive, that is, $f(u) \to +\infty$ as $\|u\| \to \infty$, if and only if f^* is continuous at the origin.

Problem 5.33. Let X be a Banach space and $f \in \Gamma_0(X)$. Show that f is continuous at the origin if and only if f^* has w*-compact sublevel sets $I_\lambda = \{x \in X : f(x) \leq \lambda\}$ with $\lambda \in \mathbb{R}$.

Problem 5.34. Let X be a Banach space and let $f: X \to \overline{\mathbb{R}} = \mathbb{R} \cup \{+\infty\}$ be a proper function for which $f'_+(x; \cdot)$ exists for all $x \in \operatorname{dom} f$. Assume that

$$f'_+(x; x-u) - f'_+(u; x-u) \geq 0 \quad \text{for all } x, u \in \operatorname{dom} f .$$

Show that f is convex.

Problem 5.35. Let X, Y be Banach spaces, $A \in L(X, Y)$, $f \in \Gamma_0(Y)$, $x_0 \in A^{-1}(\operatorname{dom} f)$, and $\varepsilon > 0$. Show that $\partial_\varepsilon (f \circ A)(x_0) = \overline{A^*(\partial_\varepsilon f(A(x_0)))}^{w^*}$.

Problem 5.36. Let X, Y be Banach spaces, $A \in L(X, Y)$, $f \in \Gamma_0(X)$, and $x_0 \in A^{-1}(\operatorname{dom} f)$. Show that

(a) $\partial(f \circ A)(x_0) = \overline{\bigcap_{\varepsilon > 0} A^*(\partial_\varepsilon f(A(u_0)))}^{w^*}$;

(b) if X is reflexive, then $\partial(f \circ A)(x_0) = \overline{\bigcap_{\varepsilon > 0} A^*(\partial_\varepsilon f(A(x_0)))}$.

Problem 5.37. Let X be a Banach space, $f, h \in \Gamma_0(X)$, and $x_0 \in \operatorname{dom} f \cap \operatorname{dom} h$. Show that

(a) $\partial(f + h)(x_0) = \overline{\bigcap_{\varepsilon > 0} [\partial_\varepsilon f(x_0) + \partial_\varepsilon h(x_0)]}^{w^*}$;

(b) if X is reflexive, then $\partial(f + h)(x_0) = \overline{\bigcap_{\varepsilon > 0} [\partial_\varepsilon f(x_0) + \partial_\varepsilon h(x_0)]}$.

Problem 5.38. Find the conjugate function of $f: \mathbb{R} \to \mathbb{R}$, when

(a) $f(x) = |x|^p / p$ with $p > 1$;

(b) $f(x) = e^x$;

(c) $f(x) = -\sqrt{|x|} + i_{\mathbb{R}_+}$;

(d) $f(x) = \sqrt{1 + x^2}$.

Problem 5.39. Let H be a Hilbert space and let $C \subseteq H$ be a nonempty, closed, and convex set. We set $f(x) = d(x, C) = \inf[\|x - u\| : u \in C]$. Show that if $x \notin C$, then f is differentiable at x and $\nabla f(x) = (x - p_C(x)) / \|x - p_C(x)\|$.

Problem 5.40. Let X be a Banach space, let $f: X \to \mathbb{R}$ be a locally Lipschitz function and let $K \subseteq X$ be a closed, convex cone and assume that $f(x) \leq f(x + y)$ for all $y \in K$. Show that

$$f^\circ(x;h) \le \sigma(h;-K^\circ) \quad \text{for all } h \in X\,,$$

where $K^\circ = K^- = \{x^* \in X^* : \langle x^*, v\rangle \le 0 \text{ for all } v \in K\}$.

Problem 5.41. Let X be a paracompact space, Y a Banach space, $C \in P_f(X)$, and $F: X \to P_{f_c}(Y)$ a lsc multifunction. Show that any continuous selection of $F|_C$ can be extended on all X to a continuous selection of F.

Problem 5.42. Let X, Y be two Banach spaces and let $A \in L(X,Y)$ be surjective. Show that there exists a continuous map $\hat{f}: Y \to X$ such that $A(f(y)) = y$ for all $y \in Y$.

Problem 5.43. Let X be a compact topological space, Y a locally convex space, and $F: X \to 2^Y \setminus \{\emptyset\}$ a multifunction with convex values. Suppose that $F^-(\{y\}) = \{x \in X : y \in F(x)\}$ is open for every $y \in Y$. Show that F admits a continuous selection.

Problem 5.44. Let X be a compact topological space and let $F: X \to P_f(X)$ be an usc multifunction. Show that there exists a nonempty, closed subset $C \subseteq X$ such that $F(C) = C$.

Problem 5.45. Let (Ω, Σ, μ) be a σ-finite, nonatomic measure space, X a Banach space, and $D \subseteq L^p(\Omega, X)$ with $1 \le p < \infty$ be nonempty, w-closed, and decomposable. Show that D is convex.

Problem 5.46. Let (Ω, Σ, μ) be a σ-finite, nonatomic measure space, X a separable Banach space, and $F: \Omega \to 2^X \setminus \{\emptyset\}$ a graph measurable multifunction such that $S_F^p \neq \emptyset$ with $1 \le p < +\infty$. Show that $\overline{S_F^p}^{w^*} = S^p_{\operatorname{conv} F}$.

Problem 5.47. Let $\Omega \subseteq \mathbb{R}^N$ be a bounded open set with C^1-boundary $\partial\Omega$ and $1 < p < \infty$. Consider the functional $\varphi: L^p(\Omega) \to \overline{\mathbb{R}} = \mathbb{R} \cup \{+\infty\}$ defined by

$$\varphi(u) = \begin{cases} \|\nabla u\|_p^p & \text{if } u \in W^{1,p}(\Omega)\,, \\ +\infty & \text{otherwise}\,. \end{cases}$$

Show that φ is lower semicontinuous.

Problem 5.48. Let $\Omega \subseteq \mathbb{R}^N$ be a bounded open set with C^1-boundary $\partial\Omega$, $1 < p < \infty$, $f: \mathbb{R} \to \mathbb{R}$ a continuous, convex function such that $|f(x)| \le c(1 + |x|^{p-1})$ for all $x \in \mathbb{R}$ and with some $c > 0$, and let $\varphi: W^{1,p}(\Omega) \to \mathbb{R}$ be the integral functional defined by

$$\varphi(u) = \int_{\partial\Omega} f(\gamma_0(u(z)))\, d\sigma(z)$$

with γ_0 being the trace operator and σ being the surface measure on $\partial\Omega$. Show that φ is continuous, convex, and $u^* \in \partial\varphi(u)$ if and only if there exists $h^* \in L^{p'}(\partial\Omega)$ with

$1/p + 1/p' = 1$ such that $h^*(z) \in \partial f(\gamma_0(u)(z))$ σ-a. e. on $\partial\Omega$ and

$$\langle u^*, u\rangle = \int_{\partial\Omega} h^*(z)\gamma_0(u)(z)\,d\sigma \quad \text{for all } u \in W^{1,p}(\Omega)\,.$$

Problem 5.49. Show that l^1 is not an Asplund space.

Problem 5.50. Suppose that X is a reflexive Banach space, $f \in \Gamma_0(X)$, and $f(x) - \langle x^*, x\rangle \to +\infty$ as $\|x\| \to \infty$ for all $x^* \in X^*$. Show that $\partial\varphi(x) = X^*$, that is, the subdifferential of f is surjective.

Problem 5.51. Let (Ω, Σ, μ) be a finite complete measure space and let $\varphi\colon \mathbb{R} \to \overline{\mathbb{R}} = \mathbb{R} \cup \{+\infty\}$ be a proper, convex function such that $\lim_{x\to\pm\infty} f(x)/|x| = +\infty$. Consider the integral functional

$$I_f(u) = \int_\Omega f(u(w))\,d\mu \quad \text{for all } u \in L^1(\Omega)\,.$$

Show that I_f has weakly compact sublevel sets.

6 Nonlinear Analysis

In this chapter we deal with some topics of "Nonlinear Analysis", which are used in the study of boundary value problems and in problems of calculus of variations, optimization, and optimal control. We start with an examination of operators of monotone type from a Banach space X into its dual space X^*. These maps are a natural generalization of increasing functions $f: \mathbb{R} \to \mathbb{R}$, but their definition does not require any order structure on X. Operators of monotone type were introduced to overcome the limitations of compact operators and they exhibit remarkable surjectivity properties. Maximal monotone and pseudomonotone maps are the two subclasses of operators of monotone type that have the strongest properties. "Degree Theory" emerged as an effective tool in the study of the solution set of an abstract equation of the form $\varphi(u) = y_0$. Degree theory provides information about the existence of solutions, their multiplicity, and their nature. We develop both the finite dimensional theory in terms of the Brouwer degree and the infinite dimensional theory in the form of the Leray–Schauder degree. Degree theory is closely related to fixed point theory. We present both the "Metric Fixed Point Theory" and the "Topological Fixed Point Theory". Next we present some important variational principles that have interesting applications. We start with the Lax–Milgram Theorem and continue with the Ekeland Variational Principle, which has important applications. We also show that it is equivalent to some other results of nonlinear analysis. Finally we deal with a mode of convergence of functions known as Γ-convergence or epigraphical convergence, which is designed in such a way that it is the suitable notion for examining the stability/sensitivity properties of variational problems.

6.1 Operators of Monotone Type

Monotone operators were introduced in the early 1960s in order to provide a framework of analysis broader than that of compact operators. Their introduction marked the advent of nonlinear functional analysis. The starting point was the observation that the Gateaux derivative of a convex function is monotone. In fact the theory developed in parallel with the theory of set-valued maps called multifunctions and the two theories interacted strongly.

The mathematical setting is the following. Let X be a Banach space and let X^* be its topological dual space. By $\langle \cdot, \cdot \rangle$ we denote the duality brackets for the pair (X^*, X). Consider a general multivalued map $A: X \to 2^{X^*}$. We introduce the following sets:

$$\begin{aligned} D(A) &= \{u \in X : A(u) \neq \emptyset\}, \\ \operatorname{Gr} A &= \{(u, u^*) \in X \times X^* : u^* \in A(u)\}, \\ A^{-1}(u^*) &= \{u \in X : (u, u^*) \in \operatorname{Gr} A\} \quad \text{for all } u \in D(A). \end{aligned}$$

Here, we call $D(A)$ the domain of A, $\operatorname{Gr} A$ the graph of A, and A^{-1} the inverse operator. Note that A^{-1} is always defined and is a multivalued map.

https://doi.org/10.1515/9783111286952-006

Definition 6.1.1. Let $A: X \to 2^{X^*}$. We define the following notions:

(a) A is **monotone** if

$$\langle u^* - v^*, u - v \rangle \geq 0$$

for all $(u, u^*), (v, v^*) \in \operatorname{Gr} A$;

(b) A is **strictly monotone** if it is monotone and

$$\langle u^* - v^*, u - v \rangle > 0$$

for all $u, v \in D(A)$ with $u \neq v$ and $u^* \in A(u)$ as well as $v^* \in A(v)$;

(c) A is **strongly monotone** if

$$\langle u^* - v^*, u - v \rangle \geq c\|u - v\|^2$$

for all $(u, u^*), (v, v^*) \in \operatorname{Gr} A$ and some $c > 0$;

(d) A is **uniformly monotone** if there exists a function $\vartheta: \mathbb{R}_+ \to \mathbb{R}_+$ which is continuous, strictly increasing, $\vartheta(0) = 0$, $\vartheta(r) \to +\infty$ as $r \to +\infty$ and

$$\langle u^* - v^*, u - v \rangle \geq \vartheta(\|u - v\|)\|u - v\|$$

for all $(u, u^*), (v, v^*) \in \operatorname{Gr} A$;

(e) A is **coercive** if $D(A)$ is bounded or $D(A)$ is unbounded and

$$\inf[\|u^*\|_*: u^* \in A(u)] \to +\infty \quad \text{as } \|u\| \to +\infty \text{ with } u \in D(A)\,.$$

A is **strongly coercive** if $D(A)$ is bounded or $D(A)$ is unbounded and

$$\frac{\inf[\langle u^*, u\rangle: u^* \in A(u)]}{\|u\|} \to +\infty \quad \text{as } \|u\| \to +\infty \text{ with } u \in D(A)\,.$$

Remark 6.1.2. From the definitions above it is clear that we always have the following implications:

$$\begin{aligned} \text{strongly monotone} \quad &\Longrightarrow \quad \text{uniformly monotone} \\ &\Longrightarrow \quad \text{strictly monotone} \\ &\Longrightarrow \quad \text{monotone} \end{aligned}$$

and strongly coercive implies coercive.

Proposition 6.1.3. *If H is a Hilbert space identified with its dual, that is, $H = H^*$, and $A: H \to 2^H$, then the following statements are equivalent:*

(a) *A is monotone;*

(b) *$\|u - v + t(u^* - v^*)\| \geq \|u - v\|$ for all $(u, u^*), (v, v^*) \in \operatorname{Gr} A$ and for all $t \in [0, 1]$;*

(c) *$\|v - u^*\|^2 + \|u - v^*\|^2 \geq \|u - u^*\|^2 + \|v - v^*\|^2$ for all $(u, u^*), (v, v^*) \in \operatorname{Gr} A$.*

Proof. (a) $\Longleftrightarrow$ (b): We have

$$\begin{aligned}&\|u - v + t(u^* - v^*)\|^2 - \|u - v\|^2 \\ &= (u - v + t(u^* - v^*), u - v + t(u^* - v^*)) - (u - v, u - v) \\ &= t[t\|u^* - v^*\|^2 + 2(u - v, u^* - v^*)] .\end{aligned} \tag{6.1.1}$$

From (6.1.1) we see that A is monotone if and only if (b) holds.

(c) $\Longleftrightarrow$ (a): We have

$$\begin{aligned}\|v - u^*\|^2 &= \|v\|^2 - 2(v, u^*) + \|u^*\|^2 , \\ \|u - v^*\|^2 &= \|u\|^2 - 2(u, v^*) + \|v^*\|^2\end{aligned} \tag{6.1.2}$$

and

$$\begin{aligned}\|u - u^*\|^2 &= \|u\|^2 - 2(u, u^*) + \|u^*\|^2 , \\ \|v - v^*\|^2 &= \|v\|^2 - 2(v, v^*) + \|v^*\|^2 .\end{aligned} \tag{6.1.3}$$

From (6.1.2) and (6.1.3) we easily conclude the equivalence of the two statements. □

Definition 6.1.4. A monotone map $A: X \to 2^{X^*}$ is said to be **maximal monotone** if

$$\langle u^* - v^*, u - v\rangle \geq 0 \quad \text{for all } (v, v^*) \in \operatorname{Gr} A$$

implies $(u, u^*) \in \operatorname{Gr} A$.

Remark 6.1.5. According to this definition, the map $A: X \to 2^{X^*}$ is maximal monotone if and only if its graph $\operatorname{Gr} A$ is not properly contained in the graph of another monotone operator, that is, $\operatorname{Gr} A$ is maximal with respect to inclusion among all monotone graphs. An increasing function $f: \mathbb{R} \to \mathbb{R}$ is monotone, but need not be maximal monotone since we can have a monotone extension by filling in the jumps at the discontinuity points. Of course if f is continuous, then f is maximal monotone. From this example it is clear that in order to have a complete theory we need to consider multivalued maps.

Proposition 6.1.6. *Every monotone map $A: X \to 2^{X^*}$ admits a maximal monotone extension, that is, there exists a maximal monotone map $\hat{A}: X \to 2^{X^*}$ such that* $\operatorname{Gr} A \subseteq \operatorname{Gr} \hat{A}$.

Proof. Clearly we may assume that $\operatorname{Gr} A \neq \emptyset$. Let

$$\mathcal{D} = \{K: X \to 2^{X^*} : K \text{ is monotone and } \operatorname{Gr} A \subseteq \operatorname{Gr} K\} .$$

Then $\mathcal{D}$ is nonempty and partially ordered by

$$K_1 \prec K_2 \quad \text{with } K_1, K_2 \in \mathcal{D} \quad \text{if and only if} \quad \operatorname{Gr} K_1 \subseteq \operatorname{Gr} K_2 .$$

Let $\mathcal{C}$ be a chain in $\mathcal{D}$, that is, a linearly ordered subset of $\mathcal{D}$. We consider the operator $S: X \to 2^{X^*}$ such that $\operatorname{Gr} S = \bigcup_{C \in \mathcal{C}} \operatorname{Gr} C$. This is an upper bound of $\mathcal{C}$. Invoking Zorn's Lemma, there exists a maximal element $\hat{A} \in \mathcal{D}$. Then $\hat{A}$ is the desired maximal monotone extension of A. □

The next proposition is an immediate consequence of Definition 6.1.4.

Proposition 6.1.7. *$A: X \to 2^{X^*}$ is maximal monotone if and only if $A^{-1}: X^* \to 2^X$ is maximal monotone.*

Proposition 6.1.8. *If $A: X \to 2^{X^*}$ is maximal monotone, then $A(u)$ is nonempty, convex, and w^*-closed for every $u \in D(A)$.*

Proof. Since $u \in D(A)$, we obtain $A(u) \neq \emptyset$. Let $u^*, v^* \in A(u)$ and define

$$y_t^* = (1-t)u^* + tv^* \quad \text{with } t \in [0,1] .$$

Given any $(x, x^*) \in \operatorname{Gr} A$ one has

$$\langle y_t^* - x^*, u - x\rangle = (1-t)\langle u^* - x^*, u - x\rangle + t\langle v^* - x^*, u - x\rangle \geq 0 .$$

Hence, $y_t^* \in A(u)$ since A is maximal monotone. This proves that A has convex values. Finally, let $\{u_n^*\}_{n\geq 1} \subseteq A(u)$ and assume that $u_n^* \xrightarrow{\mathrm{w}^*} u^*$ in X^*. Then

$$0 \leq \langle u_n^* - x^*, u - x\rangle \quad \text{for all } (x, x^*) \in \operatorname{Gr} A \text{ and for all } n \in \mathbb{N} .$$

This shows that $0 \leq \langle u^* - x^*, u - x\rangle$ for all $(x, x^*) \in \operatorname{Gr} A$. Thus, $u^* \in A(u)$ since A is maximal monotone. This proves that A has w^*-closed values. □

Definition 6.1.9. (a) A multivalued map $A: X \to 2^{X^*}$ is said to be **locally bounded at** $u \in D(A)$ if there exists $U \in \mathcal{N}(u)$ such that $A(U) \subseteq X^*$ is bounded. We say that A is **locally bounded** if it is locally bounded at every $u \in D(A)$.
(b) If $C \subseteq X$ and $u \in C$, then we say that u is an **absorbing point** of C if the set $C - u$ is absorbing, that is, $X = \bigcup_{t>0} t(C-u)$; see also Definition 3.1.55.

Remark 6.1.10. If $\operatorname{int} C \neq \emptyset$, then every $u \in \operatorname{int} C$ is an absorbing point of C. However, a set C can have absorbing points even if $\operatorname{int} C = \emptyset$. For example, let $C = \partial B_1 \cup \{0\}$ with $\partial B_1 = \{x \in X : \|x\| = 1\}$. Then $\operatorname{int} C = \emptyset$, but 0 is an absorbing point of C.

Proposition 6.1.11. *If $A: X \to 2^{X^*}$ is monotone and $v \in D(A)$ is an absorbing point of $D(A)$, then A is locally bounded at v.*

Proof. Let $v^* \in A(v)$ and let $A_1(u) = A(u+v) - v^*$. Evidently, A_1 is maximal monotone as well. So, without any loss of generality, we may assume that $v = 0$ and $(0,0) \in \operatorname{Gr} A$. We set

$$\varphi(u) = \sup[\langle y^*, u - y\rangle : y \in D(A), \|y\| \leq 1, y^* \in A(y)] \tag{6.1.4}$$

for all $u \in X$ and $L^1_\varphi = \{u \in X : \varphi(u) \le 1\}$. The function φ is lower semicontinuous and convex as the supremum of affine continuous functions. Hence, $L^1_\varphi \subseteq X$ is closed and convex. Since $(0,0) \in \operatorname{Gr} A$ it is clear from (6.1.4) that $\varphi \ge 0$. Moreover, if $(y, y^*) \in \operatorname{Gr} A$, then $0 \le \langle y^*, y\rangle$, recall that $(0,0) \in \operatorname{Gr} A$. Hence, $\varphi(0) = 0$ and so $0 \in L^1_\varphi$.

By hypothesis, $D(A) \subseteq X$ is absorbing. So, if $u \in X$, then there exists $t > 0$ such that $tu \in D(A)$, that is, $A(tu) \neq \emptyset$. Let $u^* \in A(tu)$. By the monotonicity of A we obtain

$$\langle y^*, tu - y\rangle \le \langle u^*, tu - y\rangle \quad \text{for all } (y, y^*) \in \operatorname{Gr} A\,.$$

This implies

$$\varphi(tu) \le \sup[\langle u^*, tu - y\rangle : y \in D(A), \|y\| \le 1] \le \langle u^*, tu\rangle + \|u^*\|_* < +\infty\,.$$

We choose $\vartheta \in (0,1)$ such that $\vartheta\varphi(tu) < 1$. Exploiting the convexity of φ, we get

$$\varphi(\vartheta tu) \le \vartheta\varphi(tu) + (1 - \vartheta)\varphi(0) = \vartheta\varphi(tu) < 1\,.$$

Hence, $\vartheta tu \in L^1_\varphi$ and this implies that L^1_φ is absorbing. Then $C = L^1_\varphi \cap (-L^1_\varphi)$ is nonempty since $0 \in C$, closed, convex, symmetric, and absorbing. Therefore, $C \in \mathcal{N}(0)$, that is, $0 \in \operatorname{int} L^1_\varphi$. We can find $\delta > 0$ such that $\varphi(u) \le 1$ for all $u \in X$ with $\|u\| \le 2\delta$. Hence

$$\langle y^*, u\rangle \le \langle y^*, y\rangle + 1$$

for all $y \in D(A)$ with $\|y\| \le 1$ and for all $y^* \in A(y)$ and with $\|u\| \le 2\delta$. This gives

$$2\delta\|y^*\|_* \le \|y^*\|_* + 1 \le \delta\|y^*\|_* + 1$$

for all $y \in D(A) \cap \overline{B}_\delta(0)$ and for all $y^* \in A(y)$. Hence, $\|y^*\|_* \le 1/\delta$ for all $y^* \in A(D(A) \cap \overline{B}_\delta(0))$. □

Proposition 6.1.12. *If $A : X \to 2^{X^*}$ is maximal monotone and $u \in \operatorname{int} D(A)$, then $A(u) \subseteq X^*$ is nonempty, convex, and w^*-compact.*

Proof. The proof follows by combining Propositions 6.1.8 and 6.1.11. □

Proposition 6.1.13. *If $A : X \to 2^{X^*}$ is maximal monotone and $\operatorname{int} D(A) \neq \emptyset$, then $A|_{\operatorname{int} D(A)}$ is usc from X with the norm topology into X^* with the w^*-topology denoted by $X^*_{\mathrm{w}^*}$.*

Proof. On account of Propositions 6.1.11, 6.1.12, and 5.6.11, it suffices to show that $\operatorname{Gr}(A|_{\operatorname{int} D(A)}) \subseteq X \times X^*_{\mathrm{w}^*}$ is closed. So, let $\{(u_\alpha, u^*_\alpha)\}_{\alpha \in I} \subseteq \operatorname{Gr}(A|_{\operatorname{int} D(A)})$ be a net such that $u_\alpha \to u \in \operatorname{int} D(A)$ in X and $u^*_\alpha \xrightarrow{\mathrm{w}^*} u^*$ in X^*. For all $(y, y^*) \in \operatorname{Gr} A$ and for all $\alpha \in I$ we obtain $0 \le \langle u^*_\alpha - y^*, u_\alpha - y\rangle$, which implies $0 \le \langle u^* - y^*, u - y\rangle$. Hence, $(u, u^*) \in \operatorname{Gr} A(|_{\operatorname{int} D(A)})$ since A is maximal monotone. Therefore, $A|_{\operatorname{int} D(A)}$ is usc from X into $X^*_{\mathrm{w}^*}$. □

From the proof above, we infer that the following result is true.

Proposition 6.1.14. *If $A: X \to 2^{X^*}$ is maximal monotone, then $\operatorname{Gr} A \subseteq X_w \times X^*$ and $\operatorname{Gr} A \subseteq X \times X^*_{w^*}$ are both closed, where X_w denotes the space X equipped with the weak topology and as we already mentioned, $X^*_{w^*}$ denotes the space X^* furnished with the weak* topology.*

Definition 6.1.15. Let V, Y be two Banach spaces and let $G: V \to 2^Y$ be a multifunction.
(a) We say that G is **demicontinuous** if it is usc from V with the norm topology into Y with the weak topology.
(b) We say that G is **hemicontinuous** if the multifunction $[0,1] \ni t \to G((1-t)v + tw)$ is usc from $[0,1]$ into Y with the weak topology for all $v, w \in V$.
(c) We say that G is **bounded** if it maps bounded sets in V into bounded sets in Y.

Remark 6.1.16. From the definitions above it is clear that demicontinuity implies hemicontinuity. If $A: X \to 2^{X^*}$ is monotone, hemicontinuous, and $\operatorname{int} D(A) \neq \emptyset$, then $A|_{\operatorname{int} D(A)}$ is demicontinuous. So, if $D(A) = X$, then monotonicity together with hemicontinuity imply demicontinuity.

Proposition 6.1.17. *If $A: X \to 2^{X^*}$ is monotone, hemicontinuous with $D(A) = X$, and $A(u) \subseteq X^*$ is convex and w^*-closed for every $u \in X$, then A is maximal monotone.*

Proof. From Proposition 6.1.6, we know that there exists a maximal monotone extension $\hat{A}: X \to 2^{X^*} \setminus \{\emptyset\}$ of A. We will show that $\hat{A} = A$. So, let $(u_0, u_0^*) \in \operatorname{Gr} \hat{A}$ and suppose that $u_0^* \notin A(u_0)$. Then, by the Strong Separation Theorem (see Theorem 3.1.60), there exists $v \in X \setminus \{\emptyset\}$ such that

$$\langle u^*, v\rangle < \langle u_0^*, v\rangle \quad \text{for all } u^* \in A(u_0)\,. \tag{6.1.5}$$

Let $\lambda > 0$ and set $u_\lambda = u_0 + \lambda v$. The monotonicity of $\hat{A}$ implies that $\lambda\langle u_\lambda^* - u_0^*, v\rangle \geq 0$ for all $u_\lambda^* \in A(u_\lambda)$. Hence,

$$\langle u_\lambda^* - u_0^*, v\rangle \geq 0 \quad \text{for all } u_\lambda^* \in A(u_\lambda)\,. \tag{6.1.6}$$

The hemicontinuity of A implies that $u_\lambda^* \xrightarrow{w^*} u^*$ in X^* with $u^* \in A(u_0)$. Therefore, due to (6.1.6), we see that $\langle u^* - u_0^*, v\rangle \geq 0$. But this last inequality contradicts (6.1.5). Thus, $A = \hat{A}$, and so A is maximal monotone. □

Suppose that $X = H$ is a Hilbert space with inner product $(\cdot,\cdot)$. We assume that $H = H^*$, which is possible according to the Riesz–Fréchet Representation Theorem; see Theorem 3.5.21. Clearly, $A \in L(H)$ is monotone if and only if $(A(u), u) \geq 0$ for all $u \in H$. In the literature, such operators are called **positive** and we write $A \geq 0$. In this way we can give a partial order "$\leq$" on $L(H)$ by setting $A \leq B$ if and only if $B - A \geq 0$. In a complex Hilbert space, every positive operator is self-adjoint. Indeed, if $(A(u), u) = (u, A(u))$ for all $u \in H$, then it follows, by the polarization identity (see Proposition 3.5.6 (b)) that $(A(u), v) = (u, A(v))$ for all $u, v \in H$. This is not true in real Hilbert spaces, since in that case it is not possible to recover $(u, A(v))$ by knowing $(u, A(u))$ for all $u \in H$. The following result is standard in operator theory and can be found in Rudin [284, Theorem 12.33, p. 314].

Proposition 6.1.18. *If $A \in L(H)$ and $A \geq 0$, then there exists a unique $B \in L(H)$ with $B \geq 0$ such that $B^2 = A$ and B commutes with every element in $L(H)$, which commutes with A. We write $B = A^{1/2}$ and call it the* ***square root of*** *A.*

Now we present examples of monotone and maximal monotone maps.

Example 6.1.19. (a) Let $f: X \to \overline{\mathbb{R}}$ be proper and convex. Then ∂f is monotone. In order to see this let $(u, u^*), (v, v^*) \in \operatorname{Gr} \partial f$. Then

$$\langle u^*, v - u\rangle + f(u) \leq f(v) \quad \text{and} \quad \langle v^*, u - v\rangle + f(v) \leq f(u) . \tag{6.1.7}$$

Adding the two inequalities in (6.1.7) yields $\langle u^* - v^*, u - v\rangle \geq 0$, which proves the monotonicity of ∂f. In fact, later we will show that if $f \in \Gamma_0(X)$, then ∂f is maximal monotone.

(b) Let $X = H$ be a Hilbert space and assume that $H = H^*$. Let $g: H \to H$ be a nonexpansive map, that is, $\|g(u) - g(v)\| \leq \|u - v\|$ for all $u, v \in H$. Then the map $A(u) = u + tg(u)$ is maximal monotone for any $t \in [-1, 1]$ and for all $u \in H$. Indeed, for all $u, v \in H$, we get

$$\begin{aligned}(A(u) - A(v), u - v) &= \|u - v\|^2 + t(g(u) - g(v), u - v)\\ &\geq \|u - v\|^2 - |t|\|g(u) - g(v)\|\|u - v\|\\ &\geq \|u - v\|^2[1 - |t|] \geq 0 .\end{aligned}$$

So, A is monotone, continuous, and thus maximal monotone; see Proposition 6.1.17.

(c) Let $T = [0, b]$, let (X, H, H^*) be an evolution triple (see Definition 4.2.39), and let

$$\begin{aligned}L_1(u) &= u' \quad \text{for all } u \in D(L_1) = \{u \in W_p(0, b): u(0) = 0\} ,\\ L_2(u) &= u' \quad \text{for all } u \in D(L_2) = \{u \in W_p(0, b): u(0) = u(b)\} .\end{aligned}$$

Then $L_k: D(L_k) \subseteq L^p(T, X) \to L^{p'}(T, X^*)$ with $1/p + 1/p' = 1$ is linear and from the integration by parts formula (see Corollary 4.2.47), we obtain

$$((L_k(u), u)) = \frac{1}{2}[|u(b)|^2 - |u(0)|^2] \geq 0 \quad \text{for all } u \in D(L_k) , \tag{6.1.8}$$

where $((\cdot, \cdot))$ denotes the duality brackets for the pair $(L^{p'}(T, X^*), L^p(T, X))$, that is

$$((g, h)) = \int_0^b \langle g(t), h(t)\rangle \, dt \quad \text{for all } h \in L^p(T, X), \text{ for all } g \in L^{p'}(T, X^*) .$$

So, L_k is monotone for $k = 1, 2$. Next we show the maximality of L_1 and L_2. To this end, let $(v, v^*) \in L^p(T, X) \times L^{p'}(T, X^*)$ and assume that

$$0 \leq ((v^* - L_k(u), v - u)) \quad \text{for all } u \in D(L_k)\,. \tag{6.1.9}$$

Choose $u = \vartheta x$ with $\vartheta \in C_c^\infty(0, b)$ and $x \in X$. Then $u' = \vartheta' x$ and $u \in D(L_k)$. Moreover, $((L_k(u), u)) = 0$; see (6.1.8). From (6.1.9) it follows that

$$0 \leq ((v^*, v)) - \int_0^b \langle \vartheta'(t)v(t) + \vartheta(t)v^*(t), x\rangle\, dt \quad \text{for all } x \in X\,. \tag{6.1.10}$$

Since

$$\int_0^b \langle \vartheta'(t)v(t) + \vartheta(t)v^*(t), x\rangle\, dt = \left\langle \int_0^b (\vartheta'(t)v(t) + \vartheta(t)v^*(t))\, dt, x \right\rangle$$

for all $x \in X$ (see Proposition 4.2.13), from (6.1.10) we infer that

$$\int_0^b (\vartheta'(t)v(t) + \vartheta(t)v^*(t))\, dt = 0 \quad \text{for all } \vartheta \in C_c^\infty(0, b)\,.$$

Hence, $v' = v^*$ by integration by parts. Therefore, because $v^* \in L^{p'}(T, X^*)$, we conclude that $v \in W_p(0, b)$. Using once again the integration by parts formula (see Corollary 4.2.47), we obtain

$$0 \leq ((v' - u', v - u)) = \frac{1}{2}[|v(b) - u(b)|^2 - |v(0) - u(0)|^2]\,. \tag{6.1.11}$$

For L_1 we choose $\{x_n\}_{n\geq 1} \subseteq X$ with $x_n \to 1/bv(b)$ in H as $n \to \infty$. We set $u(t) = tx_n$. Then $u \in D(L_1)$ and from (6.1.11) we get

$$0 \leq |v(b)|^2 - |v(0)|^2 + 2(v(0) - v(b), u(0)) \quad \text{for all } u \in D(L_2)\,.$$

Since u can be any constant function in X, it follows that $v(0) = v(b)$, that is, $v \in D(L_2)$ since $X \hookrightarrow H$ densely. In fact the maximal monotonicity of L_1 remains valid even if in the definition of $D(L_1)$, $u(0) = u_0 \in H$; see Roubíček [281, Lemma 8.93, p. 289].

(d) Let $X = H$ be a complex Hilbert space with inner product $(\cdot, \cdot)$ and assume that $H = H^*$. Suppose that $A, T \geq 0$ and $AT = TA$. Then $AT \geq 0$. In order to see this, let $u \in H$. Using Proposition 6.1.8 we have

$$(AT(u), u) = (AT^{\frac{1}{2}}T^{\frac{1}{2}}(u), u) = (T^{\frac{1}{2}}AT^{\frac{1}{2}}(u), u) = (AT^{\frac{1}{2}}(u), T^{\frac{1}{2}}(u)) \geq 0\,.$$

Hence, $AT \geq 0$. Moreover we can easily see that if $A \in L(H)$, then $A \geq 0$ if and only if $A + A^* \geq 0$ if and only if $A^* \geq 0$. In addition, A^*A, AA^*, $A - A^*$, and $A^* - A$ are all positive, that is, monotone.

Another very important maximal monotone map is given in the next definition.

Definition 6.1.20. Let X be a Banach space. We define the map $\mathcal{F}: X \to 2^{X^*}$ given by

$$\mathcal{F}(u) = \{u^* \in X^* : \langle u^*, u \rangle = \|u\|^2 = \|u^*\|_*^2\}\,.$$

On account of Proposition 3.1.50 we see that $\mathcal{F}(u) \neq \emptyset$ for all $u \in X$.

Remark 6.1.21. We can state a more general notion. Let $\xi: \mathbb{R}_+ \to \mathbb{R}$ be an increasing, continuous function such that $\xi(0) = 0$ and $\xi(t) \to +\infty$ as $t \to +\infty$. The map $\mathcal{F}_\xi: X \to 2^{X^*} \setminus \{\emptyset\}$ defined by

$$\mathcal{F}_\xi(u) = \{u^* \in X^* : \langle u^*, u \rangle = \xi(\|u\|)\|u\|, \|u^*\|_* = \xi(\|u\|)\}$$

is called **the duality map with gauge function** ξ. If $\xi(t) = t$ for all $t \geq 0$, then we recover Definition 6.1.20. In what follows we focus on the duality map $\mathcal{F}$ also known as the **normalized duality map**. The duality map $\mathcal{F}$ is essentially dependent on the norm of the space. More precisely, if $\|\cdot\|_1$ and $\|\cdot\|_2$ are two equivalent norms on X and $\mathcal{F}_1$, $\mathcal{F}_2$ are the corresponding duality maps, then we do not necessarily have $\mathcal{F}_1 = \mathcal{F}_2$. In fact the results that follow show a close connection between the properties of $\mathcal{F}$ and the geometry of the Banach spaces X and X^*. Finally, note that $\mathcal{F}$ is monotone. In order to see this, let $(u, u^*), (v, v^*) \in \operatorname{Gr} \mathcal{F}$. Then we obtain

$$\langle u^* - v^*, u - v \rangle \geq \|u\|^2 - 2\|u\|\|v\| + \|v\|^2 = (\|u\| - \|v\|)^2 \geq 0\,;$$

see Definition 6.1.20.

Proposition 6.1.22. *If X is a reflexive Banach space with X^* being strictly convex, then the duality map $\mathcal{F}: X \to X^*$ is single-valued, odd, demicontinuous, maximal monotone, strongly coercive, and bounded.*

Proof. Let $u_1^*, u_2^* \in \mathcal{F}(u)$. We have $\langle u_k^*, u \rangle = \|u\|^2 = \|u_k^*\|_*^2$ for $k = 1, 2$. This implies

$$2\|u_1^*\|_*\|u\| \leq \|u_1^*\|_*^2 + \|u_2^*\|_*^2 = \langle u_1^* + u_2^*, u \rangle \leq \|u_1^* + u_2^*\|_*\|u\|\,.$$

Hence, $\|u_1^*\|_* \leq 1/2\|u_1^* + u_2^*\|_*$ and so, $u_1^* = u_2^*$ due to the strict convexity of X^*. Therefore, $\mathcal{F}$ is single-valued.

From Definition 6.1.20 it is clear that $\mathcal{F}(-u) = -\mathcal{F}(u)$ for all $u \in X$. Hence, $\mathcal{F}$ is odd.

Next we show the demicontinuity of $\mathcal{F}$. So, suppose that $u_n \to u$ in X. Then we have

$$\|\mathcal{F}(u_n)\|_* = \|u_n\| \to \|u\| \quad \text{as } n \to \infty, \tag{6.1.12}$$

which gives

$$\{\mathcal{F}(u_n)\}_{n\geq 1} \subseteq X^* \quad \text{is bounded}. \tag{6.1.13}$$

Since X^* is reflexive, we may assume that

$$\mathcal{F}(u_n) \xrightarrow{\mathrm{w}} u^* \quad \text{in } X^* \text{ as } n \to \infty$$

at least for a subsequence. Then we obtain for all $v \in X$ that

$$\langle u^*, v\rangle = \lim_{n\to\infty} \langle \mathcal{F}(u_n), v\rangle \leq \lim_{n\to\infty} \|\mathcal{F}(u_n)\|_* \|v\| = \|u\|\|v\| ; \tag{6.1.14}$$

see (6.1.12). Taking (6.1.13) into account gives, for all $v \in X$,

$$\begin{aligned}\langle u^*, v\rangle &= \lim_{n\to\infty} \langle \mathcal{F}(u_n), v\rangle = \lim_{n\to\infty} [\langle \mathcal{F}(u_n), u - u_n\rangle + \langle \mathcal{F}(u_n), u_n\rangle] \\ &= \lim_{n\to\infty} \|u_n\|^2 = \|u\|^2 .\end{aligned} \tag{6.1.15}$$

From (6.1.14) we see that $\|u^*\|_* \leq \|u\|$ while from (6.1.15) we infer that $\|u\| \leq \|u^*\|_*$. Therefore, we conclude that $\|u^*\|_* = \|u\|$. This means that $u^* = \mathcal{F}(u)$ and so we have that $\mathcal{F}$ is demicontinuous. Monotonicity (see Remark 6.1.21) and demicontinuity of $\mathcal{F}$ imply the maximal monotonicity of $\mathcal{F}$; see Proposition 6.1.17. We have $\langle \mathcal{F}(u), u\rangle = \|u\|^2$ for all $u \in X$. Hence, $\mathcal{F}$ is strongly coercive.

Finally, from Definition 6.1.20, it is clear that $\mathcal{F}$ is bounded. □

The duality map is in fact a subdifferential map. More precisely, we can state the following result.

Proposition 6.1.23. *If X is a Banach space and $\mathcal{F}: X \to 2^{X^*} \setminus \{\emptyset\}$ is the duality map, then $\mathcal{F}(u) = \partial\varphi(u)$ for all $u \in X$ with $\varphi(u) = 1/2\|u\|^2$ for all $u \in X$.*

Proof. Let $u^* \in \mathcal{F}(u)$. Then, for all $v \in X$, we have

$$\begin{aligned}\langle u^*, v - u\rangle &\leq \|u^*\|_* \|v\| - \|u\|^2 \leq \frac{1}{2}[\|u^*\|_*^2 + \|v\|^2] - \|u\|^2 \\ &= \frac{1}{2}[\|u\|^2 + \|v\|^2] - \|u\|^2 = \frac{1}{2}\|v\|^2 - \frac{1}{2}\|u\|^2 = \varphi(v) - \varphi(u) .\end{aligned}$$

Hence, $u^* \in \partial\varphi(u)$ and so

$$\mathcal{F}(u) \subseteq \partial\varphi(u) . \tag{6.1.16}$$

On the other hand, let $\psi(u) = \|u\|$ for all $u \in X$. Then we obtain

$$\begin{aligned}\psi'(u; h)\|u\| &= \lim_{\lambda\searrow 0} \frac{\|u + \lambda h\| - \|u\|}{\lambda}\|u\| \\ &\leq \lim_{\lambda\searrow 0} \frac{1}{2}\frac{\|u + \lambda h\|^2 - \|u\|^2}{\lambda} = \varphi'(u; h)\end{aligned} \tag{6.1.17}$$

and

$$\begin{aligned}\varphi'(u;h) &= \lim_{\lambda\searrow 0} \frac{1}{2}\frac{\|u+\lambda h\|^2 - \|u\|^2}{\lambda} \\ &= \lim_{\lambda\searrow 0}\left[\frac{1}{2}\frac{\|u+\lambda h\| - \|u\|}{\lambda}(\|u+\lambda h\| + \|u\|)\right] = \psi'(u;h)\|u\| .\end{aligned} \tag{6.1.18}$$

From (6.1.17) and (6.1.18) it follows that $\varphi'(u;h) = \psi(u;h)\|u\|$. We know that

$$u^* \in \partial\varphi(u) \quad \text{if and only if} \quad \langle u^*, h\rangle \le \varphi'(u;h) = \psi'(u;h)\|u\| \tag{6.1.19}$$

for all $h \in X$. Then, if $u \neq 0$, we obtain

$$\left\langle \frac{u^*}{\|u\|}, h \right\rangle \le \psi'(u;h) \le \psi(u+h) - \psi(u) \le \|h\| \quad \text{for all } h \in X . \tag{6.1.20}$$

Hence,

$$\|u^*\|_* \le \|u\| . \tag{6.1.21}$$

Moreover, from (6.1.20) we infer that $u^*/\|u\| \in \partial\psi(u)$, which gives

$$\left\langle \frac{u^*}{\|u\|}, v - u \right\rangle \le \psi(v) - \psi(u) \quad \text{for all } v \in X .$$

For $v = 0$ the last inequality yields

$$\left\langle \frac{u^*}{\|u\|}, u \right\rangle \ge \psi(u) .$$

Hence,

$$\|u\| \le \|u^*\|_* . \tag{6.1.22}$$

From (6.1.21) and (6.1.22), we see that $\|u\| = \|u^*\|_*$. Therefore, $u^* \in \mathcal{F}(u)$ and so

$$\partial\varphi(u) \subseteq \mathcal{F}(u) . \tag{6.1.23}$$

If $u = 0$, then $\partial\varphi(0) = \{0\}$ (see (6.1.19)), and so $\partial\varphi(0) = \{0\} = \mathcal{F}(0)$. Hence, (6.1.23) is still valid. From (6.1.16) and (6.1.23), we conclude that $\partial\varphi(u) = \mathcal{F}(u)$ for all $u \in X$. □

Remark 6.1.24. From Propositions 6.1.22 and 6.1.23, we see that if X is reflexive with X^* to be strictly convex, then φ is Gateaux differentiable and $\varphi'(u) = \mathcal{F}(u)$ for all $u \in X$. Moreover, ψ is Gateaux differentiable at every $u \neq 0$ and

$$\psi'(u) = \frac{\mathcal{F}(u)}{\|u\|} \quad \text{for all } u \in X \setminus \{0\} .$$

So, when X^* is strictly and convex, then X has a Gateaux differentiable norm, that is, X is Gateaux smooth.

Proposition 6.1.25. *If X is a reflexive Banach space and both X and X^* are strictly convex, then the duality map $\mathcal{F}: X \to X^*$ is strictly monotone and bijective, and $\mathcal{F}^{-1}$ is the duality map of X^*.*

Proof. Suppose that

$$\langle \mathcal{F}(u) - \mathcal{F}(v), u - v \rangle = 0 . \tag{6.1.24}$$

Note that

$$\langle \mathcal{F}(u) - \mathcal{F}(v), u - v \rangle \geq \|u\|^2 - 2\|u\|\|v\| + \|v\|^2 = \left[\|u\| - \|v\|\right]^2 . \tag{6.1.25}$$

Then combining (6.1.24) and (6.1.25) yields

$$\begin{aligned} 0 &= \left\langle \mathcal{F}(u) - \mathcal{F}\left(\frac{u+v}{2}\right), \frac{u-v}{2} \right\rangle + \left\langle \mathcal{F}\left(\frac{u+v}{2}\right) - \mathcal{F}(v), \frac{u-v}{2} \right\rangle \\ &\geq \left[\|u\| - \left\|\frac{u+v}{2}\right\|\right]^2 + \left[\left\|\frac{u+v}{2}\right\| - \|v\|\right]^2 . \end{aligned}$$

This shows that $\|u\| = \|(u+v)/2\| = \|v\|$. The strict convexity of X implies that $u = v$. Therefore, $\mathcal{F}$ is strictly monotone. In particular, $\mathcal{F}$ is injective as well. Moreover, we know from Proposition 6.1.22 that $\mathcal{F}$ is maximal monotone and coercive. As we will show later in this section in Corollary 6.1.33, these two properties imply surjectivity of $\mathcal{F}$. Hence, $\mathcal{F}$ is a bijection and clearly, $\mathcal{F}^{-1}$ is the duality map for X^*. □

Proposition 6.1.26. *If X is a reflexive Banach space and X^* is locally uniformly convex, then the duality map $\mathcal{F}: X \to X^*$ is continuous.*

Proof. Suppose that $u_n \to u$ in X. Then

$$\|\mathcal{F}(u_n)\|_* \to \|\mathcal{F}(u)\|_* . \tag{6.1.26}$$

From Proposition 6.1.22 we know that $\mathcal{F}$ is demicontinuous. Hence,

$$\mathcal{F}(u_n) \xrightarrow{w} \mathcal{F}(u) \quad \text{in } X^* . \tag{6.1.27}$$

According to Proposition 3.4.32, X^* has the Kadec–Klee Property. Therefore, from (6.1.26) and (6.1.27), it follows that $\mathcal{F}(u_n) \to \mathcal{F}(u)$ in X^*. Hence, $\mathcal{F}$ is continuous. □

Remark 6.1.27. Under the hypotheses of the proposition above, the function $u \to \psi(u) = \|u\|$ is Gateaux differentiable at every $u \neq 0$ and $\psi'(u) = \mathcal{F}(u)/\|u\|$. In fact, using the result of Proposition 6.1.26, we get that $u \to \psi'(u)$ is continuous on $X \setminus \{0\}$, thus it is Fréchet differentiable at every $u \neq 0$. Therefore, when X^* is locally uniformly convex, then X has a Fréchet differentiable norm, that is, X is Fréchet smooth.

From Propositions 6.1.25 and 6.1.26 we deduce at once the following result.

Proposition 6.1.28. *If X is a reflexive Banach space and both X and X^* are locally uniformly convex, then the duality map $\mathcal{F}: X \to X^*$ is a homeomorphism.*

Proposition 6.1.29. *If X is a reflexive Banach space and X^* is uniformly convex, then the duality map $\mathcal{F}: X \to X^*$ is uniformly continuous on bounded subsets of X.*

Proof. We start by showing that $\mathcal{F}$ is uniformly continuous on $\partial B_1 = \{x \in X : \|x\| = 1\}$. Arguing by contradiction, suppose that $\mathcal{F}|_{\partial B_1}$ is not uniformly continuous. Then there exist $\varepsilon > 0$ and two sequences $\{u_n\}_{n\geq 1}, \{v_n\}_{n\geq 1} \subseteq \partial B_1$ such that

$$\|u_n - v_n\| \to 0 \quad \text{as } n \to \infty \quad \text{and} \quad \|\mathcal{F}(u_n) - \mathcal{F}(v_n)\| \geq \varepsilon \quad \text{for all } n \in \mathbb{N}\,. \tag{6.1.28}$$

For all $y, w \in X$, we have

$$\begin{aligned} \|\mathcal{F}(y) + \mathcal{F}(w)\|_* \|y\| &\geq \langle \mathcal{F}(y) + \mathcal{F}(w), y\rangle \\ &= \langle \mathcal{F}(y), y\rangle + \langle \mathcal{F}(w), w\rangle + \langle \mathcal{F}(w), y - w\rangle \\ &\geq \|y\|^2 + \|w\|^2 - \|w\|\|y - w\|\,. \end{aligned} \tag{6.1.29}$$

We choose $y = u_n$ and $w = v_n$ for all $n \in \mathbb{N}$ in (6.1.29). Then

$$\|\mathcal{F}(u_n) + \mathcal{F}(v_n)\|_* = 2 - \|u_n - v_n\| \quad \text{for all } n \in \mathbb{N}\,.$$

Hence, $\|\mathcal{F}(u_n) + \mathcal{F}(v_n)\|_* \to 2$ as $n \to \infty$. Then the uniform convexity of X^* implies that $\|\mathcal{F}(u_n) - \mathcal{F}(v_n)\|_* \to 0$ as $n \to \infty$ (see Definition 3.4.21 (b)) and this contradicts (6.1.28). So, indeed $\mathcal{F}|_{\partial B_1}$ is uniformly continuous. We know that $\mathcal{F}(\lambda u) = \lambda \mathcal{F}(u)$ for all $\lambda > 0$ and for all $u \in X$. This fact implies for $u, v \in X \setminus \{0\}$ that

$$\begin{aligned} &\|\mathcal{F}(u) - \mathcal{F}(y)\|_* \\ &= \left\| \|u\| \mathcal{F}\left(\frac{u}{\|u\|}\right) - \|v\| \mathcal{F}\left(\frac{v}{\|v\|}\right)\right\|_* \\ &\leq \|u\| \left\| \mathcal{F}\left(\frac{u}{\|u\|}\right) - \mathcal{F}\left(\frac{v}{\|v\|}\right)\right\|_* + \|u - v\| \left\|\mathcal{F}\left(\frac{v}{\|v\|}\right)\right\|_*\,. \end{aligned} \tag{6.1.30}$$

From (6.1.30) and the uniform continuity of $\mathcal{F}|_{\partial B_1}$ we conclude that $\mathcal{F}$ is uniformly continuous on bounded subsets of X. ☐

Next we prove an auxiliary result, which is fundamental in the study of operator equations involving maximal monotone maps. The result is known as the "Debrunner–Flor Lemma".

Lemma 6.1.30 (Debrunner–Flor Lemma). *If X is a Banach space, $K \subseteq X$ is nonempty, compact, convex, $A: K \to 2^{X^*}$ is a monotone map, $L: K \to X^*$ is a continuous map, and $h^* \in X^*$, then there exists $u_0 \in K$ such that*

$$\langle u^* + L(u_0) - h^*, u - u_0 \rangle \ge 0 \quad \textit{for all } (u, u^*) \in \operatorname{Gr} A\,. \tag{6.1.31}$$

Proof. Set $A_1(u) = A(u) - h^*$. We see that A_1 is still monotone and so, without any loss of generality, we may take $h^* = 0$. Arguing by contradiction, suppose that (6.1.31) is not true. Then, for every $u_0 \in K$, there exists a pair $(u, u^*) \in \operatorname{Gr} A$ such that $\langle u^* + L(u_0), u - u_0 \rangle < 0$. For each $(u, u^*) \in \operatorname{Gr} A$, let

$$N(u, u^*) = \{y \in K : \langle u^* + L(y), u - y \rangle < 0\} \neq \emptyset\,.$$

Then $\{N(u, u^*)\}_{(u,u^*) \in \operatorname{Gr} A}$ is an open cover of K. So, by compactness, there exists a finite family $\{N(u_k, u_k^*)\}_{k=1}^m$ such that

$$K = \bigcup_{k=1}^{m} N(u_k, u_k^*)\,.$$

Let $\{\xi_k\}_{k=1}^m$ be a corresponding continuous partition of unity. We introduce the maps $g_1 : K \to X$ and $g_1^* : K \to X^*$ defined by

$$g_1(y) = \sum_{k=1}^{m} \xi_k(y) u_k \quad \text{and} \quad g_1^*(y) = \sum_{k=1}^{m} \xi_k(y) u_k^* \quad \text{for all } y \in K\,.$$

The convexity of K implies $g_1 : K \to K$ and that it is continuous. Then, by the Schauder Fixed Point Theorem (see Theorem 6.3.21), there exists $y_0 \in K$ such that $g_1(y_0) = y_0$.

Let $\tau(y) = \langle g_1(y) + L(y), g_1(y) - y \rangle = \tau_1(y) + \tau_2(y)$ with

$$\tau_1(y) = \sum_{k=1}^{m} \xi_k(y) \langle u_k^* + L(y), u_k - y \rangle$$

and

$$\tau_2(y) = \sum_{1 \le k < i \le m} \xi_k(y) \xi_i(y) \left[\langle u_k^* + L(y), u_i - y \rangle + \langle u_i^* + L(y), u_k - y \rangle \right]. \tag{6.1.32}$$

For each $y \in K$, there exists at least one $j \in \mathbb{N}$ with $1 \le j \le m$ such that $\xi_j(y) \neq 0$ and $y \in N(u_j, u_j^*)$. Therefore, $\langle u_j^* + L(y), u_j - y \rangle < 0$ and so

$$\tau_1(y) < 0 \quad \text{for all } y \in K\,. \tag{6.1.33}$$

From the monotonicity of A we obtain, for all $y \in N(u_k, u_k^*) \cap N(u_i, u_i^*)$, that $\xi_k(y)\xi_i(y) > 0$ and

$$\begin{aligned} &\langle u_k^* + L(y), u_i - y \rangle + \langle u_i^* + L(y), u_k - y \rangle \\ &= \langle u_k^* + L(y), u_k - y \rangle + \langle u_i^* + L(y), u_i - y \rangle + \langle u_k^* - u_i^*, u_i - u_k \rangle < 0\,. \end{aligned} \tag{6.1.34}$$

Therefore, from (6.1.32), (6.1.33) and (6.1.34) we conclude that $\tau(y) < 0$ for all $y \in K$. On the other hand

$$\tau(y_0) = \langle g_1^*(y_0) + L(y_0), g_1(y_0) - y_0 \rangle = 0 ,$$

a contradiction. □

Using this lemma, we can prove the following fundamental theorem that characterizes maximal monotone maps.

Theorem 6.1.31. *If X is a reflexive Banach space and both X and X^* are strictly convex and $A: X \to 2^{X^*}$ is a monotone map, then A is maximal monotone if and only if $R(A+\lambda\mathcal{F}) = X^*$ for all $\lambda > 0$ (resp. for some $\lambda > 0$).*

Proof. $\Longrightarrow$: Evidently it suffices to consider the case $\lambda = 1$ and we will show that $0 \in R(A + \mathcal{F})$. Let $V_n \subseteq X$ be a n-dimensional subspace of X and let $i_n: V_n \to X$ be the corresponding embedding map. Then $i_n^*: X^* \to V_n^*$ and we set $L = i_n^* \circ \mathcal{F} \circ i_n$, $K = \{u \in V_n: \|u\| \le r\}$ with $r > 0$ large enough such that $B_r(0) \cap D(A) \ne \emptyset$. By Lemma 6.1.30 with $h^* = 0$ there exist $u_n^r \in K$ and $y_n^r = L(u_n^r)$ such that

$$\langle u^* + y_n^r, u - u_n^r \rangle \ge 0 \quad \text{for all } (u, u^*) \in \operatorname{Gr}((i_n^* \circ \mathcal{F} \circ i_n) \cap K) \times V_n^* = G_n . \tag{6.1.35}$$

Hence,

$$\|u_n^r\|^2 \le \|u^*\|_* \|u\| + \|u_n^r\|[\|u\| + \|u^*\|_*] \quad \text{for all } (u, u^*) \in G_n .$$

Hence, $\{u_n^r\}_{r>0} \subseteq V_n$ is bounded. So, we have $u_n^r \to y_n \in V_n$ as $r \to \infty$. Due to Proposition 6.1.26, we obtain $y_n^r = L(u_n^r) \to L(y_n) = y_n^*$. From (6.1.35) it follows that

$$\langle u^* + y_n^*, u - y_n \rangle \ge 0 \quad \text{for all } (u, u^*) \in \operatorname{Gr} A \text{ with } u \in V_n . \tag{6.1.36}$$

The coercivity of $\mathcal{F}$ (see Proposition 6.1.22) and (6.1.36) imply that $\{y_n\}_{n\ge1} \subseteq X$ and $\{y_n^*\}_{n\ge1} \subseteq X^*$ are bounded. Therefore, there exist $\hat{u} \in K$ and $\hat{u}^* \in X^*$ such that

$$y_n \xrightarrow{w} \hat{u} \quad \text{in } X \quad \text{and} \quad y_n^* \xrightarrow{w} \hat{u}^* \quad \text{in } X^* .$$

Let $f(x) = 1/2\|x\|^2$ for $x \in X$. Taking Proposition 5.3.19 into account gives

$$\langle \hat{u}^*, \hat{u} \rangle \le f(\hat{u}) + f^*(\hat{u}^*) \le \liminf_{n\to\infty} f(y_n) + \liminf_{n\to\infty} f^*(y_n^*) = \liminf_{n\to\infty} \langle y_n^*, y_n \rangle . \tag{6.1.37}$$

Moreover, from (6.1.36), we derive

$$\langle y_n^*, y_n \rangle \le \langle u^*, u \rangle + \langle y_n^*, u \rangle - \langle u^*, y_n \rangle \quad \text{with } n \in \mathbb{N} .$$

This implies

$$\limsup_{n\to\infty}\langle y_n^*, y_n\rangle \le \langle u^*, u\rangle + \langle \hat{u}^*, u\rangle - \langle u^*, \hat{u}\rangle \tag{6.1.38}$$

for all $(u, u^*) \in \operatorname{Gr} A$ and for all $u \in V_n$. But $\overline{\bigcup_{n\ge1} V_n} = X$. Therefore, (6.1.38) holds for all $(u, u^*) \in \operatorname{Gr} A$ and for all $u \in X$. Then, in view of (6.1.37), we obtain

$$\langle -\hat{u}^* - u^*, \hat{u} - u\rangle \ge 0 \quad \text{for all } (u, u^*) \in \operatorname{Gr} A\,,$$

which due to the maximal monotonicity of A results in

$$-\hat{u}^* \in A(\hat{u})\,. \tag{6.1.39}$$

We choose $u = \hat{u}$ and $u^* = \hat{u}^*$ in (6.1.38). Then

$$\limsup_{n\to\infty}\langle y_n^*, y_n\rangle \le \langle \hat{u}^*, \hat{u}\rangle\,. \tag{6.1.40}$$

From (6.1.37) and (6.1.40) we deduce that $\langle \hat{u}^*, \hat{u}\rangle = f(\hat{u}) + f^*(\hat{u}^*)$ and so, by Proposition 5.3.19, $\hat{u}^* \in \partial f(\hat{u})$. Then Proposition 6.1.23 implies that $\hat{u}^* = \mathcal{F}(\hat{u})$ and due to (6.1.39) we obtain $0 \in (A + \mathcal{F})(\hat{u})$.

$\Longleftarrow$: We may assume that $R(A + \mathcal{F}) = X^*$. Suppose that for $y \in X$ and $y^* \in X^*$ we have

$$\langle u^* - y^*, u - y\rangle \ge 0 \quad \text{for all } (u, u^*) \in \operatorname{Gr} A\,. \tag{6.1.41}$$

There exists $u_1 \in D(A)$ such that

$$u_1^* + \mathcal{F}(u_1) = y^* + \mathcal{F}(y) \quad \text{with } u_1^* \in A(u_1)\,. \tag{6.1.42}$$

We choose $(u_1, u_1^*) \in \operatorname{Gr} A$ in (6.1.41). Then

$$0 \le \langle y^* + \mathcal{F}(y) - \mathcal{F}(u_1) - y^*, u_1 - y\rangle = \langle \mathcal{F}(y) - \mathcal{F}(u_1), u_1 - y\rangle\,,$$

which gives $y = u_1 \in D(A)$ and $y^* = u_1^*$; see (6.1.42). Hence, A is maximal monotone. □

Theorem 6.1.32. *If X is a reflexive Banach space and $A\colon X \to 2^{X^*}$ is maximal monotone, then $R(A) = X^*$ if and only if A^{-1} is locally bounded.*

Proof. $\Longrightarrow$: The maximal monotonicity of A implies that A^{-1} is maximal monotone as well. We have $D(A^{-1}) = R(A) = X^*$. Hence, A^{-1} is locally bounded; see Proposition 6.1.11.

$\Longleftarrow$: It suffices to show that $R(A) \subseteq X^*$ is clopen. First we show that $R(A)$ is closed. So, let $\{u_n^*\}_{n\ge1} \subseteq R(A)$ and assume that $u_n^* \to u^*$ in X^*. We can find $u_n \in D(A)$ such that $u_n^* \in A(u_n)$. We have

$$\langle u_n^* - v^*, u_n - v\rangle \ge 0 \quad \text{for all } (v, v^*) \in \operatorname{Gr} A \text{ and for all } n \in \mathbb{N}\,. \tag{6.1.43}$$

Since by hypothesis A^{-1} is locally bounded, it follows that $\{u_n\}_{n\ge1} \subseteq X$ is bounded and so by passing to a suitable subsequence if necessary we may assume that $u_n \xrightarrow{w} u$ in X.

Letting $n \to \infty$ in (6.1.43) gives

$$\langle u^* - v^*, u - v\rangle \geq 0 \quad \text{for all } (v, v^*) \in \operatorname{Gr} A\,.$$

Due to the maximal monotonicity of A this implies $(u, u^*) \in \operatorname{Gr} A$. We may assume that $u = 0$ otherwise we can work with $\hat{A}(v) = A(u + v)$. Let $r > 0$ be such that $A^{-1}|_{B_r(u^*)}$ is bounded where $B_r(u^*) = \{v^* \in X^* : \|v^* - u^*\|_* < r\}$. Using the Troyanski Renorming Theorem (see Theorem 5.8.8) without any loss of generality we may assume that both X and X^* are locally uniformly convex. Let $v^* \in B_{r/2}(u^*)$. Then, according to Theorem 6.1.31, the operator equation

$$u_\lambda^* + \lambda\mathcal{F}(u_\lambda) = v^* \quad \text{with } u_\lambda^* \in A(u_\lambda) \tag{6.1.44}$$

has a solution u_λ for every $\lambda > 0$. The maximal monotonicity of A implies $\langle v^* - \lambda\mathcal{F}(u_\lambda) - u^*, u_\lambda\rangle \geq 0$; recall that $x = 0$. Hence, $\|v^* - u^*\|_* \|u_\lambda\| \geq \lambda\|u_\lambda\|^2$, which gives

$$\lambda\|u_\lambda\| < \frac{r}{2} \quad \text{for all } \lambda > 0\,. \tag{6.1.45}$$

From (6.1.44) and (6.1.45) we see that

$$\|v^* - u_\lambda^*\|_* = \lambda\|\mathcal{F}(u_\lambda)\|_* = \lambda\|u_\lambda\| < \frac{r}{2}\,. \tag{6.1.46}$$

Hence, $\|u_\lambda^* - u^*\|_* < r$ for all $\lambda > 0$. But $A^{-1}|_{B_r(u^*)}$ is bounded. So, $\{u_\lambda\}_{\lambda>0} \subseteq X$ is bounded and from (6.1.46) we see that $u_\lambda^* \to v^*$ in X^* as $\lambda \searrow 0$. Recall that $R(A) \subseteq X^*$ is closed. Hence, $v^* \in R(A)$ and so $B_{r/2} \subseteq R(A)$. Therefore, $R(A)$ is clopen, and thus $R(A) = X^*$. □

Corollary 6.1.33. *If X is a reflexive Banach space and $A: X \to 2^{X^*}$ is maximal monotone and coercive, then A is surjective, that is, $R(A) = X^*$.*

Proof. The coercivity of A implies that A^{-1} is locally bounded. Applying Theorem 6.1.32, we conclude that $R(A) = X^*$. □

Corollary 6.1.34. *If X is a reflexive Banach space and $A: X \to X^*$ is monotone, hemicontinuous, and coercive with $D(A) = X$, then A is surjective, that is, $R(A) = X^*$.*

A characteristic example of a maximal monotone map is the convex subdifferential.

Theorem 6.1.35. *If X is a reflexive Banach space and $f \in \Gamma_0(X)$, then $\partial f: X \to 2^{X^*}$ is a maximal monotone map.*

Proof. From Troyanski's Renorming Theorem (see Theorem 5.8.8), we may assume that both X and X^* are locally uniformly convex. Then Proposition 6.1.28 implies that the duality map $\mathcal{F}: X \to X^*$ is a homeomorphism.

From Example 6.1.19 (a) we know that ∂f is monotone. Then, according to Theorem 6.1.31, in order to show the maximal monotonicity of ∂f, it suffices to show that $R(\partial f + \mathcal{F}) = X^*$.

So, let $u^* \in X^*$ and consider the function $\psi: X \to \overline{\mathbb{R}} = \mathbb{R} \cup \{+\infty\}$ defined by

$$\psi(u) = \frac{1}{2}\|u\|^2 + \varphi(u) - \langle u^*, u\rangle \quad \text{for all } u \in X .$$

Evidently, $\psi \in \Gamma_0(X)$ and $\psi(u) \to +\infty$ as $\|u\| \to \infty$. Hence, the reflexivity of X implies that there exists $u_0 \in \operatorname{dom}\psi$ such that

$$\psi(u_0) = \inf[\psi(u): u \in \operatorname{dom}\psi] .$$

Due to Proposition 5.3.18 we obtain $0 \in \partial\psi(u_0)$ and because of Remark 6.1.24, we know that $u^* \in \partial f(u_0) + \mathcal{F}(u_0)$. Since $u^* \in X^*$ is arbitrary, we infer that $R(\partial f + \mathcal{F}) = X^*$. Finally, Theorem 6.1.31 implies that ∂f is maximal monotone. □

Remark 6.1.36. The result is true for any Banach space X not necessarily reflexive; see Rockafellar [274] and Phelps [260, Theorem 3.25, p. 59].

In fact the convex subdifferential is a special kind of monotone operator.

Definition 6.1.37. Let X be a Banach space, $A: X \to 2^{X^*}$, and let $n \in \mathbb{N}$ be such that $n \geq 2$. We say that A is ***n*-cyclically monotone** if, for every $\{u_k\}_{k=1}^{n+1} \subseteq X$ and $\{u_k^*\}_{k=1}^{n} \subseteq X^*$,

$$(u_k, u_k^*) \in \operatorname{Gr} A \quad \text{for all } k = 1, \ldots, n \text{ and } u_{n+1} = u_1$$

implies

$$\sum_{k=1}^{n} \langle u_k^*, u_{k+1} - u_k\rangle \leq 0 .$$

If A is n-cyclically monotone for every integer $n \geq 2$, then we say that A is **cyclically monotone.** If A is cyclically monotone and $\operatorname{Gr} A$ is not properly contained in the graph of another cyclically monotone map, then we say that A is **maximal cyclically monotone.**

Remark 6.1.38. Clearly, monotonicity and 2-cyclic monotonicity coincide. Moreover, a maximal monotone map that is also cyclically monotone is maximal cyclically monotone. It is easy to see that if $f: X \to \overline{\mathbb{R}} = \mathbb{R} \cup \{+\infty\}$ is proper and convex, then $\partial f: X \to 2^{X^*}$ is cyclically monotone.

Example 6.1.39. Let $X = \mathbb{R}^2$ and let $A = \left(\begin{smallmatrix} 0 & -1 \\ 1 & 0 \end{smallmatrix}\right)$. Then A is maximal monotone but not 3-cyclically monotone. Indeed, let $u_1 \in \mathbb{R}^2 \setminus \{0\}$ and define $u_2 = A(u_1)$, $u_3 = A(u_2) = A^2(u_1)$ and $u_4 = u_1$. Then, since $\|Au\| = \|u\|$ and $(A(u), u) = 0$ for all $u \in \mathbb{R}^2$,

$$\begin{aligned}\sum_{k=1}^{3} \langle A(u_k), u_{k+1} - u_k\rangle &= \langle A(u_1), u_2\rangle + \langle A(u_2), u_3\rangle + \langle A(u_3), u_1\rangle \\ &= \|A(u_1)\|^2 + \langle -u_1, -u_1\rangle + \langle -A(u_1), u_1\rangle = 2\|u_1\|^2 > 0 .\end{aligned}$$

Hence, A is not 3-cyclically monotone. Note that this operator performs rotation by $\pi/2$.

Next we show that convex subdifferentials are the only maximal cyclically monotone maps.

Theorem 6.1.40. *If X is a Banach space, then $A: X \to 2^{X^*}$ is maximal cyclically monotone if and only if there exists $f \in \Gamma_0(X)$ such that $\partial f = A$.*

Proof. $\Longrightarrow$: We have $\operatorname{Gr} A \neq \emptyset$. Let $(u_0, u_0^*) \in \operatorname{Gr} A$ and define $f: X \to \overline{\mathbb{R}} = \mathbb{R} \cup \{+\infty\}$ by

$$f(u) = \sup_{n \in \mathbb{N}} \sup_{\{(u_k, u_k^*)\}_{k=1}^n \subseteq \operatorname{Gr} A} \left[\langle u_n^*, u - u_n \rangle + \sum_{k=0}^{n-1} \langle u_k^*, u_{k+1} - u_k \rangle \right] \tag{6.1.47}$$

with $n \geq 2$. Then f being the supremum of affine continuous functions is itself convex and lower semicontinuous. On account of the cyclical monotonicity of A, we obtain

$$\sum_{k=0}^{n} \langle u_k^*, u_{k+1} - u_k \rangle \leq 0 ;$$

see Definition 6.1.37. Thus, $f(u_0) \leq 0$ and so $f \in \Gamma_0(X)$.

Suppose that $(u, u^*) \in \operatorname{Gr} A$ and let $v \in X$. Equation (6.1.47) gives

$$f(v) \geq \sum_{k=0}^{n-1} \langle u_k^*, u_{k+1} - u_k \rangle + \langle u_n^*, u - u_n \rangle + \langle u^*, v - u \rangle ,$$

which implies $f(v) \geq f(u) + \langle u^*, v - u \rangle$ for all $v \in X$. Hence, $u^* \in \partial f(u)$ and so $\operatorname{Gr} A \subseteq \operatorname{Gr} \partial f$. But the cyclical monotonicity of ∂f and the maximality of A imply that $\operatorname{Gr} A = \operatorname{Gr} \partial f$, hence $A = \partial f$.

$\Longleftarrow$: This follows from Remarks 6.1.36 and 6.1.38. □

Remark 6.1.41. In fact, $f \in \Gamma_0(X)$ such that $A = \partial f$ is unique up to an additive constant. In $\mathbb{R}$ the situation simplifies and $A: \mathbb{R} \to 2^{\mathbb{R}}$ with $\operatorname{Gr} A \neq \emptyset$ is monotone if and only if it is cyclically monotone. In particular, then A is maximal monotone if and only if $A = \partial f$ for some $f \in \Gamma_0(\mathbb{R})$.

Continuing with maximal monotone maps we can state the following result concerning their domain.

Proposition 6.1.42. *If X is a reflexive Banach space and $A: X \to 2^{X^*}$ is maximal monotone, then $\overline{D(A)} \subseteq X$ is convex.*

Proof. As before without any loss of generality we may assume that both X and X^* are locally uniformly convex. For any $u_0 \in X$ and $\eta > 0$, the inclusion

$$0 \in \eta A(u) + \mathcal{F}(u - u_0) \tag{6.1.48}$$

has a solution $u_\eta \in D(A)$ (see Theorem 6.1.31), and this solution is unique since A is monotone and $\mathcal{F}$ is strictly monotone; see Proposition 6.1.25. So, there exists $u_\eta^* \in A(u_\eta)$ such that

$$\eta u_\eta^* + \mathcal{F}(u_\eta - u_0) = 0 \quad \text{for all } \eta > 0 . \tag{6.1.49}$$

Then, by applying (6.1.49) and since A is monotone, we obtain, for $(v, v^*) \in \operatorname{Gr} A$, that

$$\begin{aligned}
\|u_\eta - u_0\|^2 &= \langle \mathcal{F}(u_\eta - u_0), u_\eta - u_0 \rangle \\
&= \langle \mathcal{F}(u_\eta - u_0), u_\eta - v \rangle + \langle \mathcal{F}(u_\eta - u_0), v - u_0 \rangle \\
&= \eta \langle v^* - u_\eta^*, u_\eta - v \rangle + \eta \langle v^*, v - u_\eta \rangle + \langle \mathcal{F}(u_\eta - u_0), v - u_0 \rangle \\
&\le \eta \langle v^*, v - u_\eta \rangle + \langle \mathcal{F}(u_\eta - u_0), v - u_0 \rangle \\
&\le \eta \|v^*\|_* \|v - u_\eta\| + \|u_\eta - u_0\| \|v - u_0\| .
\end{aligned} \tag{6.1.50}$$

Hence, $\{\mathcal{F}(u_\eta - u_0)\}_{\eta > 0} \subseteq X^*$ and $\{u_\eta\}_{\eta>0} \subseteq X$ are bounded. So, there exists $\eta_n \to 0^+$ such that $\mathcal{F}(u_{\eta_n} - u_0) \xrightarrow{\mathrm{w}} \hat{y}$ in X^*. This implies

$$\limsup_{n\to\infty} \|u_{\eta_n} - u_0\|^2 \le \langle \hat{y}, v - u_0 \rangle \quad \text{for all } v \in D(A) ; \tag{6.1.51}$$

see (6.1.50). Evidently, (6.1.51) holds for all $v \in \overline{D(A)}$. If $u_0 \in \overline{D(A)}$, then from (6.1.51) we get $u_{\eta_n} \to u_0$ in X as $n \to \infty$.

Now let $u_0^1, u_0^2 \in \overline{D(A)}$ and let $u_0^\lambda = (1-\lambda)u_0^1 + \lambda u_0^2$ with $\lambda \in [0, 1]$. Let $u_\eta^\lambda \in D(A)$ be a solution of (6.1.48) with u_0^λ instead of u_0. Then from (6.1.51) we see that $u_\eta^\lambda \to u_0^\lambda$ and $u_\eta^\lambda \in D(A)$. Hence, $u_0^\lambda \in \overline{D(A)}$ and so $\overline{D(A)}$ is convex. □

Corollary 6.1.43. *If X is a reflexive Banach space and $A\colon X \to 2^{X^*}$ is maximal monotone, then $\overline{R(A)}$ is convex.*

Proof. Since A^{-1} is maximal monotone and $R(A) = D(A^{-1})$, the result follows from Proposition 6.1.42. □

Now we restrict ourselves to a Hilbert space H. We assume that $H = H^*$ and consider a maximal monotone map $A\colon H \to 2^H$. For such a map we can define some useful single-valued, regular approximations.

Definition 6.1.44. Let $A\colon H \to 2^H$ be a maximal monotone map. For every $\lambda > 0$ we define

$$J_\lambda = (\mathrm{id} + \lambda A)^{-1} \quad \text{and} \quad A_\lambda = \frac{1}{\lambda}[\mathrm{id} - J_\lambda] ,$$

where J_λ is called the **resolvent operator** and A_λ stands for the **Yosida approximation**.

Remark 6.1.45. From Theorem 6.1.31 we see that $D(J_\lambda) = D(A_\lambda) = H$ for all $\lambda > 0$. Moreover, it is easy to see that J_λ is single-valued, and hence so is A_λ.

Theorem 6.1.46. *If $A: X \to 2^H$ is a maximal monotone map, then the following hold for every $\lambda > 0$:*
(a) *J_λ is nonexpansive, that is, J_λ is 1-Lipschitz;*
(b) *$A_\lambda(u) \in A(J_\lambda(u))$ for all $u \in H$;*
(c) *A_λ is monotone and $1/\lambda$-Lipschitz, thus maximal monotone as well;*
(d) *$\|A_\lambda(u)\| \le \|A^\circ(u)\|$ for every $u \in D(A)$ with $A^\circ(u) = p_{A(u)}(0)$ (see Proposition 6.1.12);*
(e) *$\lim_{\lambda \searrow 0} A_\lambda(u) = A^\circ(u)$ for all $u \in D(A)$;*
(f) *$\overline{D(A)}$ is convex and $\lim_{\lambda \searrow 0} J_\lambda(u) = p_{\overline{D(A)}}(u)$ for all $u \in H$.*

Proof. (a) Let $u, v \in H$. From Definition 6.1.44 we know that

$$u - v \in J_\lambda(u) - J_\lambda(v) + \lambda(A(J_\lambda(u)) - A(J_\lambda(v))) .$$

Taking the inner product with $J_\lambda(u) - J_\lambda(v)$ and using the monotonicity of A we obtain

$$\|J_\lambda(u) - J_\lambda(v)\|^2 \le \|u - v\| \|J_\lambda(u) - J_\lambda(v)\| .$$

Hence, $\|J_\lambda(u) - J_\lambda(v)\| \le \|u - v\|$.

(b) From Definition 6.1.44 we see that

$$(u, u^*) \in \operatorname{Gr} A_\lambda \quad \text{if and only if} \quad (u - \lambda u^*, u^*) \in \operatorname{Gr} A . \tag{6.1.52}$$

From this it follows that $A_\lambda(u) \in A(J_\lambda(u))$ for all $u \in H$.

(c) From (a) we know that J_λ is nonexpansive. Then Example 6.1.19 (b) implies that $\mathrm{id} - J_\lambda$ is monotone and this shows that A_λ is monotone. For all $u, v \in H$ we have

$$u - v = J_\lambda(u) - J_\lambda(v) + \lambda(A_\lambda(u) - A_\lambda(v)) ,$$

which gives

$$\begin{aligned}&(u - v, A_\lambda(u) - A_\lambda(v))\\ &= (J_\lambda(u) - J_\lambda(v), A_\lambda(u) - A_\lambda(v)) + \lambda\|A_\lambda(u) - A_\lambda(v)\|^2 .\end{aligned} \tag{6.1.53}$$

From (b) and the monotonicity of A we get

$$(J_\lambda(u) - J_\lambda(v), A_\lambda(u) - A_\lambda(v)) \ge 0 . \tag{6.1.54}$$

Combining (6.1.54) and (6.1.53) yields

$$\|A_\lambda(u) - A_\lambda(v)\|^2 \le \frac{1}{\lambda}\|u - v\| \|A_\lambda(u) - A_\lambda(v)\| .$$

Therefore,

$$\|A_\lambda(u) - A_\lambda(v)\| \le \frac{1}{\lambda}\|u - v\| .$$

(d) From part (b) and the monotonicity of A we conclude that

$$(A^\circ(u) - A_\lambda(u), u - J_\lambda(u)) \ge 0 \quad \text{for all } u \in D(A) \text{ and for all } \lambda > 0 .$$

Since $\lambda > 0$, this implies

$$(A^\circ(u) - A_\lambda(u), A_\lambda(u)) \ge 0 . \tag{6.1.55}$$

Hence, $\|A_\lambda(u)\|^2 \le \|A^\circ(u)\|\|A_\lambda(u)\|$ and so $\|A_\lambda(u)\| \le \|A^\circ(u)\|$ for all $u \in D(A)$.

(e) Applying (6.1.52) we see that $(A_\lambda)_\mu = A_{\lambda+\mu}$ for all $\lambda, \mu > 0$. Then from (6.1.55) we derive

$$\|A_{\lambda+\mu}(u)\|^2 \le (A_{\lambda+\mu}(u), A_\lambda(u)) \quad \text{for all } u \in H \text{ and for all } \lambda, \mu > 0 . \tag{6.1.56}$$

Then (6.1.56) implies

$$\begin{aligned}\|A_{\lambda+\mu}(u) - A_\lambda(u)\|^2 &= \|A_{\lambda+\mu}(u)\|^2 + \|A_\lambda(u)\|^2 - 2(A_{\lambda+\mu}(u), A_\lambda(u)) \\ &\le \|A_\lambda(u)\|^2 - \|A_{\lambda+\mu}(u)\|^2\end{aligned}$$

for all $u \in H$ and for all $\lambda, \mu > 0$. Hence, $\{A_\lambda(u)\}_{\lambda>0}$ is a Cauchy sequence. It follows that

$$A_\lambda(u) \to y \quad \text{in } H \text{ as } \lambda \searrow 0 . \tag{6.1.57}$$

Since $u - J_\lambda(u) = \lambda A_\lambda(u)$ for all $u \in H$ and for all $\lambda > 0$, we infer that

$$J_\lambda(u) \to u \quad \text{as } \lambda \searrow 0 . \tag{6.1.58}$$

From (b), (6.1.57) and (6.1.58) it follows that

$$(u, y) \in \operatorname{Gr} A . \tag{6.1.59}$$

But $\|A_\lambda(u)\| \le \|A^\circ(u)\|$ for all $\lambda > 0$; see (d). Hence, $\|y\| \le \|A^\circ(u)\|$ and so, by (6.1.59), $y = A^\circ(u)$.

(f) From Proposition 6.1.42 we already know that $\overline{D(A)}$ is convex. Also note that

$$\|J_\lambda(u)\| \le \|u\| + \|J_\lambda(0)\| \quad \text{for all } u \in H \text{ and for all } \lambda > 0$$

and $J_\lambda(0) = \lambda A_\lambda(0)$ for all $\lambda > 0$. It follows that $\|J_\lambda(u)\| \le \|u\| + \lambda\|A_\lambda(0)\|$. Hence, $\{J_\lambda(u)\}_{\lambda>0} \subseteq H$ is bounded for all sufficiently small $\lambda > 0$; see (6.1.57). Therefore, there exists a sequence $\{\lambda_n\}_{n\geq 1} \subseteq (0, +\infty)$ such that

$$J_{\lambda_n}(u) \xrightarrow{w} y \quad \text{in } H \text{ as } n \to \infty . \tag{6.1.60}$$

Since $J_{\lambda_n}(u) \in D(A)$ for all $n \in \mathbb{N}$ and $\overline{D(A)}$ is convex, from (6.1.60) it follows that $y \in \overline{D(A)}$. From (b) we obtain

$$(u - J_{\lambda_n}(u) - \lambda_n v^*, J_{\lambda_n}(u) - v) \geq 0 \quad \text{for all } (v, v^*) \in \operatorname{Gr} A \text{ and for all } n \in \mathbb{N} .$$

This implies

$$\|J_{\lambda_n}(u)\|^2 \leq (u - \lambda_n v^*, J_{\lambda_n}(u) - v) + (J_{\lambda_n}(u), v) \quad \text{for all } n \in \mathbb{N} .$$

Hence, $\|y\|^2 \leq (u, y - v) + (y, v)$ and so $(u - y, y - v) \geq 0$ for all $v \in \overline{D(A)}$. Then Proposition 3.5.20 shows that $y = p_{\overline{D(A)}}(u)$. Therefore, we conclude that $J_\lambda(u) \to p_{\overline{D(A)}}(u)$ for all $u \in H$ as $\lambda \searrow 0$. □

Remark 6.1.47. From the proof of (e) in Theorem 6.1.46 we see that

$$\text{if } u \notin D(A), \text{ then } \|A_\lambda(u)\| \to +\infty \text{ as } \lambda \searrow 0 .$$

In general the sum of two maximal monotone maps need not be maximal monotone.

Example 6.1.48. Let $H = \mathbb{R}^2$ and $e = (1, 0) \in \mathbb{R}^2$. We consider the sets $C_1 = \overline{B}_1(e)$ and $C_2 = \overline{B}_1(-e)$. Let $f_1 = i_{C_1}$ and $f_2 = i_{C_2}$ where i_{C_j} is the indicator function of $C_j, j = 1, 2$. Then $A_1 = \partial f_1$ and $A_2 = \partial f_2$ are both maximal monotone; see Theorem 6.1.35. Moreover, from Remark 5.7.42, we know that $A_1 = N_{C_1}$ and $A_2 = N_{C_2}$. We have $D(A_1) \cap D(A_2) \neq \emptyset$ and it is bounded. So, if $A_1 + A_2$ is maximal monotone, then we should have that $A_1 + A_2$ is surjective. But $R(A_1 + A_2) = \mathbb{R} \times \{0\}$ and so $A_1 + A_2$ cannot be maximal.

Theorem 6.1.49. *If X is a reflexive Banach space and $A_1, A_2 : X \to 2^{X^*}$ are maximal monotone maps such that $D(A_1) \cap \operatorname{int} D(A_2) \neq \emptyset$, then $A_1 + A_2$ is maximal monotone.*

Proof. Translating things if necessary, we may assume without any loss of generality that $0 \in D(A_1) \cap \operatorname{int} D(A_2)$ and $0 \in A_1(0), A_2(0)$. Moreover, as before, we can suppose that both X and X^* are locally uniformly convex.

First assume that $D(A_1)$ is bounded. Let $\mathcal{F}: X \to X^*$ be the duality map. According to Theorem 6.1.31, it suffices to show that $R(A_1 + A_2 + \mathcal{F}) = X^*$. So, for any given $h^* \in X^*$, we need to show that the operator inclusion $h^* \in A_1(u) + A_2(u) + \mathcal{F}(u)$ has a solution. Replacing A_2 with $A_2 - h^*$, we see that it suffices to show that $0 \in A_1(u) + A_2(u) + \mathcal{F}(u)$ has a solution. So, we need to find $(v, v^*) \in X \times X^*$ such that

$$-v^* \in \left(A_1 + \frac{1}{2}\mathcal{F}\right)(v) \quad \text{and} \quad v^* \in \left(A_2 + \frac{1}{2}\mathcal{F}\right)(v) .$$

Let

$$K(v^*) = -\left(A_1 + \frac{1}{2}\mathcal{F}\right)^{-1}(-v^*) \quad \text{and} \quad L(v^*) = \left(A_2 + \frac{1}{2}\mathcal{F}\right)^{-1}(v^*) .$$

Then $K, L: X^* \to X$ are both demicontinuous and monotone. We have $R(L) = D(A_2 + 1/2\mathcal{F}) = D(A_2)$ and so,

$$R(L) \text{ is bounded and } 0 \in \operatorname{int} D(L) . \tag{6.1.61}$$

We have to find $v^* \in X^*$ such that

$$K(v^*) + L(v^*) = 0 . \tag{6.1.62}$$

Evidently, $K + L$ is monotone and demicontinuous with $D(K + L) = X^*$. Hence, $K + L$ is maximal monotone; see Proposition 6.1.17. Moreover, we see that $0 \in A(0)$ and $\mathcal{F}(0) = 0$. Therefore, $K(0) = 0$. Since K is monotone, one gets

$$\langle K(u^*), u^* \rangle \geq 0 \quad \text{for all } u^* \in X^* . \tag{6.1.63}$$

Exploiting the monotonicity of L gives

$$\langle L(u^*), u^* \rangle \geq \langle L(v^*), u^* \rangle + \langle L(u^*) - L(v^*), v^* \rangle \quad \text{for all } u^*, v^* \in X^* . \tag{6.1.64}$$

Since $R(L)$ is bounded (see (6.1.61)), there exists $M > 0$ such that

$$|\langle L(u^*) - L(v^*), v^* \rangle| \leq M \|v^*\|_* \quad \text{for all } u^*, v^* \in X^* .$$

Since $0 \in \operatorname{int} D(A_2)$, it follows that $0 \in \operatorname{int} R(L)$ and so Proposition 6.1.11 yields that L^{-1} is locally bounded at $u^* = 0$. Hence, there exist $\delta, \varepsilon > 0$ such that

$$\|L(v^*)\| \leq \delta \quad \text{implies} \quad \|v^*\|_* \leq \varepsilon .$$

Recall that $0 \in \operatorname{int} R(L)$. So, we can find $\delta' \in (0, \delta]$ such that for each $y^* \in X^*$ with $\|y^*\|_* \leq \delta'$, the operator equation $y^* = L(v^*)$ has a solution. Then, due to (6.1.63) and (6.1.64),

$$\langle L(u^*), u^* \rangle \geq \sup[\langle y^*, u^* \rangle + \langle L(u^*) - L(v^*), v^* \rangle : \|y^*\|_* = \delta'] \geq \delta \|u^*\|_* - M\varepsilon .$$

It follows that there exists $\rho > 0$ such that

$$\langle L(u^*), u^* \rangle > 0 \quad \text{for all } u^* \in X^* \text{ with } \|u^*\|_* > \rho . \tag{6.1.65}$$

From (6.1.63) and (6.1.65) we infer that $K + L$ is coercive and so, with view to Corollary 6.1.33, $K + L$ is surjective. This means that (6.1.62) has a solution.

Now we assume that $D(A_2)$ is unbounded. Let $i_r = i_{\overline{B}_r}$ be the indicator function of $\overline{B}_r = \{u \in X : \|u\| \leq r\}$. We know that

$$\partial i_r(u) = \begin{cases} \{0\} & \text{if } \|u\| < r\,, \\ \{\lambda\mathcal{F}(u) : \lambda > 0\} & \text{if } \|u\| = r\,, \\ \emptyset & \text{if } \|u\| > r\,. \end{cases} \tag{6.1.66}$$

Note that $u^* \in \partial i_r(u)$ if and only if $u^* = \lambda\mathcal{F}(u)$ for some $\lambda \geq 0$. From the first part of the proof, we know that $u \to A_2(u) + \partial i_r(u)$ is maximal monotone and $0 \in D(A_1) \cap \operatorname{int} D(A_2 + \partial i_r)$. In addition, the first part of the proof gives that $u \to A_1(u)+A_2(u)+\partial i_r(u)$ is maximal monotone for every $\lambda > 0$. Taking Theorem 6.1.31 into account yields $R(A_1+A_2+\partial i_r+\mathcal{F}) = X^*$ for every $r > 0$.

Let $h^* \in X^*$ and choose $r > \|h^*\|_*$. Then there exists $u \in X$ with $\|u\| \leq r$ such that $h^* \in (A_1 + A_2)(u) + (1 + \lambda)\mathcal{F}(u)$; see (6.1.66). This implies $h^* = u^* + (1 - \lambda)\mathcal{F}(u)$ with $u^* \in (A_1 + A_2)(u)$ and so $\langle h^*, u\rangle = \langle u^*, u\rangle + (1 + \lambda)\langle \mathcal{F}(u), u\rangle$. Since $0 \in (A_1 + A_2)(0)$, we obtain $\langle h^*, u\rangle \geq (1+\lambda)\|u\|^2$ and this gives $(1+\lambda)\|u\| \leq \|h^*\|_* < r$. Thus, $\|u\| < r$. Therefore, we conclude that $R(A_1 + A_2 + \mathcal{F}) = X^*$ and this means that $A_1 + A_2$ is maximal monotone; see Theorem 6.1.31. □

Next we introduce some extensions of maximal monotone maps. These extensions are useful in the study of nonlinear boundary value problems.

The setting remains the same as before, namely, X is a reflexive Banach space, X^* is its topological dual space, and $A: X \to 2^{X^*}$ is a multivalued map.

Definition 6.1.50. (a) We say that A is **pseudomonotone** if the following hold:
- (i) $A(u) \in P_{\mathrm{wkc}}(X^*)$ for every $u \in X$;
- (ii) $A|_V$ is usc into X^*_{w} for every finite dimensional subspace $V \subseteq X$;
- (iii) if $\{(u_n, u_n^*)\}_{n\geq 1} \subseteq \operatorname{Gr} A$ with $u_n \xrightarrow{\mathrm{w}} u$ in X and $\limsup_{n\to\infty}\langle u_n^*, u_n - u\rangle \leq 0$, then for every $v \in X$ there exists $y^*(v) \in A(u)$ such that

$$\langle y^*(v), u - v\rangle \leq \liminf_{n\to\infty}\langle u_n^*, u_n - v\rangle\,.$$

(b) We say that A is **generalized pseudomonotone** if for any sequence $\{(u_n, u_n^*)\}_{n\geq 1} \subseteq \operatorname{Gr} A$ with $u_n \xrightarrow{\mathrm{w}} u$ in X, $u_n^* \xrightarrow{\mathrm{w}} u^*$ in X^* and $\limsup_{n\to\infty}\langle u_n^*, u_n - u\rangle \leq 0$, then $u^* \in A(u)$ and $\langle u_n^*, u_n\rangle \to \langle u^*, u\rangle$.

From the definition above we deduce the following result.

Proposition 6.1.51. *The map $A: X \to 2^{X^*}$ is generalized pseudomonotone if and only if $A^{-1}: X^* \to 2^X$ is generalized pseudomonotone.*

Proposition 6.1.52. *If $A: X \to 2^{X^*}$ is pseudomonotone, then A is generalized pseudomonotone.*

Proof. Let $\{(u_n, u_n^*)\}_{n\geq 1} \subseteq \operatorname{Gr} A$ and assume that

$$u_n \xrightarrow{\mathrm{w}} u \quad \text{in } X, \quad u_n^* \xrightarrow{\mathrm{w}} u^* \quad \text{in } X^* \quad \text{and} \quad \limsup_{n\to\infty}\langle u_n^*, u_n - u\rangle \leq 0\,. \tag{6.1.67}$$

Since A is pseudomonotone, given any $v \in X$, there exists $y^*(v) \in A(x)$ such that

$$\langle y^*(v), u - v\rangle \le \liminf_{n\to\infty}\langle u_n^*, u_n - v\rangle . \tag{6.1.68}$$

We may assume that $\langle u_n^*, u_n\rangle \to \eta \in \mathbb{R}$. Then we obtain

$$\limsup_{n\to\infty}\langle u_n^*, u_n - u\rangle = \eta - \langle u^*, u\rangle \le 0 ; \tag{6.1.69}$$

see (6.1.67). Moreover, (6.1.68) implies

$$\eta - \langle u^*, v\rangle \ge \liminf_{n\to\infty}\langle u_n^*, u_n - v\rangle \ge \langle y^*(v), u - v\rangle ,$$

which, because of (6.1.69), leads to

$$\langle u^*, u - v\rangle \ge \langle y^*(v), u - v\rangle \quad \text{for all } v \in X . \tag{6.1.70}$$

Suppose that $u^* \notin A(u)$. Since $A(u) \in P_{\mathrm{wkc}}(X^*)$ there exists $y \in X$ such that

$$\langle u^*, y\rangle < \inf[\langle v^*, y\rangle : v^* \in A(u)] . \tag{6.1.71}$$

We choose $v = u - y$ in (6.1.70) to get

$$\langle y^*(v), y\rangle \le \langle u^*, y\rangle \quad \text{and} \quad y^*(v) \in A(u) . \tag{6.1.72}$$

Comparing (6.1.71) and (6.1.72) gives a contradiction. Therefore, $u^* \in A(u)$.

Next, let $v = u \in X$ in (6.1.68). This yields $0 \le \liminf_{n\to\infty}\langle u_n^*, u_n - u\rangle$ and so

$$\langle u^*, u\rangle \le \liminf_{n\to\infty}\langle u_n^*, u_n\rangle . \tag{6.1.73}$$

On the other hand, from (6.1.69) we obtain

$$\limsup_{n\to\infty}\langle u_n^*, u_n\rangle \le \langle u^*, u\rangle . \tag{6.1.74}$$

From (6.1.73) and (6.1.74) we infer that $\langle u_n^*, u_n\rangle \to \langle u^*, u\rangle$. Hence, A is generalized pseudomonotone. □

Proposition 6.1.53. *If $A: X \to 2^{X^*}$ is a bounded, generalized pseudomonotone map and $A(u) \in P_{\mathrm{wkc}}(X^*)$ for every $u \in X$, then A is pseudomonotone.*

Proof. Suppose that $\{(u_n, u_n^*)\}_{n\ge1} \subseteq \operatorname{Gr} A$ with $u_n \xrightarrow{w} u$ in X and

$$\limsup_{n\to\infty}\langle u_n^*, u_n - u\rangle \le 0 .$$

Let $v \in X$. Arguing by contradiction, suppose that A is not pseudomonotone. Then there exists $v \in X$ such that

$$\liminf_{n\to\infty}\langle u_n^*, u_n - v\rangle < \inf[\langle y^*, u - v\rangle : y^* \in A(u)] .$$

By passing to a suitable subsequence if necessary we can suppose that

$$\lim_{n\to\infty}\langle u_n^*, u_n - v\rangle < \inf[\langle y^*, u - v\rangle : y^* \in A(u)] . \tag{6.1.75}$$

Since A is bounded and $\{u_n\}_{n\geq 1} \subseteq X$ is bounded, it follows that $\{u_n^*\}_{n\geq 1} \subseteq X^*$ is bounded. Therefore we may assume that $u_n^* \xrightarrow{w} u^*$ in X^*. Since A is generalized pseudomonotone it follows that

$$u^* \in A(u) \quad \text{and} \quad \langle u_n^*, u_n\rangle \to \langle u^*, u\rangle . \tag{6.1.76}$$

Therefore, thanks to (6.1.75),

$$\lim_{n\to\infty}\langle u_n^*, u_n - v\rangle = \langle u^*, u - v\rangle < \inf[\langle y^*, u - v\rangle : y^* \in A(u)]$$

and this is a contradiction since $u^* \in A(u)$; see (6.1.76). Therefore, requirement (iii) in Definition 6.1.50 is satisfied.

Next we show that A is usc from X into X^*_w. Since A is bounded, on account of Proposition 5.6.11 and the Eberlein–Smulian Theorem, it suffices to show that $\operatorname{Gr} A \subseteq X \times X^*_w$ is sequentially closed. So, let $\{(u_n, u_n^*)\}_{n\geq 1} \subseteq \operatorname{Gr} A$ and assume that $u_n \to u$ in X and $u_n^* \xrightarrow{w} u^*$ in X^*. Since $\langle u_n^*, u_n - u\rangle \to 0$ and A is generalized pseudomonotone, we see that $u^* \in A(u)$ and so A is usc from X into X^*_w. Therefore we conclude that A is pseudomonotone; see Definition 6.1.50 (a). □

The next result shows that the family of generalized pseudomonotone maps contains maximal monotone ones.

Proposition 6.1.54. *If $A: X \to 2^{X^*}$ is maximal monotone, then A is generalized pseudomonotone.*

Proof. Consider a sequence $\{(u_n, u_n^*)\}_{n\geq 1} \subseteq \operatorname{Gr} A$ such that

$$u_n \xrightarrow{w} u \quad \text{in } X, \quad u_n^* \xrightarrow{w} u^* \quad \text{in } X^* \quad \text{and} \quad \limsup_{n\to\infty}\langle u_n^*, u_n - u\rangle \leq 0 . \tag{6.1.77}$$

Let $(v, v^*) \in \operatorname{Gr} A$. Then by the monotonicity of A we have

$$\langle u_n^* - v^*, u_n - v\rangle \geq 0 \quad \text{for all } n \in \mathbb{N} . \tag{6.1.78}$$

Moreover, we obtain

$$\langle u_n^*, u_n\rangle = \langle u_n^* - v^*, u_n - v\rangle + \langle u_n^*, v\rangle + \langle v^*, v\rangle \quad \text{for all } n \in \mathbb{N} .$$

Then, (6.1.77) and (6.1.78) give

$$\langle u^*, u\rangle \geq \limsup_{n\to\infty}\langle u_n^*, u_n\rangle \geq \langle u^*, v\rangle + \langle v^*, u\rangle - \langle v^*, v\rangle .$$

Hence,

$$\langle u^* - v^*, u - v\rangle \geq 0 . \tag{6.1.79}$$

Since $(v, v^*) \in \operatorname{Gr} A$ is arbitrary, (6.1.79) along with the maximal monotonicity of A lead to $(u, u^*) \in \operatorname{Gr} A$. Hence,

$$\langle u_n^* - u^*, u_n - u\rangle \geq 0 \quad \text{for all } n \in \mathbb{N}$$

and so $\liminf_{n\to\infty}\langle u_n^*, u_n\rangle \geq \langle u^*, u\rangle$. Using (6.1.77) gives $\langle u_n^*, u_n\rangle \to \langle u^*, u\rangle$. Therefore, A is generalized pseudomonotone. □

Corollary 6.1.55. *If $A: X \to 2^{X^*}$ is maximal monotone with $D(A) = X$, then A is pseudomonotone.*

Similar to maximal monotone maps (see Corollary 6.1.33), pseudomonotone maps exhibit remarkable surjectivity properties. The proof of the surjectivity result uses Galerkin approximations. For this reason, we first prove a finite dimensional surjectivity result.

Proposition 6.1.56. *If V is a finite dimensional Banach space and $F: V \to P_{\mathrm{kc}}(V^*)$ is usc and strongly coercive, then F is surjective.*

Proof. For every $v^* \in V^*$, the multifunction $F_{v^*}(v) = F(v) - v^*$ has the same properties as F. Therefore, in order to prove the proposition, it suffices to show that $0 \in R(F)$.

Arguing by contradiction, suppose that $0 \notin R(F)$. Then, for every $v \in V$, there exists $y(v) \in X \setminus \{0\}$ such that

$$0 < \inf[\langle v^*, y(v)\rangle : v^* \in F(v)] .$$

The strong coercivity of F implies that for a given $M > 0$ there exists $r = r(M) > 0$ such that

$$\frac{\langle v^*, v\rangle}{\|v\|} \geq M \quad \text{for all } v^* \in F(v) \text{ and for all } \|v\| \geq r .$$

This yields

$$\langle v^*, v\rangle \geq Mr \quad \text{for all } v^* \in F(v) \text{ and for all } \|v\| = r .$$

For such $v \in X$ we can take $y(v) = v$ and for $v \in V \setminus \{0\}$ we define

$$U(v) = \{y \in V : \inf[\langle y^*, v\rangle : y^* \in F(y)] > 0\} .$$

The upper semicontinuity of F implies that

$$y \to \inf[\langle y^*, v\rangle : y^* \in F(y)]$$

is lower semicontinuous. So, $U(v) \subseteq V$ is open. Then $\{U(v)\}_{v \in V \setminus \{0\}}$ is an open cover of V. Since V is finite dimensional there exists an open cover $\{W_k\}_{k=1}^m$ of $\overline{B}_r = \{v \in V : \|u\| \leq r\}$ such that for each $k \in \{1, \ldots, m\}$ there is $v_k \in V$ such that $W_k \subseteq U(v_k)$ and

$$W_k \cap \partial B_r \neq \emptyset \quad \text{implies} \quad v_k \in W_k \cap \partial B_r \quad \text{and} \quad \operatorname{diam} W_k < \frac{r}{2} . \tag{6.1.80}$$

Let $\{\xi_k\}_{k=1}^m$ be a continuous partition of unity subordinate to the cover $\{W_k\}_{k=1}^m$. We define $f : \overline{B}_r \to V$ by setting

$$f(v) = \sum_{k=1}^m \xi_k(v) v_k .$$

Then f is continuous and for each $k \in \mathbb{N}$ such that $\xi_k(v) > 0$ and each $v^* \in F(v)$ we obtain $\langle v^*, v_k\rangle > 0$ since $v \in W_k \subseteq U(v_k)$. So, for every $v \in \overline{B}_r$ and every $v^* \in F(v)$ one has

$$\langle v^*, f(v)\rangle = \sum_{k=1}^m \xi_k(v)\langle v^*, v_k\rangle > 0 .$$

Hence, $f(v) \neq 0$ for all $v \in \overline{B}_r$ and so

$$d_B(f, B_r, 0) = 0 , \tag{6.1.81}$$

where $d_B(f, B_r, 0)$ denotes the Brouwer degree of f on B_r with respect to zero; see Section 6.2.

On the other hand, if $v \in \partial B_r$, $f(v)$ is a convex combination of the points $v_k \in \partial B_r$ with $k = 1, \ldots, m$ and $\|v_k - v\| < r/2$ for all $k \in \{1, \ldots, m\}$; see (6.1.80). Thus, $\|f(v) - v\| \leq r/2$ for all $v \in \partial B_r$. Therefore, f is homotopic to the identity and so

$$d_B(f, B_r, 0) = 1 . \tag{6.1.82}$$

Comparing (6.1.81) and (6.1.82) leads to a contradiction. Hence, $0 \in R(F)$ and so F is surjective. □

We will use Proposition 6.1.56 and Galerkin approximations to prove the surjectivity result for pseudomonotone maps.

Theorem 6.1.57. *If $A : X \to 2^{X^*}$ is pseudomonotone and strongly coercive, then A is surjective, that is, $R(A) = X^*$.*

Proof. As we already mentioned, we will use Galerkin approximations to prove this theorem. So, let $\mathcal{L}$ be the family of finite dimensional subspaces partially ordered by inclusion. Given $V \in \mathcal{L}$, let $i_V: V \to X$ be the embedding operator. The adjoint $i_V^*: X^* \to V^*$ is the projection operator onto V^*. We set $A_V = i_V^* \circ A \circ i_V$. Then A_V has values in $P_{\mathrm{kc}}(V^*)$ and is usc. Moreover, for every $u_V^* \in A_V(u)$, we see that $u_V^* = i_V^* u^*$ for some $u^* \in A(u)$. Hence,

$$\langle u_V^*, u\rangle_V = \langle i_V^* u^*, u\rangle_V = \langle u^*, i_V(u)\rangle .$$

This shows that A_V is strongly coercive as well.

In order to prove the theorem, it suffices to show that $0 \in R(A)$. On account of Proposition 6.1.56, for every $V \in \mathcal{L}$, there exists $u_V \in V$ such that $0 \in A_V(u_V)$, which gives

$$0 = i_V^* u_V^* \quad \text{for some } u_V^* \in A(u_V) . \tag{6.1.83}$$

The strong coercivity of A implies that $\{u_V\}_{V\in\mathcal{L}} \subseteq X$ is bounded. For $V \in \mathcal{L}$, let

$$E_V = \bigcup_{\substack{V' \in \mathcal{L} \\ V' \supseteq V}} \{u_{V'}\} .$$

Then, $E_V \subseteq \overline{B}_M$ for some $M > 0$. The reflexivity of X implies that $\overline{B}_M$ is w-compact. Then the finite intersection property says

$$\bigcap_{V\in\mathcal{L}} \overline{E}_V^{\mathrm{w}} \neq \emptyset .$$

Let $u_0 \in \bigcap_{V\in\mathcal{L}} \overline{E}_V^{\mathrm{w}}$ and $y \in X$. We choose $V \in \mathcal{L}$ such that $\{u_0, y\} \subseteq V$. Let $\{u_{V_k}\}_{k\geq 1} \subseteq E_V$ be such that

$$u_{V_k} \xrightarrow{\mathrm{w}} u_0 \quad \text{in } X \text{ as } k \to \infty .$$

From (6.1.83) we know that $0 = i_{V_k}^* u_{V_k}^*$ with $u_{V_k}^* \in A(u_{V_k})$. Thus,

$$\langle u_{V_k}^*, u_{V_k} - u_0\rangle = 0 \quad \text{for all } k \in \mathbb{N} .$$

The pseudomonotonicity of A implies that there exists $v^*(y) \in A(u_0)$ such that

$$\langle v^*(y), u_0 - y\rangle \leq \liminf_{k\to+\infty} \langle u_{V_k}^*, u_{V_k} - y\rangle = 0 \quad \text{for all } y \in X . \tag{6.1.84}$$

Suppose that $0 \notin A(u_0)$. Then by the Strong Separation Theorem, we find $y \in X$ such that

$$0 < \inf[\langle u^*, u_0 - y\rangle : u^* \in A(u_0)] . \tag{6.1.85}$$

Comparing (6.1.84) and (6.1.85), we reach a contradiction. So, A is surjective. □

Definition 6.1.58. Let $C \subseteq X$ be a nonempty set and let $A: C \to X^*$. We say that A is an **(S)$_+$-map** if for every sequence $\{u_n\}_{n\geq 1} \subseteq C$ such that

$$u_n \xrightarrow{w} u \quad \text{in } X \quad \text{and} \quad \limsup_{n\to\infty} \langle A(u_n), u_n - u \rangle \leq 0\,, \tag{6.1.86}$$

it follows that $u_n \to u$ in X.

Remark 6.1.59. The prototype of a (S)$_+$-map is a uniformly monotone map.

Let $\Omega \subseteq \mathbb{R}^N$ be a bounded domain with Lipschitz boundary, $1 < p < \infty$, and consider the map $A: W^{1,p}(\Omega) \to W^{1,p}(\Omega)^*$ defined by

$$\langle A(u), h \rangle = \int_\Omega |Du|^{p-2}(Du, Dh)_{\mathbb{R}^N}\, dz \quad \text{for all } u, h \in W^{1,p}(\Omega)\,. \tag{6.1.87}$$

Proposition 6.1.60. *The map $A: W^{1,p}(\Omega) \to W^{1,p}(\Omega)^*$ defined by* (6.1.87) *is of type* (S)$_+$.

Proof. Clearly, A is monotone and continuous. Hence, A is maximal monotone; see Proposition 6.1.17. Then Proposition 6.1.54 implies that

$$A \text{ is generalized pseudomonotone}\,. \tag{6.1.88}$$

Suppose $\{u_n\}_{n\geq 1} \subseteq X$ satisfies (6.1.86). Then from (6.1.88) we obtain that $\langle A(u_n), u_n \rangle \to \langle A(u), u \rangle$. Hence,

$$\|Du_n\|_p \to \|Du\|_p\,.$$

Recall that $W^{1,p}(\Omega)$ is uniformly convex for $1 < p < \infty$ and $u_n \xrightarrow{w} u$ in $W^{1,p}(\Omega)$. Then from the Kadec–Klee Property we have $u_n \to u$ in $W^{1,p}(\Omega)$. Hence, A is of (S)$_+$-type. □

6.2 Brouwer Degree

In this section we develop the finite dimensional degree theory, that is, Brouwer's degree. In what follows, let $\Omega \subseteq \mathbb{R}^N$ be a bounded open set.

Definition 6.2.1. Let $\varphi \in C^1(\Omega, \mathbb{R}^N) \cap C(\overline{\Omega}, \mathbb{R}^N)$. We say that $u_0 \in \Omega$ is a **critical point of** φ if $J_\varphi(u_0) = \det \varphi'(u_0) = 0$, where $J_\varphi(u_0)$ is called the Jacobian of φ at u_0. Let

$$K_\varphi(\Omega) = K_\varphi = \{u \in \Omega: J_\varphi(u) = 0\}\,,$$

which is called the **critical or singular set** of φ. The set $\varphi(K_\varphi(\Omega))$ is the set of **critical values** (or the **crease**) of φ. If $h \notin \varphi(K_\varphi)$, then we say that h is a **regular value** of φ.

The next theorem known as "Sard's Theorem" says that there are only a few critical values.

Theorem 6.2.2 (Sard's Theorem). *If $\varphi \in C^1(\Omega, \mathbb{R}^N)$, then $\lambda^N(\varphi(K_\varphi)) = 0$ with λ^N being the Lebesgue measure on $\mathbb{R}^N$.*

Proof. Let $C \subseteq \Omega$ be a closed cube of side a. The uniform continuity of φ' on C implies for a given $\varepsilon > 0$ the existence of $k \in \mathbb{N}$ such that

$$|\varphi'(u) - \varphi'(v)| \le \varepsilon \quad \text{for all } u, v \in C \text{ with } |u - v| \le \frac{\sqrt{N}a}{k} = \delta . \tag{6.2.1}$$

From (6.2.1), for all $u, v \in C$ with $|u - v| \le (\sqrt{N}a)/k = \delta$, we obtain

$$|\varphi(u) - \varphi(v) - \varphi'(v)(u - v)| \le \int_0^1 |\varphi'(v + t(u - v)) - \varphi'(v)||u - v| \, dt \le \varepsilon |u - v| .$$

We subdivide C into $\{C_i\}_{i=1}^{k^N}$ cubes of side a/k. We choose $u \in C_i \cap K_\varphi(\Omega)$. Then

$$\varphi(u + v) = \varphi(u) - \varphi'(u)v + r(u + v, u) \quad \text{for all } v \in C_i - u ,$$

with $|r(u + v, u)| \le \varepsilon(\sqrt{N}a)/k$. Therefore, we get

$$\varphi(C_i) = \varphi(u) + \varphi'(u)(C_i - u) + r(u + v, C_i) .$$

But $\det \varphi'(u) = 0$. So, $\varphi'(u)(C_i - u)$ is contained in an $(N-1)$-dimensional subspace of $\mathbb{R}^N$ and this implies that $\lambda^N(\varphi'(v)(C_i - u)) = 0$. So, we have

$$\lambda^N(\varphi(C_i)) \le 2^N \varepsilon^N \left(\frac{\sqrt{N}a}{k} \right)^N . \tag{6.2.2}$$

Note that

$$\varphi(K_\varphi(C)) \subseteq \bigcup_{i=1}^{k^N} \varphi(C_i) .$$

Therefore, thanks to (6.2.2), one obtains

$$\lambda^N(K_\varphi(C)) \le 2^N \varepsilon^N (\sqrt{N}a)^N .$$

Letting $\varepsilon \to 0^+$ yields

$$\lambda^N(K_\varphi(C)) = \lambda^N(\varphi(K_\varphi \cap C)) = 0 .$$

Since K_φ is a countable union of sets of the form $K_\varphi \cap C$, we conclude that $\lambda^N(\varphi(K_\varphi))= 0$. □

Remark 6.2.3. There is a more general version of the theorem above that says that if $\varphi \in C^1(\Omega, \mathbb{R}^N)$ and $A \subseteq \Omega$ is measurable, then $\varphi(A) \subseteq \mathbb{R}^N$ is measurable and

$$\lambda^N(\varphi(A)) \le \int_A |J_\varphi(z)|\, dz\,.$$

We will define the Brouwer degree by successively extending the range of φ and of the reference point $h \in \mathbb{R}^N$. We start with the following definition.

Definition 6.2.4. Suppose that $\varphi \in C^1(\Omega, \mathbb{R}^N) \cap C(\overline{\Omega}, \mathbb{R}^N)$ and let $h \notin \varphi(\partial\Omega \cup K_\varphi)$. Then the **degree** of φ at h with respect to Ω is defined by

$$d_B(\varphi, \Omega, h) = \sum_{u \in \varphi^{-1}(h)} \operatorname{sgn} J_\varphi(u)\,, \tag{6.2.3}$$

where the sign function sgn is defined by

$$\operatorname{sgn}(t) = \begin{cases} -1 & \text{if } t < 0\,, \\ 0 & \text{if } t = 0\,, \\ 1 & \text{if } 0 < t\,. \end{cases}$$

When $\varphi^{-1}(h) = \emptyset$, then we set $d_B(\varphi, \Omega, h) = 0$.

Remark 6.2.5. The Inverse Function Theorem implies that $\varphi^{-1}(h)$ is finite and so the sum in (6.2.3) is well-defined.

Consider a mollifier ϑ, that is, $\vartheta \in C(\mathbb{R}^N, \mathbb{R})$ such that

$$\operatorname{supp} \vartheta \subseteq B_1(0) = \{x \in \mathbb{R}^N : |x| < 1\} \quad \text{and} \quad \int_{\mathbb{R}^N} \vartheta(x)\, dx = 1\,.$$

For $\varepsilon > 0$ we define

$$\vartheta_\varepsilon(x) = \frac{1}{\varepsilon^N} \vartheta\left(\frac{x}{\varepsilon}\right) \quad \text{for all } x \in \mathbb{R}^N\,.$$

Evidently, $\operatorname{supp} \vartheta_\varepsilon \subseteq B_\varepsilon(0) = \varepsilon B_1(0)$.

Proposition 6.2.6. *If (Ω, φ, h) are as in Definition 6.2.4 and $\{\vartheta_\varepsilon\}_{\varepsilon>0}$ is as above, then there exists $\varepsilon_0 = \varepsilon_0(\varphi, h) > 0$ such that*

$$d_B(\varphi, \Omega, h) = \int_\Omega \vartheta_\varepsilon(\varphi(x) - h) J_\varphi(x)\, dx \quad \textit{for all } 0 < \varepsilon \le \varepsilon_0\,.$$

Proof. The case $\varphi^{-1}(h) = \emptyset$ is trivial since then $\vartheta_\varepsilon(\varphi(x) - h) = 0$ for $\varepsilon < d = d(h, \varphi(\overline{\Omega}))$. So, we assume that $\varphi^{-1}(h) = \{u_k\}_{k=1}^m$. By the Inverse Function Theorem, there exist disjoint balls $B_r(u_k)$ and $U_k \in \mathcal{N}(h)$ for $k = 1, \dots, m$ such that $\varphi\colon B_r(u_k) \to U_k$ is a homeomorphism. Then $\operatorname{sgn} J_\varphi(u) = \operatorname{sgn} J_\varphi(u_k)$ on $\overline{B}_r(u_k)$. Let $r_0 > 0$ be such that $B_{r_0}(h) \subseteq \bigcap_{k=1}^m U_k \in \mathcal{N}(h)$. We set $V_k = B_r(u_k) \cap \varphi^{-1}(B_{r_0}(h))$. Then

$$|\varphi(u) - h| \geq \eta \quad \text{for all } u \in \overline{\Omega} \setminus \bigcup_{k=1}^m V_k \text{ for some } \eta > 0\,.$$

Then, $\varepsilon < \eta$ implies that

$$\int_\Omega \vartheta_\varepsilon(\varphi(x) - h) J_\varphi(x)\,dx = \sum_{k=1}^m \operatorname{sgn} J_\varphi(u_k) \int_{V_k} \vartheta_\varepsilon(\varphi(x) - h)|J_\varphi(x)|\,dx \tag{6.2.4}$$

Since $J_\varphi = J_{\varphi - h}$ and $\varphi(V_k) - h = B_{r_0}(h)$, we obtain

$$\int_{V_k} \vartheta_\varepsilon(\varphi(x) - h)|J_{\varphi-h}(x)|\,dx = \int_{B_{r_0}} \vartheta_\varepsilon(x)\,dx = 1 \quad \text{for } \varepsilon < \min\{\eta, r_0\}\,. \tag{6.2.5}$$

Finally, using (6.2.5) in (6.2.4), we conclude that

$$\int_\Omega \vartheta_\varepsilon(\varphi(x) - h) J_\varphi(x)\,dx = \sum_{k=1}^m \operatorname{sgn} J_\varphi(u_k) = d_B(\varphi, \Omega, h)\,;$$

see Definition 6.2.4. □

Using this expression for the degree we can pass from regular to singular values, that is, remove the restriction $h \notin \varphi(K_\varphi)$.

So, let $\varphi \in C^2(\Omega, \mathbb{R}^N) \cap C(\overline{\Omega}, \mathbb{R}^N)$ and assume that $h \notin \varphi(\partial\Omega)$. Then $d = d(h, \varphi(\partial\Omega)) > 0$. Suppose that $h_1, h_2 \in B_d(h)$ are two regular values of φ, and the existence of these values follows from Theorem 6.2.2. Let $\delta = d - \max\{|h_k - h| : k = 1, 2\} > 0$. Invoking Proposition 6.2.6, there exists $\varepsilon \in (0, \delta)$ such that

$$d_B(\varphi, \Omega, h_k) = \int_\Omega \vartheta_\varepsilon(\varphi(x) - h) J_\varphi(x)\,dx \quad \text{for } k = 1, 2\,. \tag{6.2.6}$$

Note that

$$\vartheta_\varepsilon(x - h_1) - \vartheta_\varepsilon(x - h_2) = \operatorname{div} g(x)\,, \tag{6.2.7}$$

where

$$g(x) = \left[\int_0^t \vartheta_\varepsilon(x - h_1 + t(h_1 - h_2))\,dt\right](h_1 - h_2)\,.$$

We show that there exists $f \in C^1(\mathbb{R}^N)$ such that $\operatorname{supp} f \subseteq \Omega$ and

$$|\vartheta_\varepsilon(\varphi(x) - h_2) - \vartheta_\varepsilon(\varphi(x) - h_1)|J_\varphi(x) = \operatorname{div} f(x) \quad \text{for all } x \in \Omega\,. \tag{6.2.8}$$

Proposition 6.2.7. *If $\Omega \subseteq \mathbb{R}^N$ is an open set, $\varphi \in C^2(\Omega)$, and $\hat{d}_{ij}(x)$ is the co-factor of $(\partial\varphi_i/\partial x_j)$ in $J_\varphi(x)$, that is, $d_{ij}(x) = (-1)^{i+j}\det A_{ij}$, where A_{ij} is the matrix obtained from $(\partial\varphi_i/\partial x_j)$ by removing the $i\,\underline{\underline{th}}$-row and the $j\,\underline{\underline{th}}$-column, and*

$$f_i(x) = \begin{cases} \sum_{j=1}^N g_j(\varphi(x))d_{ij}(x) & \text{if } x \in \overline{\Omega}\,, \\ 0 & \text{otherwise}\,, \end{cases}$$

then $f = (f_i)_{i=1}^N$ satisfies $\operatorname{div} f(x) = \operatorname{div} g(\varphi(x))J_\varphi(x)$.

Proof. Note that $\operatorname{supp} g \subseteq \overline{B}_r(h)$ with $r = d - (\delta - \varepsilon) < d$ where $\operatorname{supp}\vartheta_\varepsilon = \overline{B}_\varepsilon(0)$. So, we have $\operatorname{supp} f \subseteq \Omega$ and

$$\partial_i f_i(x) = \sum_{j,k=1}^N d_{jk}\partial_k g_j(\varphi(x))\partial_i\varphi_k(x) + \sum_{j=1}^N g_j(\varphi(x))\partial_i d_{ij}(x)\,,$$

where $\partial_i = \partial/(\partial x_i)$. We claim that

$$\sum_{i=1}^N \partial_i d_{ij}(x) = 0 \quad \text{for all } j = 1, \ldots, N\,.$$

For every j, let $\hat{\varphi}_{x_k}$ be the column $(\partial_k\varphi_l : l = 1, \ldots, j-1, j+1, \ldots, N)$. It holds that

$$d_{ij}(x) = (-1)^{i+j}\det(\hat{\varphi}_{x_1}, \ldots, \hat{\varphi}_{x_{i-1}}, \hat{\varphi}_{x_{i+1}}, \ldots, \hat{\varphi}_{x_N})\,.$$

Since the determinant is linear in each column we infer that

$$\partial_i d_{ij}(x) = (-1)^{i+j}\sum_{k=1}^N \det(\hat{\varphi}_{x_1}, \ldots, \hat{\varphi}_{x_{k-1}}, \partial_i\hat{\varphi}_{x_k}, \hat{\varphi}_{x_{k+1}}, \ldots, \hat{\varphi}_{x_N})\,.$$

Set

$$c_{ki} = \det(\partial_i\hat{\varphi}_{x_k}, \hat{\varphi}_{x_1}, \ldots, \hat{\varphi}_{x_{i-1}}, \hat{\varphi}_{x_{i+1}}, \ldots, \hat{\varphi}_{x_{k-1}}, \hat{\varphi}_{x_{k+1}}, \ldots, \hat{\varphi}_{x_N})\,.$$

As $\varphi \in C^2(\Omega)$ it is clear that $c_{ki} = c_{ik}$ and since the sign of the determinant changes upon permutation of two adjacent columns, we have

$$(-1)^{i+j}\partial_i d_{ij}(x) = \sum_{i,k=1}^N (-1)^{k-1}c_{ki} + \sum_{k>i}(-1)^{k-2}c_{ki} = -\sum_{i,k=1}^N (-1)^{k-1+i}\eta_{ki}c_{ki}\,,$$

where $\eta_{ki} = 1$ for $k < i$, $\eta_{ii} = 0$ and $\eta_{ki} = -\eta_{ik}$ for $i, k = 1, 2, \ldots, N$. Therefore,

$$(-1)^j \sum_{k=1}^{N} \partial_i d_{ij}(x) = \sum_{i,k=1}^{N} (-1)^{k-1+i} \eta_{ki} c_{ki} = \sum_{k,i=1}^{N} (-1)^{i-1+k} \eta_{ik} c_{ik}$$
$$= -\sum_{i,k=1}^{N} (-1)^{k-1+i} \eta_{ki} c_{ki} = 0 .$$

It follows that

$$\partial_i f_i(x) = \sum_{j,k=1}^{N} d_{ij} \partial_k g_j(\varphi(x)) \partial_i \varphi_k(x) + \sum_{j=1}^{N} g_j(\varphi(x)) \partial_i d_{ij}(x) .$$

On the other hand, we see that

$$\sum_{i=1}^{N} d_{ij} \partial_i \varphi_k(x) = \delta_{jk} J_\varphi(x)$$

with the Kronecker symbol δ_{jk}, that is, $\delta_{jk} = 1$ if $j = k$ and $\delta_{jk} = 0$ if $j \neq k$. So, finally we derive

$$\operatorname{div} f(x) = \sum_{k,j=1}^{N} \partial_k g_j(\varphi(x)) \delta_{jk} J_\varphi(x) = \operatorname{div} g(\varphi(x)) J_\varphi(x) .$$ □

Consider a cube $Q = [-u_0, u_0]^N$ such that $\Omega \subseteq Q$. From (6.2.6), (6.2.7) and (6.2.8), we infer that

$$\begin{aligned} & d_B(\varphi, \Omega, h_2) - d_B(\varphi, \Omega, h_1) \\ &= \int_\Omega \operatorname{div} f(x)\, \mathrm{d}x = \int_Q \operatorname{div} f(x)\, \mathrm{d}x \\ &= \sum_{i=1}^{N} \int_{-u_0}^{u_0} \cdots \int_{-u_0}^{u_0} \left(\int_{-a}^{a} \partial_i f_i \, \mathrm{d}x_i \right) \mathrm{d}x_1 \ldots \mathrm{d}x_{i-1} \mathrm{d}x_{i+1} \ldots \mathrm{d}x_N = 0 . \end{aligned}$$

This means that the degrees in (6.2.6) are equal and we may define $d_B(\varphi, \Omega, h)$ as $d_B(\varphi, \Omega, h_1)$.

Definition 6.2.8. Suppose that $\varphi \in C^2(\Omega, \mathbb{R}^N) \cap C(\overline{\Omega}, \mathbb{R}^N)$ and assume that $h \notin \varphi(\partial\Omega)$. Then we define $d_B(\varphi, \Omega, h) = d_B(\varphi, \Omega, h_1)$ with a regular value h_1 of φ such that $|h_1 - h| < d(h, \varphi(\partial\Omega))$ and $d_B(\varphi, \Omega, h)$ is given by Definition 6.2.4.

As a final step in the definition of the Brouwer degree we pass from functions in $C^2(\Omega, \mathbb{R}^N) \cap C(\overline{\Omega}, \mathbb{R}^N)$ to functions in $C(\overline{\Omega}, \mathbb{R}^N)$. We will show that the definition is independent of the mollifier ϑ.

Lemma 6.2.9. *If* $\varphi \in C^1(\overline{\Omega}, \mathbb{R}^N) \cap C(\overline{\Omega}, \mathbb{R}^N)$, $0 \notin \varphi(\partial\Omega)$, $\vartheta_0 \colon \mathbb{R}_+ \to \mathbb{R}$ *is continuous,* $\operatorname{supp} \vartheta_0 \subseteq [0, \varepsilon]$ *with* $0 < \varepsilon < d(0, \varphi(\partial\Omega))$, *and* $\int_0^\infty r^{N-1} \vartheta_0(r)\, dr = 0$, *then*

$$\int_\Omega \vartheta_0(|\varphi(x)|) J_\varphi(x)\, dx = 0 .$$

Proof. Regularizing φ if necessary, we may assume that $\varphi \in C^2(\Omega, \mathbb{R}^N)$. Let $\eta \colon \mathbb{R}_+ \to \mathbb{R}$ be the function defined by

$$\eta(0) = 0 \quad \text{and} \quad \eta(r) = \frac{1}{r^N} \int_0^r t^{N-1} \vartheta(t)\, dt \quad \text{if } r > 0 .$$

Evidently, $\eta \in C^1(0, \infty)$ and it has compact support. Moreover,

$$r\eta' + N\eta = \vartheta . \tag{6.2.9}$$

Let $\xi(x) = \eta(|x|)x$ for all $x \in \mathbb{R}^N$. Then

$$\operatorname{div} \xi(x) = |x|\eta'(|x|) + N\eta(|x|) = \vartheta_0(|x|) ;$$

see (6.2.9). For $x \in \Omega$, we obtain

$$\begin{aligned}
&\sum_{k=1}^N \partial_k \sum_{i=1}^N d_{ki}(x) \xi_k(\varphi(x)) \\
&= \sum_{k=1}^N \sum_{i=1}^N (\partial_k d_{ki}(x) \xi_k(\varphi(x))) + \sum_{k=1}^N \sum_{i=1}^N (d_{ki}(x) \partial_k \xi_k(\varphi(x))) \\
&= \sum_{k=1}^N \sum_{i=1}^N d_{ki}(x) \sum_{m=1}^N \partial_k \varphi_m(x) \frac{\partial \xi_i}{\partial v_m}(\varphi(x)) \\
&= \sum_{i=1}^N \sum_{m=1}^N \left(\sum_{k=1}^N d_{ki}(x) \partial_k \varphi_m(x) \right) \frac{\partial \xi_i}{\partial v_m}(\varphi(x)) .
\end{aligned} \tag{6.2.10}$$

Applying Cramer's rule gives

$$\sum_{k=1}^N d_{ki}(x) \partial_k \varphi_m(x) = \delta_{mi} J_\varphi(x) , \tag{6.2.11}$$

where δ_{mi} denotes the Kronecker symbol as before. Using (6.2.11) in (6.2.10) yields

$$\sum_{k=1}^N \partial_k \left(\sum_{i=1}^N d_{ki}(x) \xi_i(\varphi(x)) \right) = J_\varphi(x) \operatorname{div} \xi(\varphi(x)) \quad \text{for all } x \in \Omega . \tag{6.2.12}$$

Therefore, taking (6.2.9) and (6.2.12) into account, gives

$$\begin{aligned}\vartheta_0(|\varphi(x)|)J_\varphi(x) &= J_\varphi(x)[r\eta' + N\eta]_{r=\varphi(x)} = J_\varphi(x)\operatorname{div}\xi(\varphi(x))\\ &= \sum_{k=1}^{N}\partial_k\left(\sum_{i=1}^{N} d_{ki}(x)\xi_i(\varphi(x))\right),\end{aligned}$$

which implies

$$\int_\Omega \vartheta_0(|\varphi(x)|)J_\varphi(x)\,dx = \sum_{k=1}^{N}\int_\Omega \partial_k\left(\sum_{i=1}^{N} d_{ki}(x)\xi_i(\varphi(x))\right)dx.$$

Finally, we obtain $\xi_i(\varphi(x)) = 0$ for all x in a neighborhood for $\partial\Omega$. So, via integration by parts, we conclude that

$$\int_\Omega \vartheta_0(|\varphi(x)|)J_\varphi(x)\,dx = 0. \qquad \square$$

Proposition 6.2.10. *Definition 6.2.8 is independent of $\varepsilon > 0$ and of the mollifier ϑ, provided $\varepsilon < d(h, \varphi(\partial\Omega))$.*

Proof. Let $\varepsilon_0 = d(h, \varphi(\partial\Omega))$ and let $\varepsilon_1, \varepsilon_2 \in (0, \varepsilon_0)$. Suppose ϑ_1, ϑ_2 are two mollifiers such that $\operatorname{supp}\vartheta_1 \subseteq B_{\varepsilon_1}(0)$, $\operatorname{supp}\vartheta_2 \subseteq B_{\varepsilon_2}(0)$ and $\int_{\mathbb{R}^N}\vartheta_1(x)\,dx = \int_{\mathbb{R}^N}\vartheta_2(x)\,dx = 1$. Let $\xi = \vartheta_1 - \vartheta_2$ and let $\hat{\xi}(|x|) = \xi(x)$ for all $x \in \mathbb{R}^N$. Applying Lemma 6.2.9 on $\hat{\xi}$ and the function $x \to \varphi(x) - h$ gives

$$0 = \int_\Omega \hat{\xi}(|\varphi(x) - h|)J_\varphi(x)\,dx = \int_\Omega \xi(\varphi(x) - h)J_\varphi(x)\,dx. \qquad \square$$

Next we will show the stability of Definition 6.2.8 with respect to φ.

Proposition 6.2.11. *If $\varphi_1, \varphi_2 \in C^2(\Omega, \mathbb{R}^N) \cap C(\overline{\Omega}, \mathbb{R}^N)$ and $h \notin \varphi_1(\partial\Omega) \cup \varphi_2(\partial\Omega)$, then, for $\varepsilon < 1/4d(h, \varphi_1(\partial\Omega) \cup \varphi_2(\partial\Omega))$ and for $\|\varphi_1 - \varphi_2\|_\infty < \varepsilon$, we have*

$$d_B(\varphi_1, \Omega, h) = d_B(\varphi_2, \Omega, h).$$

Proof. Without any loss of generality we may assume that $h = 0$; otherwise we replace φ_i by $\varphi_i - h$ and recall that $d_B(\varphi_i, \Omega, h) = d_B(\varphi_i - h, \Omega, 0)$ for $i = 1, 2$. Let ξ be a C^∞-function such that

$$\xi(t) = \begin{cases} 1 & \text{if } t \in [0, \varepsilon], \\ 0 & \text{if } t \geq 2\varepsilon. \end{cases}$$

Let $\varphi_3(x) = (1 - \xi(|\varphi_1(x)|))\varphi_1(x) + \xi(|\varphi_1(x)|)\varphi_2(x)$. Then $\|\varphi_i - \varphi_j\|_\infty < 3\varepsilon$ for $1 \leq i, j \leq 3$, $|\varphi_i(x)| > 3\varepsilon$ for all $x \in \partial\Omega$, and for all $i \in \{1, 2, 3\}$ and

$$\varphi_3(x) = \begin{cases} \varphi_2(x) & \text{if } |\varphi_1(x)| \le \varepsilon\,, \\ \varphi_1(x) & \text{if } |\varphi_1(x)| \ge 2\varepsilon\,. \end{cases}$$

Consider mollifiers ϑ_1, ϑ_2 such that

$$\operatorname{supp}\vartheta_1 \subseteq B_{3\varepsilon}(0) \setminus \overline{B}_{2\varepsilon}(0), \qquad \int_{\mathbb{R}^N} \vartheta_1(x)\,dx = 1\,,$$

$$\operatorname{supp}\vartheta_2 \subseteq B_{\varepsilon}(0), \qquad \int_{\mathbb{R}^N} \vartheta_2(x)\,dx = 1\,.$$

It follows that

$$\vartheta_1(\varphi_3(x))J_\varphi(x) = \vartheta_1(\varphi_1(x))J_\varphi(x)\,,$$
$$\vartheta_2(\varphi_3(x))J_\varphi(x) = \vartheta_2(\varphi_2(x))J_\varphi(x)\,.$$

From Definition 6.2.8 and Proposition 6.2.6 we infer that

$$d_B(\varphi_3, \Omega, 0) = d_B(\varphi_1, \Omega, 0) \quad \text{and} \quad d_B(\varphi_3, \Omega, 0) = d_B(\varphi_2, \Omega, 0)\,. \qquad \square$$

This proposition allows passage to functions that are only continuous on $\overline{\Omega}$. Indeed, let $\varphi \in C(\overline{\Omega}, \mathbb{R}^N)$ and consider a sequence $\{\varphi_n\}_{n\ge1} \subseteq C^1(\Omega, \mathbb{R}^N) \cap C(\overline{\Omega}, \mathbb{R}^N)$ such that $\|\varphi_n - \varphi\|_\infty \to 0$ as $n \to \infty$. Suppose $h \notin \varphi(\partial\Omega)$. Then there exists $n_0 \in \mathbb{N}$ such that $h \notin \varphi_n(\partial\Omega)$ for all $n \ge n_0$. Hence, $d_B(\varphi_n, \Omega, h)$ is well-defined for all $n \ge n_0$. Moreover, if $n_0 \in \mathbb{N}$ is such that

$$\|\varphi_n - \varphi_m\|_\infty < \frac{1}{4} d(h, \varphi_n(\partial\Omega) \cup \varphi_m(\partial\Omega)) \quad \text{for all } n, m \ge n_0\,,$$

then the degree stabilizes with respect to φ_n; see Proposition 6.2.11. Therefore, we can state the following definition of Brouwer's degree for functions $\varphi \in C(\overline{\Omega}, \mathbb{R}^N)$.

Definition 6.2.12. Suppose $\varphi \in C(\overline{\Omega}, \mathbb{R}^N)$ and $h \notin \varphi(\partial\Omega)$. Then

$$d_B(\varphi, \Omega, h) = \lim_{n\to\infty} d_B(\varphi_n, \Omega, h)\,,$$

where $\{\varphi_n\}_{n\in\mathbb{N}} \subseteq C^2(\Omega, \mathbb{R}^N) \cap C(\overline{\Omega}, \mathbb{R}^N)$ and $\|\varphi_n - \varphi\| \to 0$. It is clear that this definition is independent of the approximating sequence $\{\varphi_n\}_{n\in\mathbb{N}}$. This degree is also known as **Brouwer's degree**.

Now we will prove the main properties of the degree function.

Proposition 6.2.13. *If $\varphi \in C(\overline{\Omega}, \mathbb{R}^N)$, then $d_B(\varphi, \Omega, \cdot)$ is constant on the connected components of $\mathbb{R}^N \setminus \varphi(\partial\Omega)$, that is, if h_1, h_2 belong to the same connected component of $\mathbb{R}^N \setminus \varphi(\partial\Omega)$, then $d_B(\varphi, \Omega, h_1) = d_B(\varphi, \Omega, h_2)$.*

Proof. Recall that $d_B(\varphi, \Omega, h) = d_B(\varphi - h, \Omega, 0)$. So, if $\varepsilon < 1/4d(h, \varphi(\partial\Omega))$, then for $|h_1 - h_2| < \varepsilon$, we see that $\|(\varphi - h_1) - (\varphi - h_2)\|_\infty < \varepsilon$ and so for h_1, h_2 in the same connected component of $\mathbb{R}^N \setminus \varphi(\partial\Omega)$ we obtain

$$d_B(\varphi, \Omega, h_1) = d_B(\varphi - h_1, \Omega, 0) = d_B(\varphi - h_2, \Omega, 0) = d_B(\varphi, \Omega, h_2)\,. \qquad \square$$

The next result is an immediate consequence of Definition 6.2.12.

Proposition 6.2.14. *If $\varphi \in C(\overline{\Omega}, \mathbb{R}^N)$, $h \notin \varphi(\partial\Omega)$ and $y \in \mathbb{R}^N$, then $d_B(\varphi - y, \Omega, h - y) = d_B(\varphi, \Omega, h)$.*

We have an additivity property of the degree with respect to the domain.

Proposition 6.2.15. *If $\Omega_1, \Omega_2 \subseteq \mathbb{R}^N$ are bounded, open, disjoint sets, $\Omega = \Omega_1 \cup \Omega_2$, $\varphi \in C(\overline{\Omega}, \mathbb{R}^N)$, and $h \notin \varphi(\partial\Omega_1) \cup \varphi(\partial\Omega_2) = \varphi(\partial\Omega)$, then $d_B(\varphi, \Omega, h) = d_B(\varphi, \Omega_1, h) + d_B(\varphi, \Omega_2, h)$.*

Proof. Let $d = d(h, \varphi(\partial\Omega)) > 0$. Then $d \le d(h, \varphi(\partial\Omega_i))$ for $i = 1, 2$. Let $\psi \in C^1(\overline{\Omega}, \mathbb{R}^N)$ such that $\|\psi - \varphi\|_\infty < d/2$. Then

$$d(\psi, \Omega, h) = d(\varphi, \Omega, h) \quad \text{and} \quad d(\psi, \Omega_i, h) = d(\varphi, \Omega_i, h) \text{ for } i = 1, 2\,. \tag{6.2.13}$$

Consider the open ball $B_\rho(h)$ with $\rho = 1/2d(h, \varphi(\partial\Omega))$. By Sard's Theorem (see Theorem 6.2.2), there exists $\xi \notin \psi(K_\psi)$. Then $h, \xi \in \mathbb{R}^N$ are in the same connected component of $\mathbb{R}^N \setminus \psi(\partial\Omega)$ and of $\mathbb{R}^N \setminus \psi(\partial\Omega_i)$ for $i = 1, 2$. Proposition 6.2.13 implies that

$$d_B(\psi, \Omega, h) = d_B(\psi, \Omega, \xi) \quad \text{and} \quad d_B(\psi, \Omega_i, h) = d_B(\psi, \Omega_i, \xi) \tag{6.2.14}$$

for $i = 1, 2$. Since $\xi \in \mathbb{R}^N$ is a regular value of ψ, Definition 6.2.4 gives

$$d_B(\psi, \Omega, \xi) = \sum_{u \in \psi^{-1}(\xi)} \operatorname{sgn} J_\psi(u) = \sum_{u \in \psi^{-1}(\xi) \cap \Omega_1} \operatorname{sgn} J_\psi(u) + \sum_{u \in \psi^{-1}(\xi) \cap \Omega_2} \operatorname{sgn} J_\psi(u)\,.$$

Hence, $d_B(\psi, \Omega, \xi) = d_B(\psi, \Omega_1, \xi) + d_B(\psi, \Omega_2, \xi)$. Then, from (6.2.13) and (6.2.14) we conclude that $d_B(\varphi, \Omega, h) = d_B(\varphi, \Omega_1, h) + d_B(\varphi, \Omega_2, h)$. $\square$

As a consequence of this additivity property, we obtain the so-called excision property.

Corollary 6.2.16. *If $\varphi \in C(\overline{\Omega}, \mathbb{R}^N)$, $K \subseteq \overline{\Omega}$ is compact and $h \notin \varphi(K) \cup \varphi(\partial\Omega)$, then $d_B(\varphi, \Omega, h) = d_B(\varphi, \Omega \setminus K, h)$.*

The next result is used to show the existence of a solution for a nonlinear equation in $\mathbb{R}^N$. For this reason this property of the degree is known as the existence property.

Proposition 6.2.17. *If $\varphi \in C(\overline{\Omega}, \mathbb{R}^N)$, $h \notin \varphi(\partial\Omega)$, and $d_B(\varphi, \Omega, h) \neq 0$, then the equation $\varphi(u) = h$ admits at least one solution.*

Proof. Evidently we need to show that if $\varphi^{-1}(h) = \emptyset$, then $d_B(\varphi, \Omega, h) = 0$. On account of Definition 6.2.12 we may assume that $\varphi \in C^1(\overline{\Omega}, \mathbb{R}^N)$. Let $0 < \varepsilon < d(h, \varphi(\overline{\Omega}))$ and consider a mollifier ϑ_ε. Then from Definition 6.2.4 we obtain $d_B(\varphi, \Omega, h) = 0$. □

Remark 6.2.18. From the proof above, it is worth remembering the following implication:

$$h \notin \varphi(\overline{\Omega}) \quad \text{implies} \quad d_B(\varphi, \Omega, h) = 0 . \tag{6.2.15}$$

An interesting consequence of Proposition 6.2.17 and (6.2.15) is the following result.

Proposition 6.2.19. *If* id *is the identity function on* $\overline{\Omega}$ *and* $h \in \mathbb{R}^N$, *then*

$$d(\text{id}, \Omega, h) = \begin{cases} 1 & \text{if } h \in \Omega , \\ 0 & \text{if } h \notin \overline{\Omega} \end{cases}, \quad d(-\text{id}, \Omega, h) = \begin{cases} (-1)^N & \text{if } h \in \Omega , \\ 0 & \text{if } h \notin \overline{\Omega} . \end{cases}$$

Proof. Applying Proposition 6.2.17 and (6.2.15) on id and on − id yields the assertion of the proposition. □

Finally we present the most important property of the degree, which is the so-called homotopy invariance property. Using this property, we can transfer the computation of the degree of a function to that of another function that is simpler and whose degree is easier to compute.

Proposition 6.2.20. *If* $h\colon [0,1] \times \overline{\Omega} \to \mathbb{R}^N$ *is continuous and* $\xi \notin h([0,1] \times \partial\Omega)$, *then* $d_B(h(t,\cdot), \Omega, \xi) = d_B(h(0,\cdot), \Omega, \xi)$ *for all* $t \in [0,1]$.

Proof. Let $\varepsilon = 1/4d(\xi, h([0,1] \times \partial\Omega))$ and note that h is uniformly continuous on $[0,1] \times \overline{\Omega}$. So, there exists $\delta = \delta(\varepsilon) > 0$ such that

$$|t_1 - t_2| < \delta \quad \text{implies} \quad |h(t_1, u) - h(t_2, u)| < \varepsilon \quad \text{for all } u \in \mathbb{R}^N .$$

Then from Definition 6.2.12, we have

$$d_B(h(t_1, \cdot), \Omega, \xi) = d_B(h(t_2, \cdot), \Omega, \xi) \tag{6.2.16}$$

for all $t_1, t_2 \in [0,1]$ with $|t_1 - t_2| < \delta$. The compactness of $[0,1]$ implies that we can cover it by a finite number of subintervals of length $\delta > 0$. Then the homotopy invariance property of the degree follows from (6.2.16). □

Using the homotopy invariance property we can show that what matters in the computation of the degree, is the structure of the function on the boundary $\partial\Omega$.

Proposition 6.2.21. *If* $\varphi, \psi \in C(\overline{\Omega}, \mathbb{R}^N)$, $\varphi|_{\partial\Omega} = \psi|_{\partial\Omega}$, *and* $\xi \notin \varphi(\partial\Omega) = \psi(\partial\Omega)$, *then* $d_B(\varphi, \Omega, \xi) = d_B(\psi, \Omega, \xi)$.

Proof. Let $h\colon [0,1] \times \mathbb{R}^N \to \mathbb{R}^N$ be the homotopy defined by $h(t,u) = (1-t)\varphi(u) + t\psi(u)$ for all $t \in [0,1]$ and for all $u \in \mathbb{R}^N$. Since $\varphi|_{\partial\Omega} = \psi|_{\partial\Omega}$ and $\xi \notin \varphi(\partial\Omega) = \psi(\partial\Omega)$, we see that $\xi \notin h([0,1] \times \partial\Omega)$. Then Proposition 6.2.20 implies that

$$d_B(\varphi,\Omega,\xi) = d_B(h(0,\cdot),\Omega,\xi) = d_B(h(1,\cdot),\Omega,\xi) = d_B(\psi,\Omega,\xi)\,. \qquad \square$$

The next theorem summarizes the main properties of the Brouwer degree. In what follows, let

$$\mathcal{L} = \{(\varphi,\Omega,h)\colon \Omega \subseteq \mathbb{R}^N \text{ is bounded, open, } \varphi \in C(\overline{\Omega},\mathbb{R}^N), h \notin \varphi(\partial\Omega)\}\,.$$

Theorem 6.2.22. *There exists a function $d_B\colon \mathcal{L} \to \mathbb{Z}$ known as the* **Brouwer degree** *such that the following hold:*

(a) *Normalization property: $d_B(\mathrm{id},\Omega,h) = 1$;*

(b) *Domain additivity property: for all disjoint open $\Omega_1,\Omega_2 \subseteq \Omega$ and $h \notin \varphi(\overline{\Omega} \setminus (\Omega_1 \cup \Omega_2))$, we have*

$$d_B(\varphi,\Omega_1,h) + d_B(\varphi,\Omega_2,h) = d_B(\varphi,\Omega,h)\,;$$

(c) *Homotopy invariance property: if $\hat{h} \in C([0,1] \times \overline{\Omega},\mathbb{R}^N)$, $h \in C([0,1],\mathbb{R}^N)$ and $h(t) \notin \hat{h}(t,\partial\Omega)$ for all $t \in [0,1]$, then we have that*

$$d_B(\hat{h}(t,\cdot),\Omega,h(t)) \text{ is independent of } t \in (0,1]\,;$$

(d) *Solution property: $d_B(\varphi,\Omega,h) \neq 0$ implies that $\varphi^{-1}(h) \neq \emptyset$;*

(e) *Dependence on the boundary values: if $(\varphi,\Omega,h), (\psi,\Omega,h) \in \mathcal{L}$ and $\varphi|_{\partial\Omega} = \psi|_{\partial\Omega}$, then $d_B(\varphi,\Omega,h) = d_B(\psi,\Omega,h)$;*

(f) *Excision property: $d_B(\varphi,\Omega,h) = d_B(\varphi,\Omega_1,h)$ for every open $\Omega_1 \subseteq \Omega$ such that $h \notin \varphi(\overline{\Omega} \setminus \Omega_1)$;*

(g) *Continuity in (φ,h): $\varphi \to d_B(\varphi,\Omega,h)$ is constant on $B_\varepsilon^{C(\overline{\Omega})}(\varphi) = \{\psi \in C(\overline{\Omega},\mathbb{R}^N)\colon \|\psi - \varphi\|_\infty < \varepsilon\}$ with $\varepsilon = d(h,\varphi(\partial\Omega))$ and $h \to d_B(\varphi,\Omega,h)$ is constant on every connected component of $\mathbb{R}^N \setminus \varphi(\partial\Omega)$.*

Remark 6.2.23. The homotopy invariance property was stated here in a slightly more general form since the reference point is also t-dependent; compare with Proposition 6.2.20. However this generalization is immediate if we recall that

$$d_B(\varphi,\Omega,h(t)) = d_B(\varphi - h(t),\Omega,0) \quad \text{for all } t \in [0,1]\,.$$

Moreover, in the construction of the degree, we used the natural basis $\{e_k\}_{k=1}^N$ of $\mathbb{R}^N$ where $e_k = (e_{ki})_{i=1}^N = (\delta_{ki})_{i=1}^N \in \mathbb{R}^N$. This basis is ordered. We obtain the same degree function if we consider instead a different ordered basis. This can be easily verified by using the transition matrix corresponding to this change of bases. In a similar way we can also replace $\mathbb{R}^N$ with an N-dimensional Banach space X. Finally, suppose that X is an N-dimensional Banach space and Y is an m-dimensional subspace of X, that is, $m < n$. Let $\Omega \subseteq X$ be bounded open and $\varphi \in C(\overline{\Omega}, Y), f = \mathrm{id} - \varphi$ and $h \notin \varphi(\partial\Omega)$. Then

$$d_B(f, \Omega, h) = d_B(f_m, \Omega_m, h)\,,$$

where $\Omega_m = \Omega \cap Y$ and $f_m = f|_{\overline{\Omega}_m}$.

Next we present some applications of the Brouwer degree. In what follows, let $\overline{B}_1^N$ and $\partial B_1 = S^{N-1}$ be the closed unit balls in $\mathbb{R}^N$ and the unit sphere in $\mathbb{R}^N$, respectively, that is

$$\overline{B}_1^N = \{u \in \mathbb{R}^N : |u| \le 1\}, \quad \text{and} \quad \partial B_1 = S^{N-1} = \{u \in \mathbb{R}^N : |u| = 1\}\,.$$

Proposition 6.2.24. *S^{N-1} is not a retract of $\overline{B}_1^N$; see Definition* 1.7.11.

Proof. Arguing by contradiction, suppose that S^{N-1} is a retract of $\overline{B}_1^N$, and let φ be the corresponding retraction map. Then $\varphi|_{S^{N-1}} = \mathrm{id}\,|_{S^{N-1}}$ and so by Theorem 6.2.22 (e) and Proposition 6.2.19, we obtain

$$d_B(\varphi, B_1^N, 0) = d_B(\mathrm{id}, B_1^N, 0) = 1\,.$$

So, from the solution property (see Theorem 6.2.22 (d)), we infer that there exists $u \in B_1^N$ such that $\varphi(u) = 0$, a contradiction since $\operatorname{im} \varphi = S^{N-1}$. □

Remark 6.2.25. This is in contrast to the infinite dimensional case where the unit sphere is a retract of $\overline{B}_1 = \{u \in X : \|u\| \le 1\}$.

The next result is the celebrated "Brouwer Fixed Point Theorem".

Theorem 6.2.26 (Brouwer Fixed Point Theorem). *If $\overline{B}_1^N = \{u \in \mathbb{R}^N : |u| \le 1\}$ and $\varphi : \overline{B}_1^N \to \overline{B}_1^N$ is continuous, then there exists $u_0 \in \overline{B}_1^N$ such that $\varphi(u_0) = u_0$, that is, φ has a fixed point.*

Proof. If there is $u \in \partial B_1^N = S^{N-1} = \{x \in \mathbb{R}^N : |x| = 1\}$ such that $\varphi(u) = u$, then there is nothing to prove. So, we assume that $u - \varphi(u) \neq 0$ for all $u \in \partial B_1^N$. Consider the homotopy $h(t, u) = u - t\varphi(u)$ for all $t \in [0, 1]$ and for all $u \in \overline{B}_1^N$. Evidently, $0 \notin h(1, \partial B_1^N)$ and $0 \notin h(0, \partial B_1^N)$. Suppose that we have $u = t\varphi(u)$ for some $u \in \partial B_1^N$ and for some $t \in (0, 1)$. Hence, $1 = t|\varphi(u)| \le t < 1$, a contradiction. Therefore, $0 \notin h(t, \partial \overline{B}_1^N)$ for all $t \in [0, 1]$ and so from Theorem 6.2.22 (c) and Proposition 6.2.19, it follows that

$$d_B(\mathrm{id} - \varphi, B_1^N, 0) = d_B(\mathrm{id}, B_1^N, 0) = 1\,.$$

Then according to Theorem 6.2.22 (d) there exists $u_0 \in B_1^N$ such that $u_0 = \varphi(u_0)$. □

Remark 6.2.27. In Section 6.4 we will generalize this theorem by replacing $\overline{B}_1^N$ with any compact, convex subset of $\mathbb{R}^N$.

In fact Proposition 6.2.24 and Theorem 6.2.26 are equivalent results.

Proposition 6.2.28. *Proposition* 6.2.24 *holds if and only if Theorem* 6.2.26 *holds.*

Proof. $\Longrightarrow$: Arguing by contradiction, suppose that there exists a continuous function $\varphi: \overline{B}_1^N \to \overline{B}_1^N$ such that $\varphi(u) \neq u$ for all $u \in \overline{B}_1^N$. Given $u \in \overline{B}_1^N$, let $t(u) > 0$ be such that

$$(1 - t(u))\varphi(u) + t(u)u \in \partial B_1^N .$$

Evidently, t is continuous on $\overline{B}_1^N$. Let

$$\psi(u) = \varphi(u) + t(u)(u - \varphi(u)) \quad \text{for all } u \in \overline{B}_1^N .$$

Then $\psi \in C(\overline{B}_1^N, \partial B_1^N)$ and if $u \in \partial B_1^N$, then $t(u) = 1$ and so $\psi|_{\partial B_1^N} = \mathrm{id}\,|_{\partial B_1^N}$. This means that ∂B_1 is a retract of $\overline{B}_1^N$, a contradiction.

$\Longleftarrow$: Arguing by contradiction, suppose that ∂B_1^N is a retract of $\overline{B}_1^N$. Then there exists a retraction $r: \overline{B}_1^N \to \partial B_1^N$. Let $\varphi(u) = -r(u)$ for all $u \in \overline{B}_1^N$. Then $\varphi: \overline{B}_1^N \to \overline{B}_1^N$ is continuous but has no fixed point, a contradiction. □

Proposition 6.2.29. *If $d(\varphi, \Omega, h) \neq 0$, then $\varphi(\Omega)$ is a neighborhood of h.*

Proof. From the solution property of the degree (see Theorem 6.2.22 (d)), we know that there exists $u_0 \in \Omega$ such that $\varphi(u_0) = h$. Let U_h be the connected component of $\mathbb{R}^N \setminus \varphi(\partial\Omega)$ which contains h. Then from Theorem 6.2.22 (g), we get

$$0 \neq d_B(\varphi, \Omega, h) = d_B(\varphi, \Omega, \xi) \quad \text{for all } \xi \in U_h .$$

Then, Theorem 6.2.22 (d) implies that $U_\xi \subseteq \varphi(\Omega)$ and so, $\varphi(\Omega)$ is a neighborhood of h. □

Corollary 6.2.30. *If $\varphi(\Omega)$ is contained in a proper subspace of $\mathbb{R}^N$, then $d_B(\varphi, \Omega, h) = 0$.*

The solution property (see Theorem 6.2.22 (d)) requires that $d_B(\varphi, \Omega, h) \neq 0$. The next theorem gives a situation where this is true. The result is known as "Borsuk's Theorem".

Theorem 6.2.31 (Borsuk's Theorem). *If $\Omega \subseteq \mathbb{R}^N$ is nonempty, bounded, open, symmetric, that is, $\Omega = -\Omega$, $0 \in \Omega$, $\varphi \in C(\overline{\Omega}, \mathbb{R}^N)$ is odd and $0 \notin \varphi(\partial\Omega)$, then $d_B(\varphi, \Omega, 0)$ is odd. In particular, $d_B(\varphi, \Omega, 0) \neq 0$.*

Proof. Without any loss of generality we may assume that $\varphi \in C^1(\Omega, \mathbb{R}^N) \cap C(\overline{\Omega}, \mathbb{R}^N)$ and that $J_\varphi(0) \neq 0$. Indeed, we approximate φ by $\psi \in C^1(\Omega, \mathbb{R}^N) \cap C(\overline{\Omega}, \mathbb{R}^N)$ in $C(\overline{\Omega}, \mathbb{R}^N)$ and then consider the odd part ψ_0 of ψ, that is, $\psi_0(u) = 1/2[\psi(u) - \psi(-u)]$. Let $\lambda \in \mathbb{R}$, which is not an eigenvalue of $\psi_0'(0)$ and set $\hat{\psi} = \psi_0 - \lambda\,\mathrm{id} \in C^1(\Omega, \mathbb{R}^N) \cap C(\overline{\Omega}, \mathbb{R}^N)$. Evidently, $\hat{\psi}$ is odd with $J_{\hat{\psi}}(0) \neq 0$ and $\|\hat{\psi} - \varphi\|_\infty$ is small provided $\lambda \in \mathbb{R}$ and $\|\psi - \varphi\|_\infty$ are both small. Taking Proposition 6.2.21 into account gives $d_B(\varphi, \Omega, 0) = d_B(\hat{\psi}, \Omega, 0)$.

So, we assume that $\varphi \in C^1(\Omega, \mathbb{R}^N) \cap C(\overline{\Omega}, \mathbb{R}^N)$ and $J_\varphi(0) \neq 0$. We will produce an odd map $\psi \in C^1(\Omega, \mathbb{R}^N) \cap C(\overline{\Omega}, \mathbb{R}^N)$ close to φ in the $C(\overline{\Omega}, \mathbb{R}^N)$-norm such that $0 \notin \psi(K_\psi)$. This will be done by induction. So, let

$$\Omega_k = \{u \in \Omega \colon u_i \neq 0 \text{ for some } i \leq k\}$$

and let $\eta \in C^1(\mathbb{R}, \mathbb{R})$ be odd such that $\eta'(0) = 0$ and $\eta(t) = 0$ if and only if $t = 0$. Let $\hat{\varphi}(u) = \varphi(u)/\eta(u_1)$ for all $u \in \Omega_1$. By Sard's Theorem (see Theorem 6.2.2), there exists $v^1 \notin \hat{\varphi}(K_{\hat{\varphi}})$ with $|v^1|$ as small as necessary for what follows. Let $\psi_1(u) = \varphi(u) - \eta(u_1)v^1$ for all $u \in \Omega_1$. Since $\psi_1'(u) = \eta(u_1)\hat{\varphi}'(u)$ for all $u \in \Omega_1$ with $\psi_1(u) = 0$ we see that 0 is a regular value of ψ_1. Suppose that we already have an odd function $\psi_k \in C^1(\Omega, \mathbb{R}^N) \cap C(\overline{\Omega}, \mathbb{R}^N)$, which is close to φ in the $C(\overline{\Omega}, \mathbb{R}^N)$-norm such that $0 \notin \psi_k(K_{\psi_k})$ for some $k < N$. We set

$$\psi_{k+1}(u) = \psi_k(u) - \eta(u_{k+1})v^{k+1}$$

with $|v^{k+1}|$ sufficiently small such that 0 is a regular value of ψ_{k+1} on the set $\{u \in \Omega \colon u_{k+1} \neq 0\}$ where $u = (u_k)_{k=1}^N \in \mathbb{R}^N$. Note that $\psi_{k+1} \in C^1(\Omega, \mathbb{R}^N) \cap C(\overline{\Omega}, \mathbb{R}^N)$ is odd and $C(\overline{\Omega}, \mathbb{R}^N)$-close to φ. If $u \in \Omega_{k+1}$ and $u_{k+1} = 0$, then $u \in \Omega_k$, $\psi_{k+1}(u) = \psi_k(u)$ and $\psi_{k+1}'(u) = \psi_k'(u)$, hence $J_{\psi_{k+1}}(u) \neq 0$. So, $0 \notin \psi_{k+1}(K_{\psi_{k+1}}(\Omega_{k+1}))$. Therefore, $\psi = \psi_N$ is odd, $C(\overline{\Omega}, \mathbb{R}^N)$-close to φ and $0 \notin \psi(K_\psi(\Omega \setminus \{0\}))$. Note that $\Omega_N = \Omega \setminus \{0\}$. From the induction hypothesis, we obtain $\psi'(0) = \psi_1'(0) = \varphi'(0)$, and hence, $0 \notin \psi(K_\psi)$.

Therefore, we have reduced the problem to the case where

$$\varphi \in C^1(\Omega, \mathbb{R}^N) \cap C(\overline{\Omega}, \mathbb{R}^N), \quad J_\varphi(0) \neq 0 \quad \text{and} \quad \varphi \text{ is odd}.$$

We have

$$\varphi^{-1}(0) = \{0\} \quad \text{or} \quad \varphi^{-1}(0) = \{0\} \cup \{(u_k, -u_k)\}_{k=1}^m \quad \text{with } u_k \in \Omega \setminus \{0\}.$$

In the first case we see that $d_B(\varphi, \Omega, 0) = \pm 1$ while in the second case we obtain

$$d_B(\varphi, \Omega, 0) = \pm 1 + \sum_{k=1}^N [\operatorname{sgn} J_\varphi(u_k) + \operatorname{sgn} J_\varphi(-u_k)] = \pm 1 + 2\sum_{k=1}^N \operatorname{sgn} J_\varphi(u_k)$$

since J_φ is even. In both cases $d_B(\varphi, \Omega, 0)$ is odd. □

Remark 6.2.32. Instead of requiring that φ be odd on $\overline{\Omega}$, it is enough to assume that $\varphi|_{\partial\Omega}$ is odd. Then use the Tietze Extension Theorem (see Theorem 1.2.44) to extend on all of $\overline{\Omega}$ and then use the odd part of this extension. Even more generally, we may require only that

$$\frac{\varphi(u)}{|\varphi(u)|} \neq \frac{\varphi(-u)}{|\varphi(-u)|} \quad \text{for all } u \in \partial\Omega. \tag{6.2.17}$$

The geometric meaning of (6.2.17) is that for $u \in \partial\Omega$, the vectors $\varphi(u)$ and $\varphi(-u)$ are never colinear. If φ is odd, then (6.2.17) follows from the requirement that $0 \notin \varphi(\partial\Omega)$.

An important consequence of Borsuk's Theorem (see Theorem 6.2.31) is the so-called "Borsuk–Ulam Theorem", which we state below.

Theorem 6.2.33 (Borsuk–Ulam Theorem). *If $\Omega \subseteq \mathbb{R}^N$ is nonempty, bounded, open symmetric, $0 \in \Omega$, $\varphi \in C(\partial\Omega, \mathbb{R}^N)$ is odd and $\varphi(\partial\Omega)$ is contained in a hyperplane H of $\mathbb{R}^N$, then there exists $\hat{u} \in \partial\Omega$ such that $\varphi(\hat{u}) = 0$.*

Proof. Arguing by contradiction, suppose that $0 \notin \varphi(\partial\Omega)$. According to Remark 6.2.32, we can apply Theorem 6.2.31 and have that $d_B(\varphi, \Omega, h) \neq 0$. This contradicts Corollary 6.2.30. □

Corollary 6.2.34. *If $\Omega \subseteq \mathbb{R}^N$ is nonempty, bounded, open, symmetric, $0 \in \Omega$, $\varphi \in C(\partial\Omega, \mathbb{R}^N)$, and $\varphi(\partial\Omega)$ is contained in a hyperplane H of $\mathbb{R}^N$, then there exists $\hat{u} \in \partial\Omega$ such that $\varphi(\hat{u}) = \varphi(-\hat{u})$.*

Proof. Let $\psi(u) = (\varphi(u) - \varphi(-u))/2$ for all $u \in \overline{\Omega}$. Then ψ is odd. We apply Theorem 6.2.33 and produce $\hat{u} \in \partial\Omega$ such that $\psi(\hat{u}) = 0$, hence, $\varphi(\hat{u}) = \varphi(-\hat{u})$. □

Remark 6.2.35. In the literature we often find a meteorological interpretation of this corollary. Namely that at two opposite ends on the earth we have the same weather.

Corollary 6.2.36. *If $N > m$, then there is no continuous, odd map from $\partial B_1^N = \{u \in \mathbb{R}^N : |u| = 1\}$ into $\partial B_1^m = \{u \in \mathbb{R}^m : |u| = 1\}$.*

Proposition 6.2.37. *If $\Omega \subseteq \mathbb{R}^N$ is nonempty open and $\varphi\colon \Omega \to \mathbb{R}^N$ is continuous and locally injective, then $\varphi(\Omega) \subseteq \mathbb{R}^N$ is open.*

Proof. We need to show that for any $u_0 \in \Omega$ there exists $r > 0$ such that

$$B_r(\varphi(u_0)) = \{v \in \mathbb{R}^N : |v - \varphi(u_0)| < r\} \subseteq \varphi(\Omega)\,.$$

We replace Ω by $\Omega - u_0$, φ by $\hat{\varphi}(u) = \varphi(u_0 + u) - \varphi(u_0)$ defined on $\Omega - u_0$, and we see that without any loss of generality we may assume that $u_0 = 0$ and that $\varphi(0) = 0$.

We choose $r > 0$ such that $\varphi|_{\overline{B_r(0)}}$ is injective and consider the homotopy

$$h(t,u) = \varphi\left(\frac{u}{1+t}\right) - \varphi\left(-\frac{tu}{1+t}\right) \quad \text{for all } t \in [0,1] \text{ and for all } u \in \overline{B}_r(0)\,.$$

Evidently, h is continuous and $h(0,\cdot) = \varphi(\cdot)$ as well as $h(1,u) = \varphi(1/2u) - \varphi(-1/2u) = \psi(u)$ is an odd function. Suppose that $0 \in \partial\varphi([0,1] \times \partial B_r(0))$. Then there exist $t \in [0,1]$ and $u \in \partial B_r(0)$ such that

$$\varphi\left(\frac{u}{1+t}\right) = \varphi\left(-\frac{tu}{1+t}\right).$$

Since $\varphi|_{\overline{B}_r(0)}$ is injective, we see that $u = -tu$ and so $u = 0$, a contradiction. Then using the homotopy invariance property of the degree (see Theorem 6.2.22 (c)), we obtain

$$d_B(\varphi, B_r(0), 0) = d_B(\psi, B_r(0), 0) \neq 0\,;$$

see also Theorem 6.2.31. Hence, $d_B(\varphi, B_r(0), h) \neq 0$ for all $h \in B_r(0)$ and so $B_r(0) \subseteq \varphi(B_r(0)) \subseteq \varphi(\Omega)$. This proves that $\varphi(\Omega)$ is open. ☐

The next result essentially says that we cannot comb a coconut without leaving tufts or whorls.

Proposition 6.2.38. *If $\Omega \subseteq \mathbb{R}^N$ is bounded open with $0 \in \Omega$, $\varphi \in C(\partial\Omega, \mathbb{R}^N \setminus \{0\})$ and N is odd, then there exist $u \in \partial\Omega$ and $\lambda \neq 0$ such that $\varphi(u) = \lambda u$.*

Proof. Using the Tietze Extension Theorem (see Theorem 1.2.44), we may assume that $\varphi \in C(\overline{\Omega}, \mathbb{R}^N)$. From Proposition 6.2.19 and since N is odd, we obtain $d_B(-\operatorname{id}, \Omega, 0) = -1$.

If $d_B(\varphi, \Omega, 0) \neq -1$, then the homotopy

$$h(t,u) = (1-t)\varphi(u) - tu \quad \text{for all } t \in [0,1] \text{ and for all } u \in \overline{\Omega}$$

must satisfy $h(t_0, u_0) = 0$ for some $(t_0, u_0) \in (0,1) \times \partial\Omega$. Hence,

$$\varphi(u_0) = \frac{t_0}{1-t_0} u_0\,.$$

Therefore, the result holds with $\lambda > t_0/(1-t_0)$.

If $d_B(\varphi, \Omega, 0) = -1$, then we can apply the same argument to the homotopy

$$h(t,u) = (1-t)\varphi(u) + tu \quad \text{for all } t \in [0,1] \text{ and for all } u \in \overline{\Omega}\,.$$ ☐

Remark 6.2.39. In case $\Omega = B_1(0) = \{u \in \mathbb{R}^N : |u| < 1\}$ being the coconut, there is no continuous vector field $\varphi\colon S^{N-1} \to \mathbb{R}^N$ such that $\varphi(x) \neq 0$ and $(\varphi(x), x)_{\mathbb{R}^N} = 0$ on S^{N-1}. If $N = 3$, then we have a real coconut that cannot be combed.

The next result concerns coverings of $\partial\Omega$. It is called the "Ljusternik–Schnirelmann–Borsuk Theorem".

Theorem 6.2.40 (Ljusternik–Schnirelmann–Borsuk Theorem). *If $\Omega \subseteq \mathbb{R}^N$ is a bounded, open symmetric set with $0 \in \Omega$ and $\{C_k\}_{k=1}^m$ is a closed cover of $\partial\Omega$, that is, $C_k \subseteq \partial\Omega$ for all $k = 1, \ldots, m$ such that $C_k \cap (-C_k) = \emptyset$ for all $k = 1, \ldots, m$, then $m \geq N+1$.*

Proof. Suppose that $m \leq N$ and define φ_k by

$$\varphi_k|_{C_k} = 1, \quad \varphi_k|_{-C_k} = -1 \quad \text{for } k = 1, \ldots, m-1 \quad \text{and} \quad \varphi_k|_{\overline{\Omega}} = 1 \tag{6.2.18}$$

for $k = m, \ldots, N$. For $k = 1, \ldots, m-1$ we extend φ_k continuously on $\overline{\Omega}$. Let $\varphi = (\varphi_k)_{k=1}^N$. Then $\varphi \in C(\overline{\Omega}, \mathbb{R}^N)$. We claim that

$$\varphi(-u) \neq \lambda\varphi(u) \quad \text{for all } u \in \partial\Omega \text{ and for all } \lambda \geq 0\,. \tag{6.2.19}$$

Indeed, if (6.2.19) is not true, then there exist $u_0 \in \partial\Omega$ and $\lambda_0 > 0$ such that $\varphi(-u_0) = \lambda_0\varphi(u_0)$. Due to (6.2.18), we see that $u_0 \notin C_k \cup (-C_k)$ for all $k = 1, \ldots, m-1$. Therefore, $u_0 \in C_m$. Then $u_0 \notin -C_m$ and so $-u_0 \in C_k$ for some $k \in \{1, \ldots, m-1\}$. Hence, $u_0 \in -C_k$, a contradiction. This shows that (6.2.19) is true.

We consider the homotopy

$$h(t,u) = \varphi(u) - t\varphi(-u) \quad \text{for all } (t,u) \in [0,1] \times \overline{\Omega}\,.$$

The homotopy invariance property of the degree (see Theorem 6.2.22 (c), which can be used because of (6.2.19)) implies that

$$d_B(\varphi, \Omega, 0) = d_B(\varphi_0, \Omega, 0) \tag{6.2.20}$$

with $\varphi_0(u) = \varphi(u) - \varphi(-u)$, which is odd. So, Theorem 6.2.31 gives $d_B(\varphi_0, \Omega, 0) \neq 0$ and so, with view to (6.2.20), we obtain $d_B(\varphi, \Omega, 0) \neq 0$. Then the solution property of the degree (see Theorem 6.2.22 (d)) implies that $\varphi(u) = 0$ for some $u \in \Omega$, a contradiction to (6.2.18). ☐

Finally we introduce a notion that is useful in determining the spectral properties of certain nonlinear homogeneous differential operators such as the p-Laplacian.

Definition 6.2.41. Let X be a Banach space and $\mathcal{L}(X) = \{A \subseteq X \setminus \{0\} : A \text{ is closed and symmetric}\}$. The **genus** $\gamma\colon \mathcal{L}(X) \to \mathbb{N}_0 \cup \{+\infty\}$ is defined by $\gamma(\emptyset) = 0$ and

$$\gamma(A) = \begin{cases} \min\{k \in \mathbb{N}\} & \text{there exists an odd } f \in C(A, \mathbb{R}^k \setminus \{0\})\,, \\ +\infty & \text{otherwise}\,. \end{cases}$$

The genus generalizes the notion of the dimension of a linear space.

Proposition 6.2.42. *If X is a Banach space and $\partial B_1 = \{u \in X : \|u\| = 1\}$, then $\gamma(\partial B_1) = \dim X$.*

Proof. Let $0 < \dim X < \infty$. Then if $k = \dim X$, the identity map satisfies the requirement of Definition 6.2.41. Moreover, Theorem 6.2.33 says that we cannot go below $\dim X$. Hence, $\gamma(\partial B_1) = \dim X$.

Next let $\dim X = +\infty$ and let X_n be an n-dimensional subspace of X. Directly from Definition 6.2.41, we have $\gamma(\partial B_1 \cap X_n) \leq \gamma(\partial B_1)$ and so the first part of the proof gives $n \leq \gamma(\partial B_1)$. Therefore, $\gamma(\partial B_1) = \infty = \dim X$. ☐

Proposition 6.2.43. *If $\Omega \subseteq \mathbb{R}^N$ is bounded, open, symmetric with $0 \in \Omega$, then $\gamma(\partial\Omega) = N$.*

Proof. Using the identity map, we see that $\gamma(\partial\Omega) \leq N$. If $\gamma(\partial\Omega) \leq N-1$, then according to Definition 6.2.41, there exists an odd function $f \in C(\partial\Omega, \mathbb{R}^{N-1} \setminus \{0\})$. Applying

Theorem 6.2.33, there exists $\hat{u} \in \partial\Omega$ such that $f(\hat{u}) = 0$, a contradiction. Therefore, $\gamma(\partial\Omega) = N$. ☐

Proposition 6.2.44. *If X is a Banach space and $A, B \in \mathcal{L}(X)$, then the following hold:*
(a) *if $\varphi: A \to B$ is continuous and odd, then $\gamma(A) \le \gamma(B)$;*
(b) *if $A \subseteq B$, then $\gamma(A) \le \gamma(B)$;*
(c) *if there is an odd homeomorphism $f: A \to B$, then $\gamma(A) = \gamma(B)$.*

Proof. (a) We assume that $\gamma(B) < +\infty$ or otherwise the result holds trivially. Let $m = \gamma(B)$. Then there exists an odd function $f \in C(B, \mathbb{R}^m \setminus \{0\})$. Then $f \circ \varphi \in C(A, \mathbb{R}^m \setminus \{0\})$ and it is odd. Therefore, $\gamma(A) \le m = \gamma(B)$.

(b) This follows from (a) with $\varphi = \mathrm{id}$.

(c) This follows from (a) applied twice with φ and with φ^{-1}. ☐

6.3 Leray–Schauder Degree

At this point a natural question arises: Is it possible to extend the Brouwer degree to all continuous functions defined on an infinite dimensional Banach space? This is important because in most applications the ambient space is infinite dimensional. The next example shows that such an extension is not possible.

Example 6.3.1. Let $H = l^2$ and let $\overline{B}_1$ be the closed unit ball of H. Consider the function $\varphi: l^2 \to l^2$ defined by $\varphi(u) = (\sqrt{1 - \|u\|^2}, u_1, u_2, \ldots)$ for all $u = (u_n)_{n\ge1} \in l^2$. Evidently, φ is continuous and $\varphi(\overline{B}_1) \subseteq \overline{B}_1$. If we could extend the Brouwer degree to infinite dimensional spaces, then according to Theorem 6.2.26, we would have had $\tilde{u} \in \overline{B}_1$ such that

$$\varphi(\tilde{u}) = \tilde{u}\,. \tag{6.3.1}$$

Then $\tilde{u}_1 = \sqrt{1 - \|\tilde{u}\|^2}$ and $\tilde{u}_{n+1} = \tilde{u}_n$ for $n \in \mathbb{N}$. Moreover, (6.3.1) gives

$$1 = \|\varphi(\tilde{u})\| = \|\tilde{u}\|\,. \tag{6.3.2}$$

This shows that $\tilde{u}_1 = 0$ and so $\tilde{u}_n = 0$ for all $n \in \mathbb{N}$, that is, $\tilde{u} = 0$, a contradiction to (6.3.2).

This examples suggests that we have to restrict the family of continuous maps. So, we limit ourselves to maps of the form

$$\varphi = \mathrm{id} - f \tag{6.3.3}$$

with f being a compact map; see Definition 3.7.1. So, the extension of the Brouwer degree that is known as the **Leray–Schauder degree** will be defined on the set

$$\hat{\mathcal{L}} = \{(\varphi, \Omega, h)\colon \Omega \subseteq X \text{ is bounded, open, } \varphi = \mathrm{id} - f\colon \overline{\Omega} \to X \\ \text{with } f \in K(\overline{\Omega}, X) \text{ and } h \notin \varphi(\partial\Omega)\} .$$

Since we are in an infinite dimensional setting, some clarifications are necessary. First note that $\overline{\Omega}$ is not compact. In fact, $\overline{\Omega} \subseteq X$ is compact if and only if X is finite dimensional. So, it is not immediately clear that $r = d(h, \varphi(\partial\Omega)) > 0$. If $r = 0$, then there exists a sequence $\{u_n\}_{n\ge1} \subseteq \partial\Omega$ such that $\|h - \varphi(u_n)\| \to 0$. The compactness of f (see (6.3.3)) implies that $\{f(u_n)\}_{n\ge1} \subseteq X$ is relatively compact. We may assume that $f(u_n) \to y$ in X as $n \to \infty$. Then $y \in f(\overline{\Omega})$ and $u_n = \varphi(u_n) + f(u_n) \to h + y$ in X as $n \to \infty$. Since $\partial\Omega \subseteq X$ is closed and $u_n \in \partial\Omega$ for $n \in \mathbb{N}$, we obtain $h + y \in \partial\Omega$. Then, the continuity of φ implies $\varphi(h + y) = h$, a contradiction to the hypothesis that $h \notin \varphi(\partial\Omega)$. Therefore, $r = d(h, \varphi(\partial\Omega)) > 0$.

In order to define the Leray–Schauder degree, we will exploit the fact that compact maps can be approximated uniformly on $\overline{\Omega}$ by finite rank maps; see Theorem 3.7.10. So, for a given $\varepsilon \in (0, r)$, let $f_\varepsilon\colon \overline{\Omega} \to X$ be a finite rank map such that $\|f(x) - f_\varepsilon(x)\| < \varepsilon$ for all $x \in \overline{\Omega}$.

Let $X_\varepsilon = \operatorname{span}\{f_\varepsilon(\overline{\Omega}), h\}$, $\Omega_\varepsilon = \Omega \cap X_\varepsilon$ and $\varphi_\varepsilon(u) = u - f_\varepsilon(u)$ for all $u \in \overline{\Omega}$. The set $\Omega_\varepsilon \subseteq X_\varepsilon$ is bounded and open, $\partial_{X_\varepsilon}\Omega_\varepsilon \subseteq \partial\Omega$, $\varphi_\varepsilon(\overline{\Omega}_\varepsilon) \subseteq X_\varepsilon$ and for all $u \in \partial\Omega$ we have

$$\|\varphi_\varepsilon(u) - h\| \ge \|u - f(u) - h\| - \|f(u) - f_\varepsilon(u)\| > r - \varepsilon > 0 .$$

Therefore, $d_B(\varphi_\varepsilon, \Omega_\varepsilon, h)$ is well-defined. If $\Omega_\varepsilon = \emptyset$, then $d_B(\varphi_\varepsilon, \Omega_\varepsilon, h) = 0$.

We show that this degree stabilizes for small $\varepsilon > 0$.

Lemma 6.3.2. *For $\varepsilon \in (0, r)$, $d_B(\varphi_\varepsilon, \Omega_\varepsilon, h)$ is independent of ε.*

Proof. Let $\varepsilon, \delta \in (0, r)$. Let $\hat{X} = \operatorname{span}\{X_\varepsilon, X_\delta\}$ and set $\hat{\Omega} = \Omega \cap \hat{X}$. From Remark 6.2.23, we see that

$$d_B(\varphi_\varepsilon, \Omega_\varepsilon, h) = d_B(\varphi_\varepsilon, \hat{\Omega}, h) \quad \text{and} \quad d(\varphi_\delta, \Omega_\delta, h) = d(\varphi_\delta, \hat{\Omega}, h) . \tag{6.3.4}$$

We consider the homotopy

$$\hat{h}(t, u) = (1 - t)\varphi_\varepsilon(u) + t\varphi_\delta(u) \quad \text{for all } t \in [0, 1] \text{ and for all } u \in \overline{\hat{\Omega}} .$$

Moreover,

$$\begin{aligned} \|\hat{h}(t, u) - \varphi(u)\| &\le (1 - t)\|\varphi_\varepsilon(u) - \varphi(u)\| + t\|\varphi_\delta(u) - \varphi(u)\| \\ &< (1 - t)\varepsilon + t\delta < r . \end{aligned} \tag{6.3.5}$$

From (6.3.5) we see that, for $u \in \partial\hat{\Omega}$, it follows that

$$\|\hat{h}(t, u) - h\| \ge \|\varphi(u) - h\| - \|\hat{h}(t, u) - \varphi(u)\| > r - r = 0 .$$

From the homotopy invariance of the Brouwer degree (see Theorem 6.2.22 (d)), we obtain

$$d_B(\varphi_\varepsilon, \hat{\Omega}, h) = d_B(\varphi_\delta, \hat{\Omega}, h)\,. \tag{6.3.6}$$

From (6.3.4) and (6.3.6), we have proven the claim of the lemma. □

Then, for any $\varepsilon \in (0, r)$ and for any finite dimensional space $V \subseteq X$ such that $X_\varepsilon \subseteq V$, let $\Omega_V = \Omega \cap V$. We have $d(\varphi_\varepsilon, \Omega_V, h) = d(\varphi_\varepsilon, \Omega, h)$; see Remark 6.2.23. This leads to the following definition.

Definition 6.3.3. Let $(\varphi, \Omega, h) \in \hat{\mathcal{L}}$ and let $\hat{f}\colon \overline{\Omega} \to X$ be a finite rank map such that

$$\|f(u) - \hat{f}(u)\| < d(h, \varphi(\partial\Omega)) \quad \text{for all } u \in \overline{\Omega}\,.$$

We choose a finite dimensional subspace $V \subseteq X$ of X such that $\hat{f}(\overline{\Omega}), \{h\} \subseteq V$. We set $\Omega_V = \Omega \cap V$ and define the **Leray–Schauder degree** of $(\varphi, \Omega, h) \in \hat{\mathcal{L}}$ to be

$$d_{\mathrm{LS}}(\varphi, \Omega, h) = d_B(\hat{\varphi}, \Omega_V, h)\,,$$

where $\hat{\varphi} = \mathrm{id}_V - \hat{f}$.

Remark 6.3.4. From the definition above it is clear that $\Omega \subseteq X$ need not be bounded. It is enough to require that for any finite dimensional subspace $V \subseteq X$ the intersection $\Omega \cap V$ is bounded. Such sets are said to be **finitely bounded**. So, the whole Leray–Schauder degree theory can be formulated using finitely bounded sets.

In what follows we always suppose that $(\varphi, \Omega, h) \in \hat{\mathcal{L}}$.

Proposition 6.3.5. *$d_{\mathrm{LS}}(\mathrm{id}, \Omega, h) = 1$ if $h \in \Omega$ and $d_{\mathrm{LS}}(\varphi, \Omega, h) = 0$ if $h \notin \overline{\Omega}$.*

Proof. Let $f_\varepsilon(u) = 0$ for all $u \in \overline{\Omega}$, $X_\varepsilon = \mathbb{R}h$ and $\Omega_\varepsilon = \Omega \cap X_\varepsilon$. Then, by Definition 6.3.3, $d_{\mathrm{LS}}(\mathrm{id}, \Omega, h) = d_B(\mathrm{id}_{X_\varepsilon}, \Omega_\varepsilon, h)$ and so, by Theorem 6.2.22 (a), we derive $d_{\mathrm{LS}}(\mathrm{id}, \Omega, h) = 1$. Similarly, we show this if $h \notin \overline{\Omega}$. □

Next we prove the following solution property.

Proposition 6.3.6. *If $d_{\mathrm{LS}}(\varphi, \Omega, h) \neq 0$, then there exists $u_0 \in \Omega$ such that $\varphi(u_0) = h$.*

Proof. Let $n > d(h, \varphi(\partial\Omega))^{-1} > 0$. We can find a finite rank map $f_n\colon \overline{\Omega} \to X$ such that

$$\|f_n(u) - f(u)\| < \frac{1}{n} \quad \text{for all } u \in \overline{\Omega}\,. \tag{6.3.7}$$

Let $X_n = \mathrm{span}\{\varphi_n(\overline{\Omega}), h\}$ where $\varphi_n = \mathrm{id} - f_n$ for all $n \in \mathbb{N}$. We set $\Omega_n = \Omega \cap X_n$ and get

$$0 \neq d_{\mathrm{LS}}(\varphi, \Omega, h) = d_B(\varphi_n, \Omega_n, h) \quad \text{for all } n \in \mathbb{N}\,.$$

Theorem 6.2.22 (d) implies that for every $n \in \mathbb{N}$, there exists $u_n \in \Omega_n$ such that

$$u_n - f_n(u_n) = h\,. \tag{6.3.8}$$

The compactness of f implies that we may assume that $f(u_n) \to y$ in X as $n \to \infty$ and due to (6.3.7) this gives $f_n(u_n) \to y$ in X as $n \to \infty$. From (6.3.8) we see that $u_n \to h + y$ in X as $n \to \infty$. Finally, combining (6.3.7) and (6.3.8), it follows that $\varphi(u_n) \to \varphi(h+y) = h$.

Since $h \notin \varphi(\partial\Omega)$ we must have $h + y \in \Omega$. Therefore, the equation $\varphi(u) = \xi$ has a solution $u_0 = h + y \in \Omega$. □

The homotopy invariance property requires that we specify the admissible homotopies.

Definition 6.3.7. We say that a family $\{h_t\}_{t\in[0,1]} \subseteq K(\overline{\Omega}, X)$ is a **homotopy of compact maps** on $\overline{\Omega}$ if for a given $\varepsilon > 0$ there exists $\delta = \delta(\varepsilon) > 0$ such that

$$\|h_t(x) - h_s(x)\| < \varepsilon \quad \text{for all } |t-s| < \delta \text{ and for all } x \in \overline{\Omega}\,.$$

Proposition 6.3.8. *If $\{\hat{h}_t\}_{t\in[0,1]}$ is a homotopy of compact maps on $\overline{\Omega}$, $\varphi_t = \mathrm{id} - \hat{h}_t$ for all $t \in [0,1]$ and $h \notin \varphi_t(\partial\Omega)$ for all $t \in [0,1]$, then $d_{\mathrm{LS}}(\varphi_t, \Omega, h)$ is independent of $t \in [0,1]$.*

Proof. First we show that there exists $\vartheta > 0$ such that

$$\|\varphi_t(u) - h\| \geq \vartheta \quad \text{for all } u \in \partial\Omega \text{ and for all } t \in [0,1]\,. \tag{6.3.9}$$

Arguing indirectly, suppose that (6.3.9) is not true. Then we can find sequences $\{t_n\}_{n\geq1} \subseteq [0,1]$ and $\{u_n\}_{n\geq1} \subseteq \partial\Omega$ such that

$$\|\varphi_{t_n}(u_n) - h\| < \frac{1}{n} \quad \text{for all } n \in \mathbb{N}\,. \tag{6.3.10}$$

We assume that $t_n \to t$. Since $\hat{h}_t \in K(\overline{\Omega}, X)$, we can say that at least for a subsequence, we have

$$\hat{h}_t(u_n) \to y \quad \text{in } X \text{ as } n \to \infty\,. \tag{6.3.11}$$

Definition 6.3.7 implies that

$$\|\hat{h}_t(u_n) - \hat{h}_{t_n}(u_n)\| \to 0 \quad \text{as } n \to \infty\,. \tag{6.3.12}$$

Taking (6.3.11) and (6.3.12) into account yields

$$\|\hat{h}_{t_n}(u_n) - y\| \leq \|\hat{h}_{t_n}(u_n) - \hat{h}_t(u_n)\| + \|\hat{h}_t(u_n) - y\| \to 0 \quad \text{as } n \to \infty\,.$$

Then, (6.3.10) implies

$$u_n = \varphi_{t_n}(u_n) - h + \hat{h}_{t_n}(u_n) + h \to y + h \quad \text{in } X \text{ as } n \to \infty\,.$$

Since $u_n \in \partial\Omega$ for all $n \in \mathbb{N}$ and because $\partial\Omega \subseteq X$ is closed, we obtain $y+h \in \partial\Omega$. Moreover, it holds

$$\varphi_t(y+h) = y+h - \hat{h}_t(y+h) = y+h - \lim_{n\to\infty} \hat{h}_{t_n}(u_n) = h\,,$$

which gives $h \in \varphi_t(\partial\Omega)$, a contradiction. Therefore, (6.3.9) holds.

On $[0,1]$ we introduce the relation $\sim$ defined by

$$t \sim s \quad \text{if and only if} \quad d_{\mathrm{LS}}(\varphi_t, \Omega, h) = d_{\mathrm{LS}}(\varphi_s, \Omega, h)\,.$$

Clearly, $\sim$ is an equivalence relation, that is, $\sim$ is reflexive, symmetric, and transitive. We show that the equivalence classes are open in $[0,1]$. So, let $s \in [0,1]$ and $\varepsilon \in (0, \vartheta/4)$; see (6.3.9). Let $K_f(\overline{\Omega}, X)$ be the set of finite rank maps from $\overline{\Omega}$ into X. Of course, $K_f(\overline{\Omega}, X) \subseteq K(\overline{\Omega}, X)$ and according to Theorem 3.7.10 there exists $\hat{h}_s^\varepsilon \in K_f(\overline{\Omega}, X)$ such that

$$\|\hat{h}_s^\varepsilon(u) - \hat{h}_s(u)\| < \frac{\vartheta}{4} \quad \text{for all } u \in \overline{\Omega}\,. \tag{6.3.13}$$

By Definition 6.3.7, there exists $\delta > 0$ such that

$$|t-s| < \delta \quad \text{implies} \quad \|\hat{h}_t(u) - \hat{h}_s(u)\| < \frac{\vartheta}{4} \quad \text{for all } u \in \overline{\Omega}\,. \tag{6.3.14}$$

From (6.3.13) and (6.3.14) we obtain

$$\|\hat{h}_s^\varepsilon(u) - \hat{h}_t(u)\| < \frac{\vartheta}{2} \quad \text{for all } |s-t| < \delta \text{ and for all } u \in \overline{\Omega}\,.$$

Hence,

$$d_{\mathrm{LS}}(\varphi_t, \Omega, h) = d_B(\mathrm{id}_V - \hat{h}_s^\varepsilon, \Omega \cap V, h) = d_{\mathrm{LS}}(\varphi_s, \Omega, h)\,,$$

where $V \subseteq X$ is a finite dimensional subspace of X such that $\hat{h}_s^\varepsilon(\overline{\Omega}) \subseteq V$. Hence, $|t-s| < \delta$ implies $t \sim s$. Therefore, the equivalence classes of $\sim$ are open sets in $[0,1]$. This implies that there is one equivalence class, namely the unit interval $[0,1]$. We conclude that $d_{\mathrm{LS}}(\varphi_t, \Omega, h)$ is independent of t. □

Proposition 6.3.9. *If $\varphi = \mathrm{id} - f$ and $\psi = \mathrm{id} - g$ with $f, g \in K(\overline{\Omega}, X)$ and*

$$f|_{\partial\Omega} = g|_{\partial\Omega} \quad \textit{and} \quad h \notin \varphi(\partial\Omega) = \psi(\partial\Omega)\,,$$

then $d_{\mathrm{LS}}(\varphi, \Omega, h) = d_{\mathrm{LS}}(\psi, \Omega, h)$.

Proof. We consider the homotopy

$$\hat{h}_t(u) = \hat{h}(t,u) = (1-t)\varphi(u) + t\psi(u) \quad \text{for all } t \in [0,1] \text{ and for all } u \in \overline{\Omega}\,.$$

Then $\{\hat{h}_t\}_{t\in[0,1]}$ is a homotopy of compact maps on $\overline{\Omega}$ and $h \notin (\mathrm{id} - \hat{h}_t)(\partial\Omega)$ for all $t \in [0,1]$. So, Proposition 6.3.8 implies that

$$d_{\mathrm{LS}}(\varphi, \Omega, h) = d_{\mathrm{LS}}(\hat{h}_0, \Omega, h) = d_{\mathrm{LS}}(\hat{h}_1, \Omega, h) = d_{\mathrm{LS}}(\psi, \Omega, h) . \qquad \square$$

Proposition 6.3.10. *If $(\varphi, \Omega, h) \in \hat{\mathcal{L}}$ and $\xi \in X$, then $d_{\mathrm{LS}}(\varphi, \Omega, h) = d_{\mathrm{LS}}(\varphi - \xi, \Omega, h - \xi)$.*

Proof. Note that $\varphi - \xi = \mathrm{id} - (f + \xi)$ and $f + \xi \in K(\overline{\Omega}, X)$. Let $g \in K_f(\overline{\Omega}, X)$ and $\psi = \mathrm{id} - g$ such that

$$d_{\mathrm{LS}}(\varphi, \Omega, h) = d_B(\psi, \Omega \cap V, h) , \tag{6.3.15}$$

where $V \subseteq X$ is a finite dimensional subspace of X such that $h \in V$ and $g(\overline{\Omega}) \subseteq V$; see Definition 6.3.3. Let $\hat{g} = g + \xi \in K_f(\overline{\Omega}, X)$. Then

$$\|\hat{g}(u) - (f + \xi)(u)\| < r = d(h - \xi, (\mathrm{id} - (f + \xi))(\partial\Omega)) .$$

Hence, due to Proposition 6.2.14 and (6.3.15),

$$d_{\mathrm{LS}}(\varphi - \xi, \Omega, h - \xi) = d_B(\psi - \xi, \Omega \cap V, h - \xi) = d_B(\psi, \Omega \cap V, h) = d_{\mathrm{LS}}(\varphi, \Omega, h) . \qquad \square$$

Next we establish the continuity of the Leray–Schauder degree with respect to the function φ.

Proposition 6.3.11. *If $\varphi = \mathrm{id} - f$, $\psi = \mathrm{id} - g$ with $f, g \in K(\overline{\Omega}, X)$, $h \notin \varphi(\partial\Omega)$, and*

$$\|\varphi(u) - \psi(u)\| < d(h, \varphi(\partial\Omega)) = r \quad \text{for all } u \in \overline{\Omega} ,$$

then $h \notin \psi(\partial\Omega)$ and $d_{\mathrm{LS}}(\varphi, \Omega, h) = d_{\mathrm{LS}}(\psi, \Omega, h)$.

Proof. We consider the homotopy

$$\hat{h}_t(u) = \hat{h}(t, u) = (1 - t)f(u) + tg(u) \quad \text{for all } t \in [0,1] \text{ and for all } u \in \overline{\Omega} .$$

This is a homotopy of compact maps on $\overline{\Omega}$. We set

$$\beta_t(u) = u - \hat{h}_t(u) \quad \text{for all } t \in [0,1] \text{ and for all } u \in \overline{\Omega} .$$

We get

$$\begin{aligned} \|h - \beta_t(u)\| &\geq \|h - \varphi(u)\| - t\|\varphi(u) - \psi(u)\| > \|h - \varphi(u)\| - tr \\ &\geq (1 - t)r \quad \text{for all } u \in \partial\Omega . \end{aligned}$$

Hence, $h \notin \beta_t(\partial\Omega)$ for all $t \in [0,1]$. Then, Proposition 6.3.8 implies that $d_{\mathrm{LS}}(\varphi, \Omega, h) = d_{\mathrm{LS}}(\psi, \Omega, h)$. $\square$

Proposition 6.3.12. *If* $\varphi = \mathrm{id} - f$ *with* $f \in K(\overline{\Omega}, X)$, *then* $d_{\mathrm{LS}}(\varphi, \Omega, \cdot)$ *is constant on every connected component of* $X \setminus \varphi(\partial\Omega)$.

Proof. Let U be a connected component of $X \setminus \varphi(\partial\Omega)$ and let $h \in U$. Taking Proposition 6.3.10 into account gives, for all $\xi \in U$,

$$d_{\mathrm{LS}}(\varphi, \Omega, \xi) = d_{\mathrm{LS}}(\varphi - (\xi - h), \Omega, \xi - (\xi - h)) . \tag{6.3.16}$$

Let $r = d(h, \varphi(\partial\Omega)) > 0$ and let $\varphi_\xi : \overline{\Omega} \to X$ be defined by $\varphi_\xi(u) = \varphi(u) - (\xi - h)$ for all $u \in \overline{\Omega}$. From Proposition 6.3.11, we obtain, if $\|h - \xi\| < r$, that

$$d_{\mathrm{LS}}(\varphi, \Omega, h) = d_{\mathrm{LS}}(\varphi_\varepsilon, \Omega, h) . \tag{6.3.17}$$

From (6.3.16) and (6.3.17) we see that

$$\|h - \xi\| < r \quad \text{implies} \quad d_{\mathrm{LS}}(\varphi, \Omega, h) = d_{\mathrm{LS}}(\varphi, \Omega, \xi) .$$

Since d_{LS} is $\mathbb{Z}$-valued, the conclusion of the proposition follows. □

We also can show the additivity of domain property.

Proposition 6.3.13. *If* $\Omega_1, \Omega_2 \subseteq \mathbb{R}^N$ *are bounded, disjoint open sets,* $\varphi = \mathrm{id} - f$ *with* $f \in K(\overline{\Omega}_1 \cup \overline{\Omega}_2, X)$ *and* $h \notin \varphi(\partial\Omega_1 \cup \partial\Omega_2)$, *then* $d_{\mathrm{LS}}(\varphi, \Omega_1 \cup \Omega_2, h) = d_{\mathrm{LS}}(\varphi, \Omega_1, h) + d_{\mathrm{LS}}(\varphi, \Omega_2, h)$.

Proof. Let $g \in K_f(\overline{\Omega}_1 \cup \overline{\Omega}_2, X)$ such that

$$\|f(u) - g(u)\| < d(h, \varphi(\partial\Omega_1 \cup \partial\Omega_2)) \quad \text{for all } u \in \overline{\Omega}_1 \cup \overline{\Omega}_2 ;$$

see Theorem 3.7.10. We set $\psi = \mathrm{id} - g$. Then

$$d_{\mathrm{LS}}(\varphi, \Omega_1 \cup \Omega_2, h) = d_B(\psi, (\Omega_1 \cup \Omega_2) \cap V, h) , \tag{6.3.18}$$

where $V \subseteq X$ is a finite dimensional subspace of X such that $g(\overline{\Omega}_1 \cup \overline{\Omega}_2) \subseteq V$ and $h \in V$; see Definition 6.3.3. From Theorem 6.2.22 (d) and Definition 6.3.3, we see that

$$\begin{aligned} d_B(\psi, (\Omega_1 \cup \Omega_2) \cap V, h) &= d_B(\psi, \Omega_1 \cap V, h) + d_B(\psi, \Omega_2 \cap V, h) \\ &= d_{\mathrm{LS}}(\varphi, \Omega_1, h) + d_{\mathrm{LS}}(\varphi, \Omega_2, h) . \end{aligned}$$

Hence, $d_{\mathrm{LS}}(\varphi, \Omega_1 \cup \Omega_2, h) = d_{\mathrm{LS}}(\varphi, \Omega_1, h) + d_{\mathrm{LS}}(\varphi, \Omega_2, h)$; see (6.3.18). □

Similarly, we also have the excision property.

Proposition 6.3.14. *If* $C \subseteq X$ *is compact,* $\varphi = \mathrm{id} - f$ *with* $f \in K(\overline{\Omega}, X)$ *and* $h \notin \varphi(C)$, *then* $d_{\mathrm{LS}}(\varphi, \Omega, h) = d_{\mathrm{LS}}(\varphi, \Omega \setminus C, h)$.

Summarizing the properties of the Leray–Schauder degree, we can state the following theorem; see also Theorem 6.2.22 for the corresponding result for the Brouwer degree.

Theorem 6.3.15. *If $\hat{\mathcal{L}} = \{(\varphi, \Omega, h) : \Omega \subseteq X$ is bounded, open, $\varphi = \mathrm{id} - f$ with $f \in K(\overline{\Omega}, X)$ and $h \notin \varphi(\partial\Omega)\}$, then there exists a map $d_{\mathrm{LS}} : \hat{\mathcal{L}} \to \mathbb{Z}$ known as the* ***Leray–Schauder degree*** *such that the following hold:*

(a) *Normalization property: $d_{\mathrm{LS}}(\mathrm{id}, \Omega, h) = 1$ if $h \in \Omega$ and $d_{\mathrm{LS}}(\mathrm{id}, \Omega, h) = 0$ if $h \notin \overline{\Omega}$;*

(b) *Domain additivity property: for all disjoint open sets $\Omega_1, \Omega_2 \subseteq \Omega$ and $h \notin \varphi(\overline{\Omega} \setminus (\Omega_1 \cup \Omega_2))$, we have*

$$d_{\mathrm{LS}}(\varphi, \Omega, h) = d_{\mathrm{LS}}(\varphi, \Omega_1, h) + d_{\mathrm{LS}}(\varphi, \Omega_2, h) ;$$

(c) *Homotopy invariance property: if $\{\hat{h}_t\}_{t\in[0,1]}$ is a homotopy of compact maps on $\overline{\Omega}$, $h \in C([0,1], X)$ and $h(t) \notin (\mathrm{id} - \hat{h}_t)(\partial\Omega)$ for all $t \in [0,1]$, then we have that*

$$d_{\mathrm{LS}}(\mathrm{id} - \hat{h}_t, \Omega, h(t)) \text{ is independent of } t \in (0,1] ;$$

(d) *Solution property: $d_{\mathrm{LS}}(\varphi, \Omega, h) \neq 0$ implies that $\varphi^{-1}(h) \neq \emptyset$;*

(e) *Dependence on the boundary values: if $(\varphi, \Omega, h), (\psi, \Omega, h) \in \hat{\mathcal{L}}$ and $\varphi|_{\partial\Omega} = \psi|_{\partial\Omega}$, then $d_{\mathrm{LS}}(\varphi, \Omega, h) = d_{\mathrm{LS}}(\psi, \Omega, h)$;*

(f) *Excision property: $d_{\mathrm{LS}}(\varphi, \Omega, h) = d_{\mathrm{LS}}(\varphi, \Omega_1, h)$ for every open set $\Omega_1 \subseteq \Omega$ such that $h \notin \varphi(\overline{\Omega} \setminus \Omega_1)$;*

(g) *Continuity in (φ, h): $d_{\mathrm{LS}}(\varphi, \Omega, h) = d_{\mathrm{LS}}(\psi, \Omega, h)$ for $(\varphi, \Omega, h) \in \hat{\mathcal{L}}$ and $\psi = \mathrm{id} - g$ with*

$$\|f(u) - g(u)\| < d(h, \varphi(\partial\Omega)) \quad \text{for all } u \in \overline{\Omega}$$

and $d_{\mathrm{LS}}(\varphi, \Omega, \cdot)$ is constant on every connected component of $X \setminus \varphi(\partial\Omega)$.

Next we will present some applications of the Leray–Schauder degree. We start with "Borsuk's Theorem".

Theorem 6.3.16 (Borsuk's Theorem). *If $\Omega \subseteq X$ is bounded, open, symmetric, $0 \in \Omega$, $(\varphi, \Omega, h) \in \hat{\mathcal{L}}$ and φ is odd on $\partial\Omega$, then $d_{\mathrm{LS}}(\varphi, \Omega, 0)$ is odd. In particular, $d_{\mathrm{LS}}(\varphi, \Omega, 0) \neq 0$.*

Proof. Using the Tietze Extension Theorem (see Theorem 1.2.44), we extend φ on $\overline{\Omega}$ and then take the odd part of the extension. So, we can assume that φ is odd on $\overline{\Omega}$. Then $f \in K(\overline{\Omega}, X)$ is odd since $\varphi = \mathrm{id} - f$ and $K = \overline{f(\overline{\Omega})}$ is compact and symmetric. Let $V \subseteq X$ be a finite dimensional subspace of X and let $g \in C(K, V)$ be such that

$$\|u - g(u)\| \leq \frac{r}{2} \quad \text{for all } u \in K \text{ with } r = d(0, \varphi(\partial\Omega)) .$$

Let $g_0(u) = 1/2[g(u) - g(-u)]$ for all $u \in K$. Then $\|u - g_0(u)\| \leq r/2$ for all $u \in K$. We set $\varphi_0 = \mathrm{id} - g_0 \circ f$. Evidently, φ_0 is odd and

$$d_{\mathrm{LS}}(\varphi, \Omega, 0) = d_B(\varphi_0, \Omega \cap V, 0)$$

is odd; see Theorem 6.2.31. ☐

Remark 6.3.17. As before, we can assume instead that

$$\frac{\varphi(u)}{\|\varphi(u)\|} \neq \frac{\varphi(-u)}{\|\varphi(-u)\|} \quad \text{for all } u \in \partial\Omega\,. \tag{6.3.19}$$

Evidently, if $\varphi|_{\partial\Omega}$ is odd, then (6.3.19) holds.

Minor changes in the proof of Proposition 6.2.37 lead to the following result.

Proposition 6.3.18. *If $\Omega \subseteq X$ is open and $f \in K(\overline{\Omega}, X)$ is locally injective, then $\varphi = \mathrm{id} - f$ is open.*

Corollary 6.3.19. *If $\Omega \subseteq X$ is bounded open, $\varphi = \mathrm{id} - f$ with $f \in K(\overline{\Omega}, X)$, φ is injective and $h \in \varphi(\Omega)$, then $d_{\mathrm{LS}}(\varphi, \Omega, h) = \pm 1$.*

In Remark 6.2.25 we mentioned that in an infinite dimensional Banach space X, the boundary $\partial B_1 = \{u \in X : \|u\| = 1\}$ is a retract of $\overline{B}_1 = \{u \in X : \|x\| \leq 1\}$. However, we show that the retraction cannot be of the form $\mathrm{id} - f$ with a compact f.

Proposition 6.3.20. *There is no $\varphi \in C(\overline{B}_1, \partial B_1)$ of the form $\varphi = \mathrm{id} - f$ with $f \in K(\overline{B}_1, X)$ such that $\varphi|_{\partial B_1} = \mathrm{id}|_{\partial B_1}$.*

Proof. Arguing by contradiction, suppose that such φ exists. Then, Theorem 6.3.15 (a), (e) imply that

$$d_{\mathrm{LS}}(\varphi, B_1, 0) = 1\,, \tag{6.3.20}$$

where $B_1 = \{u \in X : \|u\| < 1\}$. From (6.3.20) and Theorem 6.3.15 (d) we infer that there exists $\hat{u} \in B_1$ such that $\varphi(\hat{u}) = 0$, a contradiction since φ is ∂B_1-valued. □

We conclude this section with the infinite dimensional analog of the Brouwer Fixed Point Theorem; see Theorem 6.2.26. The result is known as the "Schauder Fixed Point Theorem". More fixed point theorems will be proven in the next section.

Theorem 6.3.21 (Schauder Fixed Point Theorem). *If X is a Banach space, $C \subseteq X$ is nonempty, bounded, closed, convex, and $f : C \to C$ is compact, then there exists $\hat{u} \in C$ such that $f(\hat{u}) = \hat{u}$.*

Proof. Let $r > 0$ be such that $C \subseteq B_r = \{u \in X : \|u\| < r\}$. By Theorem 1.7.29, there is a compact map $\hat{f} : \overline{B}_r \to C \subseteq \overline{B}_r$ such that $\hat{f}|_C = f$. Moreover, using again Theorem 1.7.29, we see that $\overline{B}_1$ is a retract of X. Let $r : X \to \overline{B}_r$ be the retraction map. We set

$$\hat{h}_t(u) = t(\hat{f} \circ r)(u) \quad \text{for all } t \in [0,1] \text{ and for all } u \in X\,. \tag{6.3.21}$$

Then $\{\hat{h}_t\}_{t \in [0,1]}$ is a homotopy of compact maps. We may assume that $\hat{f}(u) \neq u$ for all $u \in \partial B_r$ or otherwise there is nothing to prove. Then from (6.3.21) it follows that

$$\hat{h}_t(u) \neq u \quad \text{for all } t \in [0,1] \text{ and for all } u \in \partial B_r\,.$$

Then Theorem 6.3.15 (c) gives

$$d_{LS}(\mathrm{id} - \hat{f}, B_r, 0) = d_{LS}(\mathrm{id}, B_r, 0) = 1\,;$$

see also Theorem 6.3.15 (a). So, the solution property (see Theorem 6.3.15 (d)) implies that there exists $\hat{u} \in B_r$ such that $\hat{u} = \hat{f}(\hat{u}) \in C$. □

6.4 Fixed Point Theory

Let X be a topological space and let $\varphi: X \to X$ be a map. Fixed point theory deals with conditions on X and φ that guarantee the existence of a point $u_0 \in X$ such that $\varphi(u_0) = u_0$, which is called a fixed point. We have already encountered two such well-known results as applications of degree theory. These are the Brouwer Fixed Point Theorem (see Theorem 6.2.26) and the Schauder Fixed Point Theorem (see Theorem 6.3.21). There is an informal classification of fixed point theorems into three categories: (a) metric fixed points, (b) topological fixed points, and (c) order fixed points. Here we focus on the first two categories. Order fixed points are investigated in the next section in connection with the study of the fixed point index. In metric fixed point theory, the emphasis is on the metric point structure of the ambient space X and on the metric properties of the map φ. The main representative in this category is the celebrated "Banach Fixed Point Theorem ". In topological fixed point theory the emphasis is on the topological properties of the space X and of the map φ. The typical results in this definition are the Brouwer and Schauder fixed point theorems mentioned above.

6.4.1 Metric Fixed Point Theory

Definition 6.4.1. Let (X, d) be a metric space and let $\varphi: C \subseteq X \to C$ be a map.

(a) We say that φ is a k-**contraction** if it is k-Lipschitz with $k \in [0, 1)$, that is,

$$d(\varphi(u), \varphi(v)) \le k d(u, v) \quad \text{for all } u, v \in C \text{ and for all } 0 \le k < 1\,.$$

(b) We say that φ is **nonexpansive** if

$$d(\varphi(u), \varphi(v)) \le d(u, v) \quad \text{for all } u, v \in X\,,$$

that is, φ is 1-Lipschitz.

(c) We say that φ is **contractive** if

$$d(\varphi(u), \varphi(v)) < d(u, v) \quad \text{for all } u, v \in X \text{ with } u \neq v\,.$$

Remark 6.4.2. We mention that if $\varphi, \psi: X \to X$ are Lipschitz functions with Lipschitz constants $k(\varphi)$ and $k(\psi)$, respectively, then $k(\psi \circ \varphi) \le k(\psi)k(\varphi)$. In particular, if $\varphi^{(n)} =$

$\varphi \circ \cdots \circ \varphi$ n-times, then $k(\varphi^{(n)}) \leq k(\varphi)^n$. Moreover, if X is a linear space, then $k(\psi + \varphi) \leq k(\psi) + k(\varphi)$ and $k(\lambda\varphi) = \lambda k(\varphi)$ for all $\lambda \geq 0$.

The first fixed point theorem that we prove is the celebrated “Banach Fixed Point Theorem”. The importance of this result derives from the fact that in addition to the existence of a fixed point, it provides additional valuable information such as the uniqueness of the fixed point, stability under small perturbation of the equation, a constructive method to generate the unique fixed point, a priori error estimates, and rates of convergence of the approximation method.

Theorem 6.4.3 (Banach Fixed Point Theorem). *If (X, d) is a complete metric space and $\varphi: X \to X$ is a k-contraction, then φ has a unique fixed point.*

Proof. Let $u_0 \in X$ and consider the sequence $\{u_{n+1}\}_{n\geq 0} = \{\varphi(u_n)\}_{n\geq 0}$. We easily see that

$$d(u_n, u_{n+1}) = d(\varphi^{(n)}(u_0), \varphi^{(n+1)}(u_0)) \leq kd(\varphi^{(n-1)}(u_0), \varphi^{(n)}(u_0)) .$$

By induction we obtain

$$d(u_n, u_{n+1}) \leq k^n d(u_0, u_1) \quad \text{for all } n \in \mathbb{N} . \tag{6.4.1}$$

For $m > n$, we have, by the triangle inequality and (6.4.1),

$$\begin{aligned} d(\varphi^{(n)}(u_0), \varphi^{(m)}(u_0)) &\leq \sum_{i=0}^{m-n} d(\varphi^{(n+i)}(u_0), \varphi^{(n+i+1)}(u_0)) \\ &\leq \sum_{i=0}^{m-n} k^{n+i} d(u_0, u_1) \leq \frac{k^n}{1-k} d(u_0, u_1) \to 0 \end{aligned} \tag{6.4.2}$$

as $n \to \infty$. Hence, $\{u_n\}_{n\geq 1} \subseteq X$ is a Cauchy sequence and since X is complete, it follows that $u_n \to \hat{u}$ in X. Then $\varphi(u_n) \to \varphi(\hat{u})$ and $u_{n+1} = \varphi(u_n) = \varphi(\varphi^{(n)}(u_0)) \to \varphi(\hat{u})$. Therefore, $\hat{u} = \varphi(\hat{u})$.

Suppose $\hat{v} \in X$ is another fixed point of φ. Then

$$d(\hat{u}, \hat{v}) = d(\varphi(\hat{u}), \varphi(\hat{v})) \leq kd(\hat{u}, \hat{v}) \quad \text{with } 0 \leq k < 1 .$$

Thus, $\hat{u} = \hat{v}$, that is, the fixed point of φ is unique. □

From the theorem above and its proof, we extract some important information.

Proposition 6.4.4. *If (X, d) is a complete metric space and $\varphi: X \to X$ is a k-contraction, then the following hold:*

(a) *the unique fixed point $\hat{u}$ satisfies $\hat{u} = \lim_{n\to\infty} \varphi^{(n)}(u_0)$ for any $u_0 \in X$;*
(b) *$d(\hat{u}, \varphi^{(n)}(u_0)) \leq k^n/(1-k) d(u_0, \varphi(u_0))$ for all $n \in \mathbb{N}_0$;*
(c) *$d(u_0, \hat{u}) \leq 1/(1-k) d(u_0, \varphi(u_0))$ for any $u_0 \in X$;*
(d) *$d(u_{n+1}, \hat{u}) \leq kd(u_n, \hat{u})$ with $u_n = \varphi^{(n)}(u_0)$ for all $n \geq 0$.*

Proof. (a) From the proof of Theorem 6.4.3 we know that $\varphi^{(n)}(u_0) = u_n \to \hat{u}$ in X.

(b) This follows from (6.4.2) by letting $m \to \infty$.

(c) This follows from (b) by taking $n = 0$.

(d) It holds that

$$d(u_{n+1}, \hat{u}) = d(\varphi(u_n), \varphi(\hat{u})) \le kd(u_n, \hat{u}) \quad \text{for all } n \ge 0 .$$

□

Remark 6.4.5. Part (b) of Proposition 6.4.4 gives an a priori error estimate for the successive approximations while part (d) determines the rate of convergence of these successive approximations.

Theorem 6.4.3 admits a local version as seen in the following proposition.

Proposition 6.4.6. *If (X, d) is a complete metric space, $u_0 \in X$, $r > 0$, $B_r(u_0) = \{u \in X : d(u, u_0) < r\}$, and $\varphi: B_r(u_0) \to X$ is a k-contraction such that $d(\varphi(u_0), u_0) < (1-k)r$, then φ has a fixed point.*

Proof. Let $\eta \in (0, r)$ such that $d(\varphi(u_0), u_0) \le (1-k)\eta < (1-k)r$ and let $\overline{B}_\eta(u_0) = \{u \in X : d(u, u_0) \le \eta\}$. Let $u \in \overline{B}_\eta(u_0)$. Then

$$d(\varphi(u), u_0) \le d(\varphi(u), \varphi(u_0)) + d(\varphi(u_0), u_0) \le kd(u, u_0) + (1-k)\eta \le \eta .$$

Therefore, $\varphi: \overline{B}_\eta(u_0) \to \overline{B}_\eta(u_0)$ and we can apply Theorem 6.4.3 to obtain a fixed point for φ. □

Moreover, there is a parametric version of Theorem 6.4.3. This result illustrates the stability of the successive approximations method.

Proposition 6.4.7. *If (X, d) is a complete metric space, (T, e) is another metric space called the parameter space, for every $t \in T$, $\varphi_t: X \to X$ is a k-contraction with $k \in [0, 1)$ independent of $t \in T$ and $t \to \varphi_t(u)$ is continuous for each $u \in X$, then, for each $t \in T$, φ_t has a unique fixed point $u_t \in X$ and $t \to u_t$ is continuous from T into X.*

Proof. Theorem 6.4.3 gives a unique fixed point $u_t \in X$ for every $t \in T$. Suppose that $t_n \to t$. Then we get

$$\begin{aligned} d(u_{t_n}, u_t) = d(\varphi_{t_n}(u_{t_n}), \varphi_t(u_t)) &\le d(\varphi_{t_n}(u_{t_n}), \varphi_{t_n}(u_t)) + d(\varphi_{t_n}(u_t), \varphi_t(u_t)) \\ &\le kd(u_{t_n}, u_t) + d(\varphi_{t_n}(u_t), \varphi_t(u_t)) \quad \text{for all } n \in \mathbb{N} . \end{aligned}$$

Hence, $d(u_{t_n}, u_t) \le 1/(1-k) d(\varphi_{t_n}(u_t), \varphi_t(u_t)) \to 0$ as $n \to \infty$. Therefore, $t \to u_t$ is continuous from T into X. □

Next we examine what can be said about contractive maps.

Theorem 6.4.8. *If (X, d) is a compact metric space and $\varphi: X \to X$ is a contractive map, then φ has a unique fixed point $\hat{u} \in X$ and for each $v \in X$, $\varphi^{(n)}(v) \to \hat{u}$ in X.*

Proof. Let $\xi: X \to \mathbb{R}_+$ be defined by $\xi(u) = d(u, \varphi(u))$. Evidently, ξ is continuous. So, there exists $\hat{u} \in X$ such that $\xi(\hat{u}) = \inf[\xi(u) : u \in X]$. We claim that $\hat{u} \in X$ is a fixed point of φ. Arguing by contradiction, suppose that $\hat{u} \neq \varphi(\hat{u})$. Then

$$\xi(\varphi(\hat{u})) = d(\varphi(\hat{u}), \varphi^{(2)}(\hat{u})) < d(\hat{u}, \varphi(\hat{u})) = \xi(\hat{u}) ,$$

a contradiction. So, $\hat{u} = \varphi(\hat{u})$ and clearly this fixed point is unique.

Now, let $v \in X$ and $\eta_n = d(\varphi^{(n)}(v), \hat{u})$ with $n \geq 0$. Then

$$\eta_{n+1} = d(\varphi^{(n+1)}(v), \varphi(\hat{u})) \leq d(\varphi^{(n)}(v), \hat{u}) = \eta_n \quad \text{for all } n \geq 0 .$$

Hence, $\{\eta_n\}_{n\geq 0} \subseteq \mathbb{R}_+$ is decreasing. Therefore, we have $\eta_n \to \eta \geq 0$. Moreover, since X is compact, there exists a subsequence $\{\varphi^{(n_k)}(v)\}_{k\geq 1}$ of $\{\varphi^{(n)}(v)\}_{n\geq 0}$ such that $\varphi^{(n_k)}(v) \to y$ in X as $n \to \infty$. Then we obtain

$$d(y, \hat{u}) = \eta . \tag{6.4.3}$$

If $\eta > 0$, then (6.4.3) gives

$$\eta = \lim_{k\to\infty} d(\varphi^{(n_k+1)}(v), \hat{u}) = d(\varphi(y), \hat{u}) = d(\varphi(y), \varphi(\hat{u})) < d(y, \hat{u}) . \tag{6.4.4}$$

From (6.4.4) and (6.4.5) we have a contradiction. Therefore, $\eta = 0$ and so $y = \hat{u}$. So, for the original sequence we have $\varphi^{(n)}(v) \to \hat{u}$ in X. □

Example 6.4.9. We cannot drop the compactness requirement on X in the theorem above. In order to see this let

$$X = \{u \in C[0,1] : 0 = u(0) \leq u(t) \leq u(1) = 1 \text{ for all } t \in [0,1]\} . \tag{6.4.5}$$

Evidently, this set is closed, convex, and bounded in $C[0,1]$. So, X equipped with the supremum metric, that is, $d(u, v) = \|u - v\|_{C[0,1]}$ for all $u, v \in X$, is a complete metric space but it is not compact. Let $\varphi: X \to X$ be defined by $\varphi(u)(t) = tu(t)$ for all $t \in [0,1]$. Then,

$$\|\varphi(u) - \varphi(v)\|_{C[0,1]} < \|u - v\|_{C[0,1]} \quad \text{for all } u, v \in X ;$$

see (6.4.5), but clearly it does not have a fixed point.

In Theorem 6.4.3 we may replace the assumption that φ is a k-contraction, by the requirement that for some $n \in \mathbb{N}$, $\varphi^{(n)}$ is a contraction.

Theorem 6.4.10. *If (X, d) is a complete metric space and $\varphi: X \to X$ is a map such that $\varphi^{(n)}$ is a contraction for some $n \in \mathbb{N}$, then φ has a unique fixed point.*

Proof. Since $\varphi^{(n)}$ is a contraction, we can apply Theorem 6.4.3 and find a unique $\hat{u} \in X$ such that $\varphi^{(n)}(\hat{u}) = \hat{u}$. Then

$$\varphi(\hat{u}) = \varphi(\varphi^{(n)}(\hat{u})) = \varphi^{(n+1)}(\hat{u}) = \varphi^{(n)}(\varphi(\hat{u})) .$$

Hence, $\varphi(\hat{u}) = \hat{u}$ since $\hat{u} \in X$ is the unique fixed point of $\varphi^{(n)}$.

Next, suppose that $\hat{v} \in X$ is another fixed point of φ. Then

$$\hat{v} = \varphi(\hat{v}) = \varphi(\varphi(\hat{v})) = \cdots \varphi^{(n)}(\varphi(\hat{v})) ,$$

which shows that $\hat{v}$ is also a fixed point of $\varphi^{(n)}$. Therefore, $\hat{v} = \hat{u}$. □

Remark 6.4.11. A function $\varphi: X \to X$ for which $\varphi^{(n)}$ is a contraction for some $n \in \mathbb{N}$ need not be continuous. In order to see this, consider the function $\varphi: \mathbb{R} \to \mathbb{R}$ defined by

$$\varphi(u) = \begin{cases} 1 & \text{if } u \text{ is rational} , \\ 0 & \text{if } u \text{ is irrational} . \end{cases}$$

Clearly, φ is not continuous but $\varphi^{(2)} \equiv 1$.

Example 6.4.12. In this example, we produce a continuous function φ, which is not a contraction, but $\varphi^{(n)}$ is one for some $n \geq 2$. In order to see this, let $\varphi: C[0,2] \to C[0,2]$ be defined by

$$\varphi(u)(t) = \int_0^t u(s)\,ds \quad \text{for all } t \in [0,2] .$$

Then

$$\varphi^{(n)}(u)(t) = \frac{1}{(n-1)!} \int_0^t (t-s)^{n-1} u(s)\,ds \quad \text{for } n \in \mathbb{N} .$$

Note that φ is not a contraction but $\varphi^{(n)}$ is one for $n \geq 2$.

The Banach Fixed Point Theorem can be used to produce results in analysis. One such result with important applications is the so-called "Invariance of Domain" result.

Theorem 6.4.13 (Invariance of Domain). *If X is a Banach space, $U \subseteq X$ is nonempty open, $\varphi: U \to X$ is a k-contraction and $f(u) = u - \varphi(u)$ for all $u \in U$, then the following hold:*
(a) *f is an open map, so $f(U) \subseteq X$ is open;*
(b) *$f: U \to f(U)$ is a homeomorphism.*

Proof. (a) We show that for any $u \in U$, if $B_r(u) \subseteq U$, then $B_{(1-k)r}(f(u)) \subseteq f(B_r(u))$. For this purpose let $u_0 \in B_{(1-k)r}(f(u))$ and consider the k-contraction $g: B_r(u) \to X$ defined by $g(v) = u_0 + \varphi(v)$. We have

$$\|g(u) - u\| = \|u_0 + \varphi(u) - u\| = \|u_0 - f(u)\| < (1-k)r .$$

Invoking Proposition 6.4.6 there exists $v_0 \in B_r(u)$ such that $v_0 = g(v_0) = u_0 + \varphi(v_0)$, thus $f(v_0) = u_0$. Then $B_{(1-k)r}(f(u)) \subseteq f(B_r(u))$, that is, f is open.

(b) Let $u, v \in U$. Then

$$\|f(u) - f(v)\| \geq \|u - v\| - \|\varphi(u) - \varphi(v)\| \geq (1-k)\|u - v\| ,$$

which shows that f is injective. Therefore, $f: U \to f(U)$ is a continuous, open bijection, and thus a homeomorphism; see Proposition 1.1.42. □

Corollary 6.4.14. *If X is a Banach space and $\varphi: X \to X$ is a k-contraction, then $f = \mathrm{id} - \varphi$ is a homeomorphism.*

Proof. According to Theorem 6.4.13 we only need to show that f is surjective. To this end, let $u_0 \in X$ and let $g(u) = u_0 + \varphi(u)$ for all $u \in X$. Evidently, g is a k-contraction and so by Theorem 6.4.3 there exists a unique $\hat{u} \in X$ such that

$$\hat{u} = g(\hat{u}) = u_0 + \varphi(\hat{u}) .$$

Hence, $f(\hat{u}) = u_0$ and so f is surjective. □

Remark 6.4.15. Sometimes $u \to f(u) = u - \varphi(u)$ is called **k-contraction vector field.**

Next we pass to nonexpansive maps. In general, such maps are fixed point free.

Example 6.4.16. Let $X = c_0 = \{\hat{u} = (u_n)_{n\in\mathbb{N}}: u_n \to 0\}$ with norm $\|\hat{u}\| = \sup_{n\geq 1} |u_n|$. Consider the map $\varphi: X \to X$ defined by

$$\varphi(\hat{u})_1 = \frac{1}{2}[1 + \|\hat{u}\|] \quad \text{and} \quad (\varphi(\hat{u}))_{n+1} = u_n \text{ for } n \geq 2 . \tag{6.4.6}$$

Here $\varphi(\hat{u}) = (\varphi(\hat{u})_n)_{n\in\mathbb{N}} \in X = c_0$. Note that

$$\|\varphi(\hat{u}) - \varphi(\hat{v})\| = \|\hat{u} - \hat{v}\| .$$

Hence, φ is nonexpansive. But φ is fixed point free. Indeed, if $\varphi(\hat{u}) = \hat{u}$, then

$$u_n = \frac{1}{2}[1 + \|\hat{u}\|] \quad \text{for all } n \in \mathbb{N} ;$$

see (6.4.6). This gives $\hat{u} \notin c_0$. Therefore, we need additional conditions in order to guarantee that a nonexpansive map has a fixed point.

We start with a simple observation.

Lemma 6.4.17. *If X is a Banach space, $C \subseteq X$ is nonempty, bounded, closed, convex, and $\varphi: C \to C$ is nonexpansive, then* $\inf[\|u - \varphi(u)\|: u \in C] = 0$.

Proof. Let $u_0 \in C$ and $t \in (0, 1)$. We consider the map $\varphi_t: C \to C$ defined by

$$\varphi_t(u) = (1-t)\varphi(u) + tu_0 .$$

For $u, v \in C$ we have

$$\|\varphi_t(u) - \varphi_t(v)\| = (1-t)\|\varphi(u) - \varphi(v)\| \le (1-t)\|u - v\| .$$

Hence, φ_t is a $(1-t)$-contraction for $t \in (0,1)$. Then Theorem 6.4.3 implies that there exists a unique $u_t \in C$ such that $\varphi_t(u_t) = u_t$. Since C is bounded, we easily see that

$$\begin{aligned}\|u_t - \varphi(u_t)\| &= \|\varphi_t(u_t) - \varphi(u_t)\| = \|(1-t)\varphi(u_t) + tu_0 - \varphi(u_t)\| \\ &= t\|u_0 - \varphi(u_t)\| \le t \operatorname{diam} C .\end{aligned} \tag{6.4.7}$$

Since $t \in (0,1)$ is arbitrary, we conclude from (6.4.7) that $\inf[\|u - \varphi(u)\| : u \in C] = 0$. □

This lemma leads to the first fixed point theorem for nonexpansive maps.

Theorem 6.4.18. *If X is a Banach space, $C \subseteq X$ is nonempty, compact, convex, and $\varphi: C \to C$ is nonexpansive, then φ has a fixed point.*

Proof. Let $\{u_n\}_{n\ge1} \subseteq C$ such that $\|u_n - \varphi(u_n)\| \searrow 0$; see Lemma 6.4.17. The compactness of C implies that we may assume, at least for a subsequence, that $u_n \to \hat{u} \in C$ in X. Then

$$\|u_n - \varphi(u_n)\| \to \|\hat{u} - \varphi(\hat{u})\| \quad \text{as } n \to \infty .$$

Hence, $\|\hat{u} - \varphi(\hat{u})\| = 0$ and so $\hat{u} = \varphi(\hat{u})$. □

Remark 6.4.19. Of course, the theorem above can be deduced from the Schauder Fixed Point Theorem; see Theorem 6.3.21. Here, we have given a proof that is not degree theoretic.

We have another fixed point theorem for nonexpansive maps. The proof is again based on Lemma 6.4.17.

Theorem 6.4.20. *If X is a Banach space, $C \subseteq X$ is nonempty, bounded, closed, convex, $\varphi: C \to C$ is nonexpansive, and $(\mathrm{id} - \varphi)(C) \subseteq X$ is closed, then φ has a fixed point.*

Proof. Taking Lemma 6.4.17 into account shows $0 \in \overline{(\mathrm{id} - \varphi)(C)} = (\mathrm{id} - \varphi)(C)$ by hypothesis. So, there exists $\hat{u} \in C$ such that $0 = \hat{u} - \varphi(\hat{u})$, and hence, $\varphi(\hat{u}) = \hat{u}$. □

This theorem leads to another fixed point result for nonexpansive maps on a Hilbert space.

Theorem 6.4.21. *If H is a Hilbert space, $C \subseteq H$ is nonempty, bounded, closed, convex, and $\varphi: C \to C$ is nonexpansive, then φ has a fixed point.*

Proof. Let $p_C: H \to C$ be the metric projection on C. Then $p_C|_C = \mathrm{id}|_C$ and it is nonexpansive. Hence, $\varphi \circ p_C$ is nonexpansive as well and by Example 6.1.19 (b) we know that $\mathrm{id} - \varphi \circ p_C$ is maximal monotone. Then, using Proposition 6.1.14, we see that $(\mathrm{id} - \varphi \circ p_C)(C) = (\mathrm{id} - \varphi)(C)$ is closed. So, we can apply Theorem 6.4.20 and obtain that φ has a fixed point. □

We can extend this theorem to uniformly convex Banach spaces. Recall that a Hilbert space is uniformly convex because of the parallelogram law.

Theorem 6.4.22. *If X is a uniformly convex Banach space, $C \subseteq X$ is nonempty, bounded, closed, convex, and $\varphi\colon C \to C$ is nonexpansive, then φ has a fixed point.*

Proof. Let $\mathcal{L} = \{K \subseteq C : K$ is nonempty, closed, convex, and $\varphi(K) \subseteq K\}$. Evidently, $C \in \mathcal{L}$ and so $\mathcal{L} \neq 0$. We introduce a partial order on $\mathcal{L}$ using the reverse inclusion, that is

$$K_1 \preceq K_2 \quad \text{if and only if} \quad K_2 \subseteq K_1 .$$

Let $\mathcal{E} \subseteq \mathcal{L}$ be a chain in $(\mathcal{L}, \preceq)$, that is, $\mathcal{E}$ is totally ordered with respect to $\preceq$. The set $\bigcap_{K\in\mathcal{E}} K$ is closed, convex, bounded, and φ-invariant, that is, $\varphi(\bigcap_{K\in\mathcal{E}} K) \subseteq \bigcap_{K\in\mathcal{E}} K$. Since C is w-compact, by the finite intersection property, $\bigcap_{K\in\mathcal{E}} K \neq \emptyset$. Hence, $\bigcap_{K\in\mathcal{E}} K$ is an upper bound for $\mathcal{E}$. Using Zorn's Lemma, we see that $\mathcal{L}$ has a maximal element $\tilde{K}$. The maximality of $\tilde{K}$ implies that $\tilde{K} = \overline{\operatorname{conv}}\,\varphi(\tilde{K})$. We will be done if we can show that $\tilde{K}$ is a singleton. Suppose that $\tilde{K}$ is not a singleton. Then $r = \operatorname{diam} \tilde{K} > 0$. We choose $u_1, u_2 \in \tilde{K}$ such that $\|u_2 - u_1\| \geq r/2$. Let u be the midpoint of the line segment connecting u_1 and u_2. Let $y \in \tilde{K}$ with $\|y - u_1\| \leq r$ and $\|y - u_2\| \leq r$. Then $u - y$ is the midpoint of the line segment connecting $u_1 - y$ and $u_2 - y$. Since X is uniformly convex, there is $t \in (0, 1)$ such that $\|u - y\| \leq (1 - t)r = r_0 < r$. We set

$$K_0 = \bigcap_{y\in\tilde{K}} \{v \in \tilde{K} : \|v - y\| \leq r_0\} .$$

Then K_0 is nonempty, closed, convex, $u \in K_0$, and it is a strict subset of $\tilde{K}$ since $r_0 < r = \operatorname{diam} \tilde{K}$. We show that $\tilde{K}$ is φ-invariant. So, let $v \in K_0$ and $y \in \tilde{K}$. Given $\varepsilon > 0$ there exist $\{z_k\}_{k=1}^N \subseteq \tilde{K}$ and $\{\lambda_k\}_{k=1}^N \subseteq [0, 1]$ such that

$$\Big\| y - \sum_{k=1}^N \lambda_k \varphi(z_k) \Big\| < \varepsilon \quad \text{and} \quad \sum_{k=1}^N \lambda_k = 1 .$$

For $v \in K_0$ we have

$$\begin{aligned} \|\varphi(v) - y\| &\leq \Big\| \varphi(v) - \sum_{k=1}^N \lambda_k \varphi(z_k) \Big\| + \varepsilon \leq \sum_{k=1}^N \lambda_k \|\varphi(v) - \varphi(z_k)\| + \varepsilon \\ &\leq \sum_{k=1}^N \lambda_k r_0 + \varepsilon = r_0 + \varepsilon . \end{aligned}$$

Since $\varepsilon > 0$ is arbitrary, we let $\varepsilon \searrow 0$ and obtain

$$\|\varphi(v) - y\| \leq r_0 \quad \text{for all } y \in \tilde{K} .$$

Hence, $\varphi(v) \in K_0$ and so, K_0 is φ-invariant. But $K_0 \subseteq \tilde{K}$ and so $\tilde{K} \preceq K_0$ with $\tilde{K} \neq K_0$, a contradiction to the maximality of $\tilde{K}$. This proves that $\tilde{K} = \{\tilde{u}\}$ and so $\varphi(\tilde{u}) = \tilde{u}$. ☐

This proof suggests the introduction of the following geometric notions.

Definition 6.4.23. Let X be a Banach space and let $C \subseteq X$ be nonempty.
(a) We say that $u \in C$ is a **diametral point** of C if

$$\sup[\|v - u\| : v \in C] = \operatorname{diam} C .$$

(b) We say that a convex set $C \subseteq X$ has **normal structure** if every bounded, convex set $K \subseteq C$ with $\operatorname{diam} K > 0$ has nondiametral points.
(c) We define

$$\begin{aligned} r_u(C) &= \sup[\|u - v\| : v \in C] , \\ r(C) &= \inf[r_u(C) : u \in C] , \\ C_0 &= \{u \in C : r_u(c) = r(C)\} . \end{aligned}$$

Here, $r_u(C)$ is called the **radius of C relative to** $u \in X$, $r(C)$ is the **Chebyshev radius** of C and C_0 is said to be the **Chebyshev center** of C.

Remark 6.4.24. If $u \in C_0$, then $B_{r_u(C)}(u)$ contains C. Any other ball centered at a point of C with a smaller radius does not have this property. Moreover, it holds that

$$r(C) \leq r_u(C) \leq \operatorname{diam} C \quad \text{for all } u \in C . \tag{6.4.8}$$

Evidently, $u \in C$ is a diametral point if $r_u(C) = \operatorname{diam} C$. Sets with normal structure have no convex subsets K that consist entirely of diametral points except singletons, that is, $r(K) < \operatorname{diam} K$.

Proposition 6.4.25. *If X is a Banach space and $C \subseteq X$ is compact and convex, then C has normal structure.*

Proof. Without any loss of generality, we assume that $\operatorname{diam} C > 0$. Suppose that C does not have normal structure. Then we may assume that for given $u_1 \in C$ there exists $u_2 \in C$ such that $\operatorname{diam} C = \|u_1 - u_2\|$. Due to the convexity of C, it holds that $1/2(u_1 + u_2) \in C$. We can find $u_3 \in C$ such that $\operatorname{diam} C = \|u_3 - 1/2(u_1 + u_2)\|$. Inductively, we produce a sequence $\{u_n\}_{n\geq 1} \subseteq C$ such that

$$\operatorname{diam} C = \left\| u_{n+1} - \frac{1}{n} \sum_{k=1}^{n} u_k \right\| \quad \text{for all } n \geq 2 .$$

Hence,

$$\operatorname{diam} C \leq \frac{1}{n} \sum_{k=1}^{n} \|u_{n+1} - u_k\| \leq \operatorname{diam} C .$$

This implies $\operatorname{diam} C = \|u_{n+1} - u_k\|$ for all $k = 1, \ldots, n$. So, $\{u_n\}_{n\geq 1} \subseteq C$ has no convergent subsequence, a contradiction to the compactness of C. ☐

Lemma 6.4.26. *If X is a reflexive Banach space and $C \subseteq X$ is nonempty, bounded, closed, convex, then C_0 is nonempty and convex.*

Proof. For $u \in C$ and $n \in \mathbb{N}$, we define

$$C_n(u) = \left\{ v \in C : \|v - u\| \leq r(C) + \frac{1}{n} \right\} .$$

We set $\hat{C}_n = \bigcap_{u \in C} C_n(u)$ with $n \in \mathbb{N}$. Then, $\hat{C}_n$ is nonempty, bounded, closed, convex, and $\{\hat{C}_n\}_{n\geq 1}$ is decreasing. From the reflexivity of X and the finite intersection property we conclude that $C_0 = \bigcap_{n\in\mathbb{N}} \hat{C}_n$ is nonempty, w-compact, and convex. □

Lemma 6.4.27. *If X is a Banach space and $C \subseteq X$ is nonempty, bounded, closed, convex, and has normal structure, then* $\operatorname{diam} C_0 < \operatorname{diam} C$.

Proof. Since C has normal structure, there exists $u \in C$ such that $r_u(C) < \operatorname{diam} C$; see Remark 6.4.24. If $v, y \in C_0$, then

$$\|y - v\| \leq r_v(C) = r(C) \leq r_u(C) < \operatorname{diam} C ;$$

see (6.4.8). Hence, $\operatorname{diam} C_0 < \operatorname{diam} C$. □

Now we can state the main fixed point theorem for nonexpansive maps.

Theorem 6.4.28. *If X is a reflexive Banach space, $C \subseteq X$ is nonempty, bounded, closed, convex, and has normal structure, and $\varphi: C \to C$ is nonexpansive, then φ has a fixed point.*

Proof. As in the proof of Theorem 6.4.22, let

$$\mathcal{L} = \{K \subseteq C : K \text{ is nonempty, closed, convex, and } \varphi(K) \subseteq K\} .$$

We partially order $\mathcal{L}$ by

$$K_1 \preceq K_2 \quad \text{if and only if} \quad K_2 \subseteq K_1 ,$$

which is the reverse inclusion partial ordering. Using Zorn's Lemma we produce a maximal element $\tilde{K}$; see the proof of Theorem 6.4.22. Suppose that $\operatorname{diam} \tilde{K} > 0$ and let $u \in \tilde{K}_0$; see Definition 6.4.23 (c). Then

$$\|\varphi(u) - \varphi(v)\| \leq \|u - v\| \leq r(\tilde{K}) \quad \text{for all } v \in \tilde{K} .$$

Hence, $\varphi(\tilde{K}) \subseteq \overline{B}_{r(\tilde{K})}(\varphi(u)) = \overline{B}_*$. Then $\varphi(\tilde{K} \cap \overline{B}_*) \subseteq \tilde{K} \cap \overline{B}_*$ and so from the maximality of $\tilde{K}$ we infer that $\tilde{K} \subseteq \overline{B}_*$. Therefore,

$$\varphi(\tilde{K}_0) \subseteq \tilde{K}_0 . \tag{6.4.9}$$

From Lemma 6.4.26 and (6.4.9) we see that $\tilde{K}_0 \in \mathcal{L}$. Since $\operatorname{diam}\tilde{K} > 0$, Lemma 6.4.27 implies that $\tilde{K}_0$ is a proper subset of $\tilde{K}$, contradiction the maximality of $\tilde{K}$. Therefore, $\tilde{K}$ is a singleton and so φ has a fixed point. □

A careful reading of the proofs of Lemma 6.4.26 and Theorem 6.4.28 reveals that essentially using the same proofs, we can have the following version of Theorem 6.4.28.

Theorem 6.4.29. *If X is a Banach space and $C \subseteq X$ is nonempty, w-compact, convex, and has normal structure and $\varphi: X \to X$ is nonexpansive, then φ has a fixed point.*

Remark 6.4.30. One can show that in a uniformly convex Banach space every nonempty bounded, closed, convex set has normal structure. Recall that a uniformly convex Banach space is reflexive; see the Milman–Pettis Theorem stated as Theorem 3.4.28.

Theorem 6.4.31. *If H is a Hilbert space, $\overline{B}_r = \{u \in H : \|u\| \leq r\}$ and $\varphi: \overline{B}_r \to H$ is nonexpansive, then one of the following statements hold:*
(a) *φ has a fixed point;*
(b) *there exist $\hat{u} \in \partial B_r$ and $\lambda \in (0,1)$ such that $\hat{u} = \lambda\varphi(\hat{u})$.*

Proof. Let $p_{\overline{B}_r} : H \to \overline{B}_r$ be the metric projection on $\overline{B}_r$, that is,

$$p_{\overline{B}_r}(u) = \begin{cases} u & \text{if } \|u\| \leq r\,, \\ r\frac{u}{\|u\|} & \text{if } \|u\| > r\,. \end{cases} \tag{6.4.10}$$

We know that $p_{\overline{B}_r}$ is nonexpansive; see Proposition 3.5.20 (d). Hence, $p_{\overline{B}_r} \circ \varphi: \overline{B}_r \to \overline{B}_r$ is nonexpansive as well. So, we can apply Theorem 6.4.21 and obtain $\hat{u} \in \overline{B}_r$ such that $\hat{u} = p_{\overline{B}_r}(\varphi(\hat{u}))$. If $\varphi(\hat{u}) \in \overline{B}_r$, then $\hat{u} = \varphi(\hat{u})$. If $\varphi(\hat{u}) \in H \setminus \overline{B}_r$, then $\hat{u} = r\varphi(\hat{u})/\|\varphi(\hat{u})\|$; see (6.4.10). So, if $\lambda = r/\|\varphi(\hat{u})\| \in (0,1)$, then $\hat{u} = \lambda\varphi(\hat{u})$ and $\|\hat{u}\| = r$. □

According to the theorem above, in order to guarantee that a nonexpansive map $\varphi: \overline{B}_r \to H$ has a fixed point, we need to impose conditions which exclude (b). Such conditions are provided by the next corollary.

Corollary 6.4.32. *If H is a Hilbert space, $\overline{B}_r = \{u \in H : \|u\| \leq r\}$, $\varphi: \overline{B}_r \to H$ is nonexpansive and for all $u \in \partial B_r$, one of the following conditions hold:*
(a) $\|\varphi(u)\| \leq \|u\|$;
(b) $\|\varphi(u)\| \leq \|u - \varphi(u)\|$;
(c) $\|\varphi(u)\|^2 \leq \|u\|^2 + \|u - \varphi(u)\|^2$;
(d) $(\varphi(u), u) \geq \|u\|^2$;
(e) $\varphi(u) = -\varphi(-u)$,

then φ has a fixed point.

6.4.2 Topological Fixed Point Theory

In this subsection we present some of the main results from the topological fixed point theory. Now, the emphasis is on the topological properties of the space X and of the map φ. We have already encountered two such results. The finite dimensional Brouwer Fixed Point Theorem (see Theorem 6.2.26) and the infinite dimensional Schauder Fixed Point Theorem; see Theorem 6.3.21. Both were proven using degree theory.

We start with an extension of the Brouwer Fixed Point Theorem.

Proposition 6.4.33. *If $C \subseteq \mathbb{R}^N$ is homeomorphic to the closed unit ball $\overline{B}_1^N \subseteq \mathbb{R}^N$ and $\varphi: C \to C$ is continuous, then φ has a fixed point.*

Proof. Let $h: C \to \overline{B}_1^N$ be the homeomorphism and consider the map $g = h \circ \varphi \circ h^{-1}: \overline{B}_1^N \to \overline{B}_1^N$. Evidently, g is continuous and so Theorem 6.2.26 gives $\hat{v} \in \overline{B}_1^N$ such that

$$g(\hat{v}) = h(\varphi(h^{-1}(\hat{v}))) = \hat{v} . \tag{6.4.11}$$

Let $\hat{u} = h^{-1}(\hat{v})$. Then from (6.4.11) we obtain $\varphi(\hat{u}) = \hat{u}$. □

Remark 6.4.34. A compact convex set $C \subseteq \mathbb{R}^N$ is homeomorphic to $\overline{B}_1^m$ for some $m \leq N$. If $\operatorname{int} C \neq \emptyset$, then $m = N$. So, every continuous $\varphi: C \to C$ has a fixed point; see also Theorem 6.3.21.

We present an interesting consequence of the Brouwer Fixed Point Theorem. The result is known as the "Perron–Frobenius Theorem".

Theorem 6.4.35 (Perron–Frobenius Theorem). *If $A = (a_{ij})_{i,j=1}^N$ is a nonnegative $N \times N$-matrix, that is, $a_{ij} \geq 0$ for all $i,j = 1,\dots,N$ or equivalently $(Au, u)_{\mathbb{R}^N} \geq 0$ for all $u \in \mathbb{R}^N$, then A has a nonnegative eigenvalue $\hat{\lambda} \geq 0$ and a corresponding nonnegative eigenvector $\hat{u} \in \mathbb{R}^N$, that is, $\hat{u} = (\hat{u}_k)_{k=1}^N \in \mathbb{R}^N$ satisfies $\hat{u}_k \geq 0$ for all $k = 1,\dots,N$.*

Proof. Let

$$C = \left\{ u = (u_k)_{k=1}^N \in \mathbb{R}^N : u_k \geq 0 \text{ for all } k = 1,\dots,N, \sum_{k=1}^N u_k = 1 \right\} .$$

If for some $\hat{u} \in C$, we have $A\hat{u} = 0$, then the desired eigenpair is $(0, \hat{u})$.

On the other hand, if $Au \neq 0$ for all $u \in C$, then there exists $c > 0$ such that

$$\sum_{k=1}^N (Au)_k \geq c \quad \text{for all } u \in C .$$

Consider the map $\varphi: C \to C$ defined by

$$\varphi(u) = \frac{1}{\sum_{k=1}^N (Au)_k} Au \quad \text{for all } u \in C .$$

Evidently, φ is continuous and so it has a fixed point. Hence, there exists $\tilde{u} = (\tilde{u}_k)_{k=1}^N \in C$ such that $\varphi(\tilde{u}) = \tilde{u}$; see Remark 6.4.34. Therefore, $A\tilde{u} = \lambda\tilde{u}$ with $\lambda = \sum_{k=1}^N (A\tilde{u})_k \geq 0$. □

Proposition 6.4.36. *If $\varphi\colon \mathbb{R}^N \to \mathbb{R}^N$ is continuous and there exists $r > 0$ such that*

$$(\varphi(u), u)_{\mathbb{R}^N} \geq 0 \quad \text{for all } u \in \overline{B}_r^N\,, \tag{6.4.12}$$

then there exists $\hat{u} \in \overline{B}_r^N$ such that $\varphi(\hat{u}) = 0$.

Proof. We argue by contradiction. So suppose that $\|\varphi(u)\| > 0$ for all $u \in \overline{B}_r^N$. We can define the map $\psi\colon \overline{B}_r^N \to \partial\overline{B}_r^N$ by setting

$$\psi(u) = -r\frac{\varphi(u)}{\|\varphi(u)\|} \quad \text{for all } u \in \overline{B}_r^N\,.$$

Clearly, ψ is continuous. So, there exists $\tilde{u} \in \overline{B}_r^N$ such that

$$\psi(\tilde{u}) = -r\frac{\varphi(\tilde{u})}{\|\varphi(\tilde{u})\|} = \tilde{u} \in \partial B_r^N\,.$$

Hence, $r^2 = \|\tilde{u}\|^2 = -r/\|\varphi(\tilde{u})\|(\varphi(\tilde{u}), \tilde{u})_{\mathbb{R}^N} \leq 0$; see (6.4.12), a contradiction. So, there exists $\hat{u} \in \overline{B}_r^N$ such that $\varphi(\hat{u}) = 0$. □

We have another fixed point theorem that is proven using degree theory.

Theorem 6.4.37. *If $\Omega \subseteq \mathbb{R}^N$ is bounded and open, $\varphi \in C(\overline{\Omega}, \mathbb{R}^N)$ and there is $y \in \Omega$ such that*

$$\varphi(u) - y \neq \lambda(u - y) \quad \text{for all } u \in \partial\Omega \text{ and for all } \lambda > 1\,, \tag{6.4.13}$$

then φ has a fixed point in $\overline{\Omega}$.

Proof. We assume that $0 \notin (\mathrm{id} - \varphi)(\partial\Omega)$ or otherwise we already have a fixed point. We consider the homotopy $h\colon [0,1] \times \overline{\Omega} \to \mathbb{R}$ defined by

$$h(t,u) = u - y - t(\varphi(u) - y) \quad \text{for all } t \in [0,1] \text{ and for all } u \in \overline{\Omega}\,.$$

We claim that $0 \notin h(t, \partial\Omega)$ for all $t \in (0,1)$. Indeed, if $h(t,u) = 0$ for some $t \in (0,1)$ and some $u \in \partial\Omega$, then

$$\varphi(u) - y = \frac{1}{t}(u - y) \quad \text{for } \frac{1}{t} > 1\,,$$

which contradicts (6.4.13). So, taking the homotopy invariance of the Brouwer degree (see Theorem 6.4.22 (c)) into account gives $d_B(\mathrm{id} - y, \Omega, 0) = d_B(\mathrm{id} - \varphi, \Omega, 0)$. Then, Propositions 6.2.14 and 6.2.19 imply that $d_b(\mathrm{id}, \Omega, y) = 1 = d_B(\mathrm{id} - \varphi, \Omega, 0)$. Hence, from the solution property of the Brouwer degree (see Theorem 6.4.22 (d)), we get that there exists $\hat{u} \in \Omega$ such that $0 = \hat{u} - \varphi(\hat{u})$, which gives $\varphi(\hat{u}) = \hat{u}$. □

Remark 6.4.38. The geometric interpretation of (6.4.13) is that, for some $y \in \Omega$ and for $u \in \partial\Omega$, does $\varphi(u)$ lie on the continuation of the line segment $[y, u]$ beyond u.

The next theorem shows that Brouwer's Fixed Point Theorem is in fact equivalent to some other important geometrical statements.

Theorem 6.4.39. *The following statements are equivalent:*
(a) $\partial B_1^N = \{u \in \mathbb{R}^N : |u| = 1\}$ *is not contractible in itself (see Definition* 1.7.19 *(b));*
(b) *every* $\varphi\colon \overline{B}_1^N = \{u \in \mathbb{R}^N : |u| \le 1\} \to \mathbb{R}^N$ *that is continuous has one of the following properties:*
   (b1) φ *has a fixed point;*
   (b2) *there exists* $u \in \partial\overline{B}_1^N$ *and* $\lambda \in (0,1)$ *such that* $u = \lambda\varphi(u)$*;*
(c) *every* $\varphi\colon \overline{B}_1^N \to \overline{B}_1^N$ *that is continuous has a fixed point;*
(d) ∂B_1^N *is not a retract of* $\overline{B}_1^N$.

Proof. (a) ⟹ (b): We argue indirectly. So, we suppose that $\varphi(u) \ne u$ for all $u \in \overline{B}_1$ and $y \ne \lambda\varphi(u)$ for all $y \in \partial B_1^N$ and for all $\lambda \in (0,1)$. Note that $y \ne \lambda\varphi(y)$ for all $y \in \partial B_1^N$ and for all $\lambda \in \{0,1\}$.

Let $r\colon \mathbb{R}^N \setminus \{0\} \to \partial B_1^N$ be defined by $r(u) = u/|u|$, which is a radial retraction. We consider the deformation $h\colon [0,1] \times \partial B_1^N \to \partial B_1^N$ defined by

$$h(\lambda, y) = \begin{cases} r(y - 2\lambda\varphi(y)) & \text{if } 0 \le \lambda \le \frac{1}{2}, \\ r((2-2\lambda)y - \varphi((2-2\lambda)y)) & \text{if } \frac{1}{2} \le \lambda \le 1. \end{cases}$$

Then $h(0,\cdot) = \mathrm{id}$ and $h(1,\cdot) = r(-\varphi(0))$. Hence, ∂B_1^N is contractible in itself, a contradiction.

(b) ⟹ (c): Evidently, (b2) cannot occur since if $u \in \partial B_1^N$ and $u = \lambda\varphi(u)$ with $\lambda \in (0,1)$, then $1 = |u| = \lambda|\varphi(u)| < 1$, a contradiction. So (b1) holds and φ has a fixed point.

(c) ⟹ (d): This implication follows from Proposition 6.2.28.

(d) ⟹ (a): We argue by contradiction. So, suppose that $\mathrm{id} \simeq 0$; see Definitions 1.7.16 and 1.7.19. Then there exists a deformation $h\colon [0,1] \times \partial B_1^N \to \partial B_1^N$ such that $h(0, \partial B_1^N) = u_0 \in \partial B_1$ and $h(1,\cdot) = \mathrm{id}$. We consider the map $r\colon \overline{B}_1^N \to \partial B_1^N$ defined by

$$r(u) = \begin{cases} u_0 & \text{if } |u| \le \frac{1}{2}, \\ h(2|u| - 1, \frac{u}{|u|}) & \text{if } \frac{1}{2} \le |u| \le 1. \end{cases}$$

Then r is continuous and $r|_{\partial B_1^N} = \mathrm{id}|_{\partial B_1^N}$, that is, r is a retraction and this contradicts (d). □

A more topological version of the result above is given in the next theorem.

Theorem 6.4.40. *The following statements are equivalent:*
(a) *The Ljusternik–Schnirelmann–Borsuk Theorem holds; see Theorem* 6.2.40.
(b) *There is no continuous odd map* $\varphi\colon \partial B_1^{N+1} = \{u \in \mathbb{R}^{N+1} : |u| = 1\} \to \partial B_1^N$.
(c) *A continuous odd map* $\varphi\colon \partial B_1^N \to \partial B_1^N$ *is not nullhomotopic; see Definition* 1.7.19 *(a).*
(d) *For every continuous map* $\varphi\colon \partial B_1^N \to \mathbb{R}^{N-1}$ *there exists* $\hat{u} \in \partial B_1^N$ *such that* $\varphi(\hat{u}) = \varphi(-\hat{u})$.

Proof. (a) $\Longrightarrow$ (b): We argue by contradiction and assume that there exists a continuous odd map $\varphi\colon \partial B_1^{N+1} \to \partial B_1^N$. We consider an N-simplex centered at the origin. Its boundary is homeomorphic to ∂B_1^N. Let $\{C_n\}_{k=1}^{N+1}$ be the images of the $(N-1)$-faces. Then each C_k does not contain antipodal points. Let $D_k = \varphi^{-1}(C_k)$ for $k = 1, \ldots, N+1$. These are closed sets that cover ∂B_1^{N+1}. Then Theorem 6.2.40 implies that there exists $k \in \{1, \ldots, N+1\}$ and $u \in D_k \cap (-D_k)$. Since φ is odd it follows that $\varphi(-u), -\varphi(u) \in C_k$, a contradiction.

(b) $\Longrightarrow$ (c): We argue again by contradiction. So, we suppose there exists a continuous odd map $\varphi\colon \partial B_1^N \to \partial B_1^N$, which is nullhomotopic. So, according to Definition 1.7.19 (a), we can find a homotopy h deforming φ to a constant map. We set

$$\hat{\varphi}(u) = \begin{cases} h(0, \partial B_1^N) & \text{if } 0 \le |u| \le \frac{1}{2}\,, \\ h(2|u| - 1, \frac{u}{|u|}) & \text{if } \frac{1}{2} \le u \le 1\,. \end{cases}$$

We see that $\varphi\colon \overline{B}_1^N \to \partial B_1^N$ is continuous and $\hat{\varphi}|_{\partial B_1^N} = \varphi$. We introduce the two hemispheres of ∂B_1^{N+1}, namely the sets

$$\begin{aligned} \partial B_{1,+}^{N+1} &= \{u = (u_k)_{k=1}^{N+1} \in \partial B_1^{N+1} : u_{N+1} \ge 0\}\,, \\ \partial B_{1,-}^{N+1} &= \{u = (u_k)_{k=1}^{N+1} \in \partial B_1^{N+1} : u_{N+1} \le 0\}\,. \end{aligned}$$

We know that $\overline{B}_1^N$ is homeomorphic to each of the sets above by using the stereographic projection. So, we can define the map $f\colon \partial B_1^{N+1} \to \partial B_1^N$ by

$$f(u) = \begin{cases} \hat{\varphi}(u) & \text{if } u \in \partial B_{1,+}^{N+1}\,, \\ -\hat{\varphi}(u) & \text{if } u \in \partial B_{1,-}^{N+1}\,. \end{cases}$$

Then f is continuous and odd, contradicting (b).

(c) $\Longrightarrow$ (d): Once again we argue indirectly. So, suppose that there exists a continuous map $\varphi\colon \partial B_1^N \to \mathbb{R}^{N-1}$ such that $\varphi(u) \ne \varphi(-u)$ for all $u \in \partial B_1^N$. Let $\psi\colon \partial B_1^N \to \partial B_1^{N-1}$ be defined by

$$\psi(u) = \frac{\varphi(u) - \varphi(-u)}{|\varphi(u) - \varphi(-u)|} \quad \text{for all } u \in \partial B_1^N\,.$$

Then $\psi|_{\partial B_1^{N-1}}\colon \partial B_1^{N-1} \to \partial B_1^{N-1}$ is odd and since $\psi|_{\partial B_{1,+}^{N+1}}$ is an extension over $\overline{B}_1^N$, $\psi|_{\partial B_1^{N-1}}$ is nullhomotopic by using the homotopy $h(t,u) = \psi|_{\partial B_{1,+}^{N+1}}(tu)$, a contradiction.

(d) $\Longrightarrow$ (a): We consider a closed cover $\{C_k\}_{k=1}^{N+1}$ of ∂B_1^{N+1} with the property that $C_k \cap (-C_k) = \emptyset$ for all $k = 1, \dots, N+1$. Using Urysohn's Lemma (see Theorem 1.2.17), we can find a continuous function $f_k : \partial B_1^{N+1} \to [0,1]$ such that

$$f_k|_{C_k} = 0 \quad \text{and} \quad f_k|_{-C_k} = 1 \quad \text{for } k \in \{1, \dots, N\}.$$

We consider the map $\hat{f} : \partial B_1^{N+1} \to \mathbb{R}^N$ defined by $\hat{f}(u) = (f_k(u))_{k=1}^N$. This is continuous and so by hypothesis there exists $\hat{u} \in \partial B_1^{N+1}$ such that $\hat{f}(\hat{u}) = \hat{f}(-\hat{u})$. Then

$$f_k(\hat{u}) = f_k(-\hat{u}) \quad \text{for all } k = 1, \dots, N.$$

Hence,

$$\hat{u} \in \partial B_1^{N+1} \setminus \left[\bigcup_{k=1}^{N} C_k \cup \bigcup_{k=1}^{N} (-C_k) \right].$$

The families $\{C_k\}_{k=1}^{N+1}$ and $\{-C_k\}_{k=1}^{N+1}$ are closed covers of ∂B_1^{N+1}. So, we must have $\hat{u} \in C_{N+1} \cap (-C_{N+1})$, a contradiction. ☐

The next proposition shows that in statement (c) of the theorem above, we can relax the oddness condition.

Proposition 6.4.41. *If $\varphi : \partial B_1^N \to \partial B_1^N$ is continuous and $\varphi(u) \neq \varphi(-u)$ for all $u \in \partial B_1^N$, then φ is not nullhomotopic.*

Proof. Consider the map $\psi : \partial B_1^N \to \partial B_1^N$ defined by

$$\psi(u) = \frac{\varphi(u) - \varphi(-u)}{|\varphi(u) - \varphi(-u)|} \quad \text{for all } u \in \partial B_1^N.$$

Clearly, ψ is continuous and odd. Suppose that we could find $y \in \partial B_1^N$ such that $\psi(y) = -\varphi(y)$, that is, ψ and φ are antipodal at the point $y \in \partial B_1^N$. It holds that

$$\left[1 + |\varphi(y) - \varphi(-y)|\right] \varphi(y) = \varphi(-y),$$

which shows that

$$1 + |\varphi(y) - \varphi(-y)| = 1,$$

since $|\varphi(y)| = |\varphi(-y)| = 1$. Hence, $\varphi(y) = \varphi(-y)$, a contradiction. So, φ and ψ are never antipodal. From the homotopy

$$h(t,u) = \frac{(1-t)\varphi(u) + t\psi(u)}{|(1-t)\varphi(u) + t\psi(u)|} \quad \text{for all } t \in [0,1] \text{ and for all } u \in \partial B_1^N$$

we see that $\varphi \simeq \psi$. Theorem 6.4.40 (c) says that ψ is not nullhomotopic. Since "$\simeq$" is an equivalence relation (see Proposition 1.7.18), we conclude that φ is not nullhomotopic as well. ☐

The next result is known as "Borsuk's Fixed Point Theorem".

Theorem 6.4.42 (Borsuk's Fixed Point Theorem). *If U is a bounded, open, convex, symmetric set with $0 \in U$, $\varphi: \overline{U} \to \mathbb{R}^N$ is continuous, and $\varphi|_{\partial\Omega}$ is odd, then φ has a fixed point.*

Proof. Let $\hat{p}_U: \mathbb{R}^N \to \mathbb{R}_+$ be the Minkowski (gauge) functional of U; see Definition 3.1.37. Then $|\cdot|' = \hat{p}_U(\cdot)$ is a norm on $\mathbb{R}^N$; see Proposition 3.1.39. Let $E^N = (\mathbb{R}^N, |\cdot|')$. Then the identity map $e: \mathbb{R}^N \to E^N$ is a homeomorphism mapping $\overline{U}$ to $\overline{B}_1^{|\cdot|'} = \{u \in E^N: |u|' \le 1\}$. Let $\psi = e \circ \varphi \circ e^{-1}: \overline{B}_1^{|\cdot|'} \to E^N$. Evidently, ψ is continuous and $\psi|_{\partial \overline{B}_1^{|\cdot|'}}$ is odd. Suppose that $\psi(u) \neq u$ for all $\overline{B}_1^{|\cdot|'}$. Let $\sigma: \overline{B}_1^{|\cdot|'} \to \partial B_1^{|\cdot|'}$ be the map defined by

$$\sigma(u) = \frac{\psi(u) - u}{|\psi(u) - u|} \quad \text{for all } u \in \overline{B}_1^{|\cdot|'} .$$

Then σ is continuous and so $\sigma|_{\overline{B}_1^{|\cdot|'}}$ is nullhomotopic, which can be seen by using the homotopy $h(t, u) = \sigma((1 - t)u)$. But $\sigma|_{\partial B_1^{|\cdot|'}}$ is odd. This contradicts Theorem 6.4.40 (c). So, we can find $\hat{u} \in \overline{B}_1^{|\cdot|'}$ such that $\psi(\hat{u}) = \hat{u}$. Hence, $(e \circ \varphi e^{-1})(\hat{u}) = \hat{u}$ and so, $\varphi(e^{-1}(\hat{u})) = e^{-1}(\hat{u})$. ☐

In Example 6.3.1 we saw that Brouwer's Fixed Point Theorem fails in infinite dimensional Banach spaces. In fact we can show that Brouwer's Fixed Point Theorem (see Proposition 6.4.33), is valid if and only the ambient space is finite dimensional.

Theorem 6.4.43. *If X is a normed space and $\overline{B}_1 = \{x \in X: \|x\| \le 1\}$, then every continuous map $\varphi: \overline{B}_1 \to \overline{B}_1$ has a fixed point if and only if X is finite dimensional.*

Proof. $\Longleftarrow$: This is Brouwer's Fixed Point Theorem.

$\Longrightarrow$: Suppose that X is infinite dimensional. Then according to Theorem 1.7.31, $X \setminus \{0\}$ and X are homeomorphic. Let $h: X \to X \setminus \{0\}$ be such a homeomorphism. For each $u \in \partial B_1$, map the line segment $[0, u]$ linearly to the line segment $[0, h^{-1}(u)]$. This way we have a continuous map $g: \overline{B}_1 \to X$ such that $g(0) = 0$. Let $\eta: \overline{B}_1 \to \partial B_1$ be defined by

$$\eta(u) = \frac{(h \circ g)(u)}{\|(h \circ g)(u)\|} \quad \text{for all } u \in \overline{B}_1 .$$

This is well defined since $h(g(u)) \neq 0$ for all $u \in \overline{B}_1$. If $u \in \partial B_1$, then $h(g(u)) = h(h^{-1}(u)) = u$. Hence, $\eta|_{\partial B_1} = \mathrm{id}$, that is, η is a retraction. Therefore, $u \to -\eta(u)$ is fixed point free from $\overline{B}_1$ into $\overline{B}_1$, a contradiction. ☐

Now we pass to infinite dimensional Banach spaces and prove an alternative theorem analogous to Theorem 6.4.31 for nonexpansive maps.

Theorem 6.4.44. *If X is a Banach space, $\Omega \subseteq X$ is bounded and open with $0 \in \Omega$, and $\varphi\colon \overline{\Omega} \to X$ is compact, then one of the following statements is true:*
(a) *φ has a fixed point.*
(b) *There exist $\hat{u} \in \partial\Omega$ and $t \in (0,1)$ such that $\hat{u} = t\varphi(\hat{u})$.*

Proof. Consider the compact homotopy $h(t,u) = t\varphi(u)$ for all $t \in [0,1]$ and for all $u \in \overline{\Omega}$. If $h(t,u) \neq u$ for all $t \in [0,1]$ and for all $u \in \overline{\Omega}$, then we can use the homotopy invariance property of the Leray–Schauder degree (see Theorem 6.3.15 (c)), and have

$$d_{LS}(\mathrm{id} - \varphi, \Omega, 0) = d_{LS}(\mathrm{id}, \Omega, 0) = 1 .$$

Then Theorem 6.3.15 (d) implies $\varphi(\hat{u}) = \hat{u}$ for some $\hat{u} \in \Omega$.

If $h(1,\hat{u}) = \hat{u}$ for some $\hat{u} \in \partial\Omega$, then $\varphi(\hat{u}) = \hat{u}$ and so φ has a fixed point. If $h(t,\hat{u}) = \hat{u}$ for some $t \in (0,1)$ and some $\hat{u} \in \partial\Omega$, then $\hat{u} = t\varphi(\hat{u})$. ☐

As a consequence of this theorem, we have the following corollary known as the "Leray–Schauder Alternative Principle" or "Schaefer's Fixed Point Theorem".

Corollary 6.4.45 (Leray–Schauder Alternative Principle). *If X is a Banach space, $\varphi\colon X \to X$ is compact and*

$$S(\varphi) = \{u \in X\colon \text{there exists } \lambda \in (0,1) \text{ such that } u = \lambda\varphi(u)\} ,$$

then either $S(\varphi)$ is unbounded or φ has a fixed point.

Remark 6.4.46. Roughly speaking when using this corollary on boundary value problems, the result says that existence of a priori bounds imply the existence of solutions.

Finally we have an infinite dimensional version of Borsuk's Fixed Point Theorem; see Theorem 6.4.42.

Theorem 6.4.47. *If X is a Banach space, $\Omega \subseteq X$ is bounded, open, convex, and symmetric with $0 \in \Omega$, and $\varphi\colon \overline{\Omega} \to X$ is compact and odd such that $0 \notin \varphi(\partial\Omega)$, then φ has a fixed point.*

Proof. Theorem 6.3.16 says that $d_{LS}(\mathrm{id} - \varphi, \Omega, 0) \neq 0$. Hence, there exists $\hat{u} \in \Omega$ such that $\varphi(\hat{u}) = \hat{u}$; see Theorem 6.3.15 (d). ☐

6.5 Fixed Point Index

In the previous sections, we have used degree theory to find fixed points of maps. However, in several cases of interest, we cannot use the whole Banach space. The formulation leads to a map on a closed convex subset that need not be a linear subspace. For such a situation we have an extension of the Leray–Schauder degree known as the fixed point index. Using it, we can develop fixed point theorems for settings in which the order structure of the ambient space plays the first role.

The starting point is the fact that in a Banach space every closed and convex set is a retract; see Corollary 1.7.30. Moreover, we know that if X is infinite dimensional, then $\partial B_1 = \{u \in X : \|u\| = 1\}$ is a retract of X and of $\overline{B}_1 = \{u \in X : \|u\| \leq 1\}$. In what follows when dealing with subsets of a retract $C \subseteq X$, we consider them equipped with the relative subspace topology, which is induced by the norm on C.

Let X be a Banach space. Our analysis will be done on the following family of triplets

$$\mathcal{L}_i = \{(f, \Omega, C) : C \subseteq X \text{ is a retract}, \Omega \subseteq C \text{ is a bounded, (relatively) open set}, \\ f : \overline{\Omega} \to X \text{ is compact}, \operatorname{Fix}(f) \cap \partial\Omega = \emptyset\},$$

where $\operatorname{Fix}(f) = \{u \in \overline{\Omega} : f(u) = u\}$ is the fixed point set of f.

So, let $(f, \Omega, C) \in \mathcal{L}_i$ and let $r : X \to K$ be a retraction.

Definition 6.5.1. The **fixed point index** $i(f, \Omega, C)$ is defined by

$$i(f, \Omega, C) = d_{\mathrm{LS}}(\mathrm{id} - f \circ r, B_R \cap r^{-1}(U), 0) \quad \text{with } \Omega \subseteq B_R . \tag{6.5.1}$$

This definition requires further justification. First, let us check that the Leray–Schauder degree involved in (6.5.1) is legitimate. Since f is compact, there exists a compact subset K of C such that $f(\overline{\Omega}) \subseteq K$. Then $(f \circ r)(\overline{r^{-1}(\Omega)}) \subseteq K$ and so $f \circ r : \overline{B_R \cap r^{-1}(\Omega)} \to X$ is a compact map. Moreover,

$$\overline{B_R \cap r^{-1}(\Omega)} \subseteq \overline{r^{-1}(\Omega)} \subseteq r^{-1}(\overline{\Omega}) .$$

Note that

$$u_0 \in r^{-1}(\overline{\Omega}) \quad \text{and} \quad (f \circ r)(u_0) = u_0 \quad \text{imply} \quad u_0 \in \Omega \quad \text{and} \quad f(u_0) = u_0 . \tag{6.5.2}$$

So, we can define the Leray–Schauder degree for $(\mathrm{id} - f \circ r, B_R \cap r^{-1}(\Omega), 0) \in \hat{\mathcal{L}}$. However, we need to show that (6.5.1) is independent of $R > 0$ and since there are many ways to retract X on C, we need to show that the definition is independent of the retraction r.

Proposition 6.5.2. *The definition of the fixed point index in* (6.5.1) *is independent of* R *and of the retraction* r.

Proof. Let $R < R'$. It holds that

$$\Omega \subseteq B_R \cap r^{-1}(\Omega) \subseteq B_{R'} \cap r^{-1}(\Omega) .$$

From (6.5.2) we see that $f \circ r$ has no fixed points in $\overline{B_{R'} \cap r^{-1}(\Omega)} \setminus (B_R \cap r^{-1}(\Omega))$. Then by the excision property of the Leray–Schauder degree (see Theorem 6.3.15 (f)), we obtain

$$d_{\mathrm{LS}}(\mathrm{id} - f \circ r, B_{R'} \cap r^{-1}(\Omega), 0) = d_{\mathrm{LS}}(\mathrm{id} - f \circ r, B_R \cap r^{-1}(\Omega), 0) ,$$

which implies that $i(f,\Omega,C)$ in (6.5.1) is independent of $R > 0$ such that $\Omega \subseteq B_R$.

Now we show that (6.5.1) does not depend on the choice of the retraction r. So, let $r' : X \to C$ be another retraction and set $D = B_R \cap r^{-1}(\Omega) \cap (r')^{-1}(\Omega)$. Clearly, D is a bounded open set in X and $U \subseteq D$. From (6.5.2) we know that $f \circ r$ has no fixed points in $\overline{B_R \cap r^{-1}(\Omega)} \setminus D$ and $f \circ r'$ has no fixed points in $\overline{B_R \cap (r')^{-1}(\Omega)} \setminus D$. Therefore,

$$d_{\mathrm{LS}} = (\mathrm{id} - f \circ r, B_R \cap r^{-1}(\Omega), 0) = d_{\mathrm{LS}}(\mathrm{id} - f \circ r, D, 0)\,, \tag{6.5.3}$$

$$d_{\mathrm{LS}} = (\mathrm{id} - f \circ r', B_R \cap (r')^{-1}(\Omega), 0) = d_{\mathrm{LS}}(\mathrm{id} - f \circ r', D, 0)\,,; \tag{6.5.4}$$

see Theorem 6.3.15 (f). We consider the compact homotopy

$$h(t,u) = r \circ ((1-t)(f \circ r)(u) + t(f \circ r')(u))\,.$$

Suppose that we can find $t_0 \in [0,1]$ and $u_0 \in \partial D$ such that $h(t_0,u_0) = u_0$. This would give $u_0 \in C$, and hence, $r(u_0) = u_0$, $r'(u_0) = u_0$, and $f(u_0) = u_0$. It follows that $u_0 \in U \subseteq D$ (see (6.5.2)), which contradicts the fact that $u_0 \in \partial D$. Therefore, we have proven

$$0 \notin (\mathrm{id} - h(t, \partial\Omega)) \quad \text{for all } t \in [0,1]\,.$$

Invoking the homotopy invariance property of the Leray–Schauder degree (see Theorem 6.3.15 (c)) yields

$$d_{\mathrm{LS}}(\mathrm{id} - f \circ r', D, 0) = d_{\mathrm{LS}}(\mathrm{id} - f \circ r, D, 0)\,. \tag{6.5.5}$$

Combining (6.5.3), (6.5.4), and (6.5.5) gives

$$d_{\mathrm{LS}}(i - f \circ r, B_R \cap r^{-1}(\Omega), 0) = d_{\mathrm{LS}}(i - f \circ r', B_R \cap (r')^{-1}(\Omega), 0)\,.$$

Thus, $i(f,\Omega,C)$ in (6.5.1) is independent of the choice of the retraction. □

Using Definition 6.5.1 we can translate all the properties of the Leray–Schauder degree into the language of the fixed point index.

Theorem 6.5.3. *There exists a map* $: \mathcal{L}_i \to \mathbb{Z}$ *called the* ***fixed point index****, which has the following properties:*

(a) *Normalization property:* $i(f,\Omega,C) = 1$ *if* $f(u) = h_0 \in \Omega$ *for all* $u \in \overline{\Omega}$;
(b) *Domain additivity property: for all disjoint open sets* $\Omega_1, \Omega_2 \subseteq \Omega$ *such that* f *has no fixed points on* $\overline{\Omega} \setminus (\Omega_1 \cup \Omega_2)$, *it holds that*

$$i(f,\Omega,C) = i(f,\Omega_1,C) + i(f,\Omega_2,C)\,;$$

(c) *Homotopy invariance property: if* $\{h_t\}_{t\in[0,1]}$ *is a homotopy of compact maps and* $h_t(u) \neq u$ *for all* $t \in [0,1]$ *and for all* $u \in \partial\Omega$, *then* $i(h_t,\Omega,C)$ *is independent of* $t \in [0,1]$;

(d) *Fixed point property:* $i(f, \Omega, C) \neq 0$ *implies that* f *has a fixed point in* Ω;
(e) *Excision property:* $i(f, \Omega, C) = i(f, \Omega_1, C)$ *for every open set* $\Omega_1 \subseteq \Omega$ *such that* f *has no fixed point in* $\overline{\Omega} \setminus \Omega_1$;
(f) *Permanence property:* $i(f, \Omega, C) = i(f, \Omega \cap K, K)$ *if* K *is a retract of* C *and* $f(\overline{\Omega}) \subseteq K$.

Next we will see how we can use the fixed point index to prove fixed point theorems exploiting the order structure of the space.

Definition 6.5.4. Let X be a Banach space. A nonempty, closed, and convex set $K \subseteq X$ is said to be a **cone** if it satisfies the following conditions:
(a) if $u \in K$ and $\lambda \geq 0$, then $\lambda u \in K$, that is, $\lambda K \subseteq K$ for all $\lambda \geq 0$;
(b) if $u, -u \in K$, then $u = 0$, that is, $K \cap (-K) = \{0\}$.

Remark 6.5.5. A cone K induces a partial order "$\leq$" on X by

$$u \leq v \quad \text{if and only if} \quad v - u \in K\,.$$

Then for a sequence $\{u_n\}_{n\geq 1} \subseteq X$ we say that it is **increasing** (resp. **decreasing**) if $u_n \leq u_{n+1}$ (resp. $u_n \geq u_{n+1}$) for all $n \in \mathbb{N}$. A set $C \subseteq X$ is **bounded above** (resp. **bounded below**) if there exists $h \in X$ such that $u \leq h$ (resp. $u \geq h$) for all $u \in C$.

By $\sup C$ (resp. $\inf C$) we denote the least upper bound of C (resp. greatest lower bound of C) if it exists. The cone K is called the **order cone** of X and its elements are said to be **positive**. A Banach space X with an order cone K is said to be an **ordered Banach space** (OBS for short).

Definition 6.5.6. Let X be an OBS with order cone K. Then the **dual cone** of K is defined by

$$K^* = \{u^* \in X^* : \langle u^*, u\rangle \geq 0 \text{ for all } u \in K\}\,.$$

The elements of K^* are the positive linear functionals on X.

Remark 6.5.7. Note that $K^* \subseteq X^*$ is closed, convex, and satisfies (a) in Definition 6.5.4. However, we need not have $K^* \cap (-K^*) = \{0\}$. If $X = K - K$, then $K^* \cap (-K^*) = \{0\}$ and so K^* is also a cone in the sense of Definition 6.5.4.

Proposition 6.5.8. *If* X *is an* OBS *with order cone* K *and* $K^* \subseteq X^*$ *is its dual cone, then the following hold:*
(a) $u \in K$ *if and only if* $\langle u^*, u\rangle \geq 0$ *for all* $u^* \in K^*$;
(b) *if* $u \in K \setminus \{0\}$, *then there exists* $u^* \in K^*$ *such that* $\langle u^*, u\rangle > 0$;
(c) *if* $y \notin K$, *then there exists* $u^* \in K^*$ *such that* $\langle u^*, y\rangle < 0$;
(d) *if* $\operatorname{int} K \neq \emptyset$, *then* $u \in \operatorname{int} K$ *if and only if* $\langle u^*, u\rangle > 0$ *for all* $u^* \in K^* \setminus \{0\}$;
(e) *if* X *is separable, then there exists* $u^* \in K^*$ *such that*

$$\langle u^*, u\rangle > 0 \quad \textit{for all } u \in K \setminus \{0\}\,.$$

Proof. (a) Let $y \notin K$. Then by the Strong Separation Theorem (see Corollary 3.1.61), there exist $u^* \in X^* \setminus \{0\}$ and $\varepsilon > 0$ such that

$$\langle u^*, y\rangle + \varepsilon \le \langle u^*, u\rangle \quad \text{for all } u \in K\,.$$

Hence,

$$\langle u^*, y\rangle \le -\varepsilon \quad \text{and} \quad u^* \in K^*\,. \tag{6.5.6}$$

Then $K^* = \{u \in X\colon \langle u^*, u\rangle \ge 0 \text{ for all } u \in K\}$.

(b) If $u \in K\setminus\{0\}$, then $-u \notin K$. So, again according to the Strong Separation Theorem, there exists $u^* \in X^*$ and $\varepsilon > 0$ such that $\langle u^*, -u\rangle + \varepsilon \le \langle u^*, v\rangle$ for all $v \in K$, which implies that $u^* \in K^*$ and $\varepsilon \le \langle u^*, u\rangle$.

(c) This follows from (6.5.6).

(d) Let $u_0 \in \operatorname{int} K$. Then there exists $\delta > 0$ such that $\overline{B}_\delta(u_0) \subseteq K$. Hence, $u_0 \pm \delta h \ge 0$ for all $h \in X$ with $\|h\| \le 1$. Then $\langle u^*, u_0 \pm \delta h\rangle \ge 0$ for all $u^* \in K^*$ and so $\langle u^*, u_0\rangle \ge \delta\|u^*\|_* > 0$.

Next suppose that $y \notin \operatorname{int} K$ and let $K_0 = \mathbb{R}_+ y$. Then K_0 is a cone and $K_0 \cap \operatorname{int} K = \emptyset$. So, according to the Weak Separation Theorem (see Theorem 3.1.59), there exists $u^* \in X^* \setminus \{0\}$ such that

$$\langle u^*, \lambda y\rangle \le \langle u^*, u\rangle \quad \text{for all } \lambda \ge 0 \text{ and for all } u \in K\,.$$

Hence, $\langle u^*, y\rangle \le 0$.

(e) Since X is separable, $(\overline{B}_1^{X^*}, \mathrm{w}^*)$ is compact metrizable; see Theorem 3.4.12. So, it is separable as well. Let $\{u_n^*\}_{n\ge1}$ be w*-dense in $K^* \cap \overline{B}_1^{X^*}$ and set $\hat{u}^* = \sum_{n\ge1} 1/n^2 u_n^*$. Then $\hat{u}^* \in K^*$ and $\langle \hat{u}^*, u\rangle = 0$ for some $u \in K$ implies $\langle u^*, u\rangle = 0$ for all $u^* \in K^*$. Hence, $u = 0$; see (b). □

Now let X be an OBS with order cone K. Recall that K is a retract of X. Let $\Omega \subseteq X$ be bounded and open. Then $\Omega \cap K$ is bounded and relatively open in K. We have

$$\partial(\Omega \cap K) = \partial\Omega \cap K \quad \text{and} \quad \overline{\Omega \cap K} = \overline{\Omega} \cap K\,.$$

Proposition 6.5.9. *If $0 \in \Omega$, $f\colon \overline{\Omega} \cap K \to K$ is compact and $f(u) \neq \lambda u$ for all $u \in \partial\Omega \cap K$ and for all $\lambda \ge 1$, then $i(f, \Omega \cap K, K) = 1$.*

Proof. Consider the compact homotopy $h_t(u) = tf(u)$ for all $t \in [0,1]$ and for all $u \in \overline{\Omega} \cap K$. Then

$$h_t(u) \neq u \quad \text{for all } (t,u) \in [0,1] \times (\partial\Omega \cap K)\,.$$

So, Theorem 6.5.3 (c) and (a) imply that $i(f, \Omega \cap K, K) = i(0, \Omega \cap K, K) = 1$. □

Proposition 6.5.10. *If $f: \overline{\Omega} \cap K \to K$ and $e: \partial\Omega \cap K \to K$ are both compact maps and*
(i) $\inf[\|e(u)\| : u \in \partial\Omega \cap K] > 0$;
(ii) $u - f(u) \neq \lambda e(u)$ *for all* $u \in \partial\Omega \cap K$ *and for all* $\lambda \geq 0$,

then $i(f, \Omega \cap K, K) = 0$.

Proof. According to Theorem 1.7.29, there exists a continuous map $\hat{e}: \overline{\Omega} \cap K \to K$ such that $\hat{e}|_{\partial\Omega \cap K} = e$ and $\hat{e}(\overline{\Omega} \cap K) \subseteq \overline{\operatorname{conv}}\, e(\partial\Omega \cap K)$ and the latter set is compact and convex. Let $D = e(\partial\Omega \cap K)$, which is a compact subset of X. We claim that

$$\inf[\|h\| : h \in \overline{\operatorname{conv}}\, D] = \eta > 0\,. \tag{6.5.7}$$

Let $V = \overline{\operatorname{span}} D$. The compactness of D implies that V is a separable Banach subspace of X. Let $K_0 = K \cap V$. Clearly, K_0 is a cone in V and $\overline{\operatorname{conv}}\, D \subseteq K_0$. From Proposition 6.5.8 (e) we know that there exists $\hat{u}^* \in K_0^*$ such that $\langle \hat{u}^*, u \rangle > 0$ for all $u \in K_0 \setminus \{0\}$. We claim that

$$\inf[\langle \hat{u}^*, u \rangle : u \in D] = \hat{\eta} > 0\,.$$

If $\hat{\eta} = 0$, then there exists a sequence $\{u_n\}_{n \geq 1} \subseteq D$ such that $\langle \hat{u}^*, u \rangle \to 0$. Since $D \subseteq X$ is compact, we may assume that $u_n \to \hat{u}$ in X. Then $\langle \hat{u}^*, u \rangle \to \langle \hat{u}^*, \hat{u} \rangle = 0$, and so $\hat{u} = 0$. Hence, $u_n \to 0$ in X, a contradiction to hypothesis (i). Therefore, $\hat{\eta} > 0$. Let $h \in \operatorname{conv} D$. Then

$$h = \sum_{k=1}^{n} \lambda_k v_k \quad \text{with } v_k \in D,\ \lambda_k \geq 0,\ \sum_{k=1}^{n} \lambda_k = 1 \text{ with } n \in \mathbb{N}\,.$$

We have

$$\langle \hat{u}^*, h \rangle = \sum_{k=1}^{n} \lambda_k \langle \hat{u}^*, v_k \rangle \geq \sum_{k=1}^{n} \lambda_k \hat{\eta} = \hat{\eta} > 0\,.$$

Hence,

$$\langle \hat{u}^*, u \rangle \geq \hat{\eta} > 0 \quad \text{for all } u \in \overline{\operatorname{conv}}\, D\,. \tag{6.5.8}$$

The set $\overline{\operatorname{conv}}\, D$ is compact. So, there exists $\hat{h} \in \overline{\operatorname{conv}}\, D$ such that

$$\inf[\|h\| : h \in \overline{\operatorname{conv}}\, D] = \|\hat{h}\|\,. \tag{6.5.9}$$

From (6.5.8) and (6.5.9) we conclude that (6.5.7) holds.

Now suppose that $i(f, \Omega \cap K, K) \neq 0$. Then from hypothesis (ii) and Theorem 6.5.3 (c) we obtain

$$i(f + t\hat{e}, \Omega \cap K, K) = i(f, \Omega \cap K, K) \neq 0 \quad \text{for all } t > 0\,. \tag{6.5.10}$$

Let $\eta_1 = \sup[\|u\| : u \in \overline{\Omega} \cap K]$ and $\eta_2 = \sup[\|f(u)\| : u \in \overline{\Omega} \cap K]$. Choosing $t_0 > 1/\hat{\eta}[\eta_1 + \eta_2]$, from (6.5.10) we get $i(f + t_0\hat{e}, \Omega \cap K, K) \neq 0$. Then, by Theorem 6.5.3 (d), one has $f(\hat{u}) +$

$t_0\hat{e}(\hat{u}) = \hat{u}$ for some $\hat{u} \in \Omega \cap K$. This gives

$$t_0 = \frac{\|\hat{u} - f(\hat{u})\|}{\|\hat{e}(\hat{u})\|} \leq \frac{1}{\eta}[\eta_1 + \eta_2],$$

a contradiction to the choice of t_0. □

Corollary 6.5.11. *If $f: \overline{\Omega} \cap K \to K$ is compact, $u_0 \in K \setminus \{0\}$ and*

$$u - f(u) \neq \lambda u_0 \quad \textit{for all } u \in \partial\Omega \cap K \textit{ and for all } \lambda \geq 0,$$

then $i(f, \Omega \cap K, K) = 0$.

Proof. Applying Proposition 6.5.10 with $e(u) = u_0$ for all $u \in \partial\Omega \cap K$ yields the assertion. □

Next we will use the fixed point index to prove fixed point theorems of expansion and compression type. These are fixed point theorems on sets of the form $(\Omega_2 \setminus \overline{\Omega}_1) \cap K$ with bounded, open sets Ω_1, Ω_2 such that $\overline{\Omega}_1 \subseteq \Omega_2$. Such sets are also known as **conical shells**. Of special interest are conical sets created by $\Omega_1 = B_r = \{u \in X: \|u\| < r\}$, $\Omega_2 = B_\rho = \{u \in X: \|u\| < \rho\}$ and $r < \rho$.

Theorem 6.5.12. *If $\Omega_1, \Omega_2 \subseteq X$ are bounded open sets, $0 \in \Omega_1$, $\overline{\Omega}_1 \subseteq \Omega_2$, $f: \overline{\Omega}_2 \cap K \to K$ is a compact map and*

(i) *$f(u) \neq \lambda u$ for all $u \in \partial\Omega_2$ with $\lambda > 1$;*
(ii) *there exists $u_0 \in K \setminus \{0\}$ such that $u - f(u) \neq \lambda u_0$ for all $u \in \partial\Omega_2$ and for all $\lambda > 0$,*

then φ has a fixed point in $(\overline{\Omega}_2 \setminus \Omega_1) \cap K$.

Proof. From Proposition 6.5.9 we have

$$i(f, \Omega_2 \cap K, K) = 1. \tag{6.5.11}$$

Moreover, Corollary 6.5.11 implies

$$i(f, \Omega_1 \cap K, K) = 0. \tag{6.5.12}$$

We assume that f has no fixed points on $\partial\Omega_2 \cap K$ and on $\partial\Omega_1 \cap K$. Otherwise we already have a fixed point of φ in $(\overline{\Omega}_2 \setminus \Omega_1) \cap K$. Then, from the domain additivity property of the fixed point index, see Theorem 6.5.3 (b), we obtain, taking (6.5.11) and (6.5.12) into account, that

$$i(f, (\Omega_2 \setminus \overline{\Omega}_1) \cap K, K) = i(f, \Omega_2 \cap K, K) - i(f, \Omega_1 \cap K, K) = 1 - 0 = 1. \tag{6.5.13}$$

Then from (6.5.13) and the solution property of the fixed point index (see Theorem 6.5.3 (d)), we conclude that there exists $\hat{u} \in (\overline{\Omega}_2 \setminus \Omega_1) \cap K$ such that $f(\hat{u}) = \hat{u}$. □

Corollary 6.5.13. *If* $\Omega_1, \Omega_2 \subseteq X$ *are bounded open sets,* $0 \in \Omega_1$, $\overline{\Omega}_1 \subseteq \Omega_2$, $f\colon \overline{\Omega}_2 \cap K \to K$ *is compact, and*

(i) $f(u) - u \notin K$ *for all* $u \in \partial\Omega_2$*;*

(ii) $u - f(u) \notin K$ *for all* $u \in \partial\Omega_1$,

then f *has a fixed point in* $(\overline{\Omega}_2 \setminus \Omega_1) \cap K$.

Remark 6.5.14. The corollary above remains valid if we reverse the conditions on the two boundaries. Namely, we assume that $u - f(u) \notin K$ for all $u \in \partial\Omega_2$ and $f(u) - u \notin K$ for all $u \in \partial\Omega_1$. In this case $i(f, \Omega_2 \cap K, K) = 0$ and $i(f, \Omega_1 \cap K, K) = 1$.

Another fixed point theorem of this type is the following one, which justifies the characterization of fixed points of expansion and compression type.

Theorem 6.5.15. *If* $\Omega_1, \Omega_2 \subseteq X$ *are bounded open sets,* $0 \in \Omega_1$, $\overline{\Omega}_1 \subseteq \Omega_2$, $f\colon \overline{\Omega}_2 \cap K \to K$ *is compact and*

(i) $\|f(u)\| \geq \|u\|$ *for all* $u \in \partial\Omega_2 \cap K$ *and* $\|f(u)\| \leq \|u\|$ *for all* $u \in \partial\Omega_1 \cap K$*;*

or

(ii) $\|f(u)\| \leq \|u\|$ *for all* $u \in \partial\Omega_2 \cap K$ *and* $\|f(u)\| \geq \|u\|$ *for all* $u \in \partial\Omega_1 \cap K$*;*

then f *has a fixed point in* $(\overline{\Omega}_2 \setminus \Omega_1) \cap K$.

Proof. We do the proof when (i) is in effect. The proof is similar if (ii) holds. We claim that

$$f(u) \neq \lambda u \quad \text{for all } u \in \partial\Omega_2 \cap K \text{ and for all } \lambda \in (0,1)\,. \tag{6.5.14}$$

If (6.5.14) is not true, then there exist $u_0 \in \partial\Omega_2 \cap K$ and $0 < \lambda_0 < 1$ such that $f(u_0) = \lambda_0 u_0$. Then $\|f(u_0)\| < \|u_0\|$, a contradiction to the hypothesis. Similarly we show that

$$f(u) = \eta u \quad \text{for all } u \in \partial\Omega_1 \cap K \text{ and for all } \eta > 1\,. \tag{6.5.15}$$

Moreover, we have

$$\inf[\|f(u)\| : u \in \partial\Omega_2 \cap K] \geq \inf[\|u\| : u \in \partial\Omega_2 \cap K] > 0\,. \tag{6.5.16}$$

Then, from (6.5.14), (6.5.16), and Proposition 6.5.10 we obtain

$$i(f, \Omega_2 \cap K, K) = 0\,. \tag{6.5.17}$$

Suppose that $f(u) \neq u$ for all $u \in \partial\Omega_1$. Otherwise we already have a fixed point for f in $(\overline{\Omega}_2 \setminus \Omega_1) \cap K$. Then (6.5.15) and Proposition 6.5.9 imply that

$$i(f, \Omega_1 \cap K, K) = 1\,. \tag{6.5.18}$$

Using (6.5.17), (6.5.18) and the domain additivity property of the fixed point index (see Theorem 6.5.3 (b)), we get $i(f, (\overline{\Omega}_2 \backslash \Omega_1) \cap K, K) = 1$. Hence, f has a fixed point in $(\overline{\Omega}_2 \backslash \Omega_1) \cap K$; see Theorem 6.5.3 (d). □

For the next fixed point theorem on conical shells, we will need the following lemma.

Lemma 6.5.16. *If X is an OBS with order cone $K \subseteq X$, $C \subseteq K$ is compact, and $0 \notin C$, then $0 \notin \operatorname{conv} C$.*

Proof. If $0 \in \operatorname{conv} C$, then we can find $u, x \in C$ with $u \neq 0$, $x \neq 0$ since $0 \notin C$ such that $1/2[u + x] = 0$. Hence, $u = -x$, contradicting Definition 6.5.4 (b). □

Theorem 6.5.17. *If $\Omega_1, \Omega_2 \subseteq X$ are bounded open sets, $0 \in \Omega_1$, $\overline{\Omega}_1 \subseteq \Omega_2$, $f : \overline{\Omega}_2 \cap K \to K$ is compact and*
(i) $f(u) \neq \lambda u$ for all $u \in \partial\Omega_2 \cap K$ and for all $\lambda > 1$;
(ii) $f(u) \neq \eta u$ for all $u \in \partial\Omega_1 \cap K$ and for all $\eta \in (0, 1)$;
(iii) $\inf[\|f(u)\| : u \in \partial\Omega_1 \cap K] > 0$,

then f has a fixed point in $(\Omega_2 \setminus \overline{\Omega}_1) \cap K$.

Proof. We assume that $0 \notin (\mathrm{id} - f)(\partial\Omega_2 \cap K)$, otherwise we already have a fixed point in $(\overline{\Omega}_2 \setminus \Omega_1) \cap K$. Then, through Proposition 6.5.9, we have $i(f, \Omega_2 \cap K, K) = 1$, which by the domain additivity property of the fixed point index results in

$$i(f, (\Omega_2 \setminus \overline{\Omega}_1) \cap K, K) = 1 - i(f, \Omega_1 \cap K, K) . \tag{6.5.19}$$

Using Theorem 1.7.29 and Lemma 6.5.16 there exists a compact map $\hat{f} : X \to \operatorname{conv}\{f(u) : u \in \partial\Omega_1 \cap K\}$ such that $\hat{f}|_{\partial\Omega_1 \cap K} = f$ and $\inf[\|\hat{f}(u)\| : u \in X] = \eta > 0$. Consider the compact homotopy

$$h_t(u) = (1 - t)f(u) + tm\hat{f}(u) \quad \text{for all } t \in [0, 1] \text{ and for all } u \in \overline{\Omega}_2 \text{ with } m > 1 .$$

Since $m > 1$ we see that $0 \notin h_t(\partial\Omega_1 \cap K)$ for all $t \in [0, 1]$. So, the homotopy invariance property of the fixed point index (see Theorem 6.5.3 (c)) implies

$$i(f, \Omega_1 \cap K, K) = i(m\hat{f}, \Omega_1 \cap K, K) . \tag{6.5.20}$$

If $u = m\hat{f}(u)$, then $m \leq \eta_0$ for some $\eta_0 > 0$. Therefore if $m > \max\{1, \eta_0\}$, we obtain

$$i(m\hat{f}, \Omega_1 \cap K, K) = 0 ; \tag{6.5.21}$$

see Theorem 6.5.3 (d). From (6.5.19), (6.5.20), and (6.5.21) we conclude that $i(f, (\Omega_2 \setminus \overline{\Omega}_1) \cap K, K) = 1$ and so, by Theorem 6.5.3 (d), $f(\hat{u}) = \hat{u}$ for some $\hat{u} \in (\Omega_2 \setminus \overline{\Omega}_1) \cap K$. □

We can also state theorems for multiple fixed points. They follow from Corollary 6.5.13 and Theorem 6.5.15.

Theorem 6.5.18. *If* $\Omega_1, \Omega_2, \Omega_3 \subseteq X$ *are bounded open sets,* $0 \in \Omega_1$, $\overline{\Omega}_1 \subseteq \Omega_2 \subseteq \overline{\Omega}_2 \subseteq \Omega_3$, $f\colon (\overline{\Omega}_3 \setminus \Omega_1) \cap K \to K$ *is compact and*
(i) $u - f(u) \notin K$ *for all* $u \in \partial\Omega_1 \cap K$;
(ii) $f(u) - u \notin K$ *for all* $u \in \partial\Omega_2 \cap K$;
(iii) $u - f(u) \notin K$ *for all* $u \in \partial\Omega_3 \cap K$;

then f *has at least two fixed points* $\hat{u}, \tilde{u} \in (\overline{\Omega}_3 \setminus \Omega_1) \cap K$ *such that*

$$\hat{u} \in (\Omega_3 \setminus \overline{\Omega}_2) \cap K \quad \text{and} \quad \tilde{u} \in (\Omega_2 \setminus \overline{\Omega}_1) \cap K\,.$$

Theorem 6.5.19. *If* $\Omega_1, \Omega_2, \Omega_3 \subseteq X$ *are bounded open sets,* $0 \in \Omega_1$, $\overline{\Omega}_1 \subseteq \Omega_2 \subseteq \overline{\Omega}_2 \subseteq \Omega_3$, $f\colon (\overline{\Omega}_3 \setminus \Omega_1) \cap K \to K$ *is compact and*
(i) $\|f(u)\| \geq \|u\|$ *for all* $u \in \partial\Omega_1 \cap K$;
(ii) $\|f(u)\| < \|u\|$ *for all* $u \in \partial\Omega_2 \cap K$;
(iii) $\|f(u)\| \geq \|u\|$ *for all* $u \in \partial\Omega_3 \cap K$;

then f *has at least two fixed points* $\hat{u}, \tilde{u} \in (\overline{\Omega}_3 \setminus \Omega_1) \cap K$ *such that*

$$\hat{u} \in (\Omega_2 \setminus \Omega_1) \cap K \quad \text{and} \quad \tilde{u} \in (\overline{\Omega}_3 \setminus \overline{\Omega}_2) \cap K\,.$$

6.6 Variational Principles

In this section we present some abstract variational principles that are important tools in many parts of nonlinear analysis and are used in many different applications.

We start with the so-called "Lax–Milgram Theorem", which is fundamental in the study of linear elliptic equations. So, let H be a Hilbert space with inner product $(\cdot,\cdot)$ and associated norm $\|\cdot\| = (\cdot,\cdot)^{1/2}$.

Definition 6.6.1. Let $a\colon H \times H \to \mathbb{R}$ be a bilinear form. We say that
(a) $a\colon H \times H \to \mathbb{R}$ is **continuous** if there exists $c > 0$ such that

$$|a(u,v)| \leq c\|u\|\|v\| \quad \text{for all } u, v \in H\,.$$

(b) $a\colon H \times H \to \mathbb{R}$ is **coercive** if there exists $\hat{c} > 0$ such that

$$a(u,u) \geq \hat{c}\|u\|^2 \quad \text{for all } u \in H\,.$$

(c) $a\colon H \times H \to \mathbb{R}$ is **symmetric** if $a(u,v) = a(v,u)$ for all $u, v \in H$.

Remark 6.6.2. Let $H = \mathbb{R}^N$ and let A be an $\mathbb{N} \times \mathbb{N}$ matrix that is symmetric and positive definite. We consider the bilinear form $a(u, v) = (A(u), v)_{\mathbb{R}^N}$ for all $u, v \in \mathbb{R}^N$. Then $a: \mathbb{R}^N \times \mathbb{R}^N \to \mathbb{R}$ is continuous, coercive, and symmetric.

Theorem 6.6.3. *If $a: H \times H \to \mathbb{R}$ is a continuous and coercive bilinear form, $C \subseteq H$ is nonempty, closed, convex, and $h \in H$, then there exists a unique $\hat{u} \in C$ such that*

$$a(\hat{u}, y - \hat{u}) \geq (h, y - \hat{u}) \quad \textit{for all } y \in C . \tag{6.6.1}$$

Proof. Fix $u \in H$ and consider the map $y \to a(u, y)$. This is a continuous, linear functional. So, by the Riesz–Fréchet Representation Theorem (see Theorem 3.5.21), there exists a unique $A(u) \in H$ such that

$$(A(u), y) = a(u, y) \quad \text{for all } y \in H .$$

Evidently, A is linear and the continuity and coercivity of $a: H \times H \to \mathbb{R}$ imply

$$\|A(u)\| \leq c\|u\|, \quad \text{that is } A \in L(H) \quad \text{and} \quad (A(u), u) \geq \hat{c}\|u\|^2 .$$

Then we can recast (6.6.1) in terms of A. So, we seek $\hat{u} \in C$ such that

$$(A(\hat{u}), y - \hat{u}) \geq (h, y - \hat{y}) \quad \text{for all } y \in C . \tag{6.6.2}$$

Let $r > 0$ specified later in the process of the proof. Then (6.6.2) is equivalent to finding $\hat{u} \in C$ such that

$$(rh - rA(\hat{u}) + \hat{u} - \hat{u}, y - \hat{u}) \leq 0 . \tag{6.6.3}$$

If we consider the map $\xi: H \to H$ defined by $\xi(y) = p_C(rh - rA(y) + y)$ for all $y \in H$, where p_C denotes the metric projection on C, then (6.6.3) is equivalent to finding a fixed point of ξ; see Proposition 3.5.20 (c). Applying Proposition 3.5.20 (d) gives, for $y_1, y_2 \in H$,

$$\begin{aligned} \|\xi(y_1) - \xi(y_2)\| &= \|p_C(rh - rA(y_1) + y_1) - p_C(rh - rA(y_2) + y_2)\| \\ &\leq \|(y_1 - y_2) - rA(y_1 - y_2)\| . \end{aligned}$$

This implies

$$\begin{aligned} \|\xi(y_1) - \xi(y_2)\|^2 &\leq \|y_1 - y_2\|^2 - 2r(A(y_1 - y_2), y_1 - y_2) + r^2\|A(y_1 - y_2)\|^2 \\ &\leq (1 - 2r\hat{c} + r^2c^2)\|y_1 - y_2\|^2 . \end{aligned}$$

If we choose $r \in (0, (2\hat{c})/c^2)$, then $\eta^2 = 1 - 2r\hat{c} + r^2c^2 < 1$ and so $\|\xi(y_1) - \xi(y_2)\| \leq \eta\|y_1 - y_2\|$. Using the Banach Fixed Point Theorem (see Theorem 6.4.3), we conclude that there exists a unique $\hat{u} \in C$ such that $\hat{u} = \xi(\hat{u})$, which yields $(h, y - \hat{u}) \leq d(\hat{u}, y - \hat{u})$ for all $y \in C$. ☐

If $a: H \times H \to \mathbb{R}$ is also symmetric, then we can characterize the solution $\hat{u} \in C$ variationally.

Theorem 6.6.4. *If $a: H \times H \to \mathbb{R}$ is in addition symmetric, $\hat{u} \in C$ is the unique solution of* (6.6.1), *and*

$$\varphi(u) = \frac{1}{2} a(u, u) - (f, u) \quad \textit{for all } u \in H\,,$$

then $\varphi(\hat{u}) = \inf[\varphi(u): u \in C]$.

Proof. Since $a: H \times H \to \mathbb{R}$ is also symmetric,

$$\langle u, v\rangle = a(u, v) \quad \text{for all } u, v \in H$$

is an inner product on H. Let $|\cdot| = \langle \cdot, \cdot \rangle^{1/2}$ be the corresponding norm. The continuity and coercivity of $a: H \times H \to \mathbb{R}$ imply

$$\hat{c}\|u\|^2 \leq |u|^2 \leq c\|u\|^2 \quad \text{for all } u \in H\,.$$

So, the two norms $\|\cdot\|$ and $|\cdot|$ are equivalent, which means that $(H, |\cdot|)$ is still a Hilbert space. By the Riesz–Fréchet Representation Theorem, see Theorem 3.5.21, there exists a unique $\hat{h} \in H$ such that

$$a(\hat{h}, v) = \langle \hat{h}, v\rangle = (h, v) \quad \text{for all } v \in H\,. \tag{6.6.4}$$

Because of (6.6.4) and the symmetry of $a: H \times H \to \mathbb{R}$ we obtain

$$\begin{aligned} \frac{1}{2}|v - \hat{h}|^2 &= \frac{1}{2} a(v - \hat{h}, v - \hat{h}) = \frac{1}{2} a(v, v) - a(v, \hat{h}) + \frac{1}{2} a(\hat{h}, \hat{h}) \\ &= \frac{1}{2} a(v, v) - (h, v) + \frac{1}{2}|\hat{h}|^2 = \varphi(v) + \frac{1}{2}|\hat{h}|^2\,. \end{aligned}$$

So, minimizing φ over C is equivalent to minimizing $v \to 1/2|v - \hat{h}|^2$ over C. Then, invoking Proposition 3.5.20 (c), there exists a unique $\hat{u} \in C$ such that

$$\langle \hat{h} - \hat{u}, v - \hat{u}\rangle \leq 0 \quad \text{for all } v \in C,$$

which gives, due to (6.6.4), $a(\hat{u}, v - \hat{u}) \geq (h, v - \hat{u})$ for all $v \in C$. □

If $C = V$ is a closed subspace of H, then (6.6.1) becomes

$$a(\hat{u}, v) \geq (h, v) \quad \text{for all } v \in V$$

by taking $y = v + \hat{u}$ with $v \in V$. Since V is a linear subspace, we obtain $a(\hat{u}, v) = (h, v)$ for all $v \in V$.

In particular, this is true if $C = H$, that is, $V = H$. Then we have the "Lax–Milgram Theorem".

Theorem 6.6.5 (Lax–Milgram Theorem). *If $a: H \times H \to \mathbb{R}$ is a continuous and coercive bilinear form, then for given $h^* \in H^*$ there exists a unique $\hat{u} \in H$ such that $a(\hat{u}, y) = \langle h^*, y\rangle$ for all $y \in h$. Moreover, if $a: H \times H \to \mathbb{R}$ is also symmetric, then*

$$\varphi(\hat{u}) = \inf[\varphi(u): u \in H],$$

where $\varphi(u) = 1/2a(u,u) - \langle h^*, u\rangle$ for all $u \in H$.

For the direct method of the calculus of variations to work, it requires that we have some kind of compactness-type condition either on the space X on which the minimization takes place or on the functional φ that we are trying to minimize. So, for example, when the space X is a reflexive Banach space, we need φ to be coercive. Coercivity implies that the sublevel sets of φ are relatively w-compact, in fact, w-compact if φ is lower semicontinuous and convex. So, we can minimize. Without coercivity, only approximate minimizers can be found. The Ekeland Variational Principle asserts that we can generate minimizing sequences of a particular kind. Not only do they approach the minimal value of the problem, but they also satisfy the first order necessary condition up to any desired approximation. The Ekeland Variational Principle became an essential tool in many parts of nonlinear functional analysis and in its applications.

Theorem 6.6.6 (Ekeland Variational Principle). *If (X, d) is a complete metric space, $\varphi: X \to \overline{\mathbb{R}}$ is a proper, lower semicontinuous, function bounded from below, $\varepsilon > 0$, and $v_0 \in X$ satisfies $\varphi(v_0) \le \inf_X \varphi + \varepsilon$, then for a given $\lambda > 0$ there exists $u_\lambda \in X$ such that*
(a) $\varphi(u_\lambda) \le \varphi(v_0)$;
(b) $d(u_\lambda, v_0) \le \lambda$;
(c) $\varphi(u_\lambda) < \varphi(u) + \varepsilon/\lambda d(u, u_\lambda)$ *for all* $u \neq u_\lambda$.

Proof. Replacing d by $1/\lambda d$ and φ by $1/\varepsilon\varphi$, we can always assume that $\lambda = \varepsilon = 1$. On X we define a relation "≤" by

$$u \le v \quad \text{if and only if} \quad \varphi(u) \le \varphi(v) - d(u, v)\,. \tag{6.6.5}$$

Clearly "≤" is reflexive. Moreover, if $u \le v$, then $\varphi(u) \le \varphi(v) - d(u, v)$ and if $v \le y$, then $\varphi(v) \le \varphi(y) - d(v, y)$. It follows that

$$\varphi(u) \le \varphi(y) - [d(u, v) + d(v, y)] \le \varphi(y) - d(u, y)$$

by applying the triangle inequality. Hence, $u \le y$, that is, "≤" is transitive. Finally if $u \le v$ and $v \le u$, then from (6.6.5) it follows that $d(u, v) = 0$, that is, $u = v$ and so, "≤" is also antisymmetric. Hence, "≤" is a partial ordering on X.

Let $v_1 = v_0$ and define $D_1 = \{u \in X : u \leq v_1\}$. Choose $v_2 \in D_1$ such that

$$\varphi(v_2) \leq \inf_{D_1} \varphi + \frac{1}{2^2}$$

and for the induction step let

$$D_n = \{u \in X : u \leq v_n\}, \quad v_{n+1} \in D_n \quad \text{and} \quad \varphi(v_{n+1}) \leq \inf_{D_n} \varphi + \frac{1}{2^{n+1}} . \tag{6.6.6}$$

We have that $D_{n+1} \subseteq D_n$ for all $n \geq 1$ and each D_n is closed since φ is lower semicontinuous. If $u \in D_{n+1}$, then $u \leq v_{n+1} \leq v_n$ and so, due to (6.6.6) and since $u \in D_n$,

$$\begin{aligned} d(u, v_{n+1}) &\leq \varphi(v_{n+1}) - \varphi(u) \leq \inf_{D_n} \varphi + \frac{1}{2^{n+1}} - \varphi(u) \\ &\leq \varphi(u) + \frac{1}{2^{n+1}} - \varphi(u) = \frac{1}{2^{n+1}} . \end{aligned}$$

Hence, $\operatorname{diam} D_{n+1} \leq 1/2^{n+1}$ for all $n \in \mathbb{N}$, that is, $\operatorname{diam} D_n \to 0^+$ as $n \to \infty$. Invoking Cantor's Intersection Theorem (see Theorem 1.5.15), we obtain $\bigcap_{n\geq 1} D_n = \{\hat{u}\}$. Then $\hat{u} \in D_1$ and so $\varphi(\hat{u}) \leq \varphi(v_1) = \varphi(v_0)$, that is, (a) holds.

Since we have assumed that $\varepsilon = \lambda = 1$, we have

$$d(\hat{u}, v_0) \leq \varphi(v_0) - \varphi(\hat{u}) \leq \inf_X \varphi + 1 - \inf_X \varphi = 1 ,$$

that is, (b) holds. Finally, in order to show (c), we need to prove that $y \leq \hat{u}$ implies $y = \hat{u}$. Note that $y \leq \hat{u}$ gives $y \leq v_n$ for all $n \in \mathbb{N}$. Hence, $y \in \bigcap_{n\geq 1} D_n$, that is, $y = \hat{u}$. □

Remark 6.6.7. In the theorem above, conclusions (b) and (c) are somehow complementary. The choice of $\lambda > 0$ determines which of the two conclusions we want to emphasize in our application. If $\lambda > 0$ is large enough, then (b) provides little information on the whereabouts of u_λ. On the other hand (c) tells us that u_λ is close to being a minimizer of φ. Conversely, if $\lambda > 0$ is small, then (b) implies that u_λ is close to v_0 but on the other hand (c) provides little information. Two cases are usually used in applications. In the first, let $\lambda = 1$, $\varepsilon > 0$ and in the second $\lambda = \sqrt{\varepsilon}$, $\varepsilon > 0$. We state them as corollaries.

Corollary 6.6.8. *If (X, d) is a complete metric space and $\varphi: X \to \overline{\mathbb{R}}$ is proper, lower semicontinuous and bounded from below, then for every $\varepsilon > 0$ there exists $u_\varepsilon \in X$ such that*
(a) *$\varphi(u_\varepsilon) \leq \inf_X \varphi + \varepsilon$;*
(b) *$\varphi(u_\varepsilon) < \varphi(u) + \varepsilon d(u, u_\varepsilon)$ for all $u \neq u_\varepsilon$.*

Corollary 6.6.9. *If (X, d) is a complete metric space and $\varphi: X \to \overline{\mathbb{R}}$ is proper, lower semicontinuous, and bounded from below, $\varepsilon > 0$, and $v_\varepsilon \in X$ satisfies $\varphi(v_\varepsilon) \leq \inf_X \varphi + \varepsilon$, then there exists $u_\varepsilon \in X$ such that*
(a) *$\varphi(u_\varepsilon) \leq \varphi(v_\varepsilon)$;*

(b) $d(u_\varepsilon, v_\varepsilon) \le \sqrt{\varepsilon}$;
(c) $\varphi(u_\varepsilon) < \varphi(u) + \sqrt{\varepsilon} d(u, u_\varepsilon)$ *for all* $u \ne u_\varepsilon$.

Another useful corollary of Theorem 6.6.6 and of its proof is the following:

Corollary 6.6.10. *If* (X, d) *is a complete metric space and* $\varphi: X \to \overline{\mathbb{R}}$ *is proper, lower semicontinuous, and bounded from below, then for any* $\varepsilon > 0$ *and* $v_0 \in X$ *there exists* $u_\varepsilon \in X$ *such that* $\varphi(u_\varepsilon) \le \varphi(v_0) - \varepsilon d(u_\varepsilon, v_0)$.

If we introduce more structure on the space X, then we can have more information for the minimizing sequence.

Theorem 6.6.11. *If* X *is a Banach space and* $\varphi: X \to \mathbb{R}$ *is lower semicontinuous, bounded from below, and Gateaux differentiable (see Definition* 5.2.1 *(a)), then for every* $\varepsilon > 0$ *there exists* $u_\varepsilon \in X$ *such that*

$$\varphi(u_\varepsilon) \le \inf_X \varphi + \varepsilon \quad \textit{and} \quad \|\varphi'(u_\varepsilon)\|_* \le \varepsilon .$$

Proof. Applying Corollary 6.6.8 we find $u_\varepsilon \in X$ such that

$$\varphi(u_\varepsilon) \le \inf_X \varphi + \varepsilon \quad \text{and} \quad \varphi(u_\varepsilon) \le \varphi(u) + \varepsilon\|u - u_\varepsilon\| \quad \text{for all } u \in X . \tag{6.6.7}$$

Let $h \in X$ and $t > 0$ be arbitrary and set $u = u_\varepsilon + th$. Then from (6.6.7) we obtain

$$\frac{\varphi(u_\varepsilon) - \varphi(u_\varepsilon + th)}{t} \le \varepsilon\|h\| .$$

Hence, $-\langle \varphi'(u_\varepsilon), h\rangle \le \varepsilon\|h\|$ for all $h \in X$. This yields $|\langle \varphi'(u_\varepsilon), h\rangle| \le \varepsilon\|h\|$ for all $h \in X$ and so $\|\varphi'(u_\varepsilon)\|_* \le \varepsilon$. □

Corollary 6.6.12. *If* X *is a Banach space and* $\varphi: X \to \mathbb{R}$ *is lower semicontinuous, bounded from below, and Gateaux differentiable, then there exists a sequence* $\{u_n\}_{n\ge1} \subseteq X$ *such that*

$$\varphi(u_n) \searrow \inf_X \varphi \quad \textit{and} \quad \varphi'(u_n) \to 0 \quad \textit{in } X^* \textit{ as } n \to \infty .$$

Remark 6.6.13. This corollary asserts the existence of a minimizing sequence whose elements are almost critical points.

Corollary 6.6.14. *If* X *is a Banach space,* $\varphi: X \to \mathbb{R}$ *is lower semicontinuous, bounded from below, and Gateaux differentiable, then, given any minimizing sequence* $\{v_n\}_{n\ge1} \subseteq X$, *that is,* $\varphi(v_n) \searrow \inf_X(\varphi)$, *there exists another minimizing sequence* $\{u_n\}_{n\ge1} \subseteq X$ *such that*
(a) $\varphi(u_n) \le \varphi(v_n)$ *for all* $n \ge 1$;
(b) $\|u_n - v_n\| \to 0$ *as* $n \to \infty$;
(c) $\|\varphi'(u_n)\|_* \to 0$ *as* $n \to \infty$.

The following notion is related to the corollaries above and plays a central role in critical point theory.

Definition 6.6.15. Let X be a Banach space and $\varphi \in C^1(X, \mathbb{R})$. We say that φ satisfies the **Palais–Smale condition** or **PS-condition** for short if every sequence $\{u_n\}_{n\geq 1} \subseteq X$ such that $\{\varphi(u_n)\}_{n\geq 1}$ is bounded and $\varphi'(u_n) \to 0$ in X^* admits a strongly convergent subsequence.

Remark 6.6.16. This is a compactness type condition on the functional φ. It transfers the burden of compactness from the space X, which is in general infinite dimensional, and hence not locally compact, to the functional φ. A similar situation we encountered in the Leray–Schauder degree theory, where we had to limit ourselves to functionals $\mathrm{id} - f$ with f compact.

Proposition 6.6.17. *If X is a Banach space and $\varphi \in C^1(X, \mathbb{R})$ is bounded from below and satisfies the PS-condition, then there exists $\hat{u} \in X$ such that $\varphi(\hat{u}) = \inf_X \varphi$.*

Proof. Applying Theorem 6.6.11 gives the existence of a minimizing sequence $\{u_n\}_{n\geq 1} \subseteq X$ such that $\varphi(u_n) \searrow \inf_X \varphi$ and $\varphi'(u_n) \to 0$ in X^* as $n \to \infty$. Since by hypothesis φ satisfies the PS-condition, by passing to a subsequence if necessary, we may assume that $u_n \to \hat{u}$ in X as $n \to \infty$. Then

$$\varphi(\hat{u}) \leq \liminf_{n\to\infty} \varphi(u_n) = \inf_X \varphi .$$

Hence, $\varphi(\hat{u}) = \inf_X \varphi$. ☐

We can use the "Ekeland Variational Principle" to show that $\overline{\mathrm{dom}\,\varphi} = \overline{D(\partial\varphi)}$ for $\varphi \in \Gamma_0(X)$.

Proposition 6.6.18. *If X is a Banach space, $\varphi \in \Gamma_0(X)$, and $\hat{u} \in \mathrm{dom}\,\varphi$, then there exists a sequence $\{u_n\}_{n\geq 1} \subseteq D(\partial\varphi)$ such that $\|u_n - \hat{u}\| \leq 1/n$ and $\varphi(u_n) \to \varphi(\hat{u})$.*

Proof. Since $\varphi \in \Gamma_0(X)$, it admits a continuous affine minorant. So, we can find $u^* \in X^*$ and $c \in \mathbb{R}$ such that $\langle u^*, u\rangle - c < \varphi(u)$ for all $u \in X$. Let $\psi\colon X \to \overline{\mathbb{R}}$ be defined by

$$\psi(u) = \varphi(u) - \langle u^*, u\rangle + c \quad \text{for all } u \in X .$$

Evidently, $\psi \in \Gamma_0(X)$ with $\psi \geq 0$. Let $\varepsilon = \psi(\hat{u}) - \inf_X \psi > 0$ and $\lambda = 1/n$ with $n \in \mathbb{N}$. We apply Theorem 6.6.6 and obtain a sequence $\{u_n\}_{n\geq 1} \subseteq X$ such that

$$\psi(u_n) \leq \psi(\hat{u}) \quad \text{and} \quad \|u_n - \hat{u}\| \leq \frac{1}{n} \quad \text{for all } n \in \mathbb{N} \tag{6.6.8}$$

and

$$\psi(u_n) < \psi(u) + \varepsilon n\|u - u_n\| \quad \text{for all } u \neq u_n \text{ and for all } n \in \mathbb{N} . \tag{6.6.9}$$

We consider the functional $\sigma_n: X \to \overline{\mathbb{R}}$ with $n \in \mathbb{N}$ defined by

$$\sigma_n(u) = \psi(u) + \varepsilon n\|u - u_n\| \quad \text{for all } u \in X \text{ and for all } n \in \mathbb{N}\,.$$

From (6.6.9) we see that u_n is the unique global minimizer of $\sigma_n \in \Gamma_0(X)$. Hence, $0 \in \partial\sigma_n(u_n)$ and so $0 = v_n^* + \varepsilon n y_n^*$ with $v_n^* \in \partial\psi(u_n)$ and $y_n^* \in \partial\tau_n(u_n)$ for all $n \in \mathbb{N}$, when $\tau_n(u) = \varepsilon n\|u - u_n\|$. We have $v_n^* = u_n^* - u^*$ with $u_n^* \in \partial\varphi(u_n)$ for all $n \in \mathbb{N}$. Therefore, $\partial\varphi(u_n) \neq \emptyset$ for all $n \in \mathbb{N}$ and

$$\varphi(u_n) \le \varphi(\hat{u}) + \langle u^*, u_n - \hat{u}\rangle \quad \text{for all } n \in \mathbb{N}\,;$$

see (6.6.8). Since $u_n \to \hat{u}$ in X (see again (6.6.8)), we derive $\limsup_{n\to\infty} \varphi(u_n) \le \varphi(\hat{u})$. Hence, $\varphi(u_n) \to \varphi(\hat{u})$ since φ is lower semicontinuous. □

Corollary 6.6.19. *If X is a Banach space and $\varphi \in \Gamma_0(X)$, then $\overline{\operatorname{dom}\varphi} = \overline{D(\partial\varphi)}$.*

The Ekeland Variational Principle is equivalent to another remarkable result of nonlinear functional analysis known as the "Caristi Fixed Point Theorem".

Theorem 6.6.20 (Caristi Fixed Point Theorem). *If (X, d) is a complete metric space, $\varphi: X \to \overline{\mathbb{R}}$ is proper, lower semicontinuous, and bounded from below and $F: X \to 2^X \setminus \{\emptyset\}$ is a multifunction such that*

$$\varphi(y) \le \varphi(u) - d(u, y) \quad \textit{for all } u \in X \textit{ and for some } y \in F(u)\,, \tag{6.6.10}$$

then F has a fixed point, that is, there exists $\hat{u} \in X$ such that $\hat{u} \in F(\hat{u})$.

Proof. From Corollary 6.6.8 with $\varepsilon = 1$ we know that there exists $\hat{u} \in X$ such that

$$\varphi(\hat{u}) < \varphi(u) + d(u, \hat{u}) \quad \text{for all } u \neq \hat{u}\,. \tag{6.6.11}$$

Suppose that $\hat{u} \notin F(\hat{u})$. This implies that $u \neq \hat{u}$ for all $u \in F(\hat{u})$. Let $y \in F(\hat{u})$ as postulated by (6.6.10). Then $\varphi(y) \le \varphi(\hat{u}) - d(\hat{u}, y)$ and so, from (6.6.11) with $u = y \neq \hat{u}$, $d(\hat{u}, y) < d(y, \hat{u})$, a contradiction. Hence, F has a fixed point. □

Remark 6.6.21. The important feature of the fixed point theorem above is that no continuity conditions are imposed on the multifunction F. Suppose that $F = f: X \to X$ is a single-valued k-contraction and $\varphi(u) = 1/(1-k)d(u, f(u))$ for all $u \in X$. We have

$$\begin{aligned}\varphi(u) - \varphi(f(u)) &= \frac{1}{1-k}[d(u, f(u)) - d(f(u), f^{(2)}(u))] \\ &\ge \frac{1}{1-k}[d(u, f(u)) - kd(u, f(u))] = d(u, f(u))\,.\end{aligned}$$

Hence, condition (6.6.10) is satisfied and we can apply Theorem 6.6.20 and produce $\hat{u} \in X$ such that $\varphi(\hat{u}) = \hat{u}$. Therefore, the Caristi Fixed Point Theorem implies the Banach Fixed

Point Theorem. However, the Banach Fixed Point Theorem contains other important information besides the existence of a fixed point.

The Caristi Fixed Point Theorem (see Theorem 6.6.20) and the Ekeland Variational Principle in the form of Corollary 6.6.8 are equivalent.

Theorem 6.6.22. *The Ekeland Variational Principle in the form of Corollary* 6.6.8 *is in fact equivalent to the Caristi Fixed Point Theorem stated in Theorem* 6.6.20.

Proof. $\Longrightarrow$: This follows from the proof of Theorem 6.6.20.

$\Longleftarrow$: We argue indirectly. So, suppose that there exists $\varepsilon > 0$ for which we cannot find $u_\varepsilon \in X$ such that $\varphi(u_\varepsilon) < \varphi(u) + \varepsilon d(u, u_\varepsilon)$ for all $u \neq u_\varepsilon$. We define $F(u) = \{y \in X : \varphi(u) \geq \varphi(y) + \varepsilon d(y, u) \text{ and } y \neq u\}$. Then $F(u) \neq \emptyset$ and Theorem 6.6.20 (see (6.6.10)) implies that there exists $\hat{u} \in X$ such that $\hat{u} \in F(\hat{u})$, which is impossible. So, Corollary 6.6.8 holds. □

There is another variational principle related to the results above known as the "Takahashi Variational Principle".

Theorem 6.6.23 (Takahashi Variational Principle). *If (X, d) is a complete metric space, $\varphi: X \to \overline{\mathbb{R}}$ is a proper, lower semicontinuous, bounded from below, and for each $u \in X$ with $\varphi(u) > \inf_X \varphi$ there exists $v \in X$ such that $v \neq u$ and $\varphi(v) + d(v, u) \leq \varphi(u)$, then there exists $\hat{u} \in X$ such that $\varphi(\hat{u}) = \inf_X \varphi$.*

Proof. Arguing by contradiction, suppose that $\inf_X \varphi$ is not attained. Consider the multifunction $F: X \to 2^X$ defined by $F(u) = \{v \in X : \varphi(v) + d(v, u) \leq \varphi(u) \text{ and } v \neq u\}$. By hypothesis $F(u) \neq \emptyset$ for all $u \in X$. Invoking Theorem 6.6.20 (see (6.6.9)), we obtain $u_0 \in X$ such that $u_0 \in F(u_0)$, a contradiction. So, there exists $\hat{u} \in X$ such that $\varphi(\hat{u}) = \inf_X \varphi$. □

Theorem 6.6.24. *The Takahashi Variational Principle stated as Theorem* 6.6.23 *is in fact equivalent to the Caristi Fixed Point Theorem stated as Theorem* 6.6.20.

Proof. $\Longrightarrow$: We argue indirectly. So, consider φ and F as postulated by Theorem 6.6.20 and assume that F has no fixed points. So, for every $u \in X$, we can find $v \neq u$ such that $\varphi(v) + d(v, u) \leq \varphi(u)$. Invoking Theorem 6.6.23, we see that there exists $\hat{u} \in X$ such that $\varphi(\hat{u}) = \inf_X \varphi$. Let $\hat{v} \in F(\hat{u})$. Then $\hat{v} \neq \hat{u}$ and $\varphi(\hat{v}) + d(\hat{v}, \hat{u}) \leq \varphi(\hat{u}) = \inf_X \varphi$, a contradiction since $d(\hat{v}, \hat{u}) > 0$. Hence, F has a fixed point and so Theorem 6.6.20 holds.

$\Longleftarrow$: See Theorem 6.6.23 and its proof. □

Corollary 6.6.25. *The following results are equivalent:*

(a) *The Ekeland Variational Principle in the form of Corollary* 6.6.8.
(b) *The Caristi Fixed Point Theorem stated as Theorem* 6.6.20.
(c) *The Takahashi Variational Principle stated as Theorem* 6.6.23.

There is a useful generalization of Theorem 6.6.6 due to Zhong [341].

Theorem 6.6.26. *If $h: \mathbb{R}_+ \to \mathbb{R}_+$ is a continuous, nondecreasing function such that $\int_0^{+\infty} dr/(1 + h(r)) = +\infty$, (X, d) is a complete metric space, $u_0 \in X$ is fixed, $\varphi: X \to \overline{\mathbb{R}}$ is* *proper, lower semicontinuous, bounded below,* $\varepsilon > 0$, $\varphi(y) \le \inf_X \varphi + \varepsilon$, *and* $\lambda > 0$, *then there exists* $u_\lambda \in X$ *such that*

$$\varphi(u_\lambda) \le \varphi(y), \quad d(u_\lambda, u_0) \le r_0 + \bar{r},$$
$$\varphi(u_\lambda) \le \varphi(u) + \frac{\varepsilon}{\lambda(1 + h(d(u_\lambda, u_0)))} d(u_\lambda, u) \quad \textit{for all } u \in X,$$

where $r_0 = d(u_0, y)$ *and* $\bar{r} > 0$ *such that*

$$\int_0^{r_0+\bar{r}} \frac{1}{1 + h(r)} \, dr \ge \lambda .$$

Remark 6.6.27. If $h = 0$ and $u_0 = y$, then Theorem 6.6.26 reduces to Theorem 6.6.6.

Finally, as promised in Section 5.5, we will use the Ekeland Variational Principle to prove the Bishop–Phelps Theorem; see Theorem 5.5.12.

Theorem 6.6.28 (Bishop–Phelps Theorem). *If* X *is a Banach space, then the set of all functionals in* X^* *that attain their norm is dense in* X^*.

Proof. Let $u^* \in X^*$ with $\|u^*\|_* = 1$, $\varepsilon \in (0, 1/4)$ and define $\varphi(u) = \|u\|^2 - \langle u^*, u\rangle$ for all $u \in X$. Minimizing $\vartheta(t) = t^2 - t$ on $\mathbb{R}_+$, we have $\varphi(u) \ge \|u\|^2 - \|u\| \ge -1/4$. According to Corollary 6.6.8 there exists $u_0 \in X$ such that

$$\varphi(u_0) - \varepsilon\|u - u_0\| \le \varphi(u) \quad \text{for all } u \in X .$$

Hence,

$$\|u_0\|^2 - \langle u^*, u_0\rangle - \varepsilon\|u - u_0\| \le \|u\|^2 - \langle u^*, u\rangle \quad \text{for all } u \in X . \tag{6.6.12}$$

Consider the subgraph of the function on the left-hand side of (6.6.12) and the epigraph of the function on the right-hand side of (6.6.12). Both are convex sets with disjoint nonempty interiors. So, with the Weak Separation Theorem (see Theorem 3.1.59), we can find a continuous affine function $a: X \to \mathbb{R}$ such that

$$\|u_0\|^2 - \langle u^*, u_0\rangle - \varepsilon\|u - u_0\| \le a(u) \le \|u\|^2 - \langle u^*, u\rangle \quad \text{for all } u \in X . \tag{6.6.13}$$

We have $a(u_0) = \|u_0\|^2 - \langle u^*, u_0\rangle$ and so

$$-\varepsilon\|u - u_0\| \le a(u) - a(u_0) \quad \text{for all } u \in X ; \tag{6.6.14}$$

see (6.6.13). We claim that $u_0 \ne 0$. If $u_0 = 0$, then $a(u_0) = 0$ and so $-\varepsilon\|u\| \le a(u)$ for all $u \in X$; see (6.6.14). Therefore, a is continuous, linear, and $\|a\|_* \le \varepsilon$. Recall that $\|u^*\|_* = 1$ and $\varepsilon \in (0, 1/4)$. So, $u^* + a \ne 0$. Choose $h \in X$ with $\|h\| = 1$ such that $\langle u^* + a, h\rangle = \eta > 0$. Then

$$\langle u^* + a, u \rangle \le \|u\|^2 \quad \text{for all } u \in X .$$

Let $u = th$ for $t > 0$. Then $\eta t \le t^2$, and hence, $\eta < t$ for all $t > 0$, a contradiction since $\eta > 0$. So, $u_0 \ne 0$.

If $\|u\| = \|u_0\|$, then $\langle u^* + a, u\rangle \le \|u\|^2 = \|u_0\|^2 = \langle u^* + a, u_0\rangle$. So, $u^* + a$ restricted on $\{u \in X : \|u\| = \|u_0\|\}$ attains its supremum at u_0. Evidently, so does the functional $v^* = u^* + a - a(0) \in X^*$ and $\|v^* - u^*\|_* \le \varepsilon$. This proves the theorem. □

6.7 Variational Convergence

In optimization and in the calculus of variations we have a functional $\varphi: X \to \mathbb{R}$ and a constraint set $C \subseteq X$ and we deal with the minimization problem

$$m(\varphi, C) = \inf[\varphi(u) : u \in C] . \tag{6.7.1}$$

Moreover, we examine the set of solutions of (6.7.1)

$$M(\varphi, C) = \{u \in C : m(\varphi, C) = \varphi(u)\} .$$

In this section we introduce modes of convergence for both φ and C, which permit the study of the dependence of the value $m(\varphi, C)$ and of the set $M(\varphi, C)$ on the pair (φ, C)

So, let (X, τ) be a Hausdorff topological space and $\varphi_n: X \to \overline{\mathbb{R}}$ with $n \in \mathbb{N}$ is a sequence of proper functions. Recall that if $u \in X$, then by $\mathcal{N}(u)$ we denote the filter of neighborhoods of u.

Definition 6.7.1. (a) The **Γ_τ-limit inferior** of $\{\varphi_n\}_{n\ge1}$ is defined by

$$\Gamma_\tau\text{-}\liminf_{n\to\infty} \varphi_n = \sup_{U \in \mathcal{N}(u)} \liminf_{n\to\infty} \inf_{u \in U} \varphi_n(u) .$$

(b) The **Γ_τ-limit superior** of $\{\varphi_n\}_{n\ge1}$ is defined by

$$\Gamma_\tau\text{-}\limsup_{n\to\infty} \varphi_n = \sup_{U \in \mathcal{N}(u)} \limsup_{n\to\infty} \inf_{u \in U} \varphi_n(u) .$$

(c) If $\Gamma_\tau\text{-}\liminf_{n\to\infty} \varphi_n = \Gamma_\tau\text{-}\limsup_{n\to\infty} \varphi_n = \varphi$, then we say that the sequence $\{\varphi_n\}_{n\ge1}$ **Γ_τ-converges** to φ and we write $\varphi = \Gamma_\tau\text{-}\lim_{n\to\infty} \varphi_n$.

Remark 6.7.2. Clearly, we always have that $\Gamma_\tau\text{-}\liminf_{n\to\infty} \varphi_n \le \Gamma_\tau\text{-}\limsup_{n\to\infty} \varphi_n$ and both limits are lower semicontinuous functions. Moreover, in the definitions above, we can replace $\mathcal{N}(u)$ by a local basis $\mathcal{B}(u)$. When the topology τ is clearly understood and no confusion is possible, then we drop the subscript τ. Finally if $\varphi_n = \varphi$ for all $n \in \mathbb{N}$, then $\Gamma_\tau\text{-}\liminf_{n\to\infty} \varphi_n = \Gamma_\tau\text{-}\limsup_{n\to\infty} \varphi_n = \overline{\varphi}^\tau$, where $\operatorname{epi} \overline{\varphi}^\tau = \overline{\operatorname{epi} \varphi}^\tau$ is the τ-lower semicontinuous regularization of φ, that is the greatest τ-lower semicontinuous minorant of φ.

This mode of convergence is distinct from the pointwise convergence.

Example 6.7.3. (a) Let $X = \mathbb{R}$, $\varphi_n(u) = nue^{-n^2u^2}$ with $n \in \mathbb{N}$ and $\varphi(u) = -1/\sqrt{2e}$ if $u = 0$ and $\varphi(u) = 0$ if $u \neq 0$. Then we have $\Gamma\text{-}\lim_{n\to\infty} \varphi_n = \varphi$ and $\varphi_n(u) \to 0$ for all $u \in \mathbb{R}$.

(b) Let $X = \mathbb{R}$, $\varphi_n(u) = \sin(nu)$ with $n \in \mathbb{N}$ and $\varphi(u) = -1$ for all $u \in \mathbb{R}$. Then $\Gamma\text{-}\lim_{n\to\infty} \varphi_n = \varphi$ but the pointwise limit of the φ_n's does not exist.

(c) Let $X = \mathbb{R}$, $\varphi_n(u) = nue^{-n^2u^2}$ if n is even and $\varphi_n(u) = 2nue^{-n^2u^2}$ if n is odd with $n \in \mathbb{N}$. Then

$$\Big(\Gamma\text{-}\liminf_{n\to\infty} \varphi_n\Big)(u) = \begin{cases} -(\frac{2}{e})^{\frac{1}{2}} & \text{if } u = 0\,, \\ 0 & \text{if } u \neq 0\,, \end{cases}$$

$$\Big(\Gamma\text{-}\limsup_{n\to\infty} \varphi_n\Big)(u) = \begin{cases} -\frac{1}{\sqrt{2e}} & \text{if } u = 0\,, \\ 0 & \text{if } u \neq 0\,. \end{cases}$$

So, the Γ-limit does not exist, but the pointwise limit exists and is zero.

The functions are characterized by their epigraphs. So, the Γ-limits can be described using the epigraphs of the φ_n's. To this we need to introduce corresponding limits for sequences of sets.

Definition 6.7.4. Let $\{C_n\}_{n\geq 1} \subseteq 2^X$. The **$\tau$-Kuratowski limits** of C_n are defined in the following way:

(a) $K_\tau\text{-}\liminf_{n\to\infty} C_n = \bigcup_{n\geq 1} \overline{\bigcap_{k\geq n} \overline{C_k}^\tau}^\tau$ is the **τ-lower Kuratowski limit** of the C_n's.

(b) $K_\tau\text{-}\limsup_{n\to\infty} C_n = \bigcap_{n\geq 1} \overline{\bigcup_{k\geq n} C_k}^\tau$ is the **τ-upper Kuratowski limit** of the C_n's.

(c) If $K_\tau\text{-}\liminf_{n\to\infty} C_n = K_\tau\text{-}\limsup_{n\to\infty} C_n = C$, then we say that $\{C_n\}_{n\geq 1}$ converges to C in the **Kuratowski sense** and we write $C = K_\tau\text{-}\lim_{n\to\infty} C_n$.

Remark 6.7.5. Note that $u \in K_\tau\text{-}\liminf_{n\to\infty} C_n$ if and only if for every $U \in \mathcal{N}(u)$ we can find $n_0 = n_0(U) \in \mathbb{N}$ such that $U \cap C_n \neq \emptyset$ for all $n \geq n_0$. Similarly, $u \in K_\tau\text{-}\limsup_{n\to\infty} C_n$ if and only if for all $U \in \mathcal{N}(u)$ and for all $n \in \mathbb{N}$ we can find $k \geq n$ such that $U \cap C_k \neq \emptyset$. Evidently,

$$K_\tau\text{-}\liminf_{n\to\infty} C_n \subseteq K_\tau\text{-}\limsup_{n\to\infty} C_n$$

and both sets are closed.

Recall that if $C \in 2^X$, then $i_C(u) = 0$ if $u \in C$ and $i_C(u) = +\infty$ if $u \notin C$, which is the indicator function of C.

Proposition 6.7.6. *If $\{C_n\}_{n\geq 1} \subseteq 2^X\setminus\{\emptyset\}$, $C_l = K_\tau\text{-}\liminf_{n\to\infty} C_n$, and $C_u = K_\tau\text{-}\limsup_{n\to\infty} C_n$, then $i_{C_l} = \Gamma_\tau\text{-}\limsup_{n\to\infty} i_{C_n}$ and $i_{C_u} = \Gamma_\tau\text{-}\liminf_{n\to\infty} i_{C_n}$.*

Proof. We prove the first equality, the proof of the second being similar.

Let $\varphi = \Gamma_\tau\text{-}\limsup_{n\to\infty} i_{C_n}$. Evidently, range$(\varphi) = \{0, +\infty\}$. So, we need to show that $\varphi(u) = 0$ if and only if $u \in C_l$. We know from Remark 6.7.5 that $u \in C_l$ if and only if for every $U \in \mathcal{N}(u)$ we can find $n_0 \in \mathbb{N}$ such that $U \cap C_n \neq \emptyset$ for all $n \geq n_0$. This is equivalent to saying that $\inf_U i_{C_n} = 0$ for all $n \geq n_0$. Therefore, $u \in C_l$ if and only if $\limsup_{n\to\infty} \inf_U i_{C_n} = 0$ for all $U \in \mathcal{N}(u)$. We conclude that $u \in C_l$ if and only if $\varphi(u) = 0$, that is, $\varphi = i_{C_l}$. □

Now we can characterize the Γ_τ-limits in terms of the Kuratowski limits of the epigraphs.

Theorem 6.7.7. *If $\varphi_n : X \to \overline{\mathbb{R}}$ with $n \in \mathbb{N}$ is a sequence of proper functions and*

$$\varphi_l = \Gamma_\tau\text{-}\liminf_{n\to\infty} \varphi_n \quad \text{and} \quad \varphi_u = \Gamma_\tau\text{-}\limsup_{n\to\infty} \varphi_n ,$$

then $\operatorname{epi}\varphi_l = K_\tau\text{-}\limsup_{n\to\infty} \operatorname{epi}\varphi_n$ *and* $\operatorname{epi}\varphi_u = K_\tau\text{-}\liminf_{n\to\infty} \operatorname{epi}\varphi_n$.

Proof. We prove only the first equality, the proof of the second being similar.

We have $(u, \lambda) \in \operatorname{epi}\varphi_l$ if and only if $\varphi_l(u) \leq \lambda$. Moreover, by Definition 6.7.1 (a) we have $\varphi_l(u) \leq \lambda$ if and only if for every $\varepsilon > 0$ and every $U \in \mathcal{N}(u)$ we have

$$\liminf_{n\to\infty} \inf_U \varphi_n < \lambda + \varepsilon . \tag{6.7.2}$$

But (6.7.2) is equivalent to saying that for every $\varepsilon > 0$, every $U \in \mathcal{N}(u)$, and every $k \in \mathbb{N}$, there exists $n \geq k$ such that $\inf_U \varphi_n < \lambda + \varepsilon$. This is equivalent to $U \times (\lambda - \varepsilon, \lambda + \varepsilon) \cap \operatorname{epi}\varphi_n \neq \emptyset$ and with view to Remark 6.7.2 we see that $(u, \lambda) \in \operatorname{epi}\varphi_l$. Hence, $\operatorname{epi}\varphi_l = K_\tau\text{-}\limsup_{n\to\infty} \operatorname{epi}\varphi_n$. □

Remark 6.7.8. The result above is the reason why some authors call the Γ-convergence of functions **epigraphical convergence**.

When X is first countable, then we can state convenient sequential versions of Definition 6.7.1.

Proposition 6.7.9. *If X is first countable, $\varphi_l = \Gamma_\tau\text{-}\liminf_{n\to\infty} \varphi_n$, and $\varphi_u = \Gamma_\tau\text{-}\limsup_{n\to\infty} \varphi_n$, then the following hold:*

(a) *for every $u \in X$ and every sequence $u_n \to u$ in X, we have*

$$\varphi_l(u) \leq \liminf_{n\to\infty} \varphi_n(u_n)$$

and there exists a sequence $u_n \to u$ in X such that

$$\varphi_l(u) = \liminf_{n\to\infty} \varphi_n(u_n) ;$$

(b) *for every $u \in X$ and every sequence $u_n \to u$ in X, we have*

$$\varphi_u(u) \leq \limsup_{n\to\infty} \varphi_n(u_n)$$

and there exists a sequence $u_n \to u$ in X such that

$$\varphi_u(u) = \limsup_{n\to\infty} \varphi_n(u_n) ;$$

Proof. (a) Let $U \in \mathcal{N}(u)$. Then there exists $n_0 = n_0(U) \in \mathbb{N}$ such that $u_n \in U$ for all $n \geq n_0$. We see that $\inf_U \varphi_n \leq \varphi_n(u_n)$ for all $n \geq n_0$. Hence,

$$\liminf_{n\to\infty} \inf_U \varphi_n \leq \liminf_{n\to\infty} \varphi_n(u_n)$$

and so

$$\varphi_l(u) \leq \liminf_{n\to\infty} \varphi_n(u_n) ; \tag{6.7.3}$$

see Definition 6.7.1 (a). Since X is first countable, there exists at each $u \in X$ a local basis $\mathcal{B}(u) = \{U_k\}_{k\in\mathbb{N}}$ with $U_{k+1} \subseteq U_k$ for all $k \in \mathbb{N}$. Suppose that $\varphi(u) < +\infty$ and $\lambda_k \searrow \varphi_l(u)$ with $\varphi_l(u) < \lambda_k$ for all $k \in \mathbb{N}$. From (6.7.3) we infer that $\liminf_{n\to\infty} \inf_{U_k} \varphi_n < \lambda_k$ for all $k \in \mathbb{N}$. So, there exists a strictly increasing sequence $n(k) \in \mathbb{N}$ such that $\inf_{U_k} \varphi_{n(k)} < \lambda_k$ for all $k \in \mathbb{N}$. Hence,

$$\varphi_{n(k)}(v_k) < \lambda_k \quad \text{for some } v_k \in U_k \text{ and for all } k \in \mathbb{N} . \tag{6.7.4}$$

Define $u_n = v_k$ if $n = n(k)$ and $u_n = u$ if $n \neq n(k)$ for all $n \in \mathbb{N}$. Then $u_n \to u$ in X and, due to (6.7.4)

$$\varphi_l(u) = \lim_{k\to\infty} \lambda_k \geq \liminf_{k\to\infty} \varphi_{n(k)}(v_k) \geq \liminf_{n\to\infty} \varphi_n(u_n),$$

which gives $\varphi_l(u) = \liminf_{n\to\infty} \varphi_n(u_n)$; see (6.7.3).

(b) The first result follows as (6.7.3).

As in the previous case assume that $\varphi_u(u) < +\infty$ and consider $\vartheta_k \searrow \varphi_u(u)$ with $\varphi_u(u) < \vartheta_k$ for all $k \in \mathbb{N}$. Consider a strictly increasing sequence $n(k)$ such that $\inf_{U_k} \varphi_n < \vartheta_k$ for all $n \geq n(k)$, which implies $\varphi_n(v_{nk}) < \vartheta_k$ for some $v_{nk} \in U_k$ and for all $n \geq n(k)$. Define $u_n = u$ if $n < n(1)$ and $u_n = u_{nk}$ if $n(k) \leq n < n(k+1)$. Then $u_n \to u$ in X and $\varphi_u(u) = \lim_{k\to\infty} \vartheta_k \geq \limsup_{n\to\infty} \varphi_n(u_n)$. Hence, $\varphi_u(u) = \limsup_{n\to\infty} \varphi_n(u_n)$. □

Corollary 6.7.10. *If X is first countable and Γ_τ-$\lim \varphi_n = \varphi$, then the following hold:*
(a) *for every $u \in X$ and every sequence $u_n \to u$ in X, we have $\varphi(u) \leq \liminf_{n\to\infty} \varphi_n(u_n)$;*
(b) *for every $u \in X$, there is a sequence $u_n \to u$ in X such that $\varphi(u) = \lim_{n\to\infty} \varphi_n(u_n)$.*

Remark 6.7.11. According to the results above, when X is first countable, we see that

$$\left(\Gamma_\tau\text{-}\liminf_{n\to\infty} \varphi_n\right)(u) = \min\left[\liminf_{n\to\infty} \varphi_n(u_n) \colon u_n \to u \text{ in } X\right],$$

$$\left(\Gamma_\tau\text{-}\limsup_{n\to\infty} \varphi_n\right)(u) = \min\left[\limsup_{n\to\infty} \varphi_n(u_n) \colon u_n \to u \text{ in } X\right],$$

$$\left(\Gamma_\tau\text{-}\lim_{n\to\infty} \varphi_n\right)(u) = \min\left[\lim_{n\to\infty} \varphi_n(u_n) \colon u_n \to u \text{ in } X\right].$$

Similarly we get this result for the convergence of a sequence of sets; see Definition 6.7.4.

Proposition 6.7.12. *If X is first countable and $\{C_n\}_{n\geq 1} \subseteq 2^X$, then*

$$K_\tau\text{-}\liminf_{n\to\infty} C_n = \left\{u \in X : u = \lim_{n\to\infty} u_n, u_n \in C_n, n \in \mathbb{N}\right\},$$
$$K_\tau\text{-}\limsup_{n\to\infty} C_n = \left\{u \in X : u = \lim_{k\to\infty} u_{n_k}, u_{n_k} \in C_{n_k}, n_1 < n_2 < \cdots < n_k < \cdots\right\}.$$

Remark 6.7.13. So, in a first countable setting, $K_\tau\text{-}\liminf_{n\to\infty} C_n$ is formed by the limits of all convergent sequences $\{u_n\}_{n\geq 1}$ such that $u_n \in C_n$ for all $n \in \mathbb{N}$. Similarly, $K_\tau\text{-}\limsup_{n\to\infty} C_n$ is formed by the limits of all convergent subsequences $\{u_{n_k}\}_{k\in\mathbb{N}}$ such that $u_{n_k} \in C_{n_k}$ for all $k \in \mathbb{N}$. If (X, d) is a metric space, then

$$d\text{-}\liminf_{n\to\infty} C_n = \left\{u \in X : \lim_{n\to\infty} d(u, C_n) = 0\right\},$$
$$d\text{-}\limsup_{n\to\infty} C_n = \left\{u \in X : \limsup_{n\to\infty} d(u, C_n) = 0\right\}.$$

The next result is a byproduct of the characterization above and is useful in many different situations.

Proposition 6.7.14. *If X is first countable, $\{u_{mn}\}_{m,n\in\mathbb{N}} \subseteq X$, and*

$$\lim_{m\to\infty} \lim_{n\to\infty} u_{mn} = u \quad \text{in } X,$$

then there exist sequences of positive integers $\{m(n)\}_{n\in\mathbb{N}}$, $\{n(m)\}_{m\in\mathbb{N}}$ being increasing, not necessarily strictly, to $+\infty$ such that

$$u_{m(n)n} \to u \quad \text{as } n \to \infty \quad \text{and} \quad u_{mn(m)} \to u \quad \text{as } m \to \infty.$$

Proof. Let $C_n = \{u_{mn}\}_{m\in\mathbb{N}}$ and $u_m = \lim_{n\to\infty} u_{mn}$. Then $u_m \in K_\tau\text{-}\liminf_{n\to\infty} C_n$ for all $m \in \mathbb{N}$. Since $u_m \to u$ in X as $m \to \infty$ and $K_\tau\text{-}\liminf_{n\to\infty} C_n$ is closed, we know that $u \in K_\tau\text{-}\liminf_{n\to\infty} C_n$. Hence, $u = \lim_{n\to\infty} v_n$ with $v_n \in C_n$; see Proposition 6.7.12. Then $v_n = u_{m(n)n}$ and finally we derive $u = \lim_{n\to\infty} u_{m(n)n}$.

Next, let $D_m = \{u_{mn}\}_{n\in\mathbb{N}}$. Then $u_m \in \overline{D}_m$ for all $m \in \mathbb{N}$ and so $u = \lim_{m\to\infty} u_m \in K_\tau\text{-}\liminf_{m\to\infty} \overline{D}_m = K_\tau\text{-}\liminf_{m\to\infty} D_m$. Therefore, there exists $v_m \in D_m$ with $m \in \mathbb{N}$ such that $u = \lim_{m\to\infty} v_m$. We have $v_m = u_{mn(m)}$ and so $u = \lim_{m\to\infty} u_{mn(m)}$. □

Next we will establish the variational properties of the Γ-convergence.

Definition 6.7.15. Let (X, τ) be a Hausdorff topological space.

(a) A function $\varphi: X \to \mathbb{R}$ is said to be **coercive** (resp. **sequentially coercive**) if for every $\lambda \in \mathbb{R}$ the set $\overline{\{u \in X : \varphi(u) \leq \lambda\}}^\tau$ is τ-countably compact (resp. τ-sequentially compact).

(b) A sequence of functions $\varphi_n: X \to \mathbb{R}$ is said to be **equicoercive** if for all $\lambda \in \mathbb{R}$ there exists τ-closed and τ-countably compact set $K_\lambda \subseteq X$ such that

$$\{\varphi_n \le \lambda\} = \{u \in X: \varphi_n(u) \le \lambda\} \subseteq K_\lambda \quad \text{for all } n \in \mathbb{N}\,.$$

Remark 6.7.16. Clearly, sequential coercivity implies coercivity. Suppose that X is a reflexive Banach space and $\varphi: X \to \mathbb{R}$ satisfies $\varphi(u) \to +\infty$ as $\|u\| \to \infty$. Then, the functional φ is sequentially coercive for the weak topology on X in the topological sense above.

Proposition 6.7.17. *If $\varphi_n: X \to \overline{\mathbb{R}}$ is a sequence of functions such that $-\infty < \inf_{n\in\mathbb{N}} \varphi_n(u)$ for all $u \in X$, then $\{\varphi_n\}_{n\in\mathbb{N}}$ is equicoercive if and only if there is a τ-lower semicontinuous, coercive function $h: X \to \overline{\mathbb{R}}$ such that $h \le \varphi_n$ for all $n \in \mathbb{N}$.*

Proof. $\Longrightarrow$: Let $h_0(u) = \inf_{n\ge 1} \varphi_n(u)$ for all $u \in X$ and set $h = \overline{h_0}^{\tau}$, which is the τ-lower semicontinuous regularization of h_0, that is, $\operatorname{epi} h = \overline{\operatorname{epi} h_0}^{\tau}$. The equicoercivity of $\{\varphi_n\}_{n\ge 1}$ implies that, for all $\lambda \in \mathbb{R}$, there exists a τ-countably compact set $K_\lambda \subseteq X$ such that $\{\varphi_n \le \lambda\} \subseteq K_\lambda$ for all $n \in \mathbb{N}$. Let $u \in \{h \le \lambda\}$ and $\varepsilon > 0$. We can find $n_0 = n_0(\varepsilon) \in \mathbb{N}$ such that $\varphi_n(u) \le \lambda + \varepsilon$ for all $n \ge n_0$. Then $u \in K_{\lambda+\varepsilon}$ and so $\{h \le \lambda\} \subseteq \bigcap_{\varepsilon>0} K_{\lambda+\varepsilon}$ which is τ-closed and τ-countably compact. Hence, h is τ-lower semicontinuous, coercive, and $h \le \varphi_n$ for all $n \in \mathbb{N}$.

$\Longleftarrow$: For every $\lambda \in \mathbb{R}$ there exists a countably compact set $K_\lambda \subseteq X$ such that

$$\{\varphi_n \le \lambda\} \subseteq \{h \le \lambda\} \subseteq K_\lambda \quad \text{for all } n \in \mathbb{N}\,.$$

Therefore, $\{\varphi_n\}_{n\ge 1}$ is equicoercive. □

Equicoercive families exhibit nice variational stability properties.

Theorem 6.7.18. *If $\varphi_n: X \to \overline{\mathbb{R}}$ with $n \in \mathbb{N}$ is an equicoercive sequence and*

$$\varphi_l = \Gamma_\tau\text{-}\liminf_{n\to\infty} \varphi_n, \quad \text{and} \quad \varphi_u = \Gamma_\tau\text{-}\limsup_{n\to\infty} \varphi_n\,,$$

then φ_l and φ_u are both coercive and

$$\min_X \varphi_l = \liminf_{n\to\infty} \inf_X \varphi_n\,.$$

Moreover, if $\varphi = \Gamma_\tau\text{-}\lim_{n\to\infty} \varphi_n$, then φ is coercive and

$$\min_X \varphi = \lim_{n\to\infty} \inf_X \varphi_n\,.$$

Proof. On account of Proposition 6.7.17 there exists a τ-lower semicontinuous, coercive function $h: X \to \overline{\mathbb{R}}$ such that $h \le \varphi_n$ for all $n \in \mathbb{N}$. Then $h \le \varphi_l \le \varphi_u$ and so both φ_l and φ_u are coercive. Since φ_l is coercive, $\inf_X \varphi_l = \min_X \varphi_l$. Then, from Definition 6.7.1 (a) we have

$$\liminf_{n\to\infty} \inf_X \varphi_n \le \min_X \varphi_l \,. \tag{6.7.5}$$

We assume that $\liminf_{n\to\infty} \inf_X \varphi_n < \min_X \varphi_l$, otherwise by (6.7.5) there is nothing to prove. Then there exist a subsequence $\{n_k\}_{k\in\mathbb{N}}$ and $\lambda \in \mathbb{R}$ such that

$$\lim_{k\to\infty} \inf_X \varphi_{n_k} = \liminf_{n\to\infty} \inf_X \varphi_n < \lambda \,. \tag{6.7.6}$$

So, we may assume that

$$\inf_X \varphi_{n_k} < \lambda \quad \text{for all } k \in \mathbb{N} \,. \tag{6.7.7}$$

The equicoercivity of $\{\varphi_n\}_{n\in\mathbb{N}}$ implies that there exists a τ-closed and τ-countably compact set K such that $\{\varphi_{n_k} \le \lambda\} \subseteq K$ for all $k \in \mathbb{N}$. From (6.7.7) we see that the sets $\{\varphi_{n_k} \le \lambda\}$ with $k \in \mathbb{N}$ are nonempty. Then

$$\inf_X \varphi_{n_k} = \inf_K \varphi_{n_k} \quad \text{for all } k \in \mathbb{N} \,. \tag{6.7.8}$$

We can find $u_{n_k} \in K$ such that $\varphi_{n_k}(u_{n_k}) = \inf_X \varphi_{n_k}$. Evidently, $\{u_{n_k}\}_{k\in N}$ has a cluster point $\hat{u} \in K$. For every $U \in \mathcal{N}(\hat{u})$ there exists $k_0 \in \mathbb{N}$ such that $u_{n_k} \in U$ for all $k \ge k_0$. Hence, $\inf_U \varphi_{n_k} \le \varphi_{n_k}(u_{n_k})$. Combining (6.7.6) and (6.7.8) gives

$$\liminf_{k\to\infty} \inf_U \varphi_{n_k} \le \liminf_{n\to\infty} \inf_X \varphi_n \,. \tag{6.7.9}$$

Since $U \in \mathcal{N}(\hat{u})$ is arbitrary, from (6.7.9) and Definition 6.7.1 (a), we have

$$\hat{\varphi}_l(\hat{u}) \le \liminf_{n\to\infty} \inf_X \varphi_n \,,$$

where $\hat{\varphi}_l = \Gamma_\tau\text{-}\liminf_{k\to\infty} \varphi_{n_k}$. Hence,

$$\min_X \varphi_l \le \liminf_{n\to\infty} \inf_X \varphi_n \tag{6.7.10}$$

since $\varphi_l \le \hat{\varphi}_l$. From (6.7.5) and (6.7.10) we conclude that

$$\min_X \varphi_l = \liminf_{n\to\infty} \inf_X \varphi_n \,. \tag{6.7.11}$$

Finally, if $\varphi = \Gamma_\tau\text{-}\lim_{n\to\infty} \varphi_n$, then directly from Definition 6.7.1 we obtain

$$\limsup_{n\to\infty} \inf_X \varphi_n \le \inf_X \varphi \,.$$

Due to (6.7.11) this finally gives

$$\min_X \varphi = \lim_{n\to\infty} \inf_X \varphi_n \,.$$

□

In order to continue our investigation of variational stability, we make the following definition.

Definition 6.7.19. Let $\varphi\colon X \to \overline{\mathbb{R}}$ be a proper function. We introduce the following set $S(\varphi) = \{u \in X\colon \varphi(u) = \inf_X \varphi\}$. If the infimum is not realized, then we look for ε-minimizers of φ for $\varepsilon > 0$. So, given $\varepsilon > 0$, we say that $u \in X$ is an **ε-minimizer** of φ such that

$$\varphi(u) \le \max\left\{\inf_X \varphi + \varepsilon, -\frac{1}{\varepsilon}\right\}. \tag{6.7.12}$$

We denote the set of all ε-minimizers of φ on X by $S_\varepsilon(\varphi)$.

Remark 6.7.20. If $\inf_X \varphi > -\infty$, then for $\varepsilon > 0$ small, $u \in X$ is an ε-minimizer of φ if and only if $\varphi(u) \le \inf_X \varphi + \varepsilon$. So, in (6.7.12) the term $-1/\varepsilon$ appears in order to accommodate the case where $\inf_X \varphi = -\infty$. Note that $S(\varphi) = \bigcap_{\varepsilon>0} S_\varepsilon(\varphi)$. The set $S_\varepsilon(\varphi)$ is nonempty for all $\varepsilon > 0$, but $S(\varphi)$ may be empty.

Proposition 6.7.21. *If $\varphi = \Gamma_\tau\text{-}\lim_{n\to\infty} \varphi_n$, then the following hold:*
(a) $K_\tau\text{-}\limsup_{n\to\infty} S(\varphi_n) \subseteq \bigcap_{\varepsilon>0} K_\tau\text{-}\limsup_{n\to\infty} S_\varepsilon(\varphi_n) \subseteq S(\varphi)$;
(b) *if $\bigcap_{\varepsilon>0} K_\tau\text{-}\limsup_{n\to\infty} S_\varepsilon(\varphi_n) \neq \emptyset$, then $S(\varphi) \neq \emptyset$ and*

$$\min_X \varphi = \limsup_{n\to\infty} \inf_X \varphi_n\,;$$

(c) *if $\bigcap_{\varepsilon>0} K_\tau\text{-}\liminf_{n\to\infty} S_\varepsilon(\varphi_n) \neq \emptyset$, then $S(\varphi) \neq \emptyset$ and*

$$\min_X \varphi = \liminf_{n\to\infty} \inf_X \varphi_n\,.$$

Proof. (a) Since $S(\varphi_n) \subseteq S_\varepsilon(\varphi_n)$ for all $n \in \mathbb{N}$ and for all $\varepsilon > 0$, the first inclusion is obvious.

Let $u \in \bigcap_{\varepsilon>0} K_\tau\text{-}\limsup_{n\to\infty} S_\varepsilon(\varphi_n)$. Then for every $\varepsilon > 0$, for every $U \in \mathcal{N}(u)$ and for every $k \in \mathbb{N}$, there exists an index $n \in \mathbb{N}$ with $n \ge k$ such that $U \cap S_\varepsilon(\varphi_n) \neq \emptyset$. Therefore,

$$\liminf_{n\to\infty} \inf_U \varphi_n \le \max\left\{\limsup_{n\to\infty} \inf_X \varphi_n + \varepsilon, -\frac{1}{\varepsilon}\right\}.$$

Since $\varepsilon > 0$ is arbitrary, we derive

$$\varphi(u) \le \limsup_{n\to\infty} \inf_X \varphi_n\,. \tag{6.7.13}$$

On the other hand, we have

$$\limsup_{n\to\infty} \inf_X \varphi_n \le \inf_X \varphi\,, \tag{6.7.14}$$

which follows directly from Definition 6.7.1. From (6.7.13) and (6.7.14) we conclude that the second inclusion holds as well.

(b) This follows from (a); see (6.7.14).

(c) For $u \in \bigcap_{\varepsilon>0} K_\tau\text{-}\liminf_{n\to\infty} S_\varepsilon(\varphi_n)$, we infer that

$$\varphi(u) \le \liminf_{n\to\infty} \inf_X \varphi_n \le \inf_X \varphi$$

and so $\inf_X \varphi = \liminf_{n\to\infty} \inf_X \varphi_n$. This together with (b) gives (c). □

Theorem 6.7.22. *If $\varphi = \Gamma_\tau\text{-}\lim_{n\to\infty} \varphi_n$, it is not identically $+\infty$ and we consider the following statements:*

(i) $\bigcap_{\varepsilon>0} K_\tau\text{-}\limsup_{n\to\infty} S_\varepsilon(\varphi_n) \ne \emptyset$;
(ii) $S(\varphi) \ne \emptyset$ *and* $\min_X \varphi = \limsup_{n\to\infty} \inf_X \varphi_n$;
(iii) $S(\varphi) = \bigcap_{\varepsilon>0} K_\tau\text{-}\limsup_{n\to\infty} S_\varepsilon(\varphi_n)$;
(iv) $\bigcap_{\varepsilon>0} K_\tau\text{-}\liminf_{n\to\infty} S_\varepsilon(\varphi_n) \ne \emptyset$;
(v) $S(\varphi) \ne \emptyset$ *and* $\min_X \varphi = \lim_{n\to\infty} \inf_X \varphi_n$;
(vi) $S(\varphi) = \bigcap_{\varepsilon>0} K_\tau\text{-}\liminf_{n\to\infty} S_\varepsilon(\varphi_n) = \bigcap_{\varepsilon>0} K_\tau\text{-}\limsup_{n\to\infty} S_\varepsilon(\varphi_n)$,

then the following hold:
(a) *(i)* $\Longleftrightarrow$ *(ii)* $\Longrightarrow$ *(iii);*
(b) *(iv)* $\Longleftrightarrow$ *(v)* $\Longrightarrow$ *(vi).*

Proof. (a) Note that (i) $\Longrightarrow$ (ii) follows from Proposition 6.7.21 (b). Let us show that (ii) implies (i). To this end, let $u \in S(\varphi)$. Then

$$\varphi(u) = \min_X \varphi = \limsup_{n\to\infty} \inf_X \varphi_n < +\infty\,.$$

Given $\varepsilon > 0$, we have that

$$\varphi(u) - \frac{\varepsilon}{2} \le \inf_X \varphi_n \quad \text{for infinitely many } n\text{'s}\,. \tag{6.7.15}$$

From Definition 6.7.1, we obtain for all $U \in \mathcal{N}(u)$

$$\limsup_{n\to\infty} \inf_U \varphi_n < \max\left\{\varphi(u) + \frac{\varepsilon}{2}, -\frac{1}{\varepsilon}\right\}.$$

Then, for $n \ge 1$ large enough

$$\inf_U \varphi_n < \max\left\{\varphi(u) + \frac{\varepsilon}{2}, -\frac{1}{\varepsilon}\right\}.$$

and because of (6.7.15) we conclude that

$$\inf_U \varphi_n < \max\left\{\varphi(u) + \varepsilon, -\frac{1}{\varepsilon}\right\} \quad \text{for infinitely many } n\text{'s}\,.$$

So, $U \cap S_\varepsilon(\varphi_n) \neq \emptyset$ for infinitely many n's and every $\varepsilon > 0$. Therefore,

$$u \in \bigcap_{\varepsilon>0} K_\tau\text{-}\limsup_{n\to\infty} S_\varepsilon(\varphi_n)\,.$$

This shows that

$$S(\varphi) = \bigcap_{\varepsilon>0} K_\tau\text{-}\limsup_{n\to\infty} S_\varepsilon(\varphi_n) \neq \emptyset\,.$$

So we have proven (i) $\Longleftrightarrow$ (ii) $\Longrightarrow$ (iii).

(b) The proof is similar to that of (a), only now $U \cap S_\varepsilon(\varphi_n) \neq \emptyset$ for all $n \geq 1$ large enough. □

Corollary 6.7.23. *If $\varphi = \Gamma_\tau\text{-}\lim_{n\to\infty} \varphi_n$, $u_n \in S_{\varepsilon_n}(\varphi_n)$ with $\varepsilon_n \to 0^+$ and u is a cluster point of $\{u_n\}_{n\geq 1}$, then $u \in S(\varphi)$ and $\varphi(u) = \limsup_{n\to\infty} \varphi_n(u_n)$. Moreover, if $u_n \to u$ in X, then $u \in S(\varphi)$ and $\varphi(u) = \lim_{n\to\infty} \varphi_n(u_n)$.*

6.8 Nonlinear Operators, Identities and Inequalities

In this section we examine some operators that we encounter in the study of boundary value problems and prove some characteristic properties that they have. This way we see how the abstract results obtained in the previous sections can be used in concrete problems.

For the first result, let $\Omega \subseteq \mathbb{R}^N$, $N \geq 1$, be a bounded domain with a Lipschitz boundary $\partial\Omega$ and $1 < p < \infty$. Moreover, let $a\colon \mathbb{R}^N \to \mathbb{R}^N$ be a continuous map satisfying

$$|a(y)| \leq c(1 + |y|^{p-1}) \quad \text{for all } y \in \mathbb{R}^N \text{ and for some } c > 0\,. \tag{6.8.1}$$

Then we introduce the operator $A\colon W_0^{1,p}(\Omega) \to W^{-1,p'}(\Omega) = W_0^{1,p}(\Omega)^*$ with $\frac{1}{p} + \frac{1}{p'} = 1$ defined by

$$A(u) = -\operatorname{div} a(Du) \quad \text{for all } u \in W_0^{1,p}(\Omega)\,. \tag{6.8.2}$$

Note that viewing $W_0^{1,p}(\Omega)$ as a closed subspace of $L^p(\Omega, \mathbb{R}^{N+1})$ and using the Hahn–Banach Theorem as well as the Riesz Representation Theorem for the L^p-spaces, we can see that $W_0^{1,p}(\Omega)^*$ is identified with the subspace of distributions L_0 of the form

$$L_0(u) = \sum_{k=1}^N \int_\Omega f_k \frac{\partial u}{\partial z_k}\, dz$$

with $\{f_k\}_{k=1}^N \subseteq L^{p'}(\Omega, \mathbb{R}^{N+1})$. Note that $\Omega \subseteq \mathbb{R}^N$ is bounded.

Proposition 6.8.1. *The operator A given in* (6.8.2) *is bounded (that is, it maps bounded sets to bounded ones), continuous and maximal monotone.*

Proof. If $u \in W_0^{1,p}(\Omega)$, then $Du \in L^p(\Omega, \mathbb{R}^N)$ and so from (6.8.1) we have $a(Du(\cdot)) \in L^{p'}(\Omega, \mathbb{R}^N)$, that is, $\operatorname{div} a(Du(\cdot)) \in W^{-1,p'}(\Omega) = W_0^{1,p}(\Omega)^*$. Then $A\colon W_0^{1,p}(\Omega) \to W^{-1,p'}(\Omega)$ is defined by

$$\langle A(u), h\rangle = \langle -\operatorname{div} a(Du), h\rangle = \int_\Omega (a(Du), Dh)_{\mathbb{R}^N}\,dz \tag{6.8.3}$$

for all $h \in W_0^{1,p}(\Omega)$ by using integration by parts.

Suppose that $C \subseteq W_0^{1,p}(\Omega)$ is bounded. Then the set $\{a(Du)\colon u \in C\} \subseteq L^p(\Omega, \mathbb{R}^N)$ is bounded and from (6.8.3) we have

$$|\langle A(u), h\rangle| \le \int_\Omega |a(Du)||Dh|\,dz \le c_0\|Dh\|_p$$

for all $h \in W_0^{1,p}(\Omega)$ with some $c_0 > 0$. Therefore, $A(C) \subseteq W^{-1,p'}(\Omega)$ is bounded and so A is bounded.

Next we show the continuity of A. To this end, suppose that $u_n \to u$ in $W_0^{1,p}(\Omega)$. Then, by using Hölder's inequality and Poincaré's inequality, we obtain for every $h \in W_0^{1,p}(\Omega)$

$$|\langle A(u_n) - A(u), h\rangle \le \int_\Omega |a(Du_n) - a(Du)||Dh|\,dz \le \|a(Du_n) \cdot a(Du)\|_{p'}\|h\|\,.$$

This implies

$$\|A(u_n) - A(u)\|_* \le \|a(Du_n) - a(Du)\|_{p'} \to 0 \quad \text{as } n \to \infty\,.$$

Thus, A is continuous.

Finally, the monotonicity of $a\colon \mathbb{R}^N \to \mathbb{R}^N$ implies the monotonicity of A. Recall that a continuous and monotone operator is maximal monotone; see Proposition 6.1.17. □

If

$$a(y) = \begin{cases} |y|^{p-2}y & \text{if } y \ne 0, \\ 0 & \text{if } y = 0, \end{cases}$$

then a is a monotone, homeomorphism and $A = -\Delta_p$ is the p-Laplace differential operator or p-Laplacian for short.

Corollary 6.8.2. *The p-Laplacian map* $u \to -\Delta_p u$ *from* $W_0^{1,p}(\Omega)$ *into* $W_0^{1,p}(\Omega)^*$ *is bounded, continuous and maximal monotone.*

Remark 6.8.3. In fact we can easily show that the p-Laplacian is a homeomorphism.

As an easy application of Proposition 6.8.1 we consider the following Dirichlet problem

$$-\operatorname{div} a(Du(z)) = g(z) \quad \text{in } \Omega, \quad u = 0 \quad \text{on } \partial\Omega\,, \tag{6.8.4}$$

where $g \in L^{p'}(\Omega)$.

Proposition 6.8.4. *If $a\colon \mathbb{R}^N \to \mathbb{R}^N$ is as above and in addition it satisfies*

$$\hat{c}|y|^p \le (a(y), y)_{\mathbb{R}^N} \quad \textit{for all } y \in \mathbb{R}^N \textit{ with some } \hat{c} > 0\,,$$

then, for all $g \in L^{p'}(\Omega)$, problem (6.8.4) *has a solution which is unique if $a\colon \mathbb{R}^N \to \mathbb{R}^N$ is strictly monotone.*

Proof. We already know that A is maximal monotone. Also we have

$$\hat{c}\|Du\|_p^p \le \langle A(u), u\rangle \quad \text{for all } u \in W_0^{1,p}(\Omega)\,.$$

Therefore, by Poincaré's inequality, A is coercive. Then Corollary 6.1.33 implies that A is surjective and so problem (6.8.4) has a weak solution.

Suppose now that $a\colon \mathbb{R}^N \to \mathbb{R}^N$ is strictly monotone and let $u, v \in W_0^{1,p}(\Omega)$ be two weak solutions of (6.8.4). Then we obtain

$$0 = \int_\Omega (a(Du) - a(Dv), Du - Dv)_{\mathbb{R}^N}\, dz\,,$$

hence

$$0 = (a(Du(z)) - a(Dv(z)), Du(z) - Dv(z))_{\mathbb{R}^N} \quad \text{for a. a. } z \in \Omega\,.$$

Since a is strictly monotone, this implies $Du(z) = Dv(z)$ for a. a. $z \in \Omega$. Therefore, $u = v$ and so the weak solution is unique. □

Remark 6.8.5. The map

$$a(y) = \begin{cases} |y|^{p-2}y & \text{if } y \ne 0, \\ 0 & \text{if } y = 0, \end{cases}$$

with $1 < p < \infty$, corresponding to the p-Laplace differential operator is strictly monotone. At this point it is worth recalling the following elementary inequalities

$$(|u|^{p-2}u - |v|^{p-2}v, u - v)_{\mathbb{R}^N} \ge \hat{c} \begin{cases} |u-v|^2(1 + |u| + |v|)^{p-2} & \text{if } 1 < p \le 2\,, \\ |u-v|^p & \text{if } 2 < p\,, \end{cases}$$

for all $u, v \in \mathbb{R}^N$ and for some $\hat{c} > 0$. Also, we have

$$\left||u|^{p-2}u - |v|^{p-2}v\right| \le c \begin{cases} |u-v|^{p-1} & \text{if } 1 < p \le 2\,, \\ |u-v|(|u|+|v|)^{p-2} & \text{if } 2 < p\,, \end{cases}$$

for all $u, v \in \mathbb{R}^N$ and for some $c > 0$.

Now, we consider a function $G\colon \Omega \times \mathbb{R}^N \to \mathbb{R}$ which satisfies the following hypotheses:

(H) (i) For all $y \in \mathbb{R}^N$, $z \to G(z,y)$ is measurable and for a. a. $z \in \Omega$, $y \to G(z,y)$ is C^1, strictly convex and $G(z,0) = 0$;

(ii) $|\nabla_y G(z,y)| \le \hat{a}(z)(1 + |y|^{p-1})$ for a. a. $z \in \Omega$ and for all $y \in \mathbb{R}^N$ with $\hat{a} \in L^\infty(\Omega)_+$ and $1 < p < \infty$;

(iii) $(\nabla_y G(z,y), y)_{\mathbb{R}^N} \le pG(z,y)$ for a. a. $z \in \Omega$ and for all $y \in \mathbb{R}^N$;

(iv) $c_0|y|^p \le pG(z,y)$ for a. a. $z \in \Omega$, for all $y \in \mathbb{R}^N$ and for some $c_0 > 0$.

We set $a(z,y) = \nabla_y G(z,y)$ and consider the nonlinear map $A\colon W^{1,p}(\Omega) \to W^{1,p}(\Omega)^*$ defined by

$$\langle A(u), h\rangle = \int_\Omega (a(z,Du), Dh)_{\mathbb{R}^N}\, dz \quad \text{for all } u, h \in W^{1,p}(\Omega).$$

Proposition 6.8.6. *If hypotheses* (H) *hold, then the operator A is maximal monotone and of type* $(S)_+$*; see Definition* 6.1.58.

Proof. As in the proof of Proposition 6.8.1, we show that A is continuous and monotone (recall that $G(z,\cdot)$ is convex) and so it is maximal monotone. It remains to show that A is of type $(S)_+$. To this end, let $\{u_n\}_{n\in\mathbb{N}} \subseteq W^{1,p}(\Omega)$ be such that

$$u_n \xrightarrow{w} u \quad \text{in } W^{1,p}(\Omega) \quad \text{and} \quad \limsup_{n\to\infty} \langle A(u_n), u_n - u\rangle \le 0\,. \tag{6.8.5}$$

From (6.8.5) we have

$$\limsup_{n\to\infty} \langle A(u_n) - A(u), u_n - u\rangle \le 0\,. \tag{6.8.6}$$

On the other hand, from the monotonicity of A, one has

$$0 \le \liminf_{n\to\infty} \langle A(u_n) - A(u), u_n - u\rangle\,. \tag{6.8.7}$$

From (6.8.6) and (6.8.7) it follows that

$$\lim_{n\to\infty} \langle A(u_n) - A(u), u_n - u\rangle = 0\,. \tag{6.8.8}$$

Let

$$\xi_n(x) = \big(a(z, Du_n(z) - a(z, Du(z)), Du_n(z) - Du(z))\big)\,.$$

Evidently, $\xi_n \in L^1(\Omega)$ and $\xi_n(z) \geq 0$ for a. a. $z \in \Omega$ and for all $n \in \mathbb{N}$, since $a(z,\cdot)$ is monotone for a. a. $z \in \Omega$ as $G(z,\cdot)$ is convex for a. a. $z \in \Omega$. Then from (6.8.8) we have $\xi_n \to 0$ in $L^1(\Omega)$ and so, by passing to a subsequence if necessary, we can assume that

$$\xi_n(z) \to 0 \quad \text{for a. a. } z \in \Omega, \quad 0 \leq \xi_n(z) \leq \eta(z) \tag{6.8.9}$$

for a. a. $z \in \Omega$ and for all $n \in \mathbb{N}$ with $\eta \in L^1(\Omega)$.

The convexity of $G(z,\cdot)$ and equality $G(z,0) = 0$ for a. a. $z \in \Omega$ imply that

$$(a(z,y),y)_{\mathbb{R}^N} \geq G(z,y) \geq \frac{c_0}{p}|y|^p \quad \text{for a. a. } z \in \Omega \text{ and for all } y \in \mathbb{R}^N\,, \tag{6.8.10}$$

see hypothesis (H) (iv). Applying (6.8.9), (6.8.10) and hypothesis (H) (ii) gives us

$$\begin{aligned}\eta(z) \geq \xi_n(z) \geq \frac{c_0}{p}(|Du_n|^p + |Du|^p) &- \hat{a}(z)(1 + |Du|^{p-1})|Du| \\ &- \hat{a}(z)(1 + |Du_n|^{p-1})|Du|\end{aligned} \tag{6.8.11}$$

for a. a. $z \in \Omega$. From (6.8.11) it follows that for all $z \in \Omega \setminus D$ with $|D|_N = 0$, $|\cdot|_N$ being the Lebesgue measure on $\mathbb{R}^N$, $\{Du_n(z)\}_{n\in\mathbb{N}} \subseteq \mathbb{R}^N$ is bounded. So, passing to a subsequence (in general depending on $z \in \Omega$), we have

$$Du_n(z) \to y(z) \quad \text{in } \mathbb{R}^N \text{ for all } z \in \Omega \setminus D\,,$$

which implies

$$a(z, Du_n(z)) \to a(z,y(z)) \quad \text{for all } z \in \Omega \setminus D_1 \text{ with } D \subseteq D_1 \text{ and } |D_1|_N = 0\,.$$

Then, in the limit as $n \to \infty$, we have

$$(a(z,y(z)) - a(z, Du(z)), y(z) - Du(z))_{\mathbb{R}^N} = 0 \quad \text{for all } z \in \Omega \setminus D_1\,,$$

see (6.8.9). Therefore, $y(z) = Du(z)$ for all $z \in \Omega \setminus D_1$, since $a(z,\cdot)$ is strictly monotone because by hypothesis $G(z,\cdot)$ is strictly convex. Therefore, by the Urysohn criterion on the convergence of sequences, we obtain

$$Du_n(z) \to Du(z) \quad \text{in } \mathbb{R}^N \text{ for a. a. } z \in \Omega\,. \tag{6.8.12}$$

From (6.8.11) we see that $\{|Du_n|\}_{n\in\mathbb{N}} \subseteq L^p(\Omega)$ is bounded, hence it is uniformly integrable. This combined with (6.8.12) allows us to use Vitali's Convergence Theorem given in Theorem 2.3.44, and so we have

$$\|Du_n\|_p \to \|Du\|_p\,. \tag{6.8.13}$$

Recall that

$$Du_n \xrightarrow{w} Du \quad \text{in } L^p(\Omega, \mathbb{R}^N)\,, \tag{6.8.14}$$

see (6.8.5). The space $L^p(\Omega, \mathbb{R}^N)$ is uniformly convex. So, (6.8.13), (6.8.14) and the Kadec–Klee property imply

$$Du_n \to Du \quad \text{in } L^p(\Omega, \mathbb{R}^N)\,; \tag{6.8.15}$$

see Proposition 3.4.32.

From (6.8.5) and the compact embedding $W^{1,p}(\Omega) \hookrightarrow L^p(\Omega)$, it follows that

$$u_n \to u \quad \text{in } L^p(\Omega)\,. \tag{6.8.16}$$

Finally, from (6.8.15) and (6.8.16) we infer that $u_n \to u$ in $W^{1,p}(\Omega)$. Therefore, A is of type $(S)_+$. □

Note that if $a(z,y) = \vartheta(z)|y|^{p-2}y$ for all $y \in \mathbb{R}^N$ with $\vartheta \in L^\infty(\Omega)$, $\vartheta(z) \geq \hat{c} > 0$ for a. a. $z \in \Omega$, $1 < p < \infty$, then we have the weighted p-Laplacian given by

$$\Delta_p^\vartheta u = \operatorname{div}(\vartheta(z)|Du|^{p-2}Du)$$

and a satisfies hypotheses (H). Therefore, from Proposition 6.8.6, we have the following result.

Corollary 6.8.7. *The weighted p-Laplace operator $\Delta_p^\vartheta: W^{1,p}(\Omega) \to W^{1,p}(\Omega)^*$ is of type $(S)_+$.*

Now, let (Ω, Σ, μ) be a σ-finite measure space and X a reflexive Banach space. By X^* we denote the topological dual of X, $\langle \cdot, \cdot \rangle$ stands for the duality brackets of the pair (X^*, X) and $\|\cdot\|$ (resp. $\|\cdot\|_*$) for the norm of X (resp. X^*). Let $A: X \to 2^{X^*}$ be a maximal monotone map with $0 \in A(0)$. By Troyanski's Renorming Theorem, see Theorem 5.8.9, we may assume that both X and X^* are locally uniformly convex Banach spaces. Then the duality map $\mathcal{F}: X \to X^*$, see Definition 6.1.20, is a homeomorphism; see Proposition 6.1.28. Let $1 < p < \infty$ and consider the "lifting" ("realization") of A on the dual pair $(L^{p'}(\Omega, X^*), L^p(\Omega, X))$, see Theorem 4.2.26, defined by

$$\mathfrak{a}(u) = \{u^* \in L^{p'}(\Omega, X^*) : u^*(z) \in A(u(z)) \text{ for } \mu\text{-a. a. } z \in \Omega\}$$

for all $u \in D(A) = \{u \in L^p(\Omega, X) : S^{p'}_{A(u(\cdot))} \neq \emptyset\}$ with

$$S^{p'}_{A(u(\cdot))} = \{u^* \in L^{p'}(\Omega, X^*) : u^*(z) \in A(u(z)) \text{ for } \mu\text{-a. a. } z \in \Omega\}\,.$$

In what follows we denote by $((\cdot,\cdot))$ the duality brackets for the pair $(L^{p'}(\Omega, X^*), L^p(\Omega, X))$, that is,

$$((u^*, h)) = \int_\Omega \langle u^*(z), h(z)\rangle \, d\mu$$

for all $h \in L^p(\Omega, X)$ and for all $u^* \in L^{p'}(\Omega, X^*)$.

Proposition 6.8.8. *If $A: X \to 2^{X^*}$ is maximal monotone with $0 \in A(0)$, then $\mathfrak{a}: L^p(\Omega, X) \to 2^{L^{p'}(\Omega, X^*)}$ is maximal monotone as well.*

Proof. Let $\vartheta: L^p(\Omega, X) \to L^{p'}(\Omega, X^*)$ be the map defined by

$$\vartheta(u)(\cdot) = \left|u(\cdot)\right|^{p-2} \mathcal{F}(u(\cdot)) \quad \text{for all } u \in L^p(\Omega, X) .$$

It is clear the ϑ is continuous and strictly monotone, hence it is maximal monotone; see Proposition 6.1.17.

Claim: $R(\mathfrak{a} + \vartheta) = L^{p'}(\Omega, X^*)$, that is, $\mathfrak{a} + \vartheta$ is surjective.

Let $h \in L^{p'}(\Omega, X^*)$ and consider the multifunction $K: \Omega \to 2^X$ defined by

$$K(w) = \left\{x \in X: h(w) \in A(x) + |x|^{p-2}\mathcal{F}(x)\right\} .$$

The map $x \to A(x) + |x|^{p-2}\mathcal{F}(x)$ is maximal monotone due to Theorem 6.1.49 and coercive as $0 \in A(0)$. Hence, it is surjective and so $K(w) \neq \emptyset$ for all $w \in \Omega \setminus N$ with $\mu(N) = 0$. On the exceptional μ-null set N, we set $K(w) = \{c\}$. Then we have

$$\operatorname{Gr} K = \left\{(w, x) \in \Omega \times X: (x, h(w) - |x|^{p-2}\mathcal{F}(x)) \in \operatorname{Gr} A\right\} .$$

The maximal monotonicity of A implies that $\operatorname{Gr} A \subseteq X \times X^*$ is closed; see Proposition 6.1.14. Moreover, the map $k: \Omega \times X \to X \times X^*$ defined by

$$k(w, x) = (x, h(w) - |x|^{p-2}\mathcal{F}(x))$$

is a Carathéodory function and so jointly measurable because of Proposition 2.2.31. Therefore,

$$k^{-1}(\operatorname{Gr} A) = \operatorname{Gr} K \in \Sigma \otimes \mathcal{B}(X)$$

with $\mathcal{B}(X)$ being the Borel σ-field of X. Then, invoking the Yankov–von Neumann–Aumann Selection Theorem, see Theorem 2.7.25, we can find a Σ-measurable function $u: \Omega \to X$ such that $u(w) \in K(w)$ for μ-a. a. $w \in \Omega$. This implies

$$h(w) \in A(u(w)) + \left|u(w)\right|^{p-2} \mathcal{F}(u(w)) \quad \text{for } \mu\text{-a. a. } w \in \Omega .$$

We take the duality brackets with $u(w)$ and obtain $\|u(w)\|^{p-1} \le \|h(w)\|_*$ for μ-a. a. $w \in \Omega$ since $0 \in A(0)$. Therefore, $u \in L^p(\Omega, X)$. We have that $h \in \mathfrak{a}(u) + \vartheta(u)$ and since $h \in L^{p'}(\Omega, X^*)$ we conclude that $\mathfrak{a} + \vartheta$ is surjective. This proves the Claim.

Evidently, $\mathfrak{a}$ is monotone. We are going to prove that in fact it is maximal monotone. To this end, let $(v, g) \in L^p(\Omega, X) \times L^{p'}(\Omega, X^*)$ and assume that

$$((h - g, u - v)) = \int_\Omega \langle h(w) - g(w), u(w) - v(w) \rangle \, d\mu \tag{6.8.17}$$

for all $(u, h) \in \operatorname{Gr} \mathfrak{a}$. Using the Claim above, we can find $(\hat{u}, \hat{h}) \in \operatorname{Gr} \mathfrak{a}$ such that

$$\hat{h} + \vartheta(\hat{u}) = g + \vartheta(v)\,. \tag{6.8.18}$$

Then we choose $(\hat{u}, \hat{h}) \in \operatorname{Gr} A$ in (6.8.17) and obtain

$$0 \le \int_\Omega \langle \vartheta(v) - \vartheta(\hat{u}), \hat{u} - v \rangle \, dz\,.$$

The strict monotonicity of ϑ implies $v = \hat{u}$ and so $g = \hat{h}$; see (6.8.18). Therefore, $(v, g) \in \operatorname{Gr} A$ and we have the maximal monotonicity of $\mathfrak{a}$. □

We have another maximal monotonicity result which is useful in evolution equations. Let (X, H, X^*) be an evolution triple and let $\vartheta\colon D(\vartheta) \subseteq L^p(T, X) \to L^{p'}(T, X^*)$ with $1 < p < \infty$ and $T = [0, b]$ be defined by

$$\vartheta(u) = u' \quad \text{for all } u \in D(\vartheta) = \{u \in W_p(0, b)\colon u(0) = x_0\}$$

with $x_0 \in X$ and the space $W_p(0, b)$ as given in Definition 4.2.43.

Proposition 6.8.9. *The map ϑ is maximal monotone.*

Proof. Replacing u by $u - x_0$, we see that we may assume that $x_0 = 0$. Clearly, ϑ is linear. Also, using the integration by parts formula from Corollary 4.2.47, we have

$$((\vartheta(u), u)) = ((u', u)) = \int_0^b \langle u'(t), u(t) \rangle \, dt = \frac{1}{2} |u(b)|^2 \ge 0,$$

which shows that ϑ is monotone.

We need to show that ϑ is maximal monotone. Suppose that $(v, h) \in L^p(T, X) \times L^{p'}(T, X^*)$ and note that

$$0 \le ((h - \vartheta(u), v - u)) \quad \text{for all } u \in D(\vartheta)\,. \tag{6.8.19}$$

We have to show that $v \in D(\vartheta)$ and $h = \vartheta(v) = v'$. Let $u = \xi x$ with $\xi \in C_c^\infty(0, b)$ and $x \in X$. Then $u' = \xi' x$ and so $u \in D(\vartheta)$. Then we have

$$\langle \vartheta(u), u \rangle = 0\,. \tag{6.8.20}$$

From (6.8.19) and (6.8.20) we obtain

$$0 \le ((h, v)) - \int_0^b \langle \xi'(t)v(t) + \xi(t)h(t), x\rangle \, dt \quad \text{for all } x \in X.$$

This implies

$$\int_0^b (\xi'(t)v(t) + \xi(t)h(t)) \, dt = 0 \quad \text{for all } \xi \in C_c^\infty(0, b).$$

Hence, $h = v'$ and so $v \in W_p(0, b)$.

We need to show that $v \in D(\vartheta)$. From the integration by parts formula we have

$$0 \le ((v' - u', v - u)) = \frac{1}{2}\left(|v(b) - u(b)|^2 - |v(0) - u(0)|^2\right). \tag{6.8.21}$$

Let $\{x_n\}_{n\in\mathbb{N}} \subseteq X$ be a sequence such that $bx_n \to v(b)$ in H and set $u = tx_n$. Then $u \in D(\vartheta)$ and from (6.8.21) we have

$$|v(0)|^2 \le |v(b) - bx_n|^2 \to 0 \quad \text{as } n \to \infty.$$

Thus, $v(0) = 0$ and so $v \in D(\vartheta)$. □

Next, we consider the map $\hat{\vartheta}: D(\hat{\vartheta}) \subseteq L^p(T, X) \to L^{p'}(T, X^*)$ defined by

$$\hat{\vartheta}(u) = u' \quad \text{for all } u \in D(\hat{\vartheta}) = \{u \in W_p(0, b): u(0) = u(b)\}.$$

Proposition 6.8.10. *The map $\hat{\vartheta}$ is maximal monotone.*

Proof. The proof is the same as the proof of Proposition 6.8.9, just the end is a bit different. More precisely, from (6.8.21), we obtain

$$0 \le |v(b)|^2 - |v(0)|^2 + 2(v(0) - v(b), u(0)) \quad \text{for all } u \in D(\hat{\vartheta}). \tag{6.8.22}$$

Let $x \equiv y \in X$. From (6.8.22) it follows $v(0) = v(b)$ and so $v \in D(\hat{\vartheta})$. □

Proposition 6.8.11. *If X is a Banach space, $u: \mathbb{R} \to X$ is a map which is weakly differentiable almost everywhere and $t \to \|u(t)\|_X$ is almost everywhere differentiable, then*

$$\|u(t)\|_X \frac{d}{dt}\|u(t)\|_X = \langle u^*, u'(t)\rangle$$

for a. a. $t \in T$ and for all $u^* \in \mathcal{F}(u(t))$.

Proof. For every $t, s \in \mathbb{R}$ and $u^* \in \mathcal{F}(u(t))$, we have

$$\langle u^*, u(t)\rangle = \|u(t)\|_X^2 \quad \text{and} \quad \langle u^*, u(s)\rangle \le \|u(t)\|_X \|u(s)\|_X.$$

Therefore,

$$\langle u^*, u(s) - u(t)\rangle \le \|u(t)\|_X(\|u(s)\|_X - \|u(t)\|_X)\,. \tag{6.8.23}$$

Let $t \in \mathbb{R}$ be a point of differentiability of $\|u(\cdot)\|_X$ and of weak differentiability of $u(\cdot)$. Dividing (6.8.23) with $s - t$ and letting $s \to t$, we obtain

$$\begin{aligned}\langle u^*, u'(t)\rangle &\le \|u(t)\|_X \frac{\mathrm{d}}{\mathrm{d}t}\|u(t)\|_X \quad \text{as } s \to t^+\,,\\ \langle u^*, u'(t)\rangle &\ge \|u(t)\|_X \frac{\mathrm{d}}{\mathrm{d}t}\|u(t)\|_X \quad \text{as } s \to t^-\,.\end{aligned}$$

Therefore, we finally have

$$\langle u^*, u'(t)\rangle = \|u(t)\|_X \frac{\mathrm{d}}{\mathrm{d}t}\|u(t)\|_X$$

for a. a. $t \in T$ and for all $u^* \in \mathcal{F}(u(t))$. □

The next result is a version of Ehrling's inequality, see Lemma 4.2.48, which is useful in boundary value problems.

Proposition 6.8.12. *If X is a reflexive Banach space, Y, V are Banach spaces, $X \hookrightarrow V$ continuously and $K \in L_c(X, Y)$, then, for a given $\varepsilon > 0$, we can find $c_\varepsilon > 0$ such that*

$$\|K(u)\|_Y \le \varepsilon\|u\|_X + c_\varepsilon\|u\|_V \quad \textit{for all } u \in X\,.$$

Proof. We argue by contradiction. So, suppose that there exist $\varepsilon_0 > 0$ and a sequence $\{u_n\}_{n\in\mathbb{N}} \subseteq X$ such that

$$\|K(u_n)\|_Y > \varepsilon_0\|u_n\|_X + n\|u_n\|_V \quad \text{for all } n \in \mathbb{N}\,.$$

Evidently, $\|u_n\|_X \neq 0$ for all $n \in \mathbb{N}$ and so dividing by $\|u_n\|_X$ we obtain

$$\left\|K\left(\frac{u_n}{\|u_n\|_X}\right)\right\|_Y > \varepsilon_0 + n\frac{\|u_n\|_V}{\|u_n\|_X} \quad \text{for all } n \in \mathbb{N}\,. \tag{6.8.24}$$

Let $v_n = \frac{u_n}{\|u_n\|_X}$ and note that since $X \hookrightarrow V$ continuously, we have $\|u_n\|_V \le c\|u_n\|_X$ for all $n \in N$ and for some $c > 0$. Then, from (6.8.24), we have

$$\|K(v_n)\|_Y > \varepsilon_0 + n\|v_n\|_V \quad \text{for all } n \in \mathbb{N}\,. \tag{6.8.25}$$

Note that $\|v_n\|_X = 1$ and $\|v_n\|_Y \le c$ for all $n \in \mathbb{N}$. Since X is reflexive, we may assume that

$$v_n \xrightarrow{\mathrm{w}} v \quad \text{in } X\,. \tag{6.8.26}$$

Then, by Proposition 3.7.5, we have

$$K(v_n) \to K(v) \quad \text{in } Y\,. \tag{6.8.27}$$

From (6.8.25) we have

$$\|v_n\|_V \le \frac{1}{n}\|K(v_n)\|_Y \quad \text{for all } n \in \mathbb{N}\,.$$

Then, using (6.8.27), we obtain $v_n \to 0$ in V and (6.8.26) yields $v_n \xrightarrow{w} 0$ in X. Then (6.8.27) implies

$$K(v_n) \to 0 \quad \text{in } Y\,. \tag{6.8.28}$$

But from (6.8.25) we have

$$\|K(v_n)\|_Y \ge \varepsilon_0 \quad \text{for all } n \in \mathbb{N}\,,$$

which contradicts (6.8.28). □

Recalling that the trace operation is compact, see Theorem 4.5.25, we deduce the following consequence of Proposition 6.8.12.

Corollary 6.8.13. *If $\Omega \subseteq \mathbb{R}^N$ is a bounded domain with a Lipschitz boundary $\partial\Omega$, $1 < p < N$ and $q \in (1, \frac{(N-1)p}{N-p})$, then for every $\varepsilon > 0$, there exists $c_\varepsilon > 0$ such that*

$$\|u\|_{L^p(\partial\Omega)} \le \varepsilon\|Du\|_p + c_\varepsilon\|u\|_p \quad \textit{for all } u \in W^{1,p}(\Omega)\,.$$

In Definition 6.7.4 we introduced the Kuratowski limits of a sequence of sets in a Hausdorff topological space with topology τ. When the topology τ is first countable, then the Kuratowski limits have the following form

$$K_\tau\text{-}\liminf_{n\to\infty} C_n = \left\{u \in X : u = \lim_{n\to\infty} u_n,\ u_n \in C_n \text{ for all } n \in \mathbb{N}\right\},$$
$$K_\tau\text{-}\limsup_{n\to\infty} C_n = \left\{u \in X : u = \lim_{k\to\infty} u_{n_k}, u_{n_k} \in C_{n_k},\ n_1 < n_2 < \cdots < n_k \to \infty\right\}.$$

So, $K_\tau\text{-}\liminf_{n\to\infty} C_n$ consists of all limit points of the convergent sequences with elements in C_n, while $K_\tau\text{-}\limsup_{n\to\infty} C_n$ consists of all subsequential limits of sequences with elements in C_n. If X is a metric space with metric d, then

$$d\text{-}\liminf_{n\to\infty} C_n = \left\{u \in X : \lim_{n\to\infty} d(u, C_n) = 0\right\},$$
$$d\text{-}\limsup_{n\to\infty} C_n = \left\{u \in X : \liminf_{n\to\infty} d(u, C_n) = 0\right\}.$$

Definition 6.8.14. Let X be a Banach space and $\{C_n\}_{n\in\mathbb{N}} \subseteq 2^X \setminus \{\emptyset\}$. We define

$$\mathrm{s}\text{-}\liminf_{n\to\infty} C_n = \{u \in X : u_n \to \text{ in } X,\ u_n \in C_n,\ n \in \mathbb{N}\}\,,$$
$$\mathrm{w}\text{-}\limsup_{n\to\infty} C_n = \{u \in X : u_{n_k} \xrightarrow{\mathrm{w}} u \text{ in } X,\ u_{n_k} \in C_{n_k},\ n_1 < n_2 < \cdots < n_k \to \infty\}\,.$$

We say that the sequence $\{C_n\}_{n\in\mathbb{N}}$ converges to C in the **Mosco sense**, denoted by $C_n \xrightarrow{\mathrm{M}} C$, if

$$C = \mathrm{s}\text{-}\liminf_{n\to\infty} C_n = \mathrm{w}\text{-}\limsup_{n\to\infty} C_n\,.$$

Remark 6.8.15. Evidently, $C_n \xrightarrow{\mathrm{M}} C$ if and only if $C_n \xrightarrow{K_s} C$ and $C_n \xrightarrow{K_{\mathrm{wseq}}} C$.

Proposition 6.8.16. *If (X,d) is a metric space and $\{C_n\}_{n\in\mathbb{N}} \subseteq 2^X\setminus\{\emptyset\}$, then, for every $u \in X$, we have*

$$\limsup_{n\to\infty} d(u, C_n) \le d\Big(u, d\text{-}\liminf_{n\to\infty} C_n\Big)\,.$$

Proof. If $d\text{-}\liminf_{n\to\infty} C_n = \emptyset$, then $d(\cdot, d\text{-}\limsup_{n\to\infty} C_n) = +\infty$ and the assertion of the proposition is trivially true. So, we assume that $d\text{-}\liminf_{n\to\infty} C_n \neq \emptyset$. Let $v \in d\text{-}\liminf_{n\to\infty} C_n$. Then $d(v_n, v) \to 0$ with $v_n \in C_n$ for all $n \in \mathbb{N}$. We have $d(u, C_n) \le d(u, v_n)$ for all $n \in \mathbb{N}$ and so

$$\limsup_{n\to\infty} d(u, C_n) \le d(u, v)\,,$$

which gives

$$\limsup_{n\to\infty} d(u, C_n) \le d\Big(u, d\text{-}\liminf_{n\to\infty} C_n\Big)$$

since $v \in d\text{-}\liminf_{n\to\infty} C_n$ is arbitrary. □

There is a converse to the above result.

Proposition 6.8.17. *If (X,d) is a metric space, $\{C_n, C\}_{n\in\mathbb{N}} \subseteq 2^X \setminus \{\emptyset\}$ and*

$$\limsup_{n\to\infty} d(u, C_n) \le d(u, C) \quad \textit{for all } u \in X\,,$$

then $C = d\text{-}\liminf_{n\to\infty} C_n$.

Proof. Let $u \in C$. Then, by hypothesis, we have $\lim_{n\to\infty} d(u, C_n) = 0$, which yields $u \in d\text{-}\liminf_{n\to\infty} C_n$ and so $C \le d\text{-}\liminf_{n\to\infty} C_n$ as $u \in C$ was arbitrary. □

For the lim sup the situation is more complicated. We have the following counterpart of Proposition 6.8.16.

Proposition 6.8.18. *If X is a Banach space, $D \subseteq X$ is nonempty, w-closed and for every $r > 0$ we have*

$$D \cap \overline{B}_r \in P_{\mathrm{wk}}(X)$$

with $\overline{B}_r = \{u \in X : \|u\|_X \leq r\}$, and $\{C_n\}_{n\in\mathbb{N}} \subseteq 2^X \setminus \{\emptyset\}$ is a sequence such that $C_n \subseteq D$ for all $n \in \mathbb{N}$, then for every $u \in X$ we have

$$d\Big(u, \mathrm{w}\text{-}\limsup_{n\to\infty} C_n\Big) \leq \liminf_{n\to\infty} d(u, C_n) .$$

Proof. Let $u \in X$ and $r = \liminf_{n\to\infty} d(u, C_n)$. If $r = +\infty$, then the assertion of the proposition holds trivially. Therefore, we assume that $r < +\infty$. Let $u_n \in C_n$ for $n \in \mathbb{N}$ be such that

$$\|u - u_n\|_X \leq d(u, C_n) + \frac{1}{n} \quad \text{for all } n \in \mathbb{N} . \tag{6.8.29}$$

Note that $\{u_n\}_{n\in\mathbb{N}} \subseteq \overline{B}_{r+\|u\|_X+1} \in P_{\mathrm{wk}}(X)$ by hypothesis. So by the Eberlein–Smulian Theorem, see Theorem 3.4.14, we may assume that

$$u_n \xrightarrow{\mathrm{w}} \hat{u} \quad \text{and} \quad \hat{u} \in \mathrm{w}\text{-}\limsup_{n\to\infty} C_n . \tag{6.8.30}$$

Recall that the norm in a Banach space is weakly lower semicontinuous; see Proposition 3.3.13. Therefore, we have by using (6.8.29)

$$\|u - \hat{u}\|_X \leq \liminf_{n\to\infty} \|u - u_n\|_X \leq \liminf_{n\to\infty} d(u, C_n) ,$$

which by (6.8.30) gives us

$$d\Big(u, \mathrm{w}\text{-}\limsup_{n\to\infty} C_n\Big) \leq \liminf_{n\to\infty} d(u, C_n) . \qquad \square$$

This leads to the following result on the convergence of the distance functions.

Proposition 6.8.19. *If X is a reflexive Banach space, $\{C_n, C\}_{n\in\mathbb{N}} \subseteq P_{\mathrm{fc}}(X)$, $C_n \xrightarrow{\mathrm{M}} C$ and $u_n \to u$ in X, then $d(u_n, C_n) \to d(u, C)$.*

Proof. Since X is reflexive, we can find $c \in C$ such that $d(u, C) = \|u - c\|_X$. Because of $C_n \xrightarrow{\mathrm{M}} C$, we can find $c_n \in C_n$ for $n \in \mathbb{N}$ such that $c_n \to c$ in X. Then

$$d(u_n, c_n) \leq \|u_n - c_n\|_X \quad \text{for all } n \in \mathbb{N} ,$$

and so

$$\limsup_{n\to\infty} d(u_n, c_n) \leq \|u - c\|_X = d(u, c) . \tag{6.8.31}$$

On the other hand, from the triangle inequality, we have

$$d(u, c_n) \le \|u - u_n\|_X + d(u_n, c_n) \quad \text{for all } n \in \mathbb{N}\,. \tag{6.8.32}$$

Since X is reflexive and $C_n \xrightarrow{\text{M}} C$, using Proposition 6.8.18 along with Proposition 6.8.16, we obtain

$$d(u, C_n) \to d(u, C). \tag{6.8.33}$$

Passing to the limit in (6.8.32) as $n \to \infty$ and using (6.8.33), we get

$$d(u, C) \le \liminf_{n\to\infty} d(u_n, C_n)\,. \tag{6.8.34}$$

From (6.8.31) and (6.8.34) we conclude that $d(u_n, C_n) \to d(u, C)$. □

Next, we have some analogous results on sequences of sets using this time the support function. Recall that if X is a Banach space and $D \subseteq X$ is nonempty, then the support function of D, $\sigma(\cdot, D)\colon X^* \to \overline{\mathbb{R}} = \mathbb{R} \cup \{+\infty\}$ is defined by

$$\sigma(y^*, D) = \sup\{\langle y^*, y\rangle : u \in D\}$$

with $\langle\cdot,\cdot\rangle$ being the duality brackets for the pair (X^*, X); see Definition 5.1.17. We know that $\sigma(\cdot, D)$ is sublinear (that is, subadditive and positively homogeneous), w*-lower semicontinuous and is not identically $+\infty$. The First Separation Theorem for convex sets given in Theorem 3.1.57 shows that there is a one-to-one correspondence between such functions and the sets in $P_{\text{fc}}(X)$. More precisely, if $C \in P_{\text{fc}}(X)$, then

$$C = \{u \in X : \langle u^*, u\rangle \le \sigma(u^*, C) \text{ for all } u^* \in X^*\}\,.$$

Proposition 6.8.20. *If X is a Banach space, $\{C_n\} \subseteq 2^X \setminus \{\emptyset\}$ and $C_n \subseteq D \in P_{\text{wk}}(X)$ for all $n \in \mathbb{N}$, then $\text{w-}\limsup_{n\to\infty} C_n \neq \emptyset$ and for all $u^* \in X^*$*

$$\sigma\Big(u^*, \text{w-}\limsup_{n\to\infty} C_n\Big) = \limsup_{n\to\infty} \sigma(u^*, C_n)\,.$$

Proof. Since $C_n \subseteq D \in P_{\text{wk}}(X)$, the Eberlein–Smulian Theorem implies that $\text{w-}\limsup_{n\to\infty} C_n \neq \emptyset$.

Let $u \in \text{w-}\limsup_{n\to\infty} C_n$. Then we can find a subsequence $\{C_{n_k}\}_{k\in\mathbb{N}}$ of $\{C_n\}_{n\in\mathbb{N}}$ and $u_{n_k} \in C_{n_k}$ for $k \in \mathbb{N}$ such that $u_{n_k} \xrightarrow{\text{w}} u$ in X. For every $u^* \in X^*$ we have

$$\langle u^*, u_{n_k}\rangle \le \sigma(u^*, C_{n_k}) \quad \text{for all } k \in \mathbb{N}\,,$$

therefore,

$$\langle u^*, u\rangle \le \limsup_{n\to\infty} \sigma(u^*, C_n)\,.$$

This implies

$$\sigma\Big(u^*, \text{w-}\limsup_{n\to\infty} C_n\Big) \le \limsup_{n\to\infty} \sigma(u^*, C_n)\,. \tag{6.8.35}$$

Given $u^* \in X^*$, there exists $u_n \in C_n$ such that

$$\sigma(u^*, C_n) - \frac{1}{n} \le \langle u^*, u_n\rangle \quad \text{for all } n \in \mathbb{N}\,. \tag{6.8.36}$$

We have $\{u_n\}_{n\in\mathbb{N}} \subseteq D \in P_{\mathrm{wk}}(X)$. So, by the Eberlein–Smulian Theorem, we can find a subsequence $\{u_{n_k}\}_{k\in\mathbb{N}}$ of $\{u_n\}_{n\in\mathbb{N}}$ such that $u_{n_k} \xrightarrow{\text{w}} u$ in X which implies that $u \in \text{w-}\limsup_{n\to\infty} C_n$.

From (6.8.36) we have

$$\limsup_{n\to\infty} \sigma(u^*, C_n) \le \langle u^*, u\rangle \le \sigma\Big(u^*, \text{w-}\limsup_{n\to\infty} C_n\Big)\,. \tag{6.8.37}$$

Then from (6.8.35) and (6.8.37), we conclude that

$$\sigma\Big(u^*, \text{w-}\limsup_{n\to\infty} C_n\Big) = \limsup_{n\to\infty} \sigma(u^*, C_n)\,.$$

□

Proposition 6.8.21. *If X is a Banach space, $\{C_n, C\}_{n\in\mathbb{N}} \subseteq 2^X \setminus \{\emptyset\}$ and*

$$\limsup_{n\to\infty} \sigma(u^*, C_n) \le \sigma(u^*, C_n) \quad \textit{for all } u^* \in X^*\,,$$

then $\text{w-}\limsup_{n\to\infty} C_n \subseteq \overline{\operatorname{conv}}\, C$.

Proof. Let $u \in \text{w-}\limsup_{n\to\infty} C_n$. Then we can find a subsequence $\{C_{n_k}\}_{k\in\mathbb{N}}$ of $\{C_n\}_{n\in\mathbb{N}}$ and $u_{n_k} \in C_{n_k}$ with $k \in \mathbb{N}$ such that

$$\langle u^*, u_{n_k}\rangle \to \langle u^*, u\rangle \quad \text{for all } u^* \in X^*\,,$$

which implies

$$\langle u^*, u\rangle \le \limsup_{n\to\infty} \sigma(u^*, C_{n_k}) \le \sigma(u^*, C)\,.$$

Hence, $u \in \overline{\operatorname{conv}}\, C$. So, we conclude that $\text{w-}\limsup_{n\to\infty} C_n \subseteq \overline{\operatorname{conv}}\, C$. □

The next proposition is useful in many different situations. It provides information about the pointwise behavior of a weakly convergent sequence in the Lebesgue–Bochner space $L^1(\Omega, X)$.

Proposition 6.8.22. *If (Ω, Σ, μ) is a σ-finite measure space, X is a Banach space, $\{u_n\}_{n\in\mathbb{N}} \subseteq L^1(\Omega, X)$, $u_n \xrightarrow{\text{w}} u$ in $L^1(\Omega, X)$ and $u_n(w) \in W(w) \in P_{\mathrm{wk}}(X)$ for μ-a. a. $w \in \Omega$, then $u(w) \in \overline{\operatorname{conv}}\, \text{w-}\limsup_{n\to\infty}\{u_n(w)\}$ for μ-a. a. $w \in \Omega$.*

Proof. Taking Mazur's Theorem given in Theorem 3.3.18 into account yields

$$u(w) \in \overline{\operatorname{conv}} \bigcup_{n \geq k} u_n(w) \quad \text{for all } k \in \mathbb{N} \text{ and for } \mu\text{-a. a. } w \in \Omega\,.$$

Then, for $k \in \mathbb{N}$, $u^* \in X^*$ and $w \in \Omega \setminus N$ with $\mu(N) = 0$, we have

$$\langle u^*, u(w)\rangle \leq \sigma\Big(u^*, \bigcup_{n \geq k} u_n(w)\Big) = \sup_{n \geq k} \langle u^*, u_n(w)\rangle\,.$$

Using Proposition 6.8.20, it follows that

$$\langle u^*, u(w)\rangle \leq \limsup_{n \to \infty} \langle u^*, u_n(w)\rangle = \sigma\Big(u^*, \text{w-}\limsup_{n \to \infty} \{u_n(w)\}\Big)\,.$$

Therefore, we conclude that $u(w) \in \overline{\operatorname{conv}}\, \text{w-}\limsup_{n \to \infty} \{u_n(w)\}$ for μ-a. a. $w \in \Omega$. □

An interesting consequence of the above proposition is the following result.

Proposition 6.8.23. *If (Ω, Σ, μ) is a finite measure space, X is a reflexive Banach space, $\{u_n\}_{n \in \mathbb{N}} \subseteq L^p(\Omega, X)$ with $1 < p < \infty$ is bounded and*

$$u_n(w) \to u(w) \quad \textit{for } \mu\textit{-a. a. } w \in \Omega\,,$$

then $u_n \xrightarrow{\text{w}} u$ in $L^p(\Omega, X)$.

Proof. We know that $L^p(\Omega, X)$ is reflexive. Since $\{u_n\}_{n \in \mathbb{N}} \subseteq L^p(\Omega, X)$ is bounded, by the Eberlein–Smulian Theorem, there exists a subsequence $\{u_{n_k}\}_{k \in \mathbb{N}}$ of $\{u_n\}_{n \in \mathbb{N}}$ such that $u_{n_k} \xrightarrow{\text{w}} \hat{u}$ in $L^p(\Omega, X)$. Then, from the hypothesis and Proposition 6.8.22, we infer that $u = \hat{u}$. By the Urysohn criterion for convergent sequences, we have $u_n \xrightarrow{\text{w}} u$ in $L^p(\Omega, X)$. □

We already know that the support function $\sigma(\cdot, C)$ is sublinear and w*-lower semicontinuous. For w-compact sets, we can say more.

Proposition 6.8.24. *If X is a Banach space and $C \in P_{\text{wkc}}(X)$, then $\sigma(\cdot, C)\colon X \to \mathbb{R}$ is continuous.*

Proof. Let $u_n^* \to u^*$ in X^* and choose $c_n \in C$ such that

$$\sigma(u_n^*, C) = \langle u_n^*, c_n\rangle \quad \text{for all } n \in \mathbb{N}\,.$$

Since $C \in P_{\text{wkc}}(X)$, we may assume that $c_n \xrightarrow{\text{w}} c$ in X with $c \in C$. Then we have

$$\sigma(u_n^*, C) \to \langle u^*, c\rangle \leq \sigma(u^*, C)\,. \tag{6.8.38}$$

On the other hand, from the w*-lower semicontinuity of $\sigma(\cdot, C)$ it follows that

$$\sigma(u^*, C) \leq \liminf_{n \to \infty} \sigma(u_n^*, C)\,. \tag{6.8.39}$$

From (6.8.38) and (6.8.39) we infer that

$$\sigma(u_n^*, C) \to \sigma(u^*, C) \quad \text{as } n \to \infty\,,$$

which yields that $u^* \to \sigma(u^*, C)$ is continuous. □

Next, we present an interesting application of degree theory. We start with a definition.

Definition 6.8.25. Let X be a Hausdorff topological space. We say that X is **contractible**, if the identity map id_X is null-homotopic, that is, it is homotopic to a constant function, that is, there exists a continuous map $h\colon [0,1] \times X \to X$ such that $h(0,u) = u$ for all $u \in X$ and $h(1,u) = u_0$ for all $u \in X$ and for some $u_0 \in X$.

Remark 6.8.26. If a function $\varphi\colon X \to X$ is null-homotopic, we write $\varphi \simeq 0$. So, X is contractible if $\mathrm{id}_X \simeq 0$.

Proposition 6.8.27. *For every $r > 0$, $\partial B_r = \{u \in \mathbb{R}^N\colon \|u\|_2 = r\}$ is not contractible.*

Proof. We proceed by contradiction. So, suppose that ∂B_r is contractible. Then, by Definition 6.8.25, we can find a continuous function $h\colon [0,1] \times \partial B_r \to \partial B_r$ such that

$$h(0,u) = 0 \quad \text{for all } u \in \partial B_r \quad \text{and} \quad h(1,u) = u_0 \in \partial B_r\,.$$

By the Tietze Extension Theorem given in Theorem 1.2.44, there exists a continuous extension $\hat{h}\colon [0,1] \times \overline{B}_r \to \mathbb{R}^N$ of h. We set

$$\hat{\varphi}(\cdot) = \hat{h}(0,\cdot) \quad \text{and} \quad \hat{\psi}(\cdot) = \hat{h}(1,\cdot)\,.$$

The homotopy invariance property of the Brouwer degree implies that $d_B(\hat{\varphi}, B_r, 0) = d_B(\hat{\psi}, B_r, 0)$ and so, by Theorem 6.2.22,

$$d_b(\mathrm{id}, B_r, 0) = d_B(u_0, B_r, 0)\,.$$

But $d_B(\mathrm{id}, B_r, 0) \neq 0$ while $d_B(u_0, B_r, 0) = 0$, a contradiction. So ∂B_r is not contractible. □

In infinite dimensional Banach spaces the situation changes.

Proposition 6.8.28. *If X is an infinite dimensional Banach space, then for every $r > 0$, the set $\partial B_r = \{u \in X\colon \|u\|_X = r\}$ is contractible.*

Proof. From Remark 6.2.25 we know that ∂B_r is a retract of $\overline{B}_r = \{u \in X\colon \|u\|_X \le r\}$. So, let $\vartheta\colon \overline{B}_r \to \partial B_r$ be a retraction and define

$$h(t,u) = \vartheta((1-t)u) \quad \text{for all } t \in [0,1] \text{ and for all } u \in \partial B_r\,.$$

Then $h(0,u) = u$ for all $u \in \partial B_r$, $h(1,u) = \vartheta(0)$ and is continuous. So, we conclude that ∂B_r is contractible. □

Remark 6.8.29. Recall that ∂B_r is a retract of $\overline{B}_r$ if X is infinite dimensional, but not if $\dim X < \infty$; see Proposition 6.2.24 and Remark 6.2.25. So, we see that the properties of being a retract and of being contractible for ∂B_r exhibit a similar behavior. This is not a coincidence. The reason behind that similarity is that for an infinite dimensional Banach space ∂B_r is an absolute retract, AR for short, see Definition 1.7.26, and this in turn is equivalent to saying that ∂B_r is contractible; see Palais [255, Theorem 8].

Related to the above discussion is the so-called **radial retraction map** defined by

$$\eta(u) = \begin{cases} u & \text{if } \|u\|_X \leq r\,, \\ r\frac{u}{\|u\|_X} & \text{if } r < \|u\|_X\,. \end{cases} \tag{6.8.40}$$

Proposition 6.8.30. *If $X = H$ is a Hilbert space, then η given in* (6.8.40) *is nonexpansive.*

Proof. Note that η is the metric projection on $\overline{B}_r$, so the best approximation map; see Definition 5.4.1. From Proposition 3.5.20 (c), we know that

$$(u - \eta(u), \eta(v) - \eta(u)) \leq 0 \quad \text{for all } u, v \in H \tag{6.8.41}$$

with $(\cdot,\cdot)$ denoting the inner product of H. We have

$$u - v = \eta(u) - \eta(v) + u - \eta(u) + \eta(v) - v = \eta(u) - \eta(v) + h$$

with $h = u - \eta(u) + \eta(v) - v$. Then

$$\|u - v\|_H^2 = \|\eta(u) - \eta(v)\|_H^2 + 2(h, \eta(u) - \eta(v)) + \|h\|_H^2\,. \tag{6.8.42}$$

Note that

$$(h, \eta(u) - \eta(v)) = -(u - \eta(u), \eta(v) - \eta(u)) - (v - \eta(v), \eta(u) - \eta(v)) \geq 0$$

due to (6.8.41). Using this in (6.8.42), we obtain

$$\|\eta(u) - \eta(v)\|_H^2 \leq \|u - v\|_H^2\,.$$

Hence, η is nonexpansive. □

Next, we will present a result which is a useful tool in proving uniqueness of solutions in boundary value problems. The result is known as "Picone's identity" and in its original form asserted that if $u \geq 0$, $v > 0$ are differentiable functions, then

$$|Du|^2 + \frac{u^2}{v^2}|Dv|^2 - 2\frac{u}{v}(Du, Dv)_{\mathbb{R}^N} = |Du|^2 - \left(D\left(\frac{u^2}{v}\right), Dv\right)_{\mathbb{R}^N} \geq 0 \quad \text{in } \Omega .$$

Here we will prove a nonlinear extended version of this inequality. So, let $\Omega \subseteq \mathbb{R}^N$ be a bounded domain and $a : \Omega \times \mathbb{R}^N \to \mathbb{R}^N$ be a map which satisfies the following hypotheses:

(H') $a : \Omega \times \mathbb{R}^N \to \mathbb{R}^N$ is a Carathéodory function such that the following hold:

(i) for a. a. $z \in \Omega$, $y \to a(z, y)$ is maximal cyclically monotone and strictly monotone;

(ii) for a. a. $z \in \Omega$ and for all $\lambda > 0$,

$$a(z, \lambda y) = \lambda^{p-1} a(z, y) \quad \text{for all } y > 0 \text{ with } 1 < p < \infty ,$$

that is, $a(z, \cdot)$ is supposed to be $(p-1)$-positive homogeneous.

From Theorem 6.1.40 we know that there exists a measurable function $\varphi : \Omega \times \mathbb{R}^N \to \mathbb{R}$ such that for a. a. $z \in \Omega$

(i) $\varphi(z, \cdot)$ is continuously differentiable and strictly convex;

(ii) $\varphi(z, \cdot)$ is p-positive homogeneous, that is,

$$\varphi(z, \lambda y) = \lambda^p \varphi(z, y) \quad \text{for all } y \in \mathbb{R}^N \text{ and for all } \lambda > 0 .$$

(iii) $\nabla_y \varphi(z, x) = a(z, y)$ for all $y \in \mathbb{R}^N$.

We have

$$p\varphi(z, y) = (a(z, y), y)_{\mathbb{R}^N} \quad \text{for a. a. } z \in \Omega \text{ and for all } y \in \mathbb{R}^N .$$

The characteristic example is $a(y) = |y|^{p-2}y$ for all $y \in \mathbb{R}^N$ and $1 < p < \infty$, which corresponds to the p-Laplace differential operator.

Theorem 6.8.31. *If hypotheses* (H') *hold,* $u, v \in C(\Omega)$ *are differentiable at a. a.* $z \in \Omega$, $u(z) \geq 0$, $v(z) > 0$ *for all* $z \in \Omega$ *and*

$$L(u, v)(z) = p\varphi(z, Du) + p(p-1)\left(\frac{u}{v}\right)^p \varphi(z, Dv) - p\left(\frac{u}{v}\right)^{p-1} (a(z, Dv), Du)_{\mathbb{R}^N} ,$$

$$R(u, v)(z) = (a(z, Du), Du)_{\mathbb{R}^N} - \left(a(z, Dv), D\left(\frac{u^p}{v^{p-1}}\right)\right)_{\mathbb{R}^N} ,$$

then the following hold:

(a) $L(u, v)(z) = R(u, v)(z)$ *for a. a.* $z \in \Omega$;

(b) $L(u, v)(z) \geq 0$ *for a. a.* $z \in \Omega$;

(c) $L(u, v)(z) = 0$ *for a. a.* $z \in \Omega$ *if and only if* $u = \vartheta v$ *for some* $\vartheta \in \mathbb{R}$.

Proof. (a) We have

$$\begin{aligned} D\left(\frac{u^p}{v^{p-1}}\right) &= \frac{1}{(v^{p-1})^2}(pu^{p-1}v^{p-1}Du - (p-1)v^{p-2}u^p Dv) \\ &= p\left(\frac{u}{v}\right)^{p-1} Du - (p-1)\left(\frac{u}{v}\right)^p Dv\,. \end{aligned} \tag{6.8.43}$$

From (6.8.43) we obtain

$$\begin{aligned} &(a(z,Du),Du)_{\mathbb{R}^N} - \left(a(z,Dv), D\left(\frac{u^p}{v^{p-1}}\right)\right)_{\mathbb{R}^N} \\ &= (a(z,Du),Du)_{\mathbb{R}^N} - p\left(\frac{u}{v}\right)^{p-1}(a(z,Dv),Du)_{\mathbb{R}^N} + (p-1)\left(\frac{u}{v}\right)^p (a(z,Dv),Dv)_{\mathbb{R}^N} \\ &= p\varphi(z,Du) + p(p-1)\left(\frac{u}{v}\right)^p \varphi(z,Dv) - p\left(\frac{u}{v}\right)^{p-1}(a(z,Dv),Du)_{\mathbb{R}^N}\,, \end{aligned}$$

which shows $L(u,v) = R(u,v)$.

(b) Note that $\varphi(z,\cdot)$ is convex and $\nabla_y\varphi(z,y) = a(z,y)$. Then, by the homogeneity properties, one has

$$\begin{aligned} &\varphi(z,Du) + (p-1)\left(\frac{u}{v}\right)^p \varphi(z,Dv) - \left(\frac{u}{v}\right)^{p-1}(a(z,Dv),Du)_{\mathbb{R}^N} \\ &= \varphi(z,Du) + (p-1)\varphi\left(z,\frac{u}{v}Dv\right) - \left(a\left(z,\frac{u}{v}Dv\right)\right)_{\mathbb{R}^N} \\ &= \varphi(z,Du) - \varphi\left(z,\frac{u}{v}Dv\right) + \left(a\left(z,\frac{u}{v}Dv\right),\frac{u}{v}Dv\right)_{\mathbb{R}^N} - \left(a\left(z,\frac{u}{v}Dv\right),Du\right)_{\mathbb{R}^N} \\ &= \varphi(z,Du) - \varphi\left(z,\frac{u}{v}Dv\right) - \left(a\left(z,\frac{u}{v}Dv\right),Du - \frac{u}{v}Dv\right)_{\mathbb{R}^N} \geq 0\,. \end{aligned}$$

This proves (b).

(c) We have $L(u,v)(z) = 0$ for a. a. $z \in \Omega$, that is,

$$\varphi(z,Du) + (p-1)\left(\frac{u}{v}\right)^p \varphi(z,Dv) - \left(\frac{u}{v}\right)^{p-1}(a(z,Dv),Du)_{\mathbb{R}^N} = 0$$

for a. a. $z \in \Omega$. This implies

$$\varphi(z,Du) + (p-1)\varphi\left(z,\frac{u}{v}Dv\right) - \left(a\left(z,\frac{u}{v}Dv\right),Du\right)_{\mathbb{R}^N} = 0$$

for a. a. $z \in \Omega$. From this we conclude

$$\varphi(z,Du) - \varphi\left(z,\frac{u}{v}Dv\right) - \left(a\left(z,\frac{u}{v}Dv\right),Du - \frac{u}{v}Dv\right)_{\mathbb{R}^N} = 0$$

for a. a. $z \in \Omega$. Then the strict convexity of $\varphi(z, \cdot)$ implies that $Du = \frac{u}{v} Dv$ and so $vDu = uDv$. This gives $D(\frac{u}{v}) = 0$ showing that $u = \vartheta v$ for some $\vartheta \in \mathbb{R}$. □

The next result that we will prove is the so-called "Hardy's inequality". This result is a valuable tool in the study of singular elliptic problems.

Let $\Omega \subseteq \mathbb{R}^N$ be a bounded domain with Lipschitz boundary $\partial\Omega$ and let $\hat{d}(z) = d(z, \partial\Omega)$ for all $z \in \Omega$. Given $u \in L^1_{\mathrm{loc}}(\Omega)$, the Hardy–Littlewood maximal operator $M(u)(\cdot)$ of u is defined by

$$M(u)(z) = \sup_{r>0} \frac{1}{|B_r(z)|_N} \int_{B_r(z)} |u(y)| \, dy \quad \text{for all } z \in \Omega .$$

Now let $p \in C^{0,1}(\overline{\Omega})$ with $1 < p(z)$ for all $z \in \overline{\Omega}$. Then $M : L^{p(\cdot)}(\Omega) \to L^{p(\cdot)}(\Omega)$ is bounded. The following pointwise Hardy-type inequality is due to Hajłasz [150, Proposition 1].

Lemma 6.8.32. *For all $u \in C_c^\infty(\Omega)$ it holds*

$$|u(z)| \le c\hat{d}(z) M(|Du|)(z) \quad \textit{for all } z \in \Omega \textit{ and for some } c > 0 .$$

Using this lemma, we can prove Hardy's inequality.

Theorem 6.8.33. *If $p \in C^{0,1}(\overline{\Omega})$ with $1 < p_- = \min_{\overline{\Omega}} p$, then there exist $c = c(p, N) > 0$ and $a_0 = a_0(p, N) > 0$ such that*

$$\left\| \frac{u}{\hat{d}^{1-a}} \right\|_{p(\cdot)} \le c \left\| |Du| \hat{d}^a \right\|_{p(\cdot)}$$

for all $u \in W_0^{1,p(\cdot)}(\Omega)$ and for all $0 \le a < a_0$.

Proof. First let $a = 0$. Recall that $C_c^\infty(\Omega)$ is dense in $W_0^{1,p(\cdot)}(\Omega)$. Therefore, it suffices to prove the inequality for functions in $C_c^\infty(\Omega)$. From Lemma 6.8.32 we have

$$\frac{|u(z)|}{\hat{d}(z)} \le cM(|Du|)(z) \quad \text{for all } z \in \Omega ,$$

which yields

$$\left\| \frac{u}{\hat{d}} \right\|_{p(\cdot)} \le c \left\| M(|Du|) \right\|_{p(\cdot)} .$$

Then the result follows from the boundedness of the maximal operator.

Next let $0 < a < 1$ and let $v = |u|\hat{d}^a$. Then

$$|Dv| \le |Du|\hat{d}^a + a|u|\hat{d}^{a-1} .$$

Using the inequality for $a = 0$ on v, we obtain

$$\left\| \frac{u}{\hat{d}^{1-a}} \right\|_{p(\cdot)} \leq c\big(\| |Du| \hat{d}^a \|_{p(\cdot)} + a \| |u| \hat{d}^{a-1} \|_{p(\cdot)}\big) .$$

This implies

$$\left\| \frac{u}{\hat{d}^{1-a}} \right\|_{p(\cdot)} \leq \frac{c}{1 - ca} \| |Du| \hat{d}^a \|_{p(\cdot)} \quad \text{if } a < \frac{1}{c} . \qquad \square$$

Corollary 6.8.34. *It holds*

$$\left\| \frac{u}{\hat{d}} \right\|_p \leq c\|Du\|_p \quad \textit{for all } u \in W_0^{1,p}(\Omega)$$

with $1 < p < \infty$ *and for some* $c > 0$.

Remark 6.8.35. In fact a converse of this inequality is also true, namely if $u \in W^{1,p}(\Omega)$ and $\frac{u}{\hat{d}} \in L^p(\Omega)$, then $u \in W_0^{1,p}(\Omega)$.

We close this section in the way we started it; see Propositions 6.8.1 and 6.8.6. We determine the properties of the double phase differential operator

$$u \to -\operatorname{div}\big(a(z)|Du|^p + |Dv|^q\big).$$

So, let $\Omega \subseteq \mathbb{R}^N$ be a bounded domain with a Lipschitz boundary $\partial\Omega$, $a \in L^\infty(\Omega)$, $a(z) \geq 0$ for a. a. $z \in \Omega$ and $1 < q < p < N$ with $p < q^*$. Let φ be the double phase density defined by

$$\varphi(z, t) = a(z)t^p + t^q \quad \text{for all } t \geq 0 .$$

We introduce the operator $V\colon W_0^{1,\varphi}(\Omega) \to W_0^{1,\varphi}(\Omega)^*$ defined by

$$\langle V(u), h\rangle = \int_\Omega (a(z)|Du|^{p-2} + |Du|^{q-2})(Du, Dh)_{\mathbb{R}^N} \, dz$$

for all $u, h \in W_0^{1,\varphi}(\Omega)$.

For this operator we have the following result.

Proposition 6.8.36. *The operator V is bounded, continuous, strictly monotone (thus maximal monotone as well) and of type* $(S)_+$.

Proof. First we show the boundedness of V. We assume $u, h \neq 0$ and set $\vartheta_1 = \|u\| > 0$ and $\vartheta_2 = \|h\| > 0$. Then, using Hölder's and Young's inequalities along with the fact that $x \to \frac{x-1}{x}$ is increasing on $(0, \infty)$, we obtain

$$
\begin{aligned}
\left|\left\langle \frac{1}{\vartheta_1}V(u), \frac{1}{\vartheta_2}h\right\rangle\right| &= \left|\int_\Omega a(z)\left(\frac{|Du|}{\vartheta_1}\right)^{p-2}\left(\frac{1}{\vartheta_1}Du, \frac{1}{\vartheta_2}Dh\right)_{\mathbb{R}^N} dz\right. \\
&\quad \left. + \int_\Omega \left(\frac{|Du|}{\vartheta_1}\right)^{q-2}\left(\frac{1}{\vartheta_1}Du, \frac{1}{\vartheta_2}Dh\right)_{\mathbb{R}^N} dz\right| \\
&\le \left(\int_\Omega a(z)\left(\frac{|Du|}{\vartheta_1}\right)^p dz\right)^{\frac{p-1}{p}}\left(\int_\Omega a(z)\left(\frac{|Dh|}{\vartheta_2}\right)^p dz\right)^{\frac{1}{p}} \\
&\quad + \left(\int_\Omega \left(\frac{|Du|}{\vartheta_1}\right)^q dz\right)^{\frac{q-1}{q}}\left(\int_\Omega \left(\frac{|Dh|}{\vartheta_2}\right)^q dz\right)^{\frac{1}{q}} \\
&\le \frac{p-1}{p}\int_\Omega a(z)\left(\frac{|Du|}{\vartheta_1}\right)^p dz + \frac{1}{p}\int_\Omega a(z)\left(\frac{|Dh|}{\vartheta_2}\right)^p dz \\
&\quad + \frac{q-1}{q}\int_\Omega \left(\frac{|Du|}{\vartheta_1}\right)^q dz + \frac{1}{q}\int_\Omega \left(\frac{|Dh|}{\vartheta_2}\right)^q dz \\
&\le \frac{p-1}{p}\left(\int_\Omega a(z)\left(\frac{|Du|}{\vartheta_1}\right)^p dz + \int_\Omega \left(\frac{|Du|}{\vartheta_1}\right)^q dz\right) \\
&\quad + \frac{1}{q}\left(\int_\Omega a(z)\left(\frac{|Dh|}{\vartheta_2}\right)^p dz + \int_\Omega \left(\frac{|Dh|}{\vartheta_2}\right)^q dz\right) \\
&\le \frac{p-1}{p} + \frac{1}{q} \le 2\,.
\end{aligned}
$$

Hence, $\|V(u)\|_* \le 2\|u\|$, so V is bounded.

The strict monotonicity of V follows from the strict convexity of the double phase density $\varphi(z,\cdot)$.

Next, we show the continuity of V. To this end, let $u_n \to u$ in $W_0^{1,\varphi}(\Omega)$. For every $h \in W_0^{1,\varphi}(\Omega)$, we have

$$
\begin{aligned}
|\langle V(u_n) - V(u), h\rangle| &\le \int_\Omega a(z)\left||Du_n|^{p-2}Du_n - |Du|^{p-2}Du\right||Dh|\, dz \\
&\quad + \int_\Omega \left||Du_n|^{q-2}Du_n - |Du|^{q-2}Du\right||Dh|\, dz \\
&\le \left(\int_\Omega a(z)\left||Du_n|^{p-2}Du_n - |Du|^{p-2}Du\right|^{p'} dz\right)^{\frac{1}{p'}} c_1\|h\|' \\
&\quad + \left\||Du_n|^{q-2}Du_n - |Du|^{q-2}Du\right\|_q c_2\|h\|
\end{aligned}
$$

for some $c_1, c_2 > 0$, where we have used Hölder's inequality and the fact that $L^\varphi(\Omega) \hookrightarrow L^p_a(\Omega)$ continuously with $L^p_a(\Omega)$ being the weighted L^p-space with weight a, which is a seminormed space. So, we obtain

$$\|V(u_n) - V(u)\|_* \le c_3\Big(\int_\Omega a(z)\big||Du_n|^{p-2}Du_n - |Du|^{p-2}Du\big|^{p'}\,dz\Big)^{\frac{1}{p'}} + c_3\big\||Du_n|^{q-2}Du_n - |Du|^{q-2}Du\big\|_{q'}$$

for some $c_3 > 0$. Note that, at least for a subsequence, we have

$$Du_n(z) \to Du(z) \quad \text{for a. a. } z \in \Omega\,.$$

The sequences

$$\{a(\cdot)\big||Du_n|^{p-2}Du_n - |Du|^{p-2}Du\big|\}_{n\in\mathbb{N}} \subseteq L^1(\Omega)\,,$$
$$\{\big||Du_n|^{q-2}Du_n - |Du|^{q-2}Du\big|\}_{n\in\mathbb{N}} \subseteq L^1(\Omega)$$

are both uniformly integrable. So, by Vitali's Convergence Theorem given in Theorem 2.3.44, it follows that

$$\|V(u_n) - V(u)\|_* \to 0\,,$$

which shows that V is continuous. We have proved that V is strictly monotone and continuous, hence it is maximal monotone; see Proposition 6.1.17.

Finally, we show that V fulfills the $(S)_+$-property. For this purpose, let $\{u_n\}_{n\in\mathbb{N}} \subseteq W_0^{1,\varphi}(\Omega)$ be a sequence such that

$$u_n \xrightarrow{w} u \quad \text{in } W_0^{1\varphi}(\Omega) \quad \text{and} \quad \limsup_{n\to\infty}\langle V(u_n), u_n - u\rangle\,. \tag{6.8.44}$$

Since V is bounded, $\{V(u_n)\}_{n\in\mathbb{N}} \subseteq W_0^{1,\varphi}(\Omega)^*$ is bounded; see (6.8.44). The space $W_0^{1,\varphi}(\Omega)^*$ is reflexive, so by the Eberlein–Smulian Theorem we may assume that

$$V(u_n) \xrightarrow{w} u^* \quad \text{in } W_0^{1,\varphi}(\Omega)^*\,. \tag{6.8.45}$$

From (6.8.44), (6.8.45) and the fact that V is generalized pseudomonotone, see Proposition 6.1.54, it follows that

$$u^* = V(u) \quad \text{and} \quad \langle V(u_n), u_n\rangle \to \langle V(u), u\rangle\,,$$

which implies that

$$\rho_\varphi(Du_n) \to \rho_\varphi(Du)\,.$$

Hence, $u_n \to u$ in $W_0^{1,\varphi}(\Omega)$; see Corollary 4.9.63. Therefore, V is of type $(S)_+$. □

6.9 Remarks

(6.1) Initially nonlinear operator equations in infinite dimensional Banach spaces were studied in the framework of compact operators. However, it became evident that compact operators limited the class of boundary value problems that one can study. For this reason there was an effort to broaden this class by introducing more general nonlinear operators. Monotone operators, and in particular maximal monotone ones, served that purpose. These operators are rooted in the calculus of variations and their study started in the early sixties. The first significant results were obtained by Minty [236] for Hilbert spaces and Browder [60, 61] for Banach spaces. The duality map (see Definition 6.1.20) was introduced by Beurling–Livingston [36], which is an important tool in the study of evolution equations, accretive operators and of the corresponding semigroups of operators they generate; see Barbu [30], Browder [61], Showalter [295]. Theorem 6.1.31 was first proven by Minty for Hilbert spaces with $\mathcal{F} = \mathrm{id}$. Theorem 6.1.32 is due to Browder [60] and in the form of Corollary 6.1.33 illustrates why maximal monotone operators are an effective tool in the study of many boundary value problems. Theorems 6.1.35, 6.1.40, and 6.1.49 are due to Rockafellar [274, 275]. The notion of pseudomonotonicity was introduced by Brézis [50] using nets and by Browder [61] using sequences. Our presentation is based on the work of Browder–Hess [62]. More about maximal monotone operators and nonlinear operators of monotone-type can be found in the books of Barbu [30], Brézis [51], Gasiński–Papageorgiou [135], Hu–Papageorgiou [173], Pascali–Sburlan [257], Showalter [295], and Zeidler [340].

We mention a basic continuity property of a maximal monotone map.

Proposition 6.9.1. *If X is a reflexive Banach space and $A\colon X \to 2^{X^*}$ is a maximal monotone map with $\operatorname{int} D(A) \neq \emptyset$, then $A|_{\operatorname{int} D(A)}$ is usc from X with the norm topology into X^* with the weak topology.*

Remark 6.9.2. In fact if X is separable and reflexive and $A\colon X \to 2^{X^*}$ is maximal monotone with $\operatorname{int} D(A) \neq \emptyset$, then the set $S = \{u \in D(A)\colon A(u) \text{ is not a singleton}\}$ is of first category; see Kenderov [189].

(6.2) Degree theory started with the seminal work of Brouwer [58]. His approach was based on algebraic topology. Here we present a purely analytical method for the construction of the Brouwer degree due to Heinz [157]. We have seen that homotopic maps have the same Brouwer degree; see Theorem 6.2.22 (c). In fact, for spheres the converse is also true; see Granas–Dugundji [144, Theorem 8.3, p. 24].

Theorem 6.9.3. *If $\Omega = B_r(0) \subseteq \mathbb{R}^N$ with $N \geq 2$, $f, g \in C(\overline{\Omega}, \mathbb{R}^N)$ and $d_B(f, \Omega, 0) = d_B(g, \Omega, 0)$, then there exists $h \in C([0,1] \times \overline{\Omega}, \mathbb{R}^N)$ such that*

$$0 \notin h([0,1] \times \partial\Omega) \quad \text{and} \quad h(0,\cdot) = f(\cdot),\ h(1,\cdot) = g(\cdot)\,.$$

In Theorem 6.2.22, the essential properties are the normalization (a), the domain additivity (b), and the homotopy invariance (c). In fact, these three properties uniquely characterize the Brouwer degree; see Führer [134] and Amann–Weiss [11].

(6.3) Example 6.3.1 illustrating that an extension of the Brouwer degree in infinite dimensions is not possible, is due to Kakutani [185]. Leray–Schauder [212] were able to have an infinite dimensional degree theory by restricting themselves to maps that are compact perturbations of the identity. The essential tool here was the uniform approximation of compact maps by finite rank maps; see Theorem 3.7.10, which is due to Schauder [289]. This way we can transfer the properties of the Brouwer degree to the new degree function, the Leray–Schauder degree d_{LS}. As before, the main properties are the normalization, the domain additivity, and the homotopy invariance; see Theorem 6.3.15 (a), (b), and (c). These three properties define d_{LS} uniquely; see Amann–Weiss [11].

The books of Deimling [85], Denkowski–Migórski–Papageorgiou [87], Fonseca–Gangbo [126], Lloyd [219], and Papageorgiou–Kyritsi-Yiallourou [256] have detailed presentations of the Brouwer and of the Leray–Schauder degrees. There are further extensions of the degree theory to multifunctions, condensing maps, and maps of monotone type. For a presentation of these extensions we refer to Hu–Papageorgiou [173].

(6.4) Theorem 6.4.3 is due to Banach [26]. The result is essentially a remarkable abstraction of the iteration method of Picard for differential equations. Theorem 6.4.8 is due to Edelstein [108]. The first fixed point theorems for nonexpansive maps were proven by Browder [59] and Göhde [141]. Their results involved uniformly convex Banach spaces. The more general version presented in Theorem 6.4.28 is due to Kirk [191]. The notion of normal structure (see Definition 6.4.23 (b)) is due to Brodskiĭ–Mil'man [55]. The following result of Browder [61] is useful in applications.

Proposition 6.9.4. *If X is a uniformly convex Banach space, $C \subseteq X$ is nonempty, closed, convex, bounded, $\varphi\colon C \to X$ is nonexpansive, and $g(u) = u - \varphi(u)$ for all $u \in C$, then g is demiclosed on C, that is, if $u_n \xrightarrow{w} u$ in C and $g(u) \to h$ in X, then $g(u) = h$.*

We also mention an extension of the Schauder Fixed Point Theorem (see Theorem 6.3.21) to locally convex spaces. The result is known as the "Tychonoff Fixed Point Theorem".

Theorem 6.9.5 (Tychonoff Fixed Point Theorem). *If X is a locally convex space, $C \subseteq X$ is closed, convex, $\varphi\colon C \to C$ is continuous and $\overline{\varphi(C)}$ is compact, then φ has a fixed point.*

Many of the results in this section have extensions to multifunctions. For more on this subject, we refer to Hu–Papageorgiou [173].

The fixed point theory – metric and topological – is discussed in the books of Goebel–Kirk [140], Granas–Dugundji [144], and Zeidler [339].

(6.5) The fixed point index was studied in detail by Amann [9] who also used it to study various semilinear elliptic boundary value problems. More on this topic can be found in Deimling [85], Granas–Dugundji [144], and Guo–Lakshmikantham [145].

(6.6) Theorem 6.6.3 is due to Stampacchia [302] and it is useful in the study of variational inequalities; see Kinderlehrer–Stampacchia [190] and Showalter [295]. Theorem 6.6.5 is due to Lax–Milgram [203] and Theorem 6.6.6 was proven by Ekeland [111]. In Ekeland [112, 113] we can find detailed surveys of the many applications that this result has. Theorem 6.6.20 is due to Caristi [70]. His proof was based on transfinite induction and Theorem 6.6.23 is due to Takahashi [306]. Also these nonlinear analysis results can be deduced from the following general principle for partially ordered sets due to Brézis–Browder [53].

Theorem 6.9.6. *If $(X, \leq)$ is a partially ordered set, every increasing sequence $\{u_n\}_{n\geq 1} \subseteq X$ has an upper bound, that is, if $u_n \leq u_{n+1}$ for all $n \in \mathbb{N}$, there exists $h \in X$ such that $u_n \leq h$ for all $n \in \mathbb{N}$, and $\varphi: X \to \mathbb{R}$ is an increasing function which is bounded above, then there exists $\hat{u}_0 \in X$ such that $\hat{u}_0 \leq v$ and $\varphi(\hat{u}_0) = \varphi(v)$.*

Finally Theorem 6.6.28 is due to Bishop–Phelps [37].

(6.7) The Γ-convergence of functions was introduced and studied by De Giorgi–Franzoni [84]. The related K_τ-convergence of sets (see Definition 6.7.4) was studied by Kuratowski [197, 200]. The Γ-convergence is designed as a tool suitable for the sensitivity analysis of optimization problems. The theory and applications of this mode of functional convergence can be found in the books of Attouch [17], Dal Maso [80], Dontchev–Zolezzi [98], and Hu–Papageorgiou [173].

(6.8) The results on the properties of the various nonlinear differential operators are well-known for the case of the p-Laplacian. Here we prove it in a more general setting. Proposition 6.8.8 on the lifting of maximal monotone maps was proved by Aizicovici–Papageorgiou–Staicu [2] for $X = \mathbb{R}^N$. Here we extend it to reflexive Banach spaces. More on the convergence of sets can be found in Hu–Papageorgiou [173]. The Mosco convergence of sets was introduced by Mosco [245, 246]. The mix of the topologies turned out to be the right tool in order to prove the continuity of the operation of convex conjugation. The Picone identity for the p-Laplacian can be found in Allegretto–Huang [8]. For Hardy's inequality we refer to Harjulehto–Hästö–Koskenoja [153].

Problems

Problem 6.1. Let X be a Banach space and let $A: X \to X^*$ satisfy

$$\frac{\mathrm{d}}{\mathrm{d}t}\langle A(u_1 + t(u_2 - u_1)), u_1 - u_2\rangle|_{t=0} \geq 0 \quad \text{for all } u_1, u_2 \in X\,.$$

Show that A is monotone.

Problem 6.2. Let X be a finite dimensional Banach space and let $A: X \to 2^{X^*}$ be a monotone map with $D(A) = X$. Show that A is bounded; that is, it maps bounded sets to bounded sets.

Problem 6.3. Let X be a reflexive Banach space and let $A: X \to X^*$ be a monotone map such that $D(A) = X$, A has convex values and $\operatorname{Gr} A \subseteq X_w \times X^*$ is sequentially closed. Show that A is maximal monotone.

Problem 6.4. Show that the duality map of a Banach space X is linear if and only if X is a Hilbert space.

Problem 6.5. Let X be a reflexive Banach space and let $A: X \to X^*$ be a monotone, hemicontinuous map with a dense linear subspace $D(A) \subseteq X$. Show that for any $h^* \in X^*$ the operator equation $A(u) = h^*$ has a solution $u_0 \in D(A)$ if and only if

$$\langle A(u) - h^*, u - u_0 \rangle \geq 0 \quad \text{for all } u \in D(A)\,.$$

Problem 6.6. Let X be a reflexive Banach space and let $A: X \to X^*$ be a linear, demicontinuous, monotone map with $D(A) = X$. Show that $A \in L(X, X^*)$.

Problem 6.7. Let X be a reflexive Banach space, let $A: X \to 2^{X^*}$ be monotone and hemicontinuous with $D(A) = X$ and let $K \subseteq X$ be w-closed and bounded. Show that $A(K) \subseteq X^*$ is closed.

Problem 6.8. Let X be a reflexive Banach space and let $A: X \to X^*$ be uniformly monotone, hemicontinuous with $D(A) = X$. Show that A is surjective.

Problem 6.9. Let X be a reflexive Banach space and let $A: X \to 2^{X^*}$, $L: X \to X^*$ be two monotone maps with $D(L) = X$. Suppose that $u \to (A+L)(u)$ is maximal monotone. Show that A is maximal monotone.

Problem 6.10. Let $A: \mathbb{R}^N \to \mathbb{R}^N$ be a surjective, monotone map with $D(A) = \mathbb{R}^N$. Show that $\lim_{|u| \to \infty} |A(u)| = +\infty$.

Problem 6.11. Let X be a reflexive Banach space and let $A: X \to 2^{X^*}$ be a maximal monotone map. Define $m(u) = \inf[\|u^*\|_* : u^* \in A(u)]$ with the usual convention that $\inf \emptyset = +\infty$. Show that $m: X \to \overline{\mathbb{R}} = \mathbb{R} \cup \{+\infty\}$ is lower semicontinuous.

Problem 6.12. Let X be a reflexive Banach space, let $A: X \to 2^{X^*}$ be a monotone map, and let $C \subseteq X$ be a closed, convex set with $D(A) \subseteq C$. Show that there exists a maximal monotone map $\hat{A}: X \to 2^{X^*}$ such that $\operatorname{Gr} A \subseteq \operatorname{Gr} \hat{A}$ and $D(\hat{A}) \subseteq C$.

Problem 6.13. Let X be a reflexive Banach space with strictly convex dual X^*, let $\mathcal{F}: X \to X^*$ be the duality map, let $C \subseteq X$ be nonempty and convex, and let $u_0 \in C$. Show that $\|u_0\| = \inf[\|u\| : u \in C]$ if and only if $\langle \mathcal{F}(u_0), u_0 \rangle \leq \langle \mathcal{F}(u_0), u \rangle$ for all $u \in C$.

Problem 6.14. Let X be a reflexive Banach space and let $\varphi \in \Gamma_0(X)$. Show that $\partial \varphi$ is surjective if and only if $\varphi(u) - \langle u^*, u \rangle \to +\infty$ as $\|u\| \to \infty$ for all $u^* \in X^*$.

Problem 6.15. Let X be a Banach space and let $f: \mathbb{R} \to X$ be a map that is almost everywhere weakly differentiable and let $t \to \|f(t)\|$ be differentiable for a. a. $t \in \mathbb{R}$. Show that $\|f(t)\| \mathrm{d}/(\mathrm{d}t) \|f(t)\| = \langle u^*, f(t) \rangle$ for a. a. $t \in \mathbb{R}$ and for all $u^* \in \mathcal{F}(f(t))$.

Problem 6.16. Let X be a Banach space and $\mathcal{F}$ its duality map. Show that X is reflexive if and only if $R(\mathcal{F}) = X^*$.

Problem 6.17. Let X be a reflexive Banach space and let $A: X \to X^*$ be a demicontinuous, $(S)_+$-map with $D(A) = X$. Show that A is pseudomonotone.

Problem 6.18. Let H be a Hilbert space (we can assume that $H = H^*$) and let $A: H \to 2^H$ be a maximal monotone map. Show that J_λ is compact for every $\lambda > 0$ if and only if the sublevel set $L_\vartheta = \{u \in D(A): \|u\| + \|A(u)\| \leq \vartheta\}$ is relatively compact in H for every $\vartheta > 0$.

Problem 6.19. Show that if a Banach space is not reflexive, then its duality map is not surjective. Compare the result with Problem 6.16.

Problem 6.20. Let X be a Banach space and let $\mathcal{F}: X \to 2^{X^*} \setminus \{\emptyset\}$ be its duality map. Show that, for all $u, v \in X$, $\|u\| \leq \|u + \lambda v\|$ for all $\lambda > 0$ if and only if there exists $u^* \in \mathcal{F}(u)$ such that $\langle u^*, v\rangle \geq 0$.

Problem 6.21. Suppose that $\varphi \in C(\mathbb{R}^N, \mathbb{R}^N)$ is injective and $|\varphi(u)| \to +\infty$ as $|u| \to +\infty$. Show that φ is surjective.

Problem 6.22. Let $(\varphi, \Omega, h) \in \mathcal{L}$ and suppose that there exists $u \in \Omega$ such that $\varphi(u) = h$. Is it true that $d(\varphi, \Omega, h) \neq 0$? Justify your answer.

Problem 6.23. Let $\Omega \subseteq \mathbb{R}^N$ be bounded, open, $\varphi(u) = A(u) + u_0$ for all $u \in \mathbb{R}^N$ with $u_0 \in \mathbb{R}^N$, and let A be an $N \times N$-invertible matrix. Show that, for every $h \in \mathbb{R}^N \setminus \varphi(\partial\Omega)$, it holds that

$$d(\varphi, \Omega, h) = \begin{cases} \operatorname{sgn} \det A & \text{if } h \in \varphi(\Omega)\,, \\ 0 & \text{otherwise}\,. \end{cases}$$

Problem 6.24. Let $\Omega \subseteq \mathbb{R}^N$ be bounded open and $\varphi, \psi \in C(\overline{\Omega}, \mathbb{R}^N)$ satisfy $|\psi(u)| < |\varphi(u)|$ for all $u \in \partial\Omega$. Show that $d_B(\varphi + \psi, \Omega, 0) = d_B(\varphi, \Omega, 0)$.

Problem 6.25. Let $\Omega \subseteq \mathbb{R}^N$ be bounded, open, $\varphi \in C(\overline{\Omega}, \mathbb{R}^N)$, $\varphi(\overline{\Omega}) \subseteq \overline{\Omega}$, and $\varphi|_{\partial\Omega} = \mathrm{id}$. Show that $\varphi(\overline{\Omega}) = \overline{\Omega}$.

Problem 6.26. Suppose that $\varphi \in C^1(\mathbb{R}^N, \mathbb{R}^N)$, $J_\varphi(u) \neq 0$ for all $u \in \mathbb{R}^N$, and $|\varphi(u)| \to +\infty$ as $|u| \to \infty$. Show that φ is surjective.

Problem 6.27. Let $\Omega = (a, b)$ with $a < b$, $\varphi \in C([a, b])$, and $h \notin \{\varphi(a), \varphi(b)\}$. Show that $d(\varphi, \Omega, h) \in \{\pm 1, 0\}$.

Problem 6.28. Suppose that $N \in \mathbb{N}$ is odd and let $i_N: \partial B_1^{N+1} \to \partial B_1^{N+1}$ be the identity map. Show that i_N and $-i_N$ are homotopic.

Problem 6.29. Let X, Y be Banach spaces, let $\Omega \subseteq X$ be a bounded open set, let $A \in L(X, Y)$ be an isomorphism, and let $k: [0,1] \times \overline{\Omega} \to Y$ be a compact map such that $k(0, u) =$

0 for all $u \in \overline{\Omega}$. Assume that $A(u) - k(t, u) \neq h$ for all $t \in [0, 1]$ and for all $u \in \partial\Omega$. Show that there exists $u_0 \in \overline{\Omega}$ such that $A(u_0) - k(1, u_0) = h$.

Problem 6.30. Let X, Y be Banach spaces, $A \in L(X, Y)$ an isomorphism, and $k: [0, 1] \times X \to Y$ a compact map such that $k(0, u) = 0$ for all $u \in X$ and $h \in Y$. Assume that there exists $r > \|A^{-1}(h)\|_X$ such that $A(u) - k(t, u) = h$ implies $\|u\| \leq r$ for all $(t, u) \in [0, 1] \times X$. Show that there exists $u_0 \in \overline{B}_r$ satisfying $A(u_0) - k(1, u_0) = h$.

Problem 6.31. Let X, Y be Banach spaces, $A \in L(X, Y)$ an isomorphism, and $f: X \to Y$ a compact map such that $1/\|x\|f(x) \to 0$ in Y as $\|x\|_X \to \infty$. Show that for every $h \in X$, there exists $u_0 \in X$ such that $A(u_0) - f(u_0) = h$.

Problem 6.32. Let X be a Banach space, $\Omega \subseteq X$ be bounded open, $f: \overline{\Omega} \to X$ be a compact map such that $(\mathrm{id} - f)(\overline{\Omega}) \subseteq \overline{\Omega}$, and $f|_{\partial\Omega} = 0$. Show that $(\mathrm{id} - f)(\overline{\Omega}) = \overline{\Omega}$.

Problem 6.33. Suppose that X is a Banach space and $(\varphi, \Omega, 0), (\psi, \Omega, 0) \in \hat{\mathcal{L}}$ (see Theorem 6.3.15) satisfy $d_{\mathrm{LS}}(\varphi, \Omega, 0) \neq d_{\mathrm{LS}}(\psi, \Omega, 0)$. Show that there exist $\eta < 0$ and $\hat{u} \in \partial\Omega$ such that $\varphi(\hat{u}) = \eta\psi(\hat{u})$.

Problem 6.34. Let $m, n \in \mathbb{N}$ with $m < n$. Show that there is no continuous injection $\varphi: \mathbb{R}^n \to \mathbb{R}^m$.

Problem 6.35. Let $m, n \in \mathbb{N}$ with $n < m$ and assume that $\varphi: \mathbb{R}^n \to \mathbb{R}^m$ is a continuous injection. Show that the set $\mathbb{R}^m \setminus \varphi(\mathbb{R}^n)$ is dense in $\mathbb{R}^m$.

Problem 6.36. Suppose that X is a Banach space, $\varphi: X \to X$ is a k-contraction, and $f = \mathrm{id} - \varphi$. Show that f is a homeomorphism of X onto itself.

Problem 6.37. Suppose that X is a uniformly convex Banach space, $C \subseteq X$ is nonempty, bounded, closed, convex, and $\varphi: C \to C$ is nonexpansive. Show that the set $\mathrm{Fix}(\varphi) = \{u \in C: \varphi(u) = u\}$ is nonempty, w-compact, and convex.

Problem 6.38. Let (X, d) be a complete metric space and let $F: X \to P_{\mathrm{bf}}(X)$ be a multifunction such that

$$h(F(u), F(v)) \leq k d(u, v) \quad \text{for all } u, v \in X \text{ with } k \in [0, 1) .$$

Show that F has a fixed point, that is, there exists $\hat{u} \in X$ such that $\hat{u} \in F(\hat{u})$, which need not be unique.

Problem 6.39. Let X be a Banach space, $\Omega \subseteq X$ be open, $0 \in \Omega$, and $\varphi: \overline{\Omega} \to X$ be compact, satisfying

$$\|\varphi(u)\|^2 \leq \|u\|^2 + \|u - \varphi(u)\|^2 \quad \text{for all } u \in \partial\Omega .$$

Show that φ has a fixed point.

Problem 6.40. Let X be a Banach space, $\Omega \subseteq X$ be bounded open, and $\varphi \in K(\overline{\Omega}, X)$, while there exists $u_0 \in \Omega$ such that $\varphi(u) - u_0 \neq \lambda(u - u_0)$ for all $u \in \partial\Omega$ and for all $\lambda > 1$. Show that φ has a fixed point.

Problem 6.41. Suppose that X is an AR (see Definition 1.7.26), and let $\varphi: X \to X$ be a continuous map such that $\varphi(X)$ is relatively compact. Show that φ has a fixed point.

Problem 6.42. Let (X, d) be a complete metric space and $\varphi: X \to X$. Assume that there exists a right continuous function $\eta: \mathbb{R}_+ \to \mathbb{R}_+$ such that $\eta(r) < r$ for all $r > 0$ and $d(\varphi(u), \varphi(v)) \leq \eta d(u, v)$ for all $u, v \in X$. Show that φ has a unique fixed point.

Problem 6.43. Suppose X is a Banach space, $C \subseteq X$ is a nonempty, closed, convex set, and $F: C \to P_{fc}(C)$ is a lsc multifunction that maps bounded sets to relatively compact ones. Show that there exists $\hat{u} \in C$ such that $\hat{u} \in F(\hat{u})$.

Problem 6.44. Let X be a Banach space and let $\varphi: X \to \mathbb{R}$ be a lower semicontinuous and Gateaux differentiable function that satisfies $\varphi(u) \geq a\|u\| - c$ for all $u \in X$ and for some $a, c > 0$. Show that $\varphi'(X)$ is dense in $a\overline{B}_1^*$, where $\overline{B}_1^* = \{u^* \in X^*: \|u^*\|_* \leq 1\}$.

Problem 6.45. Let X be a Banach space, $\varphi: X \to \mathbb{R}$ is a lower semicontinuous and Gateaux differentiable function and there exists a continuous map $\eta: \mathbb{R}_+ \to \mathbb{R}$ such that $\eta(t)/t \to +\infty$ as $t \to +\infty$ and $\varphi(u) \geq \eta(\|u\|)$ for all $u \in X$. Show that $\varphi'(x)$ is dense in X^*.

Problem 6.46. Let $X = c_0$ and let $\overline{B}_1 = \{u \in c_0: \|u\| \leq 1\}$. Produce an isometry $\varphi: \overline{B}_1 \to \overline{B}_1$ that is fixed point free.

Problem 6.47. Let E be a compact metric space and let $\varphi: E \to E$ be a continuous map such that there exists an AR X and continuous maps $f: E \to X$, $g: X \to E$ with $\varphi = g \circ f$. Show that φ has a fixed point.

Problem 6.48. Let X be an OBS with order cone K, $\Omega \subseteq X$ is a bounded, open set, and let $\varphi: \overline{\Omega} \cap K \to X$ be compact such that

(i) $\varphi(u) \neq \lambda u$ for all $\lambda \in [0, 1]$ and for all $u \in \partial\Omega \cap K = \partial(\Omega \cap K)$;
(ii) the set

$$\left\{ \frac{\varphi(u)}{\|\varphi(u)\|} : u \in \partial\Omega \cap K \right\} \subseteq X$$

is relatively compact.

Show that $i(\varphi, \Omega \cap K, K) = 0$.

Problem 6.49. Let X be a reflexive Banach space, $C \subseteq X$ a nonempty, closed convex set and $\varphi: C \to C$ a sequentially weakly continuous map such that $\varphi(C)$ is bounded. Show that φ has a fixed point.

Problem 6.50. Let $X = \mathbb{R}$ and let $C_n = [0, 1/n] \cup [n, +\infty)$. Is $\{C_n\}_{n\geq 1}$ K-convergent? Justify your answer.

Bibliography

[1] R. A. Adams. *Sobolev Spaces*. Academic Press, New York-London, 1975.

[2] S. Aizicovici, N. S. Papageorgiou and V. Staicu. Nonlinear periodic systems with unilateral constraints. *Topol. Methods Nonlinear Anal.*, 54(2B):871–885, 2019.

[3] L. Alaoglu. Weak topologies of normed linear spaces. *Ann. Math. (2)*, 41:252–267, 1940.

[4] P. Alexandrov. Über die Metrisation der im Kleinen kompakten topologischen Räume. *Math. Ann.*, 92(3–4):294–301, 1924.

[5] P. Alexandrov. Über stetige Abbildungen kompakter Räume. *Math. Ann.*, 96(1):555–571, 1927.

[6] P. Alexandrov and P. Urysohn. Zur Theorie der topologischen Räume. *Math. Ann.*, 92(3–4):258–266, 1924.

[7] C. D. Aliprantis and K. C. Border. *Infinite Dimensional Analysis*. Springer, Berlin, third edition, 2006.

[8] W. Allegretto and Y. X. Huang. A Picone's identity for the p-Laplacian and applications. *Nonlinear Anal.*, 32(7):819–830, 1998.

[9] H. Amann. Fixed point equations and nonlinear eigenvalue problems in ordered Banach spaces. *SIAM Rev.*, 18(4):620–709, 1976.

[10] H. Amann and J. Escher. *Analysis. I*. Birkhäuser Verlag, Basel, 2005.

[11] H. Amann and S. A. Weiss. On the uniqueness of the topological degree. *Math. Z.*, 130:39–54, 1973.

[12] L. Ambrosio and P. Tilli. *Topics on Analysis in Metric Spaces*. Oxford University Press, Oxford, 2004.

[13] R. F. Arens. A topology for spaces of transformations. *Ann. Math. (2)*, 47:480–495, 1946.

[14] C. Arzelà. Un' osservazione intorno alle serie di funzioni. *Rom. Acc. L. Mem. (3)*, 18:142–159, 1882–1883.

[15] G. Ascoli. Le curve limite di una varietà data di curve. *Rom. Acc. L. Mem. (3)*, 18:521–586, 1884.

[16] R. B. Ash. *Real Analysis and Probability*. Academic Press, New York-London, 1972.

[17] H. Attouch. *Variational Convergence for Functions and Operators*. Pitman, Boston, 1984.

[18] H. Attouch, G. Buttazzo and G. Michaille. *Variational Analysis in Sobolev and BV-Spaces, MPS/SIAM Series on Optimization*. SIAM, Philadelphia, 2006.

[19] J.-P. Aubin. Un théorème de compacité. *C. R. Acad. Sci. Paris*, 256:5042–5044, 1963.

[20] J.-P. Aubin and H. Frankowska. *Set-Valued Analysis*. Birkhäuser, Boston, 1990.

[21] R. J. Aumann. Measurable utility and the measurable choice theorem. In *La Décision, 2: Agrégation et Dynamique des Ordres de Préférence (Actes Colloq. Internat., Aix-en-Provence, 1967)*, pages 15–26. Éditions du Centre Nat. Recherche Sci, Paris, 1969.

[22] W. G. Bade. Weak and strong limits of spectral operators. *Pac. J. Math.*, 4:393–413, 1954.

[23] R. Baire. Sur les fonctions de variables réelles. *Ann. Mat. Pura Appl.*, 3(1):1–123, 1899.

[24] E. J. Balder. *Lectures on Young Measure*. Cahiers de Mathematiques, CEREMADE, Université de Paris-Dauphine, 1995.

[25] E. J. Balder. Lectures on young measure theory and its applications in economics. *Rend. Ist. Mat. Univ. Trieste*, 31(suppl. 1):1–69, 2000.

[26] S. Banach. Sur les opérations dans les ensembles abstraits et leur application aux équations intégrales. *Fundam. Math.*, 3:133–181, 1922.

[27] S. Banach. Sur les fonctionnelles linéaires. *Stud. Math.*, 1:211–216, 1929.

[28] S. Banach. *Theory of Linear Operations*. North-Holland, Amsterdam, 1987.

[29] S. Banach and H. Steinhaus. Sur le principe de la condensation de singularités. *Fundam. Math.*, 9:50–61, 1927.

[30] V. Barbu. *Nonlinear Semigroups and Differential Equations in Banach Spaces*. Noordhoff International Publishing, Leiden, 1976.

[31] V. Barbu and Th. Precupanu. *Convexity and Optimization in Banach Spaces*. D. Reidel Publishing Co., Dordrecht, The Netherlands, second edition, 1986.

[32] B. Beauzamy. *Introduction to Banach Spaces and their Geometry*, volume 68 of *North-Holland Mathematics Studies*. North-Holland, Amsterdam-New York, 1982.

https://doi.org/10.1515/9783111286952-007

[33] Y. Benyamini and J. Lindenstrauss. *Geometric Nonlinear Functional Analysis. Vol. 1*, volume 48 of *American Mathematical Society Colloquium Publications*. American Mathematical Society, Providence, RI, 2000.
[34] D. P. Bertsekas and S. E. Shreve. *Stochastic Optimal Control: The Discrete Time Case*. Academic Press, New York-London, 1978.
[35] A. S. Besicovitch. A general form of the covering principle and relative differentiation of additive functions. *Proc. Camb. Philos. Soc.*, 41:103–110, 1945.
[36] A. Beurling and A. E. Livingston. A theorem on duality mappings in Banach spaces. *Ark. Mat.*, 4:405–411, 1962, (1962).
[37] E. Bishop and R. R. Phelps. The support functionals of a convex set. In *Proc. Sympos. Pure Math., Vol. VII*, pages 27–35. Amer. Math. Soc., Providence, R. I., 1963.
[38] S. Bocher. Integration von Funktionen, deren Werte die Elemente eines Vektorraumes sind. *Fundam. Math.*, 20:262–276, 1933.
[39] S. Bochner and A. E. Taylor. Linear functionals on certain spaces of abstractly-valued functions. *Ann. Math. (2)*, 39(4):913–944, 1938.
[40] V. I. Bogachev. *Measure Theory. Vol. I*. Springer-Verlag, Berlin, 2007.
[41] V. I. Bogachev. *Measure Theory. Vol. II*. Springer-Verlag, Berlin, 2007.
[42] H. F. Bohnenblust and A. Sobczyk. Extensions of functionals on complex linear spaces. *Bull. Am. Math. Soc.*, 44(2):91–93, 1938.
[43] E. Borel. *Leçons sur la Théorie des Fonctions*. Gauthier-Villars, Paris, 1898.
[44] K. Borsuk. *Theory of Retracts*. Państwowe Wydawnictwo Naukowe, Warsaw, 1967.
[45] J. M. Borwein and J. D. Vanderwerff. *Convex Functions: Constructions, Characterizations and Counterexamples*. Cambridge University Press, Cambridge, 2010.
[46] N. Bourbaki. *General Topology. Part 1*. Addison-Wesley, Reading, MA, 1966.
[47] N. Bourbaki. *Integration. Chap. IX: Integration sur les espaces topologiques séparés. Livre VI*. Hermann, Paris, 1969.
[48] N. Bourbaki. *Integration. I. Chapters 1–6*. Springer-Verlag, Berlin, 2004.
[49] N. Bourbaki. *Topologie Generale: Chapitres 5 à 10*. Springer-Verlag, Berlin, 2007.
[50] H. Brézis. Équations et inéquations non linéaires dans les espaces vectoriels en dualité. *Ann. Inst. Fourier (Grenoble)*, 18(fasc. 1):115–175, 1968.
[51] H. Brézis. *Opérateurs Maximaux Monotones et Semi-Groupes de Contractions dans les Espaces de Hilbert*, volume 5 of *North-Holland Mathematics Studies*. North-Holland Publishing Co., Amsterdam-London, 1973; American Elsevier Publishing Co., Inc., New York, Notas de Matemática (50).
[52] H. Brézis. *Functional Analysis, Sobolev Spaces and Partial Differential Equations*. Springer, New York, 2011.
[53] H. Brézis and F. E. Browder. A general principle on ordered sets in nonlinear functional analysis. *Adv. Math.*, 21(3):355–364, 1976.
[54] H. Brézis and E. Lieb. A relation between pointwise convergence of functions and convergence of functionals. *Proc. Am. Math. Soc.*, 88(3):486–490, 1983.
[55] M. S. Brodskiĭ and D. P. Mil'man. On the center of a convex set. *Doklady Akad. Nauk SSSR (N.S.)*, 59:837–840, 1948.
[56] A. Brøndsted. Conjugate convex functions in topological vector spaces. *Mat.-Fys. Medd. Danske Vid. Selsk.*, 34(2):27, 1964, pp. 1–28.
[57] J. K. Brooks and R. V. Chacon. Continuity and compactness of measures. *Adv. Math.*, 37(1):16–26, 1980.
[58] L. Brouwer. On continuous one-to-one transformations of surfaces into themselves. *Proc. K. Ned. Akad. V. Wet. Ser. A*, 11:788–798, 1909, 12:286–297, 1910, 13:767–777, 1911, 14:300–310, 1912, 15:352–360, 1913.
[59] F. E. Browder. Nonexpansive nonlinear operators in a Banach space. *Proc. Natl. Acad. Sci. USA*, 54:1041–1044, 1965.
[60] F. E. Browder. Nonlinear maximal monotone operators in Banach space. *Math. Ann.*, 175:89–113, 1968.

[61] F. E. Browder. *Nonlinear Operators and Nonlinear Equations of Evolution in Banach Spaces*, pages 1–308, 1976.
[62] F. E. Browder and P. Hess. Nonlinear mappings of monotone type in Banach spaces. *J. Funct. Anal.*, 11:251–294, 1972.
[63] L. D. Brown and R. Purves. Measurable selections of extrema. *Ann. Stat.*, 1:902–912, 1973.
[64] T. Bühler and A. Salamon. *Functional Analysis*. American Mathematical Society, Providence, RI, 2018.
[65] G. Buttazzo. *Semicontinuity, Relaxation and Integral Representation in the Calculus of Variations*. Longman Scientific & Technical, Harlow, Essex, UK, 1989.
[66] P. L. Butzer and H. Berens. *Semi-Groups of Operators and Approximation*. Springer-Verlag, New York, 1967.
[67] G. Cantor. De la puissance des ensembles parfaits de points. *Acta Math.*, 4(1):381–392, 1884.
[68] C. Carathéodory. Über das lineare Maß von Punktmengen – eine Verallgemeinerung des Längenbegriffs. *Gött. Nachr.*, 1914:404–426, 1914.
[69] C. Carathéodory. *Vorlesungen über reelle Funktionen*. Teubner, Leipzig, second edition, 1927.
[70] J. Caristi. Fixed point theorems for mappings satisfying inwardness conditions. *Trans. Am. Math. Soc.*, 215:241–251, 1976.
[71] C. Castaing and M. Valadier. *Convex Analysis and Measurable Multifunctions*, volume 580 of *Lecture Notes in Mathematics*. Springer-Verlag, Berlin-New York, 1977.
[72] G. Choquet. *Topology*. Academic Press, New York-London, 1966.
[73] J. P. R. Christensen. *Topology and Borel Structure*. North-Holland Publishing Co., Amsterdam-London, 1974.
[74] F. H. Clarke. *Optimization and Nonsmooth Analysis*. John Wiley & Sons, Inc., New York, 1983.
[75] J. A. Clarkson. Uniformly convex spaces. *Trans. Am. Math. Soc.*, 40(3):396–414, 1936.
[76] D. L. Cohn. *Measure Theory*. Birkhäuser, Boston, Mass, 1980.
[77] M. G. Crandall and T. M. Liggett. Generation of semi-groups of nonlinear transformations on general Banach spaces. *Am. J. Math.*, 93:265–298, 1971.
[78] Á. Crespo-Blanco, L. Gasiński, P. Harjulehto and P. Winkert. A new class of double phase variable exponent problems: existence and uniqueness. *J. Differ. Equ.*, 323:182–228, 2022.
[79] D. V. Cruz-Uribe and A. Fiorenza. *Variable Lebesgue Spaces*. Birkhäuser/Springer, Heidelberg, 2013.
[80] G. Dal Maso. *An Introduction to Γ-Convergence*. Birkhäuser, Boston, 1993.
[81] P. J. Daniell. A general form of integral. *Ann. Math. (2)*, 19(4):279–294, 1918.
[82] P. J. Daniell. Stieltjes derivatives. *Bull. Am. Math. Soc.*, 26(10):444–448, 1920.
[83] M. M. Day. *Normed Linear Spaces*. Springer-Verlag, New York-Heidelberg, third edition, 1973.
[84] E. De Giorgi and T. Franzoni. Su un tipo di convergenza variazionale. *Atti Accad. Naz. Lincei, Rend. Cl. Sci. Fis. Mat. Nat. (8)*, 58(6):842–850, 1975.
[85] K. Deimling. *Nonlinear Functional Analysis*. Springer-Verlag, Berlin, 1985.
[86] C. Dellacherie and P.-A. Meyer. *Probabilities and Potential*. North-Holland Publishing Co., Amsterdam, 1978.
[87] Z. Denkowski, S. Migórski and N. S. Papageorgiou. *An Introduction to Nonlinear Analysis: Theory*. Kluwer Academic Publishers, Boston, 2003.
[88] R. Deville, G. Godefroy and V. Zizler. *Smoothness and Renormings in Banach Spaces*, volume 64 of *Pitman Monographs*. Pitman, New York, 1993.
[89] L. Diening, P. Harjulehto, P. Hästö and M. Růžička. *Lebesgue and Sobolev Spaces with Variable Exponents*. Springer, Heidelberg, 2011.
[90] J. Diestel. *Sequences and Series in Banach Spaces*, volume 92 of *Graduate Texts in Mathematics*. Springer-Verlag, New York, 1984.
[91] J. Diestel and J. J. Uhl Jr. *Vector Measures*. American Mathematical Society, Providence, R. I., 1977.
[92] J. Dieudonné. Une généralisation des espaces compacts. *J. Math. Pures Appl. (9)*, 23:65–76, 1944.
[93] J. Dieudonné. Sur le théorème de Lebesgue-Nikodym. III. *Ann. Univ. Grenoble, Sect. Sci. Math. Phys. (N.S.)*, 23:25–53, 1948.

[94] J. Dieudonné. Natural homomorphisms in Banach spaces. *Proc. Am. Math. Soc.*, 1:54–59, 1950.

[95] J. Dieudonné and L. Schwartz. La dualité dans les espaces $\mathscr{F}$ et $(\mathscr{L}\mathscr{F})$. *Ann. Inst. Fourier (Grenoble)*, 1:61–101, 1949, (1950).

[96] N. Dinculeanu. *Vector Measures*. Pergamon Press, New York, 1967.

[97] N. Dinculeanu and C. Foiaş. Sur la représentation intégrale des certaines opérations linéaires. IV. Opérations linéaires sur l'espace L^p_a. *Can. J. Math.*, 13:529–556, 1961.

[98] A. L. Dontchev and T. Zolezzi. *Well-Posed Optimization Problems*, volume 1543 of *Lecture Notes in Mathematics*. Springer-Verlag, Berlin, 1993.

[99] R. M. Dudley. Convergence of Baire measures. *Stud. Math.*, 27:251–268, 1966.

[100] R. M. Dudley. On measurability over product spaces. *Bull. Am. Math. Soc.*, 77:271–274, 1971.

[101] R. M. Dudley. *Real Analysis and Probability*. Wadsworth & Brooks/Cole Advanced Books & Software, Pacific Grove, CA, 1989.

[102] J. Dugundji. *Topology*. Allyn and Bacon, Inc., Boston, Mass., 1966.

[103] N. Dunford. Uniformity in linear spaces. *Trans. Am. Math. Soc.*, 44(2):305–356, 1938.

[104] N. Dunford and B. J. Pettis. Linear operations on summable functions. *Trans. Am. Math. Soc.*, 47:323–392, 1940.

[105] N. Dunford and J. T. Schwartz. *Linear Operators. I. General Theory*. Interscience Publishers, New York, 1958.

[106] W. F. Eberlein. Weak compactness in Banach spaces. I. *Proc. Natl. Acad. Sci. USA*, 33:51–53, 1947.

[107] M. Edelheit. Zur Theorie der konvexen Mengen in linearen normierten Räumen. *Stud. Math.*, 6:104–111, 1936.

[108] M. Edelstein. On fixed and periodic points under contractive mappings. *J. Lond. Math. Soc.*, 37:74–79, 1962.

[109] D. F. Egorov. Sur les suites de fonctions mesurables. *C. R. Math. Acad. Sci. Paris*, 152(10):244–246, 1911.

[110] G. Ehrling. On a type of eigenvalue problems for certain elliptic differential operators. *Math. Scand.*, 2:267–285, 1954.

[111] I. Ekeland. On the variational principle. *J. Math. Anal. Appl.*, 47:324–353, 1974.

[112] I. Ekeland. Nonconvex minimization problems. *Bull. Am. Math. Soc. (N.S.)*, 1(3):443–474, 1979.

[113] I. Ekeland. The ϵ-variational principle revisited. In *Methods of Nonconvex Analysis (Varenna, 1989)*, volume 1446 of *Lecture Notes in Math.*, pages 1–15. Springer, Berlin, 1990. With notes by S. Terracini.

[114] I. Ekeland and R. Temam. *Convex Analysis and Variational Problems*. North-Holland Publishing Co., Amsterdam, 1976.

[115] P. Enflo. A counterexample to the approximation problem in Banach spaces. *Acta Math.*, 130:309–317, 1973.

[116] L. C. Evans and R. F. Gariepy. *Measure Theory and Fine Properties of Functions*. CRC Press, Boca Raton, FL, 1992.

[117] M. Fabian, P. Habala, P. Hájek, V. Montesinos Santalucía, J. Pelant and V. Zizler. *Functional Analysis and Infinite-Dimensional Geometry*, volume 8. Springer-Verlag, New York, 2001.

[118] K. Fan. Entfernung zweier zufälligen Grössen und die Konvergenz nach Wahrscheinlichkeit. *Math. Z.*, 49:681–683, 1944.

[119] P. Fatou. Séries trigonométriques et séries de Taylor. *Acta Math.*, 30(1):335–400, 1906.

[120] H. Federer. *Geometric Measure Theory*. Springer-Verlag, New York, 1969.

[121] W. Fenchel. *Convex Cones, Sets and Functions*. Princeton Univ. Press, Princeton, 1953.

[122] E. Fischer. Sur la convergence en moyenne. *C. R. Math. Acad. Sci. Paris*, 144:1022–1024, 1907.

[123] L. C. Florescu and C. Godet-Thobie. *Young Measures and Compactness in Measure Spaces*. De Gruyter, Berlin, 2012.

[124] K. Floret. *Weakly Compact Sets*, volume 801 of *Lecture Notes in Mathematics*. Springer, Berlin, 1980.

[125] G. B. Folland. *Real Analysis*. John Wiley & Sons, Inc., New York, second edition, 1999.

[126] I. Fonseca and W. Gangbo. *Degree Theory in Analysis and Applications*. Clarendon University Press, New York, 1995.

[127] R. H. Fox. On topologies for function spaces. *Bull. Am. Math. Soc.*, 51:429–432, 1945.
[128] M. Fréchet. *Sur Quelques Points de Calcul Fonctionnel*. Thèse–Faculté des sciences de Paris, 1906.
[129] M. Fréchet. Sur les ensembles de fonctions et les operations lineaires. *C. R. Math. Acad. Sci. Paris*, 144:1414–1416, 1907.
[130] M. Fréchet. Sur divers modes de convergence d'une suite de fonctions d'une variable. *Bull. Calcutta Math. Soc.*, 11:187–206, 1921.
[131] M. Fréchet. *Les Espaces Abstraits*. Grautier-Villars, Paris, 1928.
[132] A. Fryszkowski. *Fixed Point Theory for Decomposable Sets*. Kluwer Academic Publishers, Dordrecht, 2004.
[133] G. Fubini. Sugli integrali multipli. *Rend. Accad. Naz. Lincei*, 16:608–614, 1907.
[134] L. Führer. Ein elementarer analytischer Beweis zur Eindeutigkeit des Abbildungsgrades im $\mathbb{R}^n$. *Math. Nachr.*, 54:259–267, 1972.
[135] L. Gasiński and N. S. Papageorgiou. *Nonlinear Analysis*. Chapman & Hall/CRC, Boca Raton, FL, 2006.
[136] L. Gasiński and N. S. Papageorgiou. *Exercises in Analysis. Part 2. Nonlinear Analysis*. Springer, Cham, 2016.
[137] J. R. Giles. *Convex Analysis with Application in the Differentiation of Convex Functions*. Pitman Publishing Inc., Boston, 1982.
[138] J. R. Giles. *Introduction to the Analysis of Normed Linear Spaces*. Cambridge University Press, Cambridge, 2000.
[139] D. C. Gillespie and W. A. Hurwitz. On sequences of continuous functions having continuous limits. *Trans. Am. Math. Soc.*, 32(3):527–543, 1930.
[140] K. Goebel and W. A. Kirk. *Topics in Metric Fixed Point Theory*. Cambridge University Press, Cambridge, 1990.
[141] D. Göhde. Zum Prinzip der kontraktiven Abbildung. *Math. Nachr.*, 30:251–258, 1965.
[142] S. Goldberg. *Unbounded Linear Operators: Theory and Applications*. McGraw-Hill, New York, 1966.
[143] H. H. Goldstine. Weakly complete Banach spaces. *Duke Math. J.*, 4(1):125–131, 1938.
[144] A. Granas and J. Dugundji. *Fixed Point Theory*. Springer-Verlag, New York, 2003.
[145] D. J. Guo and V. Lakshmikantham. *Nonlinear Problems in Abstract Cones*. Academic Press, Boston, 1988.
[146] H. Hahn. Über die allgemeinste ebene Punktmenge, die stetiges Bild einer Strecke ist. *Jahresber. Dtsch. Math.-Ver.*, 23:318–322, 1914.
[147] H. Hahn. *Theorie der reellen Funktionen*. Springer-Verlag, Berlin Heidelberg, 1921.
[148] H. Hahn. Über Folgen linearer Operationen. *Monatshefte Math. Phys.*, 32(1):3–88, 1922.
[149] H. Hahn. Über lineare Gleichungssysteme in linearen Räumen. *J. Reine Angew. Math.*, 157:214–229, 1927.
[150] P. Hajłasz. Pointwise Hardy inequalities. *Proc. Am. Math. Soc.*, 127(2):417–423, 1999.
[151] O. Halmos. *Measure Theory*, volume 18 of *Graduate Texts in Math.* Springer-Verlag, New York, 1974.
[152] P. Harjulehto and P. Hästö. *Orlicz Spaces and Generalized Orlicz Spaces*. Springer, Cham, 2019.
[153] P. Harjulehto, P. Hästö and M. Koskenoja. Hardy's inequality in a variable exponent Sobolev space. *Georgian Math. J.*, 12(3):431–442, 2005.
[154] F. Hausdorff. Dimension und äußeres Maß. *Math. Ann.*, 79(1–2):157–179, 1918.
[155] F. Hausdorff. *Grundzüge der Mengenlehre*. Chelsea Publishing Company, New York, 1949.
[156] T. Hawkins. *Lebesgue's Theory of Integration: Its Origins and Development*. Chelsea Publishing Co., New York, 1975.
[157] E. Heinz. An elementary analytic theory of the degree of mapping in n-dimensional space. *J. Math. Mech.*, 8:231–247, 1959.
[158] E. Helly. Über lineare Funktionaloperatoren. *Sitzungsber. Nat. Kais. Akad. Wiss.*, 265–297, 1911/1912.
[159] H. Hermes and J. P. LaSalle. *Functional Analysis and Time Optimal Control*. Academic Press, New York-London, 1969.

[160] E. Hewitt and K. Stromberg. *Real and Abstract Analysis*, volume 18 *Graduate Texts in Math.* Springer-Verlag, New York, 1975.

[161] F. Hiai and H. Umegaki. Integrals, conditional expectations, and martingales of multivalued functions. *J. Multivar. Anal.*, 7(1):149–182, 1977.

[162] D. Hilbert. *Grundzüge einer allgemeinen Theorie der linearen Integralgleichungen*. Chelsea Publishing Company, New York, 1953.

[163] T. H. Hildebrandt. On uniform limitedness of sets of functional operations. *Bull. Am. Math. Soc.*, 29(7):309–315, 1923.

[164] E. Hille. *Functional Analysis and Semi-Groups*. American Mathematical Society, New York, 1948.

[165] E. Hille and R. S. Phillips. *Functional Analysis and Semigroups*. American Mathematical Society, Providence, RI, 1957.

[166] J.-B. Hiriart-Urruty. ε-subdifferential calculus. In *Convex Analysis and Optimization (London, 1980)*, volume 57 of *Res. Notes in Math.*, pages 43–92. Pitman, Boston, Mass.-London, 1982.

[167] J.-B. Hiriart-Urruty and C. Lemaréchal. *Convex Analysis and Minimization Algorithms. I.* Springer-Verlag, Berlin, 1993.

[168] J.-B. Hiriart-Urruty and C. Lemaréchal. *Convex Analysis and Minimization Algorithms. II.* Springer-Verlag, Berlin, 1993.

[169] J.-B. Hiriart-Urruty and R. R. Phelps. Subdifferential calculus using ϵ-subdifferentials. *J. Funct. Anal.*, 118(1):154–166, 1993.

[170] O. Hölder. Über die allgemeinste ebene Punktmenge, die stetiges Bild einer Strecke ist. *Nachr. Akad. Wiss. Gött. Math.-Phys. Kl.*, 38–47, 1889.

[171] R. B. Holmes. *Geometric Functional Analysis and its Applications*, volume 24 of *Graduate Texts in Mathematics*. Springer-Verlag, New York-Heidelberg, 1975.

[172] L. Hörmander. Sur la fonction d'appui des ensembles convexes dans un espace localement convexe. *Ark. Mat.*, 3:181–186, 1955.

[173] S. Hu and N. S. Papageorgiou. *Handbook of Multivalued Analysis. Vol. I.* Kluwer Academic Publishers, Dordrecht, 1997.

[174] S. Hu and N. S. Papageorgiou. *Handbook of Multivalued Analysis. Vol. II.* Kluwer Academic Publishers, Dordrecht, 2000.

[175] S.-T. Hu. *Theory of Retracts*. Wayne State University Press, Detroit, 1965.

[176] A. D. Ioffe. On lower semicontinuity of integral functionals. I, II. *SIAM J. Control Optim.*, 15:521–538 and 991–1000, 1977.

[177] A. D. Ioffe and V. M. Tichomirov. *Theory of Extremal Problems*. North-Holland Publishing Co., Amsterdam-New York, 1979.

[178] A. Ionescu Tulcea and C. Ionescu Tulcea. *Topics in the Theory of Lifting*. Springer-Verlag New York Inc., New York, 1969.

[179] R. C. James. Characterizations of reflexivity. *Stud. Math.*, 23:205–216, 1963/1964.

[180] R. C. James. Weakly compact sets. *Trans. Am. Math. Soc.*, 113:129–140, 1964.

[181] R. C. James. Reflexivity and the sup of linear functionals. *Isr. J. Math.*, 13:289–300, 1972, (1973).

[182] J. L. W. V. Jensen. Sur les fonctions convexes et les inégalités entre les valeurs moyennes. *Acta Math.*, 30(1):175–193, 1906.

[183] C. Jordan. Sur la série de Fourier. *C. R. Math. Acad. Sci. Paris*, 92:228–230, 1881.

[184] C. Jordan. Remarques sur les intégrales définies. 8:69–100, 1892.

[185] S. Kakutani. A generalization of Brouwer's fixed point theorem. *Duke Math. J.*, 8:457–459, 1941.

[186] T. Kato. *Perturbation Theory for Linear Operators*. Springer-Verlag, Berlin-New York, second edition, 1976.

[187] J. L. Kelley. Convergence in topology. *Duke Math. J.*, 17:277–283, 1950.

[188] J. L. Kelley. *General Topology*, volume 27 of *Graduate Texts in Mathematics*. Springer-Verlag, New York-Berlin, 1975. Reprint of the 1955 edition [Van Nostrand, Toronto, Ont.]

[189] P. S. Kenderov. The set-valued monotone mappings are almost everywhere single-valued. *C. R. Acad. Bulgare Sci.*, 27:1173–1175, 1974.

[190] D. Kinderlehrer and G. Stampacchia. *An Introduction to Variational Inequalities and their Applications*. Academic Press, New York, 1980.

[191] W. A. Kirk. A fixed point theorem for mappings which do not increase distances. *Am. Math. Mon.*, 72:1004–1006, 1965.

[192] V. L. Klee Jr. Convex sets in linear spaces. *Duke Math. J.*, 18:443–466, 1951.

[193] H.-A. Klei. A compactness criterion in $L^1(E)$ and Radon-Nikodým theorems for multimeasures. *Bull. Sci. Math. (2)*, 112(3):305–324, 1988.

[194] E. Klein and A. C. Thompson. *Theory of Correspondences*. John Wiley & Sons, Inc., New York, 1984.

[195] A. Kolmogorov. Zur Normierbarkeit eines allgemeinen topologischen Raumes. *Stud. Math.*, 5:29–33, 1935.

[196] A. Kufner, O. John and S. Fučík. *Function Spaces*. Noordhoff International Publishing, Leyden; Academia, Prague, 1977.

[197] C. Kuratowski. Les fonctions semi-continues dans l'espace des ensembles fermés. *Fundam. Math.*, 1:148–159, 1932.

[198] C. Kuratowski and W. Sierpiński. Le théorème de Borel-Lebesgue dans la théorie des ensembles abstraits. 2:172–178, 1921.

[199] K. Kuratowski. *Topology. Vol. I*. Academic Press, New York, 1966.

[200] K. Kuratowski. *Topology. Vol. II*. Academic Press, New York, 1968.

[201] K. Kuratowski and C. Ryll-Nardzewski. A general theorem on selectors. *Bull. Acad. Pol. Sci., Sér. Sci. Math. Astron. Phys.*, 13:397–403, 1965.

[202] P.-J. Laurent. *Approximation et Optimisation*. Hermann, Paris, 1972.

[203] P. D. Lax and A. N. Milgram. *Parabolic Equations*, volume 33 of *Annals of Mathematics Studies*, pages 167–190. Princeton University Press, Princeton, N. J., 1954.

[204] E. B. Leach and J. H. M. Whitfield. Differentiable functions and rough norms on Banach spaces. *Proc. Am. Math. Soc.*, 33:120–126, 1972.

[205] H. Lebesgue. Intégrale, longueur, aire. *Ann. Mat. Pura Appl. (4)*, 7:231–359, 1902.

[206] H. Lebesgue. *Lecons sur l'integration et la recherche des fonctions primitives*. Gauthier-Villars, Paris, 1904.

[207] H. Lebesgue. Sur les fonctions représentables analytiquement. *J. Math. Pures Appl. (9)*, 1:139–216, 1905.

[208] H. Lebesgue. Sur l'intégration des fonctions discontinues. *Ann. Sci. Éc. Norm. Supér. (3)*, 27:361–450, 1910.

[209] G. Lebourg. Valeur moyenne pour gradient généralisé. *C. R. Acad. Sci. Paris, Sér. A-B*, 281(19):Ai, A795–A797, 1975.

[210] S. J. Leese. Multifunctions of Souslin type. *Bull. Aust. Math. Soc.*, 11:395–411, 1974.

[211] G. Leoni. *A First Course in Sobolev Spaces*, volume 105 of *Graduate Studies in Mathematics*. American Mathematical Society, Providence, RI, 2009.

[212] J. Leray and J. Schauder. Topologie et équations fonctionnelles. *Ann. Sci. Éc. Norm. Supér. (3)*, 51:45–78, 1934.

[213] B. Levi. Sopra l'integrazione delle serie. *Ist. Lomb. Accad. Sci. Lett. Rend. A*, 2:775–780, 1906.

[214] V. L. Levin. Lebesgue decomposition for functionals on the space L_X^∞ of vector-valued functions. *Funkc. Anal. Priložen.*, 8(4):48–53, 1974.

[215] V. L. Levin. Convex integral functionals, and lifting theory. *Usp. Mat. Nauk*, 30(2(182)):115–178, 1975.

[216] E. Lindelöf. Sur quelques points de la théorie des ensembles. *C. R. Math. Acad. Sci. Paris*, 137:697–700, 1904.

[217] J. Lindenstrauss and L. Tzafriri. *Classical Banach Spaces. I. Sequence Spaces*. Springer-Verlag, Berlin-New York, 1977.

[218] J.-L. Lions. *Quelques Méthodes de Résolution des Problèmes aux Limites Non-Linéaires*. Dunod; Gauthier-Villars, Paris, 1969.

[219] N. G. Lloyd. *Degree Theory*. Cambridge University Press, Cambridge, 1978.

[220] E. R. Lorch. On a calculus of operators in reflexive vector spaces. *Trans. Am. Math. Soc.*, 45(2):217–234, 1939.

[221] A. R. Lovaglia. Locally uniformly convex Banach spaces. *Trans. Am. Math. Soc.*, 78:225–238, 1955.

[222] G. Lumer and R. S. Phillips. Dissipative operators in a Banach space. *Pac. J. Math.*, 11:679–698, 1961.

[223] N. Lusin. Sur les propriétés des fonctions mesurables. *C. R. Math. Acad. Sci. Paris*, 154:1688–1690, 1912.

[224] N. Lusin. Sur la classification de M. Baire. *C. R. Math. Acad. Sci. Paris*, 164:91–94, 1917.

[225] N. Lusin. *Leçons sur les Ensembles Analytique et Leurs Applications*. Gauthier-Villars, Paris, 1930.

[226] A. Lyapounov. Sur les fonctions-vecteurs complètement additives. *Bull. Acad. Sci. URSS. Sér. Math. [Izv. Akad. Nauk SSSR]*, 4:465–478, 1940.

[227] R. H. Martin. *Nonlinear Operators and Differential Equations in Banach Spaces*. Wiley-Interscience [John Wiley & Sons], New York-London-Sydney, 1976.

[228] S. Mazur. Über konvexe Mengen in linearen normierten Räumen. *Stud. Math.*, 4(1):70–84, 1933.

[229] E. J. McShane. Linear functionals on certain Banach spaces. *Proc. Am. Math. Soc.*, 1:402–408, 1950.

[230] R. E. Megginson. *An Introduction to Banach Space Theory*, volume 183 of *Graduate Texts in Mathematics*. Springer-Verlag, New York, 1998.

[231] E. Michael. A note on paracompact spaces. *Proc. Am. Math. Soc.*, 4:831–838, 1953.

[232] E. Michael. Continuous selections. I. *Ann. Math. (2)*, 63:361–382, 1956.

[233] E. Michael. Another note on paracompact spaces. *Proc. Am. Math. Soc.*, 8:822–828, 1957.

[234] E. Michael. Yet another note on paracompact spaces. *Proc. Am. Math. Soc.*, 10:309–314, 1959.

[235] H. Minkowski. *Diophantische Approximationen*. Teubner, Leipzig, 1907.

[236] G. J. Minty. Monotone (nonlinear) operators in a Hilbert space. *Duke Math. J.*, 29:341–346, 1962.

[237] I. Miyadera. *Nonlinear Semigroups*. American Mathematical Society, Providence, RI, 1992.

[238] E. H. Moore. Definition of limit in general integral analysis. *Proc. Natl. Acad. Sci. USA*, 1(12):628–632, December 1915. JFM:45.0426.03.

[239] E. H. Moore and H. L. Smith. A general theory of limits. *Am. J. Math.*, 44(2):102–121, 1922.

[240] E. H. Moore. *General Analysis. Part II. Number 1*. American Philosophical Society, Philadelphia, 1939.

[241] R. L. Moore. An extension of the theorem that no countable point set is perfect. *Proc. Natl. Acad. Sci.*, 10:168–170, 1924.

[242] R. L. Moore. Concerning upper semi-continuous collections of continua. *Trans. Am. Math. Soc.*, 27(4):416–428, 1925.

[243] J.-J. Moreau. Proximité et dualité dans un espace hilbertien. *Bull. Soc. Math. Fr.*, 93:273–299, 1965.

[244] J.-J. Moreau. *Fonctionnelles Convexes*. College de France, Paris, 1967. Séminaire sur les équations aux dérivées partielles.

[245] U. Mosco. Convergence of convex sets and of solutions of variational inequalities. *Adv. Math.*, 3:510–585, 1969.

[246] U. Mosco. On the continuity of the Young-Fenchel transform. *J. Math. Anal. Appl.*, 35:518–535, 1971.

[247] J. R. Munkres. *Topology: A First Course*. Prentice-Hall, Inc., Englewood Cliffs, N. J., 1975.

[248] J. Musielak. *Orlicz Spaces and Modular Spaces*. Springer-Verlag, Berlin, 1983.

[249] K. Musial. Pettis integral. In *Handbook of Measure Theory, Vol. I, II*, pages 531–586. North-Holland, Amsterdam, 2002.

[250] J.-I. Nagata. *Modern General Topology*. North-Holland Publishing Co., Amsterdam, 1968.

[251] I. P. Natanson. *Theory of Functions of a Real Variable. Vol. II*. Frederick Ungar Publishing Co., New York, 1961.

[252] D. S. Nathan. One-parameter groups of transformations in abstract vector spaces. *Duke Math. J.*, 1(4):518–526, 1935.

[253] O. M. Nikodym. Sur une généralisation des mesures de m. j. radon. *Fundam. Math.*, 15:131–179, 1930.

[254] C. Olech. Decomposability as a substitute for convexity. In *Multifunctions and Integrands (Catania, 1983)*, volume 1091 of *Lecture Notes in Math.*, pages 193–205. Springer, Berlin, 1984.
[255] R. S. Palais. Homotopy theory of infinite dimensional manifolds. *Topology*, 5:1–16, 1966.
[256] N. S. Papageorgiou and S. Th. Kyritsi-Yiallourou. *Handbook of Applied Analysis*. Springer, New York, 2009.
[257] D. Pascali and S. Sburlan. *Nonlinear Mappings of Monotone Type*. Martinus Nijhoff Publishers, The Hague, 1978. Sijthoff & Noordhoff International Publishers, Alphen aan den Rijn.
[258] P. Pedregal. *Parametrized Measures and Variational Principles*. Birkhäuser Verlag, Basel, 1997.
[259] B. J. Pettis. On integration in vector spaces. *Trans. Am. Math. Soc.*, 44(2):277–304, 1938.
[260] R. R. Phelps. *Convex Functions, Monotone Operators and Differentiability*, volume 1364 of *Lecture Notes in Mathematics*. Springer-Verlag, Berlin, 1989.
[261] R. S. Phillips. Dissipative operators and hyperbolic systems of partial differential equations. *Trans. Am. Math. Soc.*, 90:193–254, 1959.
[262] R. S. Phillips. On linear transformations. *Trans. Am. Math. Soc.*, 48:516–541, 1940.
[263] J. Radon. Theorie und Anwendungen der absolut additiven Mengenfunktionen. *Wien. Ber.*, 122:1295–1438, 1913.
[264] M. Reed and B. Simon. *Methods of Modern Mathematical Physics. I. Functional Analysis*. Academic Press, New York-London, 1972.
[265] F. Riesz. Stetigkeitsbegriff und abstrakte Mengenlehre. *CHECK*, (1):1–20.
[266] F. Riesz. Sur les systèmes orthogonaux de fonctions. *C. R. Acad. Sci. Paris*, 144:615–619, 1907.
[267] F. Riesz. Sur une espèce de géométrie analytique des systèmes de fonctions sommables. *C. R. Acad. Sci. Paris*, 144:1409–1411, 1907.
[268] F. Riesz. Sur les ensembles de fonctions. *C. R. Acad. Sci. Paris*, 143:738–741, 1908.
[269] F. Riesz. Sur les suites de fonctions mesurables. *C. R. Math. Acad. Sci. Paris*, 148:1303–1305, 1909.
[270] F. Riesz. Untersuchungen über Systeme integrierbarer Funktionen. *Math. Ann.*, 69(4):449–497, 1910.
[271] F. Riesz. über lineare Funktionalgleichungen. *Acta Math.*, 41(1):71–98, 1916.
[272] R. T. Rockafellar. Extension of Fenchel's duality theorem for convex functions. *Duke Math. J.*, 33:81–89, 1966.
[273] R. T. Rockafellar. *Convex Analysis*, volume 28 of *Princeton Mathematical Series*. Princeton University Press, Princeton, N. J., 1970.
[274] R. T. Rockafellar. On the maximal monotonicity of subdifferential mappings. *Pac. J. Math.*, 33:209–216, 1970.
[275] R. T. Rockafellar. On the maximality of sums of nonlinear monotone operators. *Trans. Am. Math. Soc.*, 149:75–88, 1970.
[276] R. T. Rockafellar. Convex integral functionals and duality. In *Contributions to Nonlinear Functional Analysis (Proc. Sympos., Math. Res. Center, Univ. Wisconsin, Madison, Wis., 1971)*, pages 215–236. Academic Press, New York, 1971.
[277] R. T. Rockafellar. *Integral Functionals, Normal Integrands and Measurable Selections*, volume 543 of *Lecture Notes in Math.*, pages 157–207, 1976.
[278] R. T. Rockafellar and R. J.-B. Wets. *Variational Analysis*, volume 317 of *Grundlehren der Mathematischen Wissenschaften [Fundamental Principles of Mathematical Sciences]*. Springer-Verlag, Berlin, 1998.
[279] L. J. Rogers. An extension of a certain theorem in inequalities. *Messenger Math.*, 17:145–150, 1888.
[280] V. A. Rohlin. On the decomposition of a dynamical system into transitive components. *Mat. Sb. (N.S.)*, 25(67):235–249, 1949.
[281] T. Roubíček. *Relaxation in Optimization Theory and Variational Calculus*. Walter de Gruyter & Co., Berlin, 1997.
[282] T. Roubíček. *Nonlinear Partial Differential Equations with Applications*. Birkhäuser/Springer Basel AG, Basel, second edition, 2013.
[283] H. L. Royden. *Real Analysis*. Macmillan Publishing Company, New York, second edition, 1968.
[284] W. Rudin. *Real and Complex Analysis*. McGraw-Hill, New York, 1966.

[285] W. Rudin. *Functional Analysis*. McGraw-Hill, New York, 1973.

[286] M.-F. Sainte-Beuve. On the extension of von Neumann-Aumann's theorem. *J. Funct. Anal.*, 17:112–129, 1974.

[287] S. Saks. *Theory of the Integral*. Dover Publications, Inc., New York, 1964.

[288] J. Schauder. Eine Eigenschaft des Haarschen Orthogonalsystems. *Math. Z.*, 28:317–320, 1928.

[289] J. Schauder. Der Fixpunktsatz in Funktionalräumen. *Stud. Math.*, 2:171–180, 1930.

[290] E. Schmidt. Zur Theorie der linearen und nichtlinearen Integralgleichungen. I. Teil: Entwicklung willkürlicher Funktionen nach Systemen vorgeschriebener. II. Teil: Auflösung der allgemeinen linearen Integralgleichung. *Math. Ann.*, 63:433–476, 1907.

[291] S. Schwabik and G. Ye. *Topics in Banach Space Integration*, volume 10 of *Series in Real Analysis*. World Scientific Publishing Co. Pte. Ltd., Hackensack, NJ, 2005.

[292] J. Schwartz. A note on the space L_p^*. *Proc. Am. Math. Soc.*, 2:270–275, 1951.

[293] L. Schwartz. *Radon Measures on Arbitrary Topological Spaces and Cylindrical Measures*, volume 6 of *Tata Institute of Fundamental Research Studies in Mathematics*. Published for the Tata Institute of Fundamental Research, Bombay by Oxford University Press, London, 1973.

[294] G. Scorza Dragoni. Un teorema sulle funzioni continue rispetto ad una e misurabili rispetto ad un'altra variabile. *Rend. Semin. Mat. Univ. Padova*, 17:102–106, 1948.

[295] R. E. Showalter. *Monotone Operators in Banach Space and Nonlinear Partial Differential Equations*, volume 49 of *Mathematical Surveys and Monographs*. American Mathematical Society, Providence, RI, 1997.

[296] W. Sierpiński. Sur l'extension des fonctions de Baire définies sur les ensembles linéaires quelconques. *Fundam. Math.*, 16:81–89, 1930.

[297] M. A. Smith. Some examples concerning rotundity in Banach spaces. *Math. Ann.*, 233(2):155–161, 1978.

[298] V. Šmulian. On the principle of inclusion in the space of the type (B). *Rec. Math. Moscou, N.S.*, 5:317–328, 1939.

[299] V. Šmulian. Über lineare topologische Räume. *Rec. Math. Moscou, N.S.*, 7:425–448, 1940.

[300] M. Souslin. Sur une définition des ensembles mesurables b sans nombres transfinis. *C. R. Acad. Sci. Paris*, 164:88–91, 1917.

[301] S. M. Srivastava. *A Course on Borel Sets*, volume 180 of *Graduate Texts in Mathematics*. Springer-Verlag, New York, 1998.

[302] G. Stampacchia. *Èquations Elliptiques du Second Ordre à Coefficients Discontinus. Séminaire de Mathématiques Supérieures, No. 16 (Été, 1965)*. Les Presses de l'Université de Montréal, Montreal, Que, 1966.

[303] A. H. Stone. Paracompactness and product spaces. *Bull. Am. Math. Soc.*, 54:977–982, 1948.

[304] M. H. Stone. *Linear Transformations in Hilbert Spaces and their Applications to Analysis*, volume XV of *Amer. Math. Soc. Colloquium Publ.*, American Mathematical Society, New York, 1932. VIII +622 p. (1932).

[305] G. A. Suchomlinov. Über Fortsetzung von linearen Funktionalen in linearen komplexen Räumen und linearen Quotientenräumen. *Mat. Sb. (N.S.)*, 3:353–358, 1938. Russin, Germany summary.

[306] W. Takahashi. Existence theorems generalizing fixed point theorems for multivalued mappings. In *Fixed Point Theory and Applications (Marseille, 1989)*, volume 252 of *Pitman Res. Notes Math. Ser.*, pages 397–406. Longman Sci. Tech., Harlow, 1991.

[307] M. Talagrand. Pettis Integral and Measure Theory. *Mem. Am. Math. Soc.*, 51(307):ix+224, 1984.

[308] M. Talagrand. Weak Cauchy sequences in $L^1(E)$. *Am. J. Math.*, 106(3):703–724, 1984.

[309] H. T. über. Funktionen, die auf einer abgeschlossenen Menge stetig sind. *J. Reine Angew. Math.*, 145:9–14, 1915.

[310] H. Tietze. Beiträge zur allgemeinen Topologie. I. *Math. Ann.*, 88(3–4):290–312, 1923.

[311] L. Tonelli. Sull'integrazione per parti. *Rend. Accad. Naz. Lincei*, 18:246–253, 1909.

[312] J. W. Tukey. Some notes on the separation of convex sets. *Port. Math.*, 3:95–102, 1942.

[313] A. Tychonoff. Über die topologische Erweiterung von Räumen. *Math. Ann.*, 102:544–561, 1930.

[314] P. Urysohn. Über die Mächtigkeit der zusammenhängenden Mengen. *Math. Ann.*, 94(1):262–295, 1925.
[315] P. Urysohn. Zum Metrisationsproblem. *Math. Ann.*, 94(1):309–315, 1925.
[316] M. Valadier. Désintégration d'une mesure sur un produit. *C. R. Acad. Sci. Paris, Sér. A-B*, 276:A33–A35, 1973.
[317] M. Valadier. Young measures. In *Methods of Nonconvex Analysis (Varenna, 1989)*, volume 1446 of *Lecture Notes in Math.*, pages 152–188. Springer, Berlin, 1990.
[318] M. Valadier. A Course on Young Measures. *Rend. Ist. Mat. Univ. Trieste*, 26(suppl.):349–394, 1994, (1995). Workshop on Measure Theory and Real Analysis (Italian) (Grado, 1993).
[319] L. Vietoris. Stetige Mengen. *Monatshefte Math. Phys.*, 31:173–204, 1921.
[320] G. Vitali. Sul problema della misura dei gruppi di punti di una retta. Nota. Bologna: Gamberini e Parmeggiani., 5 S. 8°, 1905, (1905).
[321] G. Vitali. Sui gruppi di punti e sulle funzioni di variabili reali. *Torino Atti*, 43:229–246, 1908.
[322] L. P. Vlasov. Several theorems on čebyšev sets. *Mat. Zametki*, 11:135–144, 1972.
[323] L. P. Vlasov. Approximative properties of sets in normed linear spaces. *Usp. Mat. Nauk*, 28(6(174)):3–66, 1973.
[324] G. von Alexits. Über die Erweiterung einer Baireschen Funktion. *Fundam. Math.*, 15:51–56, 1930.
[325] J. von Neumann. Allgemeine Eigenwerttheorie Hermitescher Funktionaloperatoren. *Math. Ann.*, 102(1):49–131, 1930.
[326] J. von Neumann. Zur Algebra der Funktionaloperationen und Theorie der normalen Operatoren. *Math. Ann.*, 102(1):370–427, 1930.
[327] J. von Neumann. Über adjungierte Funktionaloperatoren. *Ann. Math. (2)*, 33(2):294–310, 1932.
[328] J. von Neumann. On complete topological spaces. *Trans. Am. Math. Soc.*, 37(1):1–20, 1935.
[329] J. von Neumann. On rings of operators. III. *Ann. Math. (2)*, 41:94–161, 1940.
[330] J. von Neumann. On rings of operators. Reduction theory. *Ann. Math. (2)*, 50:401–485, 1949.
[331] J. V. Wehausen. Transformations in linear topological spaces. *Duke Math. J.*, 4(1):157–169, 1938.
[332] J. Weidmann. *Linear Operators in Hilbert Spaces*, volume 68 of *Graduate Texts in Mathematics*. Springer-Verlag, New York-Berlin, 1980. Translated from the German by Joseph Szücs.
[333] R. Whitley. Mathematical Notes: Projecting m onto c_0. *Am. Math. Mon.*, 73(3):285–286, 1966.
[334] S. Willard. *General Topology*. Addison-Wesley Publishing Co., Reading, Mass.-London-Don Mills, Ont., 1970.
[335] V. Yankov. On the uniformization of certain A-sets. *Doklady Akad. Nauk SSSR (N.S.)*, 30:591–592, 1941.
[336] K. Yosida. *Functional Analysis*. Springer-Verlag, Berlin-New York, fifth edition, 1978.
[337] K. Yosida. On the differentiability and the representation of one-parameter semi-group of linear operators. *J. Math. Soc. Jpn.*, 1:15–21, 1948.
[338] Z. Zalcwasser. Sur une propriété du champ des fonctions continues. *Ann. Soc. Pol. Math.*, 9:171, 1931.
[339] E. Zeidler. *Nonlinear Functional Analysis and its Applications. I.* Springer-Verlag, New York, 1986.
[340] E. Zeidler. *Nonlinear Functional Analysis and its Applications. II/A and II/B.* Springer-Verlag, New York, 1990.
[341] C.-K. Zhong. On Ekeland's variational principle and a minimax theorem. *J. Math. Anal. Appl.*, 205(1):239–250, 1997.
[342] W. P. Ziemer. *Weakly Differentiable Functions*, volume 120 of *Graduate Texts in Mathematics*. Springer-Verlag, New York, 1989.

Index

https://doi.org/10.1515/9783111286952-008

List of Symbols

https://doi.org/10.1515/9783111286952-009

·nt.com/pod-product-compliance
·BA/4555